International Conference on Applied Sciences (ICAS2017)

IOP Conference Series: Materials Science and Engineering
Volume 294

Hunedoara, Romania
10 – 12 May 2017

ISBN: 978-1-5108-5567-0
ISSN: 1757-8981

Printed from e-media with permission by:

Curran Associates, Inc.
57 Morehouse Lane
Red Hook, NY 12571

Some format issues inherent in the e-media version may also appear in this print version.

Printed by Curran Associates, Inc. (2018)

For permission requests, please contact the Institute of Physics
at the address below.

Institute of Physics
Dirac House, Temple Back
Bristol BS1 6BE UK

Phone: 44 1 17 929 7481
Fax: 44 1 17 920 0979

techtracking@iop.org

Additional copies of this publication are available from:

Curran Associates, Inc.
57 Morehouse Lane
Red Hook, NY 12571 USA
Phone: 845-758-0400
Fax: 845-758-2633
Email: curran@proceedings.com
Web: www.proceedings.com

Table of contents

Volume 294

International Conference on Applied Sciences (ICAS2017)

10–12 May 2017, Hunedoara, Romania

Accepted papers received: 22 December 2017
Published online: 12 January 2018

Preface

Papers

IOP Conf. Series: Materials Science and Engineering **294** (2017) 011001 doi:10.1088/1757-899X/294/1/011001

International Conference on Applied Sciences ICAS2017

Hunedoara, Romania, May 10-12, 2017

Organizers:

University Politehnica Timisoara **University of Banja Luka**

in cooperation with

Academy of Romanian Scientists
Academy of Sciences Republic of Srpska
Academy of Technical Sciences of Romania - Timisoara Branch
and
General Association of Romanian Engineers - Hunedoara Branch

The **International Conference on Applied Sciences ICAS2017** took place in Hunedoara, Romania, in the period May 10–12, 2017, at the Engineering Faculty of Hunedoara, with the aims to serve as a platform for exchange of information between various areas of applied sciences, and to promote the communication between the scientists of different nations, countries and continents.

The conference has been focused on several fields of application, operation and influence of the applied sciences and technologies on industry.

Topics of the conference covers a comprehensive spectrum of issues from:
*1. **Materials Science:*** Metallic Materials, Composite Materials, Metal Alloys, Metallurgy, Heat Transfer, and others...
*2. **Mechatronics:*** Mechanical Engineering, Robotic Systems Engineering, Control Engineering, Reliability, and others...
*3. **Electrical Engineering:*** Circuits and Systems, Signal Processing, Electric Motors, and others...
*4. **Computers Engineering:*** Modeling and Simulation, Computational Methods in Engineering, Software Engineering, Data Bases, and others...
*5. **Fundamental Sciences:*** Numerical approximation and analysis, Interdisciplinary applications of mathematics and physics, Chemistry, and others...

The conference gathered qualified researchers whose expertise can be used to develop new engineering knowledge that has applicability potential in Industry. The number of participants was almost 110 from 11 countries (Romania, Bosnia and Herzegovina, U.S.A., Austria, Italy, Netherlands, Portugal, Poland, Slovakia, Serbia and Moldova).
During the three days of the conference, 5 invited speakers and 125 oral talks were delivered.

IOP Conf. Series: Materials Science and Engineering **294** (2017) 011001 doi:10.1088/1757-899X/294/1/011001

Invited speakers
Mohsen Razzaghi, Mississippi State University
Miroslav Prsa, University of Novi Sad
Bogdan Lungu, Horia Hulubei National Institute for R&D in Physics and Nuclear Engineering
Mircea Octavian Popoviciu, Academy of Romanian Scientists
Ioan Ilca, Academy of Technical Sciences of Romania

Special thanks should be directed to the **Organizing Committee** for his tremendous efforts in organizing the conference:

Co-Presidents
Mataruga Milan, University of Banja Luka
Serban Viorel-Aurel, University Politehnica Timisoara

Honorary Co-Presidents
Borko Durić, President of the Chamber of Commerce
Harau Carmen, Parliament of Romania - Senat
Ilca Ioan, Academy of Technical Sciences of Romania
Mirjanic Dragoljub, Academy of Sciences Republic of Srpska
Petrescu Doru Ion, C.E.O. Arcelor Mittal S.A.
Predrag Damjanovic, Director of Pedagogical Institute of the Republic of Srpska
Radojičić Igor, Mayor of Banja Luka
Yang Xi, S C Eurosport DHS S.A.

Vice-Presidents
Knezevic Darko, University of Banja Luka
Panoiu Caius, University Politehnica Timisoara

Coordinators
Latinovic Tihomir, University of Banja Luka
Lemle Ludovic Dan, University Politehnica Timisoara

Program Chairman
Kiss Imre, University Politehnica Timisoara

Members
Abrudean Cristian, University Politehnica Timisoara
Alexa Vasile, University Politehnica Timisoara
Ardelean Marius, University Politehnica Timisoara
Dinis Corina, University Politehnica Timisoara
Gherman Lucian, University Politehnica Timisoara
Ratiu Sorin, University Politehnica Timisoara
Socalici Ana, University Politehnica Timisoara
Tirian Ovidiu, University Politehnica Timisoara

Secretariat
Berdie Adela, University Politehnica Timisoara
Josan Ana, University Politehnica Timisoara
Rob Raluca, University Politehnica Timisoara

Translator
Gaianu Oana, University Politehnica Timisoara

Responsible with Media Relations
Birtok-Băneasă Corneliu, University Politehnica Timisoara

Based on the works presented at the conference, 95 selected papers have been selected to be included in a new volume of *IOP Conference Series: Materials Science and Engineering*. These papers present new researches in the various fields of Materials Engineering, Mechanical Engineering, Computers Engineering and Electrical Engineering.

The members of the **Scientific Committee** who reviewed and selected the papers included in the Proceedings of ICAS2017 are:

Chairman
Lemle Ludovic Dan, University Politehnica Timisoara

Vice-Chairmen
Barz Cristian, Technical University of Cluj Napoca
Gherman Laurian, "Henri Coanda" Air Force Academy of Brasov
Jiang Yiwen, Military Economics Academy of Wuhan
Latinovic Tihomir, University of Banja Luka

Chairmen of Materials Science Session
Ardelean Erika, University Politehnica Timisoara
Zrilic Ranko, University of Banja Luka

Chairmen of Mechatronics Session
Bordeasu Ilare, University Politehnica Timisoara
Rogic Miroslav, University of Banja Luka

Chairmen of Electrical Engineering Session
Blanusa Branko, University of Banja Luka
Popa Gabriel, University Politehnica Timisoara

Chairmen of Computers Engineering Session
Dejanovic Ratko, University of Banja Luka
Panoiu Manuela, University Politehnica Timisoara

Chairmen of Fundamental Sciences Session
Bistrian Diana, University Politehnica Timisoara
Romano Daniel, University of Banja Luka

Chairmen of Video-conference Session
Knezevic Bojan, University of Banja Luka
Muscalagiu Ionel, University Politehnica Timisoara

Members
Borojević Stevo, University of Banja Luka
Cioaca Catalin, "Henri Coanda" Air Force Academy of Brasov
Constantinescu George, "Henri Coanda" Air Force Academy of Brasov
Cosma Marius, Technical University of Cluj Napoca
Erdei Zoltan, Technical University of Cluj Napoca
Fang Zhengqi, Military Economics Academy of Wuhan
Gvero Petar, University of Banja Luka

IOP Conf. Series: Materials Science and Engineering **294** (2017) 011001 doi:10.1088/1757-899X/294/1/011001

Guo Zhonghou, Military Economics Academy of Wuhan
Heput Teodor, University Politehnica Timisoara
Lesenciuc Adrian, "Henri Coanda" Air Force Academy of Brasov
Li Xuewu, Military Economics Academy of Wuhan
Luculescu Doru, "Henri Coanda" Air Force Academy of Brasov
Lung Claudiu, Technical University of Cluj Napoca
Mandru Dan, Technical University of Cluj Napoca
Milašinović Aleksandar, University of Banja Luka
Miron Liliana Ecaterina, "Henri Coanda" Air Force Academy of Brasov
Osaci Mihaela, University Politehnica Timisoara
Pinca-Bretotean Camelia,University Politehnica Timisoara
Pop Sitar Petrica, Technical University of Cluj Napoca
Rusu-Anghel Stela, University Politehnica Timisoara
Stoychitch Mihaylo, University of Banja Luka
Tanasić Zorana, University of Banja Luka
Vilceanu Lucia, University Politehnica Timisoara
Zhang Lijun, Military Economics Academy of Wuhan
Zhang Tao, Military Economics Academy of Wuhan

Event supported by

Hunedoara County Council

S.C. Micro-Mega HD S.A.

ArcelorMittal S.A.

S.C. EUROSPORT DHS S.A.

We thank the authors for their contributions and we express our gratitude to everyone who contributed to the success of our conference.

Ludovic Dan LEMLE
Editor of the Proceedings of ICAS2017

IOP Conf. Series: Materials Science and Engineering **294** (2017) 011002 doi:10.1088/1757-899X/294/1/011002

Peer review statement

All papers published in this volume of *IOP Conference Series: Materials Science and Engineering* have been peer reviewed through processes administered by the proceedings Editors. Reviews were conducted by expert referees to the professional and scientific standards expected of a proceedings journal published by IOP Publishing.

IOP Conf. Series: Materials Science and Engineering **294** (2017) 012001 doi:10.1088/1757-899X/294/1/012001

Electromagnetic fields and their impacts

M A Prša[1] and K K Kasaš-Lažetić[2]

[1]Retired from University of Novi Sad, Faculty of Technical Sciences, Department of Power, Electronics and Telecommunication Engineering, Trg Dositeja Obradovića 6, 21000 Novi Sad, R. of Serbia
[2]University of Novi Sad, Faculty of Technical Sciences, Department of Power, Electronics and Telecommunication Engineering, Trg Dositeja Obradovića 6, 21000 Novi Sad, R. of Serbia

E-mail: prsa@uns.ac.rs

Abstract. The main goal of this paper is to briefly recall some different electromagnetic field definitions, some macroscopic sources of electromagnetic fields, electromagnetic fields classification regarding time dependences, and the ways of field determination in concrete cases. After that, all the mechanisms of interaction between electromagnetic field and substance, on atomic level, are described in details. Interaction between substance and electric field is investigated separately from the substance and magnetic field interaction. It is demonstrated that, in all cases of the unique electromagnetic field, total interaction can be treated as a superposition of two separated interactions. Finally, the main electromagnetic fields surrounding us is cited and discussed.

1. Introduction

Our living and working environment is filled with all kinds of electromagnetic fields. Some of them, like the Earth magnetic field, are natural, while the large number of electromagnetic fields are artificial, created by people. Nowadays, in the era of wireless communications, the number of electromagnetic fields with different specifications is growing rapidly.

The influence of electromagnetic fields on everything and everybody inside them could be useful or harmful. For this reason, it is important to be well acquainted with all electromagnetic fields surrounding us, as well as with the mechanisms of their impact on the substances exposed to the electromagnetic fields' influence.

The aim of the first part of this paper is to briefly remind the audience about the electromagnetic fields definitions, about some macroscopic sources of electromagnetic fields and some methods for the electromagnetic field determination. The second part is dedicated to the interaction between electromagnetic fields and the substances in which electromagnetic field is established, while the third part brings the discussion on the most common electromagnetic fields in our environment.

Some electromagnetic field classifications into groups, with the characteristics of each group are also presented and discussed in this paper, together with some standards defining maximal allowed electric and magnetic field values.

2. Basic electromagnetic fields' features

2.1. Electromagnetic fields definitions

In any paper concerning electromagnetic fields description, first of all, a definition of electromagnetic field should be presented. The definition of a kind "What is an electromagnetic field" can also be found in the science world. For instance, Einstein defined a field as "regular, geometrical curvature of space". This definition is correct, because a field really is "regular", meaning that certain sources always produce the same field. The field is "geometrical" because it is always established in the space; however, it is not always easy to comprehend what "the curvature of space" is.

Actually, this could be understood if one graphically presents the space with no field as the family of parallel, plane surfaces which are not "curved". On the other hand, if the space with a field is 3D visualized, like a gravitational field shown in Figure 1 [1], or electric field of an electric dipole, presented in Figure 2 [2], the concept of "the curvature of space" could be accepted.

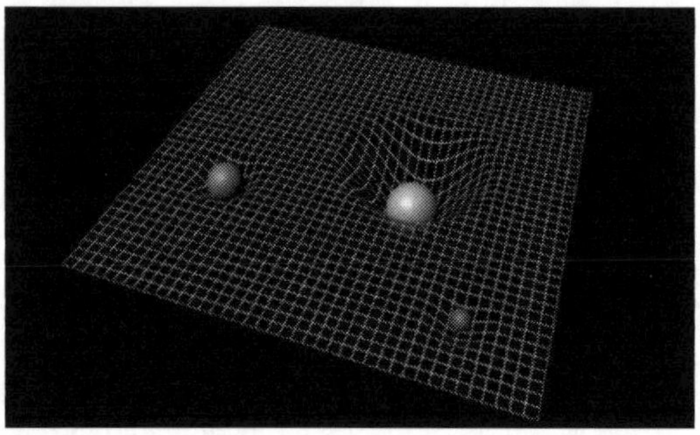

Figure 1. 3D presentation of a gravitational field

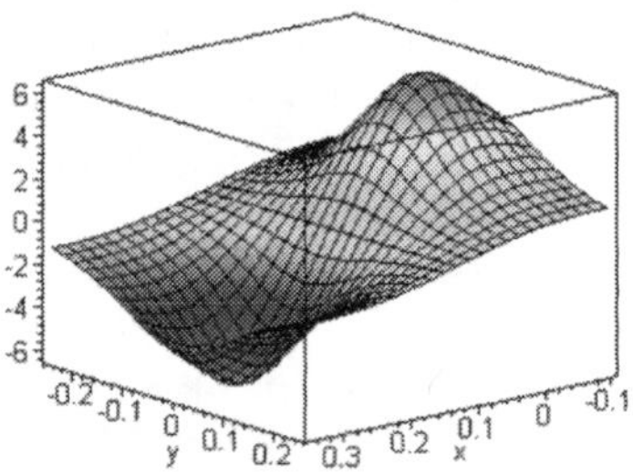

Figure 2. 3D presentation of an electric dipole field

In both images, the "curved" surfaces could be noticed and this could be a graphical explanation of the Einstein's field definition.

Nevertheless, in the engineering theory and practice, another field definition is much more in use and much easier to imagine. Namely, the most common field definition is through the field consequences, i.e. the force produced by a field [3-5].

For this reason, an electromagnetic field can be defined using the Lorenz's force expression,

$$\vec{F} = \Delta Q\vec{E} + \Delta Q\vec{v} \times \vec{B}. \tag{1}$$

The electromagnetic field definition can now be presented as follows: **If in the certain domain one puts the test charge, ΔQ and if there is a force defined by (1), acting on the test charge, it could be concluded that in this domain an electromagnetic field exists**. Obviously, the electromagnetic field, produced by some external sources, exists independently of the test charge; i.e. the test charge serves only to detect the field. It should be noticed that this implicit definition of electromagnetic field is simple and easily comprehensible.

At the same time, the expression (1) is the basic equation for the force on charges in the electric and magnetic field. In (1) $\vec{E}$ is the electric field strength vector, $\vec{B}$ is the magnetic flux density vector, while $\vec{v}$ is the velocity of moving charges. As mentioned before, ΔQ is the test charge. By its definition, the geometrical size of the test charge is very small and, at the same time, the charge itself is very small as well; actually, it is so small that the field produced by the test charge is negligible compared to the fields define by $\vec{E}$ and $\vec{B}$.

Observing (1), two more conclusions can be derived. The electric field force acts on positive charges in the direction of $\vec{E}$, adding kinetic energy to the charges and accelerating them, while the magnetic field force acts perpendicularly to the directions of moving particles, attending to change the moving direction, with no possibilities for accelerating them and adding any kinetic energy to them.

Moreover, due to relevant physical values which can be reached in nature, in the case of time constant electromagnetic fields, the first part of (1) is larger than the second part. This means that the influence of time constant magnetic field on time constant current distribution can be neglected.

2.2. Macroscopic sources of electromagnetic fields

The real sources of all macroscopic events in electrical engineering theory and practice are elementary charges, i.e. electrons and protons, considering the simple Bohr's model of atom. Actually, in macroscopic electromagnetic fields theory, the sources are defined as a simultaneous action of a large number of elementary charged particles, distributed in space. For this reason, as the macroscopic electromagnetic field sources, two macroscopic quantities are defined, in order to describe the average performances of elementary charges. These quantities are [3-5]:

- Volume charge density, $\rho(x,y,z,t)=(\mathrm{d}Q)_{\text{in dV}}/\mathrm{d}V$
- Current density vector, $\vec{J}(x,y,z,t)=NQ\vec{v}=N(-Q)(-\vec{v})=\rho\vec{v}$.

In the above expressions, $\mathrm{d}V$ is a so-called „Physically small volume" or "Elementary volume", small enough to represent a point and large enough to encounter the large number of elementary charges; N is the concentration of moving charges; Q is the macroscopic charge of each moving particle; and $\vec{v}$ is the velocity of moving charged particles.

2.3. Expression for vectors $\vec{E}$ and $\vec{B}$ produced by known electromagnetic fields sources

Let us consider the domain, with the volume v', in which all electromagnetic field sources are, and let us denote all the quantities concerning these sources with a prime, as demonstrated in Figure 3.

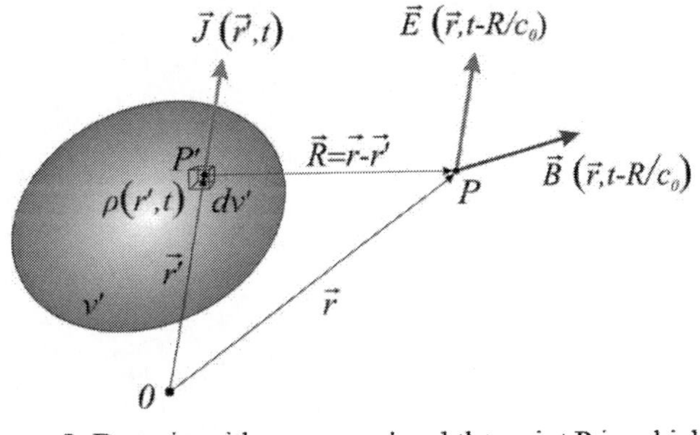

Figure 3. Domain with sources, v' and the point P in which electromagnetic field vectors are expressed

According to Figure 3, adopting the origin, denoted by 0, the position of the point P′, in which the sources $\rho(\vec{r}',t)$ and $\vec{J}(\vec{r}',t)$ are known values, is defined by the position vector $\vec{r}'$, while the position of the point P, in which the electromagnetic field vectors $\vec{E}$ and $\vec{B}$ are calculated, is defined by the position vector $\vec{r}$.

If the system presented in Figure 3 is in vacuum, the expressions for vectors $\vec{E}$ and $\vec{B}$ are as follows [3-5]:

$$\vec{E}(\vec{r},t)=\frac{1}{4\pi\varepsilon_0}\int_{v'}\frac{\rho(\vec{r}',t-R/c_0)(\vec{r}-\vec{r}')}{R^3}\mathrm{d}v'-\frac{\partial}{\partial t}\frac{\mu_0}{4\pi}\int_{v'}\frac{\vec{J}(\vec{r}',t-R/c_0)}{R}\mathrm{d}v',$$
$$\vec{B}(\vec{r},t)=\frac{\mu_0}{4\pi}\int_{v'}\frac{\vec{J}(\vec{r}',t-R/c_0)\times(\vec{r}-\vec{r}')}{R^3}\mathrm{d}v'. \tag{2}$$

In the above equations, the physical constants for vacuum are the following:
ε_0 – permittivity of vacuum, $\varepsilon_0 = (8.85419 \pm 0.00002).10^{-12}$ F/m (As/Vm),
μ_0 – permeability of vacuum, $\mu_0 = 4\pi.10^{-7}$ H/m (Vs/Am).
c_0 – light speed in vacuum, $c_0 = 1/\sqrt{\varepsilon_0\mu_0}$, $c_0 = (2.997924 \pm 0.000003).10^8$ m/s.

Another important fact, called the retardation effect, can also be noticed in (2). Due to the finite speed of the electromagnetic field establishing, it is obvious in (2) that the field vectors in the point P, at the time t, are not produced by sources at the same time, but acting at the previous time, $t - R/c_0$.

The electromagnetic field vectors are frequently defined through the potentials, mostly through the Lorenz's electric scalar potential, V and magnetic vector potential, $\vec{A}$, connected by the Lorenz's condition [3], [4],

$$\nabla\cdot\vec{A}=-\varepsilon_0\mu_0\frac{\partial V}{\partial t}. \tag{3}$$

If the electromagnetic field sources in the volume v' are known, electric scalar potential, V and magnetic vector potential, $\vec{A}$ are the solutions of partial differential equations [3-5],

$$\Delta V-\varepsilon_0\mu_0\frac{\partial^2 V}{\partial t^2}=-\frac{\rho}{\varepsilon_0},$$
$$\Delta\vec{A}-\varepsilon_0\mu_0\frac{\partial^2\vec{A}}{\partial t^2}=-\mu_0\vec{J}. \tag{4}$$

Considering the system shown in Figure 3, electric scalar potential, V and magnetic vector potential, $\vec{A}$ are retarded potentials [3], [4], [6],

$$V(r,t)=\frac{1}{4\pi\varepsilon_0}\int_{v'}\frac{\rho(\vec{r}',t-R/c_0)}{R}\mathrm{d}v',$$
$$\vec{A}(r,t)=\frac{\mu_0}{4\pi}\int_{v'}\frac{\vec{J}(\vec{r}',t-R/c_0)}{R}\mathrm{d}v'. \tag{5}$$

In the above equations, c_0 is the light speed in vacuum; $c_0 = 1/\sqrt{\varepsilon_0\mu_0}$; it is on this fact that Maxwell based the hypothesis that the light is also an electromagnetic field (wave).

Electromagnetic field vectors, electric field strength vector and magnetic flux density vector are defined by potentials as [3-5],

$$\vec{E}(\vec{r},t)=-\nabla V(\vec{r},t)-\frac{\partial\vec{A}(\vec{r},t)}{\partial t},\qquad \vec{B}(\vec{r},t)=\nabla\times\vec{A}. \tag{6}$$

From both expressions, (2) and (6), it can be concluded that, in general, the electric field, defined by the electric field strength vector, has both components, conservative and rotational, while the magnetic field expressed by the magnetic flux density vector is a thoroughly rotational field. The second part of the electric field strength vector is the so-called induced electric field strength vector.

2.4. Classification of electromagnetic fields regarding their time varying behavior

Electromagnetic fields can be either time constant or time varying. If the sources which produce electromagnetic fields are not varying in time, the electromagnetic fields themselves are also time constant. The general characteristic of any time constant electromagnetic field is that the electric field can always be separated from the magnetic field. In the case when the macroscopic charges are not moving, they produce an electrostatic field and in this case the magnetic field does not exist. In the case when the charges move uniformly, producing the time constant current, it can be shown that the influence of the time constant magnetic field on the current distribution is negligible. Consequently, in this case as well the electric field can be observed independently of the magnetic field.

Time constant electric field is always a conservative one, meaning that the work along any closed path is equal to zero, i.e. the integral of the conservative field strength vector, along any closed path, is equal to zero, too,

$$\oint_{C_{cl}} \vec{E}\cdot \mathrm{d}\vec{\ell} = 0. \tag{7}$$

A completely different situation is when the electromagnetic field sources are time varying, producing the time varying field. Due to the electromagnetic induction, the electric field strength vector obtains the rotational component, too, produced by the time varying magnetic field. On the other hand, a magnetic field can also be caused by the time varying electric field. For this reason, time varying electric and magnetic fields cannot be treated separately; they are only the components of the unique electromagnetic field. Usually, it can be defined that the time varying electric field is always followed by the time varying magnetic field and vice versa.

It is also usual that time varying electromagnetic fields are classified in two groups, depending on the field retardation. If the field retardation can be neglected, the electromagnetic field is **time slow varying**, or **quasi-static**, or a **quasi-stationary** electromagnetic field. On the other hand, if the field retardation cannot be neglected, the field is **time fast varying**, or it is an **electromagnetic wave**.

When the time variations of the sources are periodic, the frequency, f and the wave length, λ are defined and the relation between those two quantities in vacuum is

$$\lambda = \frac{c_0}{f}. \tag{8}$$

If the electromagnetic field sources vary in time as a harmonic function, i.e. sin or cosine function of time, the complex approach can be adopted and the partial differential equations (4) could be accepted as the complex ones [3], [4],

$$\begin{aligned} \Delta \underline{V} + \omega^2 \varepsilon_0 \mu_0 \underline{V} &= -\frac{\underline{\rho}}{\varepsilon_0}, \\ \Delta \underline{\vec{A}} + \omega^2 \varepsilon_0 \mu_0 \underline{\vec{A}} &= -\mu_0 \underline{\vec{J}}, \end{aligned} \tag{9}$$

defining the complex electric scalar potential and complex magnetic vector potential [3], [4], [6],

$$\begin{aligned} \underline{V}(\vec{r}) &= \frac{1}{4\pi\varepsilon_0} \int_{v'} \frac{\underline{\rho}(\vec{r}\,')\mathrm{e}^{-\mathrm{j}\beta_0 R}}{R} \mathrm{d}v', \\ \underline{\vec{A}}(\vec{r}) &= \frac{\mu_0}{4\pi} \int_{v'} \frac{\underline{\vec{J}}(\vec{r}\,')\mathrm{e}^{-\mathrm{j}\beta_0 R}}{R} \mathrm{d}v'. \end{aligned} \tag{10}$$

In the above equations, the phase coefficient for vacuum is defined as [3-5],

$$\beta_0 = \frac{\omega}{c_0} = \omega\sqrt{\varepsilon_0\mu_0}, \qquad \omega = 2\pi f. \tag{11}$$

From known values of complex potentials, the complex electric field strength vector and complex magnetic flux density vectors are [3-5],

$$\underline{\vec{E}} = -\nabla \underline{V} - j\omega\underline{\vec{A}}, \qquad \underline{\vec{B}} = \nabla \times \underline{\vec{A}}. \tag{12}$$

In the electrical engineering practice, the majority of electromagnetic fields are periodic, harmonic functions of time. For this reason, it is usual that time varying electromagnetic fields are classified according to the frequency, as it is presented in Table 1 [7].

Table 1. Classification of electromagnetic fields by sources' frequency

Number of ranges *N*	Frequency range (lower limit included, upper limit excluded)	Name of the range	Abbre viation	Wave-length range	Name of the wave-length range	Abbre viation
1.	3Hz to 30 Hz			100,000 km to 10,000 km	Deca-megameter	
2.	30 Hz to 300 Hz	Extremely Low Frequencies	ELF	10,000 km to1,000 km	Megameter	
3.	300Hz to 3kHz	Super Low Frequencies	SLF	1,000 km to 100 km	Hecto-kilometer	
4.	3 kHz to 30 kHz	Very Low Frequencies	VLF	100 km to 10 km	Myriameter	B.Mam
5.	30 kHz to 300 kHz	Low Frequencies	LF	10 km to 1 km	Kilometer	B.km
6.	300 kHz to 3 MHz	Middle Frequencies	MF	1 km to 100 m	Hectometer	B.hm
7.	3 MHz to 30 MHz	High Frequencies	HF	100 m to10 m	Decameter	B.dam
8.	30 MHz to 300 MHz	Very High Frequencies	VHF	10 m to 1 m	Meter	B.m
9.	300 MHz to 3 GHz	Ultra High Frequencies	UHF	1 m to 10 cm	Decimeter	B.dm
10.	3 GHz to 30 GHz	Super High Frequencies	SHF	10 cm to 1 cm	Centimeter	B.cm
11.	30GHz do 300GHz	Extremely High Frequencies	EHF	1 cm to 1 mm	Milimeter	B.mm
12.	300 GHz to 3,000 GHz			1 mm to 0.1 mm	Sub-milimeter	
13.	> 3,000 GHz			< 0.1 mm	Light (Optical)	

As it can be observed in Table 1, under certain conditions all time varying electromagnetic fields could be treated as electromagnetic waves. For this reason, the term "radiation", usually applied for electromagnetic waves, can conditionally be applied for all time varying electromagnetic fields.

3. Electromagnetic fields in a substance

When an electromagnetic field is present in a substance, an interaction between the field and the substance occurs. This means that the field modifies the substance, provoking the movement of free or bound charges. The modified substance produces its own electromagnetic field, which superposes the existing field and changes it. In observing these interactions, the electric field interaction can be explained separately from the magnetic field interaction.

The interaction between electric field and a substance depends on the conductive nature of the substance; the interaction differs inside the conductors with a large number of free (unbound) charges and the insulators, in which the majority of charges are bound.

3.1. Interaction of electric field and the substance

3.1.1. Electrostatic induction. In the case when a conductor is placed in an electric field, the field provokes the movement of free charges as the consequence of the first part of the Lorenz force, (1). The positive charges move in the electric field strength vector direction, while the negative charges move in the opposite direction. This movement of free charges is known as the electrostatic induction, and it is presented in Figure 4.

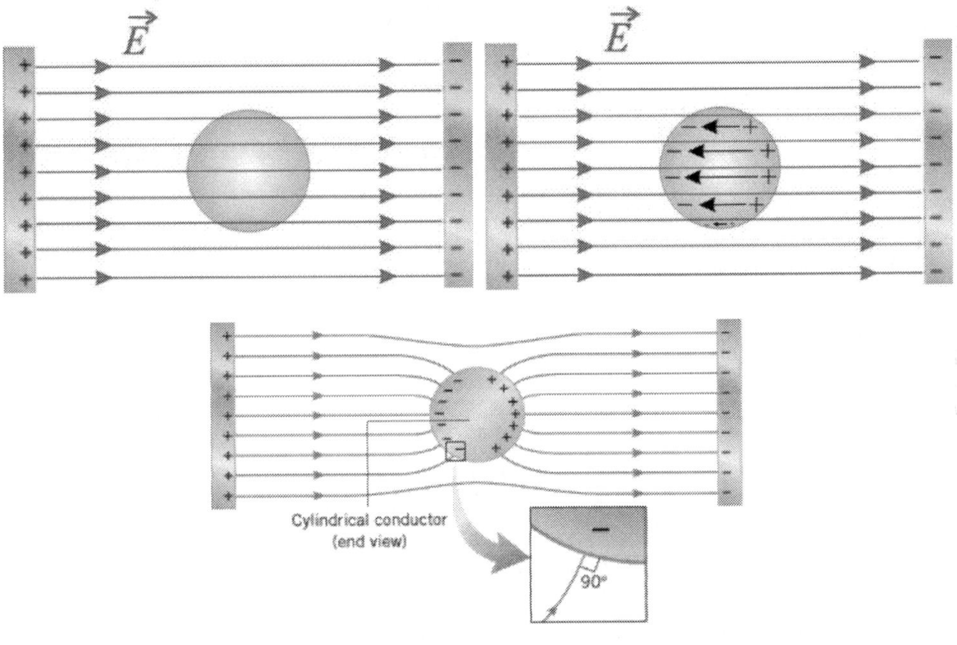

Figure 4. Conductive body inside an electric field – electrostatic induction

The free charges, positioned at the conductor body's surface, produce their own electric field, which cancels the external field inside the body. This effect is widely applied in the Faraday's cage case, determining that the inside domain is perfectly protected against the electrostatic field and very well protected against any other electric field.

3.1.2. Electric polarization. If an insulator (dielectric) is placed inside the electric field, the first part of the Lorenz force, (1), displaces the bound charges at the atomic or molecular level; this effect is referred to as an electric polarization. The electric polarization can be electronic, dipole or ionic, depending on the atomic structure of material. Inside materials with non-polar atoms, the electronic polarization occurs, as presented in Figure 5.

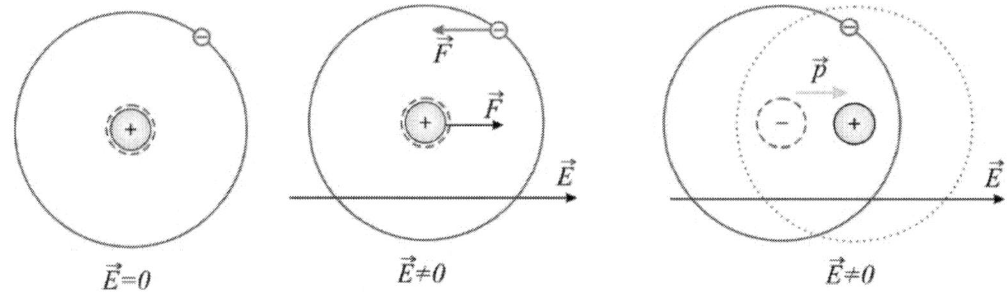

Figure 5. Electronic polarization

The first part of the Lorenz force produces the movement of positive bound charges in the direction of the electric field strength vector and the movement of negative bound charges in the opposite direction. As the result of this action, the atom is deformed and it acts as an electric dipole; hence it can be treated as a microscopic dipole, with the dipole moment rotating it in the field direction, as shown in Figure 6. There is also a force dragging it into the domain of the stronger field, presented as well in Figure 6.

Actually, Figure 6 graphically represents the dipole polarization occurring in the case of a substance with polar molecules, like, for instance, water.

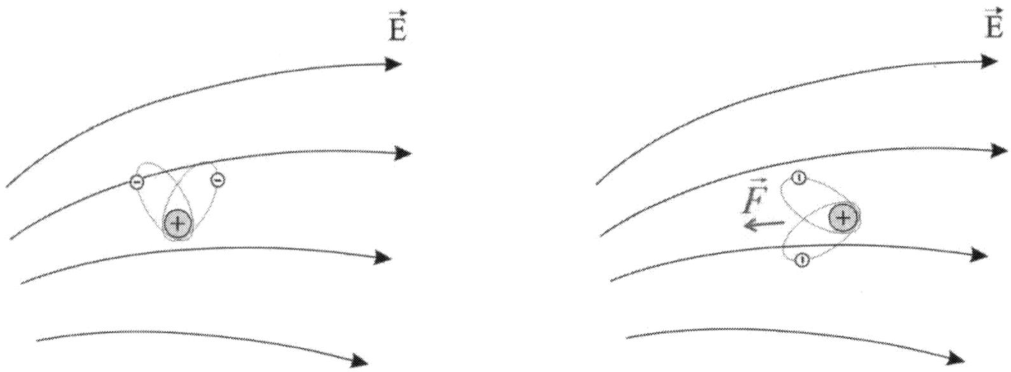

Figure 6. Dipole polarization

The electric field produces the torque on polar molecules, which rotates the molecule to the electric field strength vector direction and the force toward the domain with the higher electric field strength vector magnitudes. The third form of polarization appears inside some of the materials with crystalline structures. An external electric field deforms the structure at the atomic level and thus the deformed structure acts as a large number of microscopic dipoles.

There are several possibilities to describe polarization at the macroscopic level; the most applied is the involvement of the dielectric permittivity, as the material property. The dielectric permittivity is defined as the product of dielectric relative permittivity, and the permittivity of the vacuum,

$$\varepsilon = \varepsilon_r \varepsilon_0 . \quad (13)$$

3.2. Interaction of magnetic field and the substance

If a closed contour with the current is placed in a magnetic field, the field produces a torque, trying to rotate the contour and position it perpendicular to the magnetic flux density vector. Although there is no contour with the current at the atomic level, the orbit spin could be treated as the small, microscopic current contour on which the magnetic field produces the torque. In the absence of the magnetic field, the orbits of electrons are randomly arranged, producing no resultant magnetic field.

The exceptions are only magnetized ferromagnetic materials, in which there are strong interactions between neighboring atoms and the thorough orientation of electron's orbits inside material's domains. These domains, the so-called Weiss domains, could be even several cubic centimeters large.

When a substance is exposed to a magnetic field, the field has the intention to rotate all electrons' orbits to the plane perpendicular to the magnetic flux density vector; this situation is known as the magnetization of material. The principle of the magnetization process at the atomic level, before and after the impact of the magnetic field, is presented in Figure 7.

The Coulomb forces between electrons oppose the magnetization process and the most of materials cannot be significantly magnetized. The significant magnetization is possible only inside ferromagnetic materials, like iron, nickel, cobalt and their alloys, some rear earths like gadolinium, and ferrimagnetic materials like ferrites.

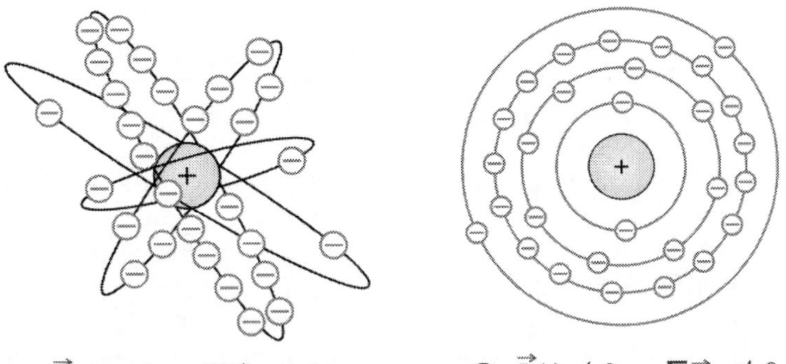

$\vec{B}(t) = 0 \quad \sum \vec{m}_{eo} = 0$ $\otimes\ \vec{B}(t) \neq 0 \quad \sum \vec{m}_{eo} \neq 0\ \otimes$

Figure 7. Principle of magnetization process

The interaction between the magnetic field and a substance can also be described in several ways; the most applied, similar to the case of the electric field, is introducing the permeability of a material as the material's characteristic.

As in the case of the interaction between the electric field and the material, the material permeability is defined as the product of material relative permeability, and the permeability of the vacuum,

$$\mu = \mu_r \mu_0 . \tag{14}$$

Comparing the electric and magnetic material characteristics, it should be emphasized that the relative permittivity is always larger than 1, while the relative permeability can be less, equal or larger than 1.

It should also be noticed that all described interactions between the material and electromagnetic fields exist in time constant or time varying electromagnetic fields, acting on steady or moving charges.

3.3. Interaction of time varying electromagnetic field and the substance

In the case of time fast varying electromagnetic fields, even the lightest charged particles, like electrons, cannot follow the electromagnetic fields changing in synchronism; however, there is always some retardation. During the moving charged particles direction changing, due to the electromagnetic field vectors direction changing, a part of energy transfers to heat. These electrical energy losses are called polarization losses and the typical example of this transfer happens inside microwave ovens.

Similar behavior of elementary charged particles, exposed to a time varying magnetic field, can be noticed in ferromagnetic or ferrimagnetic materials. The electrons' orbits also cannot follow the magnetic fields changing in synchronism, so more or less significant retardation could be expected, defining the so-called hysteresis losses.

3.4. Joule's effect - heating losses

In all conducting materials with the current, except in the superconductivity cases, the so-called Joule's effect appears. Namely, during the moving of charged particles through the conductors' structure, the successive collisions of these particles with steady (uncharged) or near-steady (heavy ions) particles appear, as shown in Figure 8.

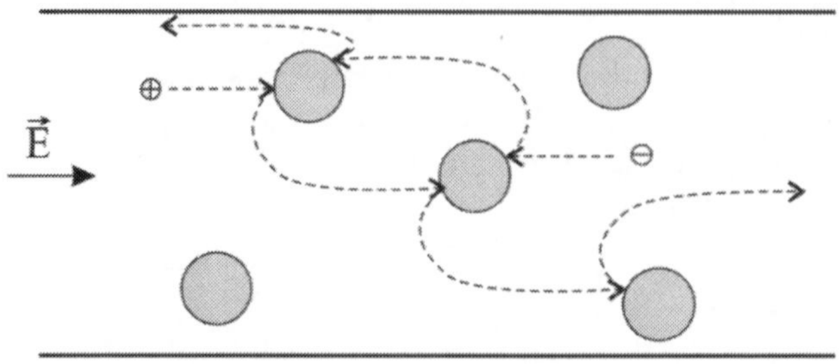

Figure 8. Principle of Joule's effect

In each collision, a part or the entire kinetic energy of the moving particle is delivered to a steady particle, transferring into additional thermal oscillations of the steady particle. The unbound particle is still in the electric field and continues to move in the electric field strength vector direction (positively charged particle) or in the opposite direction (negatively charged particle).

Considering the energy necessary for moving, the free charged particle takes it from the electric field and this is the mechanism of transferring the electromagnetic field energy into heat. This effect is present in the real, non-superconductive conductor in the cases of DC or AC currents.

This might be the right place to emphasize that the energy transferred from the electromagnetic field energy to heat is always permanently lost for the electromagnetic field.

4. Electromagnetic induction

Except for the effects described previously, and concerning the interaction between any electromagnetic field and any substance appearing in DC and AC fields, in time varying magnetic fields another effect, known as the electromagnetic induction, may appear. This effect is probably the most important effect in the entire electrical engineering theory and practice.

If any structure is placed in a time varying magnetic field, an additional electric field, denoted as the induced electric field, is produced, followed by the so-called induced electromotive force, defined as [4],

$$e_{\text{ind}}(t) = \int_C \vec{E}_{\text{ind}} \cdot \mathrm{d}\vec{\ell}. \tag{15}$$

The induced electromotive force can be the consequence of the time varying magnetic field acting on a solid, steady system. In this case, the effect is called the static electromagnetic induction, and the static induced electromotive force is produced by the induced electric field strength vector, defined as,

$$\vec{E}_{\text{ind}} = -\frac{\partial \vec{A}}{\partial t}. \tag{16}$$

The other possibility of the induced electromotive force appearing is the situation when a system moves or/and deforms in the time constant magnetic field. This case is known as the dynamic electromagnetic induction, and the dynamic induced electromotive force is produced by the induced electric field strength vector defined as,

$$\vec{E}_{\text{ind}} = \vec{v} \times \vec{B}, \tag{17}$$

where $\vec{v}$ is the velocity of the moving system.

Naturally, the induced electromotive force can be the consequence of combined effects, moving or/and deforming system in the time varying magnetic field.

The induced electromotive force has the character of voltage and it superposes to all voltages and potentials produced by other sources. This means that the induced electromotive force has an influence on all electric and electronic equipment in vicinity, as well as on all living organisms in which, as it is well known, the electric potentials exist as well.

If the time varying magnetic field and, consequently, the induced electric field and induced electromotive force appear inside the conductive material, induced currents will occur as well. These currents, known as eddy currents or Foucault's currents, superpose other existing currents, changing the previous current distribution. The non-uniform current distribution, due to eddy currents inside a single conductor, is known as the **skin effect** and the non-uniform current distribution, due to the time varying magnetic field produced by a neighboring conductor known as the **proximity effect**. Eddy currents are also followed by Joule's heating losses, depicted as **eddy current losses**. Moreover, as any other currents, eddy currents produce the magnetic field, which superposes the already existing time varying magnetic field. As the consequence, again a non-uniform magnetic field distribution appears inside any magnetic circuit; this non-uniform distribution is also called the **skin effect**.

The significance of the skin effect and proximity effects depends on how fast magnetic field time variations are (or on frequency in the case of periodical time varying fields), as well as on the conductivity and the permeability of the material in which the system is placed.

5. Electromagnetic fields we are mostly exposed to

5.1. Outdoor exposure

Most of all electromagnetic fields that we might encounter are positioned outdoor.

There are many possibilities of the exposure to ELF (Extremely Low Frequencies), mostly to the industrial frequency of 50 Hz (60 Hz), produced by the electrical energy transmission and distribution systems, including the corresponding transformers. Some photos of these examples are demonstrated in Figure 9.

Figure 9. Some examples of outdoor ELF electromagnetic fields sources

The other kinds of electromagnetic fields existing outdoor are high frequency electromagnetic fields, mostly RF (Radio Frequency) electromagnetic fields. These fields are produced by radio or television broadcasting emitting antennas and, nowadays, particularly by mobile phones base stations. Some of these sources are presented in Figure 10.

Figure 10. Some examples of outdoor RF electromagnetic fields sources

5.2. *Indoor exposure*

Inside our living and working places, we can be exposed to both ELF and RF electromagnetic fields. Considering ELF electromagnetic fields, the fields are mostly at 230 V, meaning that the magnetic field is much more significant comparing to the electric field. This implies that the fields are produced mostly by powerful apparatuses, such as all kind of heaters, as shown in Figure 11.

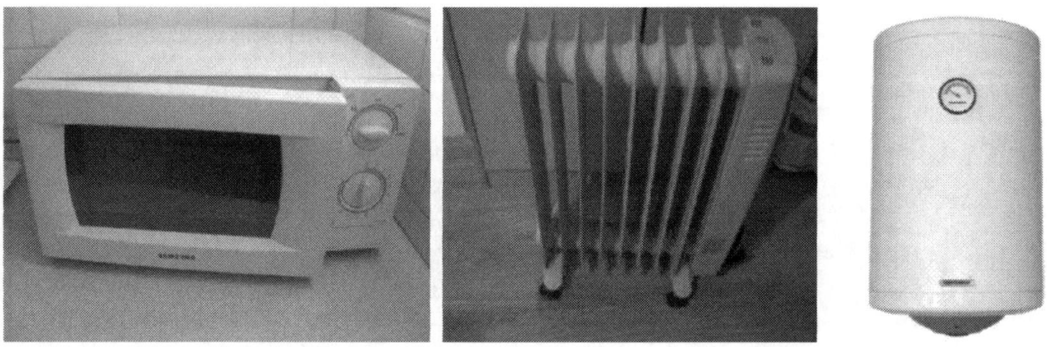

Figure 11. Some examples of indoor ELF (electro)magnetic fields sources

Some of domestic apparatuses could also produce some indoor HF (High Frequency) electromagnetic fields, like microwave ovens or CRT televisions, but only if they are not well functioning.

5.3. *Mobile (Cellular) phone*

Finally, outdoor and indoor, always present, always with us, our necessity, our mobile (cellular) phone is also always the source of HF electromagnetic field, frequently, dangerously very close to our head or our heart.

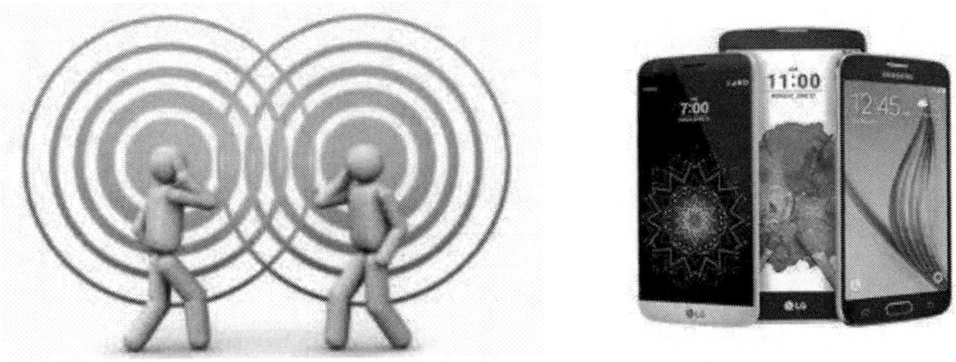

Figure 12. Always with us, our mobile (cellular) phone

Although today the mobile phone is an almost inevitable part of our lives, it must be applied with certain precaution. Namely, the electric field strength vector magnitude and the magnetic flux density vector magnitude are increasing with the decreased distance from the field source (mobile phone) and the point in which the electromagnetic field is observed (point in our brain) as $1/r^2$, while the radiated electromagnetic wave power increases as $1/r^4$. During the conversation via a mobile phone, the distance between the phone and our brain is very small and the long term consequences could be significant. For this reason, in order to avoid health risks, it is advisable that the conversation using a mobile phone does not last too long. Likewise, an application of headphones or wireless equipments, like "Bluetooth", is advisable as well.

6. Conclusion

As it could be noticed from everything mentioned above, all electromagnetic fields exert influence on everything positioned inside a domain in which a field exists. With the growing industrialization, the electric energy production and consumption increases, and the electromagnetic fields' intensities follow this increase, together with all negative effects produced by the electromagnetic fields. Moreover, nowadays some new fields, which did not exist previously, appear and are applied worldwide. The typical example of that are the application of microwave ovens and mobile phones, as well as the applications of electromagnetic fields in medical diagnostics and therapies.

The protection against undesirable electromagnetic fields, such as undesirable radiations of technical systems or the protection of certain domains has become our necessity. The part of contemporary science dealing with these problems is EMC (Electromagnetic Compatibility), developing toward an interdisciplinary approach. A closed collaboration of biologists, medical doctors, physicists and electrical engineers has become inevitable in order to explore in details all electromagnetic fields' impacts. As a solution of the investigations, the harmful impacts should be eliminated or reduced to the level as small as possible, while the useful impact should be applied in the best possible manner.

According to all this, it is very important to be well acquainted with the electromagnetic fields in our environment, as well as the effects produced by them on electrical or electronic equipments or on living beings. Electromagnetic fields' influence on electric or electronic systems should be an occupational task for electrical engineers, while the electromagnetic fields' impact on living beings should bring together the team work of the experts in different fields.

References

[1] ***Available at http://www.esa.int/spaceinimages/Images/2015/09/Spacetime_curvature
[2] ***Available at https://www.maplesoft.com/view.aspx?SI=3707/electricDipole5.gif

[3] Popović B D 1986 Electromagnetics, (in Serbian), Second Edition, Građevinska knjiga, Belgrade, Serbia
[4] Prša M 2016 *Electromagnetics*, (in Serbian), Prepared for printing, Novi Sad, Serbia
[5] Surutka J 2006 *Electromagnetics*, (in Serbian), Eighth Edition, Akademska Misao, Belgrade, Serbia
[6] Stojaković M 2001 *Mathematical Analysis 2 – Second Part*, (in Serbian), Novi Sad, Edition of Faculty of Technical Sciences, Novi Sad, Serbia
[7] Šunjarević M 2004 Outlines of radio communications with radio techniques, (in Serbian), pp 30, Studio LINE, Belgrade, Serbia

Studies on gamma irradiated rubber materials

I B Lungu[1], M D Stelescu[2] and M Cutrubinis[1]

[1]Horia Hulubei National Institute for R&D in Physics and Nuclear Engineering, IRASM department, Str. Reactorului Str, no.30, p.o.box MG-6, Magurele, Romania
[2]National R&D Institute for Textile and Leather – Leather and Footwear Research Institute, 93 Ion Minulescu St, Bucharest, Romania, tel. 040 21 323 50 60, fax. 040 21 323 52 80

E-mail: dmstelescu@yahoo.com

Abstract. Due to the increase in use and production of polymer materials, there is a constant pressure of finding a solution to more environmental friendly composites. Beside the constant effort of recycling used materials, it seems more appropriate to manufacture and use biodegradable and renewable row materials. Natural polymers like starch, cellulose, lignin etc are ideal for preparing biodegradable composites. Some of the dynamic markets that use polymer materials are the food and pharmaceutical industries. Because of their desinfastation and sometimes sterility requirements, different treatment processes are applied, one of it being radiation treatment. The scope of this paper is to analyze the mechanical behaviour of rubber based materials irradiated with gamma rays at four medium doses, 30.1 kGy, 60.6 kGy, 91 kGy and 121.8 kGy. The objectives are the following: to identify the optimum radiation dose in order to obtain a good mechanical behaviour and to identify the mechanical behaviour of the material when adding different quantities of natural filler (20 phr, 60 phr and 100 phr).

1. Introduction

The elastomeric materials have a wide spread of usage, including medical, pharmaceutical and food domain. They are used in manufacturing of seal gaskets, vibration cushioning, protection equipment, medical devices, etc. [1], [2]. Beside the constant effort of recycling used materials, it seems more appropriate to manufacture and use biodegradable and renewable row materials. Natural polymers like starch, cellulose, lignin, etc. are ideal for preparing biodegradable composites. Lately, the beneficiary market and also the producers became more interested in using biodegradable materials.

In this paper it is studied the effect of ionizing gamma radiation over natural rubber materials regarding its mechanical behaviour. Natural rubber is the most ancient and important elastomer which was the model of the first synthesis trials of synthetic rubbers. Natural rubber is widely used in tire industry, and, at least by now, it cannot be replaced entirely with other type of synthetic rubber [2], [3].

In order to obtain products, are used mixes of rubber which include batches, cross-linking agents, plasticizers, oxidants etc. The most important step in the manufacturing technology of rubber is represented by vulcanization or cross-linking. Vulcanization represents the process in which the rubber is transformed from a plastic material in an elastic material. Conventionally, it is given the term "cross-linking" (vulcanization) to all processes in which there are created bridges between macromolecules.

There are multiple possibilities of cross-linking rubber materials. Sulphur vulcanization was the first method discovered for cross-linking and by now is the most known. Other cross-linking systems, like the ones that uses peroxides, ultra-violet radiation, resins etc were discovered later and became important along with the progressive development of synthetic rubbers [2-5].

The use of gamma irradiation for cross-linking and grafting of elastomers has become more attractive along with the constant request for more cheaper, durable and resistant materials. The process is very clean, the energy consumption is small, permits high processing speeds than in other processes of curing and operates at ambient temperature [6]. Numerous polymers were studied and characterized regarding their behaviour to radiation processing. Today, this technology is very well established and used in almoust every industries. The purpose of using radiation processing is to enhance the materials mechanical, chemical and temperature resistance in different environments and also to eliminate the cross-linking agents.

Regarding the use of this materials in various industries and the global tendencies in using eco-friendly and biological materials, the scope of this study to reduce and/or eliminate toxic and carciogenic substances. So, the inorganic batch, like silicea precipitate and carbon black, which are carciogenic [7-9], were replaced with an organic, non-toxic component. Nevertheless, the material must be able to comply with requirements in the form of mechanical characteristics, both in ready to use material and also when treated with gamma radiation.

The objectives of this study are: to identify the optimum radiation dose in order to obtain a good mechanical behaviour and to identify the mechanical behaviour of the material when adding different quantities of natural filler (20 phr, 60 phr and 100 phr).

2. Experimental

2.1. Materials

The materials used in this experiment are the following: natural rubber, vulcanizing agents, batch, plasticizers, compatibilizer and other ingredients.

The following natural rubber was used: Crep from Sangtvon Rubber Ltd, in the form of white rubber sheets, Mooney viscosity 67.64 ML (1 '+ 4') 100°C, volatile matter content of 0.5%, nitrogen content 0.45%, ash content of 0.25%, impurity content of 0.026%.

The vulcanisation of NR by sulfur in presence of organic accelerators which lead to elimination of nitrosamines was replaced with another cross-linking and less toxic system, in this case, gamma radiation processing. The process was performed using low quantities of cross-linking agents. The following were used: perkadox 40 benzoyl peroxide (density 160 g/cm^3, 3.8% active oxygen content, 40% peroxide content, pH 7) and polyfunctional monomer trimethylolpropane trimethacrylate Luvomaxx TMPT DL 75 (TMPT) (22 % ash, pH 9.2, density 1.36 g/cm^3, 75 $\pm$ 3 % active ingredient).

The mixes contain the following proportions: 0 phr (parts per 100 parts of rubber), 20 phr (parts per 100 parts of rubber), 40 phr and 100 phr of organic filler.

The organic filler have special advantage in comparison to inorganic filler, but are hydrophilic. This leads to poor interface between organic filler and hydrophobic matrix of elastomer. In order to improve the interfacial interactions between rubber and organic filler, addition of compatibilizers is required [10], [11] in the case of maleated natural rubber (NR-g-AM). The compatibilizing agent was obtained by roll mixing NR with 5 phr (parts per 100 parts of rubber) of maleic anhydride for synthesis (produced by Merck KGaA, Germany, melting point 52°C) and 0,75 phr of Perkadox 40. The resulting mixture was kept at a temperature of 160°C for 30 minutes and then used as such.

Other ingredients used were the following: (a) antioxidant Richon IPPD (4010 NA) N-isopropyl - N-phenyl - phenylene diamine, 98% purity, molecular mass 493.6374; (b) plasticizer – glycerine, free acidity 0.02%, density 1.26 g/cm^3, purity 99.5%, (c) polyethylene glycol PEG 4000 (1.128 g/cm^3 density, 4-8°C melting point range), (d) zinc oxide and (f) stearic acid.

2.2. Sample preparation

Sample preparation was carried out at National R&D Institute for Textile and Leather – Leather and Footwear Research Institute –Rubber Department.

The blends were carried out using a laboratory two-roll mill machine. Formulations are presented in Table 1. First, the NR and NR-g-AM was added into two roll mills. The temperature of this machine was between 50-70°C. Afterwards, antioxidant, organic filler, plasticizer, zinc oxide, stearic acid, PEG 4000 and lastly the vulcanizing agent (peroxide and TMPT) were added into the compounding. The time taken to complete the entire process was about 18 to 30 minutes. Nip gap, the speed roll mill and the sequence of adding ingredients were kept constant for all formulations.

Table 1. Formulations

Mixture symbol	**N-20 (phr)**	**N-60 (phr)**	**N-100 (phr)**
Natural rubber	95	95	95
NR-g-AM	5	5	5
Organic filler	20	60	100
Plasticizer	10	30	50
Peroxide	2	2	2
TMPT	3	3	3
Zinc oxide	1	1	1
Stearic acid	0.1	0.1	0.1
PEG 4000	3	3	3
Antioxidant 4010	1	1	1

The test specimen sheets of all compounds were produced using compression moulding at 165°C, 5 minutes duration and pressure of 300 MPa.

2.3. Sample iradiation treatment

The radiation treatment was carried out at IRASM department using the SVST Co60 gamma irradiator. The materials were irradiated at four medium doses of 30.1 kGy, 60.6 kGy, 91 kGy and 121.8 kGy. For the first medium dose (30.1 kGy), the minimum/maximum values were 29.3 kGy respectively 30.9 kGy with a dose uniformity ratio of 1.054. For the second medium dose (60.6 kGy), the minimum/maximum values were 58.8 kGy respectively 62.3 kGy with dose uniformity ratio of 1.059. The third medium dose (91 kGy), the minimum/maximum values were 88.9 kGy respectively 93 kGy with dose uniformity ratio of 1.046. The forth medium dose (121.8 kGy), the minimum/maximum values were 119.2 kGy respectively 124.4 kGy with dose uniformity ratio of 1.043. The dose debit was approximately 2.3 kGy/hour. The radiation treatment was carried out at room temperature.

2.4. Mechanical testing

The mechanical tests were performed on a Zwick Roell Z005 testing machine equipped with a 5 kN cell force. The tests were conducted according to ISO 37 with a constant load speed of 200 mm/min and all measurements were repeated for minimum five times. The samples were cut in a bone shape specimen, type 4: 35 mm in length, 6 mm shoulder width, 12 mm gage length, 2 mm specimen width.

Tearing strength tests were carried out with a Schopper strength tester with testing speed 460 mm/min, using angular test pieces (Type II) in according to SR EN 12771/2003.

Hardness was measured by using a hardener tester according to ISO 7619-1/2011 using 6-mm thick samples.

Elasticity was evaluated with a test machine of type Schob using 6-mm thick samples, according to ISO 4662/2009.

3. Results and discussions

The following mechanical tests were carried out: tensile strength determination, elongation, hardness, elasticity and tearing strength.

3.1. Tensile strength

Figure 1 shows the dependence of the tensile strength depending on the amount of organic filler introduced in the mixture and irradiation dose. Complementary to the mean value, there are also shown the values for error of the mean.

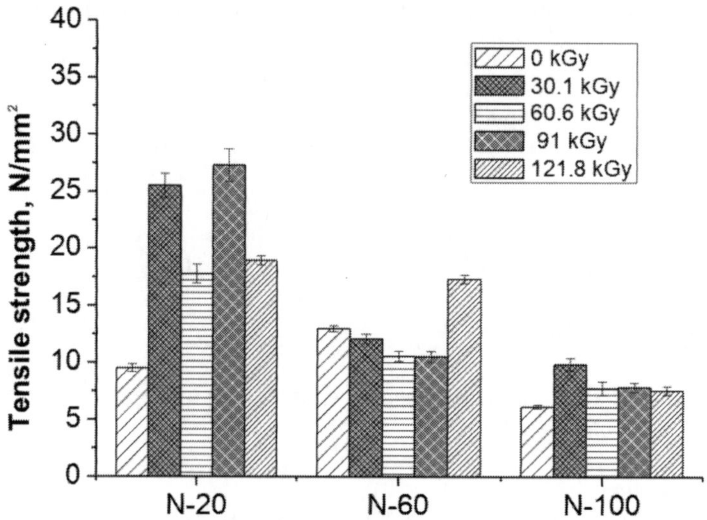

Figure 1. Tensile strength variation depending on the amount of organic filler introduced in the mixture and irradiation dose

The highest values are obtained for N-20 mix with two peaks, at 30.1 kGy and 91 kGy while N-100 obtained the lowes values, its hihest peak being obtained at 30.1 kGy. For N-60, the hihest peak value was obtained at 121.8 kGy. Per all, the tensile strength shows an increase in value to a maximum value which differs from the irradiation dose and the quantity of filler introduced in the mixes and after it decreases. This behaviour is consistent with reports from the scientific literature regarding tensile strength variation depending of the cross-linking degree [12-14].

3.2. Elongation at break

Figure 2 shows the dependence of elongation at break depending on the amount of organic filler introduced in the mixture and irradiation dose.

Elongation is most effected for N-20 mix, which scored higher at the tensile strength. For N-60 and N-100, the radiation dose does not effect significantly the elongation behaviour. This information is used when elongation is regarded as a signifiant parameter for the use of the material.

3.3. Hardness

Figure 3 shows the dependence of elongation at break depending on the amount of organic filler introduced in the mixture and irradiation dose.

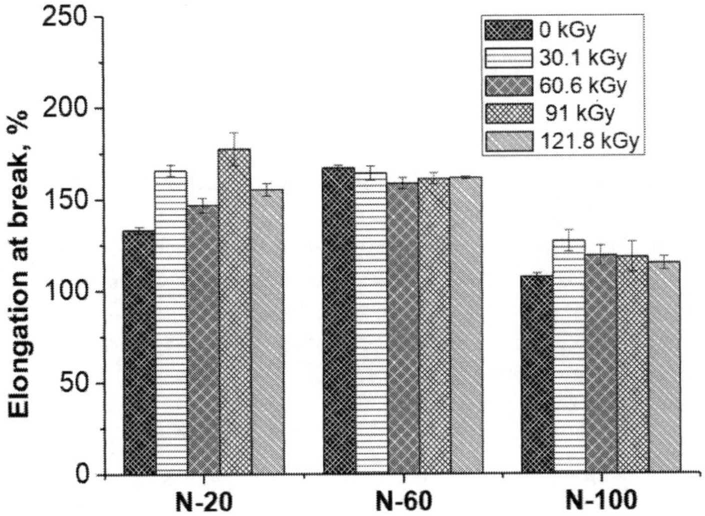

Figure 2. Elongation at break variation depending on the amount of organic filler introduced in the mixture and irradiation dose

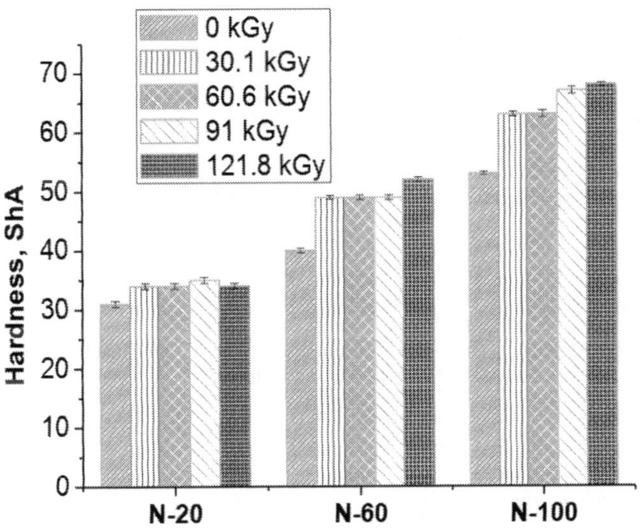

Figure 3. Hardness variation depending on the amount of organic filler introduced in the mixture and irradiation dose

It can be observed that increasing the filler quantity with 40 phr organic filler, from N-20 to N-60 leads to an increase in hardness between 29-53%, respectively 71-100% when adding 80 phr organic filler, from N-20 to N-100.

This means that organic filler leads to reinforcement of NR, therefore it can replace active fillers such as carbon black or precipitated silica which are harmful and may lead to occupational illnesses of employees of the rubber processing industry [7], [8].

Using radiation processing there is a 4-15°ShA increase in hardness due to the increase of cross-linking.

3.4. Elasticity

Elasticity variation depending on the amount of organic filler introduced in the mixture, respective irradiation dose is presented in Figure 4.

Elasticity tends to decrease along with the addition of the organic filler. An addition of 40 phr organic filler, from N-20 top N-60 decreases elasticity between 16.6-25% and an addition of 80 phr organic filler from N-20 top N-100 decreases elasticity between 18.23-38.5%.

Radiation processing using the above doses determines, per all, a decrease till 15.4% which can be due to modification of materials' surfaces or cross-linking.

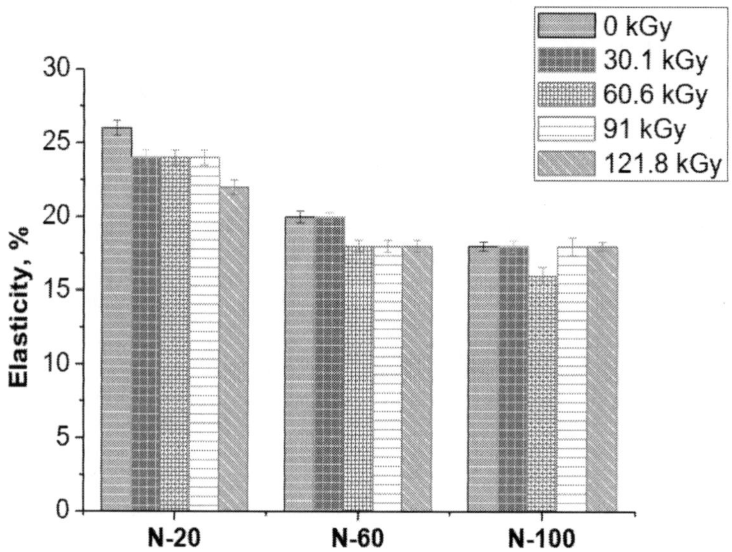

Figure 4. Elasticity variation depending on the amount of organic filler introduced in the mixture and irradiation dose

3.5. Tearing strength

Tearing strength variation depending on the amount of organic filler introduced in the mixture, respective irradiation dose is presented in Figure 5.

It can be observed that the tearing strength decreases along with the addition of 40 phr organic filler, from N-20 to N-60, between 26.9-38.5% and 23.1-61.5% with the addition of 80 phr organic filler from N-20 to N-100. Radiation processing induces an non-uniform variation in tearing strenghts, with tendences of ageing. These results follow the characteristics needed in a wide range of applications.

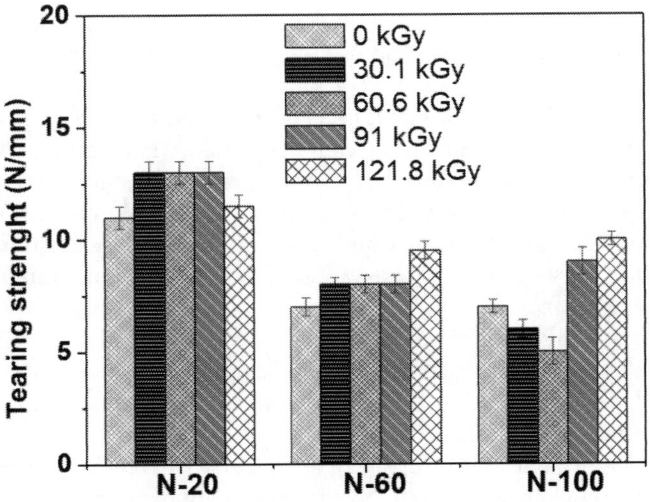

Figure 5. Tearing strength variation depending on the amount of organic filler introduced in the mixture and irradiation dose

4. Conclusions

It has been shown that gamma radiation processing for natural rubber mixed with different quantities of organic fillers has a positive effect on the material. The mechanical behaviour varies between accepted intervals of values needed for various industries, esspecialy for medical and pharmaceutical markets where radiation processing is mostly used. An important aspect showed by the results is that relatively low doses of radiation increase some of materials' characteristics like, tensile strength, elasticity and tearing strength starting from 30.1 kGy for N-20. Increase in hardness is achieved for N-100 starting also from low doses like from 30.1 kGy.

Independent of application, a good result can be considered when using a low radiation dose, which can be translated in shorter processing time and low cost and high quantity of natural material which makes the final product more eco-friendly and less toxic for manufacturing workes.

Acknowledgements

The material, sample preparation and determination of hardness, elasticity and tearing strength were financed through Nucleu Program 2016-2017 PN 16 34 01 01 supported by Romanian Ministry of Education and radiation processing, tensile strength and elongation tests were financed through Nucleu Program Nucleu Program PN 16 42 06/2016 supported by Romanian Ministry of Education.

References

[1] Ikarashi Y, Toyoda K, Ohasawa N, Uchima T, Tsuchiya T, Kaniwa M A, Sato M, Takahashi M and Nakamura A 1992 Comparative studies by cell culture and in vivo implantation test on the toxicity of natural rubber latex materials, *J. Biomed. Mater. Res.* **26** 339-356

[2] Manaila E, Craciun G, Stelescu M D, Ighigeanu D and Ficai M 2014 Radiation vulcanization of natural rubber with polyfunctional monomers, *Polym. Bull.* **71** 57-82

[3] Gonzalez L, Rodriguez A, Marcos-Fernandez A and Chamorro C 1998 A new interpretation of the crosslinking mechanism of NR with dicumyl peroxide, *Kautsch. Gummi Kunstst.* **51** 83-87

[4] Gonzalez L, Rodriguez A, Valentin J L, Marcos-Fernandez A and Posadas P 2005 Conventional

and efficient crosslinking of natural rubber: Effect of heterogeneities on the physical properties, *Kautsch. Gummi Kunstst.* **58** 638-643

[5] Stelescu M D, Manaila E, Craciun G and Dumitrascu M 2014 New green polymeric composites based on Hemp and Natural Rubber Processed by Electron Beam Irradiation, *Sci. World J.* 2014

[6] Manaila E, Stelescu M D, Craciun G and Ighigeanu D 2016 Wood Sawdust/Natural Rubber Ecocomposites Cross-Linked by Electron Beam Irradiation, *Materials* **9** 503

[7] ***IARC Monographs on the Evaluation of Carcinogenic Risks to Humans: Silica, Some Silicates, Coal Dust and Para-Aramid Fibrils, 1997 World Health Organization, International Agency for Research on Cancer, World Health Organization: Lyon, France, Vol. 68, Available online: http://monographs.iarc.fr/ENG/ Monographs/vol68/index.php (accessed on 18 September 2014)

[8] Beliczky L S and Fajen J 1998 Rubber industry. In Encyclopaedia of Occupational Health and Safety, 4th ed., Stellman JM, Ed. International Labor Office: Geneva, Switzerland, 80, pp 1

[9] Manaila E, Stelescu M D and Doroftei F 2015 Polymeric composites based on natural rubber and hemp fibers, *Iran. Polym. J.* **24** 135-148

[10] Pracella M, Haque M U and Alvarez V 2010 Functionalization, Compatibilization and Properties of Polyolefin Composites with Natural Fibers, *Polymers* **2** 554-574

[11] Yu L, Dean K and Li L 2006 Polymer blends and composites from renewable resources, *Prog. Polym. Sci.* **31** 576-602

[12] Manaila E, Craciun G, Martin D, Ighigeanu D and Zuga Stelescu D 2011 Chapter *EB and MW processing of rubber mixtures with MPFs*, pp 199-212, book *Practical Aspects and Applications of Electron Beam Irradiation, Rditors Nemteanu M, Brasoveanu M,* Published by Transworld Research Network, Kerala, India

[13] Manaila E, Stelescu M D and Craciun G 2012 *Chapter 1 Aspects Regarding Radiation Crosslinking of Elastomers", book - Advanced Elastomers – Technology, Properties and Applications*, Edited by Anna Boczkowska, Ed. In Tech., Croatia

[14] Stelescu M D, Manaila E, Martin D, Craciun G, Ighigeanu D and Alexandrescu L 2011 *New Technologies of Grafting and Cross-Linking Rubber Blends by Electron Beam and Microwave Irradiation*, Performantica Publishing House, Iasi, Romania

Aspects regarding manufacturing technologies of composite materials for brake pad application

A L Craciun, T Heput and C Pinca-Bretotean

Politehnica University of Timisoara, Engineering and Management Department, 5 Revolutiei Street, 331128 Hunedoara, Romania

E-mail: camelia.bretotean@fih.upt.ro

Abstract. Current needs in road safety, requires the development of new technical solutions for automotive braking system. Their safe operation is subject to following factors: concept design, materials used and electronic control. Among the factors previously listed, choice of materials and manufacturing processes are difficult stage but very important for achieving technical performance and getting a relatively small cost of constituting parts of brake system. The choice is based on the promotion of organic composite material, popular in areas where the weight of materials plays an important role. The brake system is composed of many different parts including brake pads, a master cylinder, wheel cylinders and a hydraulic control system. The brake pads are an important component in the braking system of automotive. These are of different types, suitable for different types of automotive and engines. Brake pads are designed for friction stability, durability, minimization of noise and vibration. The typology of the brake pads depends on the material which they are made. The aim of this paper is to presents the manufacturing technologies for ten recipes of composite material used in brake pads applications. In this work will be done: choosing the constituents of the recipes, investigation of their basic characteristics, setting the proportions of components, obtaining the composite materials in laboratory, establishing the parameters of manufacturing technology and technological analysis.

1. Introduction

Current trend in road safety impose the development of new technical solutions for vehicle „vital" systems. These systems are made of components with multiple functions inside the vehicle. The optimal operation of vehicle is ensured as part of each subassembly and the links between them. Each subassembly is made up of several parts that can be made from a wide variety of materials. Technical performance of the vehicle subassemblies and default of which they are part, depend on physical and mechanical characteristics of the materials which they are made, [1].

Many parts of vehicles contributes active and passive to increased operating performance. Safety is closely linked to the effectiveness of the braking system, which is one of the most important components of a vehicle. On the other hand, the correct operation of braking systems is essential for the safe transport of goods and people, [2]. In the case of brake components, the safety in operation is conditioned by three factors, [1]: concept design, materials used and electronic control. Among the factors previously listed, choice of materials and manufacturing processes are difficult stages to achieve technical performance of vehicles. The choice of materials is based on the promotion of low-cost, which will be readily purchased, being carried out the optimum use of their technological properties, [2]. On background of sustainable raw material resources and environmental problems caused by plastics and metals, readily degradable, automotive manufacturers are always looking for new materials with low environmental impact. These materials must provide the same performance in function, as the metals but will be produced as friendly is possible. Due to the environmental problems caused by the large volume of waste, the world is moving towards organic composite materials. These are popular in the fields in which the weight of the material plays an important role, such as automotives, [3].

Regarding manufacturing technologies of brake pads, the literature indicates that there are government rules that restrict the use of materials in the brake pads applications, [3]. In this context, current issues of research in materials science and engineering aims to develop new materials with superior performance to meet the requirements of modern industry, simultaneously with the application of technology for the processing of maximum economic efficiency.

This paper presents the manufacturing technologies for ten recipes of composites used in brake pad applications. In this work will be done: choosing the constituents of the recipes, investigation of their basic characteristics, setting the proportions of components, obtaining the composite materials in laboratory, establishing the parameters of manufacturing technology and technological analysis.

2. Technical Requirements

2.1. Design and fabrication of mold for producing composites

To produce composite materials for brake pads, it was designed and built a metallic mold. The main components of the mold were made with Autodesk Inventor and are shown in Figure 1a-e. Figure 2 shows the mold assembly designed to produce the samples in laboratory, [4].

a-Motherboard b-Nut c-Extraction disc d-Disc guide e-Rop plate

Figure 1. Components of mold for making samples

Figure 2. Mold assembly

2.2. Establishing the proportion of constituents

In the literature, the main categories of materials used in organic brake pads are: filler, abrasive, solid lubricant, binder, friction modifier and additives, [3]. The composition of organic composites for different components of vehicles, a great importance are lingo-cellulosic fibres. This are used to reinforcement composite materials and can be obtained from a wide variety of plants, such as, [5]: cereal straw stalks (wheat, barley, rice, rye); fibre plants (flax, hemp, raffia, cotton); plants with tough fibres (cane sugar, cane, reed, oilseed rape, coconut, bamboo, banana, sisal, etc.). In this paper, was chosen as a filler and reinforcement material, coconut fibre.

In order to establish prescriptions was started from a mixture of raw materials, including: aluminium, graphite, zirconium oxide, silicon carbide, titanium oxide, phenolic resin and coconut fibre. Their choice was made in accordance with the properties required for brake pads in vehicle operation. The properties of each material that is independent of the recipe are further analyzed, [6]:

- aluminium forming alloys and it has a high plasticity and low specific weight. It is used as a matrix because of its low density and high resistance to progressive oxidation;

- zirconium oxide is a material with high resistant. It is used to achieve the components which are resistant to thermomechanical stresses. It shows a very good resistance to abrasion and impact, it has high compressive strength, low density and porosity;
- silicon carbide has a crystalline structure which gives high rigidity, tensile strength and high hardness values. Add in aluminum alloys contributes to increased wear resistance, in particular to abrasive wear, but the mechanical processing will lead to intense wear of cutting tools. It has high hardness which gives a proper behavior of the composite under conditions of friction wear or abrasion.
- graphite introduced in composites in which is added silicon carbide improves the anti-friction properties, leading to a low friction coefficient. Due to the difference in specific gravity between the matrix and graphite to obtain a homogeneous mixture is required mechanical mixture.
- titanium oxide is stable and has a low density. Titanium gives superior mechanical strength, corrosion resistance in marine environments, acids, alkali, water, natural and industrial chemicals. It has high heat transfer in operating conditions and a superior oxidation resistance at ambient temperature. Titanium may be alloyed with aluminium, vanadium, molybdenum, and other elements, to produce light alloy and mechanical high strength;
- phenolic resin has excellent high temperature properties and form carbon-carbon bonds that develop resistance to high temperatures. This is thermosetting and provides unique properties of composite materials, such as: chemical resistance and superior mechanical characteristics. It is used in applications that support high operating temperatures.

On the other side, the composition of the material which contain silicon carbide and aluminium have high rigidity, resistance to wear and to high temperatures. These are used in manufacturing of braking systems components, [6].

Aluminium used in the preparation of composites was obtained by cutting the pieces presented in Figure 3a. In Figure 3b it is shown aluminium after the cutting process. Phenolic resin used in new developed composite materials is fluid, Figure 4, and coconut fibre was obtained by grinding the shells of fruit, Figure 5. In Figures 6a, b, c, d are shown other materials used to obtain composites for brake pads.

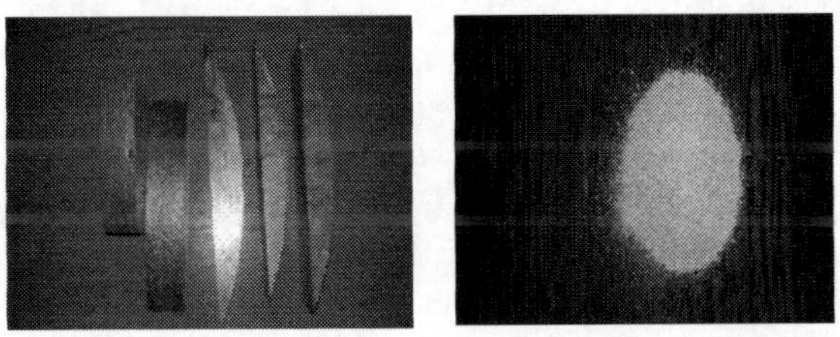

Figure 3. Aluminium used in the manufacture of composites

Figure 4. Phenolic resin

Figure 5. Coconut fibre

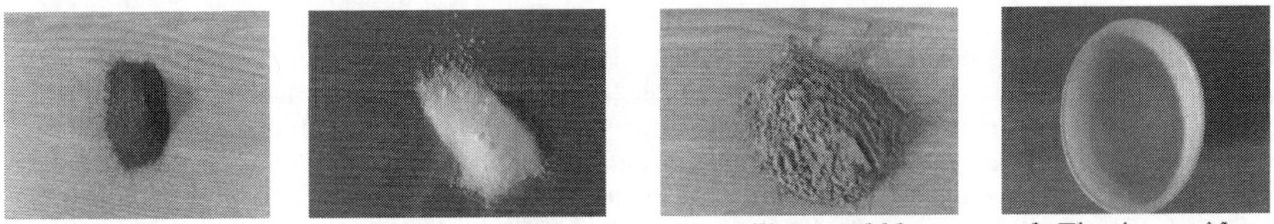

a-Graphite b- Zirconia oxide c- Silicon carbide d- Titanium oxide

Figure 6. Materials used in the manufacturing technologies of composites for automotive brake pads

Table 1 shows ten recipes for composite materials used in vehicles brake pads.

Table 1. Composite samples recipes (P1-P10)

No. of sample	Aluminium (%)	Graphite (%)	Zirconia oxide (%)	Silicon carbide (%)	Titanium oxide (%)	Phenolic resin (%)	Coconut fibre (%)
P1	45	8	2	15	13	13	2
P2	30	8	2	15	13	30	2
P3	20	10	2	15	14	40	0
P4	15	8	2	15	13	40	2
P5	20	8	2	10	10	45	2
P6	20	10	2	13	10	40	5
P7	25	10	2	10	13	40	0
P8	20	10	2	10	13	40	5
P9	15	10	2	10	13	40	10
P10	10	10	2	10	13	40	15

2.3. Experimental equipement

In the manufacturing technologies of organic composite materials was used powder metallurgy. Thus, the materials used in recipes are prepared in powder form, [7]. In this regard, they are milled in a ball mill, type PM100, which is shown in Figure 7. Technical characteristics of the mill are: ball diameter: 8 mm, the initial particle size of the material: < 10 mm; the final particle size: < 40 µm, rotor speed: 6.000 rot/min, rotor diameter: 99 mm and capacity: 900 ml.

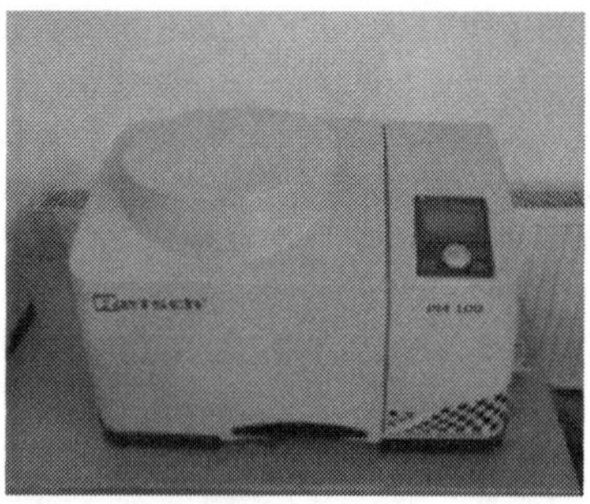

Figure 7. The ball mill machine type PM 100

Figure 8. Sartorius balance CP2202S-OCE

Figure 9. Oven type Nabertherm model L15/12/B180

After milling, the components were weighed in proportions according to the recipes. Their weighing was performed by a Sartorius balance, type CP2202S-OCE, which isshown in Figure 8, [8]. Technical characteristics are: minimum quantity of measuring 0,5g; maximum quantity of measuring of: 2200 g; precision: 0,01g. The mixture was placed in the mold and the assembly mold-sample was placed in an oven, Nabertherm mark, model L15/12/B180. Technical characteristics of the oven are: the maximum heating temperature 1200^{0}C, tension 220V, frequency 60 Hz, amperage 15.7 A, power 3.6KW.

3. Discusion and results

New composite materials developed were obtained in laboratory based on a number of ten recipes with the composition shown in Table 1. Their implementation was based on the thorough documentation in technical literature and many attempts and experiments. Manufacturing technology was improved step by step, from one sample to another, trying to solve the drawbacks and identifying the causes occurred to obtain each sample. Table 2 shows the parameters of manufacturing technology

IOP Conf. Series: Materials Science and Engineering **294** (2017) 012003 doi:10.1088/1757-899X/294/1/012003

for each sample. Technological analysis on the results obtained in terms of compactness, integrity, elasticity and appearance of the sample, [9].

Table 2. Manufacturing technology and the resulting sample

Sample	Manufacturing parameters							Appearance of sample
	Heating temperature (°C)	Heating time (min)	Retention time (min)	Pressing force		Cooling		
				Hot (KN)	Cold (KN)	Medium	Time (h)	
P1	220	15	15	-	-	Water Air	0.15 15	
P2	220	15	25	-	-	air	10	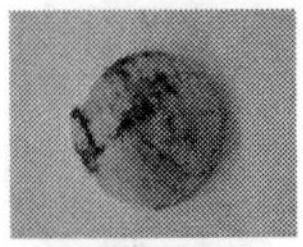
P3	200	15	25	-	-	air	10	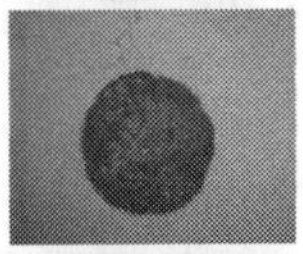
P4	180	15	30	-	-	air	10	
P5	200	15	45	-	-	air	10	
P6	200	15	45	15	15	air	10	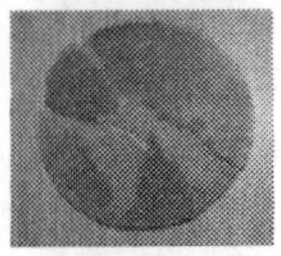
P7	200	15	45	5	5	air	10	
P8	200	15	45	5	5	air	10	
P9	200	15	45	5	5	air	10	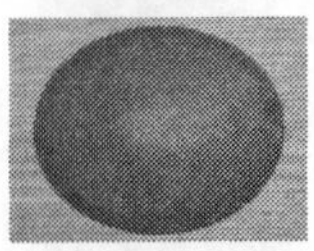
P10	200	15	45	5	5	air	10	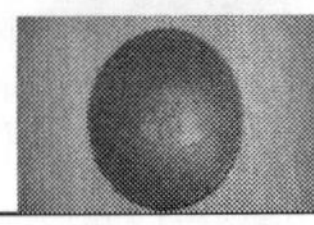

Implementation steps for P1 sample were: mixing the constituents of the ground; the phenolic resin is added; the composition is manually homogenized; the mixture is introducing into the mold; the piston of the mold is take up; the assembly sample-mold is insert in the oven.

Technological analysis of P1: the sample obtained was destroyed when removed it from the mold because of its tendency to stick to the bottom of the mold. This requires the search for a solution to protect the sample. The sample is inadequate in appearance and structure. The explanation is the small amount of phenolic resin compared with the other components of the sample. It will increase the amount of resin. On the other side, the sample is brittle because the heating-cooling regime was inadequate. So, heating temperature will decrease. In this sense, will increase the holding time in the oven and will give up water to cool the sample.

To achieve the second sample (P2) were developed certain corrections imposed by non-compliance observed at the previous sample. In order to obtain a better link of the receipe components, it has doubled the amount of phenolic resin and reducing the amount of aluminium. To remove the sample from, the mold have taken steps against sticking composite on the bottom of the mold and the active surface of its piston. Thus, these surfaces will put a layer of aluminium foil and a layer of graphite.

Implementation steps for P2 sample were: mixing the constituents of the ground at a time; the phenolic resin is added; the composition is manually homogenized; an aluminium foil is placed at the base of the mold and on the active surface of the piston; the base mold was coat with a thin layer of graphite; the piston mold is take up; the assembly sample-mold is insert in the oven.

Technological analysis of P2: this sample is inadequate in appearance and structure. It shows areas with dismantling in the structure, which is due to the small amounts of phenolic resin in the composition compared to other sample constituents. It will increase the amount of resin. The homogenization of the components was inadequate. In the structure of sample it can be observe traces of graphite non-homogenized. The sample obtained has an incompletely circular shape which is due to the mold clamping force which was not applied consistently and uniformly. It has been found that the aluminium foil in combination with graphite layer solved the problem of sample sticking.

To achieve the third sample (P3) was kept the same level of heating-cooling regime at the previous sample but it has increased the amount of phenolic resin.

Technological analysis of P3: the sample was removed from the mold without problems which means that the solution to protect the lower mold with a layer of foil and graphite is suitable. The sample obtained has good look in shape, but in terms of structure is inadequate. This presents small areas with craters due the inhomogeneity of phenolic resin, which requires remedial action: homogenization of the components of the recipe using a mini-mixer.

To achieve the sample four (P4) will increase the holding time in the oven for 30 minutes. The other parameters were kept the same to the previous sample. Another modification consists of mixing the constituents using a mini-mixer.

Technological analysis of P4: The sample obtained has a good look in shape and structure. The link between constituents is good. There is a proper mixing of the constituents which means that the adoption of mechanical mixing is suitable. There are craters due to nehomogenity of resin and will change the steps in operations. All the constituents will introduce in phenolic resin. The sample has a high porosity, to remove this shortcoming will try pressing the mixture in a hydraulic press. There will be changes to the heating-cooling regime.

To achieve the sample five (P5) the components are homogenized and then are introduced into the phenolic resin homogenizing the composition and than performs a cold pressing on a hydraulic press, type PH 40.Technical characteristics of the hydraulic press are: maximum pressing force: 100KN, the size of the table surface: 700 x 600 mm, table height: 700 mm, voltage: 380 V, power: 9.7 KW. In the oven will increase the holding time.

Implementation steps for P5 sample were: the phenolic resin is put into the mixing tank; homogenize the constituents with a mecanical mixer; constituents homogenized are introduced in the phenolic resin; the composition is homogenized with a mixer; place a layer of aluminium foil on the base of the mold and on the active surface of the piston; place a layer of graphite on the based mold;

the mixture is introducing into the mold; the piston of the mold is take up; performing a cold pressing force on a hydraulic press; the assembly sample-mold is insert in the oven.

Technological analysis of P5: the sample obtained has good look in shape but lacks toughness. Homogenization of constituents was performed properly, but it has been observed that there is excessive phenolic resin. Porosity is still high. It will explore ways to increase hardness.

To achieve the sample six (P6) will try a cold pressing of the mixture after placing it in the mold. 15 minutes after introduction the sample in the oven and will be performing a hot pressing force. It is used the same sequences of operations used in P5. Fifthineen minutes after the introduction into the oven, the mold assembly is removed from the oven and is performed a hot pressing force. After that, the sample is return in the oven and maintained for 30 minutes.

Technological analysis of P6: the sample obtained has a low porosity, but it is brittle. The force pressing is too much and it is lost a considerable amount of resin, because of this fact unrelated solid constituents remaining in the structure and the sample was cracked when extracted from the mold. It aims to reduce the pressing force.

To achieve the sample seven (P7) the pressing force is 5 KN and will performing a hot pressing force. It is used the same sequences of operations as in P6 sample.

Technological analysis of P7: the sample has an appropriate consistency and is satisfactory in shape and structure. In the central part of the sample it can be seen a crater. This is not due to the material sticking on the piston mold because its surface does not show traces of material. Crater formation is due to accumulation of gases resulting from the sintering process. Due to excessive pressing, the gases could not be released outside. This solution of pressing will cancel. This is considered a successful sample and can be used for collecting samples necessary for the characterization of the new composite.

To achieve the sample eight (P8) it is used the previous parameters and the sequence of manufacturing steps.

Technological analysis of P8: good sample as consistency, form and structure with a satisfactory hardness. It is noted that the foil bonding of the surface sample. Sample surface is quite smooth, only occasionally seen of adhesions. The test is considered a good sample.

To achieve the sample nine P9 were maintained the parameters and manufacturing technology used in P7 and P8 samples.

Technological analysis of P9: the sample obtained has a low porosity, and the hardness is satisfactory. The shape and structure of the sample are appropriate. The surface is smooth but the touch is off a fine black powder which is due to the graphite the composition.

To achieve the sample nine P10 were maintained the parameters and manufacturing technology used in P7 P8 and P9 samples.

Technological analysis of P10: The sample obtained showing good shape and structure and has a satisfactory hardness. There is a proper mixing of the constituents and a suitable solidification which shows that the order of operations is appropriate to obtain composite materials the heating-cooling is adequate pressing force applied to both hot and cold compaction performed efficiently material. Sample surfaces are smooth and these were considered good in terms of quality.

For further experiments and tests it was selected four samples made from recipes P7, P8, P9 and P10. Thus P7 has the largest amount of aluminium and 0% coconut fibre, P8 has 5% coconut fibre, P9 has 10% coconut fibre and P10 has the smallest aluminium of all and the largest proportion of coconut fibre coconut, respectively 15%. The disc samples are shown in Figure 10. From each sample will make specimens to determine the mechanical characteristics of composites materials for brake pad application.

In order to characterize the new composite material, were achieved tests to determine density, porosity, hardness, thermo-gravimetric analysis, mechanical testing, and microstructure. The samples for each test were carried out by cutting in concordance with actual standards

Figure 10. Samples selected for determining the mechanical properties of composite materials

4. Conclusions

From the results and discussion of this study it can be observed:

The laboratory phase to obtain composite materials have encountered a number of difficulties such as, [10]: failure of proper mixing of the components in the recipe; correlating the ratio of the powdered material / phenolic resin; determine the dosage of constituents in the recipes; sticking the samples to the mold`s wall samples and its piston.

Solving these shortcomings was made by:

- using a mecanic stirrer to obtain a homogenous mixture
- increasing the amount of phenolic resin to make a link between constituents;
- aluminum foil and the layer of graphite, solve the problem sample sticking on the active surface of the mold and its piston.

During the components mixing it has been observed that the factors that influence a homogenous mixture are:

- mixing rate;
- the amount of the phenolic resin and the order of constituents mixing.

Regarding, the order of mixing the components can be generalized as follows: the first mixed is the material which has low specific weight, using a lower speed mixer. These will be introduced over the materials with higher specific weight and they are mixed with a high speed mixer.

Obtaining a compact sample is influenced by:

- the heating-cooling regime;
- pressing force.

During the experiments were made the following observations:

- if the time of mixing is small it is possible to obtain inhomogeneity of constituents
- overgrowth the mixing time does not lead to further improve the homogeneity of the constituents;
- excessive tightening and pressing of the mold leads to the accumulation of gases in the middle of the sample, it solidifies accomplished with a crater.

Manufacturing technology parameters and sequence of operations that have yielded good samples are those used in obtaining P7, P8, P9 and P10.These samples are considered appropriate, both in terms of concentration of constituents and in terms of manufacturing technology. These samples may be used to make specimens to determine the mechanical characteristics of the new composite material for brake pads application.

References

[1] Ghiorghita V 2013 *Research on the use of new structures for automotive systems*, University Transilvania Brasov, Romania, Doctoral Thesis

[2] Vasile Ghe 2010 *Structures with high rigidity made of composite materials used in automotive construction*, University Transilvania Brasov, Romania, Doctoral Thesis

[3] Blau P J 2001 *Compositions, functions and testing of friction brake materials and their additives*, Contract DE-AC05-00OR22725, Oak Ridge, Tennesse, USA

[4] Craciun A L and Pinca-Bretotean C 2016 Advanced materials with natural fibred reinforced aluminium composite for automotive brake disc, *IOP Conf. Ser.:Mater. Sci. Eng.* **163** 012014

[5] Terciu O, Curtu I, Cerbu I and Stanciu M D 2012 Traction testing of lignocellulosic composites with applications in the automotive industry, *Buletinul AGIR* **1**(1) 9-16

[6] Maleque M A, Atiqah A, Talib R J and Zahurin H 2012 New natural fibre reinforced aluminium composite for automotive brake pad, *International Journal of Mechanical and Materials Engineering* **7**(2) 166-170

[7] Leman Z, Sapuan S M, Saifol S M, Maleque A M and Ahmad M M 2008 Mosture absorption behavior of sugar palm fibre reinforced apoxy composites, *International Journal of Materials and Design* **29**(8) 1666-1670

[8] Park J H, Chung J O and Kim H R 2010 Friction characteristics of brake pads with aramid fiber and acrylic fiber, *Industrial Lubrication and Tribology* **62**(2) 91–98

[9] Liew N 2009 On the Dry and Wet Sliding Performance of Potentially New Friction Brake Pad Materials for Automotive Industry, *Wear* **266**(2) 275-287

[10] Pascu L V 2014 *Research on improving the quality of rolling stock intended for brake blocks*, University Politehnica Timisoara, Romania, Doctoral Thesis

IOP Conf. Series: Materials Science and Engineering **294** (2017) 012004 doi:10.1088/1757-899X/294/1/012004

Mathematical and physical modeling of thermal stratification phenomena in steel ladles

V Putan, L Vilceanu, A Socalici and A Putan

Politehnica University of Timisoara, Engineering and Management Department, 5 Revolution Street, Hunedoara, 331128, Romania

E-mail: vasile.putan@fih.upt.ro

Abstract. By means of CFD numerical modeling, a systematic analysis of the similarity between steel ladles and hot-water model regarding natural convection phenomena was studied. The key similarity criteria we found to be dependent on the dimensionless numbers Fr and $\beta\Delta T$. These similarity criteria suggested that hot-water models with scale in the range between 1/5 and 1/3 and using hot water with temperature of 45 ^{0}C or higher are appropriate for simulating natural convection in steel ladles.

With this physical model, thermal stratification phenomena due to natural convection in steel ladles were investigated. By controlling the cooling intensity of water model to correspond to the heat loss rate of steel ladles, which is governed by Fr and $\beta\Delta T$, the temperature profiles measured in the water bath of the model were to deduce the extent of thermal stratification in liquid steel bath in the ladles. Comparisons between mathematically simulated temperature profiles in the prototype steel ladles and those physically simulated by scaling-up the measured temperatures profiles in the water model showed good agreement. This proved that it is feasible to use a 1/5 scale water model with 45 ^{0}C hot water to simulate natural convection in steel ladles. Therefore, besides mathematical CFD models, the physical hot-water model provided an additional means of studying fluid flow and heat transfer in steel ladles.

1. Introduction

Due to inevitable heat losses, natural convection is common phenomenon occurring in steel ladles during the holding period prior to casting. A typical consequence of this phenomenon is thermal stratification of the liquid steel bath. The thermal stratification phenomenon in steel ladles and its potential influence on temperature control during continuous casting are important in steelmaking. This is because the temperature of liquid steel coming from a thermally stratified melt bath held in the ladles, i.e., teeming stream temperature, will have a direct impact on the temperature of steel melt held in the tundish. Therefore, investigation concerning this impact are obviously necessary, which, in turn, presupposes a good understanding of thermal stratification phenomena inside the steel ladles.

By means of CFD numerical modeling, a systematic analysis of the similarity between steel ladles and hot-water model regarding natural convection phenomena was studied, [1-3]. The key similarity criteria we found to be dependent on the dimensionless numbers Fr and $\beta\Delta T$, [4].

These similarity criteria suggested that hot-water models with scale in the range between 1/5 and 1/3 and using hot water with temperature of 45 ^{0}C or higher are appropriate for simulating natural convection in steel ladles, [4].

The present study focuses on establishing a hot-water model to simulate natural convection phenomena in steel ladles during the holding period before casting for the following purposes:

- Realization of an experimental physical model;
- Further evaluation of the key similarity criteria previously obtained from non-isothermal physical modelling of fluid flow and heat transfer in steel ladles, i.e., Fr and $\beta\Delta T$;
- Validation of the CFD models developed for simulating natural convection in steel ladles.

2. Realization of experimental physical model

Based on the previous analysis of the similarity between natural convection phenomena in steel ladles and in hot water models [4], which suggests that the water models with size scale in the range between 1/5 and 1/3 could be appropriate for modeling steel ladles, a 1/5 scale hot-water model of mid-aged 105 t steel ladles was set up in the laboratory, [5]. Figure 1 illustrates this physical model set-up.

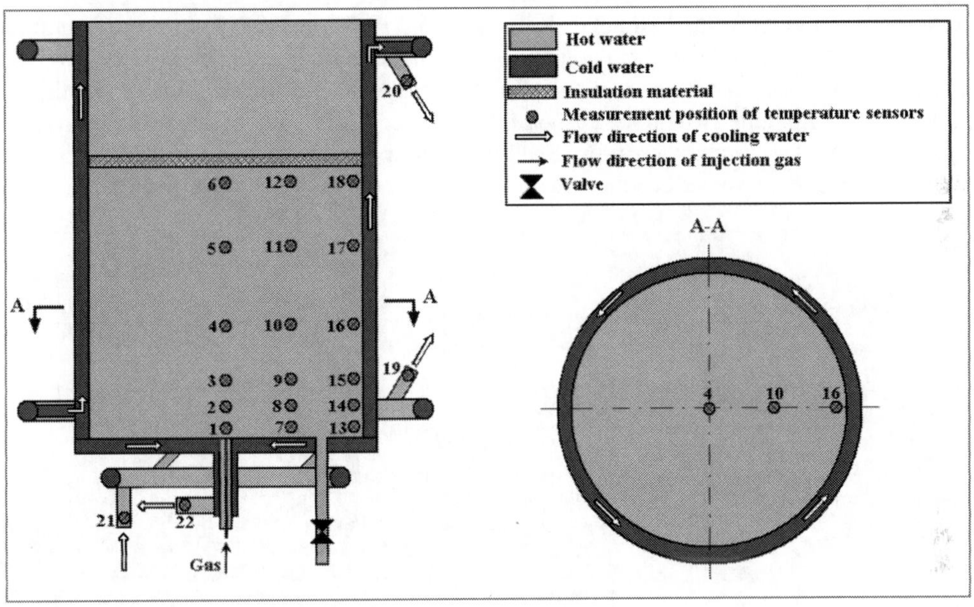

Figure 1. Hot-water model set-up, [5]

The model consists of two cooling chambers:

- The one is a cylindrical cooling chambers for simulating the ladle wall;
- The other is a flat cooling chambers for simulating ladle bottom.

The cooling chambers are made of 0,5 mm-thick stainless steel. Hot water is used as the liquid bath simulating liquid steel bath in ladles, while cold water with controllable temperatures is tangentially introduced into the cooling chambers in directions shown in Figure 1. KTY81-120-type temperature sensor were employed to get the temperature information from the water model. Principle of functioning of these sensors is based on the phenomenon of change electrical resistance with temperature and measuring range is between -50 ... 150 ^{0}C. Was preferred using these sensors in place of thermocouple elements as temperatures developed in the installation during the simulations are relatively low (max. 50 ^{0}C), and for this domain temperature sensors used have a measurement error small (± 1%) .

Figure 1 also schematically illustrates the sensor measurement position. 18 sensors were used for measuring the temperature profile in the water bath on a vertical plane bounded by sidewall and center axis. 4 sensors were used for measuring temperature of water inflows and outflows of the cooling chambers. All the temperature signals were recorded into a data logger for post processing.

To prevent heat loss from the top free surface of the water bath, the free surface was covered with a light porous plastic plate that can float on the surface. In order to homogenize the hot-water bath, if needed, pressurized air can be blown into the water bath via the tuyere located at the center of the bottom-cooling chamber.

1) A series of tests using this hot-water model was carried out with the following general procedure:
2) Supply cold water to cooling chambers at certain temperatures and flowrates;
3) Fill the model with hot water (around 50 ^{0}C) to a certain bath level;
4) Stir the hot-water bath by blowing pressurized air until the average temperature of the water bath decreases to about 45 ^{0}C;
5) Stop stiring the hot-water bath for a certain time lapse (i.e., the holding period for the development of thermal stratification);
6) Stir the hot-water bath again or drain the hot water out.

From the water model tests, temperature profiles in the hot-water bath were measured. The measured temperatures can be used to compare the predictions of the CFD model developed for the water model as an indirect verification of the CFD model developed for steel ladles. Furthermore, by controlling the cooling intensity to correspond to the heat loss rate of steel ladles, obeying the similarity criteria, the temperature profiles (exhibiting thermal stratification) obtained from the water model may be used directly to deduce the extend of thermal stratification in the prototype steel ladle.

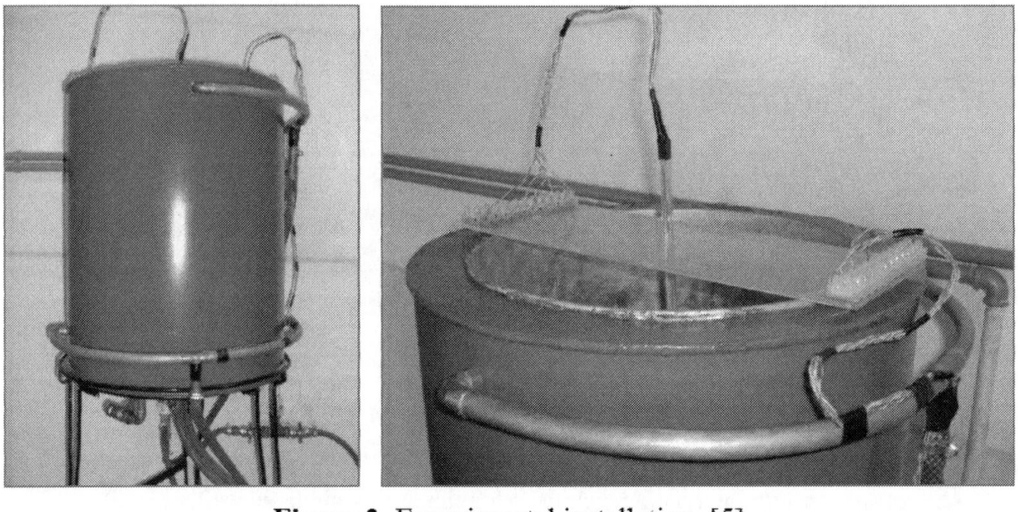

Figure 2. Experimental installation, [5]

2. Mathematical simulation

2.1. Mathematical equations that govern the phenomena

In order to use the results obtained by physical modeling with water based models for validating the CFD model to simulate the natural convection phenomenon in steel ladles, the same CFD model can also be applied to the water model. In fact, flow phenomena due to natural convection, both in liquid steel and in water, are governed by the same set of partial differential equations of turbulent flow, Navier-Stokes type, the differences occurring being due to the different thermo-physical properties of those Two fluids and different initial and contour conditions. In Cartesian coordinates, these equations have the following form:

Continuity equation:

$$\frac{\partial \rho}{\partial t}+\frac{\partial(\rho v_j)}{\partial x_j}=0 \tag{1}$$

The moment transfer equation:

$$\frac{\partial(\rho v_i)}{\partial t}+\frac{\partial(\rho v_i v_j)}{\partial x_j}=\frac{\partial}{\partial x_j}\left[v_{ef}\left(\frac{\partial v_i}{\partial x_j}+\frac{\partial v_j}{\partial x_i}\right)\right]-\frac{\partial p}{\partial x_i}-\rho_{ref}\beta\Delta T g_i \tag{2}$$

where the approximation of Boussinesq on the ascension force was made.
Energy transfer equation:

$$\frac{\partial(\rho c T)}{\partial t}+\frac{\partial(\rho v_j c T)}{\partial x_j}=\frac{\partial}{\partial x_j}\left(\lambda_{ef}\frac{\partial T}{\partial x_j}\right) \tag{3}$$

Turbulent flow equations:
To describe the effects of turbulent flow occurring in the case of natural convection phenomena in liquid steel and water, the standard k-ε model of turbulent flow was used. This is described by the following equations:

- the equation of the turbulent kinetic energy:

$$\frac{\partial(\rho k)}{\partial t}+\frac{\partial(\rho v_j k)}{\partial x_j}=\frac{\partial}{\partial x_j}\left(\frac{v_t}{\sigma_k}\frac{\partial k}{\partial x_j}\right)+v_t\left(\frac{\partial v_i}{\partial x_j}+\frac{\partial v_j}{\partial x_i}\right)\frac{\partial v_i}{\partial x_j}-\rho\varepsilon \tag{4}$$

- the equation of the speed of dissipation of turbulent energy:

$$\frac{\partial(\rho\varepsilon)}{\partial t}+\frac{\partial(\rho v_j\varepsilon)}{\partial x_j}=\frac{\partial}{\partial x_j}\left(\frac{v_t}{\sigma_\varepsilon}\frac{\partial\varepsilon}{\partial x_j}\right)+C_1 v_t\frac{\varepsilon}{k}\left(\frac{\partial v_i}{\partial x_j}+\frac{\partial v_j}{\partial x_i}\right)\frac{\partial v_i}{\partial x_j}-C_2\rho\frac{\varepsilon^2}{k} \tag{5}$$

In these equations, $\nu_{ef}=\nu_l+\nu_t$ is laminar viscosity, and ν_t turbulent viscosity. Similarly, effective thermal conductivity is defined as, $\lambda_{ef}=\lambda_l+\lambda_t$ where λ_l is laminar thermal conductivity, and λ_t is turbulent thermal conductivity. Using the turbulence model described above, ν_t and λ_t can be determined by means of relations:

$$v_t=C_v\rho\frac{k^2}{\varepsilon} \tag{6}$$

and

$$\lambda_t=\frac{v_t c}{Pr_t} \tag{7}$$

Where Pr_t is the Prandtl number of turbulent flow, this value being 0.9 for fully turbulent flows [6]. According to the recommendations of the literature [7], the constants that appear in the standard k-ε model of the turbulent flow have the following values:

C_μ = 0,09; C_1 = 1,44; C_2 = 1,92; σ_k = 1,0; σ_ε = 1,3

2.2. Mathematical simulation

In order to use the water modeling results to validate the CFD model developed for simulating natural convection phenomena in steel ladles, the same CFD model should be applied to the water model as well. In fact, the natural convection flows in liquid steel and in hot water are governed by the same set of turbulent Navier-Stokes type partial differential equations, except from the differences in thermal-physical properties of the fluids and initial and boundary conditions.

For the present study, a two-dimensional CFD model were developed. The model simulate heat transfer both in liquid and in solid, so-called conjugate-heat-transfer CFD model. This CFD model was established specially for the water model which calculates fluid flow and heat transfer both in hot-water bath and in the cold water flowing in cooling chambers including heat conduction inside the inner walls of the chambers. For this, a two-dimensional cylindrical-polar computation domain was defined for the water model as shows in Figure 3.

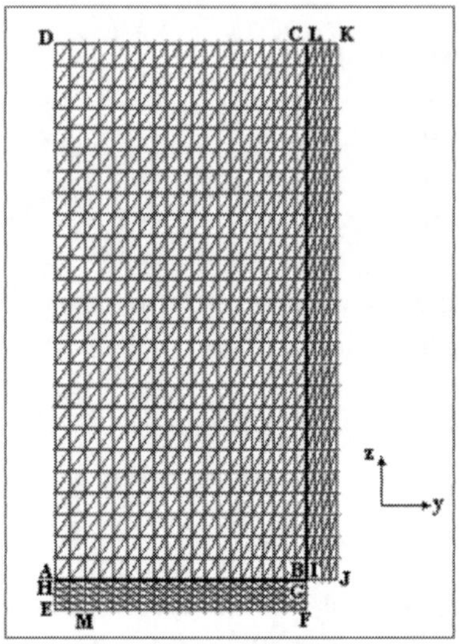

Figure 3. Computation domains and grids used for CFD simulations of water model

The reason for developing a conjugate-heat-transfer CFD model was that it was found from water model experiments and test simulations, [8], [9], that the extend of thermal stratification in the hot-water bath is dependent not only on the average magnitude of the heat loss flux but also, more crucially, on the heat flux distribution along the side wall. Therefore, it is obvious that such knowledge cannot be directly obtained from the water model by measurements. Thus, the heat loss fluxes and their distributions have to be totally calculated by means of mathematical simulations.

Table 1 list the description of various regions in the computational domains defined in Figure 3.

Table 1. Description of computation domains defined in Figure 3

Regions	Descriptions
ABCD	Hot-water bath
EFBA	Bottom-cooling chamber
EFGH	Bottom-cooling water
HGBA	Inner shell of bottom-cooling chamber
BJKC	Side-cooling chamber
IJKL	Side-cooling water
BILC	Inner shell of side-cooling chamber
FG	Inlet of bottom-cooling chamber
EM	Outlet of bottom-cooling chamber
IJ	Inlet of side-cooling chamber
LK	Outlet of side-cooling chamber
ED	Centre axis
MF	Interface between cooling water and outer shell of bottom-cooling chamber
JK	Interface between cooling water and outer shell of side-cooling chamber

3. Results and discussions

3.1. Comparison between water model and mathematical model simulation results
Figure 4 shows the comparison between the measured temperatures and those predicted using the conjugate-heat transfer CFD model for a water model simulations case.

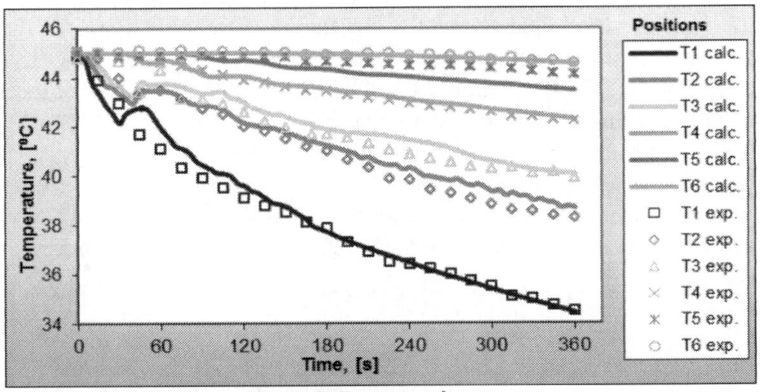

a) Centre axis

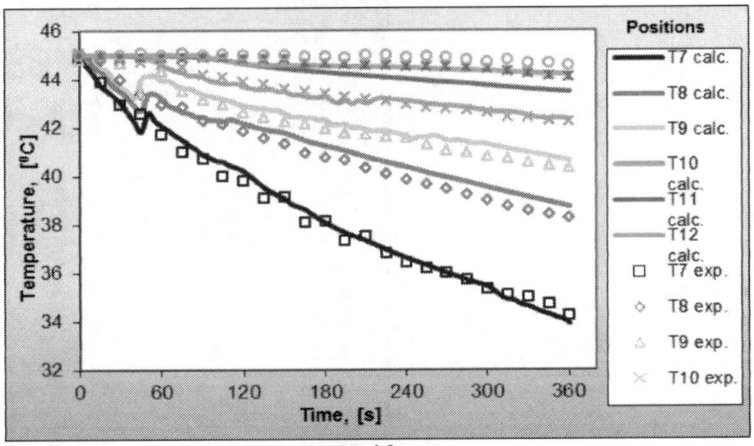

b) Half-way

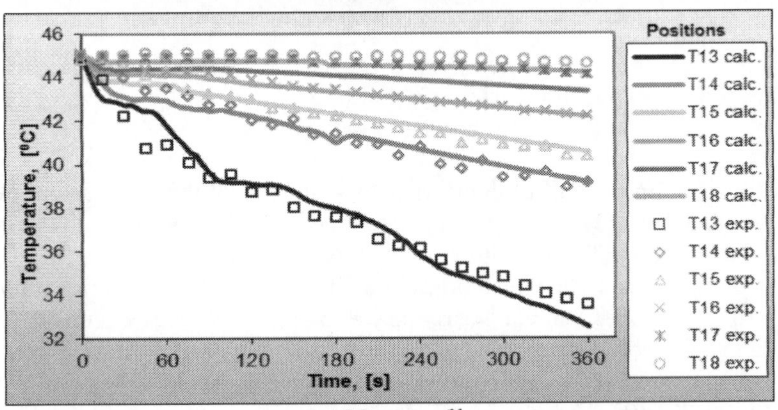

c) Near wall

Figure 4. Comparison between temperatures calculated (lines) and measured (symbols) at different position in the water model

In this simulations case, initially, after stop the gas bubbling, the hot-water bath was nearly homogeneous and had an average temperature of 45 ^{0}C. It was cooled by introducing 12 ^{0}C cold water into the side and bottom cooling chambers at flowrates of 10 liters per minute. The cooling lasted 6 minutes to allow the development of thermal stratification in the hot-water bath. It can be seen from this Figure that a generally satisfactory agreement between calculated temperatures and measurement ones was obtained, showing that the conjugate-heat-transfer CFD model developed in the present study is feasible for use. It should be noted that in the flow regions close to the top surface the calculated temperatures decrease relatively faster than the measured ones. This discrepancy could be due to the use of the standard high-Reynolds number k-ε two-equation turbulence model that may overestimate turbulent heat transfer in this flow region.

3.2. Flow and temperature fields in hot-water model

Figure 5 illustrates the flow and temperature fields in the hot-water model after 6 min. of cooling, predicted by the conjugate-heat-transfer CFD model, for the above-mentioned experimental case.

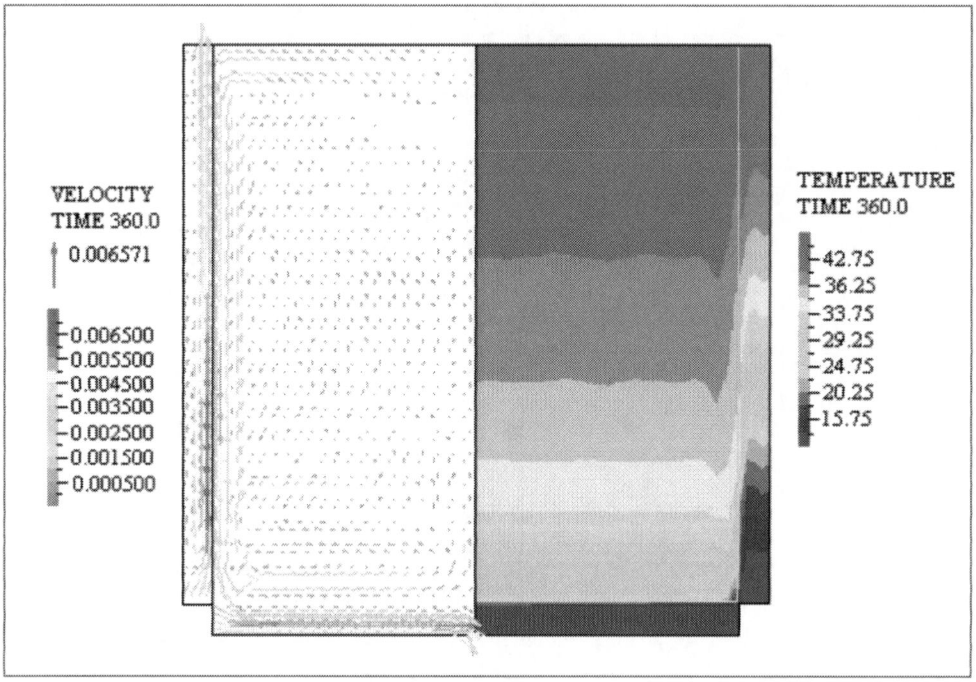

Figure 5. Predicted flow and temperature fields, after 6 min. of cooling, in a 1/5 scale hot-water model

Due to axis-symmetry, only a half of the flow field and a half of the temperature field are shown in the Figure. It is seen that the CFD model well revealed the natural convection flow pattern and thermal stratification in the hot-water bath. In addition, with help of conjugate-heat-transfer CFD model, fluid dynamics in the cooling chambers can also be examined. A well-established radial plug-flow is found in the bottom-cooling chamber with a nearly uniform velocity distribution appearing across the vertical gap inside the chamber. A nearly vertical plug-flow is found in the side-cooling chamber but with noticeable velocity gradients across the radial gap inside the chamber. It is believed that the difference in velocity distributions between bottom and side cooling chambers is caused by the buoyancy effect, which is weaker inside the bottom and side cooling chamber, but rather stronger inside the side-cooling chamber. As results, a nearly temperature distribution is revealed in the bottom-cooling chamber, while a non-uniform temperature distribution is found in the side-cooling chamber with, of course, higher temperatures appearing close to the wall of the chamber.

4. Conclusions

From these studies can be drawn following conclusions:

- The present non-isothermal water model study has confirmed the validity of the dimensionless number Fr and $\beta\Delta t$ as key criteria governing the similarity between natural convection phenomena in hot-water models and in prototype steel ladles.
- Establishing a non-isothermal water modeling system is useful for verification of CFD mathematical modelling results. In addition to mathematical methods, the hot-water model provides an alternative means of studying fluid flow and heat transfer phenomena in steel ladles.
- It is feasible to use a 1/5-scale hot-water model to simulate natural convection and resulting thermal stratification phenomena in steel ladles during the holding period before casting, if the heat loss fluxes and, especially, their distributions in steel ladles are accurately known.
- The water modeling results prove the fact that the in steel ladles development rate of thermal stratification is linearly proportional to the bulk-cooling rate, as has been reported by other authors.

References

[1] Ilegbusi O J and Szekely J 1987 Melt Stratification in Ladles, *Transactions Iron Steel Inst. Jpn.* **27** 563-569

[2] Koo Y S, Kang, T and Lee I R 1989 Steelmaking Conference Proceedings 72, Iron and Steel Society, USA 415-421

[3] Austin P R, Camplin J M, Herbertson J and Taggart I J 1992, Mathematical Modelling of Thermal Stratification and Drainage of Steel Ladles, *ISIJ Int* **32**(2) 196-200

[4] Pan Y and Bjorkman B 2002 Computational Fluid Dynamics Numerical Analysis on the Similarity between Steel Ladles and Hot-Water Models regarding Natural Convection Phenomena, *ISIJ International* **42**(1) 53-62

[5] Puţan V 2005 *Studies on the hydrodynamics and thermal work conditions of the alloy in the casting ladles*, Technical Univrsity of Cluj-Napoca, Romania, Doctoral Thesis

[6] McComb W D 1997 *Turbulenţa fluidelor*, Editura Tehnica, Bucureşti, Romania

[7] Launder B E and Spalding D B 1974 The numerical computation of turbulent flows, *Computer Methods in Applied Mechanics and Engineering* **3**(2) 269-289

[8] Xia J L and Ahokainen T 2001 Thermal Stratification in a Steel Ladle, *Canadian Metallurgical Quarterly* **40**(4) 497-487

[9] Grip C E 1998 *Temperature and Steel Flow in Ladle and Tundish*, Lulea University of Technology, Sweden, Doctoral Thesis

IOP Conf. Series: Materials Science and Engineering **294** (2017) 012005 doi:10.1088/1757-899X/294/1/012005

Research regarding the vacuuming of liquid steel on steel degassing

M Magaon[1], M Radu[1], S Şerban[1] and L Zgripcea[2]

[1]Politehnica University of Timisoara, Engineering and Management Department, Revolutiei str., no. 5, 331128 Hunedoara, Romania
[2]S.C Hydromatic Sistem S.R.L, Calea Buziasului Street, no. 11, Timisoara, Romania

E-mail: miruna magaon@yahoo.com

Abstract. When the liquid steel comes in contact with the atmosphere of the elaboration aggregates, a process of gas diffusion into the metal bath takes place on the one hand, and on the other hand a process that allows them to pass from the metal bath into the atmosphere. The meaning of these processes is determined by a number of factors as follows: the quality of raw and auxiliary materials (moisture content, oils, etc.), the boiling intensity, the evacuation duration, the properties of used slags, the values of the casting ladle processing parameters (bubbling , vacuuming, etc.). The research was carried out at an electrical steelwork, equipped with an electric arc furnace type EBT (Electric Bottom Tapping) capacity 100t, LF (Ladle-Furnace) and VD (Vacuum Degassing) facilities, establishing some correlations between the vacuuming parameters from the V.D.facility and the amounts of hydrogen and nitrogen removed from the metal bath, as well as their removal efficiency, were taken into consideration. The obtained data was processed in MATLAB calculation program, the established correlations form was presented both in analytical and graphical form. The validity of these correlations was verified in practice, being particularly useful in research.

1. Introduction

When the liquid steel comes in contact with the atmosphere of the elaboration aggregates, on one hand, a process of gas diffusion into the metal bath takes place, and, on the other hand, a process that allows them to pass from the metal bath into the atmosphere. The meaning of these processes is determined by a number of factors such as: the quality of raw and auxiliary materials (moisture content, oils, plastic materials etc.), the boiling intensity, the evacuation duration, the parameters compliance with the casting ladle processing (LF/VD), the properties of used slags, the values of the casting ladle processing parameters etc.

In steel manufacturing process, besides non-metallic inclusions, gases like hydrogen, nitrogen, and oxygen remain trapped in its structure.

The hydrogen represents an impurity for steel products, its negative influence is manifested especially in steel because: it is one of the causes of sulphides in ingots and calmed steel castings, it decreases the elasticity and toughness of the steel, it contributes to the occurrence of the defect called "flakes“ in steels alloyed with chromium and nickel, which substantially reduces fatigue strength of steel parts and affects the electrical and magnetic properties of steels. In order to produce steel with low hydrogen content, a series of measures must be applied to the manufacturing flow such as [1-4]: the cargo used must be of good quality, without moisture, oils, plastic materials, calcinated additions, freshly burnt limestone; the treatment of steel under vacuum; the adequate heating of the casting machine (the ladle and the distributor). The negative influence of nitrogen in steel is manifested in the following considerations [1], [2], [5]: it reduces the plasticity and toughness of the steel; in combination with hydrogen contributes to the appearance of sulphides in the cast steel; it causes the ageing phenomenon of the steel by depositing nitrides at the structural grains limits.

During the steel elaboration process, boiling the metal bath has a very important role in the degassing process, caused by the release of the carbon monoxide bubbles.

When these carbon monoxide bubbles are formed, the partial pressure of hydrogen and nitrogen is practically null and, therefore, the gases diffuse inside these bubbles under the action of concentration gradients, and then disposed outside the bath [4], [6].

In order for the speed of the gas removal from the steel bath to be higher than the passing speed through the slag in the metal bath and thus the nitrogen content to fall, the speed of decarburization needs to be high enough to overcome a certain critical value, in order to cause pressure in the carbon monoxide bubble above the ferrostatic pressure [5], [6].

Regarding the electrical steelworks equipped with electric arc furnaces type EBT, very good conditions to reduce gas absorption in the steel bath (high-speed boiling of the metal by the intense blowing of oxygen, the foaming of slag and the possibility of bubbling with argon in the oven) were provided [3], [5], [6].

The introduction of the so-called "Metallurgy into the pot" into the steel elaboration technological flow produced a genuine technical revolution in steel industry, some of the processes that took place in aggregates elaboration were transferred to the casting ladle, particularly in LF (Ladle-Furnace) facilities and vacuum facilities without heat input (R. H., V.D.) and with heat input (V.A.D., V.O.D. etc.) [1], [2], [5], [6].

2. Industrial experiments. Data processing

In the research conducted, the possibilities to reduce the hydrogen and the nitrogen content from steels designed for pipes manufacturing were investigated. The obtained results are presented in this paper.

The research was carried out at an electrical steelwork, equipped with an electric arc furnace type EBT (Electric Bottom Tapping) capacity 100t, LF (Ladle-Furnace) facility, VD (Vacuum Degassing) facility and a 5 wires continuous casting (ITC) facility. On the flow presented a total of 20 steel batches were followed, S 275 JOH (EN 10216-1) type steel, circular rolled with a diameter of Ø180mm. The 20 batches were monitored throughout the technological flow, stage by stage, as follows:

- Elaboration in the furnace: loads of good quality, without scrap with oil, moisture, rust under about 2%, freshly burnt limestone, calcined iron ore, adequately heated casting ladle;
- LF and VD treating facility: respecting the limits of variation of technological parameters (heating, bubbling), additions without moisture (for slag formation and alloying), mixing chemical and thermal by bubbling with argon followed by treatment in the vacuum facility without VD heat input, in the casting process the ladle and the distributor were heated adequately;
- The continuous casting compliance parameters: the casting speed correlated with the casting temperature, flow rate and pressure of cooling water in each cooling zone.

Based on the the hydrogen and nitrogen content before and after vacuuming, the quantities of removed gases were calculated (ΔH [ppm] and ΔN [ppm]), respectively their removal efficiency (ηH [%] and ηN [%]), which are considered to be the dependent parameters. The independent parameters considered are: the total duration of vacuuming Dtv [min]; the duration of ultra-high vacuuming Dva [min] and the pressure in the vacuum facility under ultra-high vacuuming Pva[mBar].

By processing the obtained data in the MATLAB calculation program, multiple correlation equations have resulted between the independent and dependent parameters, the results being presented both in the analytical and graphical form in figures 1-15.

In order to obtain these correlations (z- dependent parameter, x and y - independent parameters), the following equations forms were used:

Equation 1

z =a(1)+a(2).*x+a(3).*x.^2+a(4).*x.^3+a(5).*y+a(6).*y.^2+a(7).*y.^3+a(8).*y.^4+a(9).*y.^5 (1)

Equation 2

(z=a(1)+a(2).*log(x)+a(3).*log(x).^2+a(4).*log(x).^3+a(5)./y+a(6)./(y.^2)+a(7)./(y.^3)+ a(8)./(y.^4)+a(9)./(y.^5) (2)

In the following figures are presented the most significant correlations obtained both in terms of values for correlation coefficients, as well as for technological direction.

The graphical and analytical representation of the obtained data using the first equation:

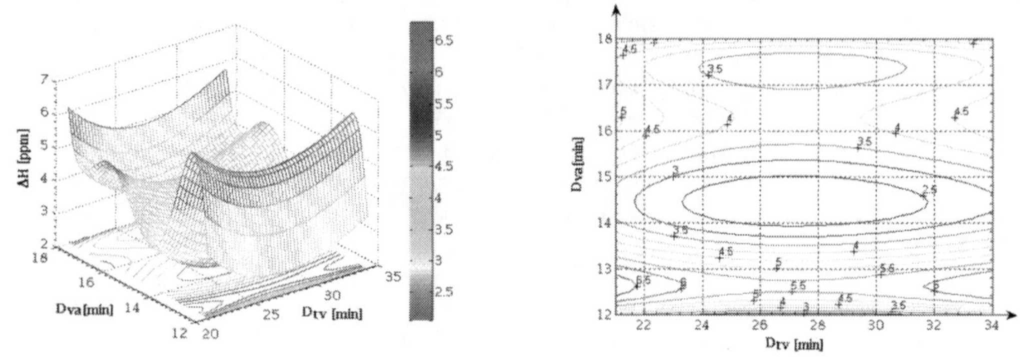

(a) (b)

D= 705116.851244; H= -99565.85323; R^2= 0.4418765; a1=-49782.926616; a2=-3.540957; a3=0.104587; a4=-0.000967; a5=16750.070210; a6=-2240.456339; a7=149.06927; a8=-4.93461; a9=0.0650288. (1)

Figure 1. Correlation ΔH =f(D_{tv}, D_{va}): a) the correlation surface; b) level curves

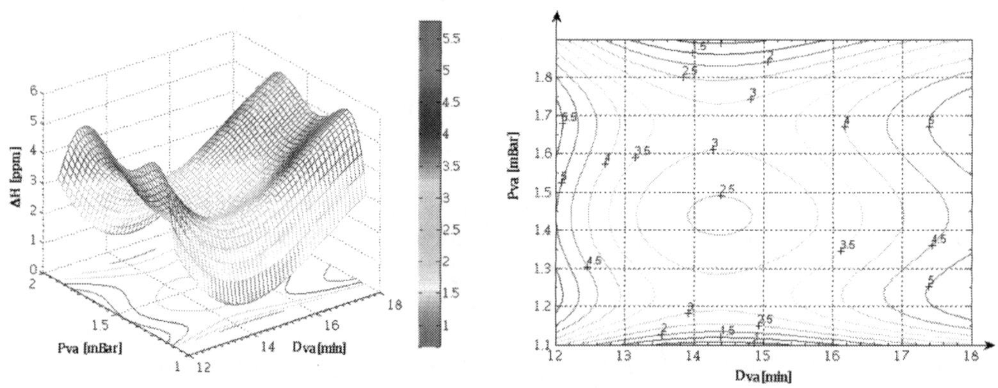

(a) (b)

D=584969.633965; H=-7208.758343; R^2=0.662044; a1=-3604.379171; a2=-40.573961; a3=2.474023;a4=-0.049264; a5=12556.088689; a6=-16305.753277; a7=10470.281392; a8=-3323.588525; a9=417.124588; y1= x2x3

Figure 2. Correlation ΔH =f($D_{va,}$ P_{va}): a) the correlation surface; b) level curves

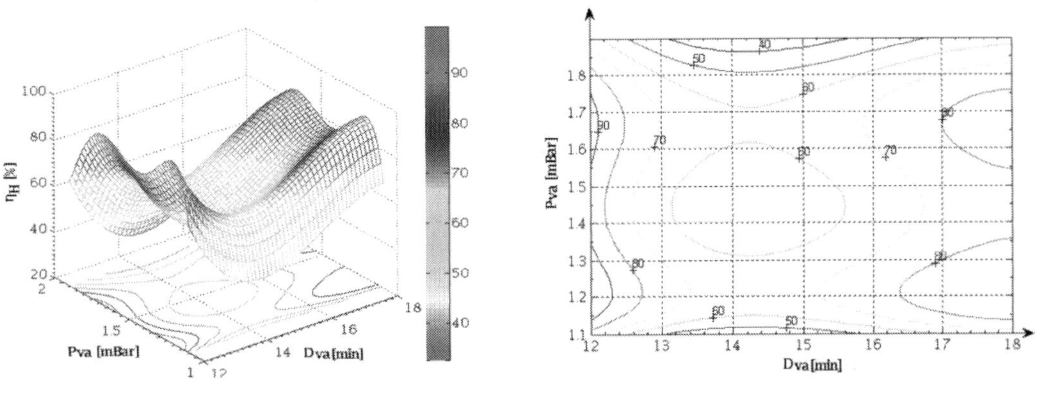

(a) (b)

D=114505539.5218; H=-90741.47209; R^2=0.6842; a1=-45370.73604; a2=-630.952437; a3=39.5767; a4=-0.8156; a5=161252.2037; a6=-211002.2892; a7=136512.1858; a8 43666.7105; a9=5524.3698.

Figure 3. Correlation $\eta_H = f(D_{va}, P_{va})$: a) the correlation surface; b) level curves

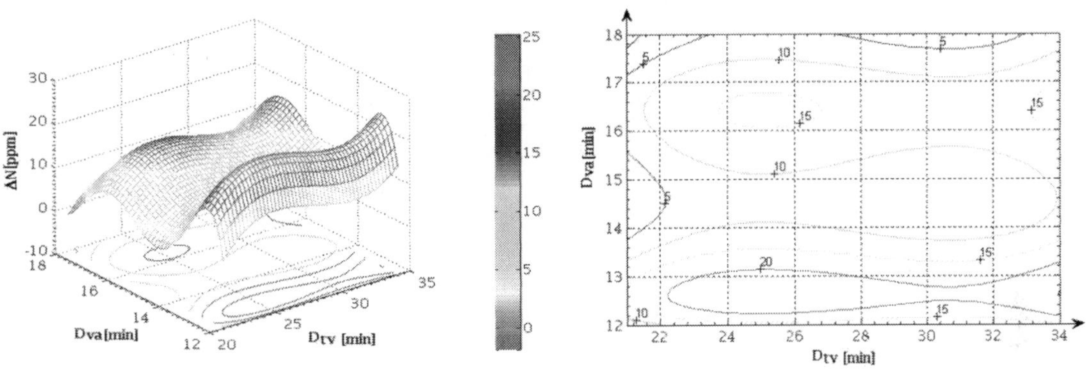

(a) (b)

D= -50936586.7253; H= -298348.3767; R^2=0.4537469; a1= -149174.1883; a2=85.36426;

a3=-3.0903; a4=0.0369; a5=49257.1978; a6=-6497.2367; a7=425.80702; a8=-13.8673; a9=0.1795.

Figure 4. Correlation $\Delta N = f(D_{tv}, D_{va})$: a) the correlation surface; b) level curves

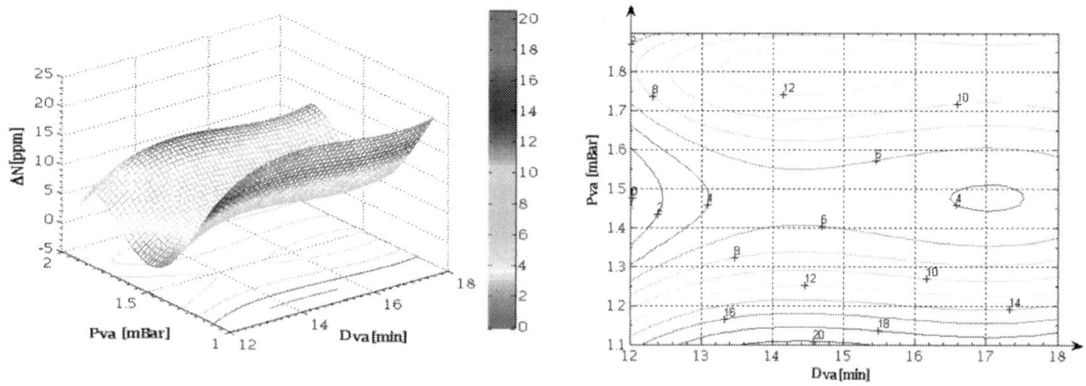

(a) (b)

D=-143127.630479; H=-648.1332002; R^2=0.772582; a1=-324.0666001; a2=110.376649; a3=-7.077384;a4=0.1501739; a5=-1530.507378; a6=3526.358771; a7=-3602.854107; a8=1676.1338907; a9=-290.3976957; Y3 = x2x3

Figure 5. Correlation ΔN =f(D_{va}, P_{va}): a) the correlation surface; b) level curves

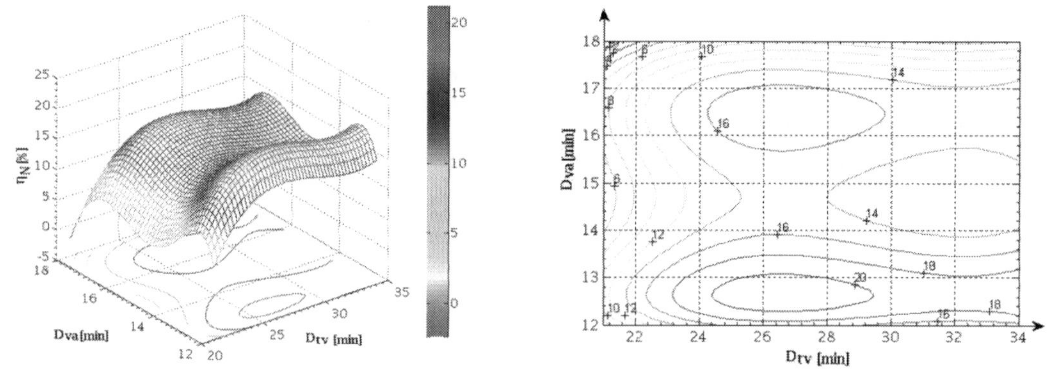

(a) (b)

D=-11520280.645924; H=-97947.685681; R^2=0.4076934; a1=-48973.842842; a2=58.808314; a3=-2.0189403; a4=0.0228829; a5=15766.0797; a6=-2035.283404; a7=130.203671; a8=-4.127074; a9=0.051839.

Figure 6. Correlation η_N =f(D_{tv}, D_{va}): a) the correlation surface; b) level curves

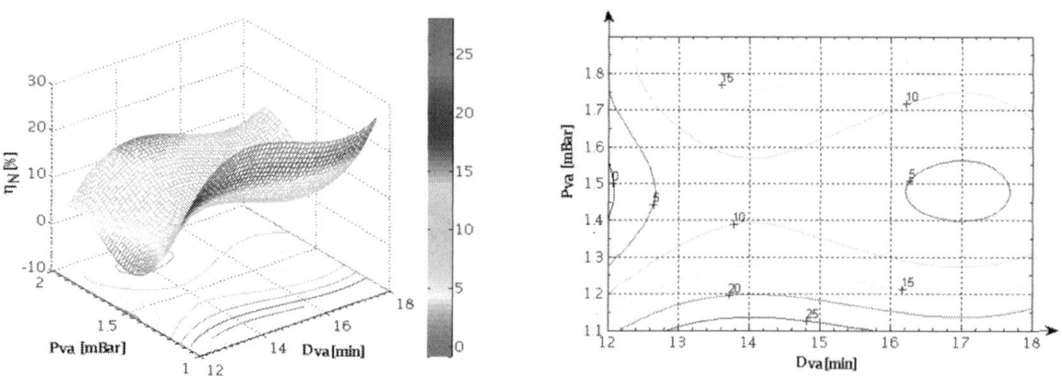

IOP Conf. Series: Materials Science and Engineering **294** (2017) 012005 doi:10.1088/1757-899X/294/1/012005

(a) (b)

D=-2271528.789341;H=-4291.142356;R^2=0.8268917; a1=-2145.571177; a2=264.641941; a3=-17.213466; a4=0.369775; a5=2922.322108; a6=-3644.849781; a7=1964.816843; a8=-426.285141; a9=20.1586318;

Figure 7. Correlation η_N =f(D_{va}, P_{va}): a) the correlation surface; b) level curves

The graphical and analytical representation of the obtained data using the second equation:

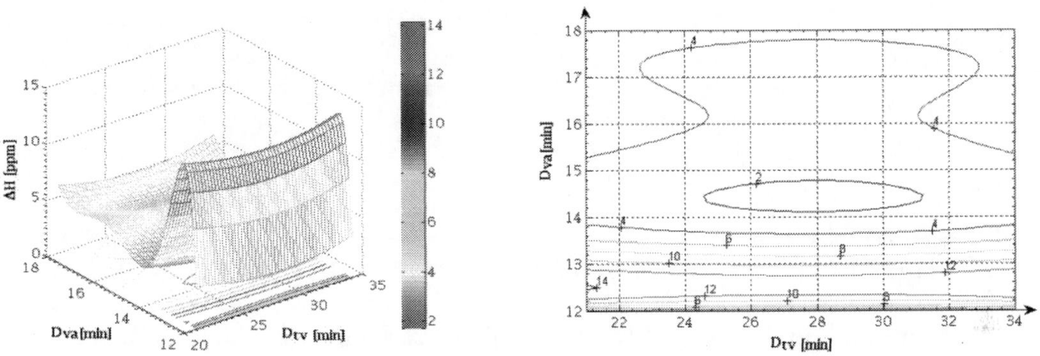

(a) (b)

D=211034332.035952; H=141201.0480609; R^2=0.4421996; a1=70600.5241; a2=747.4977; a3=-246.1921; a4=26.8182; a5=-5403329.0948; a6=162957468.5082; a7=-2445846673.9457; a8=18265542497.9899; a9=-54285159029.6510.

Figure 8. Correlation ΔH =f(D_{tv}, D_{va}): a) the correlation surface; b) level curves

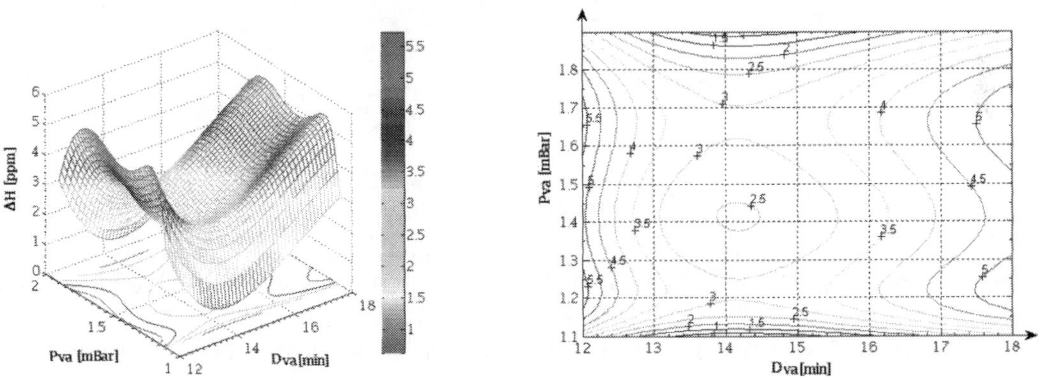

(a) (b)

D=-21656490.378032; H=3003.724125; R^2=0.664805; a1=1501.8621; a2=-3366.523469; a3=1196.776858; a4=-141.281788; a5=10429.813573; a6=-25745.12912; a7=30786.126119; a8=-17670.656116; a9=3832.083735.

Figure 9. Correlation ΔH =f(D_{va}, P_{va}): a) the correlation surface; b) level curves

IOP Conf. Series: Materials Science and Engineering **294** (2017) 012005 doi:10.1088/1757-899X/294/1/012005

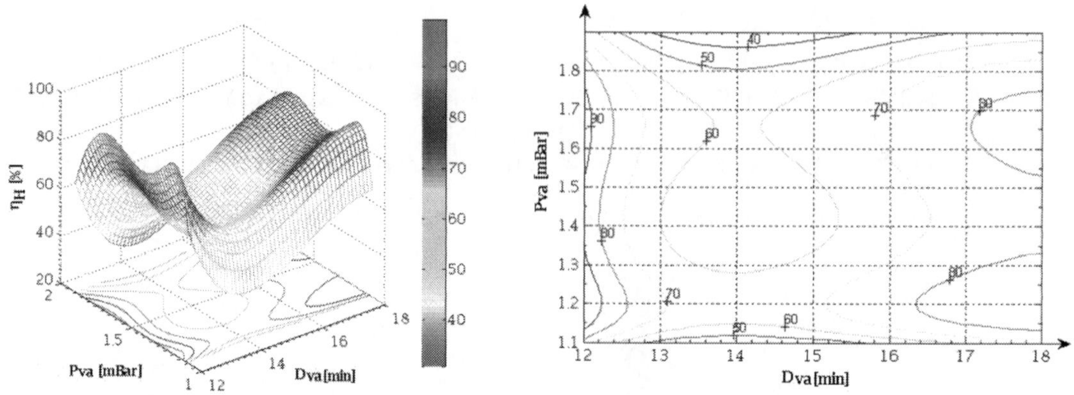

(a) (b)

R^2=0.69051; a1=38765.73730; a2=-61629.34808; a3=22304.67077; a4=-2684.798405; a5=111192.7945116; a6=-266863.828855; a7=306915.8584278; a8=-166257.084709; a9=32774.859651;

Figure 10. Correlation η_H =f(D_{va}, P_{va}): a) the correlation surface; b) level curves

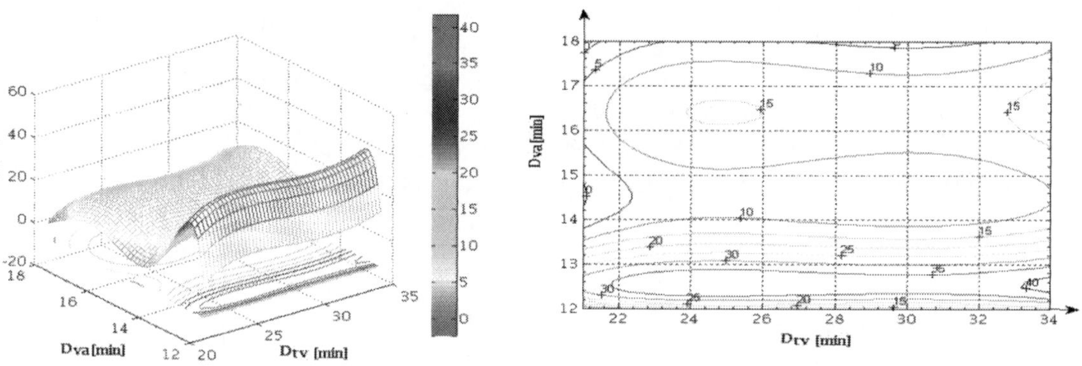

(a) (b)

R^2=0.4596777: a1=84423.7519; a2=21774.374948; a3=-6590.29533; a4=664.32948; a5=-8554613.52623; a6=267878925.313956; a7=-4159978032.23881;a8=32039478846.2628; a9=-97909697827.6924

Figure 11. Correlation ΔN =f(D_{tv}, D_{va}): a) the correlation surface; b) level curves

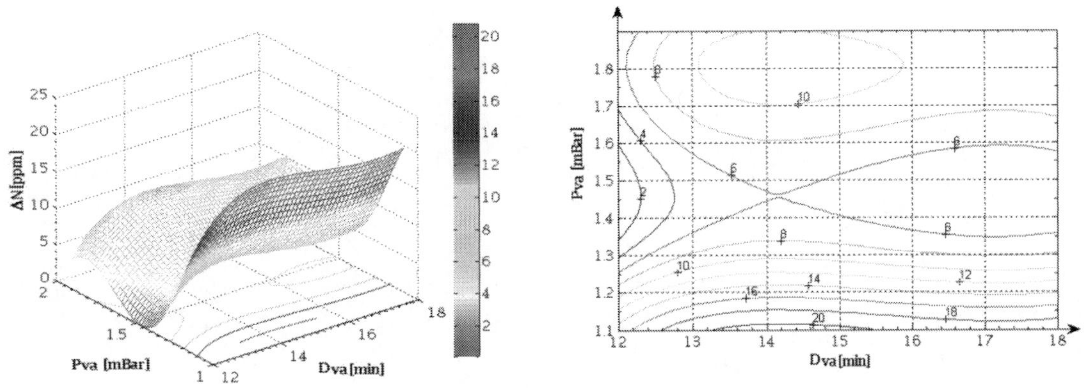

(a) (b)

R^2=0.7712696; a1=-15128.621905; a2=10627.85422007;a3=-3873.1318372; a4=469.9147357; a5=36905.8670319; a6=-98389.2377431; a7=128597.8130189; a8=-82499.246152; a9=20828.695835;

Figure 12. Correlation $\Delta N = f(D_{va}, P_{va})$: a) the correlation surface; b) level curves

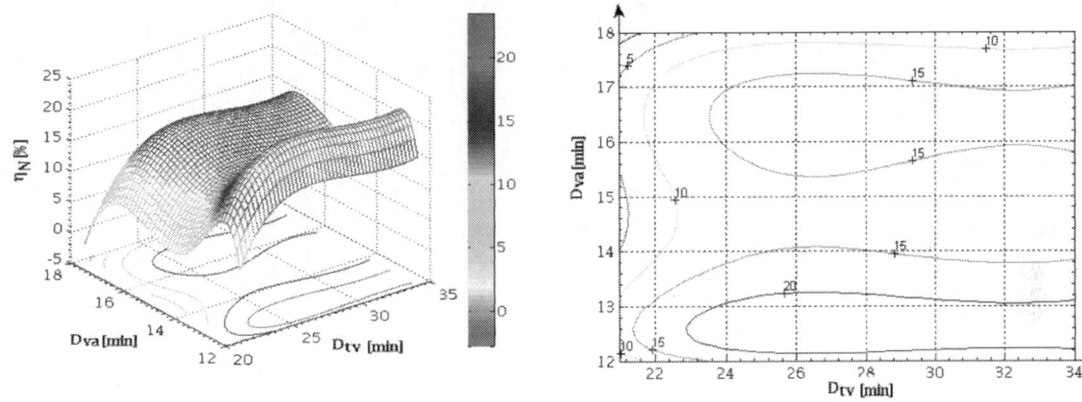

(a) (b)

D=-1459235866.36140; H=-59400.56406; R^2=0.410592; a1=-29700.28203; a2=12173.443148; a3=-3607.920043; a4=356.1214758; a5=930120.193999; a6=-19966438.3295872; a7=185770804.328888;a8=-592499637.225002;. a9=-435401682.323592

Figure 13. Correlation $\eta_N = f(D_{tv}, D_{va})$: a) the correlation surface; b) level curves

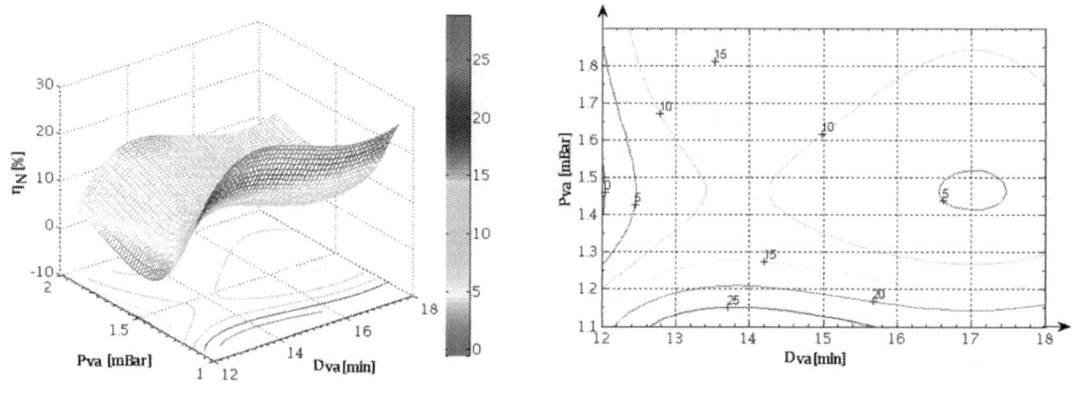

(a) (b)

D=-2753858029.17180; H=-48780.7745931; R^2=0.8272189; a1=-24390.387296; a2=27205.9563839; a3=-9980.1191003; a4=1218.552021; a5=-3175.214067; a6=13171.984069; a7=-25368.699896; a8=22854.023043; a9=-7758.362733;

Figure 14. Correlation η_N =f(D_{va}, P_{va}): a) the correlation surface; b) level curves

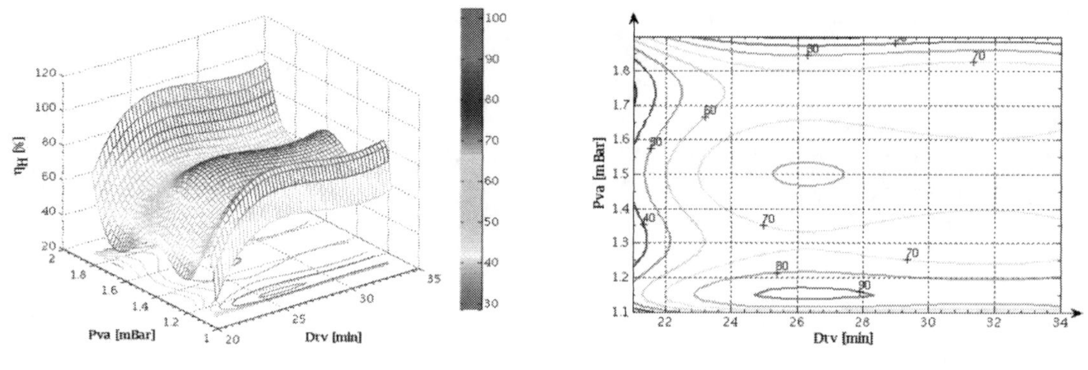

(a) (b)

R2 = 0.550274; a1 = 63736.280833; a2 = 49403.656749; a3 = -14701.351024; a4 = 1457.077641; a5 = -853705.21048; a6 = 2431071.365372; a7 = -3434032.851742; a8 = 2406387.735273; a9 = -669292.604992085.

Figure 15. Correlation η_H =f(D_{tv}, P_{va}): a) the correlation surface; b) level curves

3. Results analysis

From the results obtained in the Matlab calculation program, the correlation surfaces and the level curves in flat projection are presented in graphical form and the correlation equation in analytical form in order to achieve their technological analysis.

From figures 1, 4, 6, 8, 11 and 13, it clearly appears that the total duration of vacuuming and the duration of ultra-high vacuuming has a combined effect on the quantities of hydrogen and nitrogen removed from the metal bath during vacuuming, as well as on their removal efficiency.

Values for the total duration of vacuum of 28-32 minutes and for the duration of ultra-high vacuuming of 13-18 minutes can ensure the hydrogen reduction content of more than 3,5ppm, the average nitrogen reduction content of 12ppm, for η_H more than 70% and for η_N an average of 15%.

Analysing figures 2, 3, 5, 7, 9, 10, 12 and 14, it also appears that the duration of ultra-high vacuuming and the pressure in the vacuum facility under ultra-high vacuuming has the same combined effect as mentioned above. Values for the duration of ultra-high vacuuming of 13-18 minutes and the pressure in the vacuum facility under ultra-high vacuuming of under 1,7 mBar can ensure the hydrogen reduction content of more than 3,5ppm, the average nitrogen reduction content of 12ppm, for η_H more than 68% and for η_N an average of 16%.

The analysis of the correlation regarding the cumulative influence of the total duration of vacuuming and of the pressure in the vacuum facility under ultra-high vacuuming on the hydrogen removal efficiency (figure 15) confirms the validity of the above assertions, meaning that values for η_H of more than 70% can be obtained.

4. Conclusion

From the results obtained by processing the data in the MATLAB calculation program, one can conclude the following:

- there is a large dispersion of the values for the amounts of hydrogen (ΔH) and nitrogen (ΔN) removed from the metal bath, as well as for the hydrogen and nitrogen removal efficiency (η_H [%] and η_N [%]), depending on the vacuuming parameters (D_{tv}, D_{va} and P_{va});
- if the values for P_{va} are under 1.4 mBar the obtained values for the hydrogen and nitrogen removal efficiency are $\eta_H \geq 70\%$ and $\eta_N \geq 15\%$;
- the ultra-high vacuuming duration has a significant influence on the hydrogen removal efficiency (η_H), its desired duration is 18 even 20 min.

By using in practice the obtained diagrams, values for the vacuuming parameters can be established in order to obtain estimated values both for the removed quantity and the removal efficiency of gases.

References

[1] Geantă V 2003 *Methods and technologies for steel refining*, Printech, Bucharest, Romania.

[2] Eastham J F, Profumo F, Tenconi A, Hill-Cottingham R J, Coles P C and Gianolio G 2002 Novel Axial flux Machine for aircraft drive: design and modeling, *IEEE Transactions on Magnetics* **38**(5) 3003-3005

[3] Drăgoi F 2012 *Research on reducing the gas content of the steel treated on the technological and EBT- TC*, Politehnica University of Timişoara, Romania, Doctoral Thesis

[4] Puţan A 2013 *Research on steel refining elaborated on the flow: electric arc furnace-ladle furnace-continuous casting*, Politehnica University of Timişoara, Romania, Doctoral Thesis

[5] Ştefănoiu R 2004 *Research on the refining process of steels with inert gases*, Politehnica University of Bucharest, Romania, Doctoral Thesis

[6] Nica G, Socalici A, Ardelean E and Heput T 2003 *Technologies for improving the quality of steel*, Mirton, Timişoara, Romania

IOP Conf. Series: Materials Science and Engineering **294** (2017) 012006 doi:10.1088/1757-899X/294/1/012006

Study of materials used for the thermal protection of the intake system for internal combustion engines

C Birtok-Băneasă, S Rațiu, V Puțan and A Josan

University Politehnica Timisoara, Faculty of Engineering Hunedoara, Department of Engineering and Management, 5 Revolution Street, 331128 Hunedoara, Romania

E–mail: sorin.ratiu@fih.upt.ro

Abstract. The present paper focuses on calculation of thermal conductivity for a new materials developed by the authors, using the heat flux plate method. This experimental method consists in placing the sample of the new material in a calorimetric chamber and heating from underside. As the heat flux which passes through the sample material is constant and knowing the values of the temperatures for the both sides of sample, the sample material thermal conductivity is determined. Six types of different materials were tested. Based on the experimental data, the values of the thermal conductivity according to the material and the average temperature were calculated and plotted.

1. Why thermal protection for intake manifold of an internal combustion engine?

Because of the position in the engine compartment the intake is highly expose to thermal radiations from the cooling radiator, exhaust pipe and other hot parts of the engine which is a disadvantage, the intake air it heated and the result is a lower density, thus containing less oxygen per volume unit than cold air [1]. Recent studies show significant differences in fuel consumption and exhaust emissions depending on the temperature of the intake air [2-4], [6].

The air mass found in cylinders after closing the intake valve is the decisive factor in relation to the mechanical work produced by cyclically burning of the fresh load, with direct influence on the engine torque. That means that the higher the air mass, the higher the engine torque will be. In other words, we can say that, on the intake route, two kinds of losses are recorded: pressure losses or gasodynamic losses caused by the existence of hydraulic resistances on the suction line and heat losses or thermal losses, caused by fluid heating from the suction line walls, which determine the final temperature to be higher than the engine inlet temperature, the temperature increase resulting in density decrease and, hence, affecting the filling [5], [7-9].

One method of reducing thermal losses is to insulate the intake whit a new kind of material. The principal scope is to resource and development a new insulates material from composite, natural and organic materials based from recycling materials. As an additional property, this material must be applied easily by brush or spray on the manifold surface and adhere to it. In order to do that, it is necessary to prove that the new material has insulated properties. The present paper focuses on calculation of thermal conductivity for a new materials developed by the authors, using the heat flux plate method.

IOP Conf. Series: Materials Science and Engineering **294** (2017) 012006 doi:10.1088/1757-899X/294/1/012006

2. Basic principle of the method and devices

The single-plate method determines the thermal conductivity λ of a building material sample (Figure 1c-1) of thickness d and surface A by measuring the temperature difference and thermal flux directly. It is important for the measurement that the thermal flux crosses the building material sample homogenously and that no heat disappears by other ways. Whit the calorimetric chamber (Figure 1), this is achieved in particular by the insulating housing.

(a) (b) (c)

Figure 1. Calorimetric chamber
a – general view, b – inside view of electric heated source, c – virtual model: 1 – material sample, 2 – calorimetric chamber wall, 3 – heated source

The chamber is electrically heated on the inside (Figure 1b,c), and ice is placed directly outside the sample (Figure 1a). In the thermal equilibrium, i.e. in stationary condition, in which the temperature is constant over time at every point, the electric power exactly matches the thermal flux, i.e. the inserted electrical energy is equal the thermal energy flowing through the building material sample.

Figure 2 shows a general view of the experimental ensemble containing the following devices: transformer 2-12 V, calorimetric chamber with insulated housing, digital thermometer with 4 inputs and a laptop for data acquisition.

Figure 2. General view of the experimental ensemble

Three NiCr-Ni thermocouple sensors were used for temperatures measurements as follows: T1 – outside sample surface, toward the exterior of the calorimetric chamber, T2 –on the underside, toward the interior of the calorimetric chamber, T3 –inside the calorimetric chamber. The material sample have square surface of $150 \times 150\ mm$ and $10\ mm$ thickness and fit perfectly inside the calorimetric chamber in horizontal position (see Figures 3 and 5).

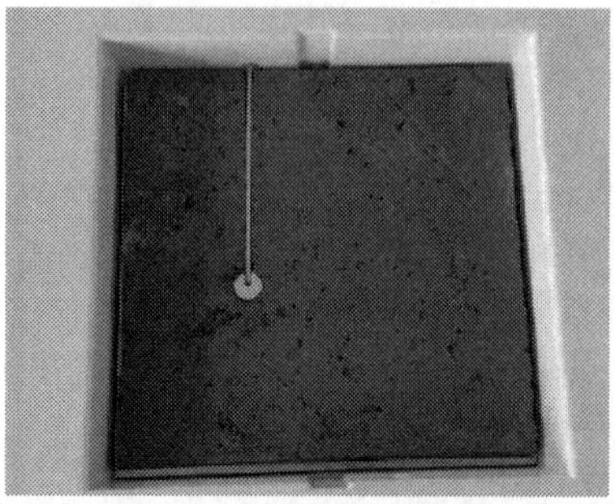

Figure 3. The way the thermocouples are mounted and the material sample are fitted inside the calorimetric chamber

Six samples of different material were developed for which thermal conductivity was determined.

Figure 4. The six samples

Material sample composition – Sample 1

Elements	Silicone	Polyurethane foam	Solvent	Cork	Pigment
Composition %	-	100	-	-	-

Material sample composition – Sample 2

Elements	Cork	Polyurethane foam	Solvent	Pigment
Composition %	8.36	68.82	19.01	3.81

Material sample composition – Sample 3

Elements	Silicone	Polyurethane foam	Solvent	Cork	Pigment
Composition %	11.38	61.20	9.64	5.69	12.09

Material sample composition – Sample 4

Elements	Silicone	Polyurethane foam	Solvent	Pigment
Composition %	21.31	69.26	6.14	3.29

Material sample composition – Sample 5

Elements	Silicone	Polyurethane foam	Solvent	Pigment
Composition %	58,1	37.28	1.73	2.89

Material sample composition – Sample 6

Elements	Silicone	Polyurethane foam	Solvent	Cork	Pigment
Composition %	32.95	21.11	3.42	1.01	2.89

Figure 5. Material samples composition

Experimental data was acquired with the help of Cassy Lab2 software. The interface of this software is presented in Figure 6.

IOP Conf. Series: Materials Science and Engineering **294** (2017) 012006 doi:10.1088/1757-899X/294/1/012006

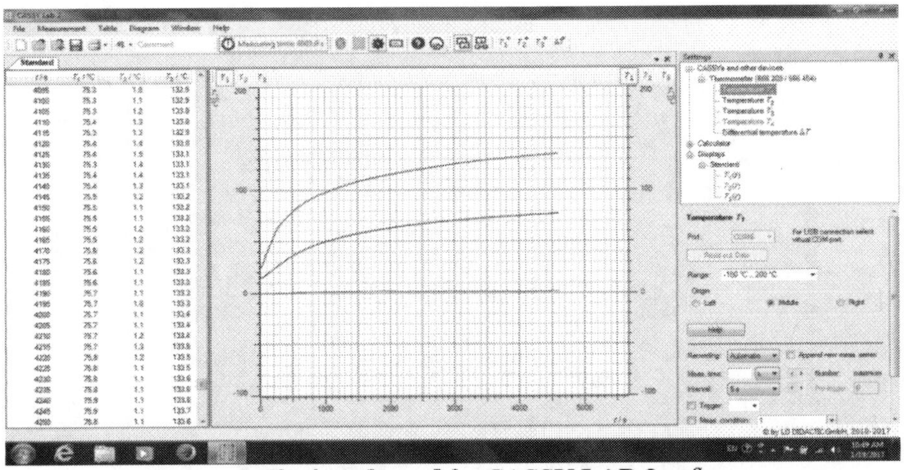

Figure 6. The interface of the CASSY LAB 2 software

The temperature values were taken from 5 to 5 seconds for 4,000 seconds, the time required to stabilize the heat transfer phenomenon.

3. Results and conclusions

Below are the results obtained from the calculations based on the experimental data taken.

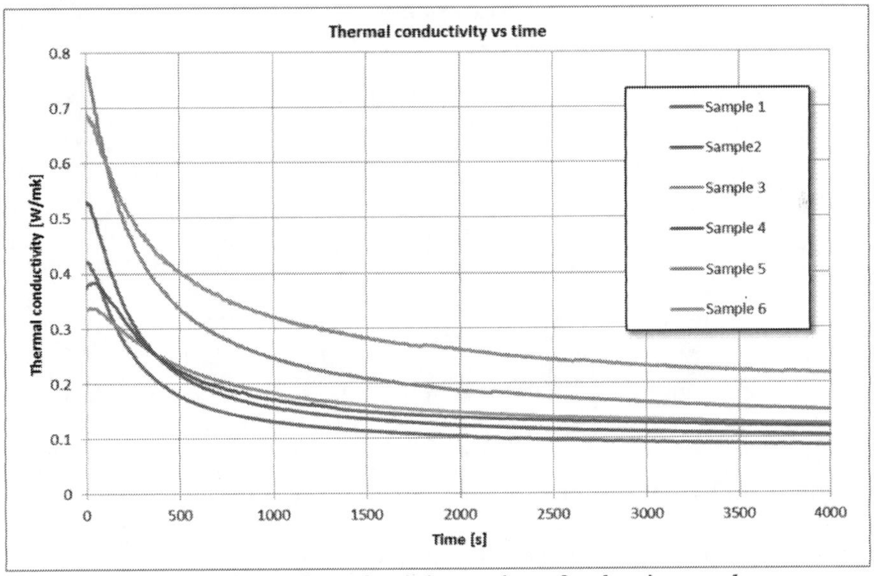

Figure 7. Thermal conductivity vs time, for the six samples

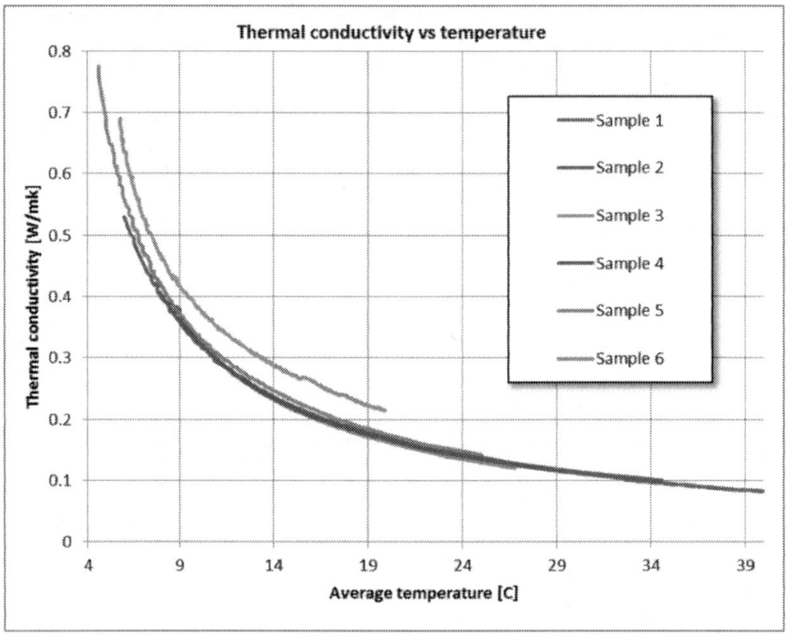

Figure 8. Thermal conductivity vs temperature, for the six samples

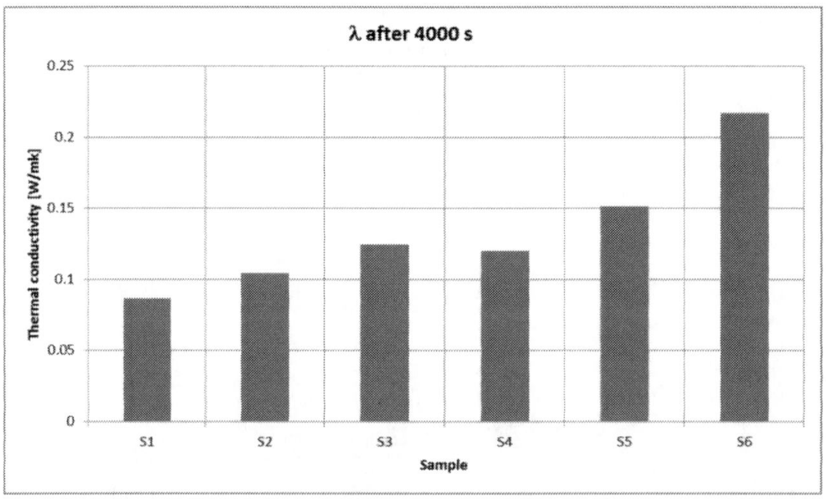

Figure 9. Comparative values of the thermal conductivity for the six samples

In conclusions, we can say that the first material sample can be used as a thermal insulation material for coating the intake manifold of an internal combustion engine.

References

[1] Birtok-Băneasă C, Raţiu S, Heput T 2017 Influence of intake air temperature on internal combustion engine operation, *IOP Conf. Ser.: Mater. Sci. Eng. Materials Science and Engineering* **163** 012039

[2] Nik R A, Hazimi I, Zeno M, Asiah A R and Hazim S 2015 Effects of air intake temperature on the fuel consumption and exhaust emissions of natural aspirated gasoline engine, *Jurnal Teknologi Sciences & Engineering* eISSN 2180–3722

[3] Mamat R, Abdullah N R, Xu H M, Wyszynski M L and Tsolakis A 2009 Effect of Fuel Temperature on Combustion and Emissions of a Common Rail Diesel Engine. SAE International Powertrains, Fuels and Lubricants. 2009-01-1896

[4] Rakesh K M and Avinash K A 2011 Experimental Investigation on the Effect of Intake Air Temperature and Air–fuel Ratio on Cycle-to-Cycle Variations of HCCI Combustion and Performance Parameters, *Journal of Applied Energy* **88** 1153-1163

[5] Raţiu S, Alexa V, Kiss I and Cioată V 2017 Experimental determination of the filling coefficient for an aspirated spark-ignition engine, *IOP Conf. Ser.: Mater. Sci. Eng.* **163** 012038

[6] Rizalman M, Ahmad F Y, Abdul A A, Amir A and Nik Ro A 2013 Effect of EGR and Air Temperature on Exhaust Emission of a Diesel Engine Operating with Biodiesel, *Journal of Biobased Materials and Bioenergy* **7** 461-463

[7] Birtok–Băneasă C and Raţiu S 2011 *Admisia aerului în motoare cu ardere internă – Filtre supraaspirante – Sisteme dinamice de transfer* [Air intake in internal combustion engines – Super absorbing filters – Dynamical Air Transfer Systems], Publisher: Politehnica, Timişoara, Romania

[8] Raţiu S 2009 *Motoare cu ardere internă pentru autovehicule rutiere – Procese şi caracteristici, Experimente de laborator* [*Internal combustion engines for road vehicles – Processes and Features – Laboratory experiments*], Publisher: Mirton, Timişoara, Romania

[9] *** 2011 *Bosch Automotive Hand–book*, 8th Edition, May 2011

IOP Conf. Series: Materials Science and Engineering **294** (2017) 012007 doi:10.1088/1757-899X/294/1/012007

Methods of silver recovery from radiographs - comparative study

L R Canda, E Ardelean and T Hepuț

Politehnica University of Timisoara, Engineering and Management Department, Revolutiei Str., no. 5, 331128 Hunedoara, Romania

E-mail: letitia.canda@fih.upt.ro

Abstract. Management and recovery of waste are activities with multiple impacts: technologically (by using waste on current production flows, thus replacing poor raw materials), economically (can substantially reduce manufacturing costs by recycling waste), social (by creating new jobs where it is necessary to process the waste in a form more suited to technological flows) and ecologically (by removing waste that is currently produced or already stored - but poses a threat to the health of the population and / or to the environment). This is also the case for medical waste, for example radiographs, which are currently produced in large quantities, for which replacement solutions are sought, but are currently stored by archiving in hospital units. The paper presents two methods used for this kind of waste management, the result being the recovery of silver, material with applications and with increasing price, but also the proper disposal of the polymeric support. This analysis aims at developing a more efficient recycling technology for medical radiographs.

1. Introduction

Silver is an important precious metal, due to its unique thermal conductivity, optical reflectivity and its photosensitivity. Its countless, medical and industrial application, have made studies on silver to be an attractive area of research related both extraction, and recovery from different waste.

Although global legislation places a growing underlining on controlling the dispersion of hazardous waste into the environment, significant amounts of waste reach it. Radiographs and effluents resulting from processing are such examples. Provided the X-rays reach patients, it is difficult to control their management, requiring a system to manage them either before reaching patients or through subsequent, controlled, safe collection, while ensuring health monitoring of patients.

The number of radiographs is rising, on one hand due to population growth and extended access to modern investigative methods (even in countries where access was formerly limited) and on the other hand, due to the increasingly intense pollution of the environment, resulting in a deterioration in the life quality, which leads to the necessity of more frequent medical investigations.

Processed radiographs contain a significant amount of silver that can be recovered. There have been several studies on its recovery from processed radiographs and their effluents [1-5], but the generally proposed methods have many disadvantages, vulnerabilities, one of which is that most of them involve high costs. Researches in this area need to reveal simple and effective techniques, both economically and ecologically. It is therefore necessary to improve these methods, as natural resources are limited and unceasingly.

In the recent years, a particular accent has been put on the purity and the form of recovered particles, although difficult to achieve, in addition to recovering silver itself, thus developing the study of nanoparticles.

The pharmaceutical industry uses silver more frequently. The shape, size, and distribution of nanoparticles depend on the methods applied in the recovery process. Electro-refining is a very important process because, in all areas, it is necessary to use silver as purely as possible. Adjusting the parameters in the electrochemical method makes it possible to control of the particle size. However, it is still difficult to establish a well-defined, quantitative relation between the reaction parameters and the resulting particulate form.

Generally, not few chemical and physical methods are used for the synthesis of silver nanoparticles. For the cluster growth, it needs an energy, at a particular size, so more silver atoms diffuse in solution and captured to the surface.

Typical anisotropic forms result in the presence of a stabilizing polymer that binds preferentially to one of the faces of the crystal faster than the other does [6].

Most part of particles obtained are spherical, pyramidal, cubic, rods, octagonal or thin sheets shaped. Spherical particles are thermodynamically the most stable because of their minimum surface at a certain volume [7]. At first, in chemical methods, usually, a strong reducing agent produces small silver particles, and then a weaker reducing agent enlarges them. For example, the following reducing agents can be used: glucose, ethylene glycol, hydrazine hydrate and citrate of sodium [8], ultrasonic irradiation as a green reducing agent [9], sodium borohydride in the presence of sodium citrate (Na_3Cit) at room temperature [10], hydroxylamine. Normally the second selective reducing agent may be add about one mole +/−5% of silver reducing agent for each mole of silver. When sodium formate is utilised as the second selective reducing agent, however, it is preferable that at least two moles of sodium formate per mole of silver should be added [11].

The research of silver recovery from exposed X-ray films has been oriented towards several directions over time. Pyrometallurgical methods are inadequate, and energy-intensive. Studies on silver recovery by enzymatic methods are in a relatively incipient phase, with inconclusive and uncertain results, assuming a long time-span of separation of the emulsion film [12], [13].

Although seemingly harmless, some of these methods can give rise to dangerous secondary products, unless there is a clear strain on processes. Therefore, it seems that the best results regarding silver recovery from radiographs are those obtained by hydrometallurgical methods.

The migration of the emulsion from the radiological film is a relatively simple process, but a particular attention should be paid to the favourable factors, which can usually be: the dissolution temperature, the concentration of the solutions and the speed of rotation if an electromagnetic stirring is used.

2. Experimental data's

In this research, base outline of the process is the separation of inorganic component from the polymer layer by solubilisation with NaOH [5] and HNO3 [3], respectively. The radiographic film contains gelatine, a protein derived from sacrificed animals leather and bones, containing glycine, proline and 4 hydroxyproline residues, in which are dispersed very fine silver bromide granules [14], [15].

Within the first experimental method, there were used aqueous NaOH solutions, to extract the polymer deposited substrate. The film was sequentially treated with NaOH solutions of various concentrations: 1M, 1.25M, 1.5M and 2M, respectively. For each 250g of solution, there were used 250g of radiographs. There were also used laboratory glassware with large base areas, so the dimensions at which the film was cut could be larger, $5x5cm^2$. To avoid the sticking to each other, radiographic films have been introduced at a time. In order to increase the leaching speed, mechanical agitation of the aqueous solution was continually carried out with a stirrer rod.

The temperatures at which the leaching took place varied between ambient temperature and 90°C. The solutions of NaOH containing black colloidal silver were decanted and filtered – Figure 1. The residue was washed with distilled water (3-4 times) and then dried – Figure 2. The silver particles were visualized at the stereo-microscope (x60) – Figure 3, Stereo-microscope from the Physical Metallurgy Laboratory within Hunedoara Engineering Faculty, Politehnica University of Timisoara.

Figure 1. Resulting solution after leaching

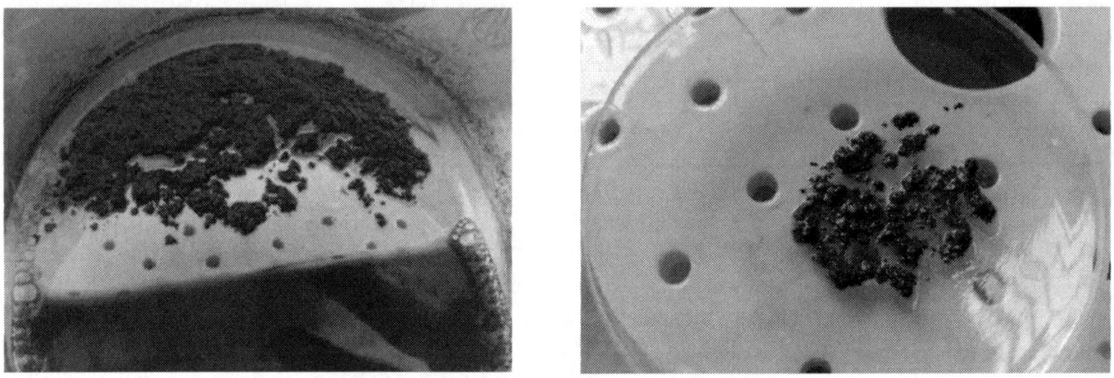

Figure 2. The solid residue before and after settling and washing

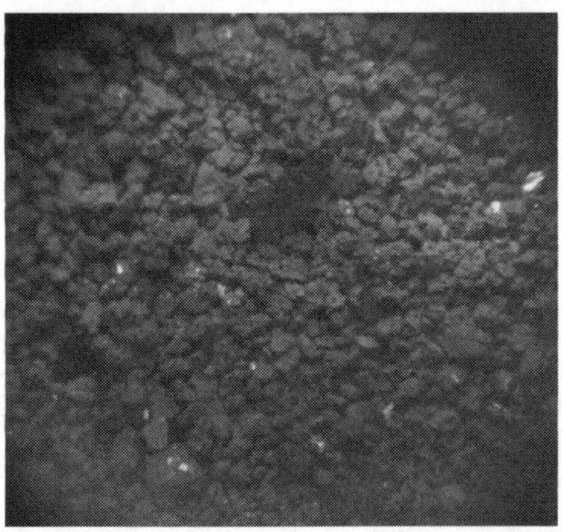

Figure 3. View by microscope after filtration and drying

The second method used involves the passing of the silver deposited with the emulsion on the polymer support into a solution of HNO_3 [16]. Three beakers, each containing 250g HNO_3 aqueous solution (68%) were used: the first beaker contained 30% HNO_3, the second 25% HNO_3 and the third 20% HNO_3 respectively. Every beaker was leached an amount of 350g film. Dissolving results in silver nitrate and impurities. During the transfer of silver to the HNO_3 solution, evaporation occurs. Thus, distilled water was added, when necessary, to the completion of the 250ml. When the leaching time increased, the solution was considered to lead to saturation.

The solutions, with different concentrations, were cooled, filtered with the Büchner funnel and subjected to several silver recovery processes by cementation.

The general chemical cementation reaction can be explained by the formula (1) [17]:

$$Ag^{+z_1} + \frac{z_1}{z_2} Me \Leftrightarrow \frac{z_1}{z_2} Me^{+z_2} + Ag \tag{1}$$

where Me is the metal used for cementation (in our case Cu, Zn, Al or Fe);

$z_1 = 1$ - valence (electro-valence);

z_2 - the valence of the metal used for cementation.

Process for cementing a metal from an acidic, aqueous solution is based on the capacity of some metals that are higher in the electromotive series than the metal to be recovered, to reduce the ions of more noble metals. Metals such as Cu, Zn, Al or Fe can cement the silver from $AgNO_3$ solutions. Copper cementation has been preferred for a better visualization of recovered silver since the initial phase. Zn and Fe tests have also been performed – Figure 4.

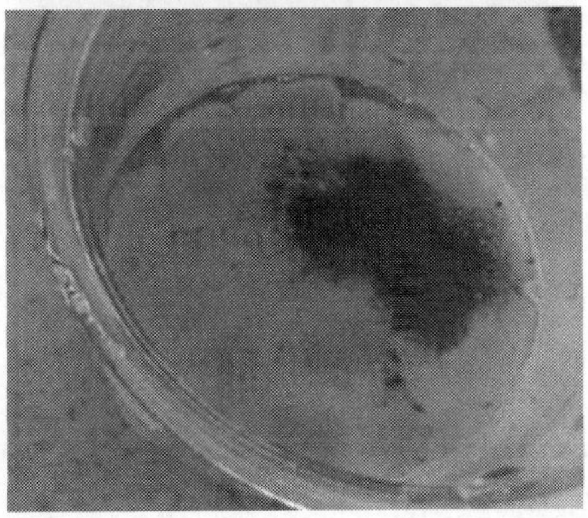

Figure 4. Copper cement (flat piece), zinc (small pieces) and iron (metal powder)

The higher initial nitric acid concentration was required to increase the silver flow rate in the solution. If the solution is not saturated with silver ions, Copper nitrate ($Cu(NO_3)_2$) will be formed concurrently with Cu – Figure 5. In the cementation process, a very small amount of copper is consumed by diluting the solution. After the copper cementation test was carried out, the solution was filtered, the silver metal remained on the filter paper – Figure 6, which was washed with distilled water to avoid further chemical reactions.

Figure 5. Complex solution of nitrate silver and copper

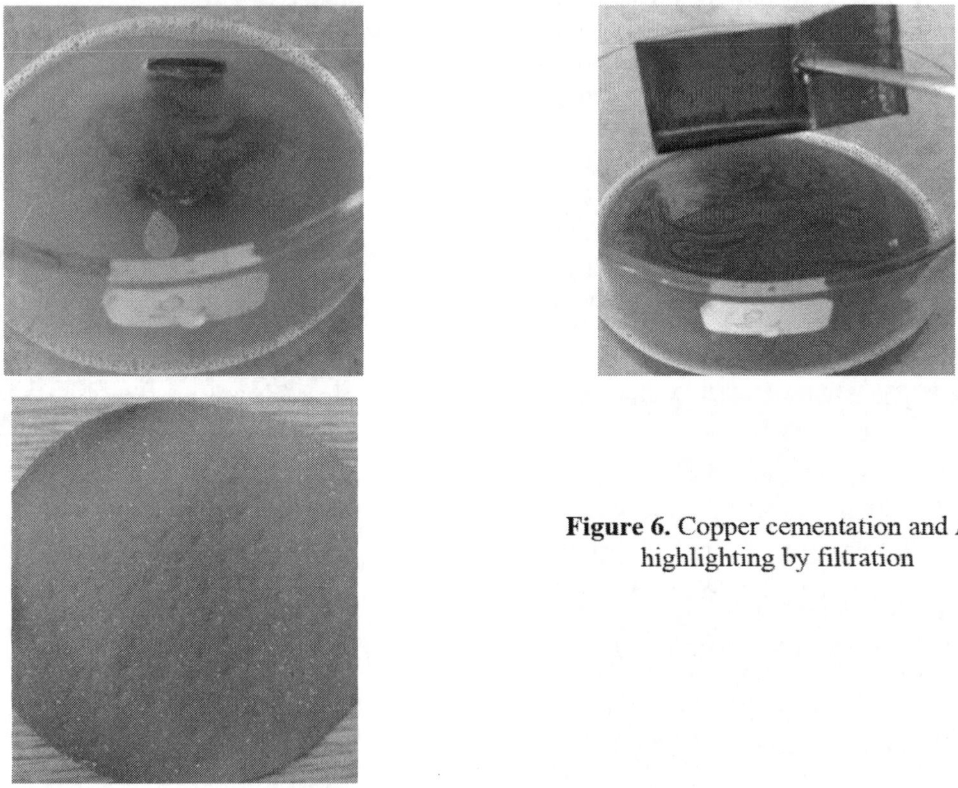

Figure 6. Copper cementation and Ag highlighting by filtration

To streamline the process, the resulting solution, after the filtration is reused in the silver leaching process, upon a new series of radiographs.

3. Results analysis

The equilibrium and kinetics of chemical reactions at the passage of silver in the solution from X-ray films depends on several factors, regardless the method used. These are:

- concentration of reagents;
- type of radiography (silver content and age after archiving);
- leaching time in the solution;
- stirring rate;
- leaching temperature;
- solid / liquid rate.

In order to study this equilibrium and kinetics of chemical reactions, the concentration of aqueous NaOH solutions was taken into account, for each experiment measuring the leaching time and the temperature. Concerning the rate of stirring, it can be considered a constant factor in the experiments carried out, considering that the mechanical agitation performed in order to decrease the leaching time. Figure 7 shows the dependence between leaching time and temperature at different concentrations of aqueous NaOH solution.

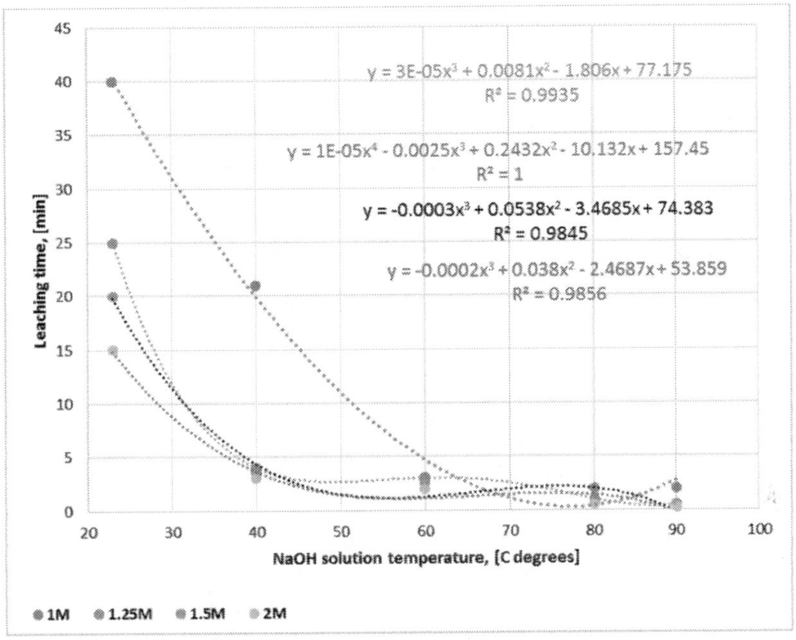

Figure 7. Dependence of leaching time and temperature at different concentrations of aqueous NaOH solution

The graph analysis in Figure 7 shows that an increase in the concentration of the solution at 2M does not lead to a significant decrease in 80-90°C the silver transition time, which is why the increase in concentration is pointless (especially from an economic and environmental point of view). The minimum leaching time, at 80-90°C, occurs at a 1.5M concentration.

A good purity silver is obtained by borax treatment and heating at high temperature, 500-600°C, by calcination [18].

Regarding the parameters characterizing the leaching process of the Ag-containing the emulsion from the radiographic films - presented in the graph of Figure 8, the following were found:

- the leaching is carried out at very low speeds, which leads to a decrease of the yield, at ambient temperature;
- the temperature has a relatively low influence, at temperatures above 40°C, which leads to the conclusion that an average temperature of 60-65°C would be sufficient for a good solubility;
- either a very high concentration of HNO_3 in the aqueous solution is not recommended and can be solubilized at the lower limit of the range, respectively a 20% HNO_3 concentration in the aqueous solution.

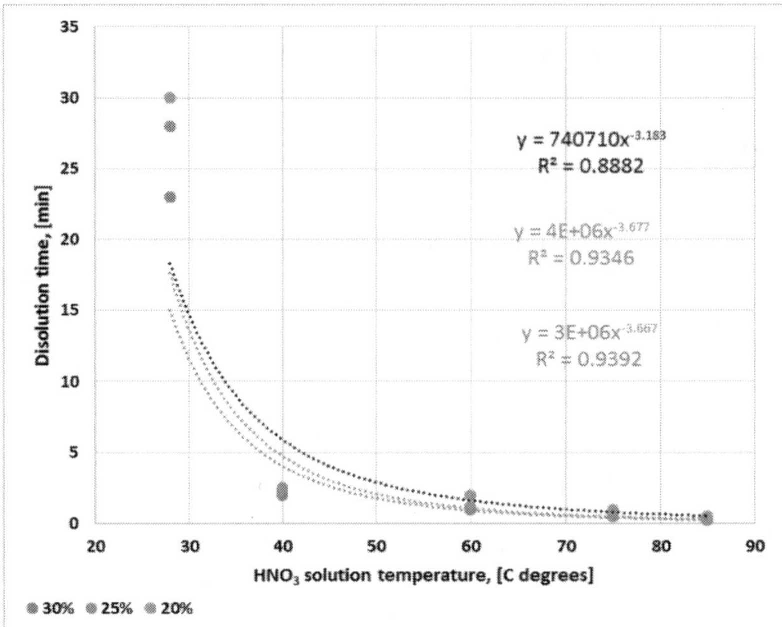

Figure 8. The dependence between dissolution time and temperature at different concentrations of HNO_3 solution

4. Conclusion

As a result of running under laboratory conditions the two different methods for recovering silver from radiographic films, the following conclusions can be drawn:

- the obtained silver has sufficiently high purity, in both experiments. Electrically refining can be performed for a high purification of silver;
- both methods are relatively simple and silver was successfully recovered with low costs;
- regarding the leaching of radiographic films in aqueous NaOH, the following process parameters are recommended: concentration of aqueous 1.5M NaOH solution, 80-90°C process interval;
- it has been found that in the method of solubilisation with HNO_3, a lower concentration of acid in the aqueous solution can also be employed; In addition, the resulting solution can be recycled for a new leaching of silver from the X-rays. Recommended parameters: concentration of 20% HNO_3 at a working temperature of max. 65°C.

Subsequent research will focus on recovering silver from X-ray films by sodium and ammonium thiosulphate [2], [19], [20] oxalic acid [1], establishing a quantitative relation between the recovered silver, solution concentration and agitation rate, as well as recovering silver from the effluents resulting from the development process [21].

References

[1] Masebinu S O and Muzenda E 2014 *Rewiew of silver recovery technics from radiographic effluents and X-ray film waste,* Proceeding of the World Congress on Engeneering and Computer Science 2014, Vol II, WCECS 2014, San Francisco, USA, 22-24 October

[2] Yazici E Y, Bas A D and Deveci H 2010 *Extraction of Silver from Waste X-ray Films by Thiosulphate Leaching,* Proceedings of the XIIth International Mineral Processing Symposium, Cappadocia-Nevşehir, Turkey, October

[3] Arslan V, Ucurum M, Vapur H and Bayat O 2011 Recovery of silver from waste radiographic films by chemical leaching, *Asian Journal of Chemistry* **23**(1) 67-70

[4] Marincovic J, Korac M, Kamperovic Z and Matic I 2006 Recycling of silver from exposed X-

ray films, *Acta Metalurgica Slovaca* **12** 262-268

[5] Nakiboğlu N, Toscali D and Nişli G 2003 A novel silver recovery method from waste photographic films with NaOH stripping, *Turkish Journal of Chemistry* **27**(1) 127-133

[6] Popa I, Nicola I M, Ceara V, Boboc C and Danec C A 2016 Obținerea și utilizarea nanoparticulelor de argint, *Electrotehnică, Electronică, Automatică (EEA)* **64**(2) 104-112

[7] Khodashenas B and Ghorbani H R 2015 Synthesis of silver nanoparticles with different shapes, *Arabian Journal of Chemistry* 1-16

[8] Guzmán M G, Dille J and Godet S 2009 Synthesis of silver nanoparticles by chemical reduction method and their antibacterial activity, *International Journal of Chemical and Biomolecular Engineering* **2:3** 104-111

[9] Randa F E, Kamyar S, Mansor B A, Nor A I and Norhazlin Z 2015 Green sonochemical synthesis of silver nanoparticles at varying concentrations of κ-carrageenan, *Nanoscale Research Letters* **10** 302

[10] Wojtysiak S and Kudelski A 2012 Influence of oxygen on the process of formation of silver nanoparticles during citrate/borohydride synthesis of silver sols, *Colloids and Surfaces A: Physicochemical and Engineering Aspects* **410** 45-51

[11] ***https://www.google.com/patents/US6290749 - Preparation of ultra-pure silver metal US 6290749 B1

[12] Shankar S, More S V and Seeta Laxman R 2010 Recovery of silver from X-ray film by alkaline protease from Conidiobolus Coronatus, *Kathmandu University Journal Of Science, Engineering And Technology* **6**(I) 60-69

[13] Li G, He D, Qian Y, Guan B, Gao S, Cui Y, Yokoyama K and Li Wang 2012 Fungus-Mediated Green Synthesis of Silver Nanoparticles Using Aspergillus terreus, *Int J Mol Sci.* **13**(1) 466–476

[14] ***http://radiologiarte.tripod.com/pelicularx.htm

[15] Jadhav U U and Hocheng H 2012 A review of recovery of metals from industrial waste, *Journal of Achievements in Materials and Manufacturing Engineering AMME* **54**(2) 159-167

[16] Aktas S, Morcali M H and Yucel O 2010 Silver recovery from waste radiographic films by cementation and reduction, *Canadian Metallurgical Quarterly The Canadian Journal of Metallurgy and Materails Science* **49**(2) 147-153

[17] Jaskuła M 2009 Cementation of Silver Ions on Metallic Copper, *Jordan Journal of Earth and Environmental Sciences* **2**(1) 84- 95

[18] Chandre D K, Surendra D A and Mishra V V 2017 Silver recovery from used X-ray film using borax compound, *IJARIIE* **3**(2) 1744-1747

[19] ***https://www.lens.org/lens/patent/US_2013_0341203_A1

[20] ***https://www.google.com/patents/US20170044644?cl=en

[21] Mahdizadeh F, Eskandarian M, Zabarjadi J, Ehsani A and Afshar A 2014 Silver recovery from radiographic film processing effluents by hydrogen peroxide: Modeling and optimization using response surface methodology, *Korean Journal of Chemical Engineering* **31**(1) 74–80

Methods for maintaining the continuity of power supply in hospitals

V V Mihut, M Panoiu, C Panoiu and I Baciu

Politehnica University of Timisoara, Department of Electrical Engineering and Industrial Informatics, 5 Revolution Street, Hunedoara, 331128, Romania

E-mail: mihut voicu@yahoo.com

Abstract. This paper presents alternative solutions to ensure power supply of critical consumer, such hospitals. In a hospital it is extremely important to have a power supply without any interruption. This is the reason of the using various solutions that ensure continuity of power supply. In this paper it is presented a solution used in a hospital in Romania. Here is used a generator controlled by microprocessor capable of taking over power supply in case there are problems with power supply from the national power grid.

1. Introduction

Knowing the current need for stability and continuity in the power supply of critiques equipment, it tends increasingly to use more alternative sources of electricity as a back-up method in power supply [1], [2]. Alternative solutions to traditional methods of producing electricity offer stability in operation of the equipment [3]. Knowing that such equipment can feed including surgery rooms, medical emergency reception centres, intensive care and resuscitation centres, etc., we can say that the implementation of these automated systems save lives [4].

Alternative methods of producing electricity must meet the following technical and economic considerations:

- reach the current and nominal power in very short time (between 0.1 to 120 seconds), depending on the type of alternative source.
- can be operated and / or put back into operation quickly by qualified staff
- provide as much as possible operational autonomy as well
- provide stability in operation from technical standpoint and electricity supplied
- have a relatively low cost of implementation and operation

From a technical standpoint we recommend the use of two types of elective alternative sources, namely:

a) Electrical sources UPS - inverter which provides a very fast coupling in load, between 0.1 - 0.5 seconds. The installed power is between 250 VA ÷ 90 kVA. The high power can provide a greater autonomy of unloading operation in both linear voltage 230V and 400V as well. This type of sources use batteries gel (mainly with nominal voltage of 12V DC) which has the following characteristics: require no maintenance operations provides maximum safety against leaks, can support up to 800 cycles of operation, and ensure longevity and reliability in operation. The advantage of this type of alternative electrical source consists in fast task switching and disadvantage group consists in using of batteries that can be discharged without possibility of supplementary charge.

b) Electric generator which provides a coupling in the load between 60-300 seconds. Coupling time depends on the type of fuel, temperature of the ambient temperature of the engine block during standby engine output heat electric power generator - mostly installed power ranges from 650 VA ÷ 300 KVA. The advantage of this type of alternative source is that we can increase the operating time as long as there is fuel in the tank. Even recommend switching to more fuel tanks, tanks that can be fed in functioning

In the electrical systems of major importance it is recommend using both types of electrical energy sources. Of course you can use other types of sources that keep mandatory features described namely: quick-coupler pregnancy, high power cut, high autonomy of operation.

Also the operation mode and the input / output to / from the load are very important. Knowing that it is difficult to balance two alternative power sources in parallel does not recommend starting simultaneously these types of equipment.

Always the first source that feed the electrical equipment must be source-in the task. Depending on the necessity it is possible to command the second source to enter in compensation with the first, but it all depends on the installed power of the first sources and power requirements of the consumer. Always has to be a balance between power generated by the power source and absorbed by the consumer.

Depending on the consumer it can be made a calculus of power of alternative sources, such as their technical characteristics and functioning and especially the installed power.

It is not recommended to use alternative power sources without knowing the entire assembly, not to endanger the optimal functioning of the equipment.

Also always recommended calibration and optimal setting of these sources (given that some have a wide range of settings), according to the type of consumer, environmental and binding according to the main system power supply.

In the following an optimal implementation in operation for generating electrical systems major is presents.

2. The electric generator group

The electric generator groups a fixed or mobile autonomous electricity production, which ensure independent power supply, emergency, if the main electrical supply interrupted [5-10]. The group is made up of a diesel engine which drives a synchronous generator capable of providing an output of at least 1.5 times the normal amount of power consumed routinely and automatically transfer the load equipment that monitors network parameters main engine and the generator ensuring command sequences automatic START - STOP.

2.1. Automatic load transfer equipment

It is the "brain" of the generator and is equipped with a microprocessor and it has multiple inputs and multiple outputs that enable it to perform the following functions:

a) Measure and display the voltage primary source. If any phase voltage main power is out of the (± 10%) U_{nom} is starting command for the generator start cycle.
b) Measure and display engine for the following parameters: battery voltage (V), engine coolant temperature (°C), lubricating oil temperature (°C), lubricating oil pressure (kPa), engine rotation speed (revolutions/minute), cumulative total number of operating hours, the number of starting attempts, fuel level in the buffer tank, the fuel level in the tank outside.
c) Measure and display synchronous generator following parameters: line and phase voltages (V), the phase currents (A), active powers (KW), total apparent power (KVA), total reactive power (KVAR), total active energy (KWh), total reactive energy

(kVARh), total power factor, the percentage of rated output (%), the level of excitation of the generator, frequency (Hz).

d) The following cases are ordered protection for stopping the engine while the onset of light signals: oil under reduced pressure, overheats coolant, packaging or excessive speed, start attempt failed, too low or to high values of frequency, too low or to high values of output voltage, generator overload, generator short circuit

e) The motor-warning for the following parameters: coolant temperature dangerously high, dangerously low coolant temperature, engine oil level too low, dangerously low engine oil pressure, dangerously low fuel level in the tank buffer, unsafe voltage batteries

f) Operating mode selection. There are three modes of operation: local manual control, local automatic control, and automatic remote control.

g) Diagnosing failures

PLC must display the diagnosis and to identify the nature of functional errors. Errors must be preserved in memory, supporting in obtaining a history of behavior of the generator, following the last or the last reset.

The main parameters or components that are aimed at the diagnosis failures: failure lubrication oil pressure, coolant temperature detection, voltage batteries, motor rotation speed detection, microprocessor or memory automatically, DC power transducers, unexplained unscheduled stoppages, failure speed regulator, failure voltage regulator, failure synchronous generator, incorrect starting sequence, failure starter, failure generator, fuel electric valve control relay, guillotine intake air control relay.

h) Generate startup sequence

i) Generate stop sequence

To explain the generation sequences mentioned, considered the case supplying a hospital. According to the figure1, KN represents mains contactor, KG emergency contactor and Kins auxiliary source realized islanding consumer class 15 for others. Islanding is necessary to use a low cost generator, with a power that permit the supply only consumers in class 15 using a backup solution from another transformer station that does not belong to the hospital with a switch manually controlled or remotely by a human operator, overload protection and short circuit.

The starting sequence is as follows:

- The main source fall after 5-10 seconds commands the starter.
- If the engine does not start, longer command the starter twice more with break between starts about 10 seconds.
- If the engine speed reaches 15% of rated speed, ordering off the starter. If this condition is not achieved, destroying the starter.

Interruptions of mains voltage distribution can last from less than a second to several minutes or longer. If interruptions between two seconds and two minutes are common, caused by minor incidents distribution system, the extended more than two minutes, are due to events such as storms, lightning, earthquakes, fault lines of high voltage switchgear (110, 230 KV) underground cable disruption due to construction etc.

2.2. Diesel Engine

Are used, four-stroke engine with direct injection and turbocharging, in particular to be able to take 100% of nominal load in a single operation, diesel fuel must have an average octane number higher than 35.

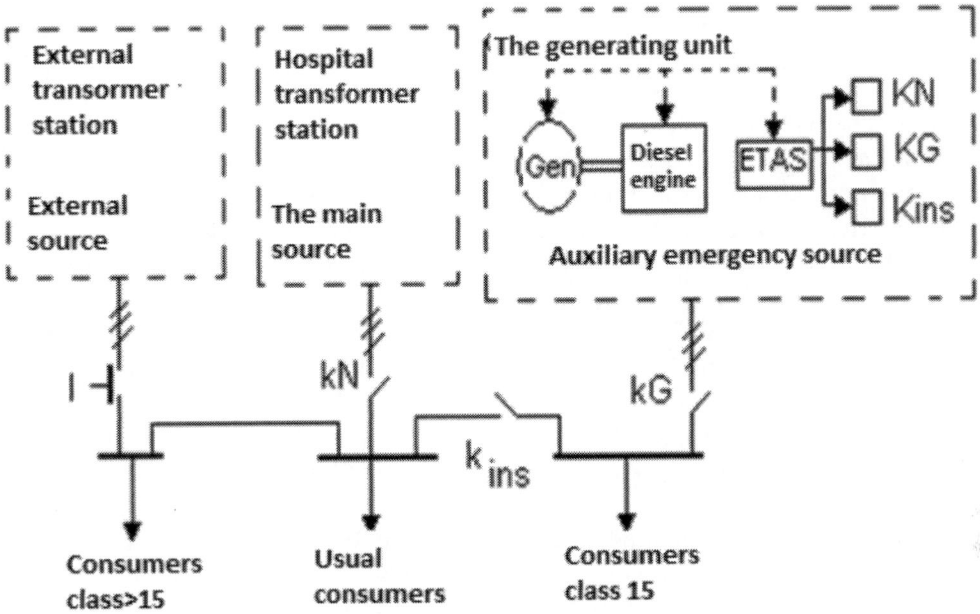

Figure 1. Power supply system for a hospital

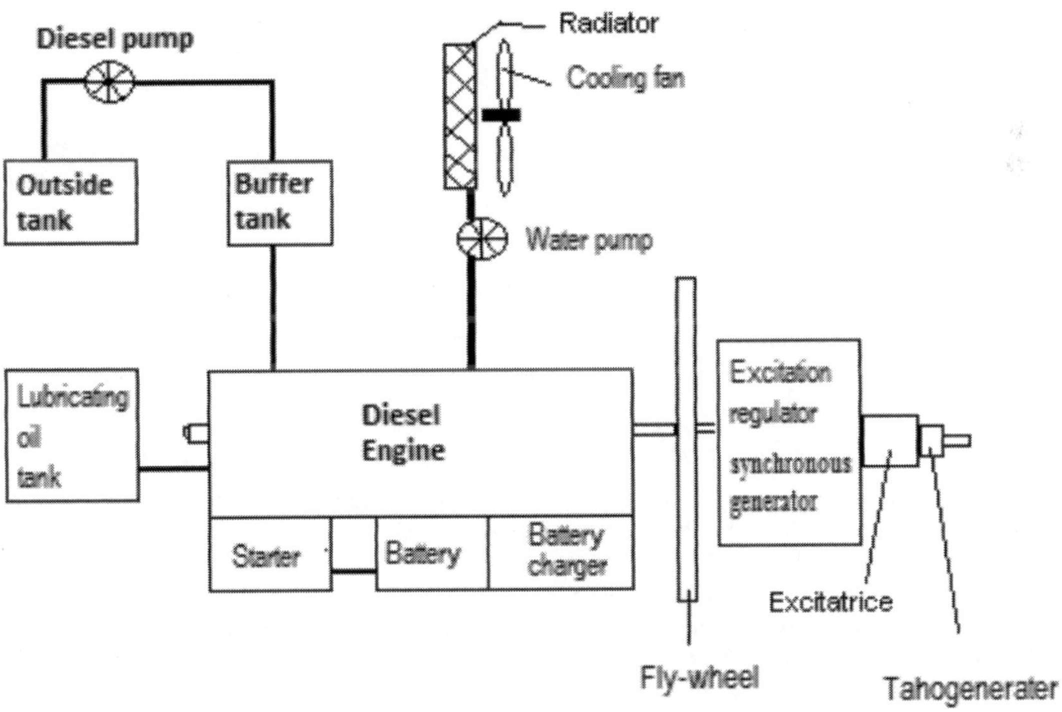

Figure 2. Schematic Diesel group - synchronous generator

The power supply circuit comprises:
- A buffer tank with a maximum capacity of 200 l,
- A tank outside, of higher capacity necessary to assure autonomy according to the group,
- An automatic pump, coupled with a manual pump, for filling the buffer tank from outside the tank.

The buffer tank is placed at a height above the engine, based on the required circuit pressure of the injection

Lubrication circuit comprises a container and a mini-pump to ensure sufficient pressure.

The cooling circuit comprises a radiator, a fan and a pump attached to the cooling water circulation. The heating standby circuit contains a heater (corded) which ensures the maintenance of internal motor temperature around 40°C. The electric starting system is composed of the group of batteries, electric starter and an independent electronic charger.

Battery capacity should allow three repeated attempts to start a current of 1000 A minimum at 0° C.

Electronic charger powered from the mains with a charging current of 10A minimum, must provide 80% of the rated load in 4 hours and 100% within 12 hours.

Speed controller includes:
- An electronic PID control to adjust motor speed by injection,
- A tachometer generator mounted on the motor shaft which produces a voltage proportional to motor speed, as input to the controller,
- A mechanical device security (guillotine for air inlet) to stop the engine immediately in case of needed.

System for noise limitation is made of specific devices for sound insulation in areas of the air supply and the special acoustic panels and panels for engine room. Allowable noise level for these areas is approximately 75 dB.

2.3. Synchronous generator

For synchronous generator driven by the diesel engine, the permissible voltage variation to variation in the slow load power provided is up 1% of nominal voltage U_{nom}. The most critical regime is characterized by varying the instantaneous voltage generator when suddenly loading or unloading the generator occurs. Loading suddenly the generator with a high current, lead to a rapid decrease of the terminal voltage.

If this drop exceeds 25% - 35% U_{nom} may occur two cases: the generator get out of synchronism, or is command stopping the group. For this regime the maximum allowable voltage variation is 5% U_{nom}. It is considered that a maximum of 1.8 $I_{nom\_generator}$ is acceptable if the number of starts is higher than 10 starts / day.

3. Conclusions

Electricity is produced and transported away from the place where we use so often, its supply interruption occurs. Continuity of power supply is important and sometimes critical for some users.

An electric generator group provides important sources of power supply, in particular when large consumers or productive units where interruptions can cause high problems. Whether banks, hospitals, hotels, guesthouses, restaurants, residential buildings, corporate headquarters, production facility or private residence, by lack of electricity can be absolutely no access to a resource required daily. Diesel generators maintenance costs are lower, are reliable, robust and resilient over time.

The electric generator groups represent modern solutions to ensure continuity of power supply regardless of the consumer who require uninterrupted electrical energy supply.

References

[1] Tutelea L N, Deaconu S I and Popa G N 2014 *Reduced cost low speed wind or hydro energy conversion system with twin stator windings induction generator*, 16th International Power Electronics and Motion Control Conference and Exposition, 21-24 sept, Antalya, Turkey, pp 317 – 324

[2] Baciu I, Cuntan C D and Gabor B 2014 *The study of the electric parameters of an autonomous supply system with photovoltaic panel*, International Conference and Exposition on Electrical and Power Engineering (EPE), 16-18 Oct, Iasi, Romania, pp 1138 – 1143

[3] Cunţan C D, Baciu I and Ghiorghioni L 2011 *DC Stabilizer With Digital Control*, 10th Aniversary International Conference on Accomplishments In Electrical and Mechanical Engineering and Information Technology, Banjaluka, Rep Srpska, pp 807-812

[4] Sărăcin C G, Sărăcin M and Zdrenţu D 2015 *Integration of residential alarm systems into electric power supplies of vital consumers*, 9th International Symposium on Advanced Topics in Electrical Engineering (ATEE), Bukarest, Romania, 7 – 9 May

[5] Chobanov V 2016 Continuity of power supply in smart grid with PV penetration, IEEE International Conference on Power Electronics and Motion Control Conference (PEMC), Varna, Bulgaria, 25-28 September

[6] Barsali S, Ceraolo M, Pelacchi P and Poli D 2002, *Control techniques of Dispersed Generators to improve the continuity of electricity supply*, IEEE Power Engineering Society Winter Meeting, New York, USA Conference Proceedings Vol. 2 pp 789 - 794

[7] Abdin E S, Osheiba A M and Khater M M 1999 Modeling and optimal controllers design for a stand-alone photovoltaic-diesel generating unit, *IEEE Transactions on Energy Conversion* **14**(3) 560-565

[8] Anderson P M and Mirheydar M 1995_Analysis of a diesel-engine driven generating unit and the possibility for voltage flicker, *IEEE Transactions on Energy Conversion* **10**(1) 37-47

[9] Lidozzi A, Solero L and Crescimbini F 2012 Adaptive Direct-Tuning Control for Variable-Speed Diesel-Electric Generating Units, *IEEE Transactions on Industrial Electronics* **59**(5) 2126-2134

[10] Shetty P K, Lewlyn L and Rodrigues R 2010 *Total Productive Maintenance of a Diesel power generating unit of a institution campus*, 2010 International Conference on Mechanical and Electrical Technology, 10 – 12 September, Singapore, Singapore, pp 68 - 71

Internal services simulation control in 220/110kV power transformer station Mintia

D Ciulica and R Rob

Politehnica University of Timisoara, Department of Electrical Engineering and Industrial Informatics, 5 Revolution Street, Hunedoara, 331128, Romania

E-mail: raluca.rob@fih.upt.ro

Abstract. The main objectives in developing the electric transport and distribution networks infrastructure are satisfying the electric energy demand, ensuring the continuity of supply to customers, minimizing electricity losses in the transmission and distribution networks of public interest. This paper presents simulations in functioning of the internal services system 400/230 V ac in the 220/110 kV power transformer station Mintia. Using simulations in Visual Basic, the following premises are taken into consideration. All the ac consumers of the 220/110 kV power transformer station Mintia will be supplied by three 400/230 V transformers for internal services which can mutual reserve. In case of damaging at one transformer, the others are able to assume the entire consumption using automatic release of reserves. The simulation program studies three variants in which the continuity of supply to customers are ensured. As well, by simulations, all the functioning situations are analyzed in detail.

1. Introduction

In order to ensure the continuity in customer supplying, reserve circuits and power suppliers are assumed. For the investment reduction, whenever is possible, the reserve circuits are used to provide power to direct circuits, taking into consideration of a reduced probability for two failures coincidence. The purpose for using the automatic release of reserves (ARR) is to serve for supplying the customers when the direct circuits suffered a fault. The measuring device detects the voltage absence on the collecting bars and transmits the command for switch releasing which ensures the customer supplying by reserve circuits.

During the functioning of an electro-energetic installation, two different situations may occur and they can be determined by the fault location:

-situation when ARR intervention is necessary;

-situation when ARR intervention is not useful.

The ARR device scheme is presented in Figure 1. The block scheme contains [1]:

-starting element which actions when the voltage is falling lower then $0.7 \cdot Un$;

-control element which conditions the functioning by the presence of the supplying voltage on the I_2 switch.

-temporization element which may deprive.

-blocking element to repeating actions.

In order the ARR installation to make difference between the situations described in this section, it has to accomplish the following conditions [2], [3]:

- ARR releasing when the voltage on the consumer bars falls below a specified value.

- ARR releasing must be accomplished only after the specified time t_{ARR};
- the reserve circuits must be released only after the direct circuit is disconnected;
-ARR must work only once, if the relay protection commands a new releasing, ARR must not response.

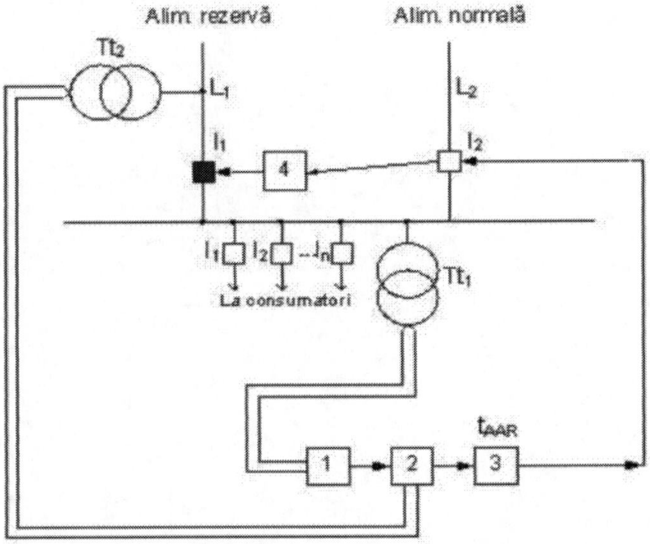

Figure 1. AAR device block scheme

2. Supplying general scheme

Power transformer station Mintia (220/110 kV) serves for transporting the electric energy to consumers which are supplied to 110kV (local power transformer stations of 110/20kV or 110/6kV). The single line scheme of the power transformer station Mintia is presented in Figure 2.

In order to supply the station internal services, three transformers are used and they can mutual reserve. So, in case of one's failure, the other two are able to assume the entire consumers using an ARR installation [4]. The supplying scheme contains simple longitudinal sectioned bar using a switch coupling. Each section has a subsection for vital consumers and one for non-vital consumers.

The scheme presented in Figure 2 is realized in Visual Basic and it permits to the user to choose the appropriate supplying way when a fault occurs. In many situation simulation is very useful to study the functioning of some electrical circuits [5-9]. In the actual situation the monitoring and control is based on SCADA, a control system architecture that uses computers, data communications and graphical user interfaces for process supervisory management [10].

In the followings, the power transformers T11 (10/0.4 kV, 800 kVA), T21 (10/0.4 kV, 800 kVA) and T31 (6/0.4kV, 600 kVA) are the supplying sources. If from various reasons, none of these sources becomes unavailable, a Diesel generator with the apparent power of 500 kVA must assume the entire consumers for up to 12 hours [10].

3. Functioning of the ARR in the internal services of power station Mintia

The 0.4 kV power station is composed by two bar sections, connected by a longitudinal couple (Q104). Section 1 is supplied by T11 transformer using Q101 switch, or by T31 transformer using Q103 switch, or by Diesel generator using Q200 switch. Section 2 is supplied by T21 transformer using Q201 switch, or by T31 transformer using Q203 switch, or by Diesel generator using Q200 switch.

In the followings, three methods for supplying consumers will be presented when fault occurs on the direct circuits.

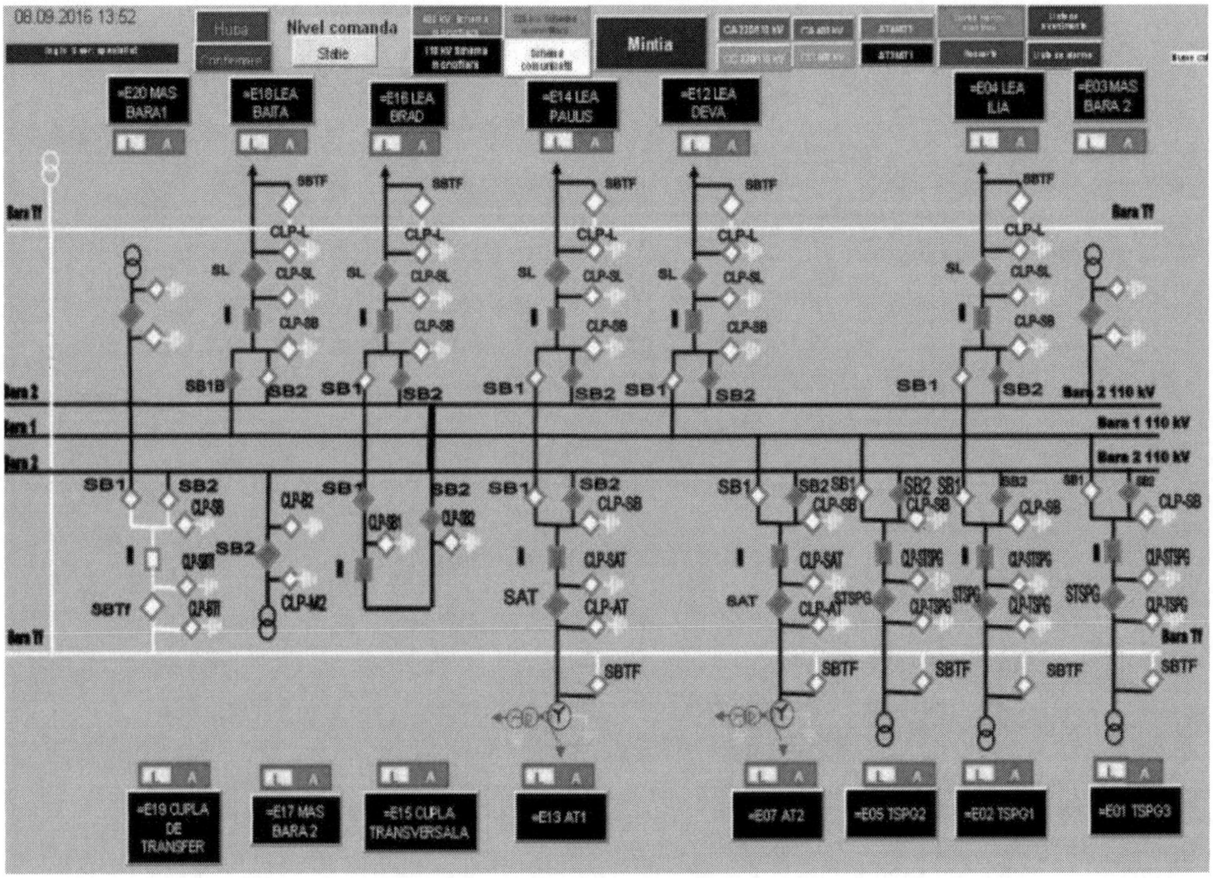

Figure 2. Power transformer station Mintia

3.1. Consumers are connected on Section 1 and Section 2 and supplied by PT1 (T11)

This is the first method for connecting the costumers when the fault occurs. From Figure 3 it can be seen that in variant 1 the Q104 and Q101 switches are connected and Q103, Q203, Q 201, Q102, and Q202 are disconnected. Elements as Diesel generator, Q200, Q105, and Q205 switches are on the 'don't care' level.

3.2. Consumers are connected on Section 1 and Section 2 and supplied by PT3 (T31)

From Figure 4 it can be seen that in variant 1A the Q104 and Q103 switches are connected and Q101, Q203, Q 201, Q102 and Q202 are disconnected. Elements as Diesel generator, Q200, Q105 and Q205 switches are on the 'don't care' level.

3.3. Consumers are connected on Section 1 and Section 2 and supplied by PT2 (T21)

From Figure 5 it can be seen that in variant 2 the Q104 and Q201 switches are connected and Q103, Q203, Q 101, Q102 and Q202 are disconnected. Elements as Diesel generator, Q200, Q105 and Q205 switches are on the 'don't care' level.

3.4. Consumers are connected on Section 1 from PT1 (T11) and Section 2 from PT2 (T21)

From Figure 6 it can be seen that in variant 3 the Q101 and Q201 switches are connected and Q103, Q104, Q203, Q102 and Q202 are disconnected. Elements as Diesel generator, Q200, Q105 and Q205 switches are on the 'don't care' level.

3.5. Consumers are connected on Section 1 from PT3 (T31) and Section 2 from PT2 (T21)

From Figure 7 it can be seen that in variant 3A the Q103 and Q201 switches are connected and Q104, Q101, Q203, Q102 and Q202 are disconnected. Elements as Diesel generator, Q200, Q105 and Q205 switches are on the 'don't care' level.

Figure 8 presents the signalling panel with the protection system state for the 400 kV cell afferent to 400/220kV autotransformer AT4.

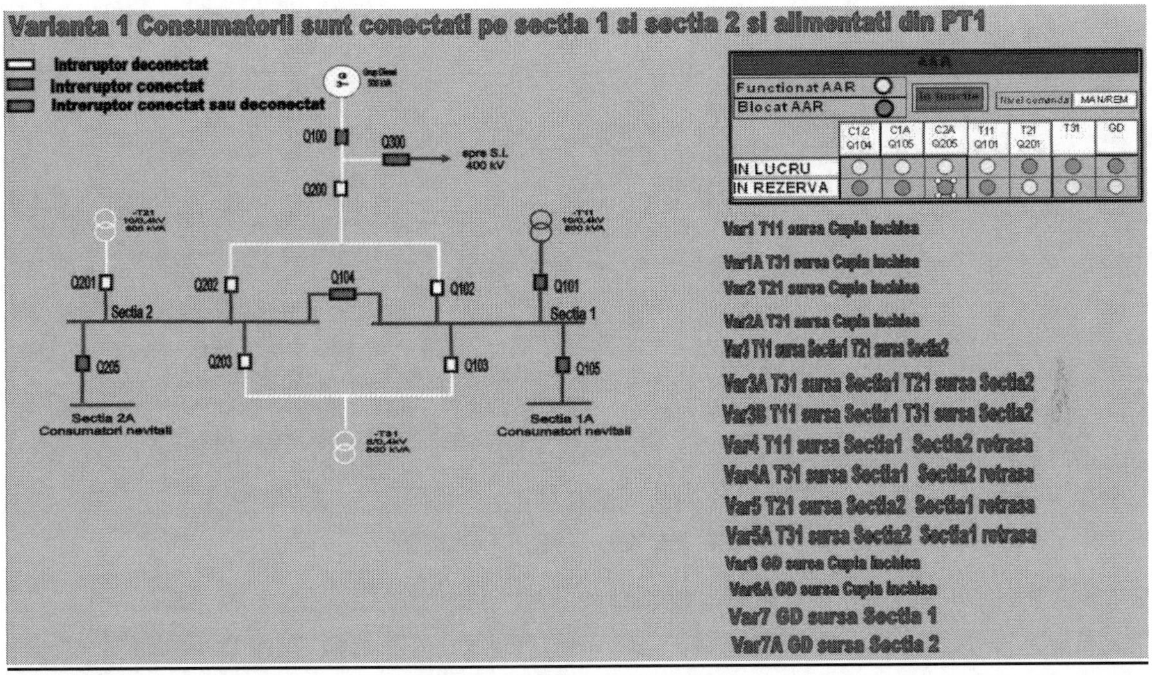

Figure 3. Scheme for Variant 1

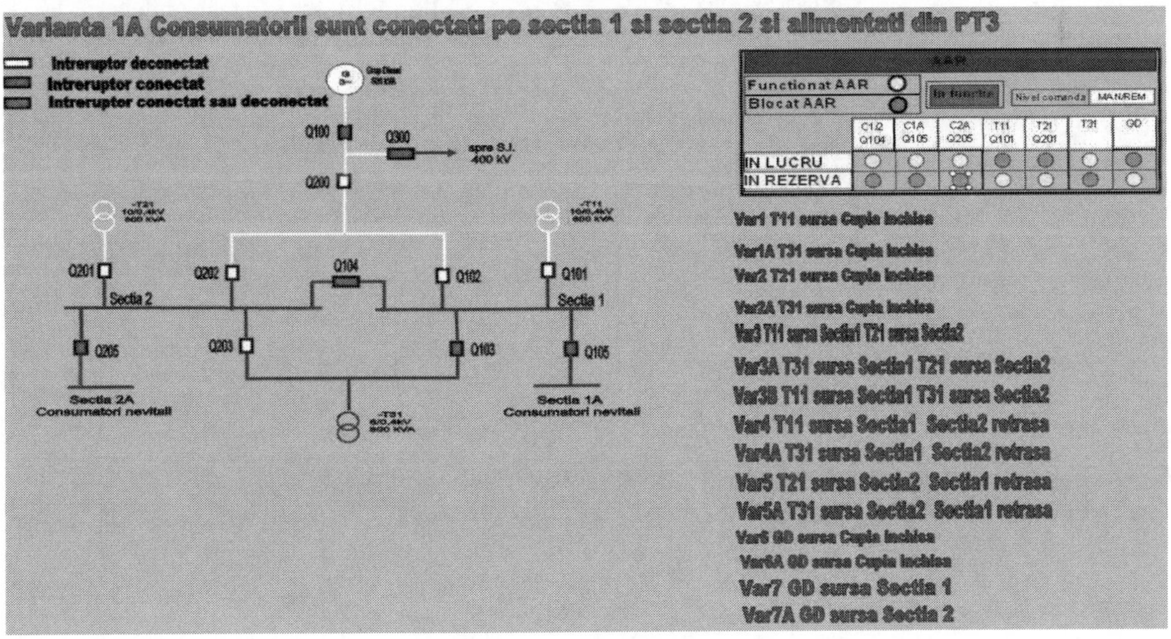

Figure 4. Scheme for Variant 1A

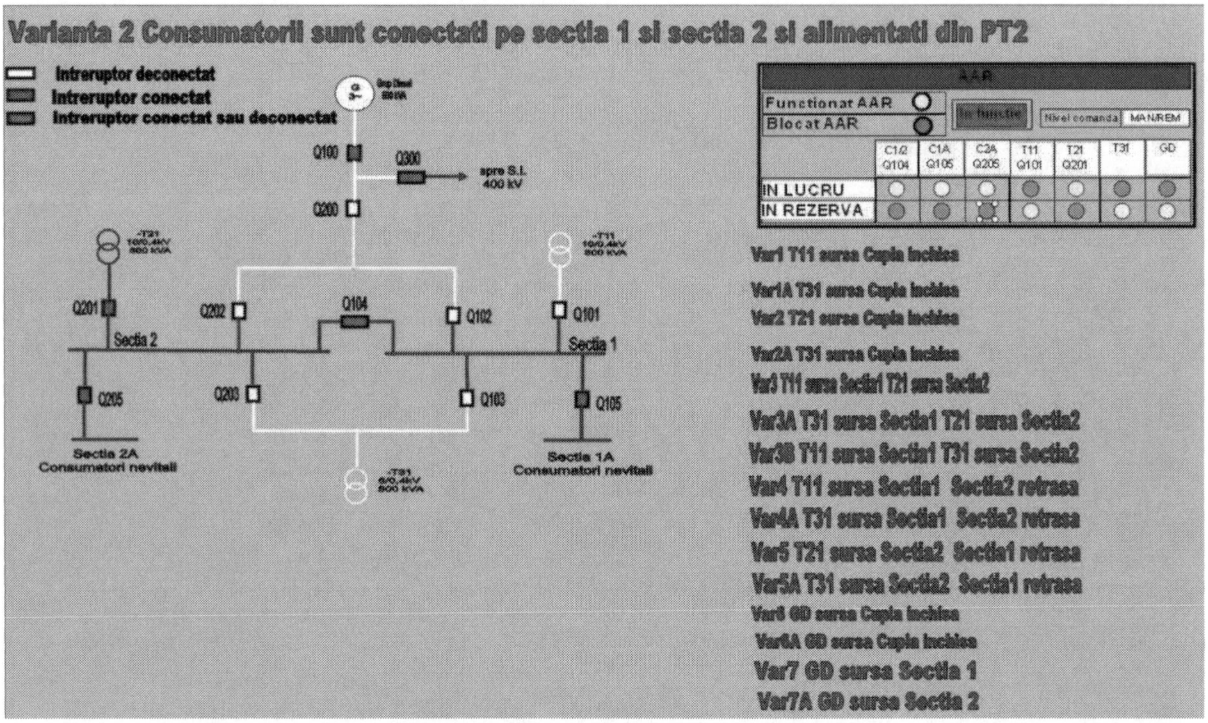

Figure 5. Scheme for Variant 2

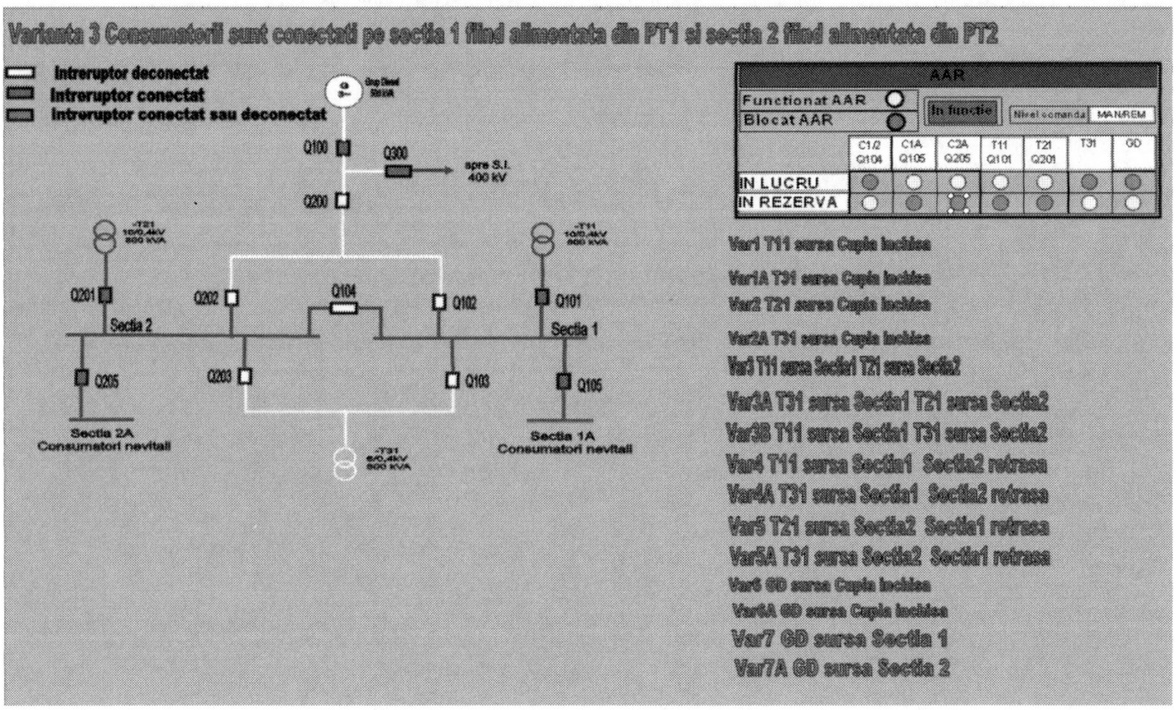

Figure 6. Scheme for Variant 3

IOP Conf. Series: Materials Science and Engineering **294** (2017) 012009 doi:10.1088/1757-899X/294/1/012009

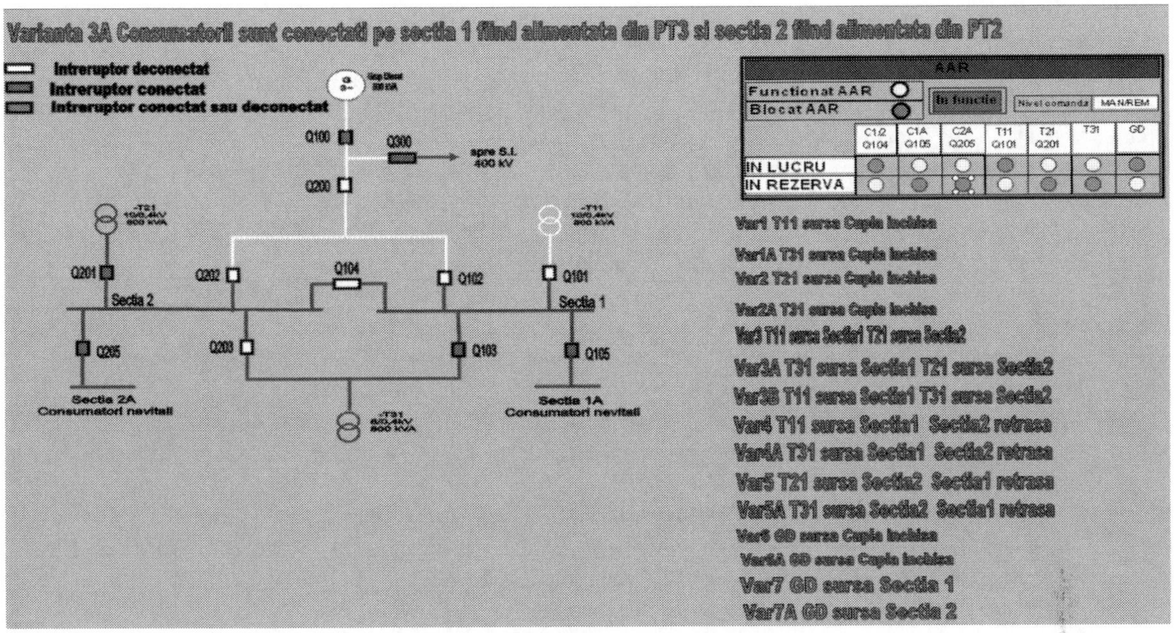

Figure 7. Scheme for Variant 3A

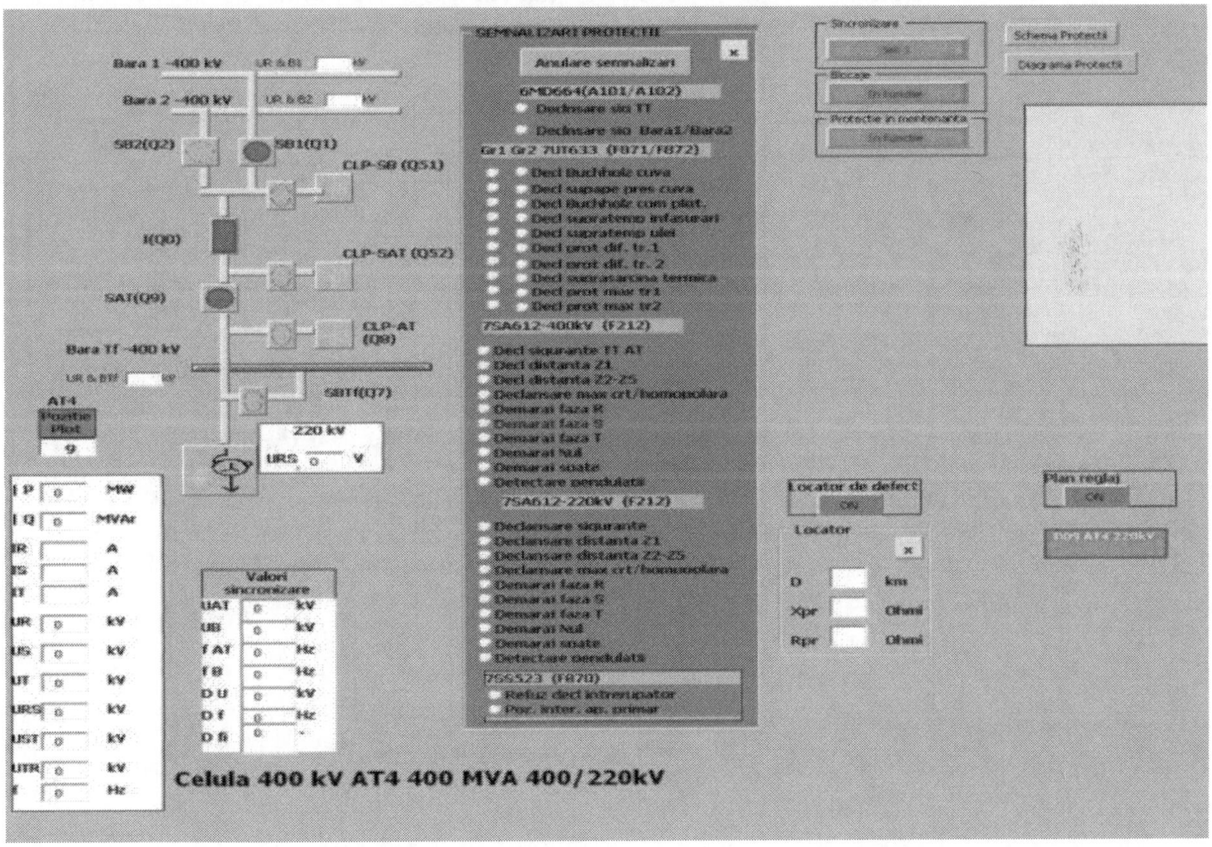

Figure 8. Signalling panel with the protection system

IOP Conf. Series: Materials Science and Engineering **294** (2017) 012009 doi:10.1088/1757-899X/294/1/012009

4. Conclusions

The states in which the installation is able to be are the followings:
1. The normal state where three situations can be distinguished:
- all the three base sources are available (T11, T21, T31). These situations were described in present paper as variants 1, 2 and 3 using ARR system;
-all the three base sources are available, but Section 2 is retired from service;
-all the three base sources are available, but Section 1 is retired from service.
2. If one of base sources in function presents abnormal parameters and one of the other sources works normally, the ARR installation will assume the entire energy consumption.

References

[1] Deshpande M V 2009 *Elements of electrical power station design*, PHI Learning publisher.

[2] Iagar A, Popa G N, Dinis C M and Moraru G 2009 *Study about Numerical Relay SEL-387 for Overcurrent and Differential Protections of 110/20 kV Transformers*, 13th WSEAS International Conference on Systems held at the 13th WSEAS CSCC Multiconference, Rhodes Isl., Greece, July 22-24, pp 265 – 270

[3] Popa G N, Deaconu S, Dinis C M and Iagar A 2009 *Implementation of a Numerical Distance Relay for the 110kV Electric Lines*, 13th WSEAS International Conference on SYSTEMS held at the 13th WSEAS CSCC Multiconference, Rhodes Isl., Greece, July 22-24, pp 271 – 276

[4] Milligan M, Donohoo P and Lew D 2010 *Operating reserves and wind power integration*, 9th Annual International Workshop on Large-Scale Integration of Wind Power into Power Systems as well as on Transmission Networks for Offshore Wind Power Plants Conference Québec, Canada, pp 1 – 19

[5] Panoiu C, Panoiu M, Muscalagiu I and Iordan A 2010 Visual Interactive Environment for Study the Power Electronics Using PSCAD-EMTDC Simulation Program, *Computer Applications in Engineering Education* **18** (3) 469-475

[6] Panoiu M, Panoiu C, Sora I and Osaci M 2007 About the possibility of power controlling in the three-phase electric arc furnaces using PSCAD EMTDC simulation program, *Advances in Electrical and Computer Engineering* **7**(1) 38-43

[7] Baciu I and Cuntan C D 2012 *Operation Analysis of a Frequency Converter with Control Realized in Labview*, IEEE International Conference On Industrial Tehnology (ICIT), 19 – 21 March, Athens Greece, pp 432 – 437

[8] Baciu I, Cuntan C D and Gabor B 2014 *The study of the electric parameters of an autonomous supply system with photovoltaic panel* , International Conference and Exposition on Electrical and Power Engineering (EPE), Iasi, Romania, October 16-18, pp 1138 – 1143

[9] Dinis C M and Popa G N 2014 Measurements in Scada System used at a Wastewater Treatment Plant, *Annals of Faculty Engineering Hunedoara* **XII**(4) 207-215

[10] Chen K, Parmee I C and Gane C R 1997 *Dual mutation strategies for mixed-integer optimisation in power station design,* IEEE International Conference on Evolutionary Computation, 13-16 April, Indianapolis, USA, pp 385 – 390

IOP Conf. Series: Materials Science and Engineering **294** (2017) 012010 doi:10.1088/1757-899X/294/1/012010

Optimizing the switching time for 400 kV SF6 circuit breakers

D Ciulica

Politehnica University of Timisoara, Electrical Engineering Department, Vasile Parvan str., no. 1-2, 300223 Timisoara, Romania

E-mail: dumitruciulica@yahoo.com

Abstract. This paper presents real-time voltage and current analysis for optimizing the wave switching point of the circuit breaker SF6. Circuit Breaker plays an important role in power systems. It provides protection for equipment in embedded stations in transport networks. SF6 Circuit Breaker is very important equipment in Power Systems, which is used for up to 400 kV due to its excellent performance. The controlled switching is used to eliminate transient modes and electrodynamic and dielectric charges in the network at manual switching of capacitor, shunt reactors and power transformers. These effects reduce the reliability and lifetime of the equipment installed on the network, or may lead to erroneous protection.

1. Introduction

SF6 has excellent dielectric and arc-quenching properties and it is used as insulating medium for high voltage distribution systems. Favorable electrotechnical, chemical and physical characteristics of gas have greatly influenced the development of switching technology. SF6 is an alternative to other conventional isolation and extinguishing media, such as e.g. oil and air. At the same time, SF6 compared to oil reduces the risk of danger (e.g. fire, explosion) to personnel and the environment. A global assessment, considering all ecological, economic, safety and technological aspects, has shown that SF6 is still an excellent choice as an insulating environment. The existing SF6 technology in power transmission and distribution is the result of decades of optimization and mainly contributes to the further development of economically efficient distribution [1].

The switching time for the circuit breakers is calculated by a program installed in the Bay Control Unit (BCU) of the control unit located in each cell of the power station and receives the command from the user to close or open the switch at the point on the voltage or current taken as a reference. The method is also called synchronous switching or controlled switching [1], [2].

2. SF6 circuit-breaker

The SF6 switch is frequently used due to its excellent performance. Due to the special properties of SF6, this extinguishing medium is used in power equipment such as high dielectric strength, arc extinguishing capability, excellent thermal stability and good thermal conductivity. The operation of the SF6 Switch is also dependent on the SF6 pressure and if the actuator is with oil depends on its pressure. In addition, the switching phenomenon relies heavily on the angles where the phases are CLOSE or OPEN. By this we understand that the sequence and angles of switching phases for a specific task are important. In this case, for capacitive load, closing is done on the peak voltage when the current is zero at phase. New applications are being researched that adapt dynamically to different angles based on the type of load. The selection of switching time of the switch for the switching point on the feature depends on the following parameters: DC voltage at the control coils, SF6 pressure, oil pressure in the actuator, load type, frequency of the reference phase in which we take the Phase R as a reference [3].

2.1. *Controlled switching*

The controlled switching is used to eliminate transient regimes and electrodynamic and dielectric charges at manual switching of capacitor batteries, shunt reactors and power transformers. An important aspect of all switching controlled applications is the accuracy obtained during arc processing and breakage. The uncontrolled closure or opening of the circuit breakers may result in wear to the equipment. Switching control provides an efficient solution taking into account the current values; the control unit optimizes the switching operations of the switch using the instantaneous voltage curve or the instantaneous current curve [4]. The operation of opening an inductive or capacitive load can result in electric arc re-ignition and overvoltage. When closing an inductive load in the case of un-optimized switching, overcurrent may occur.

2.2. *The switching point on the feature*

The switching point at zero pass characteristic is a method of eliminating the transient effects occurring in controlled time switching operations. The closing or opening commands of the circuit breaker are delayed so that the closing or opening of the contact takes place at the optimum time characteristic of the phase angle. With this method we can increase the performance and lifetime of the switch. In the following example (Figure 1) there is presented the operating principle of a system for commanding a LEA (aerial electric line). To avoid transient effects, choosing switching moment is depending on the capacitive load of the LEA and it must be at zero voltage travel. For simplicity, only one phase [5-7] is considered.

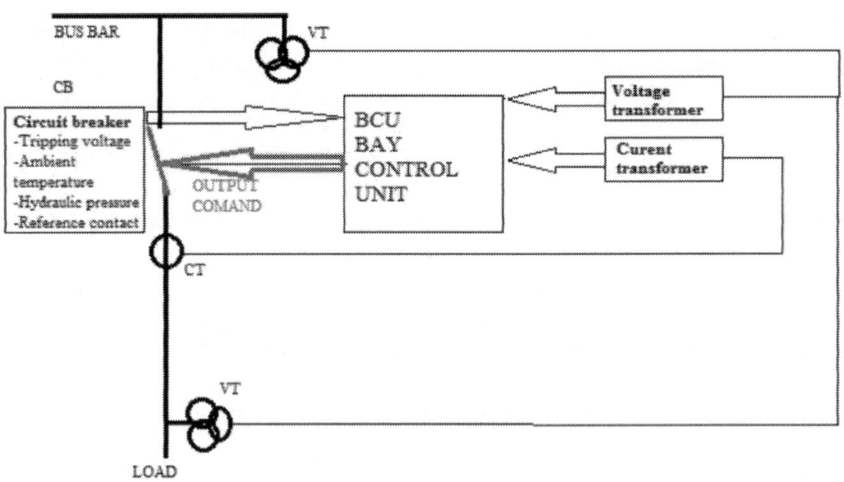

Figure 1. Switching Block Diagram

2.3. *Typical switching times*

2.3.1. Closing Time. The time from excitement of the closing coil until the active contact of the switch is closed.

2.3.2. Opening time. The time from energizing the opening coil to separating the active switch of the circuit breaker.

2.3.3. Make /Break Time. The time elapsed from excitement of the closing coil until the current starts to circulate in the main circuit.

2.3.4. Pre-arcing Time. From the start of the main circuit current until the ignition is reached.
Pre-arcing time = Closing time - Make time

2.3.5. Arcing Time. The time at which the electric arc burns at the contacts of the circuit breaker during its opening.

2.3.6. Dielectric strength decrease rate. Dielectric strength drop rate, circuit breaker characteristics that describe the voltage slope when closing a circuit breaker

2.3.7. Dielectric strength increase slope. The dielectric resistance increase slope, characteristic of the circuit breaker that describes the slope of the increase of the resistance to the opening of a circuit breaker. The dielectric resistance increase slope defines the minimum arc time required for the inductive load interruption without ignition [6-10].

2.4. The advantages of this method
The main advantage of this method is the reduction of the transient switching effects, with implicit reduction of voltages in the system and its equipment. The decrease of the transient stresses, e.g. on a power transformer, will prolong the life of the equipment [7]. In addition, the circuit-switched circuit-breaker itself will have reduced breaking currents, which will result in reduced circuit breaker wear. For capacitor, filter and reactant coil batteries, the number of load switching operations can usually be doubled before a scheduled repair, compared to uncontrolled operation [11], [12].

3. Control system

The role of the control system is to provide controlled (synchronous) switching commands to the OPEN and CLOSE coils of the switch. Successively controlled switching reduces mechanical and electromagnetic wear suffered during normal switching operations by reducing ignition currents to shut-off and re-ignition during opening. Given the precise inputs (temperature, pressure, auxiliary voltage, etc.) over the switching time, the compensation is integrated into the calculation algorithm, ensuring performance for all types of loads.

Due to phase shifts, the system is designed to calculate and send the switching commands separately at each phase of the switch. The monitoring of the ignition and re-ignition currents ensures a successful synchronous switching of the circuit breaker. The operating principle uses the zero crossing of a sinusoidal signal. A voltage signal will be used as a reference before closing and a current or voltage signal can be used before opening. When switching characteristics are identical for each phase, there is a constant time delay (T / 6) between the synchronous commands in each phase, for example in the sequence "R, T, S" (1/3 of a global cycle).

As soon as a CLOSE command is received, the command is held and a time delay is applied, waiting for the first zeroing of the voltages. If no zeroing is detected within the preset time interval, the synchronous interruption is interrupted and CLOSE is executed on the three phases simultaneously.

If there is a zero crossing in the default delay range and no other faults are detected, the CLOSE synchronous commands for each phase are emitted after the calculated time delays. The calculated time delays are measured from the zero crossing and are compensated for the variable effects and variations of switch characteristics for each phase. Auxiliary switch contacts are monitored for timing analysis, if a sequence error is detected, the synchronous command is interrupted, and a three-phase switching is performed. Figures 2 and 3 show the opening operation and the closing operation with a respective control system.

The operational principle of the OPEN operation is the same as for CLOSE with the difference that it can be used as a reference signal for a zero-pass current and that the recharged currents [13], [14] are monitored.

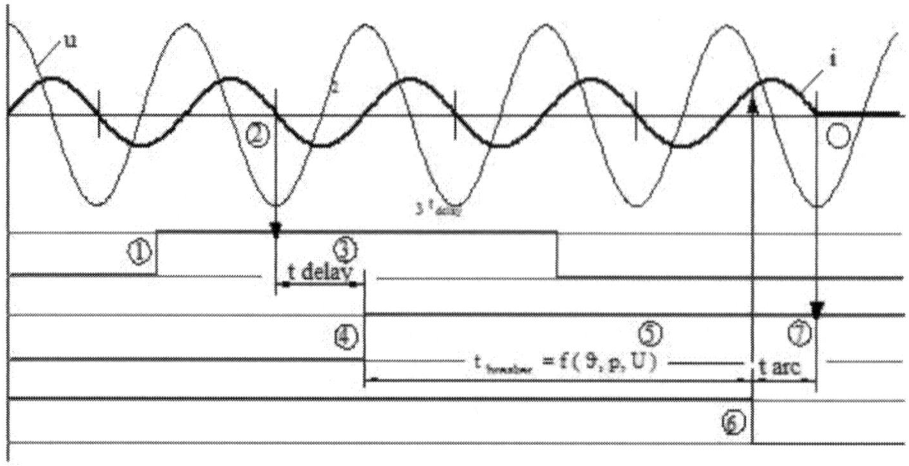

Figure 2. Opening operation of circuit breaker with a control system [6]

1 – Open command, 2 – Identify the zero point of the current, 3 – Delay time, 4 – Command to breaker opening coil, 5 – Opening time, 6 – Separating contacts, 7 – Arc time, 8 - End of current flow

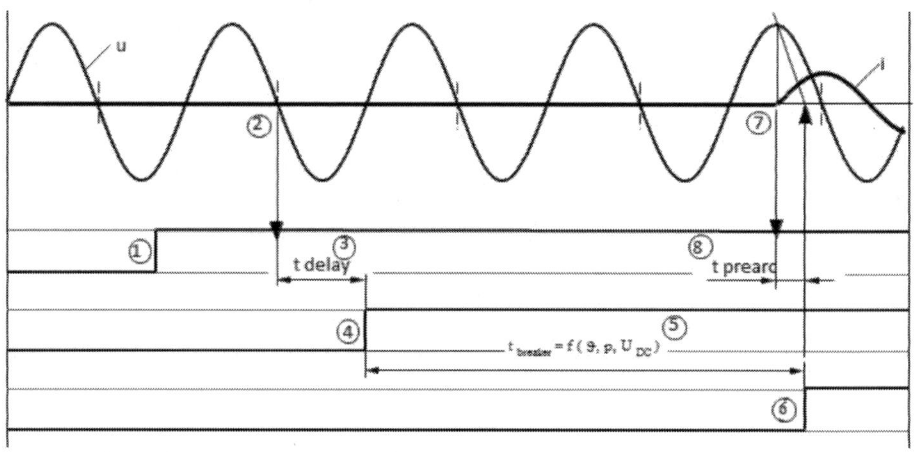

Figure 3. Closing operation of circuit breaker with a control system [6];

1 – Closing command, 2 – Identify the zero point of the voltage, 3 – Delay time, 4 – Command to the breaker closing coil, 5 – Closing time, 6 – Touch contacts, 7 – Begin of current flow, 8 – Timp Pre-arcing

4. Compensation in the control system

When the switch has a constant functioning in its behaviour, with variations in external conditions, corrections can be made for them. The verification of systematic mechanical variations is according to variations in ambient conditions, such as the influence of idle time, temperature variation, variation of coil voltage, hydraulic fluid pressure change and SF6 gas pressure change. Ambient temperature variations can be compensated by suitable transducers. The need for compensation depends on the angle variation and the actual operating conditions. For frequent operation, adaptation control may be good enough to gradually consider small variations. The system can compensate the expected operating time of the circuit breaker for variations in temperature, pressure, SF6 gas pressure, and auxiliary power supply.

5. Results

The most important variables influencing the mechanical operation time of the circuit breaker are the control voltage on the CLOSE and OPEN coil (U), the ambient temperature where the circuit breaker (T) is located, and the hydraulic pressure (p) for the actuators hydraulic.

Table 1. Limits variables for circuit-breaker

	Limits		Current values		
	min	*max*	*Pole A*	*Pole B*	*Pole C*
Frequency	48	52	50		
Control voltage [V]	187	255	242		
Trip voltage CLOSE [V]	187	255	242		
Trip voltage OPEN [V]	154	255	241		
Hydraulic pressure [bar]	0	400	335	335	341
Temperature [°C]	40	80	17.5		

Table 2 shows the measured values in the closure process of the respective circuit breaker, it is noticed that the closing times are approximately equal to the three phases.

Table 3 shows the reference contact delays and the duration of the signal given by this reference contact. Table 4 shows the set-up time according to the phase shift in degrees between the phases

Table 2. Value variables for circuit-breaker close

CLOSE			
Initial values	Pole A	Pole B	Pole C
Closing times [ms]	79.8	79.5	79.8
Hydraulic pressure [bar]	330	330	330
Trip voltage [V]	240		
Temperature [°C]	20		

Table 3. Value compensation for circuit-breaker close

CLOSE			
	Pole A	Pole B	Pole C
Delay reference contact [ms]	-18.4	-18.7	-18
Reference contact signal duration [ms]	8.9	9.5	8.9
Reference contact travel [mm]	0	0	0

Table 4. Adjusting time function phase shift\pole closing

	Pole A	Pole B	Pole C
phase shift\pole[o]	90	210	150
phase shift\breaker[o]	-10		
adjusting time [ms]	14.3	13	12.7

In Figure 4, it can be seen that the closing command is given at the passage of phase reference A by zero, the following signals and sizes are presented.

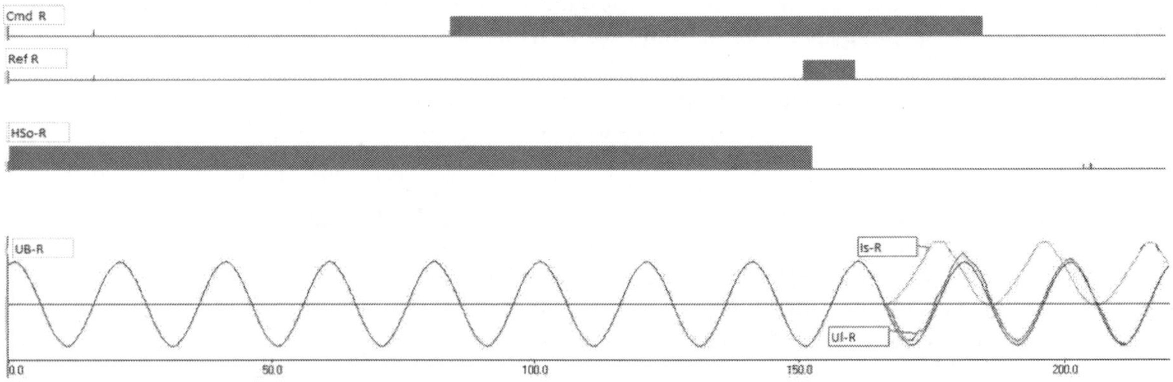

Figure 4. Closing phase A- shunt reactor

Cmd R- Displays commands transmitted by BCU to pole A
Ref R – Display of the reference contact signal at pole A
HSo-R- Display of the reference contact signal at pole A
UB-R Analogue Bar Signal Signal Signal Signal Signaling for Field Phase Switch A
Ul-R Display analog line voltages for pole A
Is-R Analog line current display for pole A

Table 5. Value variables for circuit-breaker open

CLOSE			
Initial values	Pole A	Pole B	Pole C
Closing times [ms]	21.6	21.4	21.1
Hydraulic pressure [bar]	330	330	330
Trip voltage [V]	240		
Temperature [°C]	20		

Table 6. Value compensation for circuit-breaker close

CLOSE			
	Pole A	Pole B	Pole C
Delay reference contact [ms]	6	6.2	6.1
Reference contact signal duration [ms]	3.1	3.2	3
Reference contact travel [mm]	0	0	0

Table 7. Adjusting time function phase shift\pole closing

	Pole A	Pole B	Pole C
phase shift\pole[o]	90	210	150
phase shift\breaker[o]	0		
adjusting time [ms]	7	7	7

As shown in Figure 5, the zero sweep current of any phase can be evaluated as an opening reference.

Separation of circuit breaker contacts takes place sooner than extinguishing the electric arc.

The operational opening is as follows:

- The opening command is issued
- The BCU follows the current curve crossing (zero zero in the negative range) by synchronizing the phase position of the current.
- After the opening time has elapsed, the contacts open and the flow of current flows without additional spring
- The current flow ends after the arc-electric time at the zero current flow.

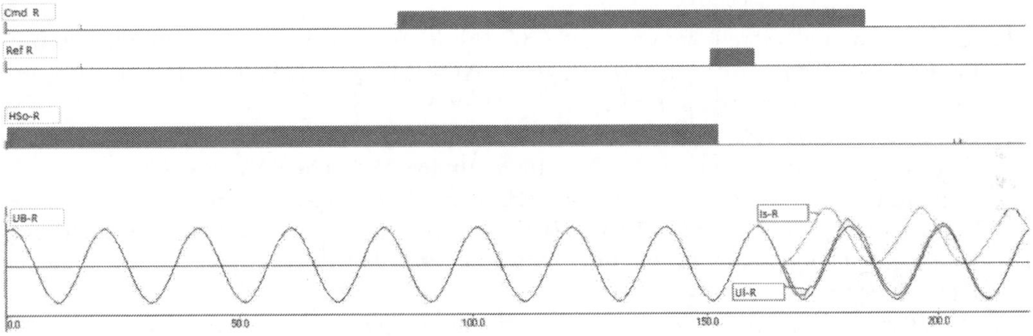

Figure 5. Opening phase A- shunt reactor

6. Conclusions

Existing SF6 switches are located outside where other equipment can be monitored for ambient temperature such as autotransformers, so the required BCU temperature can be read from the control command system (SCADA)

Considering that the last types of SF6 circuit-breakers have a spring-operated actuator so it can be removed from the hydraulic pressure sensor

Depending on the type of cell that is fed line, BC, AT is installed a program characteristic of each type of task.

Action has to be done on each independent phase

The program does not have to act when the command is issued from the protections from the cell-mounted automation

The program may send error messages if there are defects in one of the sensors that use and block the program's malfunction.

Since this program runs in parallel with the SCADA systems, it must act only after checking the interlocks in the BCU both locally and remotely.

Under normal operating conditions, operating characteristics could be constantly compensated over a long period of time by using the features of factory tests.

References

[1] Jun Liu F, Huang G M, Zhiying Ma and Geng Y 2011 *A* Novel Smart High-Voltage Circuit Breaker for Smart Grid Applications, *IEEE Transactions on Smart Grid* **2** (2) 254-264

[2] Namjoshi Y *2013 Auto-Calibration of a Sf6 Circuit Breaker for Point-on-wave Switching an Application of Signal processing to Switchgear*, International Conference on Electrical and Electronics Engineering, Kuala Lumpur, Malasya, December 4-5, pp 14-18

[3] Sharma S and Bhardwaj H 2012 How to maintain sf6 circuit breaker, *International Journal of Scientific Research Engineering &Technology* **1** 030-035

[4] Akesson U 1992 *Condition monitoring of SF6 circuit breakers* CIGRE 1992 Session, Paper 13-104

[5] Tsudata H, Hirai T, Kohyama H and Ito H 2002 *Controlled Switching System for Capacitor Bank and Transformer Switching* International Conference on Electrical Engineering, Jeju Island, Korea, July 7-11, pp 2125-2130

[6] ***ABB 2005, *Controlled Switching with Switchsync*, Presentation manual

[7] Ito H 2002 *Controlled Switching Technologies, State-of-the-Art*, Transmission and Distribution Conference and Exhibition: Asia Pacific, IEEE/PES, Vol. 2, pp 1455 – 1460

[8] *** http://www.abb.com. 2006 *Controlled Switching, Buyer's Guide, Application Guide, Catalogue publication 1HSM* 9543 22-01

[9] ***Siemens 2001, *Operating instructions PSD01 controller Order No:927 00414 174 C*, Presentation manual Edition

[10] Panoiu M, Panoiu C, Sora I and Osaci M 2007 About the possibility of power controlling in the three phase electric arc furnaces using PSCAD-EMTDC simulation program, *Advances in Electrical and Computing Engineering* **71**(1) 38-43

[11] Panoiu C, Rob R, Pănoiu M and Sora I 2012 *Measuring system for distorted regime generated by three phase nonlinear loads*, 15th International Power Electronics and Motion Control Conference and Exposition, Novi Sad, Serbia, pp LS4c.1-1-LS4c.1-6

[12] Rob R, Panoiu C and Panoiu M 2014 Voltage and Current Signals Processing in Functioning of Nonlinear Loads, *Soft Computing Applications. Advances in Intelligent Systems and Computing* **357** 1171-1184

[13] Rob R, Panoiu C and Panoiu M 2015 Study regarding the power quality improvement in functioning of nonlinear loads, *IOP Conf. Ser.: Mater. Sci. Eng.* **85** 012029

[14] Rob R, Panoiu C and Panoiu M 2013 *Harmonic analysis in functioning of nonlinear loads*, 13th International Conference on Electric Power Systems, High Voltages, Electric Machines, Chania,Greece, August 27-29, pp 126-131

The numerical modelling of mixing phenomena of nanofluids in passive micromixers

R Milotin and D Lelea

Politehnica University of Timisoara, Department for Mechanical Machines, Equipment and Transportation, B-dul Mihai Viteazu 1, 300223 Timisoara, Romania

E-mail: dorin.lelea@upt.ro

Abstract. The paper deals with the rapid mixing phenomena in micro-mixing devices with four tangential injections and converging tube, considering nanoparticles and water as the base fluid. Several parameters like Reynolds number (Re = 6 – 284) or fluid temperature are considered in order to optimize the process and obtain fundamental insight in mixing phenomena. The set of partial differential equations is considered based on conservation of momentum and species. Commercial package software Ansys-Fluent is used for solution of differential equations, based on a finite volume method. The results reveal that mixing index and mixing process is strongly dependent both on Reynolds number and heat flux. Moreover there is a certain Reynolds number when flow instabilities are generated that intensify the mixing process due to the tangential injections of the fluids.

1. Introduction

The micromixers are the good candidate for various applications in the engineering fields like bioreactors, drug-delivery devices or μ-TAS. The optimization of the channels configuration is the important task, in order to obtain the compactness, uniform concentration and temperature of the mixture and short mixing length. For passive mixing this might be obtained by magnifying the area of the fluids that come in contact, inducing instabilities or increasing the mixing path. Besides the micromixers could realize more rapid and dynamic chemical reactions. Special attention should be taken in effective mixing in low Re regime because high shear rates and some biological species are not compatible [1].

There is a large number of research papers on passive mixing in microchannels, both numerical and experimental, reported by various authors in last decade. Four different microfluidic configurations were analyzed by Jeon and Shin [2] numerically and the conclusion was that the best mixing efficiency at low Re has zig-zag. The results were also confirmed experimentally. The influence of the stream position was analyzed by Ansari et al. [3] regarding mixing phenomena in T-shape microchannels. It was found that stream position has more influence at low Re. Miranda et al. [4] investigated numerical analysis on a microfluidic device that has obstacles and a flow that alternates. Besides, the Σ-micromixer was analyzed and optimized by Tafti et al. [5]. The optimization of the micromixer with convergent-divergent sinusoidal walls taking into account both the pressure drop and the mixing efficiency was made by Afzal and Kim [6]. The optimization was made for Re of 30. The analysis showed a great variation in the ratio of throat width to depth of the convergent-divergent areas, whereas the quantitative relation of amplitude to wavelength of the sinusoidal walls maintained a mainly constant value. The microfluidic device for nano-drug delivery was numerically

IOP Conf. Series: Materials Science and Engineering **294** (2017) 012011 doi:10.1088/1757-899X/294/1/012011

explored by Kleinstreuer et al. [7]. The conclusion was that a better mixing efficiency of nanoparticles and a shorter mixing length than a simple T-shape configuration has a micromixer with baffles and injection unit. Lin and Yang [8] examined numerically the main issues on chaotic mixing. The mixing phenomena in T-type micromixer strengthened by J-shaped baffles was analyzed numerically and experimentally by Lin et al. [9]. They discovered that amount of mixing is raising as the number of baffles and Re raises. Long et al. [10] studied the numerical and experimental exploration of rapid vortex micromixers. Enhanced mixing is because of the bigger interfacial contact area and fluid convection. Review analysis on passive micromixers was published by Hardt et al [11] for μ-TAS and microreactors, Mansur et al. [12] and Kumar et al. [13]. SAR μ-reactor was analyzed by Fang and Yang [14] and Broboana et al. [15] studied unsteady diffusion process in Y- shaped micromixers. Lin et al. [16] explored quick circular micromixer with unbalanced driving force. The conclusion was that rotational flow was established even at Re = 2. Numerically planar configuration of passive micromixers with circular chambers was studied by Alam and Kim [17]. They tested four different configurations with Re = 0.1 – 100. They noticed that mixing performance is mainly dependent on micromixer configuration and Re. The micromixer they proposed has gorgeons mixing abilities at low Re but a poor behaviour at Re > 50. Dykman and Klebetsov [18] presented the review on gold nanoparticles in therapeutic and diagnosis applications.

In this paper the mixing efficiency is evaluated for nanofluid mixing in microneedle arrays.

2. Numerical details

The passive mixing phenomena was analyzed, considering the micromixer presented in Figure 1a based on the microneedle arrays [19] (Figure 1b). The geometry and the flow conditions are presented table 1. In the case of the microchannels conventional theories are appropriate to microtubes with diameters down to 100 μm accordingly to Lelea et al. [20].

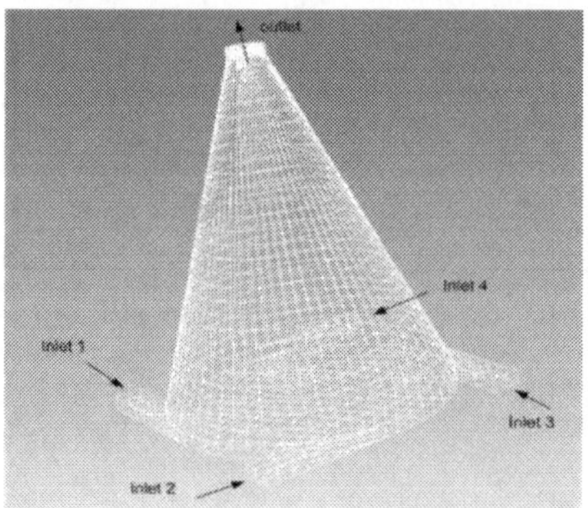

Figure 1a. The micromixer configuration

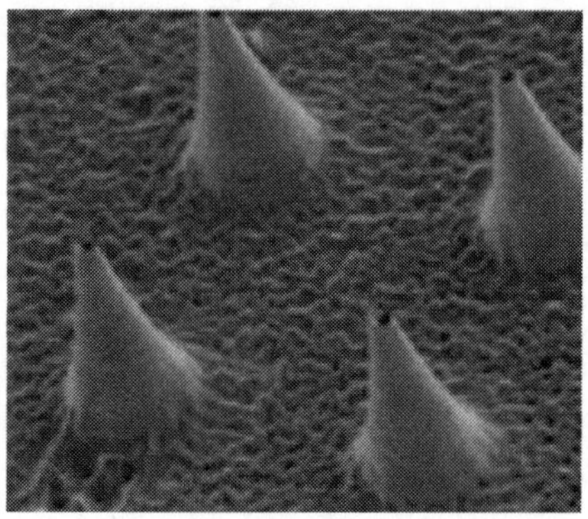

Figure 1b. The microneedle arrays [20]

IOP Conf. Series: Materials Science and Engineering **294** (2017) 012011 doi:10.1088/1757-899X/294/1/012011

Table 1. The geometry and fluid flow conditions

L, cm	a, μm	b, μm	D_{base}, μm
1	80	40	800
D_{out}, μm	t_{out}, °C	m, kg/s	Re
100	37	$10^{-7} - 5\ 10^{-6}$	6.24 - 284

So the set of the Navier-Stokes equations can be used to analyze the present phenomena, as follows:

The conservation of mass

$$\frac{\partial(\rho \cdot u_i)}{\partial x_i} = 0 \tag{1}$$

The conservation of momentum

$$\frac{\partial(u_i \cdot \rho \cdot u_j)}{\partial x_i} = -\frac{\partial p}{\partial x_j} + \frac{\partial}{\partial x_i}\left(\mu \frac{\partial u_j}{\partial x_i}\right) \tag{2}$$

The conservation of species

$$\frac{\partial(u_i \cdot C)}{\partial x_i} = \frac{\partial}{\partial x_i}\left(D \frac{\partial C}{\partial x_i}\right) \tag{3}$$

The upcoming boundary conditions are determined for the micromixers presented in this paper:

- The fluid flow is stationary, incompressible and laminar;
- The fluid properties of the nanofluid and water were found as temperature dependent with the upcoming equations:

Thermophysical properties of the water:

Dynamic viscosity:

$$\mu(t) = 2.6412018 \cdot 10^{-4} + 0.0014009 \cdot e^{-\frac{t}{31.0578605}} \tag{4}$$

Density:

$$\rho(t) = 1000.0 \cdot \left(1 - \frac{t + 288.9414}{508929.2 \cdot (t + 68.12963) \cdot (t - 3.9863)^2}\right) \tag{5}$$

Thermal conductivity:

$$k(t) = -0.58166 + 6.355 \cdot 10^{-3} \cdot T - 7.964 \cdot 10^{-6} \cdot T^2 \tag{6}$$

Specific heat:

$$c_p(t) = 8958.9 - 40535 \cdot T + 0.11243 \cdot T^2 - 1.014 \cdot 10^{-4} \cdot T^3 \tag{7}$$

Thermophysical properties of nanofluid:

The nanofluid effective thermal conductivity normalized with the based fluid thermal conductivity is given as [21]:

$$\frac{k_{eff}}{k_{bf}} = 1 + 64.7 \cdot \phi^{0.7460} \left(\frac{D_{bf}}{D_p}\right)^{0.3690} \cdot \left(\frac{k_p}{k_{bf}}\right)^{0.7476} \cdot \Pr^{0.9955} \cdot \mathrm{Re}^{1.2321} \tag{8}$$

The Pr is defined as:

$$\Pr = \frac{\mu}{\rho_{bf} \cdot \alpha} \tag{9}$$

The Re is calculated as:

$$\mathrm{Re} = \frac{\rho_{bf} \cdot k_b \cdot T}{3 \cdot \pi \cdot \mu^2 \cdot l_{bf}} \tag{10}$$

where k_b = 1.3807x10^{-23} J/K, l_{bf} = 0.17 nm and D_{bf} = 0.384 nm.

The temperature dependent viscosity of water is defined with equation (4).

The effective density is defined as:

$$\rho_{eff} = (1-\phi)\cdot \rho_f + \phi \cdot \rho_p \tag{13}$$

The effective specific heat is calculated as:

$$c_{peff} = \frac{(1-\phi)\cdot(\rho \cdot c_p)_f + \phi \cdot (\rho \cdot c_p)_p}{\rho_{eff}} \tag{14}$$

The effective viscosity is a sum of base fluid viscosity and apparent viscosity:

$$\mu_{eff} = \mu_{bf} + \mu_{app} \tag{15}$$

The base fluid viscosity (water) is calculated by the equation (4), while the evident viscosity of the nanofluid is explained by the upcoming relation [22]:

$$\mu_{app} = \frac{\rho_p \cdot V_b \cdot d_p^2}{72 \cdot C \cdot \delta} \tag{16}$$

where the Brownian velocity is defined as:

$$V_b = \frac{1}{d_p}\sqrt{\frac{18 \cdot K_b \cdot T}{\pi \cdot \rho_p \cdot d_p}} \tag{17}$$

A distance between the centers of the particles is obtained from:

$$\delta = \sqrt[3]{\frac{\pi}{6 \cdot \phi}} \cdot d_p \tag{18}$$

The correction factor is defined as:

$$C = \mu_{bf}^{-1} \cdot \left[(c_1 \cdot d_p + c_2)\cdot \phi + (c_3 \cdot d_p + c_4)\right] \tag{19}$$

where: c_1 = -1.2331·10^3, c_2 = -1.5331·10^{-6}, c_3 = 94.383 and c_4 = -4.5731·10^{-7}

- The properties of the gold particles are defined as: ρ_p = 19300 kg/m^3, c_{pp} = 1290 J/kg K, k_p = 314 W/mK.
- Because of the low flow rates, the viscous dissipation is neglected;
- The uniform velocity field and the constant temperature are imposed at the channel inlet, while at the outlet the partial derivatives of the velocity and temperature in the streamwise direction are vanishing;

The nanofluid was considered as Fluid 1 and water as Fluid 2, with inlet cross-sections presented in Figure 1a. Mass fraction is considered 1 at the nanofluid and 0 for water inlets.

The set of the partial differential equations with boundary conditions are solved using the Ansys Fluent commercial solver [23] with methods described in [24]. The Simple algorithm is used for the velocity-pressure coupling solution and QUICK scheme for discretization of the partial differential equations for momentum and concentration species. The under-relaxation factors are used for pressure field (α = 0.3), momentum conservation (α = 0.5) and species conservation (α = 0.8). The convergence criterion is defined as:

$$R^{\phi} = \frac{\sum_{cells,P}\left|\sum_{nb} a_{nb} \cdot \phi_{nb} + b - a_p \phi_p\right|}{\sum_{cells,P}\left|a_p \phi_p\right|} \tag{20}$$

The residuals for velocity components, continuity equation and concentration field were 10^{-8}.

3. Results and discussion

The mixing behavior of micromixer with four tangential inlets was evaluated in terms of mixing index defined with the following equation:

$$M = 1 - \sqrt{\frac{\sigma^2}{\sigma_{max}^2}} \tag{21}$$

where the standard deviation of the mass fraction was calculated at specific cross-sections with the following relation:

$$\sigma = \sqrt{\frac{\sum_{i=1}^{n}(c_i - c_{av})^2}{n}} \tag{22}$$

the c_{av} is the average of the mass fraction c_i at specific cross-section defined with the following relation:

$$c_{av} = \frac{\sum_{i=1}^{n} c_i}{n} \tag{23}$$

Besides, Re is calculated with the following relation:

$$\mathrm{Re} = \frac{\rho \cdot u_{av} \cdot D_i}{\mu} \tag{24}$$

The temperature dependent diffusion coefficient is defined as:

$$D = \frac{k_b \cdot T}{6 \cdot \pi \cdot \mu \cdot d_p} \tag{25}$$

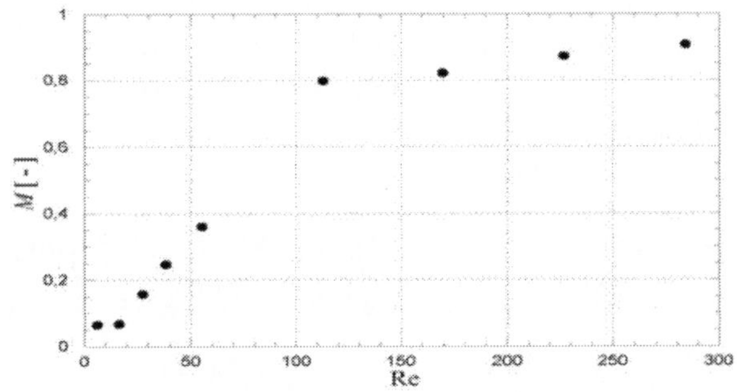

Figure 2. The mixing index versus Re

In Figure 2 the mixing versus Re is presented. It is observed that mixing index is very low for low values of Re < 20. There are two main reasons for this behaviour. The first one is the outcome of low diffusion coefficient $D = 10^{-11}$. The second one is related to stable hydrodynamic field without the swirl motion generated by tangential inlets. Besides the molecular diffusion is the main mixing mechanism in this case. As the Re increases the fluid flow became unstable and the swirl motion is initiated inside the micromixer. The outcome might be observed in the rapid increasing of the mixing index. This mixing behavior is observed for 30 < Re < 113. For Re > 113 the mixing index is slightly increasing from $M = 0.8 - 0.9$.

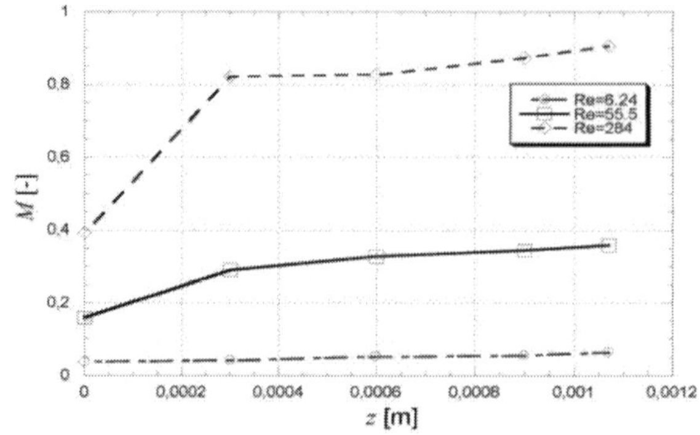

Figure 3. The axial distribution of the mixing index

In Figure 3 local distribution of the mixing index is presented for three different Re. In the case of very low Re = 6.24 the mixing index is almost constant regardless the axial position $M < 0.1$. For Re = 55.5 the mixing index slightly increases from $M = 0.2$ to $M = 0.3$ immediately after the inlet cross-sections remaining almost constant for the rest of the micromixer. This behavior indicated that the swirl motion could not be supported due to the hydrodynamic losses.

In the case of higher Re = 284, the mixing index sharply increases from M = 0.4 to M = 0.8, in the first quarter of the micromixer. For the rest of the axial distance the mixing index slightly increases to M = 0.9. In this case the swirl motion generated by four tangential inlets is fully established so the advection is responsible for better nanofluid mixing.

Figure 4. The mass fraction of nanofluids for Re = 6.24

Figure 5. The mass fraction of nanofluids for Re = 27.5

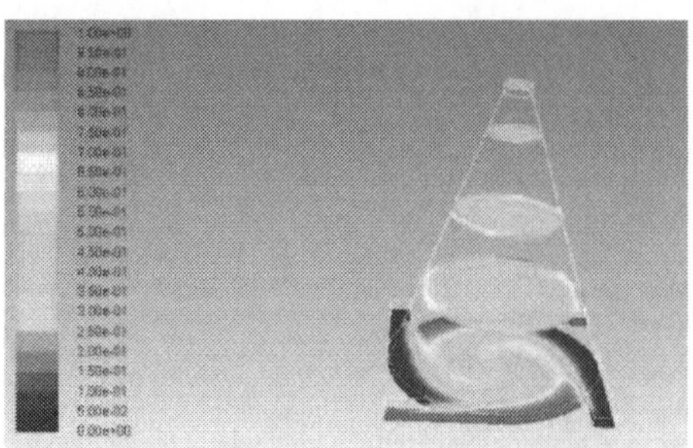

Figure 6. The mass fraction of nanofluids for Re = 170

Figure 7. The mass fraction of nanofluids for Re = 284

In Figure 4 – 5 the nanofluid mass fraction plot for Re = 6.24 and 27.5 is presented at different axial positions. It might be observed that for Re = 6.24 the mass fraction field is almost unchanged regardless the axial position. As Re = 27.5, the minor flow instabilities are observed only near the inlet cross-sections.

As the Re increases to Re = 170 (Figure 6), the swirl motion is fully established and the nanofluid mixing is almost completely done up to the outlet cross-section. The same conclusion might be outlined for Re = 284 (Figure 7).

4. Conclusions

The analysis regarding the mixing efficiency of the micromixer with four tangential inlets and converging microchannel, designed for nano-drug delivery was made. The following conclusions are outlined:

- The hydrodynamic conditions strongly influence the mixing efficiency.
- The four inlet jets enlarged the contact area and at higher Re create the swirl flow that was a result in better mixing.
- The real diffusion coefficient, based on temperature, influence the passive mixing phenomena.
- The rapid mixing influenced by advection is established for $55 < Re < 100$.
- The better mixing behavior is limited by the body temperature (t = 37 °C)
- Moreover the higher mixing index might be obtain by increasing the length of the outlet channel.

References

[1] Nguyen N T 2008 *Micromixers-Fundamentals, Design and Fabrication,* William Andrew
[2] Jeon W and Shin C B 2009 Design and simulation of passive mixing in microfluidic systems with geometric variations, *Chem. Eng. J.* **152** 575-582
[3] Ansari M A, Kim K Y and Kim S M 2010 Numerical study of the effect on mixing of the position of fluid stream interfaces in a rectangular microchannel, *Microsyst Technol* **16** 1757–1763
[4] Miranda J M, Oliveira H, Teixeira J A, Vicente A A, Correia J H and Minas G 2010 Numerical study of micromixing combining alternate flow and obstacles, *Int Commun Heat Mass* **37** 581–586
[5] Tafti E Y, Kumar R and Cho H J 2011 Diffusive mixing through velocity profile variation in microchannels, *Exp Fluids* **50** 535–545
[6] Afzal A and Kim K Y 2014 Multi-Objective Optimization of a Micromixer with Convergent-Divergent Sinusoidal Walls, *Chem Eng Commun* **202**(10) 1324-1334
[7] Kleinstreuer C, Li J and Koo J 2008 Microfluidics of nano-drug delivery, *Int J Heat Mass Tran* **51** 5590–5597
[8] Lin K W and Yang J T 2007 Chaotic mixing of fluids in a planar serpentine channel, *Int J Heat Mass Tran* **50** 1269–1277
[9] Lin Y C, Chung Y C and Wu C Y 2007 Mixing enhancement of the passive microfluidic mixe with J-shaped baffles in the tee channel, *Biomed Microdevices* **9** 215–221
[10] Long M, Sprague M A, Grimes A A, Rich B D and Khine M 2009 A simple three-dimensional vortex micromixer, *Appl Phys Lett* **94** 133501
[11] Hardt S, Drese K S, Hessel V and Schonfeld F 2005 Passive micromixers for applications in the microreactor and μ-TAS fields, *Microfluid Nanofluid* **1** 108-118
[12] Mansur E A, Mingxing Y E, Yundong W and Youyuan D 2008 A State-of-the-Art Review of Mixing in Microfluidic Mixers, *Chinese J Chem Eng* **16**(4) 503-516
[13] Kumar V, Paraschivoiu M and Nigam K D P 2011 Single-phase fluid flow and mixing in microchannels, *Chem Eng Sci* **66** 1329-1373
[14] Fang W F and Yang J T 2009 A novel microreactor with 3D rotating flow to boost fluid reaction and mixing of viscous fluids, *Sensor Actuat B-Chem* **140** 629-642

IOP Conf. Series: Materials Science and Engineering **294** (2017) 012011 doi:10.1088/1757-899X/294/1/012011

[15] Broboana D, Balan C M, Wohland T and Balan C 2011 Investigations of the unsteady diffusion process in microchannels, *Chem Eng Sci* **66** 1962-1972
[16] Lin C H, Tsai C H, Pan C W and Fu L M 2007 Rapid circular microfluidic mixer utilizing unbalanced driving force, *Biomed Microdevices* **9** 43-50
[17] Alam A and Kim K Y 2013 Mixing performance of a planar micromixer with circular chambers and crossing constriction channels, *Sensor Actuat B-Chem* **176** 639-652
[18] Dykman L A and Khlebetsov N G 2011 Gold Nanoparticles in Biology and Medicine: Recent Advances and Prospects, *Reviews – Acta Naturae* **3**(2) 34-55
[19] Wang W and Soper S 2007 *Bio-Mems: Technologies and Applications*, CRC Press, New York
[20] Lelea D, Nishio S and Takano K 2004 The experimental research on microtube heat transfe and fluid flow of distilled water, *Int J Heat Mass Tran* **47** 2817-2830
[21] Chon C H, Kihm K D, Lee S P and Choi S U S 2005 Empirical correlation finding the role of temperature and particle size for nanofluid (Al_2O_3) thermal conductivity enhancement, *Applied Physics Letters* **87** 153107-(1-3)
[22] Masoumi N, Sohrabi N and Behzadmehr A 2009 A new model for calculating the effective viscosity of nanofluids, *J. Phys. D: Appl. Phys* **42** 055501-(1-6)
[23] ***2011 *Ansys Fluent* 14.0 documentation, Ansys Inc.
[24] Patankar S V 1980 *Numerical Heat Transfer and Fluid Flow*, McGraw Hill, New York

IOP Conf. Series: Materials Science and Engineering **294** (2017) 012012 doi:10.1088/1757-899X/294/1/012012

Comparative study of oxihydrogen injection in turbocharged compression ignition engines

L Barna and D Lelea

Politehnica University of Timisoara, Department of Thermal Machines and Transportation, B-dul Mihai Viteazu, no. 1, 300222 Timisoara, Romania

E-mail: stocksproiect@yahoo.com

Abstract. This document proposes for analysis, comparative study of the turbocharged, compression-ignition engine, equipped with EGR valve, operation in case the injection in intake manifold thereof a maximum flow rate of 1l/min oxyhydrogen resulted of water electrolysis, at two different injection pressures, namely 100 Pa and 3000 Pa, from the point of view of flue gas opacity. We found a substantial reduction of flue gas opacity in both cases compared to conventional diesel operation, but in different proportions.

1. Introduction

Some main directions of study and points of view from the literature related to the studied phenomenon is presented in the introduction chapter:

Notes that in the case of naturally aspirated diesel engines, the hydrogen injection into the intake manifold can reach a maximum of 15% energy [1]. And the fact that in the case of medium load testing, an energy share of 10% hydrogen in diesel blends can reduce the smoke index by up to 50% and, in the case of maximum loading regimes, up to 17%.

In the review paper White et al. [2] stated the importance of studies on the use of internal combustion engines adapted to hydrogen-exclusive operation, both with compression ignition and spark ignition with applications in various areas of technology. They have highlighted some of the issues that require resolution to ensure that the H2ICEs engines can be safely exploited in everyday human activities. The various aspects are encountered such as detonation , wide range of flammability, combustion rate, supercharging, pollution problems with nitrogen oxides, use of flue gas recirculation valves to reduce thermodynamic cycle temperatures, maintaining the engine's specific power, achieving good filling efficiency, efficiency of Real Thermodynamic Cycle, Direct Injection of hydrogen and others.

Gomes et al. [3] conducted the experiments on the one-cylinder diesel engine and show that it is possible to operate the hydrogen-powered engine by producing an air preheat system, to obtain sufficiently high temperatures at the end of the compression phase as the air- Hydrogen mixture might self-ignites. They claimed the achievement of the satisfactory mechanical parameters, and a very low pollution level even with nitrogen oxides. The engine operated with an excess of air of 6, achieving a 45% efficiency. The increase of the final pressure in the cylinder at values exceeding those of the diesel engine operation, thus limiting the operation to high loading regimes.

Swaja et al. [4] presented the operation of diesel engines with a hydrogen-rich blend. They have shown that it is possible to operate them with 17% energy substitution rate. In the case of small substitution percentages of up to 5%, it reduces the delay time for self-ignition of the mixture, with reduction of the pressure gradient in the cylinder. It favors a smoother operation of the engine and thus a higher durability.

In the case of high power diesel engines [5], there is a decrease in total fuel consumption by approximately 12.6% at the engine supply with the 60 l/min H_2/O_2 fuel mixture and respectively a reduction in the level of pollutant compounds in the flue gas composition with 9.5% HC, 7.2% CO, 4.4% CO2, 19.3% particles, respectively an increase in the NOx level by 9.9%.

Considering the multitude of results obtained by the researchers, with conclusions that confirm the different tendencies of variation of the mechanical parameters of the diesel engines as well as the variation of the level of pollutants in the flue gases, the measurements have been made in which we can observe the variation of the opacity of flue gases according to the amount of oxyhydrogen induced in the engine mainfold at two different injection pressures and with the same amount of HHO.

2. The experimental setup

In Figure nr 2 the Block diagram of experimental installation is presented.

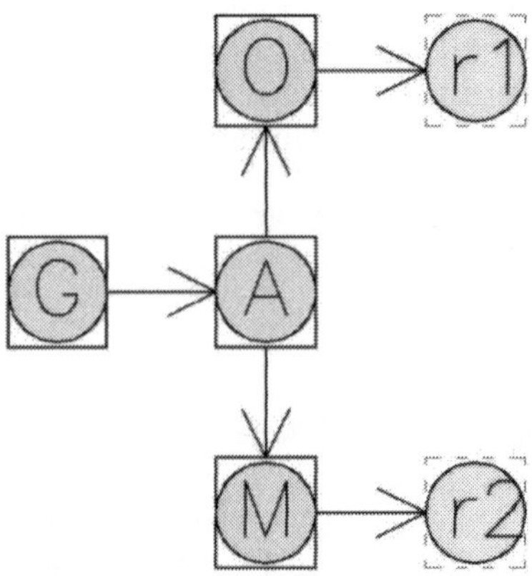

Figure 1. The schematic presentation of the experimental setup[6]:
A - Test used vehicle;
G - Oxyhydrogen gas generation equipment;
O - Laptop used for OBDII plug data acquisition;
M - Dynamometer MAHA LPS 3000 Roller type R100;
r1 - DDT2000 OBDII data acquisition (Timing angle, fuel consomption, air flow, egr psition, Load value);
R2 - MAHA LPS 3000 data acquisition (Wheel Power, Torq, smoke opacity).

Moreover the relevant technical data of the test setup used in the experiments are shown in the table 1.

Table 1. Used test stand main technical data [6]

Equipment	Description		Notifications
Dynamometer MAHA LPS 3000 R100 (M)	LxWxH: Measurement Accuracy: Max Force: Max Wheel Power: Power:	3345x1100x625mm 2% 6000N 260kW 230V	Used to achieve power and torque measurement at vehicle wheel
Oxyhydrogen gas generation plant (G)	Nominal gas flow: Voltage/Current (nominal): Electrodes configuration (fig. 3): Electrode Surface: Used electrolyte solution:	1,0l/min 13,6V/15A -NNNN+NNNN- 140x185mm NaOH, 28% mass	Electrolyzer with multiple boards, with the function to perform electrolysis of water (Figure 3)
The vehicle used for the test (A)	Max Power: Speed of Max Power: Max Torque CEE: Speed of Max Torque:	48(65)[kW/CP] 4000 [rot/min] 160[Nm] 2000[rot/min]	Dacia Logan 1.5 DCI engine K9K 790 (Euro 3), DCM12 injection unit type, soft number 004C, soft version 0071, Calibration version 4845
Laptop (O)	Used for data acquisition through OBDII, including USB/VAG - KKL cable and software		

For an easier understanding of the operation of the HHO generator (Figure nr 2), we present below a drawing of the HHO generator (Figure nr 3), and a schematic diagram of the assembly of the setup.

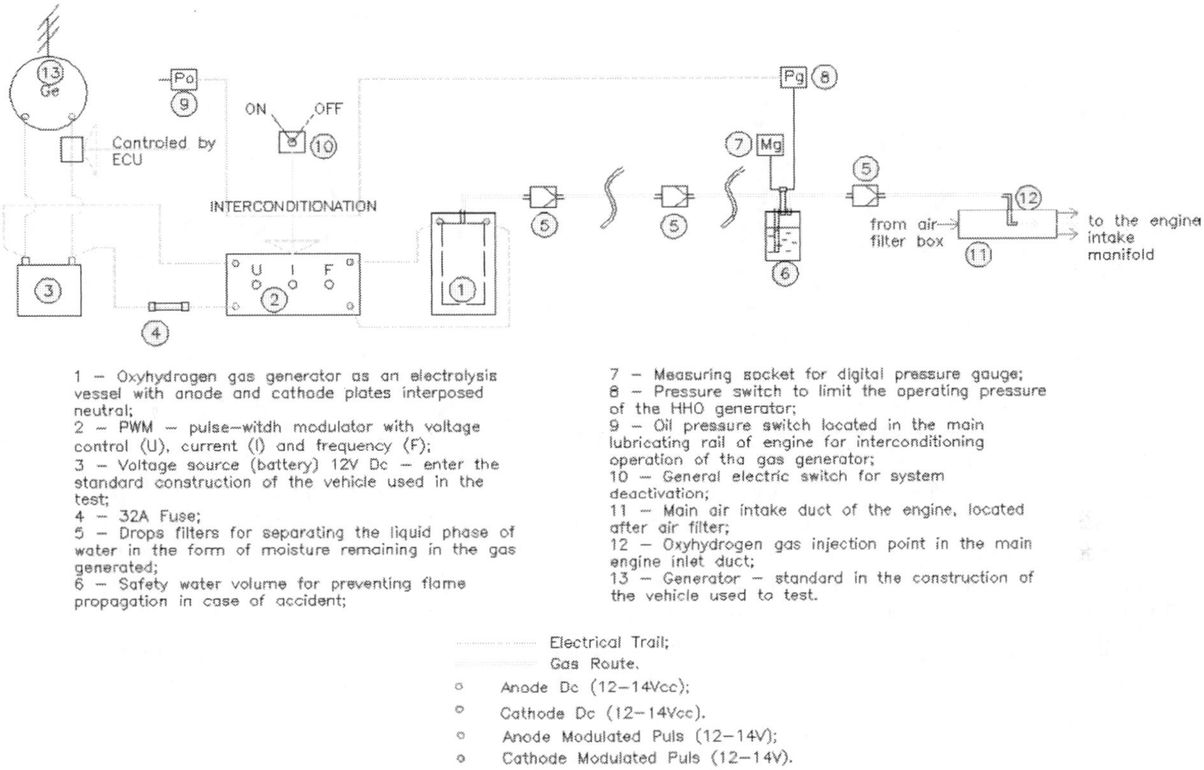

Figure 2. The general scheme of oxyhydrogen generator [6]

3. Experimental Procedure

The opacity tests have been made in 2 stage, the first stage for the standard diesel fuel mixed with HHO at the pressure p1 and in the second stage for the standard diesel fuel mixed with HHO at the pressure p2 (p2> p1), by applying the following procedure.

In the first phase, the instant but not brutal acceleration was made, up to a maximum speed limited by vehicle regulator, maintaining the speed for 2 seconds, then releases the accelerator pedal, leave a pause of a 2 seconds and repeat twice the same operation. The following phases have been employed: D_n – engine fed exclusively with diesel oil, P_i – engine fed with diesel oil and HHO mixture at 0,1 l/min, P_{i+1} – engine fed with diesel oil and HHO mixture at 0,3 l/min and P_{i+2} – engine fed with diesel oil and HHO mixture at 0,7 l/min, for Stage 1 and Stage 2.

At the same time the opacity measurements of flue gases have been made with the AVL opacimeter connected to the MAHA dynamometer and recording the available data series. The maximum supply pressure with HHO is p_1 = 80 Pa at stage 1 and p_2 = 3000 Pa at stage 2.

The recorded data are processed and the variation of the opacity of flue gases is plotted, depending on the oxyhydrogen flow rate induced in the two situations (for the pressure p_1 respectively p_2).

4. Results and Discussions

After carrying out the proposed experiments and tracing the variation of the flue gas opacity

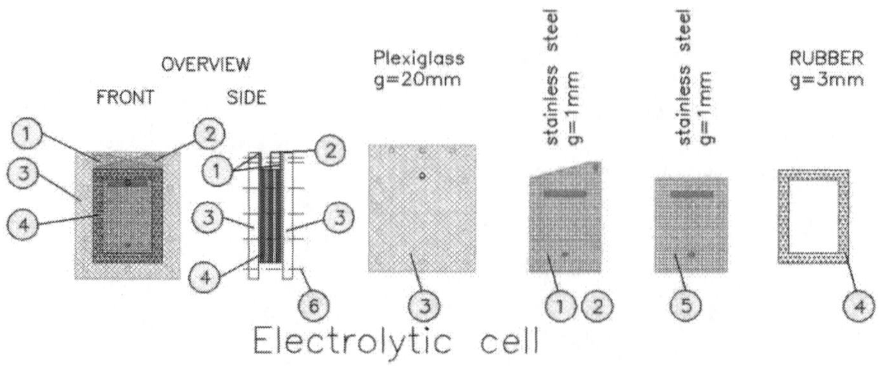

Figure 3. The components of the oxyhydrogen generator [6]

depending on the amount of HHO induced in the intake manifold, it can be clearly noticed that there is a tendency to reduce the opacity of the combustion gases for both p1 and p2 (Figure nr 4). This behavior might be explained by the reduction in the amount of diesel particles that is formed, due to a more complete burning, influenced by the presence of HHO. After the first flame cores occur, due to the self-ignition of the diesel fuel from the pilot injection, the entire mass of the mixture burns more quickly, more homogeneously and with a higher intensity. So the conditions at the beginning of the main injection will be more favorable, the combustion becomes even more homogeneous, and finally more complete.

Table 2. Measured values

	HHO feed pressure (p1, mmH2O)	HHO feed pressure (p2, mmH2O)	HHO volume flow rate (l/min)	Electrolysis Current (A)
D1	0.0	-	0.0	0.000
P1	3.5	-	0.1	2.400
P2	5.0	-	0.3	5.930
P3	8.0	-	0.7	12.66
D2	-	0.00	0.0	0.000
P4	-	167	0.1	2.400
P5	-	240	0.3	5.930
P6	-	300	0.7	12.66

5. Conclusion

Induction of oxyhydrogen in the intake manifold of turbocharged diesel engine, at low flow rates (up to 1 l/min), for 2 diferent HHO injection pressure are made. The following phenomena are observed:

The substantial reduction of flue gas opacity was found in both cases compared with conventional diesel operation;

Decrease the opacity of the flue gases with increasing amounts of added gas in both cases;

Flue gas opacity is lower in case of engine operation with diesel fuel enriched with HHO at 3000 Pa than 100 Pa.

At long-term use of an enriched fuel with HHO, the overall effect of reducing the opacity of the flue gases is observed for all cases considered.

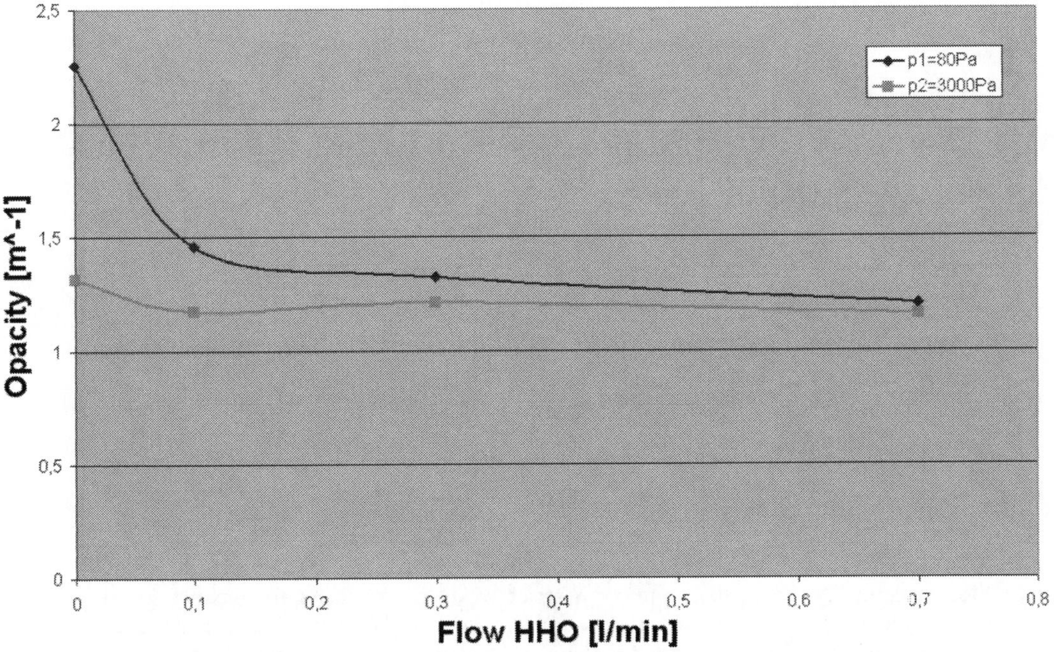

Figure 4. The smoke opacity variation for various flow of injected HHO

Acknowledgement
This work was partially supported by the strategic grant POSDRU/159/1.5/S/137070 (2014) of the Ministry of National Education, Romania, co-financed by the European Social Fund – Investing in People, within the Sectorial Operational Programme Human Resources Development 2007-2013.

References

[1] Varde K S and Frame G A 1983 Hydrogen aspiration in a direct injection type diesel engine – its effects on smoke and other engine performance parameters, *Int. J. Hydrogen Energy* **8**(7) 549-555

[2] White C M, Steeper R R and Lutz A E 2006 The hydrogen-fueled internal combustion engine: a technical review, *International Journal of Hydrogen Energy* **31** 1292-1305

[3] Gomes Antunes J M, Mikalsen R and Roskilly A P 2008 An investigation of hydrogen-fuelled HCCI engine performance and operation, *International Journal of Hydrogen Energy* **33** 5823-5828

[4] Stanislaw S and Grab-Rogalinski K 2009 Hydrogen combustion in a compression ignition diesel engine, *International Journal of Hydrogen Energy* **34**(10) 4413-4421

[5] Wang H K, Cheng C Y, Lin Y C and Chen K S 2012 Emission reductions of Air Pollutants from a Heavy-duty Diesel Engine Mixed with Various Amounts of H2/O2, *Aerosol and Air Quality Research* **12** 133-140

[6] Barna L and Lelea D 2015 Oxyhydrogen as a Catalyst of Combustion Proces in Compression Ignition Engines, *Scientific Bulletin of the Politehnica University of Timisoara – Transaction on Mechanics* **58**(72) 89–94

IOP Conf. Series: Materials Science and Engineering **294** (2017) 012013 doi:10.1088/1757-899X/294/1/012013

Diagnosis diagrams for passing signals on an automatic block signaling railway section

E Spunei[1], I Piroi[1], C P Chioncel[1] and F Piroi[2]

[1]Eftimie Murgu University of Resita, Electrical Engineering and Informatics Department, Traian Vuia str., no. 1-4, 320085 Resita, Romania
[2]Vienna University of Technology, Institute of Software Technology and Interactive Systems, Viena, Austria

E-mail: e.spunei@uem.ro

Abstract. This work presents a diagnosis method for railway traffic security installations. More specifically, the authors present a series of diagnosis charts for passing signals on a railway block equipped with an automatic block signaling installation. These charts are based on the exploitation electric schemes, and are subsequently used to develop a diagnosis software package. The thus developed software package contributes substantially to a reduction of failure detection and remedy for these types of installation faults. The use of the software package eliminates making wrong decisions in the fault detection process, decisions that may result in longer remedy times and, sometimes, to railway traffic events.

1. Introduction

The role of the railway traffic security installations is to command, control and ensure that the railroad traffic happens under completely safe conditions.

Any breakdown in such installations leads to traffic delays, and when wrong decisions are made, even to railway events. Therefore it desirable that:

- These installations have a low failure rate;
- In case of failures, the diagnosis and repair time are as short as possible;
- The train traffic during the installation malfunction is made in accordance to specific regulations, in order to avoid any traffic events;
- The fault diagnosis and remedy is done by specialized personnel, showing adequate theoretical and practical abilities.

For a number of Romanian railway junctions, the electro-dynamic centralized installations have been replaced with electronic centralized installations. Still, there are railway stations equipped with outdated traffic installations, installed in the 1970. Due to their physical wear and obsolescence, these installations fail often and can result in railway traffic events. Therefore, special attention must be given to the maintenance of these installations.

Currently, the diagnosis time for these installations depends strongly on the theoretical and practical know-how of the maintenance staff. In this respect, the most important problems faced are:

- Shortage of the maintenance and intervention staff;
- Remote intervention locations (sometimes distances as far as 80 km must be taken into account);
- Work in isolated conditions (lone walks between two railway stations);
- Difficult work conditions (night time, adverse weather, etc.).

All these put a certain pressure on the maintenance staff, such that in cases of failure their decisions may be negatively influenced.

The diagnosis charts proposed here together with the proposed software package, assist the maintenance personnel in their work by rapid and safe detection of the failure cause.

This work contributes to previous work on diagnosis chart design for all installation types for railway stations equipped with CR4 electro-dynamically centralized installations [1-7].

Worldwide, the control light signals can be done with specialized equipment [8-10], and diagnosis with the various logical systems [11-13].

2. Passing light signals

Along an Automatic Block Signaling block (ABS) lighting signal posts are placed at the end of each sector of the block. The minimal length of an ABS block is 1,200 m [13].

Each lighting signal on an ABS sector is a prior warning signal of the next lighting signal.

On all lighting signal posts white rectangular markers are installed. These markers indicate the possibilities available to a train driver to pass a signal post when the lighting signal on the post shows a red indication, an unclear indication, or the light is off.

Where ABS systems are installed, the normal indication of the passing signals (lighting signals), is either permissive or forbidding (stop), depending on the ABS block orientation [2], [3]. Figure 1 presents the unified ABS signal orientation for a double I track line (DI) and a double II track line (DII).

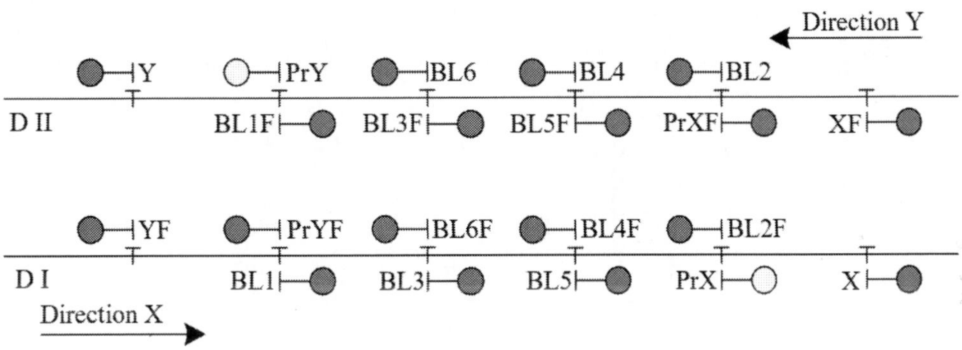

Figure 1. Signal indications on an ABL block with double line

The indications shown by the passing lighting signals of an ABS block are [14]:

- Red – "STOP, without passing the signal!" The meaning of the signal is that the first following sector of the block is occupied;
- Yellow – "Free line, travel with pre-established speed, ATTENTION! The next signal indicates stop". The meaning of the signal in this case is that the first block sector is free, but the second one is occupied;
- Green – "FREE, proceed at line speed." The meaning of the signal is that the next signal indicates free passing with the pre-established speed and that at least the next two block sectors are free;
- Blinking Yellow – "FREE with pre-established speed, the next signal is free with reduced speed." This indication is used by signals that also function as prior warning of the entry signal.

The light indications of the ABS light signals must be distinguishable, both during night and day, from at least 300 m on the lines with traffic speed of at most 120 km/h, and from at least 400 m for lines with traffic speed higher than 120 km/h.

When the train driver encounters an ABS lighting signal that is unclear, red, or off, the driver must stop the train before reaching the signal, without passing it by. After this, the driver must urgently notify the train dispatcher in the next station. If the driver is informed or sees that the next block sector is occupied he is forbidden to drive the train further. If no obstacle is identified and on the signal's post a white rectangle marker is observed, after the time the braking is off, the train driver may continue with a speed of at most 20 km/h, paying very close attention to the railroad, up to the next signal.

To remedy a faulty lighting signal on an ABS block sector, the maintenance staff must travel to the respective signal, perform measurements for pinpointing the fault and do the necessary repairing.

3. Electrical schema for passing light signals

Figure 2 presents electric schema for the BL1 and PrYF fire signals on a double track line, unified ABS block.

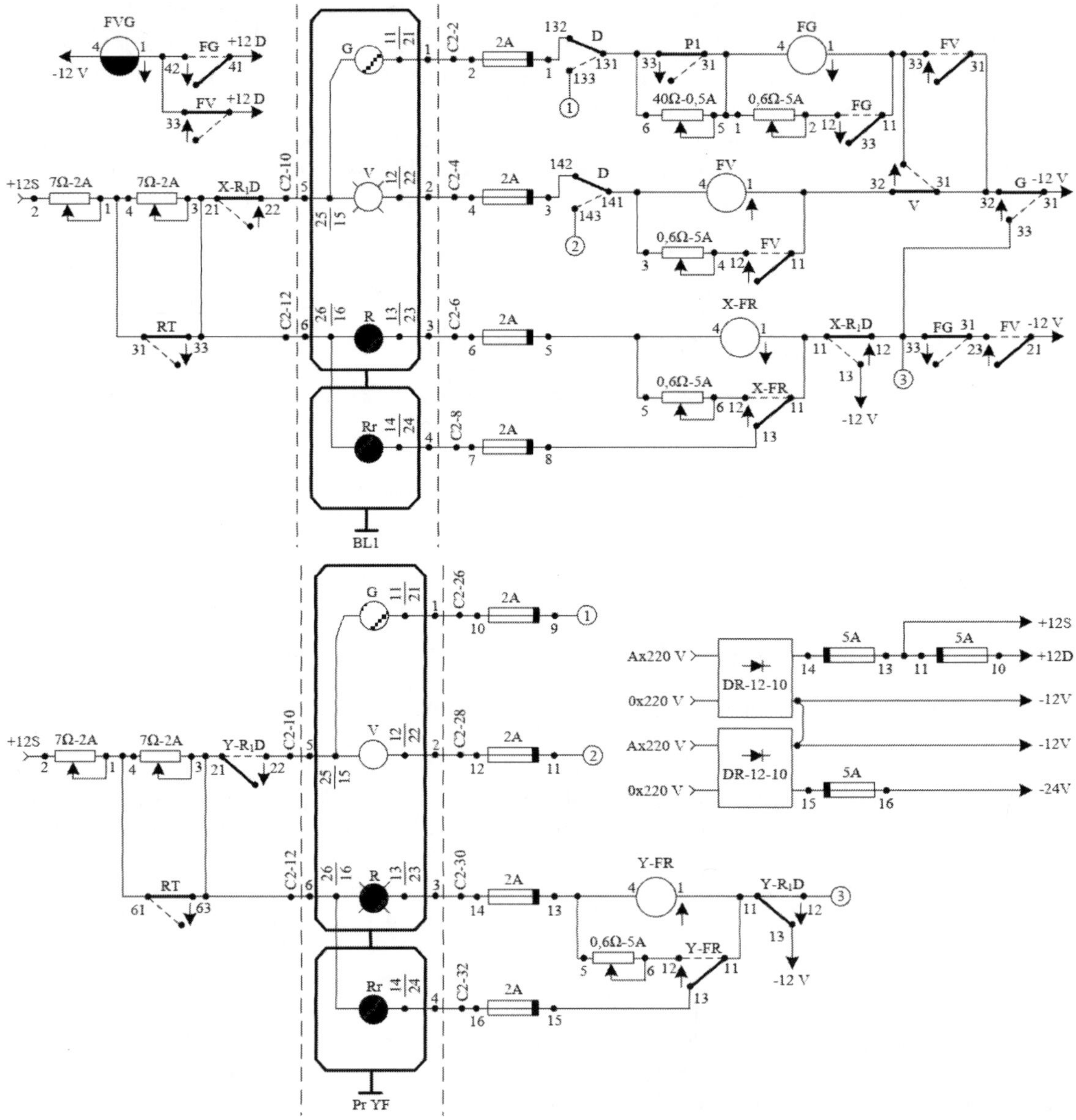

Figure 2. Electric schema for BL1 and PrYF passing signals

In this context "unified" refers to the fact that, on each track, the train traffic can happen in both directions. When the traffic runs in the Y direction on track 1, we say that the train travels on the wrong track, and the following signals will be seen having a permissive indication: BL2F, BL4F, BL6F, PrYF. The signals are 12V c.c. powered by the DR-12-10 rectifiers.

The signal fires (i.e. light bulbs) are controlled by the fire relays (green fire – FV, yellow fire – FG, red fire – FR) whose coil is serially connected with the light bulbs' filaments. Thus, whenever a light bulb is burnt, the signal changes to a restrictive or a stop indication.

When a track circuit is occupied with rolling stock or is damaged, the signal automatically changes to a stop indication by the relay G de-energization. When reversing the ABS installation's traffic orientation, the signal indication is changed through the directing relay, D, and the orientation directing relays, X-R_1D, and Y-R_1D.

For improved safety, the red indication light is doubled by a spare red light, Rr, which insures the red indication operation when the primary red light is burnt. The blinking yellow warning signal is controlled by the pulsing relay P1 contacts. The signal lights and the fire relays voltages are regulated by adjustable resistors, while the shunting protection is insured by fuses.

4. Diagnosis Charts and the Sem-Bla Software Package

The main passing signal defects that may occur are:

- All signal lights are off, no matter the ABS orientation;
- The green signal indication is off, although at least two block sectors ahead are free;
- The yellow signal indication is off, although at least one block sector ahead is free;
- The signal's red indication is off.

Figure 3 shows the diagnosis chart for situations where all light indications are off, no matter the ABS track orientation.

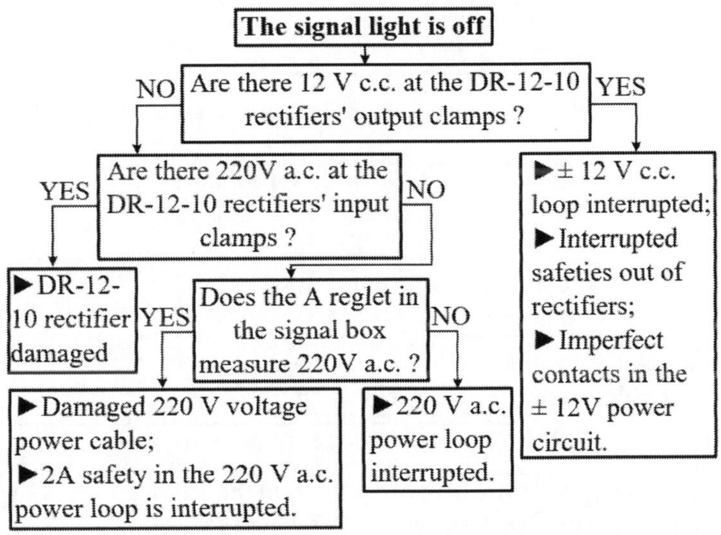

Figure 3. Diagnosis diagram for cases when the signal lights are off

Figure 4 presents the diagnosis diagram for situations where the signal cannot indicate the green light, while Figure 5 shows the diagnosis charts for situations where the yellow indication is not working.

Figure 6 shows the diagnosis charts for cases where the signal cannot indicate the red or spare red lights.

These diagnosis charts, which are created to assist the maintenance staff in their decision taking, have been used as a basis for a software package, SEMBla, written in Visual Basic.Net. The software package has a button based interface which, by its simplicity, is one of the most efficient interaction with the software.

The SEMBla software package shows a user a series of questions to which the user (i.e. the maintenance staff) answers based on the state of matters in the field. When a question was wrongly answered, the user can navigate backwards through a series of buttons in the lower area of the screen.

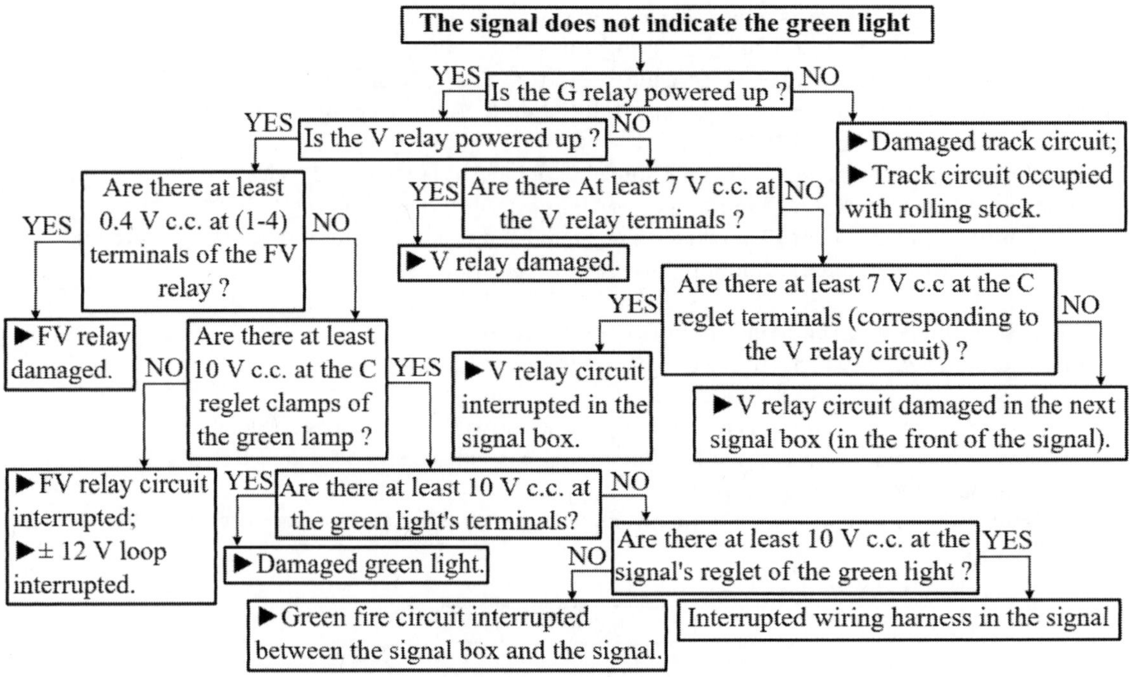

Figure 4. Diagnosis chart for cases where the green indication is not working

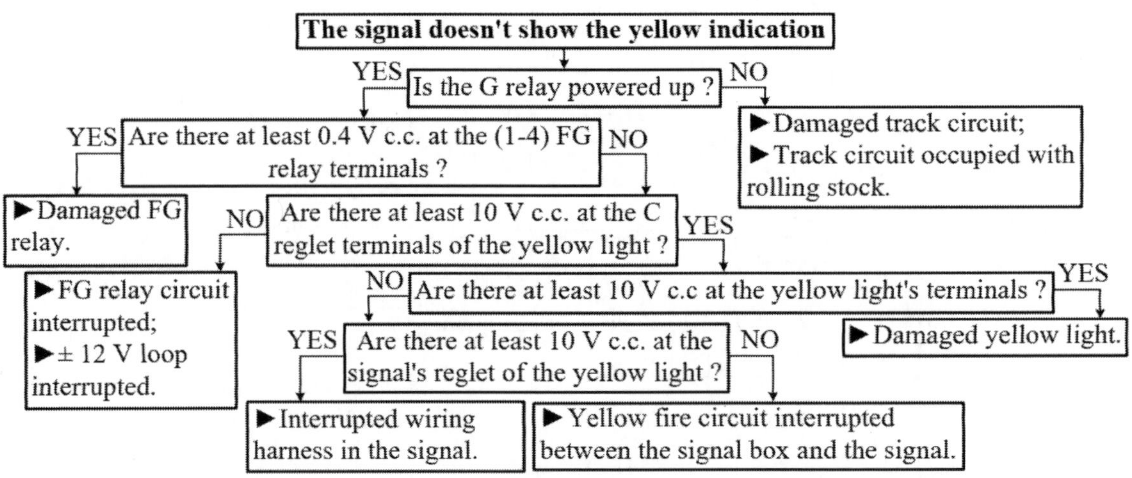

Figure 5. Diagnosis chart for cases where the yellow indication is not working

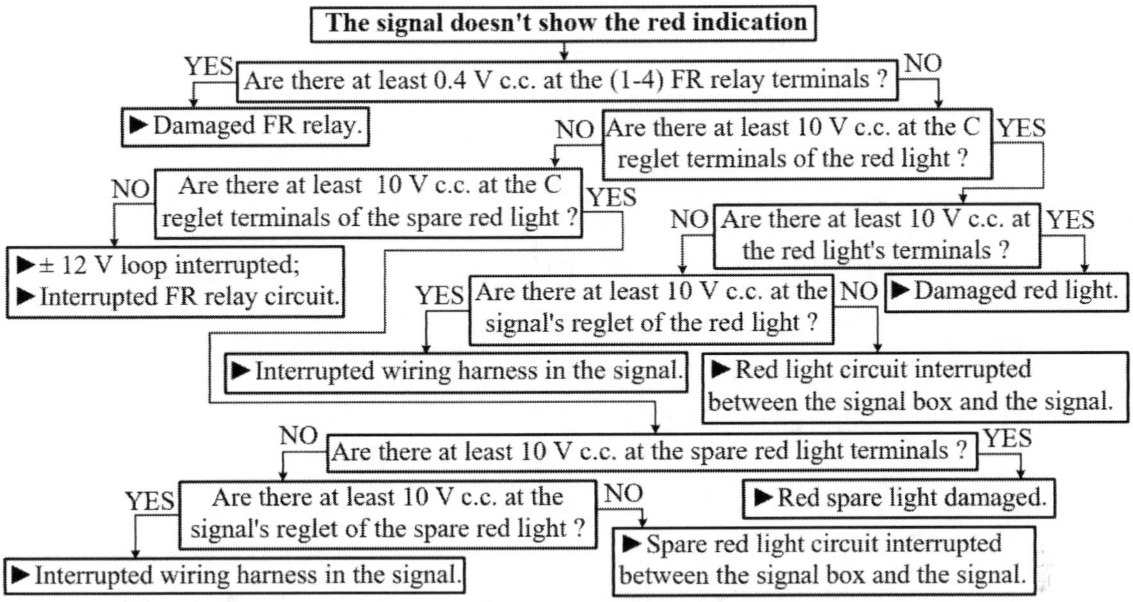

Figure 6. Diagnosis chart for situations where the red or spare red light indications are not working

Figures 7 and 8 shows an example of a SEMBla dialogue window, while Figure 9 shows the diagnosis reached based on the user answers and field measurements.

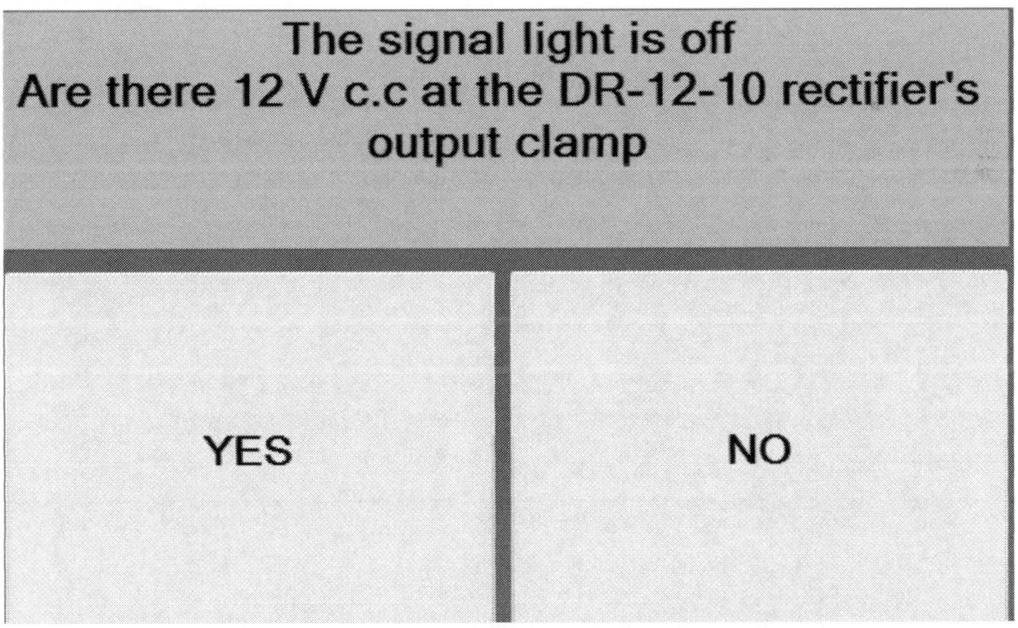

Figure 7. SEMBla dialogue window example

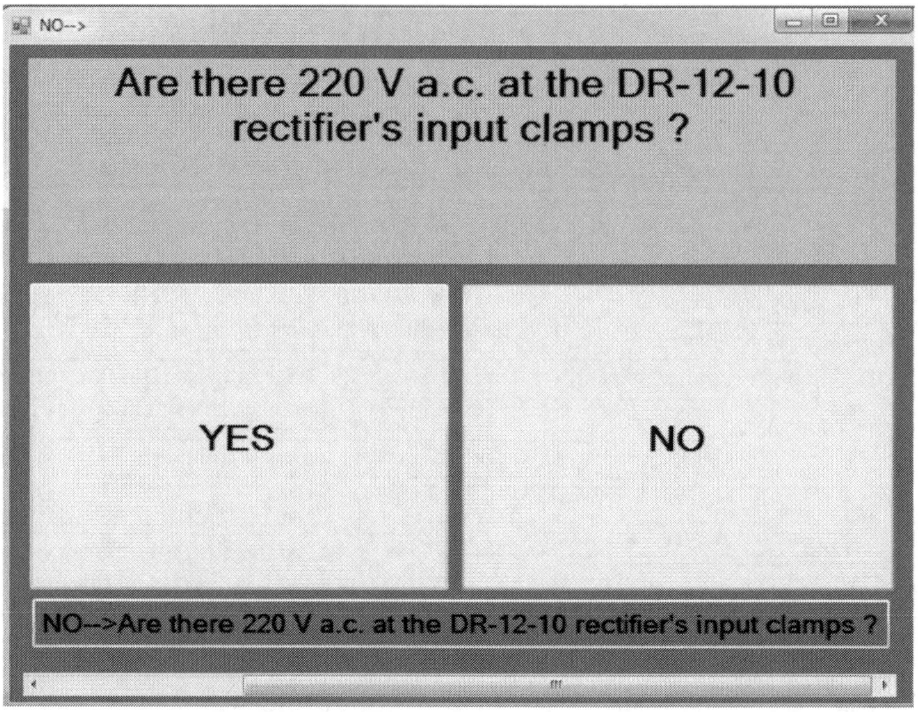

Figure 8. SEMBla dialogue window example

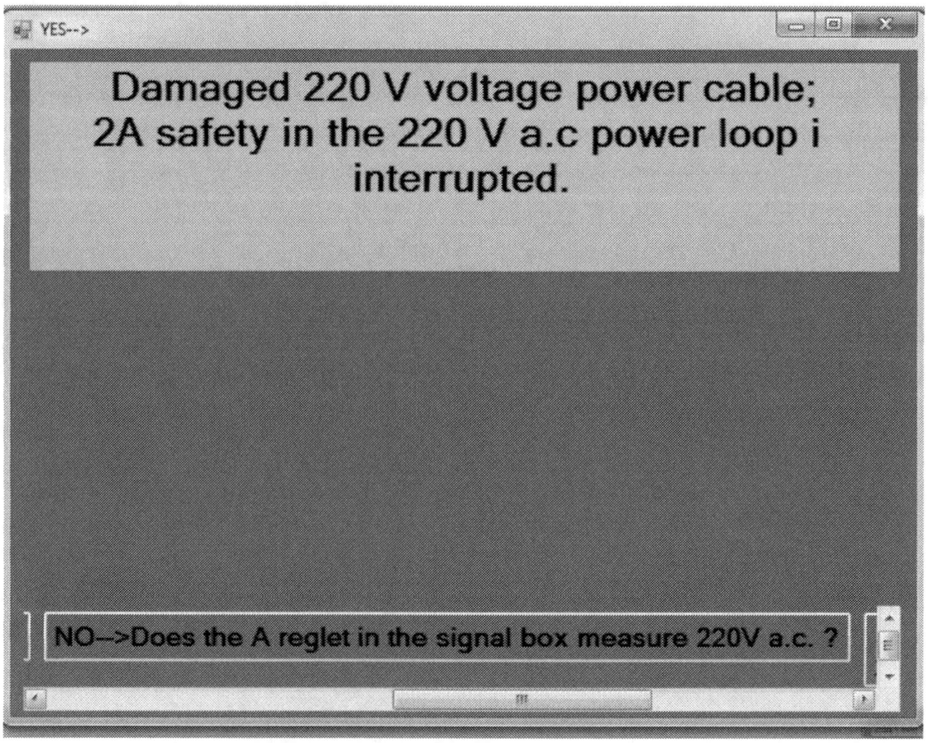

Figure 9. SEMBla dialogue window example

5. Conclusions

The realization of the SEMBla software has been made to use the diagnosis charts to detect ABS passing signal faults.

The software allows significant reduction of remedy times, and provides helpful and quick guidance to the maintenance staff using it.

The diagnosis method presented in this article does not necessitate additional infrastructure and implementation costs, being a useful assistant to the maintenance staff, both as a training method and a failure detection method.

References

[1] Spunei E, Piroi I, Muscai C and Piroi F 2014 *Software Module for Switch Electromechanics Failure Detection*, International Conference and Exposition on Electrical and Power Engineering EPE 2014, Iaşi, Romania, October 16-18, pp 1089-1091

[2] Spunei E, Piroi I, Muscai C and Piroi F 2014 *ABS Failure Diagnosis Charts for a Blocked CL*, International Conference on Applied and Theoretical Electricity ICATE 2014, Craiova, Romania, October 23-25

[3] Spunei E, Piroi I, Muscai C and Piroi F 2014 *Automatic Block Signaling Installation Failure Diagnosis with LCOBla*, International Conference on Applied and Theoretical Electricity ICATE 2014, Craiova, Romania, October 23-25

[4] Piroi I, Spunei E, Muscai C and Piroi F 2014 *Diagnosis Charts for Regular Inversion Failures of an Automatic Block Signal Installation*, International Conference on Applied and Theoretical Electricity ICATE 2014, Craiova, Romania, October 23-25

[5] Piroi I, Spunei E, Chioncel C P and Piroi F 2015 *Rapid Diagnosis of Track Circuits in a Railroad Station*, 9th International Symposium on Advanced Topics In Electrical Engineering ATEE 2015, Bucureşti, Romania, May 7-9, pp 710-715

[6] Piroi I, Spunei E, Chioncel C P and Piroi F 2016 *Computerized Diagnostic of the Red and White Entry Lighting Signal Indications*, International Conference on Applied and Theoretical Electricity ICATE 2016, Craiova, Romania, October 6-8

[7] Piroi I, Spunei E, Muscai C and Piroi F 2016 *The Diagnosis of the Entry Lighting Signal when the Free Command Cannot be Set*, International Conference on Applied and Theoretical Electricity ICATE 2016, Craiova, Romania, October 6-8

[8] Takashige T 1999 *Signalling Systems for Safe Railway Transport*, Railway Technology Today 8, Japan Railway & Transport Review 21, Japan, September [last retrieved: 25.07.2017]

[9] Ishima R, Fukuta Y, Matsumoto M, Shimizu N, Soutome H and Mori M 2008 *A New Signalling System for Automatic Block Signal between Stations Controlling through an IP Network*, WCRR, May, http://www.railway-research.org/IMG/pdf/o.3.4.2.1.pdf

[10] Hayakawa K, Miura T, Ishima R, Soutome H, Tanuma H and Yoshida Y 2010 An IP network-based signal control system for automatic block signal and its functional enhancement, *WIT Transactions on The Built Environment* **114** Computers in Railways XII 783-794 https://www.witpress.com/Secure/elibrary/ [last retrieved: 25.07.2017]

[11] Mascardi V, Briola D, Martelli M, Caccia R and Milani C 2009 *Monitoring and Diagnosing Railway Signalling with Logic-Based Distributed Agents*, Proceedings of the International Workshop on Computational Intelligence in Security for Information Systems CISIS'08, Volume 53 of the series Advances in Soft Computing, pp 108-115

[12] Durmus M S 2014 *Control and Fault Diagnosis of Railway Signaling Systems: A Discrete Event Systems Approach*, Graduate School of Engineering, Osaka University, Doctoral Thesis

[13] ***Ministry of Public Works, Transportation and Living 2001 *Regulation for Railway Technical Operation, Nr. 002,* Railroad Publishing House, (in Romanian)

[14] ***Ministry of Transportation, Constructions and Tourism 2006 *Signaling Regulation, Nr. 004,* Railroad Publishing House, (in Romanian)

IOP Conf. Series: Materials Science and Engineering **294** (2017) 012014 doi:10.1088/1757-899X/294/1/012014

Truck circuits diagnosis for railway lines equipped with an automatic block signalling system

E Spunei[1], I Piroi[1], C Muscai[1], E Răduca[1] and F Piroi[2]

[1]Eftimie Murgu University of Resita, Electrical Engineering and Informatics Department, Traian Vuia str., no. 1-4, 320085 Resita, Romania

[2]Vienna University of Technology, Institute of Software Technology and Interactive Systems, Viena, Austria

E-mail: e.spunei@uem.ro

Abstract. This work presents a diagnosis method for detecting track circuits failures on a railway traffic line equipped with an Automatic Block Signalling installation. The diagnosis method uses the installation's electrical schemas, based on which a series of diagnosis charts have been created. Further, the diagnosis charts were used to develop a software package, CDCBla, which substantially contributes to reducing the diagnosis time and human error during failure remedies. The proposed method can also be used as a training package for the maintenance staff. Since the diagnosis method here does not need signal or measurement inputs, using it does not necessitate additional IT knowledge and can be deployed on a mobile computing device (tablet, smart phone).

1. Introduction

Most failures of the railroad traffic security installations (especially traffic command and control) are due to specific component breakdowns. Other failures are due to poor maintenance work. No matter the failure cause, these lead to train traffic delays and possibly traffic events.

To identify and diagnose failures of the track circuits several methods can be applied: electric models [1], control and diagnosis equipment [2-4], neuro-fuzzy systems, neural networks [4-6].

Currently, when a track circuit is faulty, the traffic manager that notices the malfunction, notifies the maintenance staff. They, in turn, must travel to the damage location (which often is tens of kilometres away), observes the state of the installations, performs measurements and established the cause of the damage. In all this time, train traffic can only happen with reduced speed or not at all.

The failure diagnosis time significantly depends on the theoretical and practical abilities of the maintenance staff. Since they often act in stress conditions to which the weather conditions are an important factor, the diagnosis method we propose here can offer substantial help.

The method we propose in this work is based on diagnosis charts created by us, and then transposed into a software package of a diagnosis software. The user of the software does not need IT know-how, as the program has a friendly, simple, question based interface. The answers given by a user to the series of questions shown by the software lead to a quick identification of the damage cause.

2. Track circuits on an ABS block

On most railroad blocks equipped with electrodynamic centralized installations (CED in Romanian) the train traffic between two consecutive stations is commanded and controlled by an Automatic Block Signalling installation (ABS). The main components of an ABS installation are:

- Track circuits which control the rail line integrity and the occupation state of the various sectors (by rolling stock for example) [7];
- Light signals which are used to communicate traffic orders to the train operators on locomotives. These signals allow or forbid train passage from one sector to the next [8, 9];
- Control and command circuits that control the ABS installation, including its traffic orientation [10-12].

Lighting signal and ABS orientation control and command diagnosis have been presented in other works [10-12].

Track circuits power supply is done in the reverse direction to the traffic direction, such that, a circulating train will always meet the reception part of the track circuit. When the ABS installation is re-oriented, the reception part becomes the powering component, while the powering component becomes the receiving one. The installation re-orientation is realised by the orientation directing relay clamps (X-D or Y-D) powered by the directing relay, D.

Figure 1 presents the power up schema for four track circuits (Y1-AD, Y2-AD, X2-AD, X1-AD) on an ABS equipped line, when the traffic direction is in the X direction (X-oriented). Each signal box (DS on Fig. 1) is outfitted with power up components (A) for a receiving and track circuit (R) for the neighbouring track circuit.

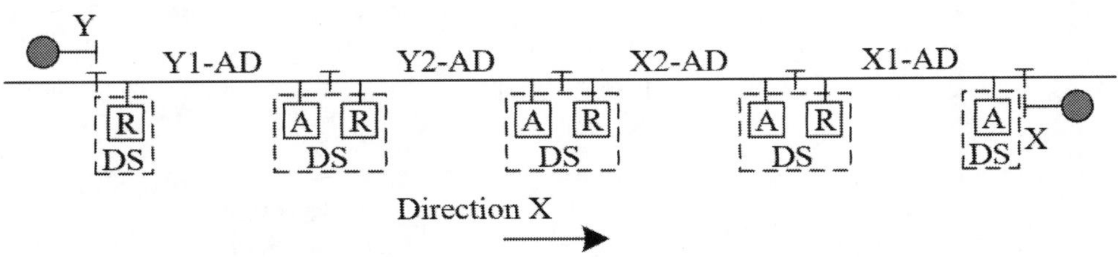

Figure 1. Powering mode for ABS track circuits

Figure 2 presents the powering schema for the track circuits on an ABS block. Each ABS track circuit is powered from the central post with two voltage levels:

- 220 V/75 Hz to power up the signal boxes and some of their equipment;
- 110 V a.c. to power up the code transmitting relays, which are placed in each signal box.

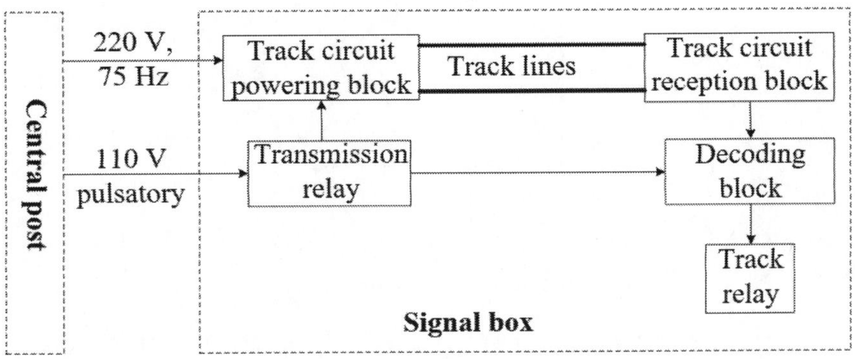

Figure 2. Power supply block scheme for the ABS track circuits

Figure 3 presents the powering scheme for the double wire X1-AD track circuit.

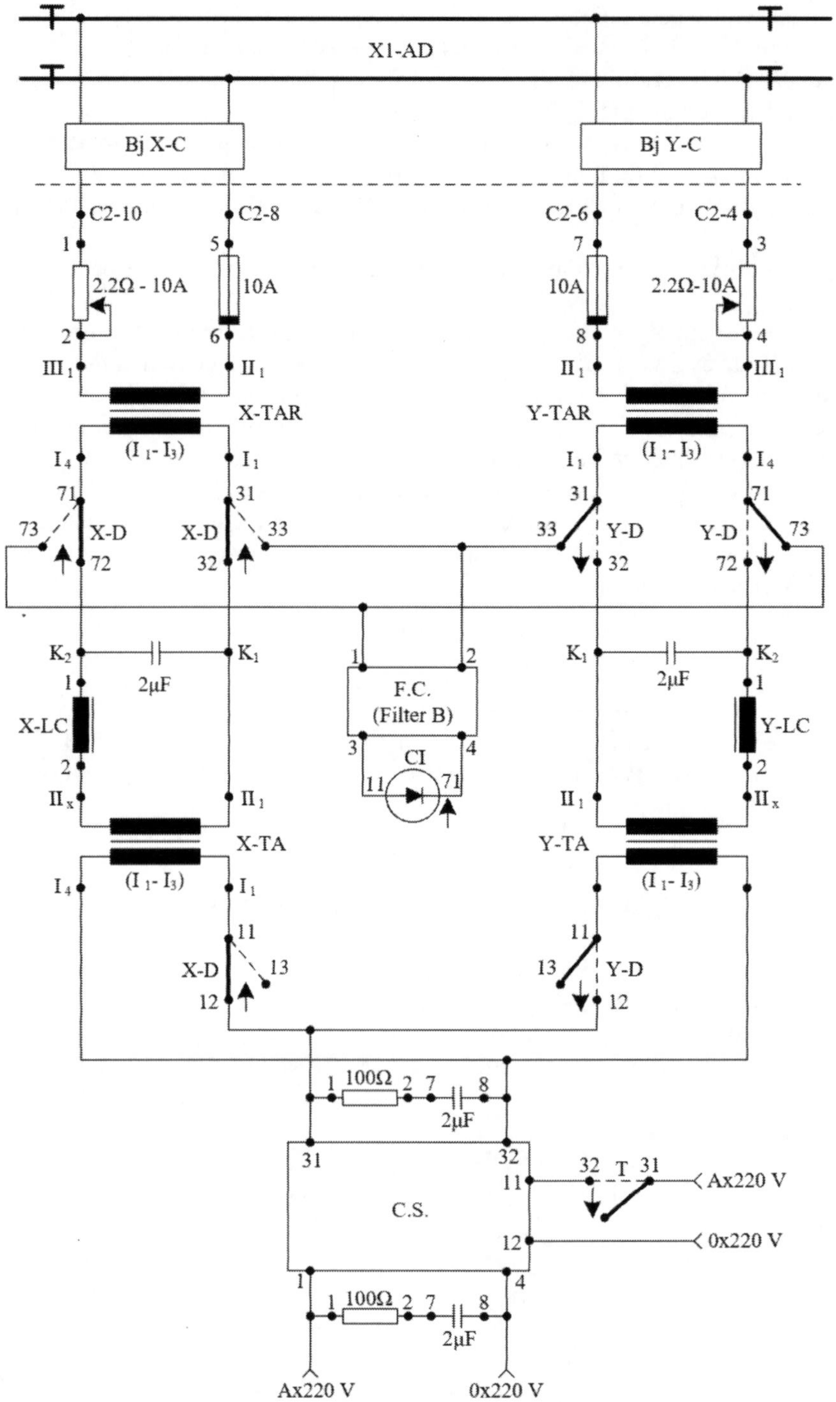

Figure 3. Power supply schema for the X1-AD double wire track circuits

To supply the power to the track circuits two codes are sent, one after the other, with a 0.22-0.23 s offset. The two codes form the sequences which successively power up each track circuit. Thus, if the

Y1AD circuit is powered up in the first sequence (S1), the Y2AD track circuit is powered in the second sequence (S2), while the X2AD track circuit is powered in the first sequence (S1). Each sequence has a 0.345 s long power impulse, followed by a 0.12 s pause [7].

The static closing switch, CS, controlled through the impulse transmitter relay, TS, powers up with pulsating 220 V the transformer's X-TA primary. The limiting coil, LC, limits the receiving circuit current when the track circuits are shunted.

Regulating the power supply voltage is done through the secondary winding plugs of the X-TA voltage reducing transformer and through the 2.2 Ω resistor on the power supply side. The voltage level is regulated depending on the ballast insulation resistance value, type of cross ties, and the track circuit length, such that the voltage measured in the track lines at the receiving end is at least 0.6 V pulsating d.c.

The role of the joint coils is to ensure that the traction current retour passes from one track circuit to a neighbouring track circuit.

When the track integrity is ensured, for the whole length of the track circuit, the voltage injected into the track lines at the power supply end is received through the joint coil and of the Y-TAR voltage step-up transformer.

Since the traction current retour voltage functions at 50 Hz we must use a track filter, F.C. This will allow the 75 Hz control voltage to pass and will stop the 50 Hz frequencies as well as their superior harmonics.

The F.C. output voltage is passed to the impulse relay, CI, which works in the received pulsing rhythm.

The integrity of the isolating coils separating the track lines of a track circuit from the neighbouring track circuits is controlled by the impulse decoding relay (Figure 4). The transmitter relay T contacts, working in a different sequence than the CI relay's sequence, power the G relay only when the isolating joints are in a corresponding state.

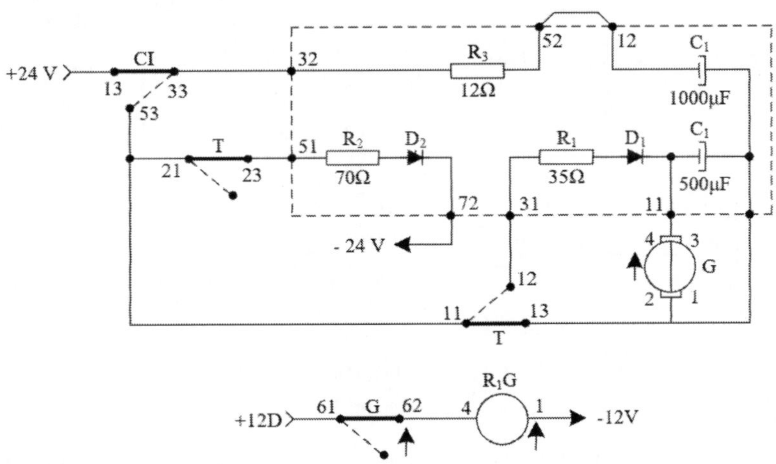

Figure 4. ABS decoding block and track circuit schema

When an isolating joint is damaged, the two sequences overlap, which disables the decoding relay and the track relay will be de-energized.

3. Diagnosis charts for ABS track circuits and the CDCBla Software

All ABS track circuits are controlled by the command apparatus in the two adjacent train stations. For the situation depicted in Figure 1 the Z1AD and Y2AD track circuits are controlled from the first station, while the X2AD and X1AD track circuits are controlled from the next station.

The main track circuit defects cases occurring are:

- One track circuit is damaged;
- All ABS track circuits are damaged;
- Half the ABS track circuits are damaged.

Figure 5 presents the diagnosis chart for the case where one track circuit is not working as expected and the ABS installation is X-oriented.

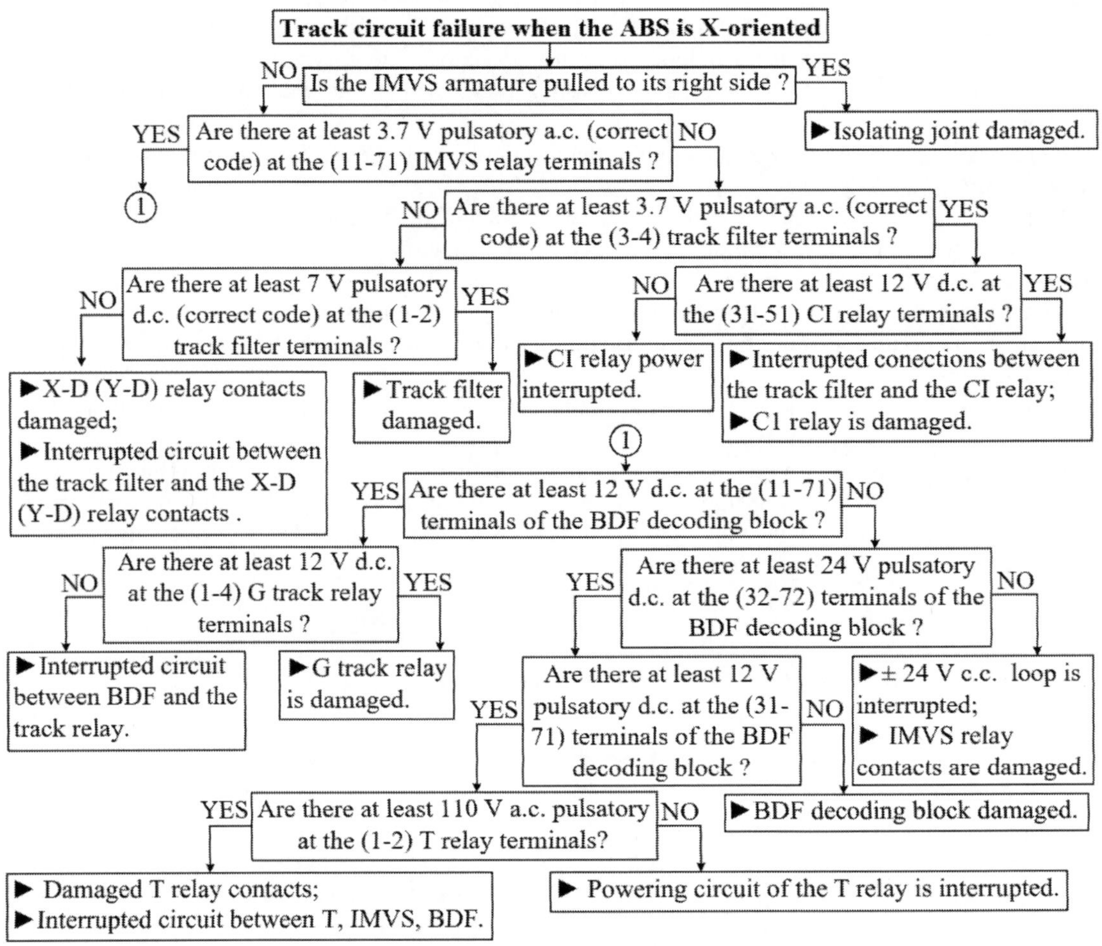

Figure 5. Diagnosis chart for cases when one ABS track circuit is damaged and the installation is X-oriented

The ABS installation that defines how the track circuits function, can be either X-oriented (train traffic away from Bucharest) or Y-oriented (train traffic towards Bucharest) [9-10]. A quick track circuit failure diagnosis can be correctly done only by explicit installation orientation and observation of the defect state depending on the ABS orientation. This is possible because the same apparatus ensures both the receiving and the power supply for different track circuits (e.g. power supply to Y1AD when the ABS installation is X-oriented and power supply for X2AD when the installation is Y-oriented).

Figure 6 presents the diagnosis chart for situations where one track circuit is malfunctioning but only when the installation is Y-oriented.

The charts presented in the previous figures were used to create the CDCBla software, with the use of Visual Basic.Net environment. Figure 7 shows a screenshot taken during a run of the software package.

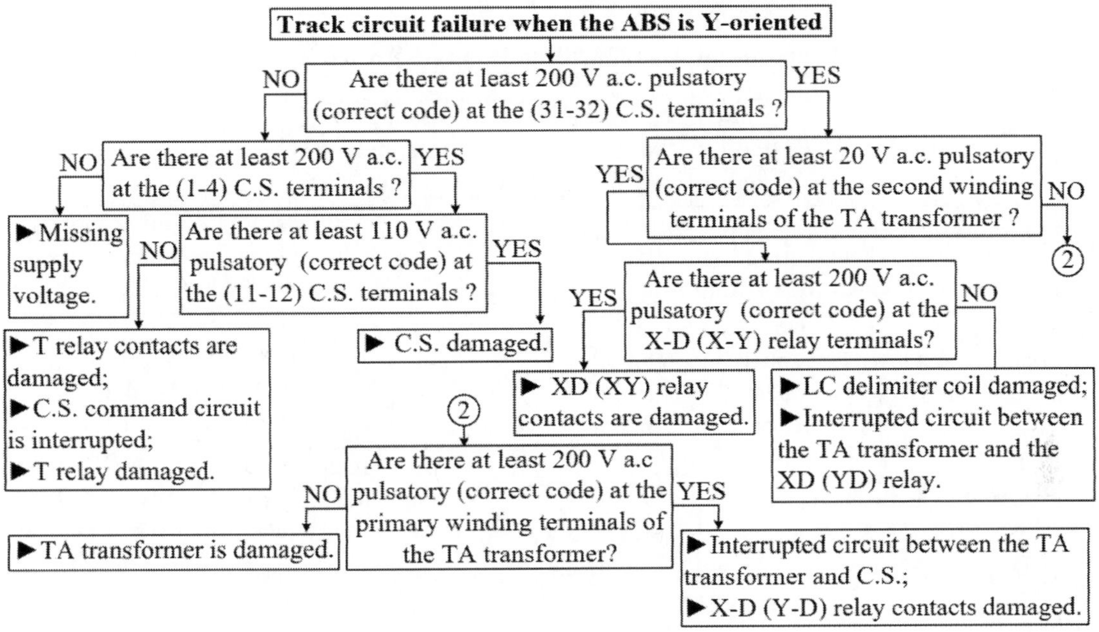

Figure 6. Diagnosis chart for cases when a track circuit malfunctions only when the ABS installation is Y-oriented

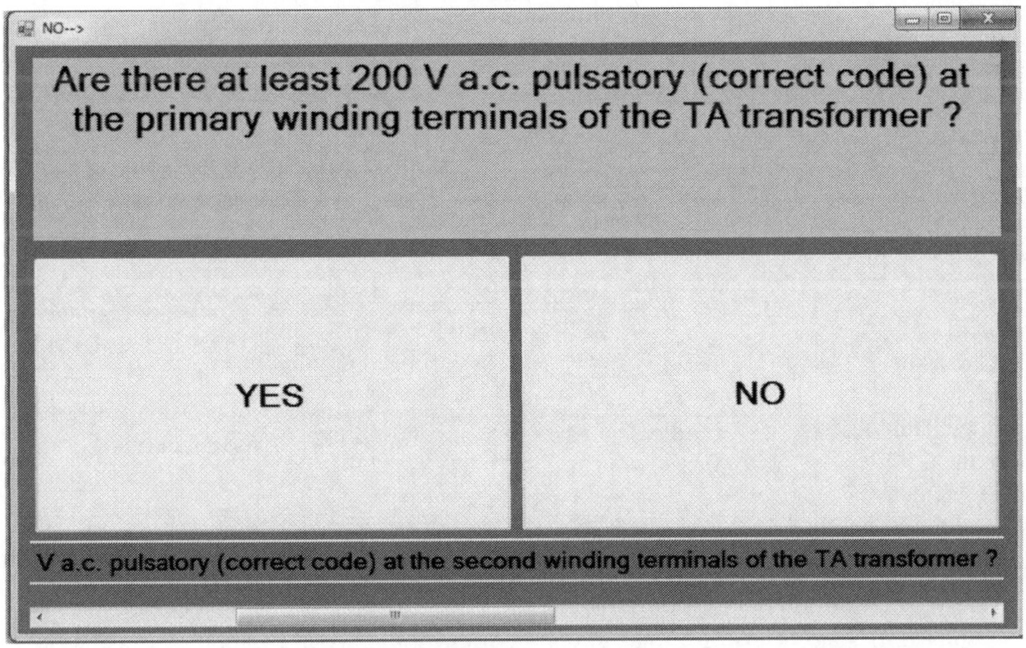

Figure 7. CDCBla software screenshot

IOP Conf. Series: Materials Science and Engineering **294** (2017) 012014 doi:10.1088/1757-899X/294/1/012014

Figure 8 depicts the diagnosis charts for the situation where, independent of the installation orientation, the same track circuit is not working. To facilitate understanding this diagram, we consider that the ABS is X-oriented, that is the orientation relay, X-D, is powered up, and the Y-D orientation relay is powered down.

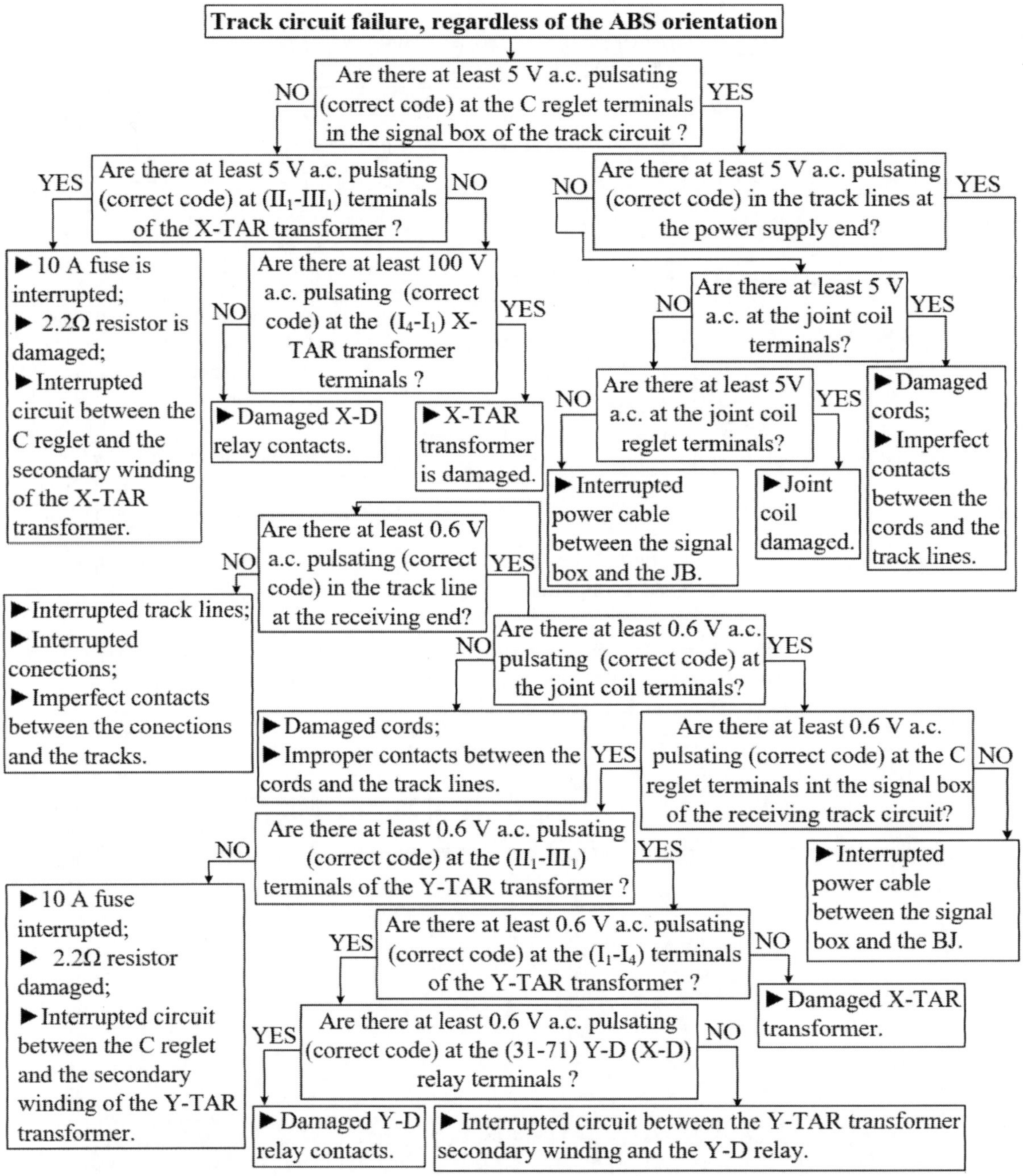

Figure 8. Diagnosis chart for track circuit failures for both ABS installation orientations

The chart presented in Figure 6 considers cases where the maintenance staff first travels to the power supply component of the damaged track circuit.

Because the latter two failures we described are caused, mainly, by failures of the power systems, we plan to analyse and describe these cases in another article.

4. Conclusions

With the design of the diagnosis charts and the CDCBla software presented in this work, a user of this software needs to answer successive questions shown by the program. At the end of the series of questions the reason for the installation failure is presented to the maintenance staff.

The work presented in this paper contributes to the diagnosis of traffic safety installations of type CED-CR2 on traffic lines equipped with ABS installation by significant reductions in diagnosis and remedy times, eliminating wrong decisions that may be made by maintenance staff in the diagnosis phase.

References

[1] Debiolles A, Oukhellou L, Aknin P and Denoeux T 2006 *Track circuit automatic diagnosis based on a local electrical modelling*, http://www.railway-research.org/IMG/pdf/492.pdf [last retrieved: 25.07.2017]

[2] Oukhellou L, Debiolles A, Denœux T and Aknin P 2010 Fault diagnosis in railway track circuits using Dempster-Shafer classifier fusion, *ScienceDirect, Engineering Applications of Artificial Intelligence* **23**(1) 117-128

[3] Ishima R, Fukuta Y, Matsumoto M, Shimizu N, Soutome H and Mori M 2008 *A New Signalling System for Automatic Block Signal between Stations Controlling through an IP Network*, http://www.railway-research.org/IMG/pdf/o.3.4.2.1.pdf [last retrieved: 25.07.2017]

[4] Gillich N, Potoceanu N, Gillich G R, Raduca M and Chioncel C P 2008 *Intelligent system for the control of ambiental parameters in hi-tech workplaces*, Proceedings of the 6 th International Conference of DAAM Baltic Industrial Engineering PTS 1 and 2, Tallinn, Estonia, pp 71-76

[5] Chen J, Roberts C and Weston P 2007 Fault detection and diagnosis for railway track circuits using neuro-fuzzy systems, *ScienceDirect, Control Engineering Practice* **16** 585-596

[6] Tirian G O, Filip I and Prostean G 2014 Adaptive control system for continuous steel casting based on neural networks and fuzzy logic, *Neurocomputing* **125**(Special Issue) 236-245

[7] Piroi I, Spunei E, Chioncel C P and Piroi F 2015 *Rapid Diagnosis of Track Circuits in a Railroad Station*, 9th International Symposium on Advanced Topics In Electrical Engineering ATEE 2015, Bucureşti, Romania, May 7-9, pp 710-715

[8] Tetsuo T 1999 Signalling systems for safe railway transport, *Japan Railway &Transport Review* September 44-50, http://www.ejrcf.or.jp/jrtr/jrtr21/pdf/F44_Technology.pdf

[9] Piroi I, Spunei E, Chioncel C P and Piroi F 2016 *Computerized Diagnostic of the Red and White Entry Lighting Signal Indications*, International Conference on Applied and Theoretical Electricity ICATE 2016, Craiova, Romania, October 6-8

[10] Spunei E, Piroi I, Muscai C and Piroi F 2014 *ABS Failure Diagnosis Charts for a Blocked CL*, International Conference on Applied and Theoretical Electricity ICATE 2014, Craiova, Romania, October 23-25

[11] Spunei E, Piroi I, Muscai C and Piroi F 2014 *Automatic Block Signaling Installation Failure Diagnosis with LCOBla*, International Conference on Applied and Theoretical Electricity ICATE 2014, Craiova, Romania, October 23-25

[12] Piroi I, Spunei E, Muscai C and Piroi F 2014 *Diagnosis Charts for Regular Inversion Failures of an Automatic Block Signal Installation*, International Conference on Applied and Theoretical Electricity ICATE 2014, Craiova, Romania, October 23-25

The comfort of the passengers

R Hasa, T Liedl and J Bucha

Slovak University of Technology in Bratislava, Institute of Transport technology and engineering design, Námestie Slobody 17, Bratislava, Slovakia

E-mail: richard.hasa@stuba.sk

Abstract. A people spending in the vehicles a lot of time. In the Slovak republic is still a lot of roads with damage road surface with holes and cracks. The idea of these paper is about increase driving comfort using additional seat foams. These foams are tested on different road surfaces at the different speed.

1. The measuring vehicle

The vehicle was loan to the Institute of Transport Technology and Engineering Design by VW Slovakia a.s. The VW e-Up! (Figure 1) is a typical city vehicle. Power drive is not the combustion engine, but by the electrical engine. The maximum driving range is around 140 km. The top speed of the vehicle is 135 km / h. At the time of testing the vehicle was 3 years old and the odometer showed around 10 000 km. At the start of each test drive the tire pressure has been checked. Vehicle has been loaded by the mandatory equipment and measuring equipment.

Figure 1. VW e-Up!

2. The measuring equipment

The measuring equipment consisted of 7 accelerometers, 12 V power supply, voltage transducer to 3,3 V, measuring card, notebook and CAN to USB convertor. The 3 axes accelerometer was connected to the network by Ethernet cable. The accelerometer recorded the data only in the Z axis with sampling frequency 100 Hz [1]. This sampling frequency made possible to create FFT analysis to 50 Hz. According to Figure 2 FFT analysis till 50 Hz is sufficient [2].

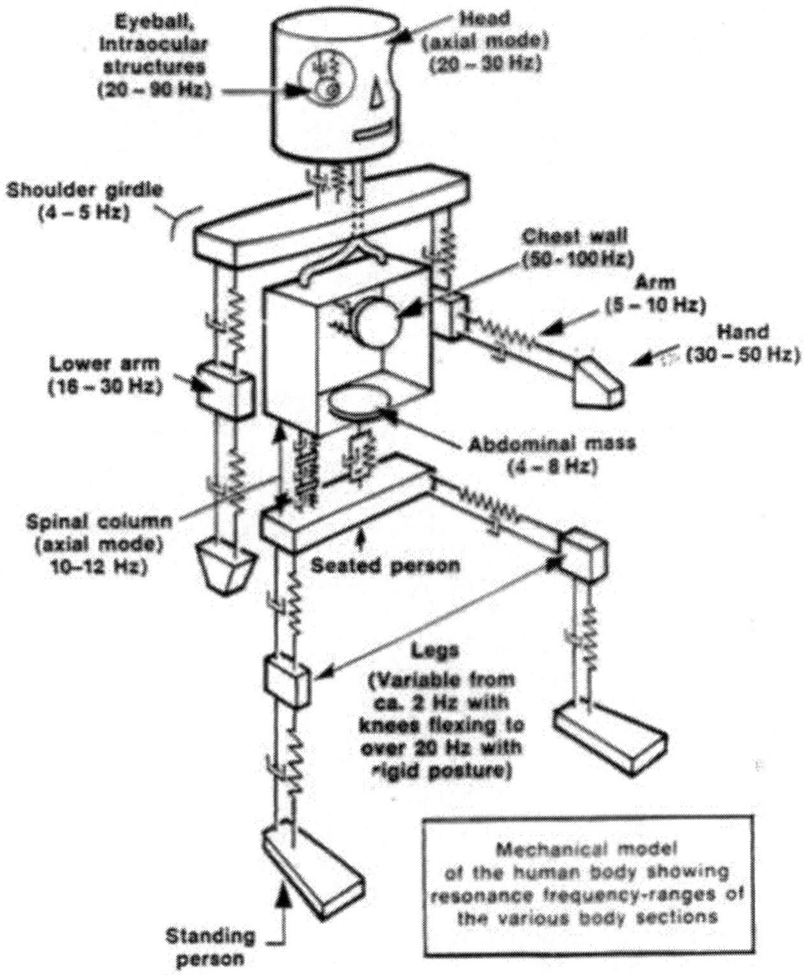

Figure 2. The human body resonance frequencies

The electrical vehicles uses CAN BUS network, such as standard vehicles with combustion engines. Using convertor that is compatible with CAN 2.0 A/B of standard ISO 11 898 data flow thru OBD/OBD II/EOED connector that convert signal from CAN to USB port. In that case was used convertor VECTOR VN1610, that communicates with Matlab-Simulink. The CAN BUS has sampling frequency 1000 Hz, but for lowering of data was sampling frequency changed to 100 Hz.

3. Sensor position

Important condition to assembly sensors was to save removability. This condition was fulfilled using fired pistol glue. Increased resistance of sensors was using electrical tape and pulling tapes. First sensor was assembly to left arm of front axle, second was assembly to right arm of front axle, third was in the middle of rear axle, fourth was in the vehicle trunk, fifth was under the front hood on chassis, sixth was on the drivers leg and seventh was on the passenger leg.

IOP Conf. Series: Materials Science and Engineering **294** (2017) 012015 doi:10.1088/1757-899X/294/1/012015

Figure 3. Sensor Positions

Under the vehicle was problem with deployment of cables. In the case of rapid weather change it has to protect sensors of short circuit. After some tests drives was made decision to remove chassis housing. This decision made easier to connect all of the sensors together with computer and cables were easily fixed to the vehicle gear (Figure 4).

Figure 4. Cable distribution

The another ends of cables were drag through the rear window. However VW e-Up! has side openable windows and in the winter month was very cold in the interior of vehicle. In that case was used tape to fill all holes among window and cables.

In the interior of the vehicle was everything original. On the rear seats were cables, measuring card and voltage transducer. During the test drives were rear seats downcast to protect measuring equipment from impact of vibrations, acceleration and deceleration. On the rear floor was 12 V power supply and in the middle of front seats was CAN to USB convertor. On the dashboard was notebook connected to measuring card and USB port.

Figure 5. Interior view

4. Testing roads

All of the testing roads were in Bratislava. The testing roads were consist of the different road surface:

1. the road consist of concrete blocks – 40 km/h, 50 km/h, 60 km/h (Figure 6)
2. standard road surface (Figure 7)
3. car park consist of concrete undulate blocks – 30 km/h, 40 km/h, 50 km/h (Figure 8)
4. the road consist of granite cubes – 40 km/h, 50 km/h, 60 km/h (Figure 9)

Figure 6. The test road No 1

Figure 7. The test road No 2

Figure 8. The test road No.3

Figure 9. The test road No 4

5. Testing materials

The choosing of seat materials was done in cooperation of Poly, s.r.o company in Žilina, that is member of Eurofoam Group. After discussion was elected 8 types of materials (Figure 10), from 3 main foam groups: Normal, Visco a Cold foams. Each foam has 2 main characteristics, first two numbers in the name of foam are bulk density (kg/m^3) and the second group of two are compression resistance (kPa).

Figure 10. Testing materials

6. Conclusion

The results after the processing and the evaluating of the measured data on the all testing road surfaces and testing materials are shown in the graph in Figures 11 and 12. The best properties of tested foams has cold foam R4036. The processing of measured data was done by Matlab. From the measured data were counted FFT analysis for each kind of foam and each test road at the different speed. From all of FFT analysis were counted abundance of each frequency. These abundance are evaluated and the frequency spectrum from 0 – 20 Hz with lowest abundance given best material for improve passenger comfort. When the values are negative the comfort decrease.

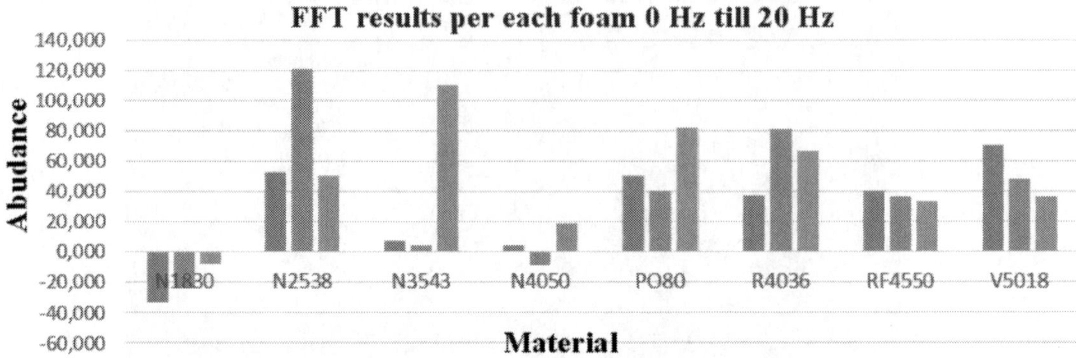

Figure 11. Results for test road No 4, blue – 40 km/h, orange – 50 km/h, grey – 60 km/h

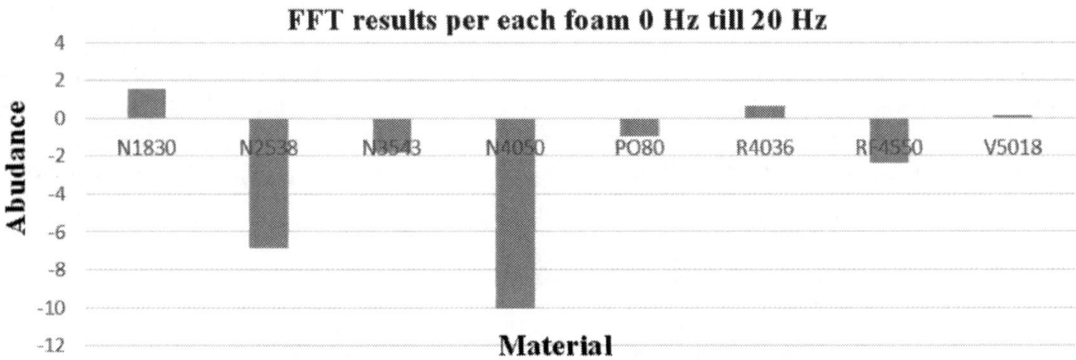

Figure 12. Results for test road No.2

Acknowledgement
We want to acknowledge to Volkswagen Slovakia a.s for car loan and Poly, s.r.o for the provision of testing materials.

References

[1] Newell G S and Mansfield N J 2006 *Influence of posture and multi-axis vibration on reaction time performance and perceived workload,* Research report
[2] Griffin M J 1990 *Handbook of human vibration,* Academic Press

IOP Conf. Series: Materials Science and Engineering **294** (2017) 012016 doi:10.1088/1757-899X/294/1/012016

Aspects regarding the capitalization of the powdery wastes in steelmaking

E C Dragna, A Ioana and N Constantin

University Politehnica of Bucharest, Engineering and Management of Metallic Materials Obtaining Department, Bucharest, Romania

E-mail: claudiadragna90@gmail.com

Abstract. The metallurgical industry is experiencing major issues which are not related to a crisis of the raw materials and energy resources, but to the stringent requirements for the environmental protection.
The small-sized and pulverous wastes, stemming mainly from the steel industry but also from the mining industry, respectively the energy industry, due to their high content of iron, manganese, carbon and various oxides, should be named by-products and be deemed as constituents of natural capital, because they can be capitalized in the steel industry.
The optimization of the steelmaking process can have radically different approaches, depending on the specific conditions of a steel plant during a specific period of time.
The production of steel is continuously increasing worldwide, due to the development of the machinery industries.
The steel manufacturers use furnaces that produce steel form various ferrous wastes, instead of traditional raw materials, because the method is effective, given that it assumes low costs.

1. Introduction

The metallurgical industry is experiencing major issues which are not related to a crisis of the raw materials and energy resources, but to the stringent requirements for the environmental protection. The development of the metallurgical industry is conditioned by the settlement of the major issues that ensue from the relationship industry-environment, being strictly targeted on the pollution control and the protection of the natural and energy resources. The small-sized and pulverous wastes, stemming mainly from the steel industry but also from the mining industry, respectively the energy industry, due to their high content of iron, manganese, carbon and various oxides, should be named by-products and be deemed as constituents of natural capital, because they can be capitalized in the steel industry [1].

The optimization of the steelmaking process can have radically different approaches, depending on the specific conditions of a steel plant during a specific period of time. When the main objective is to cut the costs, the optimization can be focused on decreasing the energy inputs and the consumption of oxygen, carbon, natural gas etc., with minimal effects on the production.

If the objective is to increase the production, the strategy must be different, other factors being taken into consideration, such as the electrical equipment, including the transformer, the reactor and the limitations of the electricity line. Usually, irrespective of the main objective, the primal goal of the optimization mission is finding the balance point of the necessary adjustments for the maximizing of benefits within each heat and the existence of a system that possesses the flexibility to adapt its operation when changes are performed on the conditions [2].

2. Aspects regarding steelmaking

The production of steel is continuously increasing worldwide, due to the development of the machinery industries [3].

Steel is the easiest metallic material to recycle worldwide, from the household waste to those resulted following the demolitions of certain buildings all steel waste can be collected and processed with a view to their re-use.

The steel manufacturers use furnaces that produce steel form various ferrous wastes, instead of traditional raw materials, because the method is effective, given that it assumes low costs.

The possibility of using waste iron as primary charge is also one of the main advantages of the electric arc furnace (EAF), which allowed its development and establishment as a steelmaking aggregate.

A long time ago, the wastes from within the plant were meeting the demand for ferrous wastes and by the steelmaking process a significant quantity of internal wastes was generated, derived from the actions as such or as a preparation that could not be sold. Nevertheless, the metallic wastes acquired in this manner did not manage to meet the entire demand. Under these circumstances, over the years, various organizations emerged and refined themselves in the preparation and establishment of metallic waste stocks for the steel plants. The majority of ferrous wastes are found in various industrial sources (initially from the rail transportation sector and installations, then from the producers of packaging and the automobile manufacturers).

Related to the development of the steelmaking technologies, the demand for metallic wastes increased robustly, which determined the development of an infrastructure of gathering, processing and sale of the wastes to steel plants and other markets, with a view to recycle.

A close interdependency, based on demand and offer, was created between the steel producers and the ferrous wastes supply units. Due to the recycling infrastructure, the steel industry receives a constant flow of ferrous wastes.

Steelmaking is performed on two major important routes (Figure 1):

1. the method “from iron ore to steel” or the traditional or “integrated” route for the so-called “crude steel” (primary) in which the point of departure is set on iron and coal and the furnace and oxygen converter are used;

2. the method “from waste to steel” or the route of “the electric arc furnace”, for the so-called “secondary steel” (recycled) within which the ferrous wastes constitute the main raw material used in the electric arc furnace.

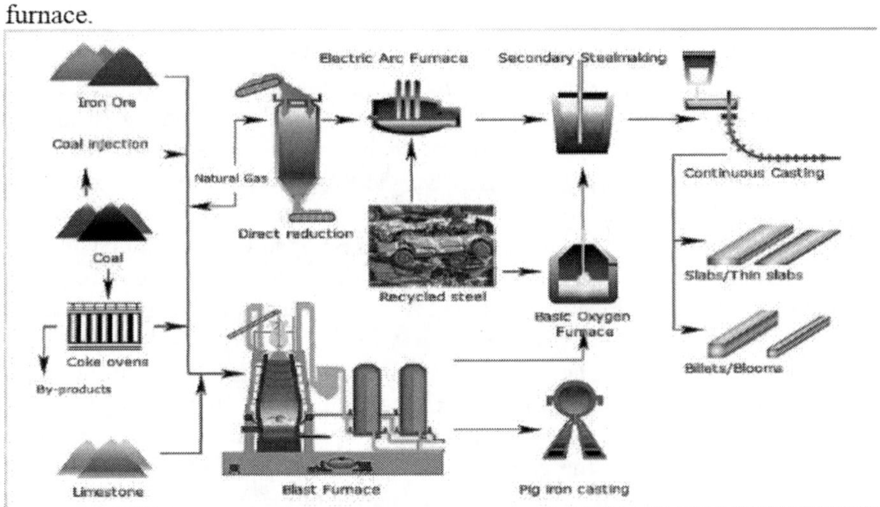

Figure 1. Routes of steelmaking [4]

The second method of steelmaking is called “from waste to steel” and, as we previously mentioned, the most suited technological aggregate for the melting of wastes is the electric arc furnace. By the

IOP Conf. Series: Materials Science and Engineering **294** (2017) 012016 doi:10.1088/1757-899X/294/1/012016

agency of the melting phase, steel wastes are converted into high quality steels, with the aid of the electric arcs.

In 2004, 1/3 of the global production of steel was achieved by the flow that uses the EAF and in the USA there was a larger quantity of steel produced from recycled wastes than from iron ores. The economic and environmental benefits are originated from the recycling of the ferrous metallic materials. Steels produced in the EAF are both low alloy steel and high alloy steel, acquired from secondary materials. Their production capacity is variable and correlated with the capacity of the furnaces (50-150 tons).

The rise in productivity and economic yield of the electric arc furnaces was possible by the implementation of some solutions, of which we can mention: the increase of the installed electrical output; the use of oxygen and additional fuels; the replacement of the refractory linings with water-cooled components; the possibility of stirring the metal heel and processing under foaming slag; the recovery of the heat from the water circulating in the cooling systems and the off-gas; the preheating of wastes; the slag-free discharge by an eccentric bottom tap-hole; the optimal management and process control by the agency of the computer [5].

The power of the transformer is determined by the capacity of the furnace and the specific energy consumption for the melting. The power of the transformer (Pt) for the electric arc furnaces is computed with the ratio [6]:

$$P_t = \frac{W}{t \cdot \eta \cdot \cos\varphi} \qquad \text{(KVA)} \qquad (1)$$

where:

W - is the total intake of power consumption during the melting phase ()kWh;

t - is the melting time (h);

η - is the yield (0,8...0,9);

cos φ - is the power factor (cca. 0,85).

3. Capitalization potentials of wastes

Figure 2 presents a an overview of the pyrometallurgical, namely hydrometallurgical proceedings applied with a view to capitalize on the pulverous wastes [7], [8].

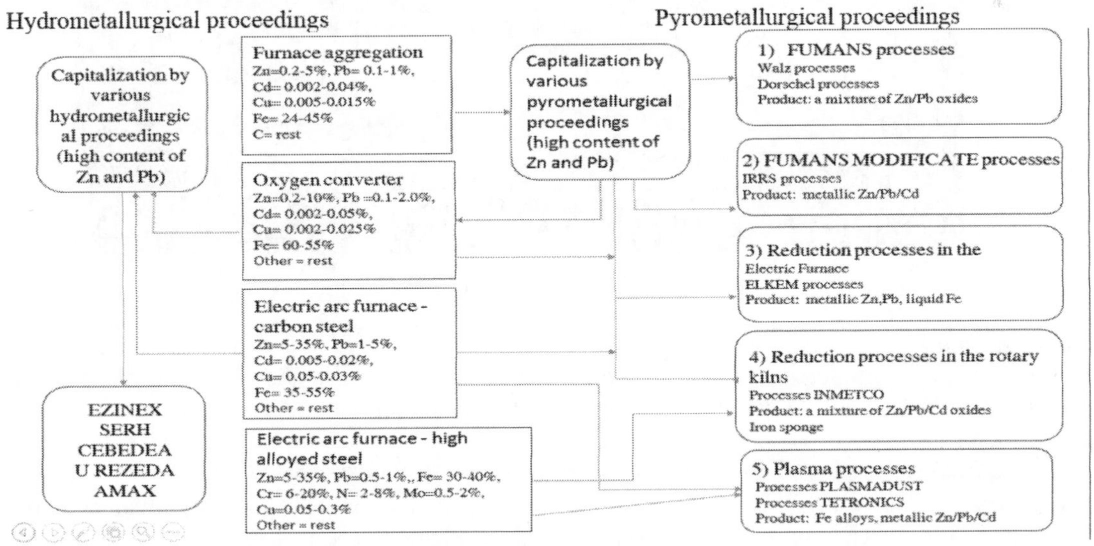

Figure 2. Pyrometallurgical and hydrometallurgical proceedings applied with a view to capitalize on the pulverous wastes generated in the siderurgical processes [6]

Albeit the EAF is the most suited technological aggregate for the conversion of steel wastes into high quality steels, these are major producers of pollutant emissions which have a negative impact on the environment.

Of all the pollutant emissions, the ones that have the strongest impact on the environment are dusts (powders) resulted during the operations of charging the raw materials and during the operations of steel's melting, stirring, alloying and discharge, that contain heavy metals (Cr. Ni, Zn, Pb, etc.), and the gasses resulted from the melting and stirring processes [5].

The improvement of dust's gathering from the process gasses within the work area and from the furnace leads to the diminution of the pollution level.

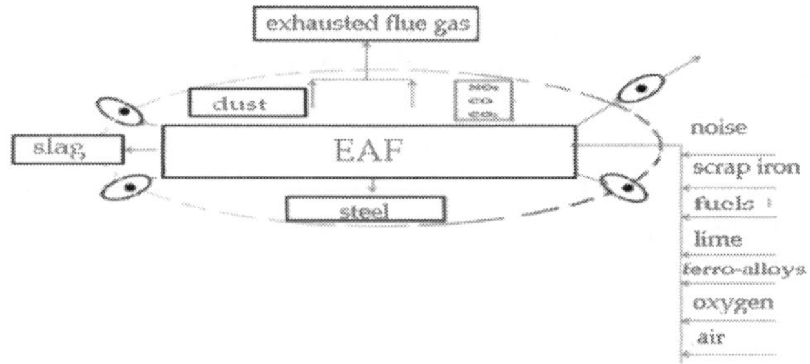

Figure 3. The main figure of environmental pollution by the agency of the EAF

Figure 4 Steelmaking routes (EAF)

4. Conclusions

Steel is the easiest material to recycle worldwide, from the household waste to those resulted following the demolitions of certain buildings all steel waste can be collected and processed with a view to their re-use.

The steelmakers employ furnaces that produce steel from various ferrous waste.

The possibility of using waste iron as primary charge is also one of the main advantages of the electric arc furnace (EAF), which allowed its development and establishment as a steelmaking aggregate.

The second method of steelmaking is called "from waste to steel" and the most suited technological aggregate for the melting of wastes is the electric arc furnace.

The economic and environmental benefits are originated from the recycling of the ferrous metallic materials.

The rise in productivity and economic yield of the electric arc furnaces was possible by the implementation of some solutions, of which we can mention: the increase of the installed electrical output; the use of oxygen and additional fuels; the replacement of the refractory linings with water-cooled components; the possibility of stirring the metal heel and processing under foaming slag; the recovery of the heat from the water circulating in the cooling systems and the off-gas; the preheating of wastes; the slag-free discharge by an eccentric bottom tap-hole; the optimal management and process control by the agency of the computer.

Albeit the EAF is the most suited technological aggregate for the conversion of steel wastes into high quality steels, these are major producers of pollutant emissions which have a negative impact on the environment.

References

[1] Socalici A V 2016 *Contribuţii privind îmbunătăţirea calităţii oţelului*, University Poltitehnica Timişoara, Habilitation Thesis
[2] Martinez F and Fernandez G 2016 *Technologies for EAF Energy Optimization*, 11th European Electric SteelMaking Conference&Expo, Venice, Italy, May 25-27
[3] Rau A, Cosma D and Ilin G H 1967 *Elaborarea Oţelului în cuptoare electrice cu arc*, Editura Tehnica, Bucureşti
[4] ***http://www.steelconstruction.info
[5] Ioana A, Nicolae A (coord), Predescu C, Sandu F, Sohaciu M and Calea G G 2002 *Conducerea Optimală a Cuptoarelor cu Arc Electric*, Editura Fair Partners, Bucureşti
[6] Martino M and Peri R 2013 Refrattari per bruciatori: caratterizzazione ed evoluzione, *La Metallurgia Italiana* **9** 37-44
[7] Heput T, Socalici A, Ardelean E, Ardelean M, Constantin N and Buzduga M 2011 *Valorificarea deşeurilor feroase mărunte şi pulverulente*, Politehnica Timişoara
[8] Heput T, Socalici A et al 2007-2013 Prevention and fighting pollution in the steel making, energetic and mining industrial areas through the recycling of small-size and powdery wastes, Proiect de cercetare nr. 31-098, Programul 4 „Parteneriate in domenii prioritare"

IOP Conf. Series: Materials Science and Engineering **294** (2017) 012017 doi:10.1088/1757-899X/294/1/012017

Methods of steel manufacturing - The electric arc furnace

E C Dragna, A Ioana and N Constantin

University Politehnica of Bucharest, Engineering and Management of Metallic Materials Obtaining Department, Bucharest, Romania

E-mail: claudiadragna90@gmail.com

Abstract. Initially, the carbon content was reduced by mixing "the iron" with metallic ingots in ceramic crucibles/melting pots, with external heat input. As time went by the puddling procedure was developed, a procedure which also assumes a mixture with oxidized iron ore. In 1856 Bessemer invented the convertor, thus demonstrating that steel can be obtained following the transition of an air stream through the liquid pig iron.
The invention of Thomas, a slightly modified basic-lined converter, fostered the desulphurization of the steel and the removal of the phosphate from it.
During the same period, in 1865, in Sireuil, the Frenchman Martin applies Siemens' heat regeneration invention and brings into service the furnace with a charge composed of iron pig, scrap iron and iron ore, that produces a high quality steel [1].
An act worthy of being highlighted within the scope of steelmaking is the start-up of the converter with oxygen injection at the upper side, as there are converters that can produce 400 tons of steel in approximately 50 minutes.
Currently, the share of the steel produced in electric arc furnaces with a charge composed of scrap iron has increased. Due to this aspect, the electric arc furnace was able to impose itself on the market.

1. Introduction:

The development of our century's technology is based, to a great extent, on the usage of steel, which must comply with conditions very diversified and, in some cases, very arduous.

Inclusions are, widely, defect which lead to the discarding of a casting or forging. These are found in the molten steel under various forms and impact in a negative manner the physicochemical properties of steels.

The production of steel has ceaselessly increased, as a result of the growing requirement for metallic materials. This increase also led to amendments regarding the quality of the steel produced and the elaboration technologies.

Constructional, technological and functional modernizations of the EAF improved this aggregate's performance and yield. The primordial objective of the modernizations was the achievement of high productivity with minimal costs. For the accomplishment of the aforementioned, the following conditions must be met:

- The rigging with a performing installation of oxy-fuel burners, for the rapid melt of the charge;
- The optimization of the charging networks, which would lead to acquiring the desired chemical composition at the end of the melting;
- High melting speed through: the charge's preheating, maximum electrical output and additional technical input;

- The abridgment of the post-melting period;
- Minimum periods of standstill. The growth related to the productivity and energo-technological yield of the elaboration of electrical steel is subordinated to the main constructional, technological and functional modernizations of the EAF.

2. The "Héroult" type Electric Arc Furnace

The electric arc furnace in which the steel is elaborated uses three-phase alternating current, the electric arc being formed between the three electrodes and the metallic charge. In view of this, it is named furnace with direct action arc

It receives the electricity from the high-voltage grid (6-110 MV), by the agency of a transformer that supplies the furnace with current at variable voltages - between 90 and 350 V [1].

The furnace (Figure 1) is composed of: the body of the furnace, the roof and the rocker. The body of the furnace, with a cylindrical shape, in the form of a spherical cap or truncated cone, is built of a shell made of thick steel plate, lined at the interior with basic refractory material (magnesite bricks or blocks of stabilized dolomite) or acid refractory material (silica bricks). To reduce the heat losses, an insulating layer is interlaid between the metallic shell and the refractory masonry. There are two holes within the the body of the furnace: the steel purging outlet, equipped with a cradle, and the furnace's operating door .The body of the furnace is fitted on a sledge (two runners) with the aid of which it can be tilted towards the steel's purging outlet with up to 45° and towards the operating door with up to 15° [2].

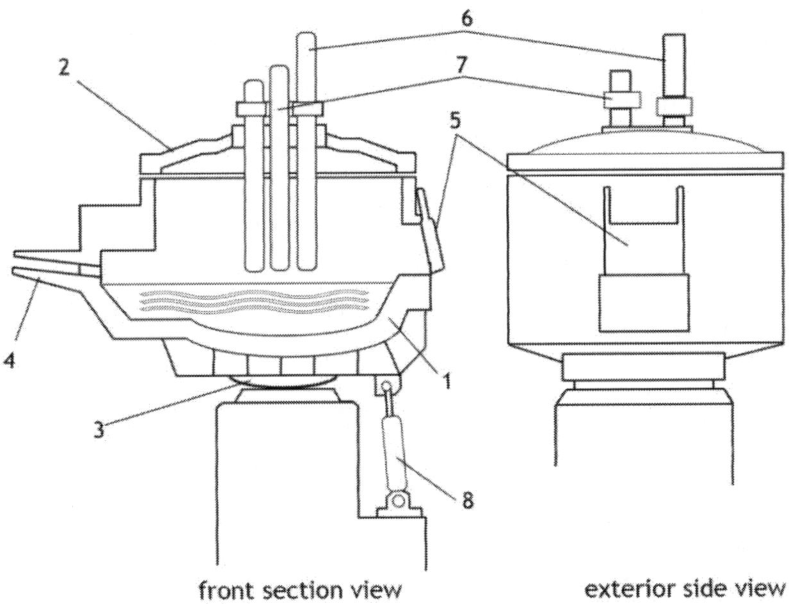

Figure 1. The drawing of the electric arc furnace: 1-the body of furnace; 2-roof; 3-sledge; 4-purging cradle; 5-operating door; 6-electrodes; 7-water-cooled copper rings; 8-roker

At the upper side of the Heroult type electric arc furnace, the body of furnace is covered by a detachable roof, made of silica bricks or chrome-magnesite bricks which are resting, by the agency of a ring, on the metal jacket. Within the furnace's roof there are three holes, disposed in the shape of an equilateral triangle and through these holes are inserted 6 electrodes that are supported by seven water-cooled copper rings. The electrodes are made of graphite or graphite carbon and have threaded holes at their ends.

The rings are positioned at the ends of some horizontal arms and are supported by vertical masts fitted on the tilting sledge in such a way that the electrodes tilt over together with the furnace. The arms that support the electrodes can move both upwards and downwards on the supporting masts, the movement being automatically controlled for the purpose of maintaining the electric arc [1]. The electric current is supplied for the transformer to the electrode's support arm with a flexible cable and, alongside the cable, with a copper rail to the electrode's clamp, which also performs the function of conveying the electricity to the electrode [1].

The Girod type electric arc furnace (Figure 2) represents a version of the Héroult type furnace and it is used for the elaboration of metals or alloys, especially for steelmaking, the electric arc being produced between the electrodes mounted within the roof and the furnace's hearth. For the single-phase alternating current the furnace was equipped with one graphite electrode fitted in the roof and two metallic water-cooled electrodes in the hearth. In the case of the three-phase alternative current are used three graphite electrodes mounted in the roof, similarly to the Héroult type electric furnace, however, with a hearth that is a good conductor of electricity. At present, these furnaces are seldom used [1].

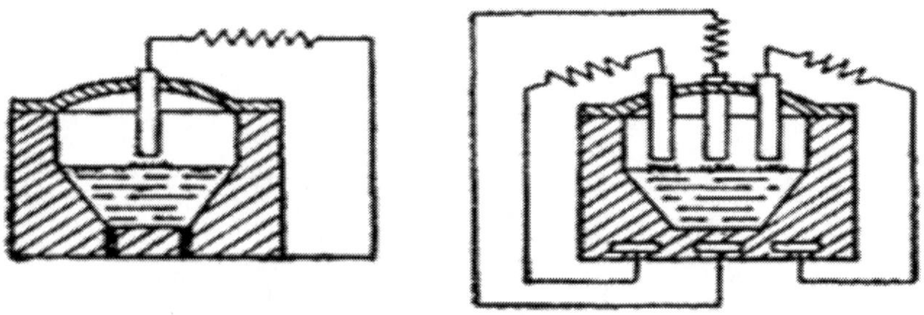

single-phase three-phase

Figure 2. Girod type electric arc furnace [2]

3. The production of steel using electricity

The steel produced by using electricity as an energy source, hence, within the Electric Arc Furnace, fills the second place of the worldwide ranking of the most important routes of steelmaking.

The steel manufacturing method by means of the electric arc furnace is the main technological route of recycling the iron scrap.

Process modelling and simulations contribute to a thorough apprehension of the energy and mass transfers during the melting process and may be applied to investigate new control strategies, possibilities of recovering the residual heat or to support the operator. The EAF off-gas especially offer development potentials, due to their energetic capacity, which represents approximately 20 - 30 % of the total absorption of energy, and the possibility of process control by means of the measurements performed on them [3].

The EAF off-gas, with its energetic capacity representing up to 30% of the total absorption of energy, is in the forefront of current and future developments, for the effective enhancement of the melting process within the EAF regarding the energy and resources [4-7].Therein, the residual energy recovery systems, by means of steam generation and the enhanced process control, by means of off-gas measurements, can contribute to the optimization of the EAF process efficiency and to remaining competitive during period of rises of the energy prices and stricter environmental protection norms.

The comprehensive process simulations of the melting process and the off-gas treatment system are capable of improving the EAF's control and to provide aidful information, for a better understanding of the energy and mass transfers. EAF's process modelling and simulation has strongly developed recently, due to its diverse possibility of application and the enhanced computing capacity. Several numeric models were developed for the improvement of the melting process, the enhancement of the process control through the analysis software programs or assistance to the plant's automatization with only limited consideration of the exhaust model [3].

Regarding the EAF a comprehensive, fast and easily adjustable model of the dedusting system can contribute, when applied, to the optimization of the control strategies, the decrease of the output energy and the improvement of the residual heat.

The development of the current century's technology is based on the use of steels, which must meet a spectrum of conditions, in some cases very difficult.

Inclusions are, widely, defect which lead to the discarding of a casting or forging. These are found in the molten steel under various forms and impact in a negative manner the physicochemical properties of steels [8], [9].

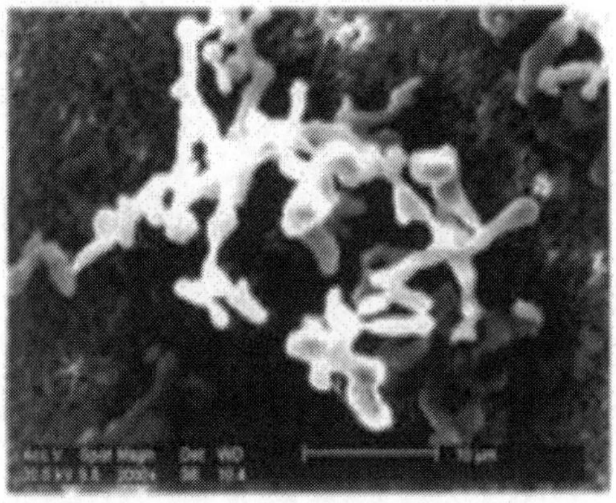

Figure 3. Al_2O_3 indogenous inclusion [8]

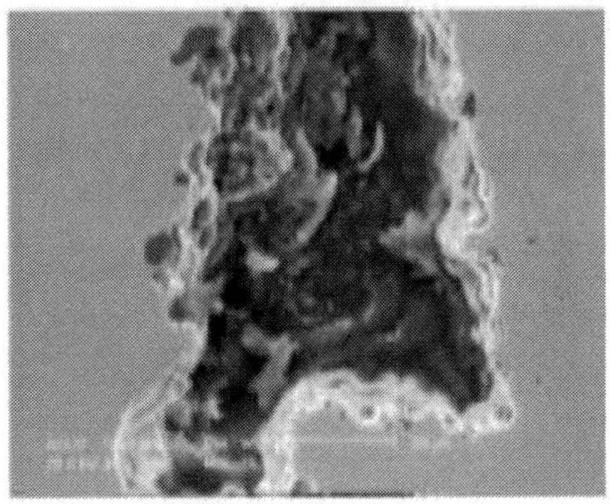

Figure 4. Al,Ca,O: Al_2O_3 • CaO exogenous inclusion [8]

Quality steels must have a high degree of purity in non-metallic inclusions and gasses to be insensitive to the formation of flakes, to overheating and to possess a high tenacity, properties that must be regularly acquired. Thus, the production needs to be uniform in time.

The amount of oxide inclusions present in each ton of steel is of the order of 1012...1015 inclusions, considering a molten steel which contains 100 ppm oxygen that was advancedly deoxidized with the aluminum, forming spheric particles of Al2O3 with ṗ=4g/cm3. Thus, depending on the diameter from Table 1, the following distribution of the number of inclusions is acquired.

Table 1. The distribution of the Al_2O_3 inclusions depending on their diameter in steel with 100 ppm and oxygen.

The inclusion's diameter in µ m	No. of inclusions/ton of steel	Steel Volume /inclusion	Minimal distance between the inclusions
1000	10^5	1,3 cm^2	1,1 cm
100	10^8	1,3 mm^2	1,1mm
10	10^{11}	$1{,}3 \cdot 10^6 \mu^7$	110 µ
1	10^{14}	$1{,}3 \cdot 10^7 \mu^7$	11 µ
10^{-1}	10^{17}	1,3 μ^7	1,1 µ
10^{-2}	10^{20}	$1{,}3 \cdot 10^{-2} \mu^2$	0,11 µ

Aside from the inclusions, significant amounts of hydrogen (derived from the dissociation of the water vapors) and nitrogen (originated from the atmosphere) are dissolved in the steel. These gases generate deficiencies, if they remain within the steel.

Within the steel, during the elaboration, excepting the non-metallic products (inclusions) resulted pursuant to the elaboration processes and the casting conditions, are dissolved substantial amounts of hydrogen (derived from the dissociation of the water vapors) and nitrogen (originated from the atmosphere).

If they remain within the steel these gasses cause defects: the hydrogen induces intercrystalline fractures, known as flakes, or structure deficiencies - shadow lines, and the nitrogen increases the steel's tendency to age.

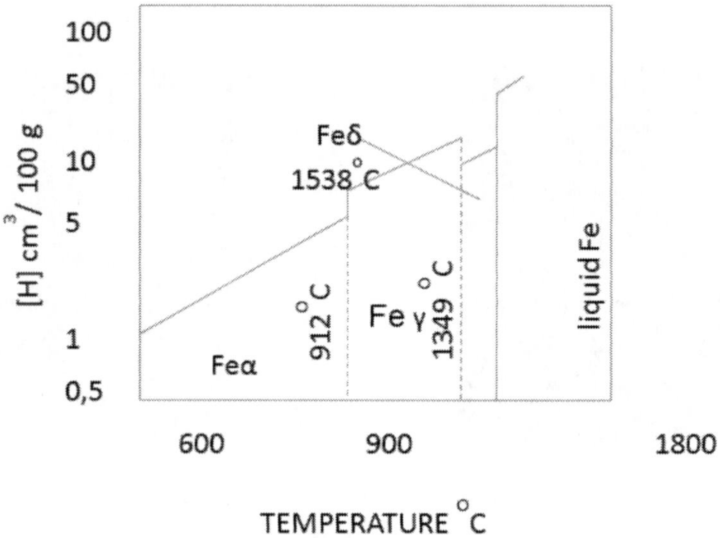

Figure 5. Hydrogen solubility in iron depending on the temperature [9]

Over the last two decades significant amendments intervened in the design and organization conception of the high capacity modern steel plants, by the emergence of the "ladle metallurgy", which consists in the fragmentation of the elaboration technology in two major phases.

During a first phase, also named as "primary elaboration", which takes place within a classic metallurgical aggregate (electric arc furnace), an intermediate liquid product is obtained (crude steel), as a result of the melting and stirring. This product is then subjected to several refining and alloying operations, which constitute the secondary elaboration phase, outside the elaboration aggregate, both in the casting ladle and some specialized aggregates.

4. Conclusion

The steel produced by using electricity as an energy source, hence, within the Electric Arc Furnace, fills the second place of the worldwide ranking of the most important routes of steelmaking.

The steel manufacturing method by means of the electric arc furnace is the main technological route of recycling the iron scrap.

The EAF off-gas is in the forefront of current and future developments, for the effective enhancement of the melting process within the EAF regarding the energy and resources.

The comprehensive process simulations of the melting process and the off-gas treatment system are capable of improving the EAF's control and to provide aidful information, for a better understanding of the energy and mass transfers.

A comprehensive model, fast and easily adaptable to the dedusting system when it is applied, can contribute to the optimization of the control strategies, the decrease of the output energy and the improvement of the EAF's residual heat.

References

[1] Carcea I, Roman C and Chelariu R 2006 *Ingineria proceselor metalurgice*, Editura Performantica, Iaşi, Romania
[2] Hotărăscu O, Iordănescu P, Rozolimo P and Tripşa I 1980 *Mică enciclopedie de metalurgie*, Editura Ştiinţifică şi Enciclopedică, Bucureşti, Romania
[3] Meier T, Hassannia Kolagar A, Echterhof T, Pfeifer H 2016 *Process modeling and simulation of an Electric Arc Furnance for comprehensive calculations of energy and mass transfers in combination with a model of the dedusting system*, University RWTH Aachen, Germany
[4] Frohling C, Hemmling P, Kempken J and Emling W H 2012 *AISTech*, Atlanta, Georgia, USA
[5] Hampel A, Fleischanderl A, Enicki G, Zauner F, Steinparzer T and Haider M 2011 *AISTech*, Indianapolis, Indiana, USA
[6] Zualiani D J and Scipolo V C 2011 *Born Milenium Steel*, Milenium Steel Publishing, London
[7] Dragomir I 1985 *Teoria Proceselor Siderurgice*, Editura Didactica şi Pedagogică, Bucuresti
[8] Radu S and Geanta V 2008 *Rafinarea oţelurilor prin injectarea gazelor inerte*, Editura BREN

IOP Conf. Series: Materials Science and Engineering **294** (2017) 012018 doi:10.1088/1757-899X/294/1/012018

Steel refining possibilities in LF

M G Dumitru[1], A Ioana[2], N Constantin[2], F Ciobanu[1] and M Pollifroni[3]

[1]SC TenarisSilcotub Calarasi, 93, Mihai Viteazul Blvd. Calarasi, 450131, Romania
[2]Politehnica University of Bucharest, Department of Engineering and Management of Metallic Materials Obtaining, 313 Spl. Independentei Street, Bucharest, 060042, Romania
[3]University of Turin, 8 Via Verdi, Turin, 10124, Italy

E-mail: adyioana@gmail.com

Abstract. This article presents the main possibilities for steel refining in Ladle Furnace (LF). These, are presented: steelmaking stages, steel refining through argon bottom stirring, online control of the bottom stirring, bottom stirring diagram during LF treatment of a heat, porous plug influence over the argon stirring, bottom stirring porous plug, analysis of porous plugs disposal on ladle bottom surface, bottom stirring simulation with ANSYS, bottom stirring simulation with Autodesk CFD.

1. Introduction

At a basic level, the refining is a purifying process of certain gaseous/solid substances. The word "refining" is commonly used in relation with raw materials. The nature and complexity of liquid steel refining depends mainly by the steel type, meaning steel chemical composition and required cleanliness [1], [2].

Refining is a stage of the secondary steel making process which is also referred as ladle metallurgy. The secondary metallurgy concept forced the reorganization of the technological flux. In this way, between primary steel making and steel casting in Continuous Casting Machine, a series of specific installations have been installed one being Ladle Furnace.

Mainly, steel refining in LF aims the following objectives [3], [4]:

- To continue steel dezoxidation started during tapping;
- To complete steel desulphurization also started during tapping;
- To homogenize steel bath thermic and chemically;
- To increase steel temperature according to required values for further events;
- To adjust the Carbon content and steel alloying;
- To eliminate inclusions in flotation in order to reach desired cleanliness;
- To modify inclusions shape in order to improve casting process at CCM;

Diversification of the refining process for steel aims to ensure a way to continue the metallurgical operations from primary aggregates in secondary aggregates, the process consisting in two stages: primary steelmaking (melting) and secondary steel making, steel treatment outside the furnace [5].

In Figure 1 we present steelmaking stages.

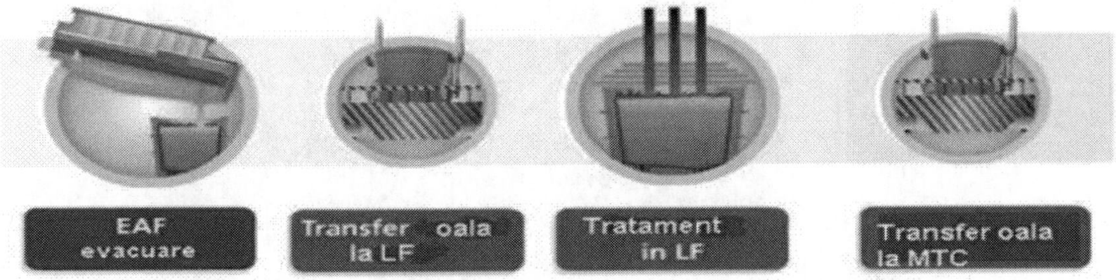

Figure 1. Steelmaking stages

The improvements done to the refining installations lately conducted to the possibility to obtain unquestionable performances regarding the quality of liquid steel which will be studied in this paper.

2. Steel refining through argon bottom stirring

During the ladle transportation process from the Electric Arc Furnace to the Ladle Furnace on the transfer car, argon (Ar) stirring must be intense in order to facilitate thermal and chemical homogenization and to prepare the slag for steel treatment. A rigorous attention is required for the steel bath level inside the ladle in order to avoid steel splashing. This has a huge impact both on company economic indicators but also on safety.

Bottom stirring effect is very complex and if the working parameters are chosen correctly and respected, the resulted alloys are high quality products. Special alloys required for cutting edge technology sometimes treated even vacuum, are treated through this procedure which has special metallurgical benefits related mainly with homogenization and refining.

The most important effects of inert gas bottom stirring are:

- Thermal homogenization of the steel bath as a result of the agitation created by the gas;
- Chemical homogenization of the steel bath as a result of intensified mass transfer processes;
- Non-metallic inclusions refining as a result of increased vertical velocity of these particles;
- Degassing through gases diffusion from the melt into the Ar bubbles;
- Interaction of the active gases with unwanted elements from the melt and formation of solid particles which no longer have the capacity to melt in the steel bath;

Refining process – intensification of physical and chemical processes thus a reduction in required time as a consequence of increased mass transfer velocity;

Gas injection into the melt is done either using a lance, protected or not with refractory material, disposed at the end with porous refractory component or, in case of the ladles, through porous plugs disposed in the bottom. The main problems of this procedure are those related to special quality of refractory materials and advanced wear rate, to purity of the injected gas, to steel melt cooling during treatment.

Steel bottom stirring start with average to high intensity generally before first heating. This intense stirring, prior to heating last for 2-3 minutes.

The main objective is to prepare the slag for an initial heating and to start the bath homogenization. The intense stirring will also facilitate the desulphurization. During this period slag deoxidizers may be added which are called "reduction mixes". In Figure 2 is presented the HMI for this online control system. The installation can use either argon or nitrogen. Nitrogen is used only for steels that require a high content.

Figure 2. Online control of the bottom stirring

Bottom stirring (Figure 3) is considered an important technological process for refining. This is a very important aspect because most of the products require a low Sulphur content. Thus, is fundamental to control the evolution of this process.

In order to obtain a good desulphurization, it is mandatory that the following conditions should be respected:

- Slag basicity;
- Deoxidized slag and steel bath;
- High temperature;
- Intense stirring;
- Fluid slag;

A very important aspect is that the intense stirring must be done at the beginning of the process. A strong stirring at the final moments of the process has repercussions for the steel refining (reoxidation, slag emulsification).

Also, too fluid slag may affect the steel refining (reoxidation due to the fact that the slag do not protect enough, emulsification, dissolution of the MgO from the refractory). This is the reason why it is fundamental to respect the slag desired composition.

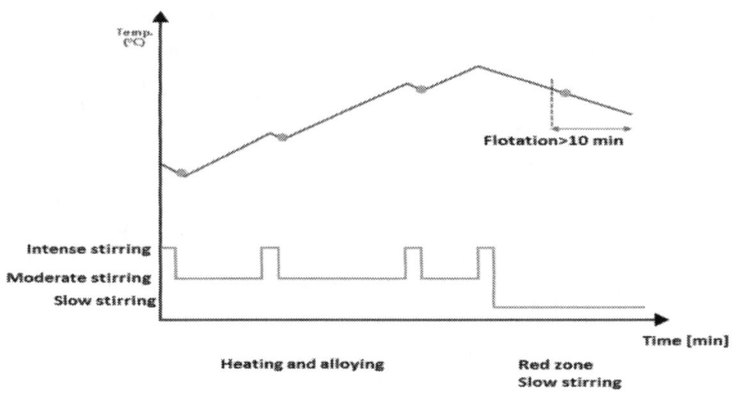

Figure 3. Bottom stirring diagram during LF treatment of a heat

The main stirring effects of the steel consist of: improving the properties through steel quality control, improving quality and productivity of continuous casting, improving overall performances of semi-products.

3. Porous plug influence over the argon stirring

A porous plug consists of: plates, refractory concrete and metallic cover. In figure 4 are presented images of a porous plug.

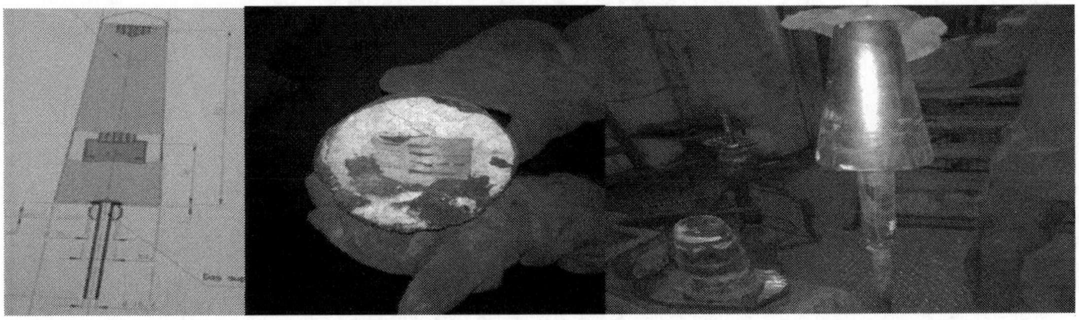

Figure 4. Bottom stirring porous plug

4. Analysis of porous plugs disposal on ladle bottom surface

Actual tendency:

In line displacement of the plugs on the same side of the ladle;
Diagonally displacement of the plugs;

4.1. In line displacement of the plugs on the same side of the ladle

This displacement method was initially designed to create a bath agitation having a circular motion from left to right (from one side to another), but on the top of the ladle in the right side (opposite side of the plugs) unassimilated ferroalloys may be observed, theory confirmed by the simulation illustrated in Figure 5. In order for the ferroalloys to be assimilated, a more dynamic stirring is required (bypass=maximum line pressure) leading to an increased argon consumption and a longer treatment time. For the plugs displaced in line, it has been observed that on the plugs side, refractory layer presented an increased wear, not uniform on the entire surface.

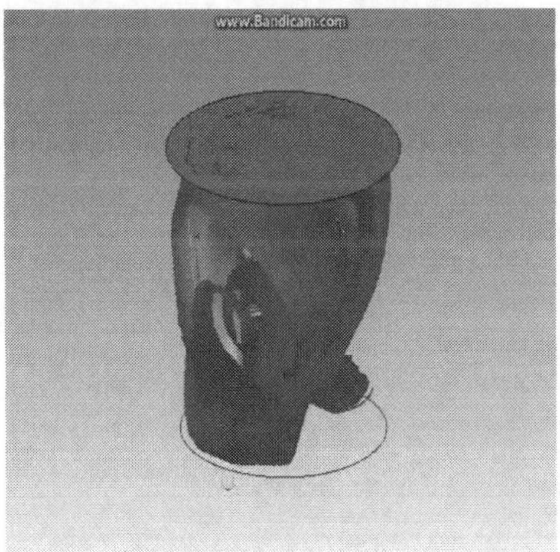

Figure 5. Bottom stirring simulation with ANSYS

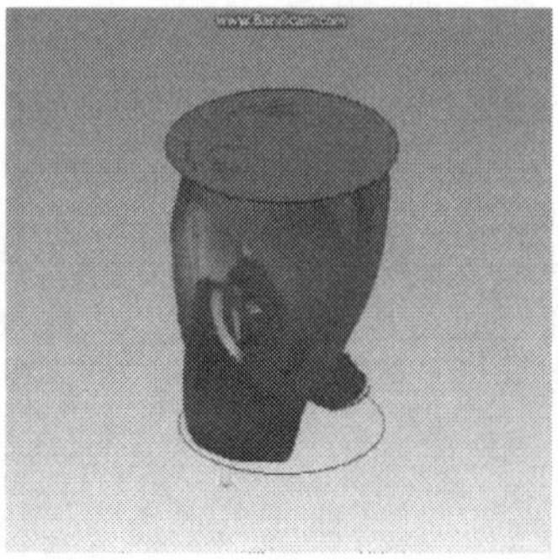

Figure 6. Bottom stirring simulation with Autodesk CFD

4.2. Diametrically porous plugs displacement

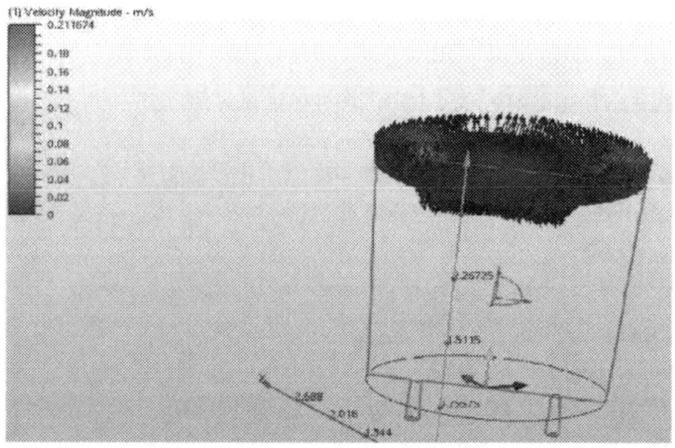

Figure 7. Bottom stirring simulation with Autodesk CFD

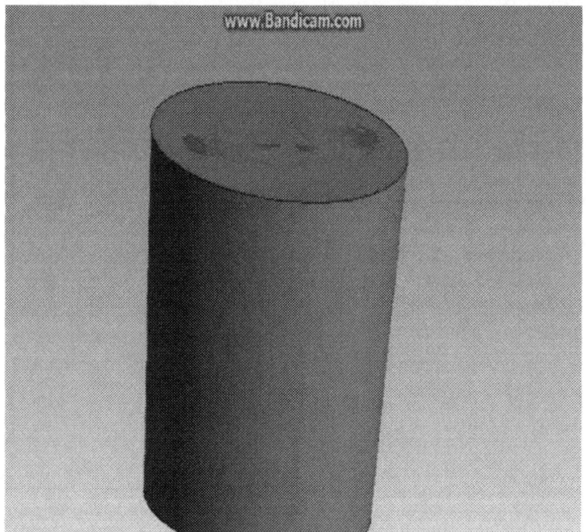

Figure 8. Bottom stirring simulation with ANSYS

Figure 9. Bottom stirring simulation with Autodesk CFD

5. Conclusions

As a result of the simulations presented above and physical experiments in the steel shop, it has been observed that diametrically plugs displacement provides better results for the process than in line plugs displacement regarding steel homogenization, ferroalloys assimilation and refractory layer uniform wear. Plugs displacement on the same side of the ladle leads to an increased argon consumption and a larger stirring time with increased temperature loss, therefore a larger amount of electrical energy is required to homogenize the steel melt both thermally and chemically.

References

[1] Ioana A, Buzatu M and Semenescu A 2016 Glance at the World, *Waste Management* **53** I-IV
[2] Ioana A 2013 Metallurgy's Impact on Public Health, *Review of Research and Social Intervention* **43** 169-179
[3] Ioana A, Constantin N and Moldovan P 2015 Constructive and functional modernization of EAF, *IOP Conf. Ser.:Mater. Sci. Eng.* **85** 012014
[4] Ioana A, Constantin N and Moldovan P 2015 About EAF and environment, *IOP Conf. Ser.:Mater. Sci. Eng.* **85** 012015
[5] Ioana A and Semenescu A 2013 Technological, Economic, and Environmental Optimization of Aluminum Recycling, *Journal of the Minerals, Metals & Materials Society* **65**(8) 951-957

Maintenance of reference standards in the field of viscosity

E A Moşulică[1], I Cîrneanu[1], N Constantin[2] and V Rucai[2]

[1]National Institute of Metrology (INM), Department of Physico-chemical quantities, Bucharest, Romania
[2]University Politehnica of Bucharest, Department of Material Science, Bucharest, Romania

E-mail: mosulica alexandra@yahoo.com

Abstract. Participation in the work of comparison in the field of viscosity, within the program conducted under the jurisdiction of ASTM (American Society for Testing and Materials), D-2 Committee, Subcommittee "Flow Properties," Newtonian Fluids) was necessary to ensure traceability of measuring unit of kinematic viscosity. Results of the comparison of the specialized participating laboratories on 4 continents, has proved annual capability of INM in the transmission unit of kinematic viscosity. Cannon Position Company in the US organizes co-operation program in the field of kinematic viscosity ASTM D 02.07. The company distributes standard substances Cannon viscosity participating laboratories and consolidate the results of the measurements.Physical-chemical laboratory has fully accepted the proposed schedule of the company Cannon. Final report of the comparison showed that in the year 2015 a number of 25 laboratories and institutes of metrology attented to the program.

1. Introduction

Since the setting up in 1938, by the man of science, inventor and Professor, Dr. Michael Cannon, firm Cannon Instrument Company had an international recognition in the field of viscosity.

The company reaches this year 79 years and has a collaboration with INM of 46 years.

Participation in the work of inter-comparison in the field of viscosity, within the framework of the program carried out under the responsibility of ASTM (D-Committee, Subcommittee 2 "Flow Properties," Newtonian fluids) has been necessary in order to ensure traceability of measuring unit of kinematic viscosity. Results of the comparison to the participating laboratories specialized on 4 continents, specialized laboratory annual capability of INM in the transmission unit of kinematic viscosity [1].

2. Presentation program ASTM

Cooperation program in the field of kinematic viscosity was discussed and adopted by the NIST (National Institute of Standards and Technology) formerly NBS (National Bureau of Standards)-Washington, in 1971. This program has been coordinated by the American society for Testing and materials (ASTM) D-02, Committee on methods of analysis in the field of petroleum, Subcommittee RASDVU (Properties, flow of Newtonian fluids). In this Position the U.S. Cannon Company uses to organize the cooperation program in the field of kinematic viscosity ASTM - D02.07.A [2].

The Cannon Company distributes standard substances to the participating laboratories and finally collect all the kinematic viscosity measurements.

2.1. Reference documents

Reference documents for participating laboratories were international standards ASTM D 445 and SR EN ISO 3104„Petroleum products. Transparent and opaque liquids. Determination of kinematic viscosity and calculation dynamic viscosity" and ISO 3105 „Glass capillary Kinematic viscometers - Specifications and operating instruction.

2.2. The participants in the ASTM D 02.07.

In the year 2015 were invited 25 laboratories from all over the world, such as:
Cannon Instrument Company – State College
Central Office of Measures – Poland
Centro Nacional de Metrologia – CENAM (Mexico)
Chevron – Richmond, CA
Evonik Oil Additives USA – Horsham, PA
INMETRO, The Nat'l Inst. Of Metrology – Brazil
Instituto Nacional De Technologia Industrial (Quimica) – Argentina
Institute for Metrology (Romania)
Kenya Bureau of Standards – Kenya
Laboratorio Dr.Lantos – Argentina
Lacomet –Costa Rica
National Institute of Metrology – China
JX Nippon Oil Co. Central Technical Research - Japan
PETROCHINA Dalian Lube Oil R&D - China
Poulten Selfe and Lee Ltd. – England
Runningland Metrology & Testing (Shanghai) – China
Slovak Institut of Metrology – Slovakia
U.S. Army Primary Standards Laboratory – Repsol, SA- Spania
ZMK – Analytik – Germany
Laboratoriuo de Lubrificantes – Uruguay [3].

2.3. Reference materials

Reference materials have been delivered to the INM by the Cannon Company in April 2015 (Table1). According to the Cannon Protocol the reporting of the results had to be done up to 1 June 2015. INM laboratory measurements proceeded in Aprilie and May 2015. The laboratory sent via fax and email the results of measurements at the end of May. The final report was received on June 22, 2015.

Table 1. Standard Materials and temperatures proposed to be measured in 2015 [3]

Code	Temperature °C
S6 (Mineral Oil)	20; 25; 40
N26 (Polialfaolefina)	40; 100
S200 (Polialfaolefina)	20; 25
S60 (Mineral Oil)	40;100
S600 (Polialfaolefina)	40
S8000 (Polibutena)	20; 25

3. Principle of Measurement

A common method for measurement the viscosity of the Newtonian fluids uses capillary extruded viscometers - Ubbelohde Type. Measurement of the viscosity of liquids shall be based on the measurement of flow time of a given volume of liquid, by a capillary tube with geometrical characteristics well established. The glass capillary viscometers, as Ubbelohde type, result in the formation of a column of liquid on which the atmospheric pressure acts on both from top to bottom (at the top level) and bottom-up (at the lower level) until the two pressures of the opposite direction shall be cancelled. The lower level of liquid formed by the breaking of the liquid column due to the atmospheric pressure transmitted through the tube connection shall make the so-called "suspended level", as shown in Figure 1 [4].

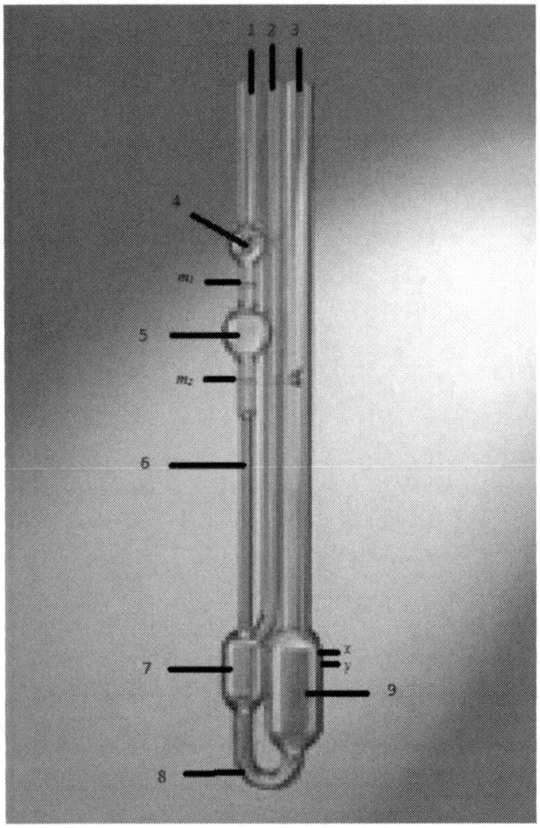

Figure 1. Capillary Viscometers with Suspended Level, Ubbelohde Type by comparison to the National Standard of Physikalisch Technische Bundesanstalt of Viscosity, PTB, Germany.[4]
1-capillary branch; 2-tube connection with the atmosphere; 3-branch with tank; 4-bubble of safety; 5-measuring Bull; 6-capillary; 7-suspended bubble level; 8-extension tubular fuel tank; 9-tank.
(On either side of the bubble of measurement shall be drawn two marks m1 and m2, which defines the volume of liquid whose capillary leakage through it is measured. The capillary tube is open in the bubble suspended level 2, which communicates with the atmosphere and the tank 9, on which are marked levels x and y, which indicates the minimum level from the viscometer and the maximum one, accordingly.)

4. The principle of calibration

The kinematic viscosity is measured by using the Ubbelohde dedicated viscometer.The flow time t, of the liquid between the two marks of a standard viscometer is directly proportional to the kinematic viscosity ν [5]:

$$\nu = k \cdot t \quad (1)$$

where: ν - kinematics viscosity of the liquid, [mm^2/s];
k - constant of the glass capillary [mm^2/s^2];
t - average time flow of liquid between the capillary marks [s].

National standard of the unit of measure of viscosity mm^2/s kinematic is a secondary standard one and consists of two sets of capillaries viscometers Ubbelohde, manufactured by the company Schott Gerate which allow to determine the kinematic viscosity of Newtonian fluids in the range of about (0.5 - 100 000 mm^2/s). Maintain traceability to the SI was achieved by calibration of two sets of capillaries at the PTB-Germany.

5. Results presentation and conclusions

According to the Protocol of the program, each participant ASTM has assigned a call sign in the final report. National Institute of Metrology has #26, respectively.

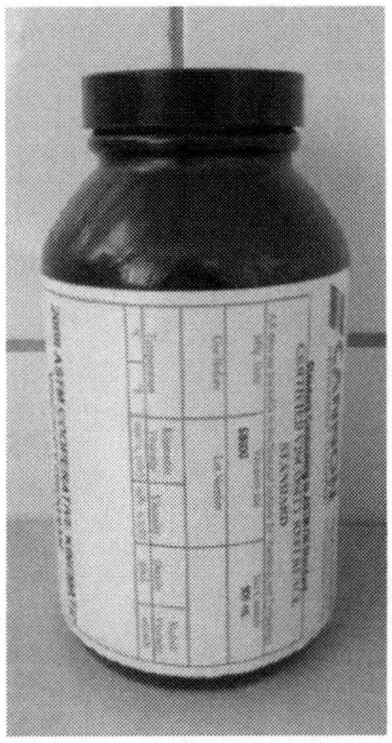

Figure 2. Reference Material Supplied to NIM by the CANNON Company, (S 8000)

In Table 2 are shown the measured reference materials (codification), temperature measurement, the value reported by the INM and the relative error of kinematic viscosity e_r versus the benchmark.

Relative error e_r represents the ratio between the absolute error of measurement (ΔX) and the real measured value of (X_e), % [6].

$$e_r = \Delta X / Xe \cdot 100 \tag{2}$$

Table 2. Measurements of Reference Materials [3]

Code(type)	T (°C)	Value reported v (mm^2/s)	Relative error e_r (%)
S6 (Mineral Oil)	20	10,77	0,13
S6 (Mineral Oil)	25	9,082	0,03
S6 (Mineral Oil)	40	5,842	0,08
N26 (Polialfaolefina)	40	25,53	0,02
N26 (Polialfaolefina)	100	5,163	0,18
S200 (Polialfaolefina)	20	543,9	0,09
S200 (Polialfaolefina)	25	403,9	-0,02
S60 (Mineral Oil)	40	54,92	0,07
S60 (Mineral Oil)	100	7,726	0,23
S600 (Polialfaolefina)	40	520,2	0,16
S8000 (Polibutena)	20	35862	0,15
S8000 (Polibutena)	25	22839	0,14

In Table 2 there are presented all 166 stock viscosity determination for each sample/set temperature and error percentage compared to the average of all values of viscosity. Measurements of viscosity values with an error > 1.36% have been removed in the calculation of the average.

In Table 3, there are presented the following issues: the measured reference materials (codification), temperature measurement, the values reported by INM, the values reported by the Cannon Company, and e_r, relative error of kinematic reported viscosities versus the benchmarks.

Table 3: Reference Material Measurement by comparison with Cannon [3]

Code (type)	T (°C)	Value reported v (mm^2/s)		Relative error e_r (%)	
		INM	Cannon	INM	Cannon
S6(Mineral Oil)	20	10,77	10,77	0,13	0,06
S6(Mineral Oil)	25	9,082	9,075	0,03	0,04
S6(Mineral Oil)	40	5,842	5,833	0,08	-0,08
N26(Polialfaolefina)	40	25,53	25,53	0,02	-0,01
N26(Polialfaolefina)	100	5,163	5,155	0,18	0,01
S200(Polialfaolefina)	20	543,9	543,0	0,09	-0,07
S200(Polialfaolefina)	25	403,9	403,6	-0,02	-0,10
S60(Mineral Oil)	40	54,92	54,70	0,07	-0,33
S60(Mineral Oil)	100	7,726	7,710	0,23	0,02
S600(Polialfaolefina)	40	520,2	518,2	0,16	-0,21
S8000(Polibutena)	20	35862	35696	0,15	-0,31
S8000(Polibutena)	25	22839	22695	0,14	-0,49

A good accordance between the averages of the values of viscosity determined by laboratory of INM vs. Cannon's finally resulted. ASTM comparison procedure is a good opportunity to prove the laboratory capability and, therefore, to maintain a good level of technical and professional competence of the personnel involved in the measurements of viscosity.

In 2015 the reported relative error is less than 0.20% out of reference amount for 73 % of results. The reported relative error is less than 0.30% out of reference amount for 80 % of results. INM reported 11 of 12 results with an average relative error of less than 0,20%, as it is shown in Figure 3.

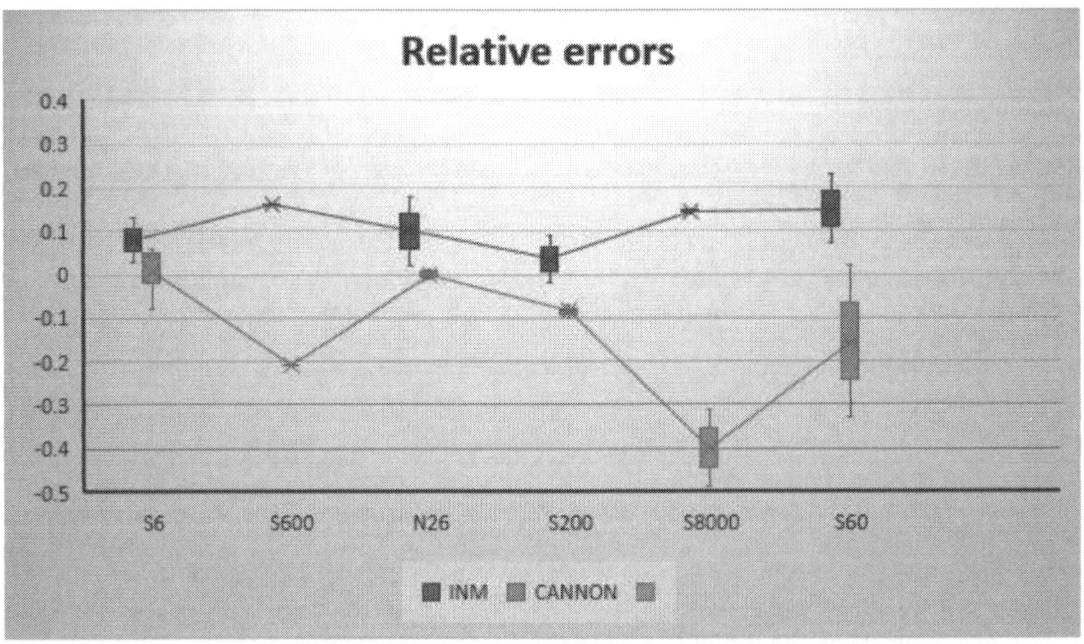

Figure 3. Relative errors of Kinematic Viscosity e_r Reported: INM vs. CANNON

References

[1] ***https://www.cannoninstrument.com/en/catalog/node/viscosity-flash-point-standards
[2] ***https://www.google.ro/url?sa=t&rct=j&q=&esrc=s&source=web&cd=3&cad=rja&uact=8&ved=0ahUKEwjzOeJxtbWAhWobZoKHckiATwQFggzMAI&url=https%3A%2F%2Fwww.researchgate.net%2Ffile.PostFileLoader.html%3Fid%3D54eb10b6d5a3f22b688b4605%26assetKey%3DAS%253A273712186626072%25401442269478565&usg=AOvVaw2RjO0oo_oeeQ8JCzfjQC04
[3] ***Report of the annual ASTM cooperative kinematic viscosity program, Spring 2015
[4] Solomon M 1958 *Viscozimetre şi elemente de teoria viscozităţii*, Tehnica, Bucuresti, Romania
[5] Oiml R 1995 Second Preliminary Draft (2CD), Procedure for the Kinematic Viscosity Measurements by Means of Standard Viscometers
[6] ***ISO/IEC Guide 98-3:2008 (JCGM/WG1/100) Preview Uncertainty of measurement Part 3: Guide to the expression of uncertainty in measurement (GUM:1995)

IOP Conf. Series: Materials Science and Engineering **294** (2017) 012020 doi:10.1088/1757-899X/294/1/012020

The use of R_z roughness parameter for evaluation of materials behavior to cavitation erosion

I Bordeasu[1], M O Popoviciu[2], C Ghera[1], L M Micu[3], L D Pirvulescu[1] and T Bena[1]

[1]Politehnica University of Timisoara, Mechanical Machines, Equipment and Transportation, Mihai Viteazul str., no. 1, 300222 Timisoara, Romania
[2]Academy of Romanian Scientists, Bd. Mihai Viteazul, no. 1, Timisoara, Romania
[3]University of Agricultural Sciences and Veterinary Medicine of Banat King Mihai I from Romania, Calea Aradului str., no.119, 300645 Timisoara, Romania

E-mail: ilarica59@gmail.com

Abstract. It is well known that the cavitation eroded surfaces have a porous appearance with a pronounced roughness. The cause is the pitting resulted from the impact with the micro jets as well as the shock waves both determined by the implosion of cavitation bubbles. The height and the shape of roughness is undoubtedly an expression of the resistance of the surface to the cavitation stresses. The paper put into evidence the possibility of using the roughness parameter R_z for estimating the material resistance to cavitation phenomena. For this purpose, the mean depth erosion penetration (MDE-parameter, recommended by the ASTM G32-2010 Standard) was compared with the roughness of three different materials (an annealed bronze, the same bronze subjected to quenching and an annealed alloyed steel), both measured at four cavitation erosion exposure (30, 75, 120 and 165 minutes). The roughness measurements were made in 18 different zones, disposed after two perpendicular diameters, along a measuring lengths of 4 mm. The results confirm the possibility of using the parameter R_z for estimating the cavitation erosion resistance of a material. The differences between the measured values of Rz and those of the characteristic parameter MDE are of the same order of magnitude as those obtained for MDE determination, using more samples of the same material.

1. Introduction

In practice, for hydraulic equipment the evaluation of the cavitation damage is done on the base of both the area and the volume eroded. The volume is frequently appreciated upon the consumed electrodes volume used for the repair work [1-3]. In research laboratories the evaluation is more complex, being used specialized devices which allow to obtain details upon the cavitation erosion evolution. In our laboratory, this device was realized upon the recommendations ASTM G32-2010 Standard and the tests are conducted following also the recommendations of this standard [4]. In parallel, many researchers are interested also by the possibility to evaluate the cavitation erosion resistance by using the roughness parameters [5-10]. Till now, there were presented in various papers the profile diagram of the roughness's [11], [9] but till the present it was not obtained a clear method to correlate directly the cavitation erosion with the roughness. The motive is the implication of different factors such as: the irregular shape of the roughness's in materials with different physical-mechanical parameters and various structures as well as the complexity of the cavitation bubble implosions [9], [11-15].

From this motive, the researchers working in the Cavitation Laboratory of Timisoara Polytechnic University have realized, in the last period, beside the cavitation erosion resistance with the well-known methods also researches about the eroded surfaces. The profile diagram was obtained using a Mitutoyo device after the final stage of cavitation erosion [8], [11]. All materials were tested in a cavitation erosion device which respect integrally the conditions imposed by the ASTM G32-2010 Standard. It was reached the conclusion that between the roughness parameter R_z and the MDE value exist a good dependence which allow to use R_z for the evaluation of materials subjected to cavitation erosion.

2. Analyzed materials

In order to obtain a reliable conclusion, for the beginning, there were chosen three different materials which were also subjected to heat treatments for increasing the resistance to cavitation erosion. These materials are:

- alloyed steel 16MnCr5 – annealed and carburized;
- CuAl10Ni5Fe2.5Mn1 - bronze subjected to volume quenching (the cavitation erosion of this steel was carefully analyzed in the doctoral degree thesis of eng. Oancă [14], and in the present work was subjected to new cavitation erosion experiments in order to measure the roughness parameters);
- alloyed steel 42CrMo4 (DIN 17200) - in annealed state (this steel is used for manufacturing some details of the hydraulic control devices).

3. Experimental procedure. Results and discussions

For each material, with the exception of steel 16MnCr5 – annealed and carburized, there were tested in the Standard Cavitation Erosion Device three specimens in agreement with the G32-2010 Standard and the laboratory customs [12], [9], [10], [16]. For one specimen the eroded surface was divide in 9 square zones (one in the central part and the others on two concentric circles) in which there were measured with the Mitutoyo device the roughness R_z on two perpendicular diameters (in total 18 measurements), in conformity with Figure 1. Those measurements were realized after (0, 30, 75, 120 and 160) minutes of cavitation exposure. Thus for each specimen were realized 90 measurements. In Table 1 are presented images with the cavitation erosion evolution, for the 5 measured times.

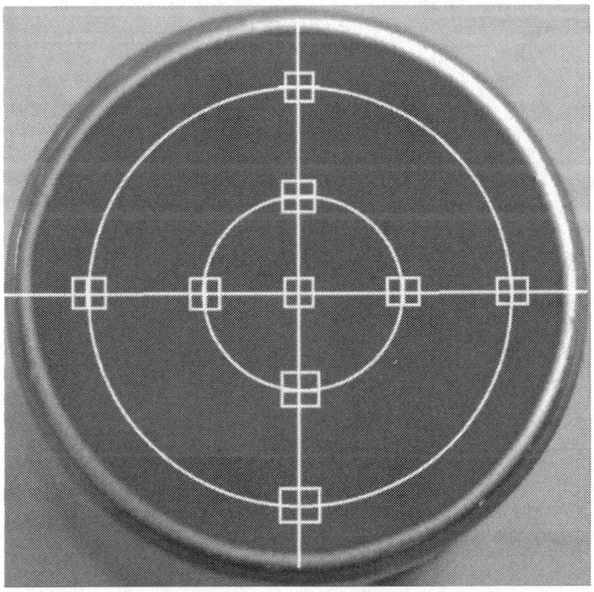

Figure 1. The 18 measuring zones for the R_z, roughness parameter

For each specimen, in Figures 2-4 is given four diagrams with the profile of the eroded area obtained with the Mitutoyo apparatus. It is presented the central area in which the greatest erosion occurs.

Table 1. Images of the cavitation erosion evolution

Cavitation exposure in minutes				
0	30	75	120	165
16MnCr5 steel – annealed and carburized				
	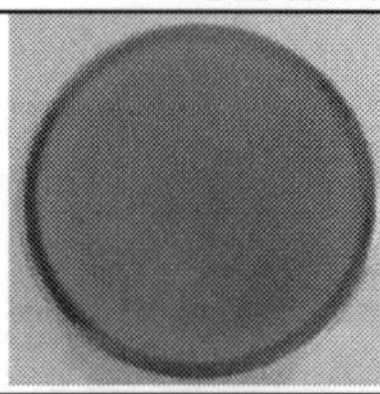			
CuAl10Ni5Fe2.5Mn1- bronze volume quenched (procedure described in [10])				
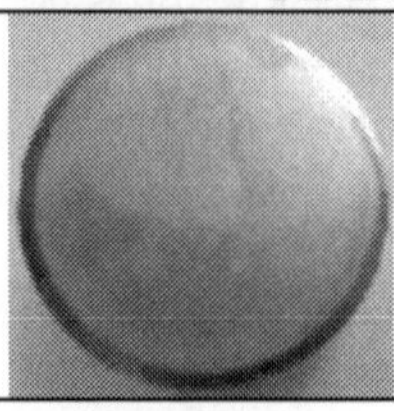	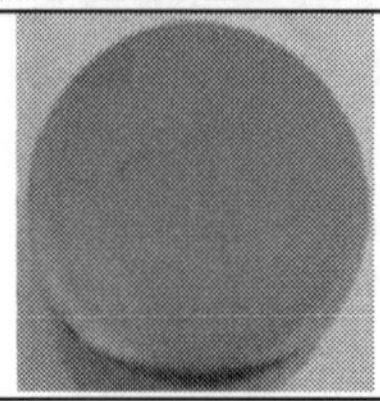			
42CrMo4 steel – annealed				
	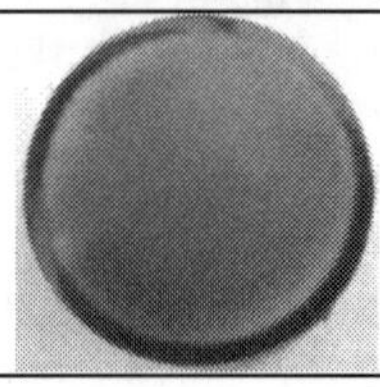			

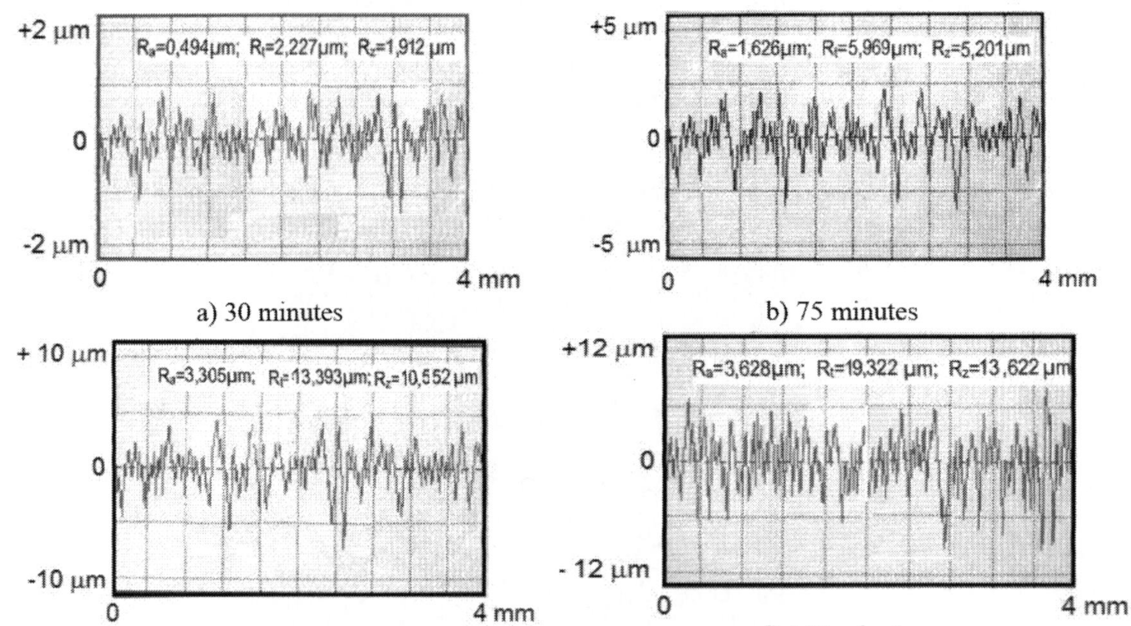

a) 30 minutes b) 75 minutes

c) 120 minutes d) 165 minutes

Figure 2. Mitutoyo roughness measurements for different exposure (Steel 16MnCr5-annealed and carburized)

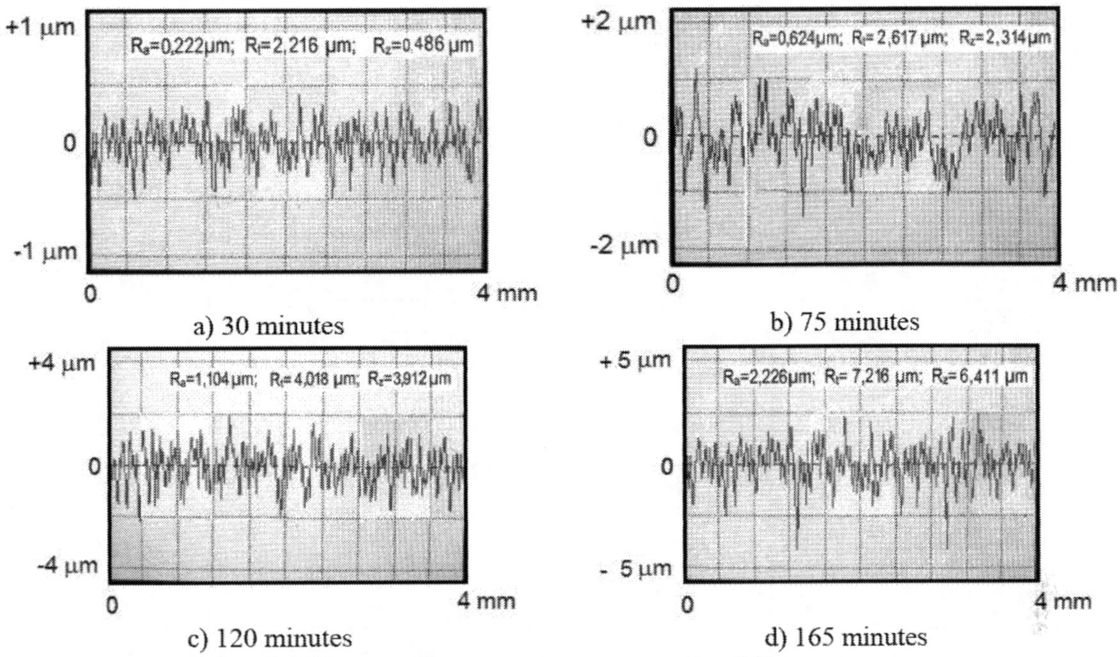

Figure 3. Mitutoyo roughness measurements for different exposure (CuAl10Ni5Fe2.5Mn1-TT bronze)

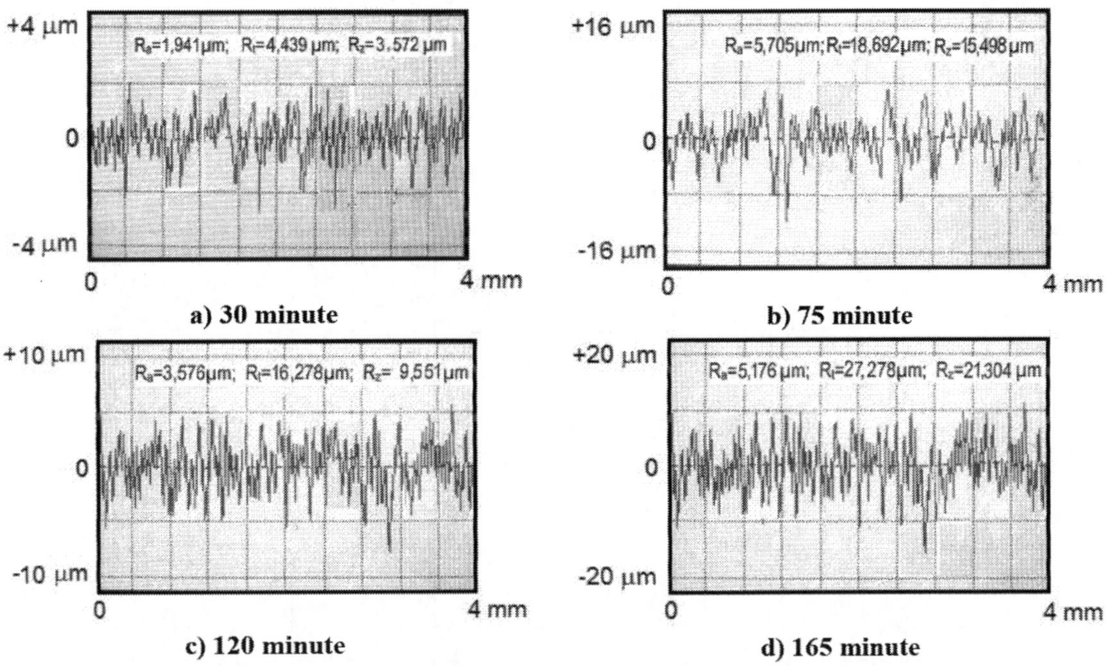

Figure 4. Mitutoyo roughness measurements for different exposure (42CrMo4 alloyed steel)

In Tables 2-4 are given the 18 values of the R_z parameter as well as the mean values for all the 9 measuring zones presented in Figure1. The parameter R_z gives the closer value to the mean depth erosion MDE [8], [15].

IOP Conf. Series: Materials Science and Engineering **294** (2017) 012020 doi:10.1088/1757-899X/294/1/012020

Specification: in MDE (t) diagrams, are presented the computed mean values of penetration for the entire area subjected to cavitation erosion.

Table 2. The values of parameter R_z [μm] (16MnCr5 steel-annealed and carburized)

0 min.	30 min.		75 min.		120 min.		165 min.	
Measured	Measured	Mean	Measured	Mean	Measured	Mean	Measured	Mean
0.02	1.488	**1.781**	5.943	**5.562**	9.925	**10.123**	13.694	**13.277**
0.02	1.721		5.672		9.841		13.589	
0.02	1.698		5.011		10.153		13.715	
0.02	1.822		5.352		9.953		13.512	
0.02	1.751		5.224		9.942		13.672	
0.02	1.771		5.932		10.281		13.883	
0.02	1.834		5.725		10.185		13.516	
0.02	1.811		5.901		9.811		13.703	
0.02	1.831		5.211		9.889		13.011	
0.02	1.792		5.867		9.921		13.785	
0.02	1.761		5.825		9.924		13.653	
0.02	1.699		5.021		9.872		13.901	
0.02	1.602		5.724		10.212		13.504	
0.02	1.912		5.201		10.552		13.622	
0.02	1.756		5.411		9.901		12.672	
0.02	1.842		5.103		9.952		13.443	
0.02	1.795		5.972		10.425		13.362	
0.02	1.587		5.895		9.898		13.198	

Table 3. The values of parameter R_z [μm] (Bronze CuAl10Ni5Fe2.5Mn1-TT)

0 min.	30 min.		75 min.		120 min.		165 min.	
Measured	Measured	Mean	Measured	Mean	Measured	Mean	Measured	Mean
0.02	0.425	**0.455**	2.233	**2.316**	4.071	**3.941**	6.208	**6.14**
0.02	0.486		2.314		3.912		6.411	
0.02	0.493		2.021		4.115		6.118	
0.02	0.488		2.225		4.071		6.189	
0.02	0.525		2.146		4.104		5.998	
0.02	0.421		2.306		3.896		6.385	
0.02	0.511		2.198		3.903		6.012	
0.02	0.412		2.187		3.845		6.214	
0.02	0.385		2.279		4.104		6.142	
0.02	0.401		2.323		4.202		6.421	
0.02	0.447		2.365		3.825		5.892	
0.02	0.411		2.402		3.925		6.141	
0.02	0.489		2.065		4.143		6.272	
0.02	0.443		2.204		4.059		6.301	
0.02	0.407		2.199		3.809		6.175	
0.02	0.491		2.285		3.998		6.007	
0.02	0.489		2.351		3.642		5.901	
0.02	0.501		2.389		4.188		6.415	

Table 4. The values of parameter R_z [μm] (42CrMo4 alloyed steel)

0 min.	30 min.		75 min.		120 min.		165 min.	
Measured	Measured	Mean	Measured	Mean	Measured	Mean	Measured	Mean
0.02	3.44		9.46		15.41		21.24	
0.02	3.375		9.375		15.521		21.272	
0.02	3.425		9.285		15.495		21.181	
0.02	3.555		9.572		15.312		21.239	
0.02	3.604		9.369		15.422		21.248	
0.02	3.402		9.522		15.495		21.421	
0.02	3.476		9.421		15.467		20.918	
0.02	3.421		9.389		15.502		21.361	
0.02	3.446	**3.469**	9.578	**9.432**	15.461	**15.435**	21.311	**21.242**
0.02	3.618		9.422		15.456		21.255	
0.02	3.572		9.551		15.498		21.304	
0.02	3.501		9.344		15.421		21.325	
0.02	3.492		9.367		15.461		21.115	
0.02	3.366		9.403		15.401		21.192	
0.02	3.381		9.298		15.431		21.236	
0.02	3.471		9.574		15.431		21.278	
0.02	3.534		9.339		15.446		21.401	
0.02	3.411		9.501		15.417		21.092	

In the histograms presented in Figures 5-7 there are compared the measured mean values of R_z, for a single tested specimen, with the mean dept erosion values for three tested specimens. It resulted the following conclusions:

- before cavitation erosion tests the roughness has evidently a given value but the mean depth erosion is absent;
- for small exposure time (30-70 minutes) the difference is enough great, the MDE indicator having enhanced values than R_z;
- for long exposure time (120-165 minutes) the differences remains but are very small.

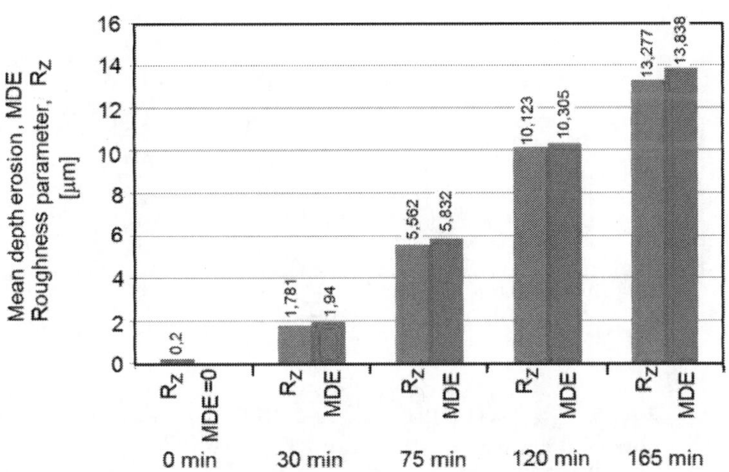

Figure 5. Comparison between the parameter MDE with the roughness R_z
(16MnCr5 Alloyed steel – annealed and carburized)

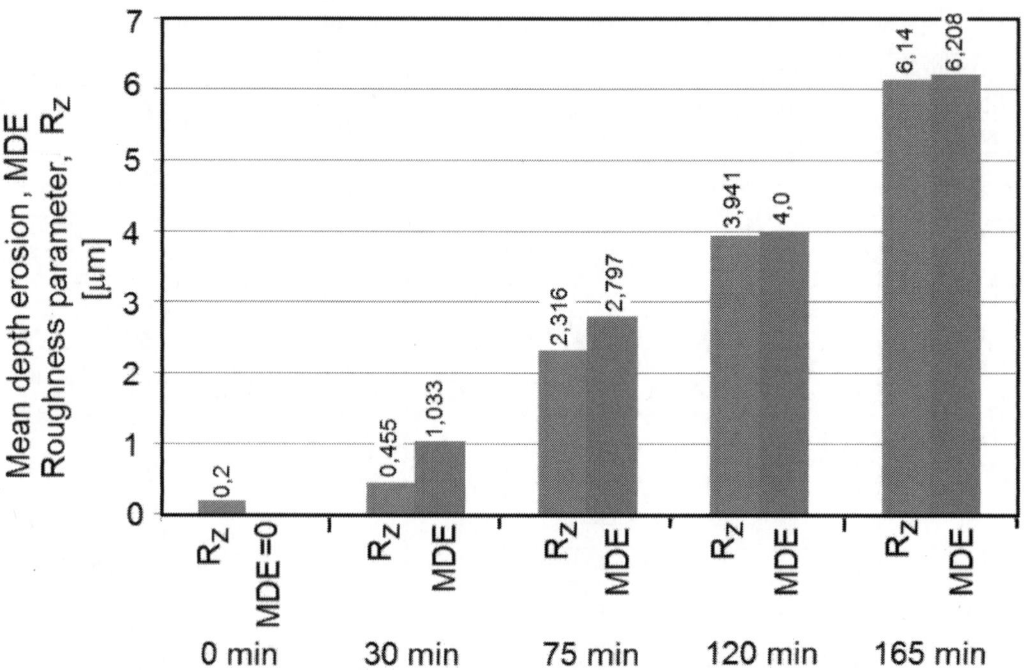

Figure 6. Comparison between the parameter MDE with the roughness R_z (Bronz CuAl10Ni5Fe2.5Mn1M45–quenching and abatement, processed after [10])

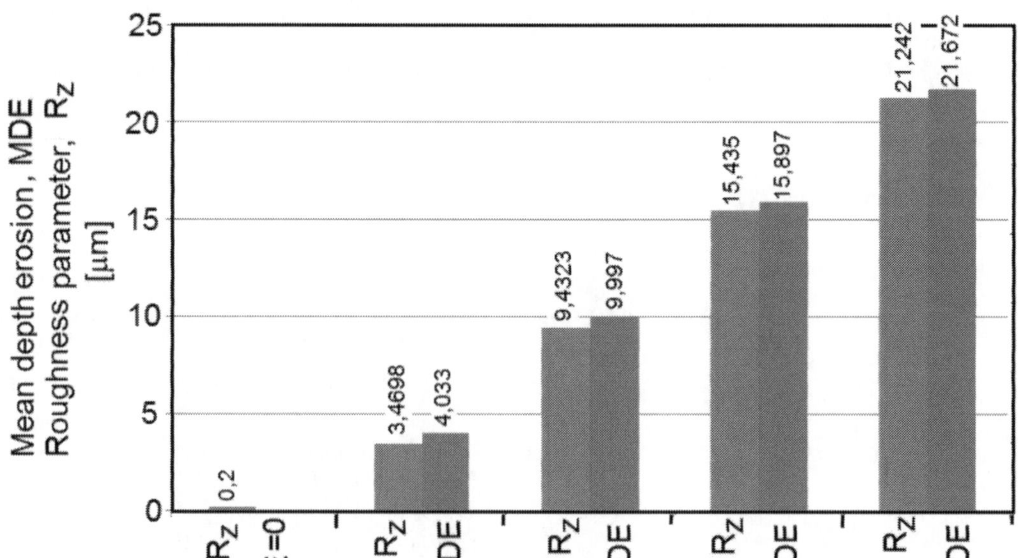

Figure 7. Comparisons between MDE and R_z roughness (42CrMo4 steel –annealed state)

These differences have the following explanations: when the cavitation phenomenon appears (first 15-30 minutes) the metallic dust resulted from the manufacturing procedure, strongly inserted between the roughnesses as well as the sharp apexes are easily removed and the volume losses being great the mean depth erosion is also very great. As the exposure time increases the influence of this important mass loses decreases sharply and the differences became without importance.

For very long exposures (120-165) minutes the losses for the same interval of exposure decreases as result of the superficial layer hardening as a result of the stresses delivered by the implosion of the cavitation bubbles [5], [12], [13], [15]. These conclusions are in agreement also with the zone III (erosion attenuation) and zone IV (erosion stabilization) which appear in the curves mean depth erosion rate [17].

In the diagrams of Figures 8-10 are presented the mean depth erosion curves, realized with the statistical relation for each material, as well as the R_z measured values. It was found out that the scatter is very small having the order of magnitude of the MDE values.

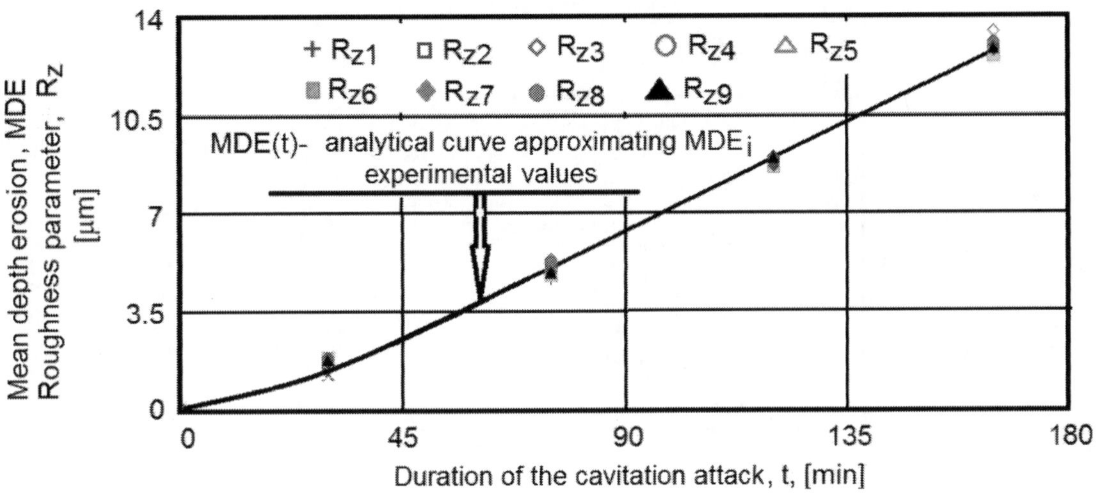

a) R_z values from 1 to 9

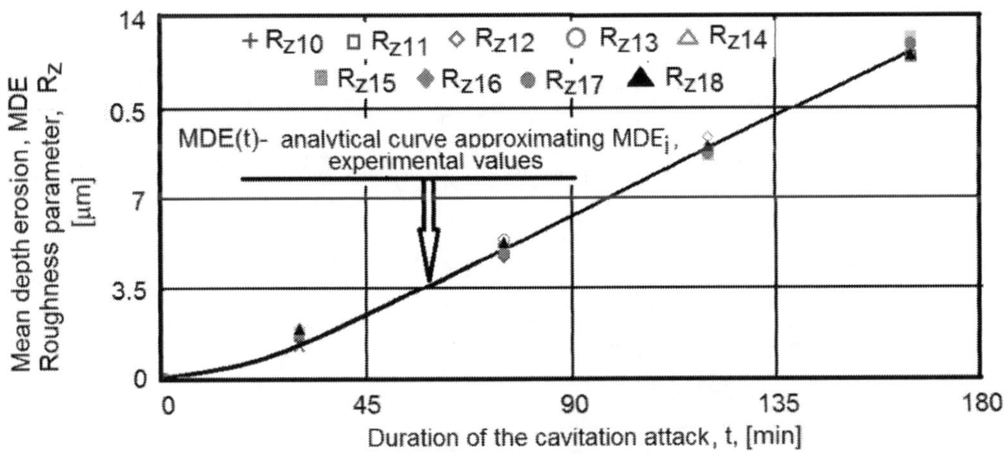

b) R_z values from 10 to 18

Figure 8. R_z parameter scatter from the curve approximating the experimental values of mean depth erosion – steel 16MnCr5-annealed and carburized

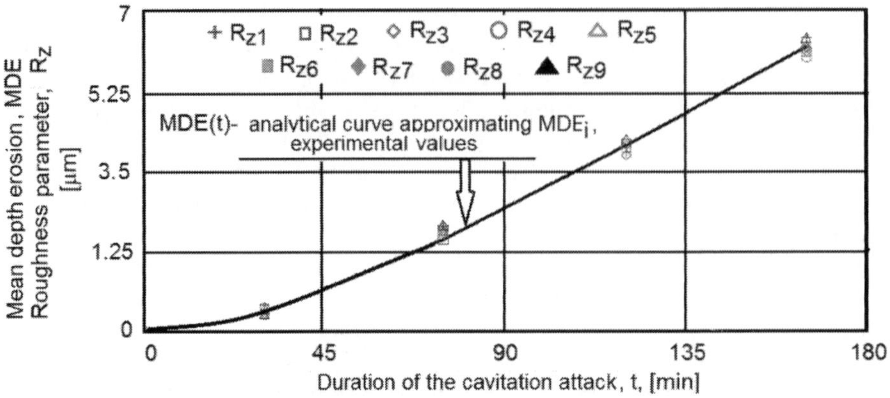

a) R_z values from 1 to 9

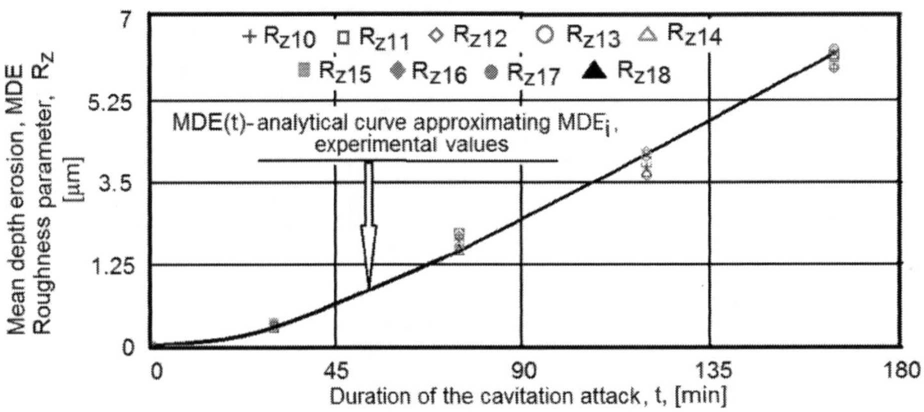

b) R_z values from 10 to 18

Figure 9. R_z parameter scatter from the curve approximating the experimental values of mean depth erosion - bronze CuAl10Ni5Fe2.5Mn1-TT

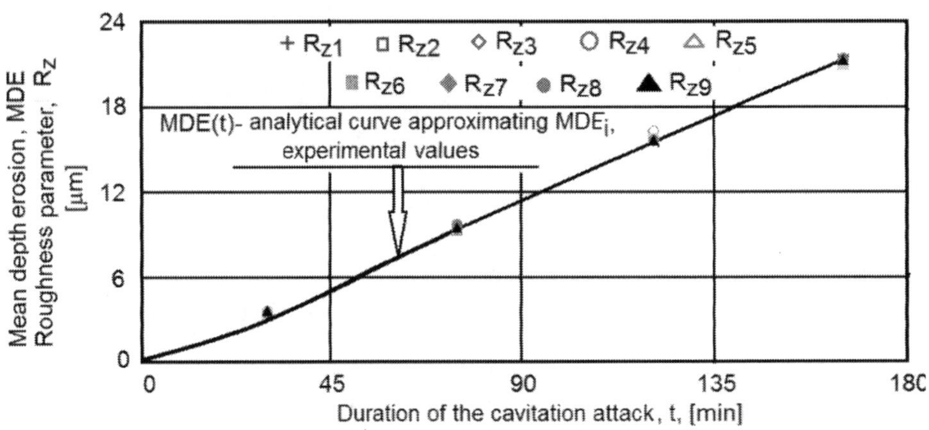

a) R_z values from 1 to 9

Figure 10a. R_z parameter scatter from the curve approximating the experimental values of mean depth erosion steel. 42CrMo4 –annealed

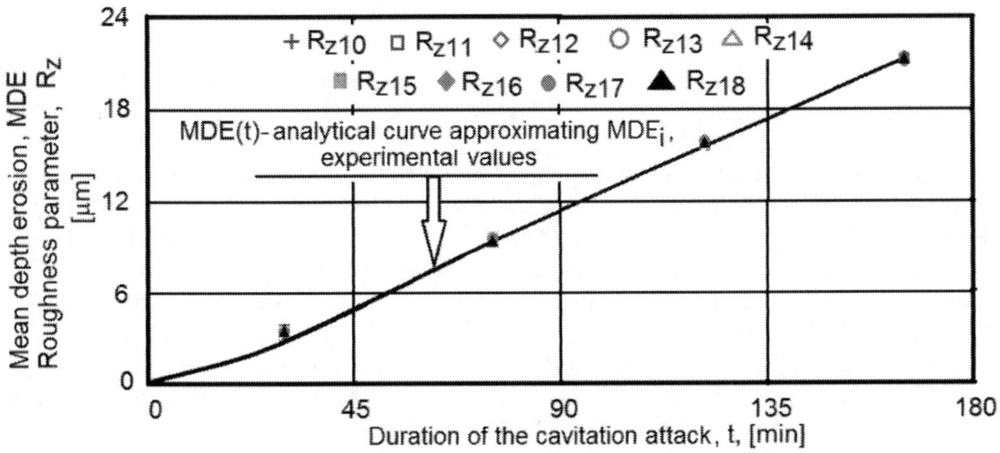

b) R_z values from 10 to 18

Figure 10b. R_z parameter scatter from the curve approximating the experimental values of mean depth erosion steel. 42CrMo4 –annealed

4. Conclusions

The profile diagrams, graphically recorded with the Mitutoyo device show that the roughness shape and level is given by the resistance of the material against cavitation stresses.

It was confirmed the possibility of using the R_z parameter to estimate the resistance to cavitation erosion. The differences obtained between the measured values of R_z and MDE (see Figures 8-10) does not exceed the differences realized by MDE test results for various samples manufactured from the same material and tested in equal conditions.

The R_z parameter has smaller values than MDE for the first minutes of attack. Those differences occur because in this time interval MDE has greater loss values by elimination of the dust remained between the roughnesses of the surface which cannot be quantified the profiles measured with the Mitutoyo device.

As a general conclusion we can say that measurements of the R_z parameter represent an excellent method to record the cavitation erosion of materials.

References

[1] Anton I 1985 *Cavitatia, vol. II*, Editura Academiei RSR, Bucureşti
[2] Bordeasu I 2006 *Eroziunea cavitaţională a materialelor*, Editura Politehnica, Timişoara
[3] Popoviciu O M and Bordeasu I 1998 *Tehnologia fabricaţiei sistemelor hidraulice*, Editura Politehnica, Timişoara, Romania
[4] ***ASTM G32-2010, *Standard test method for cavitation erosion using vibratory apparatus*, ASTM International
[5] Bordeasu I, Popoviciu M O, Ghera C, Salcianu L C, Micu L M and Podoleanu C E 2016 Cavitation erosion behavior of the steel 17CrNiMo6, *Machine Design* **8**(4) 149-154
[6] Jurchela A D 2012 *Cercetări asupa eroziunii produse prin cavitaţie vibratorie la oţelurile inoxidabile cu conţinut constant în crom şi variabil de nichel*, University Politehnica Timisoara, Romania, Doctoral Thesis
[7] Karabenciov A 2013 *Cercetări asupra eroziunii produse prin cavitaţie vibratorie la oţelurile inoxidabile cu conţinut constant în nichel şi variabil de crom*, University Politehnica Timisoara, Romania, Doctoral Thesis
[8] Katona S E 2017 *Eroziunea cavitaţională a oţelurilor inoxidabile cu transformare martensitică indirectă*, University Politehnica Timisoara, Romania, Doctoral Thesis
[9] Mitelea I, Micu L M, Bordeasu I and Craciunescu C M 2016 Cavitation erosion of sensitized

UNS S31803 Duplex Stainless Steels, *Journal of Materials engineering and performance* **25** (5) 1939 – 1944

[10] Oanca O 2013 *Tehnici de optimizare a rezistenței la eroziunea prin cavitație a unor aliaje CuAlNiFeMn destinate execuției elicelor navale*, University Politehnica Timisoara, Romania, Doctoral Thesis

[11] Mitelea I, Ghera C, Bordeasu I and Craciunescu C M 2015 Ultrasonic cavitation erosion of a duplex treated 16MnCr5 steel, *International Journal of Materials Research* **106**(4) 391-397

[12] Bordeasu I, Popoviciu M O, Mitelea I, Balasoiu V, Ghiban B and Tucu D 2007 Chemical and mechanical aspects of the cavitation phenomena, *Revista de Chimie* **58**(12) 1300-1304

[13] Franc J P, Kueny J L, Karimi A, Fruman D H, Fréchou D, Briançon-Marjollet L, Billard J Y, Belahadji B, Avellan F and Michel J M 1995 *La cavitation. Mécanismes physiques et aspects industriels*, Press Universitaires de Grenoble, Grenoble, France

[14] Franc J P, Riondet M, Karimi A and Chahine G L 2012 Material and velocity effects on cavitation erosion pitting, *Wear* **274-275** 248-259

[15] Garcia R, Hammitt F G and Nystrom R E 1960 Correlation of cavitation damage with other material and fluid properties, Erosion by Cavitation or Impingement, *ASTM, STP 408* Atlantic City, New Jersey, U.S.A

[16] Mitelea I, Bordeasu I, Pelle M and Craciunescu C M 2015 Ultrasonic cavitation erosion of nodular cast iron with ferrite-pearlite microstructure, *Ultrasonics Sonochemestry* **23** 385-390

[17] Thiruvengadam A and Preiser H S 1963 *On testing materials for cavitation damage resistence*, Report 233 – 3

IOP Conf. Series: Materials Science and Engineering **294** (2017) 012021 doi:10.1088/1757-899X/294/1/012021

Researches on the development of new composite materials complete / partially biodegradable using natural textile fibers of new vegetable origin and those recovered from textile waste

M P Todor[1], C Bulei[1], T Heput[2] and I Kiss[2]

[1]University Politehnica Timisoara, Doctoral School, Timisoara / Hunedoara, Romania
[2] University Politehnica Timisoara, Faculty of Engineering Hunedoara, Department of Engineering and Management, Hunedoara, Romania

E–mail: imre.kiss@fih.upt.ro

Abstract. The objective of the research is to develop new fully / partially biodegradable composite materials by using new natural fibers and those recovered from various wastes. Thus, the research aims to obtain some composites with matrix of various types of polymeric materials and the reinforcement phase of textile materials (of different natures, morphologies and composites) so that the resulting products to be (bio)degradable. The textile inserts used as raffle are ecological, non–toxic and biodegradable and they contain (divided or in combination) bast fibers (flax, hemp, jute) and other vegetable fibers (cotton, wool) as plain yarn or fabric, which can replace fibers of glass commonly used in polymeric composites. The main activities described in this article are carried out during the first phase of the research (phase I – initiation of research) and they are oriented towards the choice of types of textile inserts from which the composites will be obtained (the materials needed for the raffle), the choice of the types of polymers (the necessary materials for matrices) and choosing the variants of composites with different types and proportions of the constituent content (proposals and working variants) and choosing the right method for obtaining samples of composite materials (realization technology). The purpose of the research is to obtain composite materials with high structural, thermo–mechanical and / or tribological performances, according to ecological norms and international requirements in order to replace the existing classical materials, setting up current, innovative and high performance solutions, for applications in top areas such as automotive industry and not only.

1. Introductory notes

Over the past fifty years composite materials have developed rapidly [1–8]. Due to its versatility, the volume and number of areas of use of composite materials have increased constantly, developing new solutions in order to improve product quality and attractiveness for new markets. Composites are no longer the privilege of the aerospace or defence industry, or even high value goods. They have quickly become a way to achieve high structural performance at a low cost and are found all around us. Using composites provides high performance, lower costs, low weight, and compliance with the environmental and safety regulations [1–8].

The original purpose of achieving composites was to increase the competitiveness of classical materials whose strength and stiffness properties could not be improved by other means [1–6]. From this point of view it is understood that the maximum efficiency of the reinforcement of a certain material is obtained by introducing in its structure some reinforcement elements. Representing the most well–known category and marking the beginning of the industrial scale use of the new materials, the fiber composites are mainly imaginable in the form of plastics (polymer) matrices reinforced with fibers(long) [1–6].

The main reasons why polymers represent most of the matrices used for the production of composite materials (about 75% of composite material production) are the following [1–9]:

- polymers can be processed at low temperatures and thus the danger of destroying the reinforcement agent (reinforcement) does not exist
- polymers can be obtained using classical methods that do not require complex technology.

As a result of using high performance fibers, polymeric materials are increasingly replacing traditional materials in all domains. The unique properties of the polymeric materials as well as the possibility of their adaptation to the applicative needs have determined the extension of their use in all the domains of human activity [1–7].

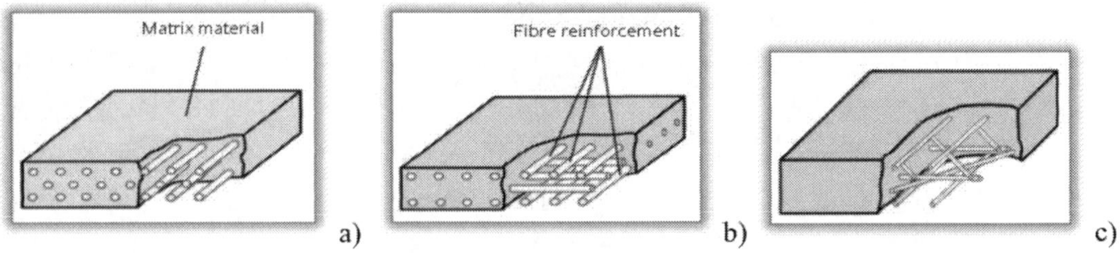

Figure 1. The general structure of long fiber reinforced composite materials.
a) same oriented fibers reinforcement, b) cross–oriented fibers reinforcement, (c) randomly oriented fiber reinforcement

Crises manifested in the whole society have also generated local or global imbalances in the material flow management, including waste. Non–renewable natural resources (coal, oil, natural gas, ferrous and non–ferrous ores) or renewable (wood, industrial plants) have become insufficient, which has required strategies to ensure the sustainable development of our society [10–28].

Thus, the return to nature, both through the processes of obtaining polymers from renewable raw materials and by inducing their end–of–life biodegradability, is the strategy that can build the sustainability of nowadays society [7–17]. The only solution to meet material needs is to obtain synthetic materials to replace natural materials and whose production can be adjusted as needed [15], [17], [22–26]. A variation of synthetic materials that has conquered all current fields of activity, through the extraordinary ability to adapt their properties to the applicative requirements are the composite polymeric materials.

2. Research directions

A sustainable development can only be ensured by proposing rational solutions for industry, designing eco–products and eco–technologies, but also recycling waste whose biodegradation has lasted for decades and hundreds of years [6], [14], [15]. A sustainable development and an intelligent use of resources in the field of industrial materials and technology require the substitution of critical materials, the development of advanced recycling technologies of secondary materials and end–of–life materials, lightweight structures and materials for substitution of heavy steel components (fiber–reinforced composites) [7], [8].

Great research efforts are being made to achieve these goals [10–28]. But even if the possibility of obtaining polymers from bio–sources (natural sources) or biotechnologies from renewable sources has been demonstrated, their price is still high and biodegradable polymers have structures and implicit properties that do not cover the whole range of needs [6], [14], [15]. At present, the most readily applicable methods for reducing the impact of polymeric materials on the environment consist in composite materials made of polymeric matrices made from common polymers for which there is production infrastructure and biodegradable natural materials (biodegradable) produced as secondary products in agriculture or even from industrial or domestic waste [9–12], [15], [19], [20]. Fibres improve or maintain the properties of the matrix polymer and induce the end–of–life biodegradability [15–17], [22–26]

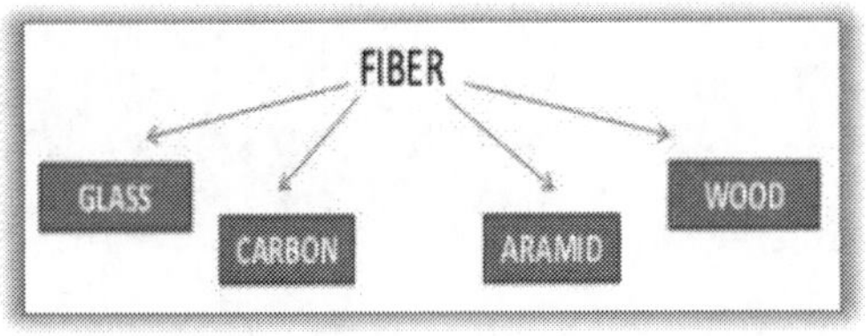

Figure 2. Most important nature of fibres

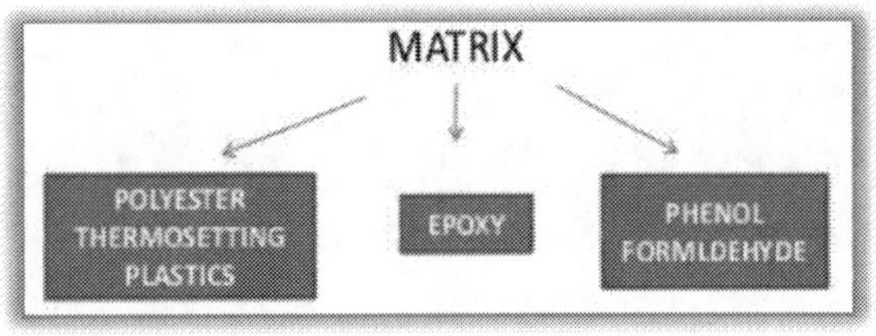

Figure 3. Most important nature of matrix

One of the research directions is related to technical textiles and textile products for strategic areas. Although the textile industry is in recession worldwide, textile technical items are on the rise, proving that manufacturers of these products are concerned about finding solutions and creating high–performance articles. Textiles or textile composites are designed to replace many of the current metal or plastic materials used in marine, aerospace, aeronautics, cars, security and defence [26–28].

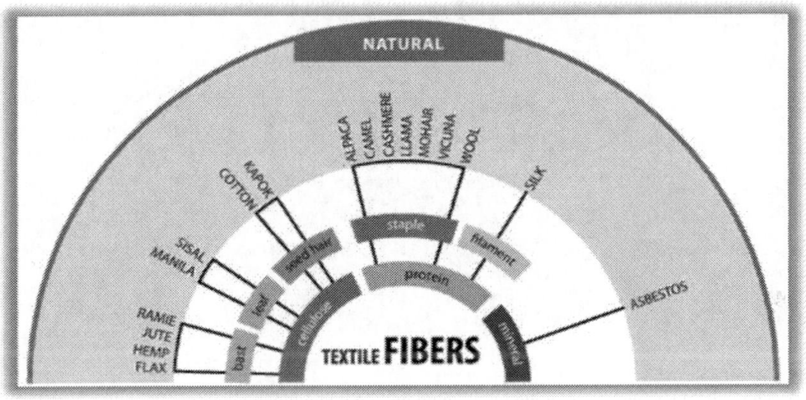

Figure 4. Classes of textile fibres

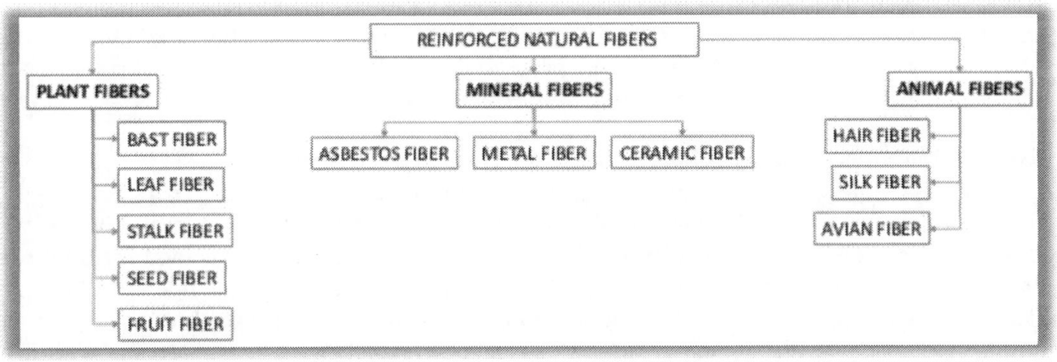

Figure 5. Reinforced natural fibers

Composite materials based on bast fibers have become the attention of researchers and manufacturers of composite materials to harness both long fibers extracted from plant stems and short fiber waste resulting from fiber extraction [15], [17], [26]. By combining this polymeric matrix plant waste there are produced polymeric biocomposite materials with a lower cost price than those based on long fibers and with mechanical characteristics comparable to them.

Figure 6. Reinforced natural fibers – plant fibers class – bastfibers

In the textiles domain, the organic concept it is also applied to the textile waste resulting from industrial activity and the management of these wastes is a process by which these products are reduced, reused, recycled, collected, stored, transported or eliminated. The development of alternative solutions for the use of textile waste in high added value products by producing composite products containing recovered textile fibers constitutes an ecological alternative for the capitalization of textile waste. They are based on a textile backing that undergoes a consolidation process.

Figure 7. Bastfibers textiles

Composites with added polymeric matrices with high adherence to the fiber reinforced textile are usable as substitutes for classical materials [7], [15], [17], [26]. Reusable textile materials, also known as textile waste, come primarily from textile processing (spinning, yarn preparation, weaving, knitting, chemical finishing), fabrication or as a result of physical and moral wear after a certain amount of time. The categories of textile waste that can be processed into yarns are: technological waste, cotton / wool type knitted scraps and fabric scraps resulting from garments.

3. Ways of inducing the biodegradability of polymeric composite materials

There have been identified several ways of designing biodegradable polymeric materials [3], [4], [6], [15], [26]. The most convenient is the use of synthetic polymers produced on a large scale to which biodegradable components are added [10–12], [19], [20], [27]. At present, this method is also the most feasible, using all the existing facilities for synthesis polymers, it uses natural fibers that are easily obtained, renewable and biodegradable, thus providing a "buffer step" between a company buried in waste non–biodegradable plastic and a partially sustainable society that uses materials with a high degree of biodegradability.

Thus, by inducing biodegradability, composite materials have been obtained in which the filler fibers are bioreinforments, most often vegetable waste (wool yarn, cotton yarn, cellulose fiber, hemp fiber, flax fiber, jute fiber, and so on) [10], [12], [17–25]. It has been demonstrated that these composites, in addition to incorporating waste, by reducing their deposited quantity, maintain or improve the properties of the polymeric matrix material (improve the thermal resistance, the irradiation and chemical resistance) and they can be recycled more favourably than the matrix [14], [17], [22–26].

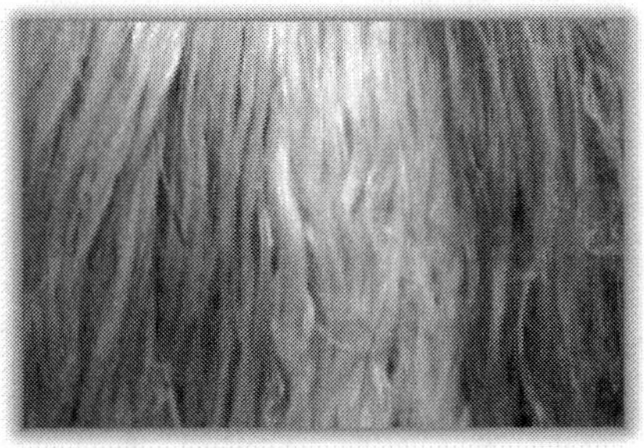

Figure 8. Hemp, flax and jute fibers

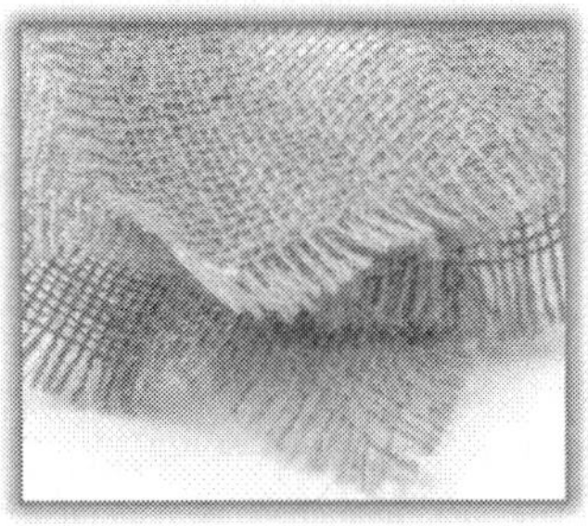

Figure 9. Bast fiber textile

Figure 10. Matrix

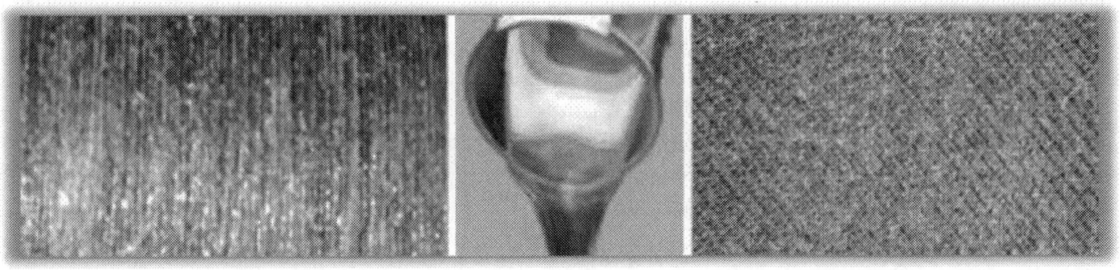

Unidirectional prepreg prepreg resin Woven prepreg

Figure 11. Technological process using the prepregs

Their properties can be modified and adapted to application requirements by reinforcing with organic or inorganic fillers (in the form of short or long fibers, filaments or yarns, knitted, woven or non–woven, or particulate powders of different sizes) or by modifying the structure and density of the composite (modification of the dispersion and branching of the reinforcement in the matrix). The surface treatments of fibers also can improve the contact at the interface with the matrix to obtain composite materials with special properties. Therefore, in recent years, the interest in composite materials based on natural fibers (bast fibers such as hemp, flax and jute, and exotic fibers such as banana, bamboo) has increased significantly [10–12], [14–17], [22], [24–26], [28]. The use of natural fibers in composites does not only improve the physical and chemical properties (low density, high chemical resistance, increased resistance to environmental conditions such as humidity, solar radiation, pH), mechanical (good mechanical strength, easy processing, good resistance to shock and fire) and material functionality (absorb vibrations better than artificial fibers, good thermal and sound insulation, low weight, low maintenance costs, a variety of forms, design dynamics and can be easily handled) as well as biodegradability by putting them in the category of organic materials [10–17], [28]

Composite materials in relation to traditional materials have important strengths and bring many functional advantages, their use contributes to increased security, and extends the life of structures due to their excellent properties. The ways of controlling the properties of polymers are virtually infinite.

All these lead to advantageous conditions for multiple applications. The main advantage of composite materials is the high ratio between their strength and their bulk density. Due to the superior mechanical properties and reduced mass, composite materials are met in various fields of activity, the industrial applications being the most important, successfully replacing traditional materials. Therefore, today, there is no field of human activity that does not gain by the benefits of using polymers.

4. Polymer composite materials with textile finishing

The problems that have arisen in trying to define the composite materials exactly as much as possible are a proof of the extremely wide field of this type of material, a field in continuous and rapid expansion. As a general definition, composite materials are systems of two or more components whose properties complement each other, resulting a material with superior properties to those specific to each component [1–7]. Thus these components will cooperate, the deficiencies of some being supplemented by the qualities of others, conferring to the assembly properties that any component can have separately. As such, fibrillated composite fiber composites are continuous or discontinuous fiber arrangements that are immersed in a polymeric matrix whose mechanical strength is much lower. Thus the composite material is a unitary assembly in which the two phases act together. In particular, the constituent–process couple is indissociable because the properties of the structure depend directly and they will only be identified in the finished product stage. The result consists in a system that includes the nature of the reinforcement, the texture and shape of the reinforcement, its strength and volume, the nature of the resin and / or the additives, the quality of the reinforcement–matrix interface, the geometry of the work piece to be analyzed, the process used and so on [1–7].

The role of the matrix in the composite material is to determine the final shape of the product, to protect the fibers both in the forming process and during use, to prevent the buckling of the fibers, to ensure the chemical and thermal compatibility with the reinforcement and to avoid rapid propagation of composite cracks. The matrix is defined as forming the continuous phase of the composite [1–7].

Reinforcement represents the discontinuous phase, evenly distributed throughout the matrix volume and which is added to the matrix to improve or modify its properties. Fiber is the element that gives the assembly the characteristics of resistance to stress. The role of the textile material in the composite material is to reinforce, thus increasing the strength and rigidity of the composite, this rise being proportional to the volumetric fiber fraction [1–7].

Bast fibers, belong to the category of natural textile fibers of vegetable origin, predominantly cellulosic, which are extracted by removing the wood from the free (stem) of plants such as flax, hemp or jute. Globally, there are over 1,000 plant species with processed bast fibers, with about 25 being profitable [10–12], [19–28].

Figure 12. Natural fibers

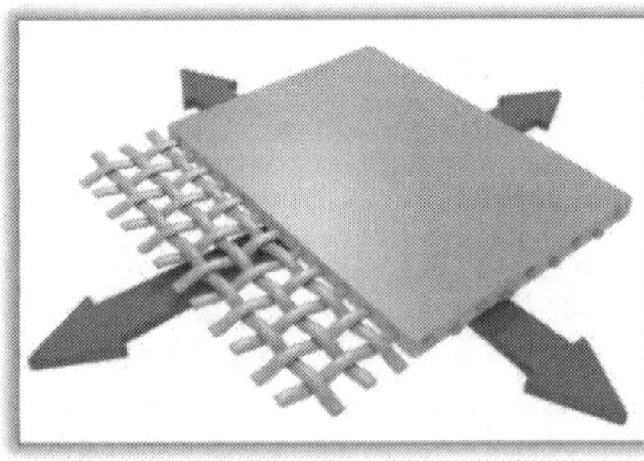

Figure 13. Fibres reinforcement

Unlike other textile fibers, bast fibers are pluricelular fibers, consisting of clusters of elementary fibers, cemented between them by the medial lamella, and which form together an ensemble called technical fiber. The destruction of the median lamellar leads to the division of technical fibers into finer technical fibers (with fewer elementary stones cemented between them), or even in elemental fibers. Thus in the twigs, the fibers can be processed in the form of technical fibers.

Soft bast fibers (such as flax) are flexible fibers for a wide range of fine and thick fabrics. Heavy bast fibers (hemp, jute) are lighter than flax and are used to produce a wide range of thick or technical fabrics. In the bast fiber industry, their own yarns are used in homogeneous mixtures based on fibers of the same origin or in heterogeneous mixtures made up of different categories of own fibers with other fiber categories and even fibers of other nature, obtained from related textile branches.

5. Reinforcing technologies

The criteria used in the choice of the technological process for the reinforcement manufacture refer to the dimensional stability, to the mechanical properties imposed, and to the rigidity properties of the reinforcement system. The variety of textile reinforcement types makes them a complex problem that requires several factors to be considered. Technical textiles may exist and can be used in various forms, such as simple fiber filament structures, to finished products with complex structures [1–7].

Reinforcing systems for composite materials comprise a wide range of preforms that can be constituted by the material obtained from all textile technologies: weaving, knitting, knitting, sewing or using non–woven fibers. To these there can be added the processes characterized by the reinforcement and the composite material in the same phase, such as filament winding and polishing.

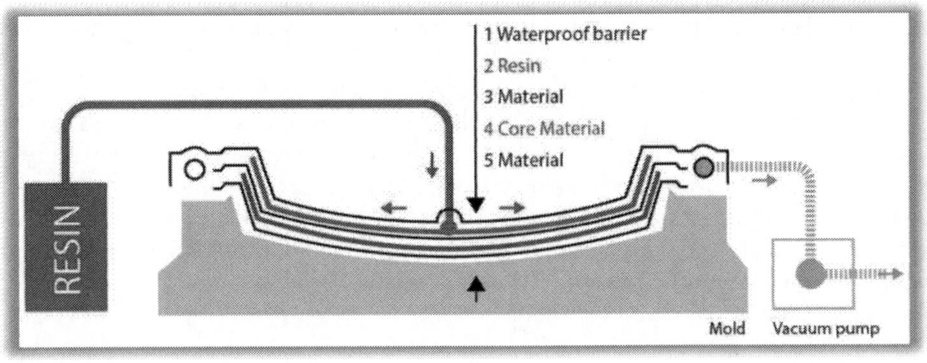

Figure 14. Reinforcing technology

In order to select the optimal preforming technology, both the strengths and weaknesses of each technology must be considered. They have the advantage of mass / strength ratio, time reduction, elimination of waste resulting from the production and processing process, better control of the final shape of the product and superior quality.

In order to obtain composite structures, different methods are used:

- embedding reinforcement material (long or short fibers, knitting material, powder) into a matrix, which may be a substance with coagulation properties;
- reinforcing the base material with curing layers, resulting in a laminate.

The criteria used in the choice of the technological process for the reinforcement manufacture refer to dimensional stability, to the imposed mechanical properties, and to the draping / forming properties of the reinforcement system.

6. Concluding remarks

Composite materials have become a more and more common solution in setting up performing structures applicable in all industrial branches. Their implementation in various fields, as advantageous alternatives to classical materials, or the acquisition of new applications, otherwise difficult or impossible to achieve, raises a series of problems due to their very complex structure and the possibilities of obtaining the behavior still insufficiently known for various requests.

In every industrialized country composite materials are a priority area, located at the forefront of the continuous process of technological innovation. Research into composite materials based on plant fibers in general and on fiber–based fibers refers to aspects related to the determination of the factors that influence the performance of the composite materials based on bast fibers and to the finding of methods for their improvement, the analysis of the physical, chemical, mechanical and technological processes, as well as finding new composite materials in order to meet the economic needs.

The public and business interests in multiple applications of these versatile plants are in a rapid growth according to the acute need for renewable alternatives for industrial resources. Bast fibers (flax, hemp, jute) are a viable good that can offer many solutions that equal or even exceed the products to be replaced.

The use of fiber composite materials has a beneficial effect in the machinery industry, which works under the pressure of reducing fuel consumption and implicitly of noxious emissions into the atmosphere. This desideratum is achievable in terms of reducing the mass of the vehicle.

Composite materials are good solutions for the mass reduction without cost increases. Synthetic fibers used as a plastic reinforcement material (carbon fibers, glass fibers, kevlar), despite offering a mass reduction of more than 50%, have the disadvantage that they are expensive. In this context, fiber–based polymer composites offer many advantages including freedom in design, mass reduction, corrosion resistance, cost reduction, and the production of complex aerodynamic and noise reduction devices. At the same time, recyclable textiles offer solutions for minimizing waste. Many technical textilesandtextile fibers minimize production costs by recycling waste, thus ensuring environmental protection.

The advantages of natural fibers are major in terms of behavior in case of an accident. Combined with a polymeric matrix, natural fibers provide lightweight flexibility and lightweight to traffic safety requirements. In the case of natural fibers, there are no punctures or sharp points, having relatively low density, low manufacturing prices, and biodegradable, unlike glass fibers that crush during processing.

Natural fibers are indicated for reinforcement or reinforcing accessories with complex shapes that stretch over wide surfaces. The widest range of applications is to decorate the interior of the doors. At the same time we can try to obtain materials for safety belts, head restraints, chairs, benches, ceiling, flooring, the back of the machine and the trunk.

References

[1] Gay D 1991 *Matériaux composites*, Editions Hermes, Paris
[2] Schwartz M M 1983 *Composite Materials Handbook*, McGraw Hill Book Co.
[3] Strong A B 1989 Fundamentals of composites manufacturing: Materials, Methods and Applications, *Society of Manufacturing Engineers*, Dearborn
[4] Stefanescu F, Neagu G and Mihai A 1996 *Materiale compozite*, Editura didactica si pedagogica, Bucuresti
[5] Ispas St 1987 *Materiale compozite*, Editura Tehnica, Bucuresti
[6] Bikiaris D and Terzopoulou Z 2014 *Green composites*, Plastics Research Online
[7] Todor M P, Bulei C and Kiss I 2016 Systematic approach on materials selection in the automotive industry for making vehicles lighter, safer and more fuel–efficient, *Applied Engineering Letters* **1**(4) 91–97
[8] Todor M P, Bulei C and Kiss I 2017 An overview on fiber-reinforced composites used in the automotive industry, *ANNALS of Faculty Engineering Hunedoara – International Journal of Engineering* **XV**(2) 181-188
[9] Panthapulakkal S and Sain M 2007 Agro–residue reinforced high–density polyethylene composites: Fiber characterization and analysis of composite properties, *Composites: Part A* **38** 1445–1454
[10] Khan J A and Khan M A 2015 *Biofiber Reinforcements in Composite Materials, Chapter 1: The use of jute fibers as reinforcements in composites*, Woodhead Publishing, Elsevier, pp. 3–34
[11] Müssig J and Haag K 2015 *Biofiber Reinforcements in Composite Materials, Chapter 2: The use of flax fibres as reinforcements in composites*, Woodhead Publishing, Elsevier, pp. 35–85
[12] Bledzki A K and Gassan J 1999 Composites reinforced with cellulose based fibres, *Progress in Polymer Science* **24**(2) 221–274
[13] Czigány T, Morlin B and Mezey Z 2007 Interfacial adhesion in fully and partially biodegradable polymer composites examined with microdroplet test and acoustic emission, *Composite Interfaces* **14**(7–9) 869–878

[14] Bartl A, Hackl A, Mihalyi B, Wistuba M and Marini I 2005 Recycling of fibre materials, *Process Safety and Environmental Protection* **83**(4) 351–358

[15] Shahzad A 2012 Hemp fiber and its composites – a review, *Journal of Composite Materials* **46**(8) 973–986

[16] Wagner P B, Little B J, Hart K R and Ray R I 1996 Biodegradation of composite materials, *International Biodeterioration & Biodegradation* 125–132

[17] Anandjiwala R D and Blouw S 2007 Composites from bast fibres—prospects and potential in the changing market environment, *Journal of Natural Fibers* **4**(2) 91–109

[18] Teodorescu H, Teodorescu F, Rosu D and Opran C 2006 The behaviour at temperature and humidity variations of fiber reinforced composite structures, *Nonconventional Technologies Review* **2** 87–92

[19] Ashori A and Nourbakhsh A 2010 Bio–based composites from waste agricultural residues, *Waste Management* **30** 680–684

[20] Cosereanu C, Curtu I, Lunguleasa A, Lica D, Porojan M, Brenci L, Cismaru I and Iacob I 2009 Influence of synthetic and natural fibers on the characteristics of wood–textile composites, *Revista Mase Plastice* **46**(3) 305–309

[21] Stanciu M D, Curtu I, Cosereanu C, Vasile O and Olarescu C 2011 Evaluation of absorption coefficient of biodegradable composite materials with textile inserts, *Romanian Journal of Acoustics & Vibration* **8**(2) 99–102

[22] Shah D U, Schubel P J and Clifford M J 2013 Can flax replace E–glass in structural composites?, *Composites Part B: Engineering* **52** 172–181

[23] Meredith J, Ebsworth R, Coles S R, Wood B M and Kirwan K 2012 Natural fibre composite energy absorption structures, *Composites Science and Technology* **72** 211–217

[24] Wambua P, Ivens J and Verpoest I 2003 Natural fibres: Can they replace glass in fibre reinforced plastics?, *Composites Science and Technology* **63**(9) 1259–1264

[25] Dahlke B, Larbig H, Scherzer H D and Poltrock R 1998 Natural fiber reinforced foams based on renewable resources for automotive interior applications, *Journal of Cellular Plastics* **34**(4) 361–378

[26] Mohanty A K, Misra M and Hinrichsen G 2000 Biofibres, biodegradable polymers and biocomposites: An overview, *Macromolecular Materials and Engineering* **27**(6)277 1–24

[27] Terciu O M, Curtu I, Stanciu M D and Cerbu C 2011 *Mechanical properties of composites reinforced with natural fiber fabrics*, Annals of DAAAM for 2011 & Proceedings of the 22th International DAAAM Symposium Intelligent Manufacturing & Automation, pp 607–608

[28] Dai S, Cunningham P R, Marshall S and Silva C 2015 Influence of fibre architecture on the tensile, compressive and flexural behaviour of 3D woven composites, *Composites Part A: Applied Science and Manufacturing* **69** 195–207

Using TELOS for the planning of the information system audit

D P Drljaca[1] and B Latinovic[2]

[1]Europrojekt centar Banja Luka, Republic of Srpska, Bosnia and Herzegovina
[2]Pan European University Banja Luka, Faculty of Information Technologies, Republic of Srpska, Bosnia and Herzegovina

E-mail: drljacad@gmail.com

Abstract. This paper intent is to analyse different aspects of information system audit and to synthesise them into the feasibility study report in order to facilitate decision making and planning of information system audit process. The TELOS methodology provides a comprehensive and holistic review for making feasibility study in general. This paper examines the use of TELOS in the identification of possible factors that may influence the decision on implementing information system audit. The research question relates to TELOS provision of sufficient information to decision makers to plan an information system audit. It was found that the TELOS methodology can be successfully applied in the process of approving and planning of information system audit. The five aspects of the feasibility study, if performed objectively, can provide sufficient information to decision makers to commission an information system audit, and also contribute better planning of the audit. Using TELOS methodology can assure evidence-based and cost-effective decision-making process and facilitate planning of the audit. The paper proposes an original approach, not examined until now. It is usual to use TELOS for different purposes and when there is a need for conveying of the feasibility study, but not in the planning of the information system audit. This gives originality to the paper and opens further research questions about evaluation of the feasibility study and possible research on comparative and complementary methodologies.

1. Introduction

The auditing of the information systems in the public administration institutions is a serious business process that requires good planning and project-oriented approach. This is important since the auditing process commits funds, resources such as personnel and time that should be managed properly in order to maximise auditing process effect, minimise occupancy of the resources and to provide as best as possible auditing material for obtaining of auditors opinion.

Feasibility study aims to objectively and rationally discover and define strengths and weaknesses of an existing business or proposed a project, as well as the opportunities and threats present in the environment, the resources needed for the implementation and chances for success [1].

It helps in the decision-making process and as such has to evaluate and predict all the possible strengths and weaknesses of the system while in the contemplative (thinking) phase, to enable savings and to prevent losses occurred without full implementation. The study provides an overview of the number of the audit aspects.

At the present, the auditing of information systems in the public administration institutions in most of the countries is voluntarily or in the frame of the internal audit process. Some institutions, such as the banks, are legally obliged to perform the audit of information system primarily to ensure security and safety of the information system, especially in the transaction aspect and even more important in the aspect of e-banking. The institutions such as stock exchange, telecoms and other institutions, which information systems collects, process and store important and sensitive data, are using the auditing of information systems on a regular basis. Having in mind that the auditing process requires access and use of different resources, and is implemented by the team of experts, it is possible to conclude that organising and implementing of the audit can be considered as the important project.

According to Avison and Fitzgerald [2], a feasibility study within SSADM (*Structured Systems Analysis and Design Method*) is implemented through four steps:

- **Prepare for the feasibility study** - a step that examines the volume of the project,
- **Define the problem** - a step that compares requirements with the present status,
- **Select feasibility option** – a step that examines available alternatives and selects one,
- **Create feasibility report** – a step of assembling the feasibility report.

Considering the auditing of information systems as a project, it is possible to follow five aspects of the project feasibility hidden behind the acronym TELOS [3]:

- **Technical feasibility** – examines the possibility of applying available technology or if new technology is needed;
- **Economic feasibility** – examines the economic effects or availability of funds to complete the project;
- **Legislative feasibility** – examines potential conflicts between conceptual system and ability of institution to discharge its legal responsibilities;
- **Operational feasibility** – examines the degree of compatibility of the existing procedures and staff skills or new procedures and training are needed; and
- **Schedule feasibility** – examines the ability of in-time implementation of the project.

When considering auditing of information system as the project, these aspects are more flexible compared to the development of information system. However, it is advisable to go through them in order to examine all strengths and weaknesses, opportunities and threats related to the audit. Feasibility study for the auditing process is done in the case of engagement of internal auditor. It is not a part of common audit frameworks but complements it.

2. TELOS Feasibility Aspects

2.1. Technical feasibility

The technical feasibility examines the possibility of applying of available technology or if new technology is needed. The answer it has to provide is – does the actual technology provides satisfactory effects, or it is time to improve, upgrade or completely change it. In the auditing process, implementation of the technology relates to:

- The tools and technologies for the implementation of the audit process, and
- The frameworks for the implementation of the audit process.

Today, IT auditors are using **Computer Aided Audit Tools and Techniques – CATT**. CAAT typically involves basic office software (eg. *MS OFFICE* package) or advanced software tools for statistical analysis (eg. *SPSS, SAS*), general audit software (eg, *ACL, Arbutus, EAS*) as well as business intelligence tools (eg. *SAP* or *Oracle* tools, *Crystal Reports, Business Objects*). In the most general sense, CAAT can refer to any computer program used to improve the automation of the audit process at any stage of extraction and analysis of data [4].

This improves the data collection process in the sense that the data collected for audit are adequate data, taken from a right place and at the right time. At the same time, these tools and techniques enable high quality and rapid analysis of collected data, thereby reducing the time required for processing, and thus the time that the auditor requires preparing a report. These tools and techniques are primarily used in financial audit for easy detection of fraud. However, with minor adjustments, these tools and techniques can be used for auditing information systems. The essence of CAAT tools is to provide a set of functions relating to the preparation, collection and analysis of data, such as:

- tools for querying,
- data stratification,
- sample extraction of data,
- identification the missing sequence,
- statistical analysis,
- calculations,
- identify redundancy, multiple and repeated transactions,
- creating pivot tables,
- cross-tabulation of data, etc.

Using CAATT, the auditor achieves multiple benefits such as:

- it simplifies the audit process,
- comes easier to data,
- the data is prepared in a predetermined format in order to facilitate their processing.

It is very important that these tools are independent of the system that is the subject of the audit. This will enable access to and extraction of data from the system without disrupting its integrity because tools are using and reading only a copy of the data (Read-only copy). Such approach protects data from potential abuse. Also, most of these tools have the possibility to test the data and to document the audit procedures that can be delivered together with an audit report.

CAATT can be classified as [5]:

- *Electronic Working Papers,*
- *Information Retrieval and Analysis*),
- *Fraud Detection,*
- *Network Security,*
- *Electronic Commerce and Internet Security,*
- *Continuous Monitoring,*
- *Audit Reporting,*
- *Database of Audit History,*
- *Computer Based Training,*
- *Time Tracking,*
- Etc.

The audit frameworks provide guidelines for the work of the auditor, guiding the implementation of the audit as systematic (qualitative and quantitative) collecting and processing of data required for the preparation of the audit report. There is a number of worldwide recognised standards and recommendations, but mostly in use are presented in Figure 1.

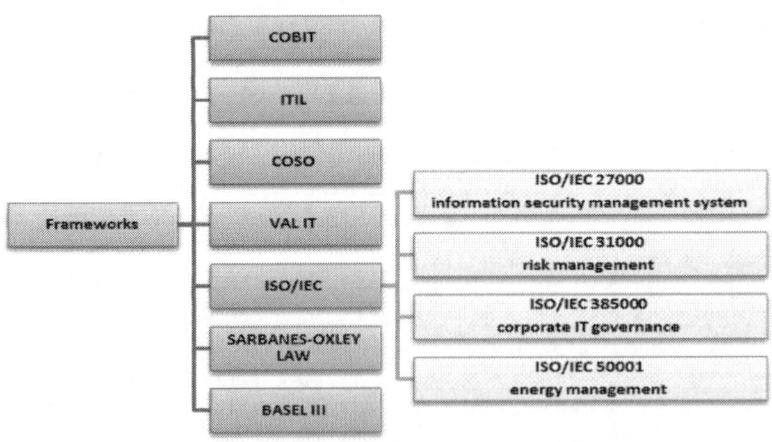

Figure 1. The most used audit frameworks and standards (author)

COBIT (*Control Objectives for Information and Related Technologies*) the most popular framework maintained by Information Systems Audit and Control Association (http://www.isaca.org) originating from IT Governance Institute (http://www.isaca.org/itgi/Pages/default.aspx) aiming to assist the governance of IT (and information systems). It was published for the first time in 1996 and actual version is number 5 published in 2012, applicable to all kind of companies and institutions [6] COBIT5 defines 7 *enablers*, with more than 300 IT controls grouped in 34 control objectives and 37 processes and 5 domains [7] These five domains of COBIT 5 are following: *Evaluate, Direct and Monitor* – EDM, *Align, Plan and Organise* – APO, *Deliver, Service and Support* – DSS, *Monitor, Evaluate and Assess* – MEA, and *Build, Acquire and Implement* – BAI.

ITIL (*IT Infrastructure Library*) provides a framework of best-practice guidance for IT service management, aiming to harmonise IT services with business needs for facilitating decision making, changes, transformation and growth [8]. The author was *Central Computer and Telecommunications Agency* - CCTA in late 1980-ies (since 2000 changed the name in *UK Office of Government Commerce* - OGC). The first version of ITIL was named GITIM (Government Information Technology Infrastructure Management); the second version was published in 2001 and last and actual version number three was published in 2007 [9]. ITIL is based on top-down approach, organised around the life cycle of services and in five processes described in five volumes [10]: *ITIL Service Strategy*, *ITIL Service Design*, *ITIL Service Transition*, *ITIL Service Operation*, and *ITIL Continual Service Improvement*. Today, this standard is maintained by AXELOS (https://www.axelos.com)

There is quite a number of ISO/IEC standards that can be related to the auditing of the information systems, but most important is ISO 27000 family of standards, so-called Information Security Management System – ISMS. The definitions and vocabulary of ISMS are given in ISO 2700:2014 (third version published in 2014). The review of ISO 27000 is presented in Table 1 below.

Val IT framework was published by IT Governance Institute (ITGI) as "*Enterprise Value: Governance of IT Investments, The Val IT Framework 2.0*" and covers three main domains: *Value Governance* – VG, *Portfolio Management* – PM and *Investment Management* - IM [12]

The main aim of Val IT is to assist the company to optimise realisation of values from investments in IT. It includes set of basic principles and number of processes that are defined with the set of key management practices. Val IT was created as a complement to COBIT from the business and financial perspective aiming to improve values coming from investments in IT.

Table 1. Review of ISO/IEC 27000 family of standards (from [11])

Standard's code	Title of the standard
ISO/IEC 27000	*Information security management systems — Overview and vocabulary*
ISO/IEC 27001	*Information security management systems — Requirements*
ISO/IEC 27002	*Code of practice for information security controls*
ISO/IEC 27003	*Information security management system implementation guidance*
ISO/IEC 27004	*Information security management — Measurement*
ISO/IEC 27005	*Information security risk management*
ISO/IEC 27006	*Requirements for bodies providing audit and certification of information security management systems*
ISO/IEC 27007	*Guidelines for information security management systems auditing*
ISO/IEC TR 27008	*Guidelines for auditors on information security controls*
ISO/IEC 27010	*Information security management for inter-sector and inter-organizational communications*
ISO/IEC 27011	*Information security management guidelines for telecommunications organisations based on ISO/IEC 27002*
ISO/IEC 27013	*Guidance on the integrated implementation of ISO/IEC 27001 and ISO/IEC 20000-1*
ISO/IEC 27014	*Governance of information security*
ISO/IEC TR 27015	*Information security management guidelines for financial services*
ISO/IEC TR 27016	*Information security management — Organisational economics*

During 1985, the USA accounting and financial associations gathered in alliance named *Committee of Sponsoring Organisations of the Treadway Commission* – COSO with the aim to fund public-private initiative named *National Commission on Fraudulent Financial Reporting* [13]. COSO framework claims that the internal control consists of five interconnected element of which for the audit of information systems the most important is the fourth one [14]: Control Environment, Risk Evaluation, Control Activities, Information and Communication and Monitoring.

Mentioned Sarbanes-Oxley Law and BASEL II and BASEL III recommendations are mostly used in the banking sector. The selection of the framework and the standards shall depend on the depth and domain of the audit, type of institution or organisation and on available funds for performing the audit.

2.2. Economic feasibility

The economic feasibility in auditing of information has aimed to examine the economic effects (positive and negative) of the auditing process without knowing or prejudicing the results of the audit. It is very important to find out the effects and consequences of the performed audit on the client. Namely, economic feasibility should offer a number of scenarios to decision makers on what to do after the audit? This aspect of the feasibility study is the most interesting for organisation's top management and may influence their decision to go into the audit or not.

The auditors, in the preparation phase of the audit and before concluding the final version of the contract, should clearly present the costs against the benefits and volume of the audit. This shall assist decision makers to have a clear picture and motivation to perform the audit. **The main questions that economic feasibility should examine are**:

- What is the level of costs for the auditing and what will be the effects of the audit? Or, what will we get with the auditing and is it beneficial to invest in the auditing?
- What are the indicators and how to make a cost-benefit analysis? How to compare the auditing costs with the corresponding effects achieved with such investment?

To be able to find answers, the economic feasibility should take into consideration following, but not exhaustive, a list of possible auditing costs:

- **Costs of external auditors service** based on to be agreed volume and domain of the audit,
- **Costs that may occur in use of resources** during the auditing process,
- **Costs that may rise in order to train the employees** for facilitation of the audit (if it is needed),
- **Costs of immediate and urgent corrections and interventions** within the system for instant improvement, and
- **Costs related to resolving other (less urgent) audit findings**.

The effects of the audit project can be quantitative or qualitative, but in general, can be divided into two categories:

- **Effects that can be measured** (the indicators are quantified and it is possible to quantify them with the numbers or trends – number of employees, decrease costs, increase of number of services, etc.), and
- **Effect impossible to measure** (the indicators are difficult to quantify or impossible to determine and to assign the number or a value; often they will not be visible immediately upon the implementation of the corrective measures but anytime in lifecycle).

For example, the audit will find that the work of an employee in a certain domain job can be automated within the system and providing the same quality of services using electronic communication between the system and clients. Savings can be calculated on the basis of the hourly rate and multiply it by the number of working hours. The employee works classic 8 work hours per day (or 40 hours working week) which is 168 hours per month. Using the automatized system, the system can improve and extend the working time and provision of services during the whole 24 hours per day, including non-working days. This means that these services will be available not 168 hours per month, but 720 hours monthly. The effective service provision is increased by more than 4 times. Multiplying with an hourly rate, it is easy to find out the direct economic effects.

But there are also some indirect effects, such as increased clients' satisfaction with services offered, operative working time is increased, increased use of services means increased the number of transactions, etc.

Also, the economic feasibility usually uses one or combination of the following methodologies:

- Cost-benefit analysis,
- Payback analysis,
- Return on investment, and
- Net present value.

2.3. Legislative feasibility

Legislative feasibility is an important aspect of TELOS. This aspect defines a legal requirement for introduction or implementation of the information system audit. The legal feasibility should provide answers on the legal basis for the audit. In order to obtain such information, the legal feasibility should provide answers to the following questions:

- Is the information system audit obligatory or voluntary?
- What is the frequency of conducting the audit – annually, bi-annually etc.?
- Is the information system audit independent or done together with the internal audit?
- Are there strict legal rules or recommendations imposed or recommended for conducting the audit?
- Are there legal consequences if the audit is not performed and how it can influence security and safety of the information system?

For example, U.S. Government Accountability Office recommends FISCAM methodology described in Federal Information System Controls Audit Manual [15]. The manual provides a methodology for assessing the effectiveness of information system controls and provides the steps in conducting information system audit. The methodology is mandatory and both auditor and audit organisations must comply with it.

International Organisation of Supreme Audit Institutions (INTOSAI) developed own standards with the set of auditing standards based on:

- ISSAI 1 - The Lima Declaration of Guidelines on Auditing Precepts (principle of independence)
- ISSAI 100-400 - INTOSAI Auditing standards (principles of auditor's work) and
- ISSAI 30 - Ethic codex (values and principles of the auditor).

These are INTOSAI key documents that should be considered in planning and conducting the audit of information systems in the public administration.

The importance of information system audit is a necessity in any public administration and should be mandatory. The Supreme Office for the Republic of Srpska Public Sector Auditing in its "Strategy of the Supreme Office for the Republic of Srpska Public Sector Auditing in period 2014-2020" stated the need for intensification of activities related to the information systems audit due to the significant and fast development of ICT [16].

The audit becomes even more important in the case of the decentralised information system of the public administration where the use of various unsecured or partially secured communication channels can jeopardise the work of the system.

Often, due to lack of qualified and skilled auditors, the information system audit is conducted as the part of the internal or financial audit. Therefore, it is necessary to educate and train at least minimal number of information system auditors to secure the quality of the conducted audit and reliable audit report findings. Another solution is contracting of auditors with adequate certification, such as Certified Information System Auditor – CISA. This certification is managed by the Information System and Control Association – ISACA.

2.4. Organisational feasibility

Each public administration institution has strict hierarchical structure. The example of the Ministry of the Government is presented in Figure 2. The top authority is the Minister and as such this position has most of authority and permissions. Depending on complexity and jurisdiction of the Ministry, it can have one or several Deputy Ministers that are in charge for one or number of departments. Each department has Head in charge and civil servants or employees.

There can be other relevant positions (such as Secretary General that can be in the same line as Deputies) in the Ministry, but this is the most common case of an organisation. Although the structure of the organisation is not a key factor for the successful information system audit, such centralised and structured organisation is less complex and enables easier follow up of the reporting and division of authorities. This facilitates check of the control points in the audit process and making conclusion where to intervene for corrective measures, an eventual increase of system efficiency or other audit purposes.

The precise definition of workflows, levels of authorities, division of responsibilities etc. is crucial in the phase of information system development. But at the same time, it is very important for the audit purposes. In that case, the auditor can easily to follow the work process, check the inconsistencies or to search for problems in security and safety.

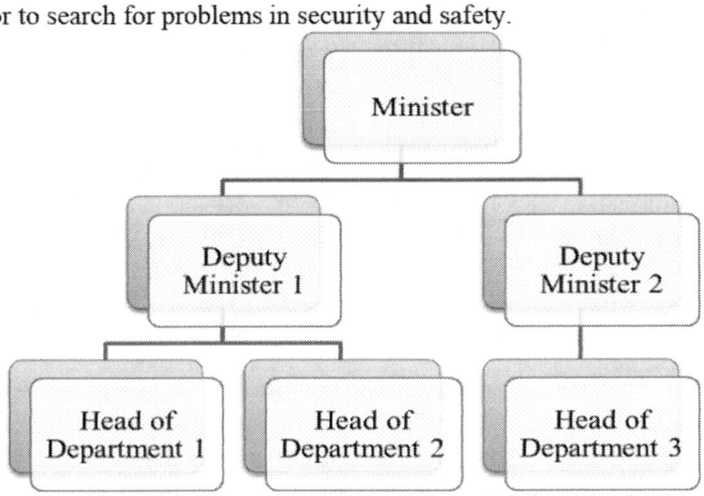

Figure 2. Common structure of the Ministry (author)

2.5. Schedule feasibility

Although it may be an important factor in assessing the overall performance of the feasibility and implementation of information systems, the schedule feasibility is also important from the aspect of information systems auditing.

There are some calculations on the time required for the implementation of each stage of an audit of information systems as shown in Table 2.

Table 2. Audit time consumption per individual phase (from [17])

The phase of the information system audit	Percentage of the total audit time
Preparation and planning	10
Analysis and documentation	10
Collecting audit evidence	25
Analysis and evaluation of the audit evidence	20
Preparation of the audit report	20
Presentation of the audit report	5
Post audit activities	10

Table 2 shows that 45% of the time (planned for the audit) should be invested in data collection and analysis (almost half of the total time), for making the presentation of the report 25% of the time, and 30% for all other audit activities. This means that it is necessary to align all stakeholders in the revision of information systems and to make available the necessary resources.

It is very important to determine when (period, time frame) the core activities of the audit will be conducted - data collection and analysis on the spot. The business processes in the institution during this period will be significantly affected and slowed down and therefore it is necessary to organise the audit in the period when it will minimise the impact of auditors' work on the business. The field work of auditor should be well elaborated and agreed in preparation and audit planning among parties in order to have the desired resources and personnel available.

This will include also the consideration of periods burdened with state or religious holidays, during vacations or in circumstances when an institution is overloaded with the business activities (eg. the organisation and conduct periodic public procurement, immediately prior to preparation of final financial documents at the end of the fiscal year, etc.).

If necessary, the time of the audit can be further extended and for the collection of data with more quality. It is possible that some functionalities of the information system are not in operation at the time of the audit, which prevents quality analysis of all necessary controls and given the possibility of latent system fault in the future. Therefore, it is sometimes necessary to simulate this functionality (if there is no real need for the work at the audit time) in order for the auditor to gain full access to the necessary controls.

3. Conclusion

Each audit of public administration institutions requires good planning and preparation. In order to prepare quality information systems audit, one of the proposed models to do it is using the TELOS methodology.

TELOS methodology recommends the creation of the feasibility study including the five main aspects of the feasibility: technical, economic, legal, organisational and schedule feasibilities.

This paper presented and discussed main questions related to using of TELOS methodology in planning information system audit in the public administration institutions. Moreover, the paper discusses positive effects implementing TELOS methodology.

The objective of the auditor is to highlight the problems of the information system functioning, its deficiencies and potential opportunities for improvement. Therefore, the institution goal should be to implement a quality audit in order to obtain more realistic results and thus managed to prevent potential damage and business losses due to improper functioning of information systems and related business processes and procedures.

In order to obtain real indicators and findings, both parties (auditor and audited institution) should work in close cooperation and with the joint aim to improve the functioning of the overall system. Therefore, it is crucial to consider an auditor more as external assistant and facilitator in the improvement process.

TELOS methodology covers important parts of the complex feasibility study that can in advance point on possible problems for the audit process. Simplicity, ease of use and adaptability to different systems are the main characteristics of this methodology.

References

[1] Justis R T and Kreigsmann B 1979 The feasibility study as a tool for venture analysis, *Business Journal of Small Business Management* **17**(1) 35-42

[2] Avison D and Fitzgerald G 2006 *Information Systems Development,* 4th Edition, McGraw-Hill Education, USA

[3] Hall J A 2011 *Information Technology Auditing and Assurance,* 3rd Edition, South-Western Cengage Learning, USA, pp 187-188

[4] Stanislav R P 2013 Informaciona bezbednost i primena CAATT alata u IT reviziji, *Pravno-*

IOP Conf. Series: Materials Science and Engineering **294** (2017) 012022 doi:10.1088/1757-899X/294/1/012022

ekonomski pogledi **2**

[5] *** Portal AuditNet 2017 https://www.auditnet.org/audit-library/computer-assisted-audit-tools-and-techniques-caatt (accessed on 12.1.2017)

[6] ***ISACA 2016 *COBIT 20th Anniversary* available at http://www.isaca.org/COBIT/Pages/COBIT-20th-Anniversary.aspx#years (accessed on 5.8.2016)

[7] ***ISACA 2012 *COBIT 5 - A Business Framework for the Governance and Management of Enterprise* IT Rolling Meadows, IL, USA

[8] ***Axelos 2016 *What is ITIL® Best practices?* available at https://www.axelos.com/best-practice-solutions/itil/what-is-itil (accessed on 1.8.2016)

[9] ***ITIL 2016 *Central History of ITIL* available at http://itsm.fwtk.org/History.htm (accessed on 1.8.2016)

[10] Arraj V 2013 *ITIL®: The basics (White paper)* The APM Group and The Stationery Office

[11] ***ISO/IEC 2008 *International standard ISO/IEC 38500:2008 (1st edition)* International Organization for Standardization Geneva Switzerland

[12] ***ITGI 2011 *ENTERPRISEVALUE: GOVERNANCE OF IT INVESTMENTS The Val IT Framework 2.0* IT Governance Institute, USA

[13] ***COSO 2016 *About us* available at http://www.coso.org/aboutus.htm, (accessed on 2.8.2016)

[14] ***Ministarstvo finansija Crne Gore 2011 *Priručnik za finansijsko upravljanje i kontrole* Podgorica (available at http://www.mf.gov.me/pretraga/107135/Prirucnik-za-finansijsko-upravljanje-i-kontrole.html accessed on 2.8.2016)

[15] ***United States Government Accountability Office 2009 *FEDERAL INFORMATION SYSTEM CONTROLS AUDIT MANUAL (FISCAM) GAO-09-232G*, Washington DC, USA (available at http://www.gao.gov/assets/80/77142.pdf accessed on 31.3.2017)

[16] ***Glavna služba za reviziju javnog sektora Republike Srpske 2013 *Strategija razvoja Glavne službe za reviziju u periodu od 2014. do 2020. godine* Banja Luka, BiH

[17] Varga M and Varga V 2011 Usporedba rezultata revizije informacijskih sustava provedenih prema CobiT okviru i uvod u CobiT 5 okvir, *Tehnički glasnik* **5**(1) 35-43

Hydroelectric power plant with variable flow on drinking water adduction

S I Deaconu[1], R Babău[2], G N Popa[1] and P L Gherman[1]

[1]Politehnica University of Timisoara, Department of Electrical Engineering and Industrial Informatics, Revolutiei street, Hunedoara, Romania
[2]Beespeed Automation Ltd Timişoara, România

E-mail: sorin.deaconu@fih.upt.ro

Abstract. The water feeding system of the urban and rural localities is mainly collected with feed pipes which can have different lengths and different levels. Before using, water must be treated. Since the treatment take place in the tanks, the pressure in the inlet of the station must be diminished. Many times the pressure must be reduced with 5- 15 Barr and this is possible using valves, cavils, and so on. The flow capacity of the water consumption is highly fluctuating during one day, depending on the season, etc. This paper presents a method to use the hydroelectric potential of the feed pipes using a hydraulic turbine instead of the classical methods for decreasing the pressure. To avoid the dissipation of water and a good behavior of the power parameters it is used an asynchronous generator (AG) which is coupled at the electrical distribution network through a static frequency converter (SFC). The turbine has a simple structure without the classical devices (used to regulate the turbine blades). The speed of rotation is variable, depending on the necessary flow capacity in the outlet of the treatment station. The most important element of the automation is the static frequency converter (SFC) which allows speeds between 0 and 1.5 of the rated speed of rotation and the flow capacity varies accordingly with it.

1. Introduction

Renewable sources will occupy an important share of energy production over the next decade.

The conversion of hydraulic energy into electricity is not polluting, involves relatively small maintenance costs, there are no problems with fuel and is a long-term solution [1-3].

The first hydroelectric power plant in Romania, Sadu I, was commissioned in 1896, having a vertical shaft turbine which was replaced in 1905 by a Francis turbine that operated until 1929 [4].

In setting up the hydropower micro-potential, both positive and negative effects on the environment must be taken into account.

Hydrotechnical constructions are non-polluting because they do not interact with the environment. On the contrary, some constructions are destined to water depollution, others increase natural self-purification capacity and others serve to dilute or annihilate some pollutants or substances.

Small hydropower plants placed on drinking water pipelines must comply with all regulations regarding pollution, and the conditions for choosing hydro-units are very strict [3], [5], [6].

The paper presents the stages of designing such plants starting from the hydraulic profile (Figures 1 and 2) and the average flows, taking into account the perspective of the increase in consumption [7], [8].

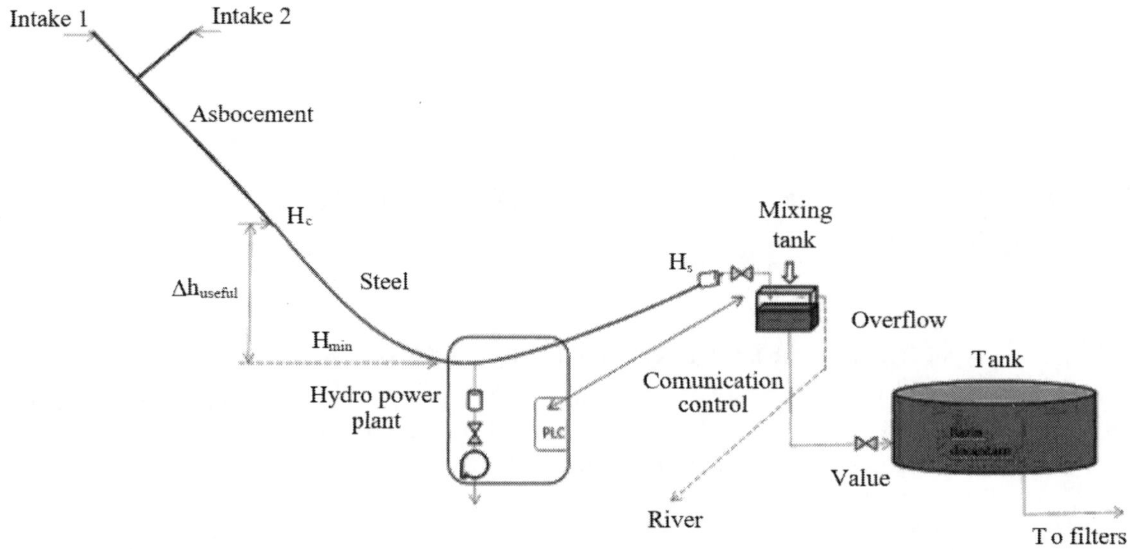

Figure 1. Hydraulic diagram of the hydroelectric power plant with oscillation of level

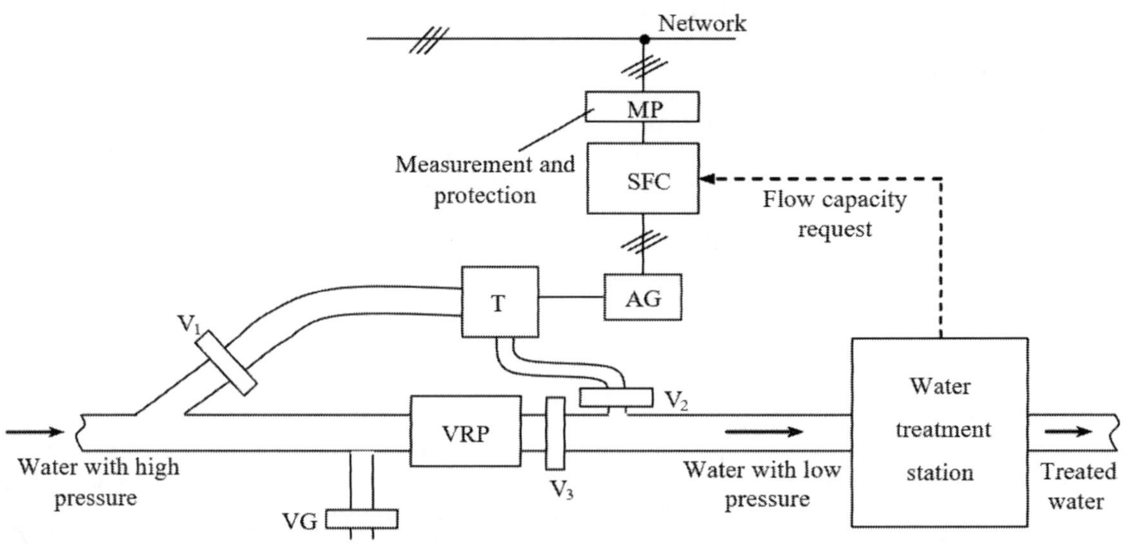

Figure 2. Hydraulic diagram of the hydroelectric power plant without oscillation of level

2. Microhydro power plant design

The hydraulic profile of the feed is shown in Figure 3 where the portion A-B is made of metal pipe and the portion B-C of the asbocement. The height difference between the capture A and the pool C of the mixture at the entrance to the treatment plant is $\Delta H_{geodezic}$= 103 m [8].

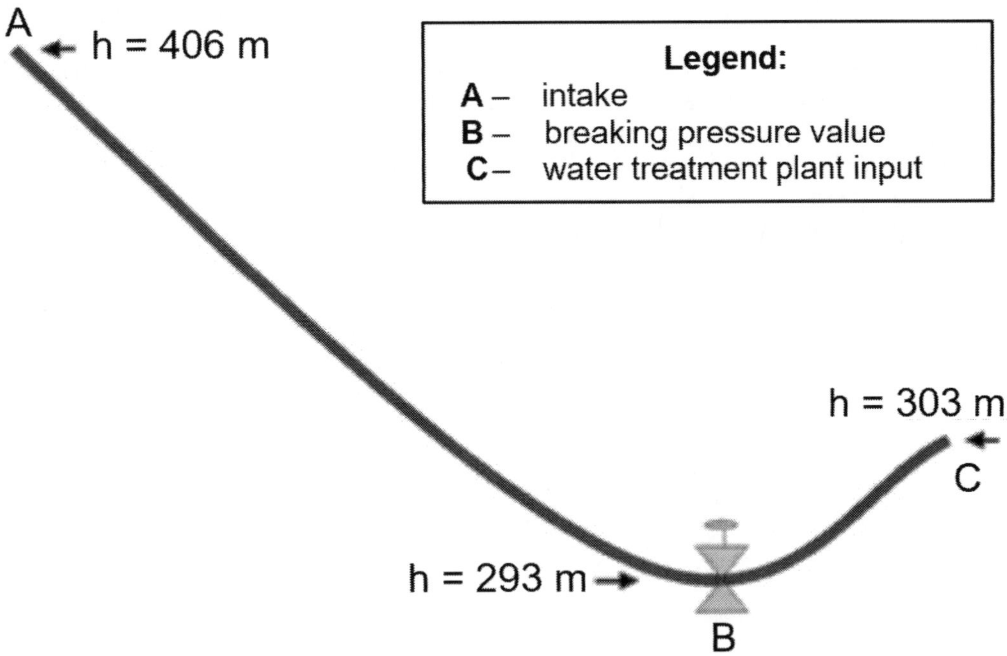

Figure 3. Hydraulic profile of crude water supply

Location of the boiler will be at point B of the minimum supply level. Currently the inlet pressure of 9.5 - 10 bar is reduced to 0.5 - 1 bar using a butterfly valve. Output pressure is required to maintain the level in the mixing tank. At a pressure difference of 8.5 bar and a flow rate of 150-210 l / s, the relationship (1) results:

$$P = \frac{Q \times h}{367}[KW], \tag{1}$$

a power dissipated on a 90-140 kW valve (proportional to flow). Table 1 gives the statistical data recorded over the period 2009-2016, indicating an average capture flow rate in the range 150-270 l / s, variable (daily or seasonal) depending on the needs in the drinking water distribution network. Based on the expansion of the number of consumers in the future the required gross water flow will reach 350-400 l / s.

Table 1. Statistical data

Flow / Year	Q_{min} [l/s]	$Q_{average}$ [l/s]	Q_{max} [l/s]
2009	165	188	206
2010	148	177	210
2011	155	180	201
2012	152	179	205
2013	160	185	215
2014	165	204	240
2015	182	216	252
2016	195	234	270

The proposed hydraulic functional scheme (which also includes the intervention pumping station) is shown in Figure 4 [8].

Taking into account the need for a variable flow rate (150-400 l / s) for the water treatment plant, a water drop of 85 m (assumed constant) requires the use of turbines that allow continuous flow control, without sensitive decrease of conversion efficiency on the required flowrate range. In this respect, it is proposed to permanently adjust the speed of hydro-aggregates so that they can generate the flow transmitted by the transducer placed at the exit of the treatment plant. In this way the reserve of drinking water in the storage tanks is always kept.

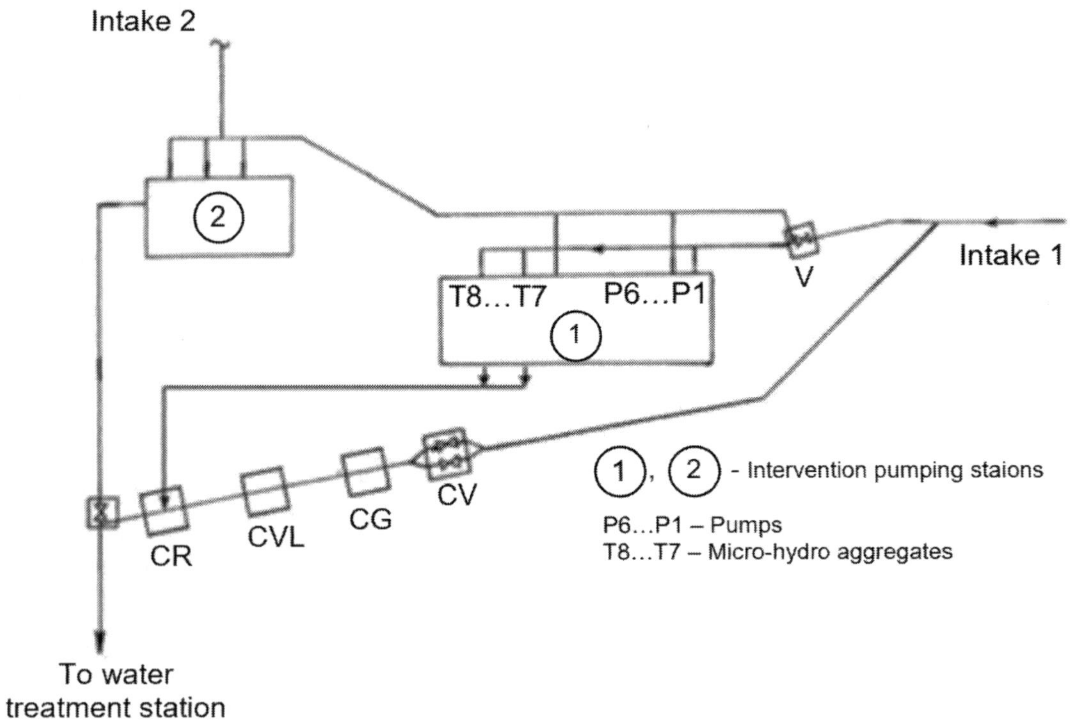

Figure 4. Functional situation with the use of micro-hydro aggregates

The generators will be asynchronous cage rotor with high efficiency. Hydraulic cutting accessories will be electrically controlled. When a turbine fault occurs, it will operate on the existing circuit without endangering the availability of the water resource necessary for the normal operation of the treatment plant [8].

All cutting elements (electrically controlled valves) will be provided with UPS (uninterruptible power supply) so that the return to the classic hydraulic circuit can also be achieved when the supply voltage drops. The switching of the flow between the two schemes must avoid hydraulic shocks (rake shot) [8].

The SCADA system (Figure 5) must ensure: controlled start-up and shutdown of the generators, generators tripping, no load and load of the generators, measure and protection of the generators, controlled output at loss of the grid, avoidance hydraulic shocks, communication with the hierarchically superior dispatcher or PLC, monitoring, controlling and controlling all dispatchable items, making real-time energy calculations for operating in an optimal area, providing screens as a communication interface with the human operator, automatic maintenance of imposed climatic parameters (temperature, ventilation) in various station compartments [8].

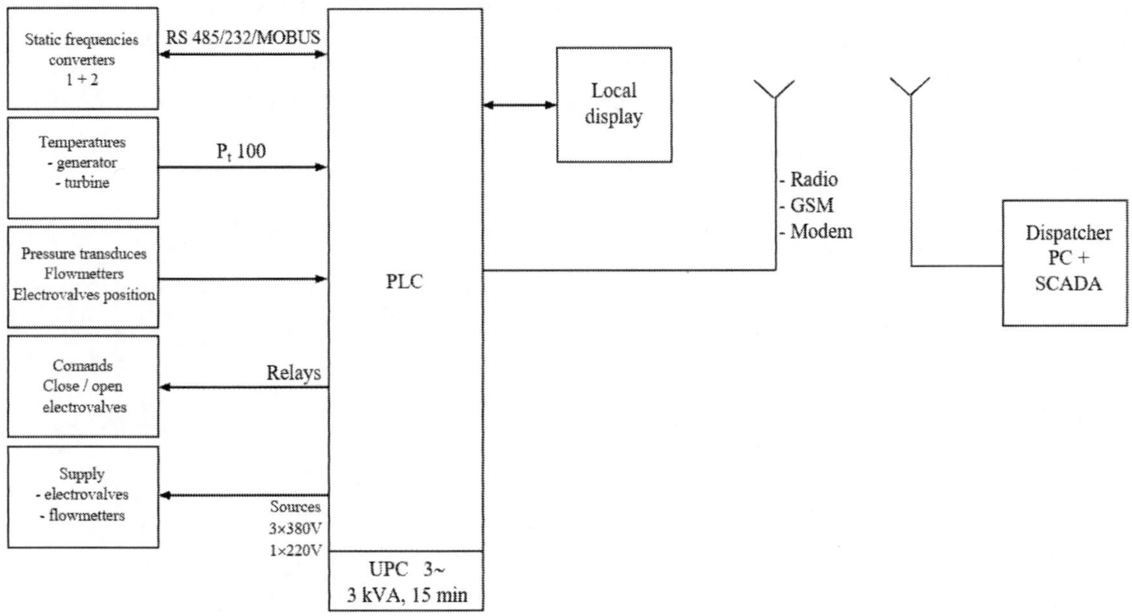

Figure 5. Block diagram of the PLC and SCADA system at micro-hydro power plant with variable speed

The yield curves are shown in Figure 6 (220 l / s and 350 l / s), and the turbine upstream pressure calculation calculates the value of 0.07 bar [7].

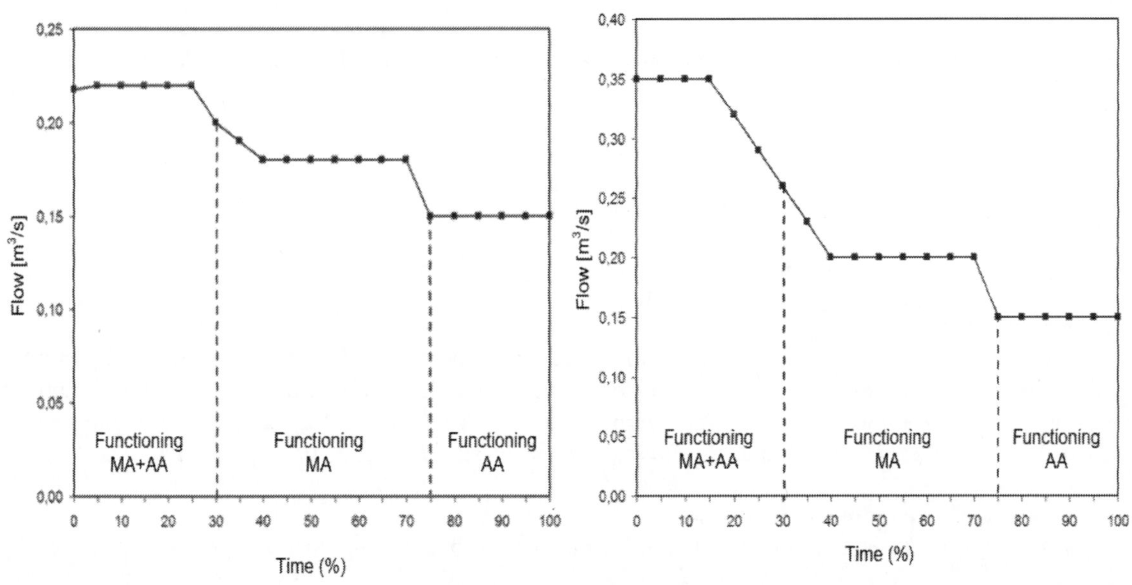

Figure 6. Adduction flow curve

Two generators, with 200 kVA power (called the main MA) and 132 kVA (called auxiliary unit AA) and the associated turbines were selected. For reasons related to the pressure drop and the limited physical space allocated, it is necessary to use irregular Francis turbines (without directing apparatus), characterized by reduced radial dimensions, the turbine speed being achieved by electric means.

Since the application requires that variable flow (prescribed online by the user) be obtained by speed adjustment, Figure 7 shows the variations in power and flow rate at constant pressure drop and different speeds [8].

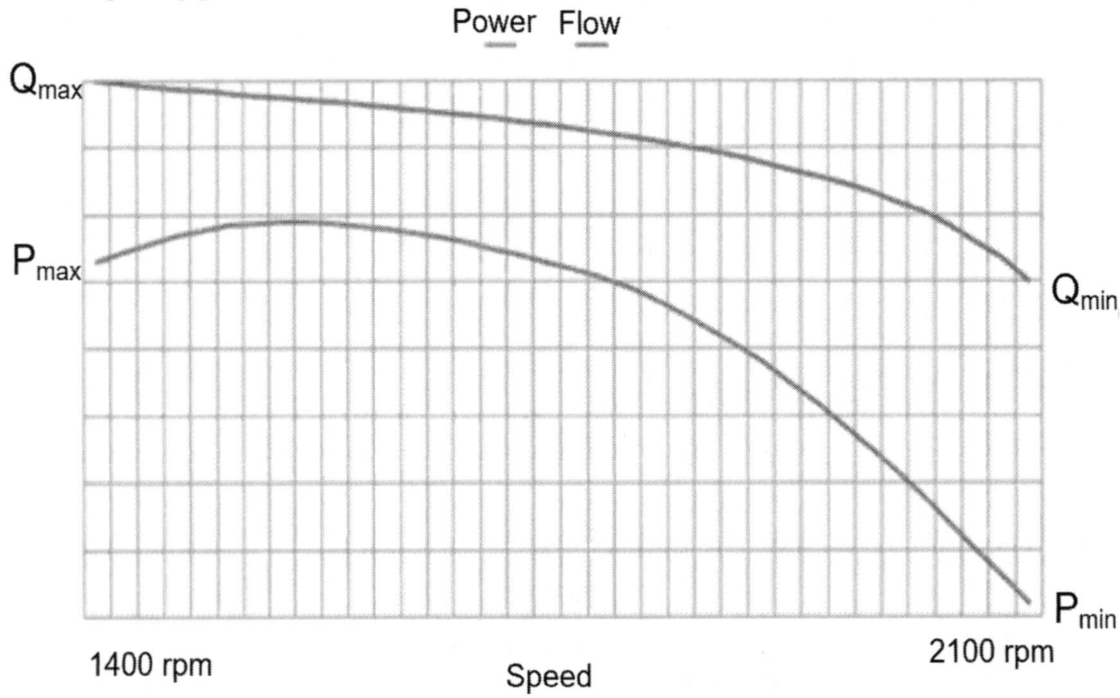

Figure 7. Variation of power and flow depending on speed, at constant drop

The failure flow rate, when dropping the load, is

$$Q_r = 0{,}7Q_n \ , \tag{2}$$

and the crash speed

$$n_r = 0{,}7n_N \ , \tag{3}$$

for the chosen turbines.

To prevent the occurrence of the hydraulic shock, it is necessary to handle the hydraulic cutting elements in a time greater than T = 9.4 s calculated for the hydraulic network characteristics. Total drainage times were chosen for 150 s (2.5 min). An overpressure valve mounted on the main duct or in the suction manifold is to be opened from 100 mca to take the calculated maximum pressure (126 mca).

3. Technical and economic calculation

With the help of a specialized program [7] in the calculation of micro-hydropower plants, based on the data previously calculated and the input quantities, the data on the value of the investment, the production of the electricity obtained and the duration of the investment recovery are obtained. The simulations have been made for the average flow of the last 3 years of 220 l / s and for the expected average flow rate of 350 l / s in the future. The results are presented in a comparative manner [7].

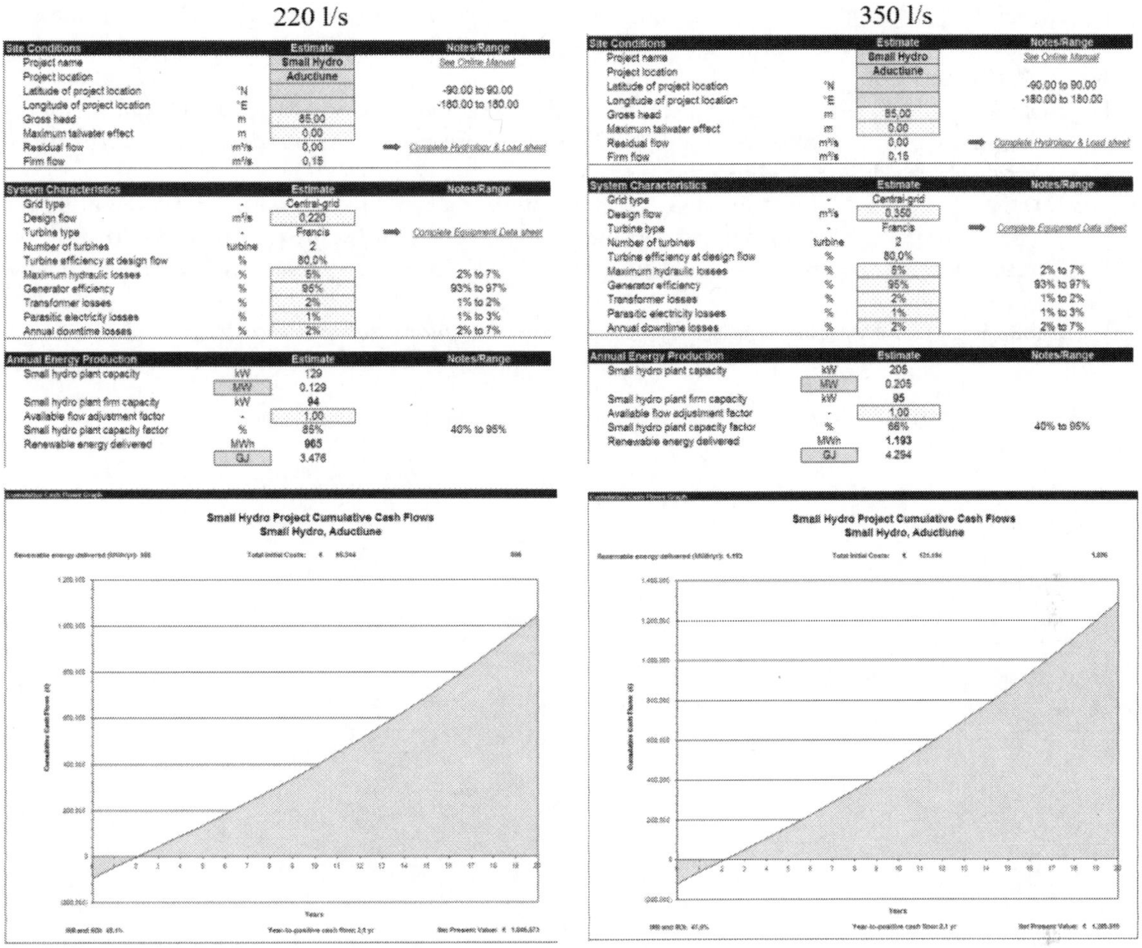

Figure 8. Calculation of micro-hydropower plants

4. Conclusions

Variable speed MHC operation implies significant advantages: there is no need for a constant speed maintenance system that involves a turbine stator directing device, the flow rate of the turbine adjusts the required flow rate, replaces the pressure reducing valve and the flow adjustment with a hydro-filling unit. From the economic calculations it is noted that due to the high pressure drop and the circulation flow, the duration of recovery of the investment of 2 years is very low, so the investment deserves to be done.

References

[1] Bolik M S 2004 *Modeling and Analysis of Variable Speed Wind Turbines with Induction Generator during Grid Fault*, Institute of Energy Technology, Aalborg University, Denmark, Doctoral Thesis

[2] Fadaeinedjad R, Moallem M and Moschopoulos G 2008 Simulation of a Wind Turbine with Doubly Fed Induction Generator by Fast and Simulink, *IEEE Transactions on Energy Conversion* **23**(2) 690-700

[3] Luu T, Abedini A and Nasiri A 2008 Power Smoothing of Doubly Fed Induction Generator Wind Turbines, *IEEE Transactions on Energy Conversion* **23**(4) 2365-2370

[4] Jayanti N G, Malabika B, Gaughan K and Conlon M F 2008 A new configuration and control of doubly fed induction generator (UPQC - WG), *IEEE Transactions on Energy Conversion* **23**(3) 2094-2099

[5] Luna A, Lima F K A, Rodriguez P, Watanabe E H and Teodorescu R 2008 Comparison of Power Control Strategies for DFIG Wind Turbines, *IEEE Transactions on Energy Conversion* **23**(3) 2131-2136

[6] Deaconu S I, Topor M and Popa G N 2009 Variable speed wind-diesel or hydro-diesel power generation systems for insulated consumers, *Automation and Instrumentation* **5-6** 18-19

[7] ****RET Screen Energy Model – Small Hydro Project,* Minister of Natural Resources Canada, 1997-2004

[8] Babău R 2017 *Technical project "Hydroelectric power plant station Petresti"*, no. 12

Particular mechanism for continuously varying the compression ratio for an internal combustion engine

S Rațiu[1], R Cătălinoiu[2], V Alexa[1], I Miklos[1] and V Cioată[1]

[1]University Politehnica Timisoara, Faculty of Engineering Hunedoara, Department of Engineering and Management, 5 Revolution Street, 331128 Hunedoara, Romania,
[2]Coramex S.R.L., Renault Service, Deva, Romania

E–mail: sorin.ratiu@fih.upt.ro

Abstract. Variable compression ratio (VCR) is a technology to adjust the compression ratio of an internal combustion engine while the engine is in operation. The paper proposes the presentation of a particular mechanism allowing the position of the top dead centre to be changed, while the position of the bottom dead centre remains fixed. The kinematics of the mechanism is studied and its trajectories are graphically represented for different positions of operation.

1. Why continuously variable compression ratio for an internal combustion engine?

An important feature of car engines is the wide variety of operating conditions. Among these, the partial load regimes play an essential role because the operation of these regimes is crucial for the engine's economy. Unfortunately, in these operating conditions the thermal efficiency is much lower compared to the one recorded at full load. In conclusion, constructive solutions that work especially in the area of these regimes should be sought, in order to determine an increase in thermal efficiency.

In the case of spark ignition engines, the quantitative load control by closing the intake channel determines not only pumping losses but also the reduction of the real compression ratio (ε_p). Because of the fixed (invariant in operation) value of the geometric compression ratio (ε_v), the lower mass of fresh load resulting from the decrease in the load level will occupy the same volume at the end of the compression stroke, so that the mixture will be more less compressed than the situation recorded at full load.

This is the main disadvantage of spark ignition engines. As a result, it becomes economical to increase the geometric compression ratio (ε_v) with the decrease of the load level, in this way the value or the real compression ratio (ε_p) being restored. In this direction, the variable compression ratio engine could be a very good alternative and its implementation is perfectly valid, taking into account existing technologies [1-3].

At the present there is a wide variety of technical solutions for obtaining the variable compression ratio that are known and presented in the literature. The present article aims to study the kinematic of a particular mechanism, conceived and designed by the authors, which allows continuous modification of the compression ratio by interposing a planar mechanism between the connecting rod and the crankshaft of the classic engine mechanism.

The value of the geometric compression ratio can be changed either by the volume of the combustion chamber or by the total volume of the cylinder, hence the displacement. The mechanism proposed for the study changes the volume of the combustion chamber by modification the position of the top dead centre (TDC), while the position of the bottom dead centre (BDC) remains fixed.

2. Analysis of plane mechanisms

Analysis of plane mechanisms has evolved over time. Thus in practices are known several methods of analysis of plane mechanisms (positional, kinematic, kinetostatic or dynamic), respectively the graphical methods, graphoanalitics and analytical each of them having advantages and disadvantages.

Graphics and graphoanalitics methods (vector equations method, projected instantaneous centre of rotation, similarity, respectively polygon method forces, etc.), have the advantage that are very easily to use, requires a relative low workload, but the accuracy of results is not always the one, because the measurement error in the graphic plane.

Analytical methods (polygonal contour method) involve writing equations contour projections and their successively derivation, respectively the equations of equilibrium, resulting linear equations systems who need to be solved for as many positions of leadership element. The workload is very high, requiring solving equation systems using programs written in different programming languages, but the accuracy of results is high. This method has the disadvantage that requires knowledge of a programming language.

With the development of CAD software (Computer Aided Design) software companies have developed software packages specific field of engineering. Thus, for modelling and simulating mechanical systems are known several applications such as Mecaplan, Simulation Mechanical, Adams, SAM, Watts & Roberts, etc. some of them doing calculations by finite element method [4]. Further in the following paragraphs will be present the modelling, analysis and simulation of mechanism with SAM 7.1 software.

3. SAM 7.1. software

SAM (Synthesis and Analysis of Mechanisms) is an interactive PC-software package for the design, analysis (motion and force) and optimization of arbitrary planar mechanisms. Mechanisms can either be generated via the design wizards or they can be assembled from basic components including beams, sliders, gears, belts, springs, dampers and friction elements. SAM integrates pre-processing, numerical analysis and post processing, such as animation and xy-plots, in an easy to-use environment offering pull-down menus, mouse support and help facilities.

The mathematical foundation of the analysis kernel, which is inspired by the well-known finite element approach, offers a large number of features and overcomes many of the problems of traditional mechanism programs. Open loop, closed loop, multiple loop and even complex planetary mechanisms can equally well be analyzed due to the finite element formulation. Even the most complex mechanisms, including planetary gear trains, can be modelled within minutes [5].

4. Presentation of the constructive solution of the studied mechanism

The new concept of the mechanism proposed by the authors implies the existence of some kinematic bonds between the connecting rod and the crankshaft, which allow continuous modification of the compression ratio of the engine (Figure 1). The mechanism is a quadrilateral plan and consists of the AM, MB and O_2B elements and the corresponding joints. AP is the connecting rod and O_1M the crankshaft arm.

Modification of the compression ratio is made by the modification by an actuator (mechanical or hydraulic) of the vertical position of the O_2 joint marked in figure 1. This causes the TDC position to change, while the position of the BDC remains fixed. The lengths of the articulated elements of the mechanism were determined by successive tests and correlated with the basic dimensions of the engine mechanism.

The kinematic analysis was performed in order to highlight the trajectory of the head of the connecting rod (point A in Figure 1) and implicitly of the kinematic bonds. Functional variants have been studied by modifying the horizontal and vertical distances between the O_2 and O_1 joints.

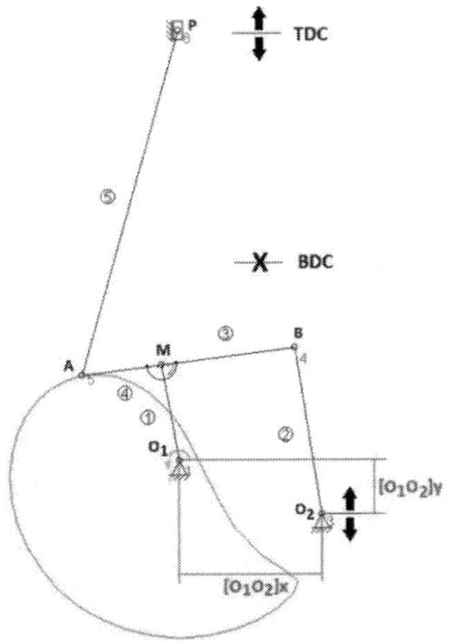

Figure 1. The mechanism

5. The cinematic study with the SAM 7.1 software

Using the constructive dimensions obtained from a preliminary analysis, the kinematic connections of the mechanism were designed through the SAM 7.1 software, simulating the movement of the mechanism and connecting rod trajectory.

The process of designing mechanisms basically can be divided into two distinct phases, namely: synthesis and analysis. After a proper specification of the demands, the first step in the design cycle consists of the synthesis phase, in which the designer attempts to find the type of mechanism and its dimensions, such that the requirements are met (as good as possible). Experience, previous designs, mechanism handbooks and the design wizards implemented in SAM can guide this creative process. Once a mechanism has been chosen, its motion and force behaviour can be analysed [5].

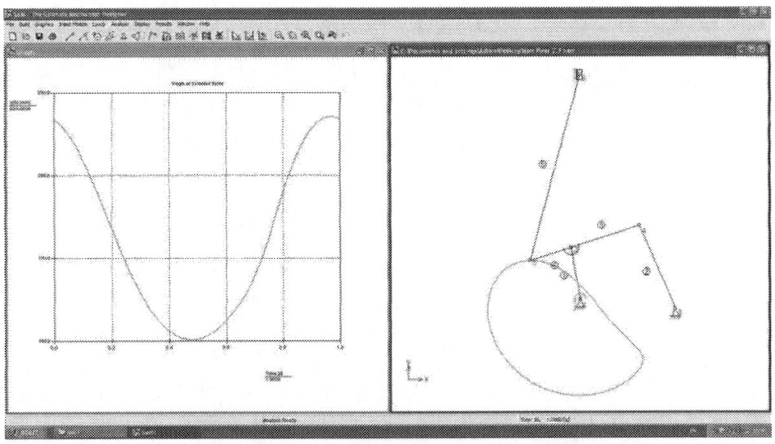

Figure 2. SAM software interface

Next we have the results of the kinematic analysis for the case where the horizontal O_1O_2 distance ($[O_1O_2]$ x - Figure 1) is 80 mm (figure 3) and 100 mm (Figure 4), and the vertical O_1O_2 distance ($[O_1O_2]$ y - Figure 1) is variable (-30, -20, -10, 0 and +10 mm), in each of the two cases.

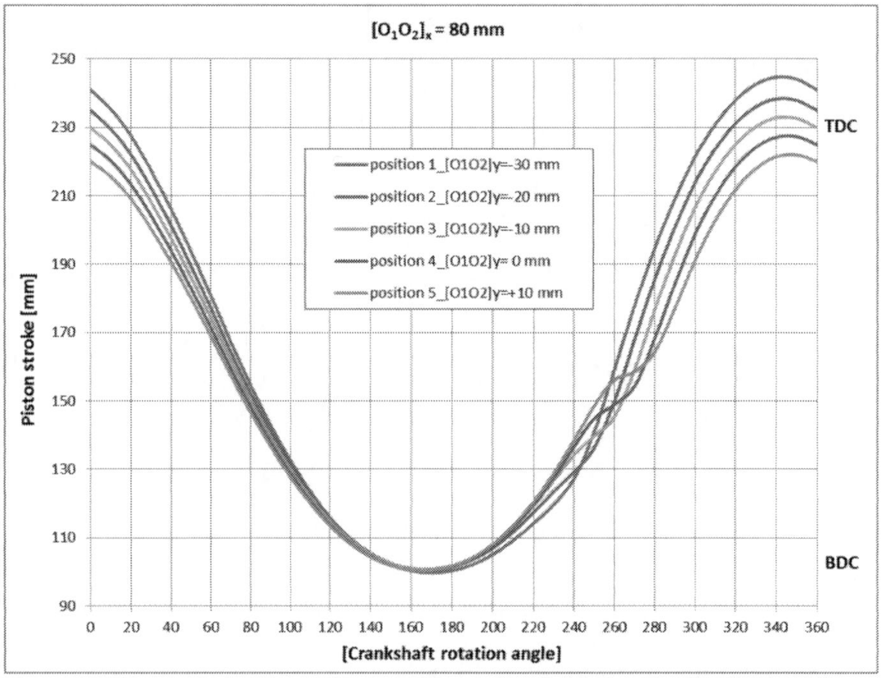

Figure 3. Piston stroke depending on the angle of rotation for $[O_1O_2]x$ = 80 mm

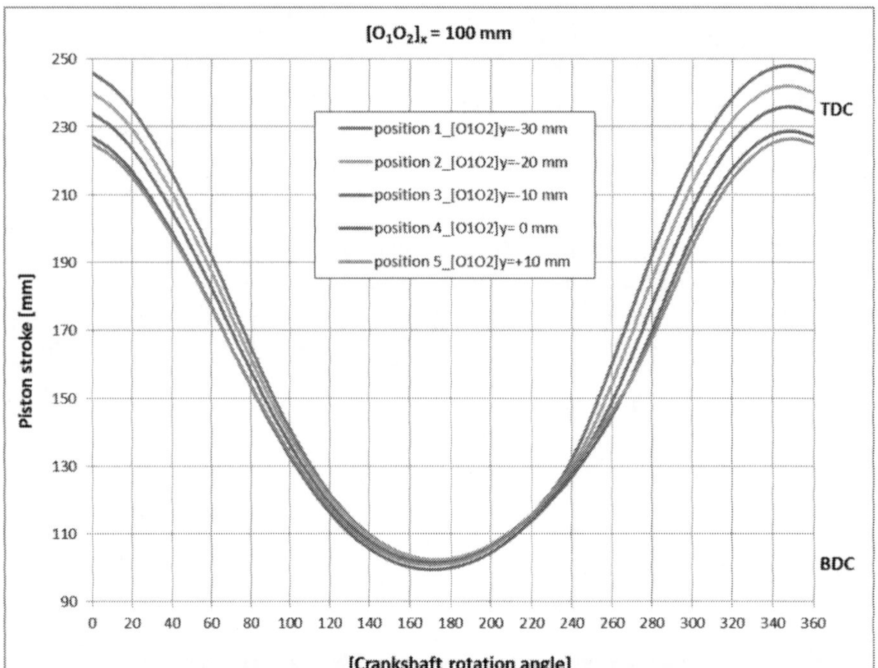

Figure 4. Piston stroke depending on the angle of rotation for $[O_1O_2]x$ = 100 mm

From Figures 3 and 4 one can notice the modification of the TDC position with the modification of the distance $[O_1O_2]y$, while the BDC position remains fixed. In the case of Figure 3, with $[O_1O_2]x$ = 80 mm, the compression ratio varies between 4.04:1 and 8.73:1. In the case of Figure 4, with $[O_1O_2]x$ =

100 mm, the compression ratio varies between 4.63:1 and 12.06:1. Next we have the comparative results in the case of the constant holding of the relative position of O_2 joint in relation to O_1 joint, in the vertical direction, in the two variants: $[O_1O_2]x = 80$ mm and $[O_1O_2]x = 100$ mm.

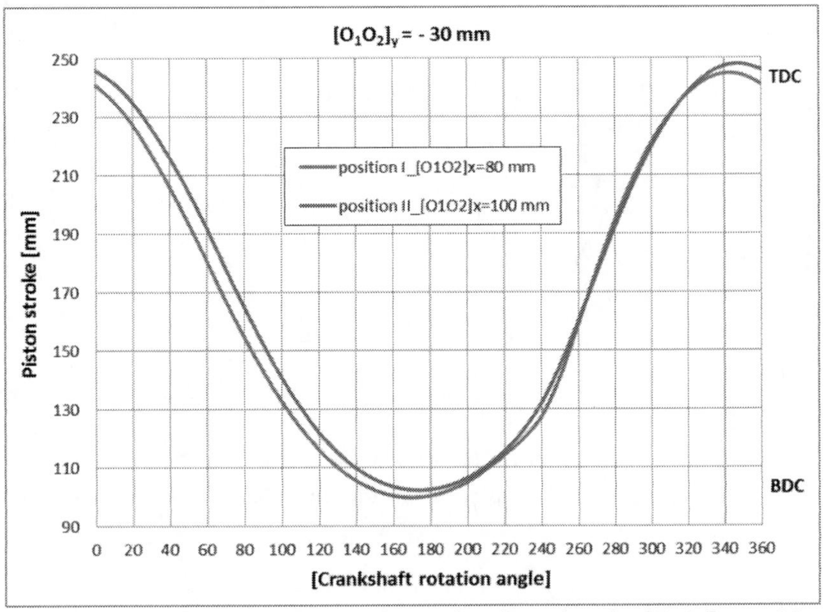

Figure 5. Piston stroke depending on the angle of rotation for $[O_1O_2]y = -30$ mm

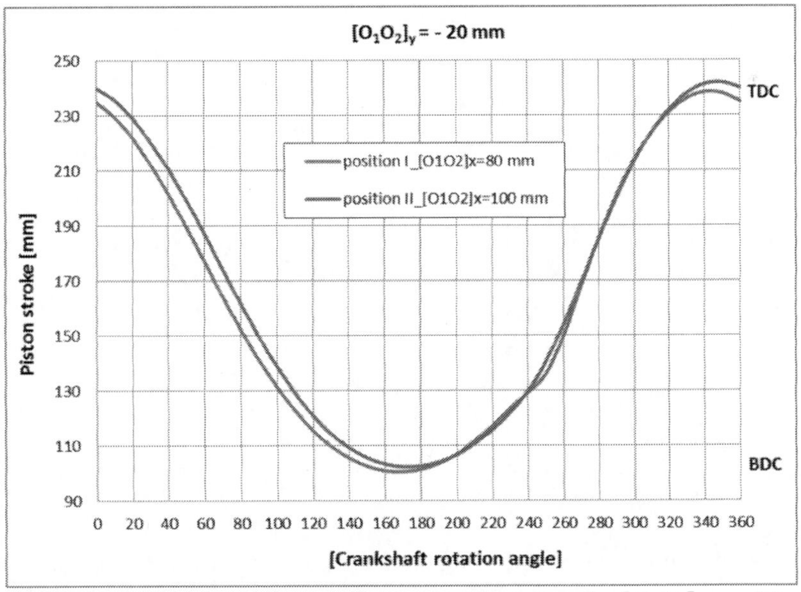

Figure 6. Piston stroke depending on the angle of rotation for $[O_1O_2]y = -20$ mm

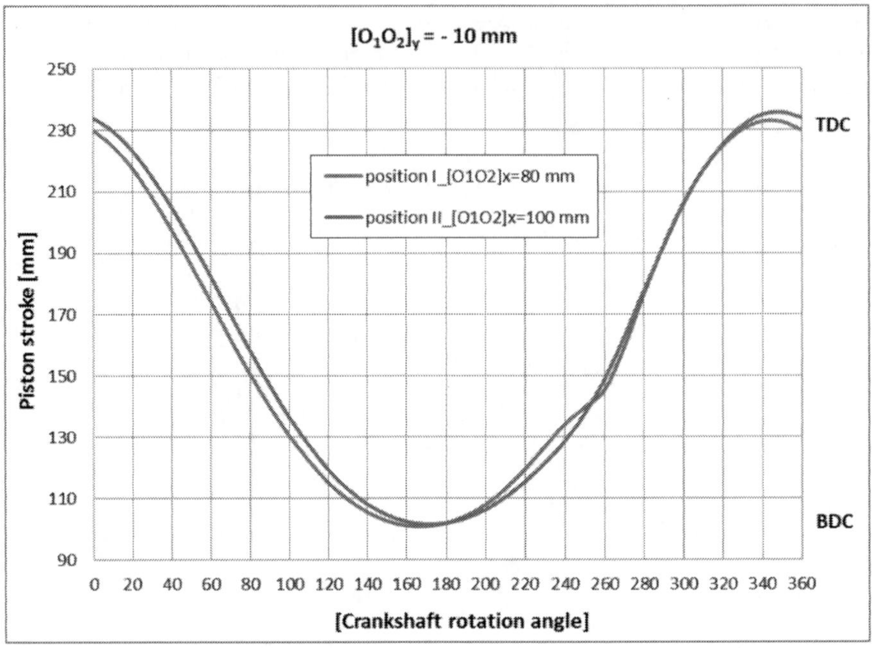

Figure 7. Piston stroke depending on the angle of rotation for $[O_1O_2]y = -10$ mm

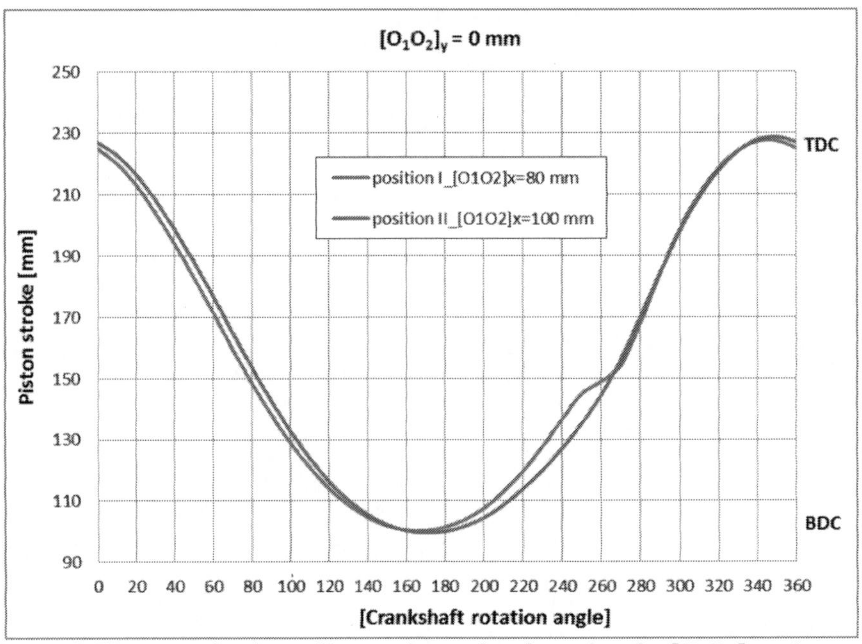

Figure 8. Piston stroke depending on the angle of rotation for $[O_1O_2]y = 0$ mm

IOP Conf. Series: Materials Science and Engineering **294** (2017) 012024 doi:10.1088/1757-899X/294/1/012024

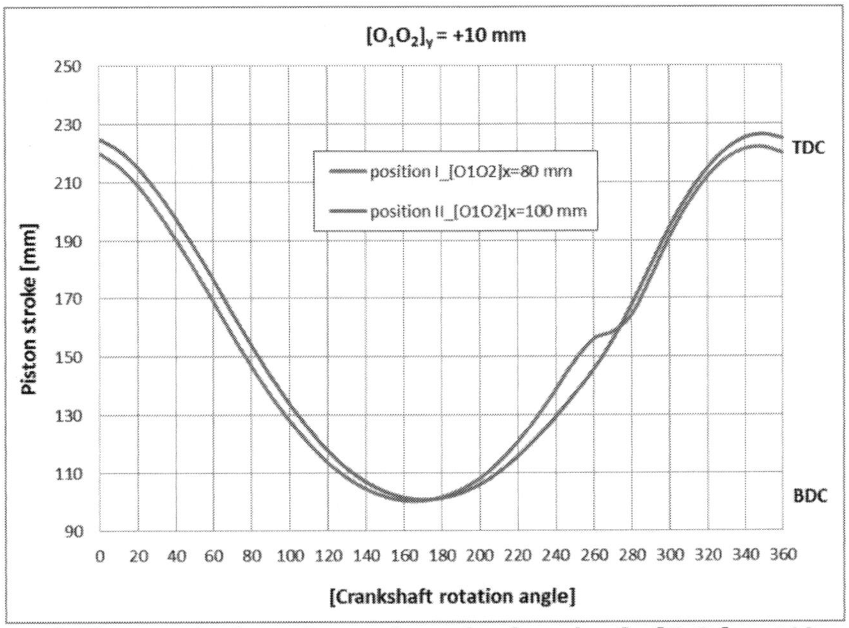

Figure 9. Piston stroke depending on the angle of rotation for $[O_1O_2]y = +10$ mm

6. 3D modeling of the mechanism

In addition, a 3D modeling of the mechanism with the AutoDesk Inventor software was performed, during which the time depending piston displacements, speeds and accelerations were determined [6-8].

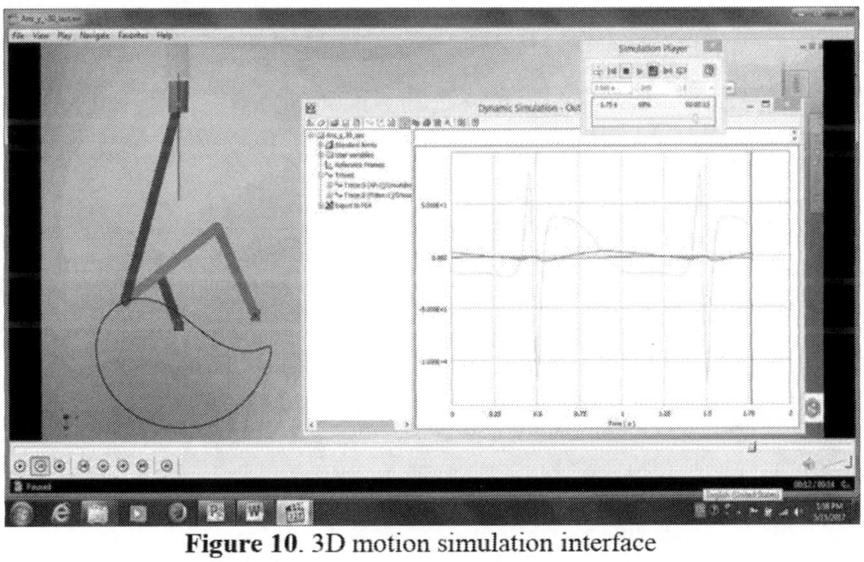

Figure 10. 3D motion simulation interface

For space considerations, only variations of the displacement, velocity and acceleration of the piston, in function of time, will be presented for the configuration of the mechanism having: $[O_1O_2]y = -30$ mm and $[O_1O_2]x = 100$ mm.

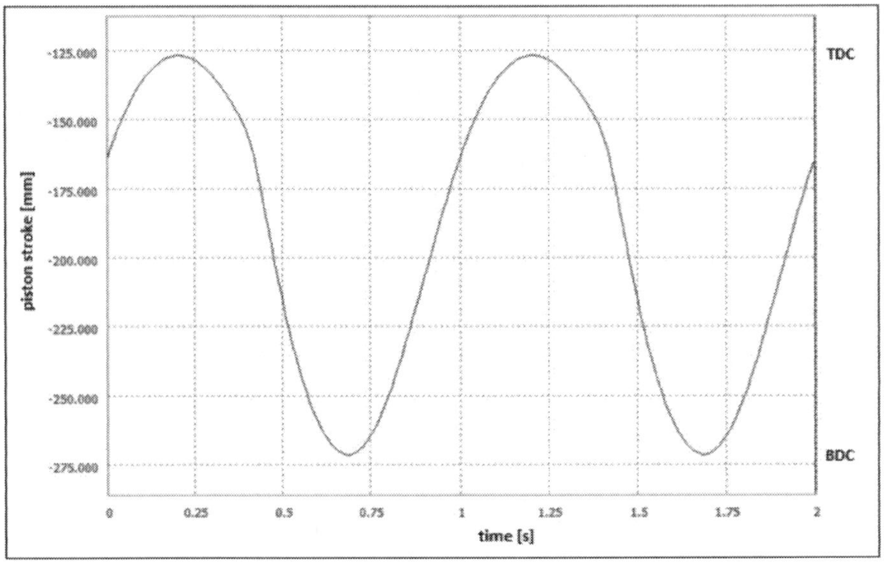

Figure 11. Piston stroke time function for $[O_1O_2]y = -30$ mm

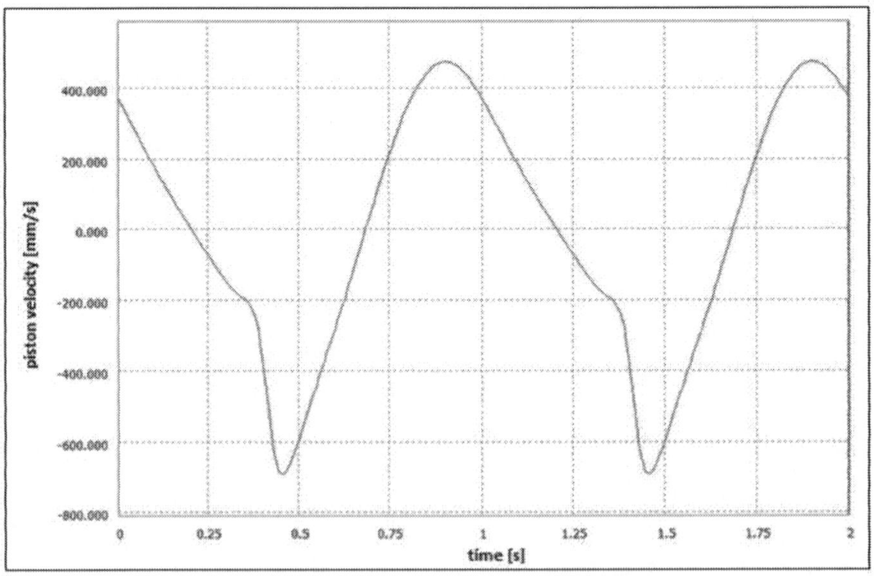

Figure 12. Piston velocity time function for $[O_1O_2]y = -30$ mm

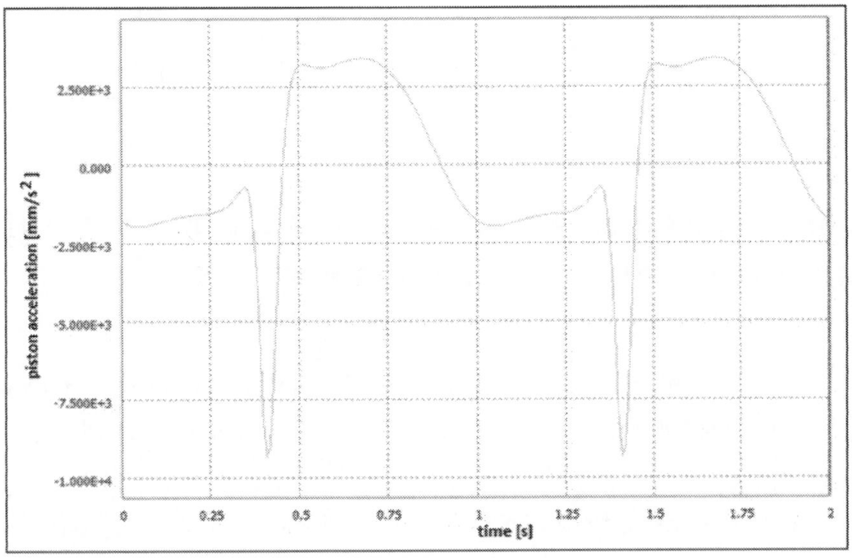

Figure 13. Piston acceleration time function for $[O_1O_2]y = -30$ mm

7. Results and conclusions

In conclusion, for the higher performances in relation to engine economy, the $[O_1O_2]x = 100$ mm variant of the mechanism can be chosen because the variation of the compression ratio is higher, ranging from 4.63:1 ÷ 12.06:1, compared to $[O_1O_2]x = 80$ mm variant, which provides a compression ratio ranging from 4.04:1 ÷ 8.73:1.

As is shown in Figure 13, at certain points of the trajectory of the mechanism, the accelerations are very high, which is why it is necessary to thoroughly study the dynamics of the motor mechanism and implicitly the calculation of the forces that occur during operation.

References

[1] Gabor A 2012 *Contribuții la determinarea răspunsului dinamic al motoarelor cu raport de comprimare variabil* [Contributions in determining the dynamic response of the variable compression ratio engines], Transilvania University of Brașov, Romania, Doctoral Thesis

[2] Rațiu S and Mihon L 2008 *Motoare cu ardere internă pentru autovehicule rutiere – Procese și caracteristici,* [Internal combustion engines for road vehicles – Processes and Features], Publisher: Mirton, Timișoara, Romania

[3] ****Bosch Automotive Hand–book*, 8th Edition, May 2011

[4] Miklos I Zs, Alic C I and Miklos C C 2010 Analysis of seat positioning mechanism in road vehicles *Acta Technica Corvinensis – Bulletin of Engineering* **III**(3) 27-30

[5] ***http://www.artas.nl/en/downloads, SAM 7.1 - The ultimate mechanism designer, © 2014 ARTAS - Engineering Software - User Guide

[6] Cioată V G and Kiss I 2017 Dynamic analysis and parametric optimisation of the connecting rod using Autodesk Inventor, *Machine Design* **9**(1) 29-34

[7] Cioată V G and Kiss I 2010 Computer Aided Design of the connecting rod, *Machine Design* **2**(1) 83-86

[8] Cioată V G and Miklos I Z 2009 *Computer Aided Design with Autodesk Inventor* (in romanian), Publisher: Mirton, Timișoara, Romania

IOP Conf. Series: Materials Science and Engineering **294** (2017) 012025 doi:10.1088/1757-899X/294/1/012025

Decoupling the structure from the ground motion during earthquakes by employing friction pendulums

G R Gillich[1], V Iancu[1], N Gillich[2], Z I Korka[1], C P Chioncel[2] and C Hatiegan[2]

[1]"Eftimie Murgu" University of Resita, Department of Mechanics and Materials Engineering, P-ta Traian Vuia 1-4, Resita, 320085, Romania
[2]"Eftimie Murgu" University of Resita, Department of Electrical Engineering and Industrial Informatics, P-ta Traian Vuia 1-4, Resita, 320085, Romania

E-mail: gr.gillich@uem.ro

Abstract. Avoiding dynamic loads on structures during earthquakes is an actual issue since seismic actions can harm or destroy the built environment. Several attempts to prevent this are possible, the essence being to decouple the structure from the ground motion during earthquakes and preventing in this way large deflections and high accelerations. A common approach is the use of friction pendulums, with cylindrical or spherical surfaces but not limited to that, inserted between the ground and the structure, respectively between the pillar and the superstructure. This type of bearings permits small pendulum motion and in this way, earthquake-induced displacements that occur in the bearings are not integrally transmitted to the structure. The consequence is that the structure is subject to greatly reduced lateral loads and shaking movements. In the experiments, conducted to prove the efficiency of the friction pendulums, we made use of an own designed and manufactured shaking table. Two types of sliding surfaces are analyzed, one polynomial of second order (i.e. circular) and one of a superior order. For both pendulum types, analytical models were developed. The results have shown that the structure is really decoupled from the ground motion and has a similar behaviour as that described by the analytic model.

1. Introduction

Earthquakes may have as a result structural damage produced in buildings, bridges and other essential infrastructure. The type of impact depends on where the earthquake is located and certainly on the capacity of the structures to withstand shaking. The seismic waves produced during earthquakes comprise a wide range of frequencies. While the energy of the waves with higher frequencies tends to be absorbed by solid rock soil, the waves with low frequency (having periods larger than one second) pass the solid rock soil without being absorbed but are eventually amplified by soft sediments. This is the reason why much worse earthquake damage is produced in areas having as base soft sediments than in the areas underlying solid rock [1].

Nowadays, seismic isolation of structures is used to protect both structural and non-structural elements, avoiding damage and minimizing economic losses due to downtime and repair costs. Seismic isolation, first implemented in 1969, is now a mature technology. From the wide variety of devices available for seismic isolation, two main categories can be identified: rubber/elastomeric bearings and frictional sliding bearings. Elastomeric isolation devices are well researched and worldwide employed in base isolation. These can be linear elastic bearings [2] which are often used with external energy dissipation systems, high damping rubber bearings [3], lead rubber bearings [4] and hybrid lead rubber bearings [5], [6]. Analytical models of these bearings are presented in [7].

The second category of seismic isolation devices consists in frictional sliding bearings. Two common types of sliding bearings are known: flat sliding bearings, used in combination with elastomeric devices since a restoration force is needed, and friction pendulum bearings.

The simplest pendulum bearing is the single friction pendulum (SFP) one, firstly proposed in [8], which consists of a spherical concave surface supporting a frictional slider. Next developments include the double friction pendulum (DFP) bearing [9] and the triple friction pendulum (TFP) bearing [10]. The behavior of DFP and TFP bearings is referred as adaptive [11], [12] because they exhibit different hysteretic properties at different displacement phases. To save space, i.e. reducing the FP diameter, multiple friction pendulum (MFP) bearings where developed [13]. As an alternative to the spherical surfaces, devices with cylindrical surfaces are now also available.

In last decades, the nature of the curves on which the sliding surfaces are built became more diverse, now being used bearings with variable curvature [14], [15]. The disadvantage of these devices consists in the need to use elastic layers [16] between the sliding surface and the slider.

This paper presents the way how SFP with surface generated by a circle respective by polynomials can decouple a structure from the ground motion induced by earthquakes, highlighting the particularities of the different surfaces and how these influence the structure isolation.

2. The principle of the single friction pendulum

Simple friction pendulum bearings are joints installed between the foundation and the structure, or in the case of bridges between the column and the superstructure. The displacement is therefore concentrated in the SFP components, the actual structure being subjected to little displacement. In fact, the structure moves like a rigid body mounted on an isolated system. In this way, the structural joint minimizes damage caused by the big inertia-induced share forces that occur at the base of the structure due to large lateral displacements observed during earthquakes.

The SFP does not absorb the earthquake energy but influences the dynamics of the system. Even if a low level of damping is beneficial for controlling an excessive displacement of the structure, the concept of isolation does not rely on damping. In contrary, excessive damping prevents displacement and reduces the effectiveness of the isolation.

The components of an SFP are presented in Figure 1. Here the bottom bearing plate has a concave surface, usually coated with special materials. If the bottom plate starts moving, due to low friction, the slider is subjected to a relative displacement versus the plate. This means that the top plate together with the superstructure is decoupled from the ground motion. An exploded view of the SFP is shown in Figure 2.

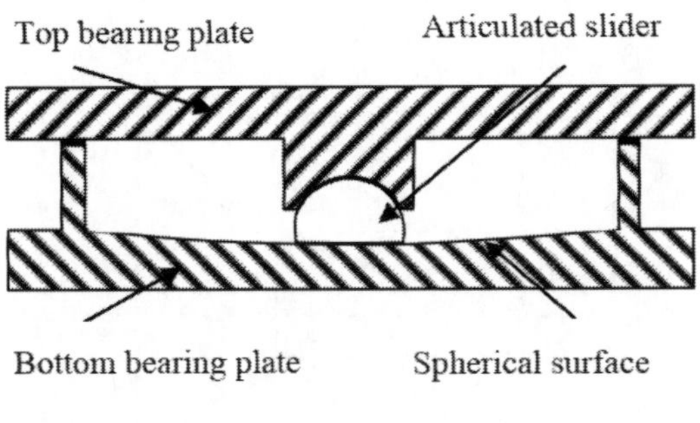

Figure 1. Cross-section of a simple friction pendulum bearing

Figure 2. Exploded view of a simple friction pendulum bearing

Versions of SFPs with the concave surface on the top plate are also known. In this case, the slider follows the ground motion, while the top plate stops moving the ground.

3. The mathematical model of a SFP with circular respectively variable curvature

Usual SFPs have spherical or cylindrical sliding surface, meaning that the surface is constructed on the basis of a circle. The scheme of such a friction pendulum having the radius R is presented in Figure 3.

Let us consider the length of chord OA being c, if the corresponding angle is α. Then, from the right triangle OAB we can deduce:

$$\sin\frac{\alpha}{2}=\frac{h}{c} . \tag{1}$$

In a similar way, from the right triangle QOC we deduce

$$\sin\frac{\alpha}{2}=\frac{c/2}{R}=\frac{c}{2R} . \tag{2}$$

From Eq.(1) and (2), naturally, result:

$$h=\frac{c^2}{2R} \tag{3}$$

Taking again the right triangle OAB, it is well-known that

$$c^2=h^2+u^2 . \tag{4}$$

Replacing Eq.(4) in Eq.(3), the relation between the vertical displacement h and the horizontal displacement u is found as:

$$h^2-2Rh+u^2=0 . \tag{5}$$

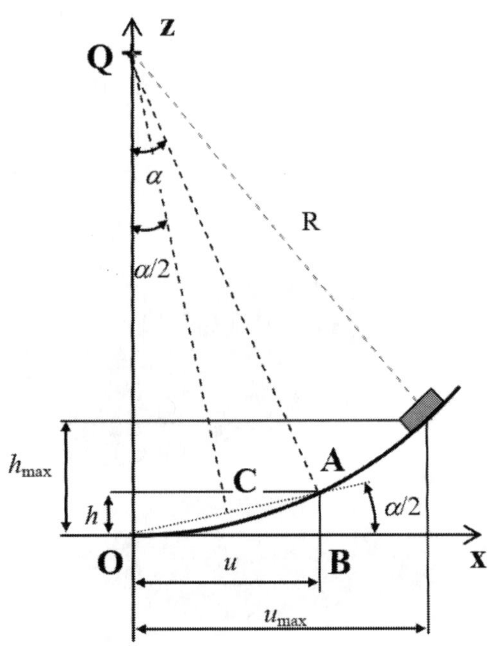

Figure 3. Friction pendulum scheme for a spherical sliding surface

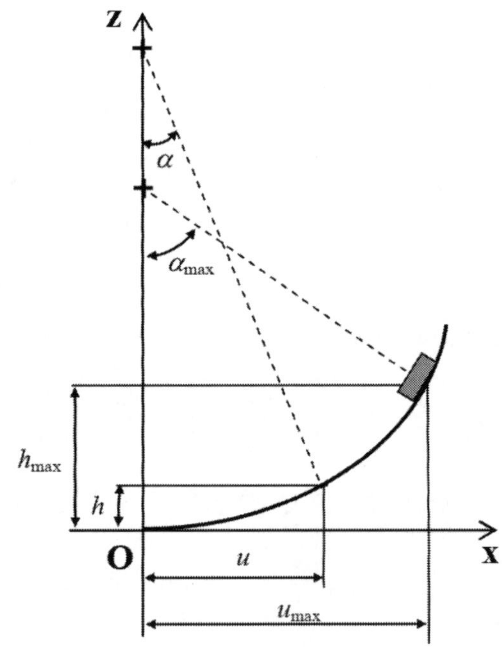

Figure 4. Friction pendulum scheme for a polynomial sliding surface

The solution of Eq.(5) takes the form:

$$h = R - \sqrt{R^2 - u^2} \ , \tag{6}$$

for the case presented in Figure 3. In this study, a spherical sliding surfaces with the radius R=5m is considered.

Three other analyzed cases base on the sliding surface generated by polynomial functions, an example of such SFP being illustrated in Figure 4. The following equation to generate the curves is employed:

$$h = a \cdot u^b \ . \tag{7}$$

For this study, we have chosen the parameter $a = 0.1$ and parameter b takes successively the values 2, 2.4 and 4. Figure 5 illustrates the arcs plotted for the equations $h = 0.1u^2$ (with solid line), $h = 0.1u^{2.4}$ (with dashed line) and $h = 0.1u^4$ (with dash-dot line). Prior research [14] has proved the similitude between the circular arc and the second order polynomial, therefore the last one is not represented in next figures.

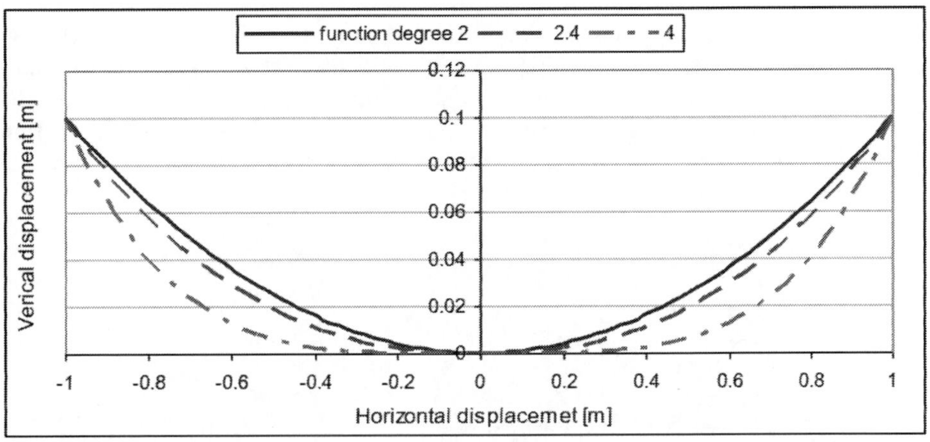

Figure 5. Curves generating the sliding surface for different function degrees

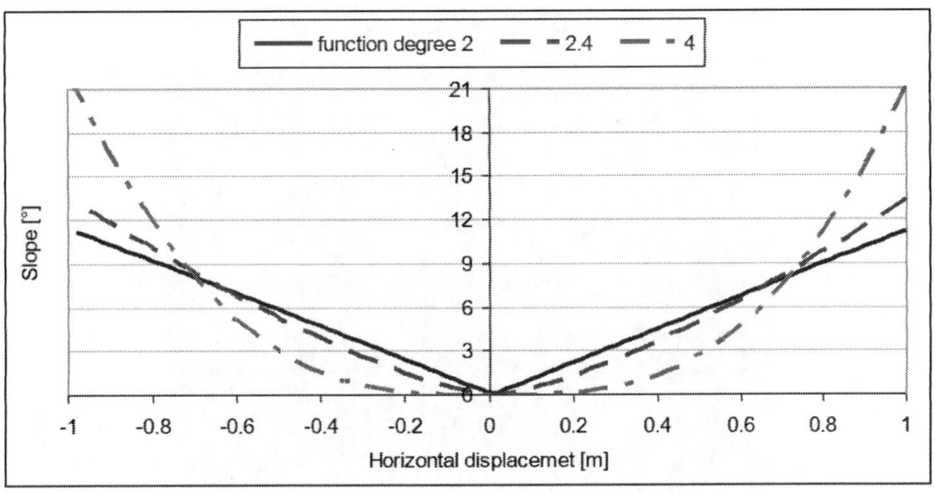

Figure 6. Slope of the curve vs. lateral displacement

Because the scales on the horizontal respectively vertical axis are different, the angle α and consequently the slope are not realistically represented in Figure 5. The evolution of angle α with the horizontal displacement is given in Figure 6. Here can be observed that for the surface constructed with the polynomial $h=0.1u^2$, or based on a circle, the evolution is linear. This is obvious since the slope is the derivative of the employed polynomial of second order. For the other two functions, $h=0.1u^{2.4}$ respectively $h=0.1u^4$, the slope evolution with the horizontal displacement is obviously polynomial.

4. Materials and methods

To simulate the ground motion, a shaking table with a mobile plate that's speed can be precisely controlled is employed. The shaking table, presented in Figure 1, was designed and realized in the *Laboratory for the study of seismic and anthropic hazards* of the "Eftimie Murgu" University Resita. It consists of a rigid structure 1, the electrical motor 2 and the mobile plate 3. The oscillation amplitude of the plate is 22 mm (11 mm in each movement sense), the movement being transmitted from the electric motor via a crank mechanism. The oscillation speed of the plate can be easily adjusted from the control panel 4 by adapting the electric motor speed. The speed value of the plate is shown on a display.

The vibration of the isolated structure 5 and the mobile plate 3 are measured with two Kistler accelerometers of type 8772, designed for fine sensing of the oscillation waves and to send an analog signal (for the horizontal oscillation direction) to the acquisition board 7. The acquisition board, consisting of the NI ENET-9163 device and the signal acquisition module NI 9234, sends binary encoding date to the computer 8. For fine processing of the vibration data, the computer is running in LabView software a specially designed application [17], [18].

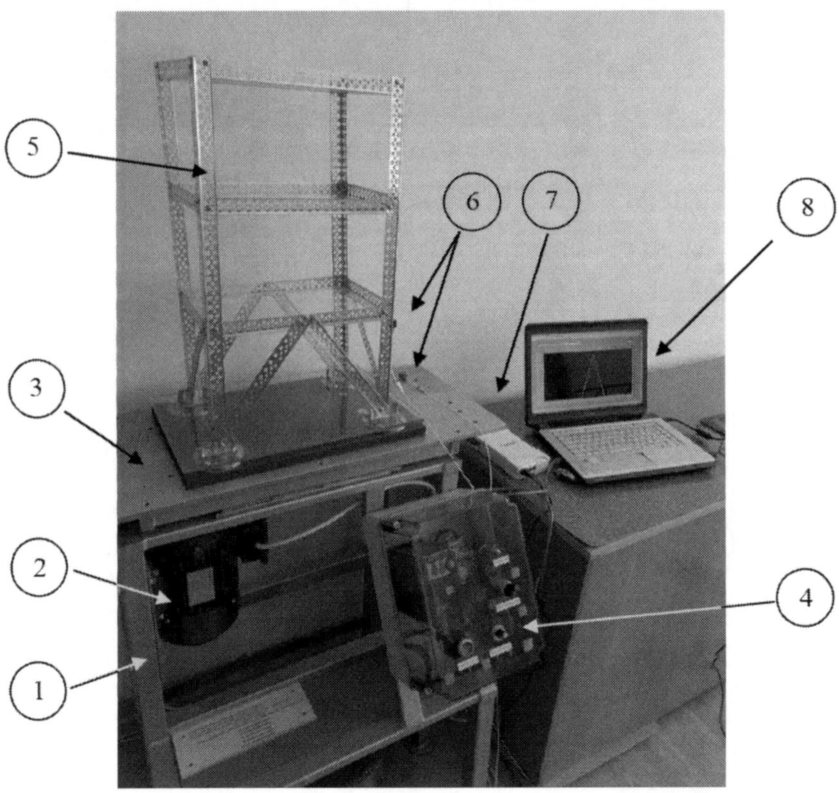

Figure 7. Shaking table

A detailed view of the isolated structure 5 is presented in Figure 8, where the positions of the two involved accelerometers 6 are also indicated. The structure 5 is supported by a glass plate 10, which has the role of the top bearing plate. The articulated slider is supported by the bottom bearing plate 9, allowing the structure 5 to move relative to the mobile plate 3. The mobile plate, in turn, oscillates in relation to the rigid structure 1. Optimal is that the structure 5 does not have a relative displacement in regard to the rigid structure 1, or it should be as small as possible.

Figure 8. Detailed view of the isolated structure

The bottom bearing plate was made of stainless steel and has a spherical sliding surface. In order to assure a good friction coefficient, the articulated slider was made of Teflon.

5. Results and discussion

From the acceleration measurements, made on the structure isolated with a SFP with the sliding surface generated with a circle, it was found that the response to a sinusoidal excitation was also sinusoidal. The two signals are presented in Figure 9; the higher-amplitude signal (marked with a black line) is the excitation and the lower-amplitude signal (marked with red line) is the response. It is worth to mention that the crank mechanism induces, beside the fundamental frequency, several harmonics. In the present study, just the fundamental frequency component obtained applying a low-pass filter was analyzed.

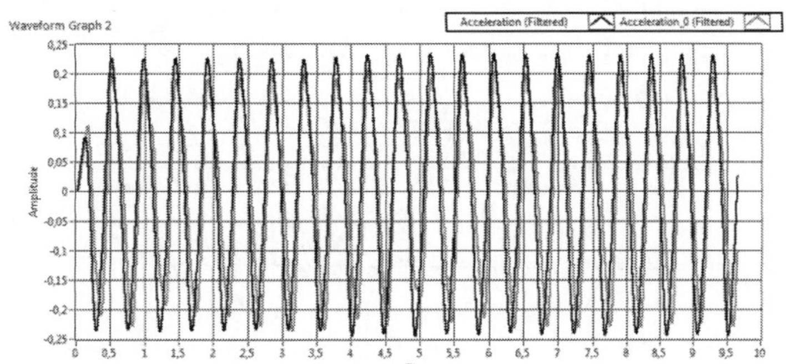

Figure 9. Filtered signal of the excitation respectively response

The results are more expressive if the analysis is performed in the frequency domain. Three cases, comprising the frequency of the excitation $f_{E1} = 2.175\,\text{Hz}$, $f_{E2} = 2.315\,\text{Hz}$ and $f_{E3} = 2.465\,\text{Hz}$, are considered in this study. The resulted spectra are illustrated in Figures 10, 11 respectively 12.

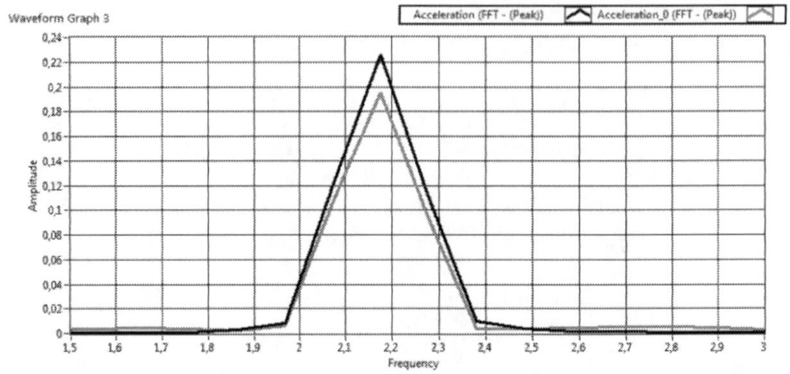

Figure 10. Response of the isolated structure for an excitation of 2.175 Hz

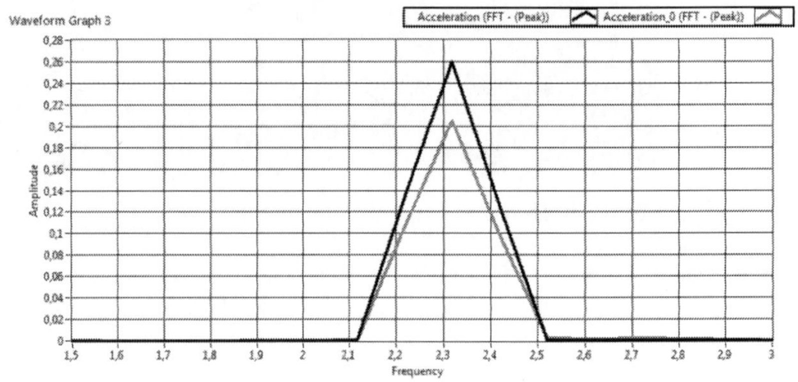

Figure 11. Response of the isolated structure for an excitation of 2.315 Hz

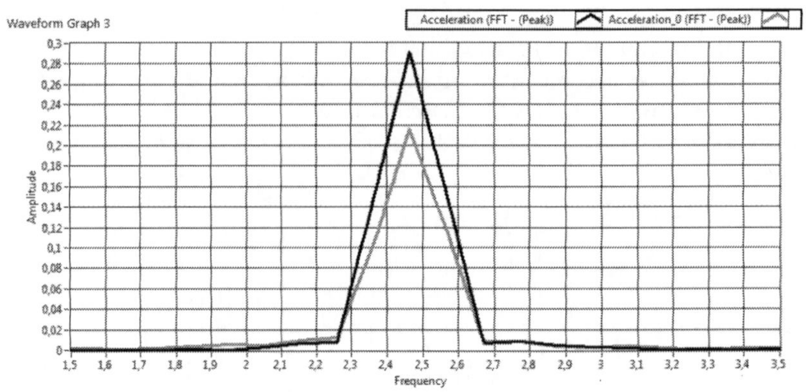

Figure 12. Response of the isolated structure for an excitation of 2.465 Hz

Having a look onto the three above-presented figures one can conclude that the accelerations transmitted to the structure are diminished, thus the SFP bearings fulfill their task. Also, it appears that the frequency of the structure was similar with that of the simulated ground motion, resulting that the decoupling was made stepwise and not continuously.

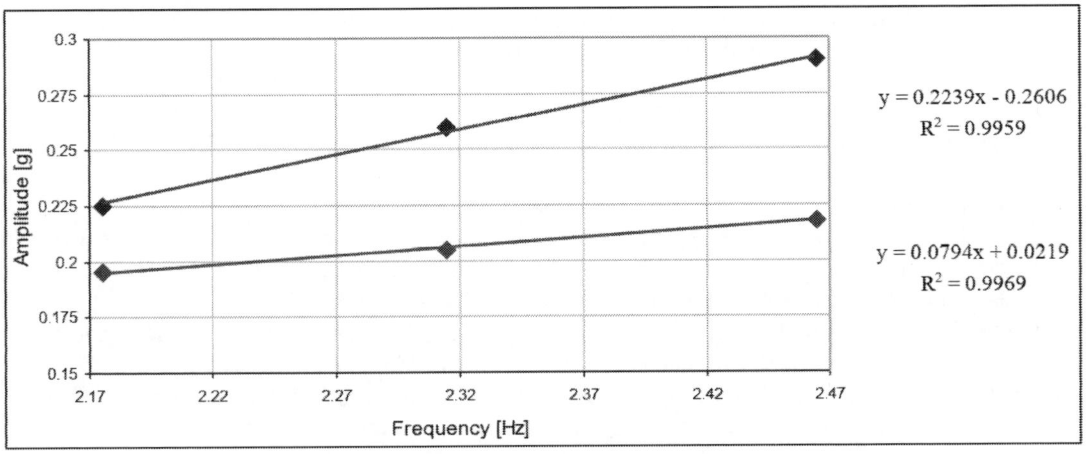

Figure 13. Evolution of acceleration amplitude vs. frequency

A precise evaluation is made in Figure 13, where the amplitudes of the simulated ground excitation respectively the structural response are depicted. From here results that, the higher the frequency, the higher the isolation rate was. This, in association with the stepwise decoupling, led us to the idea that the friction coefficient was not low enough. Actually, the intersection of the two curves presented in Figure 13 takes place if the two equations have same value, that is

$$0.2239x - 0.2606 = 0.0794x + 0.0219 \tag{8}$$

which results in $x = f_D = 1.955\,\text{Hz}$. Therefore, the frequency f_D at which the bearing decouples is near to 2 Hz. This is confirmed by the experiments since no isolation as observed before the ground motion achieved approximately 2 Hz.

6. Conclusion

The paper presents a study regarding the possibility of decoupling the superstructure from the ground motion via single friction pendulums. The theoretical approach has shown that sliding surfaces based on polynomials of a higher order as two can assure some advantages, as decoupling by lower forces due to a smaller slope, or an increase of dissipated energy for important horizontal displacements due to a dramatical slope increase.

Experimental results have shown that the excitation frequency increase leads to a better isolation. Another conclusion refers to the importance of ensuring a low friction coefficient in order to achieve decoupling for small lateral forces, i.e. for early earthquake stage. Further studies will be made with a slider made of stainless steel for concave sliding surfaces with a bigger radius in the rest position.

Acknowledgement
The work has been funded by the Sectoral Operational Programme Human Resources Development 2007-2013 of the Ministry of European Funds through the Financial Agreement POSDRU/159/1.5/S/132395.

References

[1] Earle S 2016 *Physical Geology*, CreateSpace Independent Publishing Platform, British Columbia, Canada
[2] Stanton J F and Roeder C W 1982 *Elastomeric Bearing Design, Construction and Material*, National Cooperative Highway Research Program Report 248, Transportation Research Board, National Research Council, Washington, DC

[3] Forni M, Antonucci R, Arcadi A and Occhiuzzi A 2004 *A Hybrid Seismic Isolation System Made of Rubber Bearings and Semi-Active Magneto-Rheological Dampers*, 13th World Conference on Earthquake Engineering Vancouver, B.C., Canada August 1-6, Paper No. 2177

[4] Sonawane Y N and Patil M N 2016 Base Isolation for Multistoried Buildings with Lead Rubber bearing, *International Journal of New Innovations in Engineering and Technology* **4**(4) 29-33

[5] Tanaka K, Hirasawa M, Ishiguro Y, Ohyama H and Nakamura Y 1993 *Base-Isolation System With Hybrid Lead Rubber Bearings*, Elsevier Science Publishers B.V., K 25/5, pp 381-386

[6] Iancu V, Vasile O and Gillich G R 2012 Modelling and Characterization of Hybrid Rubber-Based Earthquake Isolation Systems, *Materiale Plastice* **49**(4) 237-241

[7] Iancu V, Gillich G R, Iavornic C M and Gillich N 2013 *Some Models of Elastomeric Seismic Isolation Devices*, 12th International Symposium AVMS' 2013-Timişoara

[8] Zayas V, Low S and Mahin S A 1987 *The FPS earthquake resisting system*, Rep. No. UCB/EERC-87/01, Earthquake Eng. Res. Center, University of California at Berkeley, CA

[9] Fenz D M and Constantinou M C 2006 Behaviour of the double concave Friction Pendulum bearing, *Earthquake Engng Struct. Dyn.* **35** 1403-1424

[10] Fenz D M and Constantinou M C 2008 Modeling triple friction pendulum bearings for response-history analysis, *Earthquake Spectra* **24**(4) 1011-1028

[11] Malekzadeh M and Taghikhany T 2010 Adaptive Behavior of Double Concave Friction Pendulum Bearing and its Advantages over Friction Pendulum Systems, *Scientia Iranica Transaction A: Civil Engineering* **17**(2) 81-88

[12] Moeindarbari H and Taghikhany T 2012 *Optimum Design of Triple Friction Pendulum Bearing Subjected by Near-Field Ground Motions*, The 15th World Conference on Earthquake Engineering, Lisbon, Portugal, 24-28 September

[13] Tsai C S, Cheng C K, Chen M J and Yu S H 2005 *Experimental study of MFPS-isolated sensitive equipment*, The 2005 ASME Pressure Vessels and Piping Conference, Seismic Engineering ASME, Tsai CS (ed.), vol. 8. Denver, CO, U.S.A., pp 11–18

[14] Minda A A, Gillich G R, Iavornic C M and Minda P F 2012 *Analytical and finite element study for friction pendulum with parameterized sliding surfaces*, Proceedings of the World Congress on Engineering, vol. 3, London, UK, July 4-6, 2012, pp 1716-1721

[15] Tsai C S, Chiang T C and Chen B J 2003 Finite element formulations and theoretical study for variable curvature friction pendulum system, *Engineering Structures* **25**(14) 1719–1730

[16] Iavornic C M, Praisach Z I, Vasile O, Gillich G R and Iancu V 2011 Study of Stress and Deformation in Elastomeric Isolation Systems Using the Finite Element Method, *Recent Advances in Signal Processing, Computational Geometry and Systems Theory* - ISCGAV'11, ISTASC'11, pp 239-244

[17] Gillich G R, Frunzaverde D, Gillich N and Amariei D 2010 The use of virtual instruments in engineering education, *Procedia - Social and Behavioral Sciences* **2**(2) 3806-3810

[18] Gillich G R, Mituletu I C, Negru I, Tufoi M, Iancu V and Muntean F 2015 A Method to Enhance Frequency Readability for Early Damage Detection, *Journal of Vibration Engineering & Technologies* **3**(5) 637-652

Time series prediction in the case of nonlinear loads by using ADALINE and NAR neural networks

L Ghiormez, M Panoiu, C Panoiu and O Tirian

Politehnica University of Timisoara, Department of Electrical Engineering and Industrial Informatics, 5 Revolution Street, Hunedoara, 331128, Romania

E-mail: loredana.ghiormez@fih.upt.ro

Abstract. This paper presents a study regarding the time series prediction in the case of an electric arc furnace. The considered furnace is a three phase load and it is used to melt scrap in order to obtain liquid steel. The furnace is powered by a three-phase electrical supply and therefore has three graphite electrodes. The furnace is a nonlinear load that can influence the equipment connected to the same electrical power supply network. The nonlinearity is given by the electric arc that appears at the furnace between the graphite electrode and the scrap. Because of the disturbances caused by the electric arc furnace during the elaboration process of steel it is very useful to predict the current of the electric arc and the voltage from the measuring point in the secondary side of the furnace transformer. In order to make the predictions were used ADALINE and NAR neural networks. To train the networks and to make the predictions were used data acquired from the real technological plant.

1. Introduction

In the present the steel is used in the whole world in many domains such as: alimentary, chemically, constructions, marine and automobile industries. The steel has the main advantage that it can be melted many times and used again without losing its properties. The main equipment used to obtain liquid steel is the electric arc furnace (EAF). This kind of equipment is mainly characterized by its capacity, its power level and its productivity.

The EAF can be classified into two categories: AC or DC furnaces. Indifferent of what type is the furnace the basic operating principle is the same: the electric arc struck from one or more electrodes to a metallic charging material [1]. The charging material in the EAF's tank has different shape and dimension, so in the initial melting period the operation of the furnace cannot be in the steady state. Because this kind of EAF is a nonlinear load it produces significant electrical disturbances such as voltage fluctuations, flicker, harmonics and unbalance between phases [2], [3]. Therefore it is necessary to develop accurate and reliable models which can be voltage-current characteristic predictors. The prediction can improve the productivity of steel and also can reduce the negative effects that can influence the other equipment connected to the same power system. The voltage flicker has randomly character and also depends on both operation mode and type of the furnace [4].

In the scientific literature have been presented the following models of the electric arc:

- models based on the nonlinear and time-varying resistors [5];
- models based on the linear approximation of the voltage – current characteristic of the electric arc [6-11];

- models where the voltage and the current are related by hyperbolic functions (using the diameter or length of the arc) [8], [12-15];
- advanced models which use the artificial neural networks [1], [16-18].

In this paper is performed a study regarding the time series prediction in the case of an AC EAF. In order to make the predictions were used ADALINE (adaptive linear neuron) and NAR (nonlinear autoregressive) neural networks. To train the networks and to make the predictions were used data acquired from the real technological plant.

The networks learn to predict the voltage-current characteristic of the EAF taking into consideration the measurements used to train the networks. It can be noticed that by choosing the proper network architecture and by using a suitable learning algorithm both networks have the capability to perform one-step or multi-steps ahead prediction.

2. Development of the EAF Models

The EAF is a complex system with multi-variable process and time-varying parameters. Also, the EAF is a nonlinear load, this nonlinearity being caused by the electric arc. Therefore, the artificial neural network (ANN) fits well for the modelling of the behaviour of this installation.

The purpose of this paper is to develop a model that can be used to accurately represent the variation of the electrical parameters that characterise the EAF. In order to train the network can be used any of the following parameters: supply voltage or furnace transformer current and voltage. The behaviour of the EAF is represented within a neural network by biases, weights, transfer function and the type of network. By modifying the neural network architecture (structure of the network, number of neurons, connections, weights, biases or different learning algorithms) a network can be trained to perform a specific function.

In this paper are developed two types of networks such as ADALINE and NAR neural networks. For both of the networks the training data set contains *500 samples* of the acquired current and voltage from the secondary side of the furnace transformer. The testing data set contains also *500 samples* of the acquired current and voltage from the secondary side of the furnace transformer but this set is different by the training data set. For this new data set is tested the capability of the network to predict the voltage and current from the secondary side of the furnace transformer. The network is trained and tested for a single phase because the behaviour is the same for the other two.

3. ADALINE Neural Network

The ADALINE neural network used in this paper is similar to the perceptron, so having one input layer and one output layer, but its transfer function is linear rather than hard-limiting. In this way the output of the ADALINE network can take on any value as compared to the perceptron output that can be either 0 or 1. The ADALINE networks use the least mean squares (LMS) or Widrow-Hoff learning rule. The learning rule algorithm adjusts the weights and biases of the network.

The ADALINE network is used in this paper as an adaptive filter by adjusting the weights and biases during network training taking into consideration new input and target vectors and also a tapped delay line of the input signal.

3.1. EAF Current and Voltage Prediction

The ADALINE neural network was created by using *linearlayer* function from *Matlab*. The *input delay* is *5* and the learning rate is *0.2*. The input of the network is a current or a voltage. In order to use the adaptive training of the ADALINE is used the *adapt* function. In this case the network receives one step at a time from the sequence of the input and the weights and biases are updated after each step, before being received a new step from the sequence. In order to test the capability of the network to perform prediction of a sequence of the input, the network is simulated for a new data set containing different values as compared to the ones used for adaptive training of the network.

In Figure 1 is illustrated the adaptive training in the case of current for *500 samples*, respectively the obtained error during network training. It can be noticed that the ADALINE neural network has the capability to learn this data set. The error decreases while the network uses more steps from the training data set.

In Figure 2 is represented the response of the network when this is simulated for a new data set. In this case are used the weights and biases obtained after performing the adaptive training. It can be observed that the response of the network follow precisely the acquired current and the error has small values even smaller as compared to the training.

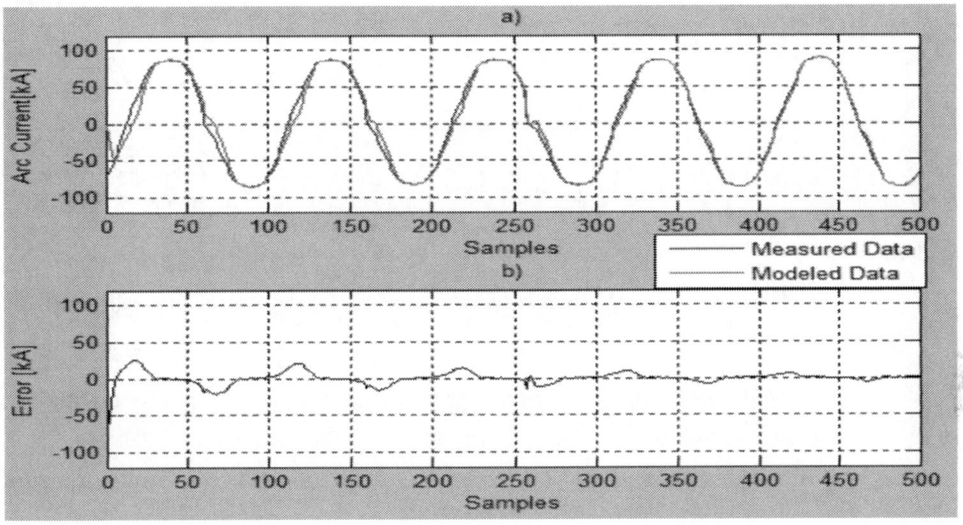

Figure 1. a) Adaptive training in the case of electric current, b) Obtained error during network training

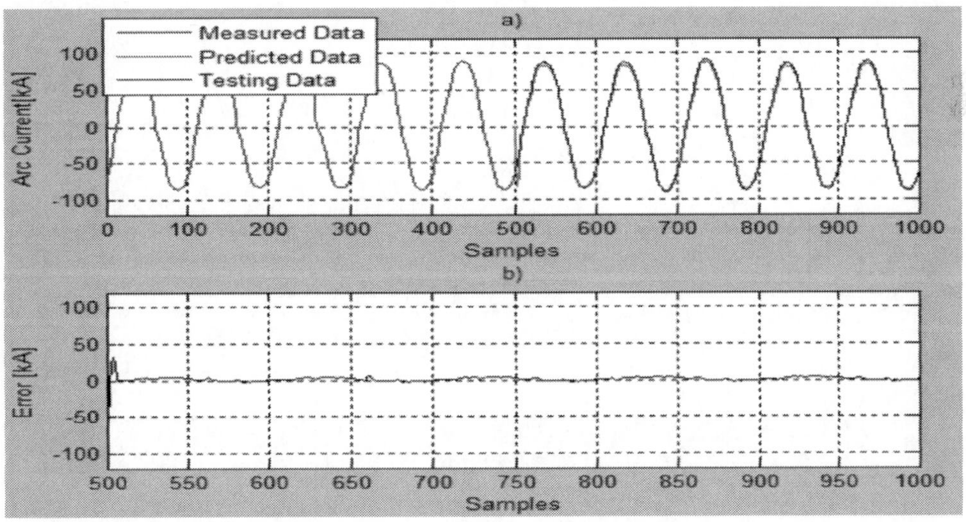

Figure 2. a) Prediction of electric current for a new data set, b) Obtained error during network prediction

Figure 3 illustrates the adaptive training in the case of voltage for a data set of *500 samples*, respectively the obtained error during network training. It can be noticed that, even for this parameter, the ADALINE neural network has the capability to learn the data set.

In Figure 4 is represented the response of the network when is simulated for a new data set that contains *500 new samples* of voltage. For this case also are used the weights and biases obtained after performing the adaptive training. It can be observed that the response of the network follow the acquired voltage but the prediction presents some errors because the voltage is more distorted as compared to the current.

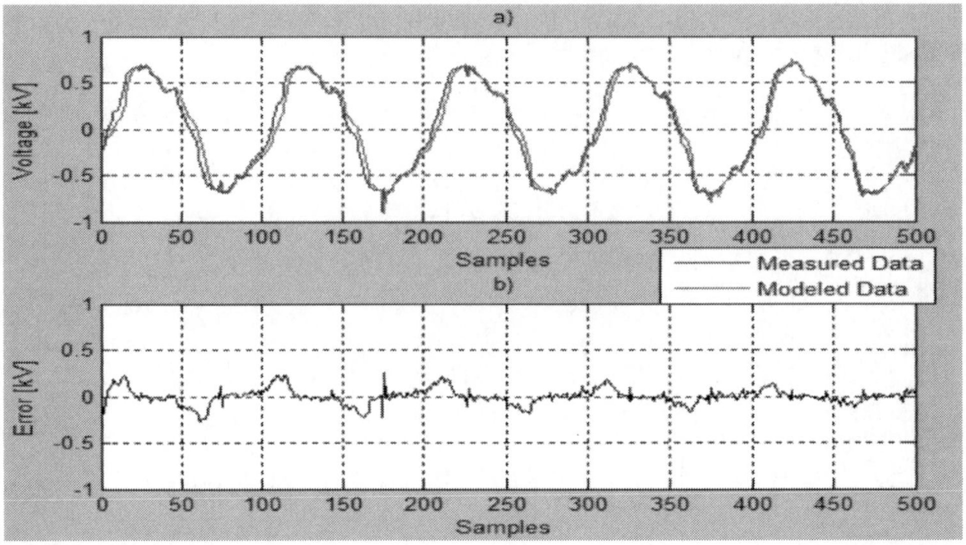

Figure 3. a) Adaptive training in the case of voltage, b) Obtained error during network training

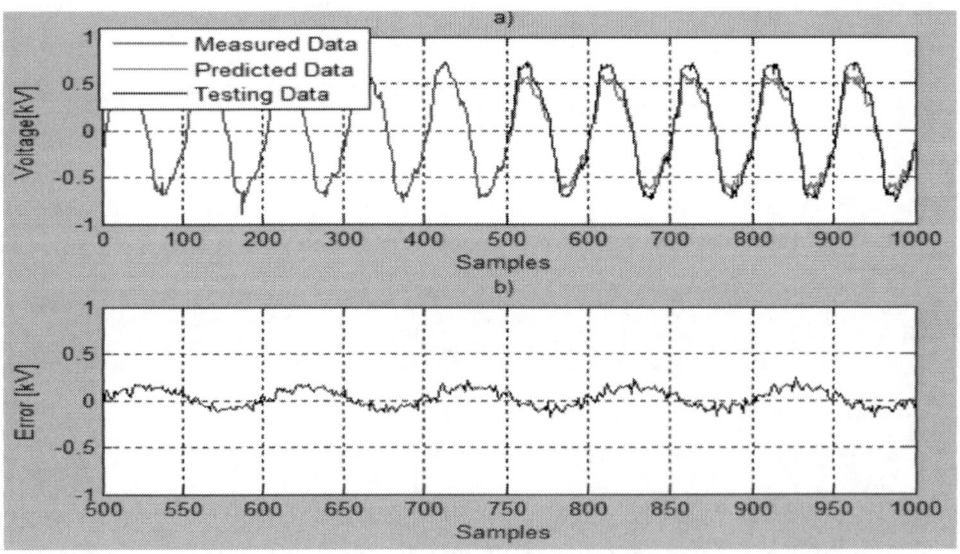

Figure 4. a) Prediction of voltage for new data set, b) Obtained error during network prediction

In Figure 5 is represented the voltage-current characteristic of the furnace that is obtained by plotting the voltage and current from the secondary side of the furnace transformer for both adaptive training of the network and prediction case taking into consideration measured and simulated data.

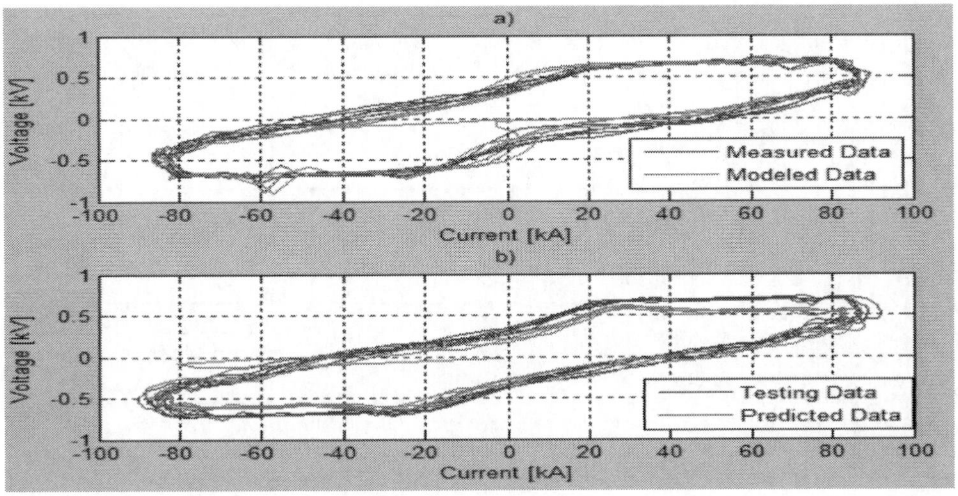

Figure 5. Voltage-current characteristic of the furnace for a) modelled data and b) predicted data

4. NAR Neural Network

The NAR neural network can be used to predict a time series from the same series past values. This network is a dynamic network with feedback that can be transformed between open-loop and close-loop.

In this paper the NAR neural network is used to simulate a network that is trained up to the present with some known values of a time-series in the open-loop mode and then is switched to the closed-loop mode in order to perform multi-steps ahead predictions. In the closed-loop the network makes predictions by using internal feedback because the external one isn't known by the network for the new data set. When the network is transformed into a closed-loop one the network is turned into a parallel configuration.

4.1. EAF Current and Voltage Prediction

The NAR neural network is created by using *narnet* function from *Matlab*. The *input delay* is *5*, the *feedback delay* is *12* and the *number of the hidden neurons* is *6*. The input of the network is a current or a voltage.

Input and target time series data is prepared for the network training by using the *preparets* function. By using this function the input and target time series are shifted as many steps are needed to fill the initial input and layer delay states. After training process the network is transformed into a closed-loop network by using *closeloop* function and the delay is removed by using *removedelay* function. In this way the network can be tested to notice the capability of the network to learn the behaviour of the input.

In order to test the capability of the network to perform prediction of a sequence of the input, the network is simulated for a new data set containing different values as compared to the ones used for the network training.

In Figure 6 is illustrated the network training in the case of current for *500 samples*, respectively the obtained error during network training. It can be noticed that the NAR neural network has the capability to learn this data set very accurately.

In Figure 7 is represented the response of the network for a new data set. It can be observed that the response of the network follow precisely the acquired current and the error has small values.

Figure 8 illustrates the training in the case of voltage for a data set of *500 samples*, respectively the obtained error during network training. It can be noticed that the NAR neural network has the capability to learn this data set very accurate.

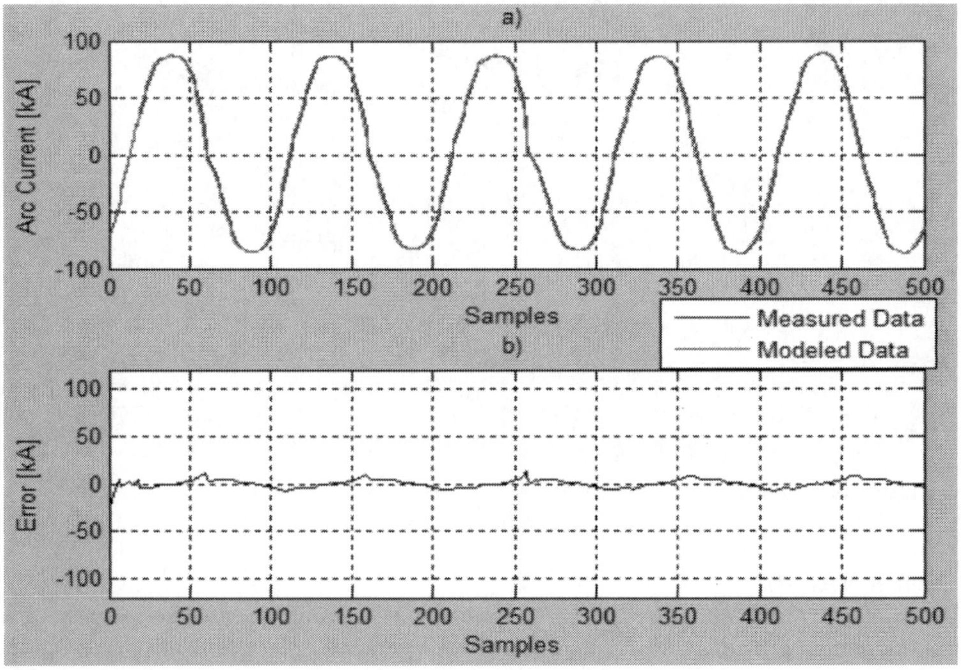

Figure 6. a) Training in the case of electric current, b) Obtained error during network training

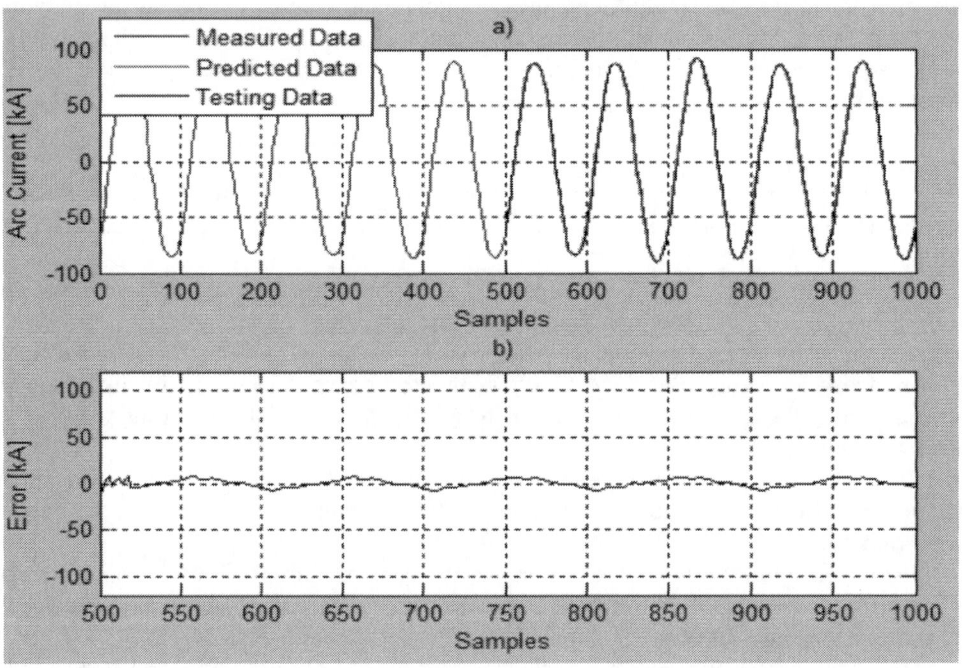

Figure 7. a) Prediction of current for new data set, b) Obtained error during network prediction

In Figure 9 is represented the response of the network when is simulated for a new data set that contains *500 samples* of voltage. It can be observed that the response of the network follow precisely the acquired voltage, the prediction being accurate.

In Figure 10 is represented the voltage-current characteristic of the furnace that is obtained by plotting the voltage and current from the secondary side of the furnace transformer for both training and prediction of the network taking into consideration measured and simulated data.

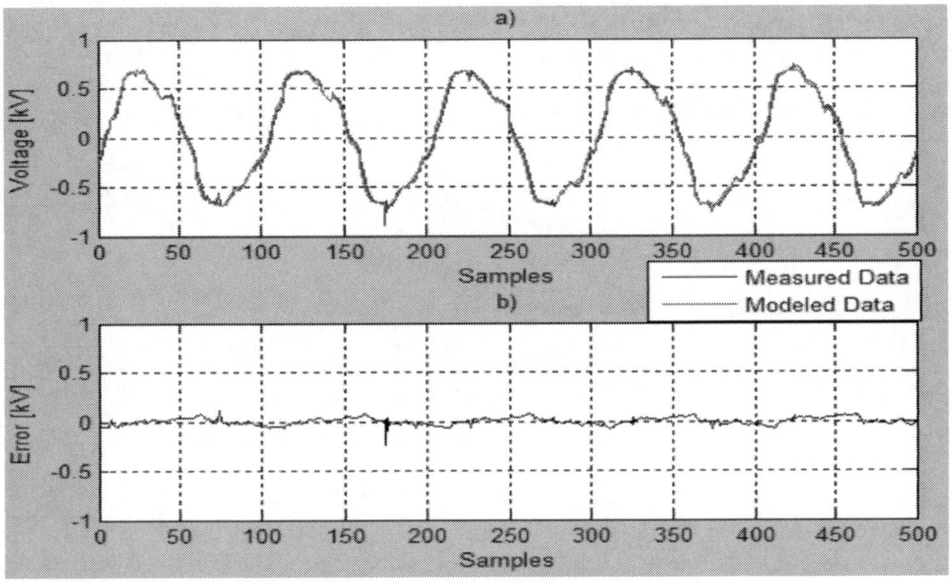

Figure 8. a) Training in the case of voltage, b) Obtained error during network training

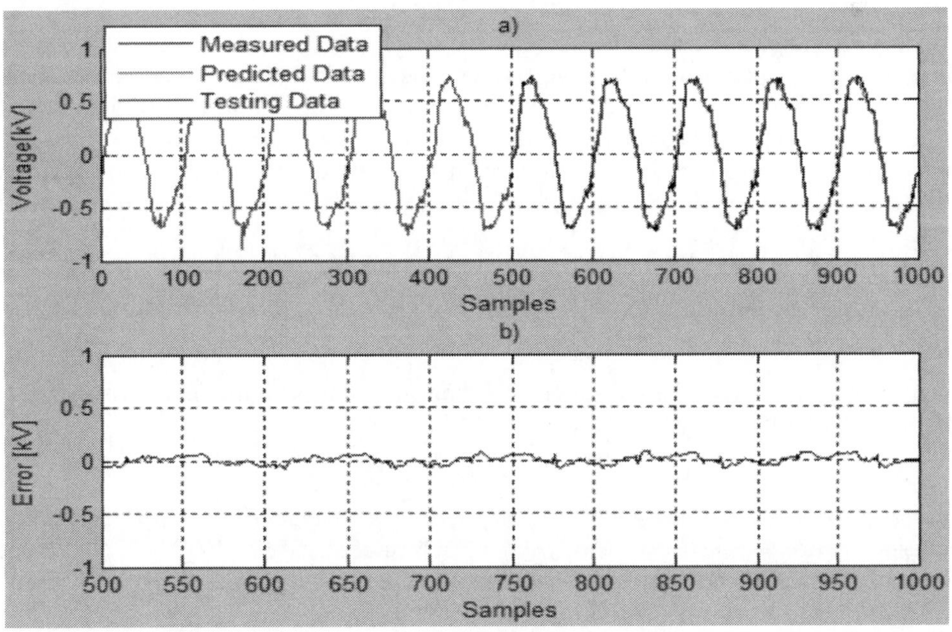

Figure 9. a) Prediction of voltage for new data set, b) Obtained error during network prediction

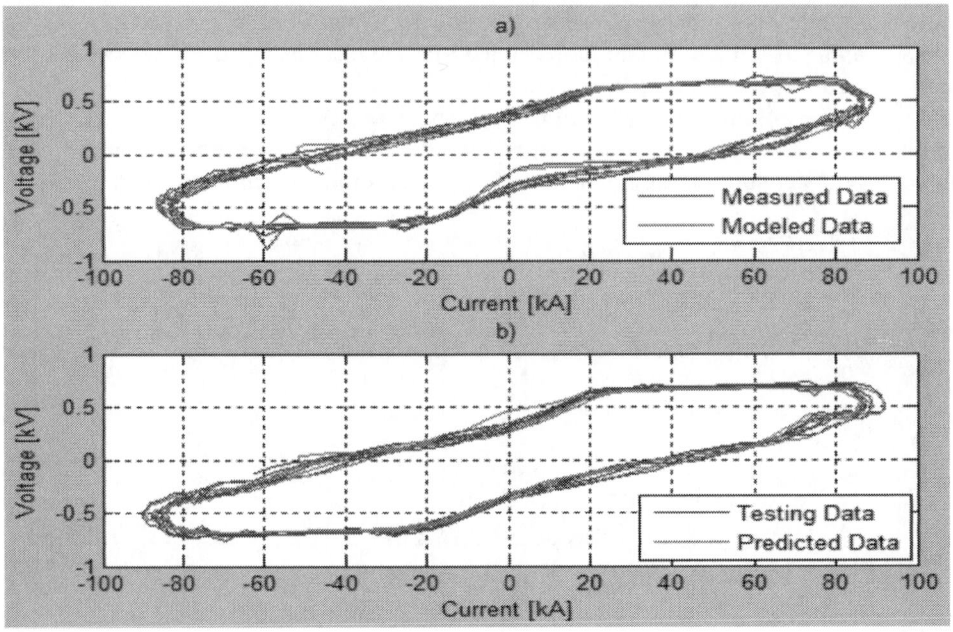

Figure 10. Voltage-current characteristic of the furnace for a) modelled data and b) predicted data

5. Conclusions

In this paper was described the development of two architecture of neural network such as ADALINE and NAR neural networks. In order to develop the networks was used Matlab 2012 environment.

The simulation results are compared with acquired data from a real electric arc furnace in order to validate the two models. The training data set was used to notice the capability of the network to learn the behavior of the electric arc furnace while the testing data set was used to obtain the capability of the network to perform generalization.

Obtained results illustrate that the NAR neural network performs a better prediction as compared to the ADALINE neural network. The results obtained from training and testing of the network can be used to estimate the voltage-current characteristic of the furnace for different meltdown stages.

These two networks illustrate the possibility of neural network to model the nonlinear and time-varying system such as electric arc furnace without using equations. Developed networks can be considered as adaptive networks.

Both networks can make prediction on multi-steps ahead and can be used to perform a particular function such as prediction or noise cancelation so can be used in signal processing and control systems.

Acknowledgments. This work was developed through the Partnerships in Priority Areas - PN II, Project No.202/01.10.2015, Code: PN-II-RU-TE-2014-4-1788.

References

[1] Sadeghian A R and Lavers J D 2000 *Application of feedforward neuro-fuzzy networks for current prediction in electric arc furnaces*, In Neural Networks, IJCNN 2000, Proceedings of the IEEE-INNS-ENNS International Joint Conference on, Como, Italy, 27-27 July, Vol. 4, pp 420-425

[2] Deaconu S I, Popa G N, Toma A and Topor M 2010 Modeling and Experimental Analysis for Modernization of 100-t EAF, *IEEE Transactions on Industry Applications* **46**(6) 2259-2266

[3] Toma A I, Popa G N, Iagar A and Deaconu S I 2010 *Experimental analysis of electric*

parameters of a 100 t UHP electric arc furnace, IEEE International Conference on Industrial Technology (ICIT), Vina del Mar, Chile, 14-17 March, pp 919 – 924

[4] Iagar A, Popa G N and Dinis C M 2015 Analysis the electrical parameters of a high-frequency coreless induction furnace, *IOP Conf. Ser.: Mater. Sci. Eng.* **85** 012013

[5] Petersen H M, Koch R G, Swart P H and Heereden R 1995 Modelling arc furnace flicker and investigating compensation techniques, *IEEE Trans Power Delivery* **8** 1733–1740

[6] Islam M and Chowdhury A H 2012 *Comparison of dynamic resistance arc furnace models for flicker study*, IEEE 7th International Conference on Electrical and Computer Engineering, Dhaka, Bangladesh, 20–22 December, pp 193–196

[7] Gomez A, Durango J M and Mejia A E 2010 *Electric arc furnace modeling for power quality analysis*, ANDESCON, the bi-annual Conference of the Andean Council of the IEEE, Bogota, Colombia, 15-17 September, pp 1–6

[8] Golkar M A, Tavakoli Bina M and Meschi S 2007 *A novel method of electrical arc furnace modeling for flicker study*, Renew Energies Power Qual **126**(7) 620–626

[9] Cano Plata E, Tacca H E 2005 *Arc furnace modeling in ATP-EMTP*, International Conference on Power Systems Transients, Montreal, Canada, 19–23 June

[10] Zheng T and Makram E B 2000 An adaptive arc furnace model, *IEEE Trans Power Delivery* **15** (3) 931–939

[11] Panoiu M, Panoiu C, Sora I and Osaci M 2007 *Using a model based on linearization of the current - voltage characteristic for electric arc simulation*, Proceedings of the 16th IASTED International Conference on Applied Simulation and Modelling, ASM 2007, pp 99 - 103

[12] Hooshmand R, Banejad M and Esfahani M T 2008 A new time domain model for electric arc furnace, *Journal of Electrical Engineering* **59**(4) 195–202

[13] Emanuel A E and Orr J A 2000 *An improved method of simulation of the arc voltage-current characteristic*, Proceedings of the Ninth International Conference on Harmonics and Quality of Power, Orlando, FL, USA, 1-4 October, pp 148–154

[14] Tang L, Kolluri S and Mark F M G 1997 Voltage flicker prediction for two simultaneously operated arc furnaces, *IEEE Trans. Power Delivery* **12**(2) 985-992

[15] Montanari G C, Loggini M, Cavallini A and Pitti L 1994 *Flicker and distortion compensation in electrical plants supplying arc-furnaces*, Ind Appl Soc Ann Meet, Denver, CO, USA, 2-6 October, pp 2249–2255

[16] Wang F, Jin Z and Zhu Z 2005 *Modeling and prediction of electric arc furnace based on neural network and chaos theory*, Advances in Neural Networks–ISNN, pp 819-826

[17] Sadeghian A R and Lavers J D 1999 *Application of radial basis function networks to model electric arc furnaces*, Neural Networks, IJCNN'99, International Joint Conference on, Washington, DC, USA, 10-16 July, Vol. 6, pp 3996-4001

[18] Chang G W, Chen C I and Teng Y F 2010 Radial-basis-function-based neural network for harmonic detection, *IEEE Transactions on Industrial Electronics* **57**(6) 2171-2179

Determination of the performance of the Kaplan hydraulic turbines through simplified procedure

I Pădureanu[1], M Jurcu[1], C V Campian[1] and C Haţiegan[2]

[1]Eftimie Murgu University of Resita, Department of Mechanical and Materials Engineering, Traian Vuia str., no. 1-4, 320085 Resita, Romania
[2]Eftimie Murgu University of Resita, Department of Electric Engineering and Informatics, Traian Vuia str., no. 1-4, 320085 Resita, Romania

E-mail: nelutupadureanu@yahoo.com

Abstract. A simplified procedure has been developed, compared to the complex one recommended by IEC 60041 (i.e. index samples), for measurement of the performance of the hydraulic turbines. The simplified procedure determines the minimum and maximum powers, the efficiency at maximum power, the evolution of powers by head and flow and to determine the correct relationship between runner/impeller blade angle and guide vane opening for most efficient operation of double-regulated machines. The simplified procedure can be used for a rapid and partial estimation of the performance of hydraulic turbines for repair and maintenance work.

1. Introduction

This paper reprezent the commissioning of the hydro-units from the Retezat downstream River, who was carried out before 1989 and was commissioned in an emergency regime, without samples that would confirm their real technical state.

The technical documentation prepared by the designer gives all the constructive and functional parameters [1], [2] that define the technical condition of each equipment, requiring experimental measuring of some these parameters at the moment of commissioning.

The simplified procedure for calculating the performance of the Kaplan's hydraulic turbines consists of: the measurements are made at the existing head of the plant; the turbine has automatically adjusted runner blades combined with automatically adjusted wicket gates, cam being the one existing in the regulator; the power of the hydro-unit increases in stages from minimum to maximum by changing the opening of the wicket gates; the efficiency is calculated only at maximum power, this being the efficiency guaranteed by the turbine supplier [3-8].

Index samples are the total tests performed in the hydroelectric plant on the hydro-aggregate to determine the flow and efficiency as relative sizes. The index efficiency is calculated from the specific hydraulic energy and the power measured by primary methods and the flow rate measured by the secondary method, according to IEC 60041 standard.

- A system for acquiring, monitoring and diagnosing the operation of hydro-aggregates has been developed, which highlights their performance before starting repairs and after repair [9].
- Through measurements and assessments made with AMDS (Automated Monitoring and Diagnostic System), an important contribution has been made to experimental research on the operation of hydraulic turbines. Complex experiments were carried out prior to the start of the

repair of the hydro-aggregates for rehabilitation and commissioning after repair, in order to monitor whether the re-assurance provided in the technical documentation was performed.

- Capital repairs, refurbishments and warranty tests require index proofing before and after repairs, in line with the recommendations of IEC 60041 standard. Upon the express request of the customer, the simplified procedure for performance determination can be used, index samples being used only when unrealized efficiency after retrofitting.

2. Methodology. Equipment using for measurement

The experimental researches have been carried out for a longer period of time, at each rehabilitation of hydro power groups, respectively before dismantling (BD) and after the commissioning (AC) [9].

At each hydro-aggregate, the table shows the results of the measurement of the measured quantities. These tables include: turbine power, turbulent flow, head and course of the wicket gates.

At each power stage, measurements are made for 3 minutes after stabilizing the hydro-aggregate operating mode. The acquisition rate of the measured quantities is 25kS/s. The values shown in the table are the average values of the measured process parameters for 3 minutes.

The power of the turbine (P_T), resulted from the active power measured with the electrical parameters analyser and process parameters, relative to the generator efficiency known in the electrical documentation. The Q flow was measured with flowmeters Winter Kennedy installation, installed in the plant. The verification of the facility and its calibration was carried out by the CCHAPT in some samples.

The head (H) resulted from the difference in the H_{am} and H_{av} levels, so is the gross head. At the hydropower plants where the measurement were made, the differences between net head and grosshead are negligible. The course of the wicket gates was measured with the linear transducer.

Turbine efficiency could be determined with relationship [9]:

$$\eta_T = \eta_H = \frac{P_T}{\rho \cdot g \cdot H_n \cdot Q_T} \quad (1)$$

where:
P_T – power of the turbine;
H_n – gross head;
Q_T – flow;
ρ – density;
g - gravitational acceleration.

The Automated Monitoring and Diagnostic System (AMDS) [10], [11] designed and built within CCHAPT Resita (Figure 1), consists of the following components: transducers for the measurement of the upstream (z_i) and downstream (z_e) level; transducer for the measurement of the course of the wicket gates (SAD); transducer for measuring the pressure difference in the spiral case (ΔWK); reducers for current and voltage measurement to determine the active power (U, I); a specialized VPA 323 equipment for the acquisition and processing of signals for the testing, monitoring and diagnosis of reactive hydraulic turbines; process interface for specialized equipment to acquire electrical and process parameters.

Transducers for the measurement of the upstream (z_i) and downstream (z_e) level, for the measurement of the course of the wicket gates (S_{AD}) and for the measurement of the pressure difference in the spiral case (ΔWK) were installed in the accessible areas of the hydroelectric plant and connected to the analogue inputs behind the specialized equipment of the electrical and process parameters.

In order to obtain the active power the voltage on the three phases (UR, US, UT) and the current on the three phases (IR, IS, IT) were measured to be connected to the specialized equipment of the electrical and process parameters on the AC inputs(a.c.) of 100V / 600V, 5A, [12].

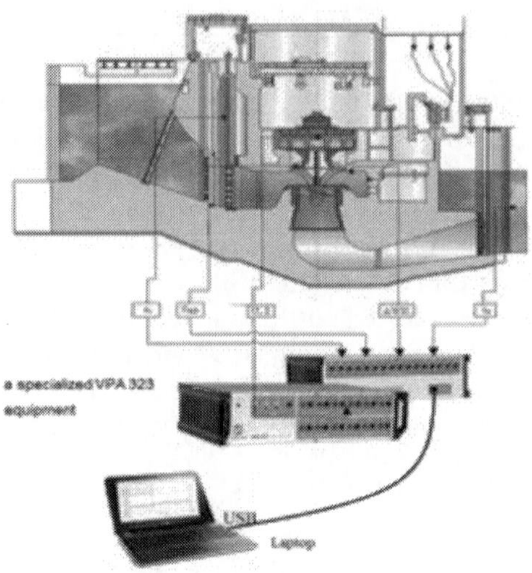

Figure 1. Automated Monitoring and Diagnostic System (AMDS)

3. Experimental results

Hydropower development The Retezat downstream (Figure 2) has a total installed capacity of 134.3 MW and provides an average annual energy output of 193.4 GWh. The arrangement located in the Hateg's depression, between the Clopotiva hydropower plant and the confluence of the Great River Strei, containtes 3 accumulations (Ostrovul Mic, Paclisa and Haţeg) carried by the dam and the hydro power plant is on retention area and 6 hydro power plant placed on the channel (Ostrovul Mare, Cârnesti I, Cârnesti II, Totesti I, Toteşti II and Orlea). All hydropower plants are equipped with two Kaplan turbines [9], [13], [14].

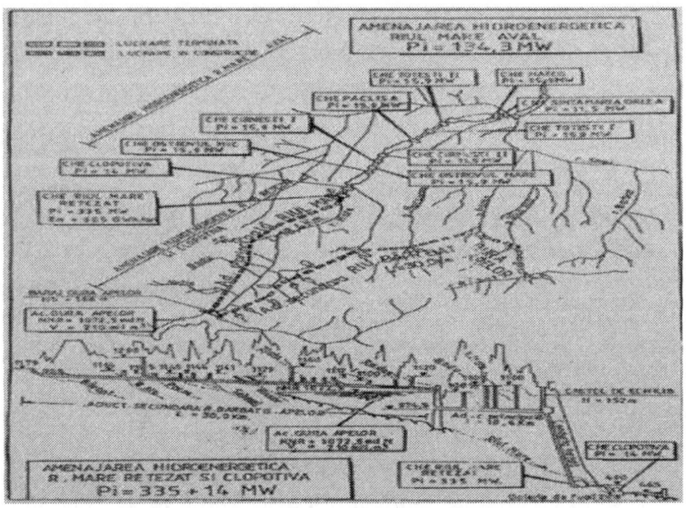

Figure 2. Great Retezat River downstream

Will be present 2 hydro aggregates before dismantling and after commissioning, together with experimental research results [15], [16].

- General data and main technical features for KVB turbine type 8.4 - 21 to HEPP Paclisa HA_2.
- Technical specifications :
  - Turbine type: 6-pallet,
  - CHE type: retention front
  - maximum net head: Hmax = 21 m.c.a

- net calculation head: Hc = 19.3 m.c.a.
- minimum net head: Hmin = 17,0 m.c.a.
- maximum coupling power: Pmax = 8400 kW
- maximum flow: Qmax = 45 m3 / sec.
- maximum suction height: Hs.max = - 2.75 m
- rotor diameter: D1 = 2750 mm
- rated speed: n = 214.3 rpm

• Maximum turbine output power corresponding to characteristic net heads:
  - Hmax: Pmaxmax = 8400 kW
  - Hc: PmaxC = 7603 kW
  - Hmin: Pmaxmin = 6027 kW

Figure 3. HEPP Paclisa HA2

Figure 3 shows the machine room at CHE Paclisa HA_2 where the measurements were made and Table 1 presents the results of the experimental investigations at this hydro-aggregate before disassembly and after commissioning.

Table 1. Process parameters measured at HEPP Paclisa HA_2

Measurements BD

P_T[MW]	Q[m3/s]	H[m]	S_{AD} [%]
2	12.6	17.41	53.16
3	20.02	17.41	67.8
4	28.02	17.41	79.3
5	36.1	17.39	93
6	39.2	17.39	99.6

Measurements AC

P_T[MW]	Q[m3/s]	H[m]	S_{AD} [%]
2	14.5	17.2	57.3
3	20.7	17	68.3
4	27.6	16.95	78.9
5	35.3	16.9	91.9
6	39.4	17.1	99.7

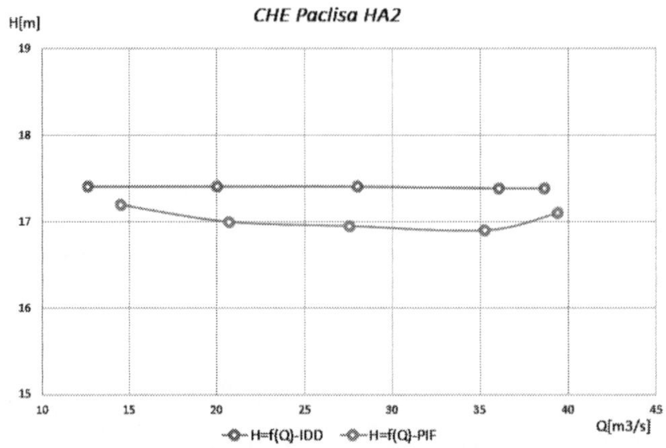

Figure 4. H=f(Q) – Head depending to flow

Figure 4 shown that both before and after repair, the turbine operates at the minimum net head, but a head stability can be observed.

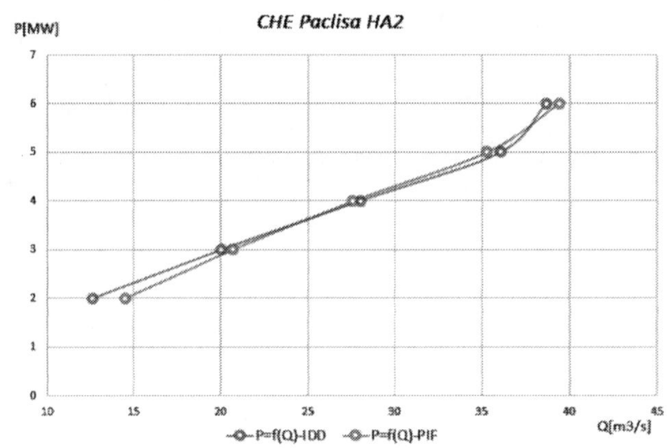

Figure 5. P=f(Q) – Power depending to flow

Figure 5 shown that active power is guaranteed at maximum net head, both before disassembly and commissioning.

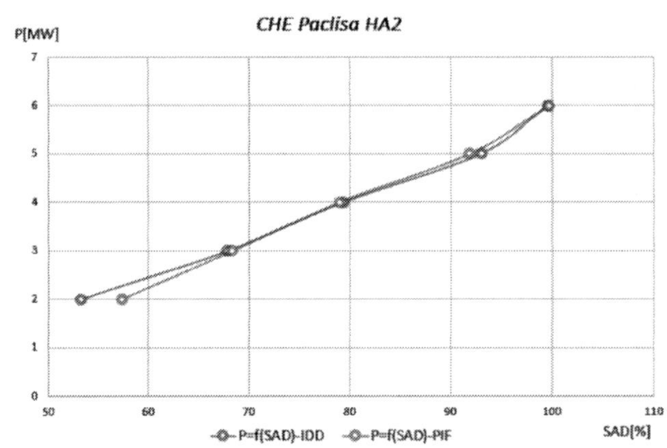

Figure 6. P=f(SAD) – Power depending to open wicket gate

Figure 6 shown that in both cases, the wicket gates is opened to the maximum, with guaranteed power at the minimum head.

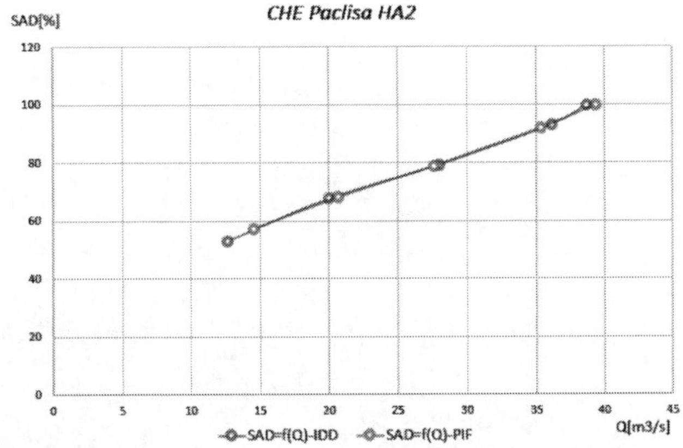

Figure 7. SAD=f(Q) – Open wicket gate depending to flow

In Figure 7 it can be seen in the SAD = f (Q) characteristics that the kinematics of the wicket gate is the same in both situations.

Knowing density ρ, gravitational acceleration g, gross head H_n, flow Q_T şi power of the turbine P_T the turbine efficiency can be determined according to therelationship (1), both before disassembly (BD) and after commissioning (AC) at the maximum measured power.

Where:

- water density at temperature T=21°C is: ρ = 998,20 kg/m^3
- gravitational acceleration is: g = 9,8054 m/s^2

- Turbine efficiency will be:

BD *Efficiency at maximum power*

$$\eta_T = \eta_H = \frac{P_T}{\rho \cdot g \cdot H_n \cdot Q_T} = 89{,}82\,\% \tag{2}$$

AC *Efficiency at maximum power*

$$\eta_T = \eta_H = \frac{P_T}{\rho \cdot g \cdot H_n \cdot Q_T} = 90{,}98\,\% \tag{3}$$

Measurement were made in the area of minimum head.

The efficiency obtained after retrofitting is the guaranteed field and is higher than before the retrofitting.

- General data and main technical features for KVB turbine type 8.4 - 21 to HEPP Cârnesti I HA_1.
- Technical specifications :
  - Turbine type: 6-pallet,
  - CHE type: retention front
  - maximum net head: Hmax = 21 m.c.a
  - net calculation head: Hc = 19.3 m.c.a.
  - minimum net head: Hmin = 17,0 m.c.a.
  - maximum coupling power: Pmax = 8400 kW
  - maximum flow: Qmax = 45 m3 / sec.
  - maximum suction height: Hs.max = - 2.75 m
  - rotor diameter: D1 = 2750 mm
  - rated speed: n = 214.3 rpm
- Maximum turbine output power corresponding to characteristic net heads:

- Hmax: Pmaxmax = 8400 kW
- Hc: PmaxC = 7603 kW
- Hmin: Pmaxmin = 6027 kW

Figure 8. HEPP Cârnesti I HA1

Figure 8 shows the machine room at HEPP Cârnesti I HA1 where the measurements were made and Table 1 presents the results of the experimental investigations at this hydro-aggregate before disassembly and after commissioning [17-19].

Table 2. Process parameters measured at HEPP Cârnesti I HA1

Measurements BD				Measurements AC			
P_T[MW]	Q[m3/s]	H[m]	S_{AD} [%]	P_T[MW]	Q[m3/s]	H[m]	S_{AD} [%]
1.5	9.2	20.24	45.4	1.41	7.4	15.3	25.66
3.1	17.8	20.14	59.84	2.7	15.8	15.1	51.78
5	28.7	20.06	75.27	3.52	23.5	15.02	66.9
6.7	37.9	19.78	90.1	4.11	28.8	15	75.54
7.9	44.4	19.55	98.9	5.25	39.6	14.7	88.8
				6.06	49.8	14	96.8

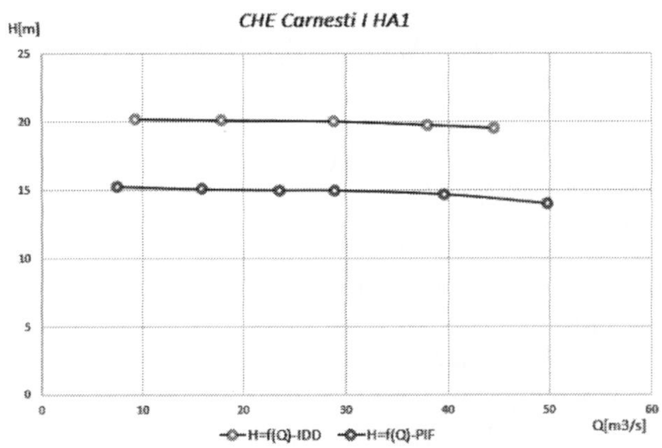

Figure 9. H=f(Q) – Head depending to flow

From Figure 9 it's shown at the BD, the turbine was operating at the net head, falling into the parameters of the technical documentation drawn up by the designer, and at the AC the head is below the minimum.

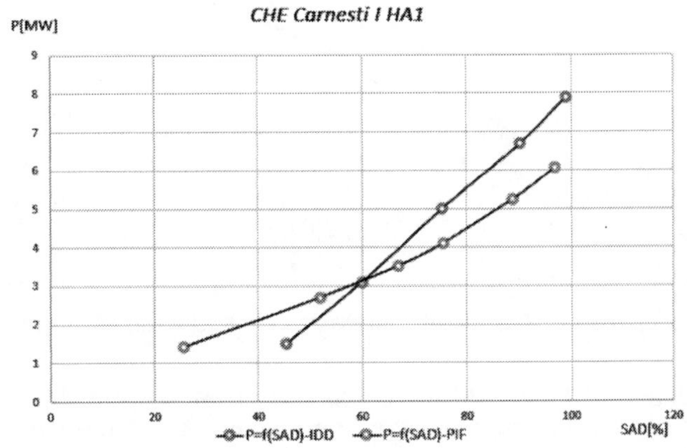

Figure 10. P=f(Q) – Power depending to flow

From Figure 10 it's shown evolution of powers from minimum to maximum is in accordance with the great difference of heads.

From Figure 11 it's shown the evolution of active powers by opening the guiding blades is similar with Figure 10.

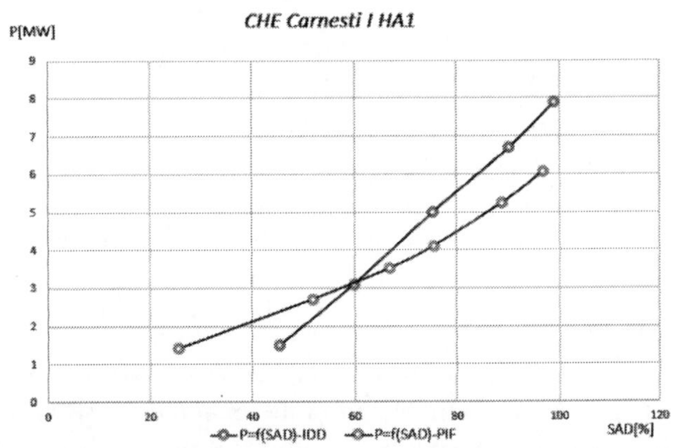

Figure 11. P=f(SAD) – Power depending to open wicket gate

From Figure 12 it's shown there is no significant difference in the kinematics of the wicket gate before and after repair.

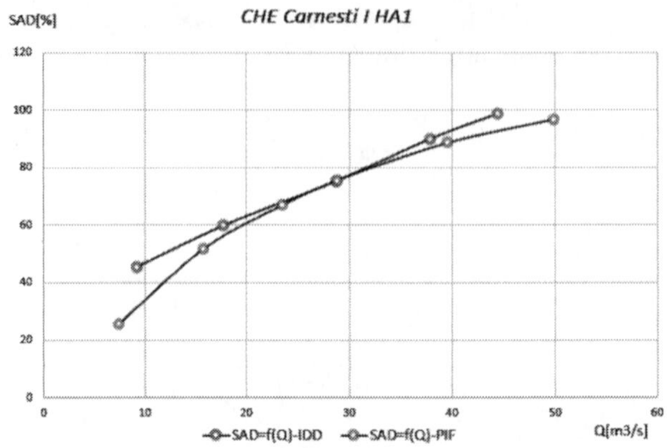

Figure 12. SAD=f(Q) – Open wicket gate depending to flow

Knowing density ρ, gravitational acceleration g, gross head H_n, flow Q_T şi power of the turbine P_T the turbine efficiency can be determined according to the relationship (1), both before disassembly (BD) and after commissioning (AC) at the maximum measured power [20-22].

Where:

- water density at temperature T=21°C is: ρ = 998,20 kg\m^3
- gravitational acceleration is: g = 9,8054 m\s^2

- Turbine efficiency will be:

BD *Efficiency at maximum power*

$$\eta_T = \eta_H = \frac{P_T}{\rho \cdot g \cdot H_n \cdot Q_T} = 92{,}98\,\% \tag{4}$$

AC *Efficiency at maximum power*

$$\eta_T = \eta_H = \frac{P_T}{\rho \cdot g \cdot H_n \cdot Q_T} 88{,}8\,\% \tag{5}$$

The two efficiency can not be compared, the measuremnts were made at totally different heads. Before to retrofitting H=19.55m, very close to the net head and after retrofitting H=14m, much less than the minimum guaranteed head.In this case after commissioning, the net head wasn't assurance.

4. Conclusions

The operating characteristics were automatically generated from the measured parameters. At HEPP Paclisa hydro power station no. 2, was measured in the minimum head area. The efficiency obtained after retrofitting is in the range guaranteed by the turbine supplier. At HEPP Cârnesti I hydro power station no. 1, efficiency can not be compared, the measurements were made at totally different heads. The functionality of this power plant need optimizations to assurance the net head, because all of this hydro power plants it works in cascade mode.

The simplified procedure can be used for a rapid and partial estimation of the performance of hydraulic turbines for repair and maintenance work. If the simplified procedure, when used after retrofitting, efficiency is lower than those guaranteed, they must be checked by complex index samples, recommended by IEC 60041 and corrected the cam combinatorial.

Complex index samples, require an experimental investigation of at least three heads, resulting in at least five days of withdrawal of the hydro-unit from the operation, namely one day of mounting measuring equipment, three days of testing and one day of dismantling the equipment.

The simplified procedure requires only two days to withdraw the hydro-unit from the operation, one day for mounting measuring equipment and one day for testing and dismantling the equipment.

Acknowledgments The work has been funded by the Sectoral Operational Programme Human Resources Development 2007-2013 of the Ministry of European Funds through the Financial Agreement POSDRU/159/ 1.5/S/132395.

References

[1] Haţiegan C, Pădureanu I, Jurcu M, Nedeloni M D, Hamat C O, Chioncel C P, Trocaru S, Vasile O, Bădescu O, Micliuc D, Filip Nedeloni L, Băra A and (Barboni) Haţiegan L 2016 Vibration analysis of a hydro generator for different operating regimes, *IOP Conf. Ser.: Mater. Sci. Eng.* **163** 012030

[2] Haţiegan C, Chioncel C P, Răduca E, Popescu C, Pădureanu I, Jurcu M R, Bordeaşu D, Trocaru S, Dilertea F, Bădescu O, Terfăloagă I M, Băra A and (Barboni) Haţiegan L 2016 Determining the operating performance through electrical measurements of a hydro generator, *IOP Conf. Ser.: Mater. Sci. Eng.* **163** 012031

[3] Yordanova S and Georgieva A 2013 Genetic algorithm based optimization of fuzzy controllers tuning in level control, *Electrotechnica and Electronics* **48**(9-10) 45-51

[4] Mostafa A Elhosseini, Ragab A El Sehiemy, Amgad H Salah and Abido M A 2017 Modeling and control of an interconnected combined cycle gas turbine using fuzzy and ANFIS controllers, *Electrical Engineering* 1–23

[5] Yordanova S 2014 Intelligent approaches for linear controllers tuning with application to temperature control, *Journal of Intelligent and Fuzzy Systems* **27**(6) 2809-2820

[6] Yordanova S and Sivchev Y 2014 Design and tuning of parallel distributed compensation-based fuzzy logic controller for temperature, *Journal of Automation and Control* **2**(3) 79-85

[7] Răduca E, Schmidt M, Ungureanu-Anghel D, Răduca M, Haţiegan C, Pop N and Novăcescu F 2014 The command of the Flour Transport Pumps in a Cereal Mill with PLC Siemens S7-1200, *Analele Universităţii "Eftimie Murgu"-Fascicula de Inginerie* **XXI**(2) 357-366

[8] Răduca E, Păun A F, Răduca M, Haţiegan C, Drăghici S, Rudolf C and Candea L 2014 Command and Running of an Experimental Model Vehicle for Covering a Labyrinth Type Runway, *Analele Universităţii "Eftimie Murgu"-Fascicula de Inginerie* **XXI**(2) 335-345

[9] Liuba Gh, Nedelea V M, Augustinov L, Pădeanu L, Pădureanu I, Jurcu M R and Pădeanu C 2012 *Documentaţie privind monitorizarea şi controlul la terminarea reparaţiei şi la punerea în funcţie a hidrogeneratorului nr. 1 din CHE Păclişa*, CCHAPT Contract 62 - 06/06/2012 Sucursala Hidrocentrale Haţeg – Hidroelectrica, (romanian language)

[10] Pădureanu I, Jurcu M, Augustinov L, Haţiegan C, Răduca E and Pădeanu L 2015 Optimisation of the Start-up and Operation Regimes of Cooling Water Pumps of a High-Power Hydro Generator, *Analele Universităţii "Eftimie Murgu"-Fascicula de Inginerie* **XXII**(1) 345-358

[11] Pădureanu I, Jurcu M, Augustinov L, Haţiegan C and Răduca E 2015 Implementation of an Automatic System for the Monitoring of Start-up and Operating Regimes of the Cooling Water Installations of a Hydro Generator, *Analele Universităţii "Eftimie Murgu"-Fascicula de Inginerie* **XXII**(1) 359-368

[12] Novăcescu F, Haţiegan C, Răduca M, Răduca E, Pop N and Nedeloni M A 2012 New Method for Testing the No-Load Work of an Electric One-Phase Transformer Using The Graphical Programming, *Scientific Bulletin of "Politehnica" University of Timişoara* **57**(71)2 54-67

[13] Jurcu M R, Padureanu I, Padeanu L, Augustinov L and Hatiegan C 2015 Tests Regarding the Transitory Regimes of Putting off Load of the Hydroagregate, *Analele Universităţii "Eftimie Murgu", Fascicula de Inginerie* **XXII**(2) 185-195

[14] Haţiegan C, Padureanu I, Jurcu M R, Biriescu M, Răduca M and Dilertea F 2016 The evaluation of the insulation performances of the stator coil for the high power vertical synchronous hydro-generators by monitoring the level of partial discharges, *Electrical Engineering* **99**(3) 1013-1020

[15] Haţiegan C, Răduca M, Frunzăverde D, Răduca E, Pop N and Gillich G R 2013 The modeling and simulation of the thermal analysis on the hydrogenerator stator winding insulation, *J*

Therm Anal Calorim **113**(3) 1217-1221

[16] Raduca M, Hatiegan C, Pop N, Raduca E and Gillich G R 2014 Finite element analysis of heat transfer in transformers from high voltage stations, *J Therm Anal Calorim* **118**(2) 1355-1360

[17] Raduca E, Nistor L, Hatiegan C, Raduca M, Padureanu I and Draghici S 2015 *Web server for command, control and monitoring of industrial equipment,* 9th International Symposium on Advanced Topics in Electrical Engineering (ATEE), 7-9 May, Bucharest

[18] Popescu C and Hatiegan C 2015 Aspects regarding the Calculation of the Dielectric Loss Angle Tangent between the Windings of a Rated 40 MVA Transformer, *Analele Universităţii "Eftimie Murgu"-Fascicula de Inginerie* **XXII**(2) 298-305

[19] Haţiegan C, Hălălae I, Popescu C, Gillich N, (Barboni) Haţiegan L, Răduca E and (Filip) Nedeloni L 2016 Evaluation insulation of the stator coil a hydro-generators through monitoring the level of partial discharges, *Annals of „Constantin Brâncuşi" University of Târgu Jiu* **3** 57-62

[20] Răduca E, Schmidt M, Haţiegan C, Hălălae I and Drăghici S 2016 An Application for the Improvement of the Transportation System of the Flour in a Grain Mill using PID Compact, *Analele Universităţii "Eftimie Murgu"-Fascicula de Inginerie* **XXIII**(1) 283-290

[21] Răduca M, Haţiegan C, Budai A M, Răduca E and Pop N 2012 *Optimization conditions of stator winding insulation of hydrogenerator,* National Symposium of Theoretical Electrical Engineering SNET'12, Vol. 3, No.1, pp 316-321

[22] Răduca M, Răduca E, Haţiegan C and Ungureanu D 2011 Fuzzy controller for adjustment of liquid level in the tank, *Annals of the University of Craiova, Mathematics and Computer Science Series* **38**(4) 33-43

Unbalance influence on the rotating assembly dynamics of a hydro

M Jurcu[1], I Pădureanu[1], C V Campian[1] and C Haţiegan[2]

[1]Eftimie Murgu University of Resita, Department of Mechanical and Materials Engineering, Traian Vuia str., no. 1-4, 320085 Resita, Romania
[2]Eftimie Murgu University of Resita, Department of Electric Engineering and Informatics, Traian Vuia str., no. 1-4, 320085 Resita, Romania

E-mail: marcelljurcu@yahoo.com

Abstract. The dynamics of the rotating parts of a hydro is characterized by the dynamic interaction between the rotor, the stator and the working fluid in order to operate the hydro.The main factors influencing the dynamics of the rotating parts of a hydro are: rotor unbalance, unbalanced magnetic pull, shaft misalignment and hydraulic flow regime.Rotor unbalanced is one of the most common factors influencing the dynamic stability of the rotating parts of a hydro.The unbalanced is determined by: uneven distribution of rotating masses, displacement of parts in the rotor during rotation, inhomogeneity of rotor component materials, expansion of the rotor due to heating, and rising speed during the transient discharge of the load.The mechanical imbalance of a rotor can lead to important forces, responsible for the vibration of the machine, which ultimately leads to a shorter operating time.Even a low unbalance can lead, in the case of high speed machines, to major unbalance forces that cause significant damage to the equipment. The unbalance forces cause additional vibrations in the bearings as well as in the foundation plate.To avoid these vibrations, it is necessary in the first stage to balance the static rotor in the construction plant and then to a dynamic rotation balancing.

1. Introduction

Considering the general tendency to reduce conventional energy resources, especially fossil fuel in the last period of time, there is a growing emphasis on the issue of increasing the efficiency of electricity production from unconventional resources [1].

The advantages of generating electricity by means of high power hydro, which typically exceed 100MW, are conditioned by the reliability of these equipments and the requirement of safe operation throughout their life [2-6].

The high power and high speed hydro power rotors have very large dimensions and weights, so they store huge amounts of kinetic energy during operation, which under certain conditions can lead to significant damage by shutting down.

Practice has shown that one of the major factors which determine the safety of operation is the rotating parts of hydro.

The dynamics of the rotating parts of a hydro is characterized by the dynamic interaction between the rotor, the stator and the working fluid in order to operate the hydro safely [7-12].

The main factors influencing the dynamics of the rotating parts of a hydro are: unbalance, unbalanced magnetic pull, shaft misalignment and hydraulic flow [13-15].

In this paper we present the factors that influence the dynamics of the rotating parts of the hydro-aggregates and the way they contribute to the instability of the position of the rotation axis of the hydro-aggregate. Also, some experimental procedures are set to highlight the causes that lead to inadequate functioning of the hydro-aggregate in terms of bearing vibrations due to the instability of the rotor dynamics.

2. Unbalance influence

Unbalance is one of the most common factors in practice, influencing the dynamic stability of the rotating parts of a hydro [16].

It is determined by: uneven distribution of rotating masses, displacement of parts in the rotor during rotation, inhomogeneity of rotor component materials, expansion of the rotor due to heating and rising speed during transient discharge from load.

The unbalance determines forces that produce radial beats and vibrations, the frequencies of which are determined by the rotor speed [16], [17]:

$$f_d = k\frac{n}{60}\ \text{(Hz)} \tag{1}$$

The centrifugal force produced by the unbalance caused by the initial eccentricity is the expression:

$$F_{em} = 11{,}2\left(\frac{n}{1000}\right)^2 G_{2l} l_0 \tag{2}$$

where:

e_o is the initial static eccentricity of the rotor shaft;

g_{2l} is the weight of the rotor between the two radial guide bearings expressed in kg.

Initial eccentricity e_o can be expressed as an amount of two eccentricities:

$$e_o = \varepsilon_1 + \varepsilon_2 \tag{3}$$

where:

ε_1 is the eccentricity of the shaft to bearings caused by bearing play, bearing wear, or imoprection of the assembly;

ε_2 is the eccentricity of the rotor axis relative to the stator axis due to ovality, deviation of the air gap, shaft line break;

The operation of a hydro with a large unbalance leads to high vibrations in the bearings or to an unsafe operation. For each machine, dynamic balancing is required in order to cancel the unbalance forces [17].

The balancing of the rotor is carried out first, statically, in the construction plant, followed by the dynamic rotation balancing [17].

To achieve proper results, dynamic balancing is required in two planes, since the distribution of masses along the length of the rotor is not uniform.

Balancing is done by placing weights on the side surface of the rotor ΔG_1 și ΔG_2.

These weights determine the centrifugal forces:

$$F_{c1} = 11{,}2\left(\frac{n}{1000}\right)\Delta G_1 \cdot r_1$$

$$F_{c2} = 11{,}2\left(\frac{n}{1000}\right)\Delta G_2 \cdot r_2 \tag{4}$$

where:

r_1 și r_2 are the rays corresponding to the points where the two weights were placed ΔG_1 și ΔG_2

The weights ΔG_1 and ΔG_2 re determined in such a way that the forces Fc_1 and Fc_2 balance the two forces of imbalance Fd_1 and Fd_2 at the two levels at which balancing is performed.

The unbalance forces Fd_1 and Fd_2 being spaced, with the length l, cause a moment that produces additional reaction forces in the radial guide bearings, which results in a further heating thereof.

Proper balancing implies:

$$F_{c1} = 11{,}2\left(\frac{n}{1000}\right)^2 \Delta G_1 \cdot r_1 = -F_{d1}$$
$$F_{c2} = 11{,}2\left(\frac{n}{1000}\right)^2 \Delta G_2 \cdot r_2 = -F_{d2} \quad (5)$$

In this way, the unbalance forces are canceled at the level of the two planes as in Figure 1, which determines that the hydro-aggregate operation is appropriate from this point of view.

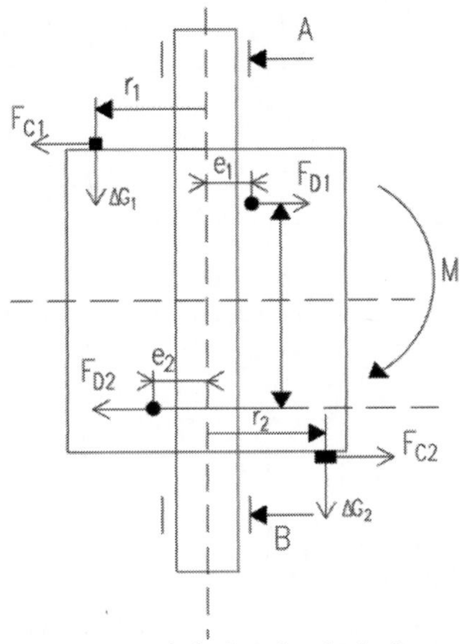

Figure 1. Dynamic balancing in 2 planes

It is worth noting that the balancing of the rotating assembly of a hydro is performed at the level of the rotor which has a very large mass and allows the balancing weights to be arranged on the rotor hub or on the fan discs [17].

3. Experimental results

Experimental tests have been carried out on different hydro of different sizes [18]:

- Hydro no.2 from Hydro power plant Cârneşti I , max power : 5.4 MW
- Hydro no.1 from Hydro power plant Măniceşti , max power : 6.1 MW
- Hydro no.1 from Hydro power plant Toteşti I , max power : 8.4 MW
- Hydro no.1 from Hydro power plant Lereşti , max power : 20 MW

The following were measured on these hydro [19]:

- Vibration in the upper generator guide bearing on the +x axis(V_LRAG+x)
- Vibration in the upper generator guide bearing on the +y axis (V_LRAG+y)
- Vibration in the lower generator guide bearing on the –x axis (V_LRI-x)
- Rotor speed (n)

3.1. Experimental results from Hydro no.2 at Hydro power plant Cârneşti

Values obtained before maintenance are given in Table 1.

Table 1. Vibration Value

No	n	Vibration [mm/s]		
	[rpm]	V_LRAG_+Y	V_LRAG_+X	V_LRI_-X

1	215.2	0.63	0.53	0.77
2	215.14	0.69	0.53	0.51
3	215.1	0.61	0.49	0.49
4	215.1	0.65	0.55	0.4
5	215.15	0.73	0.54	0.27
6	215.22	0.75	0.53	0.27
7	215.28	0.77	0.56	0.46
8	215.3	0.61	0.53	0.28
9	215.27	0.69	0.53	0.57
10	215.28	0.68	0.5	0.51
11	215.25	0.77	0.59	0.43
12	215.23	0.72	0.46	0.29
13	215.2	0.68	0.48	0.23
14	215.21	0.73	0.5	0.26
15	215.21	0.81	0.5	0.28
Average value		**0.73**	**0.51**	**0.33**

A high level of vibration is observed in upper generator guide bearing on the +y axis. To reveal the causes of these vibrations we proceed to decompose the harmonic vibration signals. Figure 2 shows the frequency spectrum of the vibration.

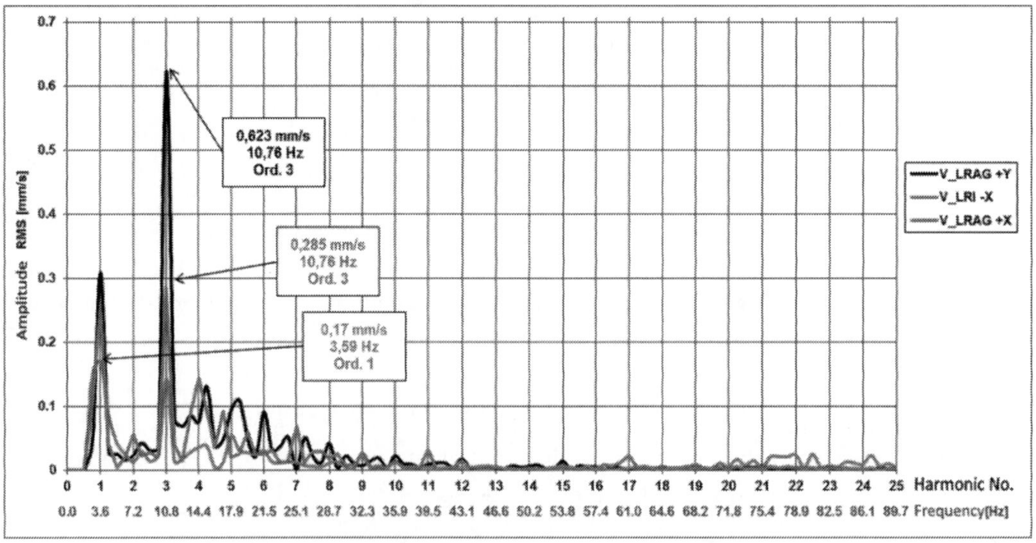

Figure 2. Spectrum frequency of vibrations

The frequency spectrum highlights a very large amplitude of the no. 3 harmonics and a very small amplitude of the fundamental, indicating a negligible mechanical imbalance, so no dynamic balancing of the rotating assembly is required. After completion of the repair work on the hydro the results are as follows:

Table 2. Vibration Value

No	n	Vibration [mm/s]		
	[rpm]	V_LRAG_+Y	V_LRAG_+X	V_LRI_-X
1	215.2	0.32	0.28	0.18
2	215.14	0.32	0.3	0.18
3	215.1	0.32	0.3	0.2

4	215.1	0.31	0.29	0.2
5	215.15	0.29	0.28	0.21
6	215.22	0.34	0.28	0.2
7	215.28	0.33	0.28	0.2
8	215.3	0.33	0.29	0.2
9	215.27	0.32	0.29	0.19
10	215.28	0.32	0.28	0.19
11	215.25	0.33	0.3	0.2
12	215.23	0.34	0.27	0.2
13	215.2	0.32	0.27	0.2
14	215.21	0.32	0.28	0.19
15	215.21	0.32	0.28	0.19
Average value		0.32	0.28	0.2

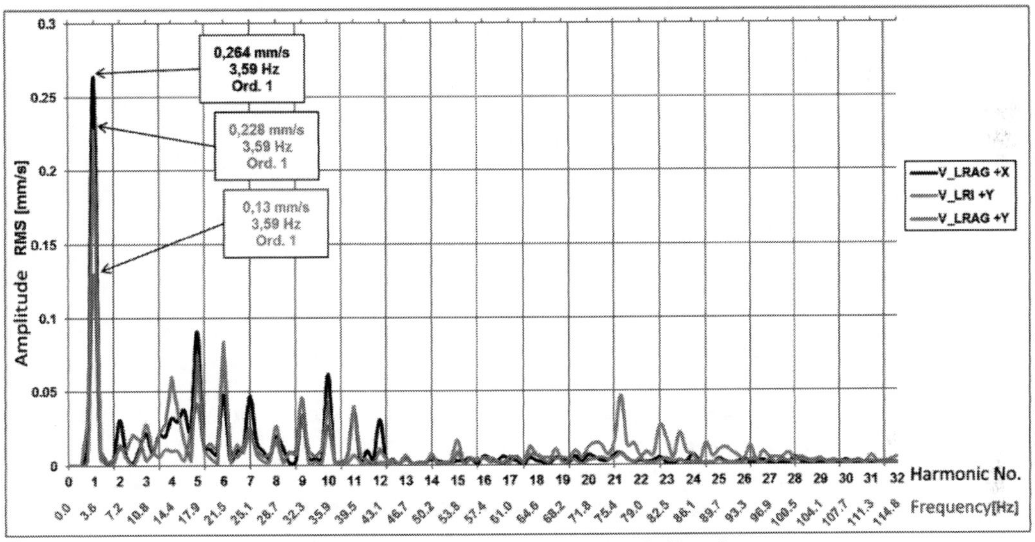

Figure 3. Spectrum frequency of vibrations

Figure 3 shows the frequency spectrum of the vibration signals, where there is an almost complete reduction of the 3rd order harmonics and some reduction of the fundamental amplitude compared to the situation before the repair.

3.2. Experimental results from Hydro no.1 at Hydro power plant Manicești

Values obtained before maintenance are given in Table 3.

Table 3. Vibration Value

Time [s]	Speed [RPM]	LRA_x_rad [mm/s]	LRA_y_rad [mm/s]	LRI_x_rad [mm/s]
0	166.7	1.87	0.39	0.11
1	166.61	2.02	0.28	0.15
2	166.69	1.67	0.39	0.12
3	166.53	2.08	0.41	0.16
4	166.53	1.93	0.4	0.13
5	166.36	1.94	0.4	0.14
6	166.61	1.7	0.33	0.21
7	166.53	1.84	0.38	0.13

8	166.69	1.74	0.34	0.14
9	166.53	1.78	0.33	0.13
10	166.36	1.67	0.34	0.13
Average		1.83	0.39	0.14

It is noted that the vibration level is very high, the maximum being in upper generator guide bearing on the +x axis, 1.83 mm / s.

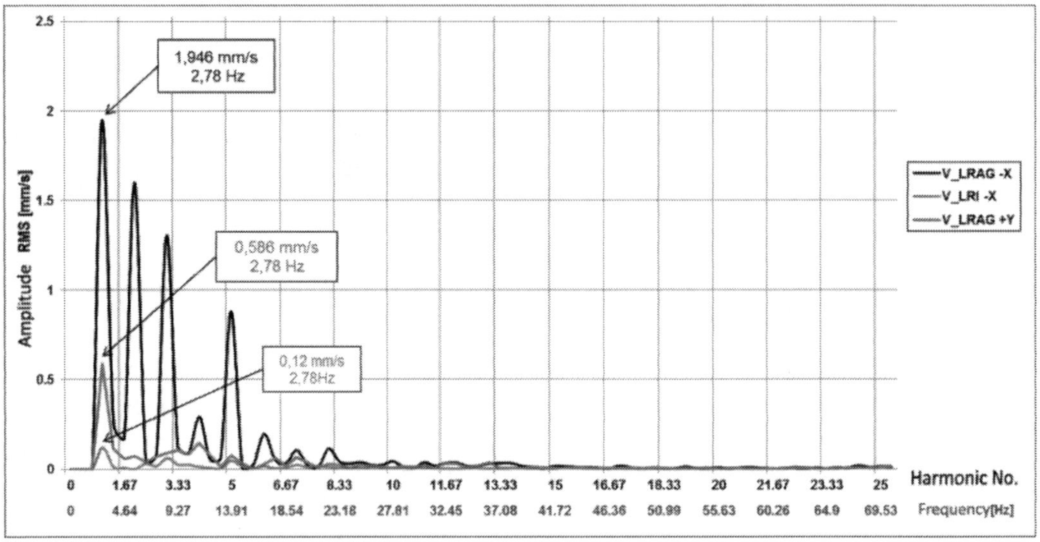

Figure 4. Spectrum frequency of vibrations

The very high amplitude of the fundamental, 1,946 mm / s indicates a pronounced unbalance, but the presence of higher harmonics with very large amplitudes of order 2, 3 and 5 indicates the presence of other factors that influence the rotor dynamics.

For this reason, it is necessary that the repair work program eliminates these factors before dynamic balancing. When checking the vibrations after balancing, it was found that the vibration level dropped a lot, the results being given in Table 4.

The maximum vibration level in radial direction + x in upper generator guide bearing on the +x axis decreased from 1.83 mm / s to 0.58 mm / s, which justifies the efficiency of maintenance work on the hydro-aggregate.

Table 4. Vibration Value

Time [s]	Speed [RPM]	LRA_x_rad [mm/s]	LRA_y_rad [mm/s]	LRI_x_rad [mm/s]
0	166.7	0,17	0,31	0,51
1	166.61	0,16	0,47	0,56
2	166.69	0,12	0,38	0,54
3	166.53	0,14	0,31	0,7
4	166.53	0,21	0,34	0,72
5	166.36	0,16	0,63	0,5
6	166.61	0,17	0,38	0,48
7	166.53	0,18	0,36	0,51
8	166.69	0,12	0,47	0,47
9	166.53	0,14	0,37	0,5
10	166.36	0,2	0,36	0,48
Average		0,18	0,38	0,58

3.3. Experimental results from Hydro no.1 at Hydro power plant Totești I

Values obtained before maintenance are given in Table 5.

Tabelul 5. Vibration Value

No	n [rpm]	Vibration [mm/s]		
		V_LRAG_+Y	V_LRAG_+X	V_LRI_-X
1	215.2	0.79	0.66	0.65
2	215.14	0.82	0.65	0.62
3	215.1	0.98	0.71	0.51
4	215.1	0.83	0.67	0.5
5	215.15	0.82	0.67	0.49
6	215.22	0.82	.067	0.56
7	215.28	0.87	0.66	0.47
8	215.3	0.8	0.7	0.54
9	215.27	0.78	0.74	0.48
10	215.28	0.84	0.72	0.61
11	215.25	0.91	0.69	0.5
12	215.23	0.82	0.69	0.44
13	215.2	0.85	0.7	0.54
14	215.21	0.72	0.64	0.55
15	215.21	0.88	0.65	.049
Average value		0.83	0.67	0.51

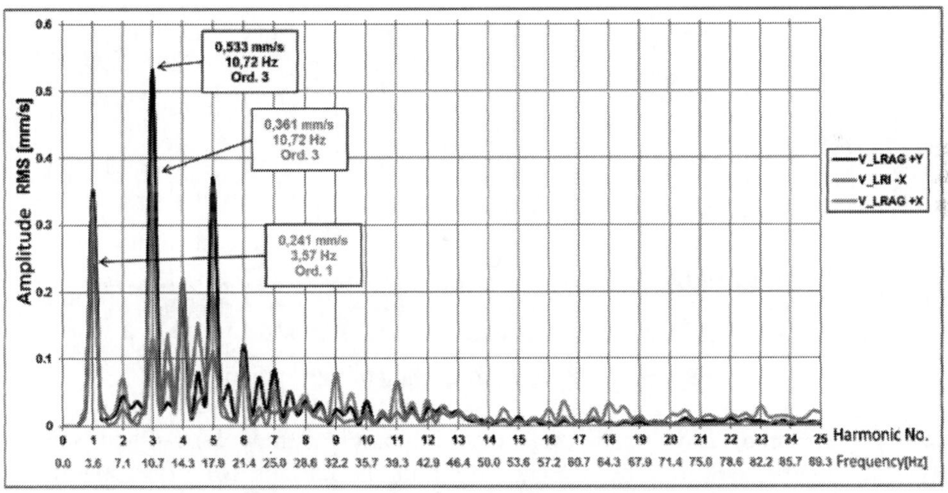

Figure 5. Spectrum frequency of vibrations

A high level of vibrations can be noticed in upper generator guide bearing on the +x axis 0.83 mm /s.

From the frequency spectrum of the vibrations shown in Figure 5, there are observed the existence of the order 3 and 5 harmonics with higher amplitudes than the fundamental.

After repair, the following results were registered:

Tabelul 6. Vibration Value

No	n [rpm]	Vibrații [mm/s]		
		V_LRAG_+Y	V_LRAG_+X	V_LRI_-X
1	212.22	0.56	0.34	0.85

2	212.68	0.56	0.34	0.76
3	213.1	0.56	0.33	0.84
4	213.31	0.51	0.43	0.77
5	213.54	0.55	0.32	0.88
6	213.66	0.59	0.37	0.97
7	213.82	0.54	0.38	0.94
8	213.95	0.59	0.35	0.86
9	214.05	0.6	0.35	0.81
10	214.17	0.58	0.4	0.86
11	214.24	0.51	0.36	0.93
12	214.26	0.63	0.31	0.78
13	214.28	0.52	0.4	1
14	214.29	0.62	0.39	0.84
15	214.31	0.52	0.39	0.84
Average value		0.56	0.36	0.86

It is noticed that after the repair the vibration level increased in lower generator guide bearing on the -x axis.

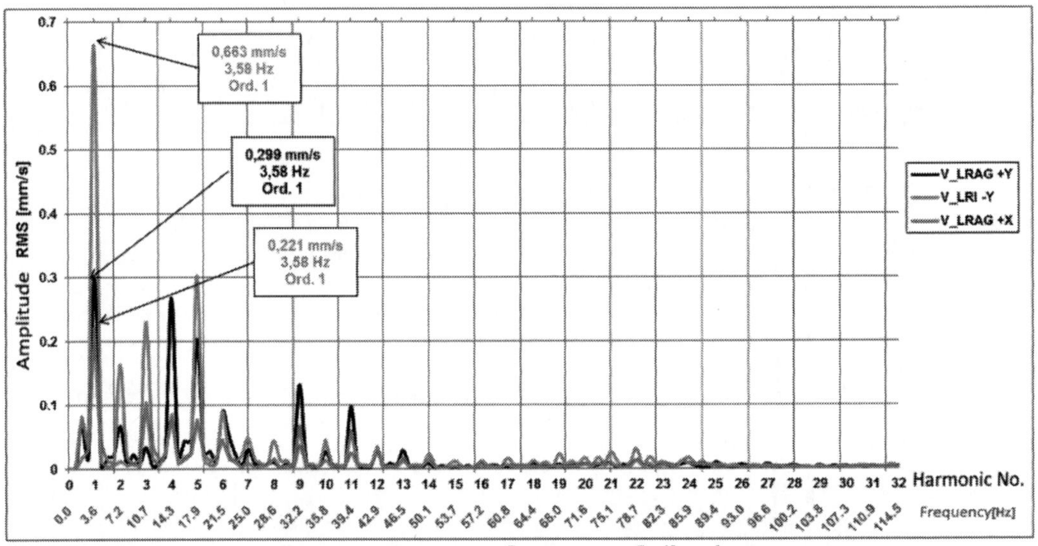

Figure 6. Spectrum frequency of vibrations

The frequency spectrum shown in Figure 6 that harmonics of the order 3,4 and 5 have not been completely reduced and indicates the presence of rotor oscillations but of lesser amplitude. The presence of these harmonics leads to the conclusion that all the conditions for proper operation have not been ensured through the maintenance work.

Since the maximum vibration level on the base is high, it is advisable to dynamically balance the rotating assembly.

3.4. Experimental results from Hydro power plant Lerești

After repair, the following results were registered in Figure 7.

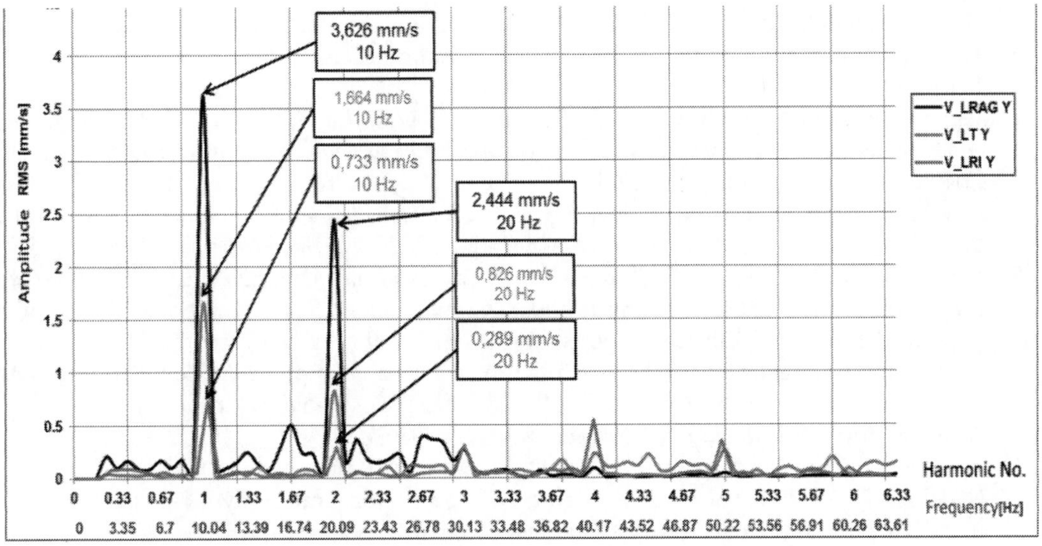

Figure 7. Spectrum frequency of vibrations

The vibration spectrum shown in Figure 7 highlights the existence of the 2nd order harmonics whose amplitude is very high, 2.44 mm / s. The existence of this harmonics reveals a shaft misalignment. Balancing is possible after shaft misalignment is resolved.

4. Conclusions

The NAR neural network can be used to predict a time series from the same series past values. This network is a dynamic network with feedback that can be transformed between open-loop and close-loop.

At hydro no.2 at Hydro power plant Cârneşti I from the measurements made before the repair showed that dynamic balancing is not necessary and after the repair the vibrations decreased from 0.73 mm / s to 0.32 mm / s, the repair works being efficient.

At hydro no.1 at Hydro power plant Măniceşti there was a mechanical imbalance, but for its correction it is necessary that the maintenance program eliminates the other factors that influence the rotor dynamics. After repair and balancing, vibrations decreased from 1.83 mm / s to 0.58 mm / s.

At hydro no.1 at Hydro power plant Toteşti I after completion of maintenance work on the frequency spectrum, a dynamic balancing is required.

At hydro no.1 at Hydro power plant Lereşti high vibrations caused by the shaft misalig in the rotary assembly prevent proper balancing.

Acknowledgment. The work has been funded by the Sectoral Operational Programme Human Resources Development 2007-2013 of the Ministry of European Funds through the Financial Agreement POSDRU/159/ 1.5/S/132395.

References

[1] Chioncel C P, Tirian G O, Gillich N, Hatiegan C and Spunei E 2016 Overview of the wind energy market and renewable energy policy in Romania, *IOP Conf. Ser.: Mater. Sci. Eng.* **163** 012009

[2] Haţiegan C, Chioncel C P, Răduca E, Popescu C, Pădureanu I, Jurcu M R, Bordeaşu D, Trocaru S, Dilertea F, Bădescu O, Terfăloagă I M, Băra A and (Barboni) Haţiegan L 2016 Determining the operating performance through electrical measurements of a hydro generator, *IOP Conf. Ser.: Mater. Sci. Eng.* **163** 012031

[3] Haţiegan C, Padureanu I, Jurcu M R, Biriescu M, Răduca M and Dilertea F 2016 The evaluation

of the insulation performances of the stator coil for the high power vertical synchronous hydro-generators by monitoring the level of partial discharges, *Electrical Engineering* **99**(3) 1013-1020

[4] Haţiegan C, Răduca M, Frunzăverde D, Răduca E, Pop N and Gillich G R 2013 The modeling and simulation of the thermal analysis on the hydrogenerator stator winding insulation, *J Therm Anal Calorim* **113**(3) 1217-1221

[5] Raduca M, Hatiegan C, Pop N, Raduca E and Gillich G R 2014 Finite element analysis of heat transfer in transformers from high voltage stations, *J Therm Anal Calorim* **118**(2) 1355-1360

[6] Raduca E, Nistor L, Hatiegan C, Raduca M, Padureanu I and Draghici S 2015 *Web server for command, control and monitoring of industrial equipment,* 9th International Symposium on Advanced Topics in Electrical Engineering (ATEE), 7-9 May, Bucharest

[7] Popescu C and Hatiegan C 2015 Aspects regarding the Calculation of the Dielectric Loss Angle Tangent between the Windings of a Rated 40 MVA Transformer, *Analele Universităţii "Eftimie Murgu"-Fascicula de Inginerie* **XXII**(2) 298-305

[8] Haţiegan C, Hălălae I, Popescu C, Gillich N, (Barboni) Haţiegan L, Răduca E and (Filip) Nedeloni L 2016 Evaluation insulation of the stator coil a hydro-generators through monitoring the level of partial discharges, *Annals of „Constantin Brâncuşi" University of Târgu Jiu* **3** 57-62

[9] Răduca M, Haţiegan C, Budai A M, Răduca E and Pop N 2012 *Optimization conditions of stator winding insulation of hydrogenerator,* National Symposium of Theoretical Electrical Engineering SNET'12, Vol. 3, No.1, pp 316-321

[10] Yordanova S and Georgieva A 2013 Genetic algorithm based optimization of fuzzy controllers tuning in level control, *Electrotechnica and Electronics* **48**(9-10) 45-51

[11] Mostafa A Elhosseini, Ragab A El Sehiemy, Amgad H Salah and Abido M A 2017 Modeling and control of an interconnected combined cycle gas turbine using fuzzy and ANFIS controllers, *Electrical Engineering* **99**(2) 1–23

[12] Răduca M, Răduca E, Haţiegan C and Ungureanu D 2011 Fuzzy controller for adjustment of liquid level in the tank, *Annals of the University of Craiova, Mathematics and Computer Science Series* **38**(4) 33-43

[13] Pădureanu I, Jurcu M, Augustinov L, Haţiegan C, Răduca E and Pădeanu L 2015 Optimisation of the Start-up and Operation Regimes of Cooling Water Pumps of a High-Power Hydro Generator, *Analele Universităţii "Eftimie Murgu"-Fascicula de Inginerie* **XXII**(1) 345-358

[14] Pădureanu I, Jurcu M, Augustinov L, Haţiegan C and Răduca E 2015 Implementation of an Automatic System for the Monitoring of Start-up and Operating Regimes of the Cooling Water Installations of a Hydro Generator, *Analele Universităţii "Eftimie Murgu"-Fascicula de Inginerie* **XXII**(1) 359-368

[15] Novăcescu F, Haţiegan C, Răduca M, Răduca E, Pop N and Nedeloni M A 2012 New Method for Testing the No-Load Work of an Electric One-Phase Transformer Using The Graphical Programming, *Scientific Bulletin of "Politehnica" University of Timişoara* **57**(71)2 54-67

[16] Saleem A, Diwakar G and Satyanarayana M R S 2012 Detection of unbalance in rotating machines using shaft deflection measurement during its operation, *IOSR Journal of Mechanical and Civil Engineering* **3**(3) 8-20

[17] Rades M 2008 *Dinamica Masinilor III,* Printech, Bucuresti, Romania

[18] Haţiegan C, Pădureanu I, Jurcu M, Nedeloni M D, Hamat C O, Chioncel C P, Trocaru S, Vasile O, Bădescu O, Micliuc D, Filip Nedeloni L, Băra A and (Barboni) Haţiegan L 2016 Vibration analysis of a hydro generator for different operating regimes, *IOP Conf. Ser.: Mater. Sci. Eng.* **163** 012030

[19] Jurcu M R, Padureanu I, Padeanu L, Augustinov L and Hatiegan C 2015 Tests Regarding the Transitory Regimes of Putting off Load of the Hydroagregate, *Analele Universităţii "Eftimie Murgu", Fascicula de Inginerie* **XXII**(2) 185-195

Possibility of using the smoothed spline functions in approximation of average course of terrain inclinations caused by underground mining exploitation conducted at medium depth

J Orwat

Silesian University of Technology, Department of Mining and Geology, Akademicka street, no. 2, 44100 Gliwice, Poland

E-mail: justyna.orwat@polsl.pl

Abstract. In this paper was presented an obtainment way of the average values of terrain inclinations caused by an exploitation of the *338/2* coal bed, conducted at medium depth by four longwalls. The inclinations were measured at sections of measuring line established over the excavations, perpendicularly to their runways, after the termination of subsequent exploitation stages. The average courses of measured inclinations were calculated on the basis of average values of measured subsidence obtained as a result of an average-square approximation done by the use of smooth splines, in reference to their theoretical values calculated via the S. Knothe's and J. Bialek's formulas. The typical values of parameters of these formulas were used. Thus it was obtained for two average courses after the ending of each exploitation period. The values of standard deviations between average and measured inclinations σ_I and variability coefficients of random scattering of inclinations M_I were calculated. Then they were compared with the values appearing in the literature and based on this the possibility evaluation of use smooth splines to determination of average course of observed inclinations of mining area was conducted.

1. Introduction

The results of geodesic measurements conducted by the use of standard methods and equipment, namely lengths and angles between measuring points based on which the situational coordinates of points are calculated, and heights differences between the points based on which the altitude coordinates of these points are calculated, are distorted by some errors. These errors are mainly binding with an accuracy of measuring devices, a human factor and the variable weather conditions.

Of course the errors appear also in case of surveys made by the use of The Global Navigation Satellite Systems or even The Unmanned Aerial Vehicles (*UAV*) [1-3]. These errors come from the additional sources like satellites constellation, troposphere parameters or gradient angle of camera lens to the terrain surface.

The above errors and the factors such as: a type and a shape of observational system (particularly the distances between measuring points); a range, a multiple [4], [5], a height and a depth of hard coal excavation; the geological structure of rock mass and its fissures [6], [7], and a random factor have an influence on the values and their uncertainties of deformation indicators of mining area determine from the results of surveys [8]. For these reasons the courses of measured deformation indicators (inter alia subsidence, inclinations) are irregular and characterized by the fluctuation (a random dispersion).

It's obvious that it should be identified the average courses of these indicators. So far the orthogonal polynomials have been mainly used to designate the average courses of measured values of subsidence, inclinations and horizontal strains. The problem with their use depends on determination of an optimal degree of a polynomial for a given indicator and it requires division of a whole profile of subsidence through into some parts [9].

Therefore in this article it has been analysed the use possibility of the smooth splines functions for designation of average courses of terrain inclinations, which were caused by an underground exploitation of hard coal bed conducted on the medium depth.

2. The courses of measured inclinations

In this section have been presented: an example of hard coal extraction from the bed located on medium depth, a description of surveys made on the points of observational line located above the exploitation field and an obtainment way of the courses of measured inclinations.

2.1. *A description of made extraction*

Some hard coal mine situated within borders of The Upper Silesian Coal Basin in Poland was running the exploitation of coal bed marked as *338/2* with four longwalls [10].

These longwalls were numbered as *001, 002, 005* and *007*. The depth of an extraction was amounted from 580 m in the northern part of region (longwall no. *001*) to 700 m in the southern part of coal bed (longwall no. *007*). The average length of longwalls was close to 250 m and their runs were amounted from 750 m (longwall no. *001*) to 1080 m (longwall no. *007*). The coal bed was exploited at a height of 2 m with the roof rocks fall to the post-mining emptiness.

It needs to be highlighted that an exploitation was conducted in the rock mass which was violated an earlier excavation. The over layers thickness amounted to 60 m. These layers were mainly formed in Quaternary and Triassic periods.

2.2. *A description of made surveys*

The deformations of mining area caused by an exploitation of four longwalls in coal bed *338/2* were observed on measuring line no. *1*. It was established perpendicularly to the runs of longwalls and consisted of 53 points. The average distance between measuring points was amounted 37 m.

The measurements of the points heights were carried out by the use of precise leveller, with an accuracy of 0.1 mm. There was used a double geometric levelling made from the middles of segments between observational points.

The sections lengths were measured using a distance-meter device, with an accuracy of 1 mm.

The points heights and segments lengths were determined in the same time, after the termination of each exploitation stage. Therefore it was obtained the four observation cycles which demonstrate the impacts coming from an exploitation of longwalls *001*; *001* and *002*; *001*, *002* and *005*; *001*, *002*, *005* and *007*.

2.3. *The obtainment way of measured inclinations courses*

On the basis of measurements of points heights before and after the termination of subsequent exploitation stages in coal bed *338/2*, there were designated the subsidence of all points from the formula (1) [11].

$$W_i^{k+1} = H_i^{k+1} - H_i^k, \tag{1}$$

where: H – height of point; W – subsidence; i – number of point; k – number of observational cycle.

The courses of measured subsidence were shown at the Figure 1.

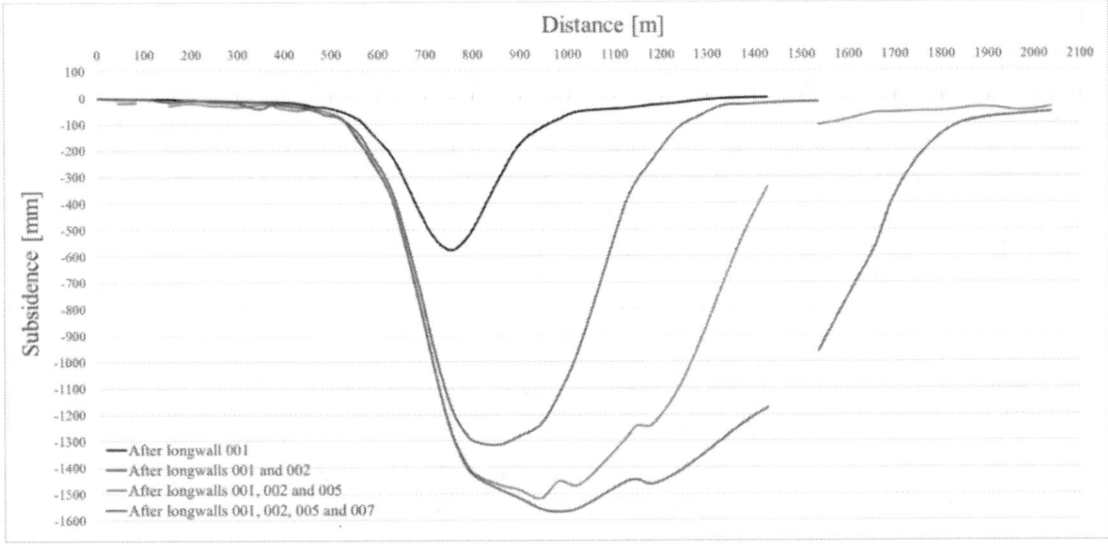

Figure 1. The courses of subsidence measured after termination of the subsequent exploitation stages in coal bed *338/2*

Based on the subsidence calculated in two neighboring points, it can be determined an inclination of segment between these points from the following formula [11]:

$$T_{i,i+1}^{k+1} = \frac{W_{i+1}^{k+1} - W_{i}^{k+1}}{l_{i,i+1}}, \tag{2}$$

where: T – inclination; W – subsidence; i – number of point; k – number of observational cycle; l – measured length of segment.

The courses of measured inclinations were shown at the Figure 2.

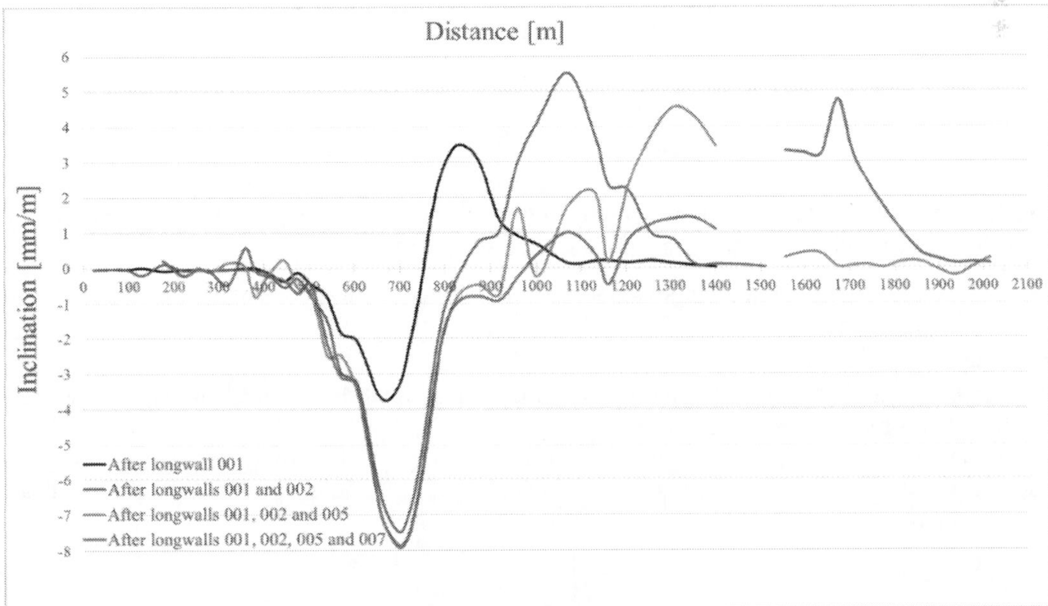

Figure 2. The courses of inclinations measured after termination of the subsequent exploitation stages in coal bed *338/2*

IOP Conf. Series: Materials Science and Engineering **294** (2017) 012029 doi:10.1088/1757-899X/294/1/012029

3. The average courses of measured inclinations

In this section have been shown how to obtain the average values of measured inclinations by the use of smooth spline functions and the basic rules of work of *smooth.spline* function from the *R* computer programme.

3.1. Calculation of average values of measured inclinations

The average values of inclinations measured after the termination of exploitation in subsequent longwalls were calculated basing on the formula (3) [12]:

$$T_{i,i+1}^{aver} = \frac{W_{i+1}^{approx} - W_{i}^{approx}}{l_{i,i+1}}, \quad (3)$$

where: T^{aver} – average value of inclination; W^{approx} – approximated value of subsidence; i – number of point; l – measured length of segment.

The average courses of measured inclinations calculated in relation to the Knothe's (circular signs) and Bialek's (square signs) formulas were shown at the Figure 3.

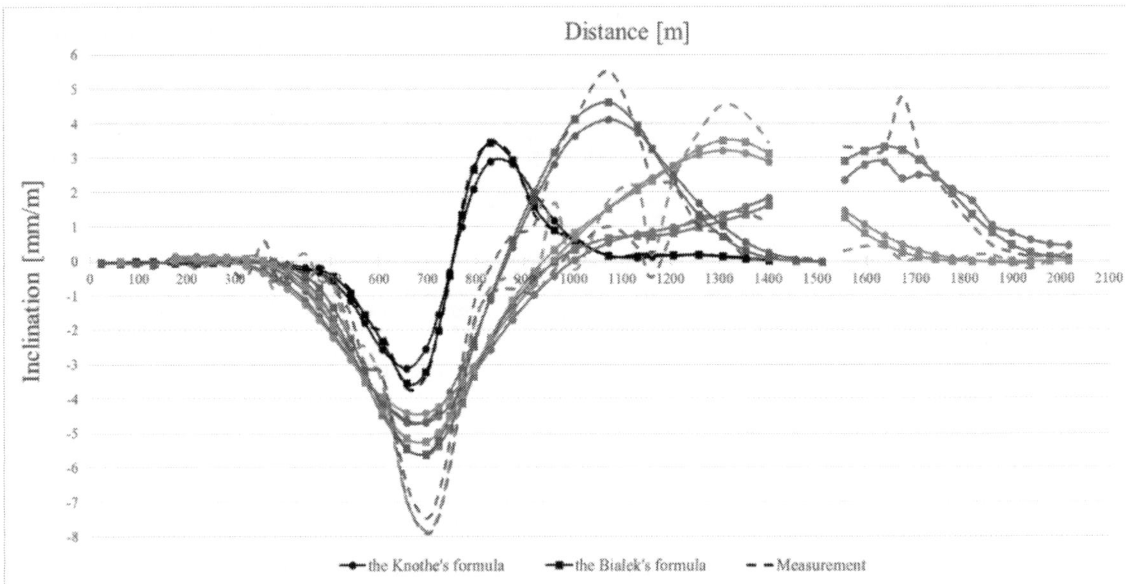

Figure 3. The average courses of measured inclinations obtained in reference to their theoretical values calculated by the use of the Knothe's and Bialek's formulas

For comparison of measured inclinations with their average values, there were calculated:

- r_T – correlation coefficient;
- σ_T – standard deviation of inclinations;
- P_T – percentages of the extreme measured values of inclinations which correspond the extreme average values of measured inclinations.

There were also calculated the values of variability coefficient of random dispersion of inclinations M_T as a standard deviation of inclinations σ_T divided by the module of average value of measured extreme inclination.

Table 1. The correlation between measured inclinations and their average values

Parameter	Reference	
	The Knothe's formula	The Bialek's formula
	After longwall 001	
r_T	0.9815	0.9978
σ_T [mm/m]	0.31	0.10
M_T[%]	10.07	2.77
P_{Tmax} [%]	83.09	98.85
P_{Tmin} [%]	84.78	96.47
	After longwalls 001 and 002	
r_T	0.9502	0.9738
σ_T [mm/m]	0.89	0.63
M_T[%]	19.04	11.31
P_{Tmax} [%]	74.64	83.70
P_{Tmin} [%]	62.70	74.87
	After longwalls 001, 002 and 005	
r_T	0.9069	0.9398
σ_T [mm/m]	1.15	0.92
M_T[%]	25.98	17.51
P_{Tmax} [%]	70.70	77.31
P_{Tmin} [%]	56.31	66.75
	After longwalls 001, 002, 005 and 007	
r_T	0.9245	0.9618
σ_T [mm/m]	1.09	0.76
M_T[%]	23.25	13.51
P_{Tmax} [%]	60.59	69.81
P_{Tmin} [%]	59.22	71.21

3.2. Approximation of average values of measured subsidence

The average values of measured subsidence were obtained as a result of least – square approximation conducted by the use of smooth spline in the *R* programme. There was used the *smooth.spline* function for that.

To the basic arguments of this function belong:

- *df* – the number of the degree of freedom assuming values from *0* to *n,* where *n* is the number of measuring points;
- *spar* – smoothing parameter of an approximating function assuming values from *0* to *1,* where value *0* denotes a total lack of smoothing and then an approximating function assumes the form of an approximated function;
- λ – penalized criterion (*PC*);
- *cv* – cross validation, when *cv* = *truth,* then we deal with *leave-one-out,* if *cv* = *false*, then we deal with *general cross validation* (*GCV*);
- *nknots* – number of knots, if *all.knots* = *false*, the value is smaller than the number of measuring points.

The least – square approximation was done in reference to the theoretical values of subsidence. There were calculated via the Knothe's formula [13] with the standard values of its parameters (a = 0.8; $tg\beta$ = 2.0; A_{obr} = 0) and via the Bialek's formula with the typical values of its parameters (a = 0.8; $tg\beta$ = 2.3; A_{obr} = 0.15). The detailed information about the reasons of such procedure were shown in the works [12], [16].

The parameters values of approximation of subsidence measured after the ending of particular exploitation stages have been juxtaposed in the Table 2.

Table 2. The parameters values of made approximation

Parameter	Reference of approximation	
	The Knothe's formula	The Bialek's formula
After longwall 001		
df	11.88	23.07
spar	0.47	0.32
λ	3.67×10^{-5}	2.02×10^{-6}
PC [mm^2]	5841	225
After longwalls 001 and 002		
df	7.24	10.54
spar	0.63	0.55
λ	3.90×10^{-4}	7.73×10^{-5}
PC [mm^2]	135434	35184
After longwalls 001, 002 and 005		
df	7.40	10.29
spar	0.68	0.61
λ	4.39×10^{-4}	1.06×10^{-4}
PC [mm^2]	254332	107365
After longwalls 001, 002, 005 and 007		
df	8.23	11.50
spar	0.66	0.58
λ	2.81×10^{-4}	6.46×10^{-5}
PC [mm^2]	205941	66549

The average courses of measured subsidence, designated in reference to the theoretical values of subsidence which were calculated by the use of the Knothe's (circular signs) and Bialek's (square signs) formulas with typical values of their parameters, were shown at the Figure 4.

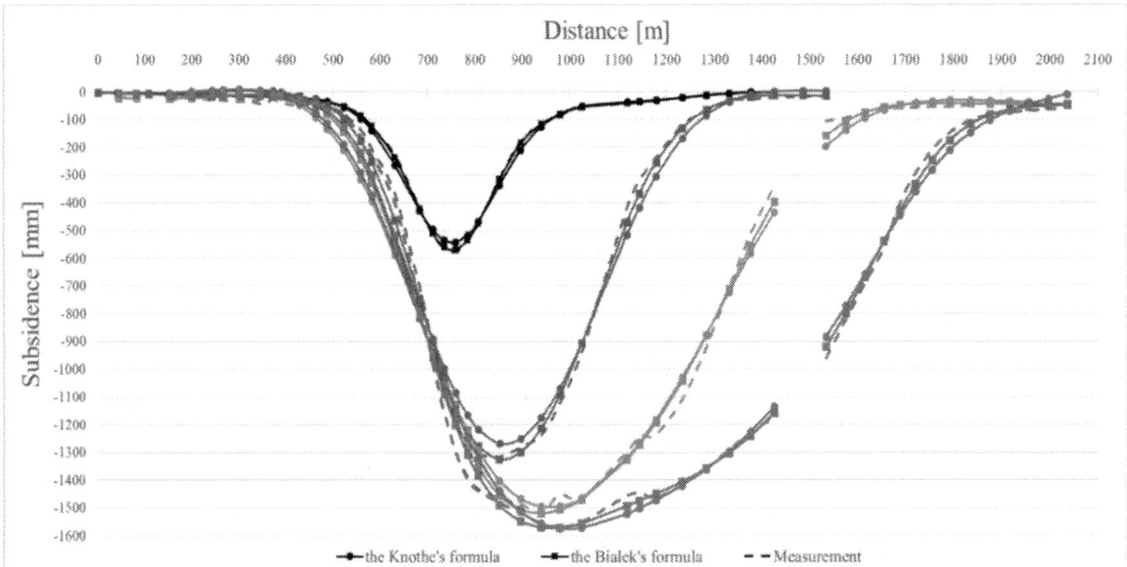

Figure 4. The approximated courses of subsidence measured after termination of the subsequent exploitation stages in coal bed *338/2*

The correlation between the measured and approximated values of subsidence has been shown by the usage of correlation coefficient r_W, standard deviation of subsidence σ_W and percentage P_W of the extreme measured value of subsidence which corresponds the extreme average value of measured subsidence. It was also calculated the value of variability coefficient of random dispersion of subsidence M_W as a standard deviation divided by the module of extreme average measured value.

Table 3. The correlation between measured and approximated subsidence

Parameter	Reference of approximation	
	The Knothe's formula	The Bialek's formula
After longwall 001		
r_W	0.9978	0.9999
σ_W [mm]	13.11	2.47
M_W[%]	2.42	0.43
P_W [%]	94.25	98.95
After longwalls 001 and 002		
r_W	0.9935	0.9982
σ_W [mm]	58.93	30.04
M_W[%]	4.64	2.27
P_W [%]	96.48	100.81
After longwalls 001, 002 and 005		
r_W	0.9923	0.9967
σ_W [mm]	73.56	46.81
M_W[%]	4.92	3.08
P_W [%]	98.47	100.06
After longwalls 001, 002, 005 and 007		
r_W	0.9944	0.9982
σ_W [mm]	64.83	36.85
M_W[%]	4.12	2.35
P_W [%]	100.28	100.18

4. Discussion of the results

The data included in the Table 2 indicate that the values of smoothing parameter *spar* of function approximating the measured courses of subsidence grow generally with the increase of exploitation range: from 0.47 after the ending of exploitation in first longwall to 0.68 after the termination of excavation in three longwalls (in case of approximation made in reference to the Knothe's formula), and from 0.32 after the first stage of exploitation to 0.61 after the third exploitation stage (in case of approximation made in reference to the Bialek's formula). It means that the approximation function reflects the measurement data in a lesser degree, when the exploitation proceeds. Above statement is confirmed by the values of penalized criterion (a residual sum of squares), which also grow together with the number of exploitation stage.

From the data juxtaposed in the Table 3 it can be seen that the values of correlation coefficient between the measured and approximated subsidence r_W decrease with the increase of exploitation range. This coefficient has reached the maximal values after first stage of extraction (0.9978 for the Knothe's formula and 0.9999 in case of the Bialek's formula). The smallest values it has reached after three longwalls (0.9923 in case of the Knothe's formula and 0.9967 for the Bialek's formula). However, its values are still high and don't come down below 0.99.

The values of subsidence standard deviation σ_W rise together with the development of exploitation in the time. It should be emphasized that for the Bialek's formula they are almost two times smaller than in the case of the Knothe's formula.

Taking into account the values of correlation coefficient of subsidence r_W and standard deviation of subsidence σ_W it can be said that the reproduction accuracy of subsidence measured courses by the approximated courses goes down with an increase of exploitation range. It's related to a random factor because its participation in the measured values of subsidence grow with the excavation development, what is confirmed by the increasing values of variability coefficient of subsidence random dispersion M_W. Its values are also two times smaller for the Bialek's formula (like in case of standard deviation).

It's surprising that the percentages of the maximal measured values of subsidence which correspond the maximal average values of measured subsidence P_W are higher for more complicated exploitation.

The correlation coefficients of inclinations r_T included in the Table 1 decrease after each exploitation stage, like in case of subsidence. Their values amount from 0.9815 to 0.9069 (for the Knothe's formula) and from 0.9978 to 0.9398 (for the Bialek's formula). As it can be seen, these values are smaller than the values of subsidence correlation coefficient r_W.

The standard deviation of inclinations σ_T increases with the exploitation development, from 0.31 mm/m after first longwall to 1.15 mm/m after three longwalls (the Knothe's formula) and from 0.10 mm/m after first longwall to 0.92 mm/m after three longwalls (the Bialek's formula). However, the differences between its values reached for the Knothe's and Bialek's formulas aren't so visible as for subsidence.

The values of variability coefficient of inclinations random dispersion M_T are higher for the Knothe's formula than for the Bialek's formula. They are almost always bigger than the values occurring in the literature. The maximal value 26 % after the third excavation stage is almost three times higher than 9.3 % identified by Kowalski in the work [8].

In case of inclinations the P_T values go down when an exploitation is developing.

5. Conclusions

It's possible to obtain the average courses of terrain inclinations measured after the termination of exploitation conducted at medium depth by the use of smoothed splines, but taking into account some conditions.

An approximating function (a smoothed spline here) should be parameterized by the use of shape parameter (*spar* here), which is responsible for fit of approximating function to the measured data.

It's recommended to do an approximation of measured courses of subsidence first and on this basis can be calculated the average courses of measured inclinations. In this way it's also possible to obtain the average courses of measured curvatures of mining area [17-19].

It should be an average – square approximation because this type of approximation ensures the greatest degree of participation elimination of random factor in the measured values.

The minimization of residual sum of squares should be referred to the theoretical courses because this criterion guarantees an obtainment of average measured courses similar to the model courses.

An accuracy of reproduction of measured courses by the approximating function is dependent on an adopted model of deformation forecasting [11], [13-15].

The reference to the Bialek's formula gives the better results than in case of the Knothe's formula. It's related to that the Bialek's formula is a three – parametric model which takes into account an existence of exploitation rim. The A_{obr} parameter is responsible for this. The Knothe's formula is only a two – parametric model [20].

Deviation of average measured courses from the measurements results is greater for the inclinations than for subsidence. Its values are increasing together with the development of exploitation in the time.

The values of variability coefficient of inclinations random dispersion M_T exceed its values occurring in the literature. It's caused by the standard values of models parameters adopted to the calculation of theoretical courses of subsidence.

To improve the results, the parameters values determined directly from the results of geodesic measurements should be used. As can be seen, the values of M_T coefficient are dependent from the parameters values of adopted model [21].

Acknowledgement
The results which have been presented in this article are the effect of work financed by the Ministry of Science and Higher Education in Poland, by means of a transfer of funds designated for the statutory activity of the Silesian University of Technology in Gliwice, the Faculty of Mining and Geology (project no. 06/050/BKM_16/0034 – BKM-558/RG5/2016).

References

[1] Janusz W, Mielimąka R, Niezabitowski M, Orwat J and Sikora P 2015 *Surface Monitoring of Water Basins Based on Use of Autonomous Flying Robots*, 20th International Conference on Methods and Models in Automation and Robotics 2015, Międzyzdroje, Poland, August 24-27, pp 376-381
[2] Klamka J, Janusz W, Maćkowski M, Niezabitowski M, Orwat J and Sikora P 2015 *UAV Application for Delivering the Internet Access During the Geodetic Measurements*, 10th Scientific Conference Internet in the Information Society 2015, Dąbrowa Górnicza, Poland, September 24-25, pp 9-25
[3] Janusz W, Niezabitowski M, Orwat J and Sikora P 2016 *The Concept of Combined Measurements of the Gliwice Radio Tower Deflection in Relation to the Speed and Direction of the Wind with Use of Quadrotor*, 17th International Carpathian Control Conference 2016, Tatranská Lomnica, Slovakia, May-June 29-01, pp 280-285
[4] Orwat J 2017 Evaluation of Forecasts of Area Deformation Caused by Multideposit Exploitation of Hard Coal Made by the Budryk-Knothe Theory – Case Study, *Conf. Proc. of SGEM 2017, Informatics, geoinformatics and remote sensing, Geodesy and mine surveying* **17**(22) 289-297
[5] Orwat J 2017 Verification of Forecasts of Subsidences, Inclinations and Curvatures of Area Caused by Multideposit Mining Conducted by Including Mining Margin – Case Analysis, *Conf. Proc. of SGEM 2017, Science and technologies in geology, exploration and mining, Exploration and mining* **17**(13) 979-986
[6] Kowalski A and Jędrzejec E 2015 Influence of Subsidence Fluctuation on the Determination of Mining Area Curvatures, *Archives of Mining Sciences* **60**(2) 487-505
[7] Piwowarski W, Strzałkowski P and Ścigała R 2017 Analysis of Rock Mass Destruction Processes with Different Activity Diagrams, *Acta Geodynamica et Geomaterialia* **14**(1) 83-92
[8] Kowalski A 2007 *Undetermined Mining Surface Deformation in Terms of the Accuracy of Forecasts*, Institute of Mining, Poland, Habilitation Thesis
[9] Stoch T, Niedojadło Z, Sopata P and Moskała S 2014 Functions Approximating the Average Course of Deformation Indicators for the Determination of the Random Dispersion Parameters of the Post-Mining Dislocations and Land Surface Deformations, *Mining Review* **70**(8) 113-123
[10] Ścigała R 2013 Mining Extraction System with Advancing Longwall as a Method of Protecting – Technical Objects and an Element of Rational Deposit Management, *Mineral Resources Management* **29**(3) 191-208
[11] Orwat J 2015 The Forecast Effectiveness of Mining Exploitation Effects on the Exploited Area Conducted with the Use of Bialek`s Formulas, *AIP Conf. Proc.* **1738** 080013-1 – 080013-4
[12] Orwat J and Mielimąka R 2017 Average Course Approximation of Measured Subsidence and Inclinations of Mining Area by Smooth Splines, *Journal of Sustainable Mining* **16**(1) 8-13
[13] Orwat J and Mielimąka R 2015 The Comparison of Measured Deformation Indicators of Mining Area with Theoretical Values Calculated Using Knothe's Formulas, *AIP Conf.*

Proc. **1738** 080014-1 – 080014-4

[14] Strzałkowski P 2015 Mathematical Model of Forecasting the Formation of Sinkhole Using Salustowicz's Theory, *Archives of Mining Sciences* **60**(1) 63-71

[15] Strzałkowski P 2001 New Mathematical Model for Description of Time-Dependent Surface Subsidence, *Transactions of the Institution of Mining and Metallurgy, Section Mining Technology* **10** A178-A182

[16] Orwat J 2017 Approximation of Average Course of Measured Subsidences of Mining Area by Smooth Splines, *AIP Conf. Proc.* **1863** 130004-1 – 130004-4

[17] Orwat J and Mielimąka R 2017 Approximation of Average Course of Measured Curvatures of Mining Area with Reference to Their Forecast Values by Bialek's Formulas, *AIP Conf. Proc.* **1863** 130003-1 – 130003-4

[18] Mielimąka R and Orwat J 2017 Approximation of Average Course of Measured Curvatures of Mining Area with Reference to Their Forecast Values by Knothe's Formulas, *AIP Conf. Proc.* **1863** 130005-1 – 130005-4

[19] Orwat J 2017 *Evaluation of Application Possibilities of Smoothed Splines to Designation of the Average Values of Terrain Curvatures Measured after the Termination of Hard Coal Exploitation Conducted at Medium Depth*, International Conference on Applied Sciences 2017, Hunedoara, Romania, May 10-12

[20] Ścigała R 2013 The Identification of Parameters of Theories Used for Prognoses of Post Mining Deformations by Means of Present Software, *Archives of Mining Sciences* **58**(4) 1347-1357

[21] Strzałkowski P 2010 Forecasts of Mine-Induced Land Deformations in Consideration of the Variability of the Parameter Describing the Process Kinematics, *Archives of Mining Sciences* **55**(4) 865-872

Appraisal of application possibilities of smoothed splines to designation of the average values of terrain curvatures measured after the termination of hard coal exploitation conducted at medium depth

J Orwat

Silesian University of Technology, Department of Mining and Geology, Akademicka street, no. 2, 44100 Gliwice, Poland

E-mail: justyna.orwat@polsl.pl

Abstract. In paper were presented results of average values calculations of terrain curvatures measured after the termination of subsequent exploitation stages in the *338/2* coal bed located at medium depth. The curvatures were measured on the neighbouring segments of measuring line No. *1* established perpendicularly to the runways of four longwalls No. *001*, *002*, *005* and *007*. The average courses of measured curvatures were designated based on average courses of measured inclinations. In turn, the average values of observed inclinations were calculated on the basis of measured subsidence average values. In turn, they were designated on the way of average-square approximation, which was done by the use of smoothed splines, in reference to the theoretical courses determined by the S. Knothe's and J. Bialek's formulas. Here were used standard parameters values of a roof rocks subsidence a, an exploitation rim A_{obr} and an angle of the main influences range β. The values of standard deviations between the average and measured curvatures σ_C and the variability coefficients of random scattering of curvatures M_C were calculated. They were compared with values appearing in the literature and based on this, a possibility appraisal of the use of smooth splines to designation of average course of observed curvatures of mining area was conducted.

1. Introduction

The influence amount of mining exploitation on terrain surface may be expressed, among others, by means of mining area curvatures. Real values of this deformation indicator are designated based on the results of geodesic measurements conducted before, during and after termination of exploitation. These surveys are carried out on the measuring points of an observational network. Generally lengths, horizontal and vertical angles, and heights of the points are measured. They can be measured by the use of standard methods and devices or the Global Navigation Satellite Systems (*GNSS*), or even the Unmanned Aerial Vehicles (*UAV*) [1-3].

To compare the measured values of curvatures with their forecasted values, it's necessary to eliminate a random factor from the measured curvatures, which is the highest from all deformation indicators [4]. It's possible thanks to an average – square approximation of measured courses of curvatures. This type of approximation can be done with the use of smoothed splines, what has been presented in this article.

2. The courses of measured curvatures

This section presents a case of hard coal exploitation, which was conducted in the *338/2* coal bed at medium depth, a characteristic of geodesic measurements done on an observational line no. *1* situated above the exploitation field and an obtainment way of curvatures measured values.

2.1. Characteristic of exploitation done in the 338/2 coal bed

The data describing the geological – mining conditions in the *338/2* coal bed and an exploitation system have been shown in the Table 1.

Table 1. The basic features of conducted exploitation

Exploitation feature	Value
Coal bed	*338/2*
Time *[years]*	1994 ÷ 1996
Exploitation system	longwall system [5]
Fulfillment of post–mining emptiness	subsidence of roof rocks [6]
Numbers of longwalls	*001, 002, 005, 007*
Interval of depth *[m]*	580 ÷ 700
Length of excavations *[m]*	250
Run of excavations *[m]*	750 ÷ 1080
Height *[m]*	2
Decline *[°]*	6 ÷ 8
Direction of coal bed inclination	N – S
Over layers thickness *[m]*	60

2.2. Characteristic of geodesic measurements

Characteristic of surveys carried out on an observational line no. *1* has been presented in the Table 2.

Table 2. The basic information about geodesic measurements

Feature	Value
Shape of observational network	line
Location in relation to the longwalls	perpendicularly to the runs
Number of measuring points	53
Average distance between points *[m]*	37
Device used for height measurement	precise leveller
Accuracy of height measurement *[mm]*	0.1
Method of height measurement	double, geometric levelling from the middle
Device used for length measurement	distance – meter
Accuracy of length measurement *[mm]*	1
Method of length measurement	double measurement

The points heights and segments lengths of measuring line were determined in the same time, after the termination of subsequent exploitation stages. Because of that, there were obtained the four curvatures courses measured after the ending of extraction in one, two, three and four longwalls.

2.3. Determination of measured curvatures courses

Based on the measurements of points heights before and after the ending of next exploitation stages in coal bed *338/2*, can be obtained the subsidence of particular points from the following formula (1) [7]:

$$W_i^{k+1} = H_i^{k+1} - H_i^k , \tag{1}$$

where: H – height of measuring point; W – measured subsidence; i – number of measuring point; k – number of measuring cycle.

The measured courses of subsidence were presented at the Figure 1.

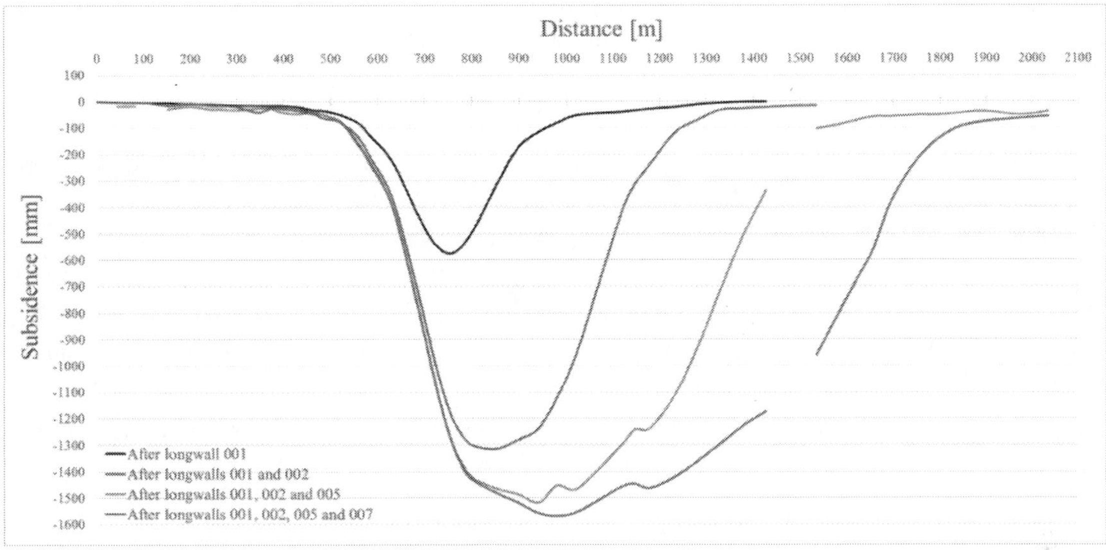

Figure 1. The subsidence courses, which were observed after the end of next excavation stages in coal bed *338/2*

On the basis of subsidence calculated in three neighboring points, it can be obtained a curvature of two segments situated between these points from the formula (2) [7]:

$$C_{i,i+1,i+2}^{k+1} = 2 \cdot \frac{W_{i+2}^{k+1} - 2W_{i+1}^{k+1} + W_{i}^{k+1}}{l_{i,i+1} + l_{i+1,i+2}}, \tag{2}$$

where: C – measured curvature; W – measured subsidence; i – number of measuring point; k – number of measuring cycle; l – measured length of segment.

The measured courses of curvatures were presented at the Figure 2.

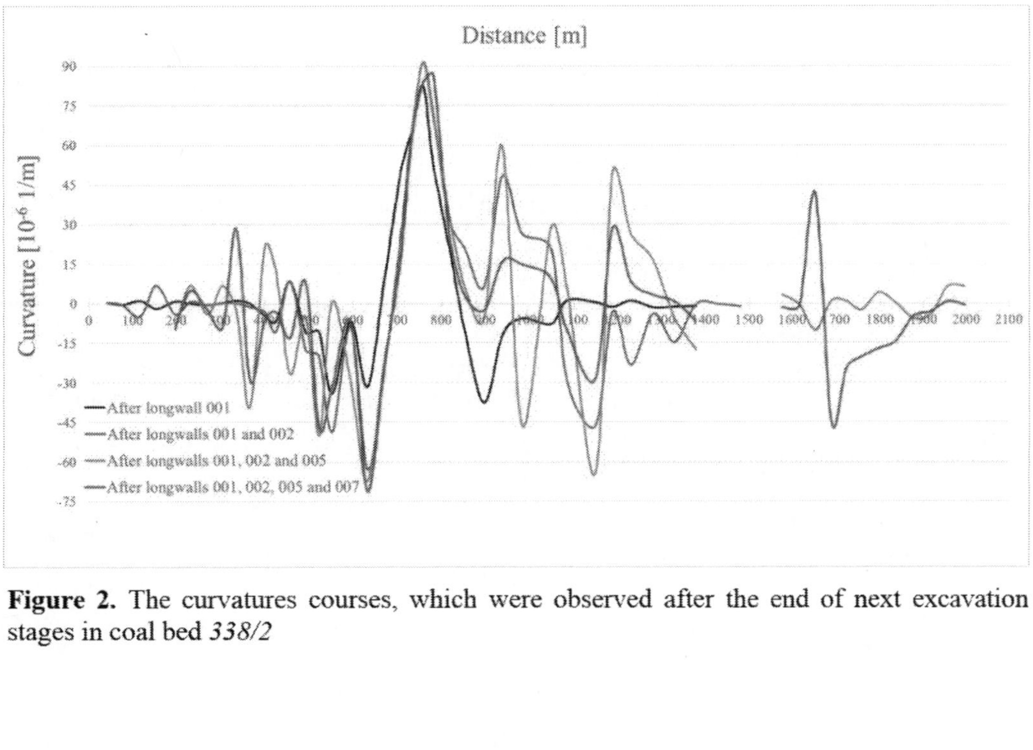

Figure 2. The curvatures courses, which were observed after the end of next excavation stages in coal bed *338/2*

3. The average courses of measured curvatures

This paragraph presents the obtainment way of average values of measured curvatures using of smoothed spline functions to that and the basic assumptions of *smooth.spline* function from the *R* programme.

3.1. Calculation of average values of measured curvatures

The average values of curvatures observed after the ending of exploitation in next longwalls were calculated based on the formula (3) [8], [9]:

$$C_{i,i+1,i+2}^{aver} = 2 \cdot \frac{W_{i+2}^{approx} - 2W_{i+1}^{approx} + W_{i}^{approx}}{l_{i,i+1} + l_{i+1,i+2}}, \tag{3}$$

where: C^{aver} – average value of curvature; W^{approx} – approximated value of subsidence; i – number of point; l – measured length of segment.

The average courses of measured curvatures calculated in reference to the Knothe's (dots) and Bialek's (squares) formulas were presented at the Figure 3.

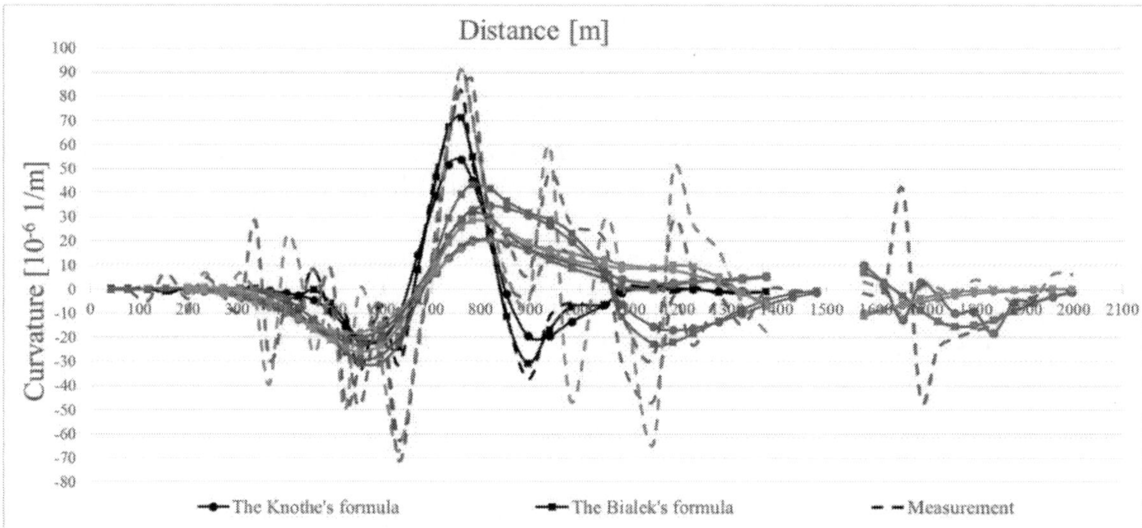

Figure 3. Observed curvatures average courses calculated regarding to their model values obtained on the basis of the Knothe's and Bialek's formulas

To compare the measured curvatures with their average values, there were calculated:

- r_C – correlation coefficient;
- σ_C – standard deviation of curvatures;
- P_C – percentages of the extreme measured values of curvatures which correspond the extreme average values of measured curvatures.

The values of variability coefficient of random dispersion of curvatures M_C were calculated as a standard deviation of curvatures σ_C divided by the module of average value of measured extreme curvature.

These values have been juxtaposed in the Table 3.

Table 3. The correlation between the measured and average values of curvatures

Parameter	Reference	
	the Knothe's formula	the Bialek's formula
	After longwall 001	
r_C	0.9498	0.9809
σ_C [10^{-6} 1/m]	9.33	4.85
M_C [%]	17.35	6.80
P_{Cmax} [%]	65.09	86.39
P_{Cmin} [%]	57.73	82.04
	After longwalls 001 and 002	
r_C	0.7843	0.8443
σ_C [10^{-6} 1/m]	20.45	17.46
M_C [%]	58.99	40.33
P_{Cmax} [%]	37.90	47.32
P_{Cmin} [%]	36.71	49.00
	After longwalls 001, 002 and 005	
r_C	0.5125	0.5962
σ_C [10^{-6} 1/m]	28.37	26.62
M_C [%]	138.40	93.17
P_{Cmax} [%]	22.51	31.37
P_{Cmin} [%]	27.03	38.58
	After longwalls 001, 002, 005 and 007	
r_C	0.5512	0.7189
σ_C [10^{-6} 1/m]	23.88	21.84
M_C [%]	114.95	68.23
P_{Cmax} [%]	23.96	36.93
P_{Cmin} [%]	28.25	41.01

3.2. Approximation of average values of measured subsidence

The average courses of observed subsidence were calculated on the way of least – square approximation, which was done by the use of smoothed splines. The *R* computer programme was used for that with the *smooth.spline* function.

The basic parameters of this function are:

- *df* – the number of freedom degree, values from the interval *<0, n>*, where: *n* – the number of observational points;
- *spar* – an approximating function smoothing parameter, values from the interval *<0, 1>*, where: *0* – a total lack of smoothing (approximating function takes the form of approximated function);
- λ – a penalized criterion (*PC*);
- *cv* – the cross validation, when: *cv = truth* it occurs the *leave-one-out* validation, if *cv = false*, then occurs the *general cross* validation (*GCV*);
- *nknots* – the number of knots, if *all.knots = false*, then the value is smaller than the number of observational points.

The least – square approximation was done regarding to the subsidence model values. These values were obtained by the use of the *EDN – OPN* computer programme and with an application of the Knothe's formula [10] with the typical values of its parameters ($a = 0.8$; $tg\beta = 2.0$; $A_{obr} = 0$) and the Bialek's formula with the typical values of its parameters ($a = 0.8$; $tg\beta = 2.3$; $A_{obr} = 0.15$). The detailed information about the reasons of such procedure were presented in the works [11], [12].

The parameters values of subsidence approximation measured after termination of the subsequent excavation stages have been juxtaposed in the Table 4.

Table 4. The values of approximation parameters

Parameter	Reference of approximation	
	the Knothe's formula	the Bialek's formula
After longwall 001		
df	11.88	23.07
spar	0.47	0.32
λ	$3.67x10^{-5}$	$2.02x10^{-6}$
PC [mm^2]	5841	225
After longwalls 001 and 002		
df	7.24	10.54
spar	0.63	0.55
λ	$3.90x10^{-4}$	$7.73x10^{-5}$
PC [mm^2]	135434	35184
After longwalls 001, 002 and 005		
df	7.40	10.29
spar	0.68	0.61
λ	$4.39x10^{-4}$	$1.06x10^{-4}$
PC [mm^2]	254332	107365
After longwalls 001, 002, 005 and 007		
df	8.23	11.50
spar	0.66	0.58
λ	$2.81x10^{-4}$	$6.46x10^{-5}$
PC [mm^2]	205941	66549

The average courses of observed subsidence, determined regarding to their model values calculated by the use of the Knothe's (dots) and Bialek's (squares) formulas with the parameters standard values, were presented at the Figure 4.

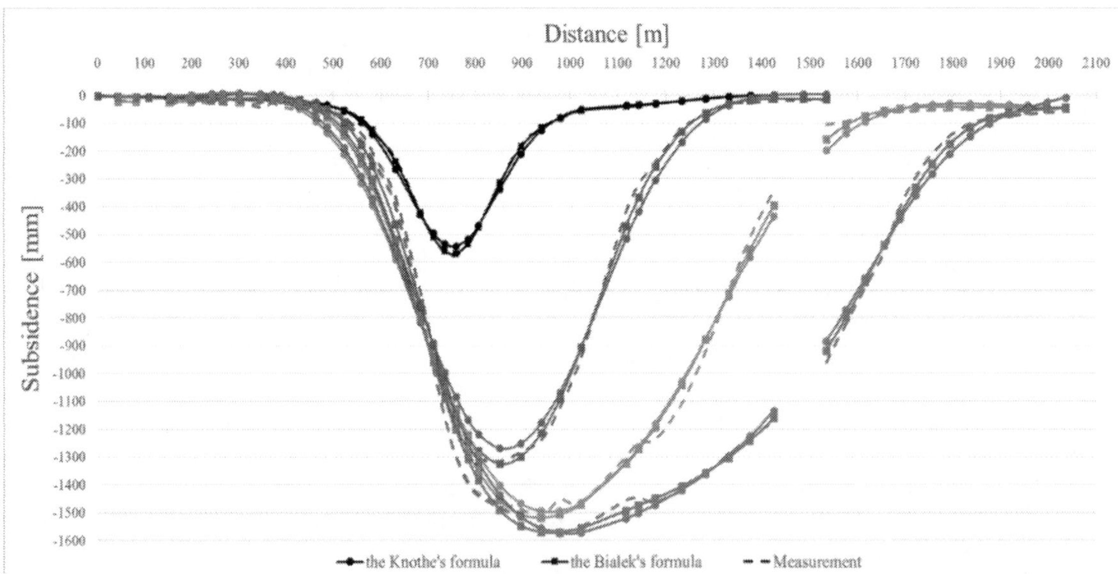

Figure 4. The average courses of subsidence observed after the end of next excavation stages in coal bed *338/2*

IOP Conf. Series: Materials Science and Engineering **294** (2017) 012030 doi:10.1088/1757-899X/294/1/012030

Dependence between the observed and average values of subsidence has been presented by: the correlation coefficient r_W, the subsidence standard deviation σ_W and the percentage of the extreme observed value of subsidence which corresponds the extreme average value of observed subsidence P_W. That was also calculated the variability coefficient of subsidence random dispersion M_W. It was obtained as the subsidence standard deviation divided by the module of the extreme average value of observed subsidence.

Table 5. The dependence between the observed and average subsidence

Parameter	Reference of approximation	
	the Knothe's formula	the Bialek's formula
	After longwall 001	
r_W	0.9978	0.9999
σ_W [mm]	13.11	2.47
M_W[%]	2.42	0.43
P_W [%]	94.25	98.95
	After longwalls 001 and 002	
r_W	0.9935	0.9982
σ_W [mm]	58.93	30.04
M_W[%]	4.64	2.27
P_W [%]	96.48	100.81
	After longwalls 001, 002 and 005	
r_W	0.9923	0.9967
σ_W [mm]	73.56	46.81
M_W[%]	4.92	3.08
P_W [%]	98.47	100.06
	After longwalls 001, 002, 005 and 007	
r_W	0.9944	0.9982
σ_W [mm]	64.83	36.85
M_W[%]	4.12	2.35
P_W [%]	100.28	100.18

4. Discussion of the results

The data, which were presented in the Table 4, show that the smoothing parameter values *spar* of the function averaging the subsidence observed courses increase together with the growth of excavation range: from 0.47 after the termination of extraction in the one longwall to 0.68 after the ending of exploitation in the three longwalls (in case of approximation done regarding to the Knothe's formula), and from 0.32 after the first period of excavation to 0.61 after the third extraction period (in case of approximation done regarding to the Bialek's formula). It can indicate that the approximating function reproduces the empirical data in a lesser degree when the excavation escalates. This thesis is confirmed by the penalized criterion values (a residual sum of squares). They also increase with the increasing of extraction period number.

The data shown in the Table 5 indicate that the correlation coefficient between the observed and average values of subsidence r_W fall down with the growth of excavation range. This coefficient has reached the maximal values after the first exploitation period (0.9978 in case of the Knothe's formula and 0.9999 for the Bialek's formula). The smallest values have been reached after the third excavation stage (0.9923 for the Knothe's formula and 0.9967 in case of the Bialek's formula). As can be seen its values are high and don't fall down below 0.99.

The excavation progress in the time causes that the values of standard deviation of subsidence σ_W grow together with the amount of exploited longwalls. Its values show that it is almost two times smaller in case of the Bialek's formula than for the Knothe's formula.

Regarding to the values of subsidence correlation coefficient r_W and subsidence standard deviation σ_W it can be seen, that the precision of mapping of observed courses of subsidence by their average courses decreases with the extraction range growth. Influence for it has got a random factor. Its participation in the subsidence observed values increases with the exploitation progress. This statement is confirmed by the growing values of variability coefficient of random dispersion of subsidence M_W. The values of M_W, as in case of σ_W, are two times smaller for the Bialek's formula than for the Knothe's formula.

As can be seen, the percentages of maximal values of observed subsidence which correspond the maximal values of average subsidence P_W are greater for more developed excavation.

The correlation coefficients between the measured courses of curvatures and their average courses r_C, juxtaposed in the Table 3, decrease after each exploitation stage (as in case of subsidence) and the decrease of these values is considerable. Their values amount from 0.9498 to 0.5125 (for the Knothe's formula) and from 0.9809 to 0.5962 (for the Bialek's formula). As can be seen, these values are definitely lower than the values of correlation coefficient of subsidence r_W.

The conventional deviation of curvatures σ_C increases with the exploitation development, from 9.33 x 10^{-6} 1/m after the first longwall to 28.37 x 10^{-6} 1/m after the three longwalls (the Knothe's formula) and from 4.85 x 10^{-6} 1/m after the first longwall to 26.62 x 10^{-6} 1/m after the three longwalls (the Bialek's formula). Notwithstanding, the differences between the values of standard deviations calculated for the curvatures in case of the Knothe's and Bialek's formulas aren't so visible like for the values of standard deviations calculated for the subsidence.

The values of variability coefficient of curvatures random dispersion M_C are higher for the Knothe's formula than for the Bialek's formula. They are much bigger than the values occurring in the literature [13]. The maximal value 138.4 % after the third excavation stage is four times bigger than 33.5 % identified by A. Kowalski.

In case of curvatures the P_C values are very small and go down when an exploitation is developing. It means that the extreme average values of measured curvatures are much smaller than extreme values of observed curvatures.

5. Conclusions

It's possible to obtain the average courses of mining area curvatures by the use of smoothed splines functions in case, when hard coal excavation was conducted at the medium depth, but taking into account below restrictions.

An approximating function (here a smoothed spline) should be parameterized by a shape parameter (here *spar*). This parameter has got an influence on fitting accuracy of approximating function to the empirical data.

The obtaining method of average courses of measured curvatures, which was presented above, requires to do an approximation of average courses of measured subsidence first. This method allows also to obtain the average courses of measured inclinations [11], [14].

In proposed manner of obtaining of average courses of measured curvatures should be use an average – square approximation. This approximation kind eliminates a random factor from the measured values of curvatures in a good degree.

The minimizing of squares residual sum should be referred to the model courses. Thanks to that, it's possible to obtain the average measured courses similar to the model courses. It's especially important in case of curvatures.

Projection accuracy of the measured courses by the average courses is dependent on an adopted model of deformation forecasting [7], [10], [15], [16].

Approximation reference to the Bialek's formula gives the better results than the Knothe's formula. It follows from that the Bialek's formula is a three – parametric model. It takes into account an excavation rim, which narrows an exploitation field [17]. For this purpose has been defined the A_{obr} parameter. It should be emphasized that the Knothe's formula is a two – parametric model [18], [19].

The values of standard deviations of subsidence and curvatures increase with the development of exploitation in the time.

The values of variability coefficient of curvatures random dispersion M_C significantly exceed its values occurring in the literature, even four times more. It's related to a high random factor which occurs in the measured values of curvatures. An average – square approximation isn't able to eliminate this factor. It's also caused by the typical values of models parameters adopted to the calculation of subsidence model courses.

The results improvement can be obtained by the use of parameters values, which have been designated directly from the geodesic measurements results [20].

Acknowledgement
The results which have been presented in this article are the effect of work financed by the Ministry of Science and Higher Education in Poland, by means of a transfer of funds designated for the statutory activity of the Silesian University of Technology in Gliwice, the Faculty of Mining and Geology (project no. 06/050/BKM_16/0034 – BKM-558/RG5/2016).

References
[1] Janusz W, Mielimąka R, Niezabitowski M, Orwat J and Sikora P 2015 *Surface Monitoring of Water Basins Based on Use of Autonomous Flying Robots*, 20th International Conference on Methods and Models in Automation and Robotics 2015, Międzyzdroje, Poland, August 24-27, pp 376-381
[2] Klamka J, Janusz W, Maćkowski M, Niezabitowski M, Orwat J and Sikora P 2015 *UAV Application for Delivering the Internet Access During the Geodetic Measurements*, 10th Scientific Conference Internet in the Information Society 2015, Dąbrowa Górnicza, Poland, September 24-25, pp 9-25
[3] Janusz W, Niezabitowski M, Orwat J and Sikora P 2016 *The Concept of Combined Measurements of the Gliwice Radio Tower Deflection in Relation to the Speed and Direction of the Wind with Use of Quadrotor*, 17th International Carpathian Control Conference 2016, Tatranská Lomnica, Slovakia, May – June 29 – 01, pp 280-285
[4] Kowalski A and Jędrzejec E 2015 Influence of Subsidence Fluctuation on the Determination of Mining Area Curvatures, *Archives of Mining Sciences* **60**(2) 487-505
[5] Ścigała R 2013 Mining Extraction System with Advancing Longwall as a Method of Protecting – Technical Objects and an Element of Rational Deposit Management, *Mineral Resources Management* **29**(3) 191-208
[6] Piwowarski W, Strzałkowski P and Ścigała R 2017 Analysis of Rock Mass Destruction Processes with Different Activity Diagrams, *Acta Geodynamica et Geomaterialia* **14**(1) 83-92
[7] Orwat J 2015 The Forecast Effectiveness of Mining Exploitation Effects on the Exploited Area Conducted with the Use of Bialek's Formulas, *AIP Conf. Proc.* **1738** 080013-1 – 080013-4
[8] Orwat J and Mielimąka R 2017 Approximation of Average Course of Measured Curvatures of Mining Area with Reference to Their Forecast Values by Bialek's Formulas, *AIP Conf. Proc.* **1863** 130003-1 – 130003-4
[9] Mielimąka R and Orwat J 2017 Approximation of Average Course of Measured Curvatures of Mining Area with Reference to Their Forecast Values by Knothe's Formulas, *AIP Conf. Proc.* **1863** 130005-1 – 130005-4
[10] Orwat J and Mielimąka R 2015 The Comparison of Measured Deformation Indicators of Mining Area with Theoretical Values Calculated Using Knothe's Formulas, *AIP Conf. Proc.* **1738** 080014-1 – 080014-4
[11] Orwat J and Mielimąka R 2017 Average Course Approximation of Measured Subsidence and Inclinations of Mining Area by Smooth Splines, *Journal of Sustainable Mining* **16**(1) 8-13
[12] Orwat J 2017 Approximation of Average Course of Measured Subsidences of Mining Area

by Smooth Splines, *AIP Conf. Proc.* **1863** 130004-1 – 130004-4

[13] Kowalski A 2007 *Undetermined Mining Surface Deformation in Terms of the Accuracy of Forecasts*, Institute of Mining, Poland, Habilitation Thesis

[14] Orwat J 2017 *Possibility of Using the Smoothed Spline Functions in Approximation of Average Course of Terrain Inclinations Caused by Underground Mining Exploitation Conducted at Medium Depth*, International Conference on Applied Sciences 2017, Hunedoara, Romania, May 10-12

[15] Strzałkowski P 2015 Mathematical Model of Forecasting the Formation of Sinkhole Using Salustowicz's Theory, *Archives of Mining Sciences* **60**(1) 63-71

[16] Strzałkowski P 2001 New Mathematical Model for Description of Time-Dependent Surface Subsidence, *Transactions of the Institution of Mining and Metallurgy, Section Mining Technology* **10** A178 – A182

[17] Orwat J 2017 Verification of Forecasts of Subsidences, Inclinations and Curvatures of Area Caused by Multideposit Mining Conducted by Including Mining Margin – Case Analysis, *Conf. Proc. of SGEM 2017, Science and technologies in geology, exploration and mining, Exploration and mining* **17**(13) 979-986

[18] Ścigała R 2013 The Identification of Parameters of Theories Used for Prognoses of Post Mining Deformations by Means of Present Software, *Archives of Mining Sciences* **58**(4) 1347-1357

[19] Orwat J 2017 Evaluation of Forecasts of Area Deformation Caused by Multideposit Exploitation of Hard Coal Made by the Budryk-Knothe Theory – Case Study, *Conf. Proc. of SGEM 2017, Informatics, geoinformatics and remote sensing, Geodesy and mine surveying* **17**(22) 289-297

[20] Strzałkowski P 2010 Forecasts of Mine-Induced Land Deformations in Consideration of the Variability of the Parameter Describing the Process Kinematics, *Archives of Mining Sciences* **55**(4) 865-872

Antropometric parameters problem solving of shoe lasts by deforming membranes with medium weight

A V Albu[1], C V Anghel Drugarin[2], E M Barla[3] and V Porav[1]

[1]University of Oradea, Department of Textile-Leather and Industrial Management, B. Ştefănescu Delavrancea str., no. 4, 4410058 Oradea, Romania
[2]"Eftimie Murgu" University of Resita, Department of Electrical Engineering and Industrial Informatics, Revolution str., no 5, 331128, Hunedoara, Romania
[3]University of Oradea, Department of Energy Engineering, Universităţii str., no 1-5 Oradea, 410087, Romania

E-mail: adina victoria@yahoo.com

Abstract. The paper presents research results into getting a virtual model of shoe last and anthropometric parameters change. The most important change occurs in the fingers region. Alternatives CAD-CAM technology for next generation is based on DELCAM software for the CAM procedure and simulation of MATLAB software. This research has led to the virtual changes of the last, anthropometric parameter - the width of the fingers (l_d) and shoe last length - (L_p) and images have been achieved with the representation in section of the shoe last changed from the original shoe lasts by FEM method (Finite element method) in MATLAB environment. The results are applied in the textile industry and in the elaboration of linings consumption or in the development of leather substitutes on fabric, knitted or woven material type.

1. Introduction

To deform shoe lasts in anthropometric parameters areas, are required specialized programs on footwear. The most important changes are in the finger area, in the fingers perimeter and in the top of the leg (shoe lasts), where you can see the direct influence of the fashion line.

Virtual modified shoe lasts are archived in a database, which can be removed and then modified at any time according to the needs in the field of footwear manufacturing and market trends in the field.

2. The experimental part

In this research we used DELCAM software, but only for CAM procedure [1]. DELCAM also has DELCAM Crispin Shoes design module [2], but it was not available due to high costs of acquisition, respectively confidentiality regime imposed by the concerned company.

That approach would certainly be the most appropriate for achieving virtual model deformation, but not for Small and Medium Factories specialized in textile and footwear, a reason for seeking alternative art technology, cheaper and easier to implement.

The proposed alternative is based on MATLAB simulation program [3], [4] which certainly will lead to positive results [1]. However, the used program will not have the comfort and convenience that is at DELCAM Crispin Shoes Design [5].

In addition, extensive work is needed (in future research) to create a reliable procedure, well-defined and in place, namely the creation of virtual databases of so generated shoe lasts.
Along the research were analyzed and implemented several iterative trials highlighting the possibility to modify anthropometric parameters of virtual shoe lasts by transforming them into linear sizes [6-14].

3. Results and discussions

As a result of the research, in Figure 1 (a, b, c, d) are images of virtual changes of the shoe last anthropometric parameter - the width of the fingers (l_d) and representing the section of the shoe lasts changed from the original shoe lasts, with values - ranging from 5 to 20 mm, by FEM method (Finite Element Method) in MATLAB environment [15].

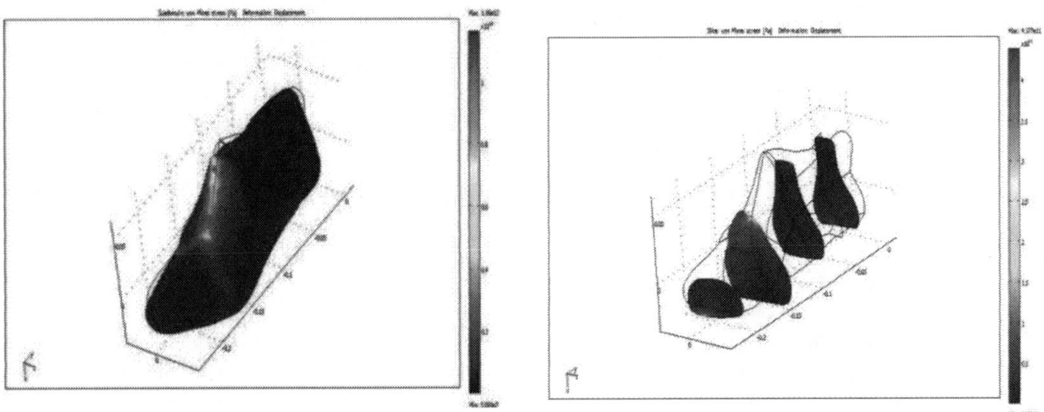

Figure 1. a – Image analysis with FEM method of the scanned and warped shoe lasts by changing the finger width of 5 mm

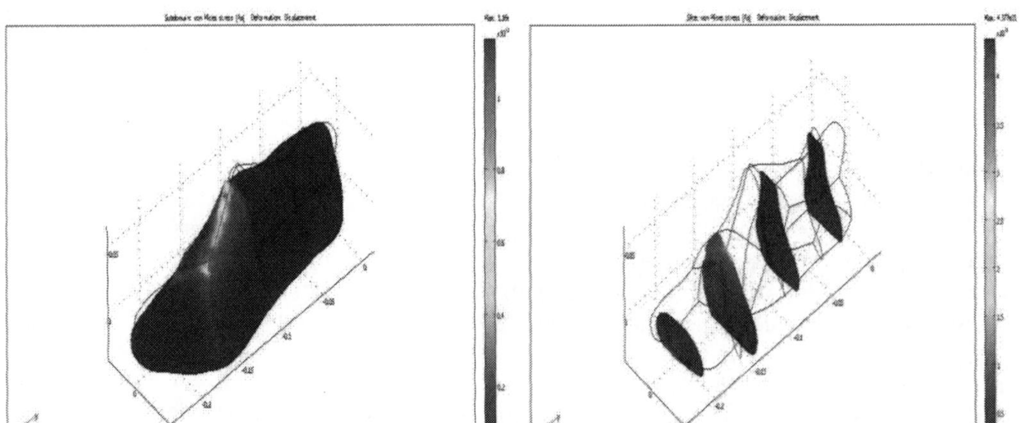

Figure 1. b – Image analysis with FEM method of the scanned and warped shoe lasts by changing the finger width of 10 mm

IOP Conf. Series: Materials Science and Engineering **294** (2017) 012031 doi:10.1088/1757-899X/294/1/012031

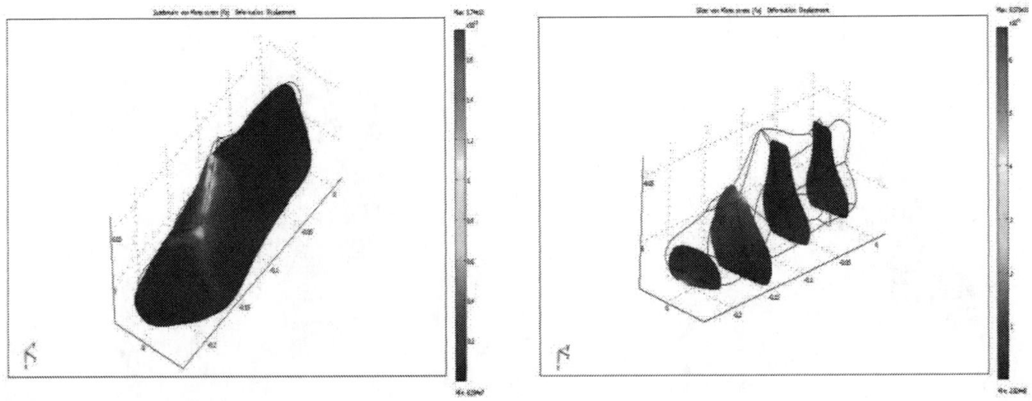

Figure 1. c – Image analysis with FEM method of the scanned and warped shoe lasts by changing the finger width of 15 mm

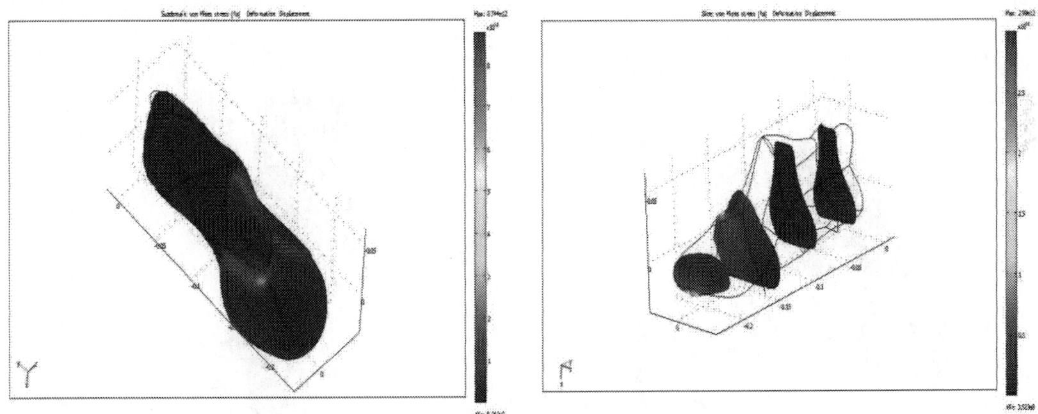

Figure 1. d – Image analysis with FEM method of the scanned and warped shoe lasts by changing the finger width of 20 mm

In Figure 2 (a, b, c) are images of virtual changes of the shoe last anthropometric parameter - shoe lasts length (Lp), as compared to the original shoe lasts, with values -ranging from 5 to 15 mm, FEM method [3], [4] in MATLAB environment [15].

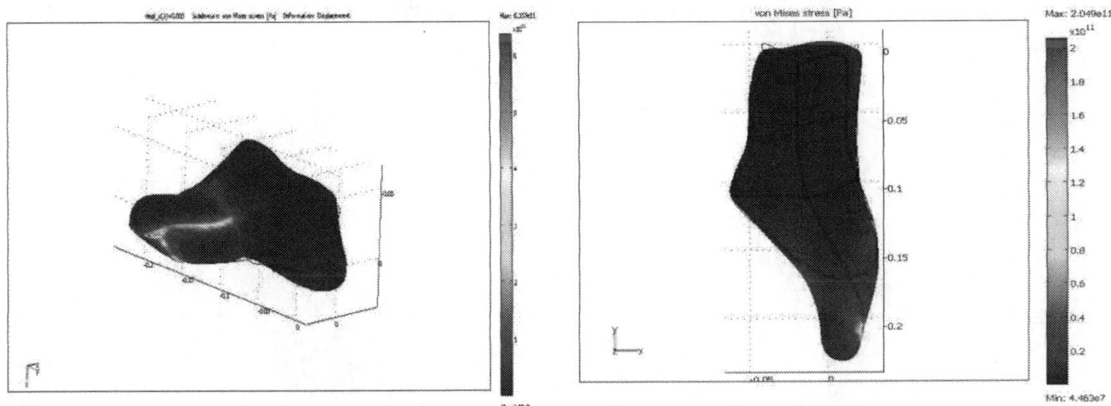

Figure 2 a. Images analysis with FEM method of shoe lasts image scanning and deformed by changing the length of 5 mm

IOP Conf. Series: Materials Science and Engineering **294** (2017) 012031 doi:10.1088/1757-899X/294/1/012031

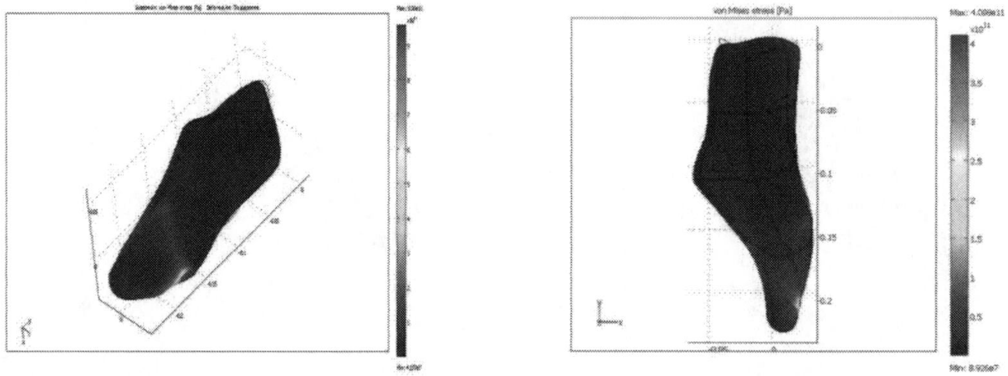

Figure 2 b. Images analysis with FEM method of shoe lasts image scanning and deformed by changing the length of 10 mm

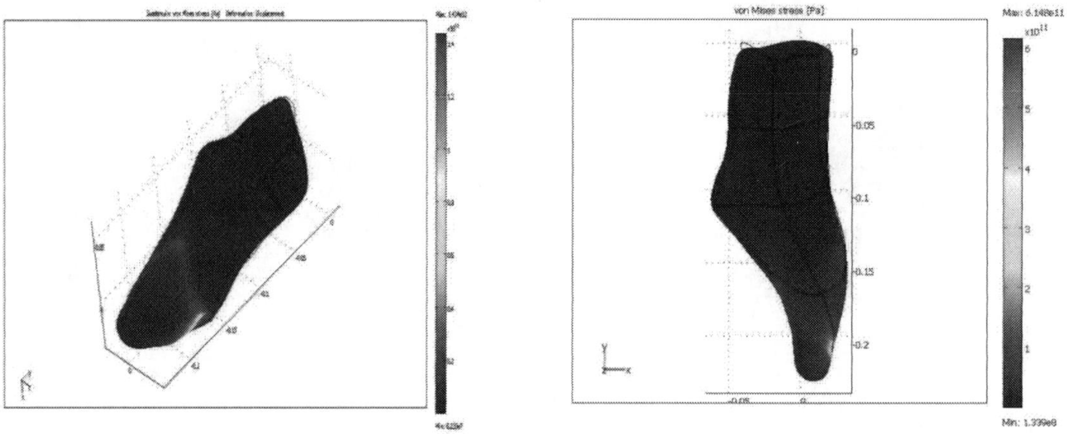

Figure 2 c. Images analysis with FEM method of shoe lasts image scanning and deformed by changing the length of 15 mm

In Figure 3 (a, b, c, d) are images of virtual changes of the shoe last, as follows: **a** - geometric pattern, **b** - network triangulation with small number of finite elements, **c** - triangulation network with a number of finite thick element; **d** - triangulation network with a number of more often finite elements.

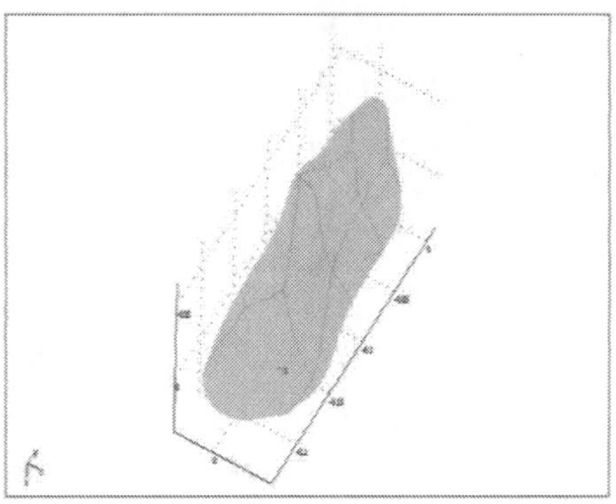

a.

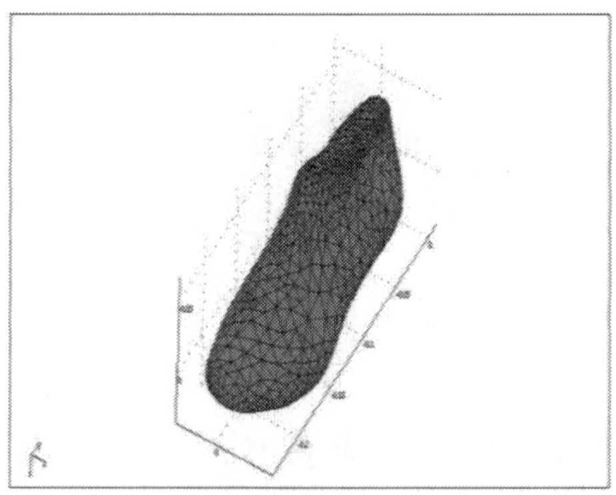

b.

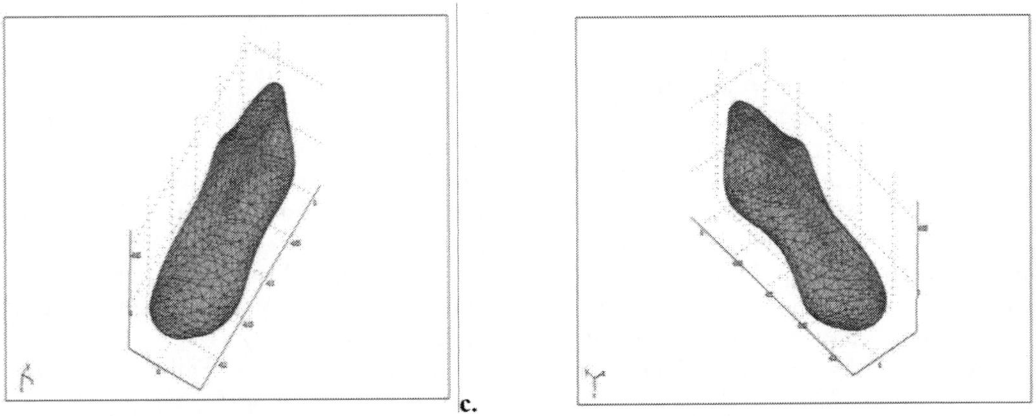

Figure 3 (a, b, c, d). The geometric triangulation networks and of virtual shoe lasts

In Table 1 it is show the Item type: Lagrange – Quadratic tetrahedral.

Table 1. Item type: Lagrange – Quadratic tetrahedral

Name of parameter value	Value
Number of freedom degrees	39711
Number of grid points	2014
Number of domain elements	7991
Number of boundary elements	2438
Number of edge elements	238
Item report / volume	0.0013

Geometric model is generated by the cloud of points and the points which are scattered were removed [1], [2]. Triangulation method increased the number of triangles (FE – finite elements) with data in Table 1 and gradual interpolations were made to reach a continuous surface needed to process of virtual shoe lasts for CAM to be physically processed on machine tool equipped with the CAM procedure by roughing with big step, then fine, and finally very fine finishing [3], [4].

4. Conclusions

Other researchers in the country are interested in mathematical modeling, to find specific equations needed to achieve shoe lasts pattern.

There is a partnership with DELCAM Crispin Shoes Design Company that can provide shoe lasts scanning and creation, and then stored in a database of different shoe lasts models [4].

The collected and stored information will have access to it and be able to develop collaboration. DELCAM Company provides post processing programs, but initially it takes 3D virtual model of the shoe lasts, which will be achieved through mathematical modeling [1].

In this paper work, the authors had attempted the development of a practical method that does not complicate the problem and allowing the development of shoe lasts without mathematics advanced knowledge of the user.

References

[1] ***www.delcam.com
[2] ***www.delcamshoes.com
[3] Moaveni S 1999 *Finite element analysis, theory and aplication with ANSYS*, Prentice Hall, New Jersey, USA

[4] Vickers G W 1999 *Numerically controlled machine tools*, University of Victoria, Canada

[5] Drişcu M and Mihai A 2008 *Proiectarea încălţămintei cu sistemul Crispin Dynamics CAD Suite - Engineer*, Editura Performantica, Iaşi, Romania

[6] Albu A, Ganea M, Bondrea I, Indrie L, Prichici M and Romocea S 2010 Processes of scanning and virtual modeling of the lasts in footwear Industry, *Annals of the Oradea University, Fascicle of Management and Technological Engineering* **IX**(XIX)

[7] Albu A 2009 *Virtual performing of the shoe last model able for CNC machining* Conferinţa Naţională Tehnico-Ştiinţifică (cu participare internaţională), Ediţia a VII-a "Tehnnologii moderne pentru mileniul III", 5-7 noiembrie, Oradea, Romania

[8] Albu A, Ganea M, Morar M and Indrie L 2009 Posibilities of virtual solid corection in the case of shoe last by using of the antropometrical parameters, *Annals of the Oradea University, Fascicle of Management and Technological Engineering* **VIII**(XVIII)

[9] Albu A, Ganea M and Ungur P 2007 *Considerations about mathematical model of the spatial surface of the lasts in the shoe industry*, Simpozionul international Inter-ing-2007, Tg. Mureş, Romania, 15-16 November

[10] Cocea M 2005 A study for the mathematic modeling of 2D irregular shapes for footwear CAD system, *Buletinul Institutului Politehnic din Iaşi - Stiinţa şi Ingineria Materialelor* **LI**(LV) 219-225

[11] Cocea M and Croitoru D F 2003 *Proceduri CAD destinate proiectării încălţămintei*, Editura Gh. Asachi, Iaşi, Romania

[12] Croitoru D F 1987 *Utilaje şi automatizări pentru industria confecţiilor din piele*, Iaşi, Romania

[13] Drişcu M 2008 *Modelarea formelor plane şi spaţiale ale încălţămintei*, Editura Pim, Iaşi, Romania

[14] Ganea M 2010 *Maşini şi echipamente tehnologice pentru prelucrarea suprafeţelor în 4 şi 5 axe CNC*, Editura Universităţii din Oradea, Romania

[15] Ganea M and Ganea C 2000 *Tehnologia prelucrării suprafeţelor curbe spaţiale*, Editura Universităţii din Oradea, Romania

IOP Conf. Series: Materials Science and Engineering **294** (2017) 012032 doi:10.1088/1757-899X/294/1/012032

General aspects of speed increaser gearboxes

A Băra[1], D Nedelcu[1], C Haţiegan[2], A C Bejinariu[3] and L (Filip) Nedeloni[1]

[1]Eftimie Murgu University of Resita, Department of Mechanical and Materials Engineering, Traian Vuia str., no.1-4, 320085 Resita
[2]Eftimie Murgu University of Resita, Department of Electric Engineering and Informatics, Traian Vuia str., no. 1-4, 320085 Resita, Romania
[3]Politehnica University of Timisoara, Faculty of Management in Production and Transportation, Remus str., no.14, 300191, Romania

E-mail: abara@resitareductoare.com

Abstract. This study aims to present the general aspects of the speed increaser gearboxes. The second part of this study will focus on the identification of the possibility to replace the special materials used for gearing execution of the speed increaser gearboxes with materials having inferior mechanical characteristics. Gearing execution from 18CrNiMo7-6 material is expensive in terms of price and applied execution technology. Also to obtain the mechanical characteristics for these materials is necessary to apply some special expensive heat treatments as carburizing heat treatment. We will replace this material with C45 steel for gear and pinion execution, we will apply an induction hardening heat treatment for getting a surface teeth hardness HRC 45 an for the increase the safety for tooth root stress coefficient we will apply a supplementary operation consisting in a tooth root shoot peening.

1. Introduction

The gears represent mechanical transmissions used in many applications due to the fact that they have high efficiencies and could cover a large scale of the speed and powers. Because the gears are executed from alloy special steel with specific heat treatments, these transmissions have small dimensions [1].

The speed increaser gearboxes are mechanical transmissions with subunit ratio. Depending on the imposed total ratio, the speed multipliers could be constructed in one or more stages.

The gears of the speed increaser gearboxes can be with cylindrical teeth or cylindrical helical teeth. We recommend cylindrical teeth when the peripheral speeds are low and the axial forces in the sleeve bearing are not accepted. When the peripheral speed are high (over 12m/s) [2] and the function is silent, we recommend using gears with cylindrical helical teeth.

Speed increaser gearboxes used in the Kaplan turbines are constructed with cylindrical helical teeth, the multipliers sleeve bearings are dimensioned so that to take the forces from the cylindrical gearing and the axial forces generated by the turbine.

2. Short description of the gears system

The gears system is classified according to the following criteria [3], [4]:

I. Axes position: A)Gears with parallel axes Figure 1.a; B)Gears with crossed axes Figure 1.d; C) Gears with intersected axes Figure 1e-f;

II. Teeth form: A) Gears with cylindrical teeth; B)Gears with cylindrical helical teeth; C) Gears with teeth in V,W,Z form; D) Gears with bevel helical teeth;
III. Relative position of the meshing surfaces: A)External meshing; B)Internal meshing
IV. Direction of rotation: A) In the same direction; B) In opposite direction;
Ratio: A) Constant ratio; B) Variable ratio;
Size of the ratio: A) Ratio i≥1-gearbox; B) Ratio i≤1-speed multipliers
Teeth profile: A) Evolving profile; B) Archimedean spiral for worm gear; C) Cycloidal profile; D) Straight line just for rack

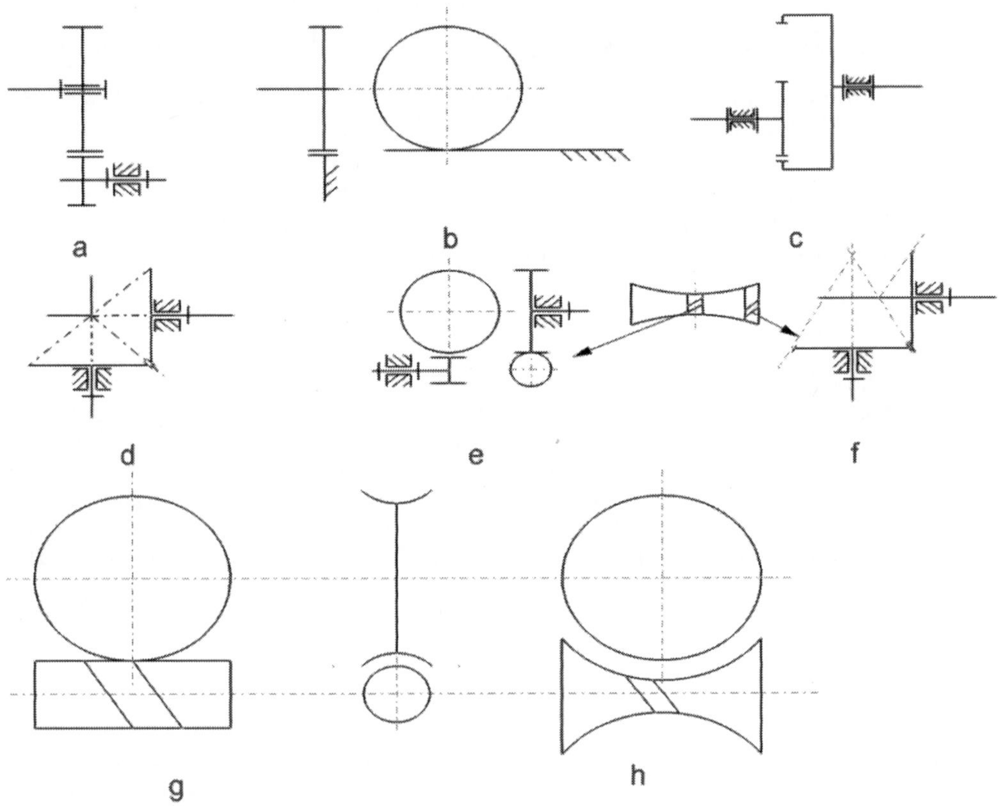

Figure 1. Types of gears

3. The geometric elements of the speed increaser gearboxes. The meshing low

The meshing low, also known as Willis theory, status the condition that has to the fulfilled by the profile curves of the teeth in contact so that the moment can be transmitted at a constant transmission ratio [3].

$$i_{12} = \frac{n_1}{n_2} = \frac{\omega_1}{\omega_2} \tag{1}$$

In order that two gears wheels to transmit the rotation moment at a constant transmission ratio it is necessary that the teeth profiles to be generated in such manner that their common normal straight line in the contact points to pass through a fixed point C (pitch point) (Figure 2) from the centre line. The conjugate profiles are those to satisfy the meshing low [3], [5].

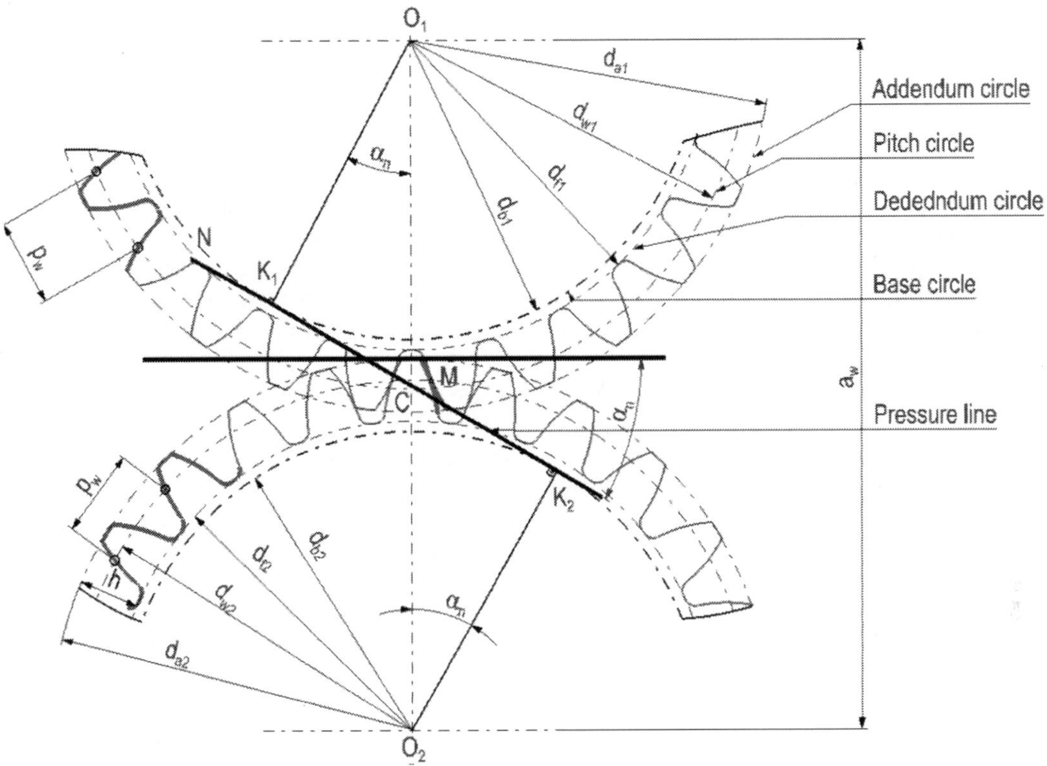

Figure 2. The geometric elements of the speed increaser gearboxes

Involute (also known as evolvent) is a curve obtained from another given curve by attaching an imaginary taut string to the given curve and tracing its free end as it is wound onto that given curve Narrow Figure/wide caption [6].

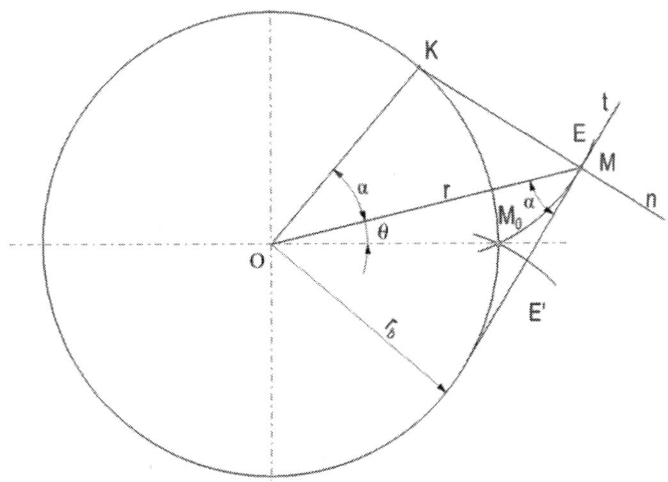

Figure 3. The geometric elements of the speed increaser gearboxes

The pressure angle α is the angle between the line of pressure (which is normal for the tooth surface) and the plane tangent to the pitch surface. The involute equation is [3]:

$$Inv\alpha = tg\alpha - \alpha \tag{2}$$

$$r = \frac{r_b}{\cos\alpha} \tag{3}$$

4. The forces acting in the cylindrical helical teeth of the speed increaser gearboxes. The stresses appearing in the cylindrical helical teeth of the speed increaser gearboxes

The normal force (F_n) which appears in gearing, decomposes in two components: the tangential force (F_t) and radial force (F_r). The tangential force introduce tooth root maximum stretching stresses in point A and maximum compression stresses in point B, and the radial force introduce the compression stresses in the ABC plan [1].

S_o, adding these stresses result that the stretching stresses in point A have a maximum value of σi-σc2 and in point B the compression stresses have a maximum value of $\sigma_{e1}+\sigma_{c2}$[1].

The tooth crack, as the result of the fatigue bending and shock stresses, appears due to the stretching stresses that occur around point A [2], [7].

We observe that an increase of the compression stress around this point would lead to decrease of the stretching stress. This can be done applying a supplementary operation consisting in a shoot peening in zone point A [2], [7].

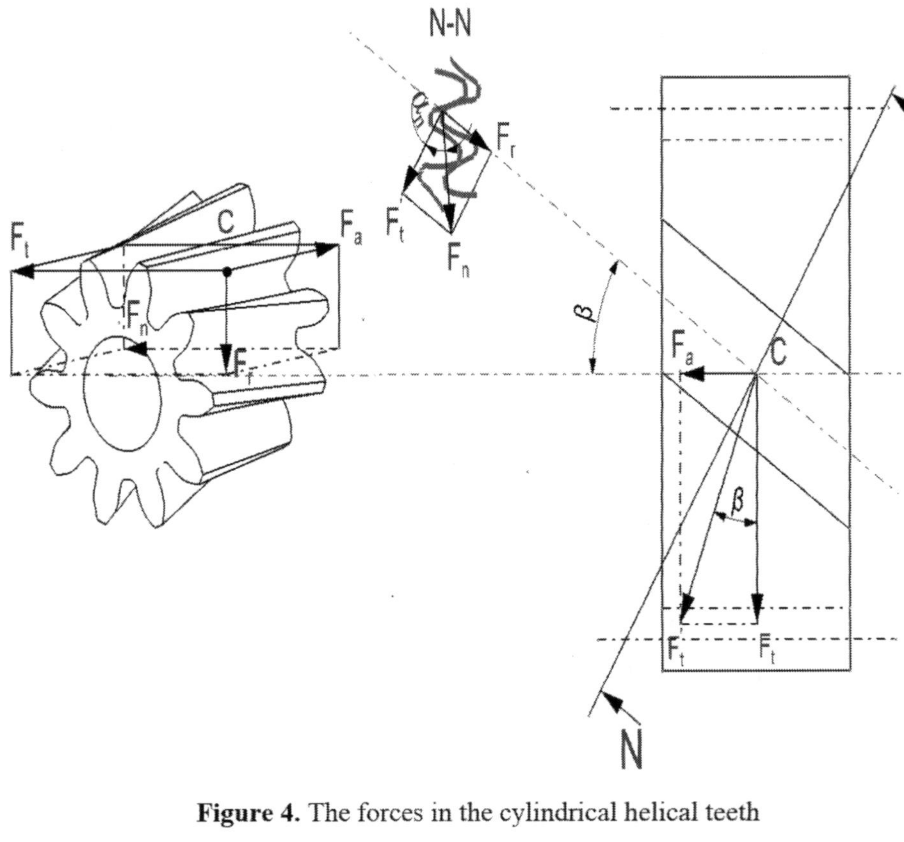

Figure 4. The forces in the cylindrical helical teeth

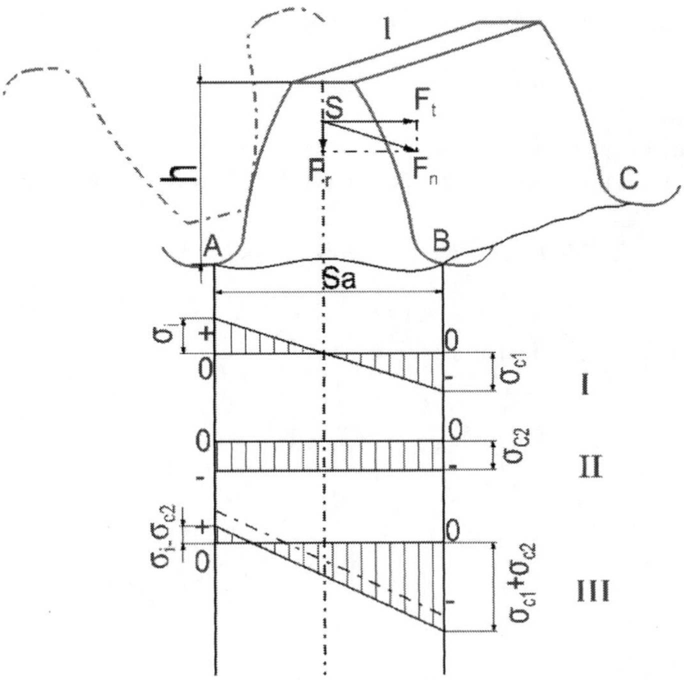

Figure 5. Tooth root stress schema

During the meshing the teeth are exposed to dynamic stressing [1].

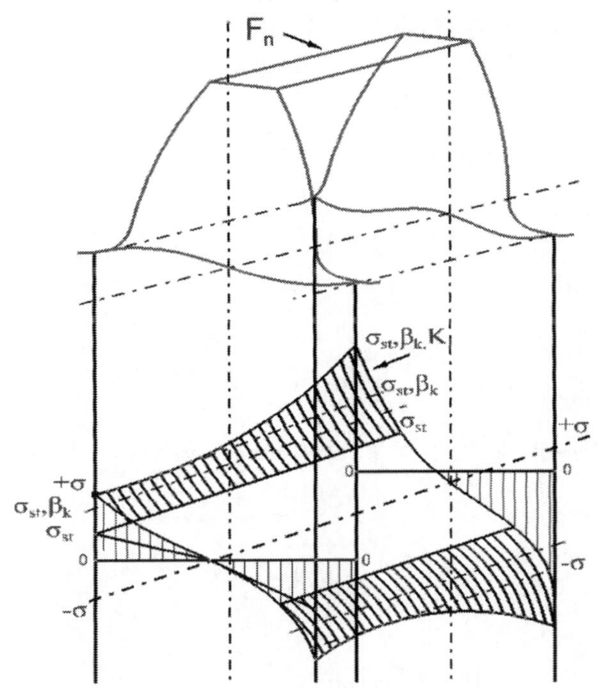

Figure 6. The geometric elements of the speed increaser gearboxes

The static stresses increase with following coefficients:
K_A – transverse load factor,
$K_{F\beta}$ – face load factor at tooth root,
K_V – dynamic factor,
$K_{F\alpha}$ – transverse load factor at tooth root.
Taking into account all these coefficients according to ISO 6336-3:2006 results that the tooth root stress is:

$$\sigma_F = \sigma_{Fo} \times K_A \times K_V \times K_{F\beta} \times K_{F\alpha} \tag{4}$$

where:
σ_{Fo} – nominal tooth root stress in static conditions;
Considering:

$$\sigma_{Fo} = \frac{F_t}{bm_n} \times Y_F \times Y_S \times Y_\beta \times Y_B \times Y_{DT} \tag{5}$$

Results:

$$\sigma_F = \frac{F_t}{bm_n} \times Y_F \times Y_S \times Y_\beta \times Y_B \times Y_{DT} \times K_A \times K_V \times K_{F\beta} \times K_{F\alpha} \tag{5}$$

where:
Y_F – tooth form factor,
Y_S – stress correction factor,
Y_β – helical load factor,
Y_B – gear rim factor,
Y_{DT} – deep tooth factor,
F_t – nominal circumferential force at pitch circle,
b – face width,
m_n – normal module.

The safety for tooth root stress coefficient according to ISO 6336-3:2006 is:

$$\sigma_F = \frac{\sigma_{FG}}{\sigma_{F1}} \geq S_{f\min} \tag{6}$$

where:
σ_{FG} – limit strength toot root;

According to ISO 6336-3:2006:

$$\sigma_{FP} = \frac{\sigma_{F\lim} \times Y_{ST} \times Y_{NT}}{S_{F\min}} Y_{\delta relT} \times Y_{RrelT} \times Y_x \tag{7}$$

Considering:

$$\sigma_{FE} = \sigma_{F\lim} \times Y_{ST} \tag{8}$$

Result:

$$\sigma_{FP} = \frac{\sigma_{FE} \times Y_{NT}}{S_{F\min}} Y_{\delta relT} \times Y_{RrelT} \times Y_x \tag{9}$$

where:

S_{fmin} – required safety,
σ_F – tooth root stress,
$Y_{\delta relT}$ – support factor,
Y_{RrelT} – surface factor,
Y_x – size coefficient,
Y_{ST} – stress correction factor.

5. Materials used for gear construction of the speed increaser gearboxes

Depending on the mechanical stresses that result from the calculation, the materials used in RRR company for the execution of the gears and pinions are:

Steel for quenching and tempering: I) 42CrMo4 according to EN 10083-3; II) 30CrNiMo8 according to EN 10083-1; III) 30MoCNi20 according to STAS 791 [8].

Case hardening steel: 18CrNiMo7-6 according to EN 10084 [9].

Carburized and heat treated this steel develops: I) Surface hardness- HRC 61; II) Tensile strength - Rm=1010-1300 [N/mm²]; III) Yield point -Rp=815-1050 [N/mm²].

For my application I will use C45 steel for gear and pinion execution with the following mechanical properties in the quenched and tempered conditions.

Tensile strength -R_{mmin}=630 [N/mm²];

Yield point -R_{pmin}=370 [N/mm²].

We will apply an induction hardening heat treatment for getting a surface teeth hardness HRC 57.

To increase the safety for tooth root stress coefficient, we will apply a supplementary operation consisting in a shot peening.

6. Choose of the constructive solution for speed increaser gearbox type AV 11.25. Geometric gears calculation of the speed increaser gearbox

Choosing the constructive variant was made starting the following input data: power P = 34.50 (kW); input speed n_1=250 (rpm); output speed n_2 = 1000 (rpm); application factor KA = 1;

The main geometrical elements of this speed increaser gearbox are [10]: center distance a = 112.500 (mm); centre distance tolerance ISO 286:2010 measure js7; normal module m_n = 2.0000 (mm); pressure angle at normal section α_n = 20.0000 (°); helix angle at reference circle β = 10.0000 (°); number of teeth z_1= 22 and z_2 = 88; face width b1= 47.50 (mm) and b_2 = 45.00 (mm); accuracy grade [Q-ISO 1328:1995], [6].

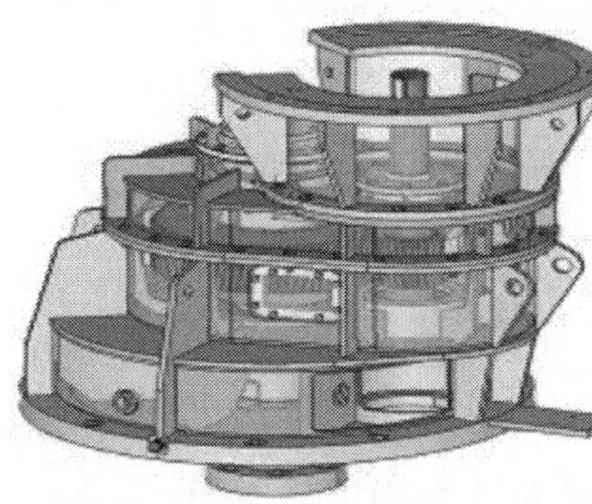

Figure 7. The geometric elements of the speed increaser gearboxes

Figure 8. Engagement between two gears

Materials used for pinion and gear wheel are [11]:

A) Gear 1: C45 , Through hardened steel, flame/induction hardened according to ISO 6336-5 Figure 11/12 (MQ)

B) Gear 2: C45 Through hardened steel, flame/induction hardened according to ISO 6336-5 Figure 11/12 (MQ)
C) Surface hardness HRC 57

The geometric gears calculation was made using the Kisssoft gear software and was made according to DIN 3960:1987, [12].

Table 1. Tooth root strength calculation

	Gear 1	Gear 2
Tooth form factor [YF]	1.79	1.53
Stress correction factor [YS]	1.81	2.15
Working angle [alfFen] (°)	21.05	21.20
Bending lever arm [h_F] (mm)	2.48	2.84
Tooth thickness at root [sFn] (mm)	4.07	4.07
Tooth root radius [roF] (mm)	1.06	0.80
Contact ratio factor [Yeps]	1.000	
Helical load factor [Ybet]	0.917	
Deep tooth factor [YDT]	1.000	
Gear rim factor [YB]	1.000	1.000
Effective facewidth [beff] (mm)	47.50	45.00
Nominal stress at tooth root [sigF0] (N/mm²)	461.84	492.75
Tooth root stress [sigF] (N/mm²)	599.17	639.26
Permissible bending stress at root of Test-gear		
Support factor [YdrelT]	0.994	1.004
Surface factor [YRrelT]	0.957	0.957
Size coefficient (Tooth root) [YX	1.000	1.000
Finite life factor [YNT]	1.000	1.000
Alternating bending coefficient [YM]	1.000	1.000
Technology factor [YT]	1.200	1.200
Stress correction factor [Yst]	2.00	
Yst*sigFlim [sigFE] (N/mm²)	740.00	740.00
Permissible tooth root stress (N/mm²)	563.13	568.69
Limit strength tooth root [sigFG]	844.69	853.03
Required safety [SFmin]	1.50	1.50
Safety for Tooth root stress	1.41	1.33

Table 2. Tooth root stresses and safeties taking into account the pretension force

	Gear 1	Gear 2
Pretension [sigmaP] (N/mm²)	45	85
Limit strength for tooth root with pretension [sigFG'] (N/mm²)	898.99	956.92
Safety for Tooth root stress [SF=sigFG'/sigF]	1.50	1050
Pretension [sigmaP] (N/mm²)	95	140
Limit strength for tooth root with pretension [sigFG'] (N/mm²)	959.32	1023.64
Safety for Tooth root stress [SF=sigFG'/sigF]	1.60	1.60
Pretension [sigmaP] (N/mm²)	145	190
Limit strength for tooth root with pretension [sigFG'] (N/mm²)	1019.66	1084.57
Safety for Tooth root stress [SF=sigFG'/sigF]	1.7	1.7

We notice that without a pretention at tooth root, the safety coefficients for tooth root stress are smaller than required safety coefficients.

7. Conclusion

The speed increaser gearboxes are mechanical transmission with subunit ratio. The pinions and the gears of the speed increaser gearboxes are executed from case hardening steel. Carburised and heat treated these materials develop high values for surface hardness and for the tensile strength. We will replace these materials with C45 steel. Appling an induction heat treatment we will get a teeth surface hardness, HRC 57.

To increase the safety factor for tooth root stress we will apply a shoot peening process. This process will be performed by bombarding the tooth root teeth of the gears with small spherical media.

By this process we will introduce the residual compressive stresses in the superficial layer of the tooth root and will improve the bending fatigue strength of the gears.

In conclusion, taking into consideration that applying an induction hardening heat treatment, we will get a surface teeth hardness HRC 57 and applying and a soot peening process for growing the safety factor for the tooth root stress it is possible to replace the special materials like 18CrNiMo7-6 or 16MnCr5 with normal steel like C45.

References

[1] Popescu N 1969 *Alegerea si Tratamentul Termic al Otelurilor pentru Roti Dintate*, Editura Tehnica, Bucuresti, Romania

[2] Anghel S and Ianici S 1993 *Proiectarea transmisiilor mecanice*, Universitatea Tehnica Timisoara, Facultatea de Inginerie, Resita

[3] Constantin V and Palade V 2005 *Organe de Masini si Mecanisme*, Editura Fundatiei universitare "Dunarea de Jos", Galati, Romania

[4] Ianici S, Ianici D and Potoceanu N 2008 *Design of Double Harmonic Transmission*, 6th International Conference of DAAAM Baltic Industrial Engineering, DAAAM Int Vienna, Tallinn, ESTONIA, April 24-26

[5] Ianici S and Ianici D 2010 *Design of mechanical systems*, Editura Eftimie Murgu, Reşiţa, Romania

[6] ***http://en.wikipedia.org

[7] Ianici D, Nedelcu D, Ianici S and Coman L 2010 *Dynamic Analysis of the Double Harmonic Transmission (DHT)*, 6th International Symposium about Forming and Design in Mechanical Engineering (KOD 2010), Palic, Serbia, September 29-30

[8] *** https://goo.gl/7w93VJ

[9] *** https://goo.gl/bHtdb1

[10] Ianici S and Ianici D 2015 Constructive Design and Dynamic Testing of the Double Harmonic Gear Transmission, *Analele Universităţii "Eftimie Murgu" Reşiţa* **22**(1) 231-238

[11] ***ISO 6336/1996 Calculation of load capacity of spur and helical gears, part 1-5

[12] ***DIN 3990/1965 Tragfahigkeisberechnung von stirnrader, Teil 1-5 Deutscher

IOP Conf. Series: Materials Science and Engineering **294** (2017) 012033 doi:10.1088/1757-899X/294/1/012033

Natural frequencies of thin rectangular plates clamped on contour using the Finite Element Method

L (Barboni) Haţiegan[1], C Haţiegan[2], G R Gillich[1], C O Hamat[1], O Vasile[3] and M D Stroia[2]

[1]Eftimie Murgu University of Resita, Department of Mechanical and Materials Engineering, Traian Vuia str., no. 1-4, 320085 Resita, Romania
[2]Eftimie Murgu University of Resita, Department of Electric Engineering and Informatics, Traian Vuia str., no. 1-4, 320085 Resita, Romania
[3]"Politehnica" University of Bucharest, Department of of Mechanics, Bucharest, Romania

E-mail: c.hatiegan@uem.ro

Abstract. This paper presents the determining of natural frequencies of plates without and with damages using the finite element method of SolidWorks program. The first thirty natural frequencies obtained for thin rectangular rectangular plates clamped on contour without and with central damages a for different dimensions. The relative variation of natural frequency was determined and the obtained results by the finite element method (FEM) respectively relative variation of natural frequency, were graphically represented according to their vibration natural modes. Finally, the obtained results were compared.

1. Introduction

A plate is a solid body which has one dimension less than its other two dimensions, called the thickness of the plate and can be regarded as "materialisation" of a surface [1-7].

It will analyze a thin rectangular plate, clamped on the contour of the structural elements, where the action is perpendicular to the median plane. The reference was chosen such that the X and Y axes in the median plane and form a right system, and the Z-axis direction to have the action of gravity.

The actual study will be carried out on two types of plates having the same dimensions, so that the first plate is no damage, the second is central plate is damage.

Damages influence the behavior of dynamic and static structures, mechanical and dynamic change, such as natural frequencies, modal shape, stiffness or flexibility, the degree of damping [8].

The damage is defined as a change in a structure that negatively affects its performance. It can also be defined as a phenomenon where weakening of the structure occurs. In another way, it could be considered as a deviation from the properties of the structure of the material or its geometry, which produce unwanted vibrations, displacements and tensions [8-12].

Study of fault is very important, since a large amount of loss property and term of life can be prevented by identifying the damage. The tendency to use structures even after deterioration, the inability to make proper renovations in a timely manner, places the study of the damage in an even more important position. Studying the damage can help specialists find more effective ways to monitor structure health and even predict possible damages. Numerous algorithms for fault detection have been performed, but many factors remain to be tested until an ideal model is developed. Due to the limits of measuring devices, sensor performance, experimental errors and complexity of structures, only a few effective fault detection methods can be called [13-19].

In recent years, important research has been carried out to develop advanced techniques for obtaining non-destructive evaluation methods to determine the damage. Regular inspection and control of engineering structures is required from the point of view of real-time damage detection, establishing the safety and reliability of the structure. Early identification of damages during the use of structures, allows scheduled maintenance, with an impact on the reduction of operating costs, or putting them back into operation and replacing the structure, to avoid breakdowns and accidents.

The development of the modern aeronautical industry, the mechanical industry and the civil industry has generated a new impetus for a more rigorous analysis of plate vibration issues and timely damage detection. The recent orientation in the development of plate theory and the detection of damages in these resistance structures is characterized by the use of performing computers using numerical methods, including the finite element method, the results comparing them with experimental data and different theoretical / analytical methods. In the literature, a large number of non-destructive control methods are treated, both local and global; these methods are not mutually exclusive, they can be used additionally [20-38].

Numeric or digital simulation uses computerized modeling and analysis to test static or dynamic behavior of stressed products under different operating conditions. Such an analysis involves, however, the assimilation of working with a specialized assisted design program as well as finite element simulation and analysis programs.

Hypotheses that underlie simulation: the system under analysis may be real or theoretical; We can build a mathematical and logical model of the analyzed system.

Performing a simulation experiment is a process that usually takes place in several stages: problem formulation; the collection and primary processing of data; formulating the simulation model; estimation of operative characteristics parameters in real data; model performance evaluation and parameter testing; building the calculation program (the simulation algorithm); program validation; planning simulation experiences; analysis of simulated data [39-42].

In this paper we determined the own frequencies of a thin rectangular plate clamped on a contour without damage, with a central damage of different sizes using the Simulation module in the SolidWorks program.

2. Analysis of the results obtained for plates with no damages clamped on all sides

The numerical method chosen for modal analysis is the finite element method (FEM) using the SolidWorks program. Modeling of the geometry to be analyzed was done in the Geometry module of the same program. To obtain its own frequencies, the static analysis module for thin rectangular plates was used under its own weight. The results obtained in this module were transferred to the modal analysis module, with the ambient temperature being 20°C. The material used in the program database, Figure 1, is steel AISI 1045, with the mechanical properties presented in Table 1.

Table 1. Material properties for AISI 1045 steel

Flow Limit $[N/mm^2]$	Breaking Resistance $[N/mm^2]$	Density $[kg/m^3]$	Elastic modulus $[N/mm^2]$	Poisson Coefficient [-]	Coefficient of thermal expansion $[K^{-1}]$
530	625	7,850	205	0.29	0.0000115

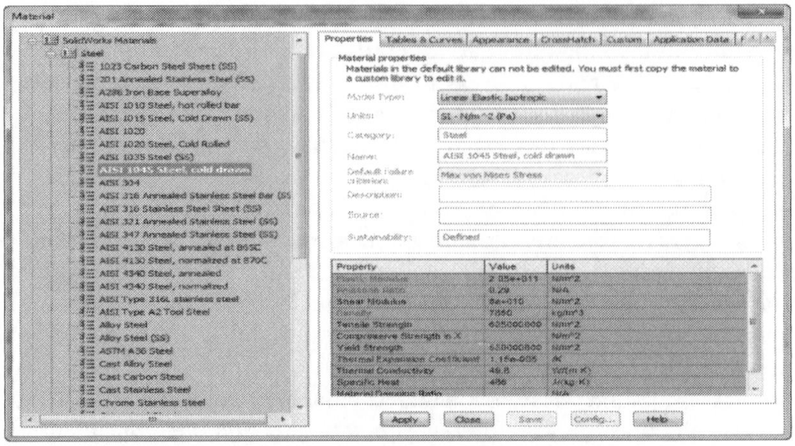

Figure 1. Select material from the SolidWorks database

The modal analysis determined the first 30 own frequencies, of which the first 20 own frequencies were taken into account. Meshing 3D model was made with tetrahedral finite elements.

The 3D model used is a real, rectangular plate with the following dimensions: length L = 950 mm; Width l = 400 mm and thickness h = 2 mm.

After creating the geometry of the plates, creating the frequency study and choosing the material, you will have to go through 4 stages, namely: applying constraints, mesh, modal analysis, and visualizing the results.

Finite element discretization was achieved through the "Solid mesh" option using the following options: High for Mesh Quality, Solid Mesh for Mesh type, Standard mesh, and Curvature based Mesh for Mesher Used. Frequency determination was performed using the Simulation module in the SolidWorks program [41-44].

The following steps are used to calculate your own frequency: create the geometry of solid-state and non-damageive plates in SolidWorks; simulation mode activation; creating simulation study of frequency type and study name; of the frequency-type study option will activate Properties, where the number of calculated frequency modes is set, in this case 30 modes, Figure 2; applying restraints; finite element mesh (mesh) and calculating analysis; view results.

Figure 2. Clamped-clamped 1,2,3 and 4 – Locked in the direction Ox, Oy si Oz

Meshing 3D model was made with tetrahedral elements the average size of 2 mm finite element. Table 2 summarizes the mesh information for the studied plate.

Table 2. Meshing information

Type of plate / bearing / size	Number of nodes	Number of items	Meshing time [min]
Clamped without damage 950x400mm	2.315.689	1.352.214	05:11

To obtain the proper frequencies of the non-damageive plates, the modal analysis module of the SolidWorks finite element analysis software was used. The first 30 vibration frequencies were determined for all plates studied.

In Figure 3 presents nine vibration modes obtained by finite element simulation for a non-damageive plate with calmped-clamped support type.

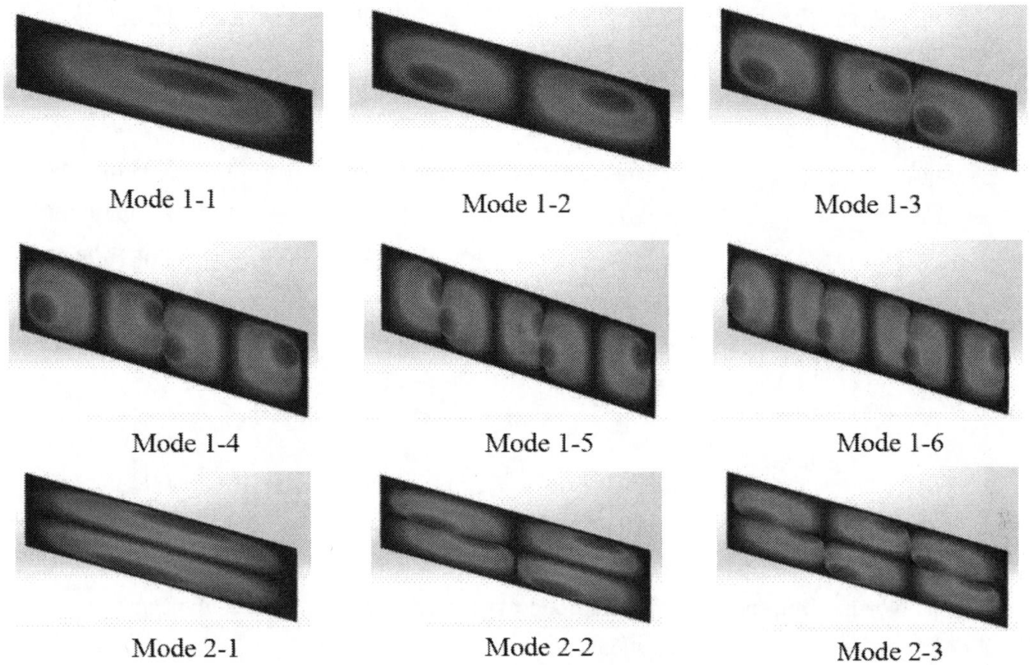

Figure 3. Examples of vibration modes determined by FEM for clamped-clamped support type

Following the numerical simulation, the frequencies for the clamped plate were obtained on all sides according to the vibration modes (Tables 3 - 6, first column). The results obtained are presented numerically in Tables 3 - 6 , 3rd column.

For the determination of the analytically determined own frequencies, the relation [43]:

$$\omega = \pi^2 \cdot \left[\left(\frac{m + \frac{1}{2}}{a} \right)^2 + \left(\frac{n + \frac{1}{2}}{b} \right)^2 \right] \cdot \sqrt{\frac{D}{\rho \cdot h}} \quad (1)$$

where:

$D = \frac{Eh^3}{12(1-\upsilon^2)}$ – the bending stiffness of the plate;

υ – cross-contractive coefficient (Poisson coefficient);

E – module of longitudinal elasticity of the plate;

h – the thickness of the plate;

ρ – density of the plate;

a, b – the length or the width of the plates;

The proprietary frequencies obtained by the FEM are compared with their analytically determined frequencies (Tables 3 - 6, column 2),comparing the percentage error with the relation [43]:

$$\varepsilon = \frac{f_A - f_{FEM}}{f_A} \cdot 100 \ [\%] \quad (2)$$

where:

ε - it represents the percent deviation or error;

f_A - the value of analytically determined own frequency;

f_{FEM} - The value of the own frequency determined by the FEM.

IOP Conf. Series: Materials Science and Engineering **294** (2017) 012033 doi:10.1088/1757-899X/294/1/012033

The values of the percentage deviation are presented in Tables 3 – 6, fourth column.

Table 3. Normal frequency for the clamped plate on all sides m=1 and n=1…10

Vibration mode m-n	Analytical [Hz]	FEM [Hz]	ε [%]
1-1	78,143606	73,047	6,5221024
1-2	92,86676	87,541	5,73483989
1-3	115,45148	113,94	1,30919067
1-4	158,89779	152,73	3,88160842
1-5	213,20566	203,54	4,53349128
1-6	268,37512	265,93	0,91108296
1-7	344,40614	339,59	1,3983897
1-8	431,29875	424,31	1,62039653
1-9	529,05293	519,99	1,71304788
1-10	637,66868	626,5	1,75148637

Table 4. Normal frequency for the clamped plate on all sides for modes m=2 and n=1…9

Vibration mode m-n	Analytical [Hz]	FEM [Hz]	ε [%]
2-1	203,67575	193,99	4,75547531
2-2	225,39891	208,78	7,37311019
2-3	257,98363	234,18	9,22679862
2-4	301,42993	279,78	7,18240886
2-5	345,73781	318,91	7,75958233
2-6	410,90726	378,66	7,84781948
2-7	466,93829	449,92	3,64465506
2-8	563,8309	532,56	5,54614868
2-9	681,58507	626,45	8,08924262

Table 5. Normal frequency for the clamped plate on all sides m=4 and n=1…4

Vibration mode m-n	Analytical [Hz]	FEM [Hz]	ε [%]
4-1	632,53827	617,59	2,36321986
4-2	654,26142	632,81	3,27872305
4-3	686,84615	658,33	4,15175218
4-4	730,29245	694,4	4,91480502

Table 6. Normal frequency for the clamped plate on all sides m=3 and n=1…7

Vibration mode m-n	Analytical [Hz]	FEM [Hz]	ε [%]
3-1	387,47398	375,67	3,04639295
3-2	409,19713	390,69	4,52279076
3-3	441,78185	416,07	5,82003312
3-4	485,22816	452,15	6,81703222
3-5	539,53603	499,21	7,4742052
3-6	604,70549	557,51	7,80470672
3-7	680,73651	627,11	7,87771909

In Figures 4, 6, 8 and 10 it's present the frequency variation according to the vibration modes for the analyzed cases and in Figures 5, 7, 9 and 11 the percentage deviation of the frequency from the analytically determined frequency is plotted, depending on the vibration modes for these cases.

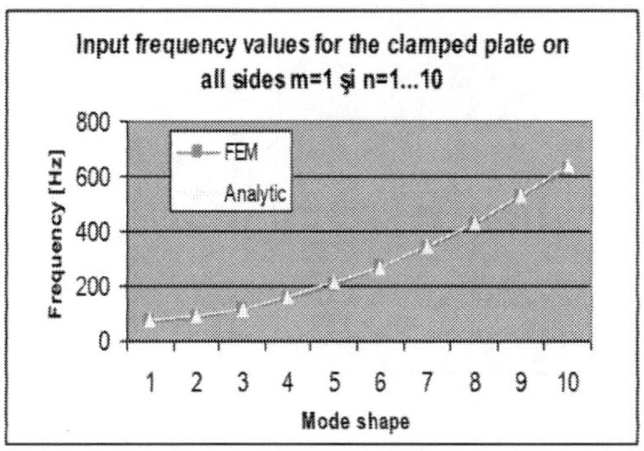

Figure 4. Frequency variation according to vibration modes for m=1 and n=1÷10

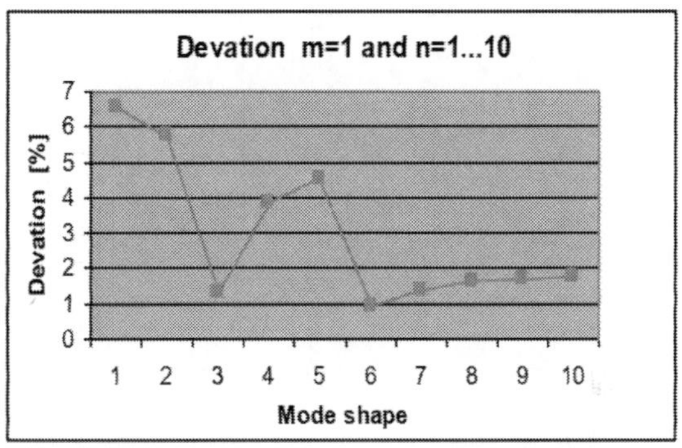

Figure 5. Percentage frequency deviation for m=1 and n=1÷10

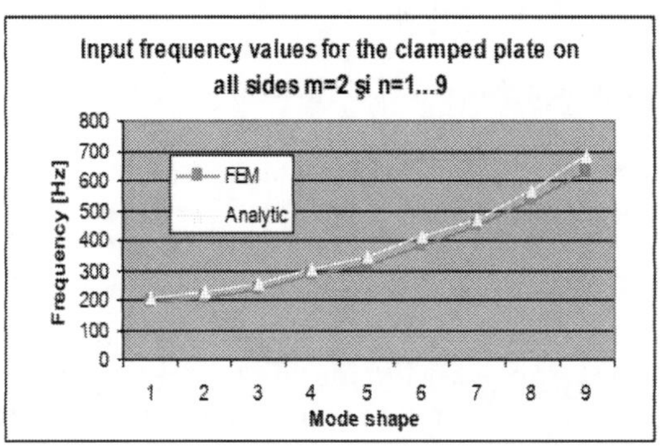

Figure 6. Frequency variation according to vibration modes for m=2 and n=1÷9

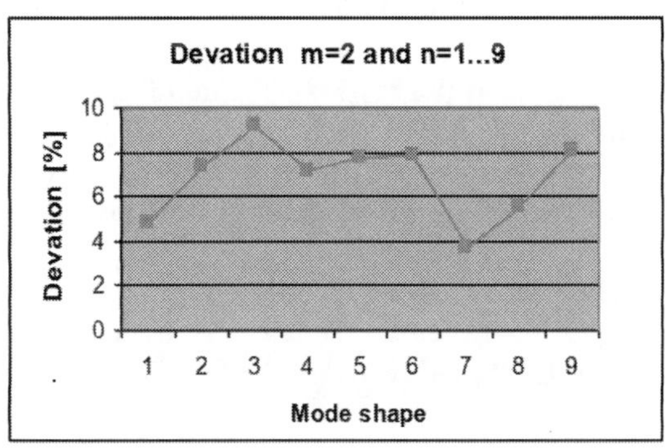

Figure 7. Percentage frequency deviation for m=2 and n=1÷9

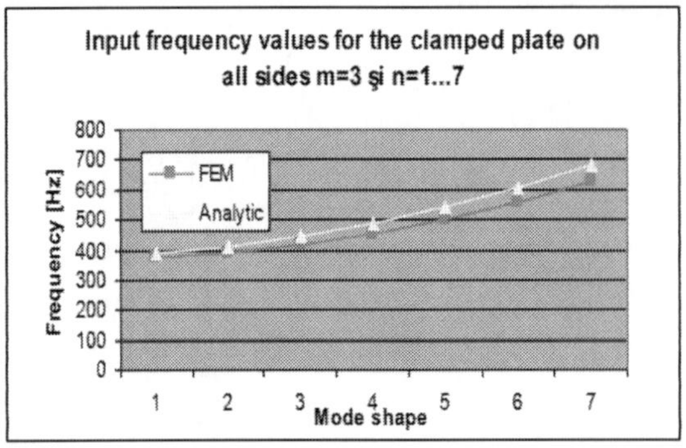

Figure 8. Frequency variation according to vibration modes for m=3 and n=1÷7

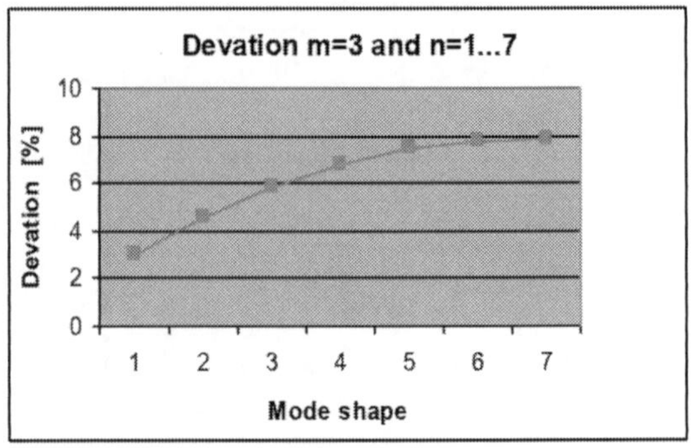

Figure 9. Percentage frequency deviation for m=3 and n=1÷7

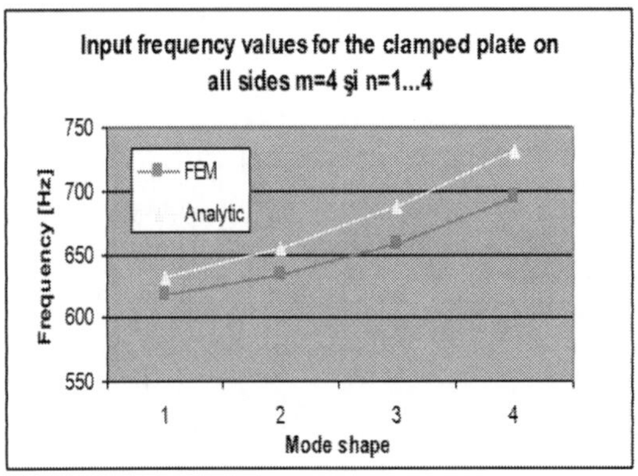

Figure 10. Frequency variation according to vibration modes for m=4 and n=1÷4

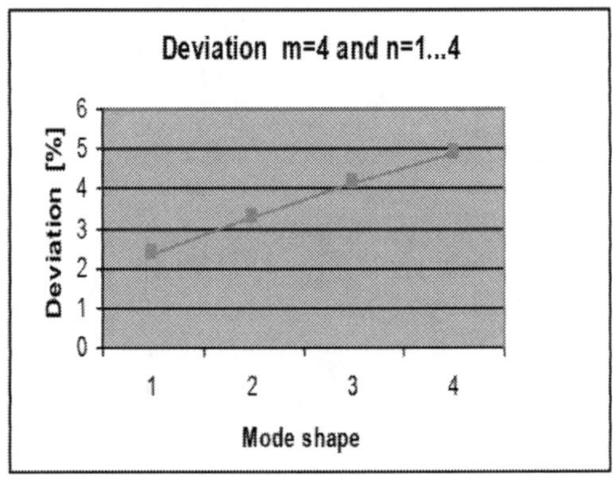

Figure 11. Percentage frequency deviation for m=4 and n=1÷4

From the results presented in Figures 5, 7, 9, 11 and Tables 3 - 6 it is observed that the maximum error is 9.3% for mode m = 2 and n = 3 it is also noted that larger deviations occur in mode m = 2.

3. Analysis of the results obtained for clamped plates on all sides with longitudinal central damage

For modal analysis of damageive plates, the Finite Element Method (FEM) was also used using the SolidWorks program. Modeling of the geometry to be analyzed was done in the Geometry module of the same program.

To obtain its own frequencies, the static analysis module for thin rectangular plates was used under its own weight. The results obtained in this module were transferred to the modal analysis module. By modal analysis the first 30 frequencies were determined. Meshing 3D model was made with 2 mm thick tetrahedron finite elements (Table 7), the material used in the program database being identical to the one used in the previous subchapter.

In Figure 12 the longitudinal central damage plate is clamped in the contour, using modal analysis with the finite element method.

Figure 12. Type of support for plate with longitudinal central damage

Table 7. Meshing information

No.	Type of plate / bearing / size	Number of nodes	Number of items	Meshing time [min]
1	No damage	2.489.930	1.476.647	04:42
2	With longitudinal central damage 5.3 × 24.8 mm	2.477.378	1.437.411	04:35
3	With longitudinal central damage 5.8 × 48.8 mm	2.517.888	1.467.061	04:38
4	With longitudinal central damage 6,05 × 75,2 mm	2.510.268	1.460.333	05:25
5	With longitudinal central damage 4.9 × 101.6 mm	2.763.438	1.648.846	05:25

In Figure 13 there are ten modes of vibration obtained by finite element simulation for a plate with longitudinal central damage clamped on the contour.

Following the numerical simulation, the actual frequencies for a real clamped plate on the contour were obtained, according to their own vibration modes with a longitudinal central damage having the dimensions: 5.3 × 24.8 mm; 5.8 × 48.8 mm; 6.05 × 75.2 mm; 4.9 × 101.6 mm. The results obtained are presented numerically in Tables 8-11 and the percent deviation values calculated by the formula (2) are presented in Tables 12 - 15, for each analyzed case.

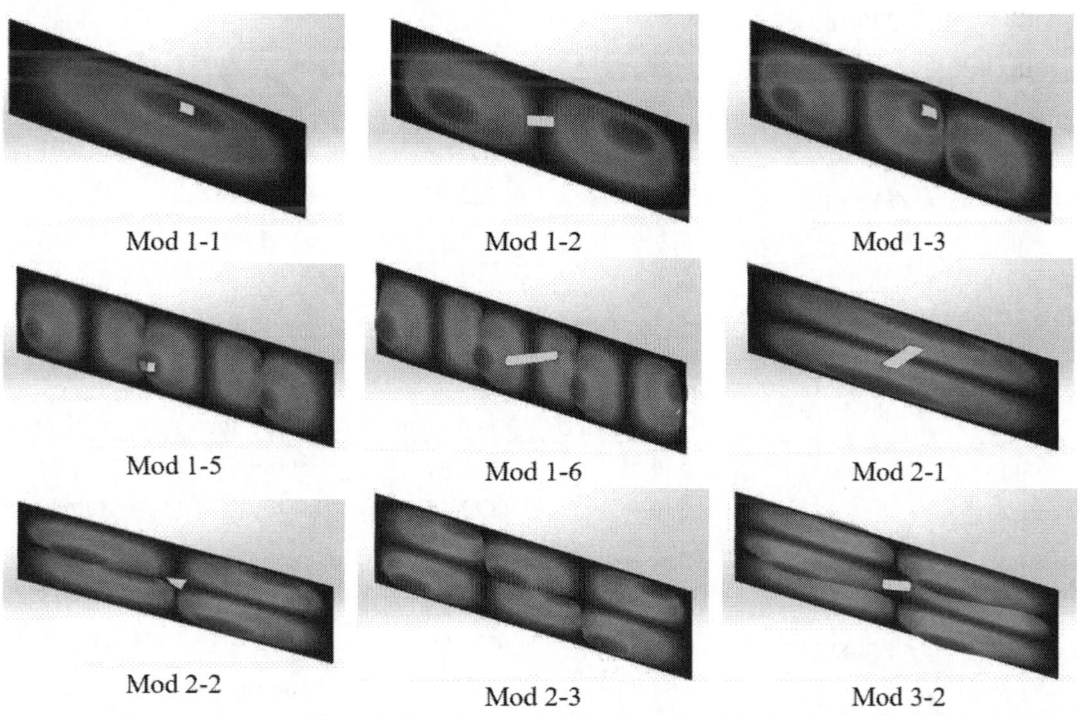

Figure 13. Examples of vibration modes for the central fault plate 50×50 mm

IOP Conf. Series: Materials Science and Engineering **294** (2017) 012033 doi:10.1088/1757-899X/294/1/012033

Table 8. Normal frequency for the clamped plate with central damage m=1 and n=1…10

Vibration mode m-n	Frequency [Hz]				
	No damage	Damage 5,3×24,8 mm	Damage 5,8×48,8mm	Damage 6,05×75,2 mm	Damage 4,9×101,6 mm
1-1	73,047	72,903	72,497	71,846	70,922
1-2	87,541	87,54	87,535	87,529	87,491
1-3	113,94	113,81	113,49	113,01	112,45
1-4	152,73	152,73	152,71	152,68	152,57
1-5	203,54	203,38	203,08	202,67	202,25
1-6	265,93	265,92	265,86	265,79	265,57
1-7	339,59	339,38	339,03	338,47	337,69
1-8	424,31	424,27	424,16	423,99	423,7
1-9	519,99	519,75	519,51	519,33	519,31
1-10	626,5	626,47	626,32	625,97	625,67

Table 9. Normal frequency for the clamped plate on all sides with central damage m=2 and n=1…9

Vibration mode m-n	Frequency [Hz]				
	No damage	Damage 5,3×24,8 mm	Damage 5,8×48,8mm	Damage 6,05×75,2mm	Damage 4,9×101,6mm
2-1	193,99	193,96	193,82	193,39	192,51
2-2	208,78	208,75	208,71	208,66	208,63
2-3	234,18	234,15	233,98	233,42	232,2
2-4	270,78	270,68	270,57	270,46	270,34
2-5	318,91	318,88	318,64	317,8	315,86
2-6	378,66	378,53	378,37	378,21	377,89
2-7	449,92	449,87	449,54	448,25	445,01
2-8	532,56	532,42	532,23	532,04	531,06
2-9	626,45	626,26	626,03	625,14	624,36

Table 10. Normal frequency for the clamped plate on all sides with central damage m=3 and n=1…7

Vibration mode m-n	Frequency [Hz]				
	No damage	Damage 5,3×24,8 mm	Damage 5,8×48,8mm	Damage 6,05×75,2mm	Damage 4,9×101,6mm
3-1	375,67	374,56	371,64	366,91	359,92
3-2	390,69	390,68	390,65	390,6	390,39
3-3	416,07	415,12	413	410,37	407,87
3-4	452,15	452,11	452	451,86	451,18
3-5	499,21	498,38	496,55	494,18	492,24
3-6	557,51	557,44	557,23	556,95	555,77
3-7	627,11	626,41	624,93	622,95	621,74

Table 11. Normal frequency for the clamped plate on all sides with central damage m=4 and n=1…4

Vibration mode m-n	Frequency [Hz]				
	No damage	Damage 5,3×24,8 mm	Damage 5,8×48,8mm	Damage 6,05×75,2mm	Damage 4,9×101,6mm
4-1	617,59	617,31	615,73	608,99	583,47
4-2	632,81	632,78	632,73	632,66	632,49
4-3	658,33	658,06	656,69	652,47	645,58
4-4	694,4	694,23	694,13	693,84	693,06

Table 12. Deviation normal frequency for the clamped plate on all sides with a longitudinal central damage m=1 and n=1…10

Vibration mode m-n	ε [%]			
	Damage 5,3×24,8 mm	Damage 5,8×48,8mm	Damage 6,05×75,2mm	Damage 4,9×101,6mm
1-1	0,197133	0,7529	1,644147	2,909086
1-2	0,001142	0,0069	0,013708	0,057116
1-3	0,114095	0,3949	0,816219	1,307706
1-4	0	0,0131	0,032738	0,10476
1-5	0,078609	0,226	0,427434	0,633782
1-6	0,00376	0,0263	0,052645	0,135374
1-7	0,061839	0,1649	0,329809	0,559498
1-8	0,009427	0,0354	0,075417	0,143763
1-9	0,046155	0,0923	0,126926	0,130772
1-10	0,004789	0,0287	0,084597	0,132482

Table 13. Deviation normal frequency for the clamped plate on all sides with a longitudinal central damage m=2 and n=1…9

Vibration mode m-n	ε [%]			
	Damage 5,3×24,8 mm	Damage 5,8×48,8mm	Damage 6,05×75,2mm	Damage 4,9×101,6mm
2-1	0,015465	0,0876	0,309294	0,762926
2-2	0,014369	0,0335	0,057477	0,071846
2-3	0,012811	0,0854	0,324537	0,845503
2-4	0,03693	0,0776	0,118177	0,162494
2-5	0,009407	0,0847	0,348061	0,956383
2-6	0,034332	0,0766	0,11884	0,203349
2-7	0,011113	0,0845	0,371177	1,091305
2-8	0,026288	0,062	0,097642	0,281658
2-9	0,03033	0,067	0,209115	0,333626

Table 14. Deviation normal frequency for the clamped plate on all sides with a longitudinal central damage m=3 and n=1…7

Vibration mode m-n	ε [%]			
	Damage 5,3×24,8 mm	Damage 5,8×48,8mm	Damage 6,05×75,2mm	Damage 4,9×101,6mm
3-1	0,295472	1,0728	2,331834	4,192509
3-2	0,00256	0,0102	0,023036	0,076787
3-3	0,228327	0,7379	1,369962	1,970822
3-4	0,008847	0,0332	0,064138	0,214531
3-5	0,166263	0,5328	1,007592	1,396206
3-6	0,012556	0,0502	0,100447	0,312102
3-7	0,111623	0,3476	0,66336	0,856309

Table 15. Deviation normal frequency for the clamped plate on all sides with a longitudinal central damage m=4 and n=1…4

Vibration mode m-n	ε [%]			
	Damage 5,3×24,8 mm	Damage 5,8×48,8mm	Damage 6,05×75,2mm	Damage 4,9×101,6mm
4-1	0,045338	0,3012	1,39251	5,524701
4-2	0,004741	0,0126	0,023704	0,050568
4-3	0,041013	0,2491	0,890131	1,936719
4-4	0,024482	0,0389	0,080645	0,192972

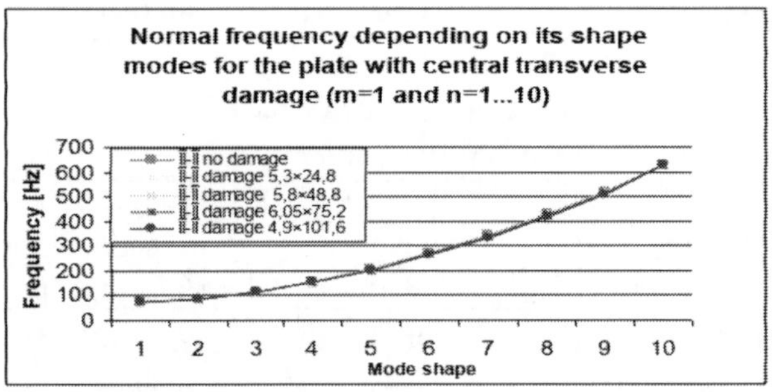

Figure 14. Frequency variation according to vibration modes for m=1 and n=1÷10

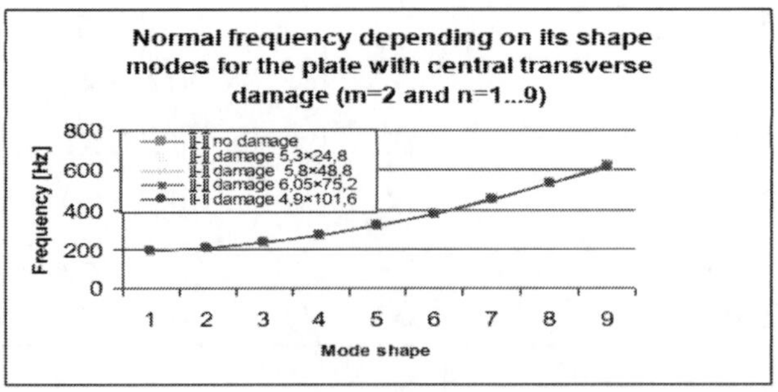

Figure 15. Frequency variation according to vibration modes for m=2 and n=1÷9

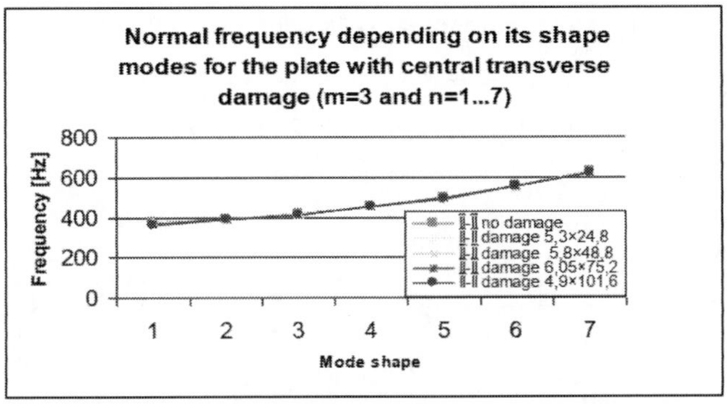

Figure 16. Frequency variation according to vibration modes for m=3 and n=1÷7

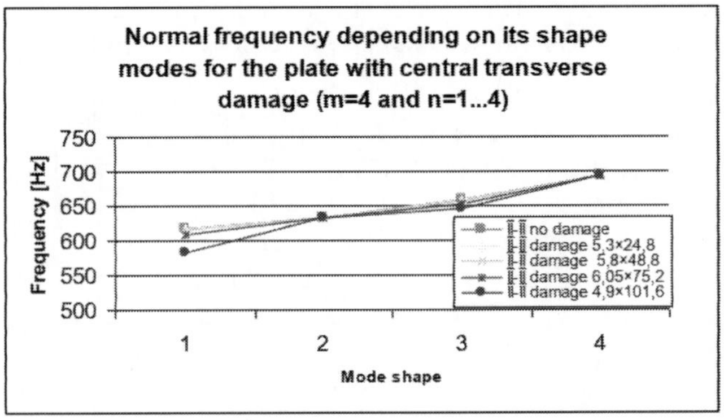

Figure 17. Frequency variation according to vibration modes for m=4 and n=1÷4

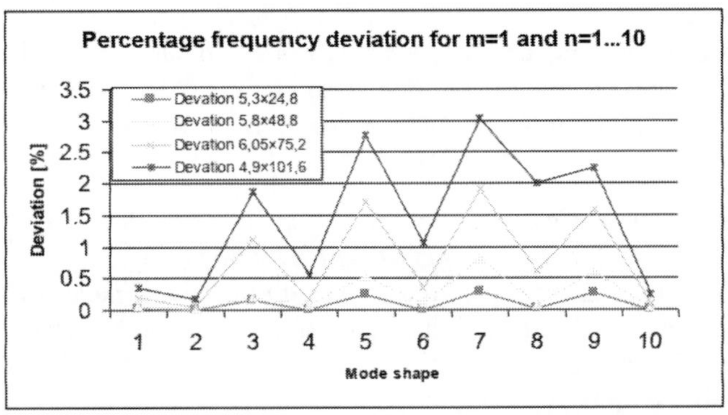

Figure 18. The relative frequency variation depending on the modes vibration for m=1 and n=1÷10

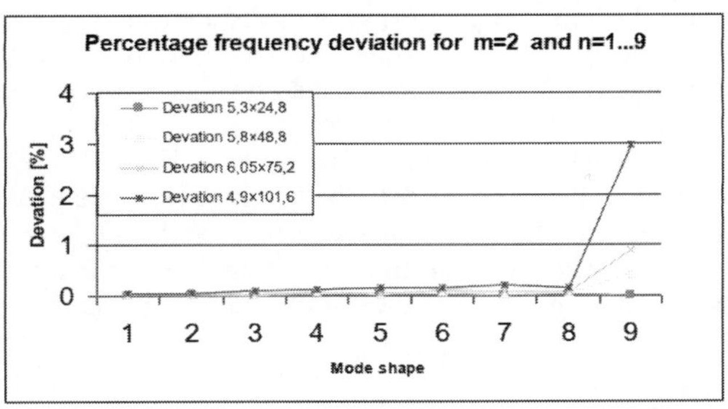

Figure 19. The relative frequency variation depending on the modes vibration for m=2 and n=1÷9

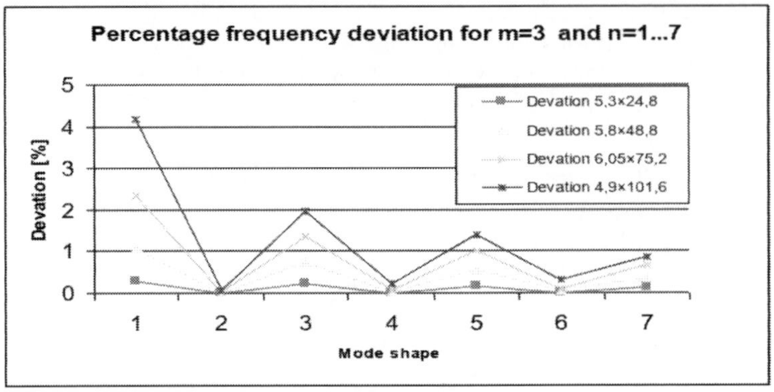

Figure 20. The relative frequency variation depending on the modes vibration for m=3 and n=1÷7

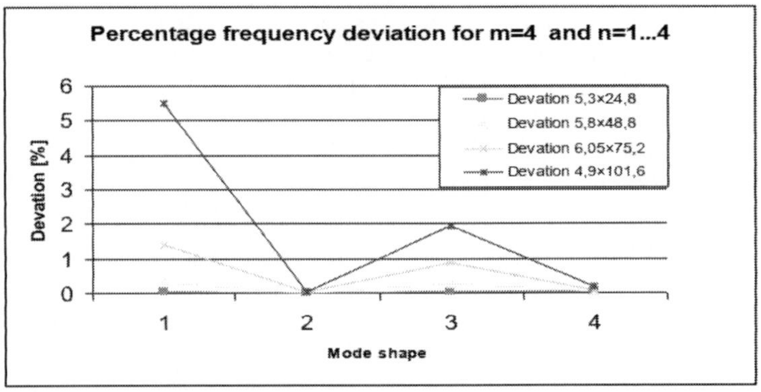

Figure 21. Relative frequency variation according to vibration modes for m=4 and n=1÷4

From the results presented in Figures 18 - 21 and tables 12 - 15 it is noted that the maximum eror is 5.52% for the m = 4 mode and n = 1 for the defective plate of 4.9 × 101.6 mm it is also noted that larger deviations occur in the defect plate of 4.9 × 101.6 mm.

4. Conclusions

From the modal analysis regarding the determination of the own frequencies and the mass participation coefficients, the following conclusions can be made:

- The meshing in finite element for all the analyzed plates was made using the same solid mesh type, resulting in total numbers of nodes and different elements for each type of plate being analyzed;
- For each analyzed plate, the value of natural frequencies gradually increases from the first mode to the last mode of vibration;
- Frequency values of the plain back plates are in all cases higher than those of the plain plates that are lengthened in width and free in length; This confirms that a flat and flawless plate has the highest values compared to other types of plates studied by the authors.

Analysis of defective contour clamped plates in a wide variety of sizes, positions, and plate orientation has highlighted the fact that each fault has its own signifi- cant signature by the relative frequency variation for a reasonable number of vibration modes m and n.

In the case of small size defects, these "signatures" are very close to those obtained in the case of bars with bigger defects, ie when changing the defect sides. In order to increase them, disturbances are introduced in the "signature" due to the essential mass change and the modal distribution.

In order to obtain credible results, an analysis of the influence of the mean sizes of the elements used in the modeling was made and chosen for various applications that lead to sufficiently precise results.

References

[1] Haţiegan C, Nedeloni M D, Tufoi M, Protocsil C and Răduca M 2013 Modal Analysis Of Natural Frequencies And Mass Participation Coefficients Of Simply Supported Thin Plates With Damages, *Constanţa Maritime University Annals* **XIV**(19) 115-120

[2] Haţiegan C, Tufoi M, Răduca E, Protocsil C and Mituleţu C 2013 Modal Analysis Through Solidworks Software Of Clamped Thin Plates With Damages, *Constanţa Maritime University Annals* **XIV**(19) 121-124

[3] Haţiegan C, Gillich G R, Gillich N, Tufoi M and Răduca M 2013 Modal Analysis of Thin Plates with Damage Simply Supported on all Edges, *Analele Universităţii "Eftimie Murgu"-Fascicula de Inginerie* **XX**(1) 117-126

[4] Haţiegan C, Gillich G R, Răduca E, Nedeloni M D and Cîndea L 2013 Equation of Motion and Determining the Vibration Mode Shapes of a Rectangular Thin Plate Simply Supported on Contour Using MATLAB, *Analele Universităţii "Eftimie Murgu"-Fascicula de Inginerie* **XX**(1) 127-138

[5] Tufoi M, Gillich G R, Praisach Z I, Ntakpe J L and Hatiegan C 2014 An Analysis of the Dynamic Behavior of Circular Plates from a Damage Detection Perspective, *Romanian Journal of Acoustics & Vibration* **X**(1) 41-46

[6] Haţiegan C, Gillich G R, Răduca M, Budai A M, Muntean F and Răduca E 2012 Finite Element nalysis Of Natural Frequencies And Mass Participation Coefficients For Thin Plates With Defects, *Scientific Bulletin of "Politehnica" University of Timişoara* **57**(71)2 43-53

[7] Leissa A W 1969 *Vibration of Shells*, NASA SP-288, Washington, DC

[8] Tufoi M, Gillich G R, Mituletu I C and Hatiegan C 2015 An Analysis of the Dynamic Behavior of Rectangular Plates from a Damage Detection Approach, *Romanian Journal of Acoustics & Vibration* **XII**(2) 146-150

[9] Tufoi M, Haţiegan C, Vasile and O Gillich G R 2013 Dynamic Analysis of Thin Plates with Defects by Experimental and FEM Methods, *Romanian Journal of Acoustics and Vibration* **X**(2) 83-88

[10] Tufoi M, Gillich G R, Haţiegan C, Mituleţu I C and Vasile I 2014 Effects of Structural Defects on Dynamic Behavior at Sandwich Composite Beams- Part I-Theoretical Approach, *Analele Universităţii "Eftimie Murgu"-Fascicula de Inginerie* **XXI**(1) 327-334

[11] Tufoi M, Gillich G, Haţiegan C, Mituleţu I C and Korka Z I 2014 Effects of Structural Defects on Dynamic Behavior at Sandwich Composite Beams- Part II- FEM Analysis, *Analele Universităţii "Eftimie Murgu"-Fascicula de Inginerie* **XXI**(1) 335-344

[12] Nicknam A, Hosseini M H and Bagheri A 2011 Damage detection and denoising in two-dimensional structures using curvelet transform by wrapping method, *Arch Appl Mech* **81** 1915-1924

[13] Choi S, Park S and Stubbs N 2005 Nondestructive damage detection in structures using changes in compliance, *International Journal of Solids and Structures* **42**(15) 4494-4513

[14] Choi S H 2002 Development of Nondestructive Damage Detection Algorithms for Plate Structures, Strength of Materials, *Journal of Civil Engineering* **6**(4) 495-501

[15] Eftekhar S A and Jafari A A 2013 A simple and accurate Ritz formulation for free vibration of thick rectangular and skew plates with general boundary conditions, *Acta Mech* **224** 193-209

[16] Nicknam A and Hosseini M H 2012 Structural damage localization and evaluation based on modal data via a new evolutionary algorithm, *Arch Appl Mech* **82** 191-203

[17] Hatiegan C, Nedeloni M D, Popescu C, Tufoi M, Pădureanu I and Rudolf C 2015 Comparative Study through Modal Analysis of Thin Trapeze Shape Plates Clamped on Contour without and with Damages, *Analele Universităţii "Eftimie Murgu"-Fascicula de Inginerie*, Anul **XXII**(2) 148-161

[18] Nedeloni M D, Haţiegan C, Muntean F, Jurcu M, Terfăloagă I M and Magheţi P 2015 A Modal Analysis Study of Thin Parallelogram Plates Clamped on the Rim with and without Defects Using the SolidWorks Program, *Analele Universităţii "Eftimie Murgu"-Fascicula de Inginerie* **XXII**(2) 273-288

[19] Gillich G R, Praisach Z I, Bobos D and Hatiegan C 2013 Assessment of corrosion damages with important loss of mass and influences on the natural frequencies of bending vibration modes,

Acoustics & Vibration of Mechanical Structures, Book Series: Applied Mechanics and Materials, vol. **430**, pp 95-100

[20] Haţiegan C, Gillich G R, Popovici G and (Barboni) Haţiegan L 2017 *The influence of temperature changes on the eigen - frequencies and mode shapes for rectangular plates clamped on the contour,* International Conference Knowledge-Based Organization, Vol. 23, Issue 3, pp 168-173

[21] Bratu P 2000 *Vibration of elastic systems*, Technical Publishing House, Bucharest, Romania

[22] Buzdugan G, Mihăilescu E and Radeş M 1979 *Măsurarea vibraţiilor*, Ed. Academiei, Bucuresti, Romania

[23] Elshafey A A, Marzouk H and Haddara M R 2011 Experimental Damage Identification Using Modified Mode Shape Difference, *J. Marine Sci. Appl.* **10** 150-155

[24] Feng G Q, Li G, Liu Z H, Niu H L and Li F C 2010 Numerically Simulating the Sandwich Plate System Structures, *J. Marine Sci. Appl.* **9** 286-291

[25] Elishakoff I and Sternberg A 1980 Vibration of Rectangular Plates with Edge-Beams, *Acta Mechanica* **36** 195-212

[26] Bratu P 2001 Structural requirements imposed to vibration systems, *International Journal of Acoustics and Vibration* **5**(2) 84-92

[27] Monfared A H 2012 Numerical Simulation of Welding Distortion in Thin Plates, *Journal of Engineering Physics and Thermophysics* **85**(1) 187-194

[28] Tufoi M, Gillich G R, Hatiegan, Gillich N and Lorenz P 2015 Some Aspects Regarding the Transition from Beam to Plate Behavior of Vibrating Structures, *Romanian Journal of Acoustics & Vibration* **XII**(1) 62-68

[29] Nedeloni M D, Haţiegan C, Vasile O, Hamat C O, Fănică C and Gillich N 2015 Numerical Study Regarding the Influence of Material Components for a Booster - Ultrasonic Horn Assembly on the Natural Frequency, *Romanian Journal of Acoustics & Vibration* **XII** (2) 155-160

[30] Haţiegan C, Pădureanu I, Jurcu M, Nedeloni M D, Hamat C O, Chioncel C P, Trocaru S, Vasile O, Bădescu O, Micliuc D, (Filip) Nedeloni L, Băra A and (Barboni) Haţiegan L 2016 Vibration analysis of a hydro generator for different operating regimes, *IOP Conf. Ser.: Mater. Sci. Eng.* **163** 012030

[31] Iancu V, Protocsil C, Gillich N, Haţiegan C and Gillich G R 2016 The Influence of the Number of Finite Elements upon the Accuracy of the Results Obtained Using Discrete Models, *Analele Universităţii "Eftimie Murgu"-Fascicula de Inginerie* **XX**(3) 189-198

[32] Hamat C O, Nedeloni M D, Haţiegan C, Ciubotariu R C and Pădureanu I 2015 Cavitation Erosion Research On C45 Carbon Steel. Part I: Multiple Tests Of 180 Minutes, *Annals of „Constantin Brâncuşi" University of Târgu Jiu* **3** 127-132

[33] Nedeloni M D, Haţiegan C, Vasile O, Hamat C O, Fănică C and Gillich N 2015 Numerical study regarding the influence of material components for a booster - ultrasonic horn assembly on the natural frequency, *Annals of „Constantin Brâncuşi" University of Târgu Jiu* **3** 100-105

[34] Muntean F, Haţiegan C, Popescu C, Gillich E V and Iancu V 2015 Study Regarding The Determining Of The Natural Frequencies And Modal Shapes Of The Column Type Structures With Additional Mass, *Annals of „Constantin Brâncuşi" University of Târgu Jiu* **3** 52-55

[35] Nedeloni M D, Haţiegan C, (Filip) Nedeloni L, (Barboni) Haţiegan L, Birtărescu E, Băra A and Pellac A 2016 Research on the cavitation erosion resistance of the x3crni13-4 stainless steel used to manufacture the runner blades of kaplan turbines, *Annals of Constantin Brâncuşi University of Târgu Jiu* **3** 70-75

[36] Pop C, Hatiegan C, Vlase T, Raduca M, Gillich G R, Vasile O and Raduca E 2014 Fermionic oscillators and their connection with the isokinetic temperature, *Romanian Reports in Physics* **66**(3) 716-722

[37] Haţiegan C, Molnar M, Trocaru S, Pădureanu I, Jurcu M R and Ilie F 2016 *Modeling and Simulation of Thermal Analysis of a Teflon Coated Plate*, International Conference Knowledge-Based Organization, Vol. 22, Issue 3, pp 639-643

[38] Nicolina P, Popov D and Haţiegan C 2013 Some Theoretical Observations Concerning the Reverberation Time for the Case of a Harmonic Emitting Source, *Analele Universităţii "Eftimie Murgu", Fascicula de Inginerie* **XX**(2) 251-258

[39] Haţiegan C, Gillich E V, Vasile O, Nedeloni M D and Pădureanu I 2015 Finite Element Analysis of thin plates clamped on the rim of different geometric forms. Part I: Simulating the Vibration Mode Shapes and Natural Frequencies, *Romanian Journal of Acoustics & Vibration* **XII**(1) 69-74

[40] Haţiegan C, Gillich E V, Vasile O, Nedeloni M D, Jurcu M and Magheţi P 2015 Finite Element Analysis of thin plates clamped on the rim of different geometric forms. Part II: The Absolute and Relative Variation of Natural Frequencies, *Romanian Journal of Acoustics & Vibration* **XII**(1) 81-86

[41] Tufoi M, Haţiegan C, Gillich G R, Protocsil C and Negru I 2013 Frequency Changes in Thin Rectangular Plates due to Geometrical Discontinuities. Part I: Finite Element Analysis, *Analele Universităţii "Eftimie Murgu"-Fascicula de Inginerie* **XX**(3) 221-232

[42] Tufoi M, Haţiegan C, Gillich G R and Vasile O 2013 Frequency Changes in Thin Rectangular Plates due to Geometrical Discontinuities. Part II: Frequency Shift Interpretation, *Analele Universităţii "Eftimie Murgu"-Fascicula de Inginerie* **XX**(3) 233-244

[43] Hatiegan C 2013 *Identification of Defects in Thin Elastic Plates by Means of Modal Analysis*, University "Eftimie Murgu" of Resita, Doctoral Thesis

[44] Haţiegan C and Suciu L 2010 *Fizică tehnologică-Teorie şi aplicaţii*, Editura „Eftimie Murgu" Reşiţa, Orizonturi Tehnice

Analysis of gear reducer housing using the finite element method

I Zs Miklos, C C Miklos, C I Alic and S Raţiu

Politehnica University of Timisoara, Department of Engineering and Management, 5 Revolution Street, Hunedoara, 331128, Romania

E-mail: imre.miklos@upt.ro

Abstract. The housing is an important component in the construction of gear reducers, having the role of fixing the relative position of the shafts and toothed wheels. At the same time, the housing takes over, via the bearings, the shaft loads resulting when the toothed wheel is engaging another toothed mechanism (i.e. power transmission through belts or chains), and conveys them to the foundation on which it is anchored. In this regard, in order to ensure the most accurate gearing, a high stiffness of the housing is required. In this paper, we present the computer-aided 3D modelling of the housing (in cast version) of a single stage cylindrical gear reducer, using the Autodesk Inventor Professional software, on the principle of constructive sizing. For the housing resistance calculation, we carried out an analysis using the Autodesk Simulation Mechanical software to apply the finite element method, based on the actual loads, as well as a comparative study of the stress and strain distribution, for several tightening values of the retaining bolts that secure the cover and the foundation housing.

1. Introduction

The gear reducers are the most common mechanical systems used to change the motion input parameters, usually provided with an AC asynchronous electric motor for reducing the speed, at the same time with the increase of the output twisting moment, in accordance with its speed transmission ratio. The toothed wheel gear reducers have the advantage of a constant transmission ratio, they have a high yield and their maintenance is easy and cheap [1].

The fixed axle toothed wheel gear reducers may have one or more stages, being equipped with cylindrical gears (with parallel axes), conical gears (with concurrent axes), worm gears (with cross axles) or with a combination of conical toothed wheels and worm gears with toothed wheels, for obtaining high transmission ratios [1].

In practice, there are also gears with conventional cylindrical or conical toothed wheels with mobile axles, known as planetary or differential, which have the advantage of a high transmission ratio under a reduced gauge size [1].

Regardless of the construction type, the gear reducers with toothed wheels have housing.

The housing is an important part of the reducer, which has the role of ensuring the relative position of the shafts and, implicitly, of the toothed wheels. At the same time, the housing takes over, via the roller bearings, the loads of the shafts resulting from gearing, and conveys them to the foundation on which it is anchored. In this regard, in order to ensure the most accurate gearing, a high stiffness of the housing is required. For easy assembly of the shafts, toothed wheels and bearings, the housing is made of two pieces: the lower housing (housing body) and upper housing (housing cover) [2].

The housing can be manufactured in welded construction, using steel sheet, for the unique and small series products, or by casting, for the large series products; in this case, the conditions imposed by the casting technology and processing economy must be observed [2]:

- Achieving a wall thickness as uniform as possible by avoiding any clumps of material; to increase the strength and rigidity, it is recommended to use ribs;
- Ensuring a minimum thickness of the walls imposed by the casting technology and the nature of the housing material;
- Gradual passing from a slightly thicker wall to a thinner one, for reducing the residual stresses after casting;
- Ensuring sufficiently large radii and wall inclination (for easy release of the model out from the mould);
- Limiting the machined surfaces to the minimum necessary by making elevations (embossments) or dents (spotfaces) for the bearing surfaces of the nuts, bolts, covers, threaded plugs, etc.

In case of welded housings, it is necessary to observe some conditions related to the welding technology [2]:

- Using easily weldable materials;
- Ensuring accessibility for the execution of welding seams;
- Choosing constructive shapes suitable for welding automation;
- Making a symmetrical construction to reduce the internal stresses and not to deform the housing;
- The thickness of the walls must be smaller than at the cast housings, but not too small for preventing their burning during welding;
- Stiffening the housings with ribs;
- De-stressing the housings before machining.

The cover is fastened to the gear housing by means of threaded assemblies (bolts and nuts), and the housing is mounted on the foundation by means of anchor bolts. An important role in the housing resistance is played by the axial forces in the bolt stems, resulting from their tightening.

2. Housing modelling

2.1. Gear reducer designing

The gear reducer designing and, therefore, the 3D modelling of the housing, were carried out by using Autodesk Inventor Professional software.

The Autodesk Inventor products provides engineers and designers professional grade design and engineering solution for 3D mechanical design, simulation, visualization, and documentation. With Inventor software, engineers can integrate 2D and 3D data into a single design environment, creating a virtual representation of the final product that enables them to validate the form, fit, and function of the product before it is ever built. Autodesk Inventor includes powerful parametric, direct edit and freeform modelling tools as well as multi-CAD translation capabilities and industry standard DWG drawings. Inventor software helps you connect more effectively, reduce errors, and deliver great product designs faster [3].

Autodesk Inventor is built for the continually evolving needs of the modern design and engineering professional.

Building on its breadth of modelling capabilities and powerful design automation tools, Autodesk Inventor 2018 marks the next step forward in professional-grade 3D mechanical engineering design. Interoperability enhancements enable distributed, multi-disciplinary teams to work together more efficiently with data from a variety of sources. The overall user experience is enhanced for more productivity and flexibility in response to direct feedback from customers around the world [3-4].

The computer-aided designing of the gear reducer was realised based on the specific input parameters, i.e. the transmitted power ($P = 10$ kW); input speed ($n1 = 600$ rot/min); transmission ratio ($i = 2{,}50$); inclination angle of the teeth ($\beta = 10$ degrees) and the required distance between axes ($A = 180$ mm).

The modelling and designing of the gear reducer components (gear, shafts, bearings, fasteners and seals, etc.) were carried out using the specific tools of the Design Module, their assembly being made by properly applying the assembly constraints. Figure 1 shows the 3D model of the gear reducer, in a simplified version, by omitting some elements such as: bearing covers, aperture cover, oil drain plug, retaining bolts, etc.

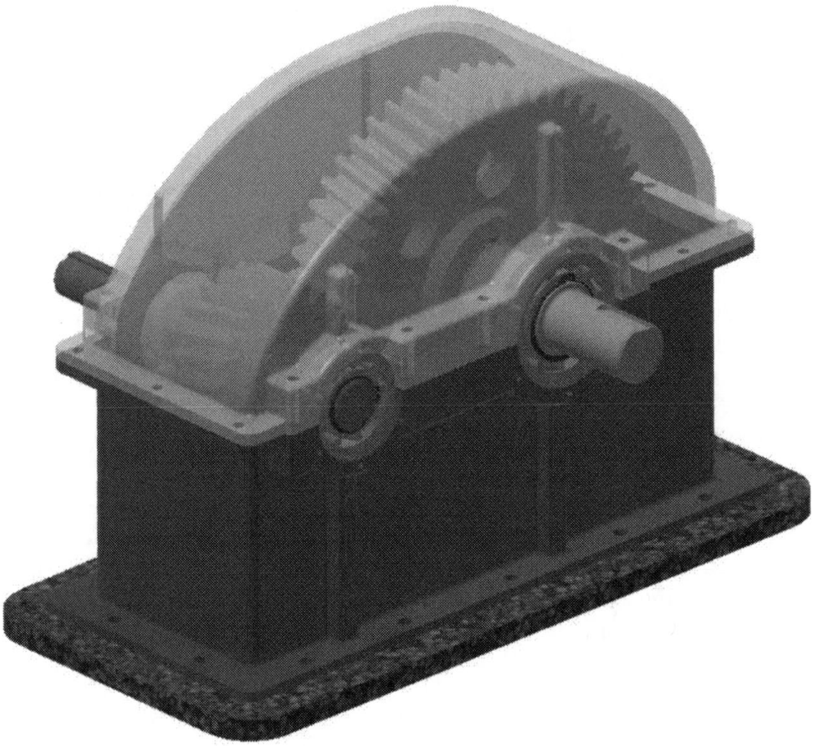

Figure 1. The 3D model of the gear reducer

2.2. Constructive sizing of the housing

When designing mechanical transmissions with toothed wheels, gear reducers, gearboxes, etc., the housing sizing is usually based on constructive relationships [5].

Thus, the thickness of the housing walls, the thickness and width of the foundation fastening shoes, cover retaining bolts and foundation retaining bolts are chosen either from tables or according to the gear size (generally given by its module) [5].

At the constructive designing of the housings, the shape and dimensions of the cooling ribs are at the discretion of the designer, without taking too much into account the strains, temperatures and their distribution in the housing walls [5].

The choice of the constructional shape of the gear housings is of particular importance, knowing that the connections, the holes for retaining bolts (in quite large number), etc., are stress concentration factors that must be also taken into account in designing [5], [6].

The anchoring of the housing and housing cover to the foundation was made using M8 bolt and nut assemblies and, respectively, M8 anchor bolts. In both cases, the threaded assemblies have been subjected to an initial tightening, which produced an axial force in the threaded rods, these ones being subjected to tensile stress.

Figure 2 shows the 3D model of the housing (with retaining bolts), obtained on the basis of the constructive sizing made by using the Autodesk Inventor Professional software.

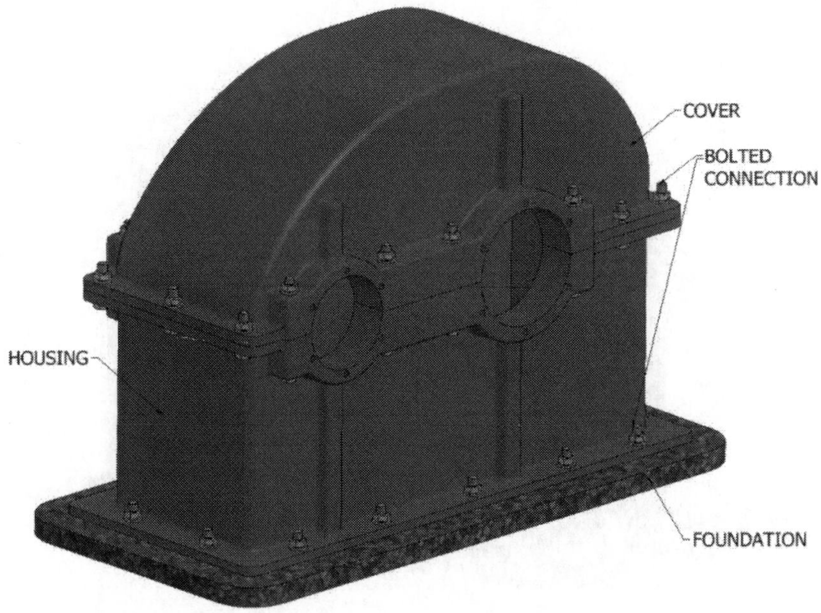

Figure 2. The 3D model of the gear reducer housing

3. Analysis of gear reducer housing using the finite element method

The finite element method is used for studying the state of stresses and strains of the resistance structures of high geometric complexity, for which the calculation is made easier if the whole is divided into simpler areas [7-9].

The finite element method is a mathematical method of numerical integration of partial differential equations put in variational form. This means that all the computational problems of the elastic structures will be reduced to a system of partial differential equations, which usually can not be solved analytically. In this respect, the assessments of the stress states in the housing walls, i.e. the sizing and verification calculations based on classical methods are quite difficult to apply to complex housings [7-9].

With the development and spreading of numerical computers and computer-aided design software packages, the problems outlined above can be solved much easier, the MEF analysis of the 3D model being solved by means of dedicated software applications, or implemented in the computer-aided design software [7-9].

For this purpose, to apply the finite element method for carrying out the housing analysis, we used the Simulation Mechanical software, an integral part of Autodesk Inventor Professional software, thus eliminating the possible geometric errors that may occur when importing the 3D model of the housing.

Because the Simulation Mechanical software enables the modelling of bolt-nut assemblies as "threaded connections", we considered a simplified 3D model of the housing for carrying out the analysis, by removing the real threaded connections (Figure 2). The simplified 3D model can be seen in Figure 3.

In order to carry out the housing analysis using the finite element method, we went through the following specific steps:

- Model digitization;
- Material selection (grey cast iron for housing and cover, concrete for foundation);
- Defining a surface contact for the 3D model elements;
- Establishing the boundary conditions, i.e. embedding on the lower surface of the foundation;

IOP Conf. Series: Materials Science and Engineering **294** (2017) 012034 doi:10.1088/1757-899X/294/1/012034

Figure 3. The simplified 3D model of the gear reducer housing

- The modelling of "threaded connections" (Figure 4) was made by specifying the thread size (M8), the diameter of the bolt head and the nut, their holding surfaces, the internal surfaces of the bolt holes and the axial force in the bolt stem resulting from its initial tightening;

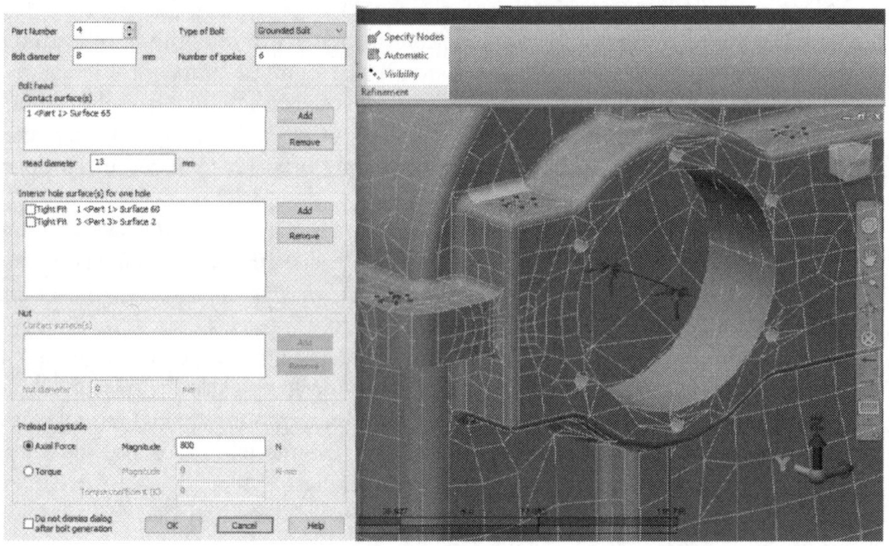

Figure 4. Modelling of the "threaded connections"

- Load definition. In addition to the axial forces in the bolt stems caused by their initial tightening (defined as shown in Figure 4), we took into account the reactions in the holders of the two gear reducer shafts obtained by designing. The values of these reactions corresponding to the hinge-type holder (fixed) and plain holder (free), as two radial forces acting in the vertical plane (along the Y-axis) and in the horizontal plane (along the X-axis), and, respectively, as axial force, are shown in the Tables 1 and 2.

Table 1. Reactions in the 1st shaft holders

Index	Type	Reaction force		
		Y (N)	X (N)	Axial Force (N)
1	Free	29	1124	
2	Fixed	1282	1972	546

Table 2. Reactions in the 2nd shaft holders

Index	Type	Reaction Force		
		Y (N)	X (N)	Axial Force (N)
1	Free	1987	-1923	
2	Fixed	1151	661	-546

These loads are defined by selecting the Bearing Load option in the symmetry axes of the bearings, as shown in the Figure 5 for the radial force in horizontal plane (along the X-axis) and, respectively, as shown in the Figure 6 for the radial force in vertical plane (along the Y-axis) and axial force.

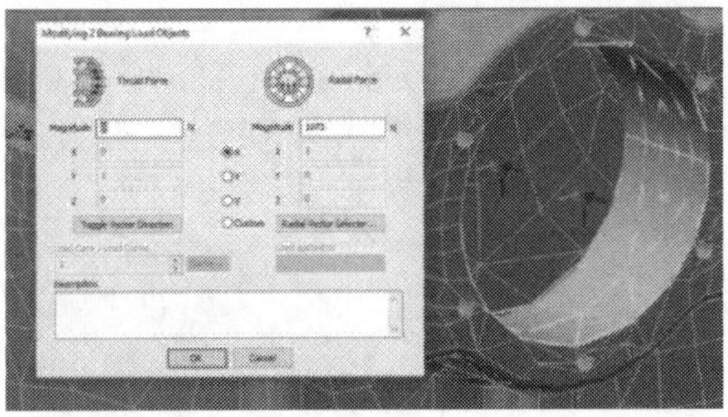

Figure 5. Defining the radial load in horizontal plane (X-axis)

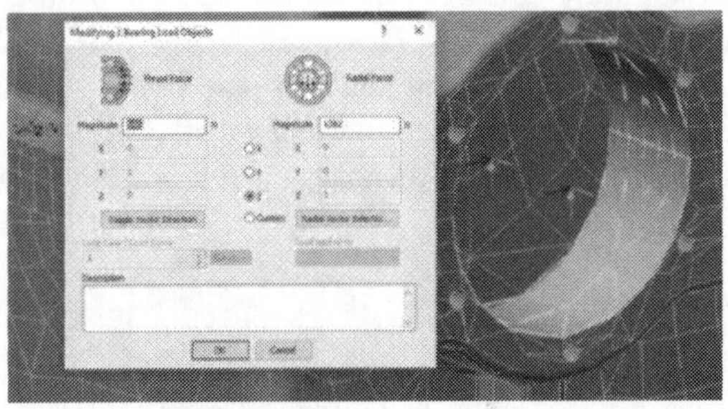

Figure 6. Defining the radial load in vertical plane (Y-axis) and axial load (Z-axis)

The 3D model of the housing, prepared for conducting the analysis according to the steps described above, can be seen in Figure 7.

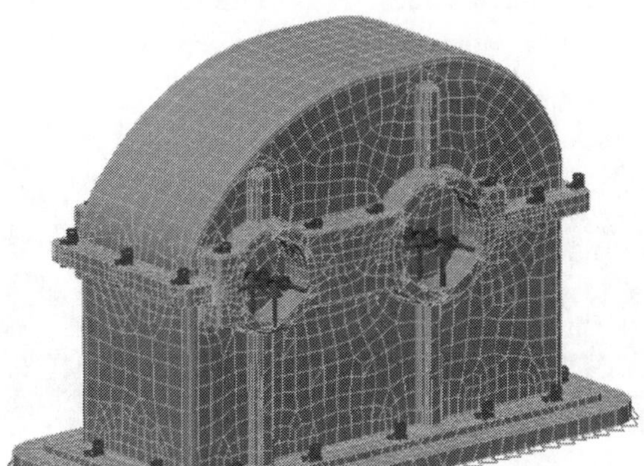

Figure 7. The 3D model prepared for analysis

IOP Conf. Series: Materials Science and Engineering **294** (2017) 012034 doi:10.1088/1757-899X/294/1/012034

The housing analysis using the finite element method was carried out for three distinct values of the initial tightening forces for the retaining bolts of the housing cover (F1) and, respectively, for the retaining bolts for anchoring the housing on the foundation (F2), according to Table 3.

Table 3. Values of the tightening forces for the three variants of analysis

	F_1 (N)	F_2 (N)
Variant 1	600	800
Variant 2	1200	1500
Variant 3	2500	3000

After performing the analyses, the results are displayed and interpreted, i.e. the nodal displacements and von Mises stress state, using the geometric model elements as shown in the Figures 8-9 for the 1st variant, Figures 10-11 for the 2nd variant, and Figures 12-13 for the 3rd variant of analysis.

It was found that, for the three variants of analysis, the stresses had maximum values (29, 31, 41 MPa) in the area of the holes made for retaining bolts, but below the permissible limits afferent to the materials used, the displacements being also small, i.e. amounting some hundredths of mm.

In the 3rd variant of analysis, the stress state in the threaded assembly elements is shown in Figure 14, where it can be seen that the mostly stressed bolts are the retaining bolts of the housing cover, next to the roller bearings, the stress value being also below the permissible limit (56 MPa)

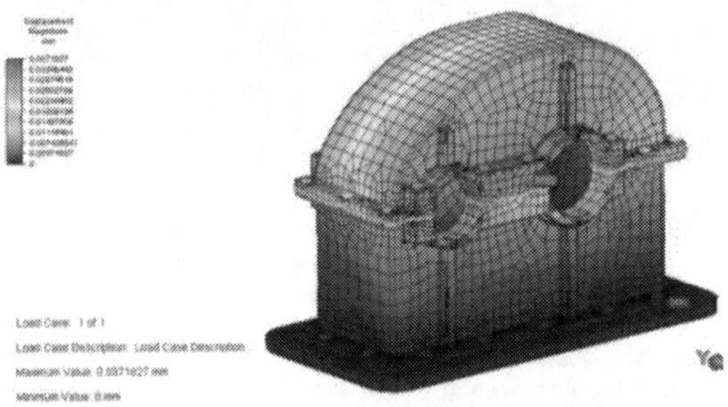

Figure 8. The nodal displacements for the 1st variant of analysis

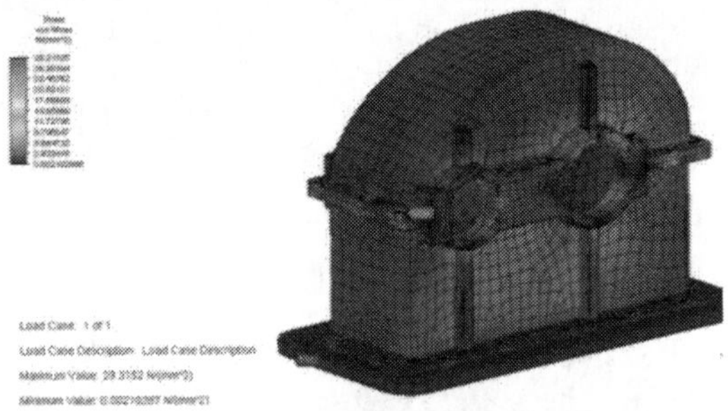

Figure 9. The von Mises stress state for the 1st variant of analysis

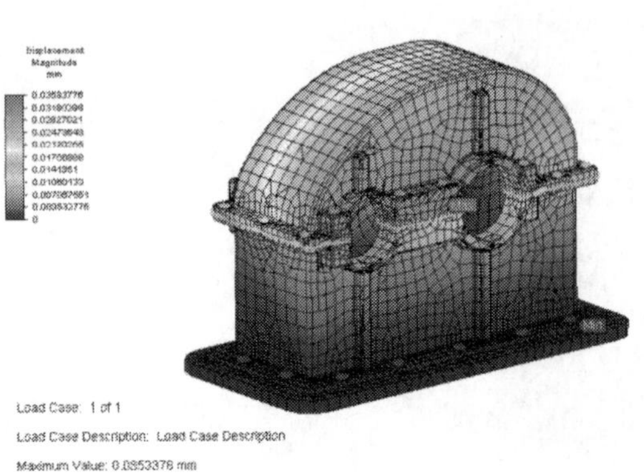

Figure 10. The nodal displacements for the 2nd variant of analysis

Figure 11. The von Mises stress state for the 2nd variant of analysis

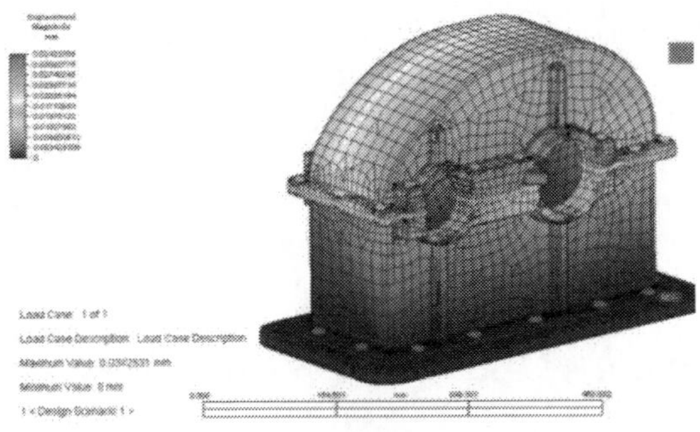

Figure 12. The nodal displacements for the 3rd variant of analysis

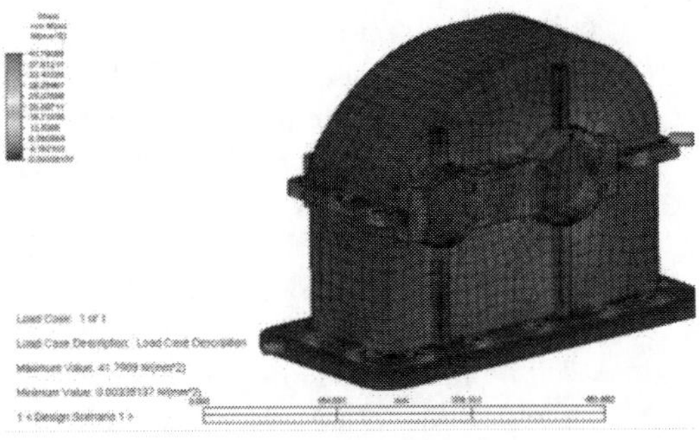

Figure 13. The von Mises stress state for the 3rd variant of analysis

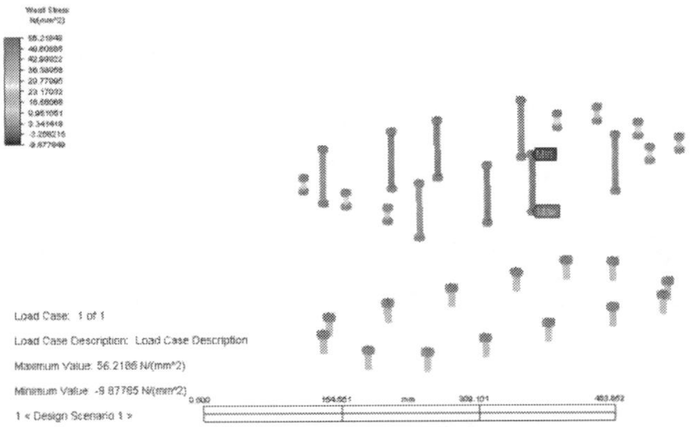

Figure 14. The stress state of the threaded elements

4. Conclusions

By applying the finite element method to analyse a gear reducer housing, we studied the influence of the retaining bolts tightening force magnitude on the stress state and nodal displacements of the housing, for the specified input parameters. It has been found that, by increasing successively the tightening forces, the nodal displacements had a very low decrease (approx. 0.001 mm), while the equivalent stress (von Mises) had a significant increase, but within the allowable stress limits of the materials from which the component elements are made.

The maximum stresses in the housing and the housing cover occur in the bolted connection areas, but in the remaining areas these stresses do not exceed 10 MPa.

The stresses in the bolted connections are also within acceptable limits.

References

[1] Crudu I, Ştefănescu I, Panţuru D and Palaghian L 1981 *Atlas Reductoare cu roţi dinţate*, Editura Didactică şi Pedagogică, Bucureşti, România

[2] Muşat M and Stoica G 2003 *Transmisii Mecanice cu reductoare într-o treaptă (Îndrumar de proiectare)*, Editura Printech, Bucureşti, România

[3] ***www.autodesk.com

[4] Dolga L, Revencu M, Maci C A and Giuchici M 2004 *Parametric and feature based modelling with applications in Catia and Inventor*, Editura Politehnica, Timişoara, România

[5] Miklos I Z, Cioată V G and Miklos I 2002 *Modelarea carcasei unui reductor cu ajutorul programului Mechanical Desktop*, Simpozionul naţional cu participare internaţională Proiectarea Asistată de Calculator, Prasic'02, Vol. III, Braşov, România, Noiembrie 7-8, pp 245-248

[6] Davis M, Mohammed Y S, Elmustafa A A, Martin P F and Ritinski C 2009 *Designing for Static and Dynamic Loading of a Gear Reducer Housing with FEA*, American Gear Manufacturers Association, Technical Ressources

[7] Spyrakos C Ch 1994 *Finite element modeling in engineering practice*, Morgantown, W. Va. West Virginia University Press

[8] Cioată V G and Kiss I 2017 Dynamic analysis and parametric optimisation of the connecting rod using Autodesk Inventor, *Machine Design* **9**(1) 29-34

[9] Cioată V G 2008 *Determining the machining error due to workpiece--fixture system deformation using the finite element method*, Annals of DAAAM & Proceedings of the 19th International DAAAM Symposium, Trnava, Slovakia, October 22-25, pp 253-255

IOP Conf. Series: Materials Science and Engineering **294** (2017) 012035 doi:10.1088/1757-899X/294/1/012035

Determination of the critical bending speeds of a multy-rotor shaft from the vibration signal analysis

I Crâştiu, E Nyaguly, S Deac, C Gozman-Pop, A Bârgău and L Bereteu

Politehnica University Timisoara, Mechanical and Strength Materials Department, Mihai Viteazul str., no. 1, 300222 Timisoara, Romania

E-mail: liviu.bereteu@yahoo.com

Abstract. The purpose of this paper is the development and validation of an impulse excitation technique to determine flexural critical speeds of a single rotor shaft and multy-rotor shaft. The experimental measurement of the vibroacoustic response is carried out by using a condenser microphone as a transducer. By the means of Modal Analysis using Finite Element Method (FEM), the natural frequencies and shape modes of one rotor and three rotor specimens are determined. The vibration responses of the specimens, in simple supported conditions, are carried out using algorithms based on Fast Fourier Transform (FFT). To validate the results of the modal parameters estimated using Finite Element Analysis (FEA) these are compared with experimental ones.

1. Introduction

Rotating machinery is seen in most industrial or domestic applications of our everyday life. During their operation rotating machines are carrying a kinetic energy of rotation and a vibratory kinetic energy. The purpose of rotor dynamics is to determine how a ratio between vibrational and rotational energy can be kept as small as possible. This can be done if the operation of the machines takes place in a speed range far away from the critical speeds.

The first reference to a limit speed over which rotating machines can not function safely was Rankine [1], which introduced the term whirling into the rotor dynamics vocabulary. The proposed model for studying the dynamics of a rotor was with two-degree system of freedom, which did not take into account the effect of Coriolis's inertial force, leading to the unlimited increase of radial deformations. This model has caused confusion amongst engineers and has delayed the development of turbo machines for nearly 50 years.

Laval was the first engineer who felt the operation of the machinery over the critical speed building a steam turbine to operate at 40 000 rpm. Dunkerly [2] and Föppl [3] have continued investigations in this area recognizing the existence of several critical speeds which in some circumstances, coincides with the natural frequencies of the rotor shaft system.

The first fundamental work in the field of rotor dynamics was that of Jeffcott [4], in which he considered a model of flexible shaft without mass and a rigid disc placed in the middle of the shaft, and which confirmed Laval and Föppl predictions. This model is named Jeffcott Laval. Important developments in rotor dynamics have been made by Stodola [5], which, among other things, has laid down some approximate methods for determining critical speeds. The graphical procedure proposed by him was widely used until the introduction of the transfer matrix method by Myklestad [6] and Prohl [7].

IOP Conf. Series: Materials Science and Engineering **294** (2017) 012035 doi:10.1088/1757-899X/294/1/012035

With the development of digital computer capacity, a series of codes have been developed to solve numerical problems in rotor dynamics. The Finite Element Method (FEM) became the strongest and most used code since the 1970s. Now, with the explosion of computer computing power, FEM techniques can be combined with model techniques to generate simulations that allow coupled behavior of flexible discs, flexible shafts, and flexible support structures into a single, massive, multidimensional model [8].

In this paper an approximate analytical / analytical determination of the critical speeds is made, for a shaft with a rotor and a three-rotor shaft. The results are compared with those obtained numerically by applying the Modal Analysis Method using the meshing technique based on the Finite Element Method. In the experimental way is developed a method for determination the critical speeds from the analysis of some vibroacoustic signals that validate the results obtained by the analytical or numerical methods.

2. Analytical analysis

Figure 1 shows a negligible mass shaft relative to the mass m of a Jeffcott rotor, placed midpoint between the rigid bearings.

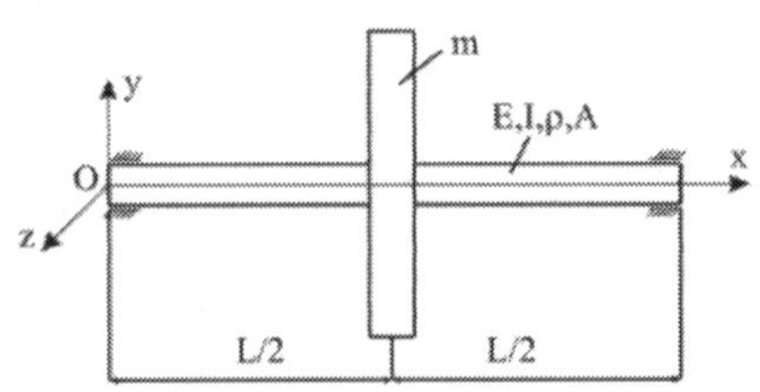

Figure 1. A Jeffcott rotor model in xoy plane

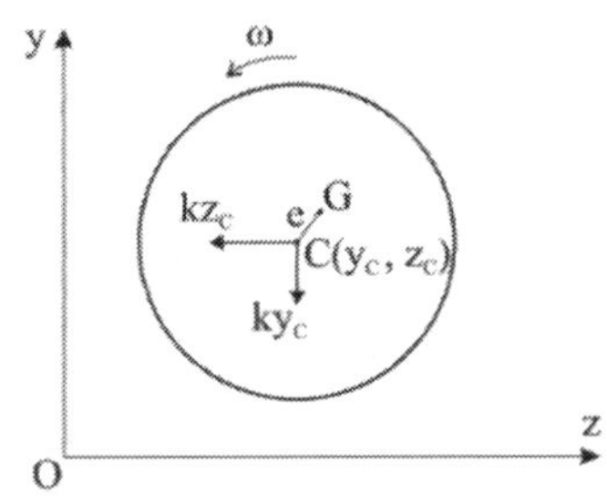

Figure 2. Free body diagram of the rotor in yoz plane

The geometric center of the disk C is located at the point (y_C, z_C) along coordinate axis defined about the bearing center line, and the disk center of mass G is located at (y_G, z_G).The vector **e** connecting the points C and G, represents the unbalance in the rotor disk. The rotating speed of the disk is given by ω. The lateral bending stiffness at the axial center of a simply supported uniform shaft is given by

$$k = \frac{48EI}{L^3}, \quad (1)$$

where E is the elastic modulus of the shaft, L is the length between the bearings, and I is the geometric inertia moment of cross area section. For a uniform cylindrical shaft with diameter d, the equation for the area moment of inertia is

$$I = \frac{\pi d^4}{64}. \quad (2)$$

The dynamic equations for the rotor are derived by applying Newton's law of motion to the rotor disk. With the assumption that the shaft is massless, the forces acting on the disk are the inertial force, the stiffness and damping forces generated by the lateral deformation of the shaft. The lateral equations of motion in the y- and z-axes, in terms of the disk geometric center as shown in Figure 2 are found to be

$$m\ddot{y}_C + c\dot{y}_C + ky_C = m\omega^2 e\cos\omega t \quad (3)$$

$$m\ddot{z}_C + c\dot{z}_C + kz_C = m\omega^2 e \sin \omega t\,, \tag{4}$$

where c is the corresponding damping.

The undamped free vibration analysis deals with the rotor vibration in the case of negligible unbalance eccentricity and damping. The equations of motion are

$$m\ddot{y}_C + ky_C = 0 \tag{5}$$

$$m\ddot{z}_C + kz_C = 0\,. \tag{6}$$

The above equations have the same undamped characteristic equation

$$ms^2 + k = 0 \tag{7}$$

Solving the above equation it is obtain the following solution:

$$s_{1,2} = \pm i\sqrt{\frac{k}{m}} = \pm i\omega_n\,, \tag{8}$$

where ω_n is the undamped natural frequency of the single rotor with the massless shaft. The undamped critical speed of the system is defined as

$$n_{cr} = \frac{30}{\pi}\sqrt{\frac{48EI}{mL^3}}\,[\mathrm{rpm}] \tag{9}$$

Because in the experimental determinations the sample has the mass of the shaft comparable to that of the disc, it will also take into account the mass of the shaft through the density ρ, the length L and the area of the section A.

Adopting Euler Bernoulli classical theory of simple supported beam the equation of motion for a uniform shaft with a rotor disk (Figure 1) can be written as [9]:

$$EI\frac{\partial^4 V(x,t)}{\partial x^4} + \rho A\frac{\partial^2 V(x,t)}{\partial t^2} + m\delta(x - L/2)\frac{\partial^2 V(x,t)}{\partial t^2} = 0 \tag{10}$$

where δ is Dirac function.

Using Ritz method [10], a series approximate analytical solution of Eq. 10 is assumed to be:

$$V(x,t) = \sum_{i=1}^{n} \Phi_i(x) q(t)_i = \{\Phi\}^T \{q\} \tag{11}$$

where $\Phi_i(x)$ are the mode shape functions assumed to satisfy geometric boundary conditions of the shaft, and $q_i(t)$ are the generalized coordinates.

The total kinetic energy of shaft whit a rotor is given by:

$$E_c = \frac{1}{2}\int_0^L \rho A\left(\frac{\partial V(x,t)}{\partial t}\right)^2 dx + \frac{1}{2}m\left(\frac{\partial V(x,t)}{\partial t}\right)^2_{x=L/2} = \frac{1}{2}\{\dot{q}\}^T\left([M_1]+[M_2]\right)\{\dot{q}\} = \frac{1}{2}\{\dot{q}\}^T[M]\{\dot{q}\} \tag{12}$$

where the matrix $[M_1]$ and $[M_2]$ are:

$$[M_1] = \rho A\int_0^L \{\Phi\}\{\Phi\}^T dx; [M_2] = m\left[\{\Phi\}\{\Phi\}^T\right]_{x=L/2} \tag{13}$$

The total potential energy of the system is given by:

$$E_p = \frac{1}{2}\int_0^L EI\left(\frac{\partial^2 V(x,t)}{\partial^2 x}\right)^2 dx = \frac{1}{2}\{q\}^T[K]\{q\} \tag{14}$$

where the stiffness matrix [K] is:

$$[K] = EI\int_0^L \frac{\partial^2}{\partial x^2}\{\Phi\}\frac{\partial^2}{\partial x^2}\{\Phi\}^T dx \tag{15}$$

It should be noted that the transverse displacement V (x, t) can take place both in the xoy plane and in the xoz plane. In the following, the two displacement functions can be written:

$$Y(x,t) = \sum_{i=1}^{n} Y_i(x)q_i(t); Z(x,t) = \sum_{i=1}^{n} Z_i(x)q_i(t) \tag{16}$$

where both the $Y_i(x,t)$ and $Z_i(x,t)$ functions are admissible functions. The number n of the chosen admissible functions determines the number of eigenfrequencies and the eigenvectors.

Applying the theorem of conservation of mechanical energy

$$E_c + E_p = \text{const} \tag{17}$$

and after derivation in respect to time, the differential equation that governs the movement of the system in matrix form is obtained:

$$[M]\{\ddot{q}\} + [K]\{q\} = \{0\}. \tag{18}$$

Assuming a modal solution

$$\{q\} = \{\mu\}e^{i\omega t}, \tag{19}$$

yields the eigenvalue problem

$$\left[-\omega^2[M] + [K]\right]\{\mu\} = \{0\}, \tag{20}$$

from which the characteristic equation, the approximate natural frequencies, eigenvectors and approximate eigenfunctions are obtained.

Given that the first vibration mode of the rotor shaft system is interested, it is necessary to choose a single admissible function for determining the critical speed. Consequently, the two functions which approximate the displacements in oxy plane and oxz plane are chosen such:

$$Y(x,t) = Z(x,t) = \sin\frac{\pi x}{L}\cos\omega t, \tag{21}$$

where ω is the fundamental circular frequency and the function $\sin \pi x/L$ is admissible function which the geometric boundary condition are exactly satisfied for simple supported shaft.

Applying this method to a shaft with a single rotor as in Figure 1, from the mechanical energy conservation theorem, for the admissible function given by equation (21), is obtained fundamental circular frequency

$$\omega^2 = \frac{\int_0^L EI\left(\frac{\pi}{L}\right)^4 \sin^2\frac{\pi x}{L}dx}{\int_0^L \rho A \sin^2\frac{\pi x}{L}dx + m\left(\sin\frac{\pi}{L}\frac{L}{2}\right)^2}, \tag{22}$$

which corresponds to the critical bending speed

$$n_{cr} = \frac{30}{\pi}\omega = \frac{30}{L}\pi\sqrt{\frac{EI}{(\rho AL + 2m)L}} \tag{23}$$

For the same rotor considering the shaft without mass, using the approximate analytical solution given by equation (21), the critical bending speed is

$$n_{cr} = \frac{30}{\pi}\omega = \frac{30}{L}\pi\sqrt{\frac{EI}{2mL}} \tag{24}$$

For a three-rotor shaft placed as in Figure 3, choosing the same admissible function as for a single rotor shaft, i.e. $\Phi(x) = \sin\pi x/L$, following the same procedure, the fundamental frequency of the bending vibrations is obtained

$$\omega^2 = \frac{\int_0^L EI(\Phi''(x))^2 dx}{\int_0^L \rho A\Phi^2(x)dx + \sum_{i=1}^{n} m_i\Phi^2(x_i)}, \tag{25}$$

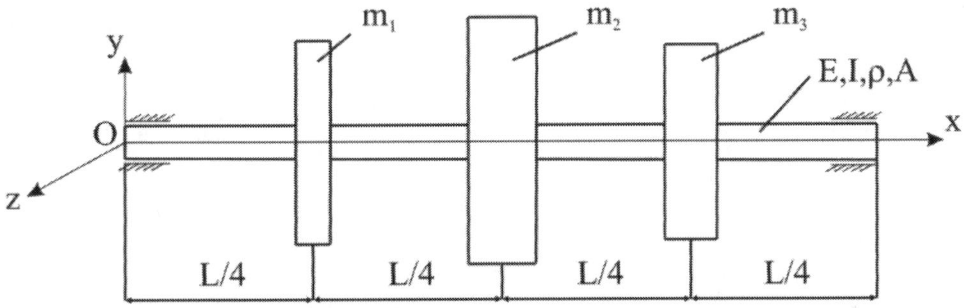

Figure 3. Three-rotor shaft

which corresponds to the critical bending speed

$$n_{cr} = \frac{30}{\pi}\omega = \frac{30}{L}\pi\sqrt{\frac{EI}{(\rho AL + m_1 + 2m_2 + m_3)L}} \tag{26}$$

3. Numerical analysis

In this paragraph are given basic steps of the numerical modal analysis simulation. The modal analysis is carried out on the steel simple supported shaft with single rotor and three-rotor simple supported shaft. The simple supported shaft is designed in the graphical environment of the ANSYS [11]. Mode shapes and natural frequencies are computed in programs ANSYS with numerical formulation of the direct solver including the block Lanczos method.

In order to determine the critical speed of the disc is required measurements of the fundamental frequency in two planes: Oxy and Oxz. Mesh of the shaft and rotor is generated automatically by ANSYS, while is used the spatial element SOLID187. The element is defined by 10 nodes while each node has three degrees of freedom. The SOLID187 has a quadratic shifting behaviour and is suitable for modelling of the finite element irregular mesh. The maximum size of the element is 5 mm. The mesh is created of 3200 elements and of 22459 nodes.

To validate the rotor shape modes of vibration, in correlation with experimental resonance frequencies test, it is necessary to do a modal analysis. For an n-degree of freedom system, by Finite Element Method the motion equation of sample can be expressed similarly as equation (18), where {q} is the vector of the displacement, and [M] and [K] are mass, and stiffness matrices. For free vibration of the shaft and rotor, the characteristic equation is:

$$\left|-\omega^2[M]+[K]\right|=0 \tag{27}$$

To solve eigenvalues problem, given by equation (14), the Block Lanczos method is use because it is efficient to extract of large number of modes in most models. It is used in complex models with mixture of solids/shells/beams etc. It offers an efficient extraction of modes in a frequency range.

The first mode shapes of the single simple supported rotor shaft in Oxy plane and in Oxz plane are shown in Figure 4 and Figure 5.

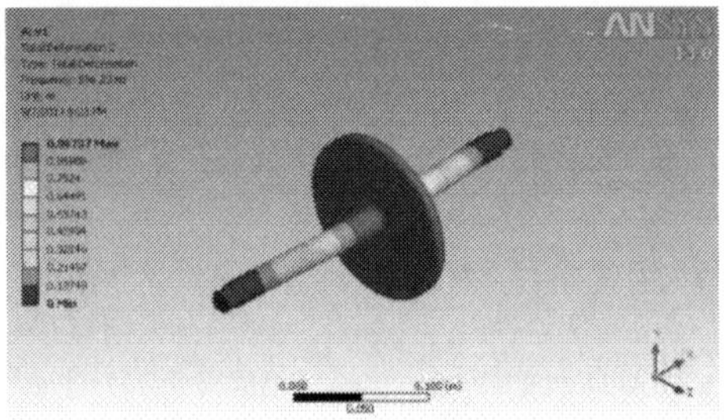

Figure 4. First bending mode in Oxy plane for single simple supported rotor shaft

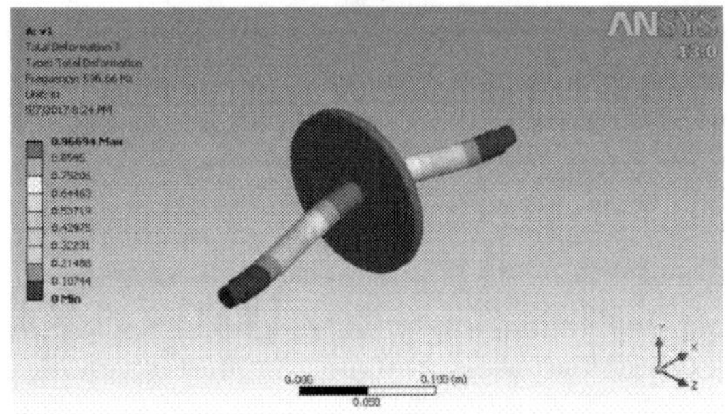

Figure 5. First bending mode in Oxz plane for single simple supported rotor shaft

The three-rotor simple supported shaft is design in the same graphical environment of the ANSYS with the same spatial element solid 187.The first mode shapes of the three-rotor simple supported shaft in Oxy plane and in Oxz plane are shown in Figure 6 and Figure 7.

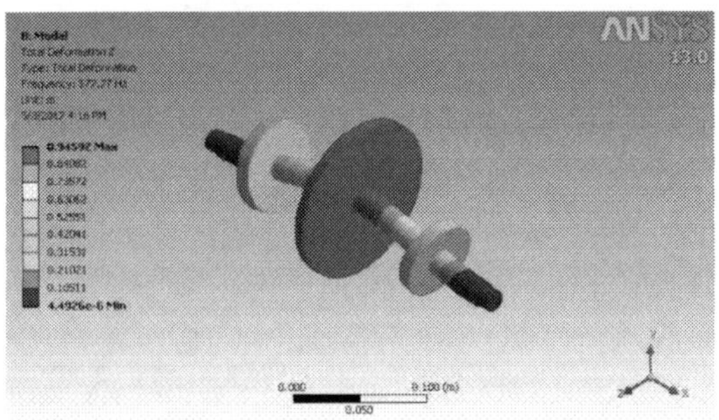

Figure 6. First bending mode in Oxy plane for three-rotor with simple supported shaft

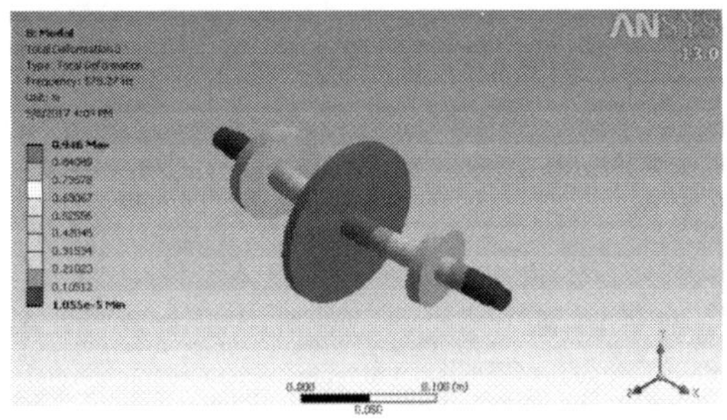

Figure 7. First bending mode in Oxz plane for three-rotor with simple supported shaft

4. Experimental analysis

The experimental stand for non-contact measurement of free vibrations of the sample having a rotor with simple supported shaft is shown in Figure 8, and it is composed by: the shaft with disk 1, which is the mechanical structure to be analyzed; impulsive mini hammer 2; brackets to support the sample 3; elastic threads for support of the structure in boundary conditions with the free ends 4; the acoustic sensor one condenser microphone 5, and the computer that has embedded and data acquisition board 6. The same stand is also used for the sample having three-rotor and simple supported shaft, as in the Figure 10. This stand or its variants used especially in the case of simulation of free- free boundary conditions was designed and used for the papers: [12], [13].

On the stands in Figures 8 and 9 vibration measurements were performed for two samples, for a single simple supported rotor shaft and for three-rotor with simple supported shaft. The vibroacoustic signal is picked up by a good quality condenser microphone. Then the signal is transmitted as an electrical voltage to a signal acquisition board, which is actually the sound card of the computer, and it is amplified by a microphone amplifier located on this board. Acquisition board is analog digital signal from the microphone, with the possibility of adjusting the sampling frequency. The signals were discretized with a sampling frequency of 44.1 kHz, thus having the possibility of spectral analysis, according to the Shannon theorem, up to a frequency of 22 kHz, i.e. in the audible spectrum.

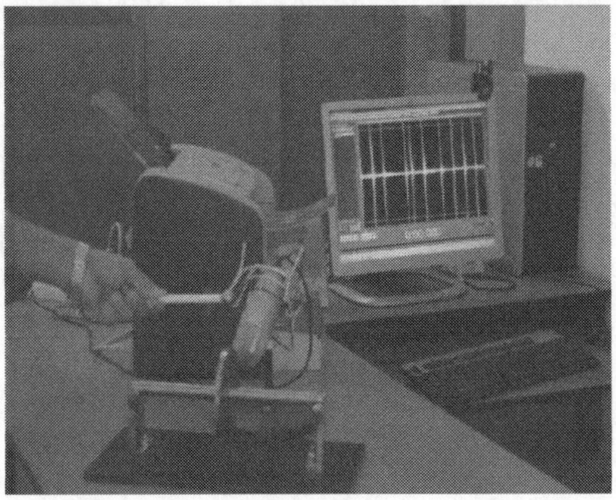

Figure 8. The experimental stand for single simple supported rotor shaft

Figure 9. The experimental stand for three-rotor with simple supported shaft

The discreet signals were purchased through the Professional CoolEdit 2 [14], software as files with the wav extension. The MATLAB [15] program was used for analysis of discreet signals stored in wap files. The Figure 10 shows the recorded signal from experimental sample of a single-rotor with simple supported shaft. For each sample, several signals were recorded for spectral analysis using Fast Fourier Transform (FFT).

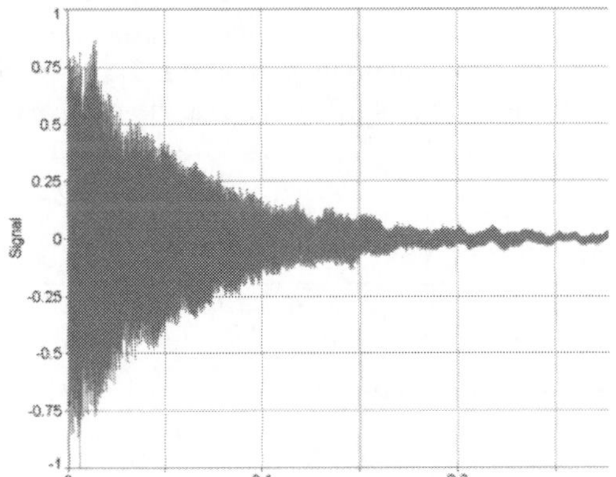

Figure 10. Signal in time domain for single simple supported rotor shaft

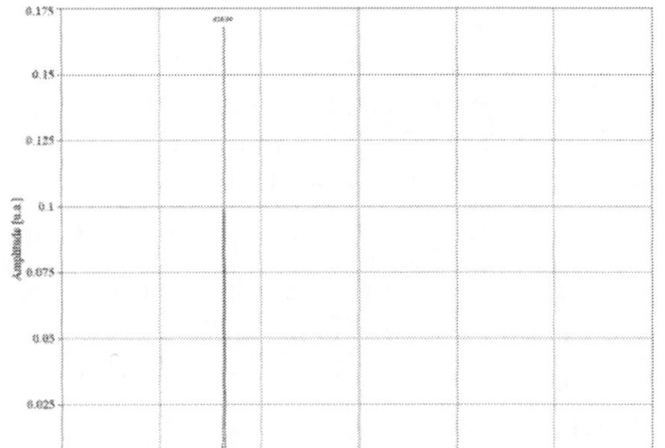

Figure 11. Signal in frequency domain for single simple supported rotor shaft

The Figure 10 shows the recorded signal from the experimental sample, consisting of a single-rotor shaft, simply supported. The spectrum of the recorded signal is shown in the Figure 11. Manual excitation was done by impulsive mini hammer consisting of a metal sphere and a plastic handle.

To retain only the component corresponding to the fundamental frequency of the vibration signal, it was filtered through a band pass filter.

Figure 12 shows the recorded signal from the experimental sample, consisting of a three-rotor shaft, simply supported. The spectrum of the recorded signal is shown in the Figure 13.

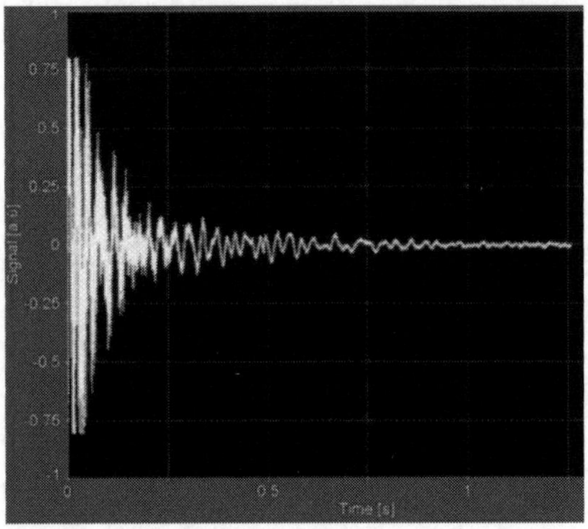

Figure 12. Signal in time domain for three-rotor simple supported shaft

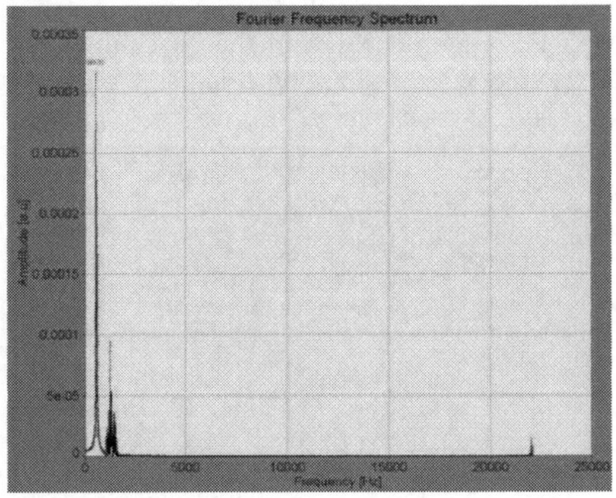

Figure 13. Fourier Frequency Spectrum Signal for three-rotor simple supported shaft

5. Results and conclusions

For experimental determinations of the critical speeds of a single rotor shaft by pulse excitation technique, the sample was made of a steel shaft having the length L=294 mm and the diameter d=20 mm. The rotor is in the form of a disc and was made of the same material having a diameter D=130 mm and a thickness h=10 mm. The material was stainless steel with Young's modulus E=210 GPa, and density ρ=7850 kg/m3. The mass of the rotor-disc and of the shaft has been measured by a precision digital balance which is accurate to one-hundredth of a gram. Average after six weighing, for the disc was $m=m_d=800.333$g, and $m_s=\rho AL=700.231$ g, for the shaft.

By making the ratio between equation (24) and equation (23), i.e. between the critical speed of the rotor in which the mass of the shaft is negligible with respect to the disc rotor, and the critical speed of the same rotor taking into account the mass of the shaft, it is ascertained that since the relative deviation of the two speeds, to be less than 1%, the ratio between the shaft mass and the disc mass must be less than 1/25. As can be seen from the mass measurement for the sample considered in the experiment, it does not answer to this requirement.

Table 1. The fundamental frequencies and critical speeds of the single rotor shaft in two planes

Type of analysis	Fundamental frequency in Oxy plane [Hz]	Critical speed in Oxy plane [rpm]	Fundamental frequency in Oxz plane [Hz]	Critical speed in Oxz plane [rpm]
Analytical analysis. Shaft massless	712.32	42739,2	712.32	42739,2
Analytical analysis. Shaft with mass	594.12	35647,2	594,12	35647,2
Numerical analysis	596,23	35773,8	596,66	35799,6
Experimental analysis	605,17	36310,2	607,03	36421,8

For experimental determinations of the critical speeds of a thre-rotor shaft by pulse excitation technique, the sample was made of a steel shaft having the same dimensions as in the first experiment. The midle rotor is noted here with m_2 and it is also the same like in first experiment. The mass of the rotor-disc and of the shaft has been measured by a precision digital balance which is accurate to one-hundredth of a gram. Average after six weighing, for the disc was $m=m_d=800.333$g, and $m_s=\rho AL=700.231$ g, for the shaft.

Table 2. The fundamental frequencies and critical speeds of the three-rotor shaft in two planes

Type of analysis	Foundamental frequency in Oxy plane [Hz]	Critical speed in Oxy plane [rpm]	Foundamental frequency in Oxz plane [Hz]	Critical speed in Oxz plane [rpm]
Analytical analysis. Shaft massless	660,34	3962,04	660,34	3962,04
Aalytical analysis.Shaft with mass	574,21	34454,6	574,21	34454,6
Numerical analysis	577,77	34666,2	578,27	34696,2
Experimental analysis	581,17	34870,2	583,34	35000,4

In conclusion, comparing the results given in Table 1 and Table 2 it can be said that the experimental method validates numerically or analytically results. The negligence of the shaft mass, when it is comparable to the rotor mass, causes large deviations in determining the critical speeds.

References

[1] Rankine W J M 1869 On the centrifugal force of rotating shaft, *The Engineer* **27** 249
[2] Dunkerley S 1895 On the whirling and vibrations of shafts, *Phil. Trans. of the Royal Soc., A*, **185**(I) 279-360
[3] Föppl A 1895 Das Problem der Lavalschen Turbinenwelle, *Der Civillingerenieur* **4** 335-342
[4] Jeffcott H H 1919 The lateral vibration of loaded shafts in neighbourhood of a whirling speed: the effect of want of balance, *Philosophical Magazine*, Ser. 6, **37** 304-314
[5] Stodola A 1924 *Dampf- und Gasturbinen*, 4, Aufl., Berlin: Springer. English translation, *Steam and Gas Turbines*, McGraw-Hill, New York
[6] Myklestad N O 1944 A new method of calculating natural modes of uncoupled bending vibrations, *Journal of. Aeronautical Science* **11**(2) 153-162
[7] Prohl M A 1945 A general method for calculating the critical speeds of flexible rotors, *Jornal of Applied Mechanics* **12**(3) 142-148
[8] Tiwari R 2014 *Theory and paractice of rotor dynamics*, NPTEL, Guwahati, Assam, India
[9] Chen Y 1963 On the vibration of beams rods carryng a concentrated mass, *Journal of Applied Mechanics* **30**(2) 310-311
[10] Ritz W 1908 Über eine neue Methode zur Lösung gewisser Variation probleme der mathematichen Physik, *Journal fur Reine und Angewandte Mathematik* **135** 1-61
[11] ***http://www.ansys.com,Simulation driven product development ANSYS, 2017
[12] Perescu A, Bereteu L, Simoiu D and Nyaguly E 2015 Nondestructive method for the determination of the elastic properties of welded aluminum plates, *Advanced Materials Research* **1111** 73-78
[13] Crastiu I, Bereteu L and Simoiu D 2015 Determination of the eccentricity of a rotor by impulse excitation technique of vibration, *Applied Mechanics and Materials* **801** 176-181
[14] ***http://oprekzone.com/download-cool-edit-pro-free-2-0-audio-editor-software
[15] ***http://download.cnet.com/Matlab/3000-2053_4-43768.html

IOP Conf. Series: Materials Science and Engineering **294** (2017) 012036 doi:10.1088/1757-899X/294/1/012036

Minimal algorithm for running an internal combustion engine

V Stoica[1], A Borborean[1], A Ciocan[1] and C Manciu[2]

[1]Politehnica University of Timisoara, Department of Mechanical Machines, Equipment and Transportation, Mihai Viteazu 1, 300222 Timisoara, Romania
[2]Continental Automotive Romania SRL, Siemens Street No. 1, 300704, Timisoara, Romania

E-mail: virgil.stoica@upt.ro

Abstract. The internal combustion engine control is a well-known topic within automotive industry and is widely used. However, in research laboratories and universities the use of a control system trading is not the best solution because of predetermined operating algorithms, and calibrations (accessible only by the manufacturer) without allowing massive intervention from outside. Laboratory solutions on the market are very expensive. Consequently, in the paper we present a minimal algorithm required to start-up and run an internal combustion engine. The presented solution can be adapted to function on performance microcontrollers available on the market at the present time and at an affordable price. The presented algorithm was implemented in LabView and runs on a CompactRIO hardware platform.

1. Introduction

There are many developed algorithms for engine control in automotive industry but, in the literature, there is no description of them, they are only mentioned [1-3].

Anand [4] presents roughly the developed algorithm but the most important parts, the engine position tracking, injector and ignition coils control, are missing. Strapko [5] provide a complete overview of the entire developed system including the electronic modules for signal conditioning, but the module for crankshaft position tracking is not clearly described and also the frequency of the FPGA loops are not provided. Busuttil [6] is still using analogic electronics for controlling the water injection within the intake manifold injection that is triggered by the fuel injector signal without the possibility to control the water injection starting moment.

Within this context we developed a minimal algorithm for starting an internal combustion engine. The presented modular algorithm can be adapted to any needs and can be completed with some other modules for controlling other injectors for water or a secondary fuel like alcohol or biogas for example.

2. Algorithm description

A spark ignition 4 stroke engine with 4 cylinders can be managed by a double spark and injection as used to be in case of EURO 3 engine. The spark is provided in the same time at two cylinders, one at desired moment (the end of the compression stroke), and the second one is provided on the evacuation stroke, and it will be unused. The same solution is applied in case of fuel injection. One injection is provided during admission stroke and the second one during detention (each time is injected half of the desired fuel mass). A schema of the injection (green), and ignition (red) timing overlapped over the engine strokes is presented in **Error! Reference source not found.**. This solution requires the crank angle tracking only in the interval 0-360 degree and not necessarily a complete engine cycle of 0-720. Therefore a second sensor mounted on the camshaft is not required.

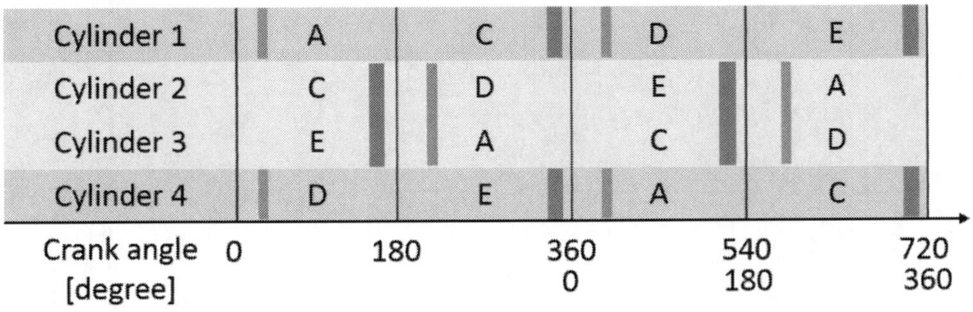

Figure 1. Ignition and fuel injection timing

The engine management program is divided in two main parts, one running on the real time module **Error! Reference source not found.**, and the second one, time critical, running on the FPGA.

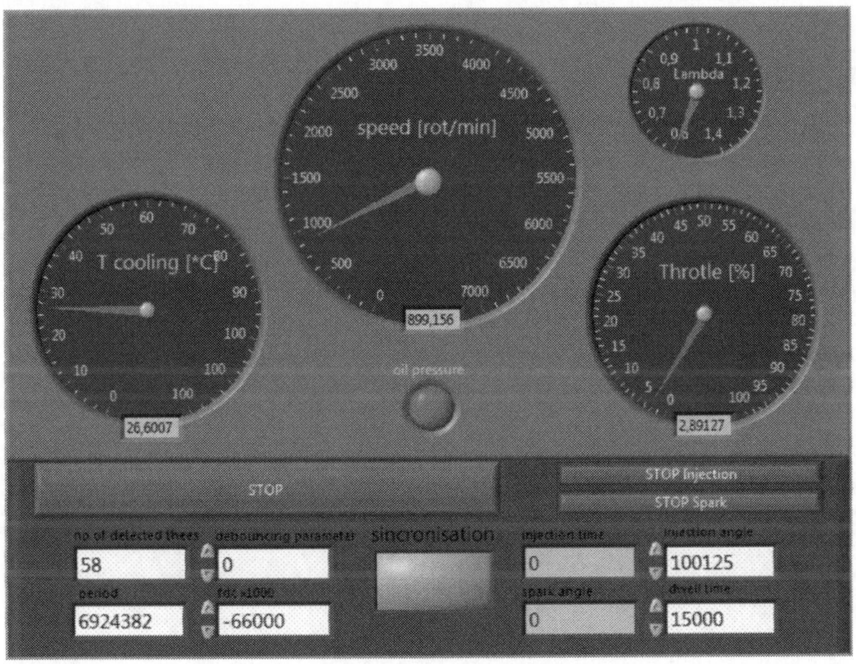

Figure 2. Program graphical user interface

An overview of the program which is running on the FPGA is presented in Figure 4. It consists of 4 parts:

1. Monitoring and establishing the crankshaft position (CRK module);
2. Commanding the injectors (Injection module);
3. Commanding the induction coil (Spark module);
4. Acquiring data from the sensors.

The first 3 parts are running in a clock cycle loop (40MHZ) and the last one, less important, running in a separate loop cycle.

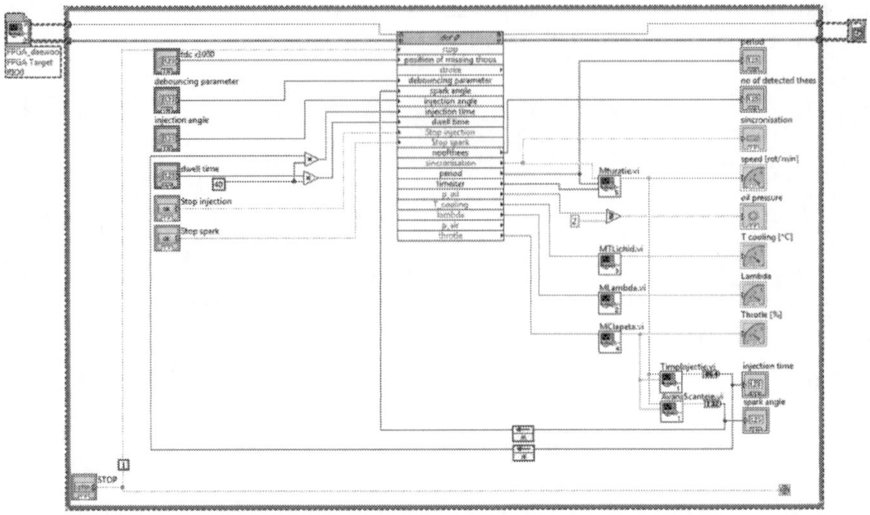

Figure 3. The main program running on the real time module

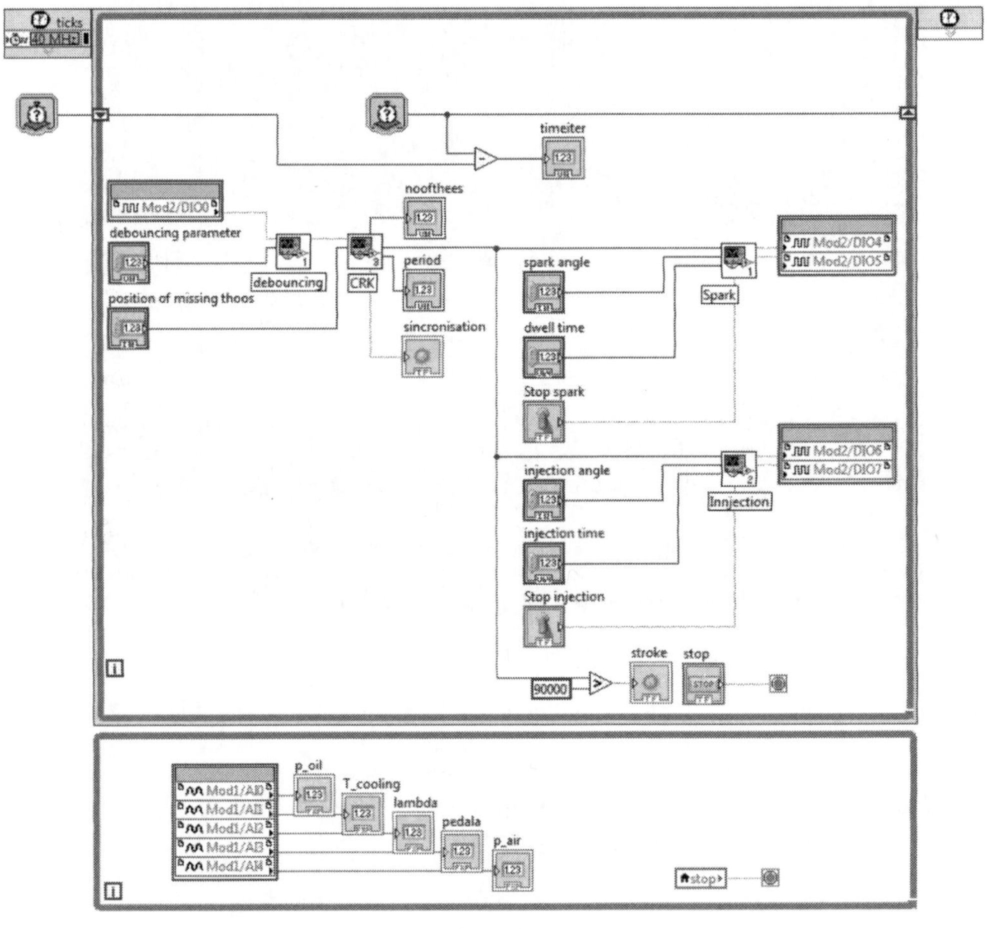

Figure 4. The overview of the module running on the FPGA

The most important parameter, that has to be monitored for an exact ignition timing, is the engine position, which in our case is done using an inductive sensor togheter with a special tooth wheel in a 60-2 configuration. The LabView program (the algorithm) is presented in Figure **Error! Reference source not found.**6. The program consists of:

a. Signal debouncing for rejecting de undesired noise;
b. Determine the angle with a precision of 6 degrees by adding 6 degrees each time a rising edge is detected. The angle variates between -90 to 270;
c. Estimating the angle in between the edges with a step of 0.375 degree by linear interpolation. For the interpolation the time between the previous consecutive edges is divided in 16 (2^4) intervals, and with a help of a timer (counter) the angle is incremented with 0.375 ($6/2^4$) at the end of each interval. As mentioned in literature other interval numbers are possible [2];
d. Detecting the missing teeth based on the fact that the time period between 2 consecutive teeth is much smaller than the time period detected in case of the two missing ones;
e. Counting the rising edges (number of teeth);
f. Counting the period of a complete revolution.

An intuitive picture of the steps b, c and d is presented in Figure 5. For a better graphical description the angle is incremented only with 1.5 degrees ($6/2^2$).

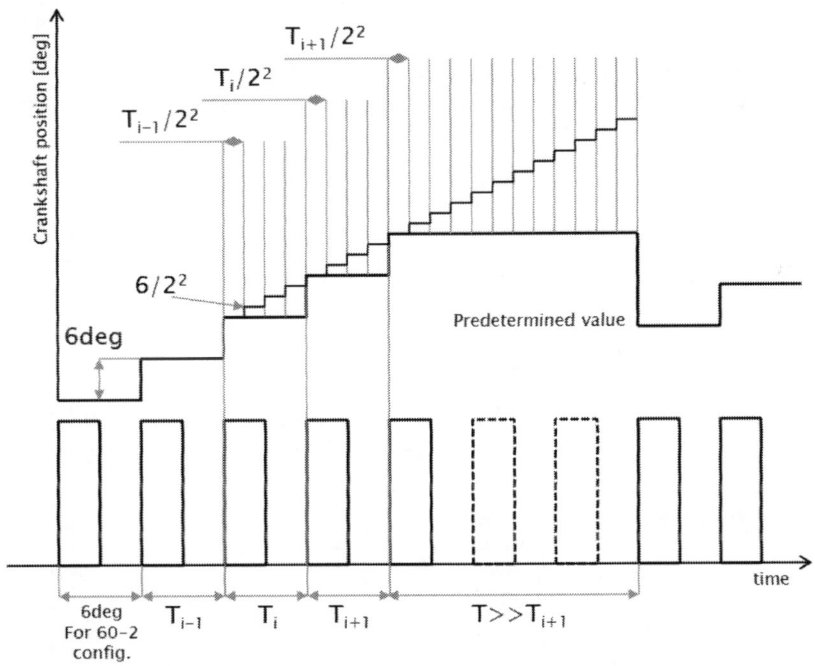

Figure 5. Crankshaft position incrementing and estimating procedure

(**Notification**: because on the FPGA is recommended to work with integer numbers and to use simple math, the angle values are multiplied by 1000).

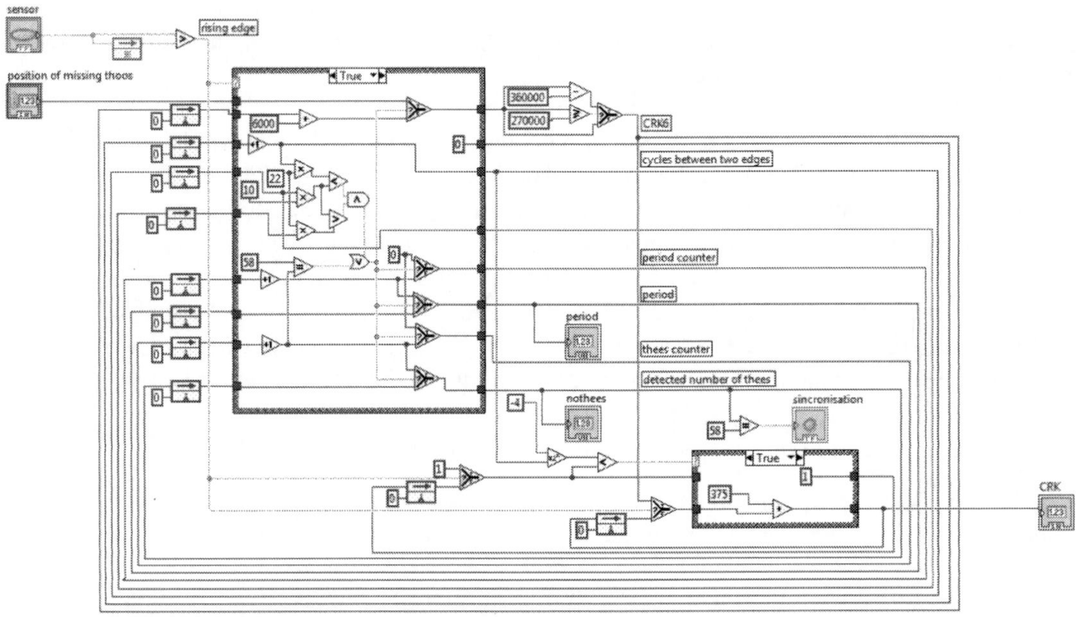

Figure 6. The module used for crankshaft position tracking

For controlling the ignition and the fuel injection mainly counters are used. The subroutine is presented in Figure 7.

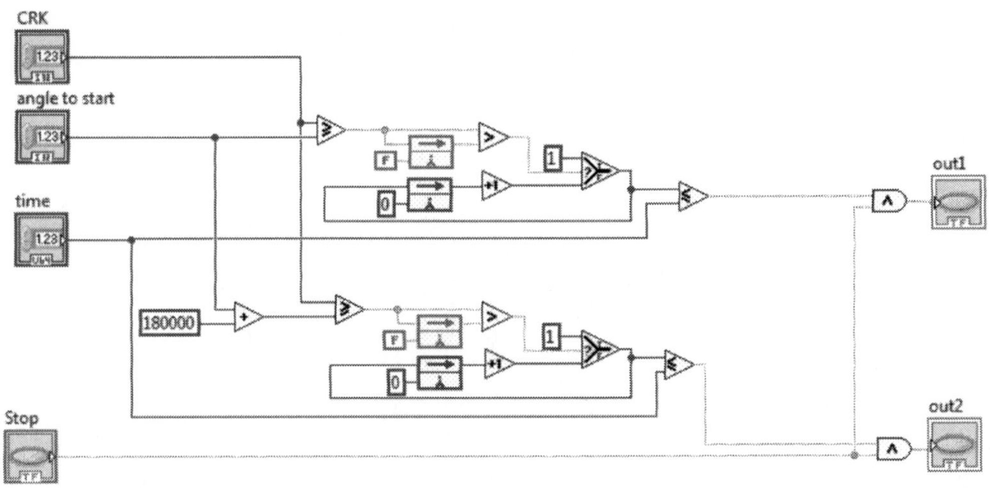

Figure 7. The module used for commanding the fuel injection and the ignition system

The counter is started at the desired crank angle and the output is maintained on 1 the prescribed time (number of cycles) which mean the injection time or the dwell angle.

3. Experimental setup

The experimental set up (Figure 8) consist of the Daewoo 1.6 liter engine 1 which is controlled by the cRIO system 2. The signals from and to the engine are passing through the signal conditioning box 3.

For monitoring the engine state, the following sensors are used: 4- intake air temperature, 5 intake air pressure, 6 throttle position, 7 cooling liquid temperature, 8- oil pressure, 9 – crank angle position, 10 – oxygen level within the exhaust gases (binary type) and cooling liquid temperature at the input 11 and output 12 of the heat exchanger. The control of the engine fueling and ignition is done with the injectors 13 and ignition coils and spark plugs 14. The fuel is brought to the common rail from the tank 15 through the filter 16 by the pump 17. The intake air passes the filter 18, the intake manifold 19 and riches the engine. The exhaust gases leave the engine and pass the catalyst 20. The engine cooling is assured with the heat exchanger 21 and the cooling fan 22. The 12 Volt power supply 23 ensure the powering of the entire system. The switches board 24 is for emergency situations and for activating or deactivating some subsystems like injection, ignition and cooling. The computer 25 is used for programing the cRIO system, for visualization of the monitored parameter and for prescribing the injection and spark timing.

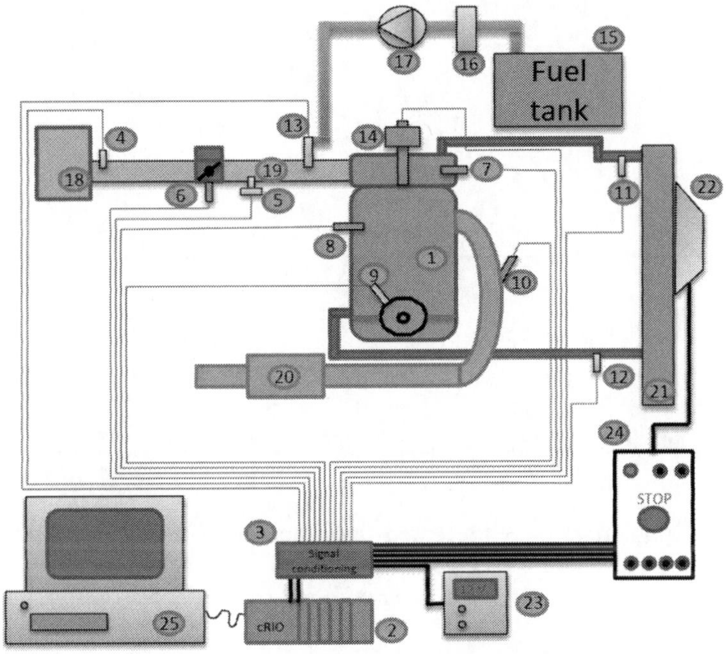

Figure 8. The experimental setup

Figure 9. The experimental test bench

The controlling system consist of a real time power PC embedded controller for Compact RIO of 533MHz, with 2Gb storage and 256 Mb DRAM, a TTL digital module NI 9401 with 8-Channels, with a response time of 100 ns, and a 16-Bit analogic input module NI 9205 with 32-Channels having the input range of ±10 V, and an acquiring frequency of 250 kS/s.

4. Results

Before implementing the algorithm, on the cRIO system, it was tested and the results are presented in. **Error! Reference source not found.**10

As mentioned before, the crankshaft angle is multiplied by 1000 therefore in figure 10 on the uperside the angle is in the range -30 to 50 degree. The spark is produce at 10 degree before top dead center at the end of the compression stroke and the injection starts at 40 degree after the top dead center during the detention stroke.

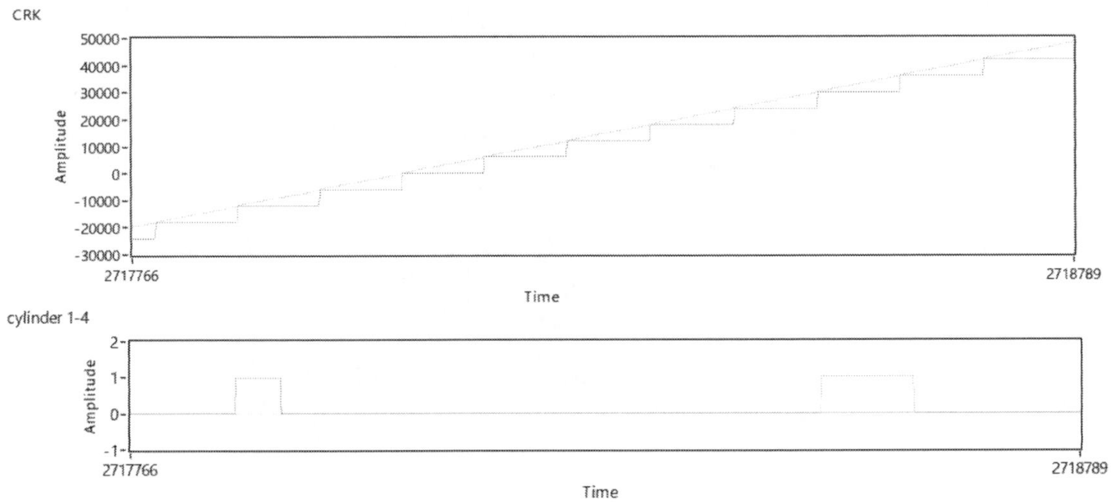

Figure 10. LabView simulation of the presented algorithm. up) The crankshaft position CRK with a 6 degree precision with black and 0.375 with red. down) The spark timing with black and injection timing with red

Similar results are obtained on the experimental bench test and they are presented in Figure 11.

Using the presented system it was possible to change, in real time, the injection timing and sparking and to monitor the most important parameters of the engine.

5. Conclusions

The modular solution of the software presented here permits in future the rapidly development of different configurations like: dual injection gas and biogas, gas and hydrogen, programs for diesel engines with common rail systems and so on. With this hardware and software solution the implementation time of different algorithm is very short. Also this solution gives to mechanical engineers the possibility to implement and test their own algorithms in a more convenient way. The presented algorithms was simulated and then implemented on a Daewoo 1.6 liters spark ignition engine. The engine was tested without load in a speed domain of 800 to 2500 rpm. During the preliminary tests the injections timing was determined for different throttle position bringing the indication of lambda sensor to 1 (stoichiometric value).

The developed system will further be used for complete calibration of the engine, after mounting a hydraulic break, and will be tested with different fuel combinations and controlling strategies.

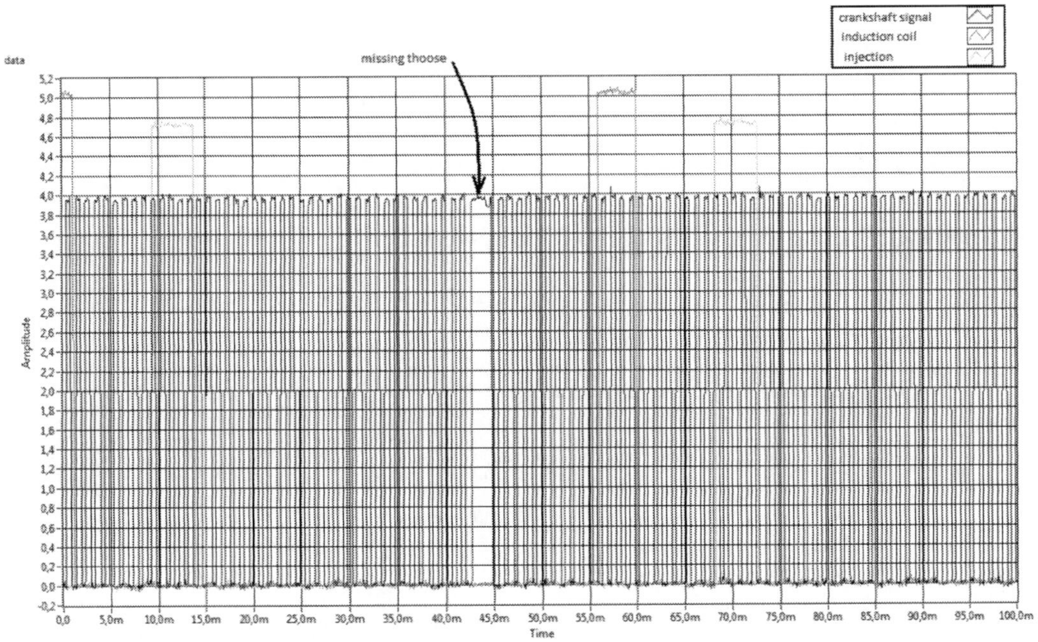

Figure 11. Measured data on the experimental test bench

References

[1] Dase C, Falcon J S and Maccleery B 2006 Motorcycle Control Prototyping Using an FPGA-Based Embedded Control System, *IEEE Control Systems Magazine*, October

[2] Viele M 2012 *Applicatons of field programmable gate arrays for engine control*, Colorado State University, Fort Collins, Colorado, USA, Doctoral thesis

[3] Irimescu A, Vasiu G and Trif-Tordai G 2014 Performance and emissions of small scale generator powered by a spark ignition engine with adaptive fuel injection control, *Applied Energy* **121** 196-206

[4] Anand B P and Saravanan C G 2012 Development of research engine control unit using FPGA based embedded control system, *Journal of KONES Powertrain and Transport* **19**(3) 9-18

[5] Strapko M and Tichánek R 2011 Engine Control Unit Based on the NI CompactRIO Platform, *Journal of Middle European Construction and Design of Cars* **9**(3)

[6] Busuttil D and Farrugia M 2015 Experimental investigtion on the effect of injecting water to the air to fuel mixture in a spark ignition engine, *MM Science Journal (March)* 585-590

Salvia officinalis used in pharmaceutics

K L Lemle

Revitalia Pharmacy, Deva, Romania

E-mail: l.katerina lucia@yahoo.com

Abstract. The paper presents some pharmaceutical properties of *Salvia officinalis*, a plant belonging the Lamiaceae family, one of the oldest medicinal plants, which play an important role in improving the state of health.

1. Introduction

Salvia Officinalis L. was discovered by Linneu in 1753, in the East-Mediterranean region and it belongs to the Lamiaceae family [1]. The plants belonging to the Lamiaceae family grow in many parts of the world, some of them being used in order to obtain various food aromas and to improve the quality of products. They also represent a basis for phytochemical products with beneficial effects for health or play an active role in improving the state of health [2].

Gender *Salvia L.* includes almost 900 species spread all over the world, some of which having a great economical value, it being used in the cosmetic industry and in perfumery [3], [4].

This plant has also been used for a long time in traditional medicine to fight fever, rheumatism, perspiration, sexual malfunction, as well as in the treatment of chronic bronchitis or of various mental diseases [5]. The best known species are *Salvia Officinalis*, *Salvia Fruticosa* and *Salvia Lavandulifolia* [6]. *Salvia officinalis L.* (Figure 1) is an evergreen plant, with wooden stems, silver leaves and flowers ranging from blue to purple and which is at present cultivated in various countries [6], [7].

Figure 1. Salvia Officinalis

Salvia is one of the oldest medicinal plants and the etymology of its Latin name suggests its medicinal properties, the word „salvia" coming from the Latin verb *salvare = to save,* which, in this context, has a medical meaning.

From times immemorial, salvia has been used on a large scale in order to add flavor to foodstuffs, as a dried herb or as en essential oil [8].

Recent studies have been done in order reveal other properties of this plant, such as its anti-inflammatory, anti-microbial, hypoglycemic, anti-diabetes, antioxidant potential, and its capacity to prevent neurovegetative diseases and to stimulate anti-tumor activities [1].

2. Chemical composition

Many studies have shown that the plants belonging to the Lamiaceae family show antioxidant and antibacterial activities, due particularly to the quantity and quality of the phenolic compounds it contains. Among these, eugenol, carvacrol and thymol, which are important components of the essential oils and are responsible for the bactericide / bacteriostatic properties. It has also been noticed that the antimicrobial effect of the extract varies from plant to plant and from one region to another, because of various factors, such as: climate, soil composition, type of solvents used in the extraction process etc.

Lately, it has been noticed that some microorganisms have become resistant to antibiotics. For instance, bacterium Staphilococcus Aureus is still responsible for the post-surgical infections, while bacterium Escherichia Coli causes infections of the urinary tract. Therefore, the researches on the biological active substances to be found in plants have been given increasing importance.

Recently, Kozlowska and collaborators have carries out a comparative study on the chemical composition and the antibacterial activity of the watery (ethanolic and methanolic) extracts from some traditional medicinal plants such as: Thymus vulgaris, Rosmarinus officinalis, Origanum Vulgare, Mentha piperita and Salvia officinalis [2].

As it is well known, the compounds with predominant medicinal properties are the monoterpenes (thujone, cineole, camphor), the diterpenes (the carnosic acid), triterpenes (the oleanolic acid, the ursolic acid) and the phenolic compounds (the rosmarinic acid) [9], [10].

Salvia Officinalis L. contains a large range of monoterpenes with carbon skeletons, including acyclic, monocyclic and bicyclic compounds. The environmental conditions, such as temperature, daytime and light influence the quantitative composition of the essential oil. The variation of these conditions leads to a seasonal metabolism of the plant, which is repeated every year. In *Salvia Officinalis L.* the monoterpenes show an accented dynamic in the period of the vegetative cycle, confirmed by studies carried out in different geographical conditions [11], [12], [13]. Grausgruber-Gröger and collaborators showed that the level of cineole decreases in the vegetation period (May - October), camphor reaches a maximum at the height of the vegetation period, while the thujone concentration grows gradually during the vegetation period [14].

It is also known that *Salvia officinalis L.* contains rosmarinic acid, tannic acid, corosolic acid, fumaric acid, nicotinic acid, etc. [15].

As it has been mentioned before, the antioxidant properties of salvia are mainly due to the presence of the phenolic acids [16]. Recent studies on the chemical composition of Salvia were mainly focused on the hydrophilic compounds deriving from the caffeic acid, over 25 derivatives being isolated and identified from the watery extracts of salvia, the A-E salvianolic acid, the rosmarinic acid and the lithospermic acid. The B – salvianolic acid, discovered in 1981, is a major component of salvia and, due to this, several intensive pharmacological studies have been done on this compound. In a recent study, a new phenolic acid was discovered, called the Y – salvianolic acid, which has the same configuration but seems to behave significantly better that the B – salvianolic acid [17].

3. Pharmaceutical use

Salvia is well known in traditional medicine as having antimicrobial properties, it being therefore used as antiseptic, anti-scabies, anti-syphilis, and anti-inflammatory agent, and frequently used in skin and eye diseases. In some regions of the Middle East, salvia was reported to be used in treating fever and some digestive diseases. The antibacterial properties of the salvia oil have been attributed to the presence of cineole, thujone and camphor [14].

The antimicrobial properties of the essential oil extracted from Salvia officinalis have drawn the attention of researchers as they can be used in developing alternatives to traditional antibiotics which prove to be less and less efficient as bacterial resistance to them increases.

It is known that Escherichia coli, Klebsiella pneumoniae and Klebsiella oxytoca are pathogen agents responsible for diseases that can be more or less serious and are transmitted orally and have been detected in various foodstuff. Moreover, all these pathogen agents have become more and more resistant to synthetic antibiotics, which leads to inefficient treatments for some patients. For this reason, the interest in discovering natural antibacterial agents has increased, and these agents can be found in the essential oils of Salvia officinalis. The main advantage of using natural agents is that they do not induce antibiotic resistance, such as it happens with the long-term use of synthetic antibiotics [18].

Also, the anti-inflammatory effects of the ursolic acid contained by Salvia Officinalis have been studied by Baricevic [19]. Salvia Officinalis tea has been traditionally used in the treatment of bronchitis, asthma, angina, mouth tissue inflammation, depressions or in various skin diseases [20], [21]. Carminative, antiseptic and antispasmodic properties of the Salvia Officinalis essential oil have been reported. Their effects in the treatment of diseases of the nervous, circulatory and respiratory systems have been brought forth by Loizzo [22]. Alizade and Shaabani explained the antimicrobial activity of the essential oil from Salvia Officinalis against Staphilococcus Aureous and Candida Albicans [23].

Salvia essential oil is used externally for inflammations and infections of the soft tissues (stomatitis, gingivitis and pharyngitis) and internally for excessive perspiration and dyspeptic symptoms. Salvia leaves and the essential oil possess carminative, antispasmodic, antiseptic and astringent properties [6].

An infusion made of this plant is used for its hemostatic, estrogenic, antiperspirant, antineuralgic antiseptic, hypoglycemic effects, as well as other therapeutic effects, while the essential oil produced from this plant is used in the large scale treatment of diseases of the nervous, circulatory, respiratory systems, as well as some digestive, metabolic or endocrine diseases [24].

A recent study shows that hydroalcoholic extracts from Salvia Officinalis are active with the promastigote forms of Leishmania major [25].

Usually, cancer therapy involves surgery, irradiation and medicine in various combinations. An efficient medicine treatment should kill the cancerous cells without affecting the activity of the healthy ones. Therefore, the identification of new cytotoxic drugs with low effects upon the immune system has developed into a new field of immune-pharmacological studies. It is known that anti-cancer agents induce an apoptosis, so that the death of the apoptotic cell reduces the sensitivity to treatment. A wide range of natural components of the anti-cancer agents show significant cytotoxic activity that operates in apoptosis. For instance, in a recent study, Zhahneh and collaborators showed that the methanol extract from Salvia Officinalis has significant benefic effects in the prophylaxis of leukemia [26].

The plant extracts used in traditional medicine have similar properties. In fact, many diterpenes isolated from some species of plants, belonging to the Salvia Gender have proved interesting pharmacological properties: antioxidant, antimicrobial, anti-inflammatory, analgesic, antipyretic, hemostatic, hypoglycemic and antitumor [26].

The antioxidant properties of the compounds isolated from Salvia officinalis have been described in [27], [28].

It is known that foodstuffs contain fatties and oils that are rich in polyunsaturated fatty acids, which suffer slight oxidation reactions that are not desired in the process of processing (heating up) and neither when stored (addition of salt, freezing). In order to prevent the loss of quality or to delay the effects of oxidation, synthetic phenolic antioxidants are currently in use (butylhydroxytoluene, butylated hydroxyanisole, third-butyl hydroquinone) [29], [30]. For this reason, recent studies have been carried out on the existence of some polyphenolic antioxidants in some plants such as salvia or rosemary.

Wu [31] used the method of oil stability in order to assess the antioxidant activities of some compounds such as: carnosol, rosmanol, epirosmanol, isorasmanol and the carnosic acid, which represented a remarkable activity comparable to that of the alpha-tocopherol. The rosmarinic acid and carnosol were the main components of all phenolic antioxidant extracts isolated from Salvia officinalis.

The increasing interest in the replacement of synthetic antioxidants with natural antioxidants showing the same efficiency could be beneficial both for the increase of foodstuff quality and for a positive influence upon the treatments of various human pathologies [32]. For instance, in [33] it has been demonstrated that the raw ethanolic extract of Salvia officinalis has antioxidant properties comparable with butylhydroxytoluene.

Salvia is a rich source of phytochemicals, including flavonoids, sesquiterpenoids, diterpenoids, sesterpenes and triterpenes. Some active components such as the rosmarinic acid and acetyl horminon have been isolated from roots of salvia officinalis. These compounds, extracts and essential oils reduce in vitro colon, breast and pulmonary cancer and leukemia and reduce the number of oncogenic cells [34], [35].

4. Conclusions

In the last years, a lot of research has been done on the aging process of the human organism. The people's interest in the physical aspect in general and that of the skin in particular has significantly developed dermatology, particularly in its part related to the aging process of the cutaneous organ [36]. A large number of plants and plant extracts are known to have a beneficial effect on the skin. Salvia officinalis is well known for the multiple and beneficial impact upon the body, as it has been used to this effect for a long time.

References

[1] Garcia C S C, Menti C, Lambert A P F et al. 2016 Pharmacological perspectives from Brazilian Salvia officinalis (Lamiaceae): antioxidant, and antitumor in mammalian cells, *Anais da Academia Brasileira de Ciências* **88** 281-292

[2] Kozłowska M, Laudy A E, Przybył J et al. 2015 Chemical composition and antibacterial activity of some medicinal plants from Lamiaceae family, *Acta Poloniae Pharmaceutica-Drug Research* **72** 757-767

[3] Velasco V and Williams P 2011 Improving meat quality through natural antioxidants, *Chil. J. Agric. Res.* **71** 313-322

[4] Hossain M B, Brunton N P, Barry-Ryan C et al. 2008 Antioxidant activity of spice extracts and phenoloics in comparison to synthetic antioxidants, *Rasayan J. Chem.* **1** 751-756

[5] Alinezhad H, Baharfar R, Zare M et al. 2012 Biochemical activities of acetone extracts of Hyssopus angustifolius, *Acta Pol. Pharm. Drug Res.* **69** 617-622

[6] Raal A, Orav A and Arak E 2007 Composition of the essential oil of Salvia officinalis L. fromvarious European countries, *Natural Product Research* **21** 406-411

[7] Mirjalili M H, Salehi P, Sonboli A et al. 2006 Essential oil variation of Salvia officinalis aerial parts during its phonological cycle, *Chemistry of Natural Compounds* **42** 19-23

[8] Perry N B, Anderson R E, Brennan N J et al. 1999 Essential oils from Dalmatian sage (Salvia officinalis L.): variations among individuals, plant parts, seasons, and sites, *Journal of Agricultural and Food Chemistry* **47** 2048-2054

[9] Lamaison J L, Petitjean-Freytet C and Carnat A, Medicinal lamiaceae with antioxidative activities, potential sources of rosmarinic acid, *Pharmaceutica Acta Helvetiae* **66** 185–188

[10] Cuvelier M E, Berset C and Richard H 1994 Antioxidant constituents in sage (*Salvia officinalis*), *Journal of Agricultural and Food Chemistry* **42** 665-669

[11] Pitarevic I, Kuftinec J, Blazevic N et al. 1984 Seasonal variation of essential oil yield and

composition of Dalmatian sage, Salvia officinalis, *Journal of Natural Products* **47** 409–412
[12] Putievsky E, Ravid U and Dudai N 1986 The influence of season and harvest frequency on essential oil and herbal yields from a pure clone of sage (*Salvia officinalis*) grown under cultivated conditions, *Journal of Natural Products* **49** 326–329
[13] Marie S, Maksimovic M and Milos M 2006 The impact of the locality altitudes and stages of development on the volatile constituents of *Salvia officinalis L.* from Bosnia and Herzegovina, *Journal of Essential Oil Research* **18** 178–180
[14] Abu-Darwish M S, Cabral C, Ferreira I V et al. 2013 Essential Oil of Common Sage (Salvia officinalis L.) from Jordan: Assessment of Safety in Mammalian Cells and Its Antifungal and Anti-Inflammatory Potential, *BioMed Research International* **2013** Article ID 538940
[15] Monsefi M, Abedian M, Azarbahram Z et al. 2015 Salvia officinalis L. induces alveolar bud growing in adult female rat mammary glands, *Avicenna J Phytomed* **5** 560-567
[16] Lu Y R and Foo L Y 2002 Polyphenolics of Salvia - A review, *Phytochemistry* **59** 117-140
[17] Gong J, Ju A, Zhou D et al. 2015 Salvianolic Acid Y: A New Protector of PC12 Cells against Hydrogen Peroxide-Induced Injury from Salvia officinalis, *Molecules* **20** 683-692
[18] Fournomiti M, Kimbaris A, Mantzourani I et al. 2015 Antimicrobial activity of essential oils of cultivated oregano (Origanum vulgare), sage (Salvia officinalis), and thyme (Thymus vulgaris) against clinical isolates of Escherichia coli, Klebsiella oxytoca, and Klebsiella pneumonia, *Microbial Ecology in Health & Disease* **26** 23289
[19] Baricevic D, Sosa S, Della Loggia R et al. 2001 Topical anti-inflammatory activity of *Salvia officinalis* L. leaves: the relevance of ursolic acid, *Ethnopharmacol* **75** 125-132
[20] Walch S, Tinzoh L, Zimmerman B et al. 2011 Antioxidant capacity and polyphenolic composition as quality indicators for aqueous infusions of Salvia officinalis L, *Front Pharmacol* **2** 29-44
[21] Khan A, Najeeb-ur-Rahman, Alkharfy K et al. 2011 Antidiarrheal andantispasmodic activities of Salvia officinalis are mediated through activation of K + channels, *J Bangladesh Pharmacol Soc* **6** 111–116
[22] Loizzo M R, Tundis R, Menichini F et al. 2007 Cytotoxic activity of essential oils from Labiatae and Lauraceae families against in vitro human tumor models, *Anticancer Res* **27** 3293–3299
[23] Alizade A and Shaabani M 2012 Essential oil composition, phenolic content, antioxidant and antimicrobial activity in (*salvia officinalis* L.) cultivated in Iran, *Advances Environment Biol* **6** 221-226
[24] Istudor V 2001 *Farmacognozie. Fitochimie. Fitoterapie*, Editura Medicala, Bucuresti, Romania
[25] Serakta M, Djerrou Z, Mansour-Djaalab H et al. Antileishmanial activity of some plants growing in Algeria: Juglans Regia, Lawsonia Inermis and Salvia Officinalis, *Afr J Tradit Complement Altern Med.* **10** 427-430
[26] Shahneh F Z, Valiyari S, Baradaran B et al. 2013 Inhibitory and Cytotoxic Activities of Salvia Officinalis L. Extract on Human Lymphoma and Leukemia Cells by Induction of Apoptosis, *Advanced Pharmaceutical Bulletin* **3** 51-55
[27] Cuvelier M E, Verset C and Richard H 1994 Antioxidant constituents in sage (*Salvia officinalis*), *J Agr Food Chem* **42** 665-669
[28] Ben H A, Gajdoova D and Have J 2006 Analysis of *Salvia offi cinalis* plant extracts by capillary electrophoresis, *J Sep* Sci **29** 1188-1192
[29] Angelo A J S, Vercellotti J, Jacks T et al. 1996 Lipid oxidation in foods, *Crit Rev Food Sci Nutr* **36** 175-224
[30] Namiki M 1990 Antioxidants/antimutagens in food, *Crit Rev Food Sci Nutr* **29** 273-300
[31] Wu Z B, Ni Z Y, Sh Q W et al. 2012 Constituents from *Salvia* species and their biological activities, *Chem Rev* **112** 5967-6026
[32] Willcox J K, Ash S L and Catignani G L 2004 Antioxidants and prevention of chronic disease, *Crit Rev Food Sci Nutr* **44** 275-295

[33] Jantova S, Hudec R, Sekretar S et al. 2014 Salvia officinalis L. extract and its new food antioxidant formulations induce apoptosis through mitochondrial/caspase pathway in leukemia L1210 cells, *Interdiscip Toxicol.* **7** 146-153
[34] Liu J, Shen H M and Ong C N 2000 *Salvia miltiorrhiza* inhibits cell growth and induces apoptosis in human hepatoma HePG2 cells, *Cancer Let* **153** 85-93
[35] Tayarani-Najarana Z, Mousavi S H, Tajfarda F et al. 2013 Cytotoxic and apoptogenic properties of three isolated diterpenoids from *Salvia chorassanica* through bioassay-guided fractionation, *Food Chem Toxicol* **57** 346–351
[36] Dragomirescu A O 2012 *Actualităţi în dermatofarmacie şi cosmetologie*, Editura Victor Babeş, Timişoara, Romania

IOP Conf. Series: Materials Science and Engineering **294** (2017) 012038 doi:10.1088/1757-899X/294/1/012038

Critical analysis of the influence of the possibilities of establishing the moulding technology on obtaining the castings

A Josan, C Pinca Bretotean and S Rațiu

Politehnica University of Timisoara, Department of Engineering and Management, 5 Revolution Street, Hunedoara, 331128, Romania

E-mail: ana.josan@fih.upt.ro

Abstract. The Romanian foundries give high priority to quality castings. Thus, the castings moulding technology decisively influences their quality. Therefore, it is necessary to study the possibilities of realization the moulds for castings. This paper presents a critical analysis of the influence of all factors of the moulding technology on the quality of steel castings. The purpose of the study is to analyze the way to establishing the separation plane, how to determine the processing allowances and technological allowances, the dimensioning of the riser, the choice of the optimal casting method of the liquid alloy. Establishing optimal moulding technology determines the optimal use of resources with direct implications for manufacturing costs.

1. Introduction

The principles of the construction of the steel castings are quite different from the known principles of the construction of cast iron castings [1]:

- First of all, the technological analysis of the steel casting construction must take into account the shrinkage factor. The large shrinkage of the cast steel during solidification and in a solid state requires measures to be taken to ensure that shrinking, tensions and cracks are produced in the casting.
- Secondly, in the case of steel castings the steps of obtaining the casting must be taken into account: moulding, cleaning, mechanical machining (which is necessary to be cheap and lead to material savings).

In this context, in the Romanian foundries it is necessary to provide technologies that will lead to a balance between all the factors of influence on the quality of the castings.

2. Critical analysis of the influence of the possibilities of establishing the moulding technology on the obtaining the castings

The castings of the *Toothed case*-type made of unalloyed cast steel, 340-550W steel grade, are very common in industrial practice of the foundries because they are used in various industries [2-4]. The chemical composition of the steel for the elaborated charges is in accordance with the standard SR ISO 3755-95 [5].

The special requirements for this type of castings imposed by beneficiaries led to the necessity of a critical analysis of the manufacturing technology for this type of casting.

With this end in view, the critical analysis of the foundry on this type of castings led to the conclusion that the percentage of the rejects is too high [6], with direct effects on the costs of foundry. Besides the registered defects due to the moulding mixtures used [6], the critical analysis has led to the idea that the possibility of establishing the moulding technology has a particular influence on the quality castings (castings without defects).

Thus, several deficiencies have been identified in the moulding technology [7-9]:

1. Inappropriate positioning of the separation plane leads to the displacement of the moulds during casting of the liquid alloy and then to rejection of the castings
2. Inappropriate choosing of the processing and technological allowances leads to an excessive increase of the labor costs; this leads to an increase of the casting price.
3. Inappropriate dimensioning and positioning of the shrink head leads to the shrinking; this defects leads to the rejected of the casting

In this context, the design of moulding technology has a prime role.

The first step in obtaining the castings quality is to establish the position of the cast in the mould and the choice of the separation surface/plan so that the casting pattern and the entire set of pattern can then be made. The initial analysis of the drawing of finished piece definitely has this role so that the designer of the technology can perfectly imagine the spatial configuration of the casting.

For these purpose it is determined [10], [11]:

a) The moulding and casting positions, respectively the separation surfaces,
b) Shrinkage allowance and processing and technological allowances,
c) The dimensions of the cores and the cores mark as well as their positioning.

All of these elements are featured on the technological drawing, in all views and sections with conventional colours and lines.

2.1. Establishing the position of the casting in the mould and chosing the separation surface

To make it possible to pattern drawing of the moulding mixture without damaging the mould it should be section with one or more separation surfaces. For each casting the possible solutions for sectioning of the pattern and mould are analyzed.

Of the possible solutions one must choose the one that ensures the following [12]:

- extracting the pattern (pattern drawing) from the moulding mixture;
- getting the mould with the fewest cores;
- easy and secure mounting of the cores in the mould;
- laying of the surfaces to be mechanically machined to the bottom or side;
- the number and dimensions of the burrs that can be formed must be minimal (the most undesirable ones are the inner ones and, therefore, the cores must not be made from two or more sealing parts except in special situations);
- the possibility of making the core correctly and easily as well as the correct and simple manufacture of the core box;
- the position of the casting in mould must be chosen so as to ensure a conduct solidification of the liquid alloy (from thin to thick parts) and to ensure minimal metal consumption and a lower cost price.

In particular, in the case of casting analysed, there are several theoretical solutions (Figure 1), which depend on the casting symmetry, the presence of the surfaces to be machined and the moulding and casting positions.

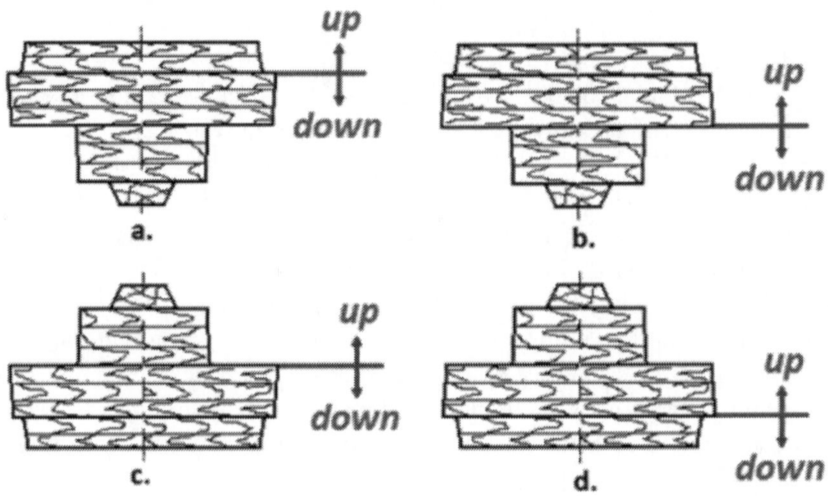

Figure 1. The possibilities of sectioning the pattern

In particular, in the case of casting analysed, there are several theoretical solutions (Figure 1), which depend on the casting symmetry, the presence of the surfaces to be machined and the moulding and casting positions.

Thus, the chosen solution must ensure the production of castings corresponding to the technical and economic requirements.

For this, it is necessary to perform the analysis of the possible solutions presented in Figure 1, with the presentation of their advantages and disadvantages:

- **case a** – has the disadvantage that the core covers the top of the casting so it is not possible to place the shrink head and it cannot be directed by the solidification process, resulting the internal shrinkages in the casting;
- **case b** – besides the disadvantage of the **case a** is added the possibility of displacement the mould parts and thus obtaining a decentralized casting;
- **case c** – present the disadvantage of the possibility of displacement in the separation plane and obtaining a decentralized casting (it is similar to **case b**);
- **case d** – is the most advantageous solution because it allows a correct positioning of the shrink head and thus directional solidification; the pouring of the liquid alloy in this case is of the cast uphill type, so pouring of liquid metal into the mould cavity is carried out without the risk of turbidity occurring (quiescent pouring).

Considering these recommendations, Figure 2 presented the mould separation plane; this surface is plane and the pouring position is horizontal.

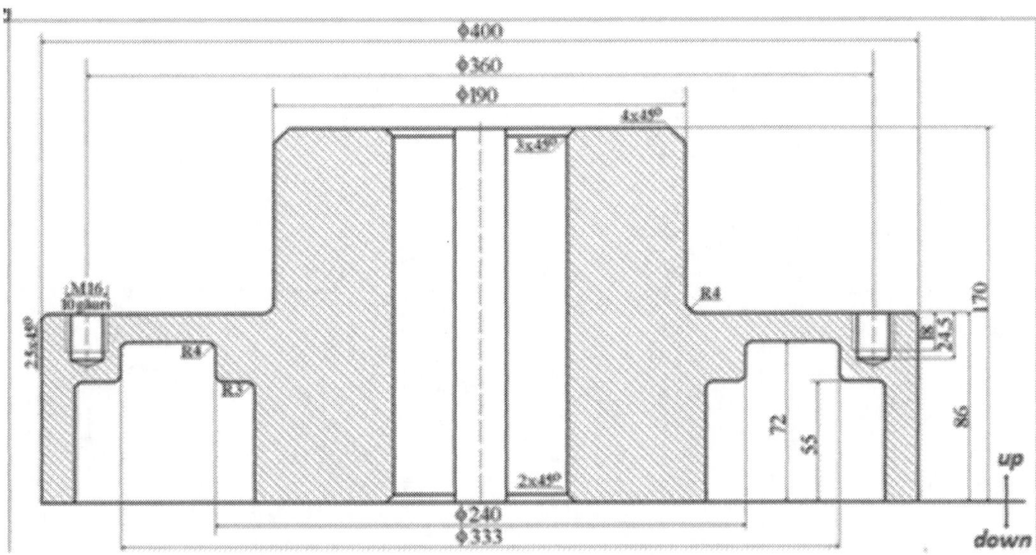

Figure 2. The separation plan of the casting studied

2.2. Establishing the processing and technological allowances

Processing allowances are additional allowances of material provided on the casting assembly surfaces. These surfaces require machining by cutting in order to obtain precise dimensions so that parts of the castings can be assembled [11], [12].

The size of processing allowances depends on several factors, namely [11], [12]:

- the type of cast alloy,
- the position of the processing surface during casting of the alloy,
- the dimensions of the casting and the surface to be processed,
- the method of moulding used etc.

Following the analysis of all these factors of influence and according to the specialty literature and industrial practice these allowances have been established for the analysed casting.

In industrial practice the processing allowances for steel castings are standardized and are deermined according to the maximum dimensions of casting [mm] and nominal dimension [mm] [11], [12]. Thus, in Table 1 are presented, for casting studied, the values for processing allowances [10], [12].

Table 1. Processing allowances according to the dimensions of casting

Maximum dimension of casting, mm	Position in mould of the surface to be processed	Nominal dimension, mm
		301…400
301…500	Up	7,0 for H = 170 mm 5,0 for h = 86 mm
	Down	5,0 for H = 170 mm
	Side	6,0 for Φ 400 mm 5,0 for Φ 190 mm 5,0 for Φ 80 mm

In Figure 3 there are presented the allowances applied to the casting as well as the final dimensions obtained on the technological drawing.

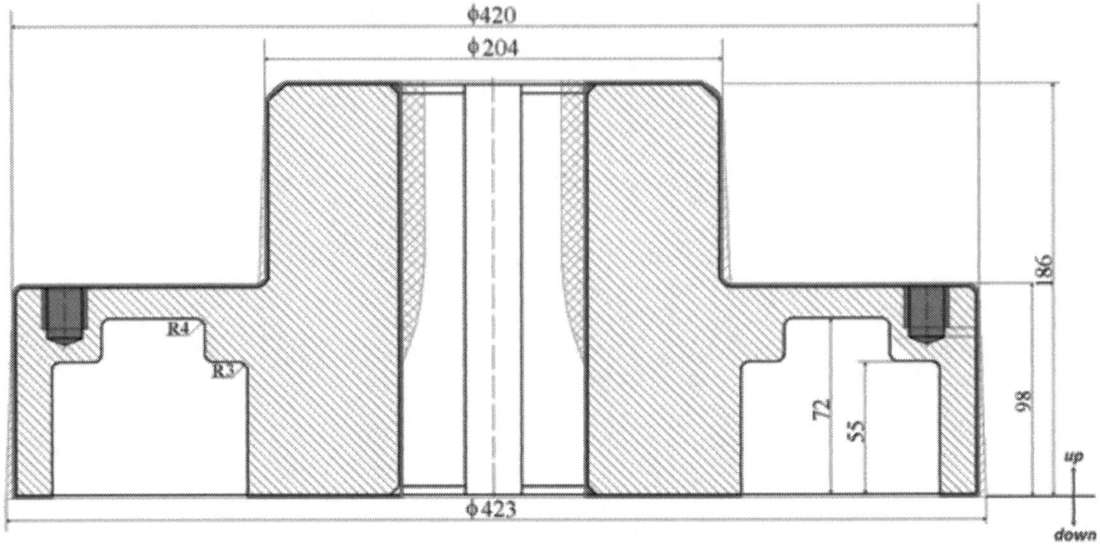

Figure 3. Presentation of the allowances for the studied casting

The elements of the pattern to be obtained are drawn on the technological drawing as far as possible according to the following rules:

- the separation surface (plane) – with a continuous blue line;
- the shrink heads – draw, color and quote in blue;
- the processing allowances and corresponding quotas – red; the graphic representation of the processing allowances, on the technological drawing, is done as follows:

- the technological allowances – for conduct solidification; the graphic representation on the technological drawing is done as follows:

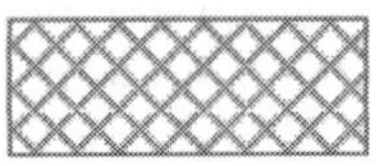

- the constructive inclinations – has the rol of preventing excessive labor and steel losses as well as ensuring that castings with precise dimensions are obtained; the size of the constructive inclinations is regulated by the standard, depending on height H of the wall to be tilted; the inclination of the walls will be perpendicular to the separation surface. The graphic representation on the technological drawing is done as follows:

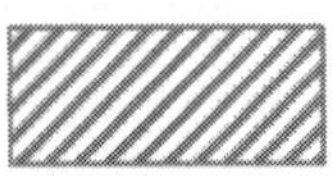

- the cores (internal or lateral cores)– with colors to show them (green e.g), on all the views and sections of the casing.

After establishing the size of shrinkage allowances, the constructive inclinations and the dimensions of the core marks, the configuration and the dimensions of the pattern can be determined.

For making the set of pattern account is taken of processing allowances, technological allowances, shrinkage allowance, respectively dimensioning the core marks [6]. It is sspecified that the set of pattern is made entirely of wood. Thus, Figure 4 presented the wooden pattern of the casting.

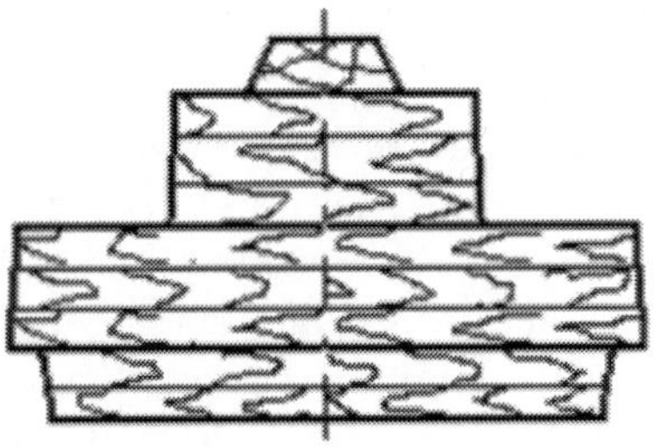

Figure 4. The wooden pattern of casting analized

Compared to other materials used in pattern shop, wood has the following advantages:

- it is cheap,
- it has a small specific weight,
- it is easy to process,
- it is easily glued by gluing and with some metallic elements.

2.3. Dimensioning the shrink head

To the solidification of some castings of alloys with a high volumetric shrinkage coefficient, at the top of the casting there is always a *concentrated shrinking*.

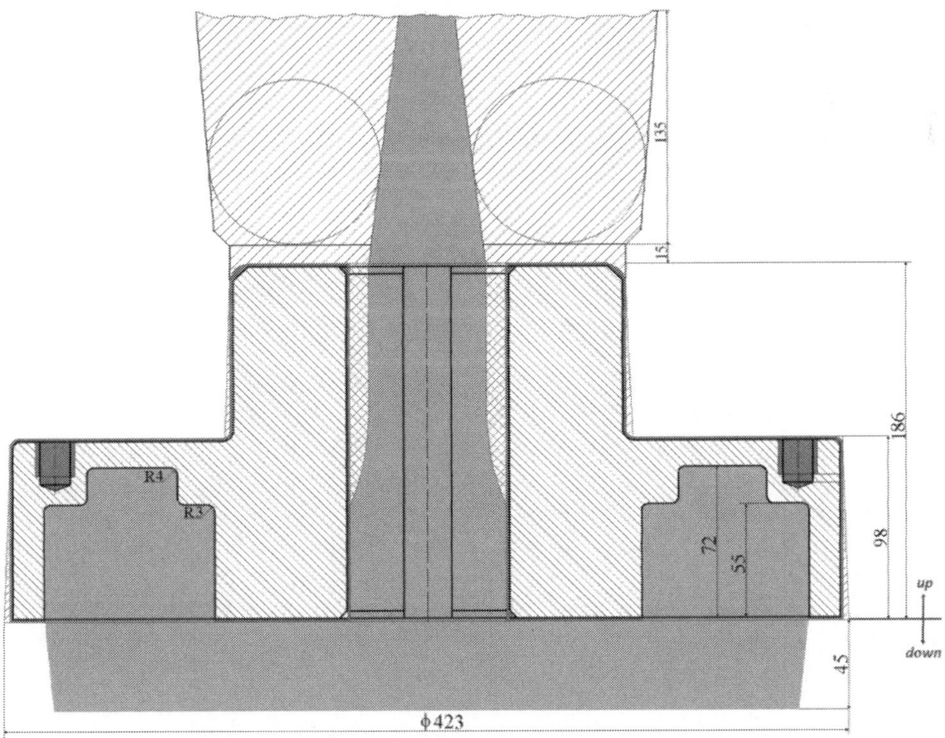

Figure 5. Technological drawing of casting analized

This occurs because during solidification, the already solidified parts decrease their volume, determining a lowering the level of liquid unsolidified [7-10].

In order to avoid the occurrence of concentrated shrinking from the top of the castings, a technological extension is made thicker than casting. In this extension will locate the shrinking so that the casting will have a compact height ("healthy") corresponding to the drawing of the finished piece. Thus, the shrinks head are "additional reservoirs" of the liquid alloy in the form of technological allowances. For calculating the dimensions the shrink head it is necessary to take into account the shrinkage coefficient β [1], [12]. This coefficient, for steel grade *340-550W* (C = 0,4...0,5%), has the value $\beta = 4,5...4,8\%$ [12], [13].

From practical experiments it has been found that for steel castings the dimensions of the casting being analysed it is best to use a circular shrink head to cover the top of the casting.

The dimensioning of the shrink head was performed using two method, respectively the graphical method of the spheres entered and expeditious dimensioning of the shrink head [10-12].

Following the correlation of the two dimensioning method of the shrink heads, the main dimensions of the shrink head are determined (Figure 5): h = 15 mm; Dm = 108 mm; Hm = 135 mm, which leads to significant material savings.

3. Conclusions

The rejected registered in the castings analyzed due to moulding technology, respectively due to positioning inadequate of the separation plane, the insufficient strength of the mould, the pouring method, represent 40% of the total rejects registered.

Thus, the position of the casting into the mould and the separation surfaces of the mould must ensure:

- required quality of the casting
- minimum consumption for mechanical machining
- minimum consumption of the metal.

The defects due to the inappropriate setting of the separation plan, depending on their severity, often lead to the increase of the cleaning work of the casting parts and even to their rejecting.

The processing and technological allowances are chosen so as to ensure the possibility of correct pattern-drawing and do not leads to an excessive increase of machining/trimming labor (this lead to an increase the price of casting).

For the studied technology, industrial practice indicates that it is best to use a circular shrink head to cover the top of the casting. Choosing this constructive option leads to a substantial material saving, compared to the current technology, which leads to a decrease in a manufacturing price of the casting.

References

[1] Socalici A, Ardelean E, Ardelean M, Heput T and Josan A 2007 *Casting and solidification of steel*, Publishing Cermi, Iaşi

[2] Josan A, Pinca Bretotean C, Ardelean E and Ardelean M 2015 Using lateral cores to casting of carbon steel parts of drive wheel type in a metallurgical enterprise, *IOP Conf. Ser.: Mater. Sci.Eng.* **85** 012016

[3] Josan A and Pinca Bretotean C 2015 Using special additions to preparation of the moulding mixture for casting steel parts of drive wheel type, *IOP Conf. Ser.: Mater. Sci.Eng.* **85** 012017

[4] Socalici A, Pascu L, Popa E and Heput T 2015 The influence of the cast iron structure upon the hardness of brake shoes meant for the rolling sock, *IOP Conf. Ser.: Mater. Sci.Eng.* **85** 012026

[5] ***SR ISO 3755-95 *Unalloyed cast steel for general mechanical applications*

[6] Josan A 2016 Studies regarding the use of molding mixtures based on chromite-sand to the casting of the steel parts, *Solid State Phenomena* **254** 243-248

[7] Josan A 2016 Improving the quality of castings by optimization of the moulding-casting technology, *Acta Technica Corviniensis – Bulletin of Engineering* **IX**(4) 75-78
[8] Josan A 2010 *The analysis of casting defects recorded in the metallurgical enterprises*, International Symposium on Advanced Engineering & Applied Management - 40th Anniversary in Higher Education (1970-2010), Hunedoara, CD
[9] ***1977 *International Atlas of casting defects*, Technical Publishing, Bucharest
[10] Stefanescu C and Cazacu I 1980 *Technologies of realization of the parts by casting*, Technical Publishing, Bucharest
[11] Josan A 2002 *The technology of moulding and casting alloys*, Editura Politehnica, Timisoara
[12] Josan A 2012 *Casting and moulding technologies of metal parts – the project guide-book*, Publishing Polytechnic, Timişoara
[13] Josan A, Pinca Bretotean C, Raţiu S, Ardelean E and Ardelean M 2017 Research on the influence of moulding-casting technology on the quality of castings, *IOP Conf. Ser.: Mater. Sci.Eng.* **200** 012010

The importance of correct specification of tribological parameters in dynamical systems modelling

S Alaci[1], F C Ciornei[1], I C Romanu[1] and M C Ciornei[2]

[1]"Stefan cel Mare" University of Suceava, Mechanics and Technologies Department, Universitatii str., no. 13, 720229 Suceava, Romania
[2]"Carol Davila" University of Medicine and Pharmacy, Department 2 Physiology I, Bd. Eroilor Sanitari, no. 8, 050474 Bucuresti, Romania

E-mail: alaci@fim.usv.ro

Abstract. When modelling the behaviour of dynamical systems, the friction phenomenon cannot be neglected. Dry and fluid friction may occur, but dry friction has more severe effects upon the behaviour of the systems, based on the fact that the introduced discontinuities are more important. In the modelling of dynamical systems, dry friction is the main cause of occurrence of the bifurcation phenomenon. These aspects become more complex if, in the case of dry friction, static and dynamic frictions are put forward. The behaviour of a simple dynamical system is studied, consisting in a prismatic body linked to the ground by a spring, placed on a conveyor belt. The theoretical model is described by a nonlinear differential equation which after numerical integration leads to the conclusion that the steady motion of the prism is an un-damped oscillatory motion. The system was qualitatively modelled using specialised software for dynamical analysis. It was impractical to obtain a steady uniform translational motion of a rigid, therefore the conveyor belt was replaced by a metallic disc in uniform rotation motion. The attempts to compare the CAD model to the theoretical model were unsuccessful because the efforts of selecting the tribological parameters directed to the conclusion that the motion of the prism is a damped oscillation. To decide which of the methods depicts reality, a test-rig was assembled and it indicated a sustained oscillation. The conclusion is that the model employed by the dynamical analysis software cannot describe the actual model and a more complex model is required in the description of the friction phenomenon.

1. Introduction

The friction phenomenon is found whenever two solid bodies come into contact [1], [2]. There are numerous effects of friction occurring in mechanical systems and describing them is not an easy task.

In various applications the presence of friction is neglected but a correct analysis requires the consideration of friction. One of the most important criteria in the classification of friction is the presence or absence of a lubricant between the boundary surfaces of the contacting solid bodies, therefore fluid friction and dry friction is found, respectively. Both types of friction have the effects of wear and energy dissipation as heat, consequences of relative velocity on the contact area, but they are completely different from a qualitative point of view. Thus, in the case of fluid friction, the magnitude of friction force depends on the size of relative velocity between the two surfaces and according to Newton's law [3-5], for a given relative velocity only one value for friction force corresponds.

However, for the case of dry friction, the magnitude of the friction force is described using inequalities, and for the same kinematical configuration of the system, there is the possibility that

IOP Conf. Series: Materials Science and Engineering **294** (2017) 012039 doi:10.1088/1757-899X/294/1/012039

several values of the friction force may exist. The behavior of mechanical systems where dry friction occurs is described by means of nonlinear differential equations. The integration of these equations, typically by numerical procedures, leads to the conclusion that in the presence of dry friction in the dynamic system may tend to different equilibrium states, arriving at bifurcation and chaos notions [6-7]. When the behaviour of dynamic systems is modelled via specialized software, the friction from kinematical pairs of the system must be accurately characterized. Regardless of the complexity of the friction model accepted by the software for dynamical modelling, this model considers simplifying hypotheses concerning friction. There are situations when the behaviour predicted by the modelling software is analogous to the behaviour of the actual model, but for other cases, there are major discrepancies between model and tangible system, the conclusion being that the simplifying hypotheses lead to an overly simplified model, and therefore a new, more complex model of friction is required. To illustrate this, a simple dynamic system whit dry friction is presented, modelled both directly and via software, and different evolutions are obtained. An experimental set-up is used and it is concluded that, although a series of simplifying hypothesis concerning friction was employed, the behaviour of the system is correctly described by the model.

2. A theoretical model with dry friction

The dynamic system form Figure 1 is considered. A prismatic body is set on an horizontal conveyor belt that has constant peripheral velocity, v_0. The body is attached to a spring with linear characteristic, having the elastic constant k. There is no lubricant between the body and the belt.

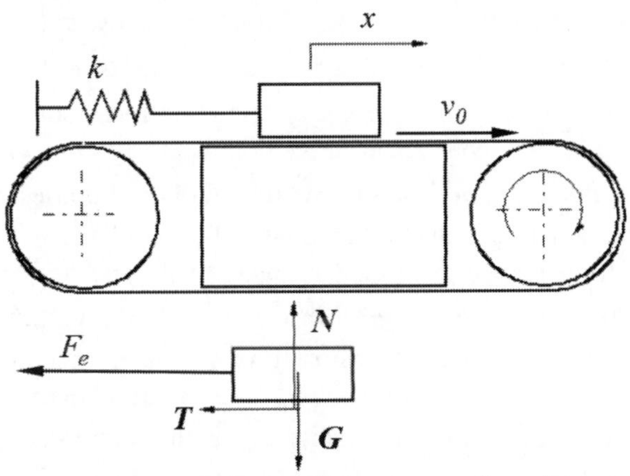

Figure 1. Scheme of the proposed model

The theorem of centre of mass is applied in order to characterize the motion of the body:

$$\begin{cases} m\ddot{x} = -T - F_e \\ N - G = 0 \end{cases} \tag{1}$$

where F_e is the elastic force:

$$F_e = k_e x \tag{2}$$

and the friction force is given by the relation, [8-11]:

$$T = -\mu N \, sgn \, v_{rel} \tag{3}$$

where μ is the coefficient of dry friction, defined as a function of the relative sliding velocity v between the belt and the body. A careful definition of the static friction force must consider the fact that, if no relative velocity exists between the contacting surfaces, the friction force may have any value between $-\mu_{st}N$ and $\mu_{st}N$, while when relative velocity exists, the friction force is according to Amonton-Coulomb law, proportional to the normal force N and oriented as to oppose to the relative displacement between the two surfaces. The dependency of the coefficient of friction on velocity can be defined by the following function:

$$\mu(v)=\begin{cases}-\mu_d, & v<0\\ -\mu_{st}\leq\mu\leq\mu_{st}, & v=0\\ \mu_d, & v>0\end{cases} \tag{4}$$

and is represented in Figure 2,

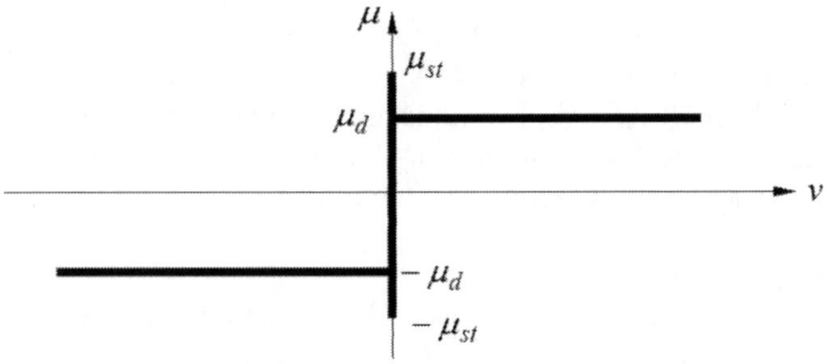

Figure 2. Theoretical dependence of dry friction coefficient versuss relative velocity

where μ_{st} and μ_d are the limit coefficients for static friction and dynamic friction, respectively. The fact that for $v=0$ the coefficient of friction takes values in the domain $[-\mu_{st},\mu_{st}]$ assumes utilizing a complicated mathematic apparatus. The same aspect occurs in the instants immediately after the exit from the equilibrium state, when the coefficient of friction suddenly jumps from μ_{st} to μ_d and from $-\mu_{st}$ to $-\mu_d$ imposes the utilization of functions with hard discontinuities whose description require the employment of distribution (generalized functions). To avoid these aspects, the hypothesis is formulated that transition from repose to motion is made in a narrow velocity domain, $[-v_{tr},v_{tr}]$ where v_{tr} is the *stiction transition velocity* used in characterisation of immobile body. The second assumption is based on accepting that the conversion from the maximum values of static friction force $\pm\mu_{st}N$ to the value of dynamic friction force $\pm\mu_d N$ is made obeying an exponential law. The dependence of the coefficient of friction on velocity under the new hypotheses is described by the function:

$$\mu(v)=\begin{cases}\mu_{st}v/v_{tr}, & |v|<v_{tr}\\ (\mu_{st}-\mu_d)exp[\alpha(v-v_{tr})]+\mu_d, & v>v_{tr}\\ -(\mu_d-\mu_{st})exp[\alpha(-v-v_{tr})]+\mu_d, & -v<-v_{tr}\end{cases} \tag{5}$$

The plot of the dry coefficient of friction versus relative velocity is shown in Figure 3.
With the stipulated relation between coefficient of friction and velocity, the equation of motion of the body is written:

$$m\ddot{x}+\mu(\dot{x}-v_0)mg+k_e x=0 \tag{6}$$

In relation 6, v_0 is the constant velocity of the belt and $\dot{x}-v_0$ is the relative velocity of the body with respect to the belt. The equation 6 is a nonlinear homogenous equation and the Runge-Kutta 4 methodology is applied in solving it [12]. Following the integration of the differential equation of motion, the elongation, velocity and acceleration graphs versus time were traced, as well as the phase diagram.

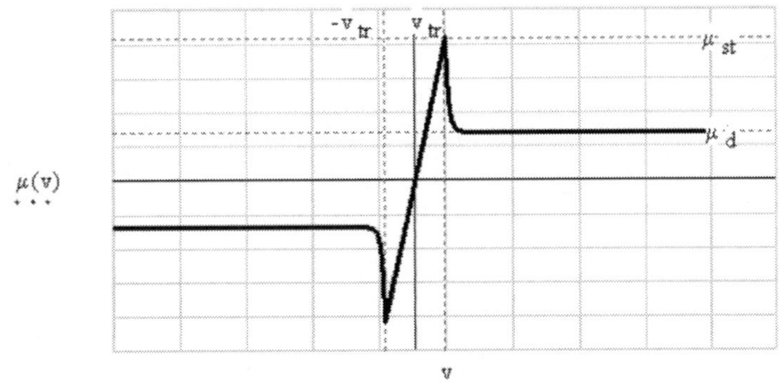

Figure 3. Approximation of dependence of the coefficient of dry friction versus relative velocity

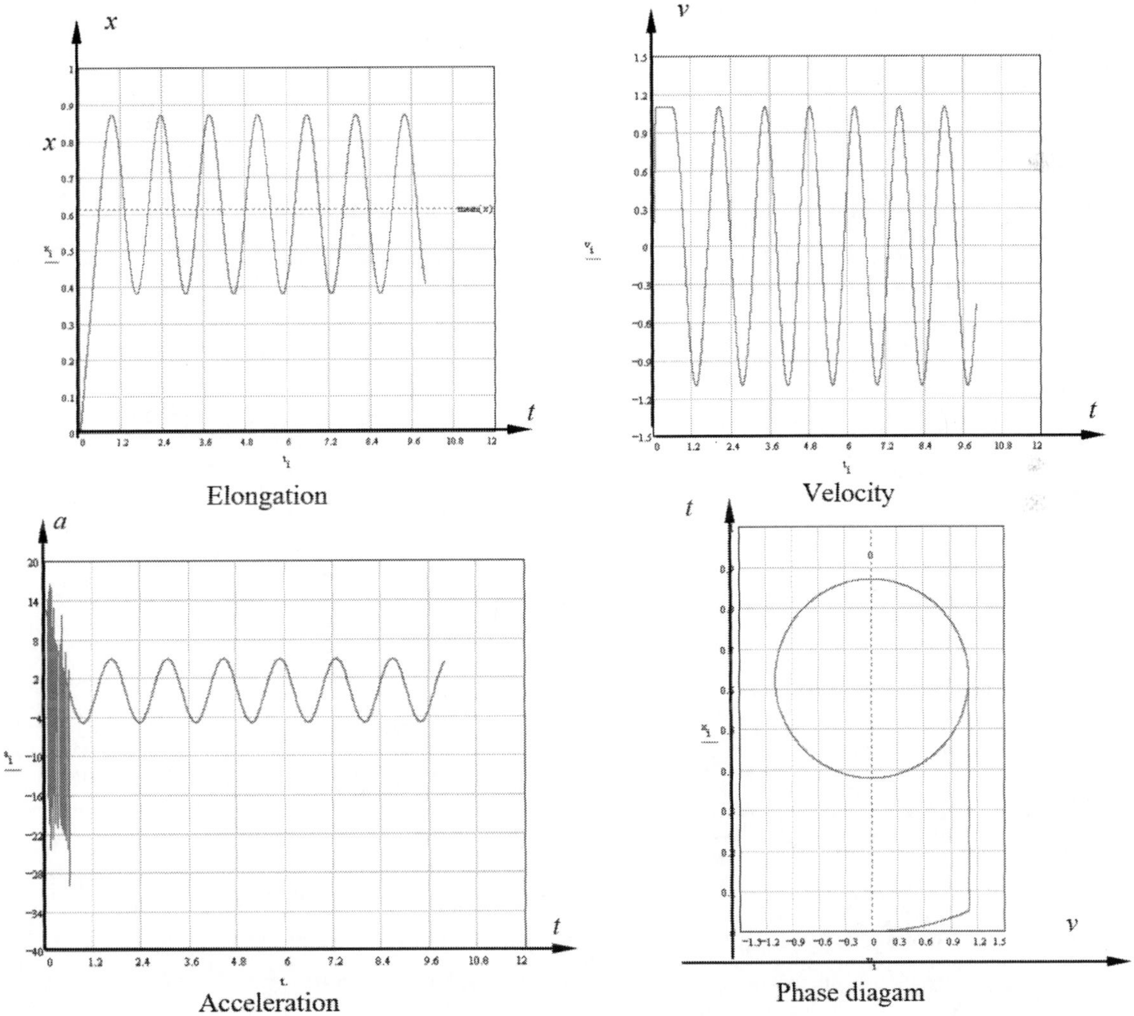

Figure 4. Elongation, velocity, acceleration and phase dyagram for $v_0 < v_{0cr}$

The integration of the differential equation for different values of belt velocity, v_0 revealed the existence of a critical value denoted v_{0cr}. The four dependencies mentioned above are presented in

Figures 4 and 5 for two values of the conveyor belt velocity. Thus, for a belt velocity smaller than the critical value, Figure 4, the body starts to move from the equilibrium position and gains an oscillatory motion, without reaching the initial position. In another situation, adopting for the belt a velocity greater than the critical one, $v_0 > v_{0cr}$, as presented in Figure 5, the body takes a periodical motion and attains the initial position. As it can be remarked from Figures 4 and 5, in both cases, the motion is un-damped since the characteristic point-in the phase diagram, moves on a circular trajectory towards the end of the analysis; for the damped case, the motion is characterized by asymptotic displacement towards a fixed point.

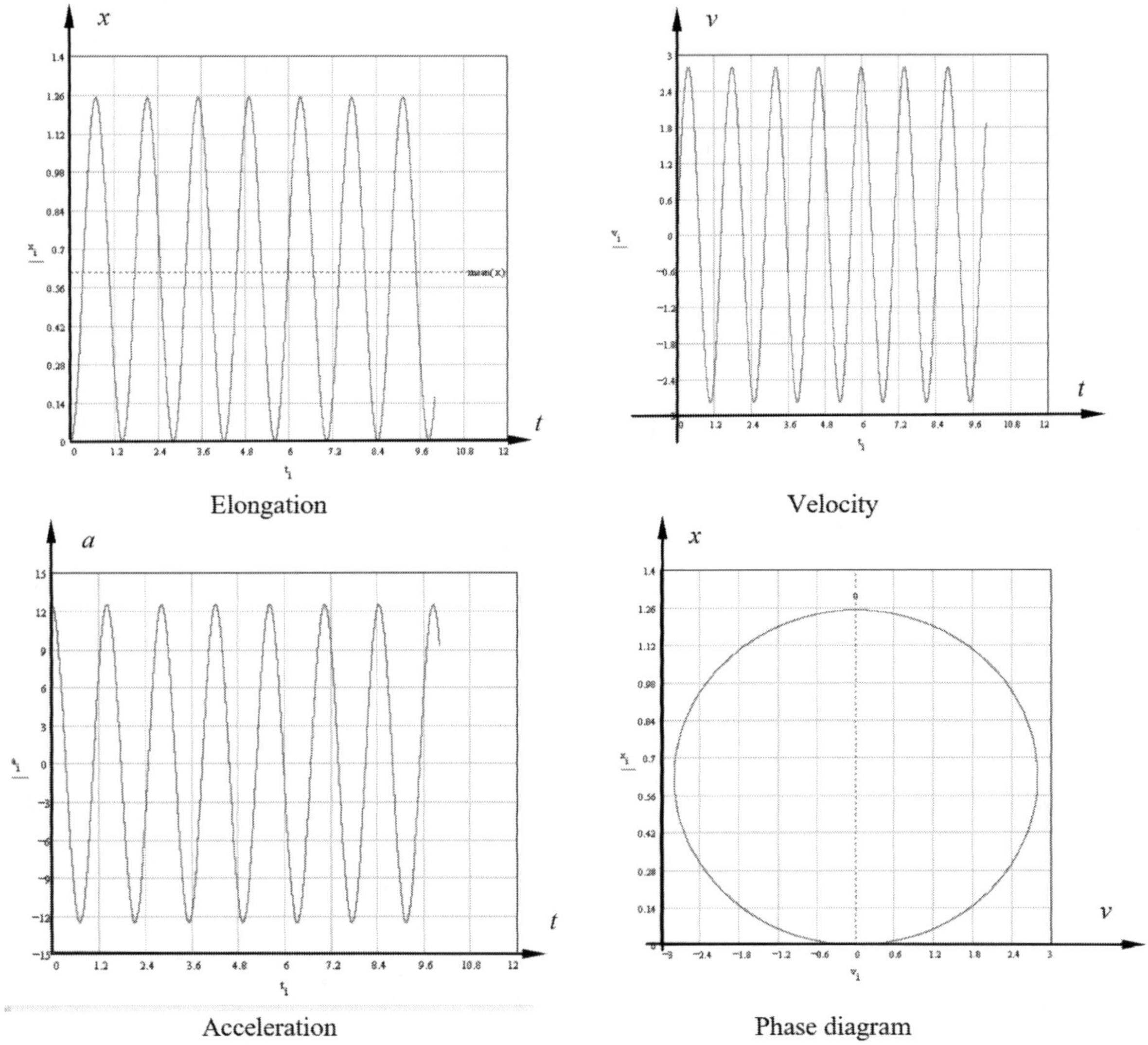

Figure 5. Elongation, velocity, acceleration and phase diagram for $v_0 > v_{0cr}$

3. Modelling the dynamic system using specialised software

For the dynamical system considered above, the dynamic behaviour was simulated using the dedicated software MSC.ADAMS [13]. A first issue which arises consists in the fact that it is impossible to model a solid body in uniform rectilinear motion, with constant velocity v_0 during a time interval of any dimension – motion ensured in the initial model by the conveyor belt. The fact that a point of the belt periodically comes into contact with the surface of the mobile body leads to the idea of replacing the belt by a disc, *1* as shown in Figure 6, which rotates about the vertical axis of the rotation pair R between the disc and the ground. A small prismatic body is placed on the frontal surface of the disc, having small dimensions compared to the radius of the disc, and situated at a distance large enough with respect to the axis of the disc, these aspects supporting the hypothesis that on the contact surface between the body and the disc the relative velocity is the same. An ideal planar parallel pair (frictionless) P was introduced in the structure of the mechanism, normal to the radius vector of the body, in order to ensure rectilinear translational motion to the prismatic body. The mobile body is also subjected to other constraints, for example C, the contact with friction between the body and the disc and *3* the helical spring ensuring the link between the ground and the prismatic body.

In order to describe the friction in the system, the tribological parameters characteristic to the contact C and to the helical spring *3* are required. The specification of the tribological parameters is not a simple task [14], [15] and to prove this, the tables required to be completed for the contact C and spring *3* are presented in Figures 7 and 8 respectively.

For the present analysis, the effect of friction from contact C is more important and therefore the adopted damping coefficient of the spring was very small, $3 \cdot 10^{-9}\, Ns/m$, that is a value corresponding to negligible damping effect. The dynamic behaviour of the system was simulated with the tribological parameters shown in Figures 7 and 8, and the same dependencies were sought as in Figures 4 and 5.

These variations are presented in Figure 9 and the *damped motion of the mobile body* may be remarked. Simulations were also made for different values of the rotation velocity of the disc and of the static and dynamic friction coefficients. The results did not differ qualitatively, since the mobile body presented a damped motion. Additionally, the manner in which the amplitudes of the oscillations decrease is an exponential function, characteristic to fluid friction. It can be easily shown that the decrease of amplitudes of oscillations for dry friction conditions is a linear function.

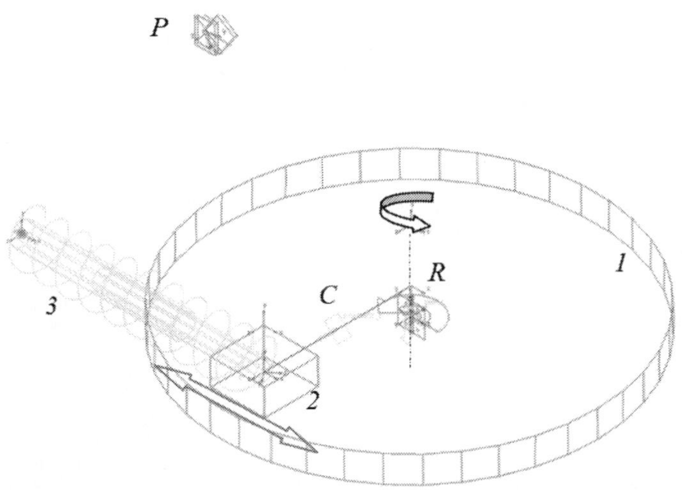

Figure 6. Dynamical simulation using specialized software

IOP Conf. Series: Materials Science and Engineering **294** (2017) 012039 doi:10.1088/1757-899X/294/1/012039

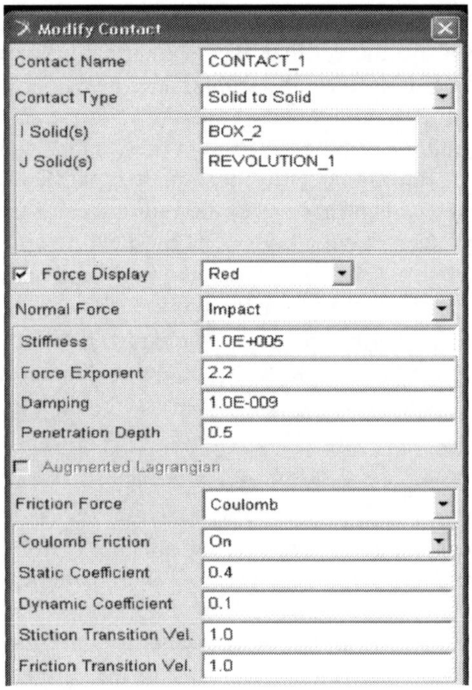

Figure 7. Tribological parameters for contact characterization

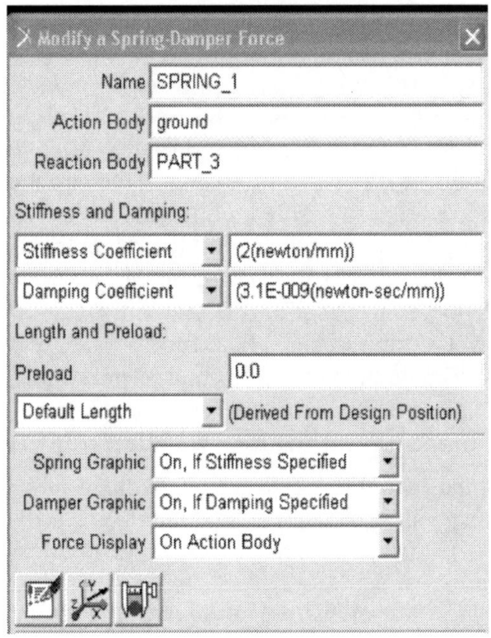

Figure 8. Characteristic parameters of helical spring

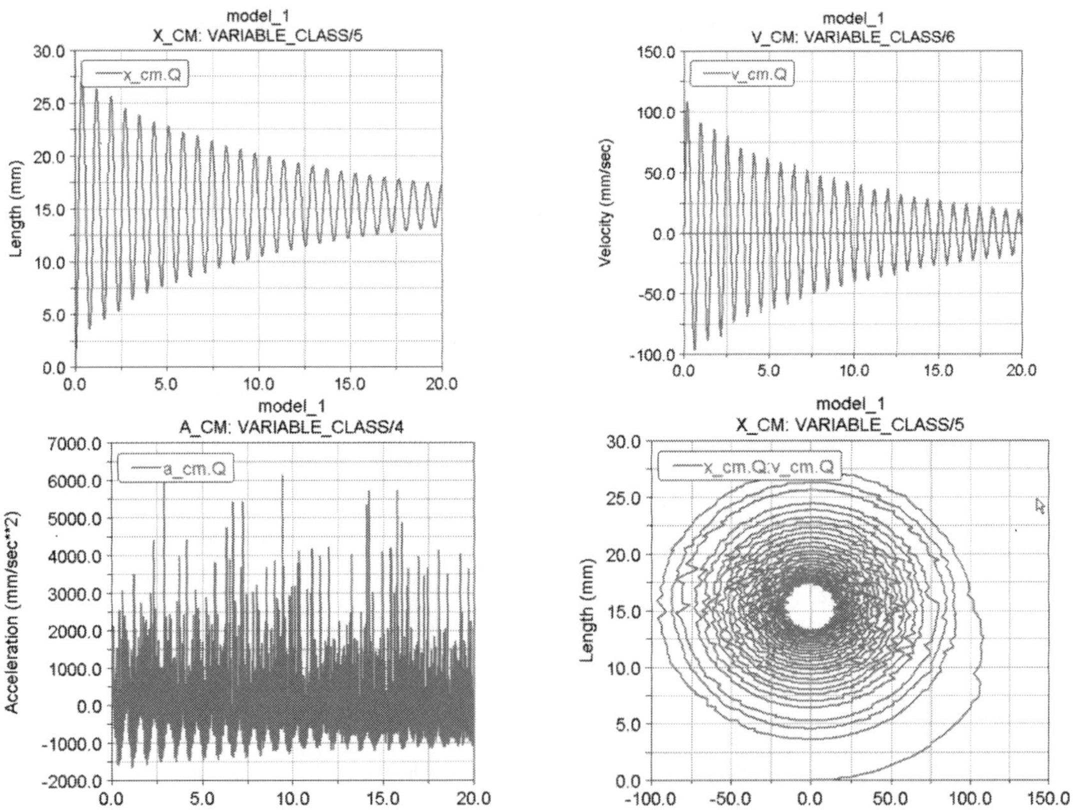

Figure 9. Elongation, velocity, acceleration and phase diagram plotted by the MSC.ADAMS software

4. Experimental validation of the correct solution

The fact that the two models, the mathematical one and the model obtained via a dynamic analysis software lead to completely different behaviours from a qualitative point of view – the mathematical model prognoses an oscillatory motion, while the dynamic simulation provides a damped motion, imposes the experimental validation of the correct model. To this purpose, an experimental device was designed and executed, consisting in a disc run into rotation via a d.c. electric motor with worm gear speed reducer. The prismatic body was replaced by a bronze cylinder placed on the horizontal surface of the disc. The need of employing a soft spring was observed after the experimental adjustments. Therefore, elastic belts of different lengths were used, Figure 10. It was observed that for a suitable selection of the length of the elastic belt and for a specified rotation velocity of the disc, the body presents periodical un-damped motion, and this confirms the correctness of the mathematical model. Because the mobile body and the disc have comparable radii, the assumption of constant relative velocity on the contact surface is not valid. A consequence resides in the occurrence of the moment of friction forces and of an oscillatory rotation motion of the cylindrical body, in addition to the translation motion.

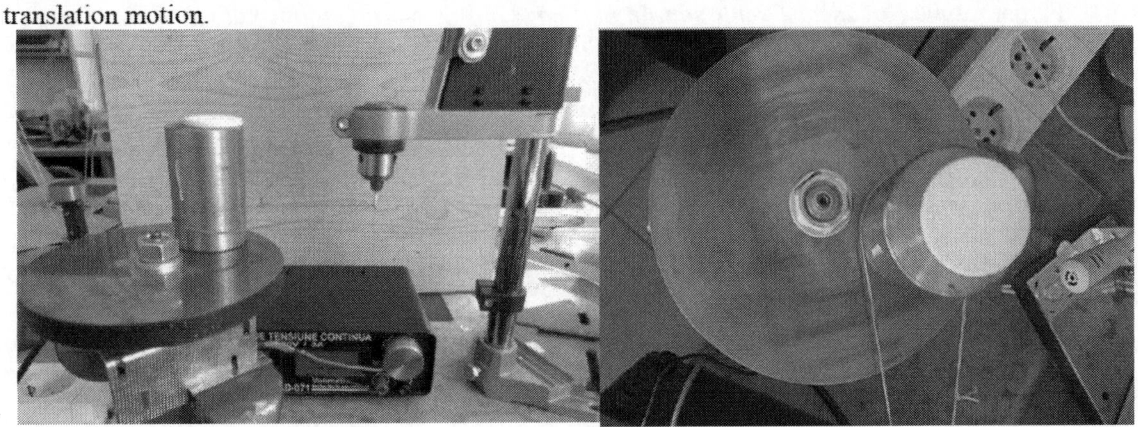

a) b)

Figure 10. Experimental test-rig: a) general view; b) top view

5. Conclusions

The main purpose of the paper consists in highlighting the importance of correct stipulation of the parameters characterizing the friction forces that occur in the pairs of a mechanical system. To this end, a simple system is considered, consisting in a horizontal conveyor belt on top of which a body - attached to the ground via a helical spring, can be in motion. Dry friction contact is assumed between the mobile body and the belt.

In the first part of the work, a law for the dependency of the coefficient of friction versus relative velocity between the belt and the body is proposed. The actual form of the function was chosen to ensure a well established value for the coefficient of friction for any velocity, but this is in opposition to the actual case when, for zero relative velocity, the friction force can take any value from a certain domain. For the proposed model of friction, the nonlinear differential equation that describes the motion of the system is obtained. After numerical integration, the elongation, velocity, and acceleration versus time are traced together to the phase diagram. The conclusion emerging is that the motion of the body is a forced oscillatory motion with constant amplitude.

The same system was then modeled utilizing specialized software, with the mention that instead of the conveyor belt, a disc in uniform rotation motion about its vertical axis was considered. This simulation, regardless of the values of the friction forces, leads to the conclusion that the body presents a damped oscillatory motion, contrary to the conclusion of the mathematical model. Moreover, the damping type of the amplitude of the motion is characteristic to fluid friction, not to dry friction.

An experimental se-up with a structure similar to the model used by the software was constructed, to determine which of the two models provides the correct solution. Experimental tests were carried out, and the motion of the mobile body proved to be an un-damped oscillatory motion, as the mathematical model predicts.

References

[1] Johnson K L 1985 *Contact Mechanics,* Cambridge University Press
[2] Popov V L 2017 *Contact Mechanics and Friction Physical Principles and Applications*, Springer
[3] Ludema K C 1996 *Friction, wear, lubrication: a textbook in tribology,* CRC Press
[4] Alaci S, Ciornei F C, Amarandei D, Irimescu L, Romanu I C and Rotar M A 2017 The effect of transport velocity upon spin torque, *IOP Conf. Ser-Mater. Sci. Eng.* **174** 012008
[5] Ciornei F C, Alaci S, Amarandei D, Irimescu L, Romanu I C and Acsinte L I 2017 Method and device for measurement of dynamic viscosity, *IOP Conf. Ser-Mater. Sci. Eng* **174** 012041
[6] Fidlin A 2006 *Nonlinear Oscillations in Mechanical Engineering*, Springer
[7] Leine R I and Nijmeijer H 2004 *Dynamics and Bifurcations of Non-Smooth Mechanical Systems,* Springer
[8] Coulomb C A 1785 Theorie des machines simples, *Mem. Math. et Phys. l'Acad. Sci.***10** 161-331
[9] Euler L 1748 Sur la diminution de la resistence du frottement *Histoire de (Academie Royale des Sciences et Belles Lettres a Berlin)*
[10] Kragel'skiĭ I V and Shchedrov V S 1956 *The Development of Science on Friction. Dry Friction,* Akad. Nauk SSSR Moscow [in Russian]
[11] Alaci S, Ciornei F C, Filote C, Romanu I C and Nastaca C P 2016 Determination of the coefficient of friction using spinning motion, *IOP Conf. Ser-Mater. Sci. Eng*. **147** 012024
[12] Hamming R 1987 *Numerical Methods for Scientists and Engineers,* Dover Publications
[13] Giesbers J 2012 *CONTACT MECHANICS IN MSC ADAMS A technical evaluation of the contact models in multibody dynamics software MSC Adams,* University of Twente, Bachelor Thesis
[14] Ciornei F C and Alaci S 2016 Aspects Regarding Evaluation of Friction Effects in Robotic Systems, *Robotica &Management* **21-1** 10-15
[15] Ciornei F C, Alaci S, Bedrule R I and Drelciuc C M 2013 The coefficient of friction, a basic parameter in design and operation of robot grasping, *Robotica & Management* **18-2** 13-16

Determination of maintainability for Dacia 1304, 1,9 D utility vehicle

A Budiul Berghian, T Vasiu and C Birtok Baneasa

Politehnica University of Timisoara, Department of Engineering and Management, 5 Revolution Street, Hunedoara, 331128, Romania

E-mail: adina.budiul@fih.upt.ro

Abstract. The study analyses the ability to be maintained or rehabilitation of Dacia 1304, 1,9D utility vehicle. The paper comprises the determination of its maintainability using the Weibull++8 specialized software.

1. Introduction

Maintainability is defined by: "The ability of a device under conditions of use to be maintained or restored to perform its specified function when maintenance is made under imposed conditions with prescribed procedures and remedies. It is the design and installation parameter, facility and safety with which maintenance can be performed.

Maintainability is a design consequence and maintenance represents the way to operate through running operations to restore the system to the optimal parameters [1]. Maintainability is related to conception activities and maintenance is located in the operation.

Maintainability as a design feature must provide:

- as an element, subsystem and system to be maintained with a prescribed probability in operation by maintenance activities;
- establishing the size and frequency of maintenance activities;
- Establishing the maintenance costs, respectively the level of expenses generated by this activity, which does not exceed the approved level on the total product.

The maintenance of appliances and systems greatly influences their usability and reliability.

Maintenance must be characterized by the requirement for the restoration time, from interrupt to functioning, to be less than the specified (predetermined) time interval.

If it is noted with τ' the time for restoration (repair) and τ'' maintenance time, then:

- the ability of repair (restoration) = $P(\tau' \leq t_r)$;
- capacity for maintenance= $P(\tau'' \leq t_i)$;

in which τ' and τ'' are random sizes and (t_r, t_i) are the value of time that the repair and maintenance must be performed under the required conditions.

Industrial experience shows that it normally follows a lognormal distribution function. Distribution laws are mathematical expressions used to model over time the phenomena of good functioning and restoration.

Their determination is based on the processing of data obtained from exploitation. Once established, the level of maintenance can be determined at any time.

IOP Conf. Series: Materials Science and Engineering **294** (2017) 012040 doi:10.1088/1757-899X/294/1/012040

Figure 1. Dacia 1304, 1,9D utility vehicle

2. Determination of Maintainability for Dacia 1304 de1,9D utility vehicle

The restoration time of the Dacia 1304 is the result of the quality characteristics under the conditions of operation, maintenance, etc. In order to establish the law on the distribution of favorable running [2] and repaired time, a statistical observation is organized [3].

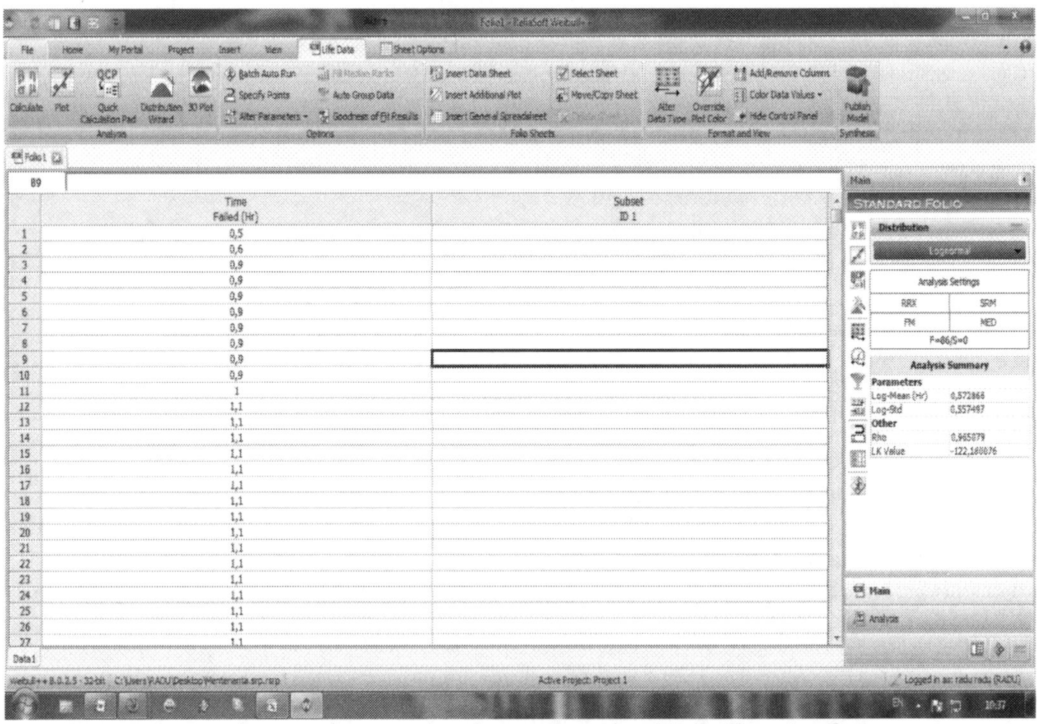

Figure 2. Interface used processing the results using Weibull ++8 software

For this, an 8-month observation program has been developed that includes data on: the times at which changes occur in the state or behavior of the entity, the timing of function after the last repair, the time of the interruption of operation, the moment of reversion operation, time of fall, moment of recommissioning after removal of causes of fall [4].

After processing the data obtained from running of Dacia 1304 utility vehicle using the Weibull ++8 specialized software, have been obtained the following results presented in graphic form in Figures 3, 4, 5 and 6.

Certification on real maintainability of the studied utility vehicle complies with the Lognormale distribution law can be seen in Figure 1, respectively the correlation coefficient are ρ=0,965079. Thus, mathematical expression of the maintainability distribution law for Dacia 1304, 1,9D is:

$$M(\tau) = 1 - \int_0^{\tau} \frac{1}{x\sigma\sqrt{2\pi}} e^{-\frac{(\lg x - \mu)^2}{2\sigma^2}} dx \tag{1}$$

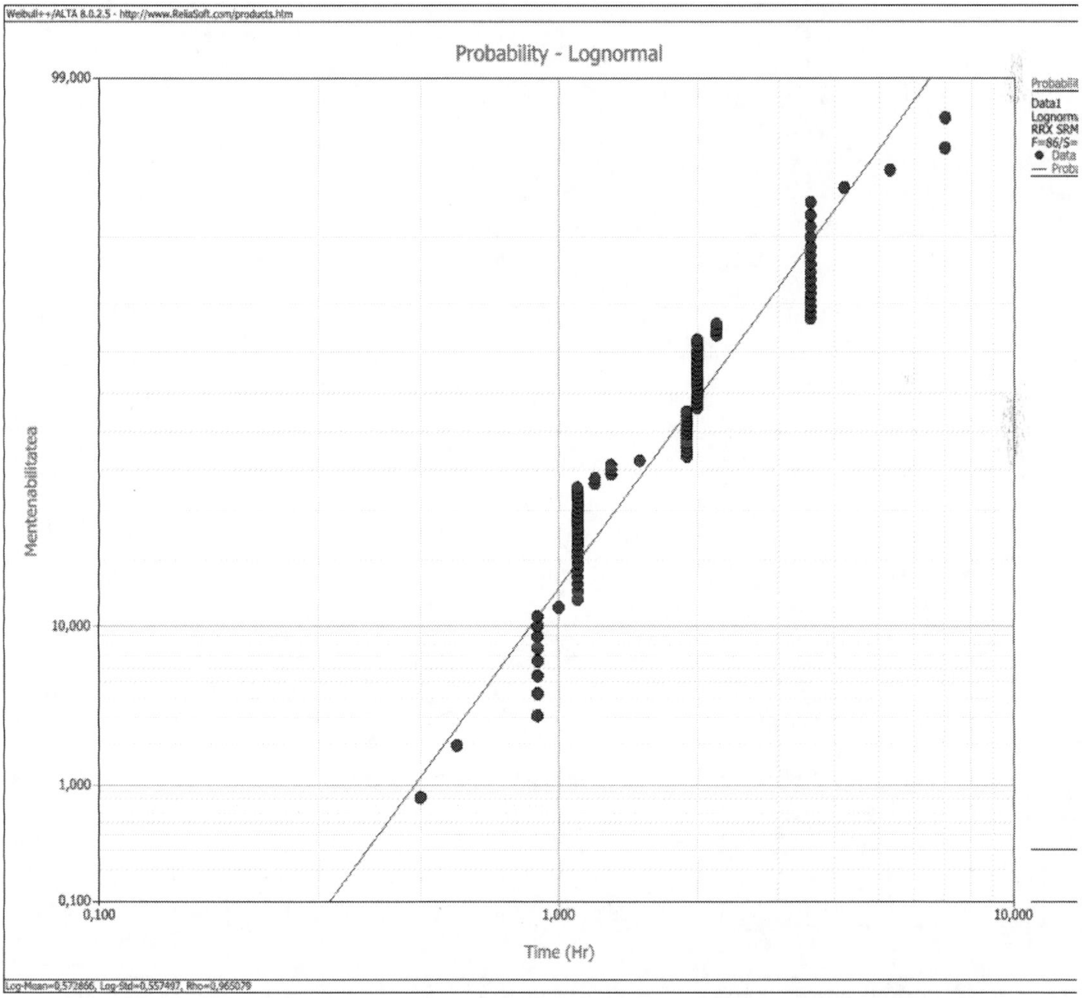

Figure 3. Graphic of maintainability depending on time in cartesian coordinates

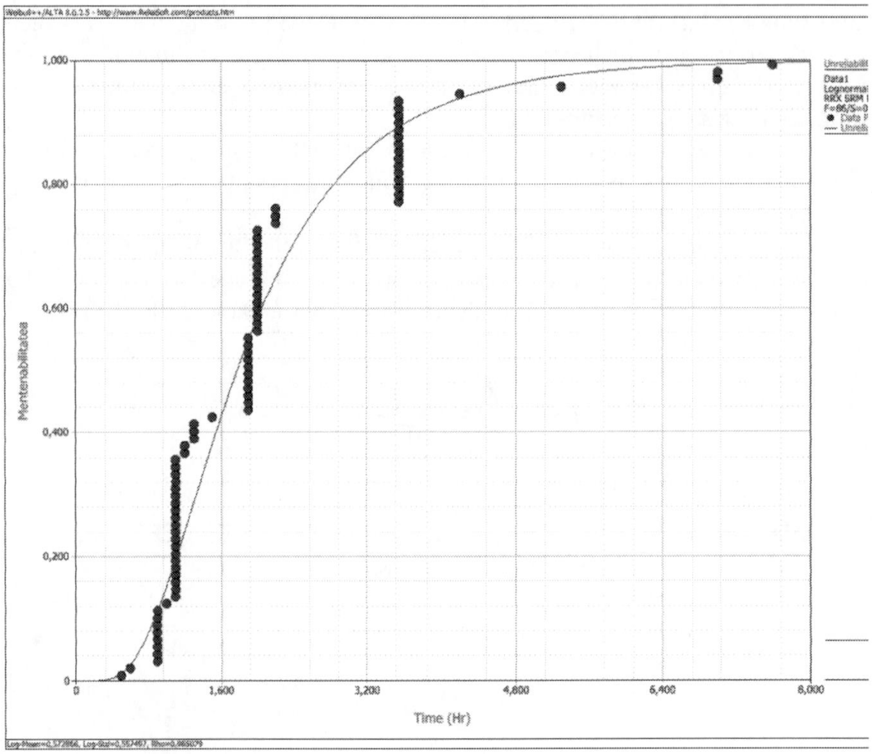

Figure 4. Graphic of maintainability depending on time

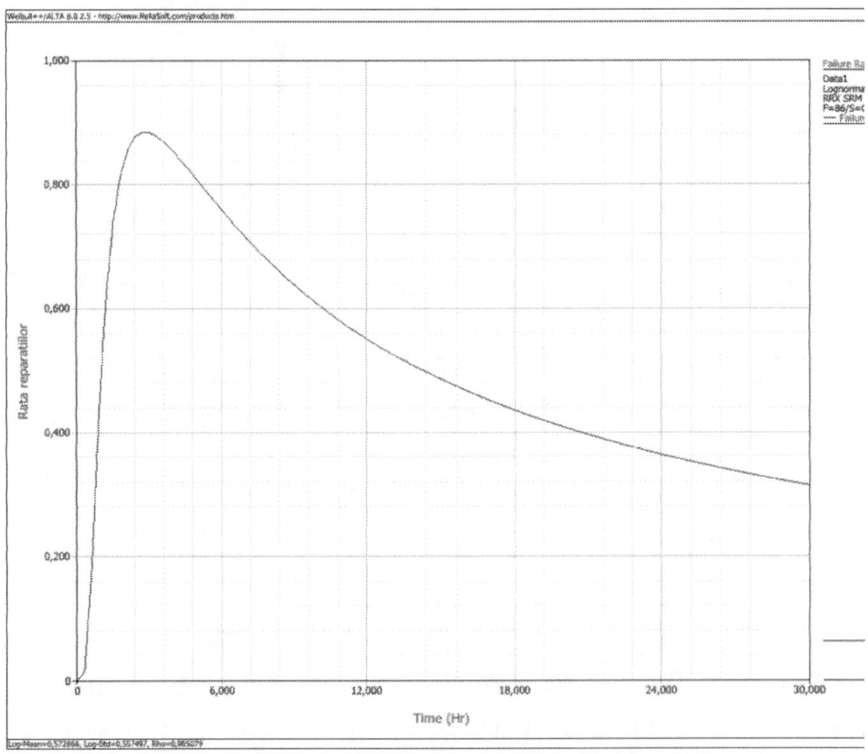

Figure 5. Graphic of rehabilitation rate depending on time

3. Conclusions

As far as *maintenance* is concerned, it is found that it complies with the lognormal distribution law. The time maintainability chart (Figure 4) shows that the service activity of the van is well established.

It is noted that the maintenance rate increases rapidly in approx. 5 hours. This means that the probability of restoring the car after the failure is approaching 1 during this time.

Regardless of mechanical or electrical problems in operation, they will be repaired within 5 hours at most.

References

[1] Roman I 1971 *Product Availability, Reliability and Maintainability in Production Quality and Metrology*

[2] Budiul Berghian A and Vasiu T 2015 Determination of Reliability for Dacia 1304, 1,9D, utility vehicule, *IOP Conf. Ser.: Mater. Sci. Eng.* **85** 012007

[3] Budiul Berghian A and Vasiu T 2008 *Reliability and maintainability of industrial entities. Theory and applications,* Publishing House: Editura "INFOMIN" Deva, Romania

[4] ***STAS 8174/1-77, *Reliability. Terminology*

Identification of exponent from load-deformation relation for soft materials from impact tests

F C Ciornei[1], S Alaci[1], I C Romanu[1], M C Ciornei[2] and G Sopon[1]

[1]"Stefan cel Mare" University of Suceava, Mechanics and Technologies Department, Universitatii str., no. 13, 720229 Suceava, Romania
[2]"Carol Davila" University of Medicine and Pharmacy, Department 2 Physiology I, Bd. Eroilor Sanitari, no. 8, 050474 Bucuresti, Romania

E-mail: florina@fim.usv.ro

Abstract. When two bodies are brought into contact, the magnitude of occurring reaction forces increase together with the amplitude of deformations. The load-deformation dependency of two contacting bodies is described by a function having the form $F = Cx^{\alpha}$. An accurate illustration of this relationship assumes finding the precise coefficient C and exponent α. This representation proved to be very useful in hardness tests, in dynamic systems modelling or in considerations upon the elastic-plastic ratio concerning a Hertzian contact. The classical method for identification of the exponent consists in finding it from quasi-static tests. The drawback of the method is the fact that the accurate estimation of the exponent supposes precise identification of the instant of contact initiation. To overcome this aspect, the following observation is exploited: during an impact process, the dissipated energy is converted into heat released by internal friction in the materials and energy for plastic deformations. The paper is based on the remark that for soft materials the hysteresis curves obtained for a static case are similar to the ones obtained for medium velocities. Furthermore, utilizing the fact that for the restitution phase the load-deformation dependency is elastic, a method for finding the α exponent for compression phase is proposed. The maximum depth of the plastic deformations obtained for a series of collisions, by launching, from different heights, a steel ball in free falling on an immobile prism made of soft material, is evaluated by laser profilometry method. The condition that the area of the hysteresis loop equals the variation of kinetical energy of the ball is imposed and two tests are required for finding the exponent. Five collisions from different launching heights of the ball were taken into account. For all the possible impact-pair cases, the values of the exponent were found and close values were obtained.

1. Introduction

The impact phenomenon is frequently encountered both in everyday life and in applied engineering, and its main characteristic is the sudden variation of the velocity of the components of a system but with a less evident change of their position. Two stages can be identified during an impact phenomenon: *the compression phase* defined as the time period between the instant when the first points of the two bodies come into contact, and the instant when the normal approach between the bodies reaches maximum, with the immediate consequence that at this moment the relative velocity between the two bodies is zero. The second stage, *the restitution phase*, starts once the compression phase ends and lasts until the instant when the last two points of the two contacting bodies detach.

In daily life the impact phenomenon happens any time when two solid bodies make direct contact. The affirmation is based on the observation that avoiding impact imposes that the two bodies are brought into contact at an infinitely small speed. This aim is practically impossible to attain and therefore when two bodies are contacting, the relative velocity is non zero and decreases rapidly, the order of impact time being in the range $10^{-3} \div 10^{-4}\,sec$ [1]. The sudden velocity variation has as consequence the development of appreciable forces characterized by accelerations within the domain $10^3 g \div 10^4 g$. In applied engineering the impact phenomenon occurs, for example, in gear boxes when the rotation sense of a shaft is changed, in all joints with clearance [2-4] etc. Two manners of approach are distinguished in technical literature concerning the impact phenomenon. The first methodology accepts the hypothesis of rigid bodies, so the method is uncomplicated but has the drawback that the impact is considered instantaneous and the impact force cannot be estimated. The second manner accepts the assumption of deformable bodies and permits the identification of all kinematical and dynamical parameters of the system. A requisite parameter in the study of impact phenomenon is e, the normal coefficient of restitution, defined as the ratio between $\boldsymbol{v}''$, the normal component of the post-impact relative velocity and $\boldsymbol{v}'$, the normal component of the pre-impact relative velocity:

$$e = -\frac{\boldsymbol{v}''\cdot\boldsymbol{n}}{\boldsymbol{v}'\cdot\boldsymbol{n}} \tag{1}$$

where $\boldsymbol{n}$ is the unit vector of the normal to the contacting surfaces.

2. Impact phenomenon modelling under the deformable bodies hypothesis

In impact modelling, besides the fundamental theorems of dynamics, a series of supplementary hypothesis must be considered for establishing the manner in which certain kinematical or dynamical parameters of the analyzed system change. The general study of the impact between two bodies of random shape is complex due to the complicate form of the equations describing the phenomena occurring during the entire process. The intricate form of differential equations makes it impossible to obtain an analytical solution and therefore numerical methods are applied in obtaining a particular solution. In order to obtain analytical solutions, many of the studies concerning the subject consider the impact between bodies having simple geometries. Timoshenko [5] applies the results from Hertzian contact theory and finds the maximum force, maximum normal approach and the impact time between two elastic spheres. Even for this very simple model, the consideration of some aspects which take it closer to an actual situation lead to a complexity difficult to surpass. Goldsmith [6] shows that in the case of impact, the kinetical energy variation is found as a work of internal friction, plastic deformation work, energy of elastic waves propagating through colliding bodies etc. Lankarani [7] proposes a nonlinear differential equation to describe the impact with internal friction between two metallic balls. To obtain the differential equation of the model, the hypothesis that the work of internal friction has the same value both for the compression phase (from contact initiation until the moment of maximum normal approach) and for the restitution phase is accepted. The hypothesis restricts the application of the model only for quasielastic bodies. Recently, Flores [8] corrected the Lankarani equation, starting from the assumption that in the phases plane, a current point moves on an elliptical trajectory, making it applicable for any type of impact. Both models, Flores and Lankarani, are based on the assumption that the impact phenomenon is not accompanied by plastic deformations. Considering the observation made by Goldsmith [9] who cites Davies, this hypothesis is hard to accept; Goldsmith shows that a $1/2inch$ diameter steel ball produces a plastic imprint on steel if the impact velocity is approximately $3in/sec \approx 0.076m/s$, corresponding to a free fall from a height of $0.3mm$. Johnson [10] confirms that the plastic deformation following the impact between a ball in free fall and an immobile horizontal plane requires a launching height of $\approx 1mm$. These experimental observations confirm the necessity of a model that considers the dissipation of impact kinetical energy as work of plastic deformation. In a recent work, Anderson [11] proposes a model for the study of impact considering that the variation of kinetical energy of the system is retrieved as work of internal friction and work of plastic deformation, and experimentally validates the presence of the two types of work for the case of an automobile impacting a fixed obstacle. The hysteresis loop for the case of impact with internal friction and plastic deformation is presented in Figure 1.

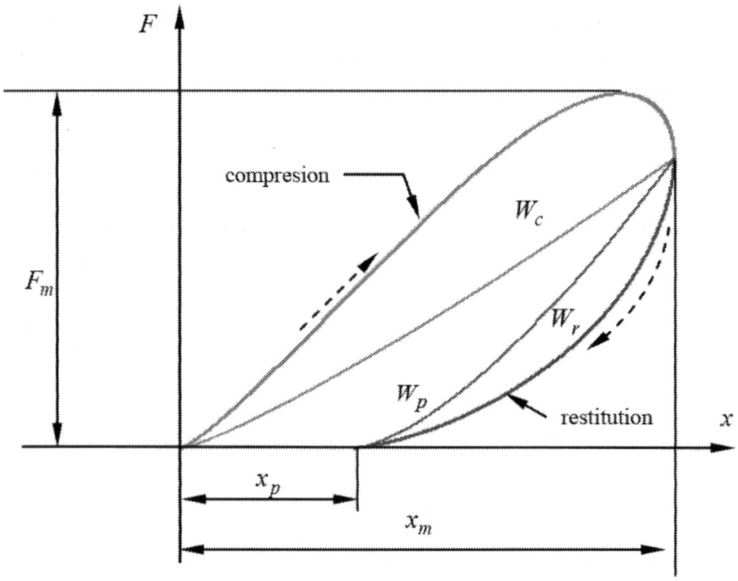

Figure 1. Hysteresis loop for the case of impact with internal friction and plastic deformation

The shape of the hysteresis curve is influenced by the ratio between the values of internal friction work $W_c + W_r$ and the plastic deformation work W_p and two limit cases can be distinguished: collision without plastic deformation, and collision only with plastic deformation. Goldsmith [6] presents the experimental hysteresis loops for several pairs of materials and identifies the cases mentioned above.

3. Theoretical considerations

A model for impact considering only plastic deformations was proposed by Lankarani [12]. More explicitly, he accepts the hypothesis that the variation of kinetical energy of a system consisting of two colliding balls is recovered as work of plastic deformation.

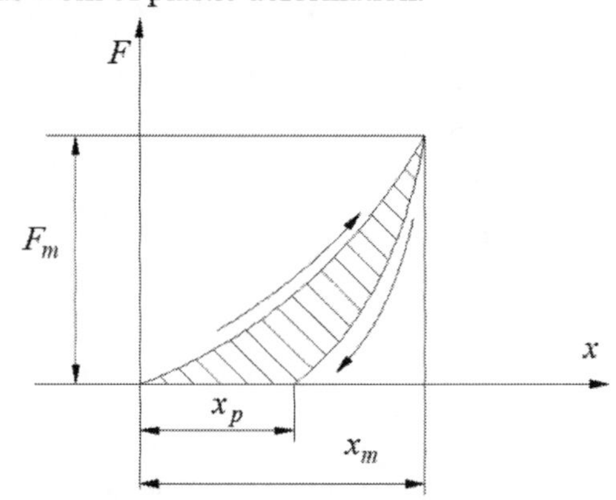

Figure 2. Impact force variation for Lankarani model

Noting by x the normal approach between the two balls and by x_m and x_p the values of maximum approach and permanent plastic deformation, respectively, Lankarani proposes for the interaction force represented in Figure 2 the expression:

$$F = \begin{cases} Cx^{3/2}, & 0 \le x \le x_m \\ C'\left(\dfrac{x - x_p}{x_m - x_p}\right)^{3/2}, & x_p \le x \le x_m \end{cases} \tag{2}$$

Assuming the geometrical and elastic characteristics of the balls are known, the initial relative velocity v' and the coefficient of restitution e, the expressions for maximum approach x_m and permanent deformation x_p can be found with the proposed model. It is obvious that for small impact velocities, the shape of hysteresis loops should be identical to the ones obtained for quasistatic contact. The hysteresis loops traced for both static and dynamic regime, presented by Goldsmith [6], confirm this hypothesis. In a recent work, Alaci [13] finds for the impact between a ball in free fall and a fixed horizontal surface, both made of steel, the coefficient of restitution and the depth of impact imprint and shows that the values of the residual imprint depth are approximately twice greater than the ones given by the Lankarani model. An analysis of the experimental results presented by Goldsmith, Figure 2, reveals that when one of the bodies is made of a soft material, the curves of impact force variation present a similar shape to the curves given by Lankarani's model. However, when both bodies are made of hard materials, the shape of the hysteresis curves changes and resembles the curves presented by Anderson [11]. In the curves given by Anderson the effect of internal friction cannot be neglected any longer. A first solution for finding the exponent from the force-deformation dependence consists in using the force-deformation curve obtained for quasistatic compression. The weak point of the method, as shown in [14], [15] is the poor accuracy in finding the instant when the contact initiation takes place. To avoid this drawback, a method is proposed for finding the law of impact force variation using hysteresis curves in the case of impact at low speeds.

4. New theoretical model

The proposed theoretical model is appropriate for the case of collision of soft bodies, characterised by reduced values of the coefficient of restitution and neglecting the effects of internal friction. The law of variation for the impact force is based on the remarks made by Goldssmith [9] and Tabor [16].

Thus, the impact force variation is according to the relation:

$$F = \begin{cases} Cx^{\alpha}, & 0 \le x \le x_m \\ C'\left(\dfrac{x - x_p}{x_m - x_p}\right)^{\beta}, & x_p \le x \le x_m \end{cases} \tag{3}$$

For the compression phase, the variation of the impact force obeys the Meyer's law and the α exponent:

$$1 \le \alpha \le 3/2 \tag{4}$$

takes the value $\alpha = 3/2$ for the case of perfect elastic materials and $\alpha = 1$ for the impact between perfect plastic materials. The assumption that during the restitution phase the plastic deformations do not occur any longer implies that:

$$\beta = 3/2 \tag{5}$$

Applying the theorem of kinetic energy for the compression phase, it results:

$$-\frac{1}{2} m_{red} v'^2 = -\int_0^{x_m} F\,dx = -\frac{C x_m^{\alpha+1}}{\alpha + 1} \tag{6}$$

For the restitution phase, the work energy theorem leads to:

$$\frac{1}{2}m_{red}(e\cdot v')^2 = -\int_{x_m}^{x_p} C'\left(\frac{x-x_p}{x_m-x_p}\right)^{\beta} dx = \frac{C'(x_m-x_p)}{\beta+1} \tag{7}$$

where m_{red} is the reduced mass of the system:

$$m_{red} = \frac{m_1 m_2}{m_1+m_2} \tag{8}$$

The condition that the maximum force should be the same for both compression and restitution phase must be added to the equations (6) and (7):

$$C x_m^{\alpha} = C' \tag{9}$$

The relations (6), (7) and (9) are the available equations for an impact test. The values of the coefficients C and C', the exponent α and the maximum normal approach x_m should be found in order to make possible the use of the relation (3). To evaluate the unknown parameters from relation (3), experimental tests were made using a bearing ball, 12.7mm diameter, as projectile, and a disc made of unhardened tool steel as target. Five launchings were made and the coefficient of restitution e was established by measuring the time elapsed between the first and the second impact of the ball with the target. The depth of the plastic imprint was measured using a laser scanner, Figure 3, and the profile of the imprint was approximated with a parabolic function as shown in Figure 3b. The results of the five tests are presented in Table 1.

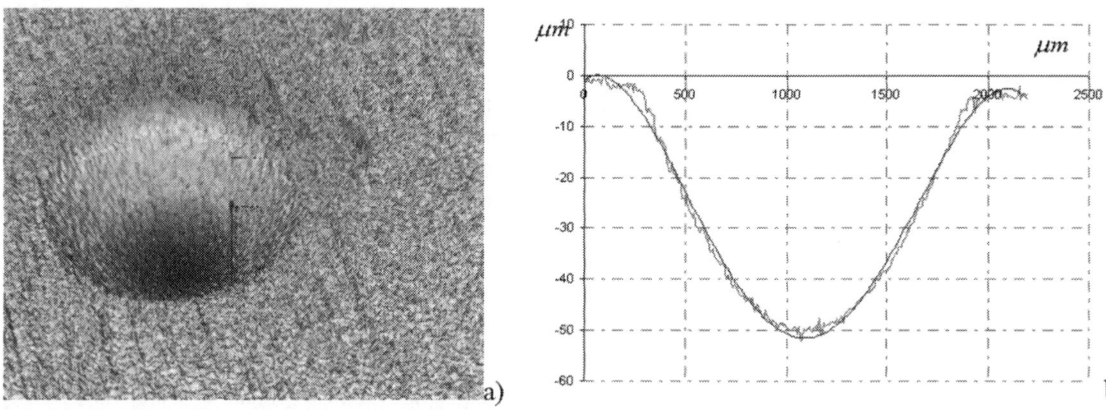

Figure 3. Evaluation of the depth of the plastic imprint

Table 1. Experimental impact parameters

Test no.	v' [m/sec]	e [−]	x_p [µm]
1.	2.334	0.530	23.5
2.	2.940	0.512	29.7
3.	4.089	0.479	41.6
4.	4.504	0.467	45.97
5.	4.993	0.453	51.03

It is assumed that for each impact test the loading is made on the same curve and the constants C and α have to be established. The unloading happens on different curves having the same exponent

$\beta = 3/2$ but with different values of the coefficient C' and depth x_{max}. Writing the equations for all $k = 1 \div 5$ collisions, the following is obtained:

$$\begin{cases} \dfrac{C x_{m_k}^{\alpha+1}}{\alpha+1} = \dfrac{1}{2} m_{red} {v'_k}^2 \\ \dfrac{C'_k (x_{mk} - x_{p_k})}{\beta+1} = \dfrac{1}{2} m_{red} (e_k \cdot v'_k)^2 \\ C x_{m_k}^{\alpha} = C'_k \end{cases} \quad (10)$$

The system has $3k$ equations and $2k+2\ (C, \alpha, x_{m_k}, C'_k)$ unknowns. For $k > 2$ the system becomes incompatible and requires finding a solution that verifies the system with minimum error. This is a difficult task considering the form of the equations and the great number of unknowns. Therefore, another method for finding the approximate solution of the system is proposed here. For $k = 2$ the system is compatible. Writing the system of equations for $k = 2$ and eliminating at each step the unknowns, finally a single transcendental equation is obtained, with unknown α:

$$\frac{\alpha+1}{\beta+1} + \frac{x_{p_1}}{x_{p_2}} \left(e_2^2 - \frac{\alpha+1}{\beta+1} \right) \left(\frac{v'_2}{v'_1} \right)^{\frac{2}{\alpha+1}} - e_1^2 = 0 \quad (11)$$

There is no difficulty in solving this equation. Once the α parameter found, the other parameters are evaluated using the relations:

$$x_{m_2} = x_{p2} \frac{1}{1 - \frac{\beta+1}{\alpha+1} e_2^2}; \quad x_{m_1} = x_{m2} \left(\frac{v_1}{v_2} \right)^{\frac{2}{\alpha+1}};$$
$$C = \frac{m_{red} v_2^2}{2} \frac{\alpha+1}{x_{m_2}^{\alpha+1}}; \quad C'_1 = C x_{m_1}^{\alpha}; \quad C'_1 = C x_{m_1}^{\alpha}. \quad (12)$$

Combining the five tests in all possible cases, a number of 10 values results for parameter α, given in Table 2. With these values, a row of points was generated (x_j, F_j) for each case with x_j taking values up to the maximum value of maximum approach found for a pair of collisions.

Table 2. Values of α parameter

1.234	1.275	1.286	1.299	1.303	1.315	1.327	1.355	1.367	1.379

All these points are interpolated using a function having the form Cx^{α}, and the final values for C and α are established. It is noticed that the value found for the α exponent by interpolation, $\alpha = 1.313$ is actually the same as the mean value of the values from Table 2, $\alpha = 1.314$. With C and α now found, the maximum approach values are found using relation (6).

The values of coefficient C'_k can be estimated either from relation (7) or from (9), using the maximum approach values. In Table 3 there are presented the values of maximum approach x_m and the coefficient C' found via both manners.

Table 3. Calculated maximum approach x_m and coefficient C'

k	1	2	3	4	5
x_m [μm]	34.08	41.61	55.34	60.16	65.77
C', (eq. 6)	1516	2009	2935	3257	3622
C', (eq. 8)	1547	2010	2924	3263	3668

The correctness of the methodology used in finding the C coefficient is validated by the practically equal values of the C' coefficient found in two different manners. The hysteresis loops were represented on the same graph for the five cases considered in Table 3, as presented in Figure 4. The curves have an identical appearance to the ones presented by Goldsmith for the impact between a steel ball with *12.7mm* diameter and the frontal face of a cylinder made of unhardened steel. The difference between the plots in Figure 4 and the loops given by Goldsmith are the values of maximum forces and maximum plastic deformations, differences explained by the values of initial impact velocity v' which were smaller in the test of present work, the launching being made via free fall, while for Goldsmith's experiments special launching devices were employed.

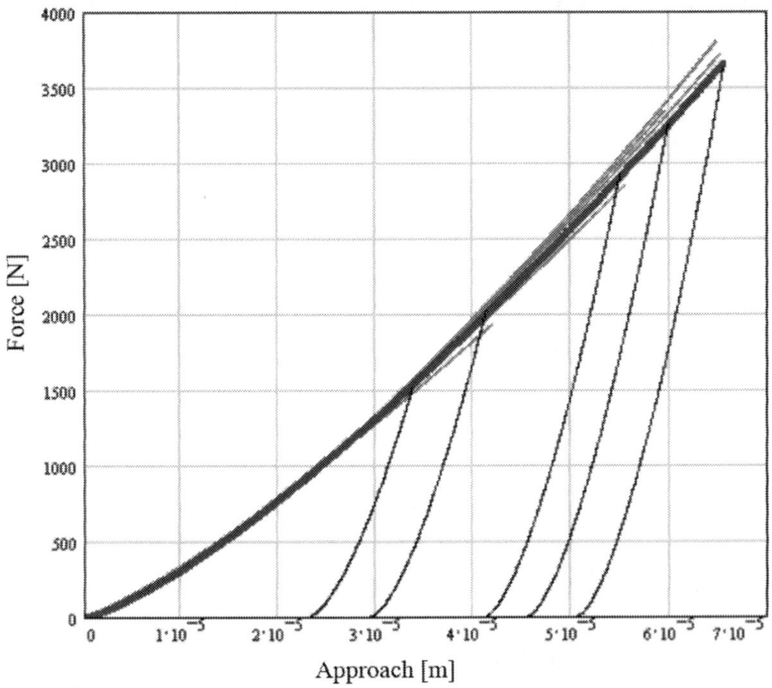

Figure 4. Generated points, the interpolation curve for compression and the restitution curves

5. Conclusions

The paper presents methodology for establishing the exponent from the force-deformation expression used for the compression phase of an impact between two bodies made of soft materials. The method is proposed as alternative to the classical method that assumes a quasistatic compression test for a nonconforming contact. The disadvantage of the classical method consists in the errors introduced due to imprecision in finding the instant of contact initiation. The proposed method is based on the remark that, for a system consisting from two metallic bodies that collide, the variation of kinetical energy of the system during impact is retrieved as work of internal friction and work of plastic deformation. A random collision should be placed between two limit models: a first model, for which the kinetical energy variation is converted into work of internal friction, and the second, for which the entire kinetical energy variation is transformed into work of plastic deformation. The second limit case is applicable when one of the impacting bodies is made of a soft material.

Admitting the hypothesis of complete transformation of kinetical energy variation into plastic deformation work, a model of impact is proposed for which the compression is obtained following a power law curve with exponent to be found, and the restitution obeys a power law curve with the exponent 3/2 that is characteristic to elastic domain. In order to find the exponent related to the compression phase, two impact tests between the bodies to be tested are required. The values of coefficient of restitution and the depth of plastic imprint must be evaluated for the two collisions. The coefficient of restitution is found via an acoustic method and the dept of plastic imprint is measured using laser profilometry. For experiments, a bearing ball made of hardened steel, with diameter of 12.7mm, falls free on the flat surface of an immobile prism made of mild steel. Five launchings of the ball were made from different heights. The values of the exponent and the coefficient from the impact force relation for restitution phase were found for all the combinations of two possible experiments. With these values, the hysteresis curves were represented on the same graph and the curves obtained have similar shapes to the experimental curves from literature.

The proposed method has the advantage of rapidity, and promptitude may be further improved if, for finding the depth of plastic imprint, the laser profilometry is replaced, and the maximum diameter of plastic imprint is measured, the depth resulting from the supposition that the residual imprint is a spherical cap.

References

[1] Alaci S, Ciornei F C, Filote C and Românu I C 2015 Analysis of Postimpact Plastic Indentation Profiles, *Applied Mechanics and Materials* **809-810** 223-228

[2] Flores P, Ambrósio J, Claro J C P and Lankarani HM 2008 *Kinematics and Dynamics of Multibody Systems with Imperfect Joints: Models and Case Studies,* vol. 34, Springer Berlin

[3] Flores P 2009 Modeling and simulation of wear in revolute clearance joints in multibody systems, *Mech. Mach. Theory* **44**(6) 1211-1222

[4] Flores P, Ambrósio J and Claro J P 2004 Dynamic analysis for planar multibody mechanical systems with lubricated joints, *Multibody Syst. Dyn.* **12** 47-74

[5] Timoshenko S 1970 *Theory of Elasticity,* McGraw-Hill College

[6] Goldsmith W 1963 *Impact at intermediate velocities involving contact phenomena,* Navweps report 8088 China Lake California

[7] Lankarani H M and Nikravesh P E 1990 A contact force model with hysteresis damping for impact analysis of multibody systems, *J. Mechanical Design.* **112** 369-376

[8] Flores P, Machado M, Silva M T and Martins J M 2011 On the continuous contact force models for soft materials in multibody dynamics, *Multibody System Dynamics* **25** 357-375

[9] Goldsmith W 1960 *Impact, The Theory and Physical Behaviour of Colliding Solids*, Edward Arnold, Sevenoaks

[10] Johnson K L 1985 *Contact Mechanics*, Cambridge University Press

[11] Anderson R W G, Long A D and Serre T 2009 Phenomenological continuous-impact modeling for multibody simulations of pedestrian-vehicle contact interactions based on experimental data, *Nonlinear Dyn.* **58** 199–208

[12] Lankarani H M and Nikravesh P E 1994 Continuous contact force models for impact analysis in multibody systems, *Nonlinear Dyn.* **5** 193–207

[13] Alaci S, Ciornei F C and Filote C 2014 Some Consideration Regarding the Models for Collisions with Plastic Indentation, *Applied Mechanics and Materials* **658** 161-166

[14] Ciornei F C and Alaci S 2016 Characterization of rubbers from spherical punch - plate indentation tests, *Robotica &Management* **21**(2) 10-14

[15] Ciornei F C, Alaci S and Besliu S 2015 Upon interpolation of contact tests experimental data from dynamical systems, *New Techn. and Prod. in Mach. Manuf. Techn.* **22** 544-548

[16] Tabor D 1951 *The Hardness of Metals,* Oxford University Press

Surface treatments for aluminium alloys

M Ardelean[1], S Lascău[2], E Ardelean[1] and A Josan[1]

[1]Politehnica University of Timisoara, Department of Engineering and Management, 5 Revolution Street, Hunedoara, 331128, Romania
[2]Anorom Hunedoara, Hunedoara Industrial Park, DJ 687 nr. 2A, 331004, Hunedoara

E-mail: marius.ardelean@fih.upt.ro

Abstract. Typically, in contact with the atmosphere, the aluminium surface is covered with an aluminium oxide layer, with a thickness of less than 1-2μm. Due to its low thickness, high porosity and low mechanical strength, this layer does not protect the metal from corrosion.
Anodizing for protective and decorative purposes is the most common method of superficial oxidation processes and is carried out through anodic oxidation.
The oxide films, resulted from anodizing, are porous, have a thickness of 20-50μm, and are heat-resistant, stable to water vapour and other corrosion agents.
Hard anodizing complies with the same obtains principles as well as decorative and protective anodization. The difference is in that hard anodizing is achieved at low temperatures and high intensity of electric current.
In the paper are presented the results of decorative and hard anodization for specimens made from several aluminium alloys in terms of the appearance of the specimens and of the thickness of the anodized.

1. Introduction

Some metals and alloys are protected against corrosion by coating with inorganic layers (films) made by chemical and electrochemical processes. In the industry are frequently applied films of oxides, phosphates and chromates.

Oxide and chromate films are used for decorative finishing of parts, carcasses of measuring and control devices. By coating with oxides and phosphates, are improved some properties of metal surfaces such as hardness, wear resistance, electro-insulating properties, etc.

Aluminium and its alloys are one of the most important categories of materials used by modern technology due to the advantages of low specific gravity, high thermal and electrical conductivity, good corrosion resistance, good mechanical features, easy machining, etc. Today, Aluminium -based alloys are the world's second metallurgical product, after steel, although the knowledge history of this metal is shorter than two centuries [1]. In the form of pure metal, aluminium is less used, but in the alloyed state it is the most used non-ferrous metal.

The aluminium trioxide film, which is formed, even at ambient temperature, is compact, waterproof and strongly adherent to metal. This makes aluminium good corrosion resistance, even in aggressive environments such as organic acids, seawater, humid atmospheres, etc. Aluminium is slightly attacked by nitric acid and sulfuric acid, instead hydrogen chloride and bases easily dissolve it. The presence of certain impurities significantly reduces the resistance of aluminium to corrosion [2].

The aluminium trioxide film, which is formed even at ambient temperature, is compact, waterproof and strongly adherent to metal. This gives to aluminium a good corrosion resistance, even in aggressive environments such as organic acids, seawater, humid atmospheres, etc. Aluminium is slightly attacked by nitric acid and sulfuric acid, instead hydrogen chloride and bases easily dissolve it.

In order to be easily recognized, Aluminium Association and the American National Standard Institute has proposed a system of symbolization of aluminium alloys that has been rapidly assimilated by all aluminium producing countries [3].

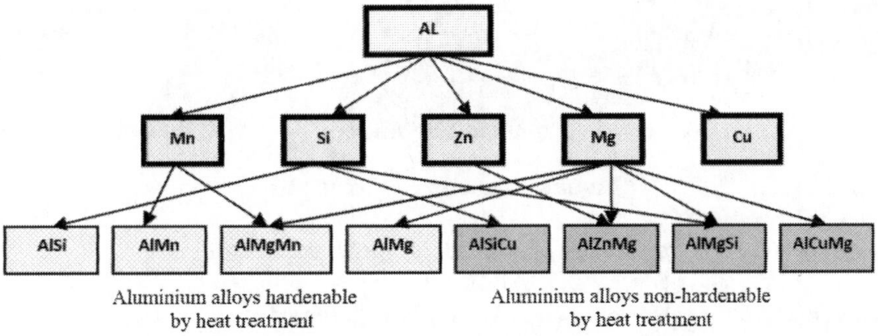

Figure 1. General classification of aluminium alloys

Another classification of aluminium alloys was also made according to the main alloying element by associating them in eight main series, each series containing a four-digit number [3], [4].

2. Surface treatment of aluminium alloys

Inorganic layers are used for the protection of metals subjected to atmospheric corrosion. Oxidation and phosphating are performed to obtain intermediate layers before applying paint layers.

Oxide and chromate films are used for decorative finishing of parts, carcasses of measuring and control devices. By coating with oxides and phosphates [5], are improved some qualities of metallic surfaces such as hardness, wear resistance, electro-insulating properties, etc.

Of the most common surface treatment processes for aluminium are anodizing, hard anodizing, chromating and bright dip (chemical polishing).

2.1. Anodizing

The process of obtaining oxide films on the surface of aluminium parts or its alloys by anodic oxidation is called anodization. Anodizing for protective and decorative purposes is the most common of superficial oxidation processes [6].

The films of oxide formed by anodization, are porous, have a thickness 20-50 microns, are heat resistant, stable to the action of water vapour and other corrosive agents. The biggest disadvantage of these films are limited elasticity and high hygroscopicity [7].

The aluminium oxide film qualities are improved by introducing the anodized pieces into the 90°C water. This process is called sealing the pores. The sealing of the pores is determined by the hydration of aluminium oxide, with the formation of crystals of type $Al_2O_3 \cdot nH_2O$, that having low specific weight and high volume, sealing the pores from this layer.

Oxide films can be impregnated with various substances, such as paraffin, insulation paints and some polymers. Impregnation reduces the hygroscopicity of the film and increases corrosion resistance [8]. Aluminium oxide is very porous, has a high absorption capacity of organic and inorganic dyes and anti-corrosion paints, thus achieving decorative effects on the surface of the pieces [9].

Colour anodized aluminium is achieved by deposition of coloured particles in the pores of the anodized layer (electrolytic staining) after the anodization phase and before the sealing phase.

Most often, as an electrolyte is used a solution of 10-20% H_2SO_4 at 15-30°C. It works at a voltage of 12-15V achieving variable current densities, from a few A/dm^2 at first, up to several hundred A/dm^2 at the end [10].

Figure 2. Anodized pieces [10]

Sulfuric acid, strongly electrolyte, dissociates according to the equation:

$$H_2SO_4 \rightarrow 2H^+ + SO_4 \quad (1)$$

Reactions that occur at the electrodes of an electrolytic cell are:

- to cathode: $2H^+ + 2e^- \rightarrow H_2\uparrow$ (2)
- to anode: $2OH^- - 2e^- \rightarrow H_2O + O$

$$SO_4^{2-} - 2e^- \rightarrow SO_3 + O \quad (3)$$
$$SO_3 + H_2O \rightarrow H_2SO_4$$
$$2Al + 3O \rightarrow Al_2O_3$$

If it is desired to obtain thick oxide films with special properties, the purity of the aluminium subjected to oxidation must be very high.

2.2. Hard anodizing

Hard anodizing follows the same training principles as decorative and protective anodizing. The difference consists in the fact that hard anodizing is done at low temperatures and high currents. The pores of the hard anodized layer are smaller than the classic anodizing.

In this case, the anodized layer is very hard, abrasion-resistant but still remain porous. Due to this very hard layer of aluminium oxide, the hard anodized parts are suitable for wide range of applications, from military applications to a multitude of other applications such as parts for the electrical industry, mechanical parts, aluminium cookware, etc [11].

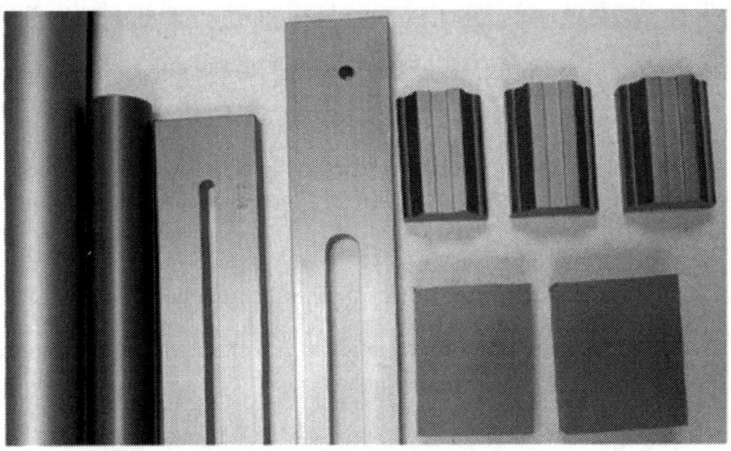

Figure 3. Different types of hard anodized parts [10]

Hard anodization is made in a sulfuric acid bath containing 180-200g/l of acid and a small amount of dissolved aluminium. The operating temperature is controlled between -5...0°C, but in some cases an acceptable oxide layer can be obtained at slightly higher temperatures.

The resulting porous oxide may then be coloured and sealed, but the final colour limitations are determined by the type of oxide and the used colour.

2.3. Chromating

Chromating is not an aluminium-specific treatment, chromating is specific to steel parts, where chromium-alloyed steels have good wear resistance, corrosion, and high temperature resistance [12]. Chromating can be applied not only to aluminium alloys but also to zinc, cadmium, copper, silver, magnesium and tin alloys.

For decades, hexavalent chromium treatment has been standard for aluminium. More recently, superficial treatments with these chromium-based products are subject to environmental, health and safety pressures.

Currently, the product used for chromating is SURTEC 650 (Tri-Crom aluminium passivation product - Best Available Technology for Aluminium Pre-treatment) [13]. The metallic surface of the aluminium piece turns into a stable barrier, mostly oxide layer. This layer prevents or slows down the corrosive attack, resistance to electric contact is small, improve the adhesion of the coatings, such as paints, sealing materials or some types of glues.

Chromating with SURTEC 650 is a passivation procedure for the surface of aluminium parts that ensure an excellent corrosion resistance up to 100°C. After treatment with SURTEC 650 the surface of the parts can be painted in electrostatic field, the chromium layer ensuring a very good adhesion of the paint.

2.4. Chemical polishing (Bright Dip)

Bright Dip is a chemical polishing process for the piece, and can be considered a preparatory operation for anodizing consisting in a chemical attack on the aluminium parts using a mixture of acids. The effect is partial and preferential dissolution of surface pieces resulting in a mirror surface of this. The method replaces electro-polishing or electroplating procedures for specialized articles such as car ornaments, picture frames, cosmetic containers and much more [14].

Problems related to aluminium cleaning are not simple, due to a sensitivity of the metal and its reactivity with many commonly used cleaning solutions. Are used an alkaline solutions, detergents or acidic cleaners, separately or together in various combinations.

Figure 4. Parts chemically polished [10]

Alkaline solutions are used to eliminate the drawings and traces of oils and glue. Even slightly alkaline cleaning solutions will attack Aluminium, especially at higher temperatures unless is used an inhibitor. An ideal cleaning cycle should include the acid cleaning as last operation.

Among the types of acids used for cleaning are:

- phosphoric acid, which leaves a clean surface without oxides and with a special quality of the surfaces of the piece;
- sulfuric acid, which requires corrosion inhibitors, without the parts would be strongly attacked;

- citric acid, which does not have great cleaning power, leaves the surface activated for the next processing steps. They are used where there are restrictions on the use of phosphoric and sulfuric acids.

After cleaning, the surface is activated is formed only from the base metal, it is not covered by any oxide and is prepared for the following operations.

3. Experiments made with aluminium alloys

Experimental tests were performed in SC ANOROM SRL consisting in colourless and black anodization versions with and without satin finish and hard anodization.

A quick review of the behaviour of the various alloys on anodization [15] is presented below:

- 1000 alloy series (pure aluminium). They respond well to anodizing treatment. The anodized layer is the more transparent as the aluminium is pure;
- 2000 alloy series (copper alloys). The copper from this alloy dissolves during the treatment, resulting after anodization an unaesthetic area (milky) [16]. The colouring of this alloy is also bad due to copper that prevents the aluminium oxide layer being uniformly deposited, that also leads to the decreasing of surface hardness and abrasion resistance;
- 3000 series alloys (manganese alloys). At concentrations above 1% Mn the anodized layer is bluish grey;
- 5000 alloys series (magnesium alloys). It performs well with anodizing. Up to a concentration of 3% Mg the anodized layer is colourless. If the magnesium concentration is greater than 3%, the anodized layer becomes grey. Sometimes a slight yellowish tendency can be observed if the alloy contains traces of manganese or chrome;
- 6000 alloys series (magnesium and silicon alloys) behave very well at anodizing and colouring, with very homogeneous anodized layer. To pieces made from this alloy can be applied satin finish to obtain a matte appearance of the surface. At thicknesses of the anodized layer over 30μm, the surface of the parts becomes dark grey;
- 7000 alloy series (zinc alloys), as the anodized layer thickness is greater as can be seen better surface appearance of the greenish yellow anodized parts. Some alloys of the 7000 series may also contain copper (eg 7075 alloy) and behave at anodization as a 2000 alloy series, the anodized surface being unaesthetic.

The samples were provided by COLOR METAL SRL through SC ANOROM SRL. These consist of specimens taken from different extruded bars from 2017 SK03, 5083 SA 02, 5083 SG 0, 6082 SF02, 7021 and 7075 aluminium alloy grade.

Figure 5. Examples of aluminium specimens, 5083 and 6082 grade

In Figure 5 shows a few specimens provided by this company. As can be seen from the Figure, the specimens are protected with a plastic foil against oxidation, a film that is removed before fitting into the mounting brackets.

In order to perform the anodization, the specimens will be fixed in special supports. They are made from titanium and will not oxidize when the pieces are treated. The fixing mode and construction of this support is shown in Figure 6.

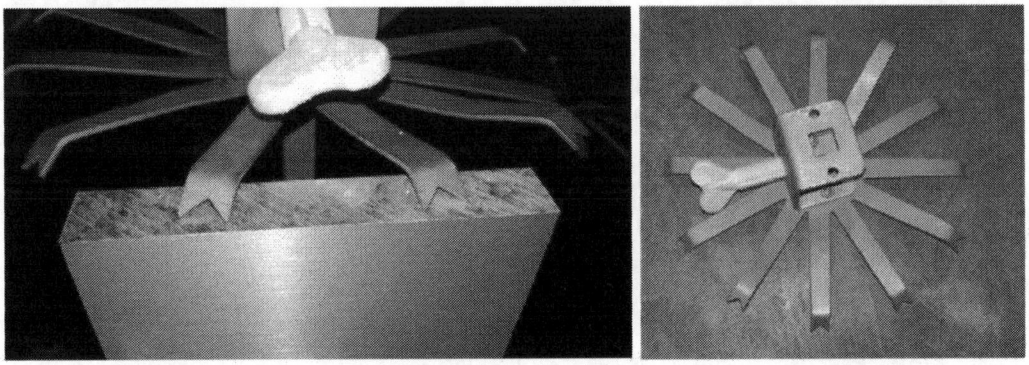

Figure 6. The fastening mode and construction of fixing support

To avoid possible defects, the electrical contact between the piece and the support should be as good as possible, in other words, the piece should be tightened as much as possible. Figure 7 shows the pieces fixed to the support before being treated.

Figure 7. Test specimens ready for anodizing operation

In order to observe the difference generated by different chemical composition (for each grade of aluminium alloy), all parts have been treated respecting the same parameters, namely alkaline degreasing, satin finishing, anodization, rinsing water, pore sealing.

All operations, parameters and technologically time are tracked and passed into a manufacturing order [10].

The supports with test specimens are handled by automatic cranes that can operate both automated and manual with operator.

It is very important to rinse the pieces when passing from one bath to another. Pieces was rinsing repeatedly until when removing from the rinse bath on the surface of the piece we have a continuous film of water.

For colourless satin pieces a technological flow involves the following steps: alkaline degreasing; current rinse; satin finishing operation; static and current rinse; introducing the test specimens into the nitric acid bath; current rinse; anodic oxidation; static and current rinse; sealing pores; current rinse – presented in Figures 8-10.

IOP Conf. Series: Materials Science and Engineering **294** (2017) 012042 doi:10.1088/1757-899X/294/1/012042

Figure 8. Alkaline degreasing

Figure 9. Satin finishing (left) and black colouring (right) of test specimens

Figure 10. Anodizing (left) and current rinse (right)

In the case of non-satin or black anodizing process, test specimens are treated with additional steps corresponding to the technological operation added or, if necessary, removed. The following Figures show aspects of the anodizing operation with satin finishing.

In the case of hard anodization the experiments was made in another periods of time, this installation is not used every day. Through this process was obtained a special values of surface hardness, which according to the literature is usually between 600...800HV. In this case the thickness of the superficial layer is very small, and hardness cannot be measured by the usual methods.

The thickness of the anodized layer is measured with a portable Foucault current device for thin layer [10]. In Figure 11 is presented a comparison between the thickness of the natural layer of aluminium oxide and the thickness of the aluminium oxide layer after anodizing.

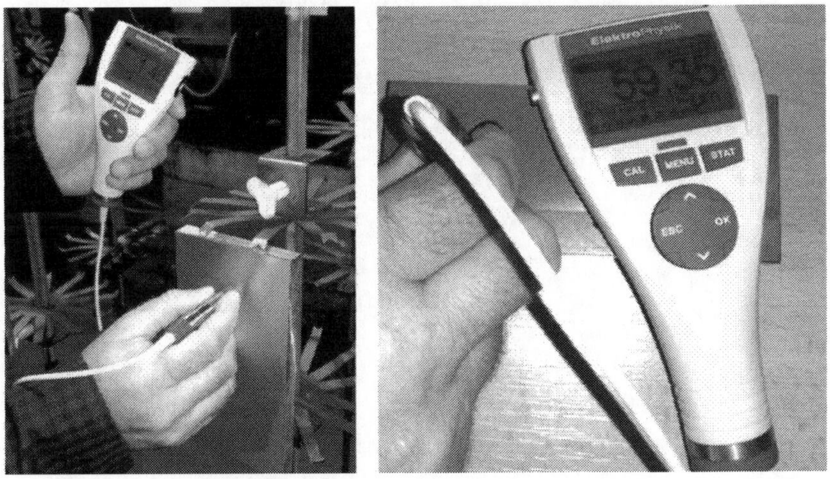

Figure 11. Comparison between the thickness of the natural layer of aluminium oxide and the thickness of oxide layer after anodizing

4. Results and discussions

After the experiments carried out on all 6 categories of materials, were obtained the following layer thicknesses, and shown in the Table 1.

Table 1. Layer thicknesses obtained from different anodizing variants

No	Material	Colourless anodizing + Satin finishing thickness of oxide layer, µm	Colourless anodizing thickness of oxide layer, µm	Black anodizing + Satin finishing thickness of oxide layer, µm	Black anodizing thickness of oxide layer, µm
1.	2017 SK03	8,64	8,44	13,26	10,80
2.	5083 SA 02	36,30	32,05	38,25	33,75
3.	5083 SG 0	39,20	37,30	45,36	43,35
4.	6082 SF02	22,36	20,24	21,10	20,58
5.	7021	43,20	33,30	45,90	36,25
6.	7075	32,75	33,60	36,70	38,45

From the data shows Table 1 results that the 2017 SK03 aluminium alloy had the worst behaviour in all anodizing variants, and the 5083 SG 0 and 7021 aluminium alloys had the best behaviour for almost all anodizing variants.

For a better analysis of the data resulting from these experiments, will be presented in the Figure 12 the same results, but in the graphical form.

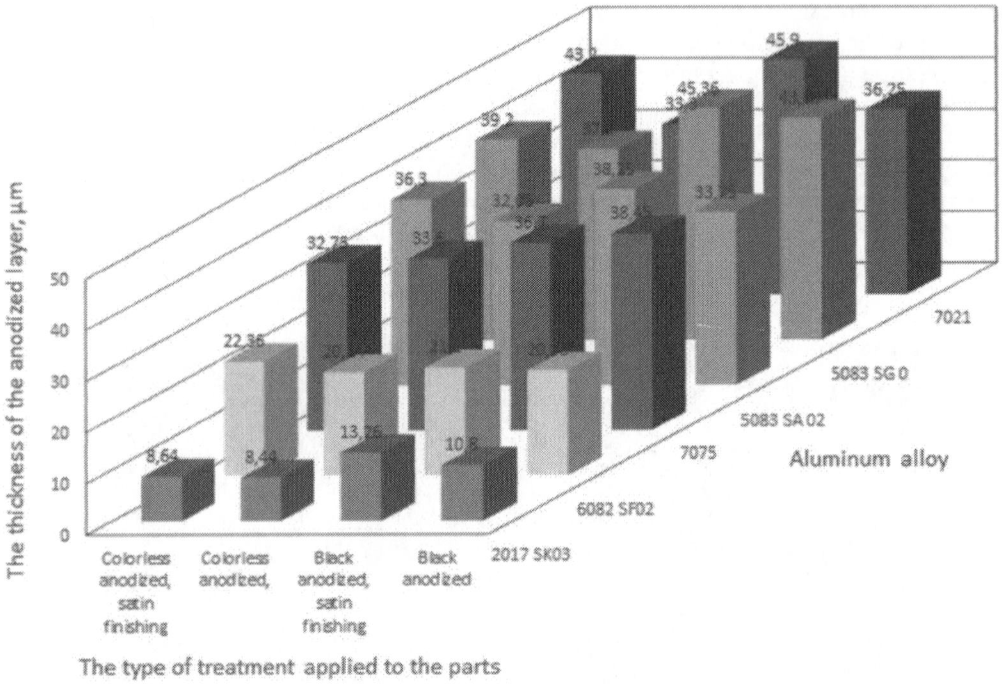

Figure 12. Layer thicknesses obtained from different anodizing variants

In Figure 13 show the colour variations for the vast majority of anodized parts in all anodizing variants.

Figure 13. Comparison between colour palette obtained at colourless anodizing (left), colourless anodizing + satin finishing (center) and hard anodizing (right), for all aluminium grade

It is interesting to look at the influence of alloying elements on the behaviour of different aluminium alloys in the different anodizing variants. Thus, depending on the average alloying element content, it is possible to represent a graphically variation of the thickness of the anodized layer for the studied alloys.

The previous graphical dependencies were made for the colourless anodizing variant with satin finishing, and in term of the chemical composition, were worked with the average values imposed by the standard.

In graphical dependence of the thickness of the anodized layer, depending on the average content of copper required by the standard, there is a tendency to decrease the thickness of the anodized layer with the increase of the copper content in the alloy for both variants.

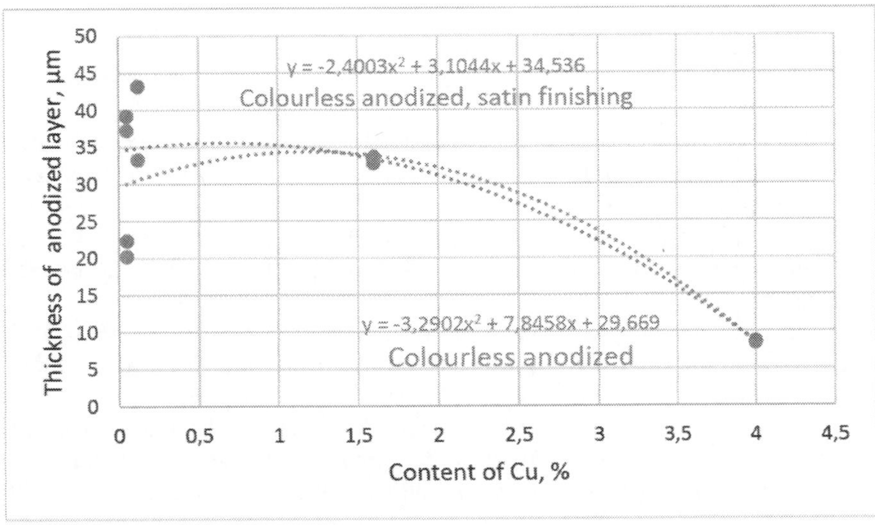

Figure 14. The variation of the layer thickness according to the average content of copper

In the case of the graphical dependence of the thickness of the anodized layer, depending on the average content of magnesium required by the standard, it observes a maximum thickness of the anodized layer thickness for 1.3...1.7% Mg. A further increase in magnesium content has adverse effects on the thickness of the anodized layer. This dependences is presented in Figure 15.

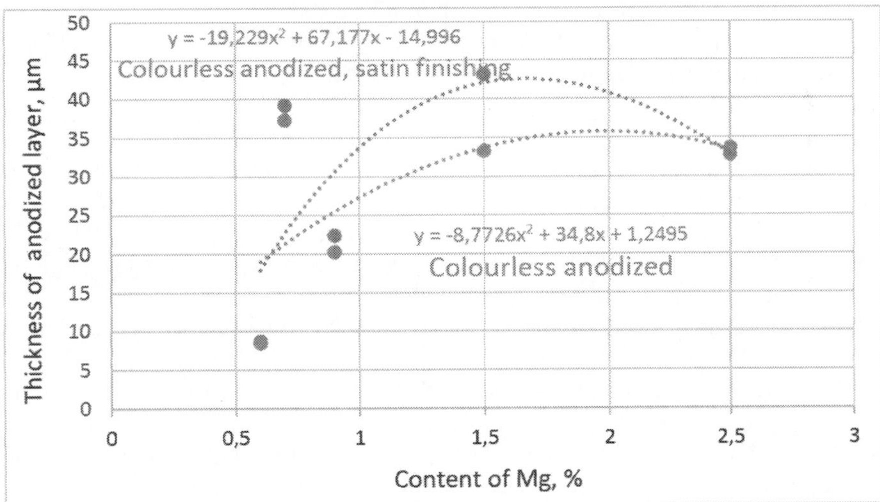

Figure 15. The variation of the layer thickness according to the average content of magnesium

Such graphical dependencies can be made for all anodizing variants and all alloying elements present in the chemical composition of the aluminium alloy.

The thickness of the anodized layer depends also on the holding time in the anodizing bath, in other words, simple or multiple dependencies can be identified depending on the holding time in the anodizing bath and the chemical composition of the anodized material.

5. Conclusions

Several conclusions can be drawn from the experiments as follows:

- the defects resulting from the production process tend to zero, they can occur mainly due to the improper fixing of the parts that will be anodized in the fixing supports;
- another possible cause of the parts with defects may be a piece made from an material different from that of the entire product group. SC ANOROM SA does not control the chemical composition of the parts just treats them superficially;
- from the data on the Table 1 we can see that the 2017 SK03 aluminium alloy has the worst behaviour in all anodizing variants, and the 5083 SG 0 and 7021 alloys has the best behaviour for almost all anodizing variants;
- the thickness of the anodized layer depends on the maintaining time in the anodizing bath;
- colour of the alloy in the case of colourless anodization with satin finish and without satin finish depends also on the maintain time in the anodizing bath;
- I think it would be of great importance to measure the hardness of the anodized layer, in which case can be made simple or multiple correlations to determine optimal areas for hardness;
- the process and stages of anodization are very well optimized, the defect rate is almost zero.

References

[1] Cioată V G and Kiss I 2008 Experimental research regarding the technological parameters of a new method of processing in the semisolid state, *Metalurgia International* **13**(6) 21-25

[2] Cioata V G and Ilca I 2009 Study of the technological parameters influence upon the mechanical characteristics of pieces obtained through die-forging in semisolid state from AlSiMg0.38, *Metalurgia International* **14**(4) 8-12

[3] Carcea I and Matei Ghe 2009 *Aliaje neferoase de turnatorie*, Editura Performantica, Iaşi, Romania

[4] ***www.makeitfrom.com/material-group/Aluminium -Alloy

[5] Sieber M, Morgenstern R, Kuhn D, Hackert-Oschätzchen M, Schubert A and Lampke T 2017 Downscaled anodic oxidation process for aluminium in oxalic acid, *IOP Conf. Ser.: Mater. Sci. Eng.* **181** 012044

[6] ***www.anodizeusa.com/anodizing-systems-hard-coat.php

[7] Aryslanova EM, Alfimov AV and Chivilikhin SA 2015 Modelling the growth process of porous Aluminium oxide film during anodization, *IOP Conf. Ser.: Journal of Physics* **643** 012008

[8] Nagaraj Chelliaha M, Saxenaa A, Sharmaa K, Singha H and Surappac M K 2017 Surface characterization of nanoporous aluminium oxide films synthesized by single-step DC and AC anodization, *Surfaces and Interfaces* **7** 139–145

[9] Dumitrascu V, Benea L and Danaila E 2017 Corrosion Behavior of Aluminium Oxide Film Growth by Controlled Anodic Oxidation, *IOP Conf. Ser.: Mater. Sci. Eng.* **209** 012016

[10] Ilango M S, Mutalikdesai A and Ramasesha S K 2016 Anodization of Aluminium using a fast two-step process, *Journal of Chemical Sciences* **128**(1) 153-158

[11] Yin A, Guico RS and Xu J 2007 Fabrication of anodic aluminium oxide templates on curved surfaces, *IOP Conf. Ser.: Nanotechnology* **18**(3) 035304

[12] Vermeşan H 2008 *Coroziune şi protecţie anticorozivă*, Editura Risoprint,Cluj-Napoca, Romania

[13] ***www.surtec.com/en/products/product-overview/aluminium-pre-treatment-50/surtec-650-268/

[14] ***www.potashcorp.com/media/bright_dip_manual.pdf

[15] ***www.anorom.ro

[16] Morgenstern R, Dietrich D, Sieber M and Lampke T 2017 Influence of the heat treatment condition of alloy AlCu4Mg1 on the microstructure and properties of anodic oxide layers, *IOP Conf. Ser.: Mater. Sci. Eng.* **181** 012043

IOP Conf. Series: Materials Science and Engineering **294** (2017) 012043 doi:10.1088/1757-899X/294/1/012043

Power quality analysis of DC arc furnace operation using the Bowman model for electric arc

P L Gherman

Politehnica University of Timisoara, Department of Electrical Engineering and Industrial Informatics, 5 Revolution Street, Hunedoara, 331128, Romania

E-mail: lucian.gherman@upt.ro

Abstract. This work is about a relatively new domain. The DC electric arc is superior to the AC electric arc and it's not used in Romania. This is why we analyzed the work functions of these furnaces by simulation and model checking of the simulation results.The conclusions are favorable, to be carried is to develop a real-time control system of steel elaboration process.

1. Introduction

An increase of steel produced quantity in electric furnaces has been noticed in the last years, mainly using DC arc furnaces. In spite of the well known advantages of this system [1], it is not used in Romania. The necessity for more powerful converters complicates the feeding diagram of the furnace, but allows an advanced degree of automation of the process, which leads in long tern to a much higher output of the DC furnaces. It is possible to determine the control algorithms only if the work conditions of the E.A. furnaces are known and can be modelled and simulated.

2. Technological considerations. The DC arc

The making of steel elaboration in an E.A. furnace involves three stages (Figure 1).

The smelting process – lasts from the moment is connected to the electric network up to the complete smelting of furnace load.

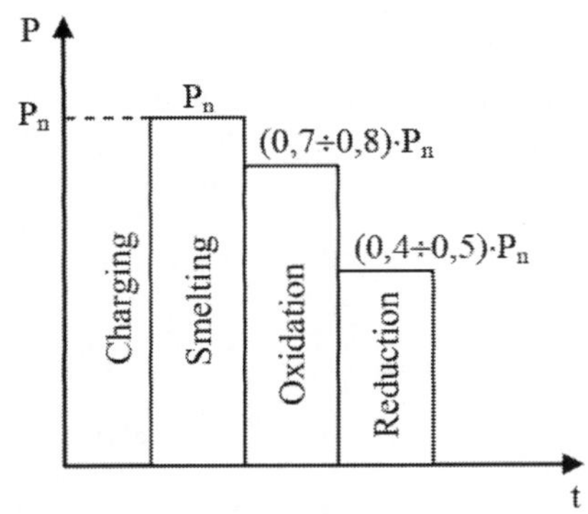

Figure 1. The stages of steel elaboration in an electric arc furnace

The oxidation process (refining) is characterized by the fact that the electric arc is steady; the temperature of the melted steel is maintained at a value required by the chemical reactions in the process. All along this stage, the furnace works at $(0.7\div 0.8)\cdot P_n$.

The reduction process. The temperature of the bath reaches 1700°C and the electric arc is stable, the power absorbed by the furnace being $(0,4\div 0,5)\cdot P_n$.

The electric arc is attached to the cathode by a small area named cathode spot, the arc reaching the smelted steel (Figure 2).

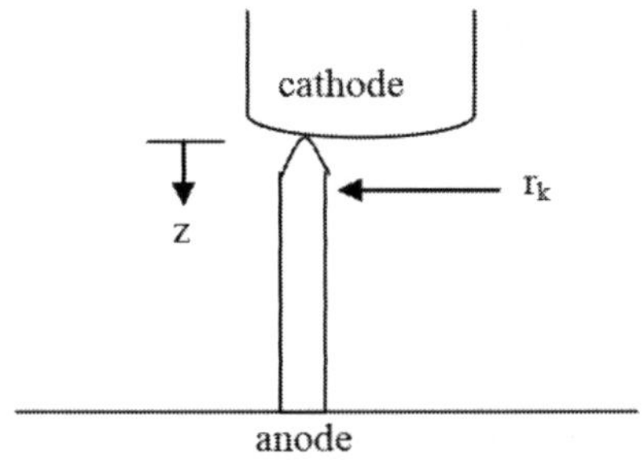

Figure 2.

The value of the current density through the cathode spot is about 3,5 KA/cm^2.

$$\frac{r_a}{r_k} = 3{,}2 - 2{,}2\cdot e^{\left(-\frac{z}{5\cdot r_k}\right)} \tag{1}$$

Relation (1) describes the shape of the conducting column of the electric arc taking in account the distance z between the metal bath and the cathode. By integration we obtain the voltage of the electric arc [2]:

$$V = 2\cdot \rho_a \cdot \sqrt{\frac{L\cdot j_k}{\pi}}\cdot \int\left(\frac{r_k}{r_a}\right)^2 \cdot dZ;\ \ Z = \frac{z}{r_k}; \tag{2}$$

where:

ρ_a – the passivity of the electric arc [$\Omega\cdot cm$];

j_k – the current density through the cathode spot, [KA/cm^2];

r_k – the radius of the cathode spot [cm];

r_a – the radius of the arc column [cm];

L – the length of the arc, [cm];

z – the axial distance between the cathode and the metal bath, [cm].

Calculations lead to a relation that links the voltage of the electric arc and its length to the current:

$$V_a = \frac{I\cdot \rho_a}{m\cdot \pi}\cdot\left[-\frac{1}{a^2 + a\cdot b} + \frac{1}{a^2 + a\cdot b\cdot e^{m\cdot L}} - \frac{\ln(a+b)}{a^2} + \frac{m\cdot L}{a^2} + \frac{\ln\left(a + b\cdot e^{m\cdot L}\right)}{a^2}\right], \tag{3}$$

where: $a = 3{,}2\cdot r_k$, $b = -2{,}2\cdot r_k$, $m = \frac{-1}{r_k}$, $r_k = \sqrt{\frac{I}{\pi\cdot 3500\left(A/cm^2\right)}}$.

The value of electric resistivity can be obtained experimentally and is strongly dependent on the composition of the atmosphere where the arc is.

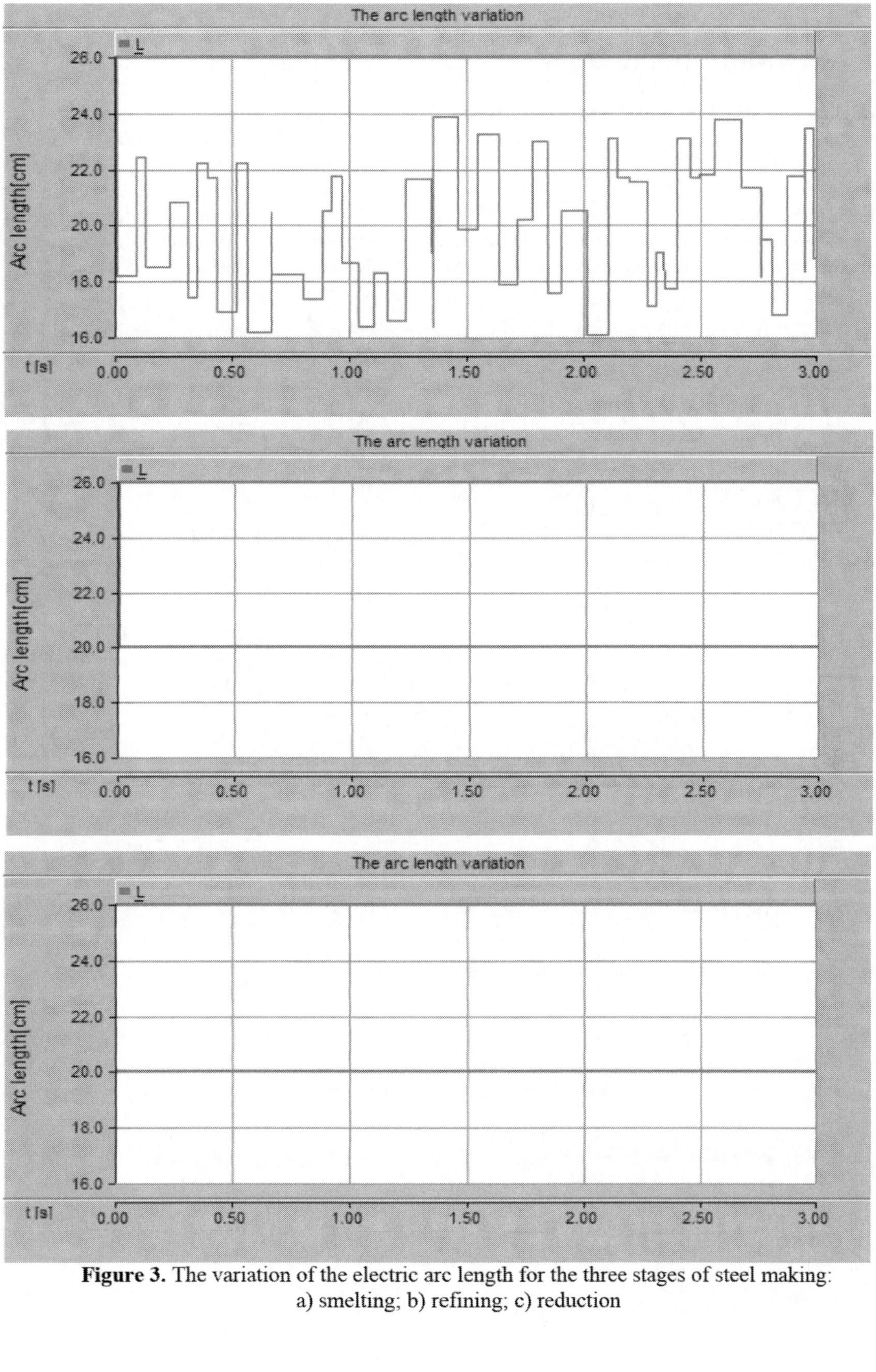

a)

b)

c)

Figure 3. The variation of the electric arc length for the three stages of steel making: a) smelting; b) refining; c) reduction

Figure 3 shows the variation of the electric arc length in the phases of smelting, refining and reduction. The average length of the electric arc is l = 20 cm [3]. During the smelting process, the arc length and its resistance vary randomly with a frequency in between 0 and 15 Hz [3].

3. Modeling and simulation of the processes in the DC arc furnace

In order to model and simulate the process of steel elaboration in the DC arc furnace, the utility program PSCAD – EMTDC 4.2 has been used. Figure 4 shows the simulation diagram that was created.

The electric arc furnace belongs to the 1st category of elements generating deforming work conditions [4]. It happens so on the one hand because of the non-linear character of the charge, respectively of the electric arc resistance, and on the other hand because of the controlled converter.

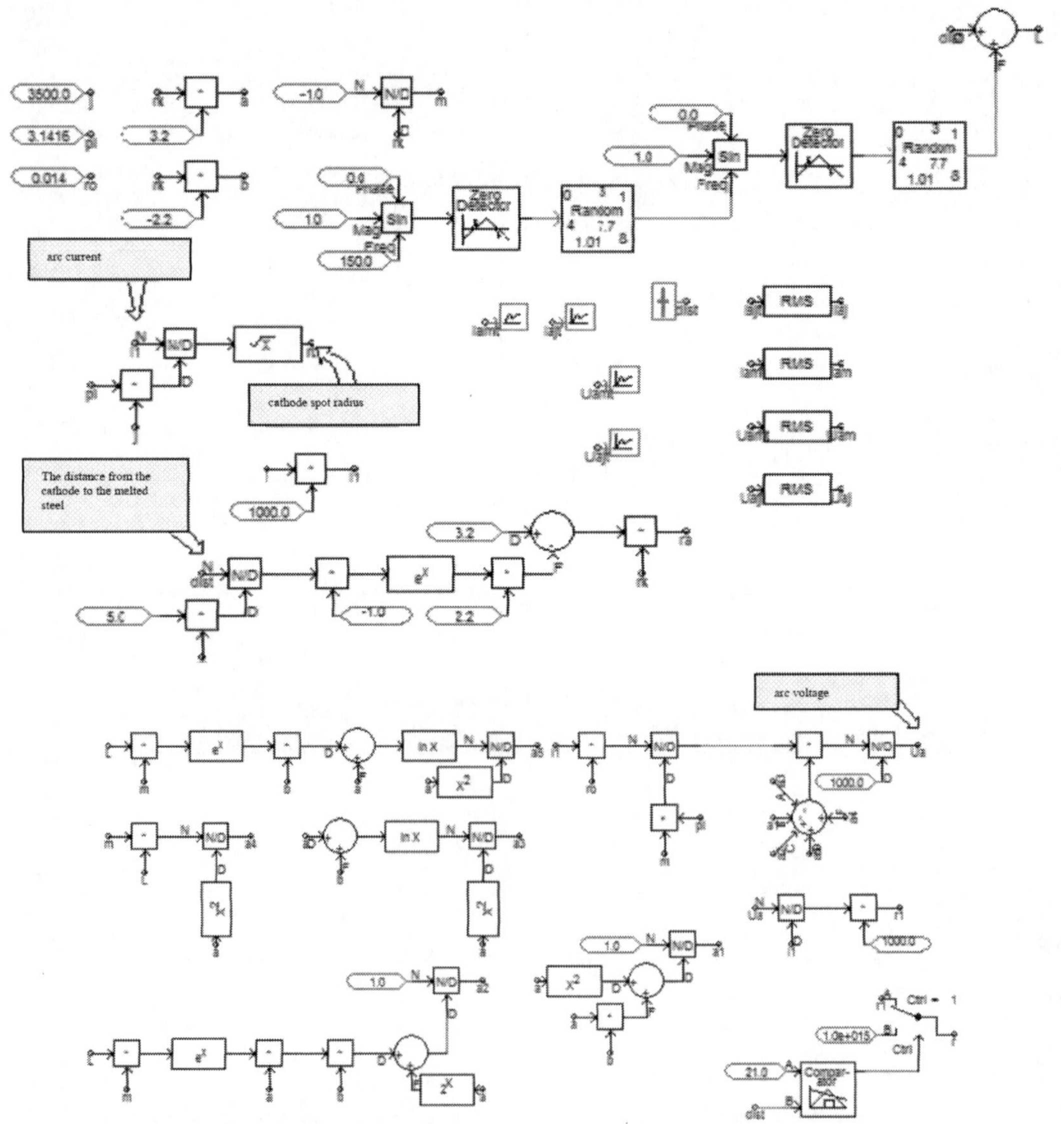

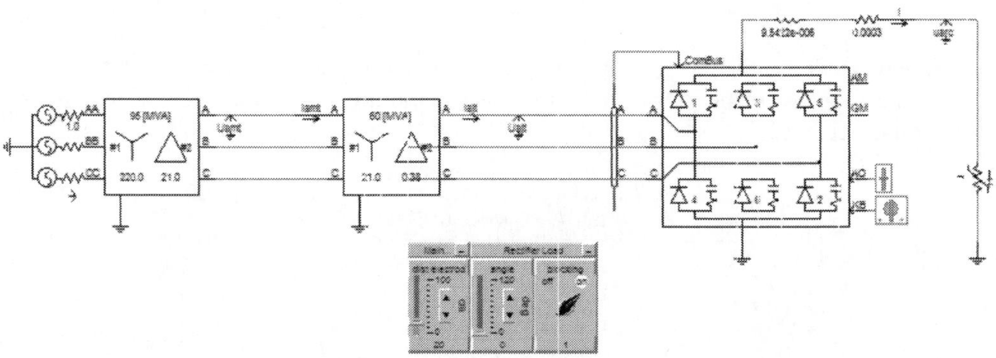

Figure 4. The simulation diagram

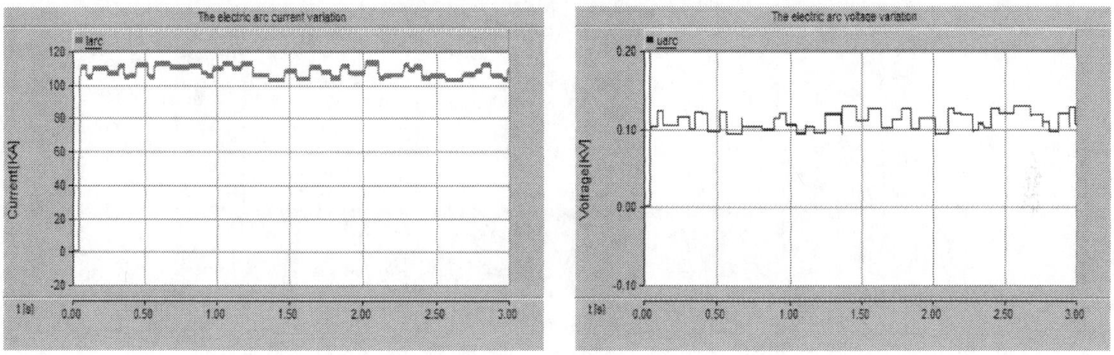

a)

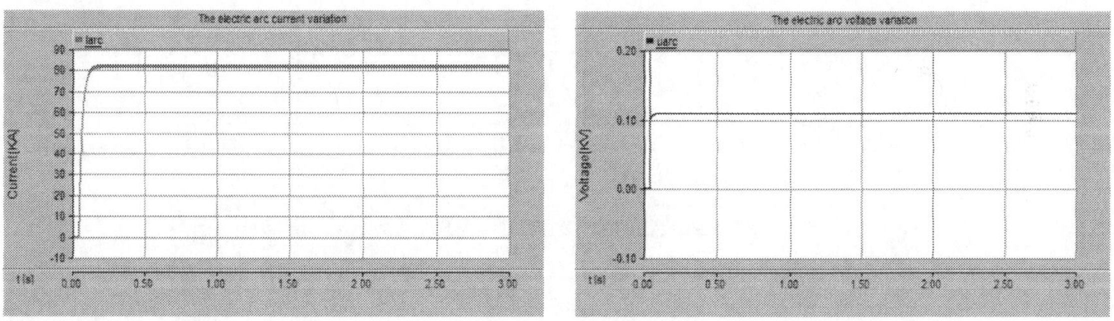

b)

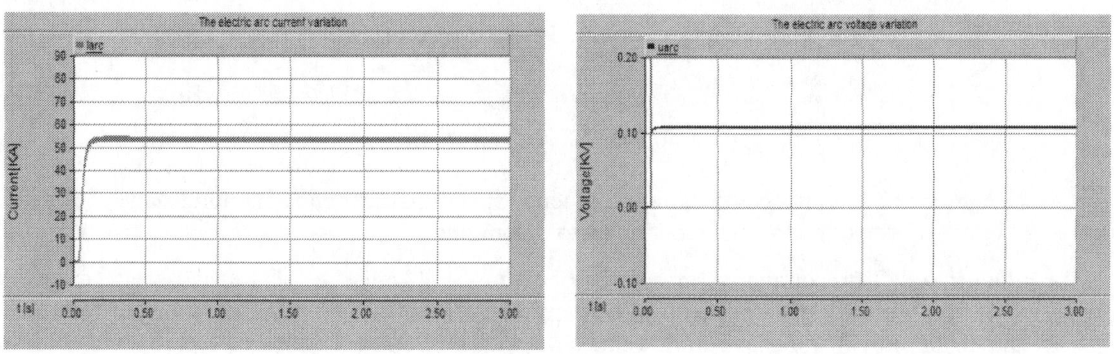

c)

Figure 5. The variation of the voltage and current of the electric arc in the phases of:
a) smelting; b) refining; c) reduction

The harmonic analysis has been done by means of a program designed by the authors under Matlab. After the analysis on the low voltage line for the smelting stage, we obtained the spectra of the current and voltage (Figure 6)

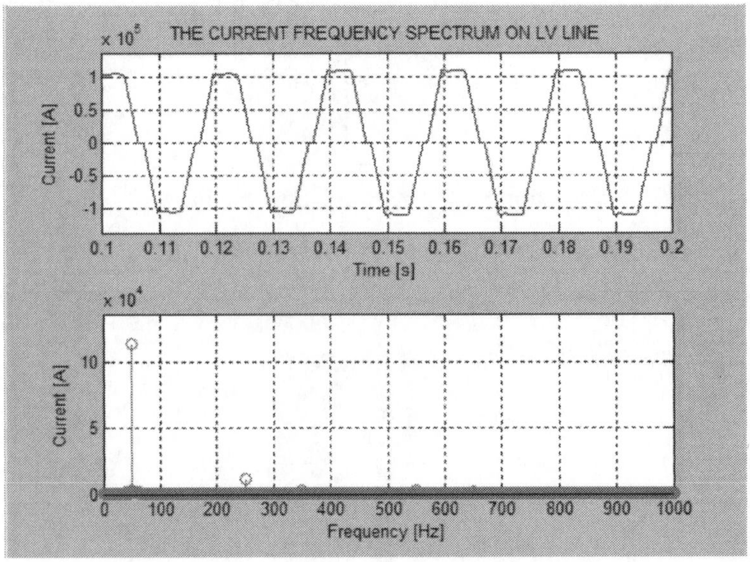

THD = 10,137% Y_d = 11590,749 A

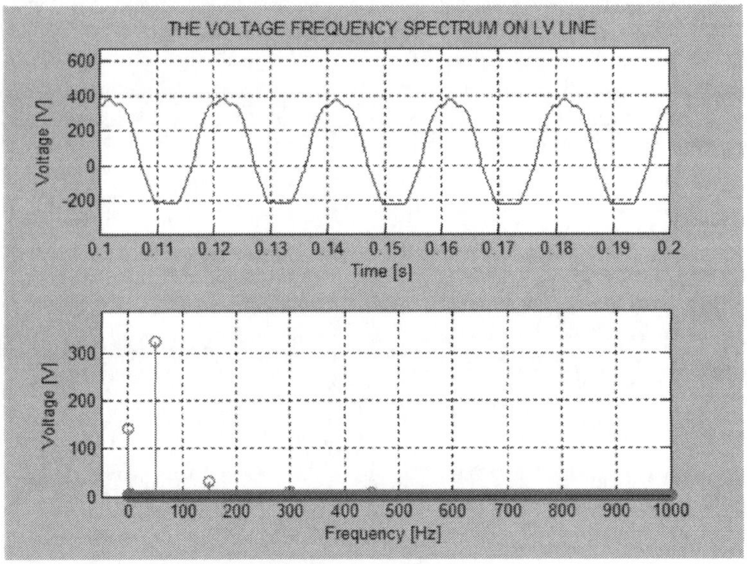

THD = 40,9757% Y_d = 145,2917 V

Figure 6. The voltage and current frequency spectrum during the smelting stage on the low voltage line

According to the CEI recommendations with respect to the 2nd class installations of electromagnetic media, included in the PE143/94 norms [4], the harmonics content of the fundamental, the harmonics 3, 9, 15 and 18 in the voltage spectrum are much larger then the limit levels of compatibility. In the current spectrum on the low voltage line, harmonic 5 with a content related to the fundamental of 9,6248% exceeds the maximal admitted value.

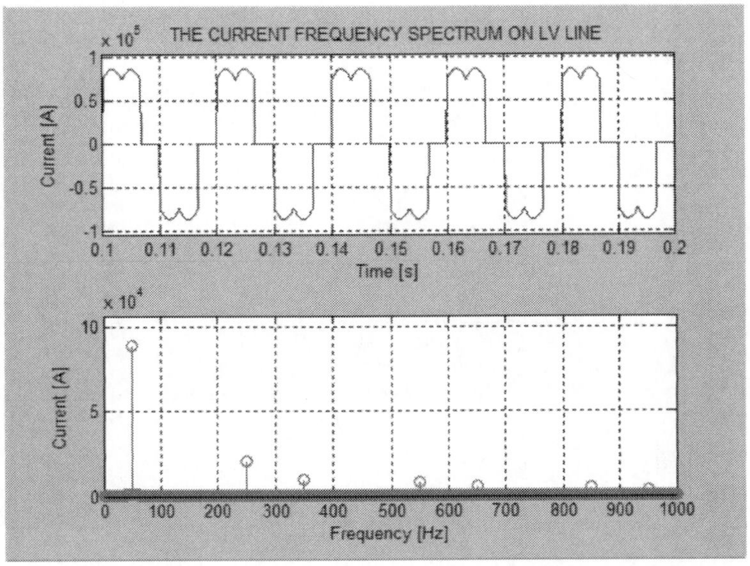

THD = 28,0481% Y_d = 25951,4773 A

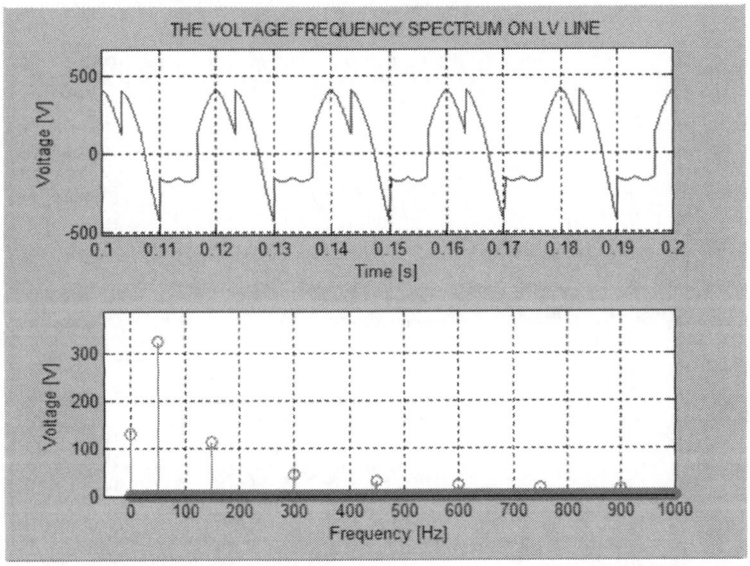

THD = 49,9998% Y_d = 186,5558 V

Figure 7. The voltage and current frequency spectrum during the refining stage on the low voltage line

Power reduction during the refining stage can be done by controlling the converter at an angle of $\alpha = 60{,}34^0$. During this stage, the voltage frequency spectrum on the low voltage line contains harmonics 0, 3, 6, 9, 12, 15 and 18 (Figure 7), all of which exceed the limit compatibility levels.

The current frequency spectrum on the low voltage line includes harmonics 5, 7, 11, 13, 17, 19, whose values exceed the maximal admitted limits.

During the reduction stage, in order to have a power $P = (0{,}4 \div 0.5) \cdot P_n$, the converter has to work at a command angle of $\alpha = 80^0$. The high value of the command angle, leads to higher levels of harmonic pollution in the feeding network.

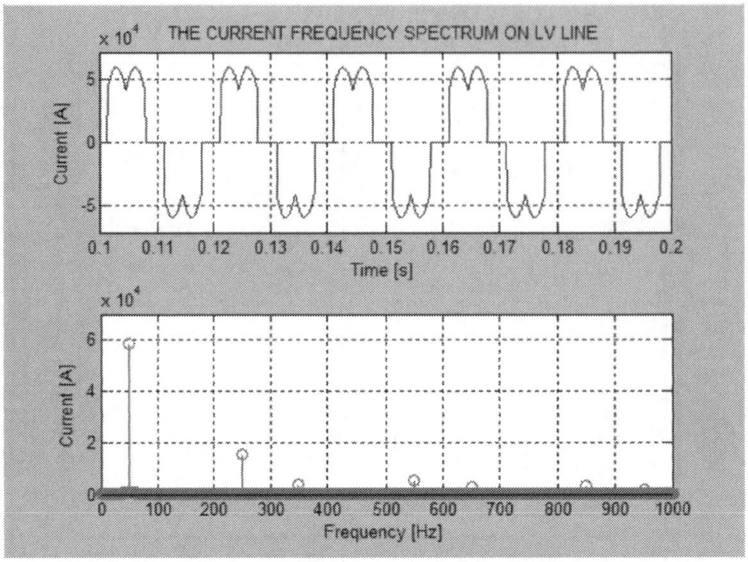

THD = 29,0364% Y_d = 17669,3544 A

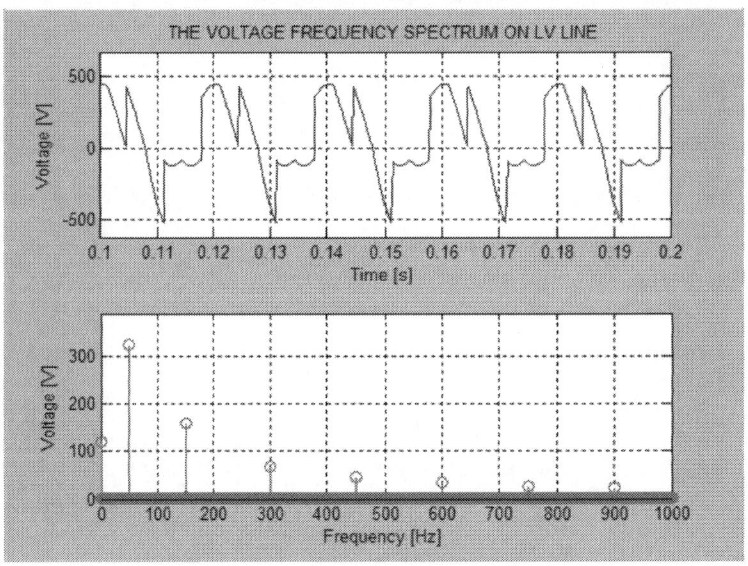

THD = 56,6584% Y_d = 223,1912 V

Figure 8. The voltage and current frequency spectrum during the reduction stage on the low voltage line

The voltage frequency spectrum on the low voltage line (Figure 8.) reveals that this is greatly polluted with harmonics and includes harmonics of 0, 3, 6, 9, 12, and 18 ranks, all of which exceed the limit values admitted. In the current spectrum on the low voltage low voltage line the harmonics of ranks 5, 7, 11, 13, 17, 19 are prevailing, their values exceeding the standardized maximal limits.

4. Conclusions

This work analyzes the performances of the smelting process in DC arc furnaces and suggests a system of controlling this process, meant to point out to the advantages of this method, minimizing its disadvantages.

References

[1] Golovanov N and Sora I 1997 *Electrotermie si electrotehnologii*, vol. I, Editura Tehnica, Bucuresti
[2] Bowman B 1994 *Properties of arcs in DC furnaces*, Electric Furnace Conference Proceedings, pp 110-120
[3] Carpinelli G and Di Manno M 1999 *AC and DC arc furnaces: a comparison on some power quality aspects*, IEEE 1999, pp 499-506
[4] Buta A, Milea L and Pana A 2000 *Impedanta armonica a retelelor sistemelor electroenergetice*, Editura Tehnica, Bucuresti

Type practical application in spectral analysis, combining Labview and open source software

C P Chioncel and C V Anghel Drugarin

"Eftimie Murgu" University of Resita, Electrical Engineering and Informatics Department, Traian Vuia Square, no. 1-4, 320085 Resita, Romania

E-mail: c.anghel@uem.ro

Abstract. The paper presents the interconnection possibility of LabVIEW with his different opportunities and Scilab, one of the successful free MatLAB clones. The interconnection between those was made possible through the LabVIEW to Scilab gateway. This tool can be applied in virtual as well as in real laboratories, representing a true assistance for self- learning, too.

1. Introduction

The development of computers technology, with high processing speed and increasing available memory, equipped with dedicated software, have turned the way in which engineers designs and work. Virtual instruments (VI) gained a well-defined place in the world of measurements and system analysis, permitting flexible customization and easy integration. In these days, virtual instruments are used as educational tools in various domains, too, [1]. On one side, those are utilized to develop and design measurement system and on the other side, are used as simulation tools in a closed virtual world in which the user can observe the behaviour of complex systems in different contexts, being of great help in understanding various phenomena or complex mathematical expression, implemented and simulated at subsystem level, [2]. This way of approaching the problems is well accepted by students in the technical field, stimulating innovation in the process of designing proper virtual instruments, mostly in the field of process control and signal processing.

The virtual instrumentations leads to discover interrelations, between mechanic and electric interconnections, and the phenomena can be better understood, theory clearly learned. Those tools can be used in the industry, in laboratories as well as home, in virtual laboratories, constituting a support for self-learning, [3].

The present paper aims to illustrate how Virtual Instruments can be used in engineering education, combining the LabVIEW environment with open source software - Scilab, with practical application in spectral analysis from the Signal Processing Laboratory of the "Eftimie Murgu" University of Resita, together with a quantitative research regarding the perception of students to this issue.

2. Virtual instruments in the learning process

Virtual Instrumentation (VI) is used as well for designing applications which simulate real-world devices and instruments, as well as for modelling and simulation of the real mechanical processes, being capable to use both, computer generated time data and/or real recorded experimental data.

Those facilities represent an important and significant characteristic of VI, which can be used to create adapted tools in the teaching process, LabVIEW software, registered trademark of National

Instrument Inc. (NI), represents a programming environment that include specific tools necessary for instrument control, data acquiring, storage, analysis, presentation and integrating of those features in a single system, [4]. It uses graphical programming and a special graphical language.

The main goal of LabVIEW represents the creation of virtual instruments. Scilab represents free and open source software for numerical computation, providing a powerful computing environment for engineering and scientific applications, ensuring multiple interoperability and interconnection through gateways in the case of LabVIEW, [5].

3. Spectral analysis - Fourier Transform Tool

In the frequency domain, signals can be analysed, synthesized, filtered or passed through different windows. This processing is performed using Fourier series and the Fourier, Laplace and Z transform, [6]. The method was, initially, based for the linear physical systems, on their remarkable properties of sinusoidal fidelity, i.e., the fact that response of a linear system to a sinusoidal signal is also a sine wave with the same frequency but different amplitude and phase, [7], [8].

The Fourier transform (FT), the transformation from the time domain to the frequency domain, and vice versa, represents one of the most important practical transforms met in the field of electrical and computer engineering, [6]. Periodic signals with period T_0, can be expressed as a sum of sinusoids using Fourier series:

$$x(t) = \frac{a_0}{2} + \sum_{k=1}^{k=\infty} (a_k \cos k\omega_0 t + b_k \sin k\omega_0 t) \tag{1}$$

where

$$a_0 = \frac{2}{T_0} \cdot \int_{t_0}^{t_0+T_0} x(t)dt \tag{2}$$

Also,

$$a_k = \frac{2}{T_0} \cdot \int_{t_0}^{t_n+T_n} x(t) * \cos(k\omega_0 t)dt \tag{3}$$

$$b_k = \frac{2}{T_0} \cdot \int_{t_0}^{t_n+T_n} x(t) * \sin(k\omega_0 t)dt \tag{4}$$

Using Euler's relation, we obtain the Fourier as a complex expression, like in relation:

$$x(t) = \sum_{k=-\infty}^{\infty} X_k e^{jk\omega_0 t} \tag{5}$$

If the repetition period T_0 increases to infinity, the sum from relation (5) turns into an integral and we obtain the Fourier transform, in terms of angular frequency $\omega = 2\pi \cdot f$, determinate by the following relations:

$$X(j\omega) = F\{x(t)\} = \int_{-\infty}^{+\infty} x(t) \cdot e^{-j\omega t} dt \tag{6}$$

$$X(t) = F^{-1}\{X(j\omega)\} = \frac{1}{2\pi} \int_{-\infty}^{+\infty} X(j\omega) \cdot e^{-j\omega} d\omega \tag{7}$$

where x(t) is the time signal and X(jω) the frequency signal.

The LabVIEW to Scilab Gateway acts like a 'script node', a text editor area, which can be moved in the LabVIEW block diagram [Sevgi]. The access to and out of the Scilab "script node" is done based on defining the input / output variables on the side of the node [9].

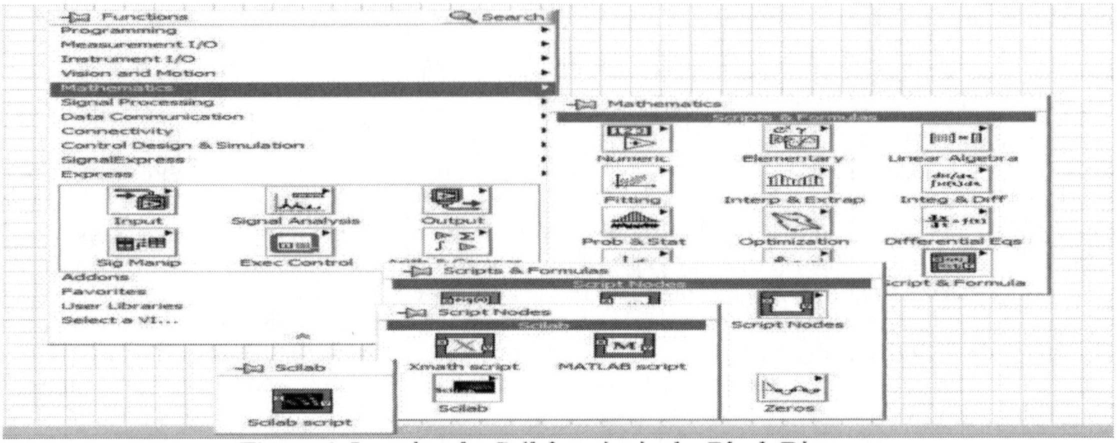

Figure 1. Inserting the Scilab script in the Block Diagram

Figure 2 presents the Block diagram window of the virtual instrument, which contains the graphical source code, designed using graphical programming, user friendly and accessible even for inexperienced users, where the analysed signals, three in number, are generated and added, the signals are viewed, the 'spectral measurements' block returns the FFT and the 'Power Spectrum' displays the signals spectrum.

In 'Scilab Script' an Scilab code sequence will compute the FFT, facilitating the use of different operations on the signal, offered by the Scilab predefined functions and / or programming language, Figure 2.

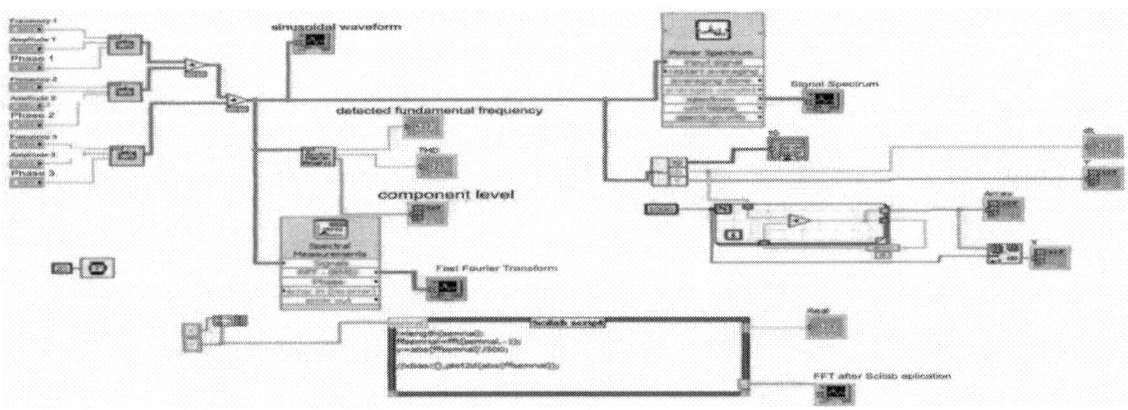

Figure 2. The complete application scheme

In Figure 3, the upper side of left hand, represents the time signal and on the upper right hand, the FFT obtained by the predefined LabVIEW block, [10] and on the bottom of the right hand, the FFT after running the Scilab application, [11-13], highlighting the three components the time represented signal, of different frequency and amplitude.

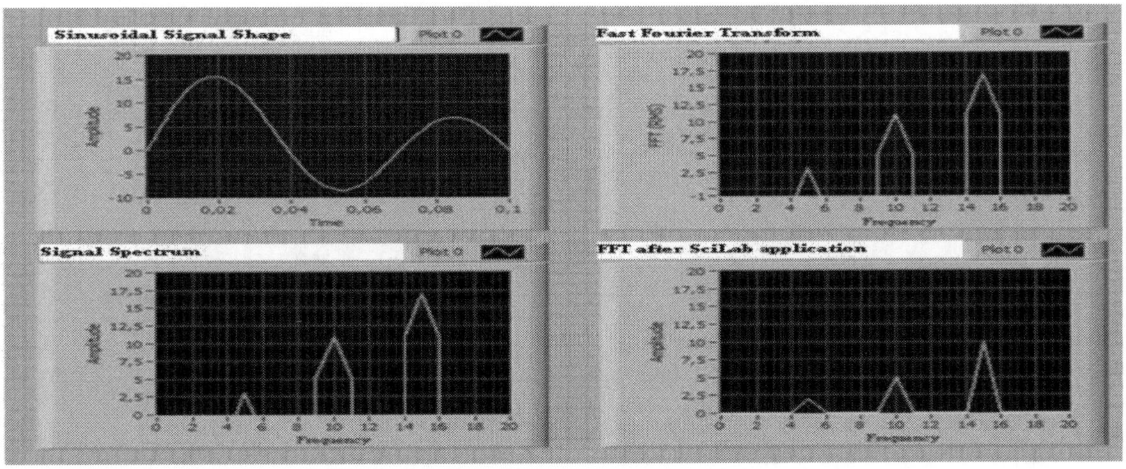

Figure 3. Results of the analysis

4. Quantitative research regarding the perception of students to use VI / Open Source software in educational process

We put the questions and the answers in the Table 1.

Table 1. Answers to Questionnaire

Statement	Answers				
	Strong reject	Reject	Neutral	Accept	Strong accept
Hardware and software resources to use VIs are available	-	-	-	16	28
It's easy to use VI's	-	-	2	19	23
VI's have been used mainly for measurements	23	17	3	1	-
Can VI's substitute seminaries and problem solving in the field of signal processing	35	4	2	3	-
Phenomena's became understandable with VI's	-	-	2	13	29
Self-learning, through VI and open source software, is easy	-	2	2	27	13

As shown in Figure 4, the availability of resources - hardware and software - in the university laboratories are appreciate among the students, so that the use of those facilities does not create an administrative issue in the educational process, [14], [15].

IOP Conf. Series: Materials Science and Engineering **294** (2017) 012044 doi:10.1088/1757-899X/294/1/012044

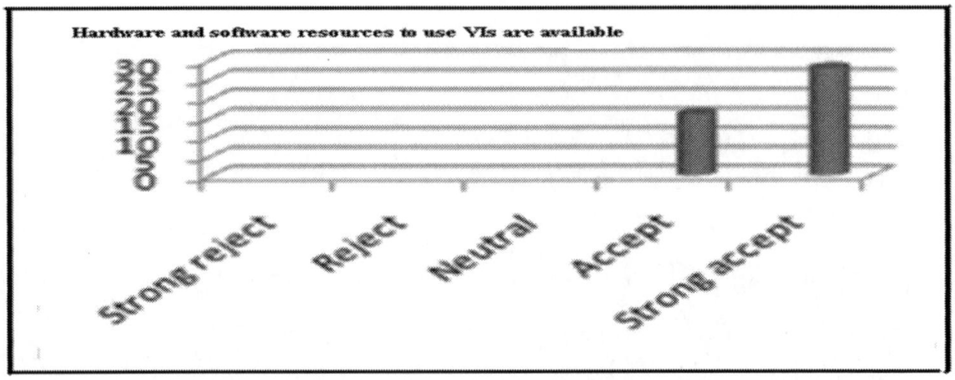

Figure 4. Availability of resources

Also, handling VI's and graphical editors to design models, simulate process and signal analysing became usual for the students, what turns out from the answers highlighted in Figure 5.

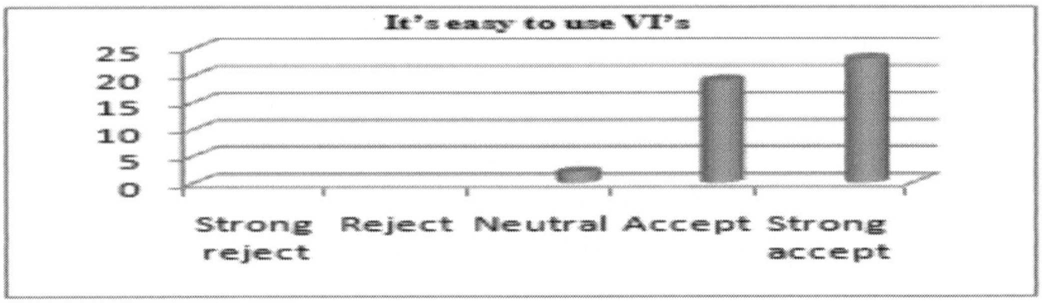

Figure 5. Easy handling

The following two questions aimed to find out the mean of using VI's. The answer, Figure 6, shows that the VI's and the related open source applications are not only used for measurements or process control, but strongly integrated in the education process, but being not able to replace problem solving and seminars, Figure 7, [15].

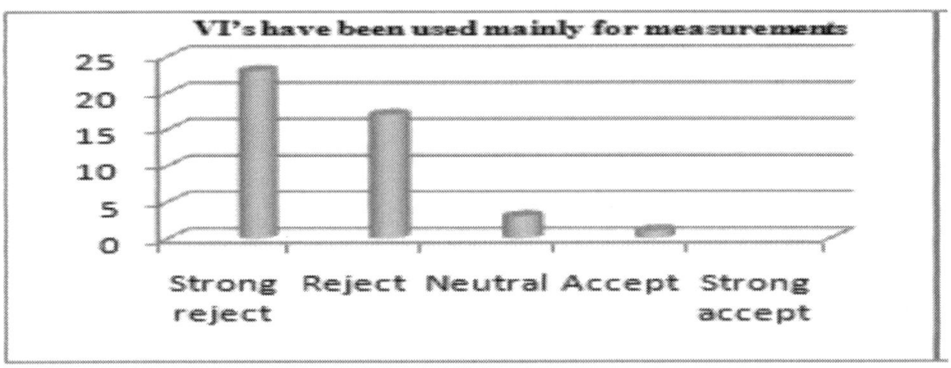

Figure 6. Uses of VI's

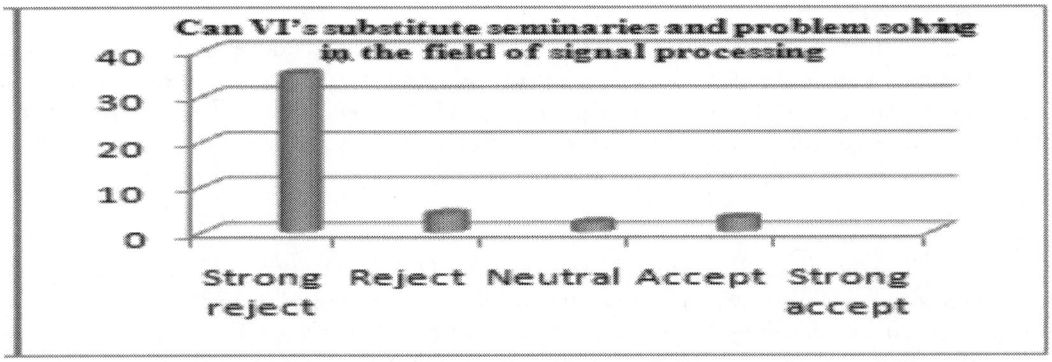

Figure 7. Substitution of seminar

Even so, virtual instruments are excellent tools to improve engineering education by providing flexibility to students and teachers too, in order to overcome the complexity of engineering study, making the understanding of the real world more clearly, [16].

This is reflected in the perception of how useful the VI together with open sources programs are among the students in assimilating information, Figure 8 and also, in Figure 9, highlighting the opportunity for individual teaching, [17].

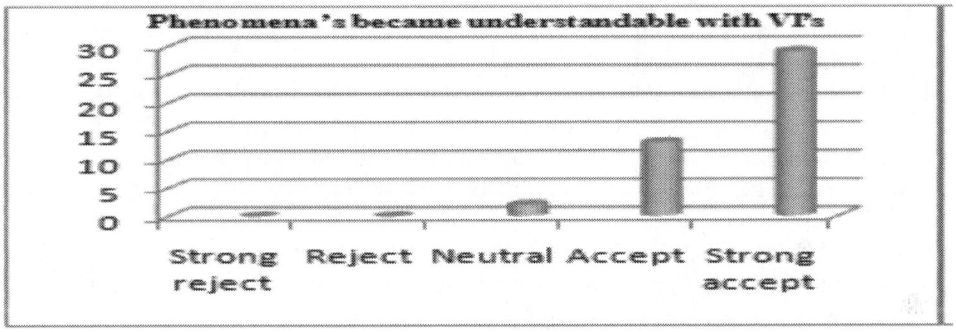

Figure 8. Uses of VI's

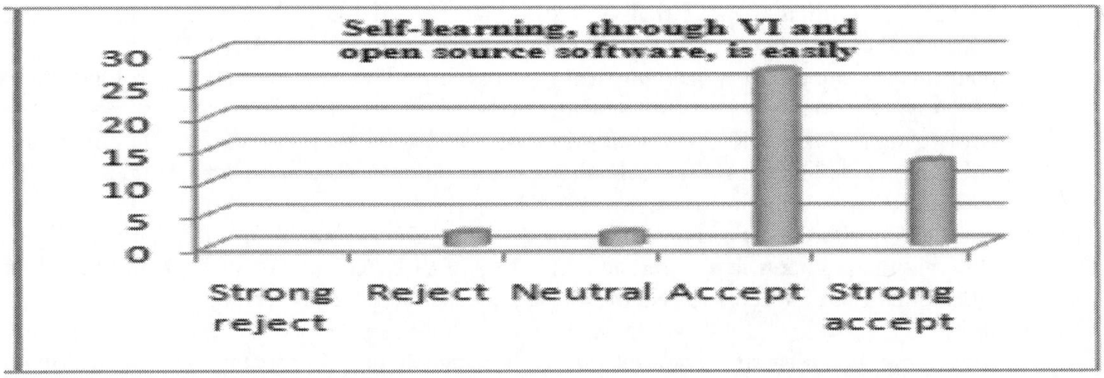

Figure 9. Substitution of seminar

Supplementary, our experience shows that virtual instruments, supported by open source application, brings more interest in engineering studies, making it more attractive and, in the same time, stimulating innovation and creativity during all involved factors.

5. Conclusion

The use of virtual instruments and open source software in experiments development and in regular lessons in the field of engineering studies makes the teaching and studying process more exciting. Especially the domain of signal processing, with one of his basic applications – the Fourier transform, in different technical directions - makes the use of virtual instruments attractive for the entire students, bringing more dynamic and improving the quality of education in the teaching process of a classroom.

References

[1] Gillich G R, Frunzaverde D, Gillich N and Amariei D 2010 The use of Virtual Instruments in engineering education, *Procedia-Social and Behavioral Sciences* **2**(2) 3806-3810

[2] Ugurlu Y 2010 *Measuring the Impact of Virtual Instrumentation for Teaching and Research*, IEEE Global Engineering Education Conference (EDUCON), pp 152-158

[3] Sevgi L and Uluisik C 2006 A Labview – Based Virtual Instrument for Engineering Education: A Numerical Fourier Transform Tool, *Turk J Elek Engin* **14**(1) 129-152

[4] Draghici S, Prostean O, Raduca E, Anghel Drugarin C V, Chioncel C P and Rudolf C 2015 The Relations Between the Mathematic Ideal, the Generating Polynome and the Matrixes G, H Used in Encodings, *Procedia-Social and Behavioural Sciences Journal* **197** 2032-2039

[5] Drăghici S and Anghel Drugărin C V 2015 The Design of The Moebius Mod-6 Counter Using Electronic Workbench Software, *Procedia-Social and Behavioural Sciences Journal* **191** 2316-2324

[6] Anghel C V 2004 Studiu comparativ al functiilor ortogonale Haar si Fourrier la transmisia semnalelor prin cablu, cu aplicatie in Matlab, *Analele Universității "Eftimie Murgu" Reşiţa, Fascicula de Inginerie* **XI**(1) 1453-7397

[7] Gillich G R and Praisach Z I 2013 *Damage-patterns based method to locate discontinuities in beams*, Proceedings of SPIE, Volume: 8695, Article Number: UNSP 869532

[8] Gillich G R and Praisach Z I 2013 Detection and Quantitative Assessment of Damages in Beam Structures using Frequency and Stiffness Changes, *Key Engineering Materials* **569-570** 1013-1020

[9] Janik Z and Zakova K 2011 *Online design of Matlab/Simulink and SciLab/Xcos block schemes,* 14th International Conference on Interactive Collaborative Learning (ICL2011), IEEE Conference Publications, pp 241-247

[10] Anghel Drugărin C V 2012 *The Dichotomy Method Programmable in LabView*, Editura Politehnica Timisoara

[11] Chioncel C P, Gillich N and Tirian G O 2011 Wigner Ville Distribution in Signal Processing, using Scilab Environment, *Annals of Eftimie Murgu Universiy* **XVIII**(2) 101-106

[12] Chioncel C P, Spunei E, Anghel Drugărin C V and Piroi I 2015 Direct Self Control Structure for the Asynchronous Machine, Implemented in Scilab /Xcos, *Acta Electrotehnica* **56**(3)

[13] Magyar Z and Zakova K 2010 *Using SciLab for building of virtual lab*, 9th International Conference on Information Technology Based Higher Education and Training (ITHET), IEEE Conference Publications, pp 280-283

[14] Anghel C V 2008 *E-Learning a new method for Environments of 21st Century*, The 14th International Scientific Conference THE KNOWLEDGE – BASED ORGANIZATION, Academia Fortelor Terestre "Nicolae Balcescu", Sibiu, 27-29 November

[15] Anghel C V, Drobot L I, Constantin A, et. al. 2009 *Aspects about information technology education in university environment,* 5th Balkan Region Conference of Engineering and Business Education, October 15-17, Sibiu, România

[16] Beirao P and Valerio D 2012 Creation of Virtual Graphic Interface Applied to a Process Control System, *Procedia Social and Behavioural Science* **46**(1) 565-569

[17] Olteanu R L, Dumitrescu C, Ghorghiu G and Gorghiu L M 2009 Related aspects to the impact of virtual instruments implementation in the teaching process, *Procedia-Social and Behavioural Science* **1** 780-784

Gamification in the context of smart cities

M R Zica[1], A C Ionica[1] and M Leba[2]

[1]University of Petrosani, Engineering and Management Department, Str. Universitatii, 20, 332006 Petrosani, Romania
[2]University of Petrosani, Computer and Electrical Engineering Department, Str. Universitatii, 20, 332006 Petrosani, Romania

E-mail: monicaleba@yahoo.com

Abstract. The recent emergence of smart cities is highly supported by the development of IT and IoT technologies. Nevertheless, a smart city needs to be built to meet the needs and requirements of its citizens. In order to build a smart city it is necessary to understand the benefits of such a city. A smart city is, beyond technology, populated by people. A smart city can be raised by its citizens' contribution, and gamification is the means to motivate them. In this paper we included gamification techniques in the stage of capturing the citizens' requirements for building a smart city. The system proposed in the paper is to create an application that allows the building of a virtual smart city customized by each user. From this virtual city, the most relevant features are extracted.

1. Introduction. Literature review. Related work

A smart city is an urban development vision to integrate information and communication technology (ICT) and Internet of things (IoT) technology in a secure fashion to manage a city's assets [1].

Its necessity is considered to be driven by the massive urban growth [2].

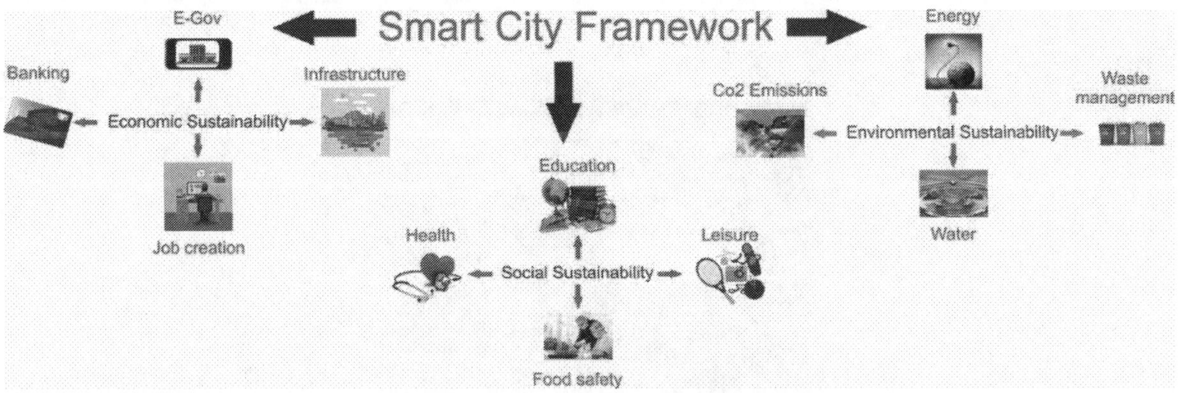

Figure 1. Smart City framework

In the context of a city and most importantly of Smart Cities, citizens became a very important piece in the city's development. In order to keep them engaged, real quality services have to be offered through the help of the application which is to be implemented [3].

Gamification is a term introduced in 2002 by Nick Pelling [4], but became popular only in 2010 [5]. Gamification represents the application of game principles and design elements in non-game contexts [6]. Gamification is about motivation, fun, rewards, competition and challenges. This concept has been applied in various fields even though it was not identified as gamification.

Connecting with the Smart City framework designed by Arup J. Paul [2], hereby are examples of the concept applied in different fields.

1.1. Economic Sustainability

Regarding the Economic Sustainability, gamification was already applied in the following fields: banking, eGovernment, Infrastructure and Job creation.

1.1.1. Banking. Gamification can motivate people to use digital channels in order to do banking operations, which leads to less operational costs for the bank. ICICI is a bank that applied gamification in the saving area of its products: deposits, savings, money management. They have designed a platform which contains a leaderboard that shows which customer/player has achieved the best score by gaining points [7].

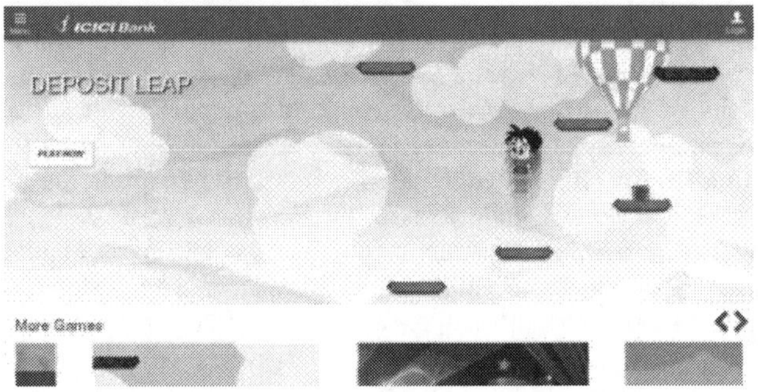

Figure 2. ICICI Bank application screen shot

1.1.2. e-Gov. Hawaii government wanted to implement a platform that can change its citizens' perception towards the online government services. The platform was launched in April 2014, with the goal of allowing citizens to access information about their state and county government interactions, services they like, paper and gasoline saving they have made and so on. The result was the online services usage increase instead of paper [8].

Figure 3. Hawaii.gov platform screen shot

1.1.3. Infrastructure. In UK and Australia gamified programs were implemented in order to encourage citizens to walk and cycle. In Australia the result was that 35% of the car trips to school were replaced with healthy transportation means. A program for public transportation was introduced in Singapore. The goal was to motivate citizens to use the public transportation in other intervals than the rush hours. They were included into a raffle and received rewards. After 6 months of testing in 2012, the result of the program was an estimate of 8% shift from the rush hour to the normal hours [9].

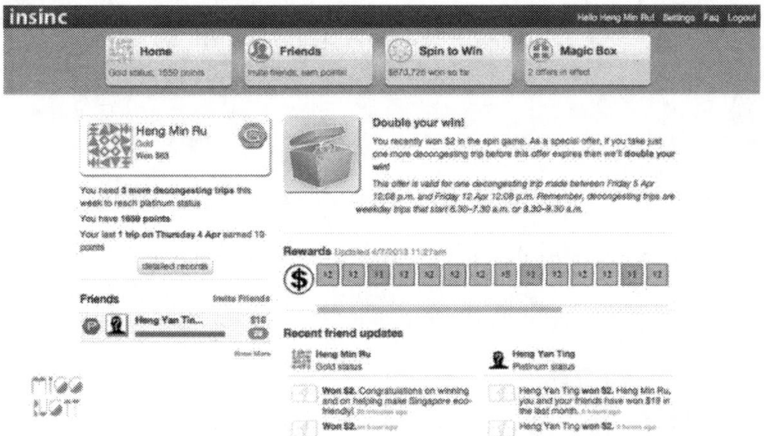

Figure 4. Insinc platform screen shot

1.1.4. Job creation. Gamification has also been applied in the field of human resources and personnel recruitment. Marriott Hotels developed an application for people who wanted to work for their company. The application, Marriott My Hotel, allows people to virtually walk inside a hotel in order to become familiar with what their potential job would imply [10].

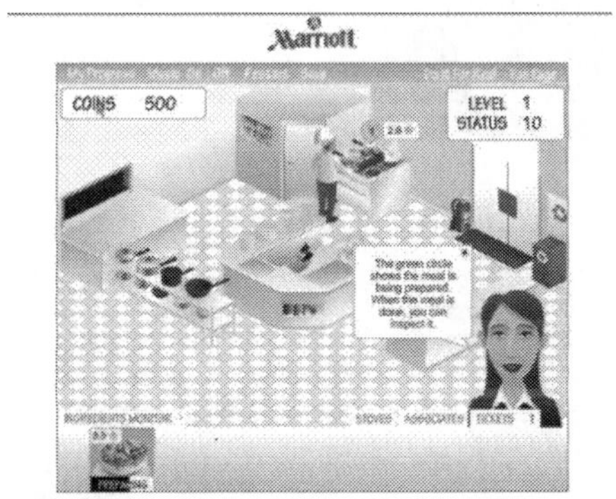

Figure 5. Marriott My Hotel application screen shot

1.2. Environmental Sustainability
Regarding the Environmental Sustainability, gamification was already applied in the following fields: CO2 emissions, Water, Energy and Waste management.

1.2.1. CO2 Emissions. VTT Technical Research Centre of Finland designed an application tested in Helsinki, Nice and Vienna. The application provides alternative plans in order to minimize the energy costs and CO2 emissions. The result of this project was the energy costs reduction by 15% and CO2 emissions by 30% [11].

Figure 6. VTT Technical Research Centre of Finland application screen shots

1.2.2. Water. Waterwise and WRc is searching for other companies to collaborate in creating a platform which has the goal to change the relationship between water companies and customers by using gamification and also reduce water consumption [12].

Figure 7. Waterwiseand WRc platform screen shot

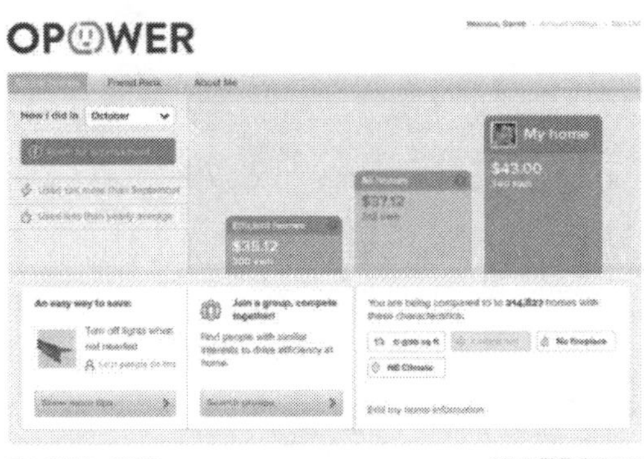

Figure 8. Opower application screen shot

1.2.3. Energy. Opower is one of the companies that implemented gamification into an application meant to encourage energy saving. 75 utility companies became partners with Opower in this project of saving the planet and making green living fun [13].

1.2.4. Waste Management. A team from University of Adelaide designed a web based game, Super Sort, which has the goal to increase awareness about the sustainable waste management system. The game was tested by 300 students and due to its success, the team is trying to also create a mobile application [14].

Figure 9. Super Sort application screen shot

1.3. Social Sustainability
Regarding the Social Sustainability, gamification was already applied in the following fields: Health, Education, Food safety and Leisure.

1.3.1. Health. In this field there are multiple examples.

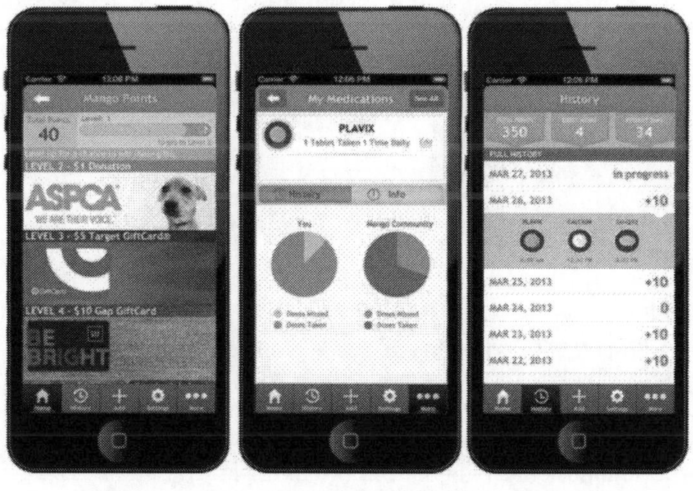

Figure 10. Mango Health application screen shots

One of them is Mango Health application which was designed to encourage people to take their medications on time. The application contains reminder alerts for the medication time. People receive gift cards for shopping at Target or GAP or they can donate money to charity [15].

1.3.2. Education. Photomath is an application which incorporates a scanner. After scanning the math problem you want to solve, the application helps you understand math, giving you the sense of companionship for learning [16].

Figure 11. Photomath application screen shot

1.3.3. Food safety. A manufacturing company needed a food safety hazards training for their employees. In order to motivate them, a virtual kitchen with potential hazards was designed. The employees had to identify the hazards and answer correctly the question about it. They received badges, trophies and certificates [17].

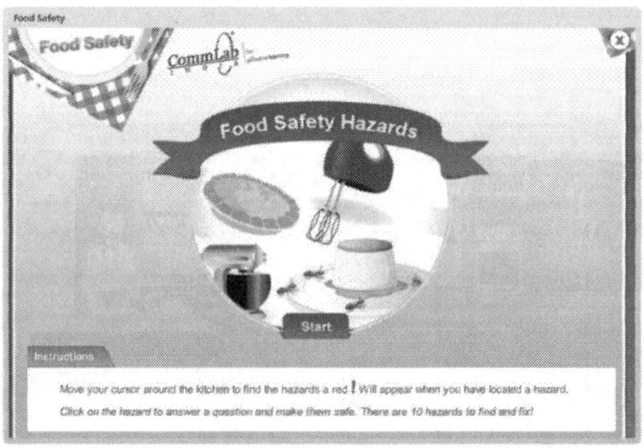

Figure 12. Food Safety Hazards platform screen shot

1.3.4. Leisure. CWT Solutions Group wanted to improve the loyalty and compliance of travelers, therefore they have designed an application called Game On. The goal is to convince travelers to be loyal not to their favorite supplier but to the offered program [18].

Figure 13. Game On application screen shot

2. The invisible wires that link a community – internet and gamification

Smart City represents the future, and it can be built with and for the people. This is emphasized by the benefits of a smart city, taking into account each sector's contribution to the real success of the smart city.

The main sectors of a city can be connected throughout the internet and citizens' contribution. The citizens can be very well motivated by using gamification techniques. The Internet of things represents the motor of a smart city, but the citizens are the fuel that makes everything act together.

We live in an era that enables the potential connection of people and increases social cohesion by IT solutions.

Not all the members of a community can be connected by smart solutions, because some of them do not have access to digital equipment or do not have the knowledge to use it. Another cause can be the pressure put on solidarity by different risks. Finally, some citizens can organize digitally connected groups which may lead to being a threat to cohesion and inclusiveness.

In order to solve these problems, the authorities and citizens should find solutions together by collecting citizens' opinion about the existing problems.

Connecting things has lately proved to be easy to achieve through the help of the Internet of things. Connecting people and citizens' involvement in a community has always been a continuous challenge. The citizens and the authorities are very important pieces in a community. By creating a real constructive relationship between these two pieces, the city will improve in all fields – infrastructure, jobs, energy saving, waste management, education, health.

Gamification can be a real solution for involving citizens in building a smart city and also in its customization so that it responds to everyone's needs.

Every citizen has his own point of view regarding the city he lives in therefore the smart city has to be designed so that each citizen can feel at home.

Gamification offers the tools to encourage the citizens to be involved, even if they do not have access or IT knowledge. Gamification represents motivation techniques specific to games. Neither all games are electronic, nor does gamification mean only IT.

Smart city represents the acceptance or not of the future. Therefore, using gamification can also mean the development of IT abilities for all citizens.

Smart city should be perceived as adding services that increase the quality of life but do not influence or eliminate traditions and originality of the place.

This also represents the subject of the research consisting in suggesting a platform which, at the beginning, allows all users to build their own virtual city. This stage represents the requirements capturing stage.

Creating an application which gives the citizen the opportunity to create their own customized city, would represent a real feedback for the authorities for what they can do for the city and its community.

3. Methodology

The suggested methodology is based on the software products life cycle stages. We chose Waterfall life cycle, because it is easier to highlight and because it is included in all the other life cycles, even in the most recent, Agile. The methodology includes also the research methods grouped in 3 stages: pre-use, during-use and post-use. Each stage is represented by specific elements of motivation based on gamification.

Table 1. Waterfall lifecycle with gamification techniques

Waterfall phases	Methods	Gamification
Requirements analysis System design	Pre-use: • Input survey • General survey • Think aloud exercise	At the beginning, surveys have to be organized in order to establish the points of interest for the community. In order to have as many surveys completed as possible, from various categories of population, people have to be motivated. At this stage, the motivation has to be a reward. The reward can be a future application badge representing that this person is a founder of the virtual community or points which can be exchanged to virtual money after the virtual city creation that can be used for real tax payments.
Implementation Verification (testing & deployment)	During-use: • Data logs • Staged usage situations • Experience sampling method	The user will be rewarded for each improvement made to his city. There will also exist competition with points, badges and leader boards between the users of the application. The application has to be fun, so mini-games of solving potential problems which appeared in the city will be implemented. The user that found the best solution will receive a badge and bonus. Sustainable solution can also be implemented in the real city.
Maintenance	Post-use: • Self-report data collection techniques • Data logs	The user who found sustainable solution for the city's problems that can be implemented into the real city will become an

	honourable citizen of the real city and will receive tax discounts from the authorities.

The architecture of the system built on the proposed methodology takes into account the steps that the application user must follow, for the first stage in which the customer requirements are identified.

Like so, the first stage consist of registering the users, stage in which data about each user involved in the study is gathered, whether we are talking about data that the user can define or pre-existing data about the user (e.g. data gathered from social media websites)

Based on big data algorithms, relevant information is extracted and used to complete the users profiles. These profiles represent the starting point for each Virtual World from the application. This is because the Initial Virtual World will contain only the elements and tools that are fit for the user based on his profile. As the application gathers more users, ad hoc groups will be formed based on the users' common interests identified by looking at their various interest fields memberships (as seen in figure 13), the completed challenges and the complementarity of the wined/used tools.

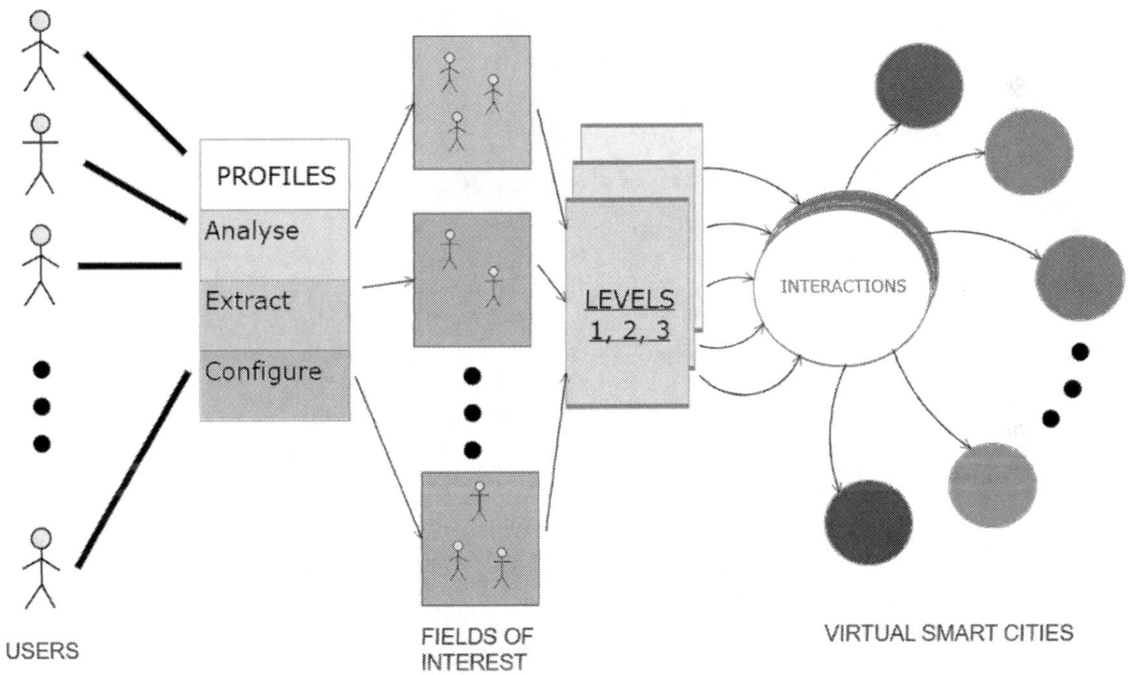

Figure 14. System architecture

This ad hoc formation is achieved by finishing the first three levels. After this, each user will be rewarded with points for the personal achievements and team work and he will be able to check his position in a leader board.

The achieved points also bring badges which will help establish the role of each user in the virtual city (mayor, hospital manager, etc.)

In the last stage, each user's achievements will be integrated, the roles will become dynamic and the participants will be able to earn cofounder badges as well as change points for virtual currency.

4. Gamification in the requirements capture stage – building the" virtual city"

The application which we want to develop is about Jiu Valley, a mono industrial area of Romania. Considering the fact that this industry, mining, has already started to disappear, the cities of this area need a real change. This change is mostly given by the citizens who, for the moment, are trying to find jobs in other towns or countries. Implementing here a smart city application can re-launch the area and keep its citizens engaged.

Figure 15. From MINE to MY City application screen shot

The application will have a log in section.

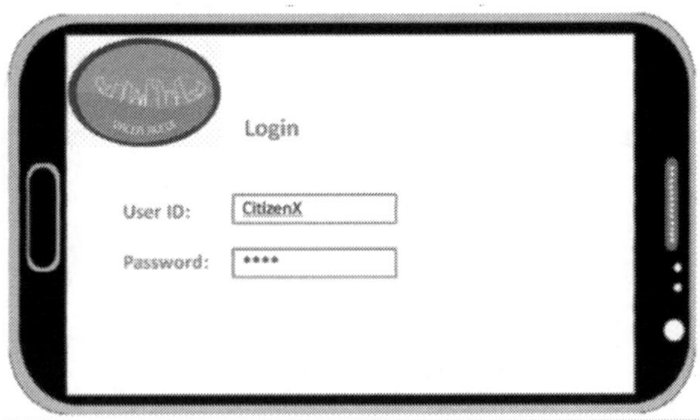

Figure 16. Log in application screen shot

After creating an account, the user has to define his profile consisting in filling some data: Age; Sex; Level of study; Income of the family; Legal status; Children; Work; Hobby.

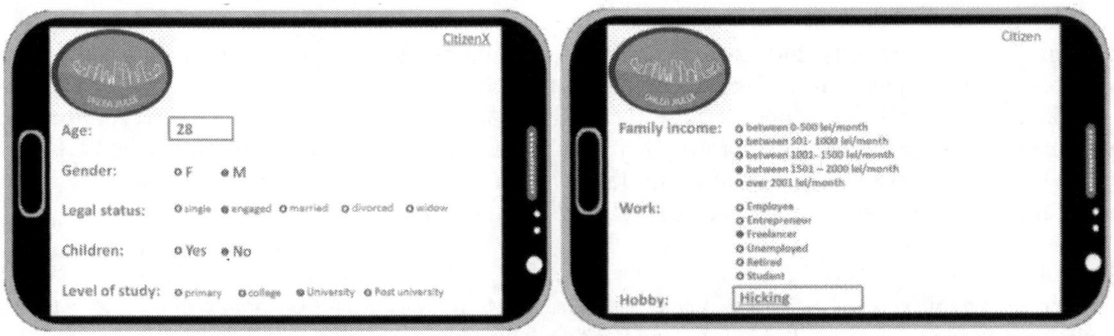

Figure 17. Profile screen shot

After completing the profile, the user starts building its own city. The first step is to choose a name for the city. In the next screen, the user will have the possibility to indicate the interest field:

Infrastructure – improve the quality of the existing roads, create new roads, and implement local transportation;

Health – improve the access to medical services, create events for encouraging a healthier lifestyle (healthy food, fitness and other competitions for the community);

Leisure – create more recreational places for every age category, create workshops with different activities (pottery, painting, dancing, handmade products, etc.), create the possibility of practicing sports for fun (riding, archery, swimming, paint ball, etc.);

Eco friendly – implementation of rewards for energy saving, water saving, selective collection of waste, participation in organized events for keeping the city and the surroundings clean;

Culture – better inform the community about the cultural events from the city, organizing more cultural events;

Personal and professional development – organize free courses for job orientation and personal development.

The elements which form a smart city will be identified (the sensors, etc.) and will become available for building the „virtual city" [19].

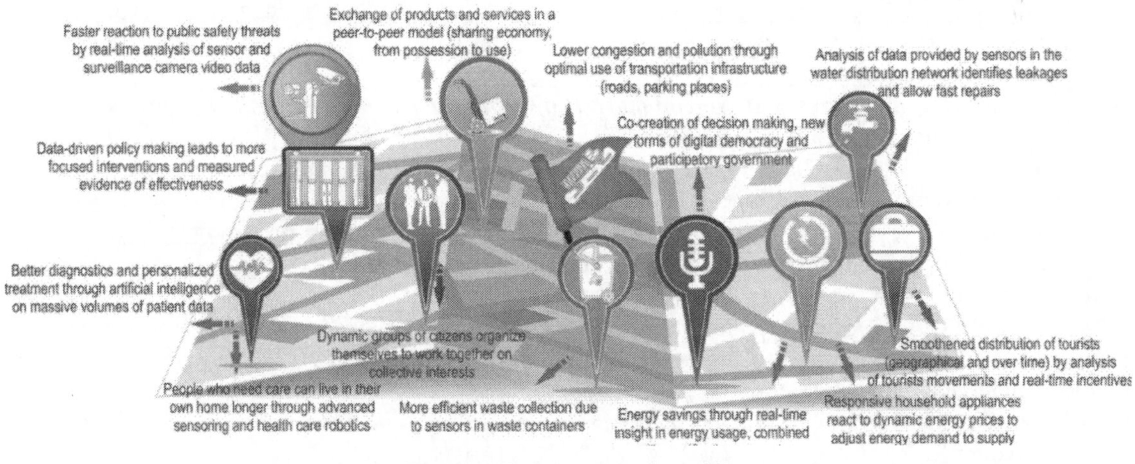

Figure 18. Smart city benefits

In the initial phase a survey will be conducted in order to determine useful user centred insights. This will be based on both direct and indirect experience. It will be important to monitor the citizens' desire for participating in this research and to use a mobile software application.

Data will be gathered in order to segment the market based on a general survey regarding the demographic issues, habits, most useful services, perceptions, beliefs, expectations, familiarity with technology. In this phase, the subjects for all the research phases are selected. Two target groups are also selected for the Think Aloud exercise. The first group is formed by citizens who are comfortable with the technology and who will project their own virtual city by using the app. The second group is formed by citizens who do not want to use the app, but their purpose is to refine the most common smart elements form the smart cities generated by the first group.

5. Conclusions

Most of the people of the IT generation can create a smart city, but not all of them will be interested in continuing its development. It is very important to keep the users engaged. This is the reason why gamification is a very important tool that can and should be used when designing a smart city application. People need motivation, fun, competition and recognition in their lives. Considering the fact that the community is the key factor in a city, the community has to be interested and motivated to really participate in the city's life.

The paper proposes a new methodology that integrates gamification into the life cycle of a software product called "From Mine to My City". It also proposes a system architecture that targets the first stage of the lifecycle that means capturing user / citizen requirements. This architecture starts from the individual user profiles and, after applying algorithms for analysing and extracting user characteristics and using gamification techniques, virtual cities are built /designed.

References

[1] ***2017 https://en.wikipedia.org/wiki/Smart_city
[2] Arup J P 2016 Smart City Sustainability Framework & Gamification, @medium.com
[3] Pinto J and Rossetti R 2016 *Fostering citizen engagement in smart cities through gamification*
[4] Marczewski A 2012 *Gamification: A simple introduction*, Amazon Digital Services Inc
[5] ***2017 https://en.wikipedia.org/wiki/Gamification
[6] Huotari K and Hamari J 2012 *Defining Gamification – A Service Marketing Perspective*, Proceeding of the 16th International Academic Mind Trek Conference, pp 18
[7] ***LTP 2016 *Gamification in Banking: From Transaction to Experiences*, https://letstalkpayments.com/gamification-in-banking-from-transactions-to-experiences/
[8] ***2015 https://www.egov.com/news/press-releases/2015/02/23/my-hawaii-gov-recognized-as-2015-harvard-ash-center-bright-idea-in-government
[9] Yen B 2016 *How 'gamification' can make transport systems and choices better for us*, Griffith University, Australia
[10] Xu F, Weber J and Buhalis D 2014 *Gamification in Tourism*, Springer International Publishing Switzerland
[11] ***2017 https://phys.org/news/2017-02-gamification-consumers-power-consumption-peaks.html, VTT Technical Research centre of Finland
[12] Snowdon C 2014 *Gamifying Water*, Waterwise Conference
[13] Shane K 2013 *3 Mobile Apps for the Gamification of Energy Conservation*, @gamification.co
[14] Babar A 2016 Smart Campus: Gamification of Disposing Waste Sustainably, @ malibabar.wordpress.com
[15] Yee C 2017 *Top Ten Gamified Healthcare Games That Will Extend Your Life*, Gamification Examples
[16] Mechelen E 2017 *Top 10 Education Gamification Examples for Lifelong Learners*, Education Gamification
[17] Reddy S 2016 *Gamification and Game based Learning for Training*, CommLab, India
[18] Carlson W 2013 *Cutting through the buzz on traveler gamification in managed travel*, CWT Vision, Sweeden
[19] Deloitte 2015 *Smart Cities*, GovLab, The Netherlands

Driver monitoring system for automotive safety

A E Lörincz[1], M N Risteiu[1], A Ionica[2] and M Leba[3]

[1]University of Petroşani, Doctoral School, Universităţii str., no. 20, 332006 Petroşani, România
[2]University of Petroşani, Management Department, Universităţii str., no. 20, 332006 Petroşani, România
[3]University of Petroşani, Computer and Electrical Engineering Department, Universităţii str., no. 20, 332006 Petroşani, România

E-mail: risteiumariusnicolae@yahoo.com

Abstract. The lifestyle of a person is a very active one from all points of view. He travels great distance every day, with car or on foot. Tiredness and stress is found in every person. These can cause major problems when driving up and driving in small or big distances by car. A system developed to prevent the dangers we are prone to in these situations is very useful. System that can be used and implemented both in the production of current cars and the use of those not equipped with this system.

1. Introduction

The systems used today by different car manufacturers include sensors or video cameras that analyze driver behavior [1].

Spanish scientists have created a system of four sensors that monitor different physical parameters and the steering position of the driver. As a result, a series of sound signals are issued in order to warn the driver. Toyota prepares projects that include, among many others, a system that contains a steering wheel sensor that analyzes the heartbeat of the driver to avoid accidents caused by heart attack while driving [2]. The engineers from Jaguar work to develop an active safety system to detect whether the driver is careless or is about to fall asleep by reading his brain waves [3]. Bosch Group has created a system that identifies signs of fatigue. They designed a function that computes the driver's fatigue level, depending on the frequency of the steering movements and other parameters, including the duration of the trip, the use of direction indicators and the time of day [4].

In recent years, there has been an increase in car accidents caused by drivers who fall asleep due to fatigue, or because of spontaneous manifestations of some disease. To prevent such accidents, there is an increased need to monitor the health of drivers, as well as to detect anomalies at an early stage. Non-contact sensors can estimate sleepiness of drivers based on pulse measurement.

Omron Corporation has announced a non-contact pulse sensor prototype. It can accurately measure heart rate inside the vehicles, where measurements are difficult to perform. The bio-detection technology prototype is able to measure the driver's pulse using radio waves and a dedicated algorithm.

The device transmits radio waves against driver and uses the return waves to detect a variation of about 100 micrometers of the blood vessels when the heart beats. The deviation is then converted into a pulse signal, which OMRON's algorithm uses to calculate heart beats. This contactless sensor allows real-time monitoring of driver health without discomfort and without affecting driving.

With this type of contactless sensor combined with other technologies such as Cut-Edge AI On-board Sensor that measures real-time driver status and Wearable Sensor (worn on the wrist) which measures blood pressure continuously, anomalies of the state of the driver will be detected with greater accuracy and certain decisions can be made to prevent problems.

OMRON combines several detection technologies that can accurately detect and estimate the risk of abnormalities before they appear. Using such a sensor inside the car can be very useful for quickly detecting driver's cardiac problems that could cause accidents if he has a crisis for some reason. OMRON wants to be the first company to create an on-board health system capable of preventing accidents caused by the driver's state of health and contributing to a safe and comfortable driving oriented society [5].

Figure 1. Contactless sensor

2. System architecture

Wearables offer great opportunities to intensify our driving experience. Wearable devices like smart glasses and smartwatches are mainly used to enhance the driver's experience and improve safety. More specifically, Mercedes, BMW and Nissan have built special applications that monitor the speed and fuel efficiency via on-board sensors and even the driver's fatigue level via heart rate sensors.

Ford S-Max Concept provides the ECG Heart Rate Monitoring Seat, which continuously monitors and records the driver's heart rate. S-Max Concept promised to enable the function of heart attack detection. Then, it will notify emergency services as well as a notification to the driver to safely come to a stop. Beyond heart monitoring, S-Max Concept is able to monitor the Blood Glucose Level. Sensors and transmitter fob are equipped to continuously measure blood sugar levels in the driver. A notification will send an alert if the level of glucose becomes dangerously low.

2.1. Selection of sensors

Sensors used to detect driver behaviour changes may be of different type and functioning principle.

The criteria for choosing these sensors are:

- Power supply voltage
- The current used
- Dimension
- Working temperature
- Case
- Mode of operation
- Output configuration
- Force sensitivity range
- Response time
- Lifetime
- Vibration / Noise Sensitivity

2.1.1. Pressure sensor FSR400. These sensors provide a simple way to sense the force that acts on it. The operation of this sensor can be compared with the operation of a resistive potentiometer that changes its proportional value to the force applied to it. It can be connected together with a resistor with a value of 10K and using the principle of the voltage divider to read the voltage drop across the resistor. Thus, the sensor changes its resistance and with this it changes the current through the circuit and also the voltage drop across the 10k resistor. This resistor's value is read by the microcontroller on an analogic input with a 10-bit resolution. The sensor has a 10% precision.

The sensor shows a decrease in resistance when the applied force increases.

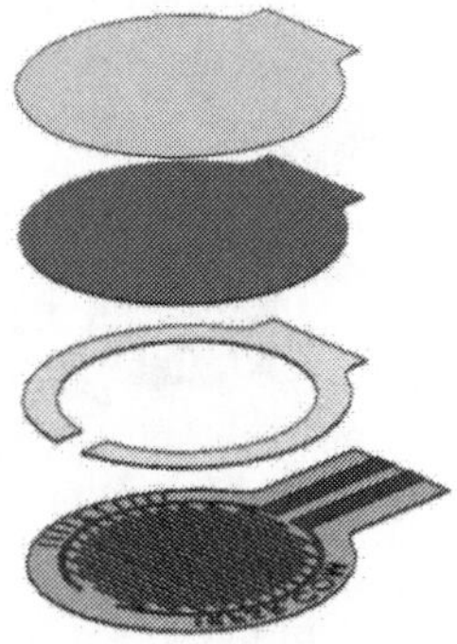

Figure 2. FSR sensor layers

Table 1. Performance specifications: [6]

Parameter	Value
Force Sensitivity Range	~0.2 to 20N
Break Force (Activation Force)	~0.2N min
Part-to-Part Force Repeatability	± 6% of established nominal
Single Part Force Repeatability	± 2% of initial reading
Hysteresis	+ 10% Average
Long Term Drift	< 5% per $\log_{10}$(time)
Force Resolution	Continuous
Stand-Off Resistance	> 10MΩ
Switch Travel	0.05mm
Device Rise Time	<3 microseconds
Maximum Current	1 mA/cm^2 of applied force
Hot Operation	-15%
Cold Operation	-5%
Hot Humid Operation	+10%
Hot or Cold Storage	-10%
Hot Humid Storage Temperature	+ 30% of established nominal resistance
Thermal Shock	± 2% typical
Tap Testing	-10%
Constant Load	-5%
Chemical Resistance	The following chemicals do not affect the operation when applied to the outside of the sensor: cola, coffee, isopropyl alcohol, soap solution, household cleaners. No others tested. Application is a single drop on the exterior of the sensor that is allowed to

	soak until evaporation and does not enter the sensor.
Sensitivity to Noise/Vibration	No effect
Positional Resolution	0.075 to 0.5 mm (0.003" to 0.02")
Positional Resolution	Better than ± 2% of full length
Force Sensitivity Range	~0.2 to 20N

2.1.2. Bending sensor FS7548. Bending sensor provides a simple way to grasp the degree of bending of an environmental element. One side of the sensor is printed with a polymer ink that has conductive particles embedded in it. When the sensor is straight, the particles give the ink a resistance of about 30k Ohms. When the sensor is bent away from the ink, the conductive particles move further apart, increasing this resistance (to about 50k-70K Ohms when the sensor is bent to 90°, as in the diagram below) [7].

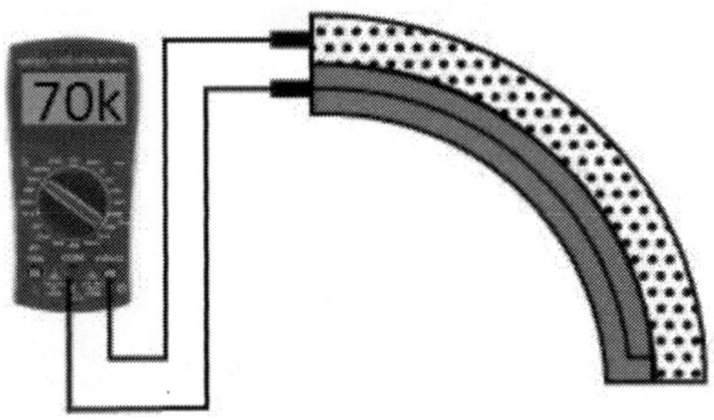

Figure 3. Conductive particles further apart – 70k [7]

When the sensor straightens out again, the resistance returns to the original value. By measuring the resistance, you can determine how much the sensor is being bent [7].

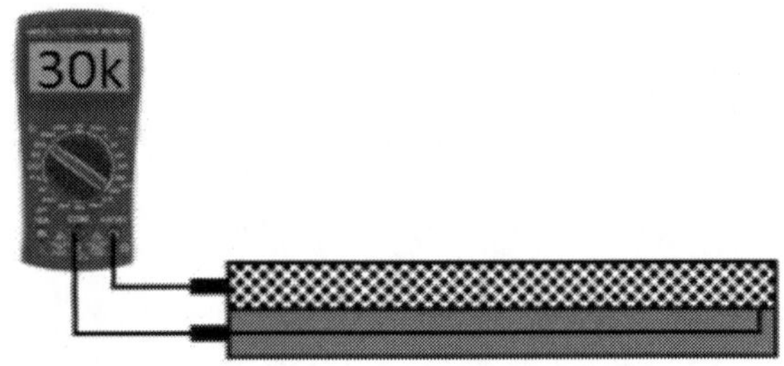

Figure 4. Conductive particles close together – 30k [7]

It changes its value depending on the degree of bending it is subjected to.

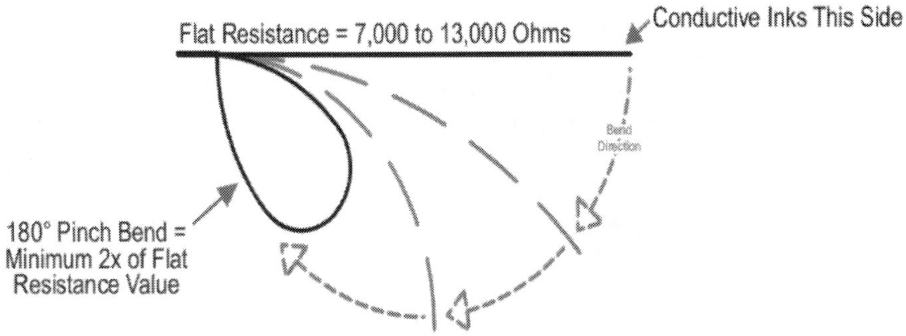

Figure 5. FS7548 principle

Table 2. Performance specifications [8]

Parameter	Value
Life Cycle:	>1 million
Height:	0.43mm (0.017")
Temperature Range	-35°C to +80°C
Flat Resistance:	10K Ohms ±30%
Bend Resistance:	minimum 2 times greater than the flat resistance at 180° pinch bend
Power Rating :	0.5 Watts continuous; 1 Watt Peak

2.2. Hardware design

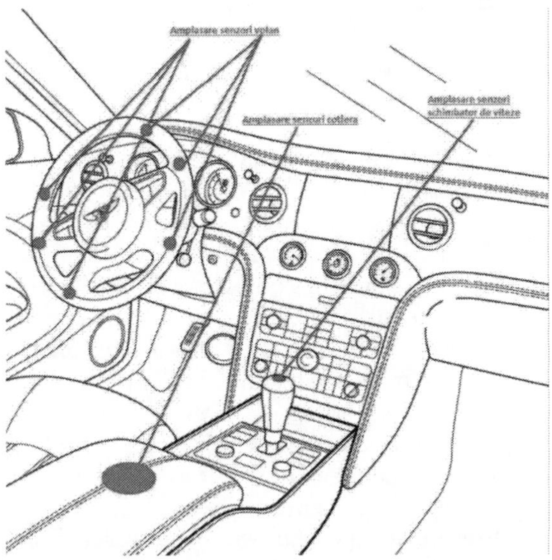

Figure 6. Sensor location in the passenger compartment

The system includes three pressure / touch sensors that are located in the vehicle, a light sensor, a microcontroller that receives sensor data, interprets and memorizes it, fills in from the vehicle's battery, and a graphical display that alerts the driver and also a buzzer.

The three pressure sensors are located on the steering wheel, gearshift and armrest. They transmit data regarding the driver's position if the positioning of both hands is on the steering wheel and if the left hand is on the steering wheel and the right hand is on the gear selector or on the armrest.

The light sensor has a very important part in the system, it detects whether the vehicle is driving at night or during the day. Driving at night is much more exhausting and driving time is lower than driving if it is daytime.

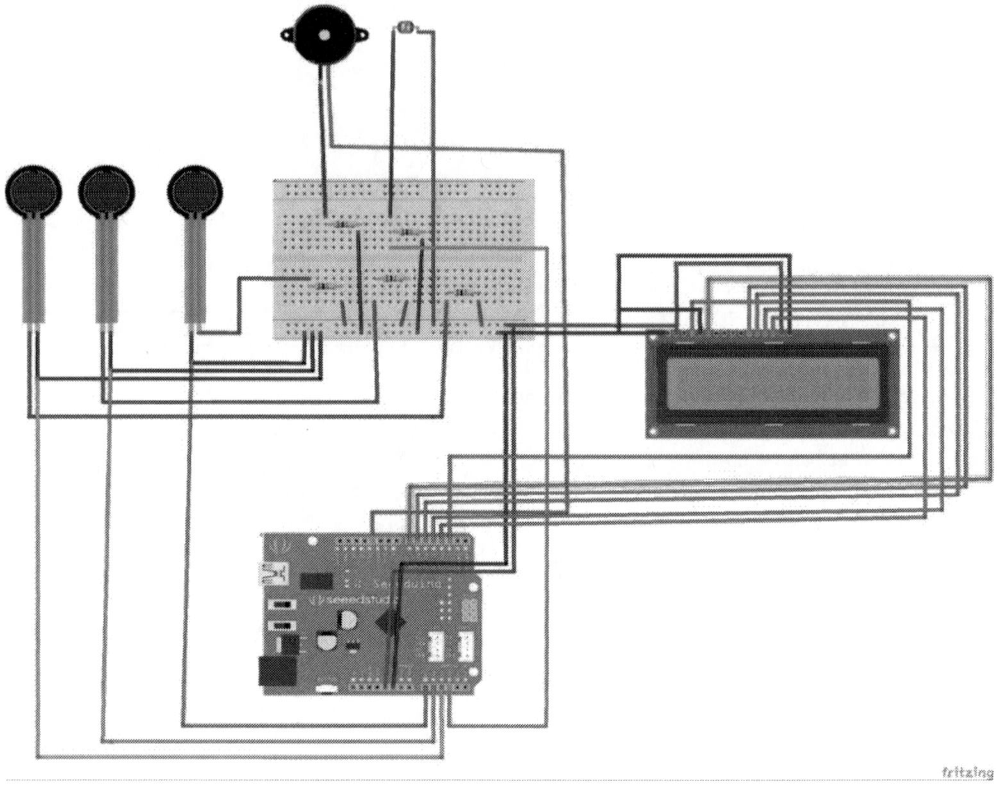

Figure 7. Circuit diagram

2.3. Software design

When the vehicle's engine is started, the system's power supply is generated and the pressure sensor, light sensor and time variable are read out. The entire driving time is counted by the microcontroller and, depending on the light sensor value, it generates decisions regarding the driver's warning. Night driving time should be shorter than daytime driving time, because the driver's sight weakening is faster when the road illumination is lower.

If no one of the sensors is active after starting the engine, the system will send only one warning after that it will consider that the engine is turned on for heating or for maintain. If the system is in this situation, driving time is not counted. The system will send again warnings only when at least one of the sensors from the steering wheel and the gearshift will be reactivated.

Time spent driving is stored by the central unit (microcontroller) and memorized before the engine stops (power off). At startup, a check is made if the driver has taken a long break or a few minutes. If the break was a few minutes then the timing continues (it adds to the previous driving time) and if the break exceeds a few hours then the variable will reset (time_driving = 0).

The pressure sensors are read every second and the driver's position is monitored if at least two sensors (one mandatory on the steering wheel) are not activated for 10 seconds then the driver is warned. This is the most common occurrence when the driver speaks on the phone, changes the radio channels, search for something in the car, etc. This means an imprudent drive and a certain degree of danger to it. If no sensor is activated then the alert is on (the driver does not hold his hands on the steering wheel). This case is monitored even when the car is passed to the automatic pilot.

During this time, the CPU registers the activity of the sensors and the average duration of their operation. The average calculated by the algorithm is the driving mode of the vehicle owner and is stored. This average is continuously compared to the driving style and when changing the driver, it can be reset or the change is made in a few hours.

The pressure sensors are read and, if the value is variable over a short period of time, then the driver is warned. Usually when the driver falls asleep at the wheel, he weakens the grip of the steering wheel, and in a short time the grip is strong (when he falls asleep). If this behavior is detected the system immediately warns the driver to stop the vehicle.

In Figure 5 is shown the entire algorithm as presented above.

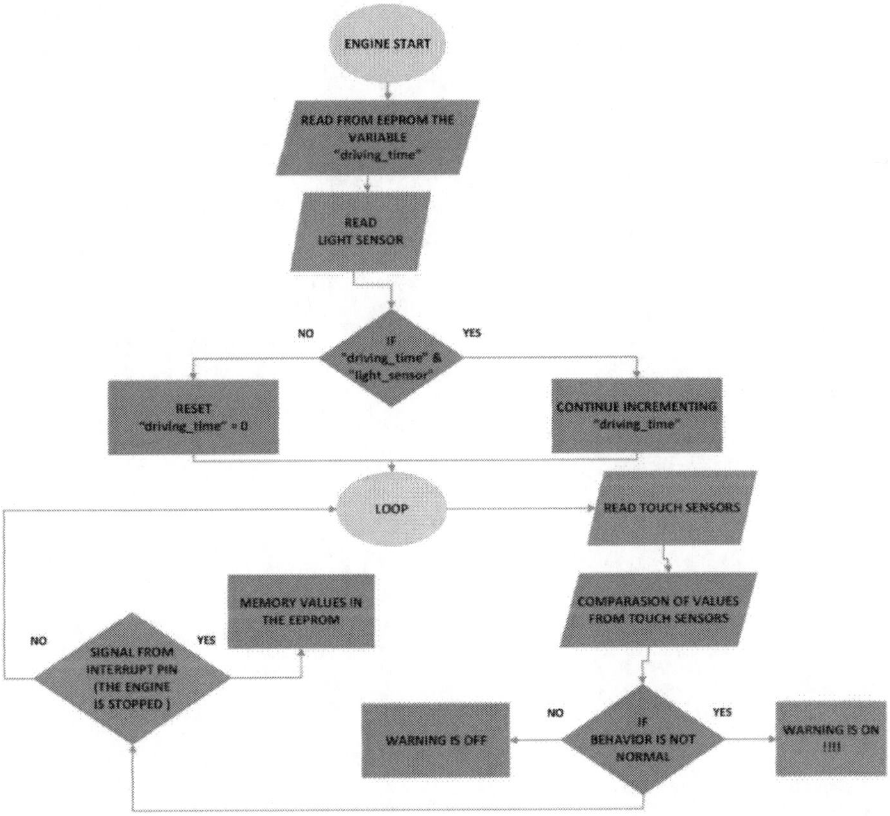

Figure 8. Algorithm diagram

3. Conclusions

Due to the fact that we are very busy every day, we get to spend a lot of time driving most of the time without realizing that we are tired and this can be a danger to us or other traffic participants. This kind of system will certainly improve traffic and life safety.

Using this system means easy mounting inside the car of several sensors together with the processing and alarm device (microcontroller and buzzer). The differences between vehicles having different manufacturers do not affect the implementation of this system, as is for the implementation of other systems.

The cost of purchasing the components/devices is very low and also the cost of implementing the system into the car is small.

The driver alert mode is adjustable to the user needs, can go from silent and only visual but it can be also loud. The buzzer used in this system produces a low frequency sound, which is not disturbing for the car passengers.

The system can detect the driver's drowsiness from the steering wheel sensors and warn the driver in this case to take a driving break. Attention to traffic, correct steering position are also monitored and, if necessary, alerted by this system.

References

[1] Purgaru D, Bălu C and Vîlceanu A 2017 *Un sistem de senzori poate detecta simptomele de oboseală ale şoferilor şi declanşează o alarmă*
[2] Mărculescu O 2011 *Tehologie Toyota - controlarea direcţiei pentru evitarea coliziunii*
[3] Toma S 2015 *Jaguar Land Rover lucrează la un sistem care să citească undele cerebrale ale şoferului*
[4] Botezatu L 2012 *Sistemul Bosch de detectare a starii de oboseala la volan salveaza vieti*
[5] ***http://www.omron.com/media/press/2016/09/c0929_2.html
[6] ***http://interlinkelectronics.com/datasheets/Datasheet_FSR.pdf
[7] ***https://learn.sparkfun.com/tutorials/flex-sensor-hookup-guide
[8] ***https://cdn-shop.adafruit.com/datasheets/SpectraFlex.pdf

IOP Conf. Series: Materials Science and Engineering **294** (2017) 012047 doi:10.1088/1757-899X/294/1/012047

A small step in VLC systems – a big step in Li-Fi implementation

S M Rîurean[1], A A Nagy[1], M Leba[1] and A C Ionica[2]

[1]University of Petroşani, Department of Computer and Electrical Engineering, Universităţii str., no. 20, 332006 Petroşani, Romania

[2]University of Petroşani, Department of Management, Universităţii str., no. 20, 332006 Petroşani, Romania

E-mail: nagyalin85@gmail.com

Abstract. Light is part of our sustainable environmental life so, using it would be the handiest and cheapest way for wireless communication. Since ever, light has been used to send messages in different ways and now, due to the high technological improvements, bits through light, at high speed on multiple paths, allow humans to communicate. Using the lighting system both for illumination and communication represents lately one of the worldwide main research issues with several implementations with real benefits. This paper presents a viable VLC system, that proves its sustainability for sending by light information not only few millimetres but meters away. This system has multiple potential applications in different areas where other communication systems are bottlenecked, too expensive, unavailable or even forbidden. Since a Li-Fi fully developed system requires bidirectional, multiple access communication, there are still some challenges towards a functional Li-Fi wireless network. Although important steps have been made, Li-Fi is still under experimental stage.

1. Introduction to VLC and Li-Fi Technologies

Into the electromagnetic spectrum, the optical frequency range starts from 300 GHz and reaches 30 PHz (Peta Hertz). The optical wireless communication (OWC) refers to infrared (Ir), visible and ultraviolet (UV) bands. This is an enormous spectral range, comparable to the 300 GHz that the radio frequency (RF) band represents.

Visual Light Communication (VLC), as part of OWC, uses the visible light medium between a frequency of 430 THz and 800 THz and distance from 750 nm to 380 nm as optical carrier both for data transmission and illumination, fast pulses of light transmitting data wirelessly [1].

Light Fidelity Li-Fi refers to a VLC system of wireless data communication that uses the conventional light emitting diode (LED) bulbs and is meant to be wireless, full-duplex (bidirectional), Multiple-Input-Multiple-Output (MIMO) network that uses the visible light medium for communication [2].

The Li-Fi technology is expected to rise significantly because of the request through different areas due to its harmless nature to human health. The market is also expected to increase significantly from the massive growth in usage of LED lights as a cleaner technology, economical and energy-efficient, over the other conventional lighting options such as high-intensity discharge (HID) and fluorescent bulbs. Some of the leading vendors operating in the global Li-Fi market are Oledcomm, LVX, PureLifi, GE, and Philips [3].

2. How Li-Fi Works

Although the VLC technology's first demonstrations began in early 1990's in countries like Japan, Germany, Korea the first implementation has been done by a Japan company, on 2003 [4]. Later, Li-Fi has been coined by Harald Haas, professor at the University of Edinburgh, UK, who demonstrated in July 2011 at Global TED Talk, its practical functionality. He explained and demonstrated" Very simple, if the LED is on, you transmit a digital 1, if it's off you transmit a 0. The LEDs can be switched on and off very quickly, which gives nice opportunities for transmitting data. So what you require at all are some LEDs and a controller that code data into those LEDs" [5]. Other researcher was able to prove that LED's could be retrofitted to send information and Haas envisioned light bulbs that could act as wireless routers. "My big idea is to turn light bulbs into broadband communication devices, so that they not only provide illumination, but an essential utility" [6].

In order to communicate data, instead of RF, Li-Fi uses visible light by sending data through a LED light bulb that fluctuates intensity faster than human eye can percept. A LED lightbulb is a semi-conductor light source. It means that the constant current of electricity supplied to an LED lightbulb can be dimmed and dipped, up and down at exceptionally high speeds, without being perceptible to the human eye. To receive the light signals, Li-Fi as VLC system, has a photo-detector and a signal processing element in order to convert the data into a content that is able to receive streams. Data is send into an LED light bulb that has incorporated the signal processing technology and is able to send data embedded in its beam to the photodiode at high speeds [2].

All these the tiny changes in the fast LED's dimming is going to be converted into electrical signal by the receiver module (Rx). This signal is then converted back into a binary data stream that is recognised as web, video and audio applications that run on internet enabled devices [7], [8].

3. Challenges for a Fully Developed Li-Fi Wireless Network

There have been lots of obstacles in front of the Li-Fi technology development and implementation, some of them have already been overcome, others are still in front of a fully developed Li-Fi network.

There are three main components of the Li-Fi's system that have to be taken into consideration: transmitter module (Tx), communication channel (CC) and receiver module Rx.

Regarding the Tx: LED types, as well as the control and modulation techniques of the lighting device are about to be improved to pass the line-of-sight (LoS) limitation, LED bandwidth, control the flickering as well as dimming level [1].

There are few type of LEDs that are suitable for a Li-Fi wireless network that use different technologies to create white light by mixing some colours: there is the phosphorescent LED emitting blue light embedded in a layer of yellow phosphor that turns some of the light into longer wavelengths, yellow and red. The human eye sees in this case a white light.

Another type is the RGB trichromatic LEDs that are merged into a distinct device to emit white light. This kind of LED enables easy color interpretation by adjusting each color individually.

Although the phosphorescent LEDs are cheaper than trichromatic LEDs, RGB LEDs are preferred for dual-use because they have a faster rise-time (high bandwidths) and the total throughput can be tripled when each of the three colors can be modulated separately. High-speed communications require either high bandwidths and/or the use of spectrally-efficient modulation schemes [9].

High bandwidths can be achieved when the signal from the LED can have the right modulation scheme. Most of the commercial phosphors LED used for lighting have rather long fluorescence lifetimes (of the order of milliseconds) and therefore, both the on-off switching and bit rate are limited.

The Colloidal Quantum Dots QDs [10], have much shorter fluorescence lifetimes, of the order of nanoseconds, and therefore, much faster switching rates for the LED device are enabled. But, long-term exposure to CdTe (cadmium telluride) quantum dots causes functional impairments in live cells [11].

Therefore, each and every one of the above mentioned LED types has its advantages and disadvantages in different scenarios and situations, for a suitable Tx transmitter data and lighting device.

As for the communication channel, the interferences from external light sources like sun light, normal bulbs, and opaque materials in the path of transmission will cause interruption in the communication or partial loss of data. On the other hand, the emitting LED light bulb can be dimmed at a level that still keep the communication available but is not disturbing during night, for example. The same situation is meet when the fog curtain becomes thick on streets, data communication can still be accessible between car bulbs and street lights [12].

The receiver must have specific properties, dedicated filters, control and synchronization.

Avalanche photodiode APS, silicon photodiode PIN diode and solar cells are used for Li-Fi systems. APS has a higher performance than a PIN but is more expensive. The Li-Fi is susceptible to interference from sunlight and indoor lights, that is why, optical filters should be designed to diminish the DC noise components present in the received signal. It is better to use a photodiode in the case of a stationary receiver; still, in case of mobility the imaging sensor is preferred instead of a photodiode because of the larger field of view FOV. Operating imaging sensors has a high energy intake and is slow. Hence, a suitable balance between the cost, speed and complexity, should be done while considering the suitable photoreceptor [13].

Another challenge facing Li-Fi's development is how the receiving device will transmit back to transmitter. There are already few implementations for this situation using IR or RF.

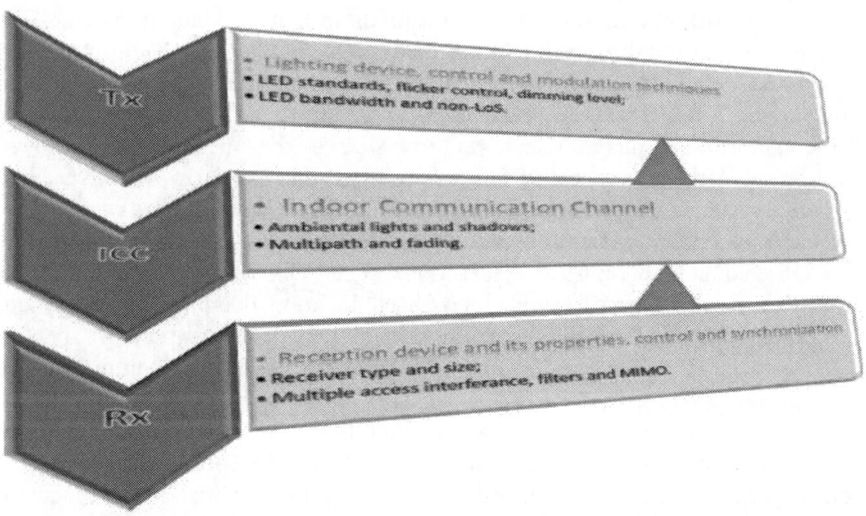

Figure 1. Tx, ICC, Rx - components of a Li-Fi system, their specific boundaries and challenges

4. Applications of Li-Fi

Any lightings device is possible to be change in such a way to be able to performed as a hotspot. It means that any LED bulb acting like an illumination device is possible to become a data communication device, as well. We refer here at ceiling lights, car lights, street lamps and so on. All of these and many others are able to spread internet connectivity using VLC, allowing in this way low cost architecture for hotspot. Hotspot covers a limited region where some devices are able to access internet connectivity.

Street lamps, for example, can act as free access points (AP) for high speed communication being able to transmit data in real time.

Traffic lights can communicate to the car and with each other. Cars have LED-based headlights, LED-based side lights, and therefore cars can communicate with each other and prevent accidents by fast exchanging of information. Industrial equipment, machines, or automobiles can be paired with light nodes to provide real-time information about the operational and functional performances, issues or any other metrics. Vehicular communication seems to be so simplified now, considering that the rear and front lights in vehicles can be used for data transfer for effective crash avoidance technologies.

Wi-Fi and many other radiation types are forbidden in many areas that are sensitive to the electromagnetic fog. On the other hand, Li-Fi communication is safe, abundant with low power consumption for all areas of these sensitive locations. These areas are power plants, petrochemical industry, nuclear plants and so on. But power plants, for example, need fast, inter-connected data systems to monitor things like demand, grid integrity and in nuclear plants core temperature need to be constantly monitored and controlled. The savings resulting from proper monitoring at a single power plant can add up to hundreds of thousands of dollars.

Other domain where the Li-Fi technology is possible to be successfully implemented is the aircrafts industry. Velmenni, which is a high-tech hardware startup based out of Delhi and Tartu, Estonia creates VLC links that can broadcast data at high speed, up to 20 meters and also work on applications in aviation. Their project is about to be applied to the Australian Airbus' BizLab innovation accelerator hubs. This would allow Airbus company to remove the looms of wiring connecting each passenger's screens to the digital content server containing all the movies, TV shows and music that are enjoyed during a flight. That wiring not only adds cost and complexity to aircraft, it adds weight which results in less fuel efficiency. It doesn't seem to mean much, but in the domain of aircraft design, every little bit of help is high valued [14].

Offices, schools, museums, metro, commercial places, hospitals in fact any indoor space can be a beneficiary of this new technology. Actually, in France for example, it has already been implemented by an important developer Oledcomm, who deployed Li-Fi products in the market, equipping in 2012 a museum in Europe (displaying a multimedia content in order to improve the visual experience of each artwork and guide the visitor in the museum according to his/her wishes/interest) and deployed in retail and Paris Metro. Carrefour with the support of Philips, installed on May 2015, in a hypermarket in Lille an indoor positioning system (IPS) based on Li-Fi technology, which allow its customers to locate the products in promotion. Philips offers a complete ISP technology. The Lille deployment is using Philips luminaires targeted at supermarket applications. The VLC implementation provides location accuracy to less than 1m. Philips offers a cloud-based location database capability and a software development kit that retailers can use to create a mobile interactive platform [15].

Another project of IPS has been presented on 2016 by the USA's company, Qualcomm [16]. Their implemented technology named Lumicast, uses LED light fixtures to broadcast positioning signals via amplitude modulation of light in a way that does not impact the fixtures' primary function of providing illumination. A mobile application that incorporates the Lumicast software framework can decode these VLC signals using those to determine the position of the mobile device. Due to its compatibility with smartphones and LED lighting infrastructure, this technology is able to support a broad range of IPS applications in commercial, office and industrial settings, and has been adopted by leading players in the LED lighting ecosystem such as Acuity Brands [17], [18] and GE Lighting [19].

As Professor Russel Griggs, the pureLiFi Chairman said on July 2016, an extension of standard Wi-Fi is coming and it's called Wi-Fi HaLow. While using less power, this new project aims to double the range of connectivity. Due to this, Wi-Fi HaLow is reportedly perfect for battery powered devices such as smartwatches, smartphones and lends itself to Internet of Things (IoT) devices such as sensors and smart applications [20].

Due to its high speed, this technology can have a big impact on the IoT too, with data transferred at much higher levels with even more devices able to connect to each other. pureLiFi company succeeded to launch on 2014 the device Li-Flame consisting of two units: Li-Flame Ceiling Unit (CU) (Figure 2) and Li-Flame Desktop Unit (DU) (Figure 3). Li-Flame technology delivered half duplex communication with a 10Mbps downlink and 10Mbps uplink up to 3 meters with standard LEDs, full mobility (portable, battery-powered desktop unit) with high data rate due to dense installation of Li-Fi access points(APs), multiple users per Li-Fi AP, supported through multiple access, while retaining high bandwidth for each user [21].

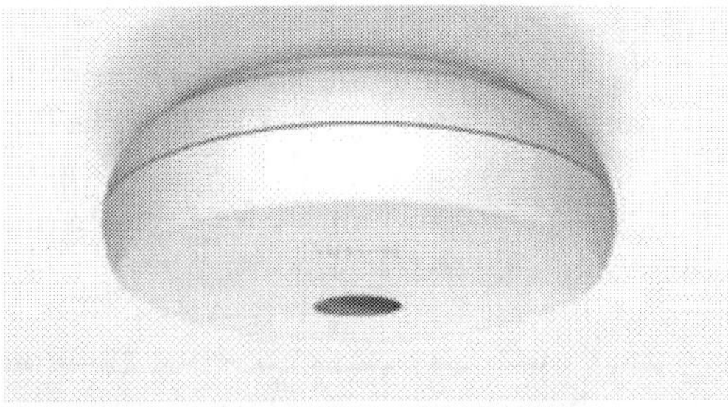

Figure 2. Li-Flame Ceiling Unit (CU)
(source http://purelifi.com/li-fire/li-flame/)

Figure 3. Li-Flame Desktop Unit (DU)
(source http://purelifi.com/li-fire/li-flame/)

The state-of-the-art device launched by pureLIFi company, named Li-Fi-X, (Figure 4) that has been publicly demonstrated along with the Li-Fi integrated Luminaire at Mobile World Congress in Barcelona on March 2017. The dongle USB-powered station has turn some of the Li-Fi's weaknesses in strengths allowing a full duplex communication with a 40Mbps downlink with a photodetector as receiver and 40Mbps uplink with infrared transmitter. The device allows full mobility being portable and is a multiple users device per Li-Fi Access Point, supported through multiple access [22].

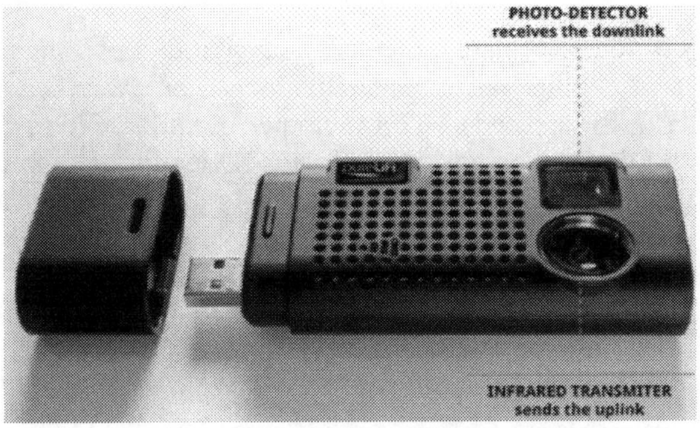

Figure 4. Li-Fi-X
(source: http://purelifi.com/Li-Fi-products/Li-Fi-x/)

Apple is also set to add Li-Fi capabilities to its future iPhones [23].

All the well-known strengths of the Li-Fi technology are supported, such as the secure wireless communications constrained by walls, eliminating the risk of signal leakage to external eyedroppers, safe wireless communication in locations where RF is forbidden and a cost-effective delivery of data as the piggyback of light on the same infrastructure [24].

There are many reasons why we should use Li-Fi in many areas nowadays. Some of these reasons are listed below:

- Safe for human eye/body;
- High security because light does not penetrate walls;
- Can provide high-speed connection;
- No interference on RF signals;
- Environmentally "green";
- Provide accurate (~1 mm) IPS;
- Both Gamma Rays and X-Rays are dangerous and have health issues;
- Ultraviolet can be used only were humans are not present;
- Infrared has heat radiation issues and using high power causes health problems.

5. The Present Scenario of RF. Comparison Between Wi-Fi and Li-Fi

Li-Fi can be thought of as a light-based Wi-Fi. It uses light instead of radio waves to transmit information. Instead of Wi-Fi modems, Li-Fi technology would use transceiver-fitted LED bulbs that can light a room as well as transmit and receive information.

WI-FI is great for general wireless coverage within buildings, and Li-Fi is ideal for high density wireless data coverage in restricted areas and for relieving radio interference issues. So the two technologies can be considered complimentary.

As we are all aware, the radio spectrum is congested and the demand doubles every year resulting in bandwidth shortage. Wireless transmission through RF is limited and expensive, insufficient spectrum for increasing data capacity. 1.8 million cellular radio stations, consume massive energy, most energy used for cooling base stations have only 5% efficiency efficiency [24]. Wi-Fi is available within the range of base stations limited, being unavailable in some aircrafts and it is less secure that Li-Fi since passes through walls. Due to VLC, the node or any terminal attach to the Li-Fi network is not visible to the host of network, thus increase communication safeness.

Li-Fi supports the broadcasting of network, it helps to share multiple thing at a single instance called broadcasting being multi user communication system.

Li-fi is a communication media in the form of light, so no matter about the frequency bandwidth problem. It does not require the any bandwidth spectrum i.e. we don't need to pay any amount for communication and license.

Although Li-Fi is not intended to replace other existing wireless communication technologies, it can be a great complement in certain places and situations. For example, nowadays, "total radio frequency (RF) wireless data traffic exceeds 11 Exabyte per month, creating a 97% gap between the traffic demands per device and the available data traffic per device in the mobile networks. This is the reason why RF wireless communication bottleneck is closely expected. Optical wireless communication (OWC), as a future reliable alternative, meets unprecedented spread in the last decade, new and interesting implementations bringing hopes for its bright future" [25].

Table 1 presents some characteristics of both technologies, Wi-Fi and Li-Fi, by comparing them.

Table 1. Comparison between Wi-Fi and Li-Fi

Characteristics	Wi-Fi	Li-Fi (VLC)
Standard	IEEE 802.11	IEEE 802.15
Frequency	2,4; 5 GHz	Hundreds of Tera Hz
Power consumption	Medium	Low
Medium of transfer	RF	Visible light
Range	100 meters	Depends on LEDs
Spectrum Range	300 Ghz	VL spectrum is 10.000 larger than RF
Data rate transfer	800 kbps – 11 Mbps	>1Gbps
Cost	Medium	High
Network topology	Point to point	Point to point
Security	Medium secure	High secure

6. VLC system with Tx and Rx PCBs

A block diagram of the VLC communication system that we have made is shown on Figure 5.

The main components of our communication system are:

- Tx - transmission circuit made by a sound device and a led driver circuit with a high brightness cold white LED has been switched on and off to a very high rate, that acts as a communication source;

- Rx - Reception circuit made by a silicon photodiode which shows good response to visible wavelength region serving as the receiving element with a photodiode.

The tests have been conducted indoor both with a regular amount of ceiling illumination and day light as well.

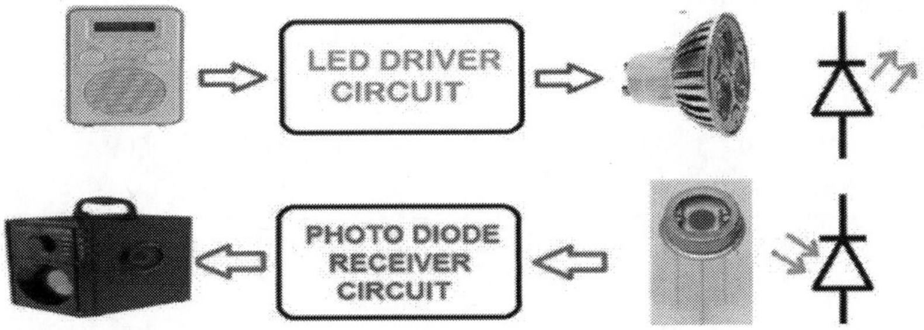

Figure 5. The block diagram of the VLC communication system made

6.1. Tx – Using 1 Watt Cold White LED

The LED driver circuit uses a 1 Watt Cold white LEDs bulb with a signal input of 10 Hz to 30 MHz and an electrical power supply of 12 V.

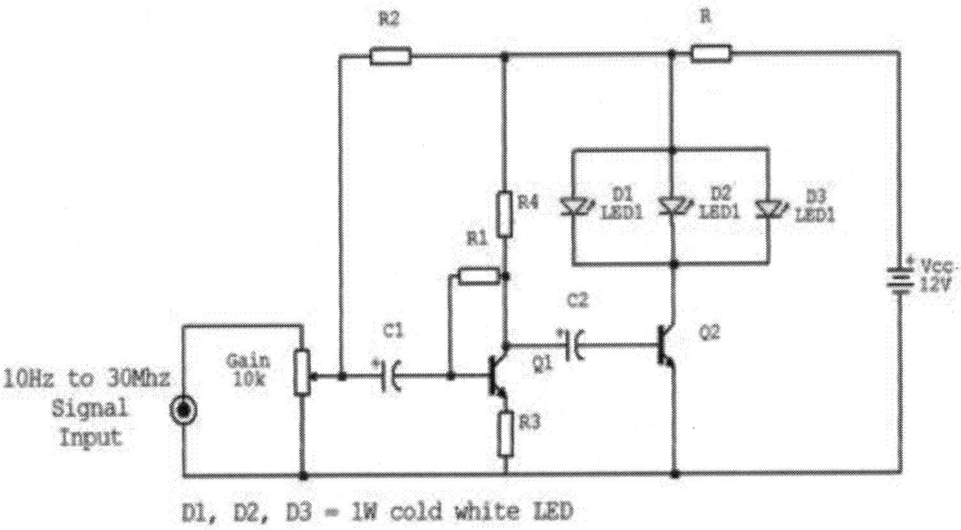

Figure 6. Tx circuit with 1 W cold white LED

TDS-P001L4Z11 Characteristics:

- Highest flux per LED family: 140-150lm/W;
- Led chip from Epistar and Bridgelux;
- Long lifespan (up to 100k hours) White: 6500K;
- Lambertian radiation pattern;
- Cool beam, safe to the touch (90-120-140degree);
- Fully dimmable;
- No UV radition.

IOP Conf. Series: Materials Science and Engineering **294** (2017) 012047 doi:10.1088/1757-899X/294/1/012047

Table 2. LED Chip TDS-P001L4Z11

LED Chip TDS-P001L4Z11	**Material**: InGaN / Sapphire	
Parameter	**Rating**	**Unit**
Power Dissipation	1365	mW
LED Junction Temperature	120	°C
Reverse Voltage	5	V
D.C. Forward Current	350	mA
Pulsed Forward Current; tp ≤ 100μs, Duty Cycle = 0.005)*1	700	mA
Operating Temperature Range	-40 to +75	°C
Electrical & Optical Characteristics:		
Parameter	**Rating**	**Unit**
Luminous Flux, IF = 350 mA	120	lm
Forward Voltage, IF = 350 mA	3.3	V
Reverse Current, Vr = 5V	5	μA
View Angle	130	deg

6.2. *Rx module*

We have used as the receiver module in two different scenarios. In the first scenario we used a photodiode and in the second one, a solar cell.

6.2.1. Rx – Using Photodiode - First scenario. BPV10 is a PIN photodiode with high speed and high radiant sensitivity in clear, T-1¾ plastic package. It is sensitive to visible and near infrared radiation.

BPV10 Photodiode characteristics are shown on Table 3.

Table 3. BPV10 Photodiode's basic characteristics

PARAMETER	TEST CONDITION	SYMBOL	TYP.	MAX.	UNIT
Forward voltage	I_F = 50 mA	V_F	1.0	1.3	V
Reverse dark current	V_R = 20 V, E =	Iro	1	5	nA
Diode capacitance	V_R = 0 V, f = 1	C_D	11		pF
	V_R = 5 V, f = 1	C_D	3.8		pF
Absolute spectral sensitivity	V_R = 5 V, λ = 950 nm	s(λ)	0.55		A/W
Angle of half sensitivity		φ	± 20		deg
Wavelength of peak sensitivity		λ p	920		nm
Range of spectral bandwidth		$\lambda_{0.1}$	380 to 1100		nm
Detectivity	VR = 20 V, λ = 950 nm	D	3 x 1012		cm√ Hz/W

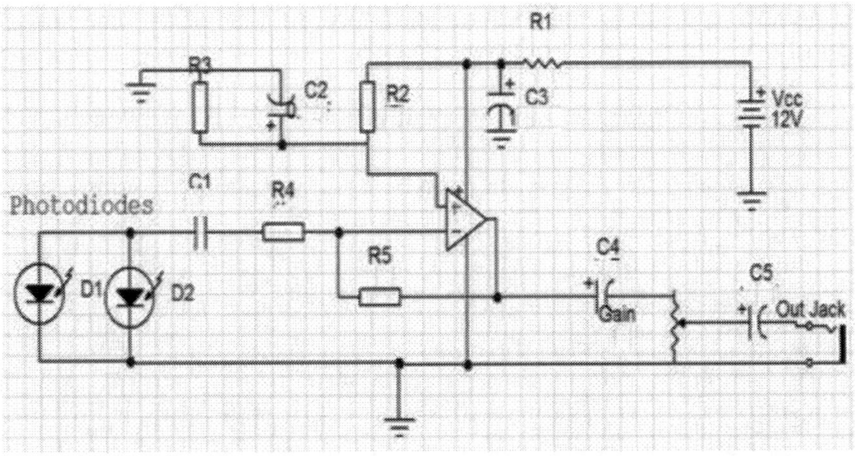

Figure 7. Rx circuit with photodiode

6.2.2. Rx – Using Solar Cell - Second scenario. And, in the second scenario, the receiver circuit contains a solar cell. The solar cell's specification are shown in Table 4.

Table 4. The solar cell's specification

Type	1 Watt
Silicon	Mono-Crystalline Silicon
Parameters	Rating/Unit
Maximum Power	1 W
Maximum Power voltage	9 V
Maximum Power current	0.11A
Open circuit voltage	10.4V
Short circuit curren	0.12A
Cells thickness	0.18mm±20μm
Number of cells	18

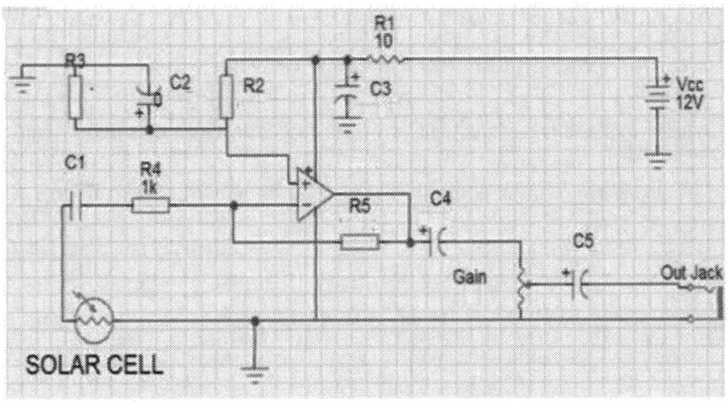

Figure 8. Rx circuit with solar cell panel

The best improvement in this scenario consists in higher distance communication achievement due to non-Line of Sight (NLOS) high capacity of the receiver.

Figure 9 reprezents a picture of the VLC system made by us both for data stream and sound communication, using the LED light bulb and one photo detector.

Figure 9. Our VLC system

The VLC system for data transmission stream has a transmission circuit designed to send data through illumination as piggyback.

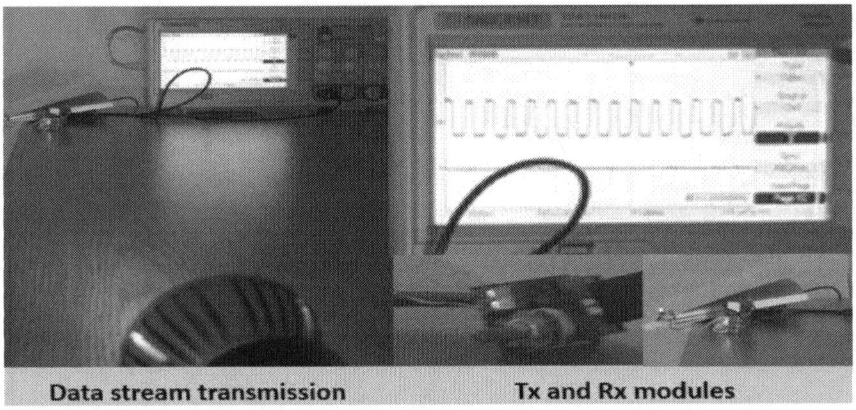

Figure 10. VLC System with the receive wave displayed on oscilloscope

The single carrier binary optical transmission is displayed on oscilloscope.
VLC system for SOUND transmission.
In case of the VLC system for sound transmission, the wave forms similarity of the two channels are dispayed on oscilloscope.

IOP Conf. Series: Materials Science and Engineering **294** (2017) 012047 doi:10.1088/1757-899X/294/1/012047

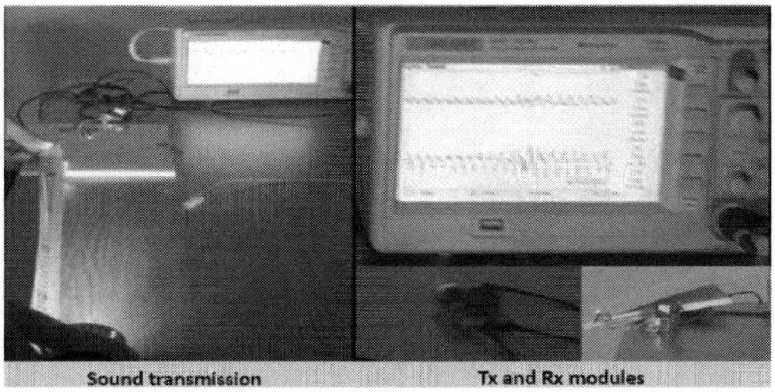

Figure 11. VLC System with the send and receive waves displayed on oscilloscope

Wave forms similarity of the two channels dispayed on osciloscope for different distances between LED and Photodiode.

We have captured the signal form of the two waves and their similarity in case that LED bulb is at 1 meter away from Rx Photodiode LoS scenario.

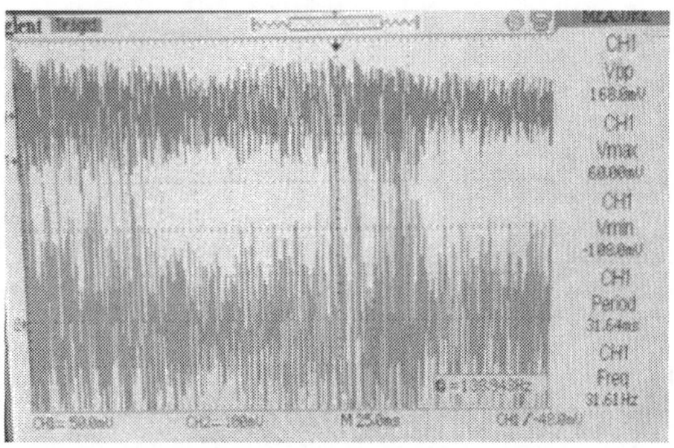

Figure 12. Tx module LED is 1.0 meter away from Rx Photodiode LoS scenario

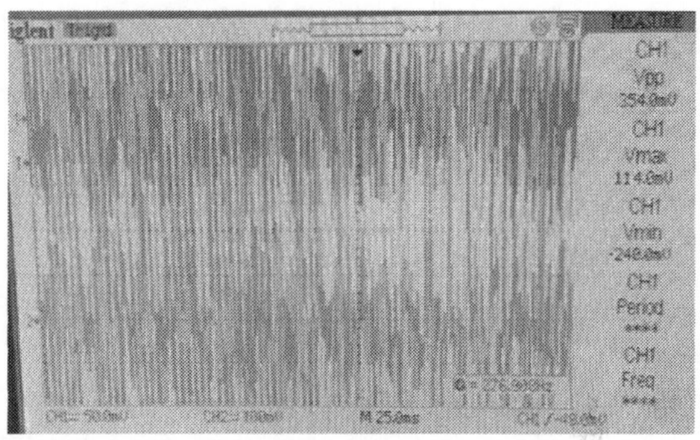

Figure 13. Tx module LED is 0.5 meter away from Rx Photodiode LoS scenario

7. Conclusions

We can every day see that more and more Wi-Fi able communication devices are hungry for data, thus the RF domainf of the electromagnetic spectrum becomes overcrowded. On the other hand, the Li-Fi technology already demonstrated to provide brighter future for wireless networks, especially where the radio frequencies are limited or even forbidden. As most of the latest developments and improvments of the technology shows, when the Li-Fi wireless network system will be fully developed, each light source can be used as a Li-Fi access point (AP), meaning where will be a LED illumination available, data communication is expected to be present, too.

This paper presents shortly the Li-Fi's state-of-the art development as well as world wide implementations realized so far. A short balance between the two technologies (Wi-Fi and Li-Fi) has been done, too.

We have made a first application of the VLC tehnology using as transmitter a 1 Watt cold white LED and as the receiver one photodiode in a fist scenario and a sollar cell in the second one. When the photodiode has been used, we have conducted indoor tests for sound communication. In a Los scenario, Rx and Tx were placed at various distances (up to 1 meter) and both the transmited wave and the received one were displayed on the oscilloscope's screen. The higher distance between the Tx and Rx, the lower sound level percept but the accuracy of music seemed to be the same.

We plan to improve and develop the system presented so far in order to make specific working implementation in different areas.

References

[1] Leba M, Riurean S and Ionica A 2017 *Li-Fi - The Path To A New Way Of Communication,* CISTI'2017 - 12ª Conferência Ibérica de Sistemas e Tecnologias de Informação

[2] Dimitrov S and Haas H 2015 *Principles of LED Light Communications Towards Networked Li-Fi*. Cambridge CB2 8BS, United Kingdom

[3] Bhisey R 2017 *Li-Fi Market to be driven due to Growing RF Spectrum Crunch,* http://www.editiontruth.com/li-fi/

[4] Sklavos N, Hübner M, Goehringer D and Kitsos P 2014 *System-Level Design Methodologies for Telecommunication*, Springer International Publishing Switzerland, pp 28, http://www.naka-lab.jp/ecsite_e/index_e.html

[5] Haas H 2011 *Wireless data from every light bulb,* Ted Talks www.ted.com/talks/harald_haas_wireless_data_from_every_light_bulb

[6] Webster G 2012 Li-Fi provides a light bulb moment for wireless web, CNN http://edition.cnn.com/2012/09/28/tech/Li-Fi-haas-innovation/index.html

[7] Haas H, Yin L and Wang Y 2015 *What is Li-Fi?* Journal of Light Wave Technology

[8] Mercer C 2016 *What is Li-Fi? How does Li-Fi work? Wi-Fi vs Li-Fi. Li-Fi news,* http://www.techworld.com/data/what-is-li-fi-everything-you-need-know-3632764/

[9] Noshad M and Brandt-Pearce M 2013 *Can Visible Light Communications Provide Gb/s Service?,*" Harvard School of Engineering and Applied Sciences Cambridge Publication arxiv

[10] Wood V and Bulović V 2010 *Colloidal quantum dot light-emitting devices* Co-Action Publishing, Vol 1

[11] Sung Ju Cho, Maysinger D, Jain M, Beate R, Hackbarth S and Winnik F M 2008 Long-Term Exposure to CdTe Quantum Dots Causes Functional Impairments in Live Cells*, Langmuir* **23**(4) 1974-1980

[12] Haas H 2015 *Forget Wi-Fi. Meet the new Li-Fi Internet,* Ted Talks https://www.ted.com/talks/harald_haas_a_breakthrough_new_kind_of_wireless_internet

[13] Aswin Kumar M, Aswin Raj G, Lokesh V and Sugacini M 2016 *Iot Enabled By Li-Fi Technology,* Proceedings of National Conference on Communication and Informatics-2016, Sri Venkateswara College of Engineering, Sriperumbudur, CompuSoft Magazine

[14] Flynn D 2016 *Airbus wants to upgrade WiFi to the speed of light,* https://www.ausbt.com.au/airbus-wants-to-upgrade-wifi-to-the-speed-of-light

[15] Wright M 2015 *Philips Lighting deploys LED-based indoor positioning in Carrefour hypermarket*, LEDs Magazine

[16] Jovicic A 2016 *A high accuracy indoor positioning system based on visible light communication,* white paper, Qualcomm® Flarion Technologies Lumicast™, https://www.qualcomm.com/media/.../lumicast-whitepaper.p.

[17] ***Press release 2016 *Qualcomm and Acuity Brands Collaborate to Commercially Deploy Qualcomm Lumicast Technology for Precise Indoor Location Services in More Than 100 Retail Locations*, https://www.qualcomm.com/news/releases/2016/03/14/qualcommand-acuity-brands-collaborate-commercially-deploy-qualcomm

[18] Lydecker S 2016 *Illuminating the In-Store Experience*, http://www.acuitybrands.com /solutions/-/media/files/acuity/solutions/ services/bytelight services indoor positioning/indoor positioning white paperrevised110315.pdf

[19] ***Press release 2015 *GE Intelligent Lighting to Transform Retail Experience through Qualcomm Collaboration*, May, http:// pressroom.gelighting.com/news/ge-intelligent-lighting-to-transform-retail-experience-through-qualcomm-collaboration

[20] Risteiu M, Dobra R, Leba M and Ionica A 2016 A Smart House in the Context of Internet of

Things, *International Journal of Internet of Thinks and Web Services* **1** 38-42

[21] ***2014 *Li-Flame* Window on Li-Fi http://www.w-o-li-fi.tk/2015/05/li-flame.html

[22] ***2016 *Researchers say Li-fi will be bigger than ENTIRE industry,* Lux Review http://luxreview.com/article/2016/06/researchers-say-li-fi-will-be-bigger-than-entire-lighting-industry

[23] Sahajune S 2016 *Can Li-Fi Drive How The World Uses Big Data And Iot?,* Yaabot Magazine. https://www.yaabot.com/23361/can-li-fi-drive-the-world-uses-big-data-and-iot/

[24] Niju S 2015 Light Fidelity. The future of Wireless Technology, https://www.slideshare.net/featured/category/technology

[25] Riurean S, Stoica R and Leba M 2017 Visible Light Communication for Audio Signals, *International Journal of Communications* **2**

Quadcopter control using a BCI

S Rosca[1], M Leba[2], A Ionica[3] and O Gamulescu[1]

[1]University of Petrosani, Doctoral School, Petrosani, Romania
[2]University of Petrosani, Department of Computer and Electrical Engineering, Petrosani, Romania
[3]University of Petrosani, Department of Management, Petrosani, Romania

E-mail: monicaleba@yahoo.com

Abstract. The paper presents how there can be interconnected two ubiquitous elements nowadays. On one hand, the drones, which are increasingly present and integrated into more and more fields of activity, beyond the military applications they come from, moving towards entertainment, real-estate, delivery and so on. On the other hand, unconventional man-machine interfaces, which are generous topics to explore now and in the future. Of these, we chose brain computer interface (BCI), which allows human-machine interaction without requiring any moving elements. The research consists of mathematical modeling and numerical simulation of a drone and a BCI. Then there is presented an application using a Parrot mini-drone and an Emotiv Insight BCI.

1. Introduction

The brain computer interface (BCI) technology make possible manipulation of an embedded system or computer using signals generated by brainwaves. A characteristic of BCI system can capture brain signals generated by neural activities, it can recognize differently neural activity patterns, and also can transform them in useful commands that can be applied to control a machine or a device. BCI are most commonly applied in prosthetic limbs, exoskeletons, robotics, autonomous vehicles, virtual keyboard and computer games [1].

From the point of view of the way the signal is acquired, a BCI system can be classified as invasive and noninvasive, depending on how the EEG biosensors are placed [2]. Noninvasive BMIs, typically named brain – computer interfaces (BCIs) are based on electroencephalography (EEG) to record the brain activities using a series of biosensors disposed on the scalp able to measure the potential generated by electrical activity of billions of cortical neurons. Our study is focused on noninvasive BCI.

Quadcopters, on the other hand, can be used in academic environment by researchers to test or to evaluate innovative new ideas from different research fields such as: real time systems, robotics, flight control theory, navigation. There is a wide range of advantages in using quadcopters as multilateral test platforms. They are relatively inexpensive, are produced in a wide range of sizes and shapes, it presents a simple mechanical design that can be understood even by persons without extensive experience in the field to build them and ensure their maintenance. Due to the interdisciplinary approach necessary to operating a quadcopter, a large number of academics from different area of science are engaged together both in the phase of elaboration and design and also in the phase of improvement the system stability and performance. Quadcopter projects are obviously based on collaboration of many researchers from the field of computer science, electrical and mechanical engineering [3].

Emotiv Insight, used to acquire the signals from the brain, is recently produced in 2015 by Emotiv and represents a novel technology in domain of brainwear device for signals acquisition by electroencephalography (EEG) and Brain-computer Interface principles.

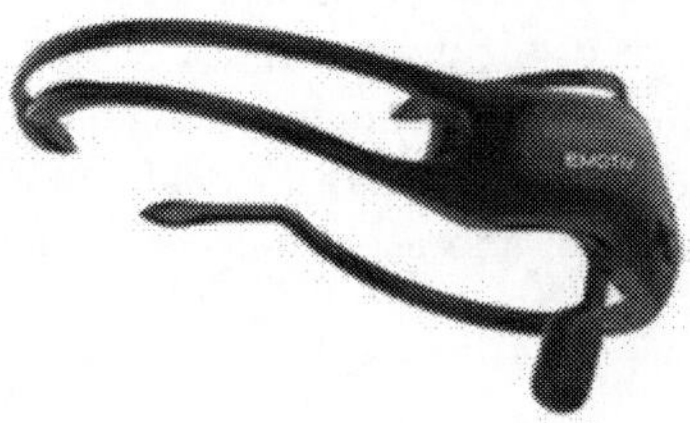

Figure 1. Emotiv Insight Headeset Overview

Emotiv Insight headset shows a number of functional features:

- ergonomic, elegant and intuitive design, is very light,
- it offers great freedom of movement as it allows wireless communication (2.4 GHz frequency) between the headset and computer or between the headset and other embedded system Bluetooth module embedded via intelligent Bluetooth module integrated (BLE 4.0),
- enables 3D visualization of the whole brain activity signal offering through 5 biosensors for electroencephalogram EEG and two reference biosensors for noise cancellation/disturbances due to movement,
- presents 9-axis motion sensors (gyroscope 3 axis motion accelerometer, 3-axis motion and 3-axis magnetometer motion) for accurate detection of the position and head movement,
- provide support for platforms and applications: Windows, OSX, Linux, Android, iOS,
- sampling rate of 128 SPS (samples/second),
- frequency response: 1-43 Hz,
- resolution: LSB 0,51uV,
- Lithium-Polymer Battery: 5V 480mAh, which provides a lifetime of up to 4 hours,
- programming languages and development environments supported: C ++, C #, VB. NET, Java, Python, Matlab Unity.

The types of measurements allowed the headset:

- Raw signal,
- Mental Commands (conscious thoughts),
- Facial expressions - facial mimicry,
- Measurements of performance of the brain.

Emotiv Insight enables:

- Measurement of key performance indices of the brain: attention, concentration, involvement, interest, excitement/enthusiasm, affinity, relaxation, stress,
- Measurements of motion sensors - emulating a virtual mouse based on the parameters obtained from sensors with 9-axis motion [4].

In this paper, remote presence was implemented to control a quadcopter Parrot Rolling Spyder by the Emotiv Insight BCI device.

1.1. Emotiv Insight BCI

The Emotiv Insight dispose to a system compose by five sensor that are projected to acquire and measure the key activity from the entire functional areas of the cortex. Emotiv Insight represent the first and only device available on the market that can covers **the entire key areas of the cerebral cortex**. The Emotiv Insight detects activity in the frontal cortex (responsible for executive functions), the parieto-temporal cortex (responsible for auditory function, spatial interaction /coordination), and the parietal-occipital cortex (responsible for visual function).

Principal key characteristics of the Emotiv Insight's Scientific design: dynamic brain-computer interface interactions with more degrees of freedom for controlling physical and virtual objects; accurate identify of mental states and emotion such as engagement, excitement, focus, meditation, relaxation and stress; possibility to build brain activity models in real-time based on spatial resolution; a deeper perspective on specific patterns of an individual's brain activity [5].

1.2. Parrot Rolling Spyder

Parrot Rolling Spider is a small sized drone with a compact design controlled by smartphone. It can flies both indoor and outdoor with a remarkable speed and stability. Due to the fact it benefit by ultra-light weight, Rolling Spider can achieve a complete and half turns in one single step; fitted with a pair of wheels, it can fly between the floor and ceiling or even on the walls; comes with a propellers cutout functionality to prevent collision with room objects and possess same flight stability as AR. Drone even if it is 6 times lighter thanks to an autopilot system based on a 3-axis gyroscope and a 3-axis accelerometer. Pressure sensor for flight altitude control. Ultrasonic sensor for precision detect flying near the ground. A vertical camera with a frame rate at 60 fps that can measure the speed based on images compared in real time. It is based on an ultra-sophisticated technology of fusing data received from sensors that makes Rolling Spider one of the most stable drone in the world. It benefit from an inedited take-off mode when the drone is in a state of free fall: the engines start at a drop detection. The Rolling Spider also allow users to take capture snapshots from the aerial views thanks the ingenious way of placing the mini-camera at the bottom of the drone [6].

2. Application

EMOTIV 3d Brain Visualizer is an application designed for Emotiv Insight BCI headset that acquire the brain electroencephalography (EEG) signals through EEG biosensors and shows on screen in real time a 3d cartography of entire neural activity according to the four major brainwave frequency bands: Delta, Theta, Alpha and Beta. User can adjust the sensibility of EEG signals acquisition through a intensity slider or can use default settings (auto-intensity) to obtain detailed information of neural stimuli in different brain regions. Delta wave (1 Hz to 4 Hz) - is the slowest frequency range. It can be measured in adults during drowsiness and deep sleep state [7]. Theta wave (4 Hz to 7 Hz) – may be seen in drowsiness, arousal or meditation [8]. Alpha wave (7 Hz to 14 Hz) – it can be measured on both side of the brain in the posterior regions of the scalp and shows a high amplitude during the relaxation and during eyes closure [8]. Beta wave (15 Hz to about 30 Hz) - it can be measured on both side of the brain but is more pronounced frontally. It is closely related to motor behavior and it shows a significant attenuation during active movements. [8].

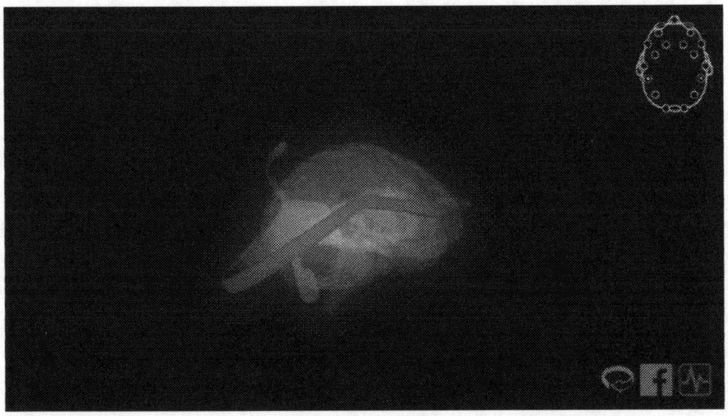

Figure 2. Overview of EMOTIV 3d Brain Visualizer

Humanoid robot controlled by BCI System could be represent a first step to develop in not far away future of an exoskeleton controlled by brain that which can be used for both military and civil purposes. Based on a BCI headset and a camera used to record a video clip of a training session for a mental task, an humanoid robot can be controlled directly by brain signals using electroencephalogram (EEG) principle by simply thinking to „turning right", „turning left" or „walking forward" [9].

Figure 3. Humanoid robot controlled by BCI System [9]

BCI for human measuring level of fatigue can be useful for drivers and aircraft pilots and for this reason it is a research line that must be promoted in the field of BCI, it is safe and effective method to prevent accidents due high workload. In order to detecting the mental state with a high accuracy up to 90% it is necessary to implement in BCI system besides electroencephalogram electrodes (EEG) both electrooculogram sensors (EOG) for estimating mental workload in tasks with high visual attention [10] and also heart rate sensors (HR) due the fact that the installation of drowsiness state is closely linked to the decrease in heart rate and the increase of blink rate [10].

Figure 4. BCI monitoring session based on EEG at flight simulator [10]

3. Software

EEGLAB is an open source Matlab toolbox for processing electroencephalography (EEG) and magnetoencephalography (MEG) signals and other electrophysiological data event-related. EEGLAB toolbox has been developed by Arnaud Delorme and Scott Makeig at the Swartz Center for Computational Neuroscience, Institute for Neural Computation at the University of California San Diego based on a set of Matlab functions for processing and visualization EEG data developed in 1997.

EEGLAB provides a standalone graphical user interface (GUI) that allows user to interact with EEG data without the need for running a Matlab syntax [11]. The user graphic interface integrates a powerful tools that provides a large number of functions placed into three individual layers for processing EEG signals (artifact rejection, filtering, epoch selection and averaging) for analysis in time or in frequency of EEG data, for independent component analysis (ICA), for event-related statistics and for data visualization (scrolling, scalp map and dipole model plotting, plus multi-trial Event-related potentials (ERP)-image plots). EEGLAB provides a programming environment that allows user to store, access, measure and manipulate the single-trial and/or averaged EEG data and display it hierarchized according to the number of EEG communication channels provided by BCI equipment [11].

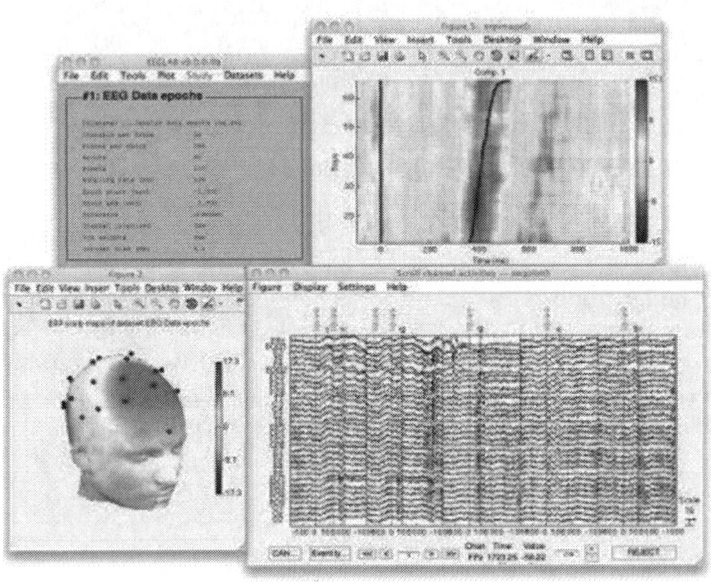

Figure 5. Overview of EEGLAB Toolbox

IOP Conf. Series: Materials Science and Engineering **294** (2017) 012048 doi:10.1088/1757-899X/294/1/012048

BCI2000 is an open source development platform for BCI systems designed to use for research or educational purposes with applicability in the field of biomedical engineering projected since 2000 by Gerwin Schalk and Dennis McFarland at the Wadsworth Center of the New York State Department of Health in Albany [12]. BCI2000 was projected in modular form using techniques of object-oriented software design in C++ and for this reason it can be implemented on any programming language and it offers support for any development environment (e.g. Matlab or LabView) [11].

BCI2000 presents a series of basic functionalities: real-time capability - acquiring and processing brain signals at high sampling rate with minimal latency, EEG data recording, removing artifacts and noise sources (e.g. interferences generate by eye blinks or eye movements), spatial filtering, offline data analyses [12].

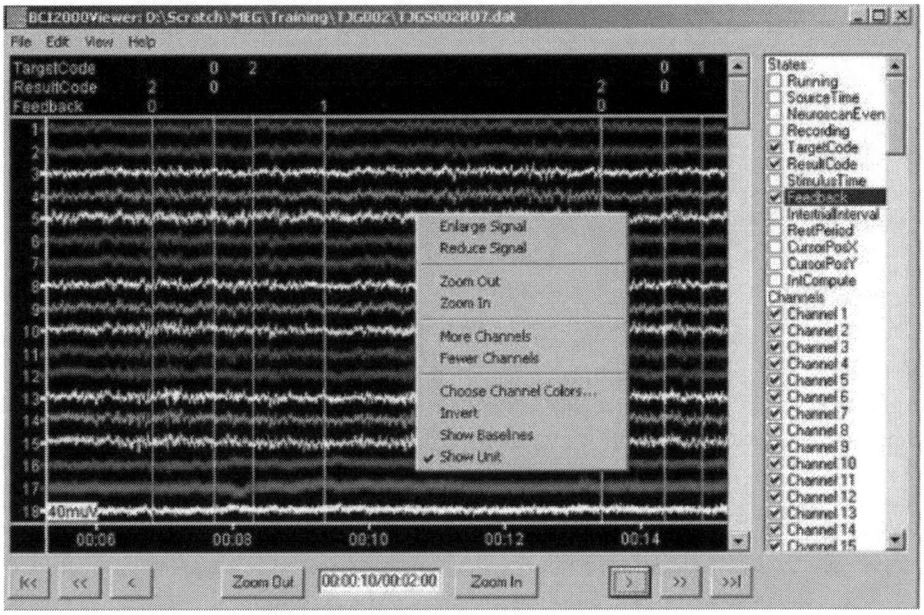

Figure 6. BCI2000 Viewer Interface [12]

Emotiv Xavier TestBench is a software development kit (SDK) designed by EMOTIV Inc., a pioneer in the field of Brainwear® Wireless EEG Technology, used to loading, displaying and analying in real time the electroencephalogram (EEG) signals aquired from EEG biosensors. Beside EEG data Emotiv Xavier TestBench can display the contact quality status with head surface for each EEG biosensors in part, an frequency analysis of the EEG signal for each communications channel based on Fast Fourier Transform (FFT) with the corespondency brainwave frequency bands (Delta, Theta, Alpha, Beta) [13] and the data received from gyroscope, accelerometer and magnetometer that are a part of Emotiv BCI headset. The user can save recorded data in European Data Format file (.edf). The EDF data previously saved can be converted into Comma-separated Value format (.csv) to be analyzed into EEGLAB toolbox.

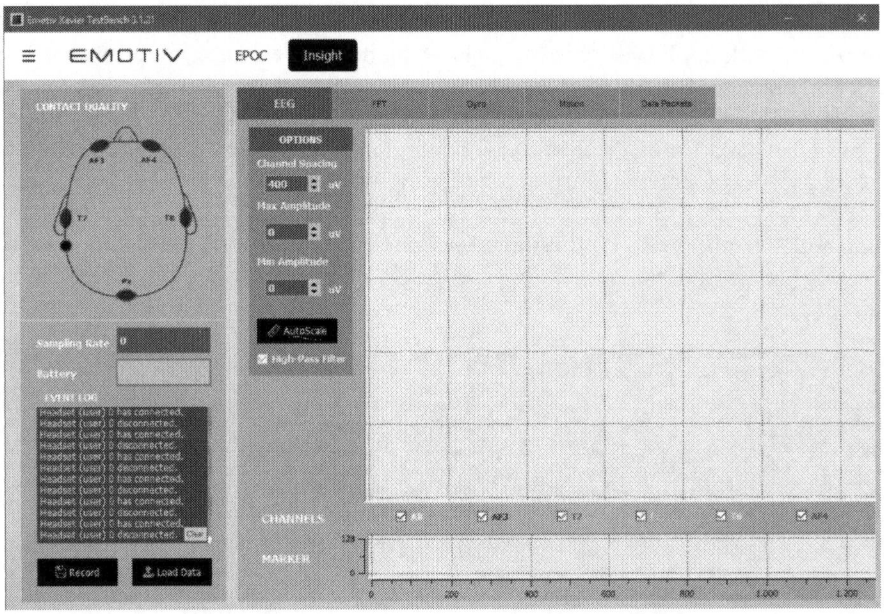

Figure 7. Overview of Emotiv Xavier TestBench SDK

Parrot mini-drone controlled by brain

The system developed in this paper aims to implement a BCI interface based on Emotiv Insight BCI headset with we can control a quadrotor Parrot mini-drone only using brain signals without requiring any moving elements.

The placement of EEG biosensors on Emotiv Insight headset corespond with 10-20 International System standardized. The number of electrodes are grouped in odd for the right hemisphere and even for the left hemisphere. The letters represent the placement of electrodes in function to lobes position: F for frontal, T for temporal, C for central (it is only for identification purposes because the central lobe not exist), P for parietal and O for occipital [14], [15].

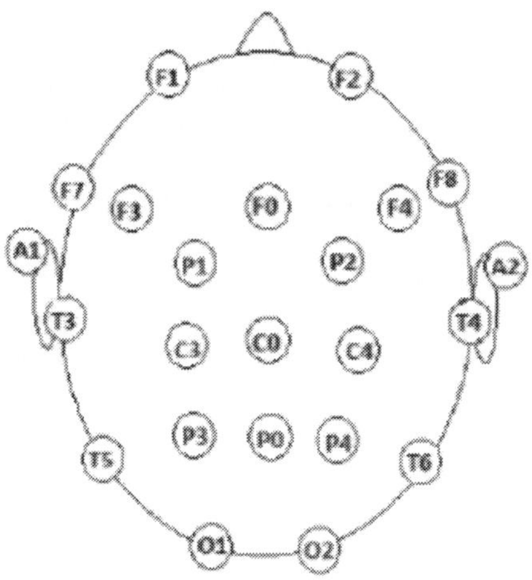

Figure 8. Electrode Placement according to the International 10-20 System [14]

The quadcopter Parrot mini-drone that we use provides 6 degree of freedom (3 degree of freedom is reserved for translational movements and other three for rotational movements) that are obtained by coupling the rotation and translation movements.

We apply the direction cousine matrix to determine quadrotor movements equation from the ground to a fixed point:

$$R_{ZXY} = \begin{bmatrix} C_\phi C_\theta & C_\phi S_\theta S_\Psi - S_\phi C_\Psi & C_\phi S_\theta C_\Psi + S_\phi S_\Psi \\ C_\phi S_\theta & S_\phi S_\theta S_\Psi + C_\phi C_\Psi & S_\phi S_\theta C_\Psi - C_\phi S_\Psi \\ -S_\theta & C_\theta S_\Psi & C_\theta C_\Psi \end{bmatrix} \#(1)$$

where:

- $S_\phi = Sin(\phi), C_\Psi = Cos(\Psi)$, etc.;
- R = matrix transformation;
- Φ = Roll angle;
- θ = Pitch angle;
- Ψ = Yaw angle.

Based on Lagrange approach we can determinated the dynamic model of the drone. Using the force and moment balance theory we can write the equations of motion as follows:

$$\left.\begin{aligned} \ddot{x} &= u_1(Cos\phi Sin\theta Cos\psi + Sin\phi Sin\psi) - \frac{K_1\dot{x}}{m} \\ \ddot{y} &= u_1(Sin\phi Sin\psi Cos\psi - Cos\phi Sin\psi) - \frac{K_2\dot{y}}{m} \\ \ddot{z} &= u_1(Cos\phi Cos\psi) - g - \frac{K_3\dot{z}}{m} \end{aligned}\right\} \#(2)$$

where:

- x: represent the forward position related to the axes of the earth;
- y: represent lateral position related to the axes of the earth;
- z: represent vertical position related to the axes of the earth
- K_i: represent the drag coefficients of the four motors

In the following, based on the fact that the drag force is insignificant at a low speed and considering that the center of gravity should be at the intersection of the axes between the four motors we assume that stability increases as angular acceleration becomes less susceptible to forces once the center of gravity moves up or down. Another way to increase the stability can be realized by targeting of rotor forces to the center of the quadrotor body frame. This will reduce the effects of pitch and roll and also the vertical thrust. For convenience, we will define the inputs to be:

$$\left.\begin{aligned} U_1 &= (Th_1 + Th_2 + Th_3 + Th_4)/m \\ U_2 &= l(-Th_1 - Th_2 + Th_3 + Th_4)/I_1 \\ U_3 &= l(-Th_1 + Th_2 + Th_3 - Th_4)/I_2 \\ U_4 &= C(Th_1 + Th_2 + Th_3 + Th_4)/I_2 \end{aligned}\right\} \#(3)$$

where:

- U_1: correspond to vertical thrust developed by the 4 rotors
- U_2: represent the pitching effect;
- U_3: represent the yawing moment;
- U_4: represent the Rolling effect;

- Th_i: represent the thrusts force developed by four rotors;
- I_i: represent the inertia intervals in relation to the axes

In this paper we also developed a simulation in real time of Parrot Rolling Spider quadcopter behavior in development environment Matlab-Simulink based on some estimation and control modules that we found in a Matlab toolbox designed by the Massachusetts Institute of Technology (MIT). Based on designed estimation and control algorithms, Matlab-Simulink can generate automatically an embedded C-code that can be load onto the drone for test it's behavior in real-life.

In next figure is presented the Simulink model of drone 's dynamics, sensor system and compensator.

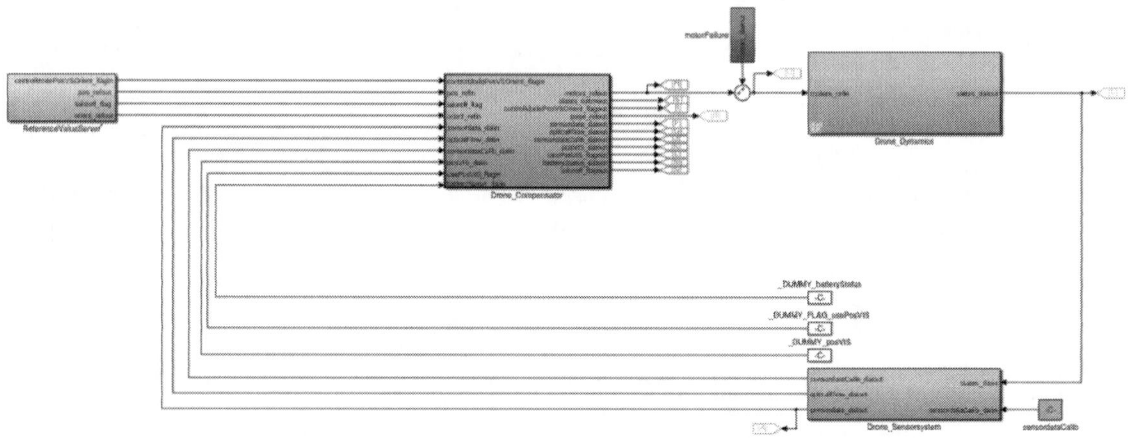

Figure 9. Simulink model of drone 's dynamics, sensor system and compensator

In Figure 10 and Figure 11 is presented the plot of data from simulated flight.

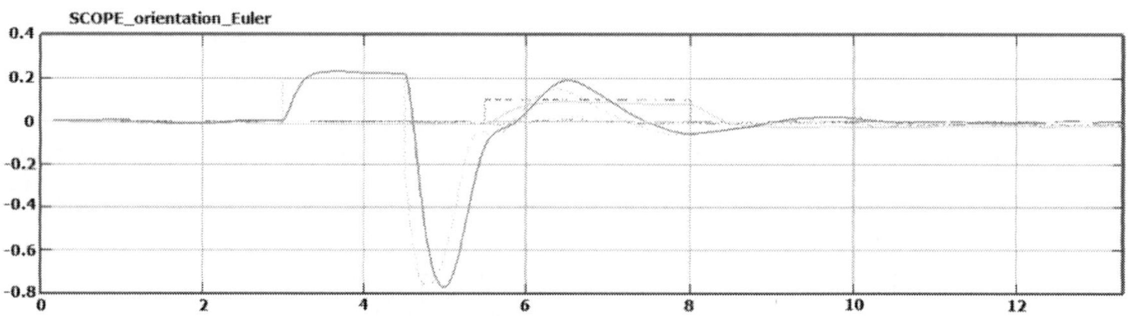

Figure 10. Simulated result for Euler Orientation

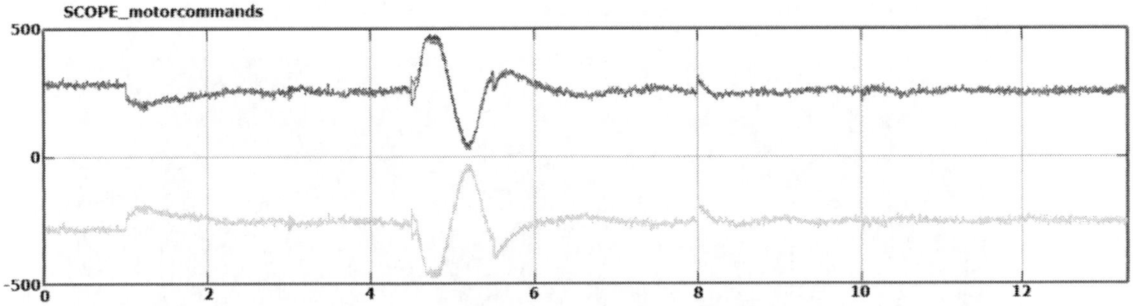

Figure 11. Simulated result for motor commands

The model developed by Simulink also gives us the opportunity to extract the actual flight data from the quadcopter and to analyze them offline, data that can then be compared to the model simulated by us.

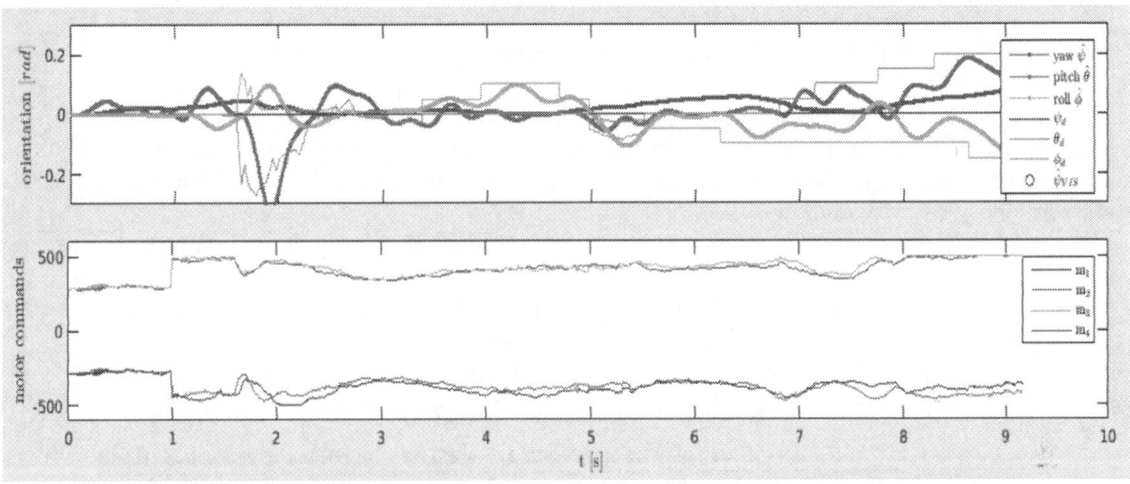

Figure 12. Result from real flight data for Euler orientation and motor commands

Based of inertial XYZ-world coordinates of the drone's center of mass and euler rotations angles (roll rotation about x axis, pitch rotation about y axis and yaw rotation about z axis) obtained by mathematical model and in accordance with simulated model, we also developed a 3d animation of the behavior of the drone in Simulink 3D Animation as presented in Figure 13.

Figure 13. Simulink 3D Animation for take-off the drone

In order to create a BCI connection between the Emotiv Insight headset and the Parrot Rolling Spider Quadcopter we use the Emotiv Xavier Control Panel interface which based on the movement of a virtual object (mental tasks), acquires the EEG signals from the BCI headset and converts them into computer binary code for computer.

We train 4 distinguish mental task one for each direction in which we propose to use the quadcopter:

- lift mental task (fig.14) for take-off the drone at 1.1 meters altitude and up to 0.2 m/s
- drop mental task for decrease altitude with 0.2 m/s
- right mental task for pitch movements with 0.05 degrees on the right side
- rotate left mental for pitch movements with 0.05 degrees on the left side

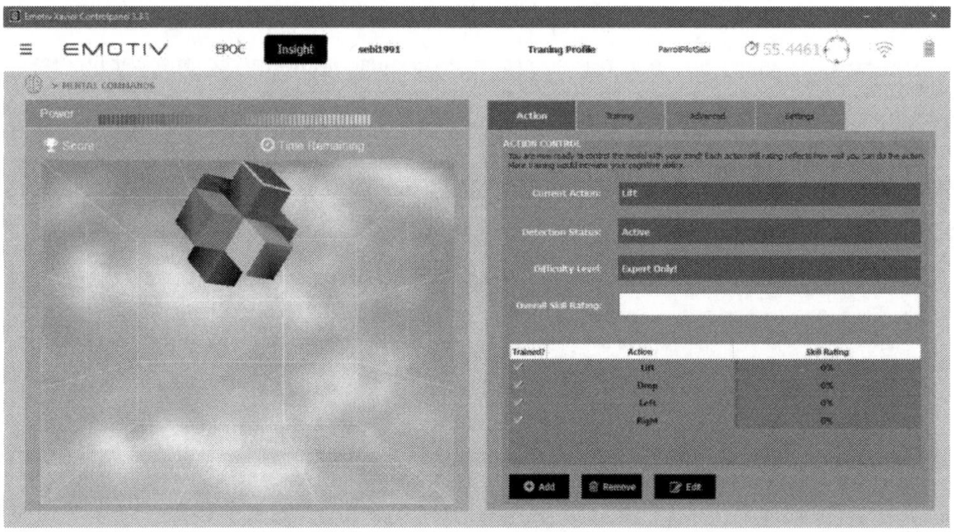

Figure 14. Emotiv Xavier Control Panel interface for Lift mental task

During on the training and execution sessions of mental task we need to maintain a good level of attention and focus key factors that correspond to a high level of beta brainwaves on frontal cortex.

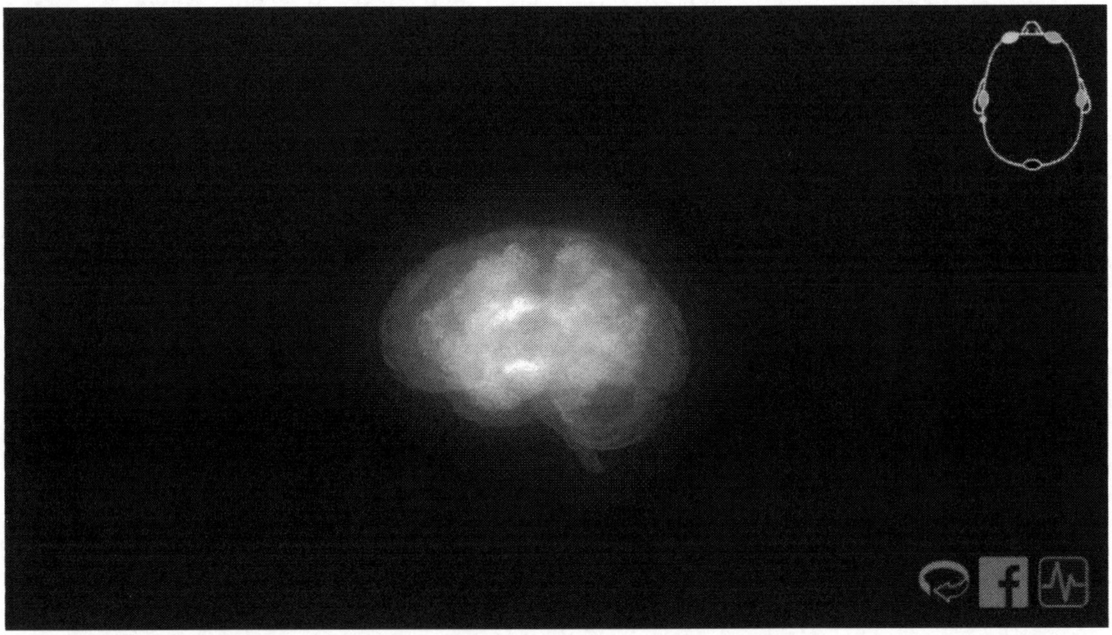

Figure 15. EMOTIV 3d Brain Visualizer Beta brainwave activation for Lift mental task

To convert the binary code into sequences of keystroke we use the Interface Emotiv Xavier EmoKey that after connect to the Interface Emotiv Xavier Control Panel based on logical rules (Mapping EmoKey) defined by us (I correspond to mental task lift, K to drop, A to right, D to left , and R to reset the angle of pich and roll).

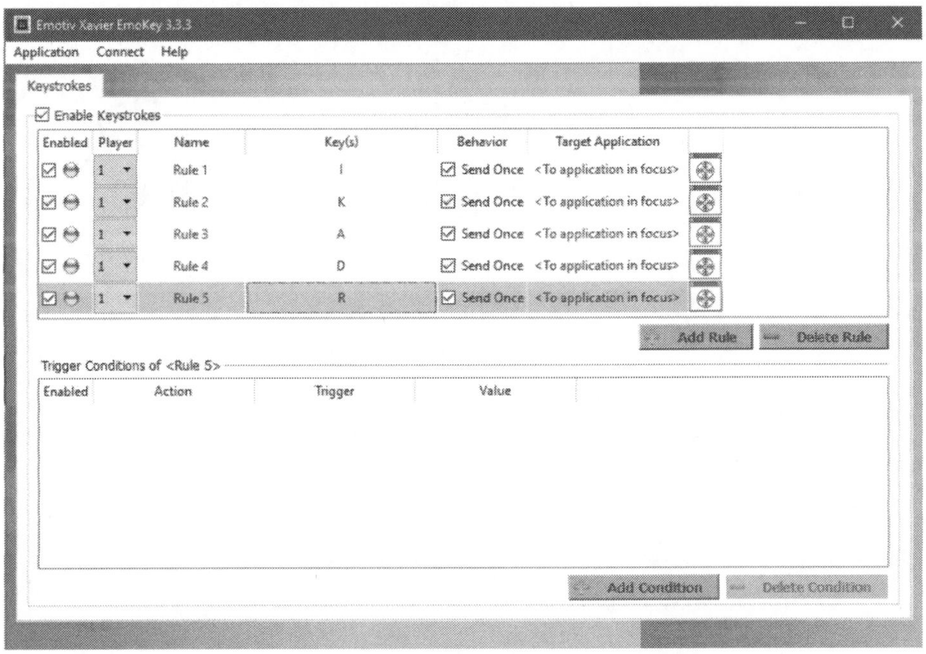

Figure 16. Interface Emotiv Xavier EmoKey

IOP Conf. Series: Materials Science and Engineering **294** (2017) 012048 doi:10.1088/1757-899X/294/1/012048

We also analyzed in development environment Matlab based on **Emotiv Xavier TestBench** development kit and EEGLAB toolbox the distribution of EEG signals for each of the five EEG channels provided by BCI headset Emotiv Insight corresponding to Lift mental task.

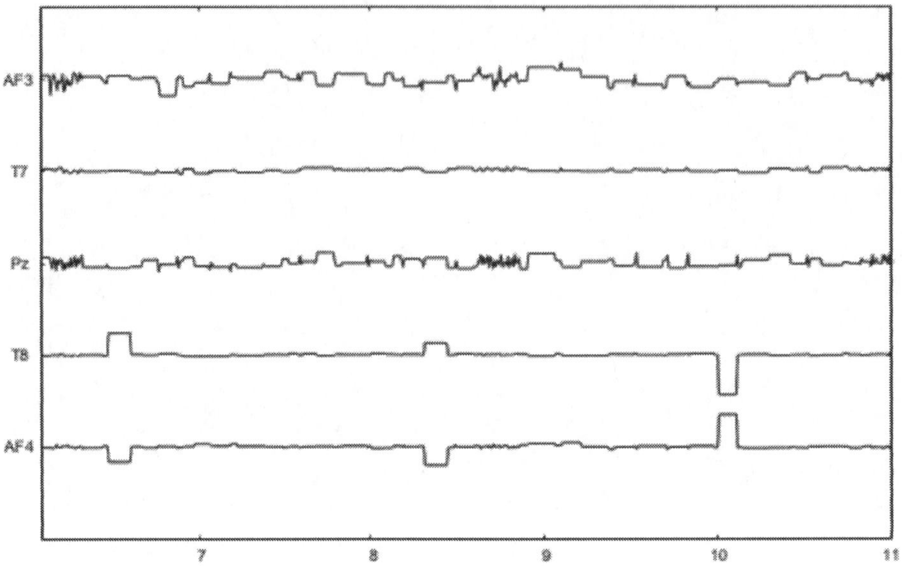

Figure 17. EEGLAB EEG Signals Plot for Lift mental task

In Figure 18 is presented the plotting of spectral analysis of the brain activity for every data channel. Each colored trace corresponds to a particular area of the brain with a distribution of power between 6Hz and 22Hz and with parieto-temporal cortex predominance.

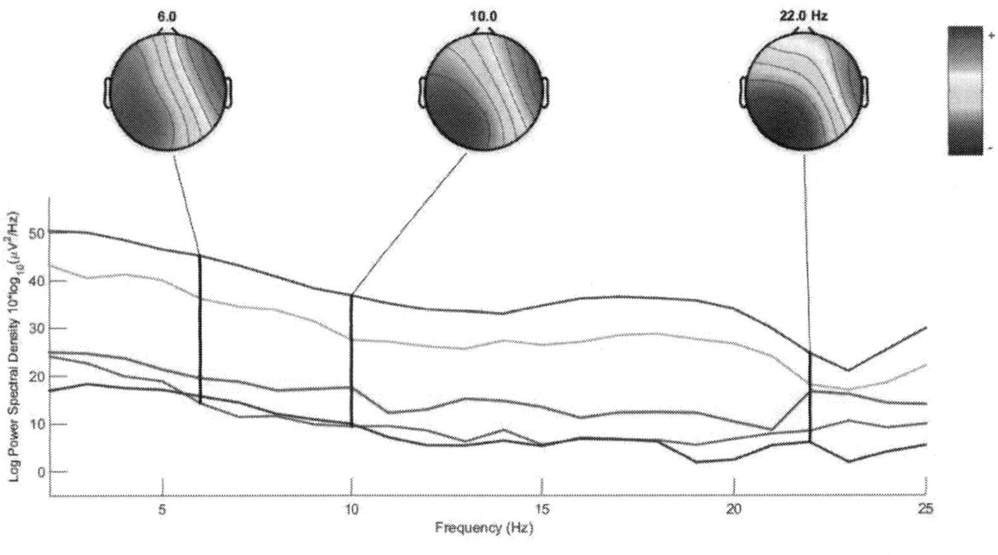

Figure 18. EEGLAB Spectral Analysis for Lift mental task

In Figure 19 is presented the diagram of brainwaves for Lift mental task obtained from analysis of AF3 data channel corresponding to the frontal cortex responsible for concentration and eye movements.

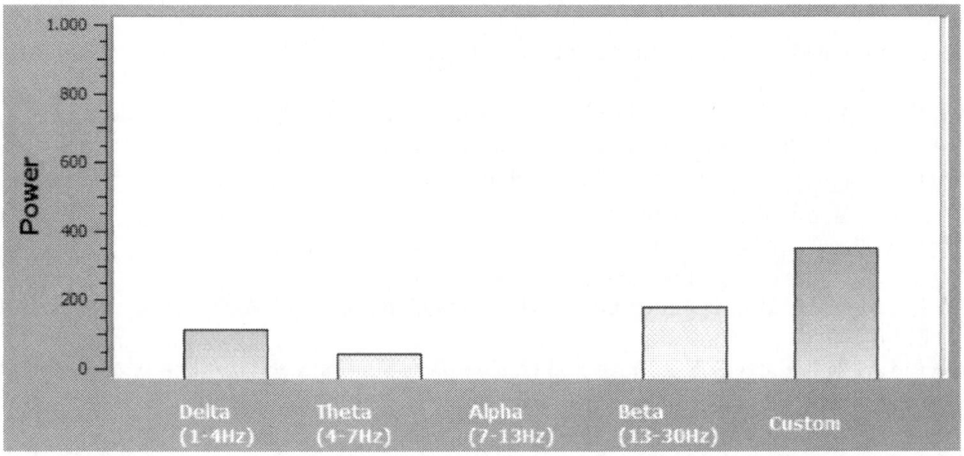

Figure 19. Emotiv Xavier TensBench SDK Brainwawes analysis of Lift mental task

4. Conclusions

Brain-computer interface represents a relatively new and reliable technology that can have appliance in major areas of activity and that can simplify human machine interaction by eliminating physical controls. Drone on the other hand also presents a wide variety of applications from monitoring and surveillance to civilian purposes.

Through this research, we proposed an unconventional method of controlling a drone directly by the brain signals acquired through a BCI system. To achieve the proposed objective we developed the mathematical model and we simulated the functioning in development environment Matlab-Simulink for the quadcopter Parrot Rolling Spider and EEGLAB toolbox for the Emotiv Insight BCI headset. We also developed a 3D animation of the quadcopter behavior in Simulink 3D animation environment based on the algorithms studied.

Further we implemented the unconventional drone control interface based on 4 mental task and we programmed the quadcopter.

Figure 20. BCI-Parrot Rolling Spider system implementation

IOP Conf. Series: Materials Science and Engineering **294** (2017) 012048 doi:10.1088/1757-899X/294/1/012048

As further developments we propose to build our drone with a 3D printed structure, together with implementation of a GPS receiver as a safety method to prevent loss of control in case of loss EEG signals due to bad detection or due to imperfect contact of the biosensors with the skin of the head.

References

[1] Rao R and Scherer R 2010 Brain-computer interfacing, *IEEE Signal Processing Magazine* **27**(4) 150-152

[2] Millan J and Carmena J 2010 Invasive or noninvasive: understanding Brain-Machine Interface Technology, *IEEE Eng. in Med. and Biol. Magazine* **29**(1) 16-22

[3] ***https://en.wikipedia.org/wiki/Quadcopter#Applications

[4] ***https://www.emotiv.com/developer/

[5] ***https://www.emotiv.com/insight/

[6] ***http://global.parrot.com/au/products/rolling-spider/

[7] Nicolas-Alonso L F and Gomez-Gil J 2012 Brain Computer Interfaces, a Review, *Sensors* **12**(2) 1211-1279

[8] Aravind M and Babu S S 2016 *Embedded implementation of brain computer interface using FPGA*, International Conference on Emerging Technological Trends (ICETT), Kollam, pp 1-5

[9] Li W, Jamarillo C and Li Y 2011 *A brain computer interface based humanoid robot control system*, In: Proceedings of the IASTED International Conference on Robotics, pp 390-391

[10] Borghini G, Astolfi L, Vecchiato G, Mattia D and Babiloni F 2012 Measuring neurophysiological signals in aircraft pilots and car drivers for the assessment of mental workload, fatigue and drowsiness, *Neurosci Biobehav Rev* **44** 58-75

[11] Schalk G, McFarland D J, Hinterberger T, Birbaumer N and Wolpaw J R 2004 BCI2000: a general-purpose brain-computer interface (BCI) system, *IEEE Trans Biomed Eng.* **51**(6) 1034–1043

[12] Schalk G and Mellinger J 2010 *A practical guide to brain–computer interfacing with BCI2000 General-purpose software for brain-computer interface research, data acquisition, stimulus presentation, and brain monitoring,* Ed. Springer Science & Business Media

[13] Szafir D J 2010 *Non-Invasive BCI through EEG*, BA, Boston College

[14] Szafir D and Signorile R 2011 An Exploration of the Utilization of Electroencephalography and Neural Nets to Control Robots, *Lecture Notes in Computer Science* **6949** 186-194

[15] ***https://en.wikipedia.org/wiki/10-20_system_(EEG)

A comparative study of the A* heuristic search algorithm used to solve efficiently a puzzle game

A E Iordan

Politehnica University of Timisoara, Department of Electrical Engineering and Industrial Informatics, 5 Revolution Street, Hunedoara, 331128, Romania

E-mail: anca.iordan@fih.upt.ro

Abstract. The puzzle game presented in this paper consists in polyhedra (prisms, pyramids or pyramidal frustums) which can be moved using the free available spaces. The problem requires to be found the minimum number of movements in order the game reaches to a goal configuration starting from an initial configuration. Because the problem is enough complex, the principal difficulty in solving it is given by dimension of search space, that leads to necessity of a heuristic search. The improving of the search method consists into determination of a strong estimation by the heuristic function which will guide the search process to the most promising side of the search tree. The comparative study is realized among Manhattan heuristic and the Hamming heuristic using A* search algorithm implemented in Java. This paper also presents the necessary stages in object oriented development of a software used to solve efficiently this puzzle game. The modelling of the software is achieved through specific UML diagrams representing the phases of analysis, design and implementation, the system thus being described in a clear and practical manner. With the purpose to confirm the theoretical results which demonstrates that Manhattan heuristic is more efficient was used space complexity criterion. The space complexity was measured by the number of generated nodes from the search tree, by the number of the expanded nodes and by the effective branching factor. From the experimental results obtained by using the Manhattan heuristic, improvements were observed regarding space complexity of A* algorithm versus Hamming heuristic.

1. Introduction

The puzzle game presented consists in a number of polyhedra of different types and colours that can be moved using the free available spaces.

The problem requires to be found the movements in order the game starts from an initial configuration and reaches to a determined configuration. The puzzle has the option for choosing from three types of polyhedra: prisms, pyramids or pyramidal frustums.

The difficulty of puzzle depends by the number of the used colours. The game is ended when on every line the polyhedra have the same colour and on every column have the same polygon as base in an increasing number of sides of the base. The game also allows the determination of the optimal solution by implementing of A* heuristic algorithm [1].

Because the problem is enough complex, the principal difficulty in solving it is given by dimension of search space, that leads to necessity of a heuristic search [2].

2. The Analysis of the Puzzle

The following examples show how to format a number of different Figure/caption combinations. From the perspective of unified modelling language, the analysis of puzzle application includes the representation of the use-case diagram.

The use-case diagram [3], [4] offers simplified and graphical representation of what the puzzle game must really do. Use-case diagram is based upon functionality and thus will focus on the "what" offers the game and not the "how" will be realized.

The use case diagram corresponding to this puzzle software, presented in Figure 1, includes:

- One actor - the user who is external entity with that the puzzle game interacts.
- Seven use-cases which describe the performance of the game.
- Relationships between user and use-cases (association relationships), and relationships between use-cases (dependency and generalization relationships).

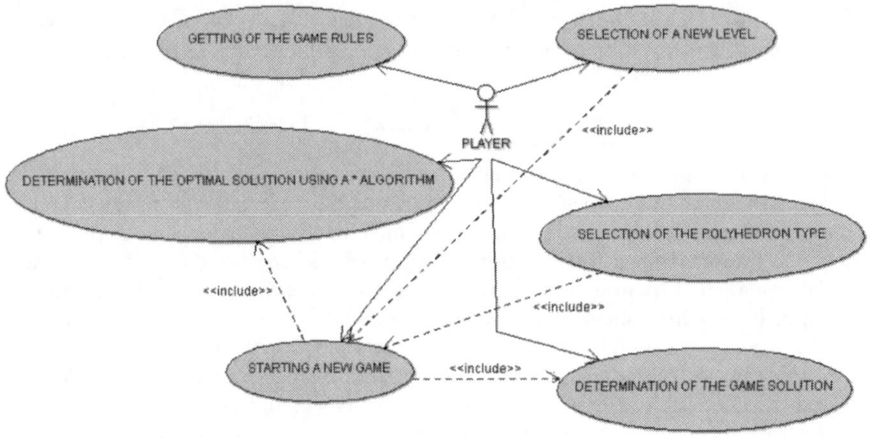

Figure 1. Use-case diagram

3. The Design of the Puzzle

Conceptual modelling allows identifying the very important elements for the interactive software [5], [6]. Class diagram is represented in Figure 2 in order to be observed the connection mode between the classes and the interfaces that are used and also the composition and aggregate relationships between instances.

After identifying the specific elements of the puzzle, they were been implemented 7 interfaces and 16 classes grouped into three packages. Package "geometry" brings together concepts that correspond to the polyhedra used by puzzle, consisting of 8 classes and one interface. Of the eight classes, three are abstract classes.

Package "algorithm" brings together concepts that correspond to the A* heuristic algorithm, consisting of 6 classes and 5 interfaces.

For memorizing a state configuration, there was implemented "State3D" class which implements "State" interface. For memorizing a node of search tree, it was implemented "Node_Puzzle" class which realizes "Node" interface and for memorizing an expanded node together with its successors, it was implemented "Node_Expanded_Puzzle" class which realizes "Node_Expanded" interface.

A* heuristic search algorithm is implemented by using "Algorithm_A_Puzzle" class. An instance of this class is composed by two instances of "Node_Puzzle" class, according to composition relationship which exists in diagram, but can also contain instances of "Node_Expanded_Puzzle" class, as it can be observed from aggregate relationship presented in diagram.

Package "gui" brings together concepts that correspond to the graphical user interface, consisting of 2 classes and one interface. For designing the graphical interface of the game, it was implemented the „Puzzle" class. It can observe that this class implements the „Runnable" interface. An instance of this class is composed by one instance of „Surface3D" class, but can also contain instances of "Algorithm_A_Puzzle" and "Node_Puzzle" classes, as it can be observed from aggregate relationship presented in diagram.

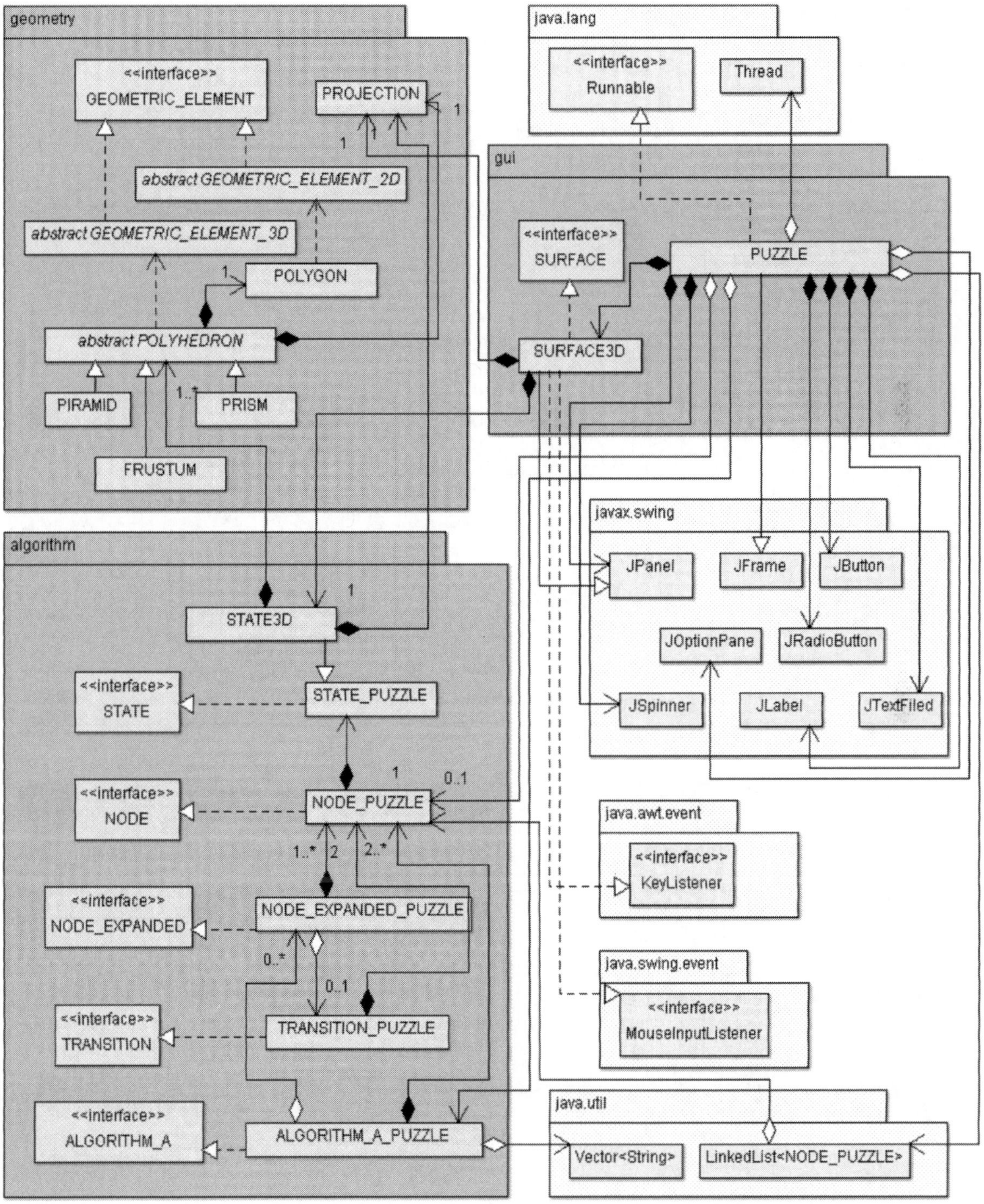

Figure 2. Class diagram

4. The Implementation of the Puzzle

The component diagram [7], [8] enables the visualization modules that compose the software and the dependencies between them. The diagram that is shown in Figure 3 describes the collection of components that all together ensure the functionality of the puzzle game.

Central component of the diagram is Puzzle.class, a component obtained by transforming by the Java compiler into executable code of the Puzzle.java component. As can be seen that component interacts directly with components Surface3D.class.

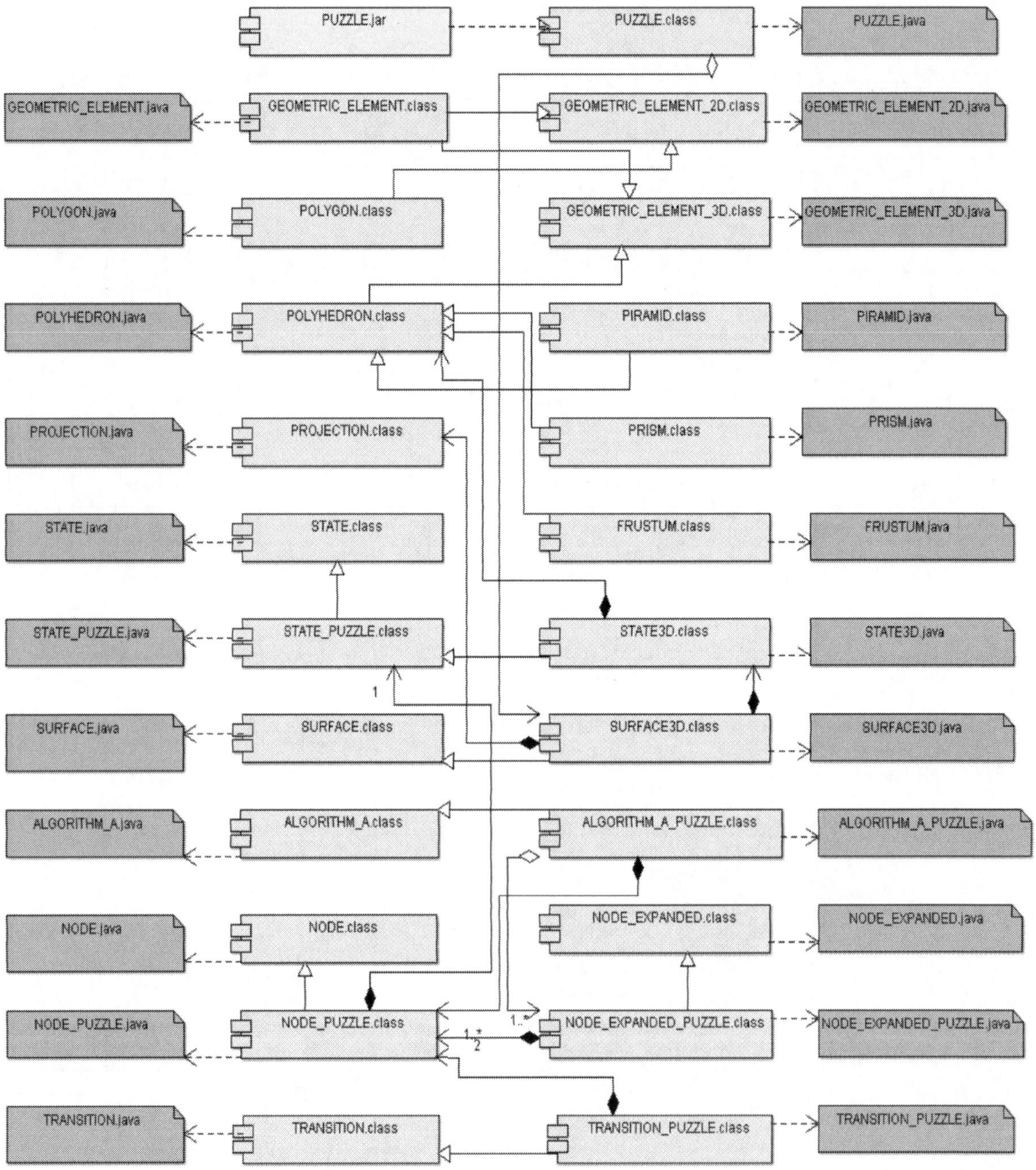

Figure 3. Component diagram

5. Graphical User Interface

The puzzle game is accomplished using the Java object-oriented programming language [9-11]. Given that specified requisites in uses cases diagram (Figure 1) it was designed graphical user interface of the puzzle. In Figures 4 and 5 are presented the five levels of difficulty. In Figure 6 is presented the corresponding final state of the third level of difficulty.

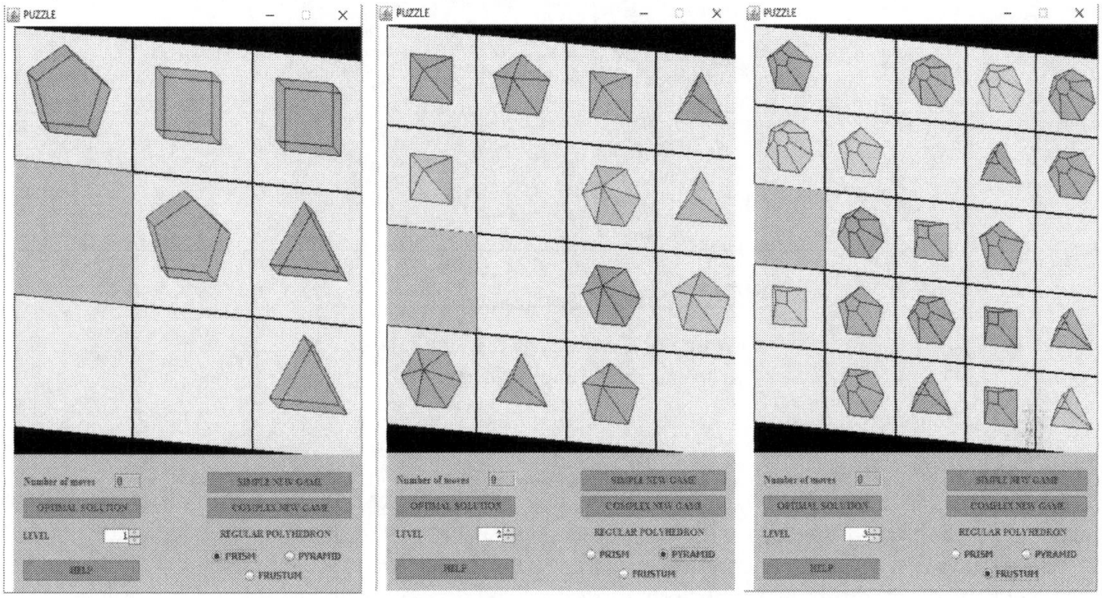

Figure 4. First three levels of difficulty

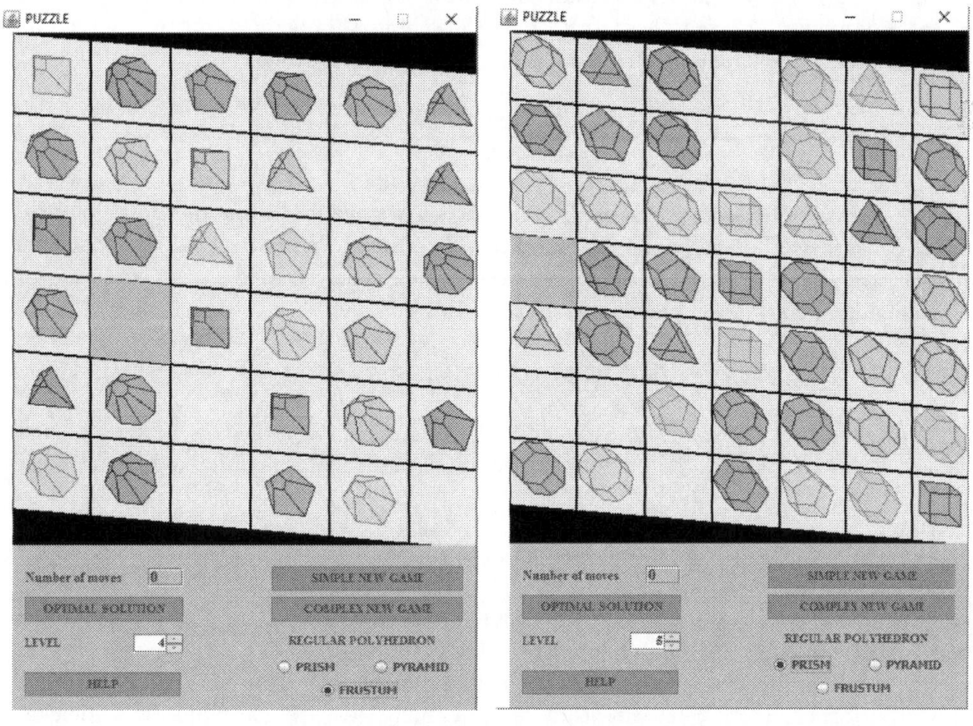

Figure 5. Last two levels of difficulty

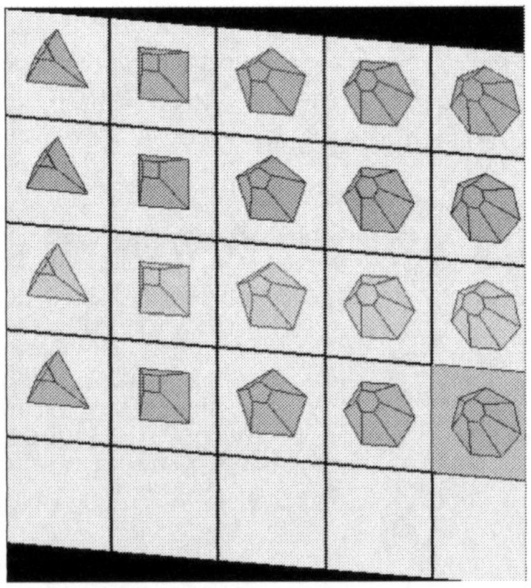

Figure 6. Corresponding final state to the third level of difficulty

6. Experimental Results

The heuristic algorithms take into consideration two pre-requisites: to find some distinction criteria which permit finding the solution with less resources that impose the uninformed algorithms and, on the other side, to determine the correct choices in order to reach the final state using the optimal path.

So, the heuristic algorithms used information from the domain of the problem to be solved. These information are usually used through an evaluation function [12] which has as argument a node of the search tree and determines a number as result, indicating the measure in which the respective node is indicated for expanding. A* search algorithm [13] combines Greedy algorithm with breadth-first search algorithm and the evaluation function has the following relation:

$$f(S) = g(S) + h(S) \tag{1}$$

where g(S) is the path cost from the start node to current node S, and h(S) is an estimation of the path cost from the current node to the goal node. As it is shown in the followings, for this puzzle game problem there was used two heuristic functions: Hamming heuristic and Manhattan heuristic. Hamming heuristic [14] is a simple heuristic function which is determined by the number of misplaced tiles. The Manhattan heuristic [15] is calculated as follows:

$$h_M(S) = \sum ManhattanDistance(k), \text{ where } k \in \{1,2,3,...,N\} \tag{2}$$

ManhattanDistance(k) represents the distance of k number position (on horizontal and also on vertical axis) in S state towards its position in the goal state which is calculated with the following relation [15]:

$$ManhattanDistance(k) = |x_k - x_{kg}| + |y_k - y_{kg}| \tag{3}$$

where (x_k, y_k) represent the coordinates of a polyhedron in current state and (x_{kg}, y_{kg}) represent the coordinates of the same polyhedron in goal state.

In order to confirm the obtained theoretical results which show that Manhattan heuristic is more efficient, one performance criterion was used: space complexity [16, 17] which was measured by the generated nodes number from the search tree and by the number of the expanded nodes.

For this puzzle game was tested a set including 10000 random configurations. This was done by running of the interactive application presented in the previous paragraph.

6.1. The graphical comparison of the results regarding the generated nodes number

For presenting the efficiency of the Manhattan heuristic from the generated nodes number point of view there are represented graphical the experimental values obtained for average value of solutions.

The comparison of the average value of solutions, presented in Figure 7, is realized for the two analyzed functions: Hamming heuristic and Manhattan heuristic.

6.2. The graphical comparison of the results regarding the expanded nodes number

For presenting the efficiency of the Manhattan heuristic from the expanded nodes number point of view there are represented graphical the experimental values obtained for average value of solutions.

The comparison of the average value of solutions, presented in Figure 8, is realized for the two analyzed functions: Hamming heuristic and Manhattan heuristic.

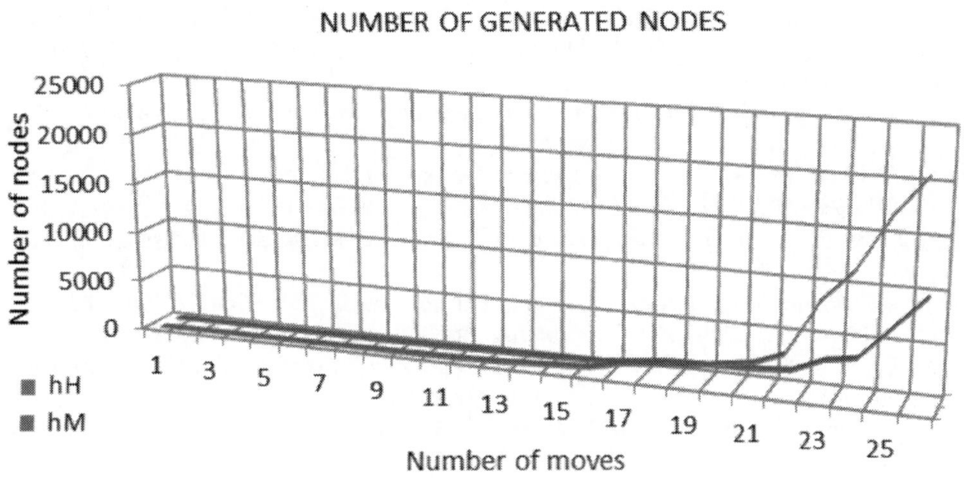

Figure 7. Number of generated nodes

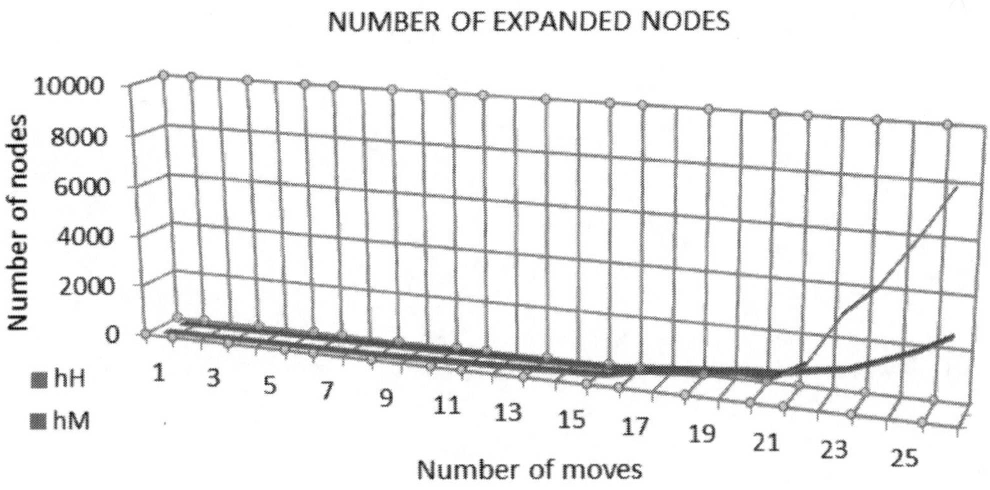

Figure 8. Number of expanded nodes

7. Conclusions

In this paper was used the Manhattan heuristic function which is more informed than Hamming heuristic. From experimental results by using the Manhattan heuristic, an improvement in spatial complexity of A* algorithm versus Hamming heuristic was observed. In two cases the generated solutions are obtained using an optimal number of movements.

Both theoretical and also experimental it demonstrated that the Manhattan heuristic is more informed than the Hamming heuristic, hence streamlines the A* heuristic search method.

References

[1] Ertel W 2011 *Introduction to Artificial Intelligence*, Springer-Verlag, Berlin
[2] Shi Z 2011 *Advanced Artificial Intelligence*, Word Scientific Publishing
[3] Jalote P 2008 *A Concise Introduction to Software Engineering*, Springer-Verlag, Berlin
[4] Fowler M and Scott K 2003 *UML Distilled: A Brief Guide to the Standard Object Modelling Language*, Pearson Education
[5] Hunt J 2014 *Guide to the Unified Process featuring UML, Java and Design Patterns*, Springer London Ltd
[6] Lunn K, Skelton J and Bennett S 2004 *Schaum's Outline of UML*, McGraw-Hill Education
[7] Bruegge B and Dutoit A 2013 *Object Oriented Software Engineering Using UML, Patterns, and Java*, Pearson Education
[8] Dennis A, Wixom B H and Tegarden D 2012 *Systems Analysis and Design with UML*, John Wiley & Sons Ltd
[9] Martin R 2012 *Java Application Architecture*, Pearson Education
[10] Bryant J 2012 *Java 7 for Absolute Beginners*, Springer, Berlin
[11] Gosling J, Joy B, Steele G, Bracha G and Buckley A 2013 *The Java Language Specification*, Oracle
[12] Bonet B and Geffner H 2001 Planning as heuristic search, *Artificial Intelligence* **23** 5-33
[13] Russel S and Norving P 2003 *Artificial intelligence – A modern approach*, Prentice Hall
[14] Rothlauf F 2011 *Design of Modern Heuristics – Principles and Application*, Springer-Verlag, Berlin
[15] Edelkamp S and Schroedl S 2011 *Heuristic Search: Theory and Applications*, Morgan Kaufmann Publishers
[16] Baase S and Van Gelder A 2000 *Computer Algorithms. Introduction to Design & Analysis*, Addison-Wesley
[17] Cormen T, Leiserson C, Rivest R and Stein C 2001 *Introduction to Algorithms*, MIT Press, Cambridge

IOP Conf. Series: Materials Science and Engineering **294** (2017) 012050 doi:10.1088/1757-899X/294/1/012050

Highlighting the harmonic regime generated by electric locomotives equipped with DC motors

I Baciu and C D Cunţan

Politehnica University of Timisoara, Department of Electrical Engineering and Industrial Informatics, 5 Revolution Street, Hunedoara, 331128, Romania

E-mail: corina.cuntan@upt.ro

Abstract. The paper presents the results of measurements made using the C.A. 8334 power quality analyzer on an electric locomotive equipped with DC motors. We carried out determinations of the current-voltage regime using a locomotive motor. The harmonic regime of the other motors being identical to the analysed one, we could easily deduce the effects caused by the entire locomotive. The data measured with the analyzer were firstly transferred into a computer system using the Qualistar software, followed by data processing in Excel, enabling therefore a graphical representation of the characteristic parameters of power quality. Based on the acquired data, we determined the power factor, as well as the active, reactive and apparent power.

The measurements revealed high values of the current harmonics, fact that required some measures to be taken for reducing the values of these harmonics. For this, we ran a simulation using the PSCAD/EMTDC software, by introducing LC filters in tune with the harmonic frequencies. The result was a significant reduction in the harmonic regime, either in the harmonics values or the power factor and reactive power.

1. Introduction

When referring to the electric power quality in a power system, we must always consider the need to supply the consumers with sinusoidal voltages whose values must fall within a well-defined range. The three-phase systems have been designed and built to operate in symmetric, balanced regimes. If one of the network or consumer elements is unbalanced, the regime becomes asymmetrical and the voltage and current systems lose their symmetry.

The most notable imbalances are cause by industrial consumers fed by high power, connected to the medium or high voltage electrical networks from which we can mention the transformation stations for the powering of electrical trains.

The number and the power of the nonlinear loads which from an electromagnetic point of view pollute the feeding electrical network grew and grow constantly. Their connection to the electrical network produces notable changes of the current, and keeping in mind the fact that their power in relation with the power of the feeding electrical network cannot be neglected, significant tension changes which deteriorates the quality of the electrical energy. Its quality depends both on the topology of the feeding network and on the quantity polluting harmonics inserted in the nonlinear loads network [1].

The sensitivity of customers to the quality of the received electrical energy varied for some categories of load. This is due to the impossibility of the loading to function correctly when fed with tensions which has variations that don't fit in the prescribed limits and the compensation measures against the abnormal function of the loadings will reflect in the costs [2].

To have a certainty in the correct functioning of the loadings from an electro energetic system there have been imposed specific norms for a series of measures that are operated in this system.

Framing the prescribed limits through rules or quality standards is not easy to realize in practical cases.

The established standards are generally different for various tension levels and most of the times, in practical work; some loadings need to produce by themselves the optimum powering tension.

This means resolving the problem of unwanted compensation for energy disturbances exterior to the charge, the result being to extend the life time of the charge [3].

The quality of electric energy can be determined with the help of qualitative indicators whose values are standardized both in our country and in Europe on all powering lines low, medium and high voltage.

Not obeying the rules or standards of some parameters inevitably leads to the dramatic decrease of the life time of the equipment or to an oversize of it to deal with the new functioning conditions [4].

2. Problem Formulation

The powering systems with electric energy in electric traction represent the vital part and their evolution allowed the evolution of technologies which are used in this domain. As presented in the previous paragraph, the types of powering systems are: in continuous current, in alternative single-phase current and three-phase alternative current [5].

The one-phase powering system with a tension of 25kV and a frequency of 50 Hz was imposed in the majority of the European states (Figure 1), through the facilities offered by the national networks and the reached travel speeds [1].

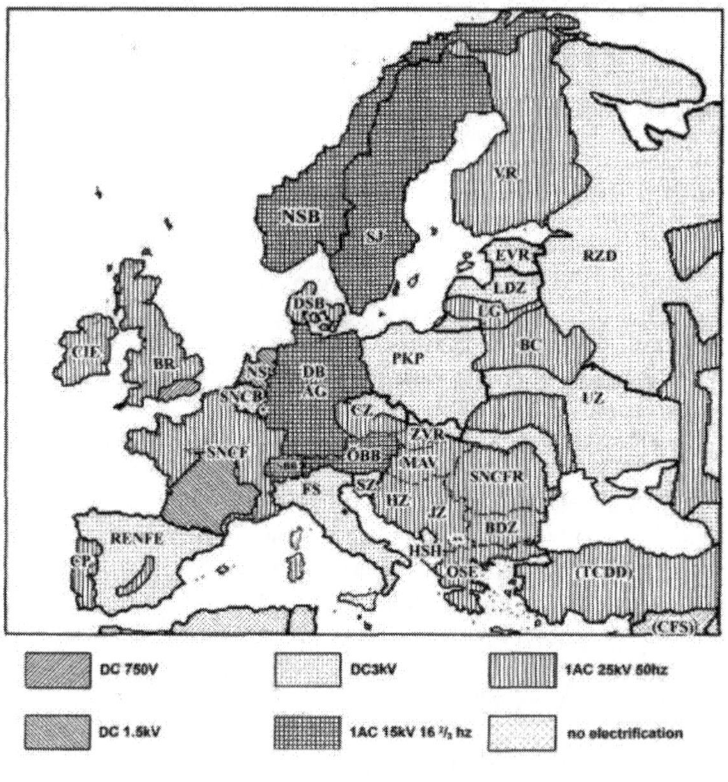

Figure 1. Powering systems with electric energy of the European railways [1]

The electric railway traction has both advantages and disadvantages in comparison to the diesel-hydraulic or diesel-electrical traction [6]. On main transport ways one can find the next advantages:

- Low cost of energy
- Using hydroelectric and coal resources
- Ecologic functioning with reduced noise
- The possibility of regaining braking energy
- The presence of a generator is not required on the vehicle's engine, obtaining a double power for the same mass (7MW for 4 axis and m_E = 85t in comparison to 3,5MW and m_E = 130t).
- A speeding cruise of over 200km/h and a bigger trailed mass.
- Possibility of the electric traction engine to overload
- Maintenance costs are reduced
- Many functioning hours with easy maintenance
- It does not require preheating of engines
- The electric engine vehicles are cheaper that the diesel ones.

As a disadvantage one can mention the initial high price of the contact way construct and of the feeding substations.

The equivalent scheme of the traction engine (Figure 2) contains the field winding, switching and compensation in series with the main winding. The field activity can be obtained in the following ways:

- The classic cars are provided with the excitation winding connected in series with the armature. It results a "soft" characteristic specific to the electric traction (Figure 2.b), with a high torque at start and acceleration and a rapid reduction with increasing speed thereof [3].

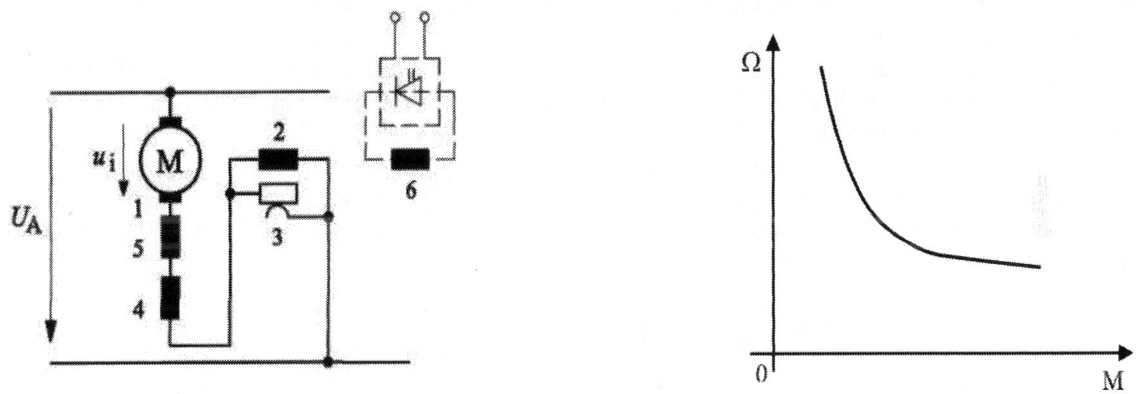

1- winding induced and collector
2- excitation winding
3- shunting
4- switching winding
5- compensation winding
6- Separate excitation powered by a chopper

Figure 2. a) The equivalent scheme of the traction engine of continuous current; b) it's mechanical characteristic

To operate at a constant power in parallel with the excitation winding resistances are connected to obtain a weaker field. For braking the reversal of the excitation field must be done with contractors.

- Branch excitation or separates is preferred in other industrial applications. If these are powered with a converter that can create load dependent current (simulating the series excitation) they can be used in traction. In this case the loosing of the field and the braking can be easily realized.
- At modern locomotives with thyristors and converters a mix system is used in which the excitation filed of the excitation winding in series in 50% of the total [7].

Traction substations with one-phase transformers connected in V/V are equipped with two identical one-phase transformers, connected to the three-phase powering system and to the contact line in the presented way (Figure 3) [8].

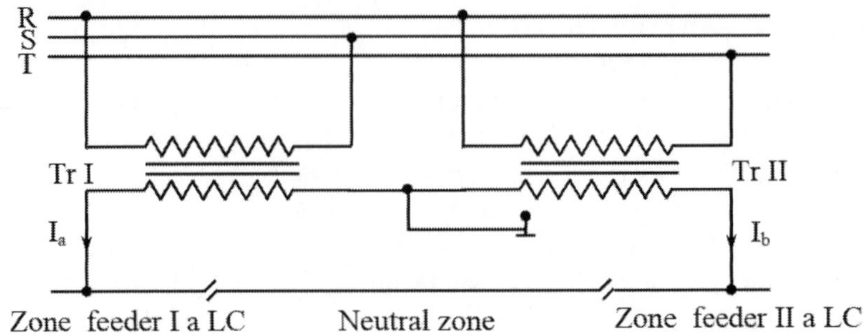

Figure 3. Traction substation with one-phase transformers connected in V/V

The scheme with two transformers tied in V/V ameliorates the tension dissymmetry which are introduced by the one-phase traction charge in the powering electric system, being in this way superior to the Y/Δ/Δ scheme [8].

By cyclic successive binding mode to the phases of the powering system, for a total of three of multiples of three traction substations, the reversed currents are notably or even entirely reduce, so the one-phase traction charges are completely equilibrated, allowing in the same time the parallel functioning of the related substation transformers and therefore the feeding from two heads of the feeder zones of the contact ways.

Also, like in the case of the Y/Δ/Δ scheme, binding the traction substations with the V/V scheme to the phases of the system lead to the notable reduction or even to the total compensation of the 3rd harmonic introduced by the locomotives with rectifiers in the systems.

The public networks of 50 Hz impose limitations regarding the recovered energy when braking and from this point of view the vehicles must be equipped with braking resistance and choppers [9].

The paper deals with some aspects regarding the electric power quality, on theoretical and practical levels, based on the measurements carried out on a locomotive motor using a C.A. 8334 power quality analyzer [10].

The front panel of the analyzer is shown in Figure 4.

Figure 4. C.A. 8334 Power quality analyzer [2]

Mathematical formulae used to compute the various parameters [10]
Half-period voltage and current RMS values.
Single RMS voltage half-period i + 1 phase:

$$\mathrm{Vdem}(i)=\sqrt{\frac{2}{N}\cdot\sum_{n=0}^{N}V(i,n)^2} \tag{1}$$

Compound RMS voltage half-period i + 1 phase:

$$\mathrm{Udem}(i)=\sqrt{\frac{2}{N}\cdot\sum_{n=0}^{N}U(i,n)^2} \tag{2}$$

RMS current half-period i + 1 phase:

$$\mathrm{Adem}(i)=\sqrt{\frac{2}{N}\cdot\sum_{n=0}^{N}A(i,n)^2} \tag{3}$$

1 sec RMS values for voltage and current:
Single RMS voltage i + 1 phase;

$$V_{rms}(i)=\sqrt{\frac{1}{N}\cdot\sum_{n=0}^{N-1}V(i,n)^2} \tag{4}$$

Compound RMS voltage i + 1 phase:

$$U_{rms}(i)=\sqrt{\frac{1}{N}\cdot\sum_{n=0}^{N-1}U(i,n)^2} \tag{5}$$

Courant efficacy phase i+1:

$$A_{rms}(i)=\sqrt{\frac{1}{N}\cdot\sum_{n=0}^{N-1}A(i,n)^2} \tag{6}$$

Neutral RMS current:

$$A_{rms}(3)=\sqrt{\frac{1}{N}\cdot\sum_{n=0}^{N-1}(A(0,n)+A(1,n)+A(2,n))^2} \tag{7}$$

N : Number of samples in a second
Calculation of the total harmonic distortion factor (THD):

$$V_{thd}(i)=\frac{\sqrt{\sum_{n=2}^{50}V_{harm}(i,n)^2}}{V_{harm}(i,1)} \tag{8}$$

$$U_{thd}(i)=\frac{\sqrt{\sum_{n=2}^{50}U_{harm}(i,n)^2}}{U_{harm}(i,1)} \tag{9}$$

$$A_{thd}(i) = \frac{\sqrt{\sum_{n=2}^{50} A_{harm}(i,n)^2}}{A_{harm}(i,1)} \tag{10}$$

i : phase (0; 1; 2) n : rang (2…50)
Different power levels 1 sec.
Active power i + 1 phase:

$$W(i) = \frac{1}{N}\sum_{n=0}^{N-1} V(i,n)\cdot A(i,n) \tag{11}$$

Apparent power i + 1 phase:

$$VA(i) = V_{rms}(i)\cdot A_{rms}(i) \tag{12}$$

Reactive power i + 1 phase:

$$VAR(i) = \frac{1}{N}\cdot\sum_{n=0}^{N-1} VF(i)\left(n - \frac{N}{4}\right)\cdot AF(i,n) \tag{13}$$

Power factor i + 1 phase:

$$PF(i) = \frac{W(i)}{VA(i)} \tag{14}$$

To emphasize with ease the effects introduced by the traction vehicles in the powering network more measures will be presented resulted from different measuring situations.

In Figure 5 we show the effective variation of the electric current in a DC motor with which the locomotive is equipped.

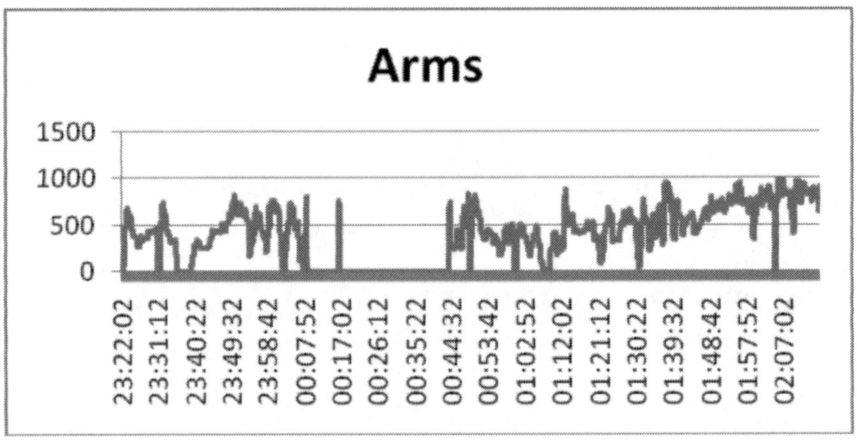

Figure 5

The periods for which we obtained zero plateau correspond to the time intervals in which the locomotive was stopped, and the high variation of the current is caused by the relief variation on the route.

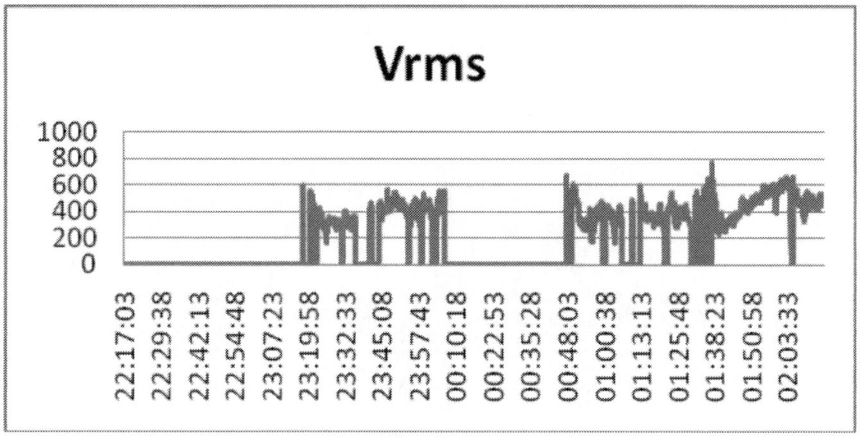

Figure 6

In Figure 6 we show how the effective voltage varies in a DC engine with which the locomotive is equipped.

There is a small variation compared to the electric current variation.

In Figure 7 we show the harmonic distortion power factor, whose high values go beyond the ones provided by the European standard.

For this reason, measures should be taken to reduce the THD factor.

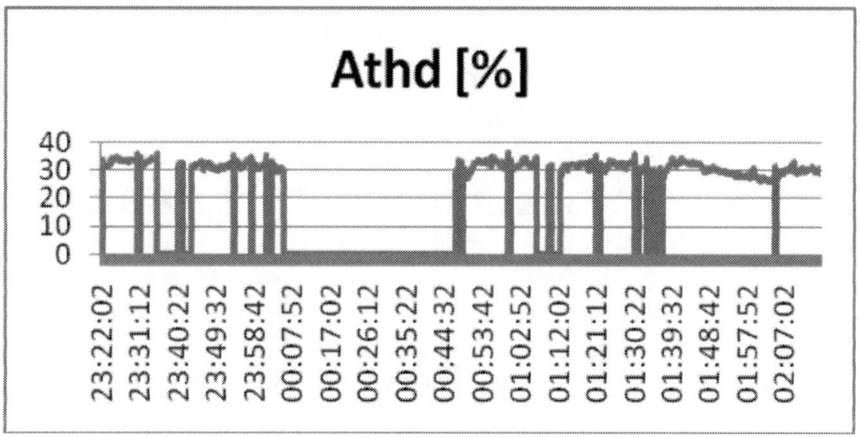

Figure 7

In Figure 8 we show the voltage harmonic distortion factor. As expected, the values are substantially lower compared to the ones obtained for the electric current, but in some cases they exceed the limits accepted by the international standards.

Based on the measured voltage and current values, we obtained the active power variation shown in the Figure 9. There is a high fluctuation of the power values throughout the entire data acquisition process.

We have also obtained the reactive power variation shown in the Figure 10. There is a high fluctuation of the power values throughout the entire data acquisition process, the values exceeding 1.4MVAR.

In Figure 11 we show the waveform for the apparent power. There is a concordance with the active and reactive power variation, the apparent power being the sum of these two types of power.

In the Figure 12 we can see a power factor value close to 0.85, with small variations around this value.

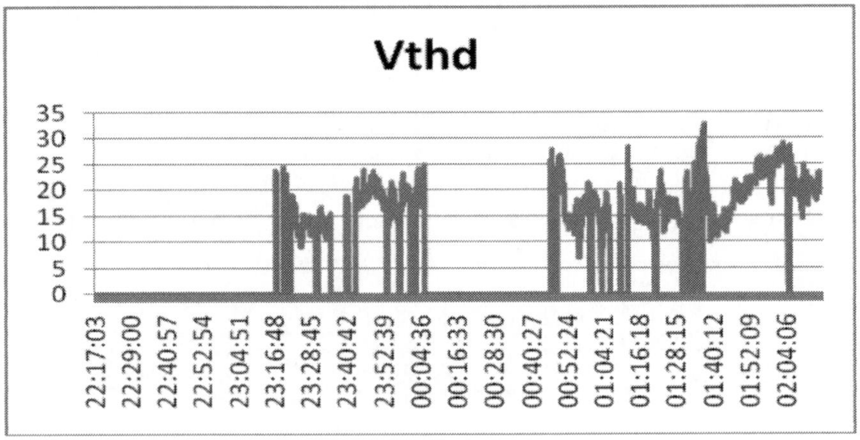

Figure 8

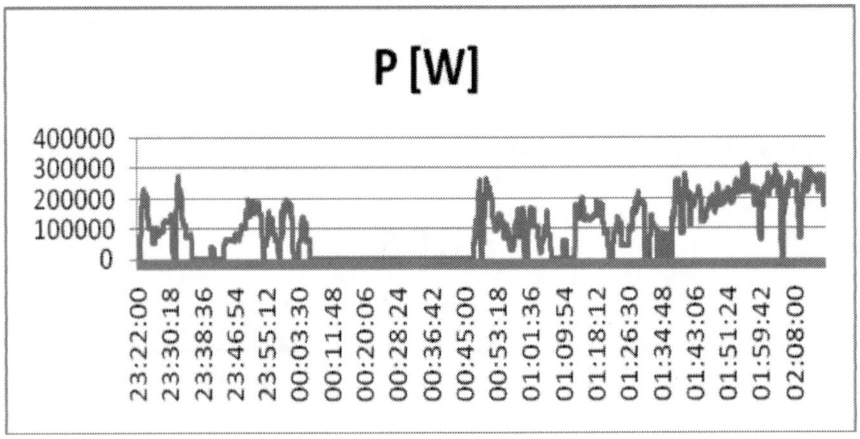

Figure 9

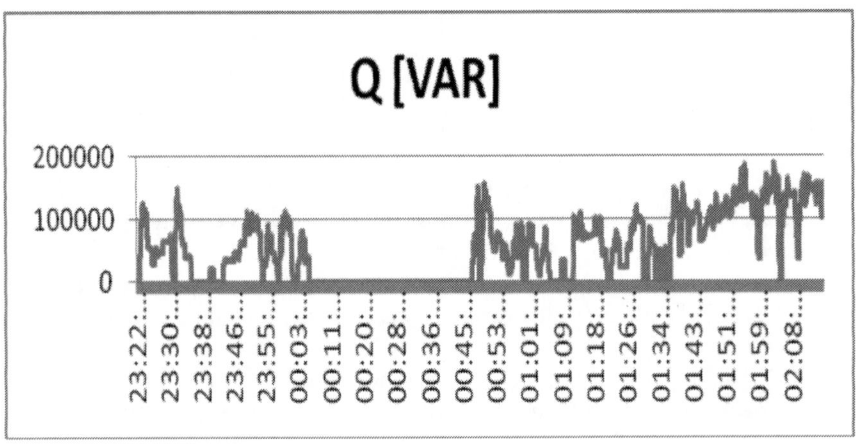

Figure 10

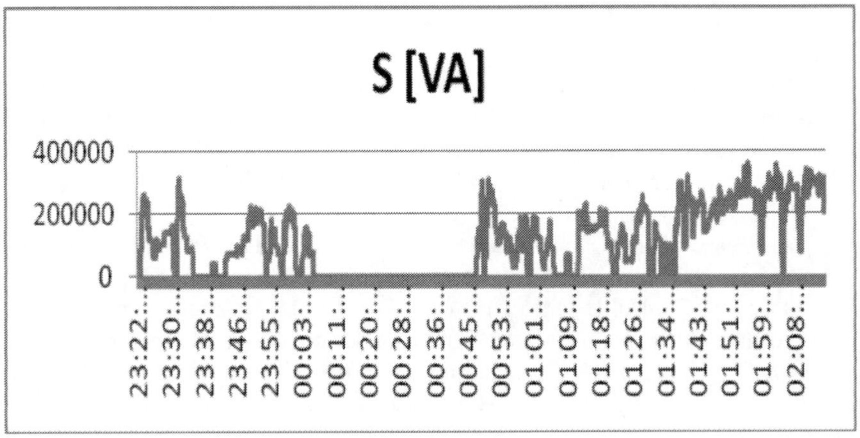

Figure 11

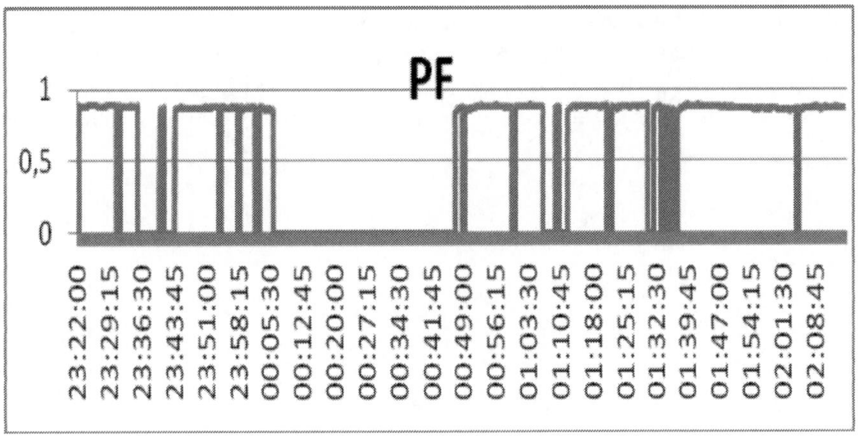

Figure 12

3. Conclusions

The experimental measurements are a real and pertinent database that has been obtained using a latest generation of high performance equipment, i.e. the C.A. 8334 power quality analyzer, France 2007.

The large amount of data, as well as the parameters determined in real time, the sampling period being set to 1 second, enables an accurate establishment of the existing deformation regime. It was found that (as we have expected) the main distorting factor is the current passing through the power supply conductors. The harmonic distortion factor of the current has values far beyond the accepted limits provided by the international standardization system.

The current variation analysis reveals a great similitude with the variation of active power and apparent power, leading to the conclusion that a highly distorting regime exists and needs to be compensated, even if the values of the voltage harmonic distortion factor falls close to the limit and, in some cases, even within the limits of compatibility.

References

[1] Steimel A 2008 *Electric Traction – Motive Power and Energy Supply Basics and practical Experience*, Oldenbourg Industrieverlag, München
[2] Popescu M, Bitoleanu A and Dobriceanu M 2008 Harmonic Current Reduction in Railway Systems, *Wseas Transactions On Systems* **7**(7) 689-698
[3] Deaconu S I, Topor M, Popa G N and Bistrian D 2010 Experimental Study and Comparative

Analysis of Transients of Induction Motor with Soft Starter Startup, *Advances in Electrical and Computer Engineering* **10**(3) 27-33

[4] Popa G N, Diniş C M and Deaconu S I 2013 Considerations on the Current Harmonics of Plate-Type Electrostatic Precipitators Power Supplies, *Elektronika ir Elektrotechnika* **19**(5) 27-32

[5] Pănoiu M, Pănoiu C, Osaci M and Muscalagiu I 2008 Simulation result about harmonics filtering using measurement of some electrical items in electrical installation on UHP EAF, *WSEAS Transactions on Circuits and Systems* **7**(1) 22-31

[6] Djeghader Y, Zellouma L, Labar H, Toufouti R and Chelli Z 2015 *Study and filtering of harmonics in a DC electrified railway system*, 7th International Conference on Modelling, Identification and Control (ICMIC), Sousse, Tunisia, 18-20 December, pp 1-6

[7] Wallace A K and Spee R 1990 The Effects of Motor Parameters on the Performance of Brushless D.C. Drives, *IEEE Transactions on Power Electronics* **5**(1) 2-8

[8] Sichenko V G and Gavrilyuk V I 2005 *The theoretical and experimental researches of electromagnetic influence from a traction electrosupply system on a railway circuits*, IEEE 6th International Symposium on Electromagnetic Compatibility and Electromagnetic Ecology, 21-24 June, Saint Petersburg, Russia, pp 41-43

[9] Toma A I, Popa G N, Iagăr A and Deaconu S I 2010 *Experimental analysis of electric parameters of a 100 t UHP electric arc furnace*, 2010 IEEE International Conference on Industrial Technology (ICIT), Vina del Mar, Chile, 14-17 March, pp 919-924

[10] ***2C.A. 8334B 2007 *Three Phase Power Quality Analyser*, Chauvin Arnaux, France

Power factor improvement in three-phase networks with unbalanced inductive loads using the Roederstein ESTAmat RPR power factor controller

C M Diniş, C D Cunţan, R O S Rob and G N Popa

Politehnica University of Timisoara, Department of Electrical Engineering and Industrial Informatics, 5 Revolution Street, Hunedoara, 331128, Romania

E-mail: corina.dinis@fih.upt.ro

Abstract. The paper presents the analysis of a power factor with capacitors banks, without series coils, used for improving power factor for a three-phase and single-phase inductive loads. In the experimental measurements, to improve the power factor, the Roederstein ESTAmat RPR power factor controller can command up to twelve capacitors banks, while experimenting using only six capacitors banks. Six delta capacitors banks with approximately equal reactive powers were used for experimentation. The experimental measurements were carried out with a three-phase power quality analyser which worked in three cases: a case without a controller with all capacitors banks permanently parallel connected with network, and two other cases with power factor controller (one with setting power factor at 0.92 and the other one at 1). When performing experiments with the power factor controller, a current transformer was used to measure the current on one phase (at a more charged or less loaded phase).

1. Introduction

The paper presents the analysis of a power factor improvement in three-phase network with unbalanced inductive loads using capacitors banks, without series coils.

Power factor (PF) is an important power engineering parameter. PF measured must be greater than neutral PF (0.92 for low voltage, and 0.95 for medium and high voltages). An artificial solution to increase the PF is to connect capacitors banks (CBs) in parallel with electric loads, which create capacitive reactive power (opposite with inductive reactive electric power), so, the overall reactive power decreases. CBs can be connected on the low or medium voltage. Capacitors values must be calculated and chosen to improve PF [1-4].

Increasing of the PF in industry can be achieved by natural means (e.g. choosing the right electric motors and transformers, reducing free running time, achieving high quality repairs of electrical equipment, changes of induction motors with synchronous motors) and/or by artificial means (using of synchronous compensators, static VAR compensators, CBs controlled by power factor controller - PFC). The use of CBs is a common solution to improve PF in low voltage electrical installations [3-6].

To improving of PF in the low voltage substation, CBs connection can be done manually or automatically (e.g. with a PFC).

To improve PF with CBs (for a single large power electric motor or for a group of electric motors), the effect of nonlinearity becomes significant and current waveforms deviate more from sine shape [4-11]. Electric waveforms (current or voltage) can be decomposed into fundamental (frequency 50

Hz) and harmonics, through Fast Fourier transform. Passive, active, or hybrid filters, can be used with CBs to improve power quality of the grid [6], [7], [9].

Experimental measurements were performed on the network with three-phase inductive loads (two three-phase induction motors) and, also single-phase inductive loads (low-pressure mercury lamps and high-pressure mercury lamp with inductive ballast) that are connected on the same phase [4], [5], [11], [12].

It is observed that the PFC controls the power factor close to the set value, for the phase on which the phase shift is measured, and on other phases, the power factor is not adjusted properly.

Operation limitations of the three-phase controller that measures the power factor on a single phase will be evaluated. It is also noted that the improvement of the power factor will increase the deformation of the phase currents.

2. Power quality and harmonics effects

The main disruptive phenomena of the power quality are:

- frequency variations;
- long-term interruptions;
- transient over-voltages;
- harmonic and inter-harmonic;
- slow variations in the effective voltage value;
- voltage shocks, short-term interruptions;
- voltage fluctuations/flicker;
- unbalances of the three-phase voltage system.

In an ideal energy system (perfectly clean), the voltage and current curves are perfectly sinusoidal. In practice, non-sinusoidal currents occur if the load is non-linear in relation to the applied voltage [13-15].

2.1. *Some indicators of power quality*

The power quality analyser CA 8334B can computes up to 50 harmonics rank [16], [17].

The RMS values of voltage and currents are:

$$V_{RMS\,i} = \sqrt{\frac{1}{N}\sum_{j=0}^{N-1}(V_{ji})^2} \tag{1}$$

$$I_{RMS\,i} = \sqrt{\frac{1}{N}\sum_{j=0}^{N-1}(I_{ji})^2} \tag{2}$$

where i is L1, L2 or L3.

Total harmonic distortion of voltages and currents:

$$V_{THD\,i} = \frac{\sqrt{\sum_{n=2}^{50}(V_{ni})^2}}{V_1}\cdot 100 \tag{3}$$

$$I_{THD\,i} = \frac{\sqrt{\sum_{n=2}^{50}(I_{ni})^2}}{I_1}\cdot 100 \tag{4}$$

Crest factor voltage and current can be compute with:

$$V_{CF\,i} = \frac{\max\left(V_{pp\ i}, \left|V_{pm\ i}\right|\right)}{\sqrt{\frac{1}{N}\cdot\sum_{n=0}^{N-1}\left(V_{ni}\right)^2}} \tag{5}$$

$$I_{CF\,i} = \frac{\max\left(I_{pp\ i}, \left|I_{pm\ i}\right|\right)}{\sqrt{\frac{1}{N}\cdot\sum_{n=0}^{N-1}\left(I_{ni}\right)^2}} \tag{6}$$

Positive voltage and negative voltage are given by relations:

$$V_{+} = \frac{V_1 + aV_2 + a^2V_3}{3} \tag{7}$$

$$V_{-} = \frac{V_1 + a^2V_2 + aV_3}{3} \tag{8}$$

The unbalance voltage and currents are calculated with:

$$V_{unb} = \frac{\left|V_{-}\right|}{\left|V_{+}\right|}\cdot 100 \tag{9}$$

$$I_{unb} = \frac{\left|I_{-}\right|}{\left|I_{+}\right|}\cdot 100 \tag{10}$$

Power factor (including harmonics) is given by:

$$PF = \frac{P}{S} \tag{11}$$

3. Power factor improving with capacitor banks

The role of capacitor banks is to provide the reactive power required for inductive loads. The goal of the capacitor banks are:

- decreasing of the phase current;
- increasing the voltage level for electric loads;
- reduction of the active and reactive power losses;
- improving the power factor of the source;
- reduction total consumption at power source.

Compensation of reactive power with capacitors banks also has the following advantages:

- the installations are static, have no wear parts, so there is no need for expensive improvement or surveillance;

- it has a relatively low gauge and can be located at any point;

- the active power losses are relatively low (2-6 W/kVAr).

4. Power factor improving in electrical networks with power factor controllers

In the experimental installation, a 12-steps ESTAmat RPR power factor controller (from VISHAY, Germany) and three phase power quality are used (Figures 1, 4, 5, 6) [12], [17].

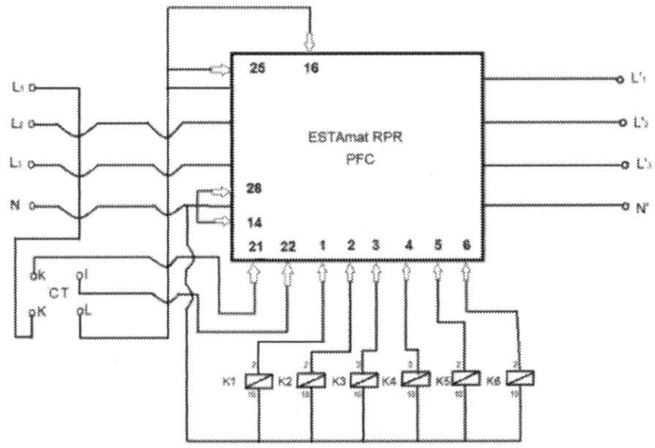

Figure 1. Electrical setup with ESTAmat RPR controller

- K1.....K6 - power relays, RI 13 type;
- CT – current transformer, with transformation ratio 15A/5A connect on the phase L1.

Inductive three-phase loads used in experiments are two induction motors:

- Motor 1 (M1): power: 0.75 kW; current: 2.1 A; efficiency: 0.72; power factor: 0.76; rotation: 1350 rpm; shaft diameter: 19 mm – Figure 2;
- Motor 2 (M2): power: 0.25 kW; current: 0,85 A; efficiency: 0,62; power factor: 0,72; rotation: 1350 rot/min; shaft diameter: 14 mm – Figure 3.

Figure 2. Induction motor M1

Figure 3. Induction motor M2

Figure 4. Power switch S1

Figure 5. Power quality analyser CA 8334 B

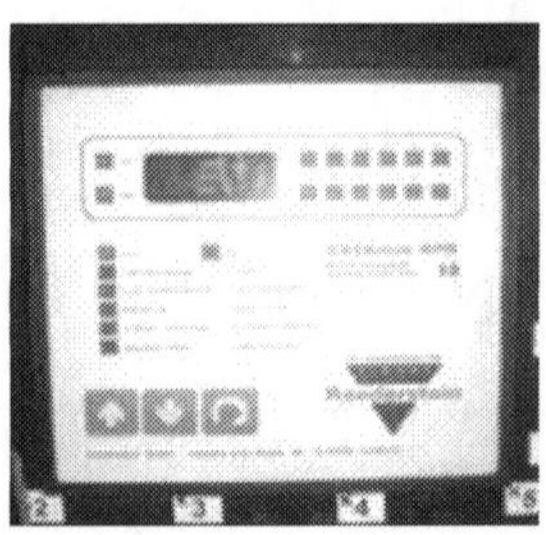

Figure 6. ESTAmat RPR controller

Single-phase inductive loads are lamps:
H1 – Hg fluorescent lamp with magnetic ballast, 2 x 20 W;
H2 – Hg high pressure lamps, 125 W;
H3 – Hg fluorescent lamp with magnetic ballast, 2 x 20 W.

In experiments, where used six capacitor banks, in delta connection (Figure 7, two capacitor banks in one box). The electrical parameters for capacitor banks, delta connection, are:

- Capacitors bank 1, B1: C1=3x3.78μF/400V, Qb1=571 VAr;
- Capacitors bank 2, B2: C2=3x4μF/400V, Qb2=616 VAr;
- Capacitors bank 3, B3: C3=3x3.71μF/400V, Qb3=560 VAr;
- Capacitors bank 4, B4: C4=3x3.9μF/400V, Qb4=588 VAr;
- Capacitors bank 5, B5: C5=3x4.12μF/400V, Qb5=622 VAr;
- Capacitors bank 6, B6: C6=3x3.6μF/400V, Qb6=542 VAr.

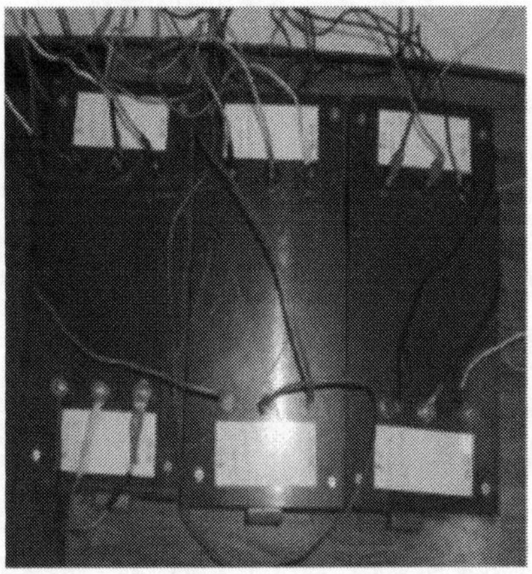

Figure 7. Capacitors banks used in experiments

5. Experimental measurements

In the following experiments with the current transformer, the current was measured on phase L1. The current transformer connected to the PFC input. The single-phase loads were connected between L1 and null, at experiments 1, 2 and 3. At experiments 1 and 2 the single phase consumers were connected between L1 and null, and at experiment 3 between L2 and null.

5.1. Experiment 1, power factor controller with DPF set=1

The stages at experiments 1 are shown in Table 1.

Table 1. Stages at first experiment DPF set = 1

Measuring stages	DPF (-)
1. Motor M1 connected	0.12 i
2. Capacitors banks B1, B2 connected	
3. Motor M2 connected	0.64 i
4. Lamp H1 connected	0.6 i
5. Capacitors bank B3 connected	0.6 i
6. Lamp H3 connected	0.91 i
7. Lamp H2 connected	0.48 i
8. B4, B5, B6 connected	0.98 c

9. Motor M1 disconnected	0.58 i
10. Capacitors banks B1, B2 disconnected	1
11. Lamps H1, H3 disconnected	0.64 i
12. B3 disconnected	0.92 i
13. B4 disconnected	0.94 c
14. Motor M2 disconnected	0.88 c
15. Lamp H2 lamp disconnected	0.06 c
16. Capacitors bank B5 disconnected	-
17. B6 disconnected	-

M2 and H1 were connected, and then PFC connected capacitors banks B1 and B2, DPF = 0.6 (Figures 8-11).

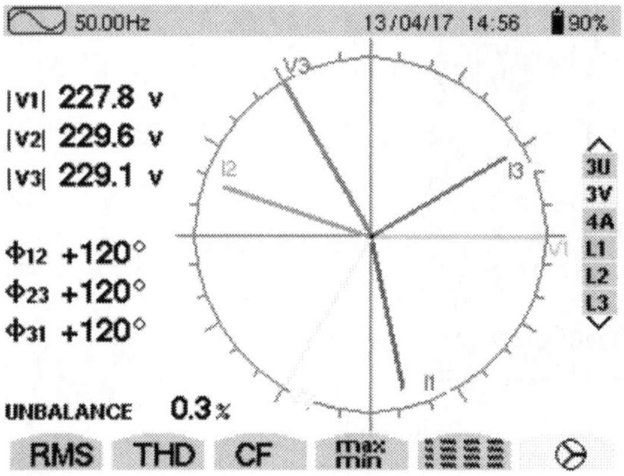

Figure 8. Fresnel diagram at connection of M1

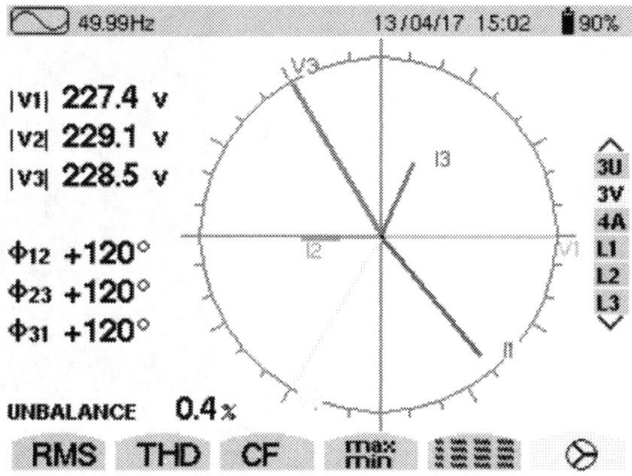

Figure 9. Fresnel diagram at connection of M1, M2 and lamp H1

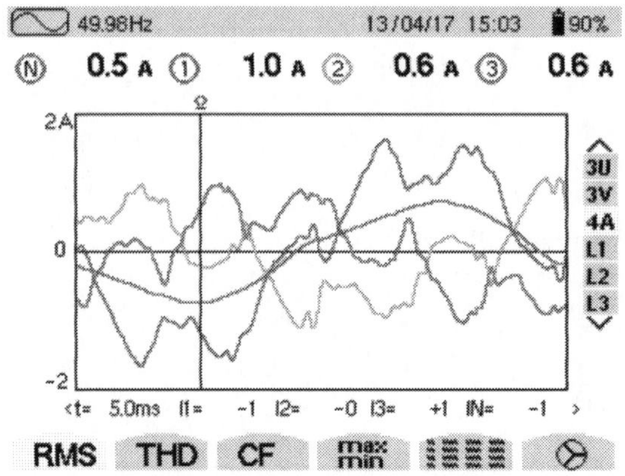

Figure 10. Phases' currents and the null current

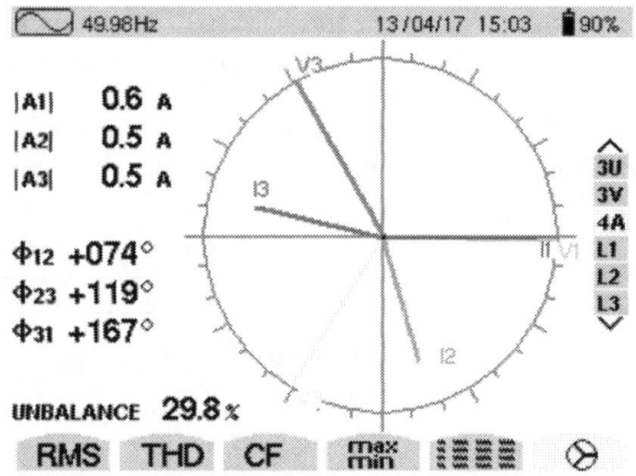

Figure 11. Fresnel diagram

An then, after a period of time, PFC was connected capacitors bank B3 (Figures 12, 13).

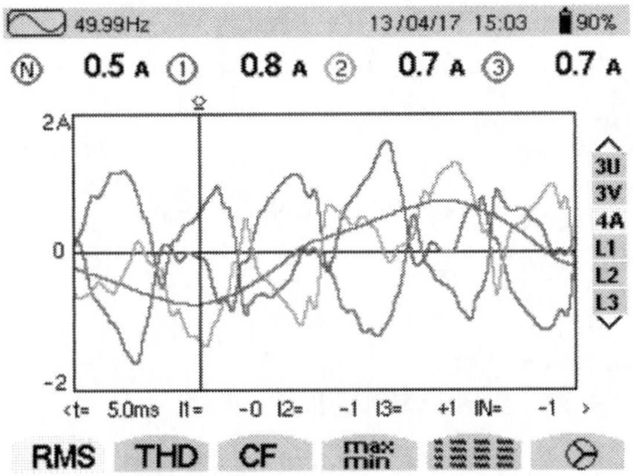

Figure 12. Fresnel diagram when was connected M2

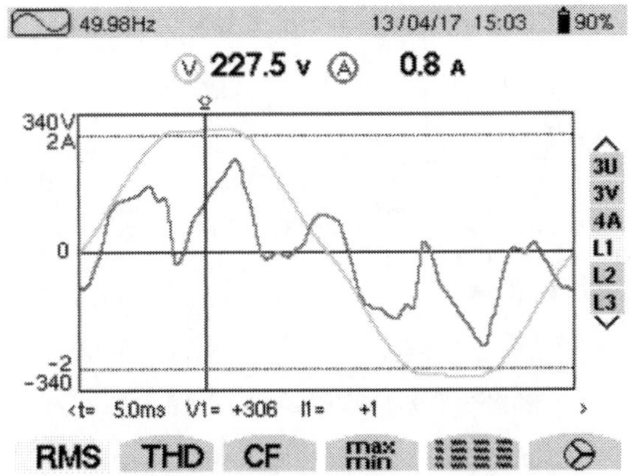

Figure 13. Voltage and current on phase L1

After a while were connected all electrical loads (M1, M2, H1, H2, H3), and the power factor controller was connected all capacitors banks: DPF = 0.98 capacitive (Figures 14, 15).

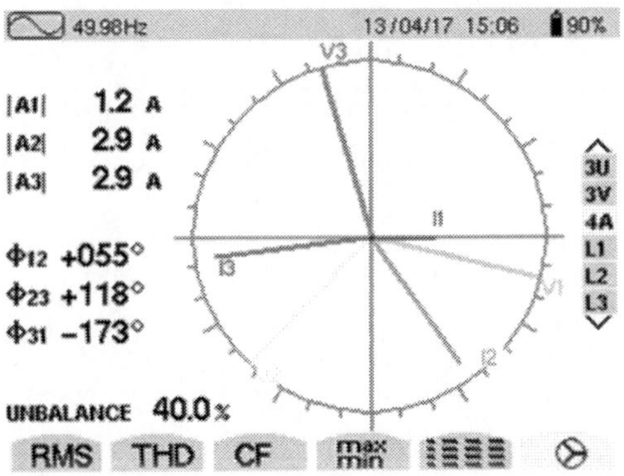

Figure 14. Fresnel diagram for currents

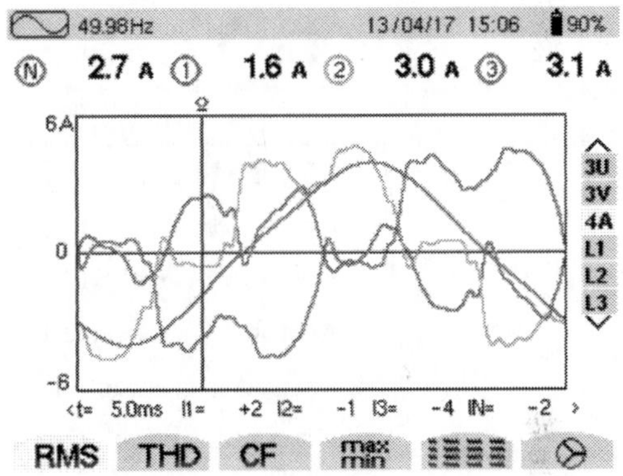

Figure 15. Phases' currents and the null current

M1 was disconnected and, after a short time, B1 and B2 were disconnected by PFC, DPF=1 (Figures 16-18).

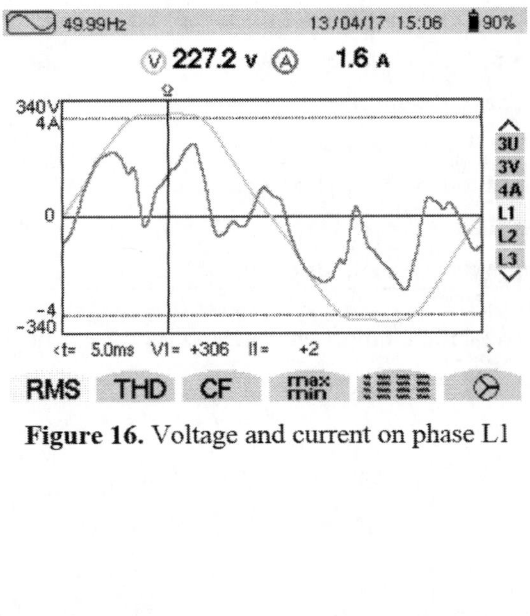

Figure 16. Voltage and current on phase L1

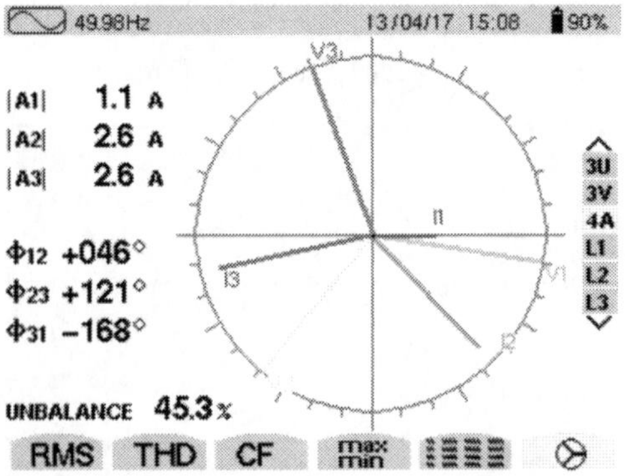

Figure 17. Fresnel diagram for currents

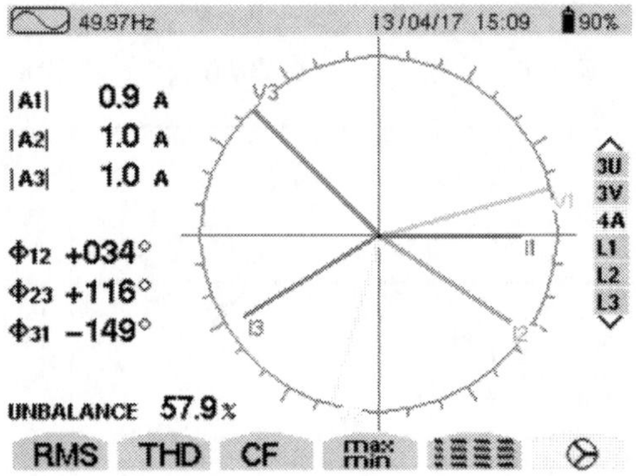

Figure 18. Fresnel diagram for currents

H1 and H3 were disconnected and, after a short time, B3 and B4 were disconnected by PFC, DPF_m= 0.63 inductive.

5.2. Experiment 2, power factor controller with DPF set =0.92
The stages at experiments 2 are shown in Table 2.

Table 2. Stages at second experiment DPF set = 0.92

Measuring stages	DPF (-)
1. Motor M1 connected	0.12 i
2. Capacitors bank B1 connected	0.29 i
3. Capacitors bank B2 connected	0.79 i
4. Motor M2 connected	0.63 i
5. Lamp H1 connected	0.59 i
6. Capacitors bank B2 connected	0.99 i
7. Lamp H3 connected	0.9 i
8. Lamp H2 connected	0.47 i
9. Capacitors banks B4, B5 connected	0.9 i
10. Motor M1 disconnected	0.74 i
11. Capacitors banks B1, B2 disconnected	0.92 i
12. Lamps H1, H3 disconnected	0.86 i
13. Capacitors bank B3 disconnected	0.95 i
14. Motor M2 disconnected	0.82 i
15. Capacitors bank B4 disconnected	0.9 i
16. Lamp H2 is connected	-
17. Capacitors bank B5 disconnected	-

M1 and M2 were connected, and in a short time B1 and B2 were connected, DPF_m = 0.63 inductive (Figures 19, 20).

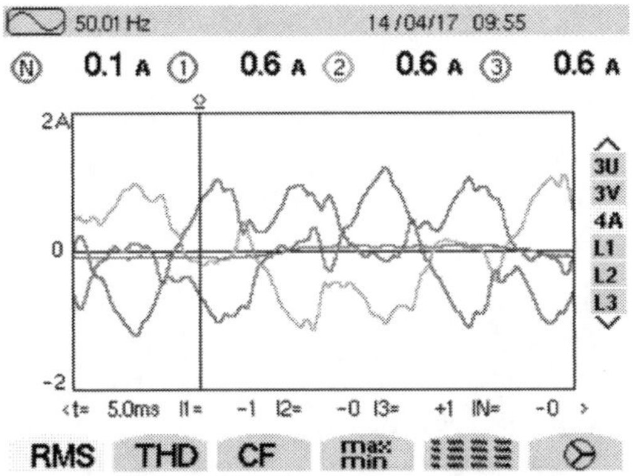

Figure 19. Phases' currents and the null current

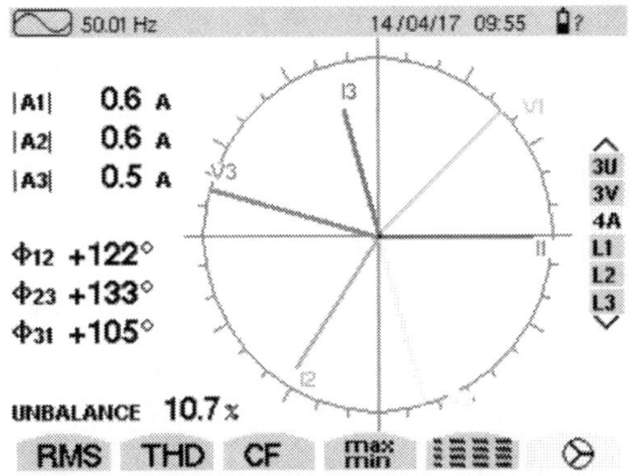

Figure 20. Fresnel diagram for currents

H1 was connected and after a short period of time, B3 was connected by PFC, DPF_m = 0.99 inductive (Figures 21-23).

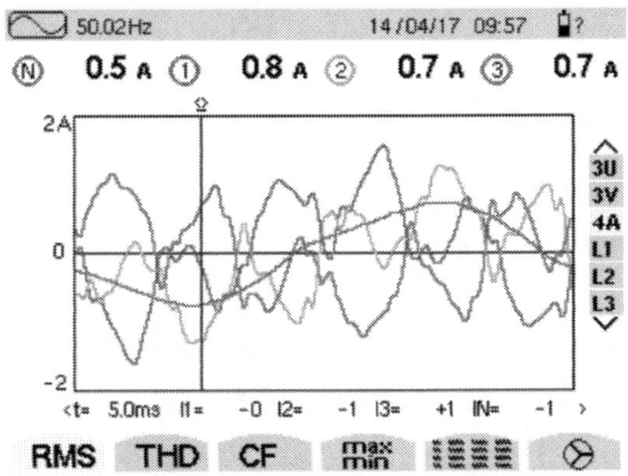

Figure 21. Phases' currents and the null current

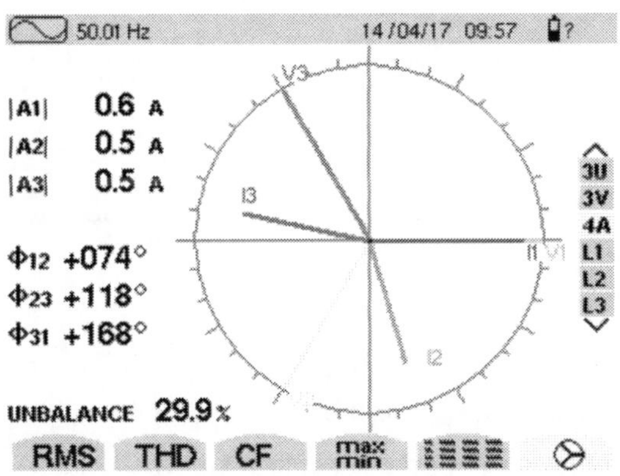

Figure 22. Fresnel diagram for currents

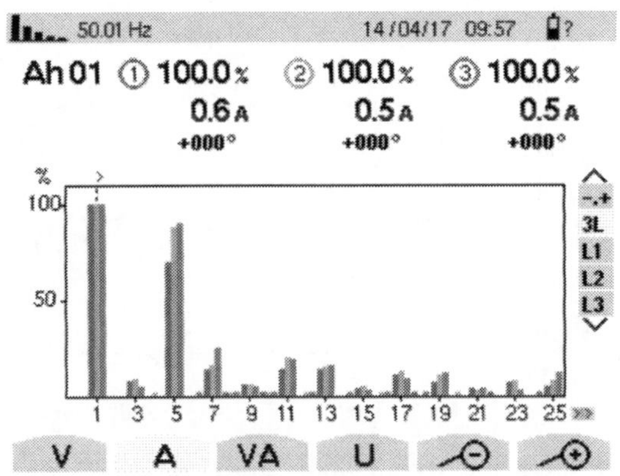

Figure 23. Harmonic spectra for currents

Then H3 and H2 were connected, and, after a short period of time, B4 and B5 were connected by PFC, DPF_m = 0.9 inductive (Figures 24-26).

IOP Conf. Series: Materials Science and Engineering **294** (2017) 012051 doi:10.1088/1757-899X/294/1/012051

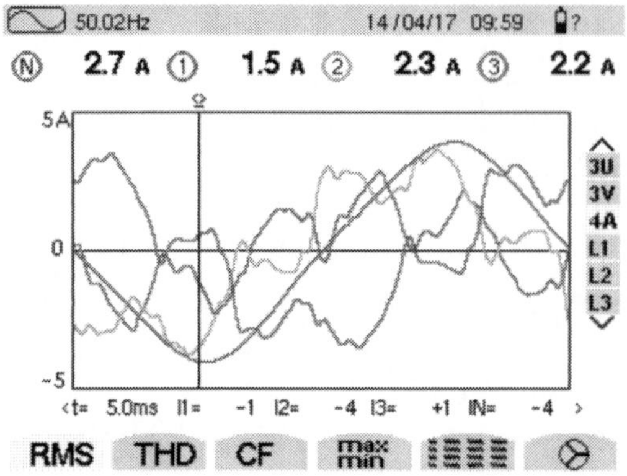

Figure 24. Phases' currents and the null current

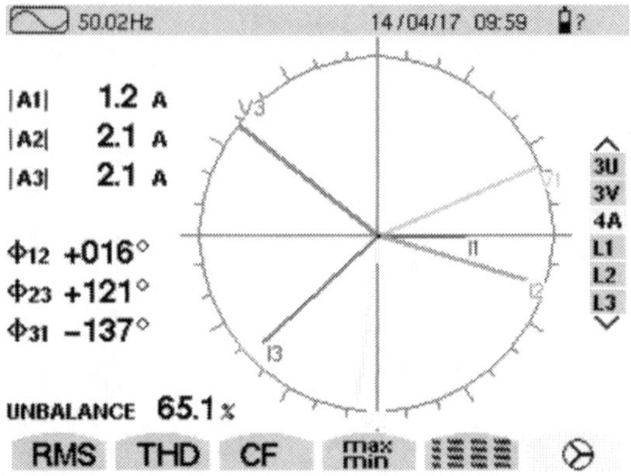

Figure 25. Fresnel diagram for currents

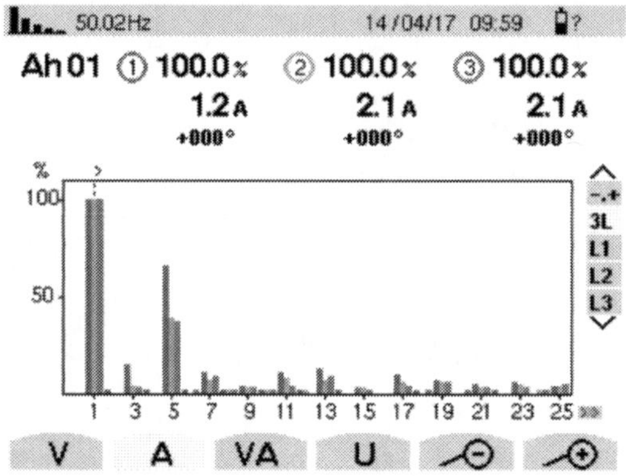

Figure 26. Harmonic spectra for currents

H1 was connected and then, immediate, B3 was connected by PFC, DPF_m = 0.99 inductive (Figures 27, 28).

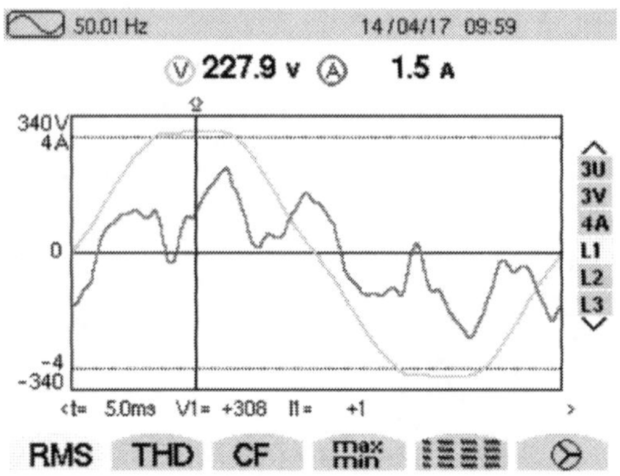

Figure 27. Voltage and current on phase L1

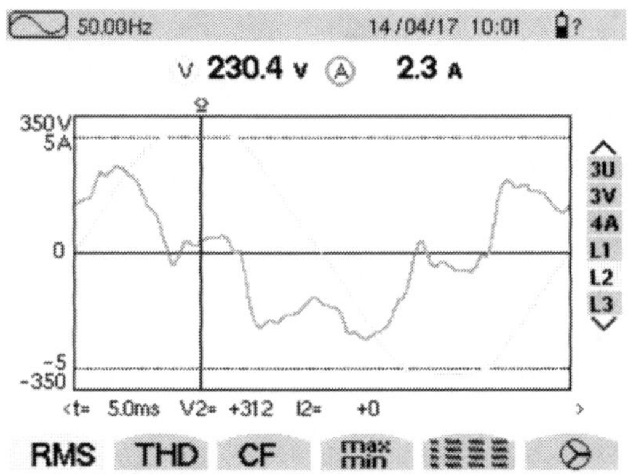

Figure 28. Voltage and current on phase L2

And then, H3 and H2 were connected, and then immediate B4 and B5 were connected by PFC, DPF_m=0.9 inductive. It was disconnected M1 and then B1 and B2 were disconnected by PFC, DPF_m=0.92 inductive. H1 and H3 were disconnected, B3 was disconnected by PFC, and DPF_m=0.95 inductive.

M2 was disconnected and B4 was disconnected by PFC, DPF_m=0.9. Finally, H2 was disconnected and B5 was disconnected by PFC.

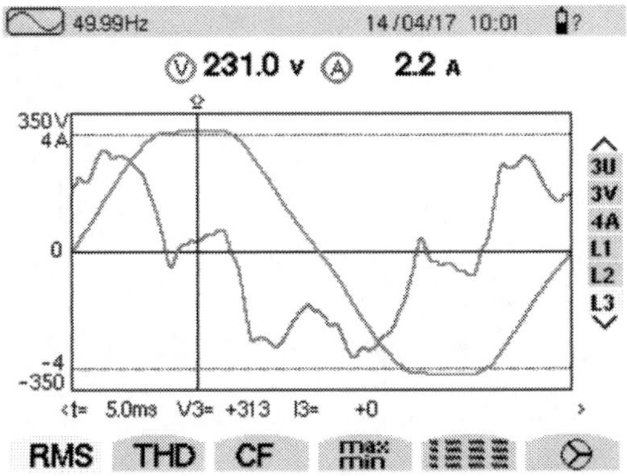

Figure 29. Voltage and current on phase L3

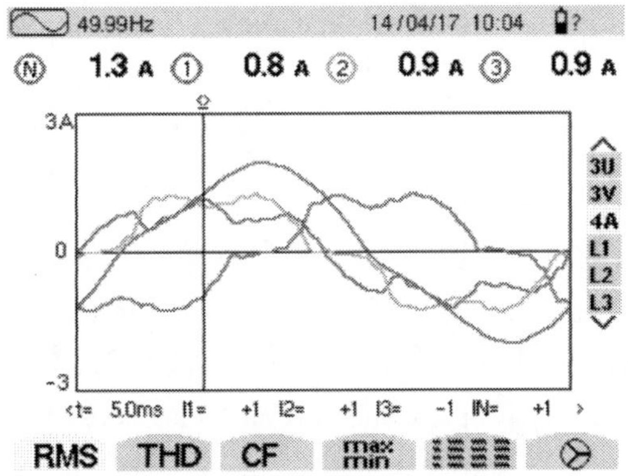

Figure 30. Phases' currents and the null current

5.3. Experiment 3, power factor controller with DPF set =0.92

At experiment 3, all single-phase loads were connected at phase L2, and the current was measure by current transformer on phase L1. The stages at experiments 3 are shown in Table 3.

Table 3. Stages at third experiment, DPF set = 0.92

Measuring stages	DPF (-)
1. Motor M1 connected	0.12 i
2. Capacitors bank B1 connected	0.29 i
3. Capacitors bank B2 connected	0.77 i
4. Motor M2 connected	0.61 i
5. Lamp H1 connected	0.62 i
6. Lamp H3 connected	0.62
7. Lamp H2 connected	0.61 i
8. Motor M1 disconnected	0.24 c
9. Capacitors bank B1 disconnected	
10. Lamps H1, H3 disconnected	Low current
11. Motor M2 disconnected	0.11 c
12. Capacitors bank B2 disconnected	-
13. Lamp H2 disconnected	-

Initially, before introducing three phase and single phase consumers DPF_m = 0.12 inductive (Figures 31, 32). Then M1 and M2 were connected, and in short time B1 and B2 were connected by PFC, DPF_m = 0.61 inductive (Figures 33-35). Then the H1, H3 and H2 lamps are connected, but the PFC will no longer connect any CB, because the single phase consumers were connected to L2 phase and not to the current measuring phase, and DPF_m = 0.61 inductive (Figures 36-38).

After disconnecting the motor M1, shortly after is disconnected the B1 by PFC, and after the motor M2 is disconnected the B2 is disconnected by PFC and DPF_m = 0.11 (Figure 39). Also, the H1 and H3 lamps were disconnected from the second phase (L2).

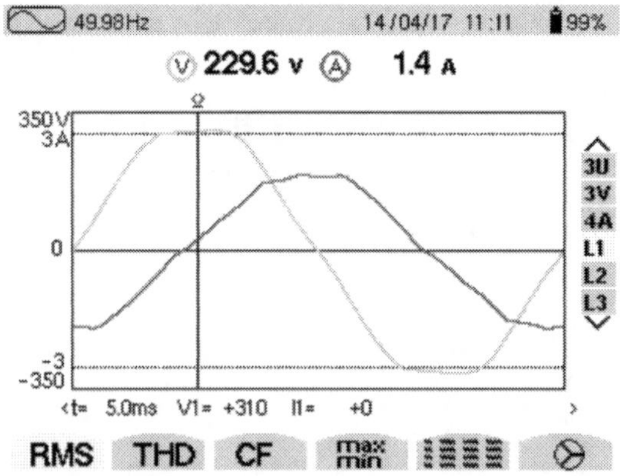

Figure 31. Voltage and current on phase L1

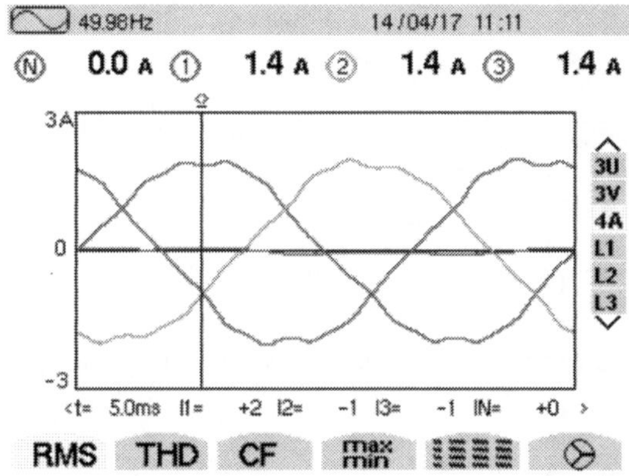

Figure 32. Phases' currents and the null current

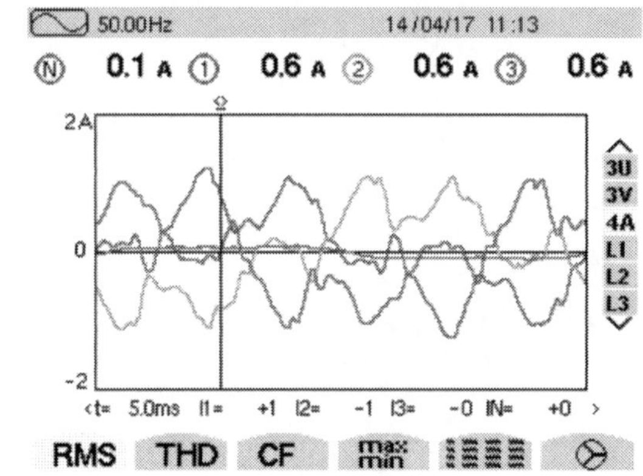

Figure 33. Phases' currents and the null current

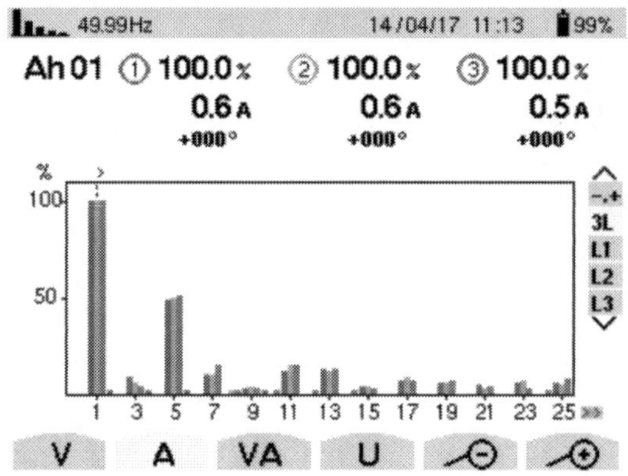

Figure 34. Harmonic spectra for currents

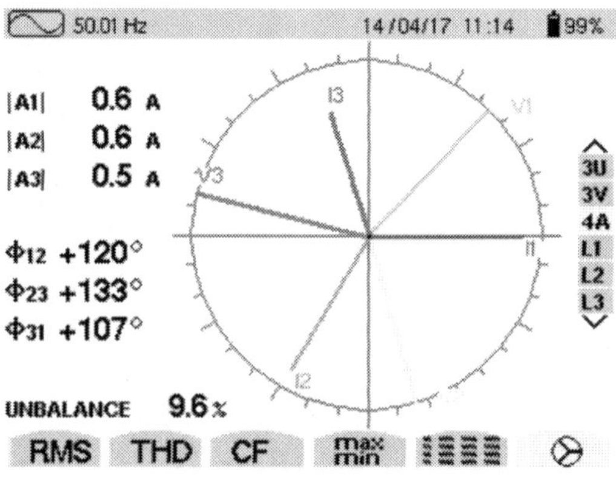

Figure 35. Fresnel diagram for currents

It can be seen from this experiment that if the single phase consumers do not mount on the same phase as the current measure (with CT), they do not influence the insertion of the capacitors banks by the power factor controller, nor will the power factor reach the value set on the regulator, only a much lower value.

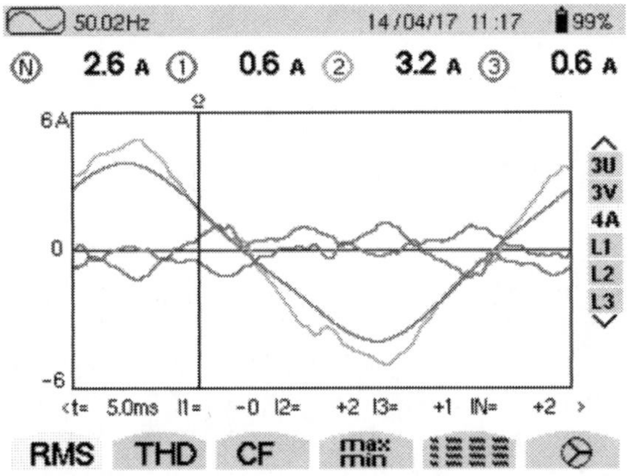

Figure 36. Phases' currents and the null current

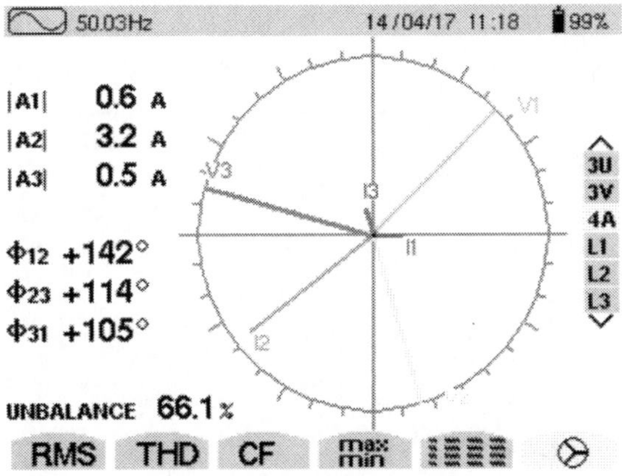

Figure 37. Fresnel diagram for currents

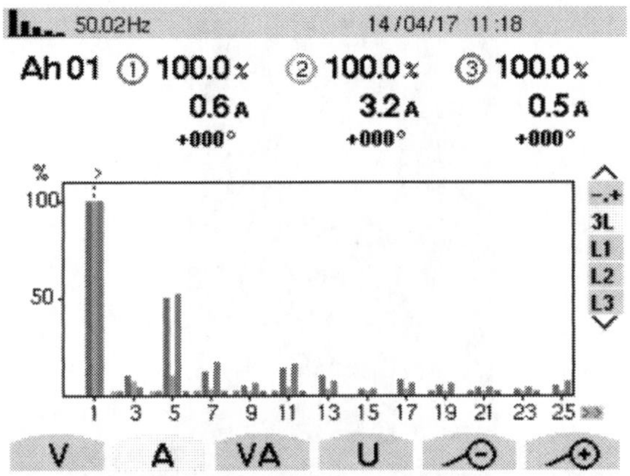

Figure 38. Harmonic spectra for currents

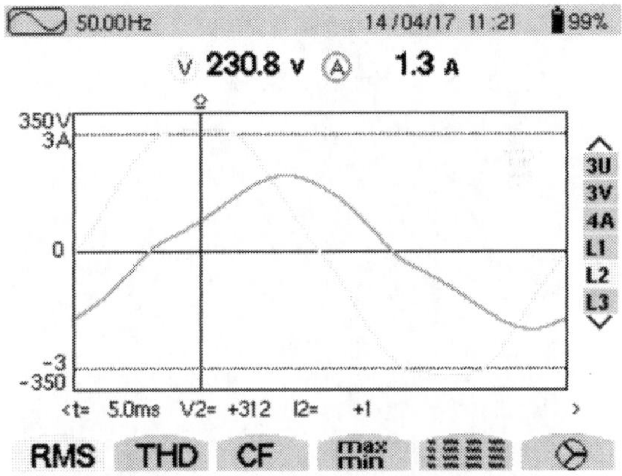

Figure 39. Voltage and current on phase L2

5.4. Experiment 4, capacitors banks permanently connected in the network

The all capacitors banks (CB1-CB6) are connected in the network and the PFC doesn't use. Stages at experiments 4 are shown in Table 4. Three-phase and single-phase consumers were connected and disconnected in the same order as the other experiments. After disconnecting the motors and lamps H1 and H3, an inductive $DPF_m = 0.37$ is obtained (Figures 40, 41).

Table 4. Stages at fourth experiment, without controller

Measuring stages	DPF (-)
1. Motor M1 connected	
2. Motor M2 connected	
3. Lamp H1 connected	
4. Lamp H3 connect	
5. Lamp H2 connect	
6. Motor M1 disconnect	0.35 i
7. Lamps H1, H3 disconnected	0.29 i
8. Motor M2 disconnected	0.37 i
9. Lamp H2 disconnected	-

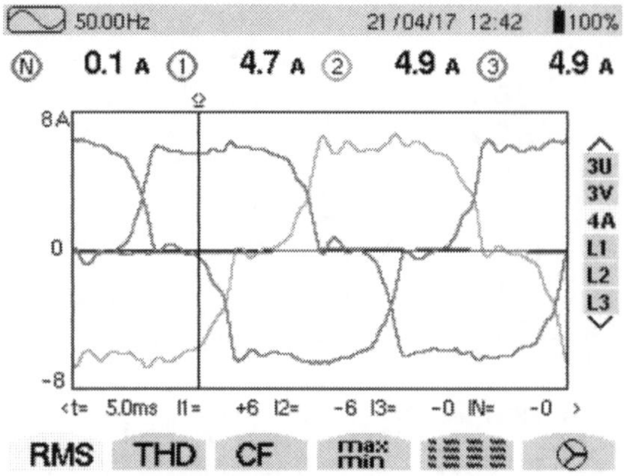

Figure 40. Phases' currents and the null current

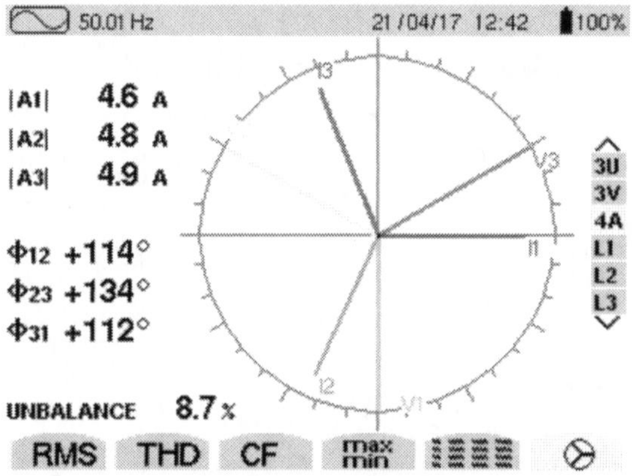

Figure 41. Fresnel diagram for currents

Some aspects from experiments are shown in Figure 42.

Figure 42. Photos from experiments

As a result of the four experiments, it is noted that the six used capacitors banks only go into operation if the DPF is set to 1. If the DPF is set to 0.92, only five (from six) capacitors banks are used. If the single phase consumers are not mounted on the same phase on which the current is measured, the ESTAmat RPR controller does not take these into account when adjusting the DPF and fails to adjust the DPF to the set value. For this case (Experiment 3), only two capacitors banks from the six existing ones are used.

It is also noted that the insertion of capacitors banks by the controller in the circuit to improve the DPF leads to a significant increase in current deformation (increase THD currents) and increase, also, unbalance.

IOP Conf. Series: Materials Science and Engineering **294** (2017) 012051 doi:10.1088/1757-899X/294/1/012051

6. Conclusions

By using a power factor correction installation, there is provided: high power factor in the consumer supply network, thus maximum energy saving; low reactive energy costs; high efficiency of network connected electrical equipment and high power transfer capacity through cables. The results of the power factor improvement with the power factor controller are as follows: eliminates penalties; reduce load in transformer and equipment; reduces losses in transformers, cables and other equipment, and helps stabilize system voltage by increasing the apparent power available.

The power factor controller continuously measures the current and voltage, only one phase (e.g. L1). The current transformer is mounted on one phase, usually the most loaded, without taking into account the charging of the other phases. Very often, sub-compensation or over-compensation, with negative effects in the electricity network, occurs, on the other phases of the low-voltage electrical network.

Most controllers take into account the capacitor operating temperature and the resonance, but do not check at the same time the harmonic distortion of voltage and current on each phase.

There is an increase in the deforming regime with undesirable effects on the electrical network. Additional losses due to harmonic current circulation, reduced electrical machine efficiency, additional charge of capacitors banks, power cables heating, additional charging of null current, additional stresses of insulators.

Usually, currents from the power station are unbalance (there are a lot of single-phase consumers) and are strongly deformed (the total harmonic distortion for current may be 25-35 %). The currents have the following higher harmonics: 5th rank (<25%), 3rd rank (<20%) and 7th (<12%), with amplitudes higher than the other harmonics, 9-40 (<10 %).

References

[1] Jankovskis J, Stepins D and Pikulins D 2010 Efficiency of PFC Operating in Spread Spectrum Mode for EMI Reduction, *Electronics and Electrical Engineering* **7**(103) 13-16

[2] Garcia O, Cobos J A, Prieto R, Alou P and Uceda J 2003 Single Phase Power Factor Correction: a Survey, *IEEE Transactions on Power Electronics* **18**(3) 749-755

[3] Popa G N, Paliciuc C and Diniş C M 2013 *On the Use of a Low Voltage Power Factor Controller in Deforming Regime*, The 4th International Symposium on Electrical and Electronics Engineering IEEE-ISEEE, Galaţi, Romania, October 11-13, pp 1-6

[4] Diniş C M, Popa G N and Iagăr A 2016 *On the Use of Low Voltage Power Factor Controller in Textile Industry*, International Conference on Applied and Theoretical Electricity ICATE 2016, Craiova, Romania, 6-8 October, pp 1-6

[5] Iagăr A, Popa G N and Diniş C M 2017 *Power Quality from Theory to Experimentations*, Politehnica Publishing House, Timişoara, Romania (in Romanian)

[6] Dell'Aquila A, Marinelli M, Giuseppe V and Yanchetta P 2004 New Power-Quality Assessment Criteria for Supply Systems Under Unbalanced and Nonsinusoidal Conditions, *IEEE Transactions on Power Delivery* **19**(3) 1284-1290

[7] *** 2011 *Capacitors and Reactors for Power Factor Correction, Manual*, Electronicon Kondensatoren, Gera Gmbh, Germany

[8] Meyer D 2004 *Switchgear for Power Factor Correction Systems, Guide*, Moller, Germany, pp 1-6

[9] Ceraolo M and Poli D 2014 *Fundamentals of Electric Power Engineering: From Electromagnetics to Power Systems*, Wiley-IEEE Press

[10] Baggini A 2008 *Handbook of Power Quality*, Ed. John Wiley & Sons Ltd., New York, USA

[11] Albert H, Gheorghe Ş, Golovanov N, Elefterescu L and Porumb R 2013 *Power Quality. Contributions. Results. Perspectives*, AGIR Publishing House, Bucureşti, Romania (in Romanian)

[12] Fărcaş C, Petreuş D and Palaghiţă N 2003 *Power Factor Improving in Modern Supplies*, Risoprint Press, Cluj-Napoca, Romania (in Romanian)

[13] Arrillaga J and Watson N R 2003 *Power System Harmonics*, John Wiley&Sons, Ltd., West Sussex, Anglia
[14] Gobbi A 2009 *Understanding Power Factor and Input Current Harmonics in Switched Mode Power Supplies*, TDI Power, pp 1-12
[15] *** 2001 *Power Quality Application Guide*, Copper Development Association, U.K.
[16] *** 2007 *Three Phase Power Quality Analyzer CA 8334B, User's Guide*, Chauvin-Arnoux, France
[17] *** 2010 *Power Factor Correction and Harmonic Filtering in Electric Plants*, Technical Applications Papers no.8, ABB, Bergamo, Italy

Automation of water supply and recirculation-filtration of water at a swimming pool using Zelio PLC

C M Diniş, G N Popa and A Iagăr

Politehnica University of Timişoara, Department of Electrical Engineering and Industrial Informatics, 5 Revolution Street, Hunedoara, 331128, Romania

E-mail: corina.dinis@fih.upt.ro

Abstract. The paper proposes the use of the Zelio PLC for the automation of the water supply and recirculation-filtration system of a swimming pool. To do this, the Zelio SR3B261BD - 24V DC with 10 digital inputs (24V DC) and 10 digital outputs (relay contacts) was used. The proposed application makes the control of the water supply pumps and the water recirculation-filtration from a swimming pool. The recirculation-filtration systems for pools and swimming pools are designed to ensure water cleaning and recirculation to achieve optimum quality and lasting service life. The water filtration process is one of the important steps in water treatment in polls and swimming pools. It consists in recirculation of the entire volume of water and begins by absorbing the water in the pool by means of a pump followed by the passing of water through the filter, disinfectant and pH dosing, and reintroducing the water back into the pool or swimming pool through the discharge holes. Filters must to work 24 hours a day to remove pollutants from pools or swimming pools users. Filtration removes suspension particles with different origins. All newly built pools and swimming pools must be fitted with water recirculation systems, and existing ones will be equipped with water recirculation and water treatment systems.

1. Introduction

A new approach to automation systems is to use programmed logic of different devices (including programmable logic controllers - PLCs). This category includes all systems that operate with a code write into an implicit memory. Automated process control systems are a conglomeration of electronic devices that provides stability, accuracy, and, not least, performance [1-9].

PLCs are structures for control of industrial process, whose implementation was used to eliminate the wired logical structures and to replace them with programmable logical structures. The majority of PLCs can be programmed via PC, but it is possible to be programmed manually with the console for programming PLCs, or in case of low capacity PLCs, using the buttons and display on the front of them. The best method of reprogramming the PLCs is using the PC, thanks to the possibility of building, saving and debugging the programs. Intelligence of the PLC consists in the ability to detect different types of sensors, input and/or output devices. Usually the most used are buttons, keys, and other types of switches at the PLCs' inputs [10-16].

Zelio Logic smart relays - Schneider Electric product - are use in small automation systems comprising 10 to 40 inputs/outputs. These relays can be used both in domestic (light installations, alarm, etc.), industrial applications (small machinery, irrigation, pumping stations, etc.), and in the commercial (management of lighting, ensuring the heating, air conditioning, etc.) [17].

Programming can be done directly using the buttons on the relays or on a PC using Zelio Soft with a programming cable, with the possibility of programming in Function Block Diagram (FBD) or in Ladder Diagram (LD) [18-21].

2. Water Supply and Recirculation-Filtration Process from a Pool

Recirculation-filtration systems for swimming pools (Figure 1) are designed to ensure water cleaning and recirculation to achieve optimal quality and lasting service life.

In addition to pipes, fittings, and valves, water recirculation and filtration systems consist of the following elements: filters, recirculation pumps and selection valves (multiport valves). When designing a swimming pool, it is necessary to consider several technological aspects that contribute to the design, such as water and hydraulic technology, civil engineering technology, electric power supply, indoor and outdoor architecture, etc.

When designing the recirculation-filtration system starts from the load capacity of a swimming pool, resulting from the number of people using the pool simultaneously and the water surface dimensions. Swimming pool load factor differs depending on the destination, for example: swimming pools, for adults, for children, etc.

Determining the filtering capacity of the plant consists of determining the amount of water recirculated per hour under the conditions of maximum use:

$$Q = N \cdot q \tag{1}$$

where:

Q = system filtering capability (m^3/h);

N = the number of people using the pool at the same time;

q = the prescribed quantity of filtered water for a person (m^3/pers. x 4).

The prescribed amount of filtered water and reintroduced into the pool for one person is: 2 m^3/ pers. x 4.

Both calculation formulas and load coefficients may differ according to the norms in force in different countries, but the final result of the calculations will differ slightly.

Figure 1. Swimming pool

The quality of the water in the pools must be closer to the drinking water qualities, that is:

- pH = 6.8 ÷ 8.5;
- transparency > 20 m;
- the iron and manganese content < 0.3 mg/l;
- acidity > 4 moll/l;
- nitrates < 40 mg/l;

• chlorinated < 100 mg/l;
• water hardness 100 ÷ 350 mg/l.

The goal of water recirculation and filtration:

• ensuring the quality of the pool water without brutal intervention and without changing the chemical properties;
• removing impurities and micro-organisms from pool water while preserving water quality;
• by filtering and disinfecting, preventing water quality changes;
• ensuring an economical and environmentally friendly process;
• permanent assurance of water quality, regardless of the weather conditions, the number of people in the pool and the amount of impurities.

The filtering process consists in recirculation of the entire volume of water and begins by absorbing the water in the pool by means of a pump followed by the passage of water through the filter, disinfect dosing and the pH, and reintroducing the water back into the pool through the discharge holes.

All newly built pools and swimming pools must be fitted with water recirculation systems, and existing ones will be equipped with water recirculation and water treatment systems. The circulation of water in the pool must be ensured so that there are no dead spaces where the water stagnates. In these spaces the water is not disinfected and is not filtered properly.

Water filtration has the purpose of removing suspensions from the water to obtain transparent and clear water (Figure 2). At the same time, the amount of disinfectants used is reduced by filtration.

Filters must to work 24 hours a day to remove pollutants from pools or swimming pools users. Filtration removes suspension particles with different origins. There are a number of types of filters available, the choice of filters will be based on certain factors, including: filtration rate, water quality, efficiency, maintenance mode, washing method. Cycle is the filter capacity typically expressed in hours and represents the length of time that all water in the swimming pool is filtered. At least one complete cycle per day is recommended.

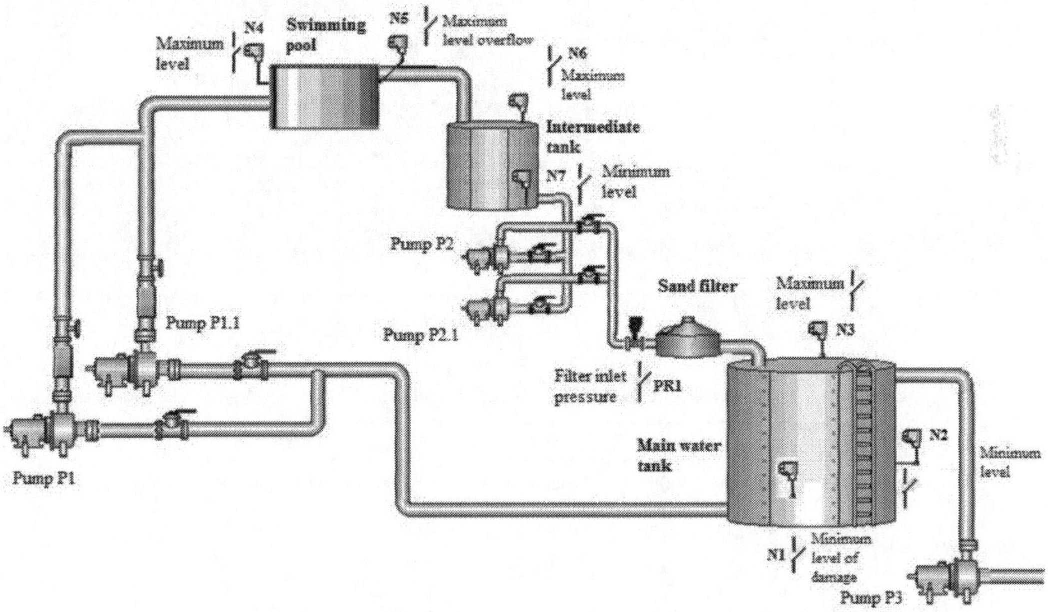

Figure 2. Schematic representation of the water supply and filtration system

Pumps P1 and P2 are working pumps, and P1.1 and P2.1 are back-up pumps. Each pump in the circuit has a normally open auxiliary contact of the contactor, which will signal to the PLC the state of the contactor (1 or 0 logic). If during main-pump operation there are faults that lead to their being switched off (disconnected of contactors), the back-up pumps will operate automatically after a period of time.

The level in the main tank during filtration cycles will remain within the maximum and minimum limits with N2 and N3 level sensors.

As long as the sensor contact N1 (fault minimum level) in the main tank is closed, the pump P1 will not have a starting condition because it would start without water on the suction line and would destroy it. From this reason, during operation of the system, the level in the main tank reaches the fault minimum level of damage, pump P1 will stop and will no longer have the starting condition until after the water supply of the basin, after opening the sensor N1 contact.

3. Automating Process of Water Supply and Recirculation-Filtration of Water from a Pool with Zelio PLC

The present application requires a large number of digital inputs and outputs, so it was chose a Zelio PLC SR3B261BD, 24V DC type with 10 digital inputs (1 logic, 24V DC) and 10 digital relay outputs.

Using PLC type it will use hardware configuration from the software (Zelio Soft 2) – Figure 3. In application it was use Zelio PLC SR3B261BD - 24V DC (Figure 3) and FBD the programming language (Figure 4).

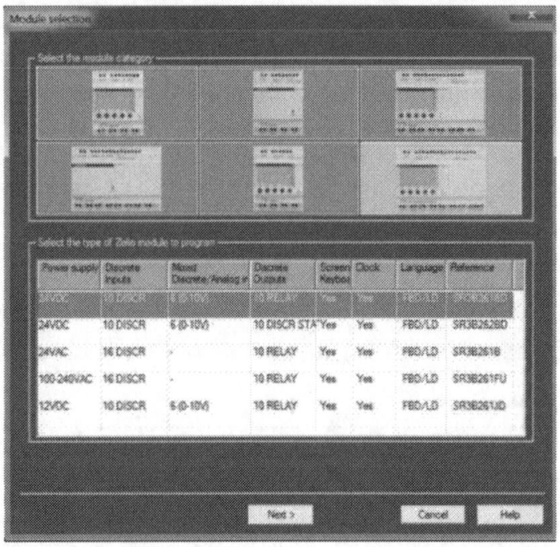

Figure 3. Configuring PLC in the software

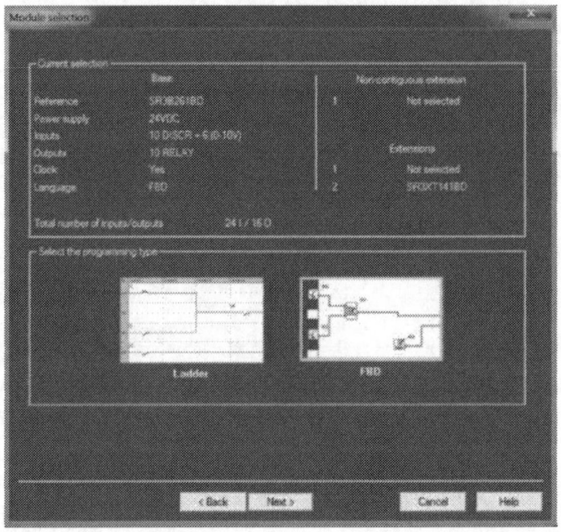

Figure 4. Information regarding final hardware configuration PLC

Table 1. The digital inputs of the PLC

Input	No	Symbol	Function	Comment
I1	B00		Pushbutton	Automatic start-up button
I2	B01		Pushbutton	Stop button in automatic mode
I3	B02		Illuminated pushbutton	Emergency stop button
I4	B03		Discrete input	Minimum damage level sensor N1 in the tank
I5	B04		Discrete input	Minimum level sensor N2 in the tank
I6	B05		Discrete input	Maximum level sensor N3 in the tank

I7	B06		Discrete input	Maximum level sensor N4 in a swimming pool
I8	B07		Discrete input	Maximum overflow sensor N5 in the swimming pool
I9	B08		Discrete input	Maximum level sensor N6 in the intermediate tank
IA	B09		Discrete input	Minimum level sensor N7 in the intermediate tank
IB	B10		Discrete input	Maximum pressure sensor PR1
IC	B11		Discrete input	Contact NO contactor P1
ID	B12		Discrete input	Contact NO contactor P1.1
IE	B13		Discrete input	Contact NO contactor P2
IF	B14		Discrete input	Contact NO contactor P2.1

In the Tables 1, 2 and 3 are shown input and output variables of the PLC, and configurable functions used in the program. In the Table 4 is shown programming functions displayed information on the PLC.

Table 2. Digital outputs of the PLC

Input	No	Symbol	Function	Comment
Q1	B16		Motor	K1 – ON P3 functioning
Q2	B17		Motor	K2 – ON P2 functioning
Q3	B18		Motor	K3 – ON P1.1 functioning
Q4	B19		Motor	K4 – ON P2 functioning
Q5	B20		Motor	K5 – ON P2.1 functioning
Q6	B21		Green indicator light	Functioning plant
Q7	B22		Red indicator light	Filter damage

Table 3. Configurable functions from the program

No	Symbol	Function	Parameters	Comment
B23		RS switching	Priority: RESET has priority	
B25		LCD display	See details below	Started
B29		LCD display	See details below	Emergency stop
B30		RS switching	Priority: RESET has priority	S/R pump P3
B34		LCD display	See details below	P3 - ON
B36		RS switching	Priority: RESET has priority	S/R pump P1

B38		Cycling timing	On time: 0H 2M 0S Off time: 0H 2M 1S Function Li Continuous flashing	
B43		RS switching	Priority: RESET has priority	
B45		LCD display	See details below	Filter damage
B46		Cycling timing	On time: 0H 0M 1S Off time: 0H 0M 1S Function Li Continuous flashing	
B49		Timer	On time: 0H 0M 2S Off time: 0H 0M 0S	
B50		LCD display	See details below	P1 - ON
B51		LCD display	See details below	P1.1 - ON
B53		Timer	On time: 0H 0M 2S Off time: 0H 0M 0S	
B55		LCD display	See details below	P2 - ON
B56		LCD display	See details below	P2.1 - ON

Table 4. Programming functions displayed information on the PLC

B25		LCD display	STARTED														
S	T	A	R	T													

B50		LCD display	P1 - ON														
						P	1	-	O	N							

B51		LCD display	P1.1 - ON														
P	1	.	1	-	O	N											

B55		LCD display	P2 - ON														
												P	2	-	O	N	

B56		LCD display	P2.1 - ON														
									P	2	.	1	-	O	N		

B34		LCD display	P3 - ON														
P	3	-	O	N													

B29	DISPLAY	LCD display	Emergency STOP														
E	M	E	R	G	E	N	C	Y		S	T	O	P				

B45	DISPLAY	LCD display	Filter Daimage														
					F	I	L	T	E	R		D	A	M	A	G	E

4. Description of the program implemented in the Zelio PLC

In view of the above requirements, it has been done using programming software program Zelio Soft 2 shown in Figure 5 and PLC implementation is shown in Figure 6.

Initially the pool and the two containers from the filtration plant are empty. The pump no. 3 is for filling the main water tank.

When the water tank is empty or has a very low level, the N2 level sensor has the "low level" contact (closed) and also the minimum level sensor N1, has the closed contact, which means that the level in the tank is at the fault minimum limit.

Pump no. 3, it will automatically stop when the level in this tank reaches the maximum level, signaled by closing the maximum N3 sensor.

Once the main water tank has been filled, the N1 low level sensor and then the N2 high level sensor will open the contacts. If this condition is achieved and the water tank is not filled with water, than the N4 peak level sensor contact is not closed, it automatically switches on the P1 water pump by transporting water from the main tank in the basin.

Because the P1 pump will consume water from the main tank, which has a much smaller volume than the basin, during the water supply of the basin, pump P3 will start several times in automatic mode, commanded by the low level sensor contact N2 and high level sensor contact N3 to resupply this tank with water. The flow rate of P3 pump is higher than the flow rates of other pumps in the plant, so the main tank will never run out of water.

After the tank has been filled, which is signaled in the PLC by the contact of the maximum level sensor N5 from swimming pool, at this moment the PLC will interrupt the basin's water supply sequence and start the water recirculation and filtration sequence from the inside of the basin, as long as the N4 sensor is on. The water recirculation and filtration sequence will be done by the on-delay time starting of the P1 pump to add water to the basin, causing the pool to overflow with water, the surplus water flowing through its overflow holes, the water being collected and transferred by falling into the intermediate tank. After a series of consecutive starts of P1 pump, the excess water stored in the intermediate tank will reach the maximum level, signaled to the PLC by closing the normally open contact of the N6 level sensor of the intermediate tank.

At this point, the water pump P2 will automatically start, which will drain the water from the intermediate tank, through the sand filter, transferring it back to the main tank.

The volume of the intermediate tank is smaller than that of the primary tank. The P2 pump will automatically switch on depending on the maximum level sensor contact N6 from the intermediate tank and will switch off depending on the level sensor contact N7 of the intermediate tank.

If the filter is partially or totally clogged, the water pressure at its inlet will increase, the pressure sensor PR1 will close the normally open contact, causing the plant to shut down and display a fault message on the PLC screen.

Also, an PLC output will be activated to supply a siren for acoustic warning about the occurrence of the fault. If the emergency stop button is pressed, all components of the system will shut down and the "Disconnect" message will be displayed on the PLC screen.

Some pictures regarding PLC operation are shown in Figures 7-10.

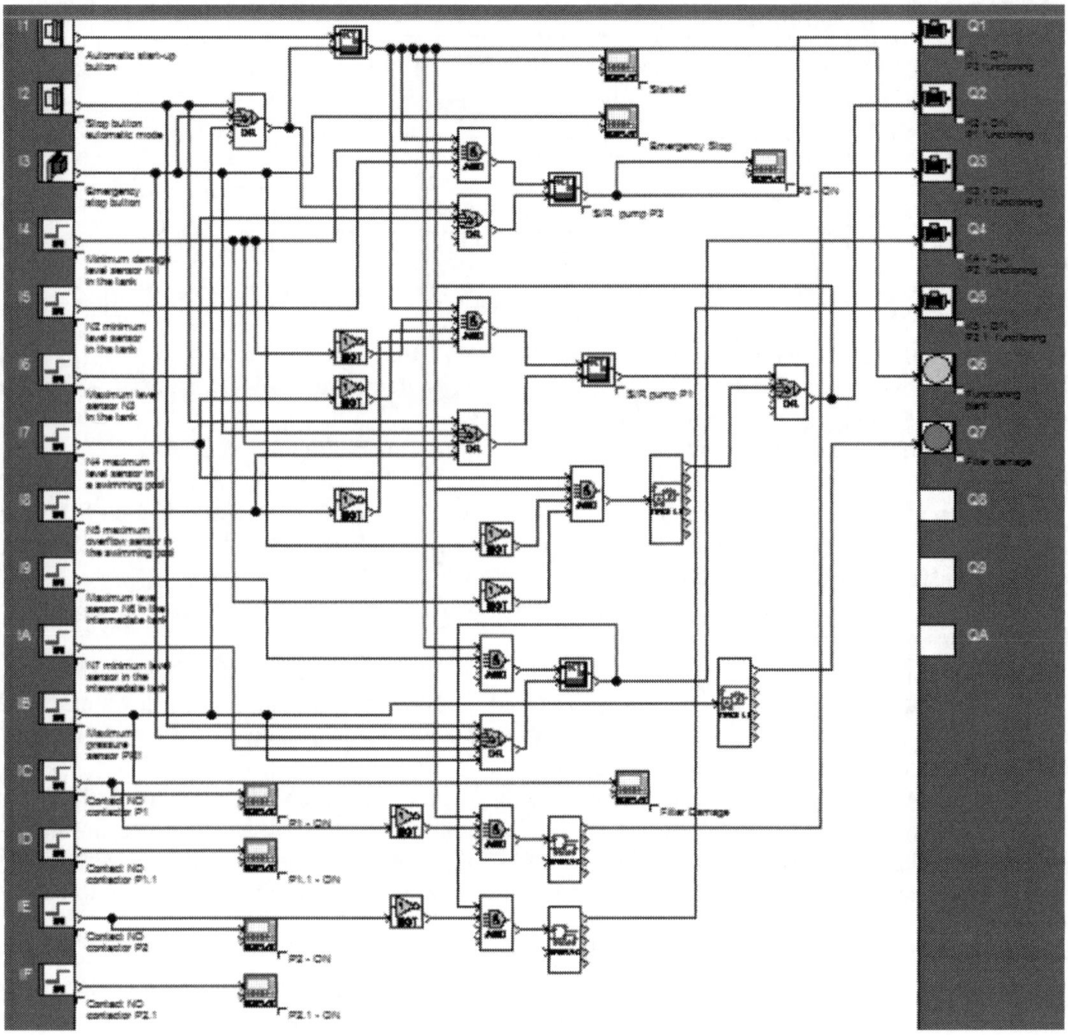

Figure 5. Program made on the PLC for swimming pool

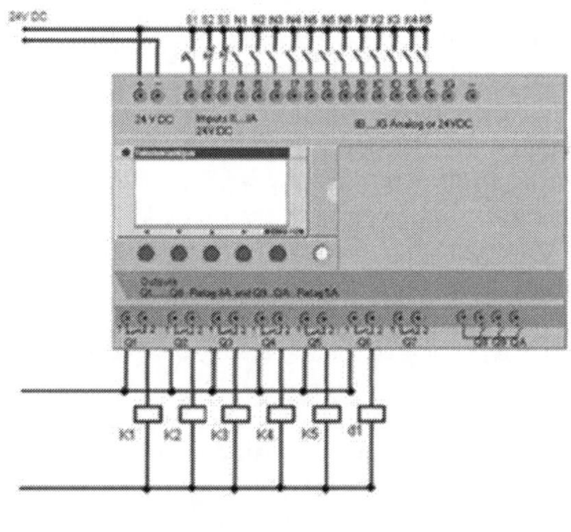

Figure 6. PLC implementation

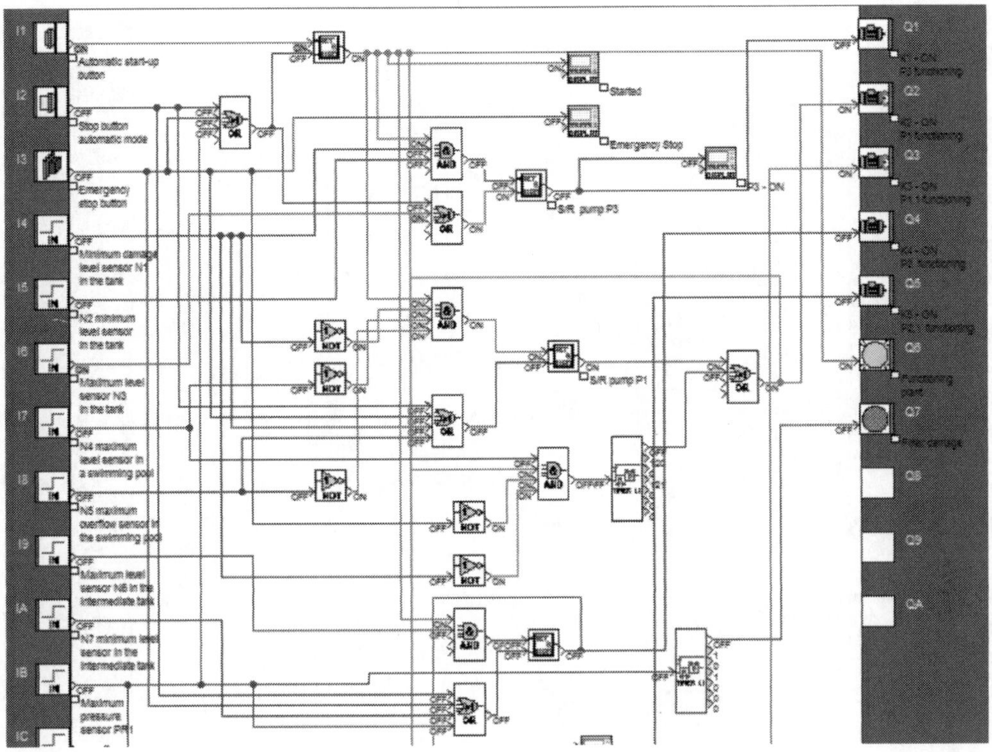

Figure 7. Sequence of PLC program - 1

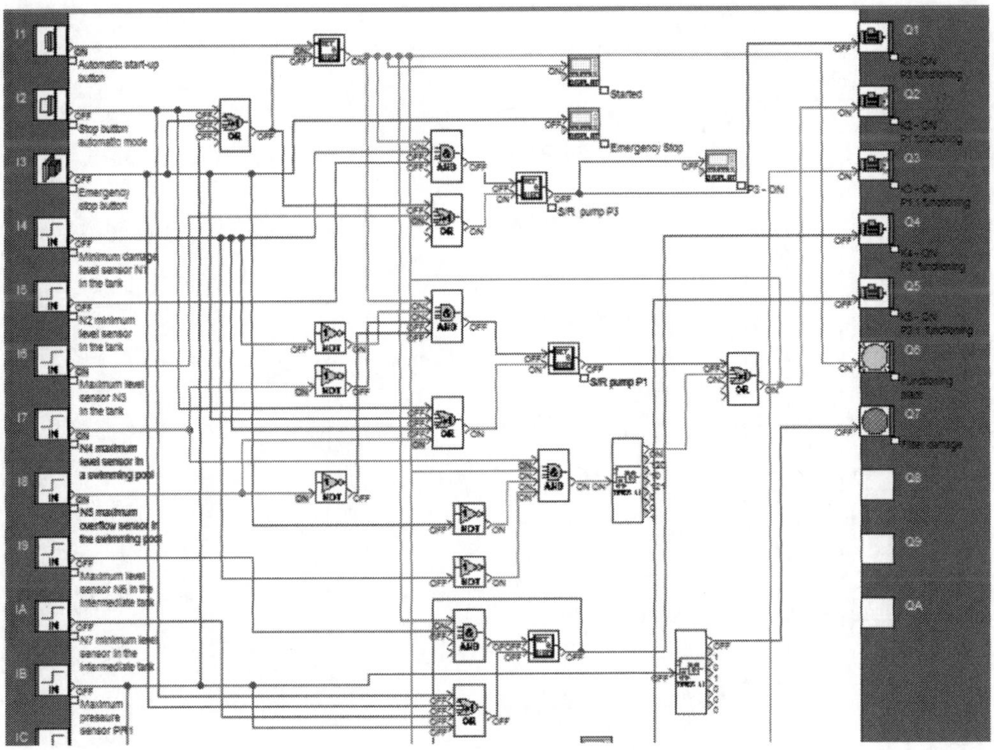

Figure 8. Sequence of PLC program - 2

IOP Conf. Series: Materials Science and Engineering **294** (2017) 012052 doi:10.1088/1757-899X/294/1/012052

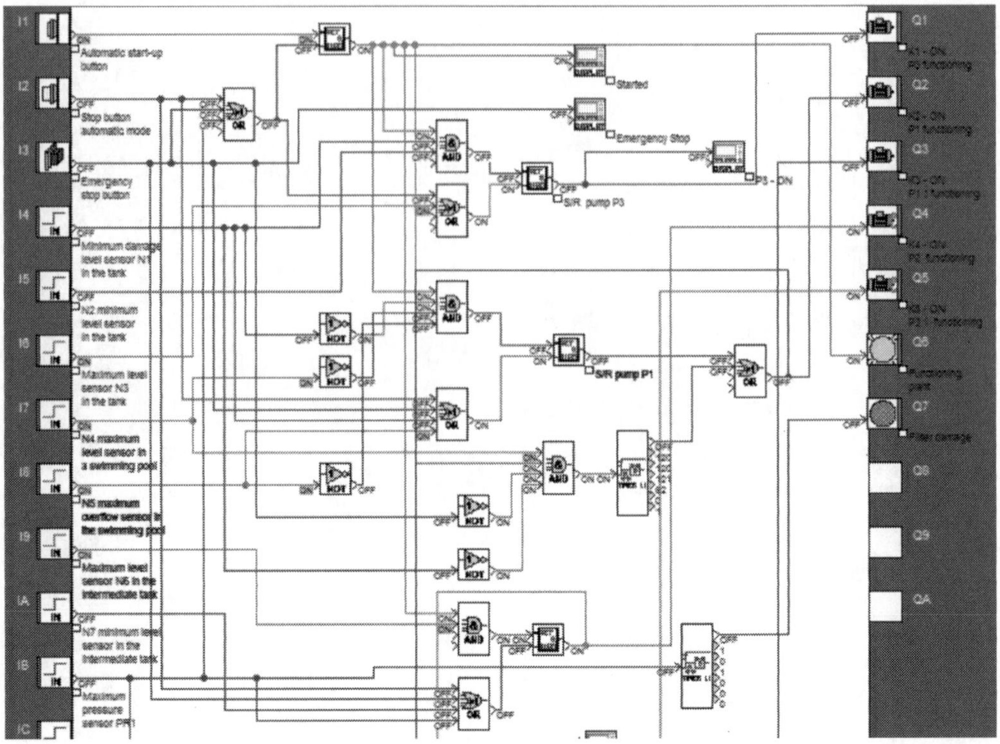

Figure 9. Sequence of PLC program - 3

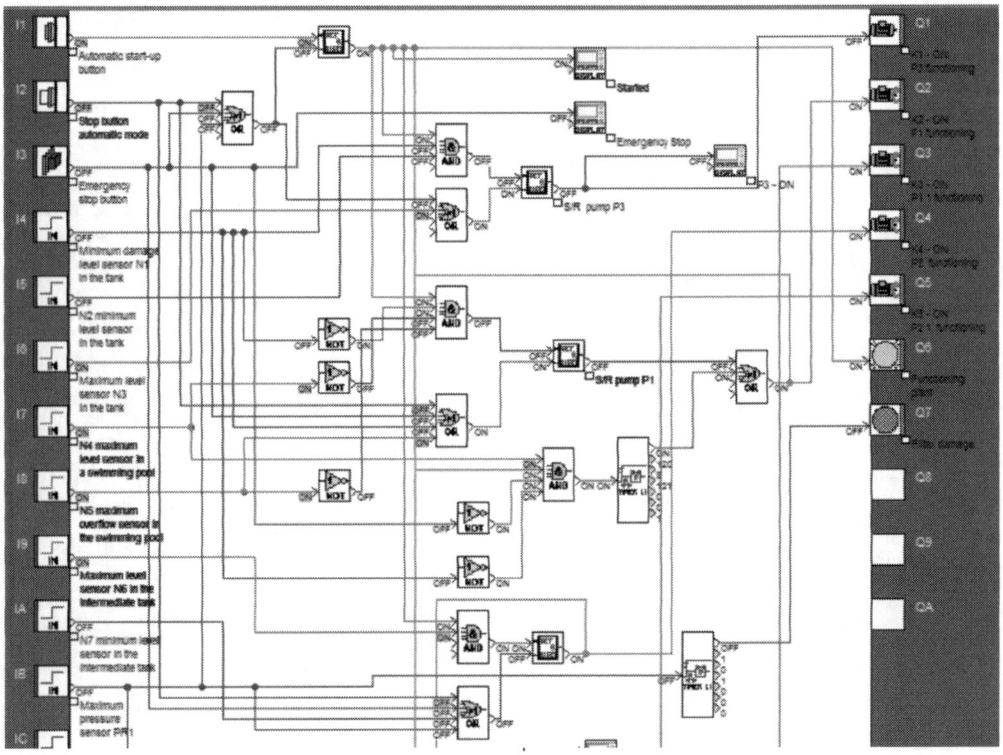

Figure 10. Sequence of PLC program - 4

5. Conclusions

Currently, many of the industrial and household applications are controlled by PLCs. There are PLCs built for easier or more complicated applications, and there are many companies that build them. Low capacity PLCs have the most common applications, both in industrial and household applications.

Compared with classical installation, by using PLC, the number of conductors used in installation are considerably reduced, and, also, the control installation simplifying. During development and debugging the PLC program, it is much easier to find a fault in the conductors, because the state of inputs and outputs of the PLC are signal on its display, and it is easily finding the fault occurred in the control installation with PLC.

Low capacity PLC can be successfully used for monitoring and control of small processes, and they have the advantages to connect of various sensors. Changing the operation mode of the electrical installation is done, in some cases, only by modifying the PLC program, and in some cases by changes, both, the program and the connections of the conductors from control installation.

References

[1] Mărgineanu I 2005 *Automate programabile*, Editura Albastră, Cluj Napoca, Romania (in Romanian)

[2] Popa G N, Popa I and Deaconu S 2006 *Automate programabile în aplicații*, Editura Mirton, Timişoara, Romania (in Romanian)

[3] Moise A 2004 *Automate programabile. Proiectare. Aplicații*, Editura MatrixRom, Bucureşti, Romania (in Romanian)

[4] Ivănescu A N, Tudorie R and Roşu A 2009 *Automate programabile*, Politehnica Press, Bucureşti, Romania (in Romanian)

[5] Nelson V P and Nagle H T 1995 *Digital Logic Circuit Analysis and Design*, Prentice Hall, NJ, USA

[6] Hugh J 2007 *Automating Manufacturing Systems with PLCs*, Version 5.0, May 4

[7] De Larminat P 2007 *Automatique appliquée*, Hermès – Lavoisier, Février, France

[8] Bhujbal K and Barve A 2013 Automation of Scrap Disposal System Using PLC, *Advance in Electronic and Electric Engineering* **3**(1) 107-112

[9] Burali Y N 2012 PLC Based Industrial Crane Automation & Monitoring, *International Journal of Engineering and Science* **1**(3) 01-04

[10] Borangiu Th and Dobrescu R 2007 *Automate programabile*, Editura Matrix Rom, Bucureşti, Romania (in Romanian)

[11] Popescu D 2005 *Automate programabile*, Matrix Rom, Bucureşti, Romania (in Romanian).

[12] Petruzella F 1996 *Programmable Logic Controllers*, Second ed., McGraw Hill, New York, USA

[13] Baciu I and Cunțan C D 2012 *Operation Analysis of a Frequency Converter with Control Realized in LabView*, IEEE International Conference on Industrial Technology, Athens, Greece, March 19-21, pp 432-437

[14] Cunțan C D, Baciu I and Osaci M 2015 Operational Study of a Frequency Converter with a Control Sequence, Utilizing Xilinx Software, *Acta Polytechnica Hungarica* **12**(6) 201-212

[15] Ali A et al. 2009 Water Pumping System with PLC and Frequency Control, *Jordan Journal of Mechanical and Industrial Engineering* **3**(3) 216-221

[16] Ioannides M G 2004 Design and Implementation of PLC-Based Monitoring Control System for Induction Motor, *IEEE Transactions on Energy Conversion* **19**(3) 469-476

[17] *** 2007 *Zelio PLC. User's manual*, Zelio Schneider Electric

[18] Sumardi S 2015 Room Temperature Control System Prototype Industry Based Programmable Logic Controller Zelio SR2 B121 BD, *Innovative Systems Design and Engineering* **6**(4) 52-58

[19] Aziz Muslim M, Goegoes Dwi N and Mahkrus A 2012 *Zelio PLC – based Automation of Coffee Roasting Process*, The 6th – Electrical Power, Electronics, Communications, Controls, and

Informatics International Seminar, Brawijaya University, Malang, Indonesia, May 30-31, pp 102-106

[20] Illes C, Popa G N and Filip I 2013 *Water Level Control System Using PLC and Wireless Sensors*, IEEE 9th International Conference on Computational Cybernetics (ICCC), Tihany, Hungary, July 8-10, pp 1-6

[21] Diniş C M, Popa G N and Iagăr A 2017 *Automation Heating and Pumping for Water Using Zelio PLC*, The 10th International Symposium on Advanced Topics in Electrical Engineering, Bucharest, Romania, March 23-25, pp 903-908

Bidirectional automatic release of reserve for low voltage network made with low capacity PLCs

I Popa, G N Popa, C M Diniş and S I Deaconu

Politehnica University of Timisoara, Department of Electrical Engineering and Industrial Informatics, 5 Revolution Street, Hunedoara, 331128, Romania

E-mail: gabriel.popa@fih.upt.ro

Abstract. The article presents the design of a bidirectional automatic release of reserve made on two types low capacity programmable logic controllers: PS-3 from Klöckner-Moeller and Zelio from Schneider. It analyses the electronic timing circuits that can be used for making the bidirectional automatic release of reserve: time-on delay circuit and time-off delay circuit (two types). In the paper are present the sequences code for timing performed on the PS-3 PLC, the logical functions for the bidirectional automatic release of reserve, the classical control electrical diagram (with contacts, relays, and time relays), the electronic control diagram (with logical gates and timing circuits), the code (in IL language) made for the PS-3 PLC, and the code (in FBD language) made for Zelio PLC. A comparative analysis will be carried out on the use of the two types of PLC and will be present the advantages of using PLCs.

1. Introduction

Electrical consumers of particular importance such as: steel casting cranes, underground mining ventilation fans, groundwater discharge pumps, electrical consumers in hospital operating rooms, etc., shall be provided with the continuity of supply with electricity. For this purpose, these consumers must be connected to at least two independent power sources with the same or different powers [1], [2].

If the power sources have equal powers and each source can provide the necessary electrical power for the proper operation of all connected consumers, the continuity of the power supply is achieved by means of bidirectional automatic release of reserve (ARR). To this goal, each source is the reserve of the other, but in order to eliminate the pervious effect of the supply between the two sources, when operational, one of them is considered as a priority, which is, usually, the source I. Eliminating the instability of the ARR system is done by using a time-on delay relay when actuated when the automaton is made with classical switching devices or with an electronic timing circuit with time-on delay, when static commutation elements are used when the control system is being used [3], [4].

In the paper there is presented a bidirectional ARR [5], which has the control system made in four variants, namely: with classical switching devices (Figure 4.a), with logical circuits of TTL or made in CMOS technologies (Figure 4.b), and timing devices with time-on and time-off delays (Figure 4.c), PLC type PS-3 and PLC type Zelio, variants have been experimented.

2. Time circuits used in control installations of ARR made on PS-3 PLC

In the ARR control system made with static commutation elements, time-on and time-off delayed electronic time circuits, made on the PLC type PS-3, which are presented below [6-9].

2.1. Time-on circuit made on PS-3 PLC

In order to achieve electronic time circuits with different time functions on the PS-3 PLC, this PLC has 32 timers (time-on delay) that have TR0 ... TR31 addresses with programmable working time in the range of 0.1 ... 6553.5 s. The time-on function is basic, because by using it all the usual time functions can be made on the PS-3 PLC. The timer TR1 has been used to achieve the timer operation. The timer setting is made with the input size I0.1. The STP (stop timing) function was not blocked, the working time t_a = 4 s (KW40) was set and the output was set to Q0.1. When the time circuit serves a control system, the input and output may, also, be internal memory signals.

For the initial time t_0, the signals I0.1 and Q0.1 have the 0 logic value (Figure 1.c), the I0.1 button is not actuated, the TR1 of the time relay is not powered, its contact in circuit 3 is open and the signal light H1 is off (Figure 1.a).

After the time t_0, push button I0.1 (I0.1 = 1) is energized with voltage the TR1, but during operation time t_a = 4 s of the timing relay, the normally open contact (n.o.c.) with time-on delay Q0.1 in circuit 3 remains in the normal state (Q0.1 = 0), after which it closes (Q0.1 = 1), and the H1 signaling light turns on. This state is maintained until I0.1 (I0.1 = 0) opens, the TR1 coil voltage is off, Q0.1 opens and H1 is turn off.

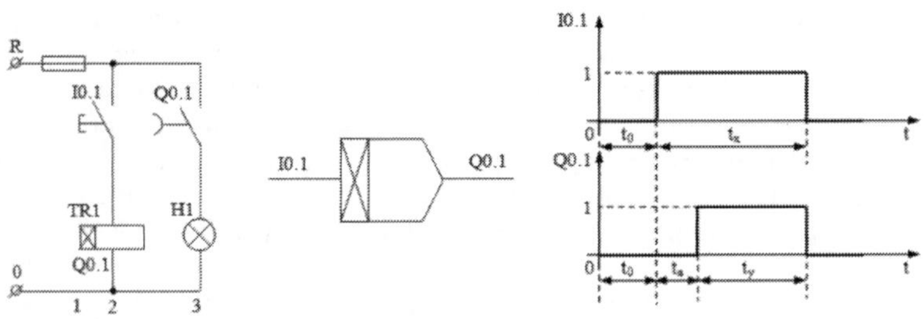

a. b. c.

Figure 1. Achieve timing function at time-on delay; a. Schematic diagram of the function using a classic timed relay; b. Block diagram of the timing circuit; c. Input signals I0.0 = $f_1(t)$ and output Q0.1 = $f_2(t)$ of the time-on delay circuit

The time-on delay program on PS-3 PLC

```
000 TR0
    TR0*S: I0.1
    TR0*STP :
    TR0*IW : KW40
    TR0*EQ : Q0.1
```

2.2. Time-off delay circuit on PS-3 PLC with the classic installation

To achieve the program on the PS-3 PLC, the schematic diagram of the classical switching elements are in Figure 2.a [6], [7]. The input and output sizes for this function are I0.1 and Q0.2. In the initial state, button I0.2 (I0.2 = 0) is not actuated during t_0 (Figure 2.c), the relay is turn off, Q0.2 (Q0.2 = 0), the signaling lamp H2 is turn off, and the signals M0.0 and M0.1 have the 0 logical value.

When pushing button I0.2 (I0.2 = 1), Q0.2 = 1 occurs, H2 lights on, Q0.2 = 1 is stored by closing the n.o.c. Q0.2 in circuit 3. Open normally close contact (n.c.c.) I0.2 of circuit 4, closes n.o.c. Q0.2 of the same circuit, so M0.0 = 0 and n.o.c. M0.0 in circuit 4 remains open, so there is no voltage supply to the TR2 timing relay coil, so M0.1 = 0 and its n.c.c. with opening delay (normally close contact with opening timing – n.c.c.o.t.) remains closed. This condition takes a while.

After time t_x, I0.2 = 0, open n.o.c. I0.2 of circuit 2, but is closed n.c.c. Q0.2 of circuit 3 (Figure 2.a). Closes n.c.c. I0.2 of circuit 4, resulting M0.0 = 1, closes n.o.c. M0.0 of circuit 5 and works TR2 which after t_r open n.c.c.o.t., M0.1 of 2 circuit, which causes the lamp H2 to turns off, M0.0 = 0,

M0.1 = 0 and return to the initial state of the n.c.c.o.t. M0.1. Using the control system with the principle diagram in Figure 2.a, program was made on PS-3 PLC.

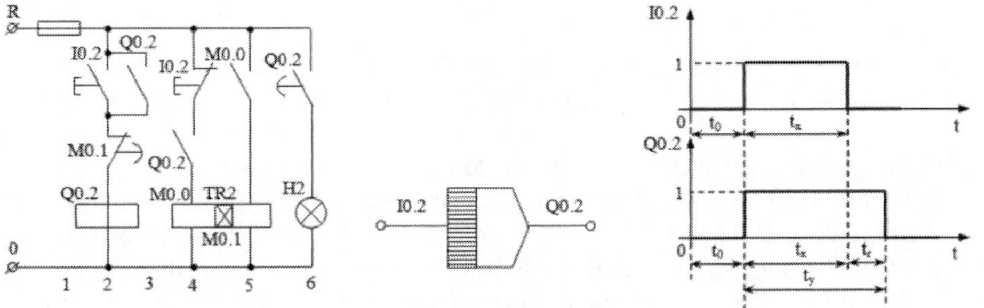

a. b. c.

Figure 2. Performing the time-off delay function: a. Scheme diagram using a classical control; b. The circuit blocks of the time-off delay circuit; c. Input signals I0.2 = $f_1(t)$ and output Q0.2 = $f_2(t)$ of the time-off delay circuit

The time-off delay program on PS-3 PLC (using Figure 2a)

```
001 L I0.2          007 = M0.0
002 O Q0.2          008 TR2
003 A NM0.1             TR2*S : M0.0
004 = Q0.2              TR2*STP :
005 L NI0.2             TR2*IW : KW60
006 A Q0.2              TR2*EQ : M0.1
```

Using this program, the t_r time-off delay is 6 s (KW60).

2.3. Time-off delay circuit with PS-3 PLC and STP command timer

The schematic diagram of this circuit is given in Figure 3 [6-9], the variations of the input signals I0.3 $f_1(t)$ and output Q0.3 = $f_2(t)$ are those shown in Figure 2.c, in which I0.2 and Q0.2 are replaced by I0.3 and Q0.3.

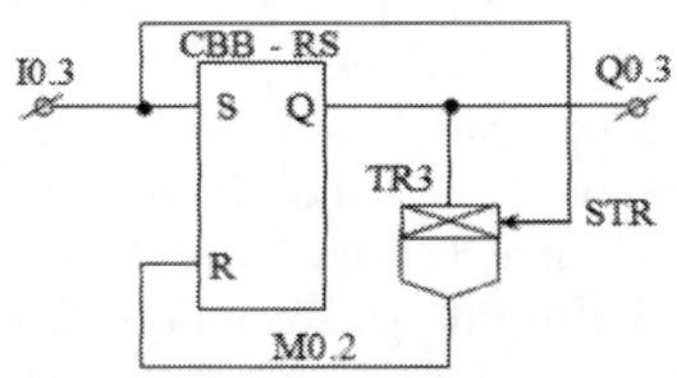

Figure 3. Schematic diagram of the electronic time-off delay with flip-flop R-S circuit, CBB-RS and timer STOP

When changing the signal value I0.3 (I0.3 = 0, S = 0), the state of the signals from the input of the flip-flop RS circuit (CBB-RS) is I0.3 = 0 (S = 0) and M0.2 = 0 (R = 0), but Q0.3 continues to have the 1 logic value due to the previous values of the S and R signals (S = 1, R = 0). Now, the TR3 timer is released which has the working time t_r, after which M0.2 = 1 (R = 1) as a result of Q0.3 = 0 (Q = 0), M0.2 = 0, so, S = 0, R = 0, and Q is maintained at the previous logical value.

The time-off delay program on PS-3 PLC (using Figure 3)

```
009TR3                  010 L I0.3
   TR3*S : Q0.3         011 S Q0.3
   TR3*STP : I0.3       012 L M0.2
   TR3*IW : KW60        013 R Q0.3
   TR3*EQ : M0.2
```

Due to the simplicity of design of the time-off delay time function, using the program, when making the bidirectional ARR, the electronic circuit in Figure 3 and the associated program are used.

3. The logical functions of the bidirectional automatic release of reserve

This ARR [5] has the diagram scheme of the main installation given in Figure 4.a. (power diagram), Figure 4.b. (classical control diagram), and Figure 4.c. (control system made with static commutation elements). Basically, the marking of the elements has been done for some of them, according to STAS 11381/83 and 88, and for others, with that recognized by PS-3 PLC.
The logical functions of the classic control device (Figure 4.b), with the marking of the switching devices, recognized by PS-3 PLC, are:

$$M0.3 = \overline{I0.3} \cdot (I0.4 + M0.3) \tag{1}$$

$$M0.0(t_{a0}) = M0.3 \tag{2}$$

$$M0.1(t_{a1}) = M0.3 \cdot \overline{I0.1} \tag{3}$$

$$M0.2(t_{a2}) = M0.3 \cdot \overline{I0.2} \tag{4}$$

$$Q0.1 = M0.3 \cdot \overline{I0.5} \cdot \overline{M0.1}(t_{a1}) \cdot \overline{M0.7}(t_{r2}) \cdot \overline{Q0.2} \tag{5}$$

$$M0.4(t_{r1}) = M0.3 \cdot Q0.1 \tag{6}$$

$$Q0.2 = M0.3 \cdot \overline{I0.6} \cdot \overline{M0.0}(t_{a0}) \cdot \overline{M0.2}(t_{a2}) \cdot \overline{M0.4}(t_{r1}) \cdot \overline{Q0.1} \tag{7}$$

$$M0.7(t_{r2}) = M0.3 \cdot Q0.2 \tag{8}$$

With these functions and taking into account the easiest configuration (Figure 3) of the electronic circuit with time-off delay, the ARR control principle was established with static commutation elements (Figure 4.c). For the TR0, TR1, TR2 electronic time-on delay circuits, output values are transmitted to the internal memories M0.0, M0.1, and M0.2. The electronic time-off circuits, with the TR3 and TR4 active elements, have outputs M0.4 and M0.7.

At the bidirectional ARR system, source I is the priority. When the source I falls, or one of the phases of the I-st source is interrupted, it is not deactivated for 2 s as the working time t_{a1} of the timer TR1 (Figure 4.c). If the supply voltage returns during this time, the consumers continue to remain connected to the source I. If after 2 s there is no voltage at the contactor terminals Q0.1 it is triggered and after the working time t_{r1} = 1 s to which the TR3 timer is set, the contactor Q0.2 is activated and the source II is activated.

When the source II is activated, the passage of the power supply from the source I goes in the same way, but in this case other logic and timing circuits occur. When passing the power supply from another source to another, if the voltage of the damaged source returns, the last voltage source remains on.

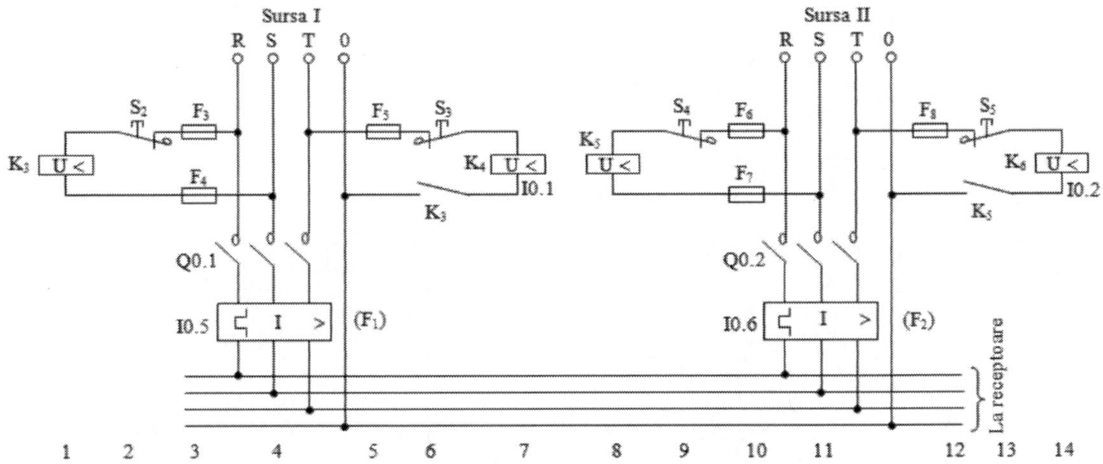
Sursa I
R S T 0
Sursa II
R S T 0
S2
F3
F4
F5
S3
K3
U <
K4
I0.1
S4
F6
F7
F8
S5
K5
K6
I0.2
Q0.1
Q0.2
I0.5
I
(F1)
I0.6
(F2)
La receptoare
1 2 3 4 5 6 7 8 9 10 11 12 13 14

a.

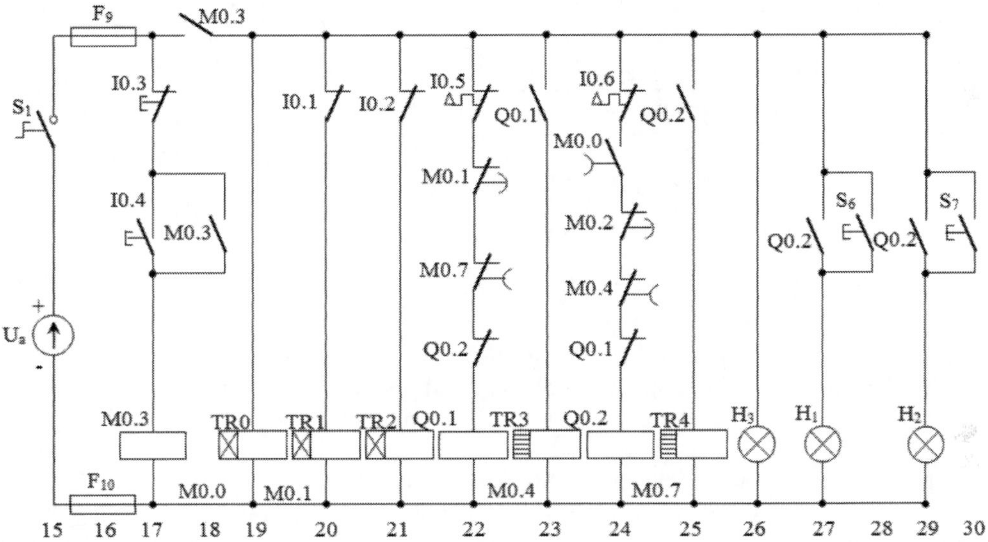
F9
M0.3
S1
I0.3
I0.4
I0.1
I0.2
I0.5
I0.6
Q0.1
Q0.2
M0.0
M0.1
M0.2
M0.7
M0.4
S6
S7
Ua
TR0
TR1
TR2
TR3
TR4
H3
H1
H2
F10
15 16 17 18 19 20 21 22 23 24 25 26 27 28 29 30

b.

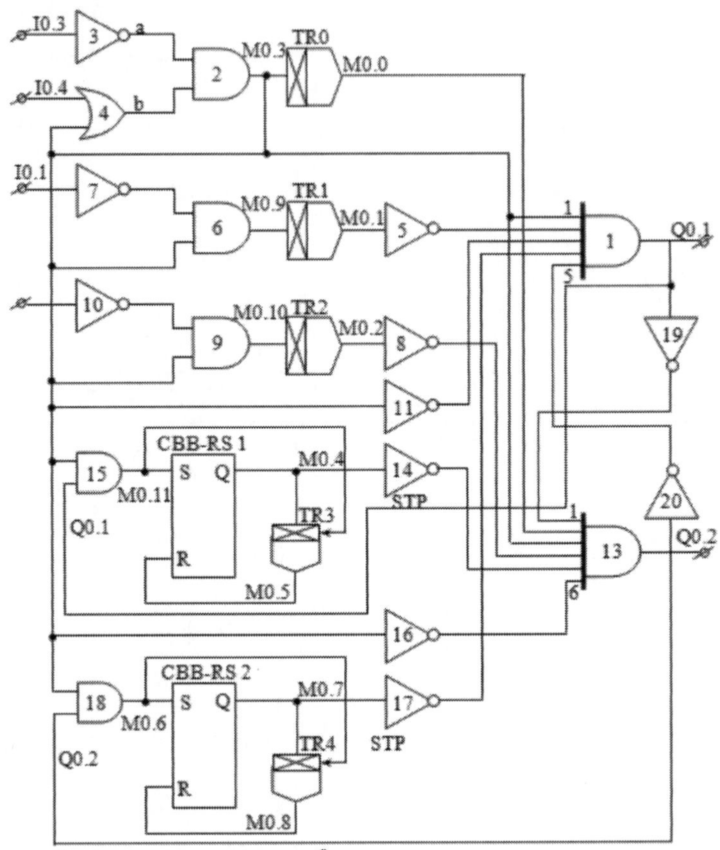

c.

Figure 4. Bidirectional ARR for low voltage electrical installations with current under 100 A: a. Scheme of the main installation; b. Classical control system diagram; c. Control system made with elements that have static commutation

4. Bidirectional ARR operation with logic integrated circuits

In the initial state, if there is voltage on each phase of sources I and II, the relays K3, K4, K5, and K6 are actuated, therefore the normally open contacts of the K4 and K6 relays are closed, the signals I0.1 and I0.2 from the inputs INVERSORS 7 and 10 have logical value 1, so M0.1 = 0, M0.2 = 0, and after INVERSORS 5 and 8, $\overline{M0.1}=1$, $\overline{M0.2}=1$. These signals are applied to the inputs 2 and 4 of the AND 1 and 13 circuits. Since the F1 and F2 protections are not actuated, the signals I0.5 and I0.6 have 0 logical value, and after INVERSORS 11 and 16, $\overline{I0.5}=1$ şi $\overline{I0.6}=1$ and the signals applied to the inputs 3 and 6 of the AND 1 and 13 circuits.

Running the ARR in operation occurs when I0.4 = 1 and I0.3 = 0, then after INVERSOR 3, a = 1, after OR 4, b = 1 and after I0.4 = 0. Signal M0. 3 = 1 unlocks the AND 1, 13, 6, 9, 15, and 18 gates. Because Q0.1 = 0 and Q0.2 = 0, after INVERSORS 19 and 20, $\overline{Q0.1}=1$ and $\overline{Q0.2}=1$. These signals are applied to the inputs 1 and 5 respectively of the AND 13 and 1 gates. After CBB-RS2, M0.7 = 0, at the INVERSOR 17 output $\overline{M0.7}=1$ applied to the input 4 of the AND circuit 1. And after CBB-RS1, M0.4 = 0 So at the output of the INVERSOR 14 $\overline{M0.4}=1$, the signal applied to the input 5 of the AND gate 13. The M0.3 = 1 validation signal of the ARR operating state is applied to the input of the timing circuit TR0, set to operate after 5 s, consequently during this time, the signal M0.0 = 0 applied to the input 2 of the AND 1 gate, blocks this gate. It is noted that on all the inputs of the circuit AND 1 the signals have the logic value 1, so Q0.1 = 1, and Q0.2 = 0 so the source I is activated by actuating the contactor Q0.1. After INVERSOR 19, a signal that blocks the gate AND 13 and after the TR0 timer works when M0.0 = 1. The TR0 timer ensures stable ARR system operation.

The signal Q0.1 = 1 also applies to the second input of the AND gate 15, so M0.11 = 1 (S = 1). Since M0.5 = 0 (R = 0) at CBB-RS1, S = 1 and R = 0, so M0.4 = 1 and M0.4 = 0 at INVERSOR 14 output and at input 5 of AND 13 gate. Specifies that the M0.4 = 1 signal does not work of the TR3 timer to operation because at the STP input M0.11 = 0. This time circuit provides a delay of 1 s when the power supply passing from the source I to the source II.

When the source I falls or the voltage drops on one of the phases, for example from phase S, the relay K3 with the coil in circuit 1 (Figure 4.a) triggers, the contact K3 from the circuit 6 is open on the relay coil circuit K4 from circuit 7, which causes the logic value of the signal I0.1 (I0.1 = 0) to be changed from the control system (Figure 4.c.), after the AND signal M0.9 = 1, the TR1 timer is activated after the time $t_{a1} = 2$ s dues M0.1 = 1, so $\overline{M0.1} = 0$ from INVERSOR 5 output and input 2 of circuit AND 1, Q0.1 = 0. The signal Q0.1 = 0 causes the contact Q0.1 to trip and the main contacts in the circuit 4 (Figure 4.a) is deactivated source I, Q0.1 = 1 at the INVERSOR 19 output and on the input 1 of the circuit AND 13. Q0.1 = 0 is applied to the input 2 of the AND circuit 15, so M0.11 = 0. Now on S and R CBB-RS1 circuit, the signals have 0 logical values, but Q = 1 due to the previous state of the signal on R and S. Since M0.11 = 0 the STP command is cancelled, the TR3 timer works and after the adjusted time $t_{r1} = 1$ s, the signal R = 1. Now S = 0 and R = 0, so Q = 0 (at CBB-RS1 input and TR3). The signal $\overline{M0.4} = 1$ at input 5 of the AND circuit 13 causes the signal value to change Q0.2 (Q0.2 = 1).

Now the power supply to the consumers is made from the source II by actuating the contactor Q2 with the main contacts in the circuit 11 (Figure 4.a). After INVERSOR 12, $\overline{Q0.2} = 0$ (interlock signal) is applied to the input 5 in the AND circuit 1 and blocks this gate. The signal Q0.2 = 1 is also applied to the input 2 of the AND circuit 18, after which M0.6 = 1 (S = 1). Because M0.8 = 0, R = 0. For this state of the signals M0.6 and M0.8 (S = 1, R = 0) at CBB-RS2 output M0.7 = 1 (Q = 1), so after INVERSOR 17, $\overline{M0.7} = 0$, which is applied to the input 4 in the AND circuit 1. This signal provides a delay of 1 s when receiving the power supply command from the source II to the source I. Operation of the timer TR4 is blocked by the signal M0.6 = 1.

When the source II falls, switching the power to the consumers from the source I takes place just as in the previous situation, but in this case other logic circuits and timing are used. When disabling a source occurs as a result of the complex protection (I0.5 or I0.6), the contactor is instantaneously switched on, after which the timed switch (1 s) takes place at the power supply from the backup source.

5. Making Bidirectional ARR Command on PS-3 PLC

Using the schematic of the bidirectional ARR control system made with integrated logic circuits and electronic timers (Figure 4.c), the corresponding program on PS-3 PLC was established. Following program experimentation, it was found that the operation of the control system made on PS-3 PLC is identical to that shown in Figure 4.c [8], [9].

IOP Conf. Series: Materials Science and Engineering **294** (2017) 012053 doi:10.1088/1757-899X/294/1/012053

The bidirectional ARR program on PS-3 PLC (using Figure 4.c)

```
000 L I0.4
001 O M0.3
002 A NI0.3
003 = M0.3
004 TR0
    TR0*S : M0.3
    TR0*STP :
    TR0*IW : KW50
    TR0*EQ : M0.0
005 L NI0.1
006 A M0.3
007 = M0.9
008 TR1
    TR1*S: M0.9
    TR1*STP :
    TR1*IW : KW20
    TR1*EQ : M0.1

009 L NI0.2
010 A M0.3
011 = M0.10
012 TR2
    TR2*S : M0.10
    TR2*STP :
    TR2*IW : KW20
    TR2*EQ : M0.2
013 L M0.3
014 A Q0.1
015 = M0.11
016 TR3
    TR3*S : M0.4
    TR3*STP : M0.11
    TR3*IW : KW10
    TR3*EQ : M0.5
017 L M0.11
018 S M0.4
019 L M0.5
020 R M0.4
021 L M0.3
022 A Q0.2
023 = M0.6
024 TR4
    TR4*S : M0.7
    TR4*STP : M0.6
    TR4*IW : KW10
    TR4*EQ : M0.8
025 L M0.6
026 S M0.7
027 L M0.8
028 R M0.7
029 L M0.3
030 A NM0.1
031 A NI0.5
032 A NM0.7
033A NQ0.2
034 = Q0.1
035 L NQ0.1
036 A M0.0
037 A NM0.2
038 A NM0.4
040 = Q0.2
```

6. Bidirectional automatic release of reserve on Zelio PLC

The PLC Zelio is a smart relay, manufactured by Schneider Electric, used for small automation with a number of inputs/outputs between 10 and 40. The Zelio PLC is available in compact formats (3 single bloc models with 10, 12, or 20 inputs/outputs with or without display) or modular (2 basic models with 10 or 26 inputs/outputs which can be extended either with Modbus or Ethernet modules or using 6, 10, or 14 individual inputs/outputs and 4 analogue extension modules for inputs/outputs). This relay has a wide range of applications in which it can be used, both in industry and construction, and in the field of services [10].

The Zelio PLC has the advantage of being simple to install and program, allowing control and monitoring of installations in any situation, either locally or remotely. It can be programmed in LD (ladder diagram) and FBD (function bock diagram). These languages are much easier to use than IL (instruction list) that is used when programming PS-3 PLC. For the design of ARR program, the FBD language was used, the program being performed after the ARR of Figure 4.c.

In Figures 5-14 (the blue connections are 0 logic, the red connections are 1 logic) the FBD program was presented during the ARR operation. To the left are the PLC's inputs (I1 presence of voltage on the three-phases of the source I, I2 presence of voltage on the three-phases of the source II, I3 is the

IOP Conf. Series: Materials Science and Engineering **294** (2017) 012053 doi:10.1088/1757-899X/294/1/012053

ARR stop button, I4 is the ARR start button, I5 is the input from protection to overloading of consumers for source I, I6 is the input from the overload protection of consumers for the source II) and the outputs (Q1 the supply of the consumers from the source I, Q2 the supply of the consumers from the source II). Inputs are digital and use mechanical locking buttons, and outputs are digital and supply coils of three-phase contactors to connect consumers to source I or source II. Both sources (I and II) used 2 s time-on delays, in which consumers are not disconnected from those sources. If the fault persists for more than 2 s, for 1 s, consumers are disconnected from damaged sources. After passing the 1 s interval, consumers are switched to the other source, and then remain connected to it even if the voltage on the first source returns the three-phases.

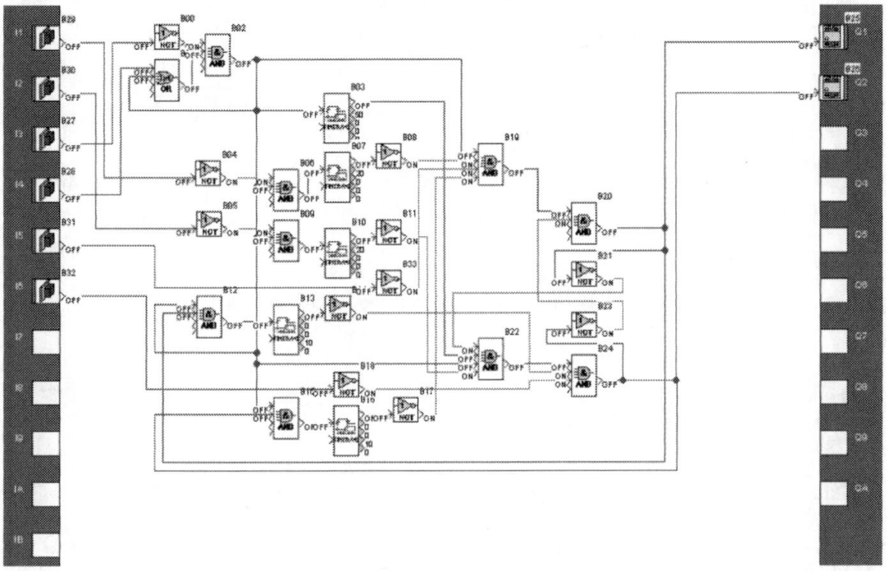

Figure 5. Principle diagram ARR

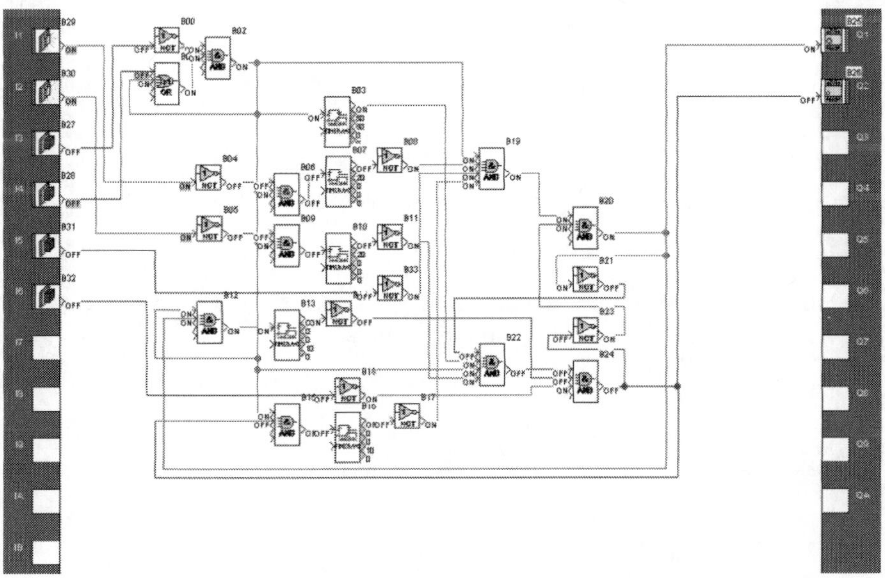

Figure 6. ARR operation, consumers are powered from source I

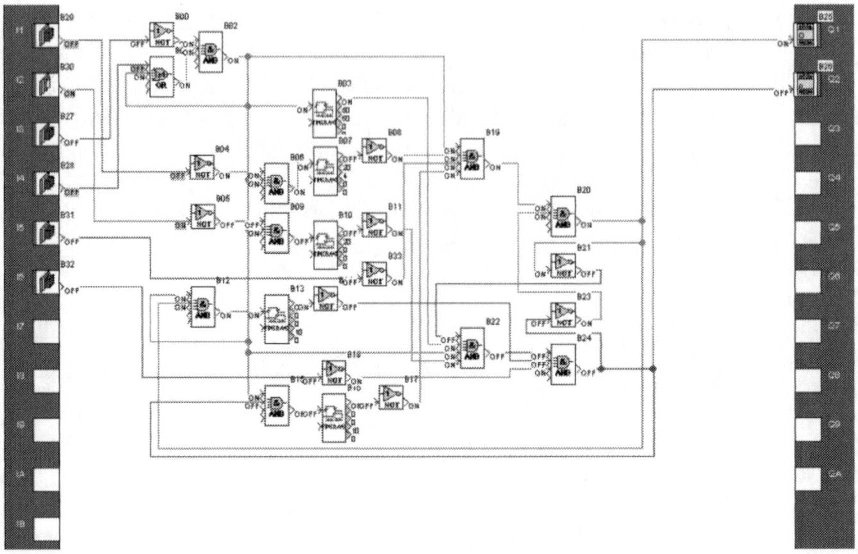

Figure 7. ARR operation, source I damaged, within 2 s

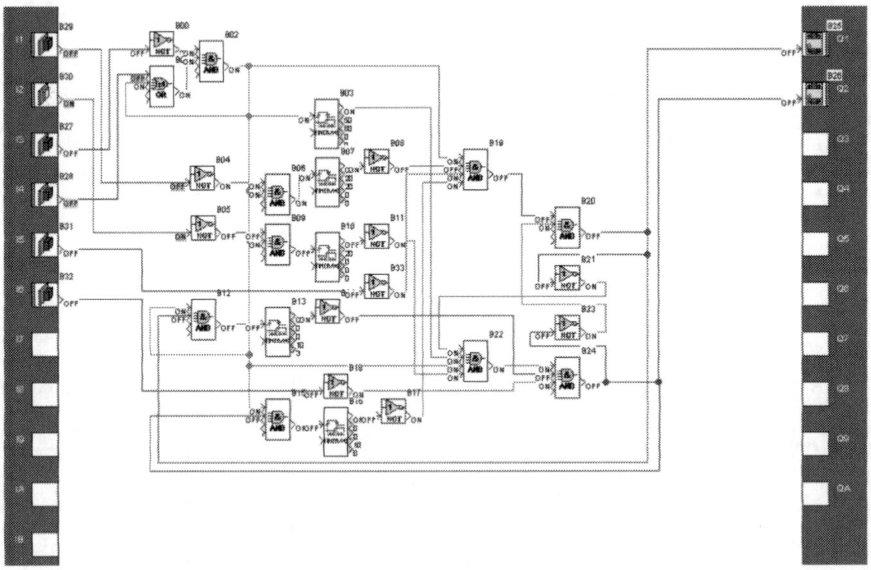

Figure 8. ARR operation, source I damaged, from 2 s to 3 s, consumers are not powered

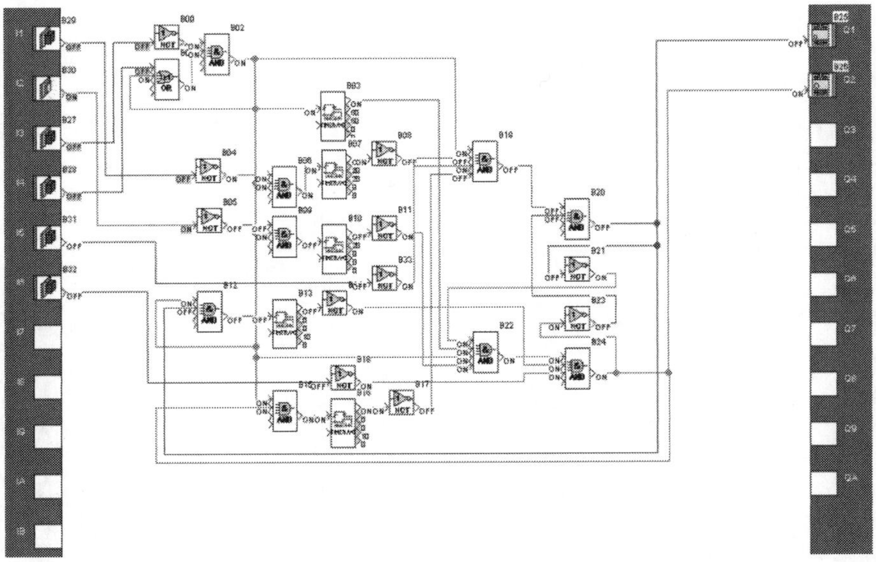

Figure 9. ARR operation, consumers are powered from source II, source I is damaged

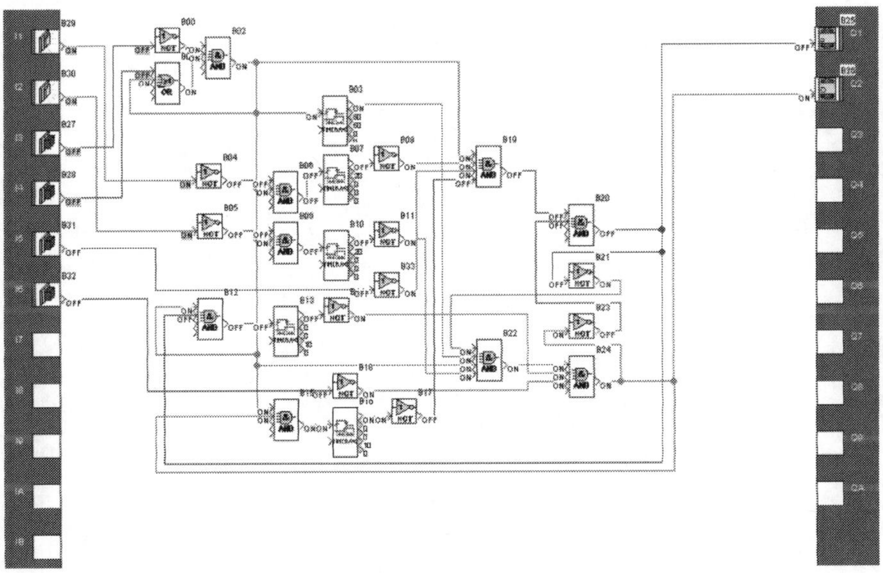

Figure 10. ARR operation, consumers are powered from source II, source I enabled

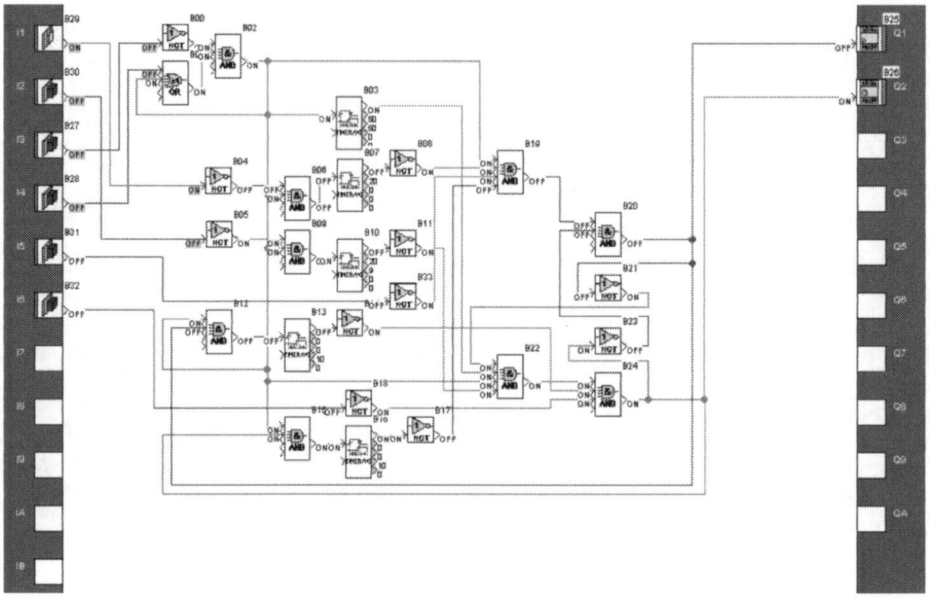

Figure 11. ARR operation, source II damaged, within 2 s

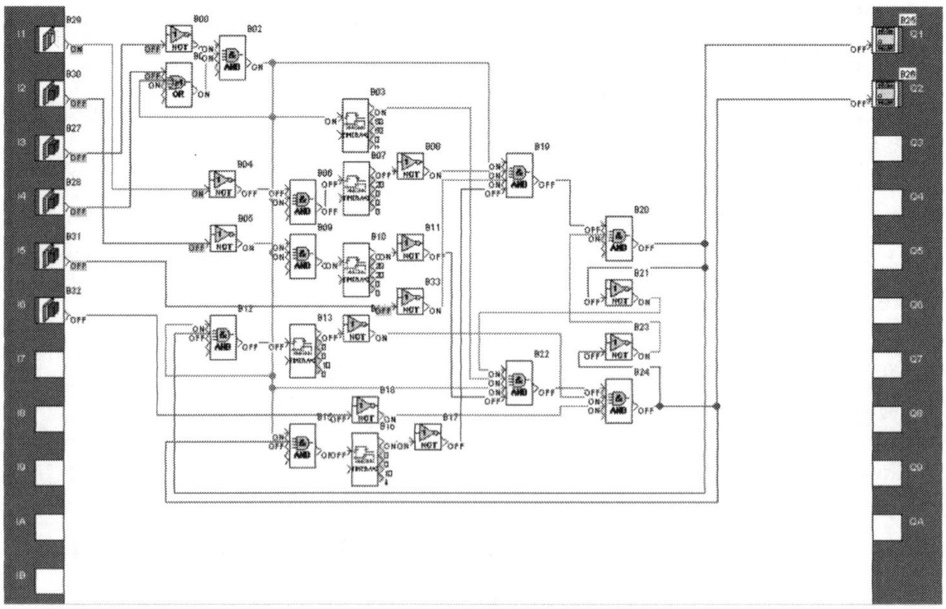

Figure 12. ARR operation, damaged source II, from 2 s to 3 s, consumers are not powered with voltage

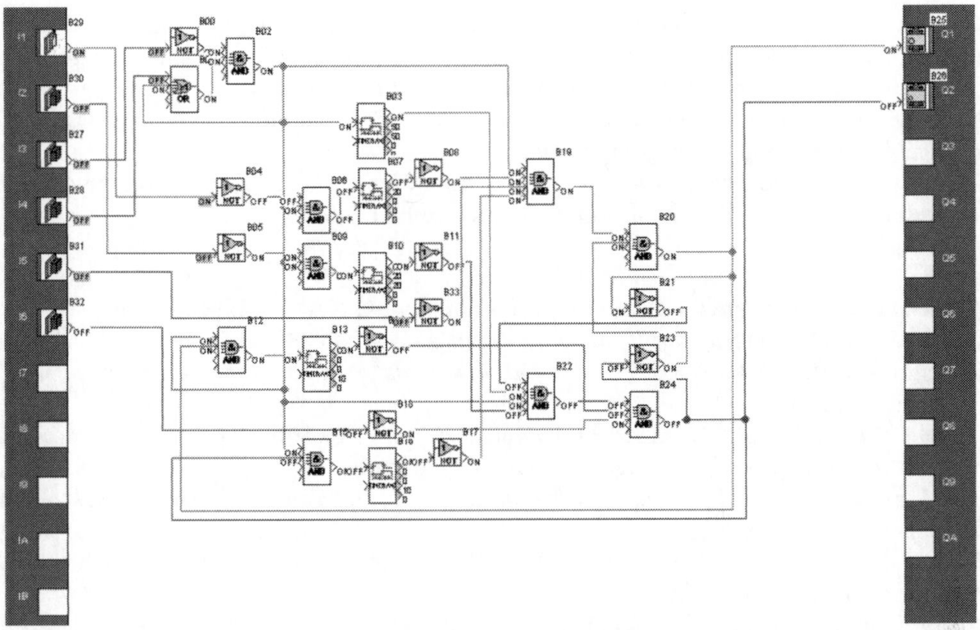

Figure 13. ARR operation, consumers are powered from source I, source II damaged

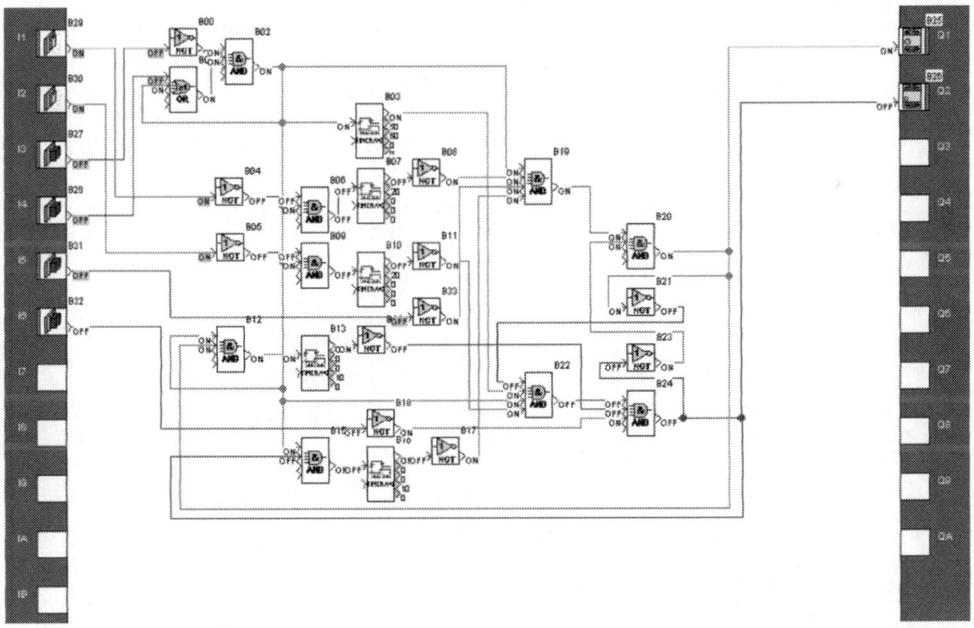

Figure 14. ARR operation, consumers are powered from source I, source II enabled

7. Conclusions

The important electrical loads require power supply from two (or even three) power sources.

The paper presents the sequences code for timing performed on the PS-3 PLC, the logical functions for the bidirectional automatic release of reserve, the classical control electrical diagram, the electronic control diagram, and the code (in IL language) made for the PS-3. The same bidirectional automatic release of reserve has been developed on Zelio PLC (in FBD language).

IOP Conf. Series: Materials Science and Engineering **294** (2017) 012053 doi:10.1088/1757-899X/294/1/012053

The control of bidirectional automatic release of reserve power supplies can be achieved using classical control installations or using PLCs (a better solution). In all type of control, the scheme of the power installation can't be replaced. PLCs can successfully replace classical control system diagram and control system made with elements that have static commutation. They can easily adapt to installations and easily debug/develop applications.

It is easiest to use modern PLCs that can be programmed in FBD (semi-graphical program) compared with PLCs can be programmed in IL (like assembler code).

References

[1] Warne D F 2005 *Electrical Power Engineer's Handbook*, Newnes, Elsevier Press

[2] Grigsby L L 2012 *The Electric Power Engineering Handbook*, CRC Press

[3] Mayergoyz I D and McAvoy P 2014 *Fundamentals of Electric Power Engineering*, World Scientific Press

[4] Ceraolo M and Poli D 2014 *Fundamentals of Electric Power Engineering: From Electromagnetics to Power Systems*, Wiley-IEEE Press

[5] Popa I and Popa G N 2000 *Electrical Installations*, vol. I, Mirton Publishing House, Timişoara, Romania (in Romanian)

[6] Popa G N, Popa I and Deaconu S 2006 *PLCs in Applications*, Mirton Publishing House, Timişoara, Romania (in Romanian)

[7] Popa I, Popa G N, Diniş C M and Deaconu S I 2014 *Upon Achievement of the Usual Functions of Time with PS-3 PLC Klöckner-Moeller*, 6th International Multidisciplinary Scientific Symposium Universitaria SIMPRO 2014, Section four, Electrical Engineering and Energetics, Petroşani, Romania, October 10, pp 283-288

[8] *** 1987 *Handbuch Teil 2: Software. Programmersprache*, Klöckner-Moeller, Germany

[9] *** 1987 *Handbuch Teil 3: Software. Programmbeispiele*, Klöckner-Moeller, Germany

[10] *** 2010 *Zelio PLC. SR3. User's Manual*, Schneider, Germany

IOP Conf. Series: Materials Science and Engineering **294** (2017) 012054 doi:10.1088/1757-899X/294/1/012054

Industrial application of low voltage bidirectional automatic release of reserve

G N Popa, C M Diniş, A Iagăr, S I Deaconu and I Popa

Politehnica University of Timisoara, Department of Electrical Engineering and Industrial Informatics, 5 Revolution Street, Hunedoara, 331128, Romania

E-mail: gabriel.popa@fih.upt.ro

Abstract. The paper presents an analysis on low voltage industrial electrical installation controlled by bidirectional automatic release of reserve. Industrial electrical installation is for removing smoke in case of fire from a textile company. The main parts of the installation of removing smoke in case of fire are: general electrical panel; reserve electrical panel; three-phase induction motors for driven fans; electrical actuators for inlet and outlet valves; clean air inlet pipe, respectively, the outlet pipe for smoke. The operation and checking of bidirectional automatic release of reserve are present in the paper.

1. Introduction

Automatic release of reserve (ARR) is the quick connection of electrical consumers to a backup power source in the event of a failure of the main power supply as a result of an interruption or disconnection imposed by the protective devices. The operation can be applied to voltage-sensitive consumers receiving two or more power supplies from different sources [1], [2].

The negative impact of voltage gaps on the power quality supplied to consumers is particularly important and depends on the type of these voltage gaps and the operating characteristics of the electrical consumers. The most important consumers which do not support the voltage gaps, materialized by accidentally interrupting of electricity, are called vital consumers. Effects caused by voltage gaps can affect human health, sometimes even life, the environment in the case of atmospheric depollution and water purification equipment, the comfort or safety of people in conference rooms, shows, and so on. To avoid these effects, choose the solution for supplying vital consumers from other available power sources in the event of accidental interruption of main power. Back-up electrical power supplies can take up all the power consumed [3], [4].

The supply of vital consumers from backup power sources in the event of an accidental interruption of the main power supply can be done by means of a ARR equipment. It is used when the two power supplies (main and back-up) have identical powers. This ensures the continuity of power supply to electrical consumers that do not support voltage gaps longer than 2 ÷ 3 seconds. The ARR determines the automatic power supply connection when the main power supply is disconnected or switches from the main power source to the backup power source [5], [6].

Smoke evacuation installations by mechanical circulation in the case of fire are an important installations that requiring a bidirectional ARR [1], [2]. Smoke evacuation installations must ensure in case of fire:

- keeping of exhaust passages and free access to smoke on the level of traffic;
- facilitating intervention operations by creating smoke-free areas;

- delaying or preventing the generalization of the fire;
- reducing the consequences and damages caused by smoke and hot gases;
- reducing efforts in building structural elements.

Mechanical removing smoke is ensured by mechanically evacuating of the smoke and naturally introducing the air, so as to ensure air circulation in the protected space and smoke evacuation. Fire safety is ensured by adapting the electrical installation to the fire resistance of the building elements, and enclosing the electrical installation in fire and explosion hazard categories. The threshold for protection of electrical appliances and equipment has been adopted for operation in the environment appropriate to their location and the purpose of the room.

2. Removing smoke installation in case of fire

Basically, the removing smoke is ensured by natural, organized or mechanical circulation, by making air circulate in the space considered and the exhaust of the smoke in relation to the introduced air or by pressure differences between the protected and the fired space depressed or by the combination of the two methods. Smoke removal is intended to remove a part of the smoke and the combustion gases from the combustion chamber in order to ensure the conditions of evacuation of users and the use of extinguishing agents as well as the limitation of fire propagation.

Smoke extraction systems must prevent the propagation of smoke according to:
- the smoke and hot gases;
- the tightness at smoke.

To ensure fire safety conditions, smoke diffusion can be limited by:
- the tightness of building compartments;
- the creation of overpressure or fresh air streams with the movement in the opposite direction to the direction of natural smoke movement.

In the building of a textile factory, the removing smoke installation is done through a mechanical system, with manual blower fans and natural circulation of air. The mechanical removing smoke is ensured by mechanically evacuating the smoke and naturally introducing the air so as to ensure air circulation in the protected space and smoke evacuation.

The technical characteristics of the fans have the following main features:
- exhaust fans are designed to operate at 400 ^{0}C for 1 hour;
- exhaust fans are electrically powered from a main power source and a back-up power source (which requires the use of a bidirectional ARR);
- air intake fans must provide at least 60% of the exhaust flow.

The main power supply line will be connected to the main power source of the plant, and the backup electrical line will be connected to the main electrical switchboard. From the ARR control system it is possible to determine, optionally, which power source is has a priority. The smoke exhaust system will be manually controlled by self-maintaining control buttons located on the escape routes.

The air intake and exhaust areas are alternately distributed, distributed evenly within the protected space, so as to ensure air circulation and smoke evacuation. Closures of the inlet or outlet areas shall be manually operated. Local manual command of the opening devices will be done mechanically or electrically.

3. The electric installation of removing smoke system

The power supply system of the removing smoke system, made with a bidirectional ARR, is shown in Figure 1. This installation is a classic one, with switching (contactors) and protection (maximum current) devices.

Three contactors are used, one for each power source and one for the electric consumers. At one time, the ARR's control system connects the electricity consumers to a single power source [7-9].

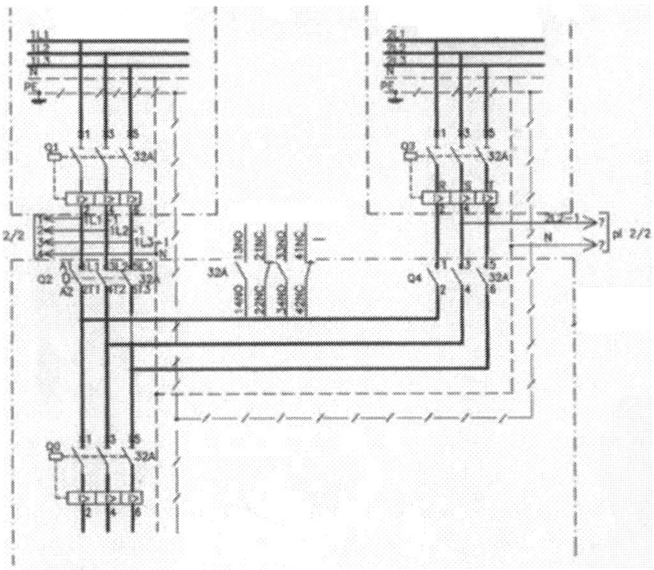

Figure 1. The power electrical installation powered by a bidirectional automatic release of reserve

The control system is a classical one (Figure 2), made with contacts, relays and time relays, which provide optional for ARR bidirectional function, with the possibility of setting the priority power source [7].

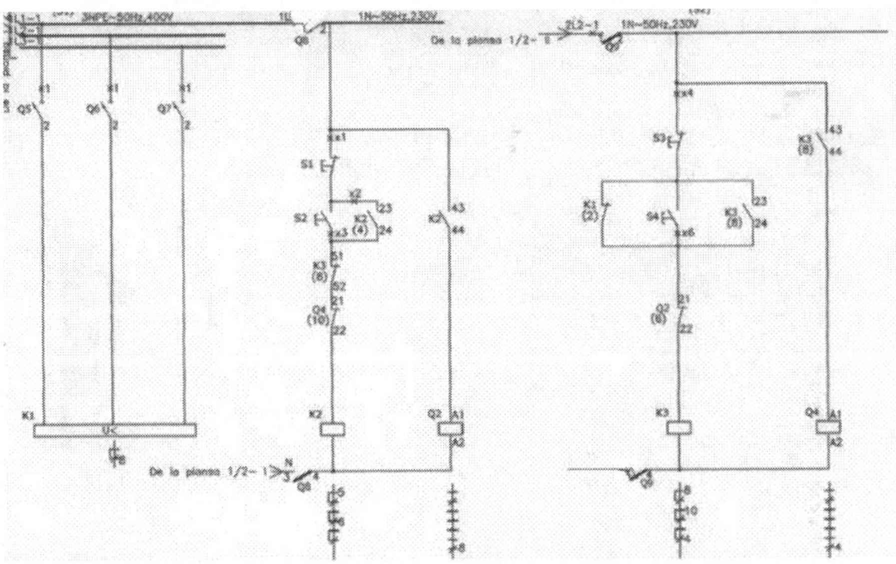

Figure 2. The electrical installation powered by a bidirectional automatic release of reserve – a part from a control installation

The power source for electric consumers (three motors that supply the fans and four actuators that regulate the air and smoke flow) is a classic installation with switching and protection devices (maximum and overload) – Figure 3.

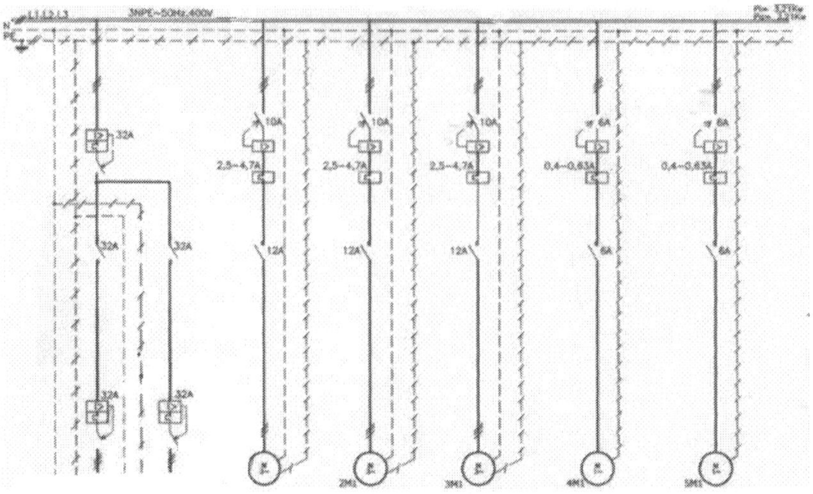

Figure 3. The electrical installation powered by a bidirectional automatic release of reserve – a part from a control installation

The removing smoke system in case of fire consists of the following elements:
- the main power source;
- the bidirectional ARR (Figure 4.a);
- the backup power source;
- three fans powered by three-phase motors (1.1 kW, 2.7A, 1430 rpm; one for air in the basement and the ground floor, and one for the smoke exhaust from the basement and the ground floor) (Figure 5.a);
- three single-phase servomotors (0.115 kW, 0.65 A) acting on the exhaust or intake valves (Figure 5.b);
- the air inlet and smoke outlet pipes (Figure 4.b).

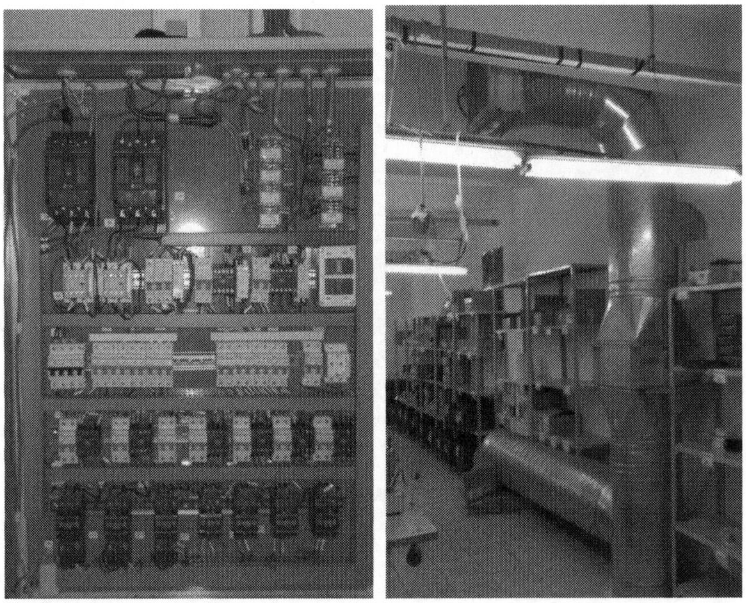

a. b.

Figure 4. a. The electric installation inside the cabinet; b. Pipes installation

IOP Conf. Series: Materials Science and Engineering **294** (2017) 012054 doi:10.1088/1757-899X/294/1/012054

a. b.

Figure 5. a. The electric drive for one fan to evacuate smoke;
b. The actuator used to control the air valve

4. Experimenting the operation of removing smoke system in case of fire

The experimental measurements (Figures 6-15) have been made for checking of the bidirectional ARR were performed using the three-phase power quality analyser (CA 8334 B).

Three phase current probes, type MN 93A (100A) were connected on phases to measure the phase currents. The four-conductor method (three phases and null) was used to measure voltages. The three-phase power quality analyser does not measure currents below 0.3 A and no voltages below 7 volts.

4.1. Measurements made in the plant when the consumers were powered from the main power source

The three fans were switched on and the air intake and smoke exhaust valves were opened with the help of servomotors. The stages of the experiment were:

- all electric consumers are connected;
- after 3 minutes, the smoke exhaust ventilator in the workshop was switched off;
- after another 1 minute the air intake fan in the basement and ground floor was switched off;
- after another 1 minute, the smoke exhaust fan was switched off in the basement and ground floor;
- after 1 minute the smoke exhaust ventilator was started in the basement and ground floor;
- after another 1 minute, the smoke exhaust ventilator in the workshop and the air intake fan in the basement and the ground floor were switched on.

The current and voltage on phases (Figures 6.a and b) were measured at a time when the consumers are approximately balanced on the phases. Currents are after voltage, on Fresnel diagram, electrical consumers have an inductive character (electric motors). The Fresnel diagram for voltages and currents are presented in Figures 7.a and b. Generally, the voltage harmonics (Figure 7.a) fall within the limits accepted by the norms regarding the limitation of the deforming and non-symmetrical regime in the electrical networks, except the 5th harmonics. The 5th harmonic current (Figure 7.b) exceeds the accepted limit of 5% [8], [10].

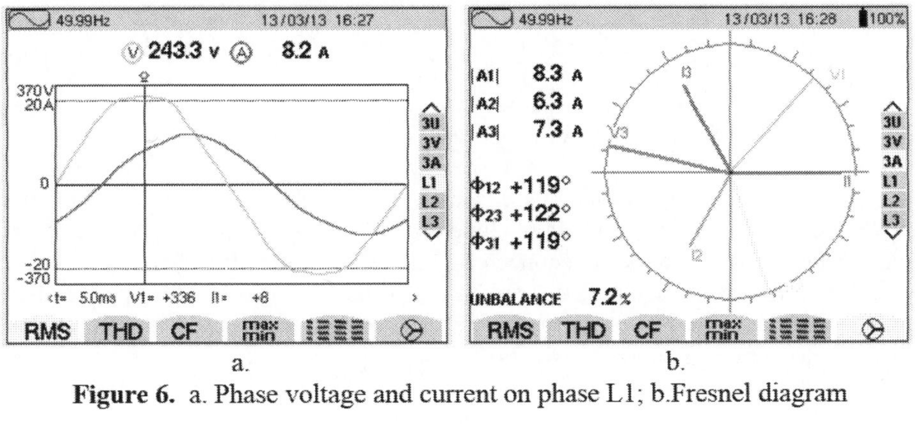

a. b.

Figure 6. a. Phase voltage and current on phase L1; b.Fresnel diagram

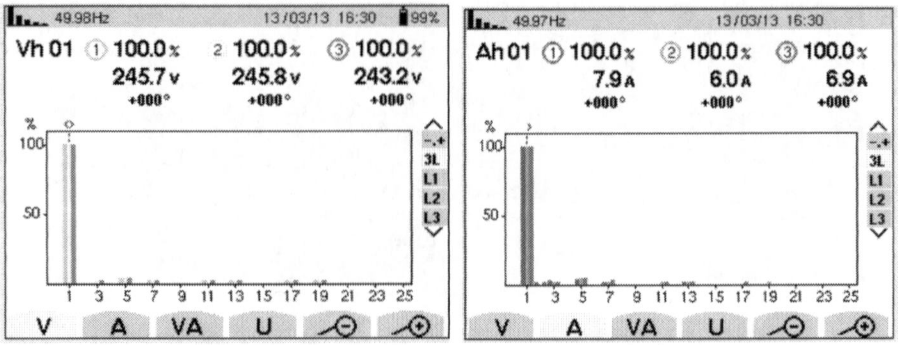

a. b.

Figure 7. a. Harmonics of phase voltages; b. Harmonics of currents

Due to the start-up of the electric consumers, voltage fluctuations (Figure 8.a) between 240 ÷ 247 V are registered. The phases are unbalanced, with a difference of approximately 3 V between the phases L1, L2 and L3.

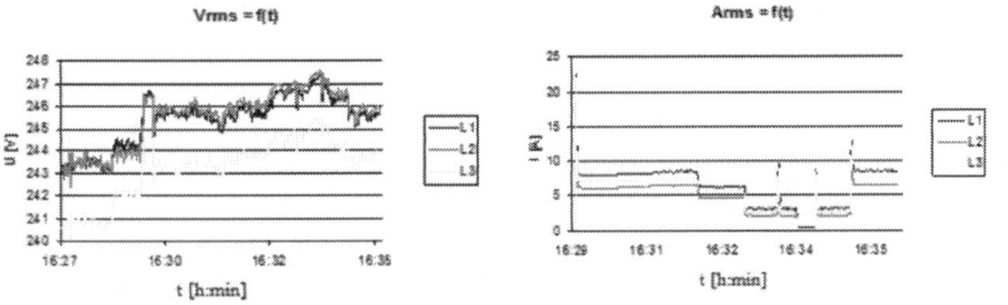

a. b.

Figure 8. a. Evolution of phase voltages; b. Evolution of currents

The phase currents (Figure 8.b) have values between 6 ÷ 8 A, the phases being slightly unbalanced. The fluctuation of the currents when coupling and disconnecting the motors is observed.

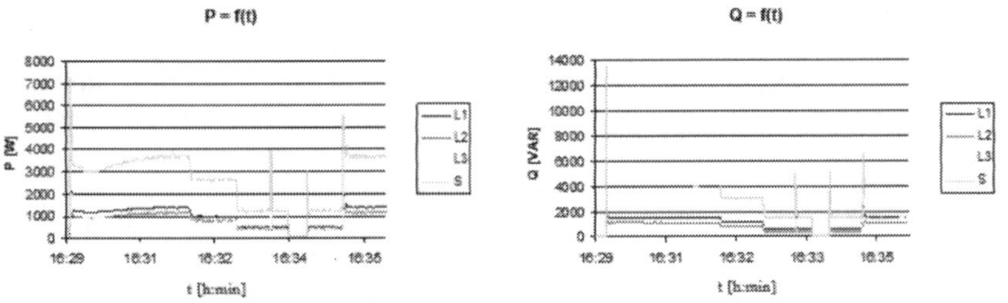

a. b.

Figure 9. a. Active powers; b. Reactive powers

The active power (Figure 9.a) on each phase is about 1 ÷ 1.5 kW, and the total power is in the 2.5 ÷ 3.8 kW range, when all engines are operating. Generally, the active power on each phase and the total power change when a new consumer is connected to the plant. As well as active power and reactive power (Figure 9.b), fluctuates when each consumer is switched on and off. The reactive power on each phase falls within the range of 0.3 ÷ 1.8 kVAR, and the total reactive power is maintained at 4 kVAR, having a linear evolution when all motors work.

Basically, apparent power (Figure 10.a) has a similar evolution to active power. It reaches a maximum of 5.5 kVA, but even higher values (8-14 kVA) are recorded during connection, disconnection of the electric consumers.

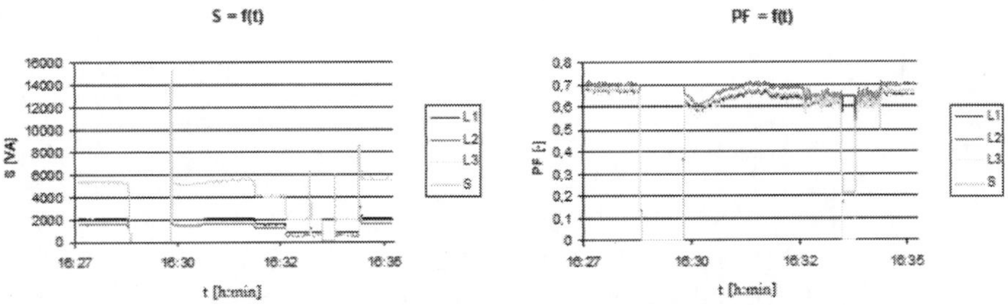

a. b.

Figure 10. a. Apparent powers; b. Power factors

The displacement power factor (Figure 10.b), on each phase, is approximately the same, and has values between 0.58 ÷ 0.72, being lower than the neutral power factor (0.92).

4.2. Measurements made at the plant when it is powered at the main power source and then switched to the backup power source

At these measurements on the ARR, on the main power source, the current was measured only on phase L1 and the voltage between phase L1 and null. On the secondary power supply, the current on phase L2 was measured, and the line voltage between phase L2 and phase L3. The four-conductor method (three phases and null) was used to measure voltages [10-12].

When checking the ARR bidirectional operation, the following steps were taken:

- after 1 minute the air inlet fan was switched on in the basement and ground floor;
- after another 1 minute the smoke exhaust ventilator was switched on in the basement and ground floor;
- after 1 minute, the smoke exhaust ventilator in the workshop was switched on;
- after about 1 minute, when the main power source voltage drops, the ARR switches between main source and back-up source.
- after 1 minute time, the smoke exhaust ventilator in the workshop was turned off;
- after another 1 minute, the air inlet fan was switched off in the basement and ground floor;
- after 1 minute the smoke exhaust fan was switched off in the basement and ground floor;
- after about 1 minute the voltage returns to the main power source;
- after 1 minute the smoke exhaust ventilator has been started in the basement and ground floor;
- after another 1 minute, the smoke exhaust ventilator in the workshop and the air intake fan in the basement and the ground floor were switched on.

Due to single-phase electric consumers, voltage fluctuations (Vrms) occur on L1 and L3, where the recorded values are approximately 240 V (Figure 11.a). Total harmonic distortion for phase voltage (Vthd) is below 3.5% (below the 8% limit) – Figure 11.b.

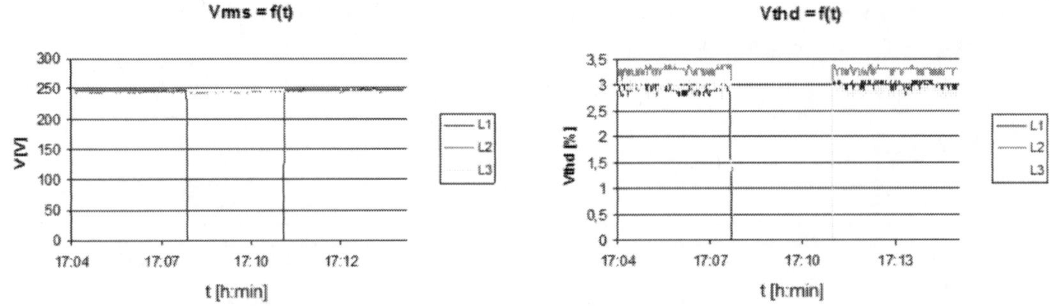

a. b.

Figure 11. a. The RMS value of phase voltages;
b. The total harmonic distortion of phase voltages

The actual measured current value (Figure 12.a) reaches up to 13.5 A when connecting motors acting on the fans of the fire-extinguishing system. The THD for current (Figure 12.b), on the L1 phase, has approximate values of 3.8%, with peaks reaching up to 10.8%.

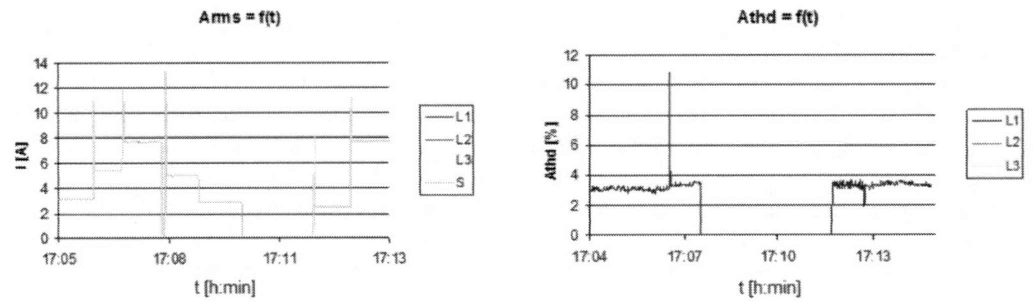

a. b.

Figure 12. a. The RMS value for current; b. The total harmonic distortion for current

Total active power generally has lower values than reactive power (Figures 13.a and b). The motors that operate the fans are oversized. Total reactive power (Figure 13.b) has values up to 2.6 kVAR, being higher than the total active power (Figure 13.a), which records values of maximum 1.9 kW. Total apparent power (Figure 14) records up to 3.2 kVA.

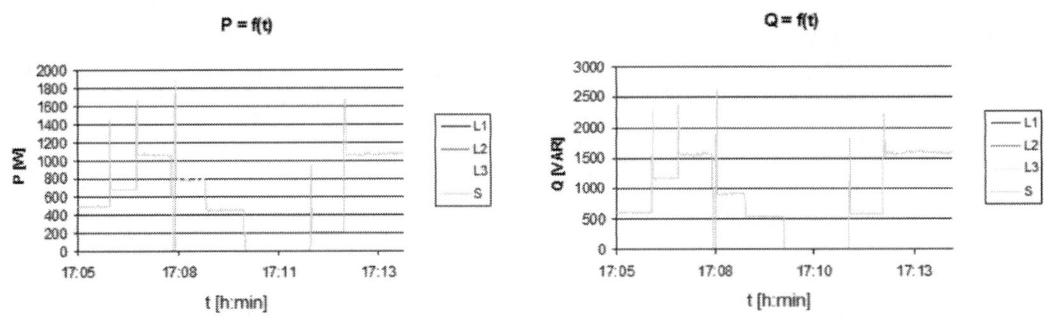

a. b.

Figure 13. a. Measured active power; b. Measured reactive power

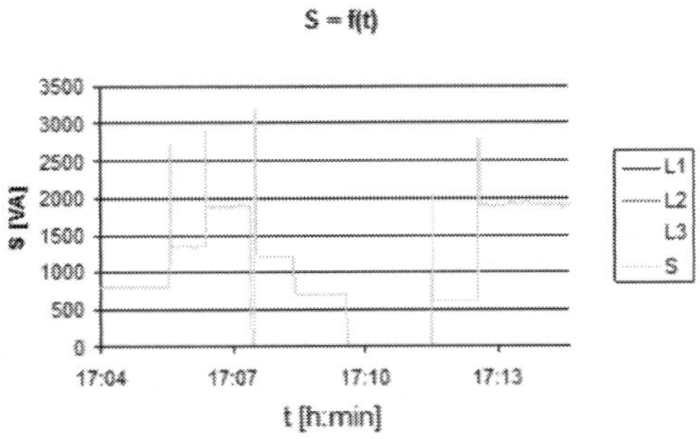

Figure 14. Measured apparent power

The deforming power factor (PF, Figure 15.a) for the L1 and L2 phases has values of 0.67, well below the required value of 0.92 (neutral power factor). As in the case of PF, the displacement power factor (Figure 15.b) has values well below the neutral power factor of 0.92.

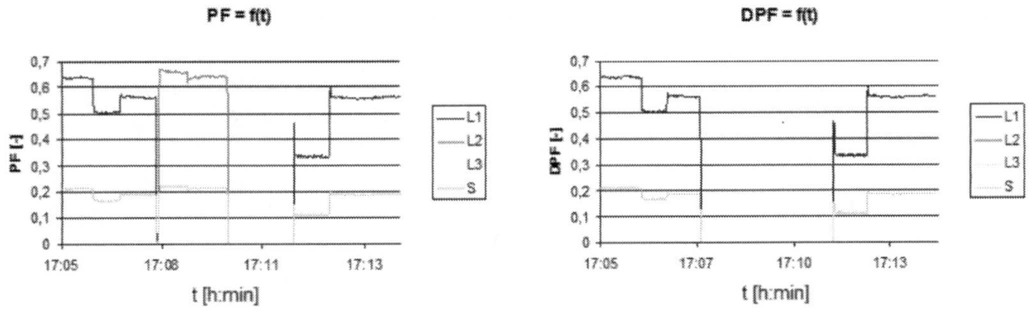

a. b.

Figure 15. a. Measured power factor; b. Measured displacement power factor

In order to improve the power factor, at the power substation, capacitors banks with power factor regulator (more expensive solution) can be used, or capacitors banks directly connected to the motors can be used to maintain the power factor above the neutral power factor.

5. Conclusions

There has been shown a smoke removing system from a textile plant. Fires often causes loss of life and serious damage. In order to prevent fires, measures must be taken in accordance with the present legislation. At industrial plant, an important component in the fight with fires are smoke removing systems. These installations must be powered by bidirectional ARR to ensure high operational reliability.

Some electric loads require electrical power from two (or even three) power sources. The control of power sources can be achieved using classical control installations or PLCs. Removing smoke system can use a third energy source (eg. internal combustion engine-electric generator). Control electrical installations must be regularly checked to prevent malfunctions. Electric drives of fans and electric actuators of valves must be regularly checked.

A three-phase portable power quality analyser is a practical solution for checking and maintaining ARR installations.

References

[1] Warne D F 2005 *Electrical Power Engineer's Handbook*, Newnes, Elsevier Press, USA
[2] Grigsby L L 2012 *The Electric Power Engineering Handbook*, CRC Press, USA
[3] Mayergoyz I D and McAvoy P 2014 *Fundamentals of Electric Power Engineering*, World

Scientific Press, USA

[4] Ceraolo M and Poli D 2014 *Fundamentals of Electric Power Engineering: From Electromagnetics to Power Systems*, Wiley-IEEE Press, USA

[5] Lopes D A R, de Jesus E G and Valee L A F 2004 *Maintaining the Continuity of Process Operation After Voltage Sag or Power Interruption*, Petroleum and Chemical Industry Technical Conference, Fifty-First Annual Conference 2004 IEEE, September 13-15, pp 81-86

[6] Parise L and Lamedica R 2015 *Service Continuity in Hospitals: Overload Risks in Operating Theaters*, 5th International Youth Conference on Energy (2015 IYCE), Pisa, Italy, May 27-30, pp 1-6

[7] Popa I and Popa G N 2000 *Electrical Installations*, vol. I, Mirton Publishing House, Timişoara, Romania, (in Romanian)

[8] Iagăr A, Popa G N and Diniş C M 2017 *Power Quality from Theory to Experimentations*, Politehnica Publishing House, Timişoara, Romania (in Romanian)

[9] Baggini A 2008 *Handbook of Power Quality*, John Wiley & Sons Ltd, USA

[10] Dell'Aquila A, Marinelli M, Giuseppe V and Yanchetta P 2004 New Power-Quality Assessment Criteria for Supply Systems Under Unbalanced and Nonsinusoidal Conditions, *IEEE Transactions on Power Delivery* **19**(3) 1284-1290

[11] Chicco G, Postolache P and Toader C 2007 Analysis of Three-Phase Systems with Neutral Under Distorted and Unbalanced Conditions in the Symmetrical Component-Based Framework, *IEEE Transactions on Power Delivery* **22**(1) 674-683

[12] Rob R, Şora I, Pănoiu C and Pănoiu M 2010 Harmonic Filters Influences Regarding the Power Quality on High Frequency Electrothermal Installation with Electromagnetic Induction, *WSEAS Transactions on Systems* **9**(1) 72-81

IOP Conf. Series: Materials Science and Engineering **294** (2017) 012055 doi:10.1088/1757-899X/294/1/012055

Generate stepper motor linear speed profile in real time

M Y Stoychitch

University of Banja Luka, Faculty of Mechanical Engineering, Vojvode S. Stepanovica 71, 78000 Banja Luka, Republic of Srpska, Bosnia and Herzegovina

E-mail: mihajlo.stojcic@mf.unibl.org

Abstract. In this paper we consider the problem of realization of linear speed profile of stepper motors in real time. We considered the general case when changes of speed in the phases of acceleration and deceleration are different. The new and practical algorithm of the trajectory planning is given. The algorithms of the real time speed control which are suitable for realization to the microcontroller and FPGA circuits are proposed. The practical realization one of these algorithms, using Arduino platform, is given also.

1. Introduction

Stepper motor is an electromechanical device that converts electrical digital pulses into mechanical shaft rotation. Many advantages are achieved using this kinds of motors, as: (*i*) precise positioning and repeatability of movement, (*ii*) the motor has full torque at standstill (if the windings are energized), (*iii*) very reliable and easy maintenance since there are no contact brushes, and (*iv*) a wide range of rotational speeds can be realized since the speed is proportional to frequency of the input pulses. Some disadvantages of these motors are: (*i*) resonance can occur if not controlled properly, and (*ii*) not easy to operate at extremely high speeds, [1], [2].

An important issue about stepper motors is that they are usually used in an open control loop. This means that the motor control system has no feedback information about the position, which eliminates expensive sensing and feedback devices.

Many systems with stepper motors need to control the speed using values of acceleration and deceleration defined in advance. Herein we will analyze the general case, when change of speed in the phase acceleration and deceleration is linear and different (ramp speed profile). In Figure 1 the relationships between acceleration $a\,[\text{rad/sec}^2]$, deceleration $d\,[\text{rad/sec}^2]$, speed $v\,[\text{rad/sec}]$ and position $s\,[\text{rad}]$ are shown. On this figure with t_a, t_v, t_d and T (all in [sec]) are labeled: the time of acceleration, time of motion of constant speed, time of deceleration and total time, respectively. Also, with N_a, N_v, N_d and N (all in [step]) are denoted number of steps on the mentioned time intervals, see Figure 1.

Since the stepper motor makes steps in discrete time (after each pulse) and the move of every step is constant, the change of speed is achieved by changing the time interval between successive steps

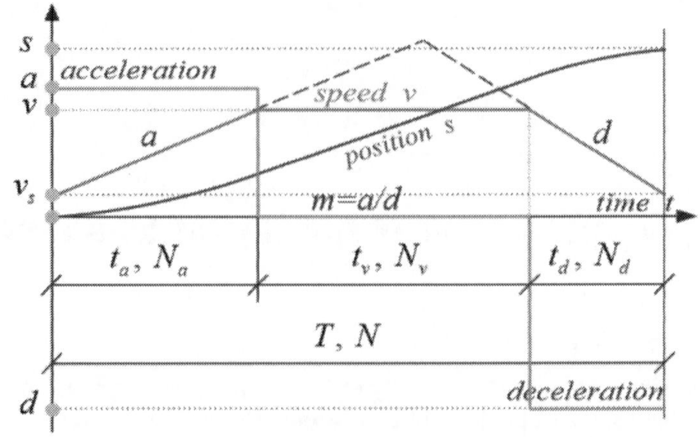

Figure 1. Change of speed, position and acceleration

(pulses). It means that the main problem of speed control is to determine instants of the time t_i (in [sec] when pulses (steps) are generated. If speed v [rad/sec] is constant $v = const.$ (independent of whether it is large or small), it is very easy to determine these instants (or equivalently, generate the pulses). In this case the time delays δt_i, between two arbitrary adjacent pulses, are the same and they are given as, $\delta t_i = \alpha / v = const.$, where α [rad] is the angle of the rotation motor shaft for every step. But, if the speed is variable, $v_i \neq const.$, it is more difficult to determine the instants when we need to generate pulses because the time delay between two successive pulses is changed, $\delta t_i = \alpha / v_i \neq const.$. In the case when acceleration/deceleration are constant, the speed is changed linearly, but the time interval δt_i between two adjacent pulses is not linear. So, we need to determine the time t_i, $i = 0,1,2,\ldots$, when pulses are generated and that ensure linear change of speed. It is shown for the acceleration phases in Figure 2.

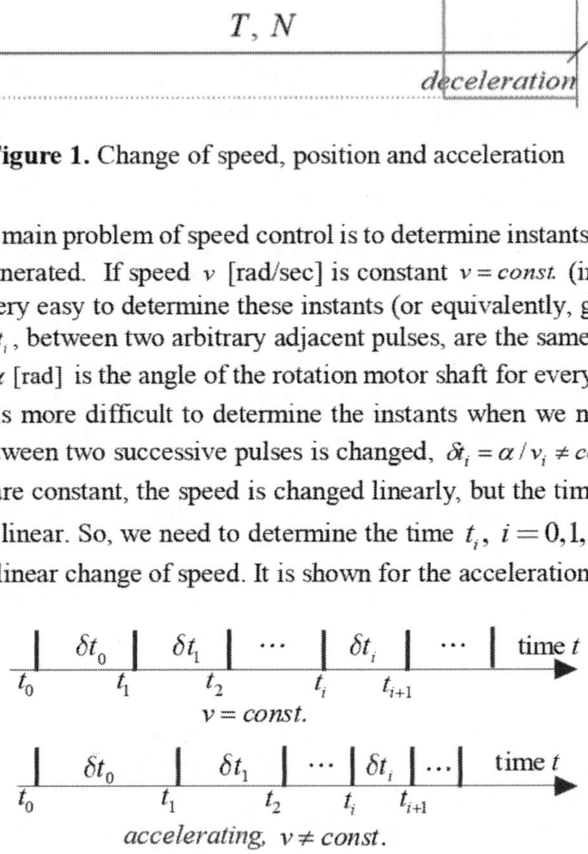

Figure 2. Constant speed and accelerating step sequences

Generally, the problem of generating appropriate speed profile to a stepper motor we can split in the two main tasks: 1) to determine a shape the speed change and 2) generate impulses in appropriate instant of time in order to achieve this speed.

And, at the end of this section, a few words about the structure of the paper. The second section is related to the trajectory planning (solving of the first task), where based on some input data about the motion and properties of the selected stepper motor we calculate characteristic points of a movement, i.e. determining a speed profile. In the third section are given two algorithms based on which to generate pulses (solving of the second task). One of these algorithms is appropriate for realization using a microcontroller and another using FPGA circuits. In the fourth section we will propose a way of realization one of these algorithms. And finally, in the fifth section, are given some conclusions, and in the appendix is given an Arduino program of practical realization one of algorithms which is proposed in the previous sections.

2. Trajectory planning

Since the change of speeds are linear, then trajectory planning means that need to determine the number of steps during which the following phases are realized: acceleration N_a, motion of constant speed N_v, deceleration N_d as well as the number of steps of the total movement N. In doing so we assume that the total movement s and the ratio $m = a/d$ between acceleration a and deceleration d are given. Usually, by the stepper motor manufacturer are specified the next parameters: number of steps per round K [step], starting speed (the stepper motor can go from 0 to this speed in no time) v_s [rad/sec], maximum speed v_M [rad/sec] and the maximum acceleration[1] a_M [rad/sec^2].

Now, our problem can be described as: we need to move a stepper motor for $N(=[s/\alpha])$ steps (here the symbol $[\circ]$ denotes the nearest integer), $\alpha = 2\pi / K$ [rad], so the start speed not higher then v_s, the acceleration a not higher then a_M, during the motion the speed v must never exceed the maximum speed v_M and the ratio acceleration/deceleration is m.

At the beginning of trajectory planning we assume that the total movement s can be achieved with a triangular speed profile, which include only two phases: acceleration and deceleration (we also assume that the acceleration a during the acceleration phase is maximum, $a = a_M$). In this case the movement s is (see Figure 3, left)

$$s = s_o + s_a + s_d = v_s T + \frac{1}{2} a t_a^2 + a t_a t_d - \frac{1}{2} d t_d^2 \tag{1}$$

where $s_o = v_s T$, $s_a = \frac{1}{2} a t_a^2$ and $s_d = a t_a t_d - \frac{1}{2} d t_d^2$ are the motions: due to the start speed v_s during the time T, due to acceleration a on the time t_a and due to deceleration d on the time t_d (we assume that the deceleration d not limited, but the time $t_d = T - t_a > 0$), respectively. Since the speed at the end of the acceleration phase is the same as the speed at the beginning of the deceleration phase, then the next equation

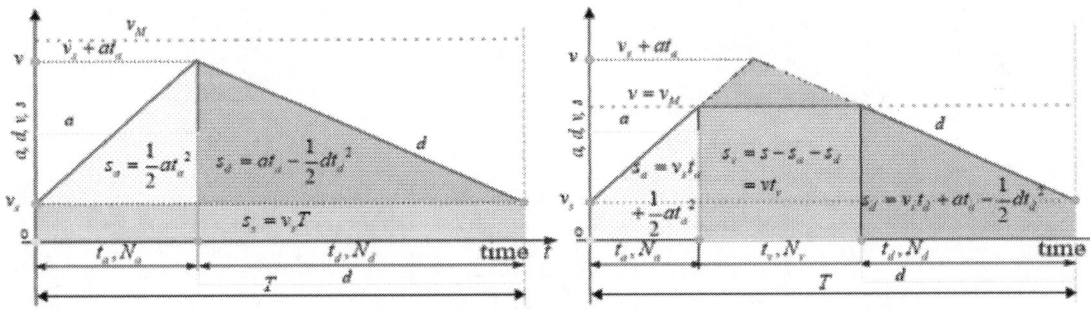

Figure 3. Triangular (left) and trapezoidal (right) speed profile

$$a t_a = d t_d = d(T - t_a) = \frac{a}{m}(T - t_a) \Rightarrow t_a = \frac{T}{1+m} \tag{2}$$

is valid. From the equations (1) and (2) it is obtained

$$2(s - v_s T) = \frac{aT^2}{1+m} \Rightarrow T^2 + \frac{2v_s(1+m)}{a} T - \frac{2s(1+m)}{a} = 0, \tag{3}$$

from where we get the solving of the above square equation per T, as

[1] If the manufacturer is specified the speed in [step/sec] and acceleration in [step/sec^2] then relation between them are: 1[step/sec] $= 2\pi / K$ [rad/sec] and 1[step/sec^2] $= 2\pi / K$ [rad/sec^2]

$$T = -\frac{v_s\ 1+m}{a} + \sqrt{\left(\frac{v_s\ 1+m}{a}\right)^2 + \frac{2s\ 1+m}{a}}. \tag{4}$$

Using this T we calculate the time of the acceleration phase $t_a = T /\ 1+m$ and checking that the speed $v = v_s + at_a$ at the end of the acceleration phase is greater than the maximum allowed speed v_M.

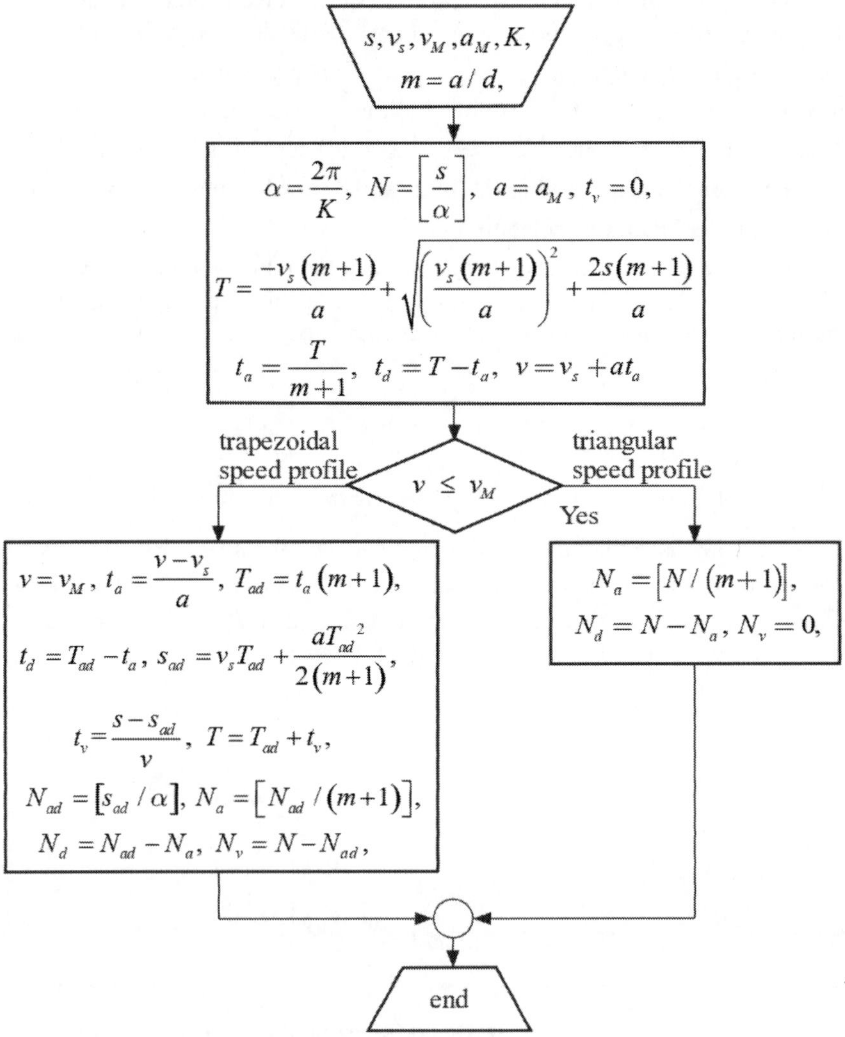

Figure 4. Flow chart of trajectory planning

If it is $v \le v_M$, then the initial assumptions is true, so that the total movement can be realized with a triangular speed profile. In this case the number of steps in the acceleration phase N_a and in the deceleration phase N_d are calculated as

$$N_a = \left[N /\ 1+m\right], N_d = N - N_a. \tag{5}$$

The flow chart of this algorithm of the trajectory planning is shown on Figure 4. However, when $v \le v_M$ is not true, then is the total movement s can be realized with trapezoidal speed profile (see Figure 3, right). In this case, we additionally need to include a third phase, the phase of moving with constant and maximum speed, $v = v_M$. Now, based the speeds v_s and v and the accelerating a we calculate the time of the acceleration phase $t_a = (v - v_s)/a$ and the total the time $T_{ad} = t_a(1+m)$ of the both phases, of acceleration and deceleration, together. Using this T_{ad} we calculate the total movement s_{ad} in the phases acceleration and deceleration (see (3)) , as

$$s_{ad} = v_s T_{ad} + \frac{a T_{ad}^{\ 2}}{2(1+m)}, \qquad (6)$$

so that the movement s_v in the phase of constant speed is $s_v = s - s_{ad}$. The time t_v during the phase of constant speed is realized and the total time T of all phases are given as

$$t_v = \frac{s - s_{ad}}{v}, \quad T = T_{ad} + t_v. \qquad (7)$$

Now, in the case of trapezoidal speed profile, using of the known movements s, s_{ad} and s_v and the step α, we can calculate the number of steps: N_a, N_d and N_v in the each phases of the motion, as:

$$N_{ad} = s_{ad}/\alpha, \; N_a = \left[N_{ad}/(1+m)\right], \; N_d = N_{ad} - N_a, \; N_v = N - N_{ad}. \qquad (8)$$

The complete algorithm of the trajectory planning for the general case of the trapezoidal speed profile is shown on Figure 4.

3. Generate of the pulses - Algorithms

There are two methods – two group of algorithms for calculating instants of time when pulses must be generate. They are named as: *"time per step"* and *"steps per time"*, that are described in [3-5], [8], [9] and partly in [6], [7], respectively. Operation mode of the first group of algorithms can be briefly described as: (*i*) calculate time period to next pulse, (*ii*) wait until that much time period elapses, (*iii*) generate next pulse and (*iv*) go back to step (*i*) and repeat until desired number of pulses is over. The another group of algorithms have an operation mode that is significantly differently, and briefly can be described as: (*i*) check the current time, (*ii*) multiply it by speed, to get expected current position, (*iii*) if difference from expected and actual current position greater than or equal one step, then to generate new pulse and (*iv*) repeat this until the final desired position is achieved.

Since in these algorithms all calculations are realized in real time (trajectory planning is not in real time), i.e. between two steps (pulses), then the basic goal is to find faster algorithms, because this allows the control of higher speed motors. Also, the difference in the speed generated through these algorithms and theoretical speeds for the given profile must be as small as possible.

The first group of algorithm is suitable for controllers that are implemented by the microcontroller, while the second group is suitable for the controllers that are implemented via FPGA circuits. In this paper are described the both groups of algorithms. From the first group is given the algorithm that is some modification of the algorithm that is proposed in [8], [9], while from the second group is proposed an algorithm that partly described in [6], [7].

3.1. Calculate times when the pulses are generated using time per steps algorithm

The first pulse (step) controller generates at the start of motion, at the start of the phase of acceleration, i.e. at the time t_0, see Figure 5. After the first pulse is generated, the controller needs to calculate the time period δt_0 until the next pulse, wait until this period has elapsed, and then generate the next pulse, at time t_1. This will go on until the desired position is achieved, or in other words, the desired number of pulses has been generated. At the start, the speed is $v_o = v_s$, and it retains its value until the moment t_1 when it becomes v_1, at the moment t_2 becomes v_2, and so on. Since after each pulse the motor makes one step for the angle α, so that is

$$\alpha = v_i \cdot \delta t_i \Rightarrow \delta t_i = \frac{\alpha}{v_i}, \qquad (9)$$

where the v_i is speed at an arbitrary instant of time t_i and δt_i the time delay between two successive instants of time, t_i and t_{i+1}. In the case when the controller is realized by using a microcontroller, the

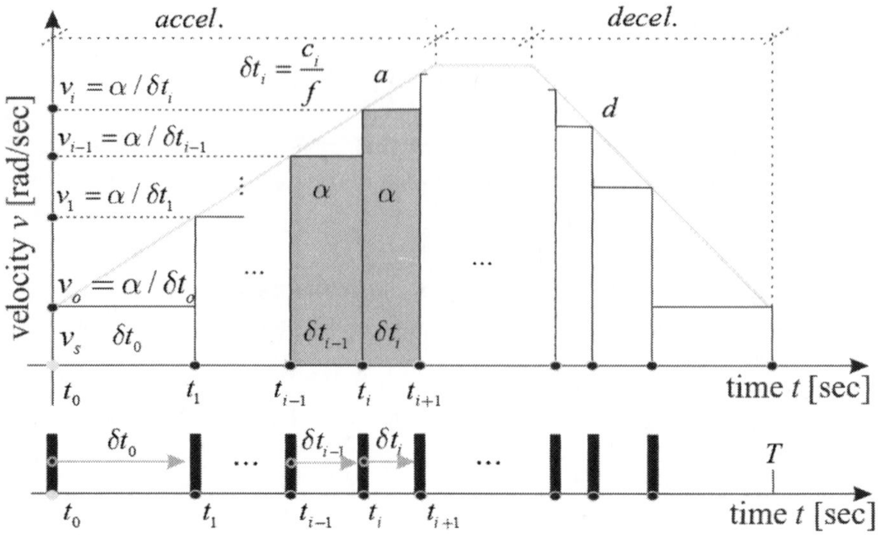

Figure 5. Calculate the time when pulses are generated

required time delay δt_i is implemented using counter c_i that counts impulses of known frequency f, so it is $\delta t_i = c_i / f$. Based on the above considerations, the speed is changed only at the discrete time t_i. But, since inertia always exists, thus we can assume that the speed v_i between two arbitrary successive instants of time t_{i-1} and t_i, $i \geq 1$, changes linearly (see bright lines in Figure 5.). Thus, the speed v_i at the arbitrary instant of time t_i, and in the phase of acceleration, becomes

$$v_i = v_{i-1} + a \cdot \delta t_{i-1},\ i \geq 1. \tag{10}$$

Using above equations, the value of the counter c_i becomes

$$c_i = \frac{\alpha f}{v_i} = \frac{\alpha f}{v_{i-1} + a \cdot \delta t_{i-1}} = \frac{\alpha f}{\frac{\alpha f}{c_{i-1}} + a\frac{c_{i-1}}{f}} \Rightarrow c_i = \frac{c_{i-1}}{1 + R_a c_{i-1}{}^2},\ R_a = \frac{a}{\alpha f^2}. \tag{11}$$

In a similar way, in the phase of deceleration, we obtain

$$c_i = \frac{c_{i-1}}{1 + R_d c_{i-1}{}^2},\ R_d = -\frac{d}{\alpha f^2}. \tag{12}$$

From (11) and (12) we can see that the time delay c_i (or equivalently δt_i) is calculated based on the previous time delay $c_{i-1}, i \geq 1$, which is already known and what is very suitable for realization. Further, it implies that it is necessary to determine the initial time delay c_o and the time delay during of the phase of constant speed (if exist) c_m. These time delays are obtained based on the start speed v_s and the maximum speed v_M, respectively, as

$$c_o = \frac{\alpha f}{v_s} \text{ and } c_m = \frac{\alpha f}{v_M}. \tag{13}$$

After detailed analysis and simulation of the above proposed algorithm, in [9] was shown that the speed profile that generated by stepper motors more exact (closer to the theoretical) if in the first and last five steps are introduced some corrections. These corrections are realized by using the next equations (i – the current number of steps)

$$c_i = c_i\left(1+\frac{0.08}{i}\right),\ 1\le i\le 5 \ \text{ and } \ c_i = c_i\left(1+\frac{0.08}{N-i}\right),\ 1\le N-i\le 5. \tag{14}$$

where are i the current number and N the total number of steps. For more details see paper [9]. On the Figure 6 is given flow chart of this algorithm (without the corrections that are given by (14)).

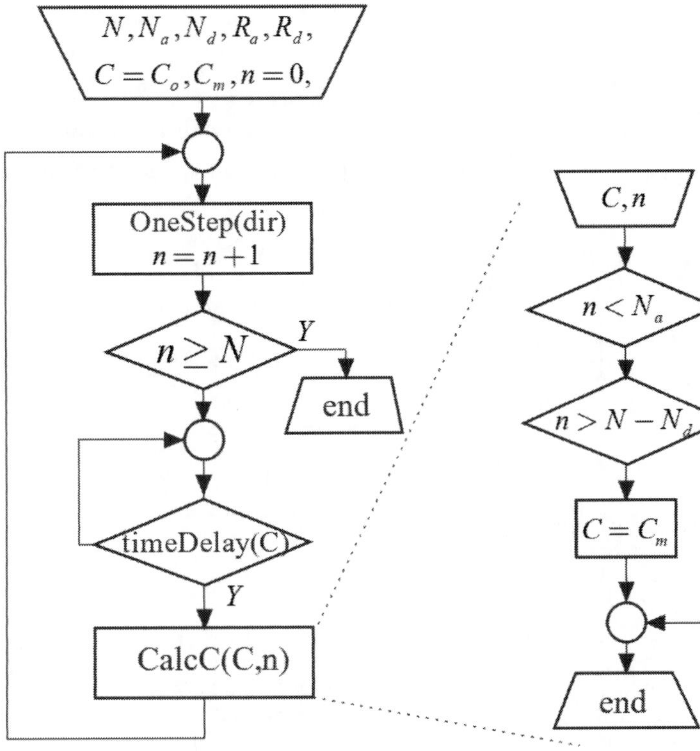

Figure 6. Flow chart of the algorithm from section 3.1

3.2. Calculate times when the pulses are generated using steps per time algorithm

Calculating the time delay between two pulses using algorithm that is given by equations (11) and (12) require two multiplication, one addition, and one division (all with float number of double precision) for each step. This algorithm it is not suitable for implementation by using FPGA circuits, because multiplication and division with them is difficult or very slow. Unlike them, the addition and subtraction with this circuits is realized very easily and very fast. Therefore, for FPGA-based controllers, it is necessary to find another algorithms, in which for all calculations in real time, are used only addition and subtraction operations.

It is known from theory that between the acceleration $a(t)$, velocity $v(t)$ and position $p(t)$, at some time interval $[0,T]$, the following relations apply:

$$v(t)=\int_0^t a(\tau)d\tau \ \text{ and } \ p(t)=\int_0^t v(\tau)d\tau,\ t\in[0,T]. \tag{15}$$

Above equations are given in a continuous domain. In a discrete domain with a constant time period T_s, so it is $T = MT_s$, where T_s is very small real number and M is integer, these equations at discrete time interval $t_k = kT_s$, $k = 0,1,2,\dots,m$, $m \le M$ become:

$$v(kT_s)=v(k)=\sum_{k=0}^{m}a(k)T_s \text{ and } p(kT_s)=p(k)=\sum_{k=0}^{m}v(k)T_s. \tag{16}$$

According equation (16), if it is dimension of the acceleration a in $\left[\text{steps}/T_s^2\right]$, then dimension of the velocity v is in $\left[\text{steps}/T_s\right]$ and dimension of the position p is in $[\text{steps}]$ (for example: if it is T_s in $[\text{ms}]$ - milliseconds, then a in $\left[\text{steps}/\text{ms}^2\right]$, v in $[\text{steps}/\text{ms}]$ and p in $[\text{steps}]$).

From (16) it can be seen that for calculating, for example $p(k)=\sum_{k=0}^{m}v(k)T_s$, necessary addition and multiplication with T_s. But if this equation it is solved at every T_s, then multiplication with T_s is not necessary, since at every instant of time T_s the new part of position $v(k)\cdot 1T_s=v(k)\cdot 1$ is added to the existing value of position.

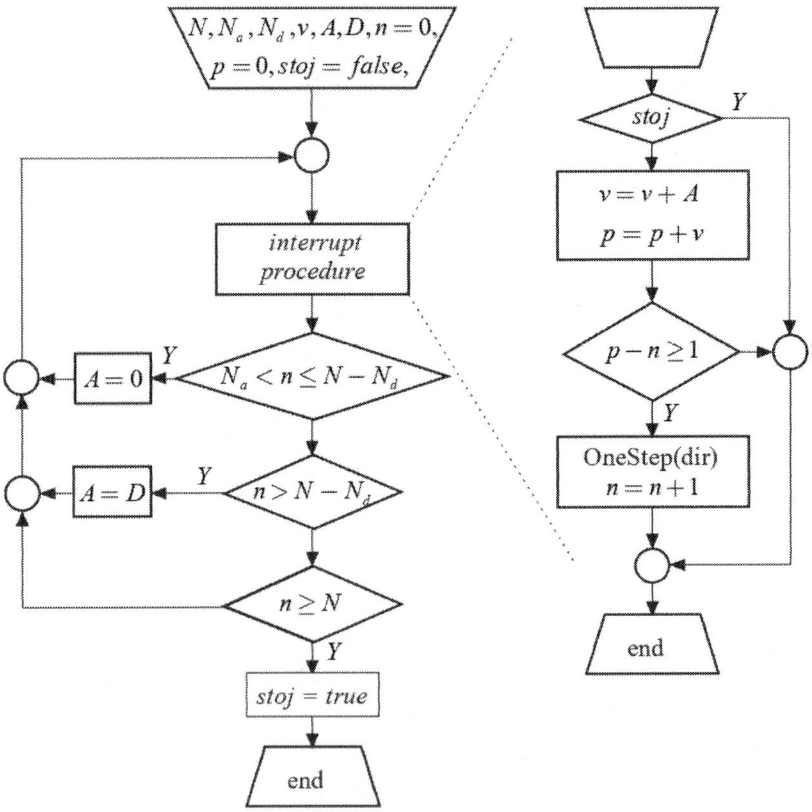

Figure 7. Flow chart of the algorithm from section 3.2

And finally, let the dimensions are: of acceleration a $\left[\text{steps}/T_s^2\right]$, of speed v $\left[\text{steps}/T_s\right]$ and of positions p $[\text{steps}]$, and let at the initial time $t_0=0=kT_s, k=0$, their values are: $acceleration=A$, $speed=V_s$, $position=0$ and number of steps $steps=0$, then by solving the next equations:

$$\begin{aligned} speed &= speed + acceleration \\ position &= position + speed \end{aligned} \tag{17}$$

at every instant of time T_s, it is equivalent to the solution of the equations (16). The value of *position* that is given from (17) it is the same to the expected current position, and value of the *steps* is the same to the actual current position. The instant of time when a new pulse (step) is generated now is determined from the next requirement: if (*position-steps* >= 1) then the new pulse is generated and the new actual current position *steps* becomes *steps=steps+1*. Flow chart of this algorithm is shown in Figure 7. In this flow chart the next label: *p – position*, *n – steps*, *v- speed* and *A-acceleration* are used.

The implementation of the procedure (17) (i.e., the solution of the equation (16)) is easiest to accomplish using an interrupt, where interrupt occurs by every T_s instant of time. The value of the time period T_s depends of the motion parameter (K, a_M, v_M) and of type and speed of a processor, which is the problem of special analysis that is not the subject of this paper.

4. Implementation of the steps per time algorithm

Based on the consideration in the sections II and III, the algorithm that is proposed in section 3.2 (step per time algorithm) is implemented by using the microcontroller ATMega328. In this implementation as a hardware we use the 28BYJ-48 stepper motor (power supply 5V-DC, 4096 steps per round, because the motor has 64 steps and a gear unit with 64:1 ratio, so that 64*64=4096), the driver based on the ULN 2003 circuit and the Arduino UNO as a platform of the ATMega328 microcontroller, see Figure 8.

Figure 8. Devices for the experiment: Arduino, driver and stepper motor

Arduino program through which, based on the entered data, the planning of the trajectory and generation of the pulses using the steps per time algorithm (section 3.2) is given in the appendix. In this program we use the interrupt frequency F = 10 000 Hz, so that the time period $T_s = 1/F = 0.1$ ms. Since the inputs data for acceleration, speed and position the next units of measure: [rad/sec^2], [rad/sec] and [deg] are used, respectively, then it must be translated into new units: [steps/T_s^2], [steps/T_s] and [steps]. This conversion are achieved using the following relations: acceleration [rad/sec^2] $\times K / 2\pi / F^2$ = acceleration[steps/T_s^2], speed[rad/sec] $\times K / 2\pi / F$ = speed[steps/T_s] and position [deg] $\times K / 360$ = position [steps].

The given program enables the input and change of all parameters of movement, as well as their displaying. After the reset, the initial values are loaded, and the current position of the stepper motor shaft is considered as a zero position. After each next entry of the new position, the program determines the direction and value of the angle from the current to the new position and plans the trajectory between these two positions.

5. Conclusion

In this paper, problem of the control of a stepper motor with linear change of the speed is considered. This problem involves solving two tasks. The first task (that is solved in section 2) is the trajectory planning so that all requirements relating to the parameters of motor and motion are fulfilled, while the second task is the real-time control of the motor motion, according to the requirements from the first task (this task is solved in section 3). The general case is analyzed, when the absolute values of the acceleration in the phases of acceleration and deceleration are different. Two completely different algorithms related to real-time control of motors are proposed. One of these algorithms is adapted for controllers based on FPGA circuits, while both algorithms are suitable for controllers that are based on a microcontroller.

All the theoretical considerations in this paper were simulated by experiment. The results of the experiment confirm the proposed theoretical considerations and show that are all proposed algorithms very fast and simple.

Appendix

Arduino program

```
#define K   4096   //number of steps per one revolution
#define F   10000  //interrupt frequency  10kHz
//
float s = TWO_PI,vs = 0.5,vM = 2,aM = 2.0, m = 2,T,ta,tv,td,ss,aa = aM,vv = vM;//a[rad/sec^2],v[rad/sec],s[rad]
double alfa = TWO_PI/K, Af = aM/alfa/F/F, Vsf = vs/alfa/F, Df = -Af/m, //Af[steps/Ts^2],Vsf[steps/Ts]
     brzina = Vsf, pozicija = 0; //
//cN-current number of steps, dN-desired number of steps
long N, Na, Nv, Nd, broj, cN, np = -1;
byte cw[] = {0b0001, //A,
             0b0011, //A,B
             0b0010, //B,
             0b0110, //B,C
             0b0100, //C,
             0b1100, //C,D
             0b1000, //D,
             0b1001}; //D,A
byte dcw = sizeof(cw), kk = 0, maska = 0xF0, ledPin = 13;
char ch, chh;
boolean smjer = true, stoj = true;
//
void setup() {
  Serial.begin(9600);
  Serial.print("CONTROL of STEPPER MOTOR\n");
  Serial.print("Commands: Pxxxx, desired position s[deg]X10 \n");
  Serial.print("       Bxxx, start speed vs [rad/sec]X10 \n");
  Serial.print("       Vxxx, max. speed  vM [rad/sec]X10 \n");
  Serial.print("       Axxx, max. acce  aM [rad/sec2]X10 \n");
  Serial.print("       Rxxx, the ratio m*10 of acce/dece \n");
  Serial.print("       S,-stop, M,- move, W-write paramet. \n");
  //
  DDRB  = 0b00001111;  //pins 8,9,10,11 are outputs
  PORTB = PORTB & 0xF0; //all outputs are zero
  pinMode(ledPin,OUTPUT); //led pin, this pin start and stop is labeled
  //settings interrupt, every 1/10[ms] the interrupt has occurred
```

```
  cli();              // disable global interrupts
  TCCR1A = 0;          // set entire TCCR1A register to 0
  TCCR1B = 0;          // same for TCCR1B
  TCNT1  = 0;          // initialize counter value to 0
  // dF-desired interrupt feq., pS- prescaler, cR-compare match register
  // cR=[16*10^6/(pS*dF)]-1
  // if dF=10^4, pS=1(no prescaler) => cR=[16*10^6/(1*10^4)]-1=1599
  // set compare match register to desired timer count:
  OCR1A = 1599;
  TCCR1B |= (1 << WGM12); // turn on CTC mode:
  TCCR1B |= (1 << CS10);  // Set CS10 bit for no prescaler:
  TIMSK1 |= (1 << OCIE1A); // enable timer compare interrupt:
  sei();               // enable global interrupts
}
// interrupt procedure, every 1/10000 [sec]=1/10[ms] program calls this procedure
ISR(TIMER1_COMPA_vect){
  //
  if (stoj) return;
  brzina  = brzina + Af;
  pozicija = pozicija + brzina;
  if (pozicija - np >= 1) {
    oneStep(smjer);
    np++;
  }
}
//motor makes one step, if right = 1, then direction = CW
void oneStep(boolean right){
  if (right) { PORTB = (PORTB & 0xF0) | cw[kk++]      ; cN++; }
  else      { PORTB = (PORTB & 0xF0) | cw[dcw-1-kk++]; cN--; }
  if (kk == dcw) kk = 0;
}
//
void loop() {
  if (np >= Na && np < N-Nd) Af = 0; //speed is constant
  if (np >= N-Nd)          Af = Df; //begin of deceleration
  if (np >= N) {
    stoj = true;
    Serial.print("...........................................\n");
    Serial.print("STEPS    = ");  Serial.print(cN); Serial.print(" [steps]\n");
    Serial.print("ANGLE    = ");  Serial.print(cN*360.0/K,2);Serial.print(" [deg]\n");
    Serial.print("POSITION = ");  Serial.print(float(cN*alfa),2);Serial.print(" [rad]\n");
    np = -1; pozicija = 0; brzina = Vsf;
    digitalWrite(ledPin,LOW);
  }
}
//this procedure enables the entry of new data
void serialEvent(){
  if (Serial.available() > 0){
    ch = toupper(Serial.read());
    if (ch == 'P' || ch == 'B' || ch == 'V' || ch == 'A'  || ch == 'W'
             || ch == 'R' || ch == 'S' || ch == 'M') chh = ch;
    if (ch >= '0' && ch <= '9') broj = broj*10 + int(ch - '0');
    //analiza komandi
```

```
    if (ch == ','){
      if (chh == 'P') {
        s    = broj/10.0;      // s[deg]
        stoj = planTrajek(s);
      }
      if (chh == 'B')  vs = broj/10.0;
      if (chh == 'V')  vM = broj/10.0;
      if (chh == 'A')  aM = broj/10.0;
      if (chh == 'R')  m  = broj/10.0;
      if (chh == 'S')  stoj = true;
      if (chh == 'M')  stoj = false;
      if (chh == 'W')  pisiSve();
      broj = 0;
    }
  }
}
//this procedure displaying current data
void pisiSve(){
 Serial.print("\n..............................................\n");
 Serial.print("Displacement   s = ");Serial.print(s,2); Serial.print("[deg]\n");
 Serial.print("Start. speed  vs = ");Serial.print(vs,2);Serial.print("[rad/sec]\n");
 Serial.print("Max.  speed   vM = ");Serial.print(vM,2);Serial.print("[rad/sec]\n");
 Serial.print("Max. accler.  aM = ");Serial.print(aM,2);Serial.print("[rad/sec^2]\n");
 Serial.print("Ratio of a/d   m = ");Serial.print(m,2); Serial.print("\n");
 Serial.print("Stop = ");Serial.print((stoj)? 1:0);
 Serial.print(", Move = ");Serial.print((stoj)? 0:1);Serial.print("\n");
 Serial.print("..............................................\n");
}
//trajectory planning procedure
boolean planTrajek(float _s){
 Serial.print("\nDATA: newP = ");Serial.print(_s,2); Serial.print("[deg], ");
 Serial.print(", curP = "); Serial.print(float(cN*360.0/K)); Serial.print("[deg]");
 //
 N = int(_s*K/360.0) - cN ; //difference from desired and current position [steps]
 if (N < 0)  smjer = false;
 else        smjer = true;
 if (N == 0) {np = -1; return true;} //
 //
 N = abs(N); float s_ = N*alfa;
 T  = sqrt( pow( (vs*(1+m)/aa),2)+2*s_*(1+m)/aa  )-vs*(1+m)/aa; //N = round(s/alfa);
 ta = T/(1+m); vv = vs + aa*ta; //td = (T-ta); tv = 0;
 if (vv <= vM){//TRIANGULAR PROFILE
   Na = int(N/(1+m)); Nd = N - Na; Nv = 0;
 }else{      //TRAPEZOIDAL PROFILE
   vv = vM; ta = (vM-vs)/aa; T = (1+m)*ta; ss = vs*T+aa*pow(T,2)/(2*(1+m));
   //ss - is tha part of movement in the phases of acceleration and deceleration only
   //np - number of steps on the movement ss
   np = int(ss/alfa); T = (1+m)*ta+(s_-ss)/vv; //tv = (s_-ss)/vv; td = T - ta - tv;
   Na = int(np/(m+1)); Nd = np - Na; Nv = N - np; //Nv - number of steps in the phase v=const.
 }
 //Vsf [steps/0.1ms] - the start speed, Af[steps/(0.1ms)^2]-acceleration, Df-deceleration
 Vsf = vs/alfa/F; Af = aa/alfa/F/F; Df = -Af/m;
 brzina = Vsf; pozicija = 0;  np = 0;
```

```
 Serial.print(", dN = "); Serial.print(N); Serial.print(", Direction = ");
 Serial.print((smjer)?"CW.\n":"CCW.\n");
 digitalWrite(ledPin, HIGH);
 return false;
 //
}
```

References

[1] *** Industrial Circuits Application Note, *Stepper Motor Basis*, http://solarbotics.net/library/pdflib/pdf/motorbas.pdf

[2] Condit R, Douglas W Jones, *Stepping Motors Fundamentals*, Microchip AN907, http://homepage.cs.uio- wa.edu/~jones /step/ an907a.pdf

[3] Austin D 2005 *Generate stepper motor speed profiles in real time,* Embedded Systems Programming, www.embedd ed.com/56800129

[4] *** Atmel Corporation, *AVR446: Linear speed control of stepper motor*, Application note, http://fab.cba.mit.edu/ classes/ MIT/961.09/prjects/i0/doc8017.pdf

[5] Eiderman A, *Real Time Stepper Motor Linear Ramping Just by Addition and Multiplication,* http://hwml.com/ LeibRamp.pdf

[6] Ranade P 2009 *Linear Motor Control Witdout the Math,* SPJ Embedded Technologies, http://www.eetimes.com/document.asp?doc_id=1276928

[7] *** *Stepping Motion Profiles in Realtime*, http://picprog.strongedge.net/step_prof/step-profile.html

[8] Stoychitch M Y 2012 *Linear Speed Control of Stepper Motor in Real Time,* XI International SAUM Conference on Systems Automatic Control and Measurements, Niš, Serbia, November 14th-16th, pp 406-409, http://www.ni.ac.rs/images/stories/events/SAUM_2012_abstracts.pdf

[9] Stoychitch M Y 2013 An Algorithm of Linear Speed Control of a Stepper Motor in Real Time, *Annals of Faculty Engineering Hunedora, International Journal Engineering* **XI** (3) 51-56, http://annals.fih.upt.ro/pdf-full/2013/ANNALS-2013-3-06.pdf

IOP Conf. Series: Materials Science and Engineering **294** (2017) 012056 doi:10.1088/1757-899X/294/1/012056

The interbelic Germans from the Banat Highland. Coal, steel, mines and forges

C Rudolf[1], D M Micliuc[2], M D Nedeloni[3], E Birtarescu[4] and A Varga[5]

[1]"Eftimie Murgu" University of Resita, Department of Social Sciences, 11Traian Vuia Square, No. 1-4, 320085, Resita, Romania
[2]"Eftimie Murgu" University of Resita, Department of Doctoral Studies, Traian Vuia Square, No. 1-4, 320085, Resita, Romania
[3]"Eftimie Murgu" University of Resita, Department of Electric Engineering and Informatics, Traian Vuia Square, No. 1-4, 320085, Resita, Romania
[4]"Eftimie Murgu" University of Resita, Department of Mechanical and Materials Engineering, Traian Vuia Square, No. 1-4, 320085, Resita, Romania
[5]"George Barițiu" Institute of History Cluj- Napoca, Str. M. Kogalniceanu, No. 12-14, Cluj- Napoca, Romania

E-mail: c.rudolf@uem.ro

Abstract. The difficulties of the reconstruction era, following World War I had been increased by the cessation of some activities in the industrial centres of the Banat Highland. For instance, the copper mines were closed in 1921, the Romanian state forbidding the extraction of this ore. Only in Ocna de Fier a special dispensation had been given. The copper mines from Moldova Nouă, Sasca Montană, Ciclova, Dognecea had also been shut down. This fact caused the acid reaction of some writers. We recall that one of the main ways for improving the material condition, embraced by the ethnic Germans, was working abroad. Many German workers of the Banat Highland had emigrated, taking up an offer of well-paid work during the crisis years: 1929-1933. The miners of the Banat Highland, especially those of German origin, travelled to the areas rich in iron ore and coal of France, namely Alsace and Loraine. Considering that German was spoken there by a significant percentage of the population, the integration into the new working environment did not represent a problem.

1. Introduction

The industrial activity in the Inter-War Banat Highland was concentrated around the urban and semi-urban centres of Resita, Anina, Bocsa, Moldova Noua and Otelu Rosu. The historian Rudolf Gräf argues in his books dedicated to S.T.G., not unjustly, that this firm, as well as its successor, U.D.R. Were the main employer and profit generator for the mountain area of that historic province. Both the secondary sources, especially the primary ones, confirm this, proving the economic and socio-cultural importance that this company had for the inhabitants of the micro-region, regardless of their ethnic, religious or linguistic affiliation. In the following we will present some information about the evolution of U.D.R. in the interwar decades [1].

The company had been named U.D.R., i.e.: the Iron Factories and Resita, and was founded by a government decision on June 8, 1920. Two days later, the first general meeting had taken place. According to the financial statement for the year 1919 the company was worth 228 million crowns,

having a debt of 85 million crowns, of which 40 million at the Boden-Kreditanstalt bank in Vienna. The remaining capital was valued at 125 million lei and divided into 250,000 shares. The first distribution of shares, 60% of the total, worth RON 74.2 million, was made for the benefit of the founding members: Prince Constantin Caragea, Victor Bontescu, C.I.C. Bratianu, Mihail Pherekide, I.G. Duca, Manole Diamandi, Ion Boamba and Adalbert Veith. The other shares, amounting to 50.8 million lei, were sold to various banks. Some of them formed the Business Finance Consortium: Marmorosch Blank and Co., Chrisoveloni, Romanian Commercial Bank, Bank of Romania, Banca Generală a Ţării Româneşti, Albina, Temeswarer Bank and Temeswarer Handels A.G. [2]
At international level, the following companies held shares in the UDR: Syndikats-Boden Kreditanstalt Wien, Vickers-Armstrong London, Continental Trust for Bank and Industrial Values Basel. After the general meeting of the shareholders of U.D.R. On February 8, 1923, the Board of Control included members of Vickers-Armstrong London, as well as representatives of local banks: Marmorosch Blank, Romanian Credit Bank, Peasant Bank, National Bank of Romania and Banca Romaneasca [3].

The means of production of U.D.R. Were made up of: Reşiţa and Anina factories, Bocşa farm tools mill, Moraviţa iron mine, coal mines from Anina, Secu, Doman, copper from Oraviţa, Ciclova, Sasca Montană and Moldova Nouă, Gold from Oraviţa and sulphur from Moldova Nouă, but also in a series of concessions. These involved the holding of resources considered important for the metallurgical production process: iron and manganese in Delineşti, iron and lignite in Armenis, lignite in Caransebes, lignite in Bozovici, copper and iron in Liupcova. The U.D.R. Also had the right of ownership: the Anina, Vasiova, Văliug and Zăvoi sawmills, the Colţani lime burning plant, the Ramna, Tyrol and Moldavian Noua vineyards and the Văliug and Anina fish ponds. In the U.D.R. We find references to 134 km of unrestricted railway, 103 km of wood transportation channels, 8 natural or artificial lakes owned as land and 27 retail outlets [4].

The Board of Directors consisted of the following members: Prince Barbu Ştirbei, Bishop Valeriu Frenţiu, Prince Mihail Cantacuzino, D. Ghica-Comăneşti, Duiliu Zamfirescu, Victor Bontescu, John Boamba, Prince Constantin Caragea, Serban Fassian, Adalbert Veith, Count Germiny, Gustave Périére, Jean Huntuere, George Günther, Anton Weiler, Victor Levy, Avram Imbroane, Constantin Ghica and Stan Vidrighin [5].

The Management Board together with the Company's President and the Control Board were headed by U.D.R. They were joined by one or two administrative delegates from the government, one or two vice-presidents and 20 shareholders' representatives, of whom 4 were appointed by the government and 16 were elected. These people were the company's first echelon. The second echelon included a Chief Executive Officer and Chief Director, as well as four Managing Directors in Banat. The head of the domains was headquartered in Oraviţa, the mines at Anina, and the factory in Resita. The UD.R. Representation in Timisoara was also headed by a director and the headquarters of the Bucharest firm had attached a character holding an identical function. The third echelon consisted of six deputy directors, four residents in Banat and two in Bucharest [5]. Among the directors of U.D.R. Who had worked in Banat before 1918, we mention Arpad Hajts, Viktor Jung, Eugen Links, Nikolaus Balint, Anton Hendrich, Karl Kneipert, Geza Halasz and Revay Karoly [6].

The position of President of the Council of Administration was ensured during the interwar years by the following persons: Prince Barbu Ştirbey (1922-1929), General Ernest Ballif (1929-1939), General Ernest Urdareanu (1939-1940) and Constantin Greceanu 1940-1941). Administrative delegates were: Adalbert Veith (1922-1929), Constantin Busila (1929-1931), Max Auschnitt (1931-1939) and Nicolae Malaxa (1939-1941). General directors had been: Adalbert Veith (1922-1924), Constantin Orghidan (1924-1927), followed by Agrippa Popescu (1927-1931). Between 1931 and 1941, the functions of Chief Executive Officer and Chief Director were confused. Among the local directors we mention Aloys Tavy (1919-1924), V. Martian (1924-1930), C. Mihălcescu (1930, 1934-1940) and Karl Rennerth (1931-1934). Directors of the mining department were: Anton Hendrich (1919-1927) and Adalbert Ferrand (1927-1932). The department of factories was led by: Karl Kneipert (1919-1925), Johann Göbl (1925-1928, 1932-1940) and C. Teodorescu (1928-1932). The direction of the domains was conducted during the interwar years by: Nikolaus Balint (1919-1934) and Alexandru Popp (1934-1941) [7]. With the militarization of state-owned enterprises, in 1939, the military staff from U.D.R. Had been led by the Colonels: Constantin Livezeanu and Panait Boitan.

2. Analysis

In the following we will describe a series of elements related to mining and metallurgical production of the under study. From the very beginning, we make it clear that the development of this company and of other industrial enterprises in the area has undergone numerous meanders, being influenced both by international events: World War I or the Great Economic Crisis, but also by internal ones: labour accidents, strikes, Reduced and incorrectly achieved give insufficient professional education of the staff employed. Regarding the number of employed workers, it is worth mentioning that it has risen sharply after 1919. This can also be attributed to the expansion of the UDR, which during this period had lucrative contracts, making important profits and thus had the possibility Employ and train more workers than in the past. However, coal production was not good: in 1913, for example, 1,400,000 tons of coal was exploited, while in 1937 only 226,830 t, of which: 176,000 t in Anina, 50,730 t in Doman and 22,100 t in Secu [8].

The annual iron ore production in Ocna de Fier was about 20000 tons in 1926. However, the transport and processing costs were enormous, three times higher than the imported ores, and in 1927 the exploitation in this locality was reduced. The following year, out of the 74,000 tons of iron ore needed in Reşiţa plants, half was imported from abroad. Technological innovations often required drastic staff restructuring. For example, in 1925, 6.194 tons of manganese ore were extracted at the Delineşti mine. On 1 December 1927 production ceased due to the modernization of the machinery. When work was restarted, only 14% of the mining personnel who were working before the modernization of the production equipment were required [9].

The UDR was the principal exploiter of 155 mine concessions, a total of 2.507 ha of carboniferous land and 633 ha of land rich in iron and manganese ore. The company belonged to the coal mines of Anina, Secu, Doman, the manganese from Delineşti, the iron from Moraviţa and Ocna de Fier, Armenis, Liupcova, Bozovici, as well as the mines closed from Oraviţa, Ciclova, Sasca Montană And Dognecea, without forgetting Anina's ammonia and coke plants, and the Bocsa, Anina and Breazova power plants [10].

The mines in Anina, seven in 1919, were named Thinfeld, Breuner, Gustav, Friedrich, Kollowrat, Ponor, and Kübeck. The first three mines had been linked through access and mining galleries and upgraded before 1918 by the installation of 4 compressed air locomotives, each of 22 hp delivered by Schwartzkopf in Berlin. In 1922 the Breuner mine was to be reopened, where in 1872 there were found coals of coal that burned underground due to the presence of high temperature and oxygen. Wilhelm Riegel, the ministry director, retired during the interwar years, attracted the attention of U.D.R. That the reopening of the mine could lead to catastrophic fires in the underground workplace. To substantiate his claims, he had presented the results of his mining prospects in 1895 and 1897 [11]. On May 15, 1923, the wooden gate that closed the Breuner mine with a concrete was doubled and the access was permanently sealed. Also in 1923 was the Ponor mine, which will not be used until 1942.

The Ferdinand mine had a depth of 644,4 m in 1919 and at 723,67 m in 1928. It was considered as the most profitable area of exploitation Carboniferous of the locality, as well as the main underground access path, because from there the access galleries to the other work and transport stations started. From the Ferdinand mine, or Nr. I, as it was called by workers, started from the horizon 4, a gallery of access to the mine Nr. II or Gustav, which had been extended in depth between 1920 and 1928 from 573.32 m to 635.92 m. The Kollowrat locality was closed in 1928 after reaching a depth of 714 m. IV, Friedrich, served as a way of venting and ventilating underground workplaces. For better air circulation, the depth of 1918 had been preserved, namely 587.56 m [12].

On January 30, 1925, the modernization of Anina began: The Kollowrat mine was equipped with a Wickers drilling machine, the Gustav mine received in 1926 an air compressor of 180 hp and in 1928 all mines in Anina were equipped with four steam boilers with a total capacity of 2,300 hp and a load of 7,800 kg. To these were added three electric drills with a total capacity of 750 hp and a load of 4,800 kg [1].

Modernization had also been done at Doman and Secu mines, but at a slower pace. In 1932, two locomotives with electric accumulators for the transport of ore were introduced at Doman. At Secu, the Albrecht mine reached a depth of 600 meters in 1923 and 800 meters in 1930. Because of the decrease in the amount of coal exploited on June 1 of that year, the mine in question was sealed. Also in that year two more mines were extended to Secu until work XV and two electric locomotives were purchased on 100 hp each [2].

The Anina coke factory had two German batteries, one produced by Bauer, and the other produced by Küpers, both imported between 1914 and 1918. These batteries consisted of 62 coal coke rooms with a total capacity of 13.2 tons. Along with coke, the plant also produced ammonium sulphate, the highest coke level being reached in 1928 by 40,370 tons, and ammonium sulphate in 1925 with 362 tons. The management of the UDR decided to build a coke factory in Resita, equipped with two batteries imported from Germany with a capacity of 22 rooms. These batteries began their work successively, between April 1, 1934 and March 1, 1935 [3].

On January 30, 1925, the modernization of Anina began: The Kollowrat Mine was equipped with a Wickers drilling machine, the Gustav mine received in 1926 an air compressor of 180 hp and in 1928 all the mines in Anina were equipped with four steam boilers with a total capacity of 2,300 hp and a load of 7,800 kg. To these were added three electric drills with a total capacity of 750 hp and a load of 4,800 kg [4].

The mines and factories of Oţelu Roşu, Ruspert and Ruschiţa belonged to the Societatea de Mine şi Uzine Călăraşi S.A., which was owned by Rimamurany S.A. In 1919, these facilities of exploitation and production were taken over by the Romanian state by being considered the assets of an enemy state [5]. The marble quarry in Ruschiţa belonged to a leading German from Banat Mountain, namely Johann Bibel, the father. As architect of this state, he made the railway construction plans: Reşiţa-Bocşa-Moraviţa, built the school and the Evangelical-Lutheran church from the last-mentioned settlement in 1871, as well as the Roman Catholic church from Anina in 1872.

Being an entrepreneur privately, he had exploited alongside the Ruschiţa career, Sviniţa and Predet, as well as the Poiana Mărului. After 1919 these societies were taken over by his son, who was also named Johann Bibel [6].

The coal mines of Lupac and Ruspert had formed the Mining Industries in Banat in 1921. Coal production increased significantly from 2,136 t in 1921 to 31,253 t in 1927. The basic capital of 150 million lei was increased successively in 1929 to 350 million ROL. Among the main shareholders we mention French bankers: Francois Sépulchre, his brother Antoine and Pierre Liedckercke with 75 million lei, Société des Minérals in Brussels with 10 million lei and Prince C. Basarab Brâncoveanu with 15 million lei. In 1928, a society called the Romanian Mining Industry was set up to manage the coal mines at Cozla. This company was funded by Marmorosch Blank and received a massive infusion of capital and mining equipment from the German company Julius Berger Tiefbohr in Berlin [7].

Regarding the post-World War, I steel production, we can state that information from primary sources has decreased significantly compared to the pre-war level. If, in 1913, S.T.E.G. Produced 115,000 tons of cast iron and 122,000 tons of steel, the steel level will only be reached in 1927, while iron production will remain during all interwar years below the level of 1913. This was probably due to the closure of a foundry section at Resita in 1921. Immediately after the First World War an engineer from Turnu-Severin, named Theodor Codreanu, rented a part of the steelworks at Oţelu Roşu and opened with them a workshop for repairing freight wagons [8].

On February 23, 1924, the Romanian Government ordered the creation of the Titan Metallurgical Plant S.A.R. With a capital of 12.4 million lei. Among the main shareholders we can mention: Max Auschnit, UDR, Titan-Nădrag-Galaţi Society and Chrissoveloni Bank. On 9 March of the same year, the newly established firm was granted the 10-year concession to redeem the shares of the Mining and Steel Company Calan S.A. Worth 135 million lei. On 1 April, the new name of the mining and steel company was established: Titan-Nădrag-Călan Metanurgical Plant Reunified. From the company's properties for 1927, we mention the plants in Bucharest, Galati, Nădrag, Găvojdia, Călan, Oţelu Roşu and the iron mines in Teliuc [9].

The locomotive factory in Resita was built between 1921 and 1923. The plans were made by Simion Avram and Georg Stanek under the supervision of the Inspector C.F.R. Johann Grycz [9]. The actual building up and installation of the machinery was entrusted to a trio of engineers: Ion Păsărică (metal skeleton), Ludwig Reisinger (foundation, walls and roof) and Bernard Mendlovici (installation of production equipment). The planned production was 1,000 locomotives per year [10]. Romanian Railways promised to send locomotives for repair but to order new models. Between 1923 and 1925, 308 C.F.R. locomotives had been repaired at Resita, and from 1926 new copies were produced, 10 in that year and 11 in the following year. The economic crisis of overproduction adversely affected the demand for locomotives and wagons, however, in 1929, 34 were produced. During 1926-1929, 113 locomotives were built in the municipality of Bârzava River and other 767 were repaired. The first locomotives produced after 1926 were made after an Austrian license by German workers sent to training in Austria and equipped with a Schmidt type heater and a Fritz-Wagner type regulator. It was not an exact copy of the Austrian model, the technicians in Resita, making some changes, such as the locomotive's ability to burn both coal and crude oil for better efficiency. Several local patents, such as the Cosmovici oil nozzle and the Alexandru Popp safety valve, contributed to the security of the process. For the local railways, under the rugged relief of Banat Montan, smaller, but more robust, more powerful locomotives were built at Resita, equipped with a Kraus-Helmholtz chassis. Resident workers sent to training courses in Germany had mastered the construction of the wheels of a mountain locomotive patented by the engineer in Munich, nominated in the previous sentence.

The difficulties of the reconstruction era that followed the first global conflagration were bolstered by the ceasing of the activities of industrial sites in the Banat Mountains. Copper mines, for example, were closed in 1921, with the Romanian state banning the exploitation of this ore. Only a special authorization was granted in Ocna de Fier. Copper exploits from Moldova Nouă, Sasca Montană, Ciclova, Dognecea were equally closed. This caused the acid reaction of some writers [11]. It is worth noting the narrative approach with satirical accents, made by Simeon Samuel Moldovan:
"Through this government moratorium, the local population has become unemployed, and the richest county of Romania (Caras-Severin) has been thrown into misery. After the beautiful industrial past of Ciclova, in which mining and ironmongery merged from the creation of a mechanical baroque workshop, until the opening of a mint, a catastrophic regression came about due to the obtuse of the governmental organs. Romanians, the ancestral population, are hungry and German industrial schoolmen have to emigrate, even the Gypsies who are privileged by the precious metal crafts carry a miserable existence only on their musical talent "[12].

We recall that one of the main ways to improve the material situation, adopted by the ethnic Germans, was to go to work abroad. Several workers of German origin in the Banat Mountains opted to emigrate based on a better paid job offer to mitigate the effects of the global overproduction crisis. Miners in the Banat Mountains, especially those of German origin, chose to move to the rich coal and iron ore areas in France, namely Alsace and Lorraine. Considering that German was spoken by a significant percentage of the population in these two regions, their integration into the new lucrative environment did not represent a problem [1].

A first group of 41 German miners from Secu emigrated to France at the end of May 1930. According to an agreement between the Romanian and French governments in June of that year, miners with 5 or more children were entitled to Leave with the whole family. The costs of transport and subsistence from Timisoara to France were borne by the employer. Workers continued to pay 10% of the monthly salary to the Romanian trade unions in which they were enrolled, while enjoying the rights under French labour law. The usual working day amounted to 8 hours, for which the miners were paid 28 francs for the coal mining and 30 francs for the iron ore exploitation. Accommodation was made in colonies built for workers, with a daily cost of between 10 and 20 francs. Daily foods cost between 10 and 12 francs [2]. The French employer undertakes to pay daily for each worker the equivalent of the cost per litter of wine.

In mid-July 1930, the committee for the recruitment of volunteer miners came to Resita. The French employer was represented by 2 specialists and the Romanian state of 1 inspector from the Ministry of Labour and Social Welfare. 120 workers were employed by individual contract. A first group of them, totalling 70 people, left on August 17 of the same year to France. Professor Franz Wagner offered French free hours to work abroad [3]. On November 7, 1930, the commission returned to Resita but had also moved to Anina, and four days later 12 miners had gone to France. The working and living conditions were not the best, as Anton Stupak, a German worker from Doman who had been employed in Luneville. The letter had been published in the Reschitzaer Zeitung. The Issuer reported the following:
"Even here our situation is not the best. We work from 7.30 to 12.30 in the morning, then we have a 45-minute meal break. In the afternoon, we work from 13.15 until 18.15. Most Luneville workers, including the natives, around 95% of them work 90 minutes over the program to earn a satisfactory salary. The average leaf is 4 francs per hour (about 24 lei, Anton Stupak's note). However, living is hard, with food being 40% to 60% more expensive than in Resita. Instead, bedding, clothes and household items are 30% to 45% cheaper. In Luneville there are 14 workers from Resita, some of whom will return home for Christmas. The others will resist the 6 months until the contract expires, then move to another city where living conditions are better. If subsistence costs are all the same we will all return after one year in Romania" [4].

3. Conclusion

These data imply unmistakable progress in demographic, but especially occupational terms, within the German minority living in the urban communities of Banat Montana. Working as vendors, barbers, blacksmiths, carpenters, tailors, butchers, tanners, plumbers, painters or as doctors, dentists, teachers, lawyers, they contributed to the material and spiritual development not only of their own minority but also of communities Where they were born and thrived [5].

Within the research undertaking there have been explained the following existence aspects of the German community from the Mountainous Banat in the timeframe between the two world wars:

1. Defining of a minority group in relation to the majority one within a national European state in the interwar period. The finality of the scientific undertaking related to one or more national minorities in Romania, generally, but also to the Germans from the Banat Highland, especially [6].
2. The impact of the ascension and recession periods during the economic evolution of the interwar Romanian state, on the German community from the Banat Mountains [7].
3. The influence on the Germans from the Banat Mountains by the main political currents and doctrines during the first half of the 20th century. Their position towards socialism social democracy, communism, fascism, and national- socialism [8].

References

[1] Ancel J 2001 *România interbelică,* Editura Universităţii „Alexandru Ioan Cuza" Iaşi, România
[2] Feneşan C, Gräf R, Zaberca V M and Popa I 1991 *Din istoria cărbunelui, Anina 200,* Editura Muzeului de Istorie al Judeţului Caraş-Severin, Reşiţa, România
[3] Gräf R 1997 *Domeniul bănăţean al STeG: 1855-1920: din istoria industrială a Banatului Montan,* Editura Banatica, Reşiţa, România
[4] Gräf R 2002 *Domeniul bănăţean al STeG: 1855-1920: minele şi uzinele aferente,* Editura Universităţii Babeş-Bolyai, Cluj-Napoca, România
[5] Lupşiasca K L 2000 *Dem Emporbringen und Aufblühen dieser Bergwerke. Eine Geschichte des Banater Berglands in der Zeitspanne 1855-1920,* ADZ- Verlag, Bucureşti, România.
[6] Lupşiasca K L 2006 *Höhepunkt ihrer Geschichte. Eine Geschichte des Banater Berglands in der Zeitspanne 1920-1948,* Editura Banatul Montan, Reşiţa, România
[7] Mărgan S 1996 *Două secole de metalurgie în Valea Bistrei,* Editura Timpul, Reşiţa, România
[8] Narai E 1998 *Evoluţia situaţiei economico-sociale în judeţul Caraş,* Editura Modus P.H., Reşiţa, România

[9] Popa I 1990 *Consideraţii asupra şomajului de la Uzinele de Fier şi Domeniile Reşiţa,* în *Banatica*, X, Reşiţa, România
[10] ****Reschitzaer Zeitung* 1930 XLIV, Nr. 27, Reşiţa, România
[11] ****Reschitzaer Zeitung* 1930 XLIV, Nr. 45, Reşiţa, România
[12] Rieß J 1935 *Deutsches Volkswerden im Banat. Reden und Aufsätze Dr. Kaspar Muths*, Graphisches Unternehmen Ideal, Timişoara, România

Industrial production of the interbelic German from the Banat Highland

C Rudolf[1], C Hatiegan[2], A Pellac[3], S L Bogdan[3], L Paduraru[3] and A Varga[4]

[1]"Eftimie Murgu" University of Resita, Social Sciences Department, Traian Vuia Square, No. 1-4, 320085, Resita, Romania
[2]"Eftimie Murgu" University of Resita, Electrical Engineering and Informatics Department, Traian Vuia Square, No. 1-4, 320085, Resita, Romania
[3]"Eftimie Murgu" University of Resita, Department of Mechanical and Materials Engineering, Traian Vuia Square, No. 1-4, 320085, Resita, Romania
[4]"George Barițiu" Institute of History Cluj- Napoca, Str. M. Kogalniceanu, No. 12-14, Cluj- Napoca, Romania

E-mail: c.rudolf@uem.ro

Abstract. The historian Fernand Braudel considered that geography dictates history because a certain region or a certain state requires a characteristic of its economic life. Commerce was often called in mercantilist period the blood that provides life of the nation. These data imply progress in terms of demographic beyond doubt, especially in the employment of the German minority living in the urban communes of the Banat Mountain. We retain the dual nature of their occupations, with farm work kept in the background, but the main plane dominated the gradual effect of industrial labour and business activities. Along with farmers, most Germans residing in villages and hamlets in the mountains of Banat were devoted to jobs in the service sector and small crafts area. They worked both as employers and employees in the stores and workshops in rural and urban areas where they resided. Working as sellers, barbers, blacksmiths, carpenters, tailors, butchers, tanners, plumbers, painters or as doctors, dentists, teachers, lawyers, they have contributed to the development of material and spiritual not only their own minority but of communities within which they were born and prospered.

1. Introduction

Realising a truthful narration about the German community of the Banat Highland in the interwar years, as it had happened, to quote the historian Theodor Mommsen, proves a necessary endeavour, especially since the works of the last decades treating the German community in Romania have done this generally, without allotting a space aside for this micro region [1].

Analysing the works that have touched, even in passing, the history of the German community in the interwar Banat Highland, we may state that many of these, of more recent or more remote date, have insisted either on the politic element of this minority, sometimes also on the economic one, without giving much importance to the social or cultural one. That which has been written so far on this subject, is in some cases correct, in some not, but in both situations still insufficient, as it has not touched the depth of the problem in all its aspects: economic, social, political and cultural. In these conditions, the endeavour of the author of this doctoral thesis was to discover much information from the primary sources, to give them back to the historiographical circuit, to analyse and interpret them [2].

Over the centuries, a series of historians, not without merit, have committed the frequent error of stating that the German population of the Banat is totally composed from Swabians, originated from the area of Würtenberg or from Alsace and Loraine. The error is explainable by the names given to the Germans from the Banat area, during the Austro- Hungarian dualism. Under the influence of some writers from Germany and Austria, such as Adam Müller Guttenbrun, the name "Swabian" received an honourable prerogative, namely that of the colonist who subdued the elements of nature. At the beginning of the 20th century, many more authors considered that the Germans from Hungary are divided in the Saxons from Transylvania and the Swabians from Banat [3].

The error of placement for these writers refers to the omission of other groups of German colonists who came from: Bohemia, Burgenland, Carinthia, Moravia, Slovakia, Styria, Tirol and Zips, and who were settled in the mountainous region of the Banat. Those who were colonised in the 18th century and were no ethnic German, underwent a germanisation process in the 19th century, ending up considering themselves as members of the said ethnic group in the 20th century. They were perceived as German by the members of the Romanian majority who did not differentiate on the region of origin their forefathers had come to the Banat Highland and did not see them as specific Austrians, Styrians, Tirolians, Bohemians, Moravians or Italians. Some authors such as: Engelmann, Bell, Herrschaft, Lupşiasca, Petri, Schöffler şi Windhager have not considered the work of the historian Ţintă, either because they wrote their own books before the period of creation of the said scholar, or because they had no access to his writings. Whatever the reason, the result found itself in the often-encountered confusion of the interwar and post war historiography, to culturally assimilate the Germans from the Banat Mountains (Austrians, Czechs and germanised Italians) with the Swabians from the plain area of the province (Würtenbergers, Alsatians, Lorrainians and Rheinlander) [4].

The scientific endeavour to identify the sources, to go through the bibliography and to write my doctoral thesis was characterised by a close cooperation between all information sources, the edited as well as the inedited ones. This way, in which the researcher uses in close cooperation the older and newer information sources bears the name Hugging Technique, was being originar from the Anglo-Saxon linguistic space, where it had been erroneously attributed to the historian Samuel P. Huntington. Of course, this scholar, famous for his work The Clash of Civilizations, did not invent the said technique, it being created by the historians of the Annales School: Lucian Febvre and Fernand Braudel, between 1922 and 1929. Concretely, the technique supposes the doubling of the bibliographic information by the archivist ones, which support and check the narration in every chapter of the paper[4]. Thus, the data from books and articles, contemporary or from the studied époque, have been completed, verified and debated. The conclusions were thus logically structured and would enjoy an increased scientific value [5].

The above-mentioned working technique was supported and completed by the research method mix, namely the micro- analytical and selective process, by which the hard points of thesis are fixed. The central research point was described, but also its surrounding environment as well as the relation: capital- province, city- village [6].

Naturally, this fact implied a rigorous filtering of the information found, both bibliographic and documentary. The narrations and descriptions, from the thesis as well as the conclusions part had to be supported by data taken from the read books, from the documentary researched funds but also from the publications of the age. The information gathered from a source have been carefully scrutinised, compared with similar ones from other sources, and, if there had been a corroboration, then they were used in the construction of the text. The used quotations, acting as examples from the publications of that age as well as from the documentary funds, if they were written in other languages than Romanian, then they were translated and inserted into the body of the text [7].

Beside books, studies and articles written by various: historians, sociologists, metallurgic and mining engineers, union leaders or journalists, the documentary funds contain inedited and interesting information, far too little researched so far. After consultations with MA Josef Wolf from the Institute for Swabian History and Homeland Studies in Tübingen, Germany, (further on abridged I.D.G.L.), the author has reached the conclusion that within the research and writing system of the paper it is necessary to put into value the informational sources from diaries of personalities from the Banat but especially from the private collections of documentary funds found in the National Archives of Romania [8]:

1. The documentary funds researched in Caransebeş (The National Archives, County Branch Caraş-Severin, further abridged A.N.D.C.S.) have revealed brand new data on the German minority as it had been described in the official papers drawn up: by the police institutions from Băile Herculane, Caransebeş, Oraviţa and Reşiţa, by the administrative activity of the Prefecture of the Caraş county, by the agricultural banks from Bocşa Montană, Caransebeş and Oraviţa, through the documents drawn up by the Commission for Application of the 1921 Land Reform at county level, without omitting, of course, the ones connected with the activity of social protection and mutual support of the Pension Institute managed by the firm U.D.R. in the interval 1918-1930 [9].
2. In parallel, the data collected in Timişoara (The National Archives, County Branch Timiş, further on abridged A.N.D.T.), has revealed information about: the activity of the ethnic Germans in the Educational Inspectorate of the county Severin as well as their level of involvement in the agricultural production of the interwar years. The information found in the Archives of the Roman-catholic Bishopric (further on abridged A.D.T.) has not been neglected, revealing brand new data regarding the correspondence of different professional and cultural associations, formed by the German inhabitants of the plain and mountainous regions of the Banat in the interval 1929-1944 [10].
3. Very interesting from the economic and politic aspect proved the information gathered because of the applicative research, carried out in the documentary centres of Bucureşti (The Central Direction of the National Historic Archives, as well as, The Archive of the National Council for the Studies of the Security's Archives, further on abridged A.N.C.D.B. and respectively A.C.N.S.A.S.). Thus, the politic evolution of the studied minority can be found in the documentary funds containing information about: The Ruling Council, the Police General Direction, the Homeland and Foreign Ministries, as well as the Secret Information Service. The ups and downs of economic life, underwent by the said minority are described via the date from the files treating: the application of the Land Reform o 1921 in the counties of the Banat Mountains, but also via the industrial and commerce legislation adopted by the Chamber of Representatives and the Senate in the interval 1923-1938. Often, the economic problems were identified as cause for the political ones, and sometimes the latter overlap the first. This is the case of the events described in the documentary pages that deal with the historic evolution of the German minority from Romania between 1929 and 1940 and with the one of the German Ethnic Group 1940-1944. The information attest to multiple interferences between those two previously mentioned plains of existence with a strong extension into the cultural one and a profound influence over the social life [11].

The German community of the interwar Banat Highland is implicitly connected to the historic concept known as: past- present. Beginning with the theoretic explanation of the concept of history carried out by Rheinhard Koselleck and of the disparity of temporal phases, the author has described the life and evolution level of the said minority in the interval 1918-1940, insisting upon the named geographic area. Applying the latter concept, formulated by the same historian the author has drawn up a description of the mental collective under the influences of the politic, economic and social phenomena, specific to the respective age [12].

The historian Fernand Braudel considered that geography dictates history because it imposes on a region or a state a characteristic of its economic life. Commerce was often called in the mercantilist era the blood that assures the life of a nation [13].

2. Analysis

Romania's industrial rebuilding after the First World War was uneven. This is obvious if we consider the following situation: in 1919, the production volume in the steel industry was 19.4% compared to 1913 although there were abundant resources. The insufficient development of the metallurgical industry meant that it could only take part in the development of the local economy, and in 1924 it covered only 46% of domestic consumption. The extractive industry has seen significant improvements in terms of production. Notable progress has also been made in the extraction and processing of salt, copper, lead and other non-ferrous metals. Within the four-year food industry: 1924- 1928, the number of enterprises increased from 842 to 1,133, and in the textile sector from 377 companies in 1924 to 570 firms in 1928. In parallel, the industry had a high degree of automation, the 73,000 hp Used at the beginning of the third decade of the 20th century, being increased to 96,000 C.P. At the beginning of the fourth decade of the same century [14].

Fundamental to organizing the exploitation of subsoil resources was the Mining Law of June 4, 1924. Using the provisions of the Constitution of March 28-29, 1923, this legislative act strengthened the hegemony of the state over the wealth found in the basement of Romania. There was also a strong degree of control of domestic capital in joint ventures [15].

The National Peasant Government issued on 28 March 1929 the Law for the Amendment of the Mines Law of June 4, 1924. This law is in line with the policy formulated by the Political Revolution. Of "Open Gates!", Different from the liberal one entitled "By ourselves!". Although both economic doctrines aimed at the general progress and the development of the Romanian society, they were different in terms of the means and methods of achieving these desiderata [16].

The years of the global economic crisis have led to an economic recession marked by the decrease of the purchasing power of the population, but also by the reduction of state orders for products. This led to a decrease in production in some of the economic branches, for example: the steel industry, the coal mining industry, the manufacturing industry, the forestry industry, the construction industry, the average index of 57% [17].

During the period 1934-1938, the Romanian economy experienced an upward trajectory due to strong government involvement. This corresponded to the neo-liberal political doctrine of involving the government in the economy and to create a state of fact conducive to the development of domestic economic forces in the banking, industrial and export fields. However, foreign investment in finance and industry continued to be a major part of the capital that has run in these economic branches. In April 1935, the Romanian government concluded a contract with the Czechoslovak Skoda Factory for the supply of weapons to the Cugir and Resita plants [18].

The legislation of the liberal national government at the end of the interwar period had beneficial effects for the national industry. By Decree Law of May 10, 1937, the regulation and control of cartels was established regarding the establishment of new factories. For such an approach, the favourable opinion of the Ministry of Industry was needed, which exemplifies state control in some industrial branches such as arms production [19].

The Banat Mountains played an important role in the strategic interests of Romania and Germany, the interest being placed on the chromium ores located in this micro-region. The mandatory military organization of the industry, starting with 1939, involved the formation of industrial offices in industry branches, the presidents of the administrative boards being appointed by the government [20].

The outbreak of the second world conflagration led to the militarization of many firms and economic companies in the industrial, banking and commercial spheres. This is the case of enterprises: Astra Romana, Concordia, Colombia, Romanian-American Society. This element affected the flow of commodities and currency in the offices of the Banat Mountains. The German Kontinental A.G. Opened in 1940 a subsidiary of Otelu Rosu to produce the necessary parts in the oil industry [21].

The conditions given by the habitats of the population have concentrated the country's demographic weight in rural areas. This led to the situation where most of the able-bodied ones were engaged in agricultural activities. The destructive effects of the First World War were cancelled until 1922, after which the agricultural production had an ascending meaning. Laws for the implementation of the agrarian reform in Banat, Crişana, Maramureş, Transylvania, Bessarabia and Bucovina had a beneficial effect on agriculture in the region under discussion. Both in the mountain area and in the Banat plain area, 24,815 families belonging to the German minority were appropriated [22]. This measure was in addition to the Law on the Establishment of the Agrarian Committee of March 1920 and the Act on the Establishment of the Municipal Slates in August 1920. For the German farmers in the Mountain Banat, who were mainly engaged in animal husbandry and fruit growing, the clear constitution and delimitation of some Grazing places had a direct beneficial effect on the evolution of their main occupation. In 1928 the cultivated areas were 20% larger than 1922.

The global economic crisis has led to the collapse of agricultural prices across the global commodity exchanges. The agricultural production, which in 1929 was worth ROL 88 billion, was only 30 billion ROL in 1933. The small property of 1 to 3 ha predominates in a proportion of 48.5%. Despite the 1921 Agrarian Reform, the world economic crisis has led to the plunder of many landowners in 1938, with 1 million peasants deprived of land. Inter-war legislation tried to help agricultural workers establish state pensions for those who reached the age of 70. We mention in this regard the Law on the Settlement of Laws of Ownership of 20 August 1929 initiated by the peasant national politician Ion Mihalache [23].

Investments in agriculture have remained relatively low and were mainly initiated by national peasant governments between 1928 and 1931. As the second world warfare was prefigured, the government took measures to boost agriculture. This was to ensure that a surplus of agri-food products would be stored within the strategic reserves of the state prepared for conflict cases [24].

One of the legislative measures referred to the organization of Chambers of Agriculture. The measure was adopted by the National Representation on July 1, 1930. The central idea of the normative act was the protection of cereal production by both the state and the private owners. The legislative act, although well intentioned, has failed to lead to increased small and medium-sized peasant production deeply affected by the crisis and increased taxes. The funds were simply insufficient to allow large-scale leasing of machinery for the selection of grain varieties and for the gradual removal of traction animals [25].

Apparent relief measures for agricultural workers did not stop, under the rule of Iorga-Argetoianu, on December 18, 1931, starting the first Law to suspend forced execution. In theory, the measure would protect small owners from the abuse of creditors and give them favourable time and conditions to pay their debts. In practice, however, the measure remained largely ineffective. The government was forced to formulate complementary solutions. One of these refers to the first Law on the conversion of agricultural debts in April 1932. The measure should, at least theoretically, provide debt repayments according to the terms and conditions of reimbursement [26].

The legislative measure of the Iorga-Argetoianu government was qualified as inadequate by the peasant national ministerial team, which is why they sought effective means to improve the material and social situation of the agricultural producers affected by the crisis. In October 1932, a first concrete element of these plans appeared, represented by a second Law on the conversion of agricultural debts. The practical applicability of this normative measure has been desirable, being completed in April 1933 with a Law on the regulation of agricultural and urban debts. The latter proposed the conversion of agricultural debts by 75% of their gross value, being an indisputable act of social justice initiated in favour of small agricultural owners who enjoyed wide support among the peasant national political base [27].

The Law on the Clearance of Agricultural Debt of 7 April 1934 deals with agricultural lending problems. Debt reliefs to the state or private institutions varied, depending on the economic level, between 50% and 70%. The rest of the borrowed amount was paid within 15 years with an annual interest rate of 3%. The measure has temporarily improved the material state for a part of the agricultural producers, but failed to fully remedy the catastrophic effects of the economic crisis on the population [28].

Our country suffered from the existence of an insufficiently developed road network. The frequent changes of governments, which affected the structure of line ministries, did not improve this situation. For example, the Autonomous House of State Roadways of the Ministry of Public Works and Communications was replaced in 1932 with the General Directorate of Roads. Between 1936 and 1938 425 km of roads were asphalted for the next two years, with another 1,800 km of roads being programmed. The railway network was performing somewhat better, in 1939 repairing 11,410 km of railway at the Reşiţa and Bucharest plants. During the liberal national government, a series of railways were built, having a secondary economic aspect of passenger and commodity traffic, but of indisputable military and strategic value. These are two routes: Ilva Mica - Vatra Dornei, Resita – Caransebes [29].

From the point of view of transports and aeronautical communications, we mention the following: The communication and shipping routes were managed by an Autonomous Directorate of Waterways Communications founded in July 1929. It will be transformed in May 1934 into an institution named Commercial Administration of Ports and Communications on Water. In March 1936, the Ministry of Air and Navigation formed the following administrative subdivisions: the Romanian Maritime Service, the Danube Naval Service and the so-called commercial administration. The merchant fleet consisted of 5 passenger ships and 7 ships and cargo ships for the Black Sea along with 12 passenger ships and 19 freight tugs on the Danube. The Romanian State Air Operations included 6 international flights with operating contracts with 7 foreign partner companies [30].

Letters, telegrams and telephone services were managed by the Directorate-General for Post and Telecommunications, founded in April 1920. This body was later renamed the Post and Telephony Registry and ranked in May 1934 in the Ministry of Public Works. On December 12, 1927, the Broadcasting Directorate was established, which became the Romanian Broadcasting Society on November 1, 1928, and in 1940 there were 4 radio stations that took over broadcasts and newsletters from similar institutions from abroad [31].

Domestic and international, wholesale and retail trade followed the sinuous moves of global financial and banking developments. The period of economic rise from 1924 to 1928 followed the years of the world economic crisis: 1929-1933, to then make sensible increases in the trading of goods and services between 1934 and 1940 [32].

The First World War negatively affected the trade balance in our country. Thus, in 1919 the figures were only 2.5% of the export and 37.7% of the average imports from 1910 to 1914. The balance of imports and exports tended to be net in favour of the former by 68.8% as against 31 ,2%. Through a series of banking measures meant to cover as much as possible the Romanian currency with precious metal in international transactions, a trade deficit was reduced from 583 million lei in 1919 to 156 million lei in 1921 [33].

From the point of view of public finances, we note that in 1922-1923 the first unitary budget was achieved for the whole of Romania. Public debt increased because of the world's first conflagration from ROL 1,400 billion in 1914 to ROL 20,000 billion in 1921. This is explained by the fact that the first amount represented the debts contracted only by the Kingdom of Romania in the territorial form 1914, while the second sum represents the debts of large Romania, a state entity existing after the acts of unification in 1918. In general, the external loans contacted during the peasant national governments attracted quite high interest rates of 8.9%. The preferred currency was US dollars along with French or Swiss francs. Overall, the interwar years have experienced significant fluctuations, in general the Romanian state is mostly indebted to both internal and external capital circles [34].

In the period immediately following World War I, anonymous banking companies increased from 487 in 1919 to 556 in 1921. Capital detained increased from 10 billion lei in 1922 to 24 billion lei in 1924. In Banat Montana, there were branches of the main institutions Banking and credit in the country. Well-represented at the micro-region level were the Romanian National Bank, an institution that controlled 25 banks and 52 industrial enterprises, the Romanian Credit Bank, a body holding the majority stake in 46 banking and credit companies, Marmorosch Blank Bank with 9 Affiliated banks and 58 industrial enterprises. They had branches at Oraviţa, Reşiţa, Caransebeş, Oţelu Roşu, Anina, Moldova Nouă, Orşova. Along with the credit institutions were represented in Banat Mountains: Banca Romaneasca and Banca Comerciala Romana [1].

Romania's situation must be integrated into the global economic situation. Like other states, the Romanian one attempted financial recovery measures. We mention in this direction the Law on Monetary Stabilization of February 7, 1929. This law regulated the issue of banknote and coin issuance of the National Bank of Romania. Also in February 1929, the national-peasant government promulgated the Law on the modification of the state of the National Bank of Romania, a measure that tried to be a social policy act meant to increase the population's trust in the country's main credit institution [2].

Over the years of the overproduction economic crisis, the Romanian government has often appealed to foreign loans. The main partner of financial engineers came from the American, French, Belgian, Dutch, German, Austrian, Czechoslovak capital. Loans were typically made for 40 years with a fixed interest rate of 7.5%. Under extreme conditions dictated by the global overproduction crisis, the government used the reserves of the Romanian National Bank to pay salaries and budget pensions. Given the financial austerity measures dictated by the circumstances, these amounts proved to be insufficient, which had a negative impact on the purchasing power of the population [3].

With the establishment of the government chaired by Gheorghe Tătărăscu, there was a change in the country's economic policy. Loyal to the liberal national motto "By ourselves!", The government has launched a series of protectionist measures to boost exports from Romania, but also to reduce imports considered unimportant. We mention in this regard the Law on the Establishment of Machinery Factories and on Important Non-Manufactured Products for the National Industry of July 31, 1937. Thus, new industrial sub-branches were created, such as the steel tubes industry, the radio lamps industry, the electric cables industry, Dyestuffs industry, aircraft industry naval instrument industry. A year earlier, the government issued a law on the mandatory use of indigenous raw materials for commodities [4].

3. Conclusion

Regarding the economic context of the evolution of the Germans in the Banat Mountain during the interwar years, there is a general ascendant course in this area registered by this community. The line of material progress is highlighted by historian Otto Gräfner. The assertions are supported by solid statistical data, which shows that in the Banat Province, more precisely within the mountain microregion, about 25% of all craft enterprises and trade companies were owned by ethnic Germans. We can observe the occupational and economic differences between the mountain and the flats in the world, the first being 56% in the interwar period in industry and mining. The economic years 1921-1926 are presented in antithesis with the economic crisis of 1929-1933, which allows a comparative analysis in social and historical aspect [5].

The bankruptcy of some renowned banks, such as Marmorosch Blank Bank, is described by Otto Gräfner through the impact of the Germans in the Banat Montan urban centres. An important role for the German peasants in the mountain region of the region, fruit growers or livestock breeders, was the creation of the cooperative selling AGRARIA in 1935. This event allowed the capitalization of the positive agricultural quotas obtained on a series of products in 1934 [6].

References

[1] Agrigoroaiei I 1983 *Mişcarea democratică din România întregită,* Editura Academiei R. S. R., Iaşi, România
[2] Ancel J 2001 *România interbelică,* Editura Universităţii „Alexandru Ioan Cuza", Iaşi, România
[3] Bell K 1926 *Das Deutschtum im Aussland. Monographiensammlung,* Deutscher Buch und Kunstverlag, Dresden, Deutschland
[4] Bolovan I 2000 *Transilvania la sfârşitul secolului al XIX-lea şi începutul secolului XX. Realităţi etno - confesionale şi politici demografice,* Editura Presa Universitară Clujeană, Cluj-Napoca, România
[5] Bolovan I and Bolovan S P 2002 *Die Deutschen in Rumänien,* Editura Fundaţia Culturală Română, Cluj-Napoca, România
[6] Bulgaru V 2003 *Reforma agrară din 1921: fundamente economice,* Editura de Vest, Timişoara
[7] Engelmann N 1959 *Das Banat. Ein Buch der Errinerung in Wort und Bild,* Pannonia Verlag, Freilassing, Deutschland

[8] Fassel H and Förster H 1999 *Kuturdialog und akzeptierte Vielfalt? Rumänien und die rumänische Sprachgebiete nach 1918,* Jan Thorbecke Verlag, Stuttgart, Deutschland
[9] Fassel H and Förster H 1997 *Das Banat als kulturelles Interferenzgebiet. Traditionen und Perspektiven,* herausgegeben vom IDGL, Tübingen, Deutschland
[10] Feneşan C, Gräf R, Zaberca V M and Popa I 1991 D*in istoria cărbunelui, Anina 200,* Editura Muzeului de Istorie al Judeţului Caraş-Severin, Reşiţa, România
[11] Gabriel J 1928 *Fünfzigjährige Geschichte der Banater Arbeiterbewegung 1870-1920,* Buchdruckerei der Schwäbischen Verlags-Aktiengeselschaft, Timişoara, România
[12] Gräf R 1997 *Domeniul bănăţean al STeG: 1855-1920: din istoria industrială a Banatului Montan,* Editura Banatica, Reşiţa, România
[13] Gräf R 2002 *Domeniul bănăţean al STeG: 1855-1920: minele şi uzinele aferente,* Editura Universităţii Babeş-Bolyai, Cluj-Napoca, România
[14] Greffner O 1994 *Populaţia şvăbească (germană) din Banat. Cine sunt, când au venit şi de unde?* s.n., Arad, România
[15] Herrschaft H 1942 *Das Banat. Ein Deutsches Siedlungsgebiet in Südosteuropa,* Grenze und Ausland Verlag, Berlin, Deutschland
[16] Hockl H W 1960 *Ungewisse Wanderungen,* Pannonia, Freilassing, Deutschland
[17] Hromadka G 1993 *Kleine Chronik des Banater Berglands,* Verlag des Südostdeutschen Kulturwerks, München, Deutschland
[18] Lupşiasca K L 2000 *Dem Emporbringen und Aufblühen dieser Bergwerke. Eine Geschichte des Banater Berglands in der Zeitspanne 1855-1920,* ADZ- Verlag, Bucureşti, România
[19] Lupşiasca K L 2006 *Höhepunkt ihrer Geschichte. Eine Geschichte des Banater Berglands in der Zeitspanne 1920-1948,* Editura Banatul Montan, Reşiţa, România
[20] Mărgan S 1996 *Două secole de metalurgie în Valea Bistrei,* Editura Timpul, Reşiţa, România.
[21] Munteanu I 2006 *Banatul istoric. 1867-1918. Vol. I.: Aşezările. Populaţia,* Editura Excelsior Art, Timişoara, România
[22] Narai E 1998 *Evoluţia situaţiei economico-sociale în judeţul Caraş,* Editura Modus P.H., Reşiţa, România
[23] Nedelea M 1991 *Prim-Miniştrii României Mari: ideile politice,* Garell Press, Bucureşti
[24] Păun N 1991 *Rolul statului în viaţa economică a României în perioada interbelică: evoluţie doctrinară şi instituţională,* s.n., Cluj-Napoca, România
[25] Petri P A 1992 *Biographisches Lexikon des Banater Deutschtums,* Th. Breit Druck-Verlag GmbH, Marquartstein, Deutschland
[26] Predescu L 1999 *Enciclopedia României: Cugetarea: material românesc: oameni şi înfăptuiri,* Editura Saeculum Vestala, Bucureşti, România
[27] Riegel W M 1994 *Das Schicksal der Deutschen in Rumänien. Eine Dokumentation,* Band IV, Weltbild Verlag, Augsburg, Deutschland
[28] Rieß J 1935 *Deutsches Volkswerden im Banat. Reden und Aufsätze Dr. Kaspar Muths,* Graphisches Unternehmen Ideal, Timişoara, România
[29] Saon S 1999 *Agricultura României în al doilea deceniu interbelic,* Editura Universităţii "Transilvania", Braşov, România
[30] Sbârnă G 2001 *Partidele politice din România: 1918-1940: programe şi orientări doctrinare,* Editura Sylvi, Bucureşti, România
[31] Schöffler J 1923 *Das Banat. Teil I: Zweihundert Jahre (1723-1923),* Schwäbische Buchdrukerei Verlag, Timişoara, România
[32] Schöffler J 1925 *Das Banat. Teil II: Deutsche Arbeit und ihr Erfolg. Mit Karten und Tabellen versehen,* Ostdeutsche Drukerei und Verlags AG, Sibiu, România
[33] Scurtu I 2008 *Civilizaţia românească interbelică: 1918-1940,* Editura Fundaţiei "România de Mâine", Bucureşti, România
[34] Şandru D 1975 *Reforma agrară din 1921 în România,* Editura Academiei R. S. R., Bucureşti, România

IOP Conf. Series: Materials Science and Engineering **294** (2017) 012058 doi:10.1088/1757-899X/294/1/012058

Electromechanical converters for electric vehicles

T Ambros[1], M Burduniuc[1], S I Deaconu[2] and N Rujanschi[1]

[1]Tehnical University of Moldova, Electrical Engineering Department, 31 August 1989 str., no. 78, MD-2012 Chisinau, Republic of Moldova
[2]Politehnica University of Timisoara, Department of Electrical Engineering and Industrial Informatics, 5 Revolution Street, Hunedoara, 331128, Romania

E-mail: tudor.ambros@ie.utm.md

Abstract. The paper presents the analysis of various constructive schemes of synchronous electromechanical converters with permanent magnets fixed on the rotor and asynchronous with the short-circuit rotor. Various electrical stator winding schemes have also been compared, demonstrating the efficiency of copper utilization in toroidal windings. The electromagnetic calculus of the axial machine has particularities compared to the cylindrical machine, in the paper is presented the method of correlating the geometry of the cylindrical and axial machines. In this case the method and recommendations used in the design of such machines may be used.

1. Introduction
Electromechanical converters - electric cars are increasingly being implemented in terrestrial and air transportation.

An attempt is made in the paper to develop an electromechanical converter designed to drive an electric or hybrid vehicle. The purpose of the research is to analyze several constructive schemes capable of efficiently achieving the electric drive of the vehicle. From the start the analysis of the converters was oriented towards the research of the machines with axial magnetic flux, excluding the classic machine with cylindrical construction.

The construction of two-rotor electromechanical converters eliminates one of their main drawbacks by compensating the axial forces of interaction between the stator and the rotors. These electric machines can be made with one winding on the stator package or with two independent flat windings.

As a result of the analysis, an experimental sample was made with cast aluminum rotor windings. In this construction, additional elements were introduced to save the electrotechnical steel and evenly distribute the magnetic induction from the air gap and the stator yokes. Analytical correlations have been established between the basic geometric dimensions of the axial and cylindrical machine.

2. Electromechanical converters of various executions
Electricity is the most efficient and convenient form of energy for transport and use. The basic element in this energy transformation process is the electric machine, capable of functioning as a generator or motor.

The accelerated development of new electrical device manufacturing technologies has also influenced parameters improvement, reduced active and constructive materials, and the appearance of the electric machine [1].

The use of computing in the design of electrical machines makes it possible to achieve a wide range of the most diverse constructions.

Electric cars are used more and more frequently as a personal electric transport. In this context, specific construction electric machines can be used to function as generator and motor [2], [3].

One of these types of constructions could be an alternating current electric motor with axial flux synchronous or asynchronous, with one or two rotors.

Figure 1 shows the synchronous electromechanical converter with an excited rotor by permanent magnets, and Figure 2 shows the asynchronous electromechanical converter with a rotor and a short-circuited winding. These converters were developed and realized within the Electromechanical Department of the Technical University of Moldova.

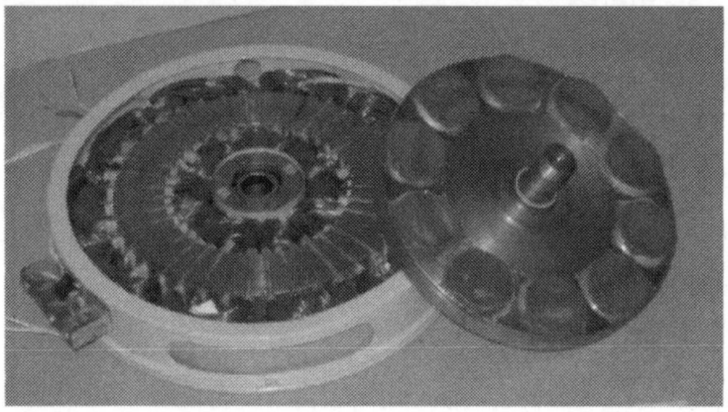

Figure 1. Synchronous axial machine with a permanent magnet rotor

Figure 2. Electromechanical asynchronous converter with a rotor

The stator winding in both cases is flat and the consumption of copper is considerable (Figure 3a). To reduce the amount of copper in axial flow electric machines, it is mounted on the toroidal winding stator (Figure 3b). In this case, it is efficient to use two rotors that are placed on both sides of the stator. The rotors are driven by the rotating magnetic field produced by the three-phase currents, which are closed through the toroidal stator winding.

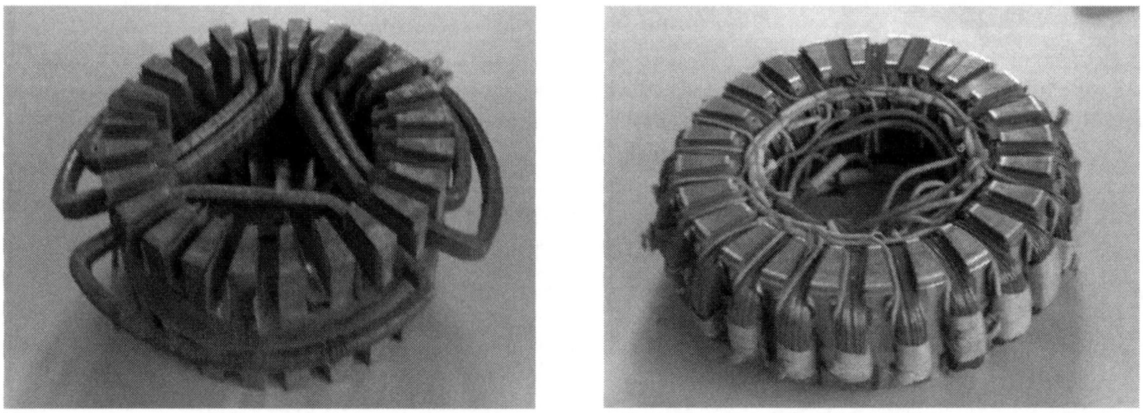

a) b)

Figure 3. Stator axial construction packages with diverse types of windings:
a) Flat winding; b) Toroidal wrapping

The large radial and small axial dimensions of the axial converter allow for a large number of poles and low angular velocity.

The priority of the two-rotor axial machine and toroidal winding consists in the possibility of changing the number of poles because the opening of the winding section does not depend on the polar step as in the case of the winding in machine of the cylindrical construction. Thus, the toroidal winding gives the possibility to realize an axial multi-speed asynchronous converter, using a single winding. This could exclude the gearbox of the vehicle by using an electric switch.

It is then proposed to construct a two-rotor three-phase asynchronous machine (Figure 4). The stator has two packages with rings and mounted on the shaft. This device together with the rotors is mounted by means of the bearings on the shafts. The shafts provide rotation of the rotors in the same direction or in different directions. In the notches of the stator packs are mounted two three-phase windings, these having $2p1 = 2p2$ or a number of different poles $2p1 \neq 2p2$. The rotor windings are short-circuited and cast from aluminum in notches or cast and further mounted in the notches of the rotor package if it has open notches, which is an advantage of this technology.

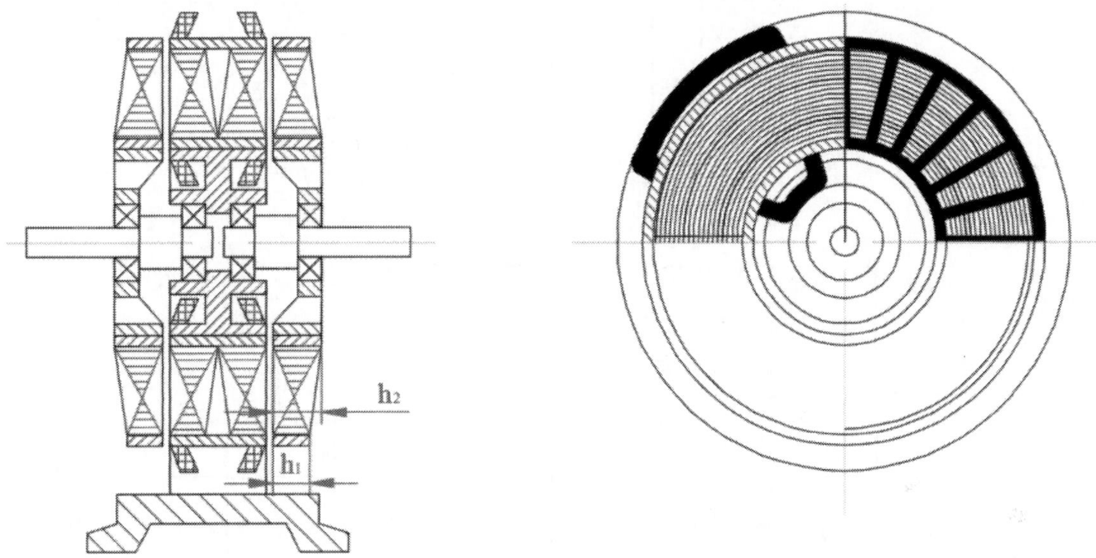

Figure 4. Electromechanical converter with flat stator windings and short-circuited rotor windings

Such a construction of the axial asynchronous machine has certain advantages compared to the cylindrical construction electric machine, being used in motor or generator modes. Firstly, the construction consists of two machines in a single block, these ensuring the reduction of the active and constructive material used in the production. Another advantage is that the machine can operate simultaneously in the motor and generator mode, thus ensuring battery charging and vehicle operation.

The proposed asynchronous machine replaces the differential in vehicle construction because each of the two stator windings can be fed at different frequencies, so the rotors can rotate at different angular speeds.

In the construction of the two-rotor axial machine, the magnetic induction in the air gap and the stator and rotor yokes are distributed unevenly (Figures 5 a, b). For uniform distribution of magnetic induction, the heights of hooks h1 and h2 are different. Thus, the electrotechnical steel is efficiently used and the magnetic induction curve of the air gap near the sinusoid is obtained. The design of the proposed asynchronous machine has reduced axial dimensions and the cooling surface of the rotors is open, therefore the current density in windings can be increased by (1.3-1.5) times.

In order to ensure the operation of the asynchronous machine in generator mode the capacitors are provided in the winding schemes [4], [5].

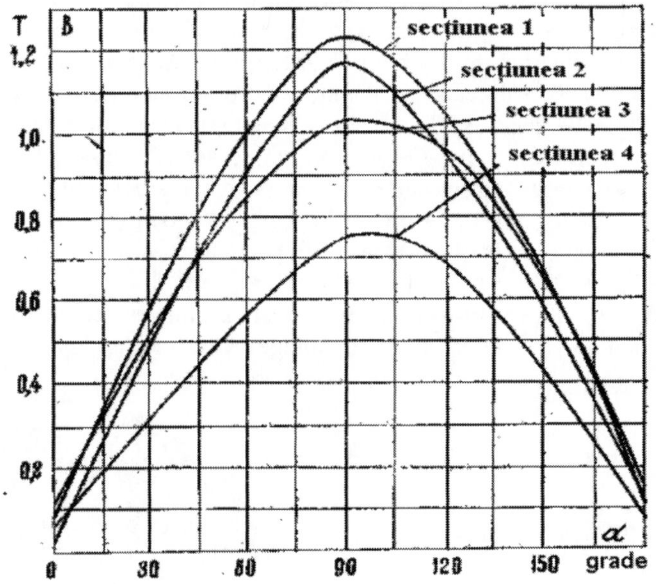

a)

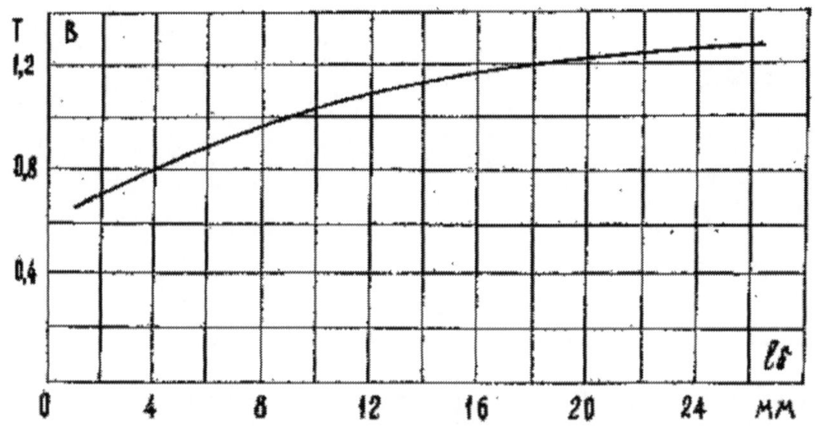

b)

Figure 5. Magnetic induction curve distribution in the rotor yoke:
a) The curves of the magnetic induction variation in the rotor yoke along the polar step;
b) The curve of the magnetic induction variation in the rotor yoke in the radial direction.

3. Correlation between the basic dimensions of cylindrical electric machines and electric machines with axial flux

In the cylindrical asynchronous machine, the magnetic induction in the tooth area is not evenly distributed.

In order to adapt the design methods of axial asynchronous machines to the design of cylindrical electric machines, further attempts are made to analytically establish correlations between the sizes and basic dimensions of these electric machines [6], [7].

Traditionally, it has been found that the basic geometric dimensions of conventional cylindrical electrical construction machines are diameter D and length ls, determined by the type of machine respectively. For synchronous, synchronous and asynchronous AC motors, D is the internal diameter of the stator, and for DC machines D equals the outer diameter of the rotor.

By accepting that some parameters of the electric machine, such as the winding factor K_{W1} and the coefficient of form of the K_B electric motor, have values close to the unit, the expression for the constant of the electric machine can be written as follows:

$$\frac{D^2 \cdot l_\delta \cdot \Omega}{P_c} = \frac{1}{\alpha_\delta \cdot A \cdot B_\delta} \tag{1}$$

Where: P_c – the calculated power of the cylindrical construction machine, A – linear load of current and, B_δ – magnetic induction in the air gap.

Considering that the shape of the magnetic curve can be considered as sinusoidal, the polar coverage coefficient:

$$\alpha_\delta = \frac{2}{\pi}$$

And the expression (1) is immediately as following:

$$\frac{D^2 \cdot l_\delta \cdot \Omega}{P_c} = \frac{\pi}{2 \cdot A \cdot B_\delta} \Rightarrow P_c = \frac{2 \cdot D^2 \cdot l_\delta \cdot \Omega \cdot A \cdot B_\delta}{\pi} \tag{2}$$

In the design of electric machines, regardless of machine construction, calculation methods are used based on the constant of the electric machine, which takes into consideration the power, geometric dimensions and electromagnetic stresses of the projected machine.

From a physical point of view, the constant of the electric machine expresses what power can be obtained from the volume of the active mass of the electric machine at a certain angular velocity Ω. This condition must also be observed for other constructions of electrical machines, such as, for example, the asynchronous machine with axial flux [8].

Figure 6 shows the basic geometric dimensions of the axial electric machine.

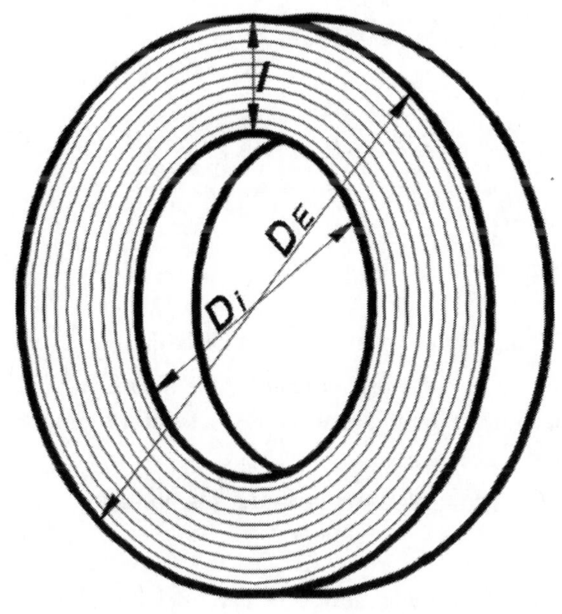

Figure 6. The basic dimensions of the stator packet

In this case the average force on a surface unit F_S acting on the side surfaces of the cylinder is given by the expression

$$F_S = \alpha_\delta \cdot B_{\delta m} \cdot A_m \tag{3}$$

Where: $B_{\delta m}$ – is the mean value in the axial machine air gap, A_m – is the linear load of current corresponding to the mean diameter D_m.

The electromagnetic torque is determined with the expression:

$$M = \frac{1}{2} \cdot \frac{D_E + D_I}{2} \cdot 2S_L \cdot F_S \tag{4}$$

Where the side surface of the active part:

$$S = \frac{\pi \cdot D_E^2}{4} - \frac{\pi \cdot D_I^2}{4} = \frac{\pi}{4}\left(D_E^2 - D_I^2\right) \tag{5}$$

Substituting (3) and (5) into (4) we obtain:

$$M = \frac{1}{2}\frac{D_E + D_I}{2} \cdot \frac{\pi}{2}(D_E^2 - D_I^2) \cdot \alpha_\delta \cdot B_{\delta m} \cdot A_m \tag{6}$$

or

$$M = D_m^2 \cdot l \cdot B_{\delta m} \cdot A_m \tag{7}$$

where

$$D_m = \frac{D_E + D_I}{2}, \quad \frac{(D_E - D_I)}{2} = l$$

The left and right sides of the expression (7) multiply at the angular velocity, so we get:

$$\Omega \cdot M = D_m^2 \cdot l \cdot B_{\delta m} \cdot A_m \cdot \Omega \tag{8}$$

Or the axial machine's calculated power

$$P_a = D_m^2 \cdot l \cdot B_{\delta m} \cdot A_m \cdot \Omega \tag{9}$$

The expression (9) can be written like follows:

$$\frac{D_m^2 \cdot l \cdot \Omega}{P_a} = \frac{1}{B_{\delta m} \cdot A_m} = C_A \tag{10}$$

Then the formula for the constant of the electric machine of axial construction is the same, only that the basic geometric dimensions D_m^2 and l do not express the real volume of the active part. The equivalent volume of the active part of the axial machine (Figure 5) can be given with the product: $D_m^2 \cdot l$, where l is the radial cylinder length of the axial flux machine.

In order to obtain the same power from the same volume of active material for both cases, it is necessary:

$$P_c = P_a \tag{11}$$

According to (2), these powers are different $\frac{2}{\pi}$ times, but the length of the axial machine package is less than the calculus length of the cylindrical machine l_δ about 0.7 times (Figure 6). The advantage of the axial machine is opportunity to save the active material.

In Figure 7, a and b are given graphs of variation $l = f(P_a)$ and $l_\delta = f(P_c)$ for different angular velocities (Ω).

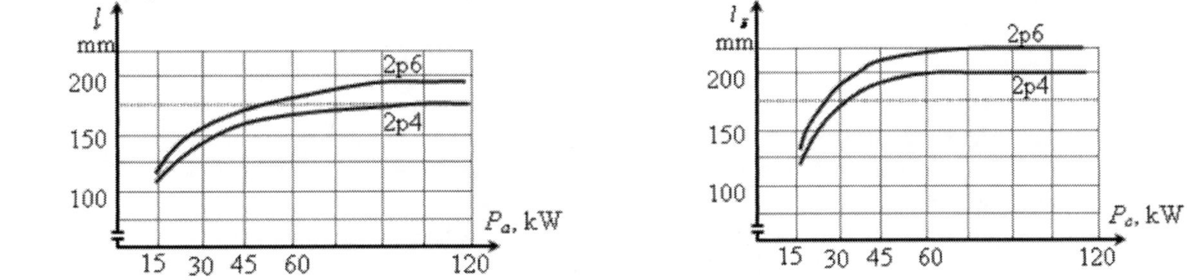

a) b)

Figure 7. The variation graphs of $l = f(P_a)$ and $l_\delta = f(P_c)$

In Figure 8, a and b show the variation graphs for $D_m = f(P_a)$ and $D = f(P_c)$ for various axial and cylindrical machines.

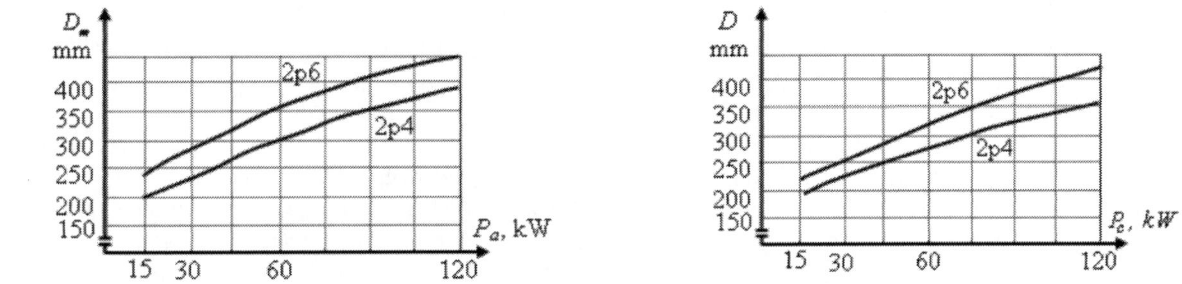

a) b)

Figure 8. The variation graphs for $D_m = f(P_a)$ and $D = f(P_c)$

Knowing the correlations between the cylindrical and axial machine diameters, the cylindrical machine recommendations can be used to design the axial machine.

Therefore, the calculation power for the asynchronous axial flux machine can be calculated with the following expression:

$$P_a^{'} = \frac{k_E P_n}{\eta_n \cos\varphi_n}$$

The coefficient of the electric motor k_E, the n_n and $cos\varphi_n$ yields in the first approximation can be considered as sizes corresponding to the cylindrical machine and the axial magnetic flux machine. Calculated length of magnetic air gap in radial direction:

$$l = \frac{P^{'}}{\alpha_\delta \cdot k_{w1} \cdot A \cdot B_\delta \cdot D_m^2 \cdot \Omega} \quad (12)$$

The outer and inner diameters of the stator packet (Figure 7) are determined, respectively, by the following relations:

$$D_E = D_m + l;$$
$$D_I = D_m - l. \qquad (13)$$

These diameters describe circles that comprise the actual active surfaces of the axial machine.

4. Conclusions

- ✓ Electric motors are used more and more frequently in personal electric transport.
- ✓ Electric axial construction machines have advantages over the cylindrical ones in terms of their use in traction on electric or hybrid vehicles.
- ✓ The proposed three-phase asynchronous machine has less weight and lower cost compared to other types of electric machines.
- ✓ The magnetic induction curve in the air gap is obtained close to the sinusoid by using yokes with different height.

References

[1] Boldea I, Topor M, Marignetti F, Deaconu S I and Tutelea L N 2010 *A Novel Single Stator Dual PM Rotor, Synchronous Machine: topology, circuit model, controlled dynamics simulation and 3D FEM Analysis of Torque Production*, 12th International Conference on Optimization of Electrical and Electronic Equipment OPTIM 2010, Brasov, Romania, May 20-22, pp 343-351

[2] Eastham J F, Profumo F, Tenconi A, Hill-Cottingham R J, Coles P C and Gianolio G 2002 *Novel Axial flux Machine for aircraft drive: design and modeling*, IEEE Transactions on Magnetics **38**(5) 3003-3005

[3] Zbant A, Livadaru L, Simion Al and Adam G 2012 Influence of permanent magnets design over the performance of axial flux synchronous motors, *Buletinul AGIR* **4** 107-112

[4] Simion Al, Munteanu A, Livadaru L and Cojan M 2007 Study on the characteristics of a new hybrid synchronous generator, *Revue Roumaine des Sciences Techniques, Serie Electrotechnique et Energetique* **52**(3) 321-330

[5] Livint Gh, Gaicinschi R and Horga V 2006 *Vehicule Electrice Hibride*, Editura Versus, Iași, Romania

[6] Ambros T, Burduniuc M and Gulpescu D 2002 *The corelation between geometric dimention of cylinder electric maschineand of the maschine with axle wave*, CNAE-2002

[7] Parviainen A 2005 *Design of AFPM low-speed Machines and Performance Comparison between Radial-Flux and Axial-Flux Machines*, Lappeenranta University of Technology, Finland, Doctoral Thesis

[8] Ambros T 2008 *Electrical and Electromechanical special converters* Editura Technica-INFO, Chisinau, Moldova

Electric machines with axial magnetic flux

I Nuca[1], T Ambros[1], M Burduniuc[1], S I Deaconu[2] and A Turcanu[1]

[1]Tehnical University of Moldova, Electrical Engineering Department, 31 August 1989 str., no. 78, MD-2012 Chisinau, Republic of Moldova
[2]Politehnica University of Timisoara, Department of Electrical Engineering and Industrial Informatics, 5 Revolution Street, Hunedoara, 331128, Romania

E-mail: ilie.nuca@ie.utm.md

Abstract. The paper contains information on the performance of axial machines compared to cylindrical ones. At the same time, various constructive schemes of synchronous electromechanical converters with permanent magnets and asynchronous with short-circuited rotor are presented. In the developed constructions, the aim is to maximize the usage of the material of the stator windings. The design elements of the axial machine magnetic system are presented. The FEMM application depicted the array of the magnetic field of an axial machine.

1. Introduction
The study of the axial electric machine, little researched in the past, is a theoretical and practical interest for their design and production in order to increase the efficiency of their use in relation to the use of the cylinder machine with radial magnetic flux.

In this context, the interest refers to the performance of the electric machine with axial magnetic flux relative to the radial magnetic flux.

At present these performances are obvious, namely: reduced axial dimensions, use of cold rolled steel with reduced magnetic losses, reduction of stamping waste, increased stator and rotor cooling surface removed from the stator package, possibility of replacing the mechanical rigid coupling between Stator and rotor with magnetic one.

Some of these performances, particularly eloquent, are seen in the use of axial magnetic flux motors to drive certain mechanisms: centrifugal electric pumps, fans, and, when used as generators, can be driven by wind and water turbine engines. When coupling the axial machines with the above-mentioned mechanisms, the rotor can be cumulated with the traction or drive mechanism, thus making it possible to couple and compensate the unilateral electromagnetic forces between the stator and the rotor. Moreover, in these machines it is possible to obtain reduced angular velocity, since the diameter is increased, allows the stamping of a large number of notches and hence a large number of poles.

The purpose of the paper is to study the construction of synchronous electric machines with axial flux with permanent magnets; the study of stator windings used in electric machines with permanent magnet rotors and the use of inexpensive axial machine manufacturing technologies.

2. Electromechanical converters of various constructions
According to the studied and analyzed sources [1], [2], it was concluded that this type of machines can be classified as follows: with a disc rotor or with multiple disk rotors; with a rotor and a stator, with flat winding on the stator (Figure1); with two rotors and a toroidal winding stator (Figure 2);

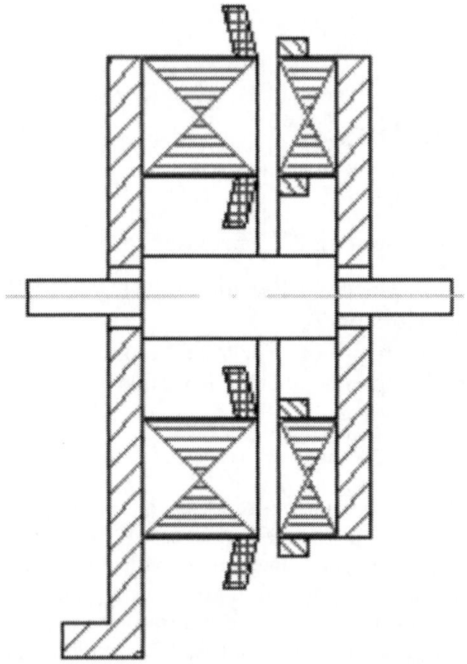

Figure 1. Axial electric machine with one rotor

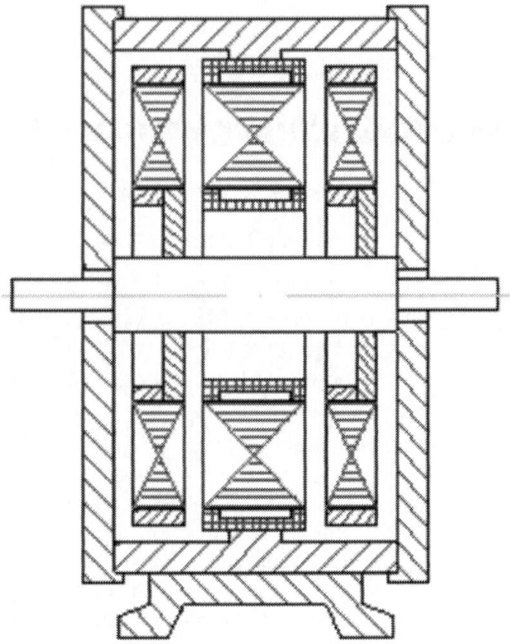

Figure 2. Axial electric machine with two rotors

In most cases, the disc rotor design schemes are proposed for use in automated systems. These motors simultaneously solve two problems: they reduce the moment of inertia and minimize the unilateral forces between the stator and the rotor.

The use of permanent magnets fixed on the rotor allows to obtain a synchronous electromechanical converter. This electric machine has high performance such as: efficient copper use or increasing the length of the active parts of the stator winding. This type of machine can be used in generator mode and in synchronous motor mode (Figure 3).

Axial synchronous generators can be used to generate electricity at wind or hydraulic micro-power plants. They can be used at low speeds and are mechanically coupled directly to the drive motor.

When manufacturing synchronous generators, the active part can be used more efficiently, thus increasing the power of the machine. This can be done by optimizing the construction of synchronous generators.

In Figures 4 and 5 there are schematically shown the constructions of two electromechanical converters with one and two rotors respectively. In both machines, three active parts of the stator windings are effectively used, thus increasing the machine's power in the same volume.

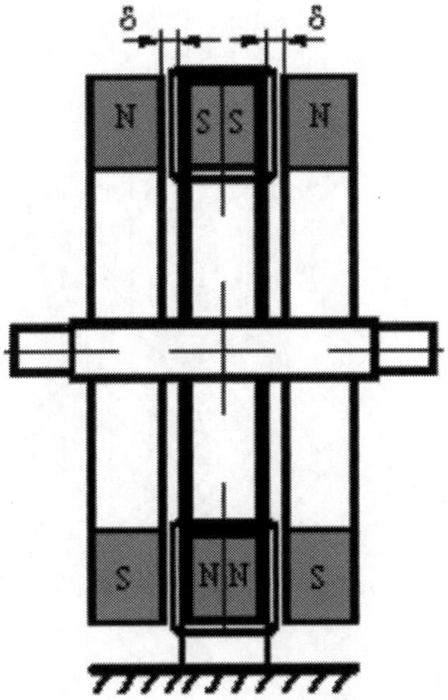

Figure 3. Axial electric machines with two stator and two rotor windings

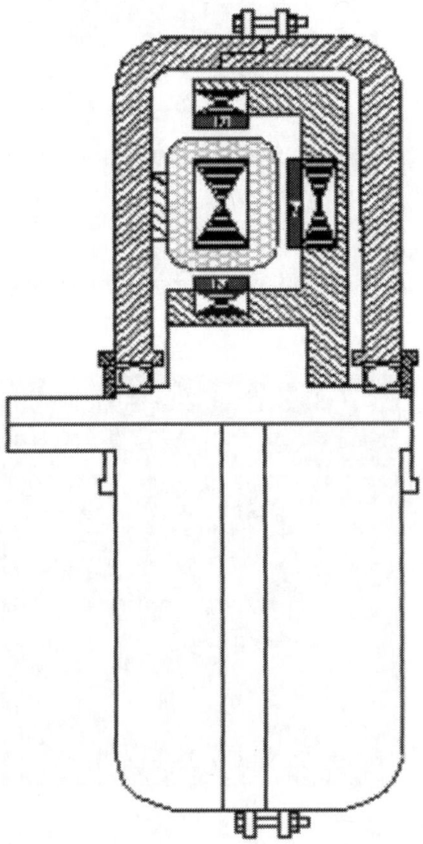

Figure 4. Synchronous axial machine with one rotor

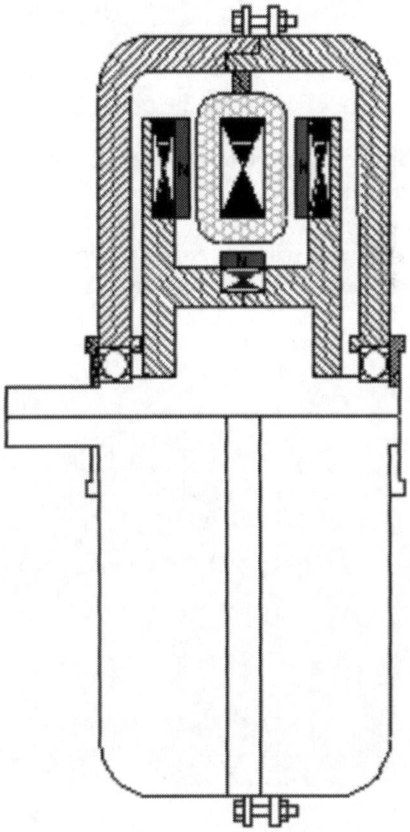

Figure 5. Synchronous axial machine with two rotors

For maximum use of the active material (stator winding), the most efficient would be the use of horseshoe magnets on the rotor (Figures 6 and 7).

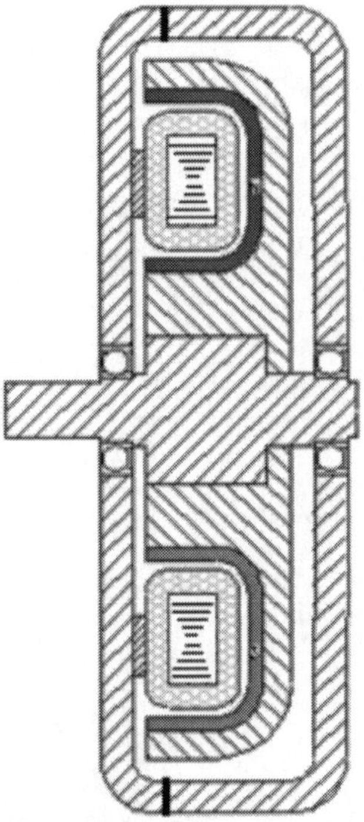

Figure 6. Synchronous axial machine with one rotor

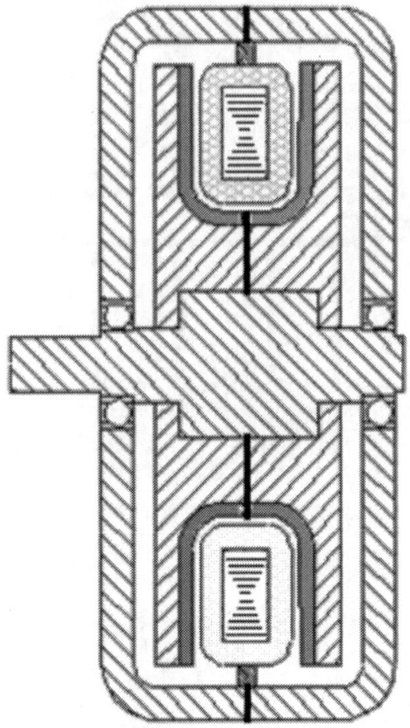

Figure 7. Synchronous axial machine with two rotor

The toroidal winding used in two-rotor axial machines is difficult to achieve from a technological point of view. This has resulted in the construction of the hybrid winding, which is easy to make (Figure 8).

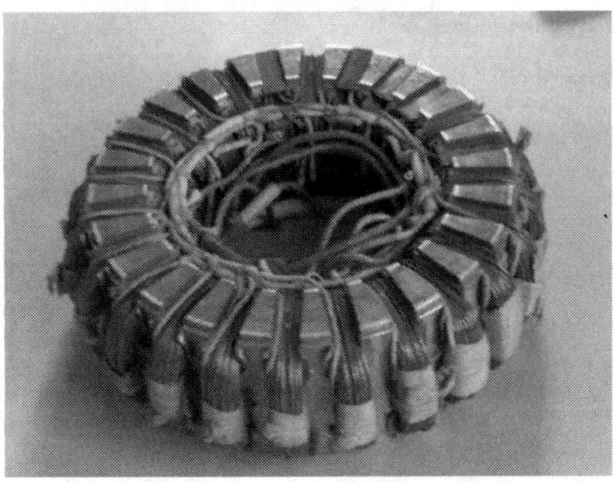

a)

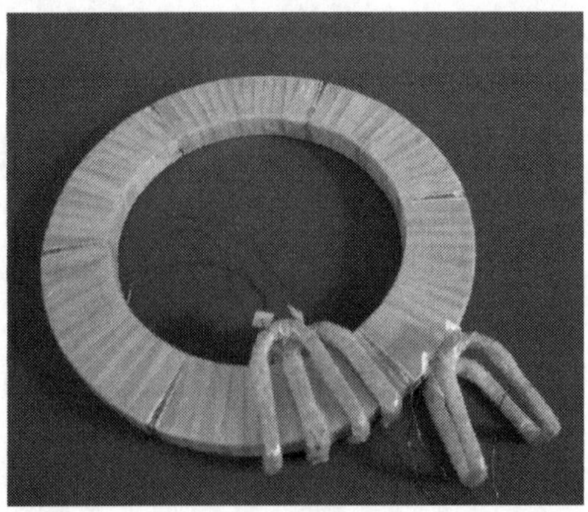

b)

Figure 8. Stator axial construction packages with different types of windings: a) toroidal winding; b) hybrid wrapping

3. Elements of electromagnetic calculation of an asynchronous axial motor with one rotor

The distinctive feature of the electromagnetic calculation for the single-phase axial motor is its equivalence with the cylindrical asynchronous machine.

This equivalence was achieved by determining the ratio between the mean diameter of the axial machine stator pack and the inner diameter of the stator pack of the cylindrical machine. This connection is made using the electric machine constant - Arnolde.

The treatment of electromagnetic computation in this way gives the possibility to use some recommendations, graphical interpretations for the calculation of the cylindrical machine. However, the axial machine electromagnetic calculus is not reduced to the direct use of straightforward recommendations for ordinary asynchronous machines, here the results of the research done in the field of axial machines were applied using classical design methods of normal and special electrical machines [3-5].

3.1. Determination of basic geometric dimensions

Below are some calculation elements of the basic dimensions of the axial asynchronous motor with one rotor and flat winding on the stator.

The calculation of power of the single-phase motor is determined by the following relationship:

$$P = \frac{k_e \cdot P_n}{\eta_n \cdot \cos\varphi_n} \tag{1}$$

where k_e – The coefficient of the electromotive voltage, proportional to the ratio $\frac{E_1}{U_1}$ [6].

In the asynchronous machine one of the basic geometric dimensions is the internal diameter of the stator. To perform the calculations for the axial single-phase motor it is necessary to equalize the single-phase motor with the classic cylindrical engine [7].

Several tests carried out within the Electromechanical Department, of the Technical University of Moldova, in this direction obtained the expression for the average diameter of the axial motor:

$$D_m = \sqrt[3]{\frac{P_1 \cdot 2p}{k_B \cdot \lambda_\delta \cdot \pi \cdot \Omega \cdot k_{W1} \cdot B_\delta \cdot A}} \tag{2}$$

where $\lambda_\delta = \frac{l_\delta}{\tau_m}$, but $\tau_m = \frac{\pi \cdot D_m}{2p}$

The length of the stator packet calculation:

$$l_\delta = \frac{P_1}{k \cdot k_B \cdot k_{W1} \cdot D_m^2 \cdot A \cdot B_\delta \cdot \Omega} \tag{3}$$

where $k = \frac{k_T}{\beta}$, β = (1,25 – 1,7) it is recommended to set the power reduction of the single-phase motor compared to the three-phase motor;

$k_T = 2$ – Indicates the efficiency of axial engine cooling

The values of the magnetic induction B_δ and the linear load of current A vary between B_δ = (0.5 - 0.7) T, and the linear current load A = (20 - 30) · 10^3 A.

The outer and inner diameter of the stator packet

$$D_E = D_m + l_\delta \quad \text{and} \quad D_i = D_m - l_\delta \tag{4}$$

IOP Conf. Series: Materials Science and Engineering **294** (2017) 012059 doi:10.1088/1757-899X/294/1/012059

Phase nominal current:

$$I_n = \frac{P}{\sqrt{2}U_n} \tag{5}$$

Number of notches for pole and phase

$$q = \frac{Z_1}{2pm_1} \tag{6}$$

and the number of conductors in a notch:

$$N_c = \frac{\pi D_m \cdot A}{Z \cdot I_n} \tag{7}$$

Number of turns in the main phase winding:

$$W_1 = \frac{N_c \cdot Z_1}{m_1 \cdot 2a_1} \tag{8}$$

where a_1 – number of current paths.

Magnetic flux closed through the engine air gap:

$$\Phi = \alpha_\delta \cdot \tau_m \cdot l_\delta \cdot B_\delta \tag{9}$$

The number of turns in the main phase is specified with the relationship:

$$W_A = \frac{k_e \cdot U_n}{4k_B \cdot f \cdot k_{W_1}} \tag{10}$$

Also, the number of conductors in the notch is verified

$$N_C = \frac{2m_1 \cdot W_A \cdot a_1}{Z_1} \tag{11}$$

The linear current load is checked:

$$A = \frac{N_C \cdot I_{1n} \cdot Z_1}{\pi \cdot D_m} \tag{12}$$

and the average value of the axial motor diameter [4]:

$$D_m = \sqrt[3]{\frac{P \cdot k_E}{k \cdot k_B \cdot k_{w1} \cdot \Omega \cdot B_\delta \cdot A \cdot \eta_n \cos\varphi_n \cdot l_\delta}} \tag{13}$$

Number of teeth corresponding to the diameters D_E, D_m, D_I - outer, medium, inner.

$$t_{Z1} = \frac{\pi D_E}{Z_1}; \quad t_{Zm1} = \frac{\pi D_m}{Z_1} \text{ şi } t_{Z2} = \frac{\pi D_I}{Z_1} \tag{14}$$

The density of the current in the stator winding is selected depending on the motor execution and varies within the limits $\Delta_1 = (5\text{-}10)\frac{\text{A}}{\text{mm}^2}$.

Cross sectional area of the phase conductor A:

$$S_A = \frac{I_{1n}}{\Delta_1} \tag{15}$$

The width of the tooth is determined by the expression:

$$b_{Z1\min} = \frac{B_\delta \cdot t_{Z1I} \cdot l_\delta}{B_{Z1\max} \cdot l_\delta \cdot k_{FI}} \tag{16}$$

then the width of the notch:

$$b_{C1} = t_{Z1I} - b_{Z1\min} \tag{17}$$

the area of the notch section occupied by the conductors is calculated, then determine the coefficient of filling of the stator groove:

$$k_U = \frac{d_{IZ}^2 \cdot N_C}{S_{C1}} \tag{18}$$

where d_{iz} is the diameter of the insulated conductor, S_{C1} - the area of the notch without insulation

4. Magnetic field analysis in synchronous machines

For determining and specifying the parameters of the proposed synchronous machine, it is necessary to determine the magnetic field produced by windings currents. In this context, the finite element method with FEMM is used [8].

Figure 9 shows the result of the magnetic field analysis of the synchronous machine with toroidal stator winding and permanent magnets on the rotor. In this case the synchronous machine with stator magnetic core was analyzed.

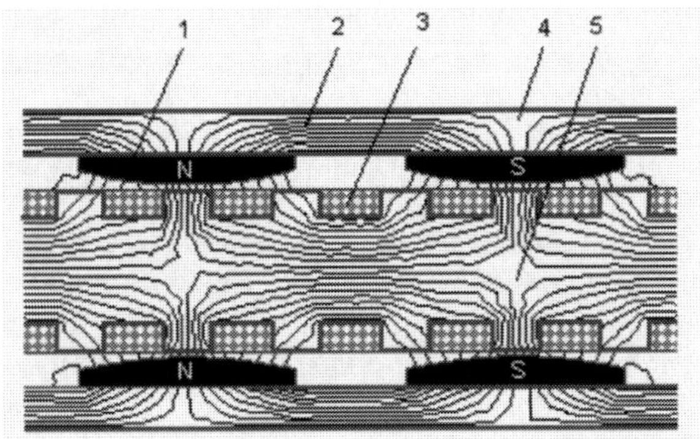

a)

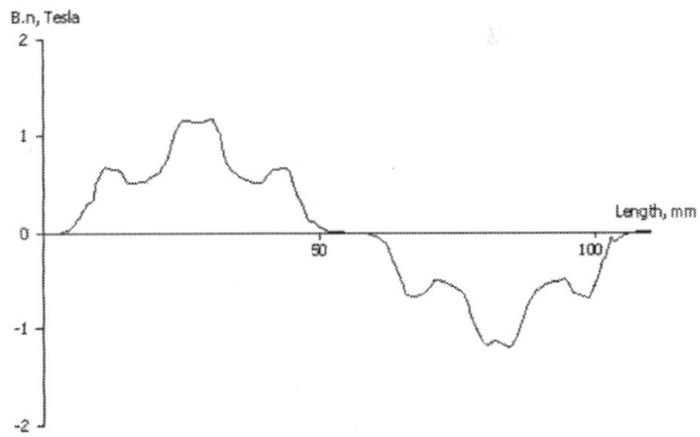

b)

Figure 9 Magnetic field analysis of the axial construction machine with notches on the stator:
a) The field of the magnetic field; b) Magnetic induction curve in the air gap

In order to minimize the action of higher order harmonics, the geometric configuration of the magnetic poles has been modified, so the magnetic induction curve of the practical sinusoidal in the air gap can be obtained, assuming that the stator is made without notches (Figure 10).

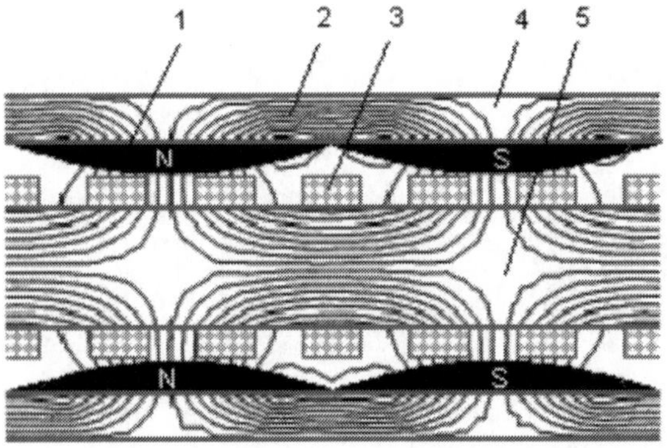

a)

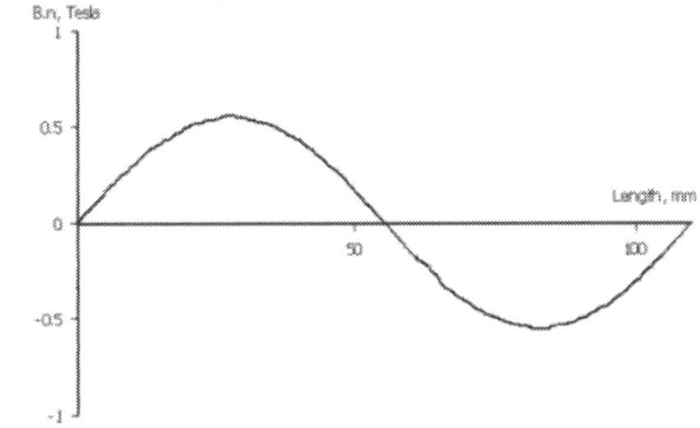

b)

Figure 10. Magnetic field analysis of the axial construction machine with the non-prong stator: a) The field of the magnetic field; b) Magnetic induction curve in the air gap

5. Conclusions

- ✓ Constructive designs schemes of axial electric machine were analyzed.
- ✓ The optimization of synchronous machines with permanent magnets with one rotor or two have been studied.
- ✓ The types of stator windings that can be used on synchronous machines with permanent magnets are described.
- ✓ The FEMM analysis showed that by modifying the construction of the pole magnets and the stator pack, the induction curve of the air gap is practically sinusoidal.

References

[1] Ambros T, Todos P and Tanasă S 1996 *Asynchronous motor with double rotor*, Development and application systems, Suceava

[2] Simion Al, Munteanu A, Livadaru L and Cojan M 2007 Study on the characteristics of a new hybrid synchronous generator, *Revue Roumaine des Sciences Techniques, Serie Electrotechnique et Energetique* **52**(3) 321-330

[3] Ambros T, Ciobanu N and Corjan A 1999 *Determinarea pasului variabil al zonei de dantură în maşinile electrice de curent alternativ cu flux axial*, A doua Conferinţa Internaţională de Sisteme Electromecanice, October 7-8

[4] Ambros T and Gulpescu D 2001 *Consideraţii privind particularităţile constructive şi de proictare ale maşinilor axiale*, Lucrările Conferinţei Internaţionale SIELMEN 2001, Chişinău

[5] Ambros T and Nuca I 1993 *Le processus transitoire o demarage de moteur axial,* ICATE'93 – Craiova, Romania

[6] Lopuhina E M and Somihina G S 1980 *Proektirovanie asinhronnîh mikrodvigatelei s polîm rotorom*, Moskva

[7] Ambros T 2008 *Electrical and Electromechanical special converters*, Editura Technica-INFO, Chisinau, Moldova

[8] Chen J, Nayar C and Xu L 2000 Design and Finite-Element Analysis of an Outer-Rotor Permanent-Magnet Generator for Directly Couple Wind Turbines, *IEEE Transaction on Magnetics* **36**(5)

Analysis of time in establishing synchronization radio communication system with expanded spectrum conditions for communication with mobile robots

T S Latinovic[1], S B Kalabic[2], C R Barz[3], P Paul Petrica[3] and A Pop-Vădean[3]

[1]Faculty of Mechanical Engineering, Banjaluka, Republic of Srpska/B&H
[2]High school Prometej, Banjaluka, Republic of Srpska/ B&H
[3]Technical University of Cluj-Napoca, North University Center of Baia Mare, Romania

E-mail: tihomir.latinovic@mf.unibl.org

Abstract. This paper analyzes the influence of the Doppler Effect on the length of time to establish synchronization pseudorandom sequences in radio communications systems with an expanded spectrum. Also, this paper explores the possibility of using secure wireless communication for modular robots. Wireless communication could be used for local and global communication. We analyzed a radio communication system integrator, including the two effects of the Doppler signal on the duration of establishing synchronization of the received and locally generated pseudorandom sequence. The effects of the impact of the variability of the phase were analyzed between the said sequences and correspondence of the phases of these signals with the interval of time of acquisition of received sequences. An analysis of these impacts is essential in the transmission of signal and protection of the transfer of information in the communication systems with an expanded range (telecommunications, mobile telephony, Global Navigation Satellite System GNSS, and wireless communication). Results show that wireless communication can provide a safety approach for communication with mobile robots.

1. Introduction

Radio communication signal with expanding spectrum was introduced to provide an essential and effective method of protection of transfer signals from interfering and interception. This method of protecting information is more reliable than the previous methods of protection by coding the radio communication signal content. Expansion of the spectrum can be realized by modulating the useful signal with a relative speed pseudorandom sequence, whose generation must be only known to certain users in order to protect information from interference and interception. To make it possible to extract the contents of NF useful signal (speech or some information) from expanded spectrum, it is necessary to generate an identical sequence in a receiver needed for modulation of the useful signal in the transmitter. It is necessary to perform receiver synchronization and locally generated PSS in the receiver. It should be noted that the sequence is pseudorandom and it is very complicated to execute their synchronization thus to enable the compression of the spectrum of the received signal and further extraction of useful information or digitized voice signals by detection which in certain circumstances must not be accessible to unauthorized persons.

These systems are used in certain communications, satellite and aviation systems, different platforms and systems for remote sensing and reconnaissance of facilities, terrain, space research, diplomacy and statesmanship of communication, in order to protect the content of that information which is important to users of communication systems with expanded spectrum. So, to be able to use these communication systems special attention must be paid to the choice of method for synchronizing of pseudorandom transmitting and received sequences, because only due to this synchronization the compression of the spectrum of expanded signal can be achieved from which useful signal can be detected. In literature, scientific and professional papers and publications in IEEE journals various methods of synchronization of PSS for radio transmission were analyzed. Systems for obtaining expanding spectrum are the following:

- Systems with direct sequence – "Direct -Sequence", whose code is DS-SS and modulation performed with fast pseudo-random sequence,
- Systems with the frequency - ("Frequency-Hopping"), whose code is FH-SS, and transfer of information is based on the elements of a standard system for digital transmission of information with incoherent demodulation signal,
- Systems with a change period ("Time-Hopping"), whose code is TH-SS in which expanding the spectrum of the information signal is performed by changing its parameters in the time domain.
- Hybrid systems, whose code is FH / DS-SS, which are obtained by combining certain elements of the three systems explained above [1].

The first radio communication system for transmitting information in the extended range was manufactured in the early 50s of the last century in Lincoln's laboratories for the purposes of the United States' Army, in communications with satellites for research NASA spacecraft, the GPS systems, asteroid belt, and then in NATO in general. By applying the techniques of expanded spectrum a high degree of resistance to impact interference signals and electronic surveillance is achieved thus enabling the system to be used by pre-planned users. The first radio communication systems operated in the band 30-88 MHz, and newer systems in the 1710-1850 MHz and higher frequencies and the width of the frequency band for signal transmission order of 22.5 MHz and data transfer speed of order 32 kb/s to 216 kb/s. Radio communication systems with the PSS are a special class of communication system, characterized by the following characteristics [2]:

- Transmitted signal (an expanded signal) occupies a much wider frequency range than the minimum required to transfer a message,
- Expansion of the frequency spectrum in the transmitter is realized using auxiliary, pseudo-random signal (PSS), which is independent of the message and is known to the receiver. PSS is in principle expressed in digital form, and in the literature it is referred to as PSS sequence or pseudorandom code sequence.

Graphic spectra of the information signal,

$$S_{u1}(f) \equiv S_{ui}(f) \equiv S_i(f)$$

and spectrum of PSS sequences $S_{PSS}(f)$ is shown in Figure 1.

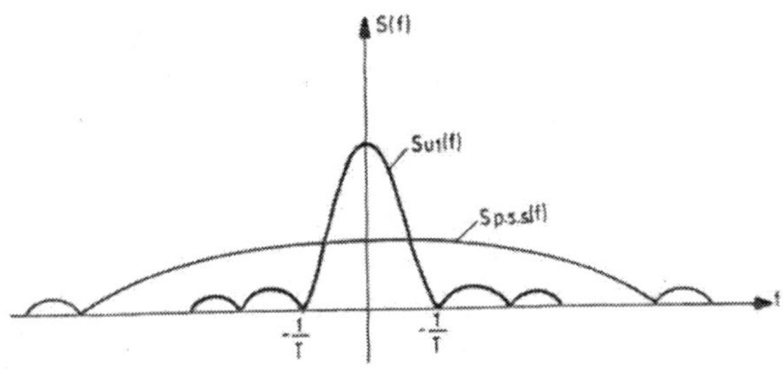

Figure 1. Graphical representation of spectra of the information signal

By applying expanding spectrum the following advantages over the standard signal transmission are achieved:

- Increased electronic reconnaissance of expanded signal,
- Disabled interception,
- The high degree of immunity to electronic interference,

- Improved quality of transmission in radio channels with multiple transfers,
- Multiplex is formed where all channels simultaneously work on the same frequency,
- Selectively addressing individual users of the system,
- The precise location of individual cells in the system and so on.

These solutions can be achieved in small communication systems in a limited space, as well as large systems involving tens of satellites.

2. The merits of signal expansion and compression

Expanding the spectrum is usually achieved by using binary pseudorandom code sequences, which can be reproduced by an authorized user [3]. This code allows the user of the communications system the identification and selective calling. With conventional radio transmitting spectral density of the useful signal $S_i(f)$ is significantly higher than the spectral density of the noise power N_0 [2].

Nonexpanded signal occupies the spectrum width $B_b \equiv n$. In systems with an expanded range $S_{PSS}(f)$, the signal occupies quite a wider range $B_{PSS} \equiv N$, wherein $B_{PSS} >> B_b$. The width ratio of the expanded and nonexpanded spectra, generates a factor of expansion or amplification process:

$$\eta = \frac{B_{PSS}}{B_b} \tag{1}$$

Since $B_{PSS} >> B_b$, this process can gain great value, and it is thus appropriate to expressed this factor in decibels:

$$\eta = 10\log\frac{B_{PSS}}{B_b} = 10\log\eta[dB] \equiv PG \tag{2}$$

In practice, it is usually $\eta = (20-60)[dB]$.

In practice, there is a number of different configurations systems with an expanded spectrum, and their principle of operation is based on a common concept. In fact, the signal spectrum $S_i(f)$, which carries the information in the transmitter where the spectrum is expanded, is transformed by a certain procedure so expansion (enlargement) of its spectrum is executed. The spectrum of the expanded signal can be represented by the relation [4]:

$$S_{PSS}(f) = \in_{PSS}[S_i(f)] \tag{3}$$

where $\in_{PSS}[S_i(f)] \equiv \in_{PSS}[\bullet]$ marks an operation for the expansion of the spectrum signal $S_i(f)$. In the radio receiver system with PSS a reverse process occurs, i.e. the compression of the spectrum of the received signal $S_{PSS}(f)$, using the same procedure $\in_{PSS}[S_i(f)] \equiv \in_{PSS}[\bullet]$, to give a spectrum signal $\hat{S}_i(f)$ from which NF or useful signal can be extracted by detection. Compression can occur only if the PSS generation law in the receiver is exactly the same as in the transmitter. Otherwise, there will be no compression of signal and the useful signal will not possible to detect. The compression of the spectrum of the received signal in the receiver can be displayed in the following form:

$$\hat{S}_i(f) = \in_{PSS}\{S_{PSS}(f)\} = \in_{PSS}\{[S_i(f)]\} = S_i(f) \tag{4}$$

From relation (4) we see that operation $\in_{PSS}\left[S_i(f)\right]\equiv\in_{PSS}[\bullet]$ is inverse of itself, that is $\in^2_{PSS}\left[S_i(f)\right]\equiv\in^2_{PSS}[\bullet]=1$.

So, after a double application of the operation for the expansion of the signal $\in_{PSS}\left[S_i(f)\right]\equiv\in_{PSS}[\bullet]$ to any signal, a signal obtains the original form, i.e. the compression of the spectrum of the received signal $S_{PSS}(f)$ from which NF useful signal can be detected only on this way. Since in the process of expanding the spectrum total energy S of the expanded signal remains unchanged, a condition must be satisfied:

$$S=S_i(f)\cdot B_b=S_{PSS}(f)\cdot B_{PSS}=\hat{S}_i(f)\cdot B_b \tag{5}$$

Both modulator and demodulator must be of the same kind, wherein they may be of any type: AM, FM, SSB, PSK, FSK, etc. PSK is usually used.

By using standard procedure of modulation of RF carrier signal with NF useful signal (e.g. a speech signal) in the transmitter a modulated signal $S(t)$ is obtained, whose spectrum is $S_i(f)$.

Multiplying the resulting modulated signal $S(t)$ with a pseudorandom sequence $C(t)$, whose range is $S(f)$ we get an expanded signal whose spectrum is $S_{PSS}(f)$. In practice, this process is implemented with a balancing modulator. The output level of the transmitter and the receiver input stage include broadband amplifiers, filter, networks for adaptation and so on. The expanded signal $S(t)\cdot C(t)$, during the current propagation can interfered by a signal of narrowband interference $J(t)$ and noise signal $N(t)$ which is time shifted during propagation τ .

In order to extract useful information from the received signal, it is necessary for the generator of PSS sequences in the receiver to generate the exact same sequence of PSS as in the transmitter and the time shifted during time propagation τ , so that by multiplying the received expanded signal, superimposed on the noise and narrow-band disturbance yields:

$$\begin{aligned}&\left[S(t-\tau)\cdot C(t-\tau)+j(t)+N(t)\right]\cdot C(t-\tau)=\\&=S(t-\tau)\cdot C^2(t-\tau)+j(t)\cdot C(t-\tau)+\\&+N(t)\cdot C(t-\tau)\end{aligned}$$

Considering that:

$$C^2(t-\tau)=1$$

This previous expression can be written as:

$$\begin{aligned}&\left[S(t-\tau)\cdot C(t-\tau)+j(t)+N(t)\right]\cdot C(t-\tau)=\\&=S(t-\tau)\cdot+j(t)\cdot C(t-\tau)++N(t)\cdot C(t-\tau)\end{aligned}$$

This process of expanding and compressing signal is illustrated in Figure 2 [4], [5].

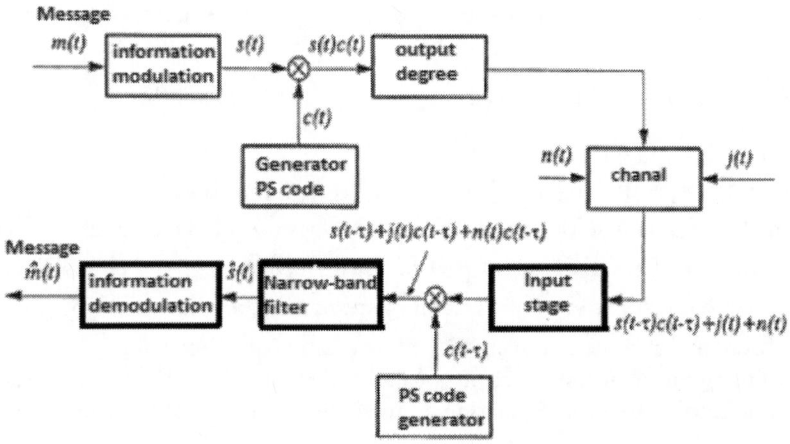

Figure 2. Block diagram of the sending and receiving part of the RK system with expanded spectrum

The total energy of the signal before and after expanding remains unchanged. These energies are defined with terms:

$\int S^2(t)dt$ - energy signal before expanding the spectrum,

$$\int [S(t)C(t)]^2 S^2(t)dt = \left| \begin{matrix} for \\ C^2(t)=1 \Rightarrow \end{matrix} \right| \equiv \int S^2(t)dt \tag{6}$$

energy signals after expanding the spectrum.
From the expression for the energy signal before and after expanding the spectrum, we see that this is the same energy. Therefore, we have the case that the level of energy of expanded signal is lower than the level of noise signal (Figure 3).

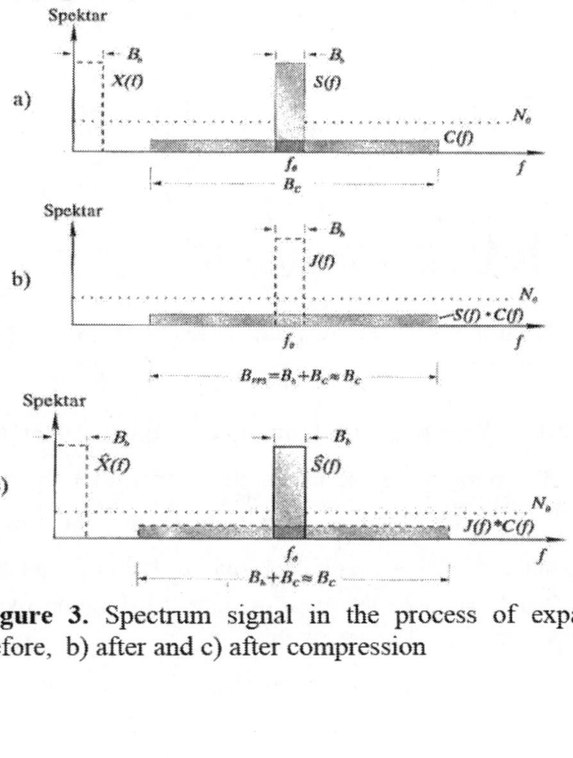

Figure 3. Spectrum signal in the process of expanding a) before, b) after and c) after compression

In Figure 4, spectra of the modulation signal, carrier signal, PSS, narrowband interference, noise and expanded signal are respectively labeled with:

$$X(f),\ S_i(f)=S(f),\ J(f), C(f),\ S(t)*C(t),$$

where convolution is marked with *.

Figure 3.c shows the spectra of waveform signal after the compression of the spectrum in the receiver, from which we see that the range of interference $J(t)$ is extended so that within the bandwidth of MF filter receiver only a small part of the disturbance energy enters. The same happens with noise $N(t)$, to which the operation of expanding spectrum is applied after compression, causing a reduction in the amount and its energy covered by narrowband MF filter.

The process of enlargement of the spectrum of the useful signal is illustrated in Figure 4 which is used in the form of a bipolar signal information (information string) and the pseudorandom sequence [6].

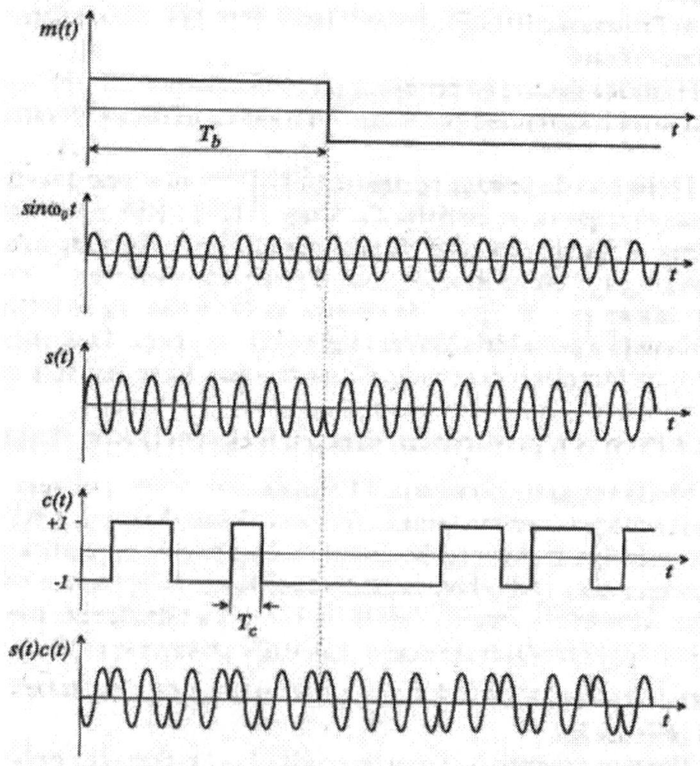

Figure 4. The process of enlargement of the spectrum of the useful signal

Usually, the minimum duration of the interval of the information signal $m(t)$ is referred to as bit, which will be denoted by T_b. The duration of the minimum interval of PSS sequences $c(t)$ will be denoted by T_c, wherein $T_c < T_b$. Usually, within an interval of the bit, the useful signal can be 1000 and more natural PSS of code intervals. Thus, it is typical that $T_b \geq 10^3 T_c$. The common name for one PSS code interval is a chip. The duration of the smallest interval duration of those signals defines bandwidth information and code PSS signals respectively:

$$B_b = \frac{1}{T_b} \qquad B_c = \frac{1}{T_c} \tag{7}$$

The analysis shows that the bandwidth of spectrum signals B_{PSS} is significantly higher than the bandwidth of the signal prior to enlargement (expanding) spectrum B_b, and it is defined by the formula:

$$B_{PSS} = B_b + B_c \tag{8}$$

The factor of expanding or extending the range is defined by the expression:

$$\mu = \frac{B_{PSS}}{B_b} = \frac{B_b + B_c}{B_b} \tag{9}$$

Bearing in mind that $B_c >> B_b$, this expression (9) can be written as:

$$\mu = \frac{B_b + B_c}{B_b} \approx \frac{B_c}{B_b} \tag{10}$$

Based on the definition of bandwidth of coding and information signal, the expression for a factor of expanding the spectrum can be written in the form:

$$\mu \approx \frac{B_c}{B_b} = \frac{T_b}{T_c} \tag{11}$$

which is approximately equal to the process gain defined with expressions (1) and (2).

Two effects occur in the receiver , namely:

- Compression of the received expanded signal $S(t-\tau)\cdot C(t-\tau)$ by re-application operations $\in_{PSS}\left[S_i(f)\right] \equiv \in_{PSS}[\bullet]$, which is practically implemented by ordinary multiplication of the modulated signal with $C(t-\tau)$, which can be seen in Figure 2
- Signal interference and noise by using operation $\in_{PSS}\left[S_i(f)\right] \equiv \in_{PSS}[\bullet]$, i.e. with multiplication with signal $C(t-\tau)$, the range of the width of the spectrum extending gets equal to $B_{J,N} = B_c$

Narrowband filter, whose bandwidth B_b, passes through the energy of the useful signal $S(t-\tau)$ and only part of the energy interference and noise. The strength of this total signal interference at the output of the narrow-band filter is reduced by a factor:

$$\mu \approx \frac{B_c}{B_b} = \frac{T_b}{T_c} \tag{12}$$

The question is how factor of expanding the spectrum is determined by the degree of protection from interference and electronic discovery of content information. For example, when the input of the receiver next to the useful signal $S_{PSS}(f)$, comes also a narrowband disturbance $J(f)$, as shown in Figure 2, then the total signal at the receiver input in the spectral domain can be written as:

$$R(f) = S_{PSS}(f) + J(f) \tag{13}$$

After the application of the operation $\in_{PSS}\left[R(f)\right] \equiv \in_{PSS}[\bullet]$ of the received signal obtains:

$$\hat{S}(f) = \in_{PSS}\{R(f)\} \equiv \\ = \in_{PSS}\{S_{PSS}(f)\} + \\ + \in\{J(f)\} = S(f) + J_{\in}(f) \tag{14}$$

where $J_{\in}(f)$ are spectrum disorders, whose range by applying operations expanding the spectrum $\in_{PSS}[R(f)] \equiv \in_{PSS}[\bullet]$ is extended to the level of the band B_{PSS}. With this operation affecting the disturbance the power of spectral density interference for expanding is reduced in the amount of the coefficient of the (enlargement) spectra. If $J(f) \equiv J$ denoted overall strength of interference, then the ratio of signal/fault within the bandwidth of the useful signal B_b after the operation of range compression is defined by the following expression:

$$\left(\frac{S}{J}\right)_{tyl} = \frac{(S)_{ul}}{\frac{B_b}{B_{PSS}}(J)_{ul}} = \mu\left(\frac{S}{J}\right)_{ul} \tag{15}$$

From expression (15) we see that the ratio of the signal and the interference in the process of expanding the spectrum improves amounting to the factor of expanding (expansion) η of the spectrum.

3. Time of the establishment of synchronization with a single integrator in the presence of Doppler signal

When the Doppler signal is present, then two aspects that influence of interference signals occur in the system for establishing synchronization between the locally generated and received PSS sequence. The first is when the phase difference between the received and locally generated PSS sequence is variable with time. This phase difference will lead to the decrease or increase of the probability of detection P_D, depending on whether the Doppler signal increases or decreases the phase difference between said sequences. The second dominant influence of the Doppler signal is on the length of the average time to establish synchronization. If the phase shift is caused by the Doppler signal in time τ_d, which is approximately the same step time of acquisition, i.e. the establishment of synchronization, then the average or median duration of time to establish synchronization decreases. An analysis of these impacts is quite a complex process, so the process of derivation of the equation for the time of establishing synchronization is quite extensive. If Doppler signal in the chip / second is characterized by Δ_{fc}, then the average length of time steps for search (acquisition) of correspondence between the received and locally generated PSS sequences is defined with the following expression [1]:

$$\mu = \frac{N_u}{q'} + \Delta_{fc} \cdot \tau_d + \Delta_{fc} \cdot \tau_d \cdot K \cdot P_{FA} = \\ = \frac{N_u}{q'} + \Delta_{fc} \cdot \tau_d (1 + K \cdot P_{FA}) \tag{16}$$

where is:

μ - step search (acquisition) cells – parts of the searching signal

$\frac{N_u}{q'}$ - the length of step for acquisitions in the areas of chip signal when there is no activity of the

Doppler signal,
$\Delta_{fc} \cdot \tau_d$ - phase shift code (sample of a sampled signal), originating due to the effects of the Doppler signal at a time,
$\Delta_{fc} \cdot \tau_d \cdot K$ - phase shift code that is created during verification of wrong signal.

If in the equation for the mean time to establish synchronization (acquisition time) when the Doppler signal is not present (execution of this term can be seen in [6] and equitation (16a)).

$$\overline{T}_{ACQ} = \frac{(2-P_D)(1+P_{FA})}{2P_D} \cdot q\tau_d \tag{16a}$$

We change $q = \frac{N_u}{\mu}$ and if the μ classified with expression (16), we get:

$$\begin{aligned}
\overline{T}_{ACQ} &= \frac{(2-P_D)(1+P_{FA})}{2P_D} \cdot q\tau_d = \\
&= \frac{(2-P_D)(1+P_{FA})}{2P_D} \cdot \frac{N_u}{\mu} .\tau_d = \\
&= \frac{(2-P_D)(1+P_{FA})}{2P_D} \cdot N_u .\tau_d \cdot \frac{1}{\mu} = \\
&= \frac{(2-P_D)(1+P_{FA})}{2P_D} \cdot N_u .\tau_d \cdot \\
&\cdot \frac{1}{\frac{N_u}{q'} + \Delta_{fc} \cdot \tau_d (1+K \cdot P_{FA})} = \\
&= \frac{(2-P_D)(1+KP_{FA})}{2P_D \left[\frac{N_u}{q'} + \Delta_{fc} \cdot \tau_d (1+K \cdot P_{FA}) \right]} \cdot N_u .\tau_d = \\
&= \frac{\overline{T}_{ACQ}}{1 + \frac{q'}{N_u} \Delta_{fc} \cdot \tau_d (1+K \cdot P_{FA})} \quad \Rightarrow
\end{aligned}$$

$$T_{ACQ} = \frac{\overline{T}_{ACQ}}{1 + \frac{q'}{N_u} \Delta_{fc} \cdot \tau_d (1+K \cdot P_{FA})}$$

Where $\overline{T}_{ACQ}$ - acquisition time without the Doppler signal.

The same procedure is carried out in the expression for the variance of the time of establishing the synchronization, shown in [6] based on the definition of the time of acquisition, and application of the term to $\overline{T}_{ACQ}$ and the condition that the number of searched phases $K << q$, for which the variance of the acquisition of the total time, can be written in the form:

$$\sigma_{ACQ}^{2}=\tau_{d}^{2}\left(1+KP_{FA}\right)^{2}\cdot q^{2}\cdot\left[\frac{1}{12}+\frac{1}{P_{D}^{2}}-\frac{1}{P_{D}}\right] \tag{17}$$

Introducing expression $q=\frac{N_u}{\mu}$ and (16) for μ we get the following expression for the variance of the total time of acquisition:

$$\begin{aligned}\sigma_{ACQ}^{2}&=\tau_{d}^{2}\left(1+KP_{FA}\right)^{2}\cdot\left[\frac{1}{12}+\frac{1}{P_{D}^{2}}-\frac{1}{P_{D}}\right]\cdot q^{2}=\\&=\tau_{d}^{2}\left(1+KP_{FA}\right)^{2}\cdot\left[\frac{1}{12}+\frac{1}{P_{D}^{2}}-\frac{1}{P_{D}}\right]\cdot\left[\frac{N_u}{\mu}\right]^{2}=\\&=\frac{\tau_{d}^{2}\left(1+KP_{FA}\right)^{2}\cdot\left[\frac{1}{12}+\frac{1}{P_{D}^{2}}-\frac{1}{P_{D}}\right]\cdot N_u^{2}}{\left[\frac{N_u}{q'}+\Delta_{fc}\cdot\tau_d\left(1+K\cdot P_{FA}\right)\right]^{2}}=\\&=\frac{\sigma_{ACQ}^{2}\quad without\ Doppler\ signal}{\left[1+\frac{q'}{N_u}\Delta_{fc}\cdot\tau_d\left(1+K\cdot P_{FA}\right)\right]^{2}}\Rightarrow\end{aligned}$$

$$\sigma_{ACQ}^{2}=\frac{\sigma_{ACQ}^{2}}{\left[1+\frac{q'}{N_u}\Delta_{fc}\cdot\tau_d\left(1+K\cdot P_{FA}\right)\right]^{2}} \tag{18}$$

Considering that Δ_{fc} may be greater or less than zero, the Doppler signal can speed up or slow down the search (acquisition) of correspondence of the received locally generated pseudorandom sequence. Usually it is in the expressions (17) and (18).

$$\Delta_{fc}\cdot\tau_d\left(1+K\cdot P_{FA}\right)<<\frac{N_u}{q'}$$

So, these equations are reduced to equations in terms of the absence of the Doppler signal. When the Doppler signal $\Delta_{fc}\left[chip/\sec\right]=0$, then the correspondence of these equations is hundred percent and $q'=q$.

Mean value of the search of correspondence is calculated by means of analogy with equation (16) by replacing q with $\frac{N_u}{\mu}$ and then the equations for the mean and variance of time are used to establish synchronization.

Based on the expression (16) we can conclude that if Doppler signal is not present, then the mean value of steps for search μ_{sred} correspondence of the received and locally generated pseudorandom sequence is time-invariant. If in addition to the useful signal of information the Doppler signal is present, then the mean value of steps for search μ_{sred} changes with time and this is a function of examined cells. Average values of searched steps μ_{sred} of any cell (segment signal) being searched is equal to the nominal step results $\frac{N_u}{q'}$ (typical $\frac{N_u}{q'} = \frac{1}{2}$ for half a chip or half-cells results) of local PN generator plus the mean change in phase of the received signal during a search of that cell. If $\mu_{sred} \equiv \mu_n$ denotes the mean steps of search of examined cells and $\Delta'_{fc}\left[\text{chip/sec}^2\right]$ denotes the value of the Doppler signal, then the Markov's model set for the time interval search of correspondence of the received and locally generated pseudorandom sequence obtains:

$$\mu_{n+1} = \frac{N_u}{q'} + \Delta_{fc} \cdot \tau_d\left(1 + K \cdot P_{FA}\right) + \\ + \frac{1}{2}\Delta_{fc} \cdot \tau_d^2\left[1 + P_{FA}\left(K^2 + 2K\right)\right] + (19) \\ + n\Delta_{fc}\left[\tau_d\left(1 + K \cdot P_{FA}\right)\right]^2$$

where $n >> 0$ is a number of examined cells.

From the expression (19) we conclude that the mean of the search step is a linear function of the number of cell search. Because of this dependence on $n >> 0$, we cannot directly replace q with $\frac{N_u}{\mu}$ in the equation for the mean time to establish synchronization (acquisition time) , when the Doppler signal is not present (execution of this term can be seen in [6] expression (16a.) in order to get the expression for median time to establish synchronization T_{ACQ} in the presence of the Doppler signal and the Doppler impact. It is more suitable to find the average mean of search steps μ_{sred}, which can be obtained by searching the average value μ_{n+1}, from equation (19) for all search cells q, i.e. [7]

$$\mu = \frac{1}{q}\sum_{n=0}^{q-1}\mu_{n+1} \tag{20}$$

and then to introduce the previously proposed changes in the expression for the mean time to establish synchronization (acquisition time) when the Doppler signal is not present. From equations (19) and (20), the following expression can be reached:

$$\mu_n = \frac{N_u}{q'} + \Delta_{fc} \cdot \tau_d\left(1 + K \cdot P_{FA}\right) + \\ + \frac{1}{2}\Delta_{fc} \cdot \tau_d^2 \cdot q \cdot \left(1 + K \cdot P_{FA}\right)^2 + \\ + K^2 \cdot P_{FA}\left(1 - P_{FA}\right) \tag{21}$$

Since the overall mean of search steps must match the total number of searched chips (cells), it obtains that:

$$q\mu = \sum_{n=0}^{q-1} \mu_{n+1} = N_u \tag{22}$$

Here it is assumed that the amounts Δ_{f_c} and Δ_{fc} are sufficiently small sizes so that the $\mu_{n+1} > 0$ for each $n = 1,2,\ldots\ldots\ldots\ldots,q-1$ and so that the search is performed only in one direction, which is defined by changing the phase of the local PSS signal replacing q with $\frac{N_u}{\mu}$ in equation (21) and after settling of the term, we get a quadratic equation with a variable μ:

$$\begin{aligned} &\mu^2 - \mu\left[\frac{N_u}{q'} + \Delta_{f_c} \cdot \tau_d\left(1 + K \cdot P_{FA}\right) + \right. \\ &\left. + \frac{1}{2}\Delta_{fc} \cdot \tau_d^2 \cdot K^2 \cdot P_{FA}\left(1 - P_{FA}\right)\right] - \\ &- \frac{1}{2}\Delta_{fc} \cdot \tau_d^2 \cdot N_u \cdot \left(1 + K \cdot P_{FA}\right)^2 = 0 \end{aligned} \tag{23}$$

If the following changes are introduced to the equation (23)

$$A = 1.$$

$$\begin{aligned} B = &\left[\frac{N_u}{q'} + \Delta_{f_c} \cdot \tau_d\left(1 + K \cdot P_{FA}\right) + \right. \\ &\left. + \frac{1}{2}\Delta_{fc} \cdot \tau_d^2 \cdot K^2 \cdot P_{FA}\left(1 - P_{FA}\right)\right] \end{aligned} \tag{24}$$

$$C = \frac{1}{2}\Delta_{fc} \cdot \tau_d^2 \cdot N_u \cdot \left(1 + K \cdot P_{FA}\right)^2$$

Then equation (23) can be written in the following form:

$$A\mu^2 - B\mu - C = 0 \tag{25}$$

The solution of the quadratic equation is

$$\mu = \frac{-b \pm \sqrt{b^2 - 4ac}}{2a} = \frac{-(-B) \pm B\sqrt{1 - \frac{4A(-C)}{(-B)^2}}}{2A} =$$

$$= \frac{B + B\sqrt{1 + \frac{4AC}{B^2}}}{2A} \Rightarrow$$

respectively

$$\mu = \frac{B}{2A}\left[1 + \sqrt{1 + \frac{4AC}{B^2}}\right] \tag{26}$$

If in the expression for the mean time to establish synchronization T_{ACQ} the expression (26) is added, and if we replace q with $\frac{N_u}{\mu}$, we obtain an expression for the time of acquisition:

$$\begin{aligned} T_{ACQ} &= \frac{(2-P_D)(1+KP_{FA})}{2P_D \cdot \mu} \cdot N_u.\tau_d = \\ &= \frac{(2-P_D)(1+KP_{FA})}{2P_D \cdot \frac{B}{2A}\left[1+\sqrt{1+\frac{4AC}{B^2}}\right]} \cdot N_u.\tau_d = \\ &= \frac{(2-P_D)(1+KP_{FA})}{P_D \cdot \frac{B}{A}\left[1+\sqrt{1+\frac{4AC}{B^2}}\right]} \cdot N_u.\tau_d \end{aligned} \qquad (27)$$

From this formula can be seen if P_D is larger, time acquisition T_{ACQ} will be less, but the amount of size P_D is affected by the Doppler signal. Analogly and for a variance of time for searching correspondence of the received and locally generated PSS sequences, we get the following expression:

$$\begin{aligned} \sigma_{ACQ}^2 &= \tau_d^2(1+KP_{FA})^2 \cdot \left[\frac{1}{12}+\frac{1}{P_D^2}-\frac{1}{P_D}\right] \cdot q^2 = \\ &= \tau_d^2(1+KP_{FA})^2 \cdot \left[\frac{1}{12}+\frac{1}{P_D^2}-\frac{1}{P_D}\right] \cdot \left[\frac{N_u}{\mu}\right]^2 = \\ &= \tau_d^2(1+KP_{FA})^2 \cdot \left[\frac{1}{12}+\frac{1}{P_D^2}-\frac{1}{P_D}\right] \cdot \\ &\cdot \left[\frac{B}{2A}\left[1+\sqrt{1+\frac{4AC}{B^2}}\right]\right]^{-2} \end{aligned} \qquad (28)$$

If $\Delta_{\dot{f}_c} = \Delta_{fC} = 0$ it obtains that the coefficients are:

$$A = \frac{N}{q}, \quad B = \frac{N}{q} \quad \text{and} \quad C = 0.$$

Based on which we can conclude that the expressions (27) and (28) are reduced the expressions for acquisition time and time variance of search of correspondence of the received and locally generated PSS sequences without the presence of the Doppler signal, respectively. The basic idea of the systems with an expanded spectrum is to reduce the time of acquisition of correspondence and variance of time of search of correspondence of the received and locally generated PSS sequences, which can be achieved by reducing the integration time within the time interval τ_d. This is achieved by reducing the probability of false alarm, increasing the possibility of correct detection of P_D, which is mathematically proven. It is shown that a lot of time of integration increases the possibility of correct detection of P_D, but if in the process of establishing synchronization false alarms appear, the average time increases to establish synchronization for repeating the acquisition process. In order to reduce this time, it is convenient to introduce a system in which the time of the first integration is very short. After the expiry of the first time of integration at the output of the assembly decision, with one integrator (Figure 5) [8].

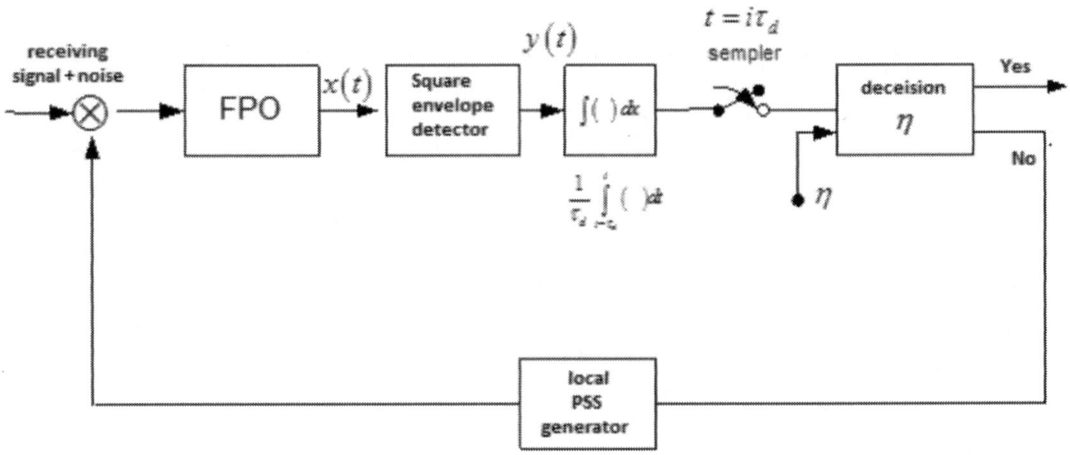

Figure 5. The circuit for establishing synchronization of PSS sequences with one integrator

or more integrators obtains a decision whether there was an establishment of synchronization or not. If synchronization is established, further search of sequence phase is terminated and the process is focused on maintaining synchronization of PSS sequences. The method of sequence detection is suitable for practical application for the establishment of synchronization, especially in the case of long sequences, when the integration time is longer, and when the probability of correct detection of P_D increases and time to establish synchronization and establishing communication between the participants in a communication system [9].

4. Radio Frequency Communication for Mobile Robots

We used the suitability of Wireless Radio Frequency (RF) inter-module communication for mobile robots. Our hypothesis is that, instead of using Infrared (IR) and wired links, RF could be used for local and global communication. Given the need to protect the contents of important information and data in the field of communications, the most reliable systems are with an expanded spectrum, which is why developing these communication systems is specially oriented for communication with mobile robots. Wi-fi Hops about every 0.1 sec, Bluetooth is hopping every 625 microseconds and as fast as 3200 hops a second when initially connecting. Hopping Benefits are:

- Bandwidth can be utilized more efficiently,
- Security level (difficult to intercept)

In order to avoid collisions with other Bluetooth transmitters in the same area, Bluetooth uses pseudo-random generation algorithm to identify the frequency hopping pattern between two devices. [10].

The hopping pattern is based on the Bluetooth Device Address information, a unique identifier that is assigned to every Bluetooth transmitter.

Bluetooth Physical Layer (RF)

Utilizes the free license 2.4GHz ISM radio band

Pseudo-random frequency-hopping scheme with 1600 frequency hops per second (FHSS) 79 carriers (f=2402+k MHz, k=0,...,78)

Performing Gaussian Frequency Shift Keying modulation

Power classes:

– I: Max output power: 100mW (20dBm) => 100m

– II: Max output power: 2.5mW (4dBm) => 20m

– III: Max Output power: 1mW(0dBm) => 10m

5. Conclusions

Given the need to protect the contents of important information and data in the field of communications in the area of dedicated communications, as well as in the field of satellite systems, military systems for communications systems for remote sensing of the universe, diplomatic and statesman communications, the most reliable systems are with an expanded spectrum, which is why special attention has been paid to the development of these communication systems for more than half a century. An important aspect of these communication systems is acquisition and the establishment of synchronization between the received and locally generated pseudorandom sequence. If specified synchronization is not achieved, communication cannot be established and the content of information cannot go through nor there can be a possibility for communication between users of these radio communication systems. Therefore, due attention is paid to the development of a system for establishing synchronization using the sliding correlator with one or more integrators. Simulation of synchronization of PN sequences in the process of establishing synchronization is performed in C++.

Comparing RF communication with the other technologies we found that the main difference between RF, Infrared or Bluetooth is that RF can provide reliable local and global communication without regard to the inter-module docking orientation, mechanism and number of modules connected. Infrared can only provide neighbor-to-neighbor communication and suffers from module misalignment issues, while Bluetooth can communicate locally and globally but requires a central module that constrains the size of the network. With Radio Frequency we did not have the above problems to communicate with mobile robots [11].

References

[1] Dixon R C 1984 *Spread Spectrum Systems*, John Wiley, New York
[2] Kalabic S 1988 *Analysis and Simulation synchronization pseudorandom sequence*, Master's thesis, Electrical Engineering, Zagreb
[3] Haykin S 1994 *Communication Systems*, John Wiley& Amp; Sons, Inc., New York
[4] Sunjeveric M 2004 *The basis of radio communications with radio Technique*, Belgrade
[5] Kalabic S 1988 Basic characteristics of transmission of information spread spectrum, *Bulletin SC The Air Force*, Sarajevo, pp 91-93
[6] Zentner E 2001 *Antennas and Radio Systems*, Zagreb, pp 281-284
[7] Stoy K, Brandt D and Christensen D J 2010 *Self-Reconfigurable Robots An introduction (Intelligent Robotics and Autonomous Agents series),* The MIT Press, pp 27-49 and 86-89
[8] Zentner E 1989 *Wireless Communications*, School Books, Zagreb, pp 199-203
[9] Dixon R C 1994 *Spread Spectrum systems (with commercial Applications)*, John Wiley & Sons, pp 101-105
[10] Yim M, Zhang Y and Duff D 2002 Modular Robots. When a task or terrain varies, reconfigurable robots can change their shape to get the job done, *IEEE Spectrum*, pp 31
[11] Yim M, Shen W M, Salemi B, Rus D, Moll M, Lipson H, Klavins E and Chirikjian G S 2007 Modular Self-Reconfigurable Robot Systems. Challenges and Opportunities for the Future, *IEEE Robotics & Automation Magazine*, pp 45

Arguments for the need of mining education continuity and development in Romania

I Bud, S Duma, I Pasca and D Gusat

Technical University of Cluj Napoca Romania, North University Center of Baia Mare, Department of Mineral Resources, Materials and Environment, 62A, Dr. V. Babeş street, 430324, Baia Mare, România

E-mail: greenminingteam@gmail.com

Abstract. Mining is considered the oldest conscious man activity. In the beginning, man searched for hard rocks in the outskirts area and used it to make weapons and ornaments. Subsequently, civilizations evolved through the development of infrastructure, buildings and monuments and finally, weapons. For all these it was necessary to have mineral raw materials obtained under increasingly difficult conditions, through increasingly evolved techniques. In this way, the art of mining and metallurgy was born, which led to the formation of scientific bases. The mining activity was equally art and science. The art and the science have been learned and taught in schools since ancient times and continue today in large universities with mining engineering, metallurgy, mining topography, mining environmental protection, and geology. Lately, in Romania, the mining high school has reached a deadlock and the middle and professional school has collapsed. The development of infrastructure, construction, etc. requires the exploitation and valorisation of mineral resources based on specialists. The paper warns against the danger of losing tradition and skills in mining engineers formation and militate for the re-establishment of professional and technical schools.

1. Introduction

The mining activity is considered to be the oldest conscious man activity, a statement supported by the fact that the great divisions of human civilization evolution have received designations related to minerals: Stone Age, Bronze Age, Iron Age, Uranium Age.

At the beginning, man was searching for hard rocks in outcrop areas and they used it for weapons and ornaments. Along the time, both rocks range and their areas of use have diversified. Copper and tin were among the first metals known and used in bronze production, approximately 4000 years ago.

The need for expansion and defense have struggling man to develop weapons and the wishes of beauty made him create jewelries, consequently he has developed the most longest-running activity, exploitation and capitalization of mineral resources. Mining activities have extended from small scale up to large and very large scale and have led to the formation of mining school which has become multi and inter disciplinary.

In Romania mining school has developed up to December 1989 Revolution, continuing in the years 90 which, at the beginning of XXI century went into decline. The paper highlights the risk of losing the frame of training of mining specialists to maximize the capitalization of mineral resources potential available in Romania.

2. Mining activity along Romanian community evolution

The precious metals, gold and silver, have been recovered since ancient time. On the Romanian territory, there are a lot of mining activity traces from the most ancient times and the period the most historically known is before the Roman occupation when the Dacians held important gold and silver reserves. These riches have attracted invaders - the most powerful empire of the tine, Roman Empire. After Daco-Roman wars, Trajan armies captured impressive amounts of precious metals which have contributed to the restoration of empire financial situation. The catch of the Roman war has been assessed by some historians to hundreds of gold and silver tons and it is certified that most Rome vestiges were built after Dacia occupation by Romans. After the war, Romans have continued - on a larger scale - the gold exploitation in Dacia. The remainders of this activity are visible today at Roşia Montană.

In Roman period, gold from Dacia was a true "El Dorado" for the populations in the Empire provinces, especially from Minor Asia and Northern Dalmatia contributing to the formation of a new population. The historians estimate that in Dacia mines were used between 15000 and 20000 people obtaining an annual production between 3000 and 5000 kg gold and a double silver production. Accordingly, in 166 years of Roman occupation, estimating a production of only 3000 kg Au/year, from Dacia went outside about 500 t Au and 950 t Ag [1].

This gold odyssey from Apuseni Mountains has continued until our days and has turned into international discord after the estimation of Roşia Montană reserves to about 300 t gold and other valuable items. Because of these disputes mining project is locked with profound consequences on the population in the area, Romanian socio-economic developments and affecting mining education.

Sliding, in time, the situation of Apuseni Mountains in Baia Mare mining area, history repeats, mineral resources capturing the interest of the great world powers. Country mineral riches have attracted along the time the interests of powerful countries, a situation which are repeated even today. The departure of these riches over the borders privates the population of its development potential, the employment places, the source of income and prosperity. Without finding occupations, most of young and adults people, with work power, have taken the foreigners path, searching for the everyday life stability and security. This condition of things was reflected in folklore by metaphors "poor country rich..." or "our mountains golden hold and we beg from door to door...".

During the communist period, the Romanian state has intensively exploited the resources, developing powerful mining sector but in the post-revolutionary period activity has entered total collapse, even in the context of raw materials prices increase and the development of mining activity in numerous European and world countries.

Extraction of solid mineral resources has an old history in Romanian territory (over 2000 years) being characterized by a variety of useful mineral capitalized. Until the end of XX century, mining industry in Romania has known a continuous and relevant development extracting different fossil fuels, ferrous and nonferrous ores, precious and radioactive ores, useful rocks for construction, ornamental rocks, mineral salts, etc.

Deposits of useful substances are spread across the whole country territory, ores being in mountain areas and those of coal and nonmetallic in hill and depressions areas.

In 1990, overall, mining industry has had a maximum production of approx. 160 million tons (coal, ores and salt), there were 278 mines and quarries in operation and 70 processing plants, of which 30 in the sector of metallic iron minerals, 34 in the sector of non-metalliferous and 6 in the coal sector. These industrial objectives have been spread in 41 mining basins located in the territory of the 23 counties [2].

Therefore, mining sector was at that time a socio-economic pole and a way of living for approximately 10 % of Romanian population and has both a well-developed infrastructure and the necessary logistics in the field, impressed by research and design units operating in cities located close to mining perimeters (Baia Mare, Petroşani, Cluj, Deva, Craiova, Bucureşti). At the same time, the training for the necessary staff in mining units (mining workers, technical staff) was ensured by mining college, technical mining schools, the two Institutes of mines in Baia Mare and Petroșani and Geological Institutes in country university centers.

In the main seven mining companies in Romania (metals, salt and coal) the employed staff January 1997 was of 175 879 persons. This statistic doesn't cover employees of the mining units specialized in useful rocks, ballast and small careers. In the same period, there were over 1 500 ballast and quarries with an annual production of more than 150 million tons of minerals aggregates and useful rocks [2].

After 1990, mining sector has entered a permanent regress process, dropping production up to approx. 70 million tons in 2005 and after 2007 when ores and a part of coal mines have close, production decrease was much more obvious.

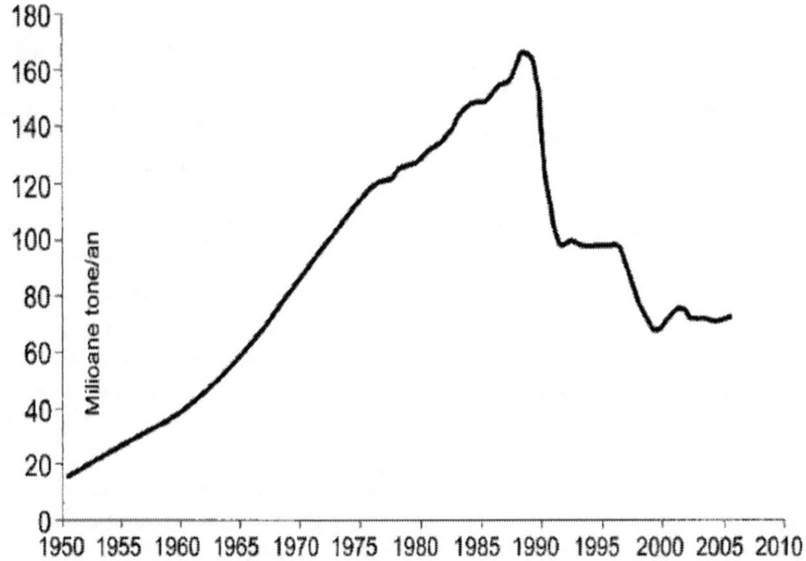

Figure 1. Evolution of total annual extraction (mining mass) 1950-2010 [2]

We had a history and a tradition in mining activity and mineral resources capitalization and we still have potential and for its valorization we need specialists trained in structured and solid mining school, supported and promoted responsible.

Mining activity during the communist period has been developed, sometimes oversized with organization deficiencies which did it - sometimes - inefficient but after 1990, in an economic and political conflict situation it has been strongly compromised: irresponsible organization, concentrates supply and sale through intermediaries, social conflicts and trade unions, etc. All these have led to the total collapse of this activity.

As an example, in comparison with Romanian mining activity evolution, in other European and world countries have elaborated development strategies for mining sector and subsequently for professional education and research, which were applied. The results obtained in these countries are spectaculars and prove the coherence and the pragmatism of their previsions.

In the context of its strategy, Sweden allow a prominent place to education and research development in mining domain. There is also a very clear structure of the economic significance of Mineral industry: operating costs, salaries and pensions, profit, taxes and contributions to the state, dividends to state, dividends to private owners, allowances for the urban transformations, allowances for environment remedies as well as the production for the main minerals and its share in the European economy.

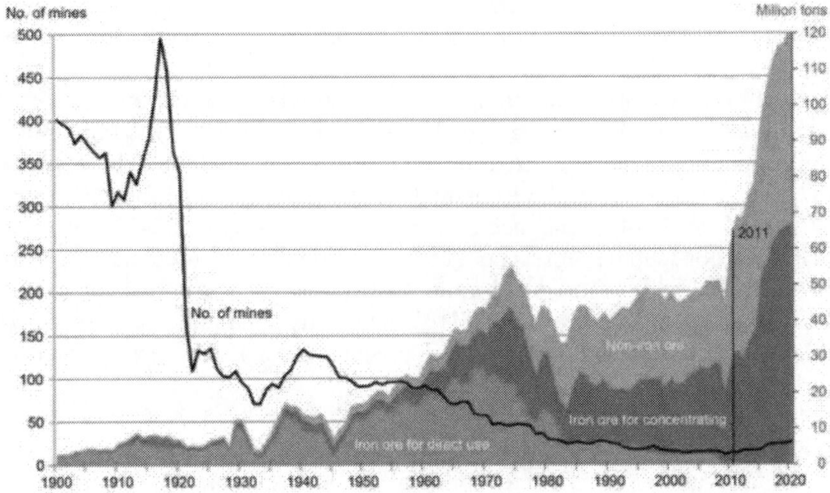

Figure 2. Evolution and prognosis of mining production 1900 – 2020 of Sweden strategy 2013 [3]

Our research team initiated a comparative analysis of Romanian mining industry strategy with development strategies of European and world countries. The conclusions of this research demarche were presented and published.

3. Actual Romanian mineral resources potential

Romania is an important and rich country from the point of view of useful mineral substances because of the 3000 minerals known in the world, 500 can be found in its underground and 35 minerals have been found and described for the first time, here. Today Romania still has over 15 billion tons of total geological reserves of various useful mineral substances which represents a potential mining wealth over 250 billion euro given by the value of the usable mining products after extraction and processing with the current technologies [2].

The total reserves which we have consist of the following: 3 billion tons lignite, 1 billion tons coal, over 750 million tons gold-silver ores, 100 million tons polymetallic ores, 1 billion tons copper ores, 4 billion tons of salt, a few tens of billions of tons of useful and ornamental rocks, etc. [2]. Mining activity continued after 2007 to a different scale and in a different form.

From the public data of NAMR page for 2013 it was recorded 1.382 proposals for new perimeters in the phase of request: 11 permits on exploration; 882 permits on exploitation; 37 licenses for exploration; 33 operating licenses. To illustrate here are a few of active permits and licenses (2014): in Maramures 53; in Cluj 29; in Bistrita 28; in Alba 44 permits + 14 licenses; Arad 38; Timis 35; Constanta 37; Gorj 37 permits + 26 licenses (19 on lignite); Bihor 4 licenses for lignite exploration; reserves of ores Titan, Zirconium, Molybdenum Tellurium, etc. are in the phase of research [4], [5].

In 2016, it were approved 1489 permits and licenses for exploitation and exploration to which are added 510 in approval procedure (940 exploitation permits, 475 operating licenses and 74 exploration licenses). These statistical data show that in Romania today there are approximately 2000 active mining perimeters [5].

Romania is an important country both in surface and population, with a remarkable geo mining potential, exploited and capitalized in historical periods. In the time of mining closure, the idea that our resources would be exhausted or too poor was driven. This statement was contradicted by the results obtained by many private companies in the research and exploration of mining sector. In the moment when somebody have expressed his intention to exploit a detrimental flow claimed that the holding of mineral reserves should be made by Romanian state. However, Romanian state has ceased operating activity at the time in which the price of metals on the Stock Exchange has reached historical level even if there were production capacities, open and/or prepared reserves, experience and qualified staff.

As an example, it may be put into question numerous projects of mineral resources exploitation in Apuseni Mountains and Baia Mare area which are tergiversated for a long time, with adverse consequences on economic, social and environmental protection situation. The exploitation of Rosia Montana is an atypical case in which the Romanian state exploit national mineral resource and sell it as raw material abroad. In 2014, it has been extracted a production of 2.37 million tons ore, after processing the obtained concentrate represents 7264 t Cu and in 2015 production was 2,66 million tons ore, and 7612 t Cu concentrate. The added value lost through the sale of copper in concentrate in comparison with finished product is triple [6], [7]. The benefits obtained by processing some resources until finished product could be used in part to solve environmental problems created by exploitation activity.

4. The need of mining education continuity in Romania

For the pursuit of activity in the field of mineral resources exploitation and capitalization, the formation of specialists at various levels is necessary: professional education, technical and engineering. In the framework of engineering there are three distinct competent authorities: the production engineer, design engineer and research engineer. To get out on the labor market with production engineers, mining engineering high school learn its graduates' knowledge that other engineering specializations do not address: development of the perforation and shooting-blasting techniques monographs, working with explosives, etc.

Technical documents to obtain exploitation permits – laborious materials including much mining specific information – are renewing on annual basis. This work can be carried out only by the specialist mining engineer attested by NAMR.

Each mining perimeter in operation requires at least a mining specialist in view of the problems complexity in this area: technical specific documents, works of deposits opening and preparation, mechanical derockment and mining blasting (only mining school learn its graduates the entire technique of mining blast), transport and climatization, specific rules of labor protection, environment and deposits protection. In addition to higher education units it must function technical schools for staff training: technical mining school, mining blasting school, geology and topography mining school.

At the moment, the lack of interest for the formation of competent staff in the field of mining is reflected in the production activity and the drafting of technical documents: design and execution mining perimeters closures (mining waste deposits, tailing ponds, underground and surface areas); salt mines operation and safety; the manner of drawing up the monographs of perforation and shooting - blasting; the quality of the exploited rocks for infrastructure works (units not complying with technical quality standards concerning characteristics of resistance and geometry).

The road and railway infrastructure development implies the use of huge quantities of hard rocks which are operated by perforation and shooting with explosives. Only mining engineering school learn these techniques.

If the Romanian school no longer forms engineers, companies coming to invest in mineral resources capitalization will be forced to bring specialists from other countries. So, Romania - which was, once, the supplier of well prepared and appreciated specialist in Europe and worldwide - will become importer of mining engineers.

5. Brief history of mining school in Romania

Baia Mare mining education has started in 1864 with 3 years mining school for mine management and organization staff. The school activity has continued up to 1873.

Between 1873 and 1918 the mining school moved on and has functioned in Baia Sprie. In 1919, it comes back to Baia Mare and has two specializations: mining and metallurgy, with Romanian teaching language. In 1937, Baia Mare mining school achieve the qualification of school for technical mining and chemical - metallurgical conductors. In troubled years between 1940 and 1945 Baia Mare mining school find refuge in Brad.

In 1945 in Baia Mare appears technical mining school which keep the same status up to 1948, in 1949 it moves in Brad achieving the degree „Institut de Ingineri" (Engineering high school). In 1948 by Romanian education reform, it was transformed in 5 years mining engineering institute.

In 1969 was set up Technical Institute of Engineers Baia Mare, under the patronage of Polytechnic Institute of Cluj Napoca.

The 1st of October 1974, it was set up the Higher Education Institute, with two faculties: Faculty of Pedagogical Education and Faculty of Technical Education. The technical domain is structured in various specializations: Mining, Technological Electromechanical Mining, Nonferrous Metallurgy and Civil constructions, training technical engineers, day and evening classes.

The 1st of October 1984, Higher Education Institute from Baia Mare go back under the patronage of Polytechnic Institute of Cluj Napoca with the name Technical Engineering Institute.

Starting with 1990, Technical Engineering Institute came back to the old denomination of Higher Education Institute of Baia Mare, which by Government Decision it is transformed in Baia mare University. This one become, in 1996, North University of Baia Mare and in its frame functions Mining Faculty (1990-1996), then the denomination changes back to Mining and Metallurgy Faculty (1996-2001), and later Mineral Resources and Environment Faculty (2002-2014).

In 2012, after the fusion with Cluj Napoca Technical University, North University become North University Centre of Baia Mare and in 2014, Mining Engineering specialization, by reorganization, belong to Engineering Faculty, Department of Mineral Resources Engineering, Materials and Environment.

Other mining schools in Romania. In 1835, it was set up in Săcărâmb, the first School (Middle Mining School) with mining specialization in Transilvania and South-East Europe with German teaching Language, up to 1881, subsequent, in Hungarian teaching language up to 1906. In 1862 this one is transformed in Technical Engineers Institute with foreign teachers.

Figure 3. Mining school between 1873 – 1918 in Baia Sprie [4]

In 1864, by the decision of Alexandru Ioan Cuza, it was set up Superior School of Bridges, Roads, Mine and Architecture and then, in 1920, by a Decision of Ferdinand the king of Romania, it was set up Polytechnic Schools in Bucureşti and Timişoara, each with one Faculty of Mining and Metallurgy.

Figure 4. Middle Mining School (in German language teaching) 1835 – Săcărâmb [4]

After the 1948 reform, in București was set up Mining Institute and in Petroșani, Coal Institute. Both concentrate after 1957 in Petroșani Mining Institute.

In conclusion (in 2017), along 153 years, we speak about Mining Education in Baia Mare (with interruptions between 1940 – 1945 and 1949 - 1969), 70 years in Săcărâmb, 93 years in București (1864 – 1957) and 69 years in Petroșani (starting in 1948).

6. Conclusions

Exploitation mining activities shall be carried out at different scales: small production capacities in which are involved small groups of people or even individuals, activities called generically " artisanal mining" up to enormous production capacities, millions and tens of millions of tons per year, carried out by large companies which allocate considerable resources, making investments of the order of billions of dollars in a single mining perimeter.

Mineral resources belong to the state and, consequently, regardless of the scale of the works, it is required a technical documentation and a legal framework for the activity pursuit. For artisanal mining Romania has no rules to encourage the small entrepreneurs to legalize their activity which generates numerous illegal activities from which, both the state and the developer are losing.

The establishment of a legal and responsible framework for the mining at any level implies the existence on the labor market of specialists trained in technical or higher schools. Those formed in technical schools will activate at the level of execution regardless the production capacity and the graduates of higher schools will involve at the level of execution in medium and large activities, up to design and research. The technical documentation to obtain the exploitation processing and recovery permit, regardless the scale of the works, presupposes the involvement of a specialist mining engineer.

In recent years, medium and vocational mining school has completely disappeared and the superior school, represented in Petroșani and Baia Mare, crossed periods particularly heavy, experiencing the lack of candidates and difficulties about work places for graduates, depriving a country with a remarkable mining potential, of the most valuable resource, the human. The number of graduates of the two schools was small and the request for mining engineers has occurred in regions distant from the university centers.

The need for mining engineers appears mainly in open pits operating by perforation and blasting. Non-availability of them has made that work performed by other specialties engineers or companies from abroad. The results of this condition of things are reflected in the technical state of mining activities and resulted mining products: bad geometry of mining perimeters, incorrect blasting with important percentages of large blocks, loads of explosive not blasted after shooting (extremely dangerous items), units not complying with quality characteristics on mechanic resistances, geometry and particle size distribution imposed by Quality Technical Normative for roads, bridges, railways, civil and industrial construction works, rehabilitations and construction of wastes deposits, etc.

Our documentary and research intention has the purpose to argue the need for continuation and development of mining school in Romania, to sensitizing and warn concerning the risk of losing the tradition and skills required by the training of mining industry specialists. If, for superior school there is a chance for continuity and development, vocational and technical school has lost - for the moment - perspective. All this is happening while in the neighboring countries and all over the world mining school develops at all levels.

References

[1] Fodor D 2005 *Mining history pages*, Publishing House INFOMIN, Deva
[2] Fodor D 2015 *Mining and Environment*, Publishing House Corvin, Deva
[3] Bud I, Duma S, Paşca I, Gusat D and Bud A 2016 The collapse of mining industry in Romania in the period of development oportunities through coherent strategies, *Scientific Bulletin Series D: Mining, Mineral Processing* **XXX**(1) 7-8
[4] Bud I, Duma S, Paşca I, Gusat D and Bud A 2015 150 years retrospective of mining education in Baia Mare, *Scientific Bulletin Series D: Mining, Mineral Processing, Non-Ferrous Metallurgy, Geology & Environmental Engineering* **XXIX**(1) 7-12
[5] ***www.namr.ro
[6] Bud I, Paşca I, Duma S and Gusat D 2016 Paradigms of Copper exploitation and capitalization in Romania, *Mining Revue/Revista Minelor* **XXII**(1) 13-15
[7] ***www.cuprumin.ro

The influence of cooling parameters on the speed of continuous steel casting

G O Tirian[1], C A Gheorghiu[2] , T Hepuţ[2] and C P Chioncel[3]

[1]Politehnica University of Timisoara, Electrotechnical Engineering and Industrial Informatics Department, 5 Revolution Street, Hunedoara, 331128, Romania
[2]Politehnica University of Timisoara, Department of Engineering and Management, 5 Revolution Street, Hunedoara, 331128, Romania
[3]Eftimie Murgu University/Electrical Engineering and Informatics, Resita, Romania

E-mail: ovidiu.tirian@fih.upt.ro

Abstract. This paper analyzes the cooling parameters of the continuous casting speed. In the researches carried out we aimed to establish some correlation equations between the parameters characterizing the continuous casting process, the temperature of the steel at the entrance to the crystallizer, the superheating of the steel and the flow of the cooling water in the crystallizer and different zones of the secondary cooling. Parallel to these parameters were also the values for the casting speed. The research was made for the casting of round ϕ270mm semi-finished steel products. The steel was developed in an electric EBT furnace with a capacity of 100t, treated in L.F. (Ladle - Furnace) and VD (Vacuum-Degassing) and poured in a 5-wire continuous casting plant. The obtained data was processed in MATLAB using three types of correlation equations. The obtained results are presented both in the analytical and graphical form, each correlation being analyzed from the technological point of view, indicating the optimal values for the independent parameters monitored. In the analysis we present a comparison between the results obtained after the three types of equations for each correlation.

1. Introduction

A very important component of the continuous casting installation is the secondary cooling zone. The secondary cooling zone has the role to continue the wire cooling after it has emerged from the crystallizing and to assure the total solidification of the semi –product. It is considered "the heart" of a continuous casting and has the role of ensuring the quality of the material, the material surface shape and has to ensure a homogeneous cooling and a uniform repartition of the water on the materials surface [1-3].

In most cases the crust doesn't offer sufficient mechanical resistance to the action of ferro static pressure. To complete solidification and guidance in good conditions of the wire it is created the secondary cooling zone. This cooling is achieved by direct spraying pressurized water, through nozzles, it is able to cross the steam layer formed by evaporation and ensure continuous water-metal contact [4-7].

The secondary cooling can be made in different cooling environments. In practice, the water is used especially as cooling agent that is sprayed through nozzles (circular conical, conical ring and

IOP Conf. Series: Materials Science and Engineering **294** (2017) 012062 doi:10.1088/1757-899X/294/1/012062

slot). In special cases it is added to the water compressed air for optimizing the automation sprayed water [7-10].

2. Data processing

Data obtained was processed in MATLAB using three types of correlation equations (z - dependent parameter; x, y - independent parameters):

$$z=a_1x^2+a_2y^2+a_3xy+a_4x+a_5y+a_6 \quad (1)$$

$$z = a_1+a_2x+a_3x^2+a_4x^3+a_5y+a_6y^2+a_7y^3+a_8y^4+a_9y^5 \quad (2)$$

$$z=a_1+a_2\log(x)+a_3\log(x)^2+a_4\log(x)^3+\frac{a_5}{y}+\frac{a_6}{y^2}+\frac{a_7}{y^3}+\frac{a_8}{y^4}+\frac{a_9}{y^5} \quad (3)$$

z - casting speed, m / min
x - steel temperature at casting [^{0}C], for all correlations;
y - cooling water flow in zone 1 [l / min], for the correlations in Figures 1, 2 and 3;
y - cooling water flow in zone 2 [l / min], for the correlations in Figures 4, 5 and 6;
y - cooling water flow in zone 3 [l / min], for the correlations in Figures 7, 8 and 9.

The obtained results are presented in graphic form, as follows: a) the correlation surface; b) in the projection plane contour; c) the spatial contour projection.

In a first analysis, it can be observed that equation 1 has the simplest analytical form (polynomial of degree 2) compared to polynomial level 5 (equation 2) and especially with equation 3 (logarithmic - polynomial combination).

3. Results

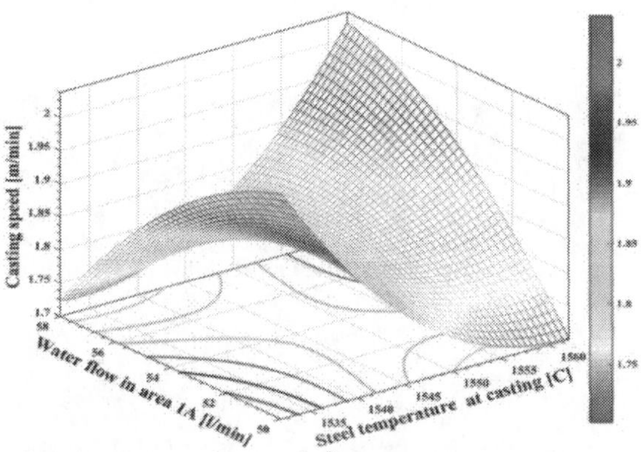

a)

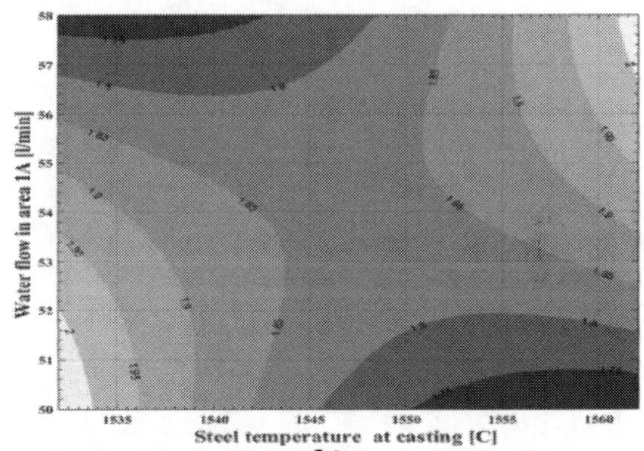

b)

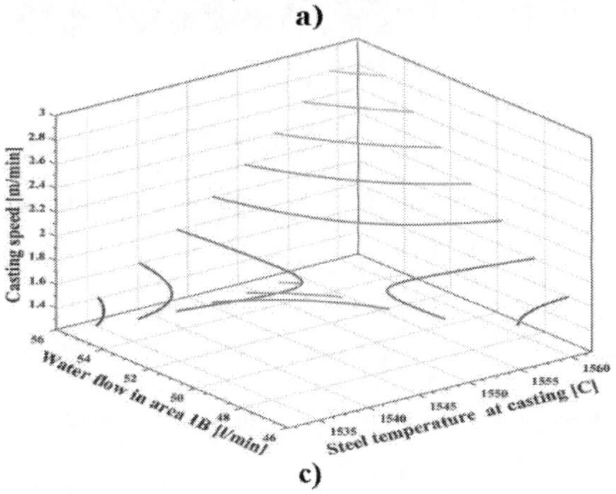

c)

Figure 1. Casting speed= f (Steel temperature at casting, Water flow in area 1, equation 1);
a) correlation surface; b) outiline in plain; c) contour in space;

Equation 1:

$$z=a_1x^2+a_2y^2+a_3xy+a_4x+a_5y+a_6$$

R_2= 0.71748;
a_1=0.00065; a_2=0.00258;
a_3=0.00925; a_4=-2.47158;
a_5=-14.56371; a_6=2274.28489;

IOP Conf. Series: Materials Science and Engineering **294** (2017) 012062 doi:10.1088/1757-899X/294/1/012062

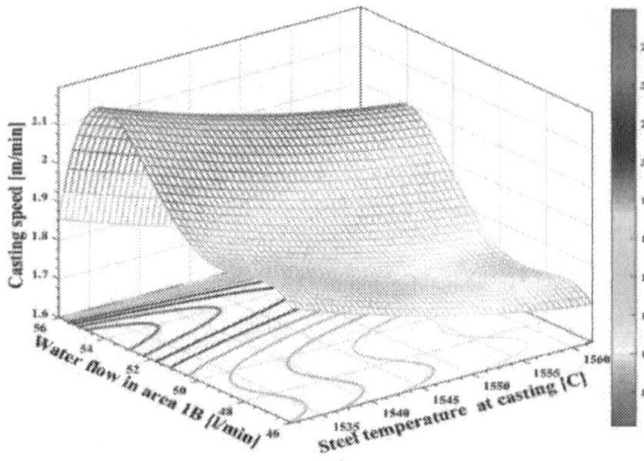

a)

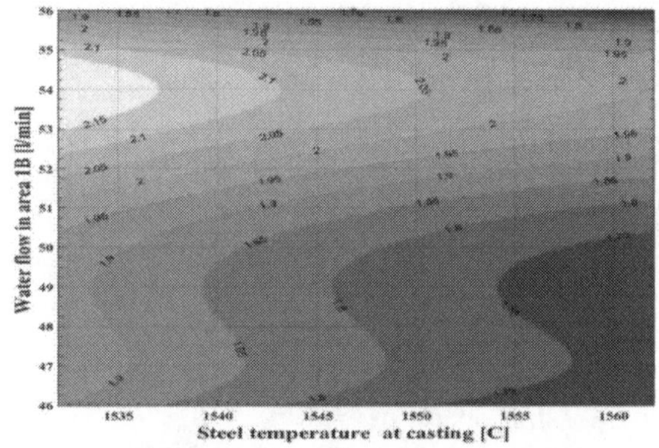

b)

Figure 2. Casting speed= f (Steel temperature at casting, Water flow in area 1, equation 2); a) correlation surface; b) outiline in plain; c) contour in space;

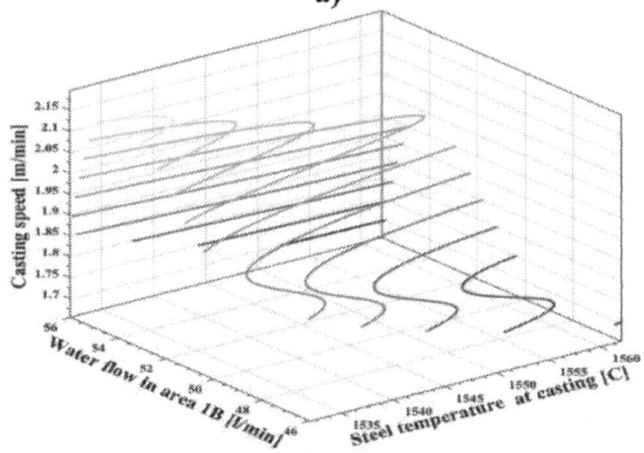

c)

Equation 2:

$$z = a_1+a_2x+a_3x^2+a_4x^3+a_5y+a_6y^2+a_7y^3+a_8y^4+a_9y^5$$

R^2 = 0.79421;

a_1=0; a_2=-1.59968;
a_3=0.00092; a_4=0;
a_5=0; a_6=3.71578;
a_7=-0.14897; a_8=0.00223;
a_9=0;

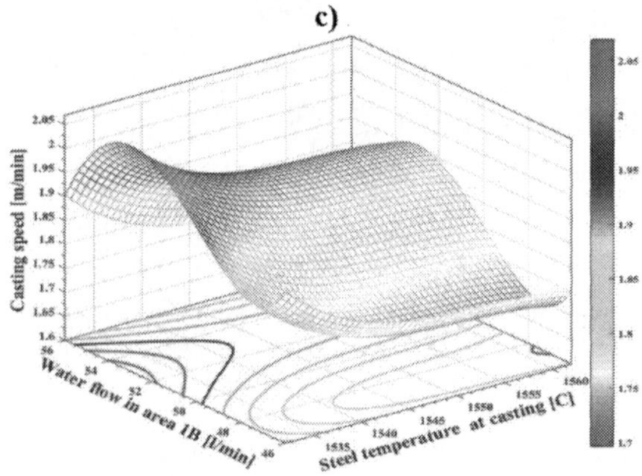

a)

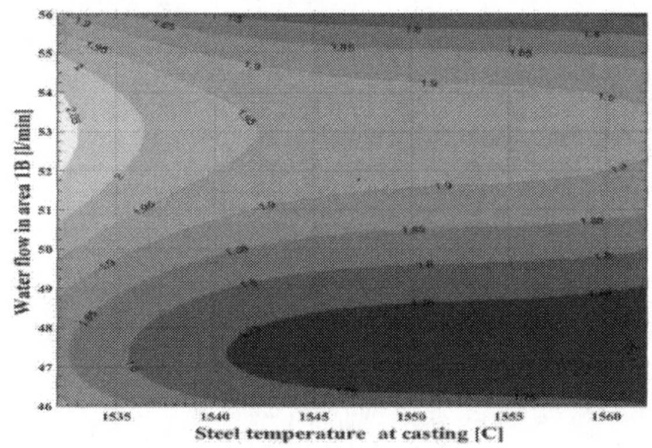

b)

Figure 3. Casting speed= f (Steel temperature at casting, Water flow in area 1, equation 3); a) correlation surface; b) outiline in plain; c) contour in space;

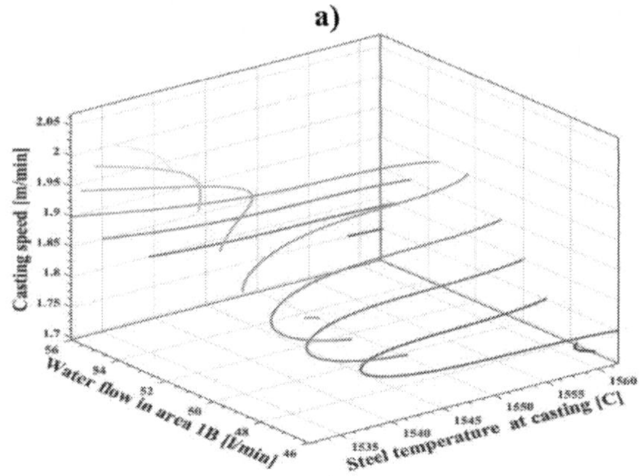

c)

Equation 3:

$$z=a_1+a_2\log(x)+a_3\log(x)^2+a_4\log(x)^3+$$
$$+\frac{a_5}{y}+\frac{a_6}{y^2}+\frac{a_7}{y^3}+\frac{a_8}{y^4}+\frac{a_9}{y^5}$$

R^2 = 0.702100744917242

a_1=21424; a_2=-87475;
a_3=11905; a_4=-54000;
a_5=44161; a_6=-22154;

a_7=36930; a_8=0; a_9=0;

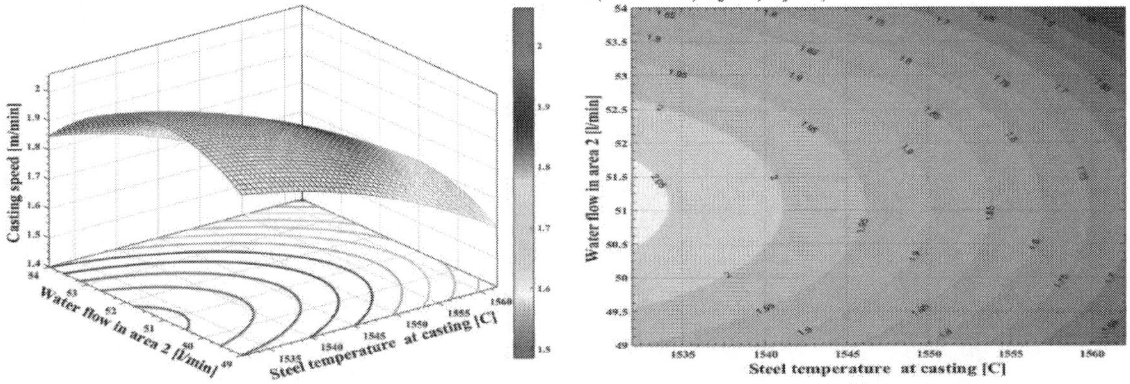

a) b)

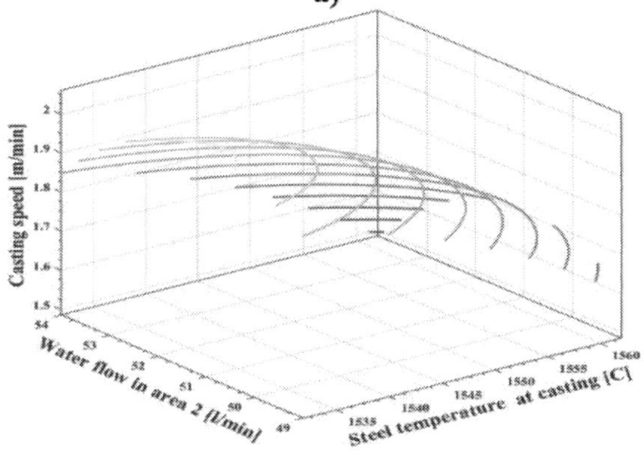

c)

Figure 4. Casting speed= f (Steel temperature at casting, Water flow in area 2, equation 1);
a) correlation surface; b) outiline in plain; c) contour in space;

Equation 1:

$$z=a_1x^2+a_2y^2+a_3xy+a_4x+a_5y+a_6$$

R_2= 0.76230;

a_1=-0.00023; a_2=-0.02534;
a_3=-0.00013; a_4=0.73284;
a_5=2.79633; a_6=-627.26823;

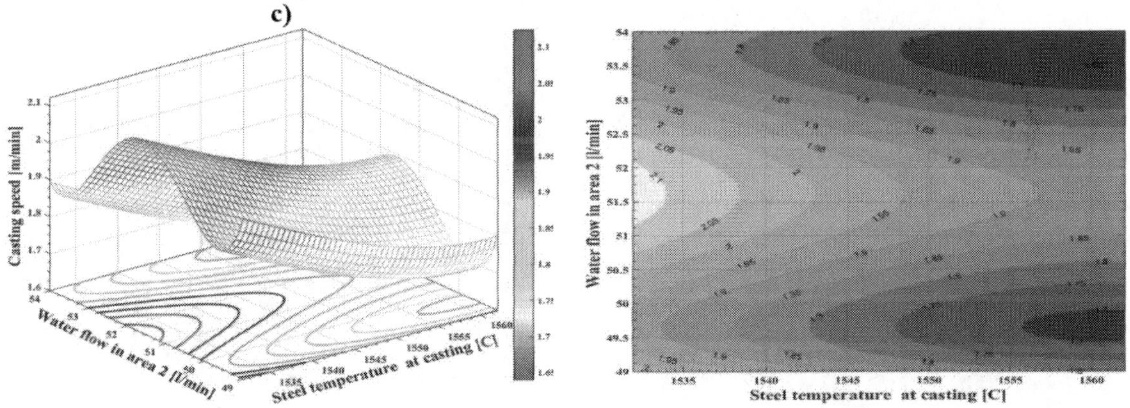

a) b)

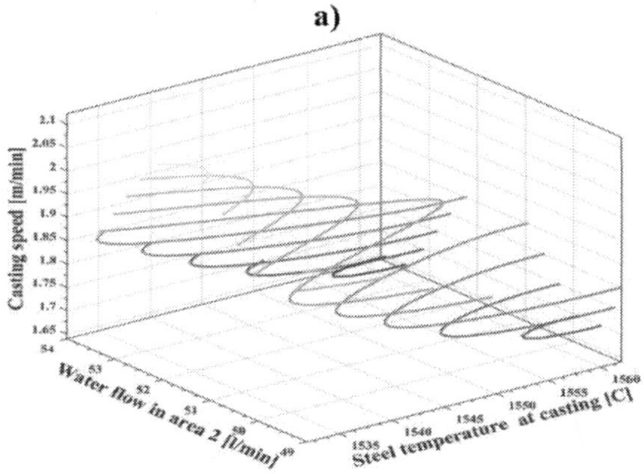

Figure 5. Casting speed= f (Steel temperature at casting, Water flow in area 2, equation 2);
a) correlation surface; b) outiline in plain; c) contour in space;

Equation 2:

$$z = a_1+a_2x+a_3x^2+a_4x^3+a_5y+a_6y^2+a_7y^3+a_8y^4+a_9y^5$$

R^2 = 0.86142;

a_1=0; a_2=36.64747;
a_3=-0.02389; a_4=0;
a_5=0; a_6=-70.41481;
a_7=2.72890; a_8=-0.03963;

c)

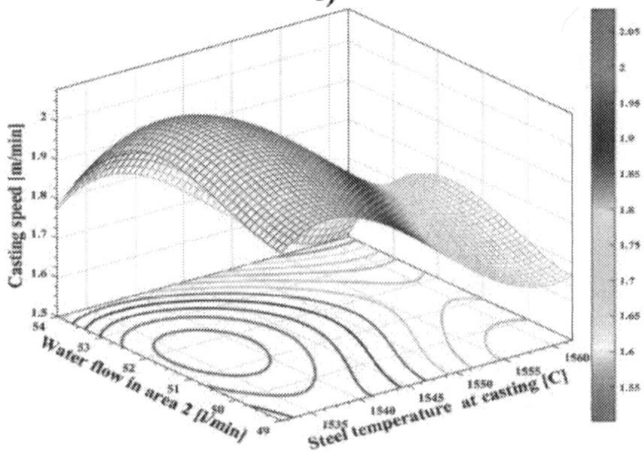

a)

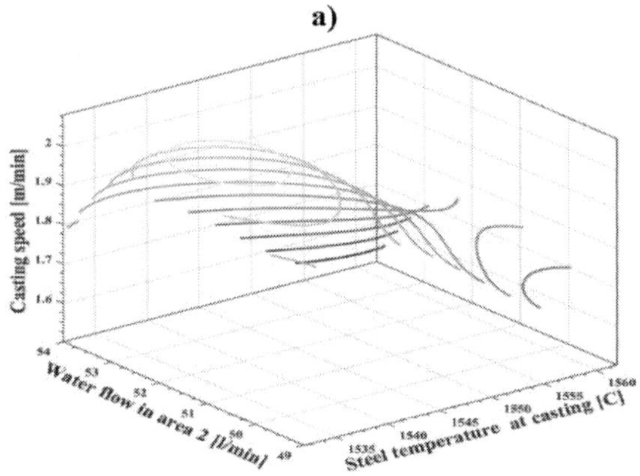

c)

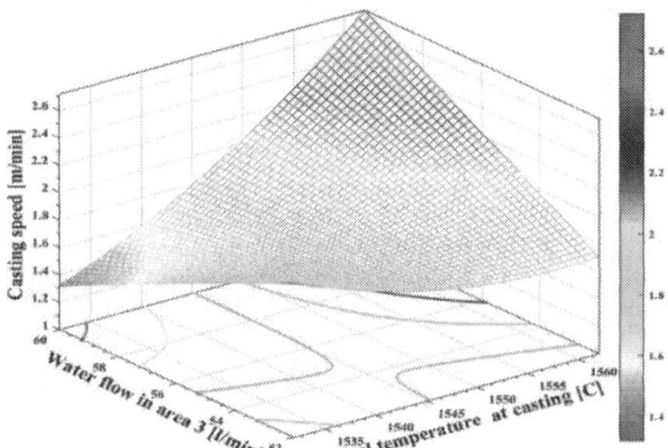

a)

a_9=0.0002;

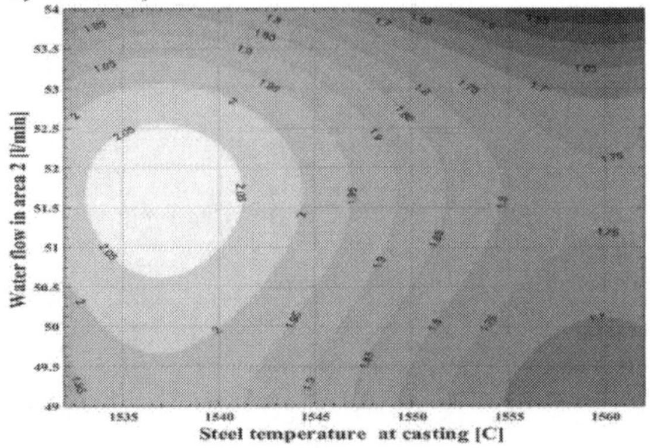

b)

Figure 6. Casting speed= f (Steel temperature at casting, Water flow in area 2, equation 3);
a) correlation surface; b) outiline in plain; c) contour in space;

Equation 3:

$$z=a_1+a_2\log(x)+a_3\log(x)^2+a_4\log(x)^3+ +\frac{a_5}{y}+\frac{a_6}{y^2}+\frac{a_7}{y^3}+\frac{a_8}{y^4}+\frac{a_9}{y^5}$$

R^2 **= 0.79973;**

a_1=-78060; a_2=31882;
a_3=-43407; a_4=197000;
a_5=17588; a_6=-88439;
a_7=14811; a_8=0; a_9=0;

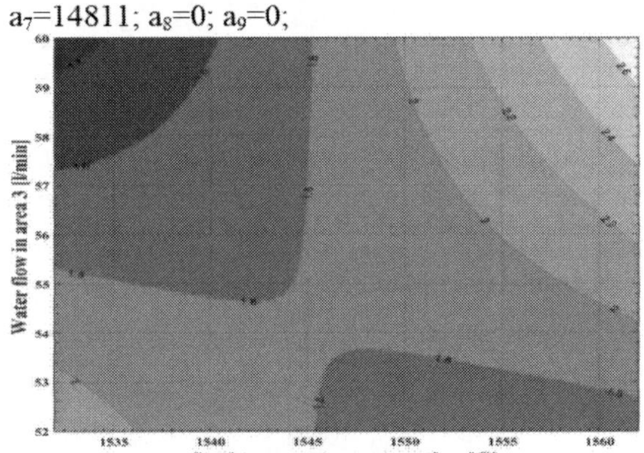

b)

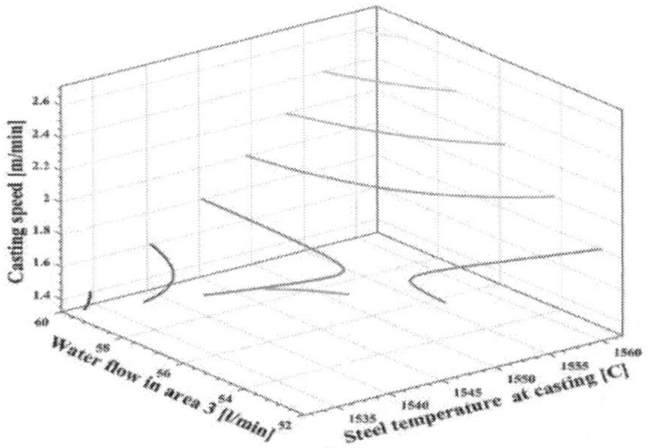

c)

Figure 7. Casting speed= f (Steel temperature at casting, Water flow in area 3, equation 1);
a) correlation surface; b) outiline in plain; c) contour in space;

Equation 1:

$$z=a_1x^2+a_2y^2+a_3xy+a_4x+a_5y+a_6$$

$\mathbf{R^2= 0.68311;}$

$a_1=0.00063$; $a_2=-0.00037$;
$a_3=0.00755$; $a_4=-2.37357$;
$a_5=-11.63264$; $a_6=2150.49806$;

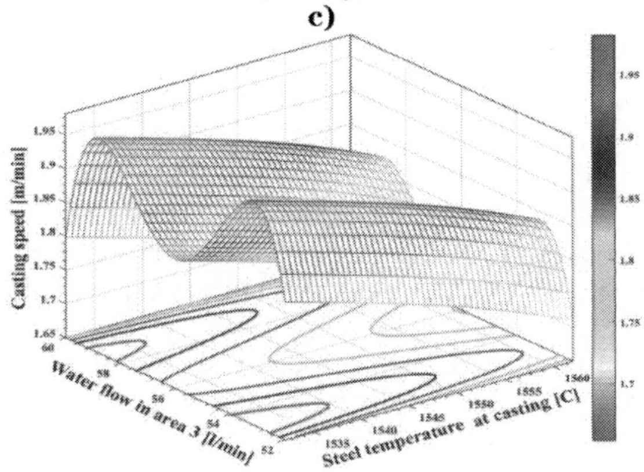

a)

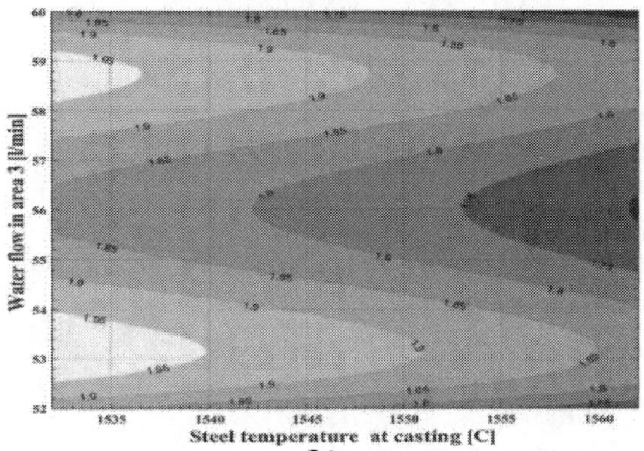

b)

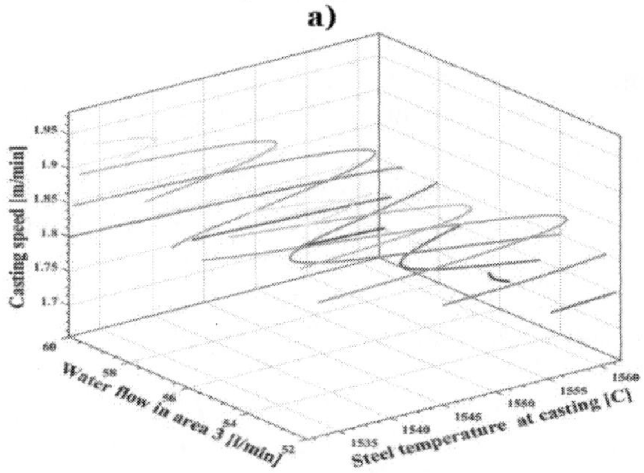

c)

Figure 8. Casting speed= f (Steel temperature at casting, Water flow in area 3, equation 2);
a) correlation surface; b) outiline in plain; c) contour in space;

Equation 2:

$$z = a_1+a_2x+a_3x^2+a_4x^3+a_5y+a_6y^2+a_7y^3+a_8y^4+a_9y^5$$

$\mathbf{R^2 = 0.62121;}$

$a_1=0$; $a_2=-8.35244$;
$a_3=0.00543$; $a_4=0$;
$a_5=0$; $a_6=13.73756$;
$a_7=-0.49166$; $a_8=0.00659$;
$a_9=0$;

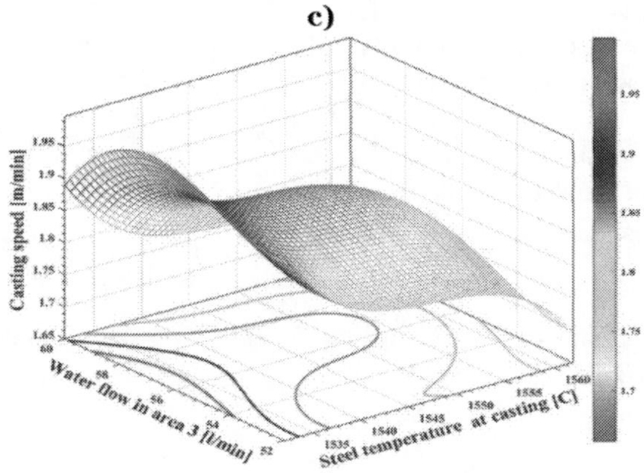

a)

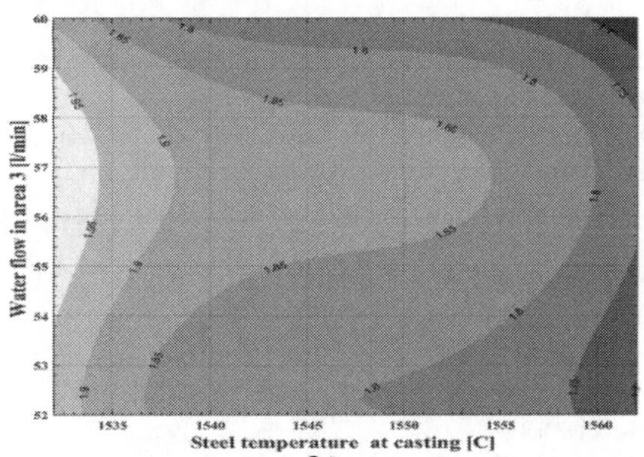

b)

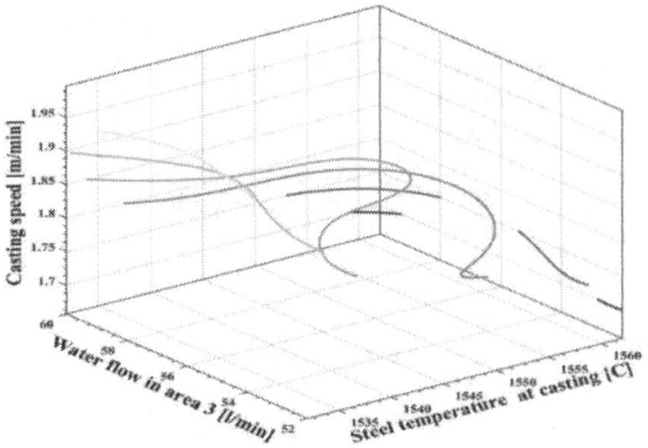

c)

Figure 9. Casting speed= f (Steel temperature at casting, Water flow in area 3, equation 3);
a) correlation surface; b) outiline in plain; c) contour in space;

Equation 3:

$$z=a_1+a_2\log(x)+a_3\log(x)^2+a_4\log(x)^3+$$
$$+\frac{a_5}{y}+\frac{a_6}{y^2}+\frac{a_7}{y^3}+\frac{a_8}{y^4}+\frac{a_9}{y^5}$$

R^2 **= 0.55463;**
a_1=44677; a_2=-18249;
a_3=24848; a_4=-113000;
a_5=38611; a_6=-21071;
a_7=38268; a_8=0;
a_9=0;

4. Technological analysis of the obtained results

Following the processing of the data collected in the industrial experiments and their processing in MATLAB, three groups of double correlations were obtained using 3 types of correlation equations. All the correlations obtained are presented in analytical and graphic form, being representative in a technological sense.

Regarding the casting temperature, the graphical representations confirm that an increase in the casting temperature results in a reduction of the casting speed, which ensures an optimal heat removal from the liquid - crystalliser steel system, in the cooling water crystalliser system, respectively secondary cooling area.

In all three correlation groups, respectively, the three types in each group, the influence of temperature has the same meaning. For example, Figures 3, 4 and 5 show that:

- Figure 3 at a steel temperature of 150 ° C and the secondary cooling rate in zone 2 of 51 l / min., The casting speed is 1.9 m / min, and in Figure 5 all at 1.9m / min;
- Figure 6 at the same values for the aforesaid parameters, the casting speed is also 1.95m / min, a difference of 2.62% compared to the two previously presented cases, therefore in unstable practice conditions;

It should be mentioned that in the case of the other correlation groups, respectively the three types of equations used in each group, similarly correlated results are obtained

5. Conclusions

On the basis of the researches conducted, respectively on the obtained results can be concluded the following:

- between the cooling parameter in the secondary area of the continuous casting installation and the continuous casting speed of the steel can be established representative technology correlations, analytically expressed and graphically;
- on the basis of the graphical representations, depending on the steel casting temperature and the flow rate of the casting water, the casting speed can be chosen;
- the results can be used in the casting of semi-finished products ∅ 180mm.

Acknowledgment

This work was supported by a grant of the Romanian National Authority for Scientific Research and Innovation, CNCS – UEFISCDI, project number PN-II-RU-TE-2014-4-1788.

References

[1] Ardelean E, Ardelean M, Socalici A and Heput T 2007 Simulation of continuous cast steel

product solidification, *Revista de Metalurgia* **43** (3) 181-187

[2] Bouhouch S, Lahreche M, Moussaoui A and Bast J 2007 Quality Monitoring Using Principal Component Analysis and Fuzzy Logic. Application in Continuous Casting Process, *American Journal of Applied Science* **4**(9) 637-644

[3] Cioată V G 2008 Determination of the moulding time of alloys processed in a semi-solid state, *Metalurgia International* **13** (12) 42-52

[4] Efimov V A 1986 *Casting and crystallizing of steel*, Ed. Technical, Bucuresti

[5] Kiflie B and Alemu D 2000 *Thermal Analysis of Continuous Casting Process*, 5th Annual Conference on Manufacturing &Process Industry, Faculty of Technology, Addis Ababa University, Ethiopia

[6] O'Conner T and Dantzig J 1994 Modeling the Thin Slab Continuous Casting Mold, *Metallurgical and Materials Transactions* **25B**(4) 443-457

[7] Pinca C and Tirian G O 2006 *The numerical analysis of the asymmetrical thermal tension from hot rolling mill cylinders*, National Conference of Metallurgy and Materials Science, Bucuresti, Romania, pp 296-303

[8] Singh J and Ganesh A 2008 Design and Analysis of GA based Neural/Fuzzy Optimum Adaptive Control, *Transactions on Systems and Control* **5** (3)

[9] Tirian G O, Gheorghiu C A, Hepuţ T and Rob R 2016 Fuzzy control strategy for secondary cooling of continuous steel casting, *IOP Conf. Ser.: Mater. Sci. Eng.* **200** 012046

[10] Tirian G O, Gheorghiu C A, Hepuţ T and Chioncel C 2016 Control system of water flow and casting speed in continuous steel casting, *IOP Conf. Ser.: Mater. Sci. Eng.* **200** 012047

Cooling water flow control realized with systems based on fuzzy mechanism

G O Tirian[1], C A Gheorghiu[2], T Heput[2] and C P Chioncel[3]

[1]Politehnica University of Timisoara, Electrotechnical Engineering and Industrial Informatics Department, 5 Revolution Street, Hunedoara, 331128, Romania
[2]Politehnica University of Timisoara, Department of Engineering and Management, 5 Revolution Street, Hunedoara, 331128, Romania
[3]Eftimie Murgu University/Electrical Engineering and Informatics, Resita, Romania

E-mail: ovidiu.tirian@fih.upt.ro

Abstract. This research proposed a solution based on fuzzy mechanisms for controlling the flow of cooling water on the first zone of the secondary cooling of steel. For this purpose, a fuzzy system with three input variables and three output variables was designed, the proposed system was tested and validated by simulation made in Matlab Simulink based on actual data collected from the continuous casting process.

1. Introduction

The necessity of manufacturing using continuous casting of some products of high quality and in the same time competitive can be made only by thorough knowledge of the phenomena and the complex processes which take place on the technological development and casting.

The method of continuous casting consists of introducing the liquid metal with a well determined temperature in a cava shape which has the walls cooled inside with water named crystallizing, and the evacuation is made on the opposite side where is obtained the solidified chord [1], [2].

A fuzzy solution is proposed that placed on the existing structure of the control system of continuous casting will reduce fissures in the secondary cooling by generating necessary value adjustments to change the water flow and the velocity of the casting.

There are no current systems in the world that can eliminate cracks if they are detected in the secondary cooling steel. Basically, the rules database was designed specifically for this purpose and it contains measures to be taken to mitigate the risk of a crack.

A very important component of the continuous casting installation is the secondary cooling zone. The secondary cooling zone has the role to continue the wire cooling after it has emerged from the crystallizing and to assure the total solidification of the semi –product. It is considered “the heart” of a continuous casting and has the role of ensuring the quality of the material, the material surface shape and has to ensure a homogeneous cooling and a uniform repartition of the water on the materials surface [3], [4].

During the continuous casting of steel, in the secondary cooling stage, the process of crust solidification continues and can develop cracks. In order to detect and remedy cracks that may occur a simulation was made using Matlab Simulink which analyzes the signals from a temperature sensor and provides the necessary corrections. Thus, compared to the old methods of tackling the problem of crust cracks in the continuous casting of steel, such as technological methods (requiring a certain speed, adjusting the composition and quantity of powders, etc.), the new method proposed and developed, brings safety and precision in this complex process.

2. Structure of the fuzzy system

Fuzzy logic allows the treatment of vague variables whose values can continually vary across any defined numerical range, making decisions based on the position of the indicator in the numerical range and predefined rules.

The applicability of fuzzy logic is varied, the metallurgical field being one in which fuzzy systems are increasingly used.

Methods based on fuzzy logic do not have very strictly defined algorithms, and they appeal largely to the experience of the specialist in the field. For the issue in question, a database was built from information gathered from technology experts. This database was supplemented with the information obtained from the mathematical model of the solidification process.

Designed fuzzy systems receive the following values from the process, depending on the casting area in which the preform is located:

- the casting speed in the cooling Z1, v1
- the cooling water flow rate in the cooling Z1, q1
- steel temperature in the cooling Z1, T1
- the casting speed in the cooling zone 2, v2
- the cooling water flow rate in the cooling zone 2, q2
- steel temperature in the cooling zone 2, T2
- the casting speed in the cooling zone 3, v3
- the cooling water flow rate in the cooling zone 3, q3
- steel temperature in the cooling zone 3, T3

Because conditions and rules change depending on the area where the blank is located, it was necessary to design different fuzzy systems for each cooling zone separately.

Depending on the casting speed and the temperature of the steel, the technological risk coefficient, which displays a percentage value and determines the risk of a crack, if the value returned is higher than 75% the system is blocked.

After the input data is processed, the necessary corrections are made to change the required values of the casting rate and the secondary cooling water flow, respectively, in order to eliminate any cracks in the blank.

It is for the first time on a national and global scale when ordering to change the six sizes (v1, v2, v3, q1, q2, q3, casting speed and cooling water flow for each cooling zone separately) to eliminate cracks in the mold continuously. The rules are designed so that the first size that changes is the flow rate of the cooling water, and then the casting speed is changed. This principle has been used to maintain the plant's productivity as far as possible (the decrease in casting speed leads to decreased productivity). IFS works with three bases of different rules, selected according to the signal received from the neural network for the detection of primary fissures. Figure 1 presents the block diagram of the IFS assembly.

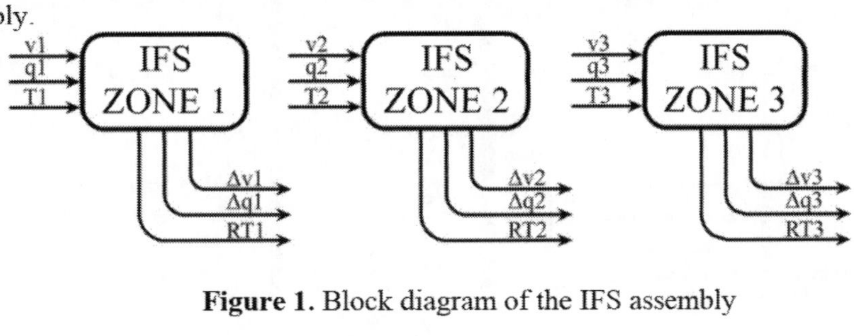

Figure 1. Block diagram of the IFS assembly

In Figure 2 is presented the block diagram of the proposed structure, which is a fuzzy decision system (IFS). V-casting speed; Q-water flow to secondary cooling; T-temperature of steel;

RT-technological risk; Two-speed pouring correction; Δq-secondary water cooling flow correction.

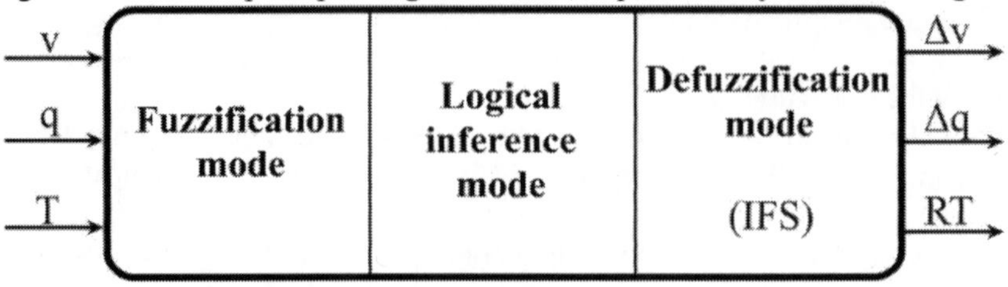

Figure 2. Block scheme of the fuzzy decision system

In this situation, IFS is making changes in casting speed and water flow from secondary cooling to eliminate the risk of a crack. It is noted that for the first time, the risk of crust cracking is taken into account, in case of a risk percentage of over 75% the thread is blocked, in the literature there is no reference to this aspect.

Establishing the bases of rules required an analysis of all possible cases. To this end, a number of technology experts with extensive experience in the exploitation of continuous casting facilities have been contacted. Also, for each individual case, an analysis of the crust solidification phenomenon was performed. Of course, the bases of rules thus obtained are not perfect, and they are also influenced by a number of factors that take into account the state of the equipment and other practical considerations. Therefore, they are to be improved when commissioning and testing the proposed scheme.

3. Design of IFS (Intelligent fuzzy system)

The actual design of IFS was done in Matlab using the fuzzy toolbox. Basically, there are three IFSs, one IFS for each secondary cooling zone, called IFS Z1(intelligent fuzzy system for cooling zone 1), IFS Z2 (intelligent fuzzy system for cooling zone 2) and IFS Z3 (intelligent fuzzy system for cooling zone 3), and have input rates for casting speed v, water flow in secondary cooling q, steel temperature T. The three Systems produce 3 outputs, namely: secondary cooling water flow correction, casting speed correction and technological risk.

For the qualitative analysis of IFS operation, it is proposed to simulate it using the Matlab-Simulink environment. Figure 3 shows the implementation of IFS-Z1 in Matlab-Simulink. The temperature reading is simulated by the data obtained from the process using a numeric data string. To start the simulation, the casting speed and water flow sizes must be entered, the program will correct these values using a loop through which the correction obtained changes the input data. The running time of the simulation is 1000 seconds, each calculation loop having 3.2 seconds.

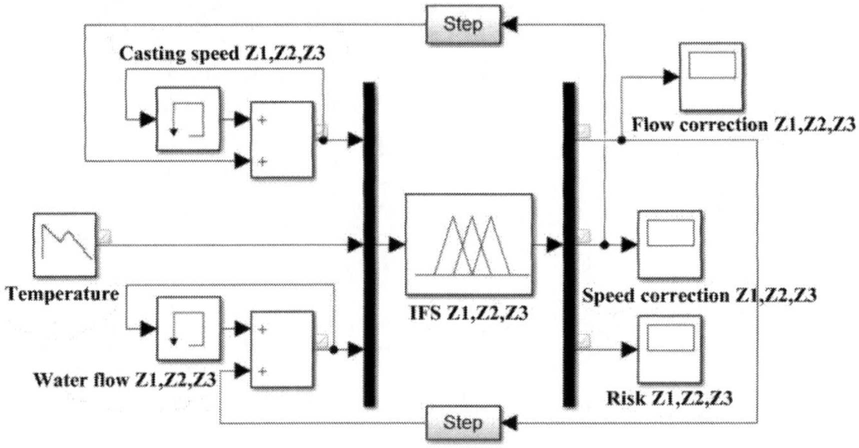

Figure 3. Implementation in Simulink of IFS-Z1,Z2,Z3

IOP Conf. Series: Materials Science and Engineering **294** (2017) 012063 doi:10.1088/1757-899X/294/1/012063

Figure 4 shows the values of the input variables. We know the temperature and the starting variables of casting, in this case we chose the simulation start speed of 1.5m/sec and 168l/min for the cooling water flow, we chose these high risk factors to highlight the risk remedy in the first 30 seconds of the simulation that can be seen in Figure 5.

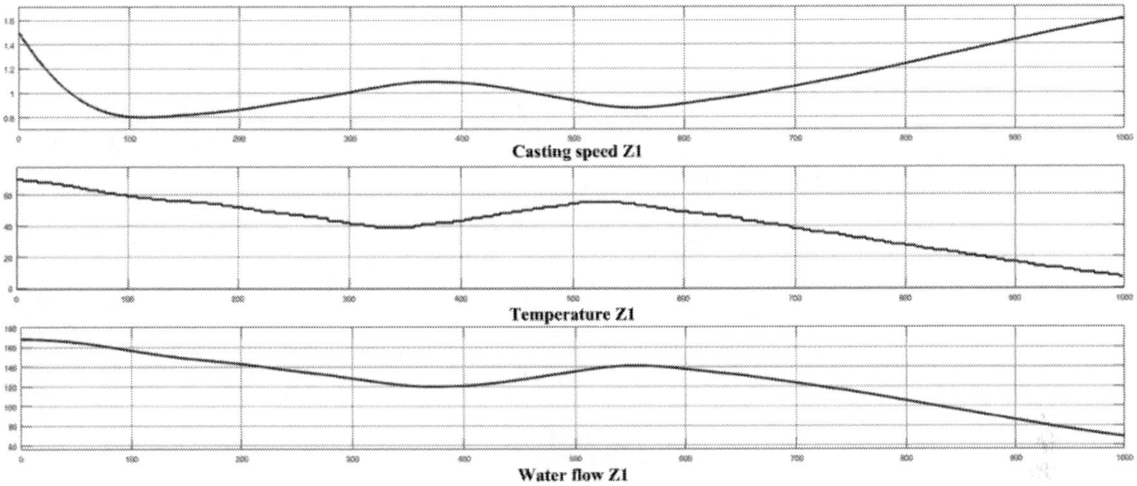

Figure 4. Values of input variables for IFS-Z1

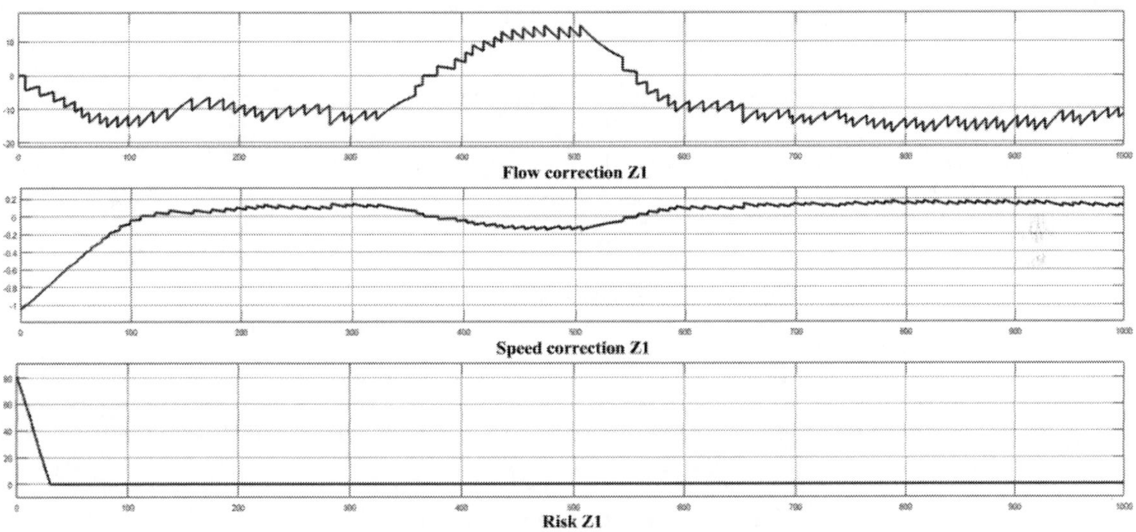

Figure 5. Corrections made by IFS-Z1

Figure 6 shows the values of the input variables. Known are the temperature and the starting variables of casting, in this case we have chosen the start value of the simulation for the casting speed of 1.4m / min and 220l / min for the cooling water flow, we have chosen these high risk factor values for To highlight the risk remedy in the first 25 seconds of the simulation that can be seen in Figure 7.

The steels heat is partially removed by heating the cooling water, but the most heat is extracted by water evaporation, even if the evaporation percentage of the sprayed water on the wires surface is in general lower than 20%. All the unevaporated water which flows from the wires surface are collected filtered and reused [5-7]. Fuzzy logic can process vague variables whose values can vary continuously in any numerical range defined by taking decisions depending on the position indicator in the numerical range and predefined rules [8-11].

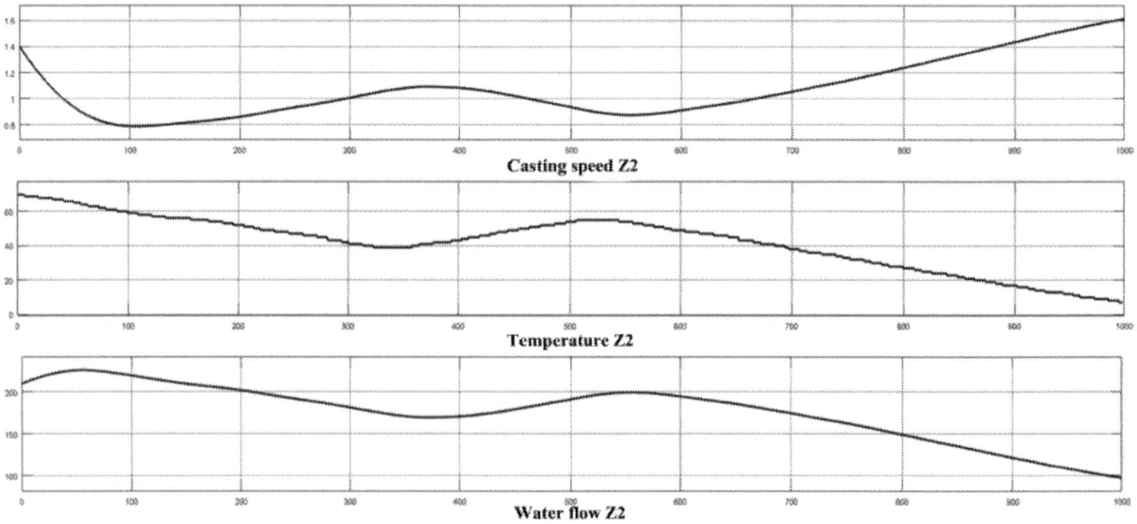

Figure 6. Values of input variables for IFS-Z2

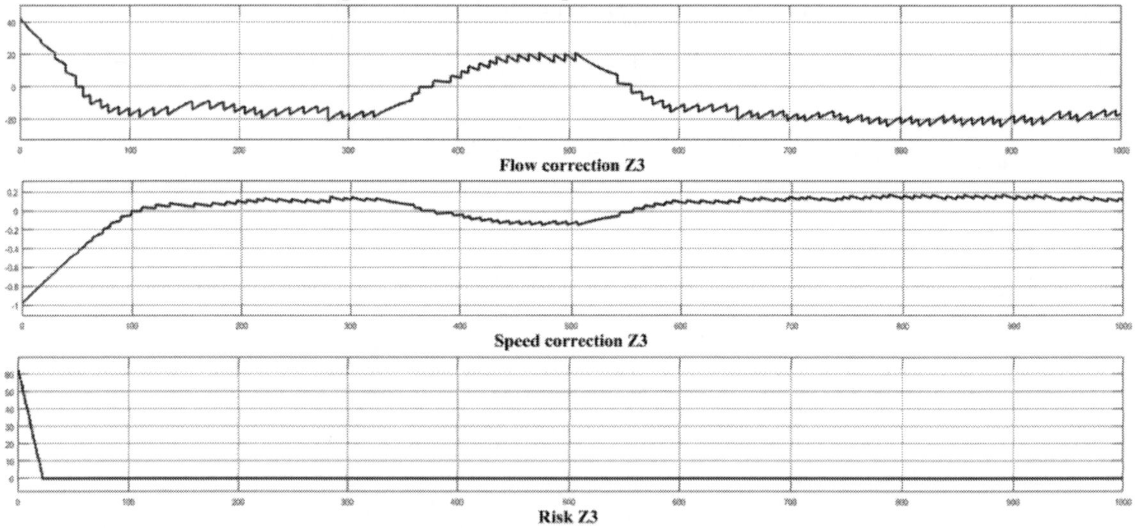

Figure 7. Corrections made by IFS-Z2

Figure 8 shows the values of the input variables. We know the temperature and the starting variables of casting, in this case we chose the start value of the simulation for the casting speed of 1.3m / sec and 280l / min for the cooling water flow, we chose these high risk factor values to Highlight the risk remedy in the first 20 seconds of the simulation that can be seen in Figure 9.

Laying the foundation of rules required an analysis of all possible outcomes. To this end, a number of expert technologists with extensive experience in the operation of the casting machines were consulted.

Also, for each case was carried out an analysis of the phenomenon of solidification of the crust. For the issue in question a database was built that consists of information gathered from technologists experts. This database was supplemented with information from the mathematical model of solidification process [7].

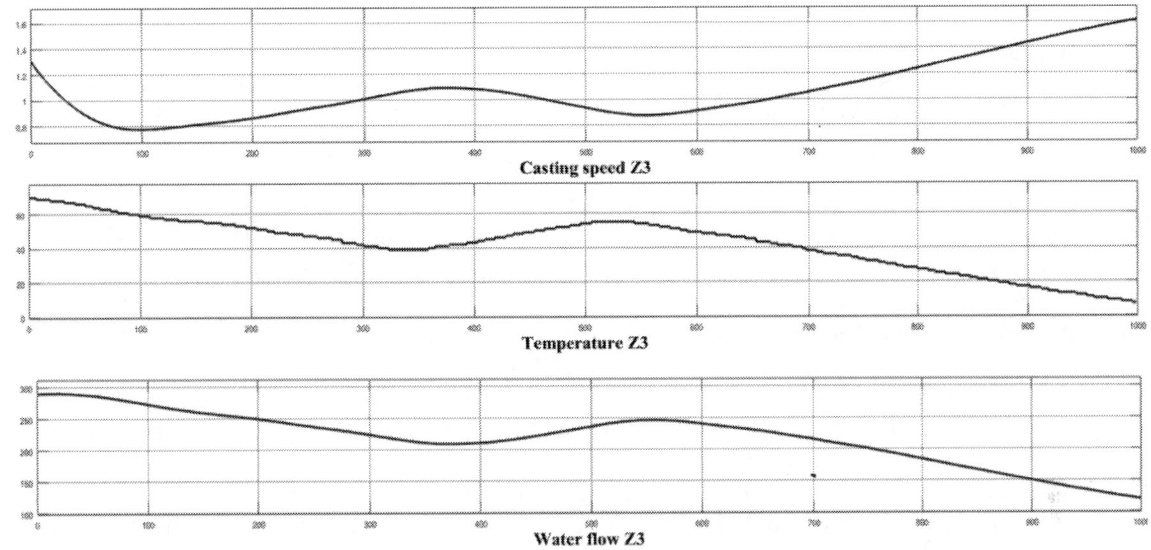

Figure 8. Values of input variables for IFS-Z3

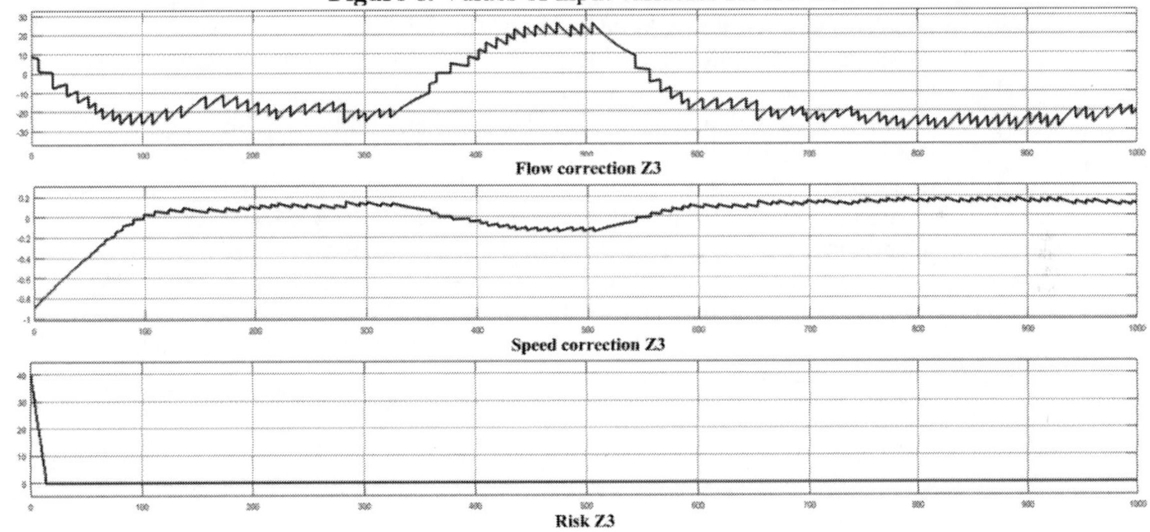

Figure 9. Corrections made by IFS-Z3

4. Conclusions

Analyzing simulation results shows that regardless of the values of the generated input, IFS develops the necessary corrections casting speed and primary cooling water flow, which confirms the validity of system operation. It is considered that, in terms of quality, use fuzzy decision system is an effective, practical and easy to implement, in order to analyze complex phenomena and nonlinear.

The goal of fuzzy system is to improve this classical adjustment system, the introduction of some adaptive components in the adjustment loops and some overall predictions over the continuous casting machine.

This paperwork introduces a new and original concept of the structure of a control system for the continuous casting.

The goal of the Simulink system is to improve this classical adjustment system, the introduction of some adaptive components in the adjustment loops and some overall predictions over the continuous casting machine.

Acknowledgment
This work was supported by a grant of the Romanian National Authority for Scientific Research and Innovation, CNCS – UEFISCDI, project number PN-II-RU-TE-2014-4-1788.

References
[1] Efimov V A 1986 *Casting and crystallizing of steel*, Ed. Technical, Bucuresti
[2] Kiflie B and Alemu D 2000 *Thermal Analysis of Continuous Casting Process*, 5th Annual Conference on Manufacturing &Process Industry, Faculty of Technology, Addis Ababa University, Ethiopia
[3] O'Conner T and Dantzig J 1994 Modeling the Thin Slab Continuous Casting Mold, *Metallurgical and Materials Transactions* **25B**(4) 443-457
[4] Pinca C and Tirian G O 2006 *The numerical analysis of the asymmetrical thermal tension from hot rolling mill cylinders*, National Conference of Metallurgy and Materials Science, Bucuresti, Romania, pp 296-303
[5] Bouhouch S, Lahreche M, Moussaoui A and Bast J 2007 Quality Monitoring Using Principal Component Analysis and Fuzzy Logic. Application in Continuous Casting Process, *American Journal of Applied Science* **4**(9) 637-644
[6] Singh J and Ganesh A 2008 Design and Analysis of GA based Neural/Fuzzy Optimum Adaptive Control, *Transactions on Systems and Control* **5** (3)
[7] Ardelean E, Ardelean M, Socalici A and Heput T 2007 Simulation of continuous cast steel product solidification, *Revista de Metalurgia* **43** (3) 181-187
[8] Lee C C 1990 Fuzzy logic in control systems: Fuzzy logic controller, *IEEE Trans. Systems, Man & Cybernetics* **20**(2) 404-435
[9] Tirian G O, Gheorghiu C A, Heputţ T and Rob R 2016 Fuzzy control strategy for secondary cooling of continuous steel casting, *IOP Conf. Ser.: Mater. Sci. Eng.* **200** 012046
[10] Tirian G O, Gheorghiu C A, Heputţ T and Chioncel C 2016 Control system of water flow and casting speed in continuous steel casting, *IOP Conf. Ser.: Mater. Sci. Eng.* **200** 012047
[11] Tirian G O, Gheorghiu C A, Heputţ T and Tirian A 2017 *Cracks detection in continuous casting processes used unconventional methods*, International Conference on Automatic Control, Modelling and Simulation, Barcelona, Spain

IOP Conf. Series: Materials Science and Engineering **294** (2017) 012064 doi:10.1088/1757-899X/294/1/012064

Directions for material recovery of used tires and their use in the production of new products intended for the industry of civil construction and pavements

C Bulei[1], M P Todor[1], T Heput[2] and I Kiss[2]

[1]University Politehnica Timisoara, Doctoral School, Timisoara / Hunedoara, Romania
[2]University Politehnica Timisoara, Faculty of Engineering Hunedoara, Department of Engineering and Management, Hunedoara, Romania

E–mail: imre.kiss@fih.upt.ro

Abstract. The management of waste from used tires is one of the major principles of recycling and reuse, which involves encouraging a high level of material recovery components, preferably by recycling. Given the current pressure on natural resources on a global scale we must fully take into account the waste in a broader framework defined by the flow of raw materials and their sustainable use. Thus, the opportunity to use various waste from used tires as raw material in order to support economic activities becomes a priority. The recycling of raw materials from waste products and their use in new production processes for their material capitalization is becoming a sustainable approach. Used tire recycling, is the process of recycling waste tires that are no longer suitable for use on vehicles due to wear or irreparable damage. These tires are a problematic source of waste, due to the large volume produced, the durability of the tires, and the components in the tire that are ecologically problematic. Although tires are usually burnt, not recycled, efforts are continuing to find value. Tires can be recycled into, among other things, typically as crumb rubber modifier in recycled pavement. The paper presents aspects of the product resulting from crushing of used tires (powder), which can be used properly valued in the street furniture field or building materials containing rubber.

1. Introductory notes

Worn tires are among the biggest and the most problematic sources of waste and because of their large shape and volume they also occupy a lot of space. However, used tires are also one of the most reusable waste. In the past, the largest share of industrial waste (including used tires) was considered unusable, the main concern for managing waste being the identification of disposal solutions and not the way we can minimize their generation or how we can reuse and valorise them. Disposal of waste tire rubber has become one of the major environmental issues in the world. Every year millions of tires are discarded, thrown away or buried all over the world, representing a very serious threat to the ecology [1]. Also, the illegal dumping of tires is a serious threat to public health through the release of chemicals and pollutants into surrounding soil, groundwater and rivers.

But nowadays the application of the principles of sustainable development implies a new approach to waste using environmentally–friendly concepts and it is necessary to re–evaluate the options for managing this waste in order to increase their use and to drastically reduce the quantities that require disposal or storage. To this end, the waste hierarchy should be applied with a focus on reuse, recycling

IOP Conf. Series: Materials Science and Engineering **294** (2017) 012064 doi:10.1088/1757-899X/294/1/012064

and recovery, while waste disposal should be interpreted as the last available option that corresponds to the highest level of resource loss and alteration [5].

Figure 1. Illegal dumping of tires and the pollution danger

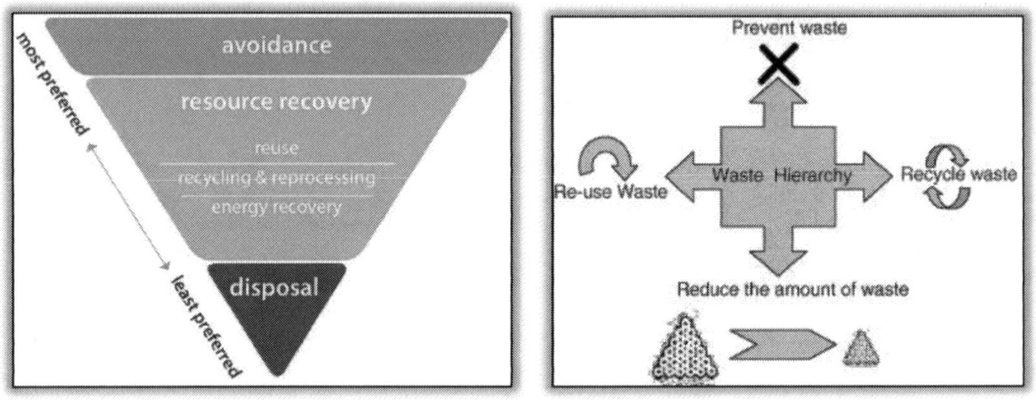

Figure 2. The waste hierarchy preferences [5]

The accentuated increase in the amount of rubber waste has led to the search for ways to combat environmental pollution by reuse, revalorize and recycling. The utilization of used rubber tires for the re–use of component materials in the industry and the release for re–farming of the land on which these tires are currently stored are a priority both worldwide and nationally. We must take into account the economic potential offered by the raw material price of waste and in this sense recycling / recovery capacity should be developed and also the encouragement of the use of raw materials that are found in waste. One of the possible solutions for the use of waste tire rubber is to incorporate into cement concrete, to replace some of the natural aggregates [1].

Among the rubber products, the tires are by far the largest share, which also explains the attention paid to the reintroduction of used tires in the economic circuit while avoiding pollution. Today, over 1 billion tires are produced annually in over 400 tire factories. Thus the potential for recycling millions of used tires is clearly very important. The main source of used tires is the automotive industry, which means cars that use tires as tread. It is estimated that less than 7% of the 300 million used tires take the form of new products and about 11% are converted into electricity. Over 77% are buried, stored or abandoned. In Romania there are about 40,000 tons / year of used tires being available for recycling, the amount being continually increasing, plus uncollected and un–recycled accumulated stock.

At the end of their life–use, tires are turned into waste that either is stored or as an alternative is recycled or reused. Good waste management practices take into accounts, excepting retreating tires or storing them, other methods of capitalizing on reusable rubber materials. In the market economy the possibility of recycling is directly related to the recovery of residual value according to the direct proportion of the willingness to make efforts for such a process. From the recycling perspective, used tires are exceptional materials, as the number of recycling without significant quality is an indefinite one [4], [8].

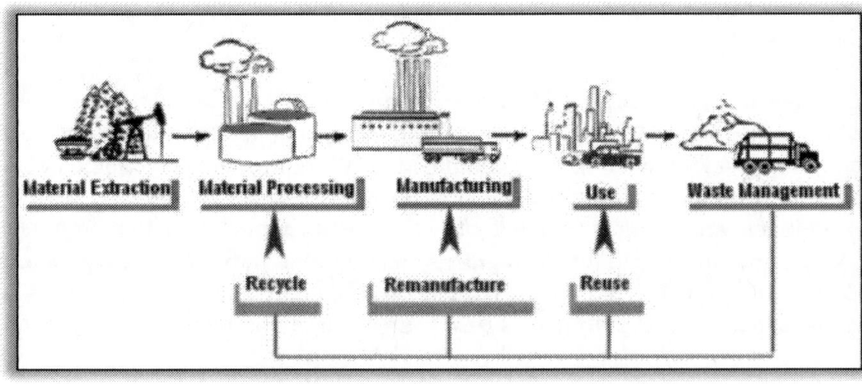

Figure 3. End of tire life–use

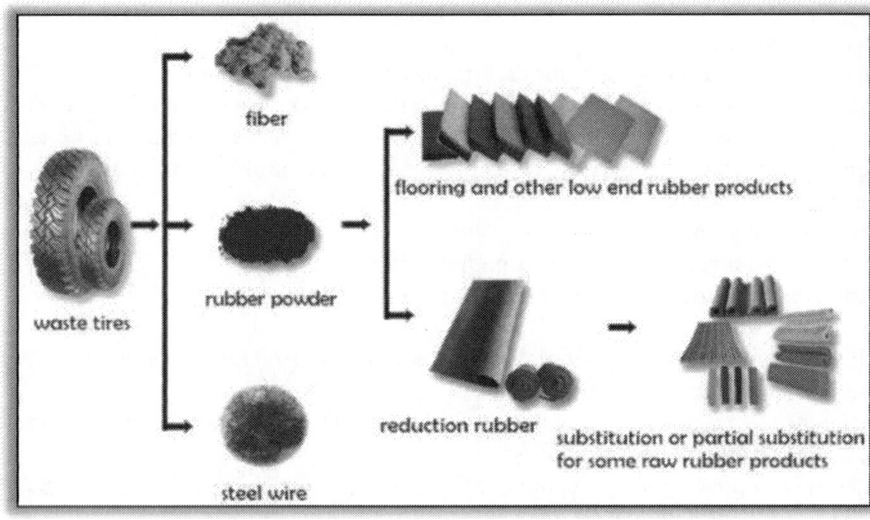

Figure 4. Waste tire to rubber powder for recycle use

Engineering recovering and recycling waste tires is an increasingly tempting area for investors and manufacturers of recycled products alike. Tire recycling allows:

- recovering precious material without reducing quality
- energy savings in relation to primary production
- reducing the amount of waste.

Currently, used tires have become the largest share of the total amount of potentially recyclable rubber. The technical solutions that are used take the aim at returning them to the economic circuit but in a limited number of cycles. The worn tire contains a number of chemicals with energy potential, but it is also a source of secondary raw material. The tires are made of flexible elastomeric rubber material, the structure of which is reinforced with textile and metallic materials. A tire is composed of several materials with a rubber content ranging from 46–48%, carbon black between 25–28%, steel inserts between 10–12%, oil and vulcanizing agents between 10–12% and inserts synthetic yarns and textiles between 3–6%. Tires contain potentially valuable constituent components, such as carbon black, organic oils and steel, extracting these materials in a cost–effective way is extremely difficult. Only a few of the technologies are viable, the others being only pilot stations, have not been used at the industrial scale yet. Thus, used tires pose problems even for the most sophisticated waste management systems. The tires are made up of complex polymeric materials with a large addition of various other components, the recovery of this added value being a dangerous operation and leading to high energy consumption.

2. Methods and directions of recovery of rubber waste

Most of the rubber waste is discarded and the natural decomposition lasts a long time because of their structure, stabilizers and additives from their composition. Worn tires are not wanted at landfills due to their high volume, which quickly consumes valuable space.

Two major problems arise: large quantities of waste are accumulated and environmental pollution is produced. Worn tire holes can grow rapidly, creating not only land use problems, but also environmental hazards, since (the dumps) can self–ignite, and causing long–lasting fires. Accumulation of increasing inventories of used tires raises particularly serious environmental problems, as:

- occupying extensive areas with environmental pollution and set aside of potentially productive land (a temporary solution may be to reduce the volume and surface occupied by using pre–slicing machines)
- being a potential fire hazard
- may become a growing environment for dangerous parasites and insects
- are not biodegradable (the decomposition period in the natural environment is over 80–100 years).

The major advantage is that rubber wastes can be reused by recycling and can be used in various fields depending on the design properties. Two solutions to these problems would be:

- recycling and reuse of waste
- recovery of waste materials.

Actions to be taken for recycling tires at a high rate concern:

- reducing the risk of pollution by temporary storage of worn tires
- increasing the utilization rate of the used tires
- reducing the consumption of raw materials, especially those from non–renewable sources, by using used tires as secondary raw materials.

The main directions of recovery of the used products and the rubber waste, grouped in descending order of the technical efficiency of the process are:

- reuse of used rubber products in the same quality as the original product by repair and retreading
- reuse for other rubber products
- recycling as material reintroduced into regenerated elastomeric compositions or rubber powder
- the use as a source of various starting materials, such as carbon black or pyrolysis oils
- use as a fuel for the production of thermal energy, for the manufacture of cement or for steel
- reuse as modification agents for various materials, especially for road construction.

Recycling abandoned tires is a difficult global problem. If real tires can be redeemed, they turn from pollution sources into raw materials. Using a well–designed tire management plan, we can focus on net benefits and used tires can be transformed from a major source of pollution into an important source of income. It is important to find ingenious and innovative solutions to store and reuse waste.

Figure 5. Rubber waste that can be used for granules

In case of used tires, they can be ecologically destroyed and the material obtained can be used for other purposes. Recycling of used tires results in three categories of products (rubber granules, steel inserts and synthetic yarn inserts and textiles) that can be recovered allowing for massive energy and raw materials savings.

These many possibilities of capitalizing on non–recoverable worn tires make them a material with great potential. The industry has discovered many qualities of worn tires: their structural strength when in use, their flexibility when turned into granules, or their calorific power when used as fuel.

Tires are mainly used in two ways:

- material capitalization, by recycling with grain conversion
- thermal (energy) recovery by co–incineration in cement plants.
- Material capitalization refers to the use of non–recoverable worn tires in various applications:
- whole tires can be used in dykes or hydrographic or anti–noise screens
- coarse chopped tires can be used to build the chassis or as a substrate for draining water for sports grounds
- fine ground tires can be used to obtain rubber granules, which can be used to produce different rubber products, heat insulating, sound absorbing or agro–touristic arrangements
- fine ground tires can be used for powder coating, which can be used in asphalt mixes, which has resulted in good results in reducing vehicle rolling noise, reducing the risk of skidding and guaranteeing a long lasting life for roads.

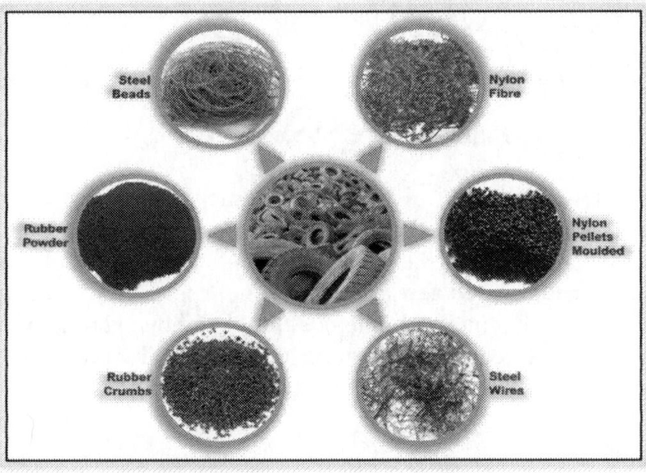

Figure 6. Tire derived materials

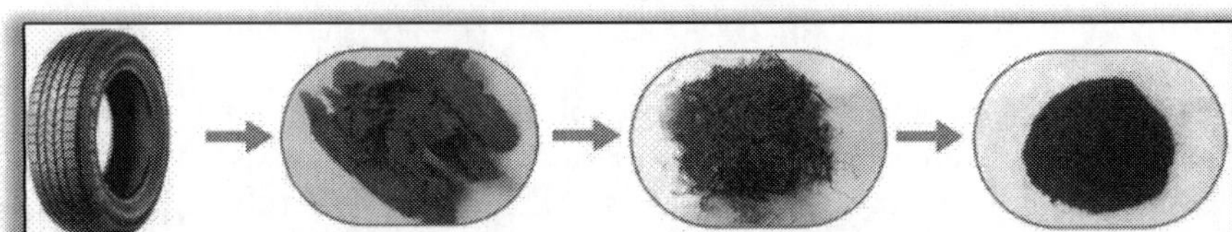

Figure 7. Tire processing

The problem of recovery of used tires cannot be considered solved at present, not even worldwide, being a generally accepted and expanded solution with technical and economic advantages. However, there have been adopted in the world various variants, the most remarkable being the milling (at ambient temperature, cryogenic or wet process) with powder–coating. Recycling of used tires implies many processes, from material recovery, cutting, grinding, granulating or other operation that causes changes in the nature or composition of the used tire by industrial processes in order to obtain raw materials. From used tires, following a grinding and sieving process, rubber and powder granules can be used to make new products.

The process of fine milling of single–material waste is an economic and perspective alternative in current practice and recycling of used tires. Thus, the rubber powder (millimeter granules) will serve to manufacture various rubber products, some of the rubber granules resulting from recycling being used for the production of products for the civil engineering and pavement industry (thermo–acoustic insulation for buildings, work floors, flooring for playgrounds, floors for sports grounds, and so on), another part of the resulting rubber powder being used for the production of technical articles made of recycled rubber.

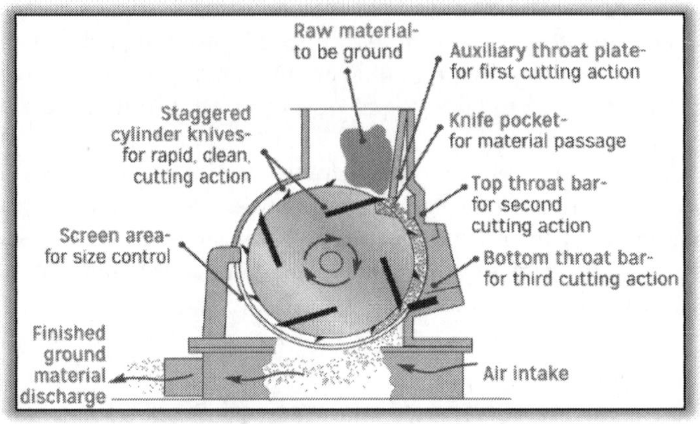

Figure 8. Tire rubber processing

In the scheme from Figure 8, there are presented some possibilities of using this material by grinding. First, from the used and rebutted tires and other rubber products there can be obtained as intermediate products (rubber sections), rubber milling (2–5 mm) and rubber powder (0.25–2 mm), which can be then dimensionally sorted and used in various application domains. Obviously, each of the transformation processes is a distinct technological process, however the prices of the intermediate products are even higher as the fineness of the material is more advanced. In the same scheme there are presented the technologies for processing the intermediate products until the final products are obtained.

3. Rubber pavers – a solution for recycling rubber waste

Due to the increasingly serious environmental problems presented by waste tires, the feasibility of using elastic and flexible tire–rubber particles as aggregate in concrete is investigated several studies [1-9].

Investigations and research into the recent use of rubber particles in concrete has been well documented. However, information on the rubber particle sizes or their distributions within concrete which may also influence the concrete properties is still limited. Tire–rubber particles composed of tire chips, crumb rubber, and a combination of tire chips and crumb rubber, were used to replace mineral aggregates in concrete. These particles were used to replace different percent of the total mineral aggregate's volume in obtained concrete [1-9]. Properties of concrete with different rubber sizes and distributions were studied by several studies [1-9]. It was concluded that there is a promising future for the use of discarded tire rubber as a partial substitute for fine aggregates in concrete, which can result in huge environmental and sustainability benefits [1-9].

A large variety of waste materials are considered feasible and even much valuable additives for concrete. Rubber obtained from scrapped tyres is considered as the most recent waste materials that have been examined because of its vital use in the construction field [9]. The recycled rubber tyre waste is a promising material in the construction industry and the sole reason for this is the

IOP Conf. Series: Materials Science and Engineering **294** (2017) 012064 doi:10.1088/1757-899X/294/1/012064

lightweight of the resulting concrete when the rubber tyre is incorporated in it as an aggregate replacement (partial or complete) [9].

The management of end-of-life tires is a great environmental challenge. Recycling of waste rubber tires in civil engineering is considered as ecological and economical solutions due to the advantages it can offer. It preserves natural resources and produces an eco-friendly material. The application of recycle waste tires into civil engineering practices, namely asphalt paving mixtures and cement based materials has been gaining ground across the world [4].

A new possibility consist in producing advanced aesthetic and functional materials by using recyclable materials (rubber, plastics) for the development of ceramic composites. In practice, this means that recycled rubber can be used as a filler in new blends for the development of new composites. Since their appearance, the composites have been used in a wide variety of applications due to their outstanding performance. Having high mechanical characteristics and superior design and manufacturing capabilities, composite materials have superior advantages over traditional materials.

Figure 9. Rubber pavers (or elastic flooring)

Rubber pavers (or elastic flooring) are an aesthetic and efficient paving solution in several areas of use where the comfort and adhesion of the surface is a priority regardless of weather conditions. The most common areas of application of rubber pads, with the proviso that they can be used for the same applications as tiles in the case of land and public playgrounds (alleys, pavements around the house, around the garden, patios, terraces, anti–slipping stuff). The careful choice of raw materials as well as state–of–the–art technology used in the manufacturing process are some important aspects that place these products first in terms of quality and differentiate it from the other traditional paving options (concrete, wood, stone, brick or other solutions). Sorting the rubber granules is very important because it helps to 'uniformize' the distribution of cohesive forces to the granules, which gives the product a very good stability in time.

Advantages of using elastic pavers:

- impact shock absorption – elasticity, comfort
- protection and safety – prevents hitting in case of a fall
- long lifetime – resistance
- instant drainage – avoiding water build–up
- impermeable surface – anti–slip
- increased resistance to environmental factors
- neat appearance – cleanliness and safety
- heavy traffic resistance.

Rubber flooring can be used both outside and inside. They are obtained from a new manufacturing technology with recycled rubber granules combined with a polyurethane binder and pigments for colouring. Thanks to the colour impregnation process even during the manufacturing process, the colour resistance is very high and the plates do not discolour no matter the cleaning process and the frequency of the operations. Being processed at high temperatures and pressures, they acquire elasticity and reliability characteristics.

Due to its properties, the rubber pavement is the wise choice of floor regarding to covering. Being anti–slipping and elastic, the risk of injuries caused by moisture is almost completely reduced. Additionally, the product made from recycled rubber granules is not absolutely toxic, being also a top heat and sound insulator. Since it carries heavy loads, the ecological pavement can support any type of furniture (street or home, indoor or outdoor).

4. Concluding remarks

Nowadays the disposal of waste from used tires represents a considerable problem. In their state of waste the tires are very difficult to degrade and occupy a substantial amount of space in the landfill. If 20–30 years ago, they were stored in huge quantities all over the world, nowadays worn tires are either used in the cement industry as an alternative fuel or converted by recycling into new products such as street furniture or materials of rubber–containing construction. The industry has discovered many qualities of used tires: their structural strength when used in their entirety, their versatility when converted to rubber granules, or the calorific power when used as fuel. These numerous possibilities of capitalizing the used tires make them a material with great potential.

One of the environmental problems is the worrying increase in the amount of such waste and it is obvious that this trend must be stopped. Their destruction by burning on open sites is not allowed due to toxic smoke. Their burning in incinerators consumes a substantial amount of energy to bring them in the right state of combustion, create large quantities of ash, and require sophisticated filtration systems for hazardous substances that are released in large quantities (heavy metals, especially zinc, fine dusts, sulphur oxides, including the danger of releasing dangerous substances). The same problems arise from their burning in cement kilns, where, in general, the injection rate of the rubber is, due to pollution, of max. 20% of batch. In addition to direct combustion, as a more environmentally friendly alternative, the development of technologies and installations for energy recovery of used tires through pyrolysis and pyrolysis followed by gasification is being investigated. That's why, obtaining fuel from used tires all over the world is an important market for recovery of tire waste. But material recovery by using worn tires in various applications is a great challenge for those who handle managing this type of waste.

In some countries used tires are used by co–incineration in cement factories. In others, the accent is placed on their recycling. Thus, recycled tires can be used as substitutes for fossil fuels under controlled conditions, or they can become completely new products. The cement factories are for Romania the priority destination of used tires for now. The trend is to focus on recycling, to obtain secondary raw material from used tires by turning them into rubber granules. There is a potential market for rubber–containing rubber products from the tire recycling process, an inadequately exploited direction yet with unquestionable potential and valences in terms of the quality and durability of these products. However, in Romania there are not many economic operators to carry out granulation and there is no sufficient development of a derivative market for the by–products that can be obtained from these granules. Unfortunately, there is no good internal demand for these granules.

Rubber pavers have become very popular in recent years in our country as well. Their popularity is due to the large variety of models, sizes, colours and thicknesses, qualities that successfully serve multiple areas of use. Thanks to the innovative technology, the elastic pavers are environmentally friendly, being granular rubber products, offering added strength and reliability, both in abrasion and heavy traffic. Ideal for both impact surfaces with playground protection and anti–skid surfaces, the elastic pavers will be a real inspiration of elegance and style wherever they are placed. They can be used for decorative purposes, but also for industrial applications due to their high durability. Elastic

IOP Conf. Series: Materials Science and Engineering **294** (2017) 012064 doi:10.1088/1757-899X/294/1/012064

pavers are the most stable rubber–based granular products and are therefore ready to handle all demands and have much more advantages over traditional paving options in concrete, wood, stone, brick or other solutions.

Rubber pavers are specially designed for areas where the risk of falling and hitting is high, and the manufacturers of such materials have developed a range of product lines suitable for many exterior or interior types designed to protect against the most diverse risks. Outside, it offers the advantage of resistance to repeated freeze–thaw cycles, and additionally the tile composition allows drainage of water, so the pavement is easy to dry.

References

[1] Thomas B S and Gupta R C 2015 Long term behaviour of cement concrete containing discarded tire rubber, *Journal of Cleaner Production* **102** 78–87

[2] Khaloo A R, Dehestani M and Rahmatabadi P 2008 Mechanical properties of concrete containing a high volume of tire–rubber particles, *Waste Management* **28**(12) 2472–2482

[3] Su H, Yang J, Ling T-C, Ghataora G S and Dirar S 2015 Properties of concrete prepared with waste tyre rubber particles of uniform and varying sizes, *Journal of Cleaner Production* **91** 288–296

[4] Yazdi M A, Yang J, Yihui L and Su H A 2015 Review on application of waste tire in concrete, *International Journal of Civil, Environmental, Structural, Construction and Architectural Engineering* **9**(12) 1648-1653

[5] Batayneh M, Marie I and Asi I 2007 Use of selected waste materials in concrete mixes, *Waste Management* **27**(12) 1870-1876

[6] Sukontasukku P, Chaikaew Ch 2006 Properties of concrete pedestrian block mixed with crumb rubber, *Construction and Building Materials* **20** 450–457

[7] Rafat S and Tarun R Naik 2004 Properties of concrete containing scrap-tire rubber – an overview, *Waste Management* **24**(6) 563–569

[8] Toutanji H A 1996 The use of rubber tire particles in concrete to replace mineral aggregates, *Journal of Cement & Concrete Composites* **18** 135–139

[9] Alam I, Mahmood U A and Khattak N 2015 Use of rubber as aggregate in concrete, *A Review International Journal of Advanced Structures and Geotechnical Engineering* **4**(2) 92–96

IPS technologies in sales improvement in companies in Banja Luka, Bosnia and Herzegovina

Lj Micic and D Preradovic

TA Business Informatics, University of Banja Luka, Faculty of Economics, Majke Jugovica 4, 78 000 Banja Luka, Bosnia and Herzegovina

E-mail: ljubisa.micic@ef.unibl.org

Abstract. Modern economy, in first order Retail sector, are introducing new and modern technologies which improve sales, communication with customers as well as marketing decision process. Retail sector is more and more oriented to technologies which improve not just communication with possible customers but also information gathering for better decision making process. Among those technologies special place is given to Indoor Positioning Systems (IPS) technologies, specialized technologies which are used for locating the people and objects but also their movement as well as for communication with clients in the store. These technologies follows customers in the store, follow their movement across, communicate with them about offers and promotions but also are base for predictive marketing of retail sector which has a goal to achieve „offer to customers what they look for and what they really need". These technologies are in limited usage in Banja Luka area but potential for their use is present. There is no enough demand from retails industry but having in mind that these technologies are raising in modern economies as well as predictions and trends it is to be expected that those technologies will be implemented in retail sector in this region too.

1. Introduction

Retail sector is in continuous demand for the new strategies of communication with clients, as well as for new technological gadgets for business intelligence improvement and finding valid information for better strategic and operational planning. Considering technological improvement in modern society, the expectations are that new technologies are going to immerge in retail sector. IPS, which contains the whole set of technologies for locating the customers in the store, their interest in offers, specialized technologies for personalized communication has a specific role in retail industry. Among IPS technologies, the great importance is given to beacon technology which is based on the small devices called beacons that transmit specific messages using Bluetooth Low Energy (BLE, which is different from the first version of Bluetooth) when the person or, more accurately, the device that has the reading compatibility of BLE broadcasting signal is on particular distance. The technologies mentioned above, with all its advantages, are becoming the basic technological gadget for obtaining the information that serve as a basis for analysis through business intelligence system, especially considering the sales analysis, heat maps analysis i.e. the map of customers movement through the store, location of products analysis as well as sales improvement which are enhancing customers' loyalty and supporting Customer Relationship Management (CRM) systems [1], [2].

The main hypothesis of this paper starts from the fact that the IPS technologies application in retail sector will increase whereas their benefits are going to bring great influence on the currently unknown changes of retail sector. These IPS technologies are likely to have the same destiny of rejection as other similar technologies from the past (such as usage of consumer card in retail sector) in our countries. However, these technologies were, subsequently, also implemented in our countries because of the benefits they have and influences from well-known leading companies.

The first part of this paper gives an overview of the main terminology and phrases used in IPS technologies with the focus on beacon technology as a most important IPS technology. Furthermore, there is a brief review on the global leading research of application of these technologies in retail sector as well as predictions for the future. In addition, there is an overview of functionality and potential benefits of application of these technologies in retail sector and on the other sectors where these technologies might be used.

The second part of this paper reveals the results of the research which was conducted in Banja Luka through anonymous questionnaire of participants who are greatly experienced popular personnel of this industry. The individuals who completed the questionnaire were the directors of retail companies and chains, stores' managers and well experienced traders with specialized skills. The purpose of this questionnaire was to identify the level of awareness that these individuals have about IPS technologies, their commitment for eventual application of these technologies in their stores, their opinion about the time frame these technologies would take to improve the retail sector, their evaluation of level of customer interest for these technologies in our countries even when the potential benefit for customers is evident.

2. Indoor Positioning Systems (IPS): definition of technologies, relationship with retail sector and global trends of implementing this technology

The easiest way to define IPS is to describe it as a GPS for indoor environment (the reason for this relation is the fact that usage of GPS in indoor environment is impossible or nearly impossible). The technologies mentioned above are frequently used for location of people and objects through the use of mobile devices, such as smart phones or tablets. IPS often relies on technologies such as beacon devices, as well as on Wi-Fi technologies using magnetic positioning, sometimes geomagnetism, video technologies, GMS/3G/4G, compass technologies, etc. However, the most frequently adopted technology is beacon technology which is based on Bluetooth low energy (BLE) that enables not only customers' location and their movement in the store, but also sending specialized messages and offers received through analysis of customers' entering and movement through the store and their earlier interests in particular products and purchases [3], [4].

The benefits of using this technology for Retail Company or chain are [5], [6]:

- Identification of entering the store
  1. Gaining passerby attention
  2. Keeping track of customers entering the store and replacement of the consumer card
- Finding the products and navigation through the object
  3. Sales improvement through finding the desirable products easier
  4. Improvement of movement through the store via linking the shopping list with products' location in the store
  5. Sales employees' efficiency improvement (receiving less customers' inquiries)
  6. Enabling customers to contact sales personnel through the use of application which brings customers' satisfaction
- Targeted advertising
  1. Additional promotional channel
  2. The possibility to offer connected products
  3. The possibility of getting the personalized coupons for individual customers
  4. The possibility of offering the replacement for products
  5. The chance of receiving the payments from particular brands
- Interaction with client
  1. Online support for customers and stores' visitors

2. Receiving the customers' impression, comments and recommendation from customers

Moreover, the potential benefits that customers and visitors in the store are receiving should also be identified and considered. These are listed below:

- The identification while entering the store
  1. Getting the information while entering the store (or in front of the store) about the current specials etc.
  2. Receiving the notification that the certain store is nearby (as customers may not notice the location of the store, so they get informed).
- Searching the product and navigation in the store
  1. Quick search through the store
  2. Navigation and movement through the store
  3. More information about the product, online recommendations etc.
  4. The chance of communicating with sales personnel via application
- Targeted advertising
  1. Adjusted targeted push-notifications
  2. More attractive personalized offers
  3. Advertising and informing the customer only about the products the which are of client's interest.
- Interaction with client
  1. Chance of getting an online support
  2. Possibilities to leave the impressions, comments and recommendations as well as to read other customers' experiences

These technologies also offer particular benefits for the brands which are:

- Having the potential of giving more information about the products
- Offering the possibility of additional advertising, especially through push-notifications
- More focused promotion (ex. advertising the customers' who expressed the interest)
- Direct communication with clients
- Collecting customers' impressions, critical views and suggestions about the brands.

These technologies are used through the particular application and they have some long lasting benefits which are presented in the tables below.

Table 1 – Positional benefits of IPS technologies implementation for different beneficiary groups (authors contribution)

Retail	**Brands**	**Customers**
Sales promotion (through the support in finding products)	Additional channels for brand promotion	Efficiency in purchasing and time saving
Customer loyalty is at higher level	Direct channel for communication with customers	More information about products and recommendations
Building more powerful channels in order to offer products, substitutes,	Possibility of providing additional information about product	Aimed offers and discounts based on customer's location

promotions and discounts		
Active channel for feedback	Tool for customer movement analysis	Channel for communication with brand or an object
Tool for measuring the effectiveness of special offers and merchandising	Tool for measuring the behavior and efficiency of marketing channel	Suitability and facilitating navigation through an object

The importance of using these technologies is greatly evident through its analysis. The application of these technologies offers certain advances for each of targeted groups. This gadget is becoming enormously useful analytical gadget as it is enabling the following:

- Customers' movements through "heat maps", display of their movement and the customers' stay
- Customers' behavior analysis and their reaction to special offers
- Informing the brands, which are on offer at the store, for the purpose of sales' analysis

Furthermore, these analyses of application are also useful for the brands as they provide the information about the customers' movement and reactions as well as about the effects of promotional campaigns that are focused on targeted customers.

These analyses allow the positive effects for sales which includes optimal organization inside the store, optimal placements of products that belong to same category as well as optimization of special offers through the observation of its efficiency. Moreover, this is one of the ways for the sales chains to offer the additional promotional channel to its customers and brands which allows the sales chains to get the special offers when purchasing the products or selling the promotional place. In addition, these brands can be offered the sale of analytical reports and customers' behavior through specific segments, effects of particular promotional campaigns as well as the information about the activities of competitive brands inside the store [7].

The application of this technology can also have positive effects on business organization through employees' efficiency analysis, reduction of excessive activities which can be replaced with more efficient and useful application as well as through the observation of employees' movements throughout the store and their adherence to the stores' policies and procedures. Ultimately, the same technology can be used for sending the messages and workplace communication which can have positive effects on costs reduction and time management.

Swirl Networks Inc. [4] report emphasizes that 73% of customers believe that their chances to buy the product would be greater if they receive the message, special offer or information about the desired product. It follows that this percentage would even increase if, through the particular system of business intelligence, these customers receive the special offers based on their previous visits to the store, their requests for information, previous purchases, movement throughout the store etc. The same report suggests that 60% of the customers would rather receive the information about the products and offers which was sent through beacon technology whereas 30% of these customers are already receiving these notifications. 61% of customers that took part in this research state that they would rather visit the store that has beacon technology system in place than the store that does not have that while 60% of customers believe that they would shop more if they receive offers through beacon technology.

Figure 1. Main results of the Swirl Report [1], graph author contribution

Cooper Smith, in the report „THE BEACONS REPORT: Growth Forecasts for The Most Important Retail Technology since the Mobile Credit Card Reader" claims that 100 leading retail chains in USA are trying out the beacon technology [8]. According to these findings, the question of the size and influences of beacon technology come into consideration. The same author points out that, beacon devices are "the fastest growing devices which are used for retail stores since credit cards" [8].

Some of his interesting findings related to development of the application of this technology are:

- The number of installed beacons is expected to be over 4.5 million until the end of 2018, from which 3.5 million are going to be installed in retail stores
- It is believed that up until the end of 2015. more than 1/3 of leading retail companies already had beacon devices installed
- It is globally believed that more than 570 million devices, whether Android or Apple OC are compatible with the use of Bluetooth low energy which certainly makes this technology extremely attractive as a tool for new communicational channel with customers.

Some of the leading retail chains, particularly in USA have a considerable usage of this technology. The examples of this usage are seen among the retail chains of Macy's stores such as Target, Lord & Taylor, Walgreen's as well as the well-known retail chain of technical equipment such as Best Buy's which has the beacon technology in place for considerable length of time for the purpose of sales improvement and better communication with clients and visitors. All of these findings are suggesting that this industry is increasing and that the percentage of those using beacon technologies is going to have considerable growth until the level when this technology will be expected in the stores. The application of beacon technology is also evident in other sectors. For instance, the art centers are using this technology to transmit messages and information about the art work exposure. There is an obvious usage of beacon technology in transport as it is used for navigation, information and promotion around the bigger places for transport including airports, harbors, vast train stations etc. Some of the health departments and educational institutions which are greatly developed are also using this technology. Furthermore, these technologies have their existence on special occasions, such as concerts, movie spectacles and music festivals etc.

3. The results of the research "Analysis of attitudes related to the implications of IPS in retail sector in Banja Luka"

In the period of March-May 2016 a research was conducted in Banja Luka on carefully chosen participants working in sales stores who completed an anonymous questionnaire. There were 34 well-known experienced individuals of this sector in 12 retail stores participating in this research. These participants were selected directors of retail companies and chains, stores' managers and well experienced traders with specialized skills.

The aims of this research were:

- Estimating the level of knowledge participants have about IPS technologies
- Evaluation of participants' interest and commitment to commence the implementation of these technologies
- Analysis of participants attitude and opinions about eventual benefits that these technologies have and their effects on sales' improvement

There is the list of questions and results summarized on the basis of participants' answers.

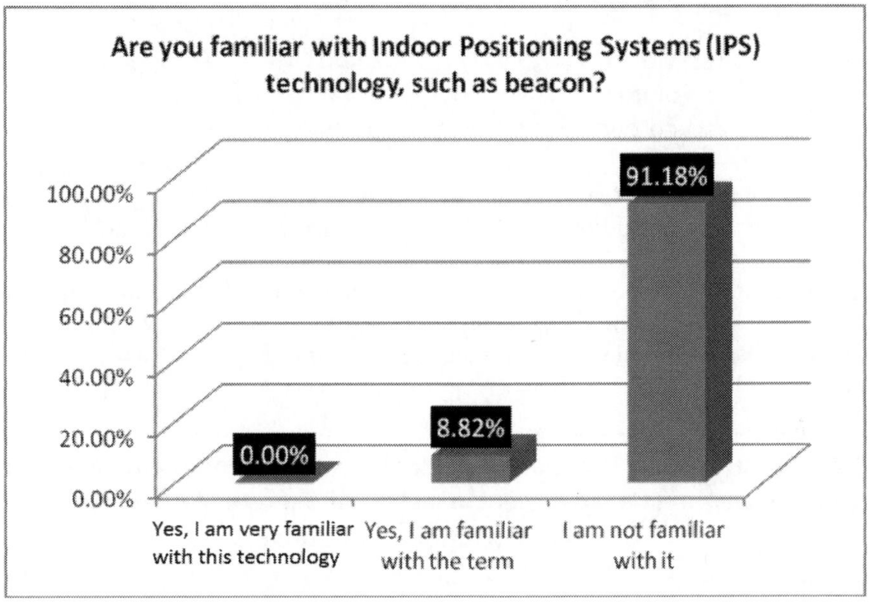

Figure 2. Answer to question "Are you familiar with Indoor Positioning Systems (IPS)?"

It is evident that majority of participants is not aware of IPS terminology which represents a foreign phrase for them. The reason we can find in a few facts. First of all is an official education in the high schools and faculties. Our education system doesn't follow the new trends in IT technologies and retail. The second one is unofficial education. Retail companies don't spend resources on training and education their employees with news in the sector. Also, employees don't want or don't have a time, money and so on to spend on education or not interest to keep track of new trends in their sector.

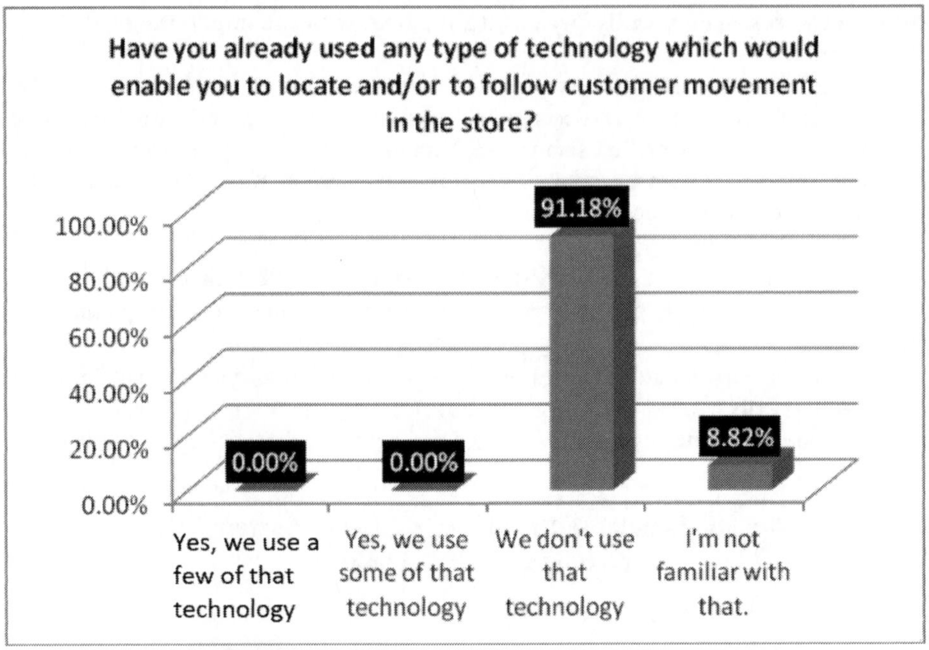

Figure 3. Answer to question "Have you already used any type of technology which would enable you to locate and/or to follow customer movement in the store?"

According to participants' responses it can be concluded that retail stores in Banja Luka is still not using any technologies for observing and locating customers in the store, which is one of the main indicators that business analysis and observing the behavior of customers in our countries is not given enough attention.

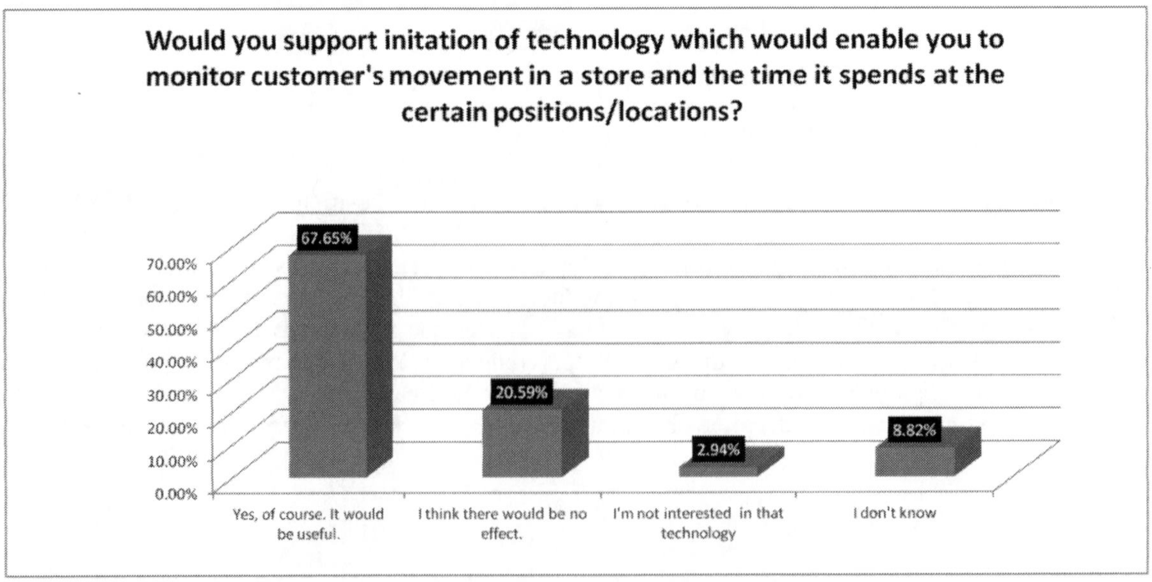

Figure 4. Answer to question "Would you support initiation of technology which would enable you to monitor customer's movement in a store and the time it spends at the certain positions/locations?"

On the other hand, 65% of participants support the idea of implementation of technology which would enable observation of customers' movement inside the store as well as their continuous stay in the particular place of the store. Even more participants, nearly 80%, would support the implementation of technology which enables sending the specialized messages to customers while in the store. This confirms the idea that participants in this study realized the fact that this way of communication would be useful for sales' improvement.

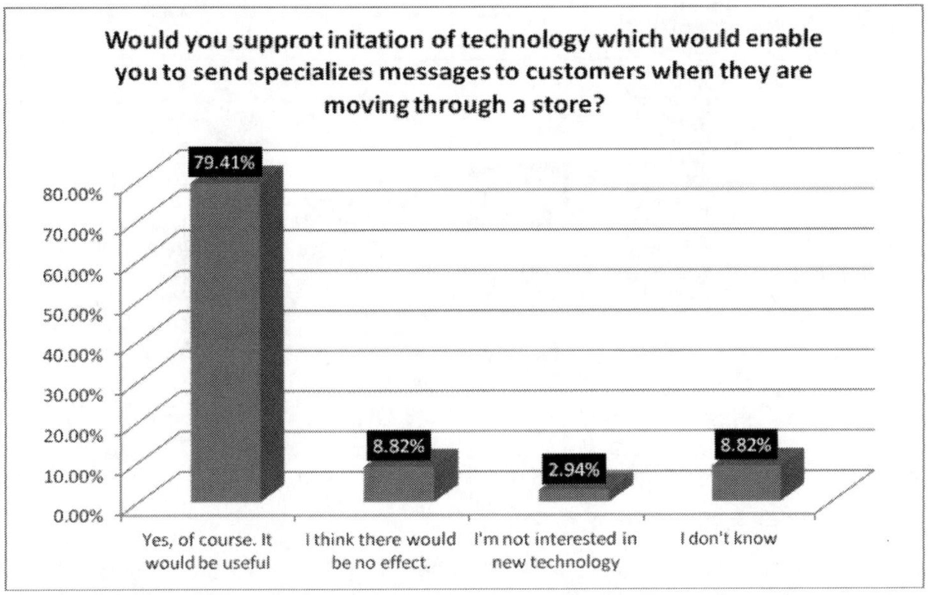

Figure 5. Answer to question "Would you support initiation of technology which would enable you to send specializes messages to customers when they are moving through a store?"

The participants also recognized that this technology would have a positive effect on increase in the sales, which is the main indicator of the success of retail store or chains. In addition, more than 60% of participants believe that these technologies would have a positive effect on enhancing customers' loyalty which is, strategically explained, significant indicator of quality of these technologies.

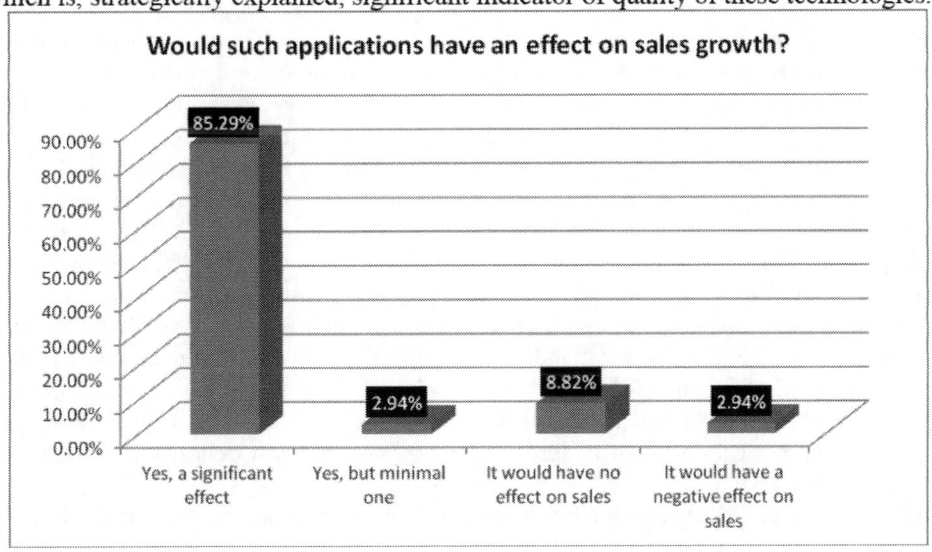

Figure 6. Answer to question "Would such applications have an effect on sales growth?"

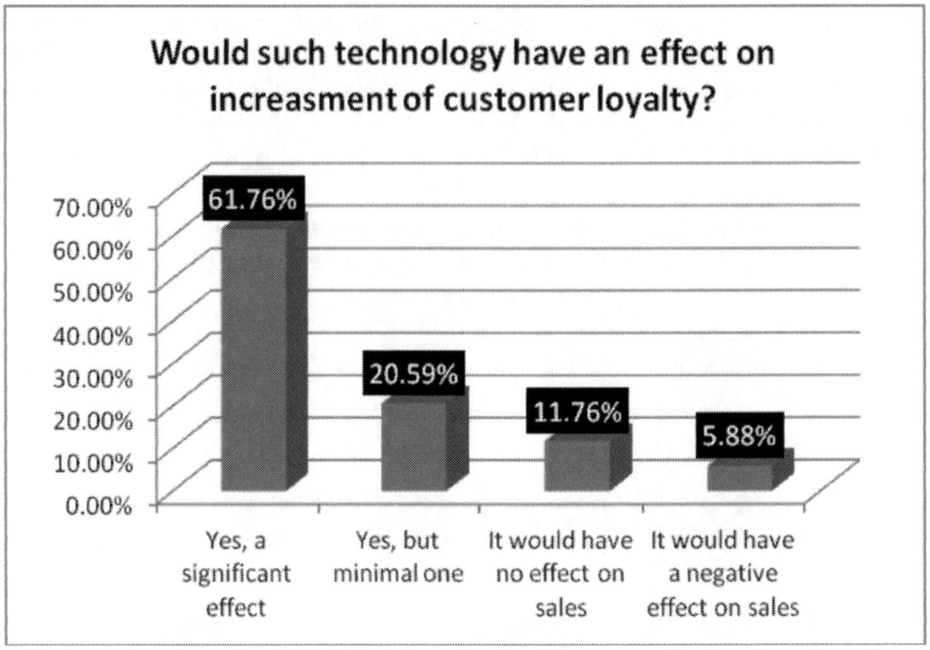

Figure 7. Answer to question "Would such technology have an effect on increase of customer loyalty?"

It is well known that people don't adapt new technologies so easily. Especially when they don't know anything about advantages or disadvantages of a new application and they not tried it. Authors are opinion this lack of information had an impact on participants answers.

According to the last population list in Bosnia and Herzegovina lives 3,53 million of people and this is not a big market. There is also a question of profitability of introducing this type of technology in so small and purchasing possibility low market.

4. Conclusion

IPS technologies, especially beacon technology is globally known by the retail sector. Their implication is constantly becoming greater and its effects are evident. There is an increasing number of retail chains, particularly in USA, which are using these technologies for decision making processes as well as for new communication processes. On the other hand, the customers are also finding the benefits which have a positive effect on their willingness to use these technologies.

This paper identifies significant benefits for retail companies of using IPS technologies. Through interaction with customers, retail companies can find out what customers want, what they look for, are they satisfied with offer etc. Based on this information retail companies can direct efforts to optimize the time, costs and efforts of customers and own employees. Lack of time for shopping today is a common feature of a significant number of customers. The most important benefits of using IPS technology for customers are spending less time and money to buy what they need. Personalized offers based on customers preferences, better customers experience, and optimized movement through the store. According to the different research, customers have a positive reaction to use IPS app. IPS technologies can make a significant impact on the brands in store, business organization, and CRM, also. IPS is found its place in a lot of different areas: in museums, warehouses, concerts, events, to find an office in buildings etc.

The local retail sector in Bosnia and Herzegovina is still not aware of the terminology related to IPS technology or the technology itself. Regardless of the fact that local retail sector is not yet implementing these technologies, there is a positive attitude towards potential benefits on sales' improvement. Nevertheless, these technologies are going to be implemented in local retail companies despite the local interest in them, whether through the local or international retail chains. Furthermore, taking into consideration growth rate of implementation of these technologies and welcoming them by the customers and retail sector, it is unquestionable that these technologies will become integral part of individual purchases and strategic marketing management of retail chains.

References

[1] ***Infsoft Inc 2017 *Indoor Positioning, Tracking and Indoor Navigation with Beacons*, Großmehring Germany, https://www.infsoft.com/technology/sensors/bluetooth-low-energy-beacons
[2] ***Geo IT Hub 2016 *What Is A Beacon?* Geo IT Hub Delhi India, http://geoithub.com/what-is-a-beacon/
[3] Senion Inc 2017 *What is Indoor Positioning Systems*, Senion Inc Linköping, Sweden, https://senion.com/indoor-positioning-system/
[4] Smith C 2014 *The Beacons Report: Growth Forecasts For The Most Important Retail Technology Since The Mobile Credit Card Reader*, Business Insider
[5] ***Acuity Brands Inc 2015 *Illuminating the In-Store Experience Indoor Positioning Services Using LED Lighting Benefit Shoppers and Retailers*, Acuity Brands Indoor Positioning Services White Paper, Atlanta, Georgia, http://www.acuitybrands.com/media/Files/Acuity/Solutions/Services/Indoor%20Positioning%20White%20Papers/indoor%20positioning%20white%20paperrevised%20110315%20pdf.pdf
[6] ***Senion Inc 2017 *Indoor positioning for retail*, Senion Inc Linköping, Sweden, https://senion.com/indoor-positioning-for-retail/
[7] Fischer J 2015 *Can IPS enhance the shopping experience?* RetailCustomerExperience.com https://www.retailcustomerexperience.com/articles/can-ips-enhance-the-shopping-experience/
[8] ***Swirl Networks Inc 2014 *Retail Store Beacon Marketing Campaigns*, Swirl Networks Inc, New York, USA

IOP Conf. Series: Materials Science and Engineering **294** (2017) 012066 doi:10.1088/1757-899X/294/1/012066

Contributions to optimization of storage and transporting industrial goods

T Babanatsas[1], R M Babanatis Merce[1], D O Glăvan[1] and A Glăvan[2]

[1]University "Aurel Vlaicu", Department of engineering, Arad, Romania
[2]Rug University of Groningen, Faculty of Mathematics and Natural Sciences, Groningen, Netherlands

E-mail: babanatsas@outlook.com

Abstract. Optimization of storage and transporting industrial goods in a factory either from a constructive, functional, or technological point of view is a determinant parameter in programming the manufacturing process, the performance of the whole process being determined by the correlation realized taking in consideration those two factors (optimization and programming the process). It is imperative to take into consideration each type of production program (range), to restrain as much as possible the area that we are using and to minimize the times of execution, all of these in order to satisfy the client's needs, to try to classify them in order to be able to define a global software (with general rules) that is expected to fulfil each client's needs.

1. Introduction

The goal of the worldwide economic and technical development that has taken place during the past years has been due to increasing social and economic needs.

The knowledge and innovation has created more flexible systems in storage and transporting industrial goods, based on information technology. The study of storage and transporting industrial goods is motivated by the needs of the people, and by the high interest in flexible manufacturing systems with high performance levels, which can replace the human factor in storage and transporting industrial processes [1-4].

Optimization of storage and transporting industrial goods in a factory are very important especially in our days when we try to protect our eco system. For this we must to use small spaces (deposits) but with as much efficiency as possible, also we must have the minimum amount possible of waste materials and to use mostly renewable energy (minimizing the use of electrical and gas power).

The aim is to improve the handling of goods in existing warehouses in Romania, which have had a wide range of products within application.

Objectives and directions are focused on analyzing, using, and capitalizing on the efficiency of storage space for materials and the entire logistics process [5].

We will analyze different product types, different storage and transportation systems, and from the results we will propose an optimal and innovative system for the storage and transportation industrial goods.

2. Current situation of storage and transportation systems

2.1. Current situation of storage and transportation systems in Romania

Most companies in Romania have not implemented an automated logistics system and especially not for the handling of goods.

The Romanian market for industrial equipment for handling, storing, and improving production flows increased by 30% in 2015 with 3,195 new units compared to 2,470 units in 2014.

Last year, on the Romanian market there were 872 forklifts sold, 568 electric forklifts and 1,755 technical vehicles for warehouse interior [6].

As a result of this study, we conclude that the current manipulation system is a classic one (man-mechanical system).

2.2. Current situation of storage and transportation systems in a society of industrial goods

A more extensive study was carried out in a society of industrial goods.

This society has as activity the trade of furniture materials and accessories, where:

- We have had a turnover of over 7,000 different landmarks in the last 5 years.
- We currently have about 4,000 different landmarks

All the landmarks in the two repositories are matrix stored, in rows and columns, each landmark having its own well-defined location, Figure 1.

- Warehouse 1 has 2,000 sqm
- Warehouse 2 has 1,500 sqm

Figure 1. Location of the warehouses

Currently, product locations are arranged by a matrix system on columns and rows, and the employee must climb the stairs to product locations Figures 2 and 3.

In this society are a very wide range of types of pieces and almost 5.000 items.

Figure 2. Matrix system on columns and rows

Figure 3. Matrix system on columns and rows

Figure 4. Small items

Figure 5. Big items

3. Experimental conditions

3.1. Objectives of experimental conditions

The main objective of this experiment is to optimize the storage and transportation systems in the warehouses of commercial enterprises.

From this the results are:

- Reduction of energy consumption;
- Reduction of investment costs;
- Reducing the consumption of labor;
- Reduce accidents, risks and human errors;
- Increased operability;
- Increase cost / benefit efficiency;
- Increase operational safety.

With this will result in efficiency of the space and of the logistics system, it is necessary to design and implement an automated system for handling goods [3], [7], [8].

Nowadays, the handling of commodity merchandise takes place in various ways both as automated systems and as classical systems serviced by the human operator, assisted by various computer programs (databases, specific software for managing goods, etc.). These systems are costly and on the one hand their efficiency can be improved [7], [9-11].

The optimization of the storage and transportation systems aims to:

- Avoid product damage due to human factor;
- Avoid serving other products and / or alter the initial position on the site;
- Rapid service;
- Higher safety of the human operator by avoiding the risk of injury;
- Better product management;
- Efficiency of work in the warehouse;
- Space efficiency.

3.2. Application in experimental conditions

As a novelty will be the creation of a coordination system (from supply to customer service) and optimization of automatic storage and transportation systems for goods, objects, a wide range of types, materials, plastic, metal (ferrous and non-ferrous), glass, etc. We will not limit to manipulating palette objects or a limited range of types.

To achieve the goal, we have proposed the creation of an automatic cargo handling system also considering the information inside the warehouses [5], [12], [13].

This system, Figure 6, will have to communicate fully with the three warehouses (reception warehouse, small size accessory storage 1 and large accessory storage 2).

This system may be a rail guided "stacker" device. Each warehouse will have such a system installed in the room and service between warehouses will be made through conveyor belts.

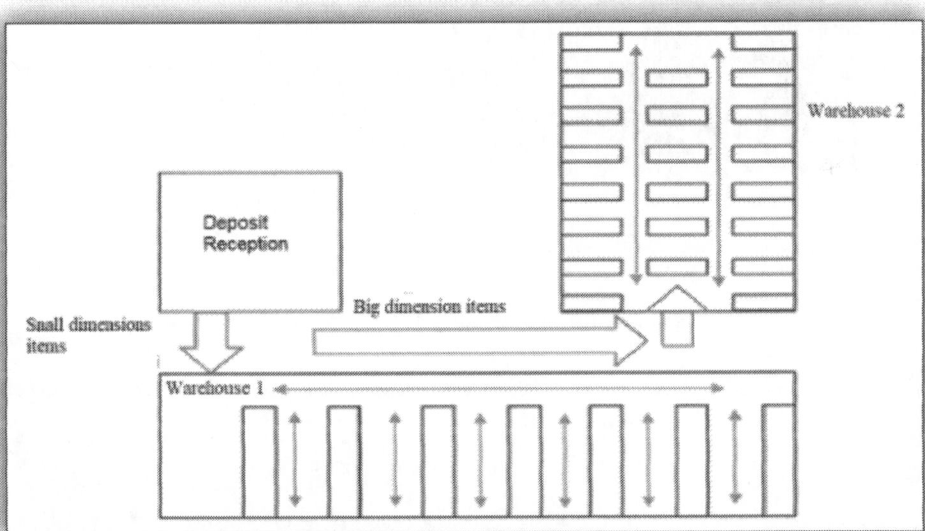

Figure 6. Logical description of the process

This system has several advantages

- Avoiding the disclosure of products due to the human factor
- Avoid serving other products and /or alter the initial position on the site
- Rapid in serving products
- Greater safety of the human operator by avoiding the risk of injury
- Better product management
- Efficiency of work in the warehouse
- Space efficiency

3.3. Creating devices for experimental conditions

For this experiment we created, in our laboratory from University "Aurel Vlaicu", engineer department, a device for recognition of objects, Figure 7.

We create this automatic and intelligent storage system prototype device that contains two departments, the factory and the retail sell department. In the next stage, the optical system must separate these items.

The device has a processor that is connected to a pc for recognition objects. In this pc we have stored 1000 photos of items. From the conveyor, the items pass in front of the camera which is responsible to detect these items. The selection device is responsible for selecting objects to be directed to the factory or to the retail branch.

The schematic structure of automatic storage system that deserve the factory and the retail sell department are exemplification in Figure 8.

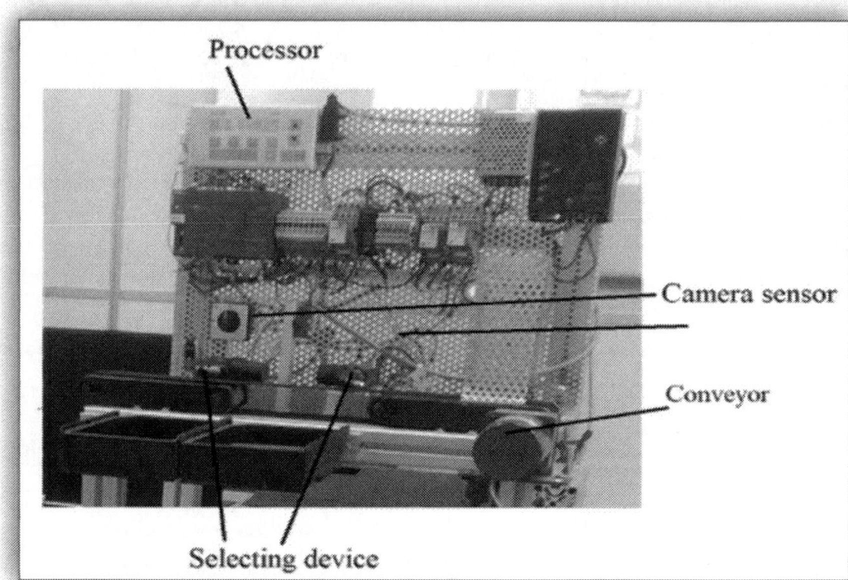

Figure 7. Logical description of the process

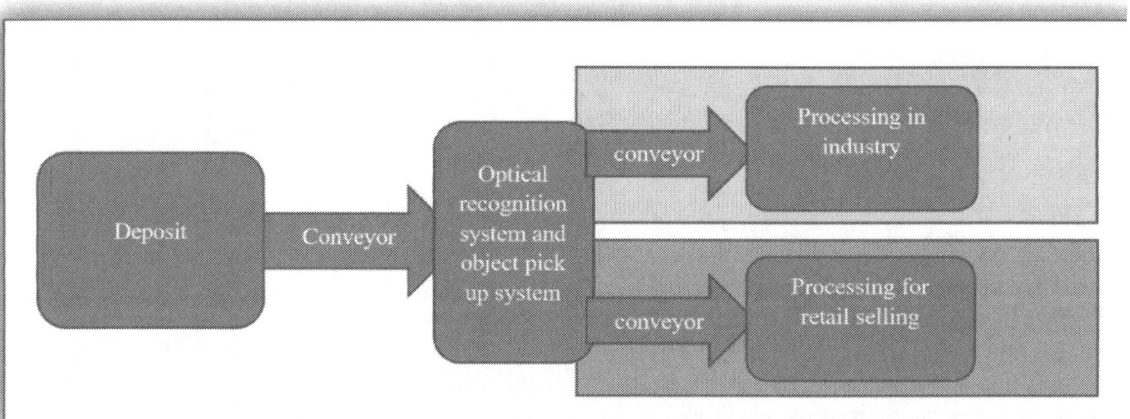

Figure 8. Schematic structure of automatic storage system

4. Results and discussions

To determine the accuracy of optical recognition software we make the tests with more image recognition software. We selected, for these tests some small metallic (Figure 9) and plastic (Figure 10) and some big metallic (Figure 11) and plastic (Figure 12) objects.

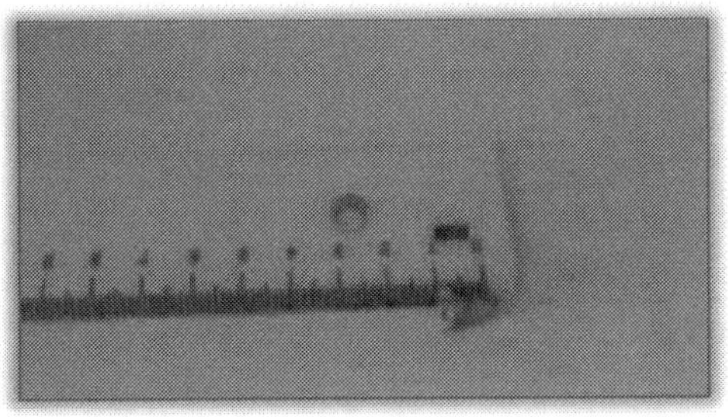

Figure 9. Small metallic items

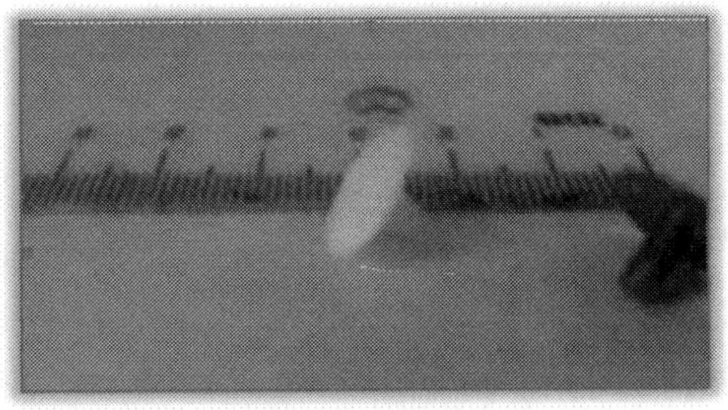

Figure 10. Small plastic items

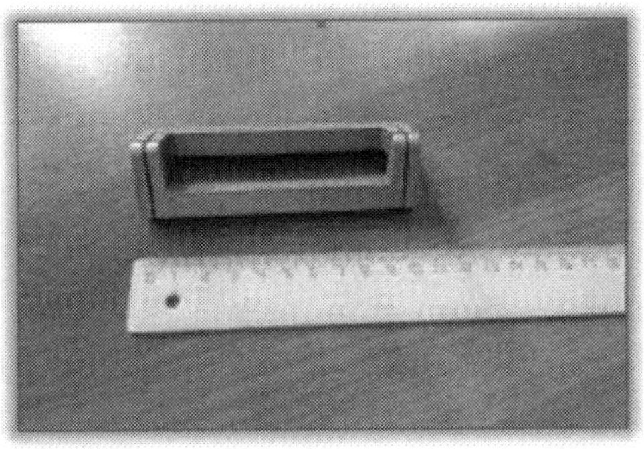

Figure 11. Big metallic items

Figure 12. Big plastic items

The experimental data, for the accuracy for each software, are presented in Table 1.

Table 1. Accuracy of recognition software

	Small plastic objects	Big plastic objects	Small metallic objects	Big metallic objects
Soft 1 SVS	96%	98%	96%	98%
Soft 2 BLIPPAR	98%	96%	98%	96%
Soft 3 GAZIRU	97%	98%	97%	97%
Soft 4 IMAGE CAPTURE PLUS	95%	96%	97%	95%
Soft 5 API	98%	97%	95%	96%

All software has no big errors when we determinate the recognition items by size and by materials in this prototype device.

Also, we calculate the average delay time from the moment when the camera detects the object and the moment when the software recognizes the item and give the command at selected device, Table 2.

Table 2. Average delay time of optical recognition software

	Maxim delay time	Minim delay time	Average delay time
Soft 1 SVS	0.092 sec	0.090 sec	0.0913 sec
Soft 2 BLIPPAR	0.121 sec	0.121 sec	0.121 sec
Soft 3 GAZIRU	0.099 sec	0.099 sec	0.099 sec
Soft 4 IMAGE CAPTURE PLUS	0.108 sec	0.108 sec	0.108 sec
Soft 5 API	0.11 sec	0.09 sec	0.1 sec

5. Conclusions

It is very important to create automatic olives storage. With this storage, we can better manage all the items in society. In this storage, we can make an automatic selection between more types of items. For this we tested more image recognition programs.

All the software has a result of 100% to determine the correct dimension of the item and over 95% accuracy of recognizing the item itself.

This result was very good and indicates some minor errors which are not important in all technological processes.

The next test was to determine the delay time between the moment when the item is in the conveyor in front of the optical reader and the moment when the selected device chooses the item.

Also, this test shows good results and indicates, in all software, a delay time between 0.09 sec and 0.12 sec, which is faster than olive movement speed on conveyor that is 0.7 sec.

References

[1] Altieri G, Di Renzo G C and Genovese F 2013 Horizontal centrifuge with screw conveyor (decanter), *Journal of food engineering* **119**(3) 561-572

[2] Gusetoriu I and Tucu D 2013 *Stress risk in management systems in metallurgical problems*. 22nd International Conference on Metallurgy And Materials, pp 1904-1908

[3] Debojit, Biswas Jain Hitesh, Arora Manoj K and Balasubramanian R 2011 Study and Implementation of a Non-Linear Support Vector Machine Classifier, *International Journal of Earth Sciences and Engineering* **4**(06) SPL

[4] Peri C 2013 *Handling, storage and transportation*, Wiley-Blackwell, Italy

[5] Tucu D 2012 *Application of life cycle cost method for willow production machinery*, Actual tasks on agricultural engineering, proceedings, Book Series: Actual Tasks on Agricultural Engineering-Zagreb, **40**, pp 549-556

[6] Cioarna A 2016 *Metodologie şi eficienţă în activitatea de cercetare ştiinţifică*, Editura Gutenberg Univers Arad, pp 21-26

[7] Tucu D, Golimba A and Slavici T 2010 *Fuzzy methods in renewable energy optimization investments*, Actual tasks on agricultural engineering, proceedings, Book Series: Actual Tasks on Agricultural Engineering-Zagreb, **38**, pp 455-462

[8] Tucu D, Rotarescu V and Bordeasu I 2006 Considerations on the relationship between product quality and milling system in plastic recycling, *Materiale Plastice* **43**(4) 308-311

[9] Glavan D 2016 *Compunerea si funcționarea Mașinilor – Unelte,* Editura Universității „Aurel Vlaicu" Arad, pp 178-179

[10] Tsarouchi P and et al 2013 *Vision System for Robotic Handling of Randomly Placed Objects*, Procedia, 2nd CIRP Global WebConference, pp 66-67

[11] Cotet G B and et al 2017 *Assessment procedure for the soft skills requested by industry 4.0*, 8th International Conference On Manufacturing Science And Education - MSE 2017 - Trends in new industrial revolution, Sibiu, Romania

[12] Popa A, Glavan D O, Bucevschi A, Mortoiu D and Gaspar M C 2015, *Solution for Industrial Process Optimization*, IManE 2015 International Conference (Innovative Manufacturing Engineering International Conference), Iasi, Romania, Applied Mechanics and Materials, **809-810**, pp 1348-1353

[13] Tucu D, Golimba A and Mnerie D 2010 *Grippers design integrated in handling systems destinated to agriculture mechanization,* Actual tasks on agricultural engineering, proceedings, Book Series: Actual Tasks on Agricultural Engineering-Zagreb, **38**, pp 447-454

IOP Conf. Series: Materials Science and Engineering **294** (2017) 012067 doi:10.1088/1757-899X/294/1/012067

Modelling in 3D the olive trees cultures in order to establish the forces (interval) needed for automatic harvesting

T Babanatsas[1], D O Glăvan[1], R M Babanatis Merce[1] and S A Maris[2]

[1]University "Aurel Vlaicu", Department of engineering, Arad, Romania
[2]University "Politehnica", Department of engineering, Timisoara, Romania

E-mail: babanatsas@outlook.com

Abstract. The purpose of this study is to bring as much as possible, close to real situation the 3D modelling for the olive trees in order to establish the necessary forces for automatic harvesting (harvesting robots). To fulfil our goal we have at our disposal different ways to do modelling very close to the real situation. One way is to use reality capture software (its results being photos) that are converted into a real 3D model, the disadvantage of the method being a mesh model that is not accurate enough. The reasonable alternative is to develop an experiment by measuring a sample orchard of olive trees (experiment who took place in Halkidiki, Greece, measuring over 120 trees). After establishing the real dimensions, we adopted as model the media that we have measured (the height of the tree, the thickness of branches, number of branches, etc.), model which we consider closer to the reality and therefor more suitable for our simulation.

1. Theoretical considerations

Common automatic harvesting is based on vibration forces to shake the trees [1]. We can establish this vibration forces with empiric mathematical calculation or we can determine this forces by 3D modelling, with specific software. In the first case, the problem is that after the calculation we need to create a real prototype for testing the forces in real conditions. In the second case, we must create a real 3D model of tree, with real dimensions and real wood characteristics.

In this situation, in order to create a 3D tree model, we can use two methods. In one method, we can create the 3D tree model from photos and the result is a 3D mesh model. In the other method, we can create the 3D tree model by measurements and drawing the tree body in CAD software. In this second case, we use the average diameter dimensions and the average density of olive tree wood in raw condition [2-5].

2. Experimental conditions

In this experiment, we were traveling in Northern Greece, in Central Macedonia, in the Region of Halkidiki where there is an olive orchard that was designed to involve planting trees.

The dimensions of this olive orchard are about 6.000 square metres with about 134 olive trees, Figure 1.

Figure 1. Olive orchard in peninsula Kassandra of Halkidiki

2.1. Creating a 3D tree model from photos

In this case we examine all the trees from orchard and we choose, for this experiment, a tree with the middle dimensions. After that we took photos in 360-degree around the olive tree, Figure 2.

Figure 2. Olive in 360-degree photo

After that we use a dedicated software for reverse engineering that can create 3D models from photos. The software is the ReMake from Autodesk and we converted the reality captured with photos into high-definition 3D meshes, Figure 3. These meshes are edited and scaled and the final result will be a 3D olive tree meshed model in 3D solid cad extensions (dwg, sat or obj), Figure 4.

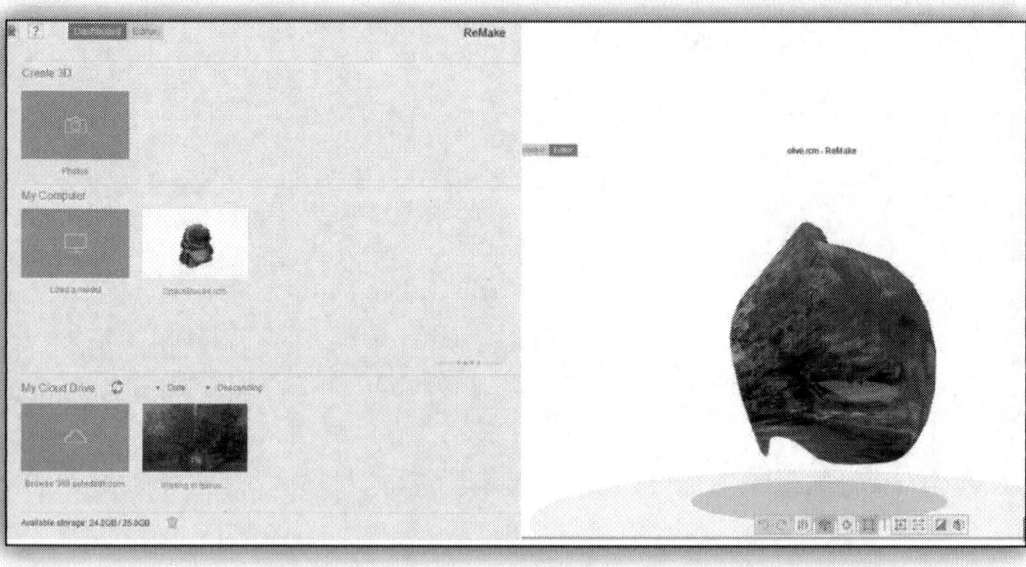

Figure 3. Olive tree mesh created from photos

Figure 4. Olive tree mesh 3D model in Autodesk Inventor

The inconvenience of the method is that the accurracy is not good, because:

- we have transformed the leaves in 3D objects with the same mass characteristic of the olive wood [6];
- it is very difficult to delete or edit every leaves;
- the tree body has different dimensions from the real one because the software has approximate dimensions;

At least one of the biggest problems is the mesh file because it isn't a real 3D object, and for this reason we can't edit the characteristics of the materials, wood density, and we cannot add the vibrating forces to see if these forces are enough or if these forces will affect the olive tree [7], [8].

2.2. Creating a 3D tree model from measurements

In this case we took measurements of all the trees from orchard. The measurements we took were the height, the thickness, the perimeter and the numbers of branches of olive trees, Figure 5 [2], [9].

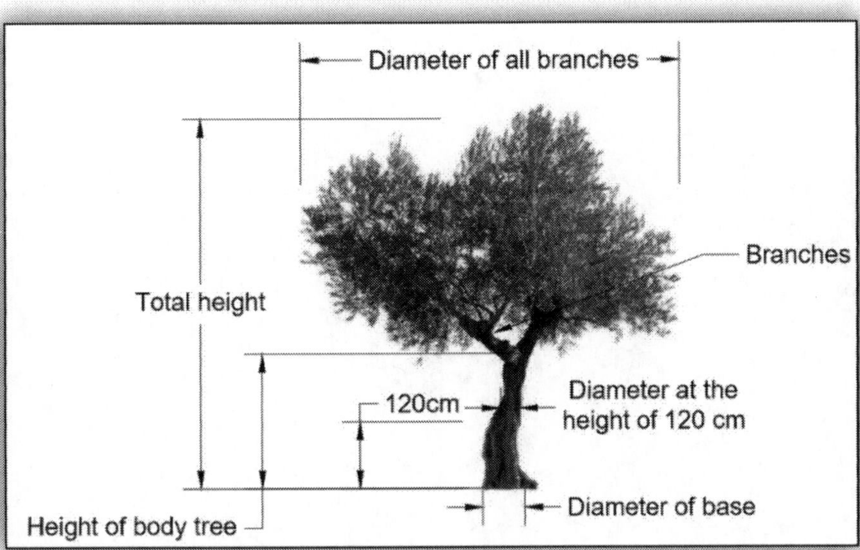

Figure 5. Olive tree measurements

All the data from the measurements was centralized in a table for processing, where the most useful data for creating the 3D model is the height of the olive and of the body tree, the perimeter of the body in different heights and the crown size of the tree.

Table 1. Measurements of olive trees

Olive tree	Height of olive tree (cm)	*Height of body tree*	Perimeter of olive tree (cm) on base	Perimeter of olive tree (cm) at high of 120 cm from sol	Number of principal branches	*Diameter of crown*
1	351	195	94,2	72,2	5	380
2	327	210	100,5	81,5	7	410
3	373	235	141,3	119,2	15	570
4	342	205	113,1	91,1	9	380
5	335	245	138,2	113,1	16	560
6	341	225	131,9	106,8	14	520
7	368	235	141,3	116,2	19	590
8	379	255	144,5	109,9	23	620
9	333	235	157,1	128,8	15	490
…	…	…	…	…	…	…
132	345	210	153,9	128,8	12	440
133	361	205	144,5	119,3	11	420
134	352	200	119,3	106,8	14	430
Average dimensions	365	225	119,4	100,5	11	485
Minimum dimensions	310	175	81,6	62,8	4	350

We used all the data from Table 1 to create a 3D model of a olive tree in Autodesk Inventor. We started to create a 3D model of olive tree, Figure 6 and we are scaling in the average and minimum dimensions.

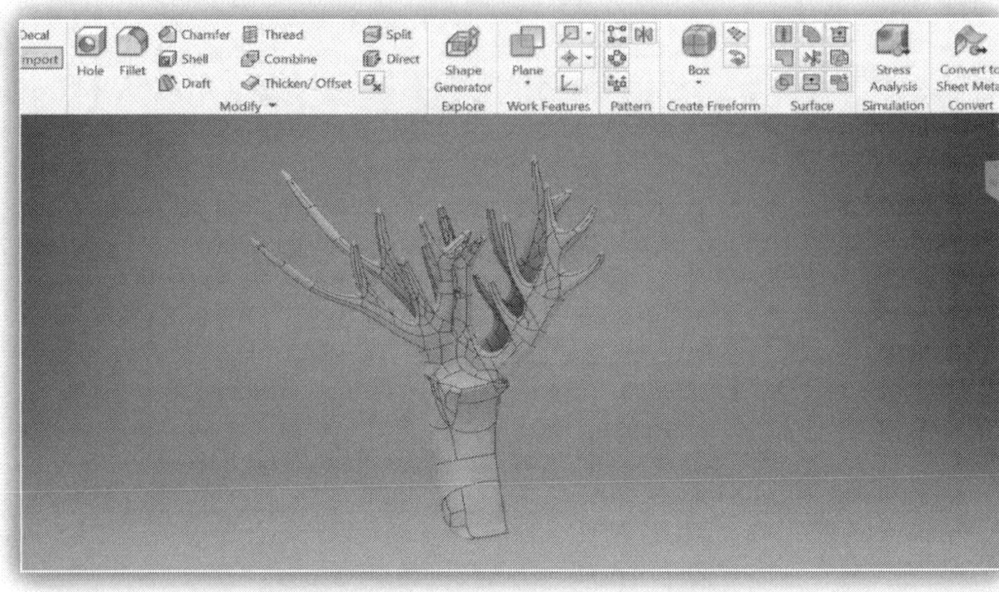

Figure 6. 3D Olive tree in Autodesk Inventor

In the next step we used the stress analysis study firm Autodesk Inventor and we simulated different vibration forces, Figure 7. The vibration forces were simulated on 3D olive model at a height of 120 cm from the ground at the positions of the shaking system [6-9].

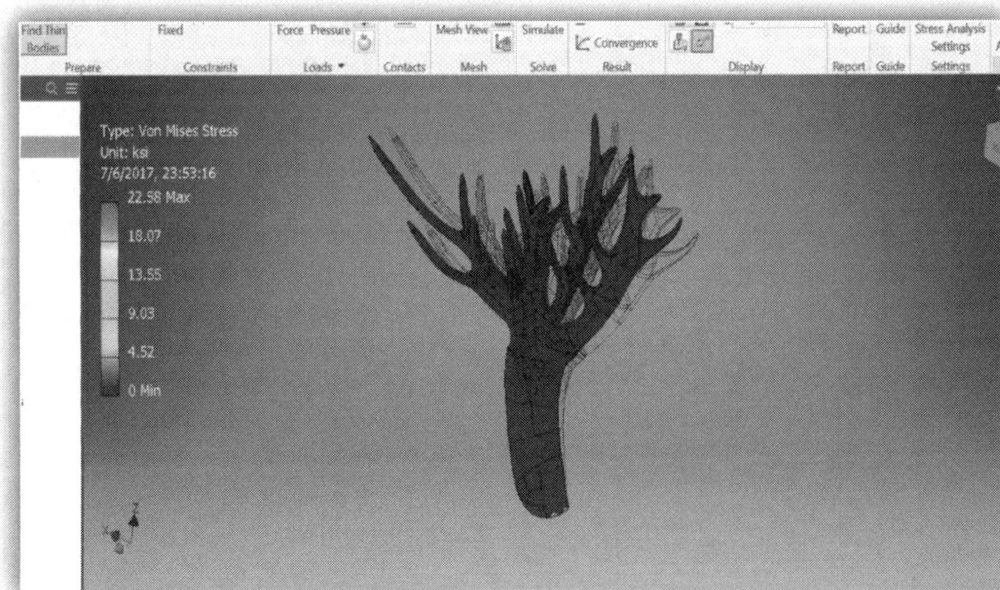

Figure 7. 3D Olive tree in Autodesk Inventor in stress analysis

3. Results and discussion

In order to determine the necessary vibration forces we can use certain software such as Inventor or Solidworks, where we can create the 3D model and simulate the forces. In this case, we used Autodesk Inventor with different vibration forces and static force, Table 2.

Table 2. Resume of Autodesk Inventor simulated study

Nucleus	Reaction Moment		Reaction Moment	
	Magnitude	Component (X,Y,Z)	Magnitude	Component (X,Y,Z)
Fixed Constraint	112.045 N	-112.045 N	96.153 N m	0 N m
		0 N		-96.153 N m
		0 N		0 N m
Fixed Constraint	223.345 N	-223.345 N	197.133 N m	0 N m
		0 N		-197.133 N m
		0 N		0 N m
Fixed Constraint	435.265 N	-435.265 N	301.223 N m	0 N m
		0 N		-301.223 N m
		0 N		0 N m

All results were satisfactory on highest forces without breaking the olive tree body or any branches. We apply these forces in x displacement (Figure 8), y displacement (Figure 9) and z displacement (Figure 10).

Figure 8. 3D Olive tree with forces distribute in x displacement

Figure 9. 3D Olive tree with forces distribute in y displacement

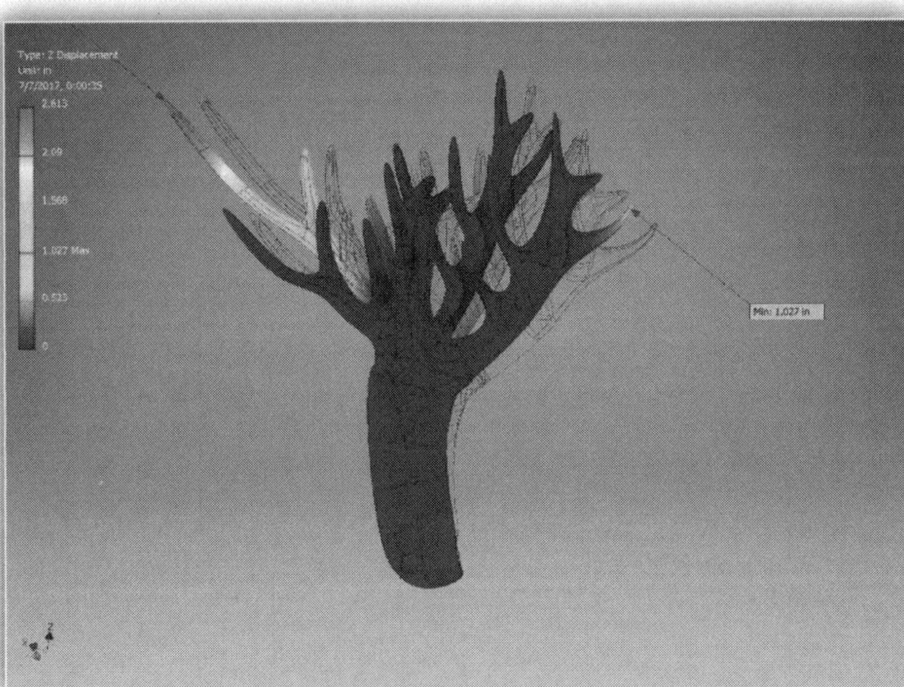

Figure 10. 3D Olive tree with forces distribute in z displacement

With this we can take a conclusion that there are more methods to create a 3D model of an olive tree, but the best solution are to create step by step the 3D model with real dimensions and with real olive wood characteristics. After we finished this 3D modelling we can use our 3D model in Inventor or in Solidworks to establish the shaking forces for automating harvesting.

References

[1] El Attar M Z, El Awady M N, Rashwan M and Genaidy M A I 2004 *Physical properties effects on shaker-model harvesting of olive-trees*, 10th MSAE conference
[2] Peri C 2014 *Olive handling, storage and transportation*, Wiley-Blackwell, Italy
[3] Tucu D, Golimba A and Mnerie D 2010 *Grippers design integrated in handling systems destinated to agriculture mechanization*, Actual tasks on agricultural engineering, proceedings. Book Series: Actual Tasks on Agricultural Engineering-Zagreb, **38**, pp 447-454
[4] Mnerie D, Tucu D, Anghel G V and Slavici T 2008 *Study about integration capacity of systems for agro-food production*, Actual tasks on agricultural engineering, proceedings, Book Series: Actual Tasks on Agricultural Engineering-Zagreb, **36**, pp 617-622
[5] Tucu D 2011 *Analyze of opportunities for willow's culture as biomass resources in banat region*, Actual tasks on agricultural engineering, proceedings, Book Series: Actual Tasks on Agricultural Engineering-Zagreb, **39**, pp 171-178
[6] Tucu D 2014 *The behavior of willow stems by cutting in nurseries*, Actual tasks on agricultural engineering, proceedings, Book Series: Actual Tasks on Agricultural Engineering-Zagreb, **42**, pp 405-413
[7] Tucu D, Golimba A and Slavici T 2010 *Fuzzy methods in renewable energy optimization investments*, Actual tasks on agricultural engineering, proceedings, Book Series: Actual Tasks on Agricultural Engineering-Zagreb, **38**, pp 455-462
[8] Peri C 2014 *The extra virgin Olive Oil Handbook*, Wiley-Blackwell, Italy
[9] Tucu D 2012 *Application of life cycle cost method for willow production machinery*, Actual tasks on agricultural engineering, proceedings, Book Series: Actual Tasks on Agricultural Engineering-Zagreb, **40**, pp 549-556

Comparative study of tool machinery sliding systems; comparison between plane and cylindrical basic shapes

D O Glăvan[1], T Babanatsas[1], R M Babanatis Merce[1] and A Glăvan[2]

[1]University "Aurel Vlaicu", Department of engineering, Arad, Romania
[2]Rug University of Groningen, Faculty of Mathematics and Natural Sciences, Groningen, Netherlands

E-mail: glavan@fortuna.com.ro

Abstract. The paper brings in attention the importance that the sliding system of a tool machinery is having in the final precision of the manufacturing. We are basically comparing two type of slides, one constructed with plane surfaces and the other one with circular cross-sections (as known as cylindrical slides), analysing each solution from the point of view of its technology of manufacturing, of the precision that the particular slides are transferring to the tool machinery, cost of production, etc.
Special attention is given to demonstrate theoretical and to confirm by experimental works what is happening with the stress distribution in the case of plane slides and cylindrical slides, both in longitudinal and in cross-over sections. Considering the results obtained for the stress distribution in the transversal and longitudinal cross sections, by composing them, we can obtain the stress distribution on the semicircular slide. Based on the results, special solutions for establishing the stress distribution between two surfaces without interact in the contact zone have been developed.

1. The influence of the shape of sliding surfaces in sliding process of the whole tool-machine

The first part of our study is referring to the case of cylindrical slides. In order to determine what is really happening in all zones of interest we have to create a mathematical model with all phenomena that are generated by the manufacturing process, all stress that is acting on the structure, the generated forces and torques, action and reaction and distributed forces.

The resultant of the forces is affecting the semicircular slide, and the formulas for acting stress has been deducted previously, is broken down using both the transversal direction of the slide, and the longitudinal direction [1], [2].

In a transversal cross section, the resultant affecting the slide, as proved previously, in a spot characterized by an angular position and by coordinates y and z, will break down in a f(x) distribution curve, which is characterized by a maximum stress value in its transversal cross section of P_{max}, and a contact length T/2, as presented in Figure 1.

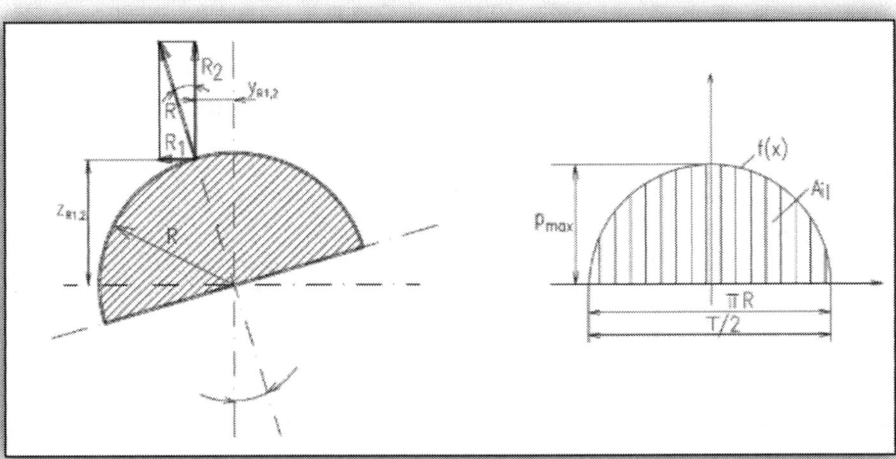

Figure 1. The resultant in a transversal cross section

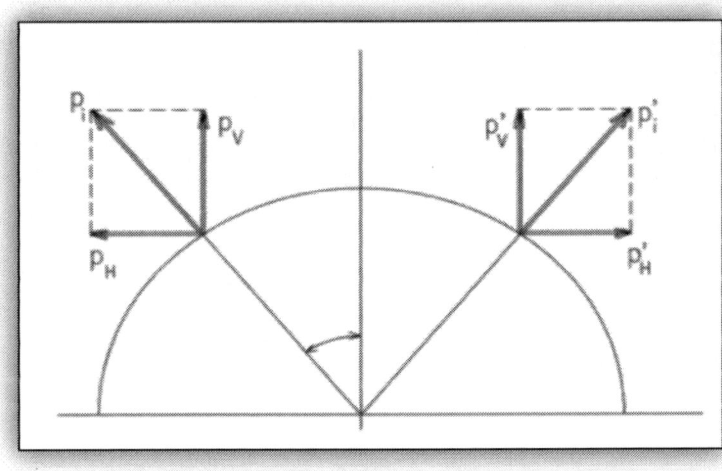

Figure 2. Reaction in the semi-circular slide

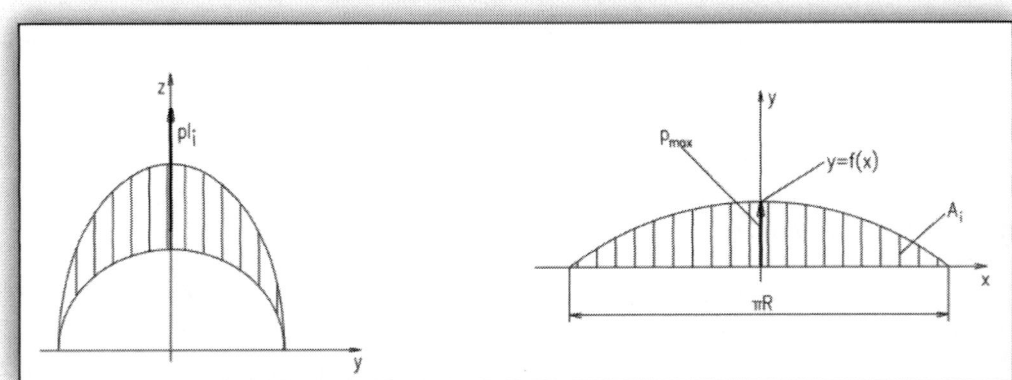

Figure 3. The broken-down stress in a transversal cross section

A specific feature of the semicircular slide is the fact that the reaction components on the horizontal axis will cancel each other out generating just a compression stress on the horizontal section of the slide, with low value fact that is not affecting at all the precision of the structure and of course the final precision of the manufacturing process. Only the vertical components will be left and they will count in the precision balance, as shown in the Figure 2 [3].

The broken-down stress values in the transversal cross section are displayed in Figure 3.

2. Experimental conditions

For this particular shape of the sliding surfaces we can write the following equations:

$$\begin{cases} p_H = p_i \sin a_i \\ p_V = p_i \cos a_i \\ \overline{p_H} + \overline{p_H} = 0 \end{cases} \quad (1)$$

The break-down in the transversal cross section allows us to anticipate the fact that the distribution curve will be of a cosine type. However, two variants will be considered in the calculation.

Two function variants which could define the stress curve in a transversal cross section were analyzed: the harmonic and the polynomial functions [2], [4].

2.1. Distribution according to a harmonic function

The following general formula was used:

$$f(x) = b \cdot cos\omega x \quad (2)$$

In order to determine the b and ω constants, the area between the f(x) function and the OX axis is equaled to the longitudinal stress corresponding to the respective I point that is defining the A_i cross section [5-8]

$$p_i l = A_i \quad (3)$$

but:

$$A_i = \int_{\frac{\pi R}{2}}^{\pi R} f(x)dx = \int_{\frac{\pi R}{2}}^{\pi R} b \cdot cos\omega x dx = 2\frac{b}{\omega} sin\omega x \begin{cases} \frac{\pi R}{2} \\ 0 \end{cases} \quad (4)$$

$$\omega = \frac{2\pi}{T} \quad (5)$$

but the T period meets the following condition, according to the diagram:

$$\pi R = \frac{T}{2} \Longrightarrow T = 2\pi R \Longrightarrow \omega = \frac{2\pi}{T} = \frac{2\pi}{2\pi R} = \frac{1}{R} \quad (6)$$

$$\Longrightarrow A_i = 2bR\,(\sin\frac{\pi R}{2} - \sin 0) = 2bR \quad (7)$$

$$\begin{cases} A_i = 2bR \\ \omega = \frac{1}{R} \end{cases} \quad (8)$$

and therefore the formula for the distribution function of the stress in the transversal cross section will be as follows:

$$\Longrightarrow f(x) = \frac{p_i l}{2R} cos\frac{1}{R}x \quad (9)$$

2.2. Distribution according to a polynomial function

The general formula of a polynomial function is as follows [9-11]:

$$y = f(x) = -ax^2 + b \quad (10)$$

$$x = 0 \Longrightarrow f(x) = b \tag{11}$$

$$A_i = 2\int_0^{\frac{\pi R}{2}} f(x)dx = 2\int_0^{\frac{\pi R}{2}} (-ax^2 + b)dx =$$

$$= 2\left(-\frac{ax^3}{3} + bx\right)\Big\{_0^{\frac{\pi R}{2}} = 2\left(-\frac{a}{3}\left(\frac{\pi R}{2}\right)^3 + \frac{b\pi R}{2}\right) \tag{12}$$

$$x = \frac{\pi R}{2} \Longrightarrow \begin{cases} f(x) = 0 \\ f(x) = -ax^2 + b \end{cases}$$

$$\Longrightarrow -a\left(\frac{\pi R}{2}\right)^2 + b = 0 \tag{13}$$

$$A_i = p_i l \Longrightarrow -\frac{a\pi^3 R^3}{12} + \frac{b\pi R}{2} = p_i l \tag{14}$$

$$\Longrightarrow b = a\frac{\pi^2 R^2}{4} \tag{15}$$

$$\Longrightarrow a = \frac{24}{\pi^3 R^3} p_i l \tag{16}$$

$$\Longrightarrow b = \frac{6}{\pi R} p_i l \tag{17}$$

Therefore, the polynomial function that could define the stress distribution in the transversal cross section is expressed as follows:

$$f(x) = -\frac{24}{\pi^3 R^3} p_i l \cdot x^2 + \frac{6}{\pi R} p_i l \tag{18}$$

Experimental studies have proved that the stress distribution in a transversal cross section occurs according to a harmonic function of cosine type, this being the closest to test results [12].

In the case of the plane sliding system it is immediate that the stress distribution is a linear one, the distribution of the stress in the cross over section being related with the value of forces and the coordinates of the pressure point where they are applied [13-15].

3. The distribution type

The distribution type of the stress in the longitudinal cross section is influenced by the slide length, by the point of application of the force and by whether there is a closing facet or not.

There are three main distribution variants: the trapezoid distribution, with its idealform being the rectangular distribution, the complete triangular distribution and the incomplete triangular distribution, the latter having several variants itself [9-12].

3.1. The trapezoid distribution variant

$$\delta = kp \tag{19}$$

$$\varphi_y = \frac{\delta_{Amax} - \delta_{Amin}}{H} = \frac{p_{Amax} - p_{Amin}}{H} \tag{20}$$

$$p_{Amin} = \frac{A}{aH} - \frac{6M_A}{aH^2} \tag{21}$$

$$p_{Amax} = \frac{A}{aH} + \frac{6M_A}{aH^2} \tag{22}$$

$$\Longrightarrow \varphi_y = \frac{12M_A k}{aH^3}\ ;\ \frac{x_A}{H} - \frac{M_A}{aH} < \frac{1}{6} \tag{23}$$

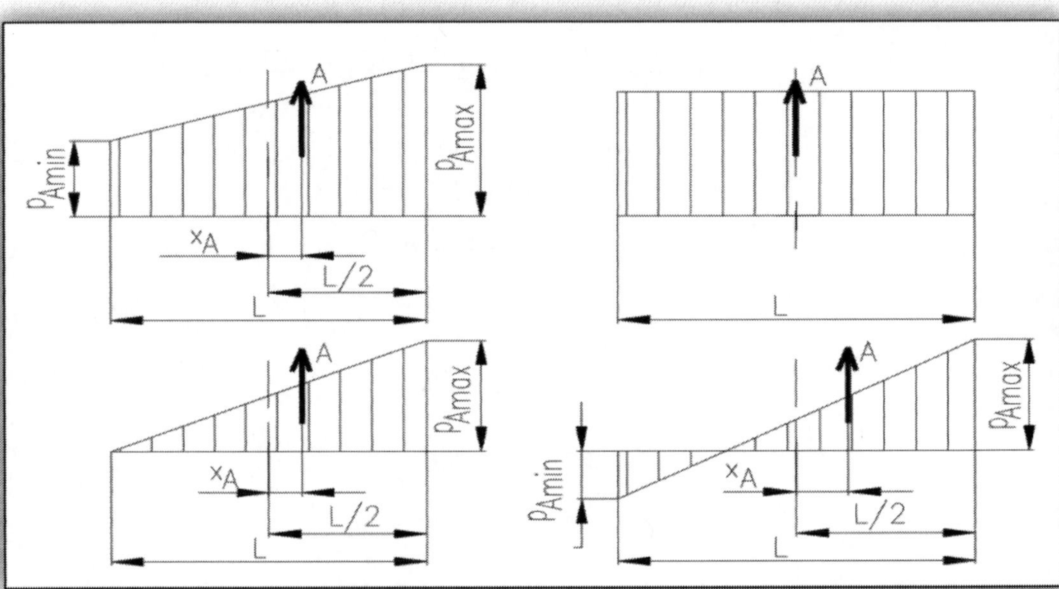

Figure 4. Trapezoid distribution of stress

3.2. *The triangular distribution variant H_m<H*

We will analyze only the variant where the closing facet is not involved, since there is no closing facet in the semicircular slide, Figures 5, 6 and 7 [11], [16].

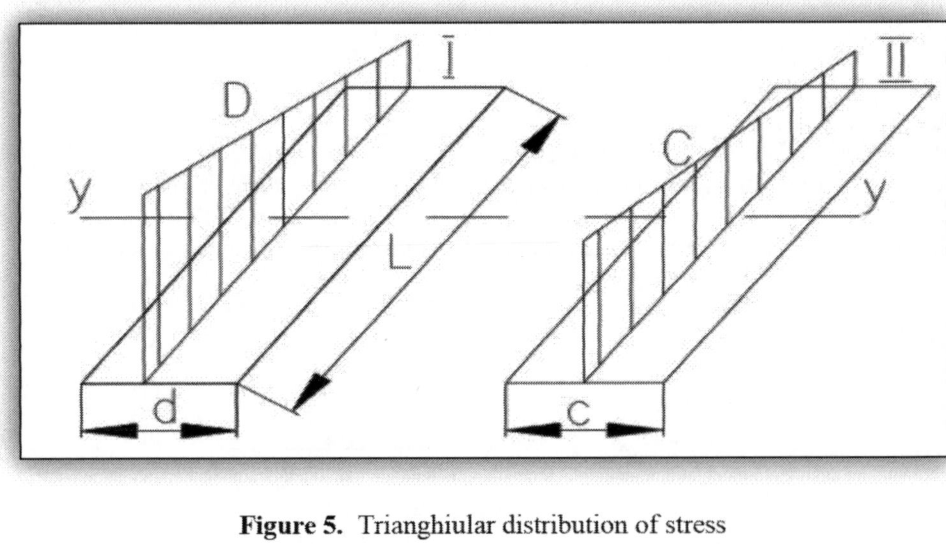

Figure 5. Trianghiular distribution of stress

$$\Delta > \varphi(H - H_m) \quad (24)$$

$$\varphi_y = \frac{\delta_{Amax}}{H_m} = k\frac{p_{Amax}}{H_m} \quad (25)$$

$$p_{Amax} = 4\frac{A^2}{3a(AH-2M_A)} \tag{26}$$

$$H_m = 3H(\frac{1}{2} - \frac{M_A}{AH}) \tag{27}$$

$$\varphi_y = \frac{12M_A k}{\lambda a H^3};\ \lambda = \frac{M_A}{AH}(1 - \frac{2M_A}{AH}) \tag{29}$$

under the conditions presented in Figures 5, 6 and 7.

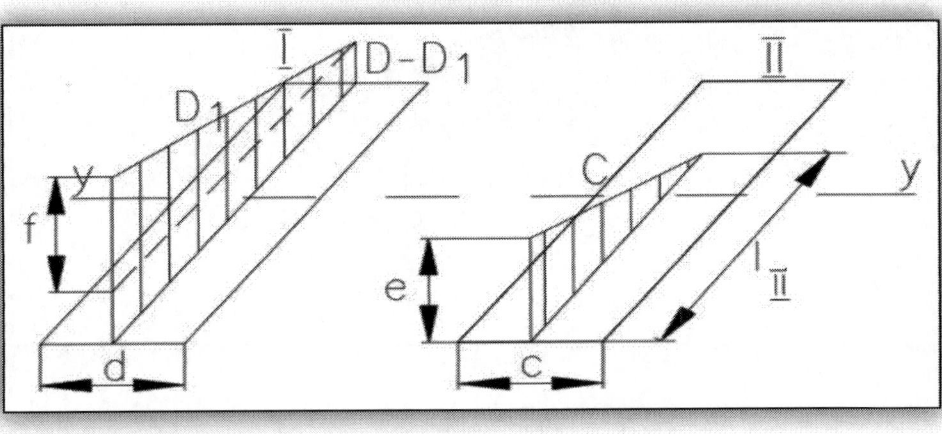

Figure 6. Trianghiular distribution of stress

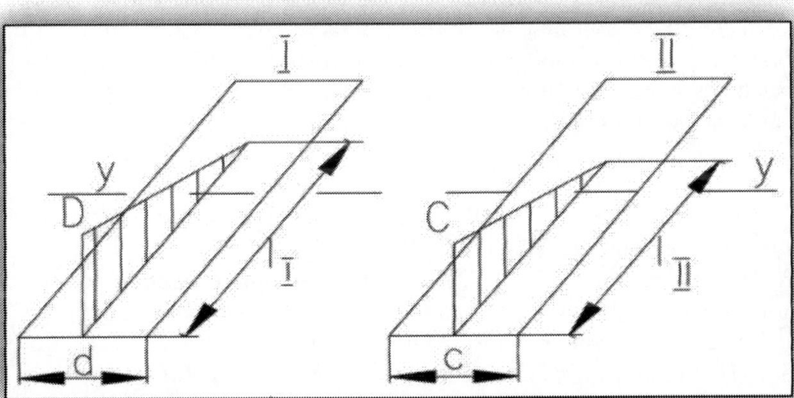

Figure 7. Trianghiular distribution of stress

4. Conclusions

Considering the results obtained for the stress distribution in the transversal and longitudinal cross sections, by composing them we can obtain the stress distribution on the semicircular slide, as shown in the Figure 8.

In the case of circular slides in the cross section, in median line of the section, we determined a value of distributed forces that is considerably higher than the values situated at the borders of the interval, fact that conduct us to the conclusion to recommend the use of this type of slides on cases of low stress manufacturing machines (basic on smaller machines).

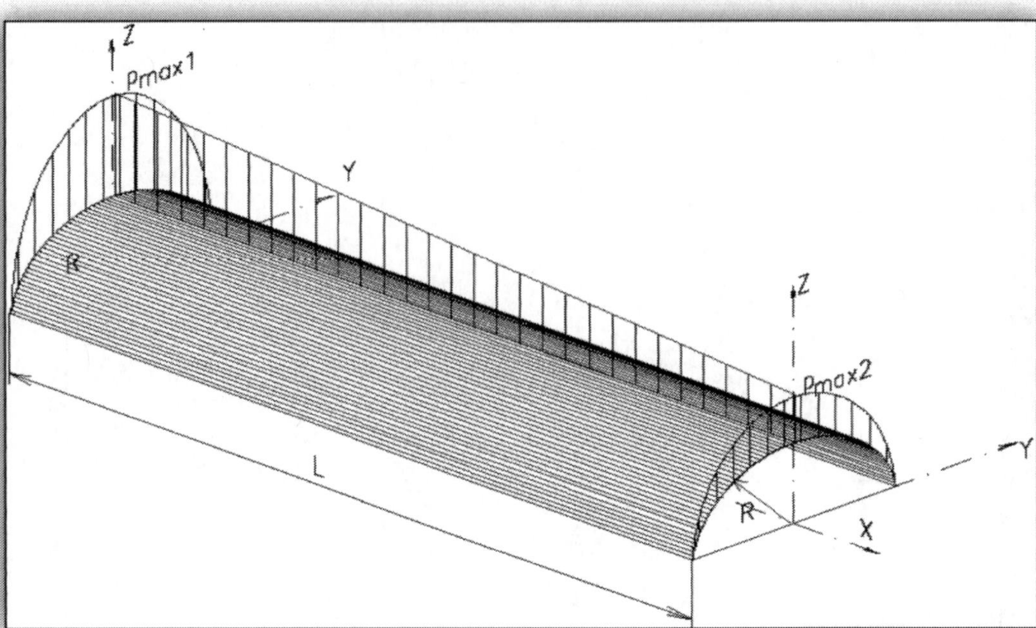

Figure 8. Triunghiular distribution of stress

The distribution in the longitudinal cross section being similar, the only determinant factor differencing the plane slides is the distribution in cross over section where we concluded that the differences between the maximum and the minimum values of pressureis much smaller than in the case of cylindrical slides, so we can recommend the plane sliding system to be used on bigger manufacturing machines where the cutting parameters and therefore the forces and torques acting on the structure are at least ten times higher.

References

[1] Meyer H, Fuchs F and Thiel K 2009 *Manufacturing Execution Systems: Optimal Design, Planning, and Deployment*, McGraw Hill, pp 105-120
[2] Kovalev V D, VasilchenkoY V and Dašiü P 2014 Adaptive optimal control of a heavy lathe operation, *Journal of Mechanics Engineering and Automation* **4**(4) 269-275
[3] Cîndea L, Hatiegan C, Gillich G R, Marta C and Nedeloni M 2012 Research Regarding Vibration Damping in Systems with a Degree of Freedom, *Scientific Bulletin of the "Politehnica" University of Timisoara, Transaction on Mathematics & Fhysics* **57**(1) 68-70
[4] Glăvan D O and Babanatsas T 2015 *Mașini unelte și agregate*, Editura Universității „Aurel Vlaicu", Arad, Romania
[5] Hakansson L 1999 *Adaptive Active Control of Machine-Tool Vibration in a Lathe – Analysis and experiments*, Lund University, Sweden
[6] Jacobson M O 1966 *Tehnologhia stankostroenia; "Maşinostroenie"*, Moscow, Rusia
[7] Pronicov A C 1957 *Iznov i dolgovecinosti stankov*, Maşghiz, Moscow, Rusia
[8] Dreucean A 1984 *Maşini unelte şi control dimensional*; Litografia UPT, Timişoara, Romania
[9] Paizi Gh, Stere N and Lazăr D 1977 *Organe de maşini şi mecanisme* Editura didactică şi pedagogică Bucureşti
[10] Sima Gh 2004 *Sisteme senzoriale utilizate la sudare*, Editura „Viata Arădeana", Arad, Romania
[11] Sima Gh, Glăvan D O, Popa A and Muncut S 2016 *Hard as a Rock or Deformation Controlled?* Soft Computing Applications. Advances in Intelligent Systems and Computing, vol **356,**

Springer, Cham

[12] Kacalak W and Tandecka K 2012 Effect of superfinishing methods kinematic features on the machined surface, *J. Machine Engineering* **4** 35-48

[13] Janecki D, Stępień K and Adamczak S 2010 Problems of measurement of barrel- and saddle-shaped elements using the radial method, *Measurement* **43**(5) 659-663

[14] Alquraana T, Kuznetsovb Yu and Tsvydb T 2016 *High-speed qlamping Mechanism of the CNC lathe with Compensation of Centrifugal Forces*, International Conference on Industrial Engineering, ICIE

[15] Dimitrova D and Szecsib T 2015 *Machining accuracy on CNC lathes under the lack of unity of the process and design data*, 48th CIRP Conference on Manufacturing Systems - CIRP CMS

[16] Radu I 2001 *Physical and mathematical pattern of a pneumatical oscilation system with two masses unlinear-D.N.4*, Magyar Tudomanyos Akademia, Kutatasi es fejlestesi tanacskozas, **25**(3), Kotet, Godolo, pp 305-309

IOP Conf. Series: Materials Science and Engineering **294** (2017) 012069 doi:10.1088/1757-899X/294/1/012069

Study on the influence of supplying compressed air channels and evicting channels on pneumatical oscillation systems for vibromooshing

D O Glăvan[1], I Radu[1], T Babanatsas[1], R M Babanatis Merce[1], I Kiss[2] and M C Gaspar[3]

[1]University "Aurel Vlaicu", Department of Engineering, Arad, Romania
[2]University Politehnica Timisoara, Faculty of Engineering, Department of Management and Engineering, Hunedoara, Romania
[3]Escola Superioar de Tehnologia, Instituto Politehnico de Castelo Branco, Castelo Branco, Portugal

E-mail: glavan@fortuna.com.ro

Abstract. The paper presents a pneumatic system with two oscillating masses. The system is composed of a cylinder (framework) with mass m_1, which has a piston with mass m_2 inside.
The cylinder (framework system) has one supplying channel for compressed air and one evicting channel for each work chamber (left and right of the piston). Functionality of the piston position comparatively with the cylinder (framework) is possible through the supplying or evicting of compressed air. The variable force that keeps the movement depends on variation of the pressure that is changing depending on the piston position according to the cylinder (framework) and to the section form that is supplying and evicting channels with compressed air.
The paper presents the physical model/pattern, the mathematical model/pattern (differential equations) and numerical solution of the differential equations in hypothesis with the section form of supplying and evicting channels with compressed air is rectangular (variation linear) or circular (variation nonlinear).

1. Introduction

The mathematical modeling of the work of some mechanic systems, frequently used in the industry, such as the vibrosmoothing device with pneumatic operation, is very important, being the first condition in achievement of the mathematical model that will be used in computer simulation. The conditions for running, execution and functioning of this device allow the formulation of some simplifying hypothesis satisfying the design. These hypotheses are defined by the vibrosmoothing device and by the elastic elements.

The vibrosmoothing device represents a perfectly centered system, meaning that the result of perturbation (elastic and dissipative) forces is oriented through the mass center of system. Therefore, all the work organs make only translational movements;

The elastic elements are considered perfectly linear and the differences between their elastic characteristics are neglected. The whole system is moving around the stable equilibrium position.

2. The mathematical calculation models

The vibrosmoothing device is represented schematically in Figure 1.

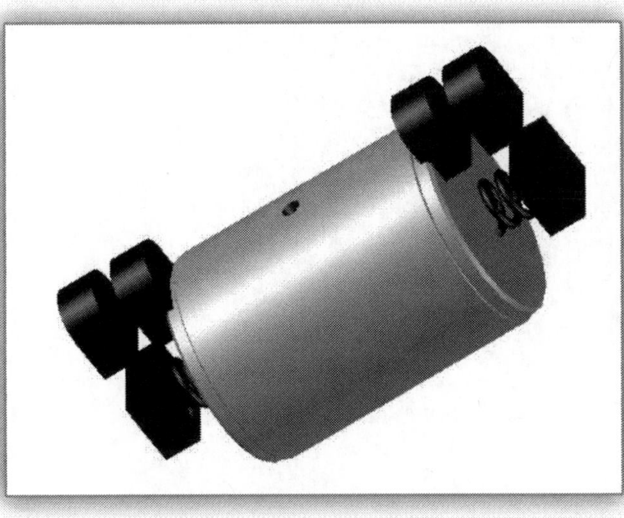

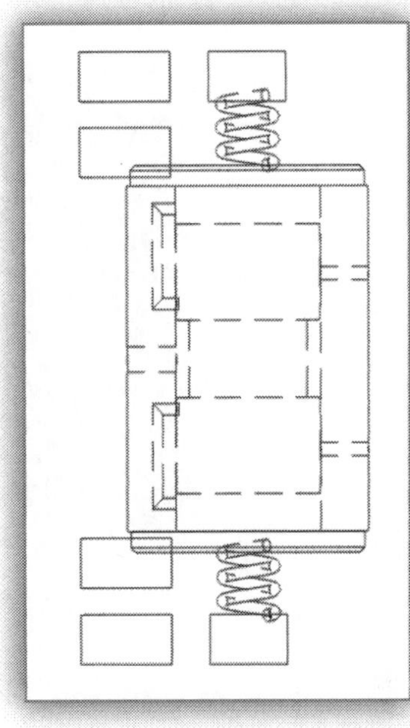

Figure 1. The classic vibrosmoothing device

The differential equations of the system movement for the pneumatical oscillation system are (Figure 2, 3 and 4):

$$\ddot{x} = \frac{K_e}{m_1} \cdot x - \frac{A}{m_1}(P_1 - P_4) - \frac{F_a}{m_1} \cdot sign\left(\dot{x}\right) \tag{1}$$

$$\ddot{u} = \frac{K_e}{m_1} \cdot x + \frac{F_a}{m_1} \cdot sign\left(\dot{x}\right) + A(P_1 - P_4) \cdot \left(\frac{1}{m_1} + \frac{1}{m_2}\right) \tag{2}$$

where:
m_1 is the framework mass;
m_2 is the piston mass;
K_e is the elastic constant of the springs mounted in parallel;
F_a is the constant horizontal force;
P_1 is the pressure on the left frontal surface;
P_4 is the pressure on the right frontal surface;
P_r is the pressure for compressed air supplying;
A is the supplying channel for compressed air;

It is imperative to take into consideration each type of production program (range), to restrain as much as possible the area that we are using and to minimize the time of execution, all of these in order to satisfy the client's needs, to try to classify them in order to be able to define a global software (with general rules) that is expect to fulfill each client's needs (Figure 5) [1-6].

Figure 2. The vibrosmoothing device

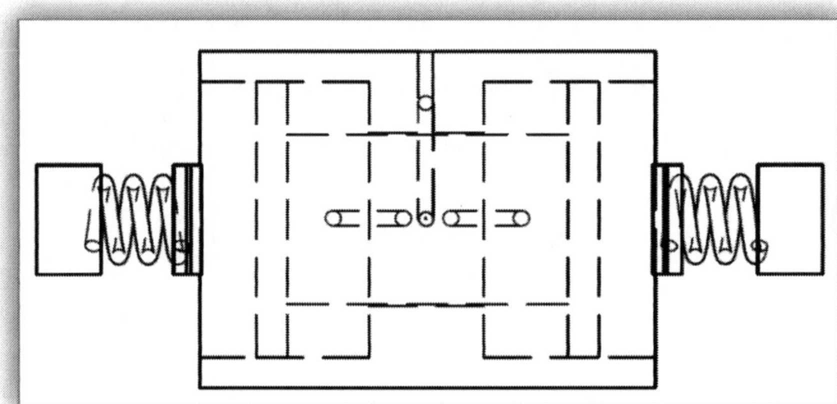

Figure 3. The upper view of vibrosmoothing device

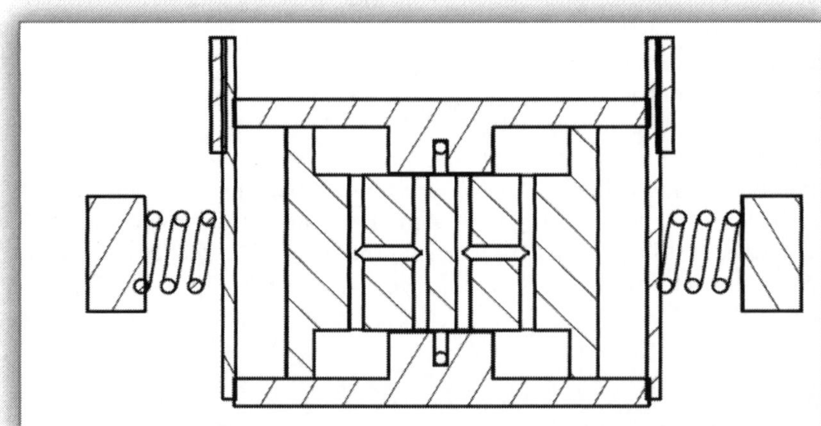

Figure 4. The front view of vibrosmoothing device

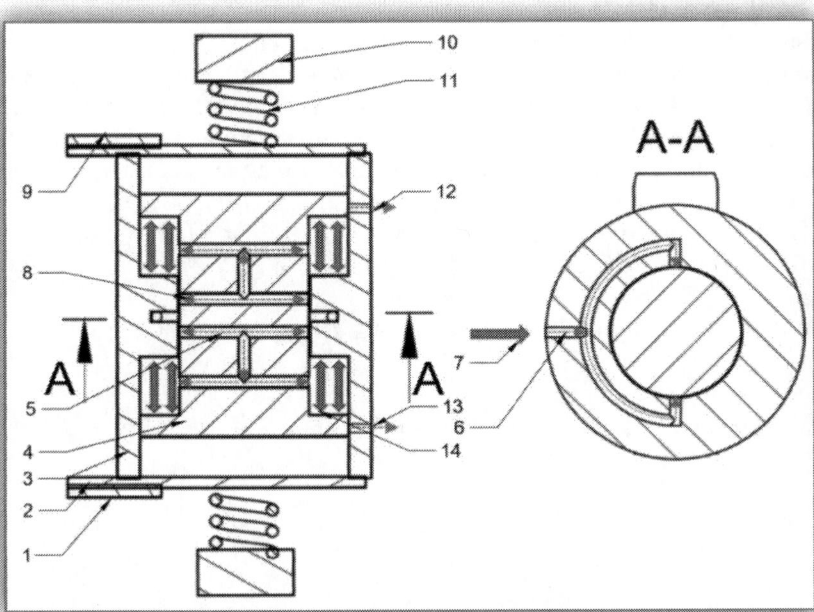

Figure 5. Schematically principle of pneumatically oscillation system

where:

1. Electromagnet
2. Cover frame of pneumatical system
3. Framework mass (m_1)
4. Piston mass (m_2), with diameter D and length l
5. The supplying channel for compressed air to the right chamber of the piston
6. The principal supplying channel for compressed air
7. Pressure for compressed air supplying mass (P_r)
8. The supplying channel for compressed air to the left chamber of the piston
9. Electromagnet
10. Fixed mass
11. Spring, u compressed distance
12. Evicting channel for compressed air from the left chamber of the piston
13. Evicting channel for compressed air from the right chamber of the piston
14. Pressure compartment (P_1, P_2)

3. The mathematical calculation model

The framework of mass m1 is executing an absolute movement in comparison with the fixed system Oxy. The relative movement of the piston of mass, m_2 is described in comparison with the system O_1uv.

The differential equations of the system movement are [1-3]:

$$\ddot{u} = \frac{K_e}{m_1} \cdot x + \frac{F_a}{m_1} \cdot \sin g(\dot{x}) + A(P_1 - P_2) \cdot (\frac{1}{m_1} + \frac{1}{m_2}) \quad (3)$$

We can see that m_1, m_2, A, K_e, F_a remain constant during the movement, but the pressures P_1 and P_2 are changing in function of the piston position in comparison with the framework.

3.1. The variation of the pressures P_1 and P_2

The variable pressures P_1 and P_2 are determined by studying the system as a pneumatical cylinder with double effect. Considering the transformation process as an adiabatic process the pressures P_1 and P_2 alternating linearly in hypothesis with the section form of supplying and evicting channels with compressed air is rectangular and the pressures P_1 and P_2 alternating nonlinearly in hypothesis with the section form of supplying and evicting channels with compressed air is circular, Figures 6 and 7 [2-5].

- if: $$u_{\min} \le u \le u_1 \Rightarrow P_1 = P_0\left(\frac{u_7-u_1}{u_7-u}\right)^{\chi};\ P_4 = P_r\left(\frac{l-l_1-u_7+u_1}{l-l_1-u_7+u}\right)^{\chi} \quad (4)$$

- if: $$u_1 \le u \le u_2 \Rightarrow P_1 = P_0;\ P_4 = P_r \quad (5)$$

- if: $$u_2 \le u \le u_3 \Rightarrow P_1 = P_0 + \left[\frac{3}{4}P_r\left(\frac{u_7-u_4}{u_7-u_3}\right)^{\chi} - P_0\right]\frac{u-u_2}{u_3-u_2};\ P_4 = P_r - \frac{1}{4}P_r\left(\frac{u-u_2}{u_3-u_2}\right) \quad (6)$$

- if: $$u_3 \le u \le u_4 \Rightarrow P_1 = \frac{3}{4}P_r\left(\frac{u_7-u_4}{u_7-u}\right)^{\chi};P_4 = \frac{3}{4}P_r\left(\frac{l-l_1-u_7+u_3}{l-l_1-u_7+u}\right)^{\chi} \quad (7)$$

- if: $$u_4 \le u \le u_5 \Rightarrow P_1 = P_r\left[\frac{3}{4}+\frac{1}{4}\left(\frac{u-u_4}{u_5-u_4}\right)^{\chi}\right];\ P_4 = \frac{3}{4}P_r\left[\left(\frac{l-l_1+u_3-u_7}{l-l_1+u_4+u_7}\right)^{\chi}+P_0\right]\frac{u-u_4}{u_5-u_4} \quad (8)$$

- if: $$u_5 \le u \le u_6 \Rightarrow P_1 = P_r;\ P_4 = P_0 \quad (9)$$

- if: $$u_4 \le u \le u_5 \Rightarrow P_1 = P_r\left[\frac{3}{4}+\frac{1}{4}\left(\frac{u-u_4}{u_5-u_4}\right)^{\chi}\right];\ P_4 = \frac{3}{4}P_r\left[\left(\frac{l-l_1+u_3-u_7}{l-l_1+u_4+u_7}\right)^{\chi}+P_0\right]\frac{u-u_4}{u_5-u_4} \quad (10)$$

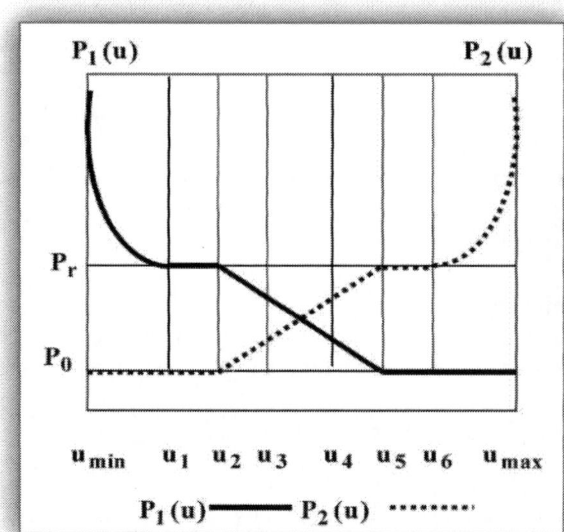

Figure 6. The resultant in a transversal cross section

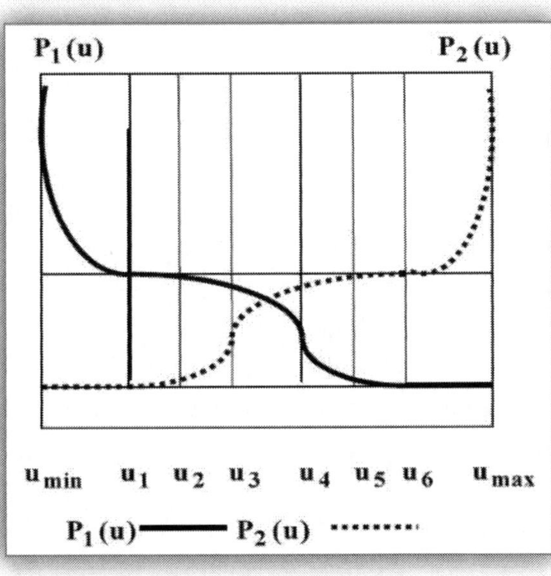

Figure 7. The resultant in a transversal cross section

4. Numerical solution of the differential equations

The differential equations system was solved numerically using the Runge-Kutta method with the use of utilitarian MathCAD. Introducing the following initial values [7-9]:

- $u_1 = 0.014$ m;
- $u_2 = 0.018$ m;
- $u_3 = 0.022$ m;
- $u_4 = 0.032$ m;
- $u_5 = 0.036$ m;
- $u_6 = 0.040$ m;
- $U_{min} = 0$ m;
- $U_{max} = 0.054$ m;
- $D = 0.050$ m;
- $l = 0.102$ m;
- $X = 1.4$;
- $P_r = 4 \cdot 10^5$ N /m^2;
- $P_0 = 1 \cdot 10^5$ N/ m^2;
- $m_1 = 4,5$ kg;
- $m_2 = 2,5$ kg;
- $K_e = 80\text{-}104$ N/ m;
- $F_a = 25$ N

We obtained the graphics of the framework displacement (Figure 8) in hypothesis with the section form of supplying and evicting channels with compressed air is rectangular and the graphics of the framework displacement (Figure 9) in hypothesis with the section form of supplying and evicting channels with compressed air is circular [10-12].

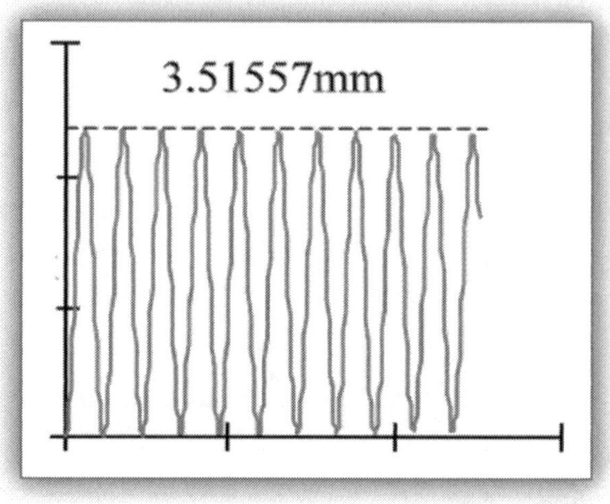

Figure 8. The graphics of the framework displacement in rectangular section

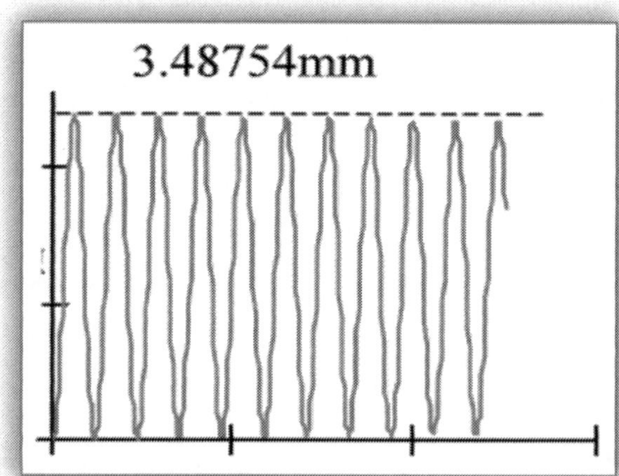

Figure 9. The graphics of the framework displacement in circular section

5. Conclusion

The physical model of the oscillation system pneumatical is used for constructing the vibrosmoothing device with pneumatical actioning of the movement. The mathematical modelling allows us to choose the constructing and running optimal parameters in order to obtain a movement with desired amplitude and frequency for the framework, satisfying the technological process demands.

References

[1] Radu I 1997 *Vibrosmoothing device with pneumatical actioning*, Romanian Academy, The annual simposyum of the institute of solid mechanics "SISOM 95", Bucharest, Romania, pp 391-394

[2] Radu I 2001 *Physical and mathematical pattern of a pneumatical oscilation system with two masses unlinear-D.N.4*, Magyar Tudomanyos Akademia, Kutatasi es fejlestesi tanacskozas, **25**(3), Kotet, Godolo, Hungary, pp 305-309

[3] Radu I 2001 *Physical and mathematical pattern of a vibrosmoothing device with two masses, unlinear,with pneumatic operation D.N. 2*, 6th International Scientific Symposium, Quality and Reliability of Machines, Nitra, Slovakia, pp 179-182

[4] Radu I and Ursu-Fischer N 1997 *Unele aspecte privind modelarea matematidi a unui dispozitiv de vibronetezire la variatia liniar-neliniara a presiunii de alimentare (partel+2)*. Conferinta ştiintifica „Tehnologii şi produse noi in constructia de maini TEHNOMUS IX", Suceava, Romania, pp 110-122

[5] Radu I and Glavan D 2002 *Study Of Line System For Vibromoothing Device D.N.6-1*, 7th International Scientific Symposium, Quality and Reliability of Machines, Nitra, Slovakia, pp 77-80

[6] Hordieievo D and Karmalita A 2014 Study of use application vibration equipment for oil-retaining holes at the neck of the trees, *Problems of Applied Sciences* **2** 95-100

[7] Spadło S, Młynarczyk P and Bańkowski D 2014 Analysis of the Effect of Processing Vibro-abrasive Finishing on the Geometric Structure Surface Scales Ammunition and Sharp Edges, *Journal of Achievements in Materials and Manufacturing Engineering* **66**(1) 39-44

[8] Spadło S and Bańkowski D 2015 *Influence Of The Smoothing Conditions In Vibro-Abrasive Finishing And Deburring Process for Geometric Structure of the Surface*, Proceedings of 24th International Conference on Metallurgy and Materials, METAL, Machine Parts Made Of Aluminum Alloys, pp 1062-1068

[9] Glavan D O and Radu I 2001 *Vibraţii mecanice în tehnică,* Editua Universitatii „Aurel Vlaicu"Arad, pp 75-92

[10] Glavan D O and Radu I 2001 *Elemente de vibraţii mecanice,* Editua Universitatii „Aurel Vlaicu"Arad, pp 29-34

[11] Kovalev V D, Vasilchenko Y V and Dašiü P 2014 Adaptive optimal control of a heavy lathe operation, *Journal of Mechanics Engineering and Automation* **4**(4) 269-275

[12] Hakansson L 1999 *Adaptive Active Control of Machine-Tool Vibration in a Lathe*, Analysis and experiments, Lund University, Sweden

IOP Conf. Series: Materials Science and Engineering **294** (2017) 012070 doi:10.1088/1757-899X/294/1/012070

Wind turbine remote control using Android devices

C L Rat[1] and M Panoiu[2]

[1]Politehnica University of Timisoara, Automation and Computers Department, Vasile Parvan str., no. 1-2, 300223 Timisoara, Romania
[2]Politehnica University of Timisoara, Department of Electrical Engineering and Industrial Informatics, 5 Revolution Street, Hunedoara, 331128, Romania

E-mail: cezara.liliana@gmail.com

Abstract. This paper describes the remote control of a wind turbine system over the internet using an Android device, namely a tablet or a smartphone. The wind turbine workstation contains a LabVIEW program which monitors the entire wind turbine energy conversion system (WECS). The Android device connects to the LabVIEW application, working as a remote interface to the wind turbine. The communication between the devices needs to be secured because it takes place over the internet. Hence, the data are encrypted before being sent through the network. The scope was the design of remote control software capable of visualizing real-time wind turbine data through a secure connection. Since the WECS is fully automated and no full-time human operator exists, unattended access to the turbine workstation is needed. Therefore the device must not require any confirmation or permission from the computer operator in order to control it. Another condition is that Android application does not have any root requirements.

1. Introduction

Distance learning laboratories have gained popularity in the academic community, supported by advances in remote monitoring technologies. This gives remote users the possibility to experiment with data monitoring, electrical measurements, remote data processing, remote storage, data mining, etc. This novel teaching tool allows students to interact and observe experimental results from a remote location using Internet connectivity. This tool can be seen as an alternative to the traditional educational methods. Remote educational laboratories can be achieved through a number of different methods [1].

Literature presents remote educational applications that use LabVIEW Remote Panels, LabVIEW Web Services, NI Data Dashboard, Web Sockets, Shared Variables or Remote Desktop Services. Here are a number of significant examples [1]. The CyberLab, the optics laboratory from Stanford University, uses Remote Panels for distance learning process [2]. Remote Panels are also used at the ALTE laboratory in the University of Hartford [3] and at the laboratory for nuclear engineering experiments from Illinois University [4]. The LabVIEW Web Services technology is used to provide real time transistor measurements in [5] and an online microelectronics device characterization laboratory at the MIT in [6]. A comparative study of both these technologies is done in [1], where a distance laboratory for solar electricity experiments is presented. In [7], Web Sockets are used instead.

This paper describes the remote control of a long distance microgrid laboratory over the internet using an Android device (a smartphone or a tablet). The microgrid laboratory simulates a grid-connected wind energy conversion system (WECS). Remote monitoring and control of the WECS is an increasingly important aspect for system improvement [8]. The WECS also has a workstation which contains a LabVIEW program that monitors and controls everything that happens in the plant simulation. The Android device connects to the LabVIEW application, working as a remote interface to the wind turbine. The purpose of the proposed system is to take remote measurements via mobile devices. Two technological solutions are proposed and analyzed, revealing their advantages and disadvantages [9].

2. Wind Energy Conversion System

Small-scale power-generation systems based on renewable energy sources have been proven to be efficient and cost effective, reducing both CO_2 emissions and primary energy use, thereby saving energy resources. Renewable energy technologies range from the well-established, such as hydropower, to the emergent, such as hybrid systems [10], [11].

Wind energy or wind power describes the process by which wind is used to generate mechanical or electric power. The wind turbine is the typical component through which wind energy is converted into mechanical energy, by means of the aerodynamically designed blades. The function of the wind turbine is to capture power from the wind, but, according to the Betz theory, the entire wind energy could not be completely absorbed by the turbine. The output power is highly dependent on the constructive features and characteristics of the turbine, being determined by factors such as power coefficient, air density, the radius of the wind wheel and wind speed [10], [12].

There are several wind turbines types proposed in literature. The two most important types are vertical axis wind turbines (VAWT) and horizontal axis wind turbines (HAWT). HAWTs are the most frequently used, being able to operate at higher wind speeds than the VAWTs [13], [14].

The wind turbine used in this paper is a 5 kW fixed pitch variable speed horizontal axis turbine; its characteristics being presented in the following table.

Table 1. The parameters of the wind turbine

Name	Value
Rated power	5.5 kW
Rated wind speed	11 m/s
Maximum speed	126 rpm
Turbine inertia	140 kg*m^2
Blade swept area	19.6 m^2
Radius of the turbine blade	2.5 m
Nominal tip-speed ratio	3
Specific density of air	1.225 kg/m^3

The wind turbine is directly coupled to a three phase permanent magnet synchronous generator (PMSG). This arrangement offers high reliability, high efficiency, low maintenance and low cost, being considered a robust solution for wind power generation. The wind turbine, being a variable speed turbine, when connected to a PMSG, permits operation over a wider range of speeds, resulting in higher power harvesting from the turbine to the generator [10].

The PMSG is a standard synchronous machine, where the DC excitation circuit is replaced with permanent magnets, placed on the rotor surface. Thus, the electrical losses in the rotor are eliminated and the thermal characteristics of the machine are improved. Also, the absence of the mechanical components such as brushes and slip rings makes the machine lighter, having a smaller physical size, a low moment of inertia leading to a higher power to weight ratio, which means a higher efficiency and reliability [15]. Due to these advantages, the permanent magnet synchronous generator is an attractive solution for wind turbine applications. The disadvantages are: high costs for the magnets and a fixed excitation. Since there is no excitation system and associated voltage controller, the generator output voltage is variable and directly influenced by the rotor speed [10].

The PMSG used in this simulator has the magnets mounted on the surface of the rotor, being a Surface mounted permanent magnet synchronous generator (SPMSG) [16], [17]. The considered PMSG has the following parameters:

Table 2. The parameters of the PMSG

Name	Value
Rated power	5 KVA
Rated current	12 A
Rated speed	120 rpm
Rated frequency	32 Hz
Pole pairs number	16
Permanent magnets flux	1.32 Wb

The PMSG output voltage and frequency is governed by the machine speed and hence needs a full rated power converter in order to interface with the AC load, which needs a fixed voltage and frequency. The conversion of the electrical energy obtained from the turbine system is based on an AC/DC/AC conversion system. The PMSG generator is connected with a PVI-Wind Interface 4000 rectifier (a three phase rectifier with overvoltage protection) to a PVI-12.5-TL-OUTD-W Wind Inverter (a dual stage transformerless wind inverter). Both these devices are produced by ABB. The ABB 4000-WIND-INTERFACE rectifies and filters the alternating currents, thus producing a direct current output to feed to the inverter. The wind interface actually controls the power output of the wind inverter through an external signal. The wind inverter uses a very precise high speed algorithm (MPPT) that maximizes the total energy harvested. Through the inverter the wind turbine simulator is connected to the main grid, which represents the AC load. The entire ensemble constitutes the wind energy conversion system (WECS). The integration of the WECS into the existing grid saves cost and improves the reliability of the power supply [10], [18], [19]. The characteristics of the equipment can be seen in Table 3.

Table 3. The characteristics of the equipment used in the conversion system

	ABB Wind Inverter	*ABB Wind Interface*
Input Voltage	DC 140-530(V)	AC 400(V)
Input Current	32 (A)	16.6 (A)
Output Voltage	AC 230(V)	DC 600(V)
Output Current	20 (A)	6 (A)
Output Power	4.2 (kW)	4 (kW)

The WECS diagram can be seen in Figure 1.

IOP Conf. Series: Materials Science and Engineering **294** (2017) 012070 doi:10.1088/1757-899X/294/1/012070

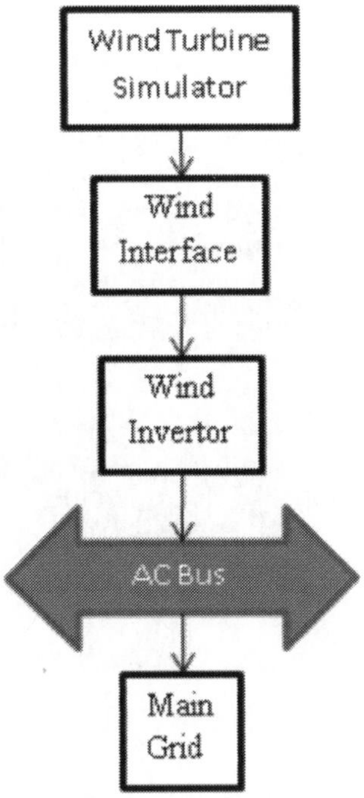

Figure 1. Wind energy conversion system block diagram

The wind interface has the possibility to drive an external diversion load resistor that can help the turbine in case of high wind or missing grid. This activates at 530 V [18], [19].

The effect the conversion has on the system is clearly visible on the power versus rotating speed curve. The Wind Inverter follows a maximum power point tracking (MPPT) algorithm, the search for the maximum power point being visible in the following Figure 2.

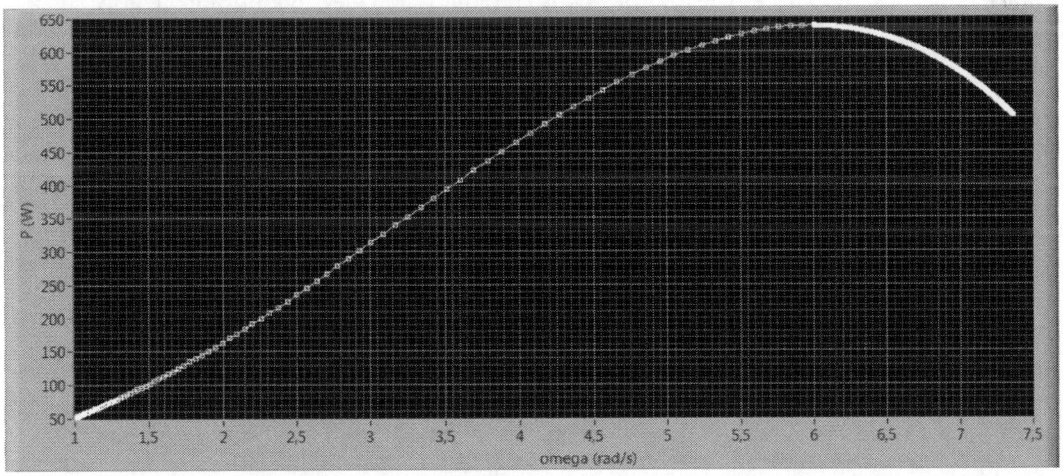

Figure 2. Power vs. angular speed characteristic for a wind speed of 5m/s

The wind turbine system is simulated in LabVIEW. The LabVIEW program also has the task of controlling a 7.5 kW three phase squirrel cage induction machine through an ABB ACS800 drive. The induction machine is connected through a gearbox to the aforementioned PMSG. The induction motor (IM) with a gearbox (GB) represent the wind turbine equivalent, driving the generator as if it were connected to the wind turbine itself. The LabVIEW program is responsible for real time acquisition and monitoring of the WECS data [11].

A view of the front panel of the LabVIEW application can be seen in Figure 3.

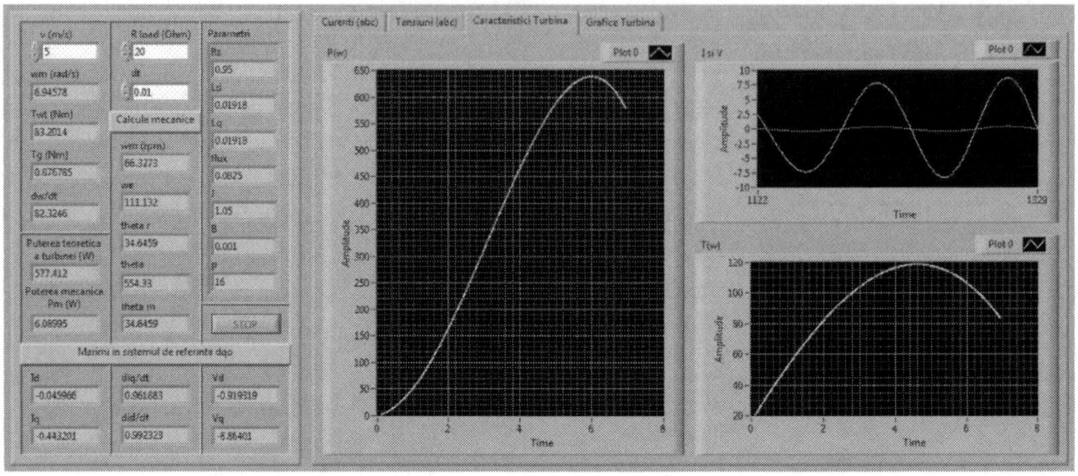

Figure 3. The front panel of the LabVIEW application

The scope of this study was the design of a remote control software capable of visualizing real-time wind turbine data through a secure connection. The interaction of the user with this system can be remote or in the laboratory [9].

3. Android Remote Control

Android is a ready-made, low-cost, customizable, mobile operating system developed by Google for such high-tech touchscreen devices as smartphones, tablets, televisions, cars, wrist watches, notebooks, game consoles, digital cameras, etc., each with a specialized user interface [20]. An Android mobile device can be a handy means of remote parameter monitoring. It also creates the possibility of remote data processing and analysis far from the microgrid laboratory [9]. The block diagram of such a system can be seen in Figure 4.

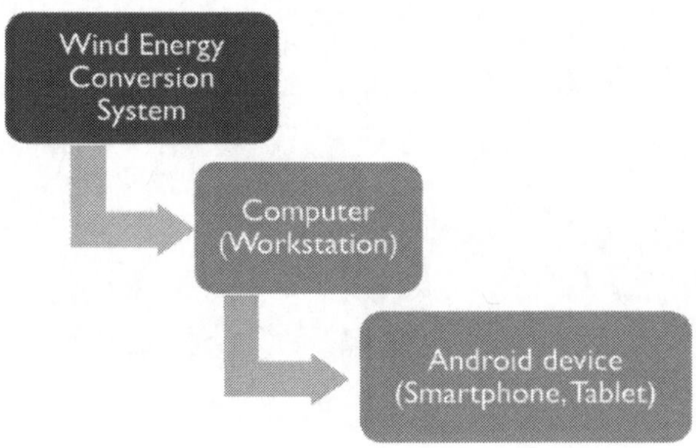

Figure 4. The structure of the Android remote control system

Two remote communication technologies have been used for this purpose: the LabVIEW Web Services and the Remote Control framework for LabVIEW.

3.1. Remote Control framework for LabVIEW

RCF (Remote Control framework) for LabVIEW from TOOLS for SMART MINDS (T4SM) is a powerful toolkit for LabVIEW that transforms an Android device, such as a smartphone or a tablet, into a remote control capable of displaying data, charts, tables, etc. RCF provides the LabVIEW application an interface towards mobile devices. Running as a background service, RCF gathers requests from connected clients, sends these requests to the LabVIEW application for processing, retrieves results and publishes them to clients. RCF is responsible for maintaining the connections, validating requests and notifying changes to the Android interface. Connections are password protected and data is transferred in encrypted format, optimized to minimize consumed band. RCF passes to the Android device only selected data that the remote viewer is allowed to see, rather than publishing the entire LabVIEW panel. This is an advantage when it comes to security, but also regarding transmission speed and Android storage capabilities. RFC also allows streaming of real-time data on Android smartphones and tablets, these devices becoming remote controls for the LabVIEW application in charge of monitoring the microgrid laboratory [21].

The schematic of the measurement system is presented in Figure 5.

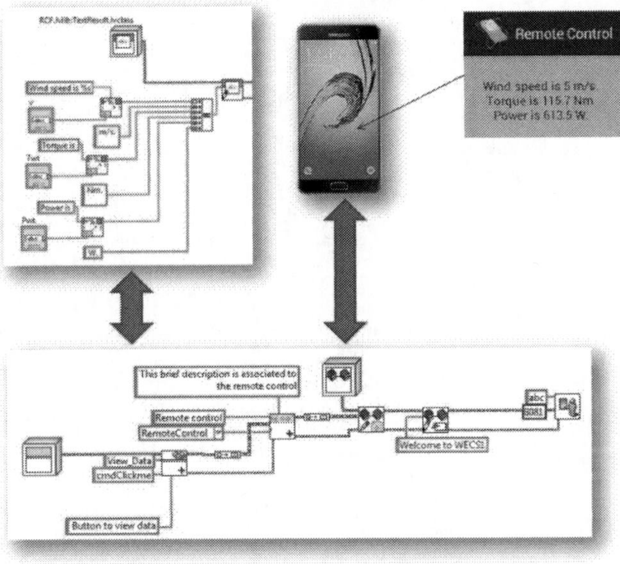

Figure 5. The diagram of the RCF system with the LabVIEW code

3.2. Data Dashboard for LabVIEW

Data Dashboard for LabVIEW 2.3. is a mobile application that enables the creation of custom views of NI LabVIEW applications. The resulting Data Dashboard application can monitor and control LabVIEW applications remotely from iOS and Android mobile devices. Users can monitor and control the LabVIEW interface from Android tablets or smartphone. They can even view real-time measurement data. Data Dashboard for LabVIEW is capable of reading/writing data or commands over the network through network published shared variables and/or a secure LabVIEW web service with the object of monitoring and control [22].

3.2.1. Shared Variables. When using shared variables, the input and output data used in the application are network-published shared variables. They can both read and write data to the host system. Single-process shared variable, as used here, are similar to the LabVIEW global variable.

Shared variables interact with the host application through the LabVIEW Shared Variable Engine (SVE) operating over TCP-based network communications.

The Data Dashboard application is updated when it receives information from the server, the updates being handled through the Publish Subscribe Protocol (NI-PSP). The NI Publish and Subscribe Protocol (NI-PSP) is a networking protocol optimized to be the transport for Network Shared Variables [8], [23] and [24].

Shared Variables support primitive data types (Numeric Doubles, Booleans and Strings) as well as arrays of those primitive data types.The disadvantages are that the limited resources of the mobile devices must be taken into consideration when selecting the data to share and the lack of means of encrypting the data being sent via shared variables. The advantages are the simplicity and robustness of the method [8]. The Block Diagram of the LabVIEW program when using shared variable can be seen in Figure 6.

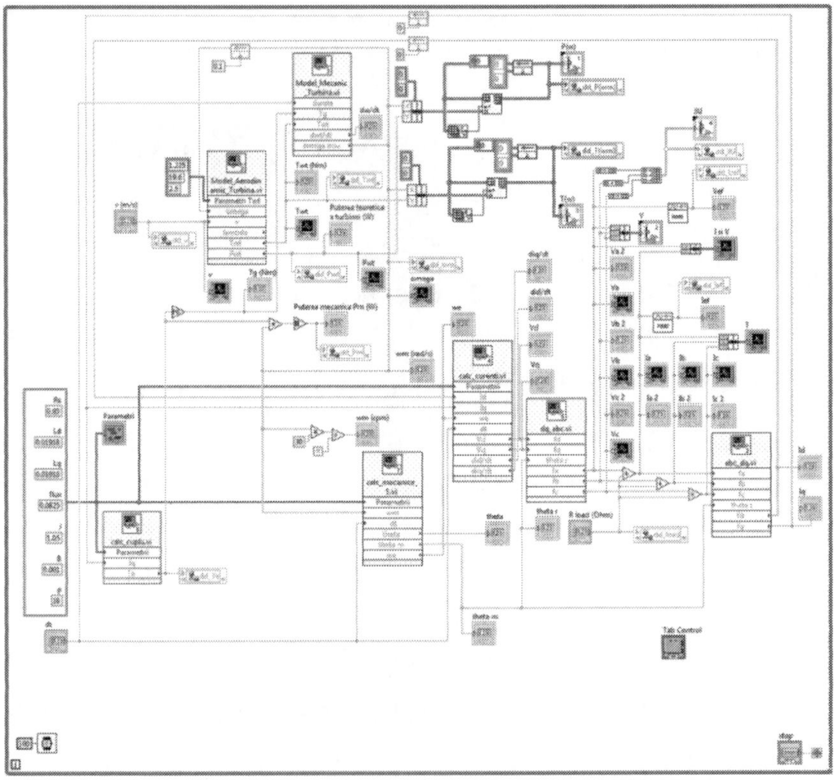

Figure 6. The Block Diagram of the LabVIEW program

3.2.2. LabVIEW Web Services. LabVIEW Web Services require the installation of Microsoft Silverlight 4.0. This tool provides the means to create a user interface. However, limitations regarding the number and type of front panel objects and their functionality are still to be evaluated [1], [9].

By using secure LabVIEW web services, dashboard users login using ID and passwords for authorization and have the digital API key to access the application. LabVIEW does not support built-in encryption, but it does use digital signatures. Secured requests between clients and LabVIEW Web services are sent unencrypted or "in-the-clear", yet "digitally-signed", with a secret key intended to be known only by authorized applications. Hence, LabVIEW Web services security is accomplished by validating the signature to verify the request was created by an application with authorization to access the requested resource. An API security key for a web service allows only requests coming with that particular key access to the data. LabVIEW's key generation algorithm creates a highly randomized string of characters. At runtime, the Web services runtime engine inspects every request as it arrives, looking for the "digital signature". If the runtime does not find a digital signature, the request is automatically rejected. If it finds a digital signature, calculates what the signature should be for that request and compares the two strings. If the two do not match, the request is rejected [22], [25].

If the communication takes place via Web Services, the remote user can only monitor designated variables. The front panel of the Data Dashboard application can be seen in the following Figure.

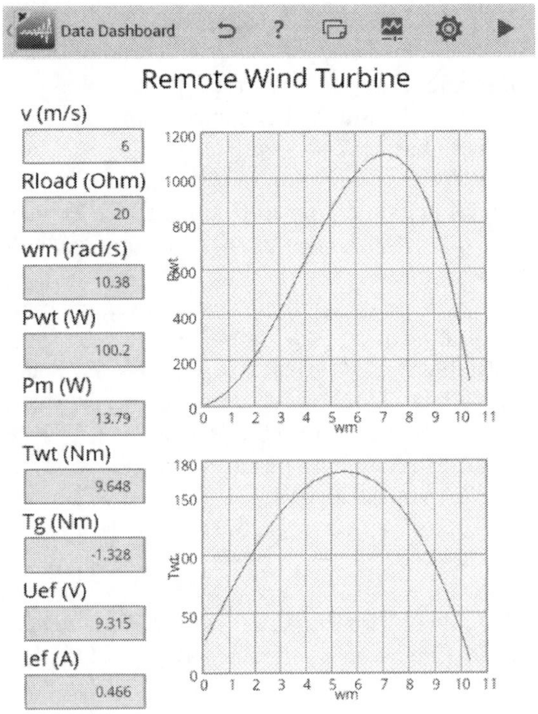

Figure 7. The front panel of the Data Dashboard application

4. Conclusions

This paper presents a means of remotely connecting to a microgrid laboratory, which consists of a wind turbine simulator directly coupled to a permanent magnet synchronous generator. The PMSG is connected to the electrical load (main grid) through a power conditioning system composed of an industrial wind turbine rectifier (Wind Interface) and a voltage source inverter (Wind Inverter). LabVIEW software is used to locally monitor and control the simulated plant. This LabVIEW application can be accessed remotely by an Android device [10].

This technology can provide a means of accessing measurement data from anywhere, a human-machine interface (HMI) for remotely viewing and controlling systems. Engineers can monitor data for trends, view historical data over periods of time, instantly receive alarms based on measurements, or remotely control an application or test sequence.

The remote control of a wind turbine system over the internet using an Android device presents, generally speaking, security concerns, but it is required in order to access to the turbine workstation. Remote Control framework for LabVIEW does cover most of the security requirements, but the solution is more complicated to implement than Data Dashboard. Data Dashboard for LabVIEW is capable of reading/writing data or commands over the network through shared variables or a LabVIEW web service. Shared variables are a simple and robust solution, but they can overload the memory of the Android device and are sent through the network unencrypted, being vulnerable to security threats. LabVIEW web services, however, can be secured using ID and passwords for authorization and a digital API key for user validation, yet secured requests between clients and LabVIEW Web services are sent unencrypted. Using these technologies requires the workstation to have a static IP address which may in itself cause security issues. None of the technologies require the Android mobile device to be rooted, which is a serious advantage.

References

[1] Bauer P and Ionel R 2013 *Labview remote panels and Web services in solar energy experiment—A comparative evaluation*, 8th International Symposium on Applied Computational Intelligence and Informatics (SACI), Timisoara, Romania, 23-25 May 2013, pp 263-268

[2] Hesselink L, Rizal D, Bjornson E, Paik S, Batra R, Catrysse P, Savage D and Wong A 2008 *Stanford cyberlab: Internet assisted laboratories*, Online and Distance Learning: Concepts, Methodologies, Tools, and Applications, IGI Global, pp 2400-2415

[3] Oruganti T, Eppes T and Schuyler P 2005 *Novel distance laboratory control panel*, Proceedings of the 2005 American Society for Engineering Education Annual Conference & Exposition

[4] Jain P, Stubbins J and Uddin R 2006 *Interactive virtual laboratory for distance education in nuclear engineering*, American Nuclear Society, 555

[5] Dutta S, Prakash S, Estrada D and Pop E 2011 A web service and interface for remote electronic device characterization, *IEEE Transactions on Education* **54**(4) 646-651

[6] Harward V J, Del Alamo J A, Lerman S R, Bailey P H, Carpenter J, DeLong K, Felknor C, Hardison J L, Harrison B, Jabbour I, Long P D, Tingting M, Naamani L, Northridge J, Schultz M, Talavera D, Varadharajan C, Wang S, Yehia K, Zbib R and Zych D 2008 The ilab shared architecture: A web services infrastructure to build communities of internet accessible laboratories, *Proceedings of the IEEE* **96**(6) 931-950

[7] Bergmans J 2010 *Extending LabVIEW to WEB using the LabSocket System*, NI LabVIEW User Group Meeting, Santa Ana, USA

[8] ***http://www.ni.com/white-paper/14689/en/

[9] Călinoiu D, Ionel R, Lascu M and Cioablă A 2014 *Arduino and LabVIEW in educational remote monitoring applications*, Frontiers in Education Conference (FIE), Madrid, Spain, 22-25 October, pp 1-5

[10] Samuel P, Chandrashekhar N and Gupta R 2010 *Wind energy conversion based on seven-level cascaded H-bridge inverter using LabVIEW FPGA*, 2010 International Conference on Power, Control and Embedded Systems (ICPCES), Allahabad, India, 29 November-1 December 2010, pp 1-6

[11] Pecen R, Salim M D and Zora A 2004 A labview based instrumentation system for a wind-solar hybrid power station, *Journal of Industrial Technology* **20**(3) 1-8

[12] Yuegang L, Peiyu X, Nailu L and Xiaoxu F 2009 *Research of VSCF wind power generation training system based on Matlab/LabVIEW*, IEEE International Conference on Automation and Logistics (ICAL'09), Shenyang, China, 5-7 August, pp 1592-1597

[13] Hui J 2008 *An Adaptive Control Algorithm for Maximum Power Point Tracking for Wind Energy Conversion Systems*, Queen's University Kingston, Ontario, Canada, Doctoral Thesis

[14] Mihet-Popa L 2007 "*Wind Turbines Using Induction Generators Connected to the Grid*", Ed. Politehnica, Timisoara

[15] Milivojevic N, Stamenkovic I, Schofield N and Emadi A 2008 *Electrical machines and power electronic drives for wind turbine applications*, Industrial Electronics, 34th Annual Conference of IEEE 2008 (IECON 2008), Orlando, USA, 10-13 November 2008, pp 2326-2331

[16] Mora M O 2009 *Sensorless vector control of PMSG for wind turbine applications*, Yüksek Lisans Tezi, Institute of Energy Technology Aalborg University, Denmark, Master Thesis

[17] Madescu G, Moţ M, Biriescu M, Greconici M and Koch C 2011 *Low speed PM generator for direct-drive wind applications*, International Conference on Computer as a Tool

(EUROCON), Lisbon, Portugal, 27-29 April 2011, pp 1-4
[18] ***http://new.abb.com/power-converters-inverters/wind-turbines/small-wind/three-phase/pvi-12-5-tl-outd-w
[19] ***http://new.abb.com/power-converters-inverters/wind-turbines/small-wind/generator-interfaces/pvi-wind-interface-4000-7200
[20] ***https://en.wikipedia.org/wiki/Android_(operating_system)
[21] ***http://www.idaq-datalogger.com/_documents/RCF%20User%20Manual.pdf
[22] ***http://www.ni.com/white-paper/14033/en/
[23] Latha N A and Murthy B R, Data Monitoring On Data Dashboard Using LIFA
[24] ***http://www.ni.com/white-paper/4679/en/
[25] ***http://www.ni.com/white-paper/7749/en/

IOP Conf. Series: Materials Science and Engineering **294** (2017) 012071 doi:10.1088/1757-899X/294/1/012071

Hardware-in-the-Loop emulator for a hydrokinetic turbine

C L Rat, O Prostean and I Filip

Politehnica University of Timisoara, Department of Automation and Applied Informatics, Vasile Parvan str., no. 1-2, 300223 Timisoara, Romania

E-mail: cezara.liliana@gmail.com

Abstract. Hydroelectric power has proven to be an efficient and reliable form of renewable energy, but its impact on the environment has long been a source of concern. Hydrokinetic turbines are an emerging class of renewable energy technology designed for deployment in small rivers and streams with minimal environmental impact on the local ecosystem. Hydrokinetic technology represents a truly clean source of energy, having the potential to become a highly efficient method of harvesting renewable energy. However, in order to achieve this goal, extensive research is necessary. This paper presents a Hardware-in-the-Loop emulator for a run-of-the-river type hydrokinetic turbine. The HIL system uses an ABB ACS800 drive to control an induction machine as a significant means of replicating the behavior of the real turbine. The induction machine is coupled to a permanent magnet synchronous generator and the corresponding load. The ACS800 drive is controlled through the software system, which comprises of the hydrokinetic turbine real-time simulation through mathematical modeling in the LabVIEW programming environment running on a NI CompactRIO (cRIO) platform. The advantages of this method are that it can provide a means for testing many control configurations without requiring the presence of the real turbine. This paper contains the basic principles of a hydrokinetic turbine, particularly the run-of-the-river configurations along with the experimental results obtained from the HIL system.

1. Introduction

The difficulties related to the global energy crisis during the past decade have pushed the boundaries of science and technology and have forced researchers to find new and innovative ways of dealing with this ever-increasing problem. The main challenge is not only extracting more energy, but doing it in a more economic and environmentally friendly fashion. One of these is generating electricity as near as possible of the consumption site, which is usually part of the urban environment and through this reducing the power inevitably lost during transmission. This solution cannot be applied for large-scale power plants due to the environmental impact and the disheartening financial investment involved. Therefore small-scale renewable energy sources can be used instead [1], [2].

Micro-hydropower plants that generate between 5 and 100 kW of power constitute such a source. These plants are reliable; low cost; small sized and can be installed to power remote areas or stand-alone loads harnessing energy from small rivers or streams nearby. Compared to other renewable energy sources, micro-hydropower plants have certain advantages such as higher efficiency and slower change rates. One such hydropower source is the hydrokinetic turbine, the study of which constitutes the object of this paper [1].

Hydrokinetic conversion systems represent an emerging class of hydropower technology that are used in free-flowing/zero-head hydro streams. The process of hydrokinetic energy conversion implies utilization of kinetic energy contained in river streams, tidal currents, or other man-made waterways for generation of electricity without significantly altering the natural pathway of the stream [3]. In this paper, we considered the hydrokinetic turbine as being in run-of-the-river configuration. This represents one of the most cost effective and environment friendly technologies, partly due to the fact that it uses no dam or water storage [4]. River turbines operate under the influence of varying volumetric flow, but have the advantage of higher resource predictability. Also, the flow of the river is unidirectional and this eliminates the requirement for rotor yawing and the complicated mechanics that goes with it [3].

However issues such as the limited energy contained in the natural flow of many rivers, the frequent presence of floating and submerged debris, the possible destruction during flash flood and corrosion make the task of implementing a functional water turbine a complicated one [5].

2. Hydrokinetic energy

The hydrokinetic method of extracting energy from flows captures the kinetic energy of the flow as it passes across a rotating device. The device is connected through a mechanical shaft to a generator either directly or through a gearbox, in a manner similar to wind turbines. The rotor turns with the current, creating rotational energy that is converted into electricity by a generator. This ensemble constitutes a hydrokinetic energy conversion system (HKECS) [6].

Although a number of novel concepts have emerged recently, hydrokinetic energy conversion has mostly seen advancements in the domain of axial (horizontal) and vertical axis turbine systems. Horizontal axis hydrokinetic turbines have rotor blades which rotate in a plane perpendicular to the axis, oriented into the direction of the flow or tidal current. These turbines are capable of high efficiency, self-starting, lack of torque fluctuation and high speed operation [3], [5], [6].

2.1. Hydrokinetic turbine modeling

2.1.1. Hydrodynamic model. The maximum output power of a hydrokinetic conversion device P_{max}, taking into account losses due to Betz law, is [7]:

$$P_{max} = \frac{1}{2}\eta_t \rho Q v^2 \tag{1}$$

where ρ is the specific density of water [kg/m^3], v is the flow rate of water [m/s], η is the total hydraulic efficiency of the turbine and Q is the volume of water flowing through the turbine [m^3/s]. Q is given by:

$$Q = A * v \tag{2}$$

where A is the cross sectional area of the turbine [m2]. Thus:

$$P_{max} = \frac{1}{2}\eta_t \rho A v^3 \tag{3}$$

If we consider η_t as being composed of the turbine efficiency $C_P(\lambda)$ and other losses η, (3) would become [7]:

$$P_{max} = \frac{1}{2}\eta C_P(\lambda_i, \theta)\rho A v^3 \tag{4}$$

The $C_P(\lambda)$ factor represents the turbine power coefficient, giving the real amount of power that the hydrokinetic turbine can be extracted from the water. This factor takes into account the efficiency of the turbine itself while other losses, such as those dissipated by the internal mechanisms of the drive train are represented through the drive train efficiency which is typically around 0.7 [8], [9].

Using (4), the output power of the hydrokinetic turbine (P_{wt}) can be calculated as:

$$P_{wt} = \frac{1}{2}C_P(\lambda_i, \theta)\rho A v^3 \tag{5}$$

This equation constitutes the aerodynamic model of the hydrokinetic turbine. It can be seen that the output power of a hydrokinetic turbine is proportional to the speed of the current cubed. Thus the available power depends primarily on the speed of the current. Hydrokinetic devices work best in locations with relatively steady flow throughout the year and without extended periods of low water level. The energy conversion factor C_p is a function of the λ_i factor and the pitch angle θ, being able to render the behavior of the modeled turbine. C_p has the following equation [6], [8], [9]:

$$C_p(\lambda_i, \theta) = 0.22\left(\frac{116}{\lambda_i} - 0.4\theta - 5\right)e^{\frac{-12.5}{\lambda_i}} \quad (6)$$

where:

$$\frac{1}{\lambda_i} = \frac{1}{\lambda + 0.08\theta} - \frac{0.035}{\theta^3 + 1} \quad (7)$$

The λ_i factor is a function of the tip-speed ratio λ, which is defined as:

$$\lambda = \omega_r R / v \quad (8)$$

where R is the turbine blade radius [m] and ωr is the rotation speed [rad/s];

With the power coefficient equation, it is possible to obtain the power coefficient curves for the turbine, which can be compared with the information provided by the turbine manufacturer. This makes it possible to reproduce the real behavior of the axial rotor, ensuring the accuracy of the turbine model [9].

2.1.2. The mechanical model. The chosen turbine is connected to a synchronous generator with permanent magnets. The mechanical part is considered to be a two-mass model. The motion equation is [8], [9]:

$$T_{ht} - T_g = J * \frac{d\omega_r}{dt} \quad (9)$$

where T_{ht} is the turbine torque [Nm], T_g is the generator torque [Nm] and J is the inertia of the hydro system [$kg \cdot m^2$]. J is composed of:

$$J = J_{ht} + J_g \quad (10)$$

where J_{ht} is the inertia of the turbine [$kg \cdot m^2$]and J_g is the generator inertia [$kg \cdot m^2$]. The power of the system can be expressed by the following relationship:

$$P_{ht} = T_{ht} * \omega_r \quad (11)$$

2.2. *HKECS Hardware-in-the-Loop System*

The hydrokinetic Hardware-in-the-Loop emulator consists of a rotational motor-generator system: a 7.5 kW/720 rpm three phase squirrel cage induction motor (IM) connected to a 5kW/120rpm permanent-magnet synchronous generator (PMSG) [10]. A 1/6 gearbox is needed in order to adjust the speed of the IM to the necessary value for the generator. The IM is controlled by a ABB ACS800 motor drive system through an National Instrument embedded controller (NI cRIO 9068). The controller simulates the hydrokinetic turbine using mathematical modeling. The mathematical model of the hydrokinetic turbine is based on the mathematical equations presented in the preceding chapter: the aerodynamic model and the mechanical model, both translated into LabVIEW, as can be seen in Figures 1 and 2.

The turbine parameters are presented in the following table:

IOP Conf. Series: Materials Science and Engineering **294** (2017) 012071 doi:10.1088/1757-899X/294/1/012071

Table 1. Parameters of the hydrokinetic turbine [10]

Parameter	Value
Rated power	$P_n = 2.5$ [kW]
Nominal water speed	$v_0 = 1.5$ [m/s]
Nominal speed	$n = 130$ [rpm]
Inertia of the turbine	$J_{wt} = 0.0018$ [kg m^2]
Inertia of the generator	$J_g = 1.05$ [kg m^2]
Surface covered by turbine blades	$A = 3.14$ [m^2]
Turbine blade radius	$R = 1$ [m]
Maximal power factor	$C_{p_{max}} = 0.4382$
Nominal speed ratio	$\lambda_0 = 6.325$
Specific water density	$\rho = 1027$ [kg/m^3]

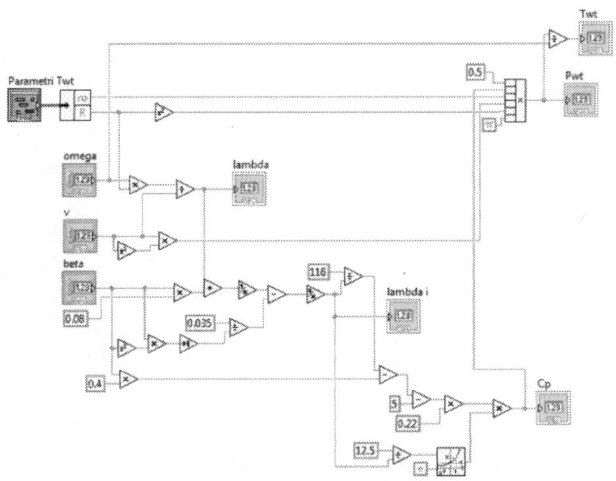

Figure 1. The hydrodynamic model implemented in LabVIEW

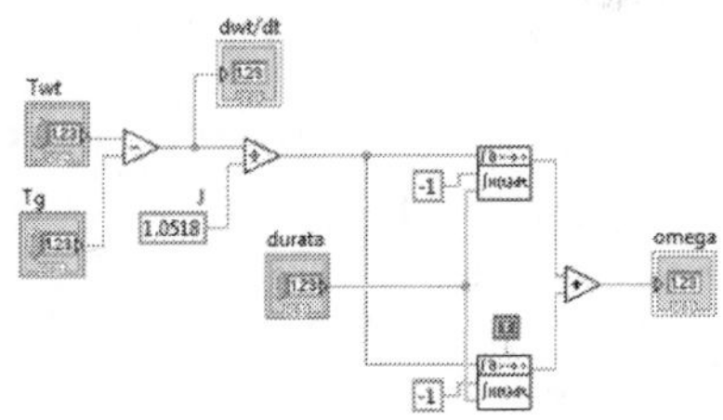

Figure 2. The mechanical model implemented in LabVIEW

The HIL emulator developed in this paper is designed to simulate the behavior of the hydrokinetic turbine in different situations. Usage of HIL enhances the quality of the testing by increasing the scope of the testing. Specifically, the NI cRIO 9068 runs a real-time mathematical model of the hydro turbine implemented in LabVIEW. These mathematical representations are referred to as the "plant simulation". The embedded system to be tested interacts with this plant simulation; the program reads the torque estimated by ACS800 and calculates the new rotating speed (ω_r) which is prescribed to the ACS800 [11-14].

The emulator hardware is depicted in Figure 3.

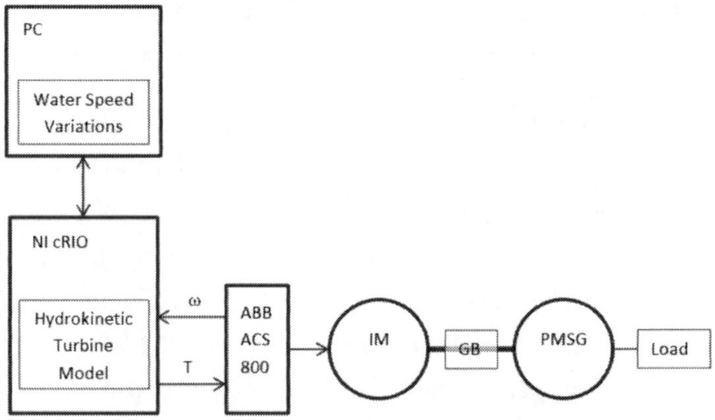

Figure 3. HIL emulator

The turbine characteristics were obtained through LabVIEW simulation varying the flow speed between 0.1 and 1.5 m/s. The performance of the turbine has been verified using these graphics. The power characteristic of the turbine can be seen in Figure 4.

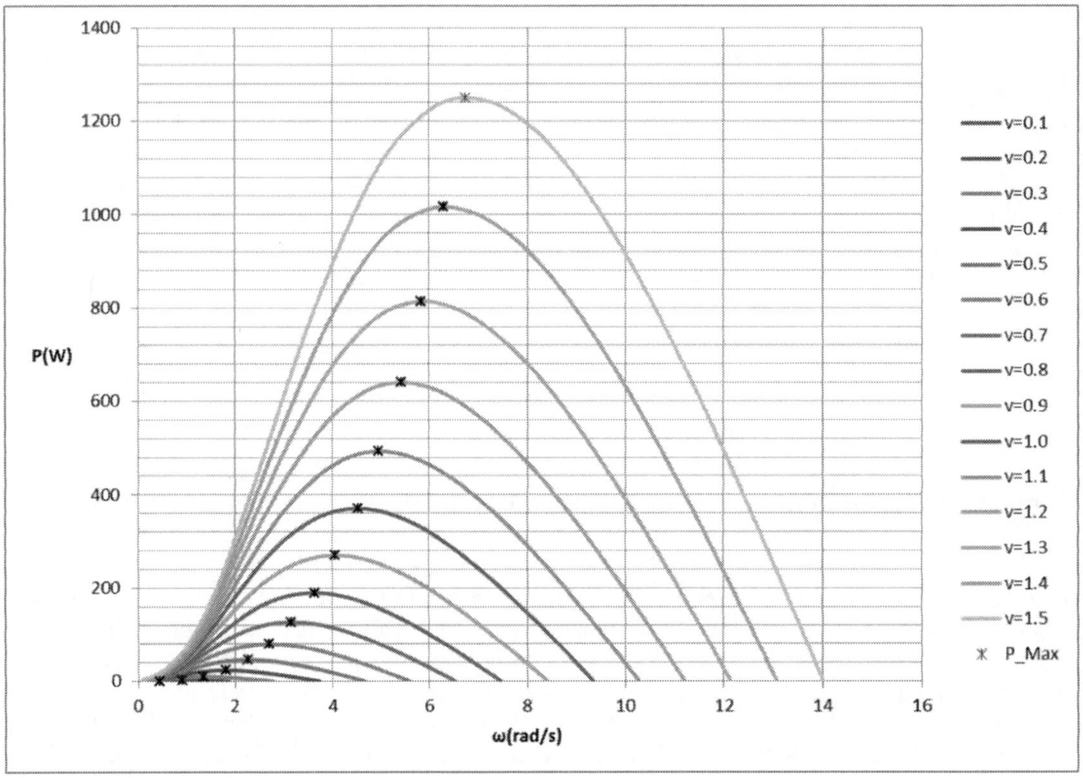

Figure 4. Power versus angular speed

Using the turbine power equation from (5), we obtain:

$$P_{ht} = 0.5A\rho C_P(\lambda) * v^3 = 0.5A\rho C_P(\lambda) * \frac{v^3}{\omega^3 R^3} * \omega^3 R^3 = 0.5\pi R^5 \rho C_P(\lambda) * \frac{1}{\lambda^3} * \omega^3$$

The maximum power for a certain θ is obtained at the optimum value of λ:

$$\lambda = \lambda_{\mathrm{opt}} = constant \text{ and } \mathrm{C_P}(\lambda_{opt}) = C_{P_{opt}} = constant \quad (12)$$

So the optimum hydrokinetic turbine power is given by:

$$\mathrm{P}_{\mathrm{ht}_{opt}} = 0.5\pi R^5 \rho \mathrm{C}_{\mathrm{P}_{opt}} * \frac{1}{\lambda_{opt}{}^3} * \omega^3 = k\omega^3 \quad (13)$$

As can be seen, the optimum power is proportional to the cube of the angular velocity (Figure 5). Hence, by measuring the hydrokinetic turbine rotor speed, it is possible to determine the optimum power. But this requires knowing the optimal power points for cross correlation [15].

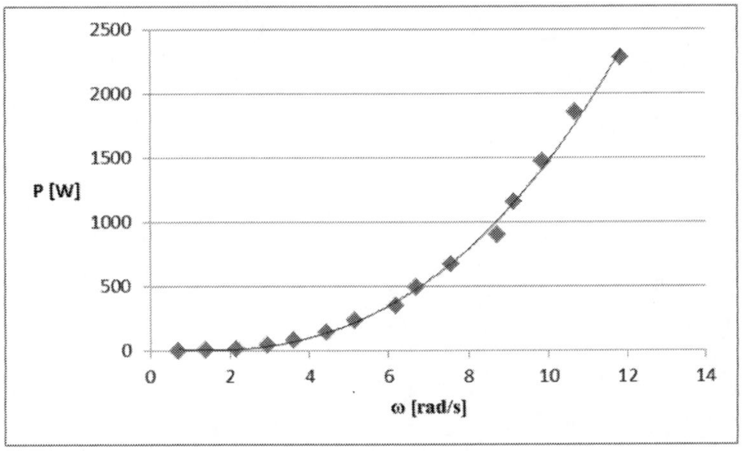

Figure 5. Optimal power versus angular speed

In (12) we have deduced that for the maximum power point to exist, there has to be a $\lambda = \lambda_{\mathrm{opt}}$ and a $\mathrm{C_P}(\lambda_{opt}) = C_{P_{opt}}$ corresponding to this point. Using the hydrodynamic equation of the turbine (5), the following relationship is obtained:

$$\mathrm{P}_{\mathrm{wt}_{opt}} = 0.5\mathrm{A}\rho \mathrm{C}_{\mathrm{P}_{opt}} * \mathrm{v}^3 = \mathrm{kv}^3 \quad (14)$$

Therefore, the optimal power is directly proportional to the cube of the water speed (Figure 6). So, by measuring the flow speed, it is possible to determine the optimum power. This result is also visible in the following Figure.

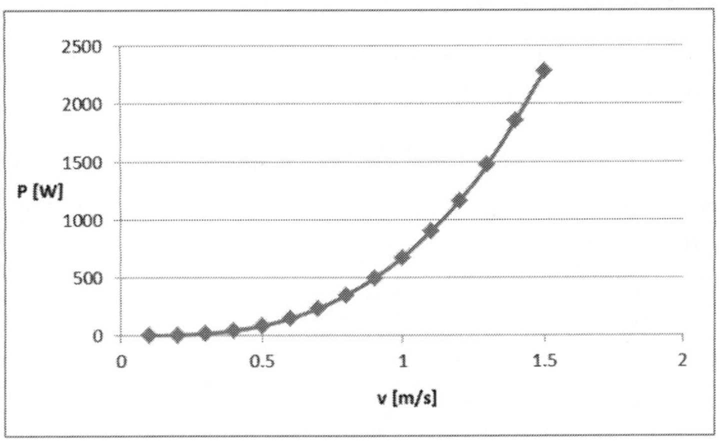

Figure 6. Optimal power versus water speed

The turbine torque relation is obtained from (11) and (5):

$$T_{wt} = 0.5A\rho C_T(\lambda) * v^2 \tag{15}$$

where C_T is the torque coefficient ($C_p = \lambda C_T$). The turbine torque characteristics obtained through varying the flow speed between 0.1 and 1.5 m/s can be seen in Figure 7.

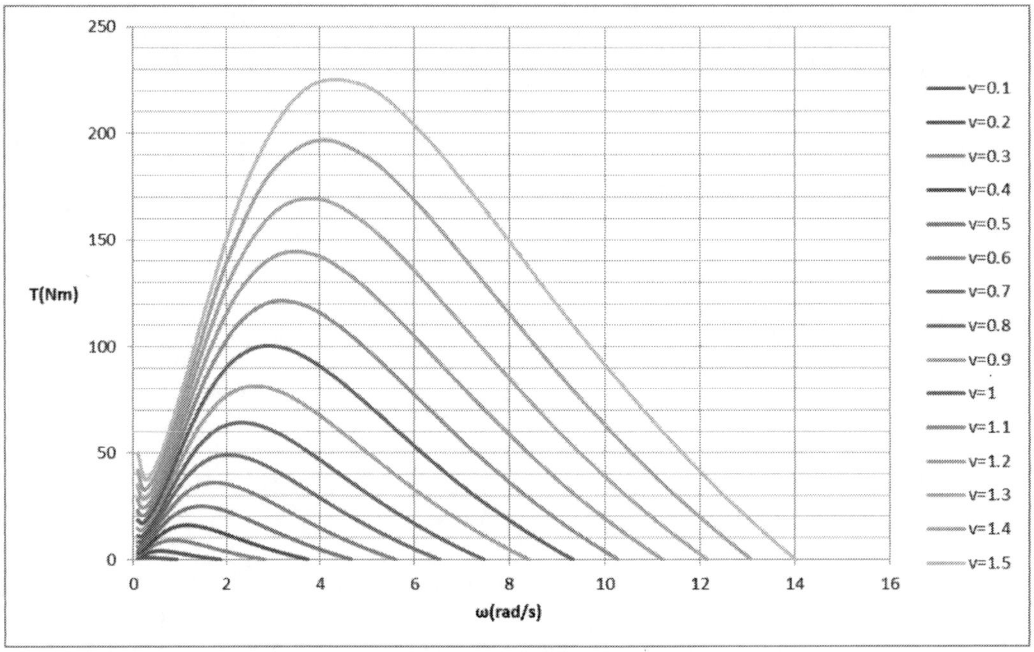

Figure 7. Mechanical torque versus angular speed

The turbine torque equation is obtained using (15):

$$T_{wt} = 0.5A\rho C_T(\lambda) * v^2 = 0.5A\rho C_T(\lambda) * \frac{v^2}{\omega^2 R^2} * \omega^2 R^2 = 0.5\pi R^4 \rho C_T(\lambda) * \frac{1}{\lambda^2} * \omega^2$$

Maximum torque is obtained at the optimal value of λ and C_T:

$$\lambda = \lambda_{opt} = constant \text{ and } C_P(\lambda_{opt}) = C_{P_{opt}} = constant \tag{16}$$

The maximum torque point does not correspond to the maximum power point. So the optimal torque of the hydrokinetic turbine is:

$$T_{wt_{opt}} = 0.5A\rho C_{T_{opt}} * \frac{1}{\lambda_{opt}{}^2} * \omega^2 = k\omega^2 \tag{17}$$

Therefore, the maximum torque is directly proportional to the square of the angular velocity, which is also visible in the following Figure:

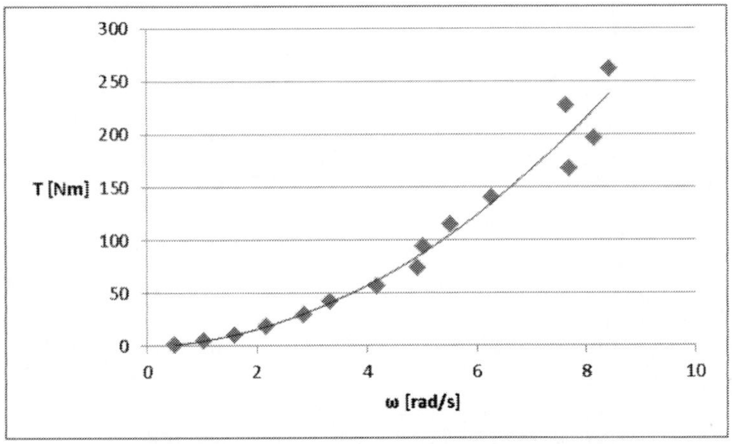

Figure 8. Optimal torque versus angular speed

In (15) we have deduced that for the maximum torque point to exist, there has to be a $\lambda = \lambda_{opt}$ and a $C_T(\lambda_{opt}) = C_{T_{opt}}$ corresponding to this point. Using the hydrodynamic equation of the turbine (5), the following relationship is obtained:

$$T_{wt_{opt}} = 0.5A\rho C_{T_{opt}} * v^2 = kv^2 \tag{18}$$

Therefore, the maximum torque is directly proportional to the square of the flow speed, which is also visible in the following Figure:

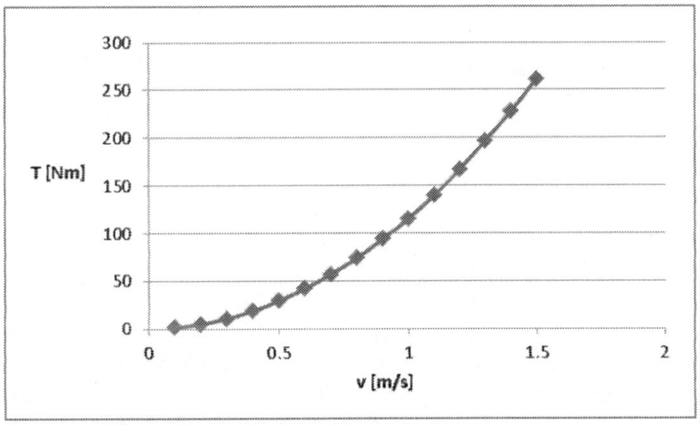

Figure 9. Optimal torque versus flow speed

The optimal power versus pitch angle is shown below. It was obtained based on experimental result and has been found to be a 2nd order polynomial function.

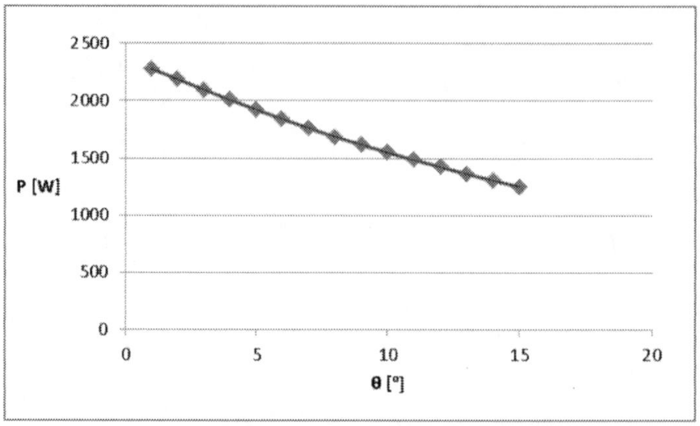

Figure 10. Optimal power versus pitch angle

The voltage versus flow speed is shown below. It was obtained based on experimental result and has been found to be a 2nd order polynomial function.

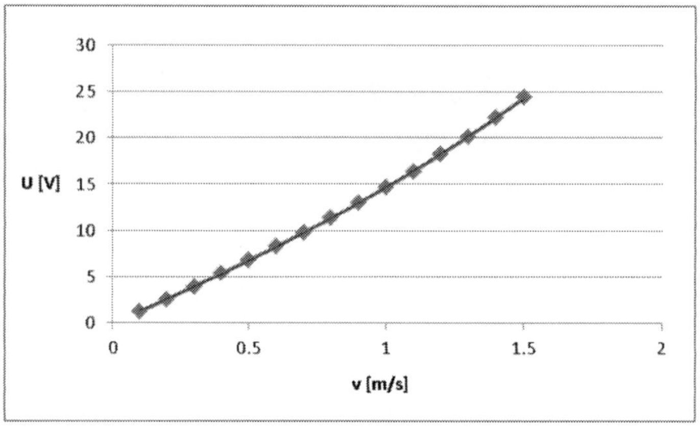

Figure 11. Actual voltage versus flow speed

3. Conclusions

The objectives of this paper were to conduct a study on a small axial river turbine in a hydrokinetic energy conversion system. The hydrokinetic converter is constructed without significantly altering the natural pathway of the water stream, specifically in run-of-the-river configuration. The power is generated by capturing the energy of naturally flowing water – stream flows, tidal flows, or wave motion – without impounding the water. This represents one of the most cost effective and environment friendly technologies. The hydrokinetic energy conversion system presented in this paper demonstrates how much use can be made of small river systems. Micro-hydropower plants are reliable, small sized; they have a low cost and can be installed to power remote areas or stand-alone loads harvesting energy from local rivers or streams [5], [6].

In this paper, the effects of the main parameters of the turbines on their operation were analyzed. There is a polynomial dependence between the angular velocity and the optimal power point, respectively the optimal torque. Dependent is 3rd order for power and 2nd order for the torque. A similar dependence exists between these parameters and water speed. The output voltage from the generator has 2nd order dependency on the variation of the flow speed. The optimal power versus pitch angle has also been found to be a 2nd order polynomial function.

References

[1] Usman A A and Abdulkadir R A 2015 Modelling and Simulation of Micro Hydro Power Plant using Matlab Simulink, *International Journal of Advanced Technology in Engineering and Science (IJATES)* 2348-7550

[2] Lago L I, Ponta F L and Chen L 2010 Advances and trends in hydrokinetic turbine systems, *Energy for Sustainable Development* **14**(4) 287-296

[3] Khan M J, Bhuyan G, Iqbal M T and Quaicoe J E 2009 Hydrokinetic energy conversion systems and assessment of horizontal and vertical axis turbines for river and tidal applications: A technology status review, *Applied energy* **86**(10) 1823-1835

[4] Temiz A 2013 *Decision making on turbine types and capacities for run-of-river hydroelectric power plants a case study on Eglence-1 HEPP*, İzmir Institute of Technology, Master Thesis

[5] Anyi M and Kirke B 2010 Evaluation of small axial flow hydrokinetic turbines for remote communities, *Energy for Sustainable Development* **14**(2) 110-116

[6] Ortega-Achury S L, McAnally W H, Davis T E and Martin J L 2010 *Hydrokinetic power review*, Prepared for the US Army Corps of Engineers Research and Development Center, Vicksburg, Mississippi

[7] Mangold E 2012 *Hydrokinetic Power: An Analysis of Its Performance and Potential in the Roza and Kittitas Canals*, Evergreen State College, USA, Doctoral Thesis

[8] Chica E, Pérez F, Rubio-Clemente A and Agudelo S 2015 Design of a hydrokinetic turbine, *WIT Trans. Ecol. Environ* **195** 137-148

[9] Sinopoli L 2012 *Marine current turbine dynamic electrical emulator*, Applied Science: School of Engineering Science, USA, Doctoral Thesis

[10] Szeidert I, Biriescu M, Mihet-Popa L and Toader D 2009 *Analysis by Numerical Simulation regarding the Stability of the Synchronous Generator Operating in Autonomous or Grid Connected Regime*, 5th International Symposium On Applied Computational Intelligence And Informatics (SACI), Timisoara, Romania

[11] Woods L D 2013 *Simulation of VAWT and Hydrokinetic Turbines with Variable Pitch Foils*, Doctoral Thesis

[12] ***https://en.wikipedia.org/wiki/Hardware-in-the-loop_simulation

[13] Neely J C, Ruehl K M, Jepsen R A, Roberts J D, Glover S F, White F E and Horry M L 2013 *Electromechanical emulation of hydrokinetic generators for renewable energy research*, Oceans-San Diego, pp 1-8

[14] Topor M, Muntean N and Sorandaru C 2015 Wind Turbine Emulator Development using NI cRIO 9068 and ABB ACS 800 Drive, *Journal of Electrical Engineering* **15**(3) 254-259

[15] Karthikeyan V, Vijayalakshmi V J and Jeyakumar P 2014 Step and Search Control Method to Track the Maximum Power in Wind Energy Conversion Systems, A Study, **1403** 5865

Study of the influence of the cutting temperature on the magnitude of the contact forces in the machining fixtures

V G Cioată[1], I Kiss[1], V Alexa[1], S A Rațiu[1] and M Racov[2]

[1]Politehnica University of Timisoara, Department of Engineering and Management, Revoluției Street, no. 5, 331128, Hunedoara, Romania
[2]University of Novi Sad, Faculty of Technical Science, Department of Mechanization and Design Engineering, Revolution Street, Novi Sad, Serbia

E-mail: vasile.cioata@.upt.ro

Abstract. In the machining process, the workpieces are installed in machining fixtures in order to establish a strictly determined position with the cutting tool or its trajectory. During the cutting process, the weight of the workpiece, the forces and moments of inertia, cutting forces and moments, clamping forces, the heat released during the cutting process determine the contact forces between the locators and the workpiece. The magnitude of these forces is important because too large value can destroy the surface of the workpiece, and a too small value can cause the workpiece to slip on the locators or even the loss of the contact with the workpiece. Both situations must be avoided. The paper presents a study, realized with CAE software, regarding the influence of the cutting temperature on the magnitude of the contact forces in a machining fixture for the milling a rectangular workpiece.

1. Introduction

The great variety of workpieces found in the industry stimulates the designers to find new solutions for machining fixtures, which meet the requirements of precision and productivity imposed by the users.

This is done shortly and at low costs if CAE type numerical simulation software, as in [1], [2], and CAD type three dimensional modeling, as in [3], are used to validate the fixtures configuration and if modern control techniques of the processes are used for the processing of their component parts, to eliminate human errors [4].

In the practice of designing machining fixtures, the designers should consider the influence of the cutting process parameters on the machining precision of the workpiece installed in the fixture. Among these, we list the cutting forces and moments, parameters of the cutting regime (cutting depth, feed, cutting speed), material of the workpiece, material of the tool, cooling-lubrication during the process, geometry and kinematic precision of the machine-tool etc.

At the same time, the chosen orientation layout, the type of locators, the nature of the contact between the locator and the workpiece and the clamping elements and the workpiece [5], the size and location of the clamping forces [6] must be taken into account.

It is a known fact that the thermal phenomena have an influence both on the behavior of the workpieces during operation [7], [8] and during machining by cutting. During cutting, the thermal phenomena influence the nature and the dynamic of some characteristic phenomena as well as the results of the process, in terms of precision and quality of the machined surface, the wear of the cutting tool, the process of the forming of the chip, vibration, etc.

The heat developed during the cutting results from the almost complete transformation of the mechanical work consumed for the cutting process. It is largely due to the plastic deformations of the material during the chip formation (Q_1), frictions at tool/chip interfaces (Q_2) and tool/cut surface (Q_3) and to a lesser extent, to additional deformation of the chips (twisting, crushing).

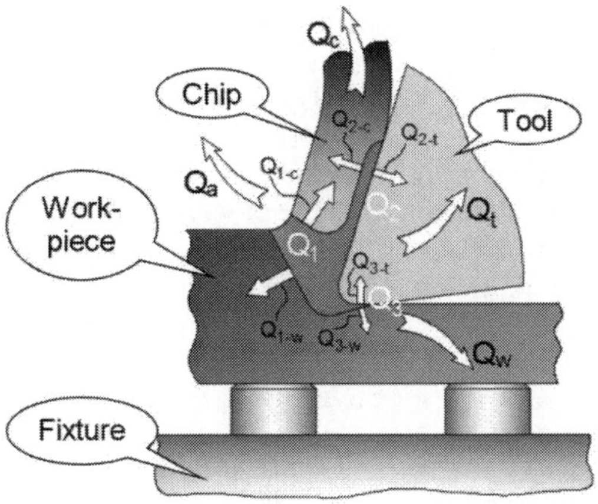

Figure 1. Main sources and distribution mode of the heat during cutting

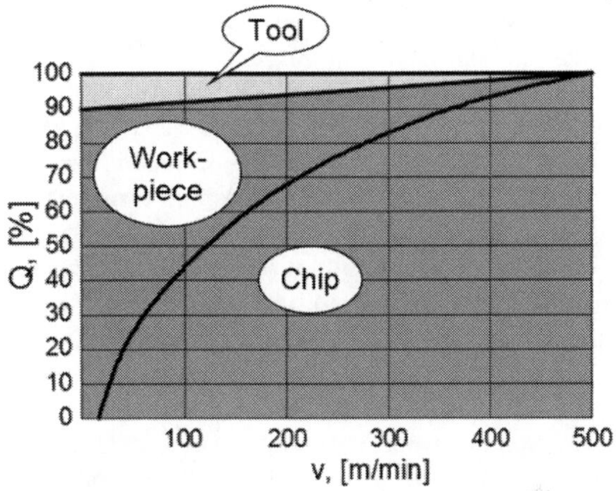

Figure 2. Percentage distribution of the heat (Q) according to the cutting speed (v)

The heat developed during the cutting process spreads from the sources where it was formed towards the cooler areas, distributing in the chip (Q_c), the tool (Q_t), the workpiece (Q_w) and the cutting environment (Q_a). Each source transmits, according to Figure 1, a certain amount of heat towards the neighboring areas.

Heat flows go in opposite directions towards cooler areas. Thus, from the source developed in the shear plane due to the deformation of the cut material (Q_1) a flow goes towards the chip (Q_{1-c}) and a much smaller flow (Q_{1-w}) goes towards the workpiece. The heat developed by the tool – chip friction (Q_2) goes towards the chip through the Q_{2-c} flow and to the tool through the Q_{2-t} flow, and the heat from the friction on the back edge (Q_3) goes towards the tool (Q_{3-t}) and towards the workpiece Q_{3-w}.

This way, the heat absorbed by the chip (Q_c) can be expressed as:

$$Q_c = Q_{1-c} + Q_{2-c} \quad (1)$$

the one absorbed by the tool (Q_t) will be:

$$Q_t = Q_{2-t} + Q_{3-t} \quad (2)$$

and the one absorbed by the workpiece (Q_w) will be:

$$Q_w = Q_{1-w} + Q_{3-w}. \quad (3)$$

The percentage distribution of heat flows depends largely on the cutting conditions, especially on the type of operation and the cutting regime. Figure 2 shows that, in general, most of the heat is absorbed by the chip, followed by the workpiece and the smallest part by the tool.

The machining precision is negatively influenced especially by the amount of heat absorbed by the workpiece and the tool, which remain in contact during the entire machining process and it is manifested especially in the field of small cutting speeds (Figure 2).

The heat absorbed by the workpiece or the cutting tool can be determined using various methods, as shown in [9], [11].

Analytically, the heat absorbed by the piece can be roughly estimated using the relationship [12]:

$$Q_w = \frac{1}{3} \cdot \lambda \cdot \frac{bl^2}{v} \nabla\theta_{max} \text{ [J]} \quad (4)$$

in which:

λ – the thermal conductivity of the material of the workpiece, W/mK;
b – is the width of the chip, in m;
l – the tool path length, in m;
v – the cutting speed, in m/s;
$\nabla\theta_{max}$ – the maximum value of the temperature gradient (K/m).

It is a known fact that the installation of the workpiece in the fixture has two functional phases: orientation and clamping. It is necessary for this system of forces to ensure the contact of the workpieces with the locators and to maintain it during machining, while also ensuring maximum rigidity of the workpiece-fixture assembly, which leads to the decrease or removal of vibrations.

At the same time, the size of the clamping forces, together with the other influences (forces and moments of cutting, weight, forces and moments of inertia, centrifugal forces, mass forces, etc.) must not cause contact deformations destroying the surfaces of the workpiece.

The heat absorbed by the workpiece is inevitably transmitted to the fixture in which it is installed, generating the expansion of its component parts, which influences the machining accuracy and the size of the contact forces between the workpiece and the locators.

The extreme values of the contact forces are very important. The maximum (fc_{max}) and minimum (fc_{min}) values of the contact forces are determined according to [13], [14].

The contact force must not exceed the maximum value fc_{max} as there is a risk of damage to the surfaces of the workpiece (indetation will appear) and the minimum contact force (fc_{min}) must prevent the loss of contact between the workpiece and the locators, and must also prevent the workpiece to slipping from the locators.

2. Model development

This paper presents a study on the influence of the heat generated during the cutting process on the size of the contact forces between the workpiece and the locators of the fixture in which the piece is installed.

For this purpose it is considered a situation in which a parallelepiped shaped workpiece (size 225x122x112 mm) is subjected to cylindrical-frontal milling. A canal, with 2 mm depth, is formed, positioned at a distance of 12 mm from the evaluated edge, which is the reference element for measuring (Figure 3). This situation has been chosen so that there are no considerable changes of rigidity in the workpiece caused by the machined canal, affecting the results of the analysis. The diameter of the cutting tool is 25 mm and has two teeth (z=2 teeth).

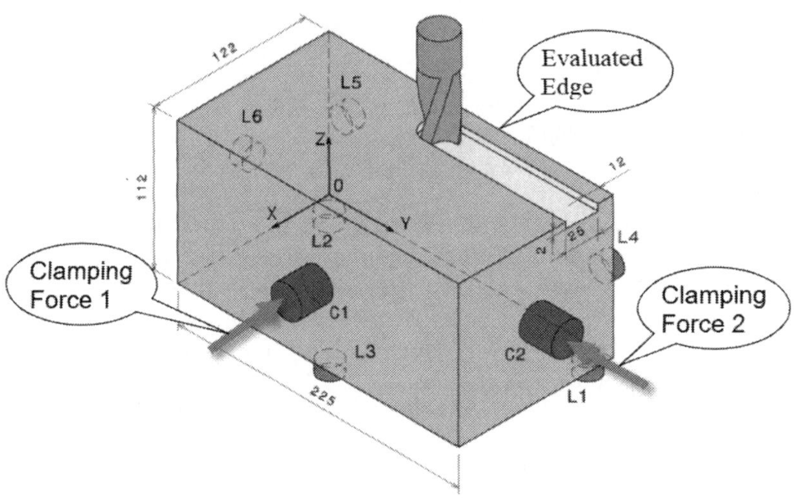

Figure 3. 3D model of the workpiece – fixture system

Table 1. Coordinates of locators and clamping elements

Coordinates	Locators						Clamps	
	L1	L2	L3	L4	L5	L6	C1	C2
X [mm]	9	9	113	0	0	61	122	61
Y [mm]	216	9	112,5	216	9	0	112,5	225
Z [mm]	0	0	0	56	56	56	56	56

The orientation layout is of type 3-2-1, locators (L1,..., L6) have a cylindrical shape, 18 mm in diameter, with a flat contact surface and the clamping elements C1 and C2, also have a cylindrical shape with a diameter of 25 mm. The values of the clamping forces are 576 N and 621 N. The contacts forces between the workpiece and the locators and between the workpiece and the clamping elements are flat and the friction coefficient at sliding is 0,1.

The locators and elements of applying clamping forces, relative to the system OXYZ, are positioned as in Figure 3 and Table 1.

The workpiece, the locators and clamping elements are made of structural steel. The properties of the material used in the simulation are presented in Table 2.

Table 2. Properties of the material

Properties	Structural Steel
Density [kg/m^3]	7850
Young's Modulus [MPa]	200000
Poisson's Ratio [-]	0,3
Tensile Yield Strength [MPa]	250
Tensile Ultimate Strength [MPa]	460

The 3D model of the workpiece-fixture system was made in Ansys Design Modeler and it was then transferred to Ansys Mechanical for the two analyzes.

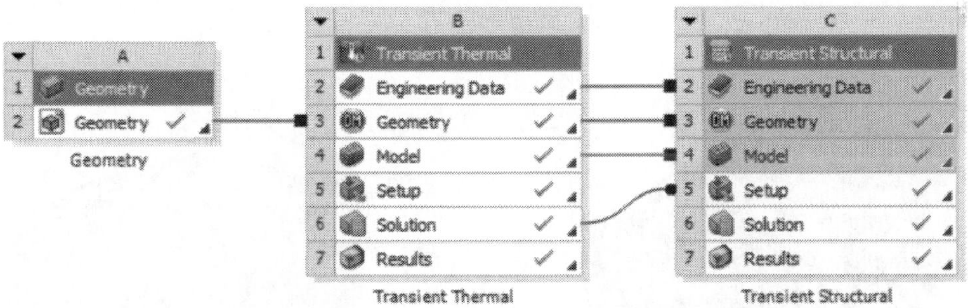

Figure 4. Project layout in Ansys

Two analyzes were carried out: a transient thermal analysis to determine the temperature in the workpiece-fixture assembly during machining and then a transient structural analysis in which the temperature field from the thermal analysis was imported. Loads and constraints specific to the structural analysis were added to this analysis, namely: forces and moment of cutting, clamping forces, binding of the degrees of freedom of the locators and of clamping elements (Figure 4).

For both analyzes, the loads were applied in 9 equidistant circular areas, 25 mm in diameter, on the upper surface of the workpiece, their centers being located on a line parallel to the evaluated edge, at a distance of 24,5 mm (Figure 5).

For the thermal analysis, a thermal flow of 2 W/mm^2 acts in the 9 circular areas, and the heat transfer between the workpiece and the environment is made by convection, the convection heat transfer coefficient being $2 \cdot 10^{-5}$ W/mm$\cdot^0$C.

For the structural analysis, the forces and the moment of cutting are considered to have the following values: Fx=131 N, Fy=232 N, Fz=55 N and M=2.77 N·m, and they also act in the 9 circular areas on the upper surface of the workpiece.

Under a cutting speed of approximately 60 m/min and a feed per tooth of 0.012 mm/z, the cutting processing time of the workpiece over the entire length of 225 mm is of approximately 81 seconds. This will also be the time used in the analysis.

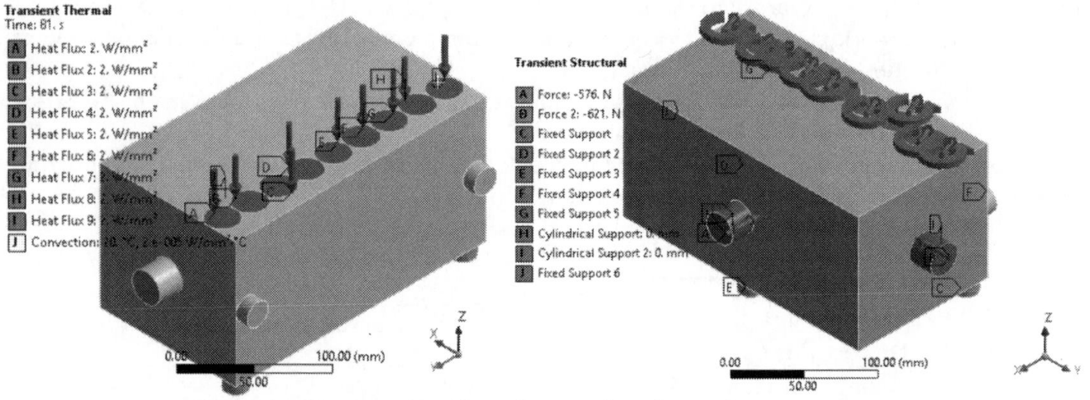

Figure 5. Loads and constaints for analyzes

3. Results

After running the thermal analysis, the field of temperatures in the workpiece is displayed for two of its surfaces: the upper and the right side of the canal, at time points 42 s and 81 s (Figure 6).

It can be seen that the contact areas between the locators and the surfaces of the workpiece get warm due to the heat released during cutting, which results in expansion of the workpiece-fixture assembly, with possible consequences on the machining precision as well as on the size of the contact forces between the locators and the surfaces of the workpiece.

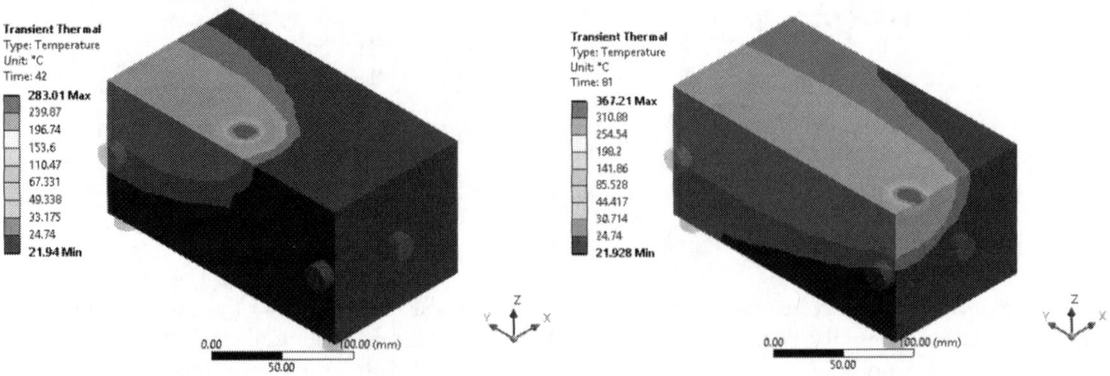

Figure 6. Temperature in workpiece at moments 42 s and 81 s

Four cases of workpiece-fixture assembly loading were considered for the study of these consequences, as follows:

- Case I: when only the clamping forces act (No T, No F);

- Case II: when only the forces and moment of cutting act (No T, With F);
- Case III: when only the cutting temperature acts (With T, No F);
- Case IV: when both forces and moment of cutting as well as cutting temperature act (With T, With F).

For each of the four considered cases, the contact forces and total deformation of the evaluated edge were determined.

As such, Figures 7…10 present the variation of the contact forces between the locators (L1,…, L6) and the workpiece in the 4 cases considered above. In Figure 7, the contact forces have constant values throughout the machining because neither the temperature which results from the cutting, nor the forces nor the moment which result from the cutting process act on the workpiece.

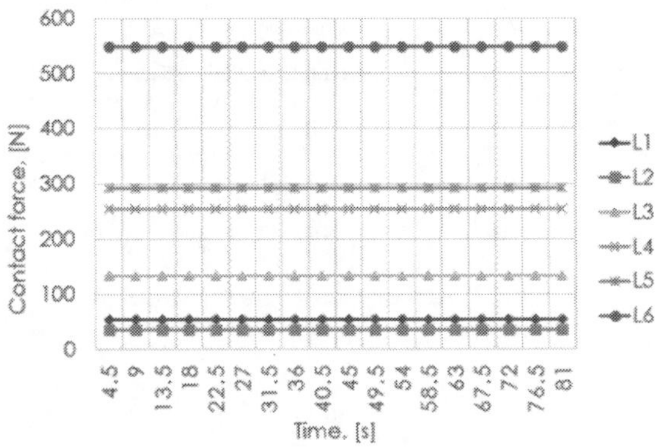

Figure 7. Contact forces for Case I (No T, No F)

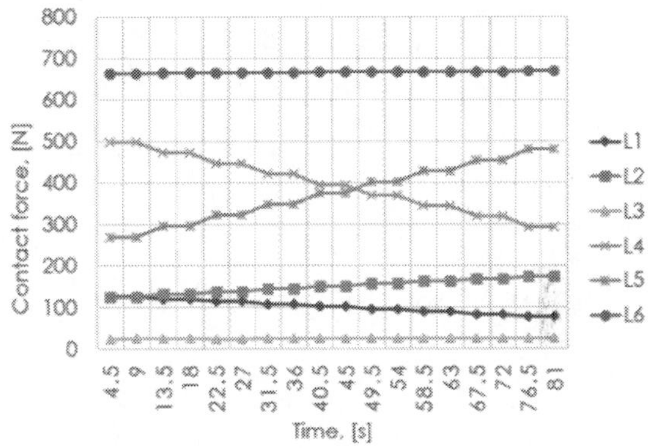

Figure 8. Contact forces for Case II (No T, With F)

For case II, when only the influence of forces and moment of cutting is taken into account, the variation of the contact forces during machining is shown in Figure 8. This is consistent with the results published in the specialty literature.

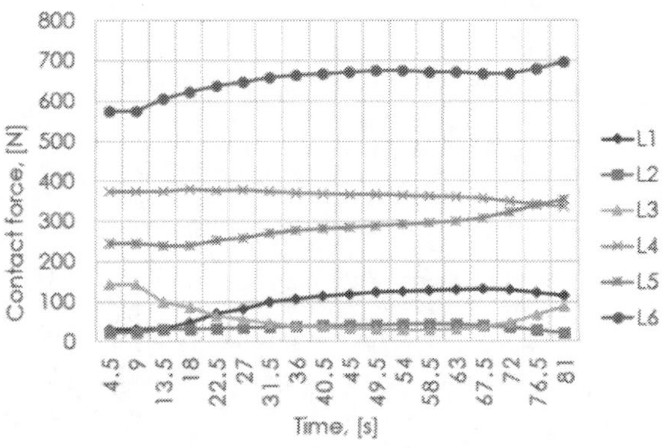

Figure 9. Contact forces for Case III (With T, No F)

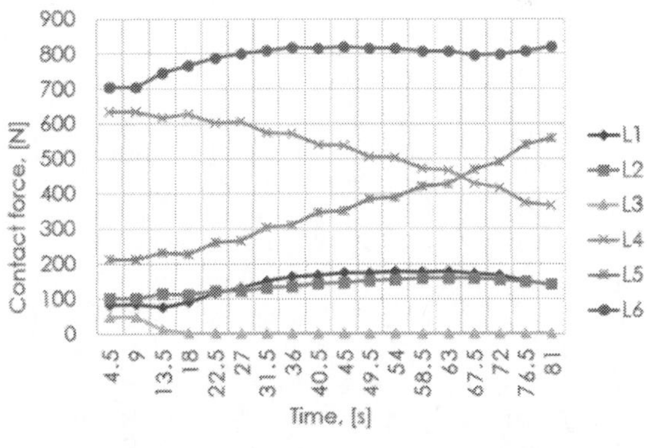

Figure 10. Contact forces for Case IV (With T, With F)

For case III, only the influence of the clamping forces and the cutting temperature on the size of the contact forces is considered. It can be seen, by comparison to the graphic in Figure 8 that the allure of the curves is changing, indicating a certain influence of the cutting temperature on the values of the contact forces.

IOP Conf. Series: Materials Science and Engineering **294** (2017) 012072 doi:10.1088/1757-899X/294/1/012072

Case IV is the one in which both loads are considered. The variation of the contact forces during machining is shown in Figure 10.

In order to highlight the variation of the contact forces in the four loading situations, for each of the locators L1,..., L6, the representations in Figures 11...16 were realized.

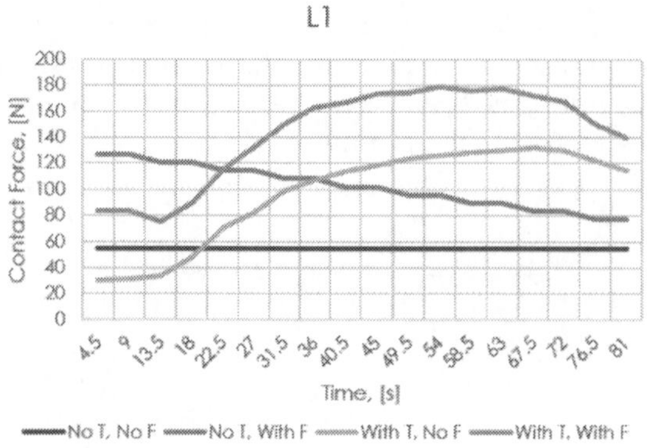

Figure 11. Contact forces for locator L1

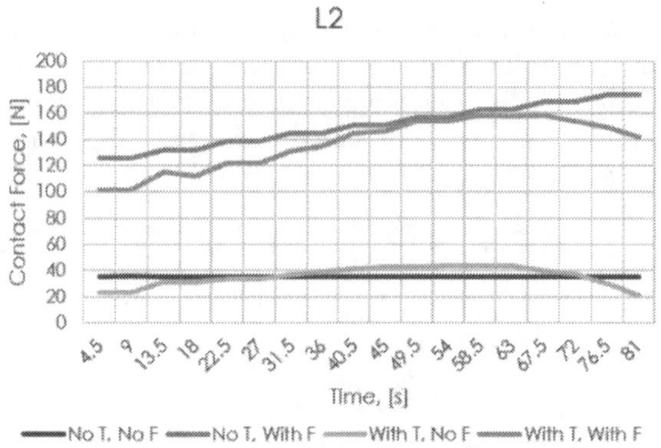

Figure 12. Contact forces for locator L2

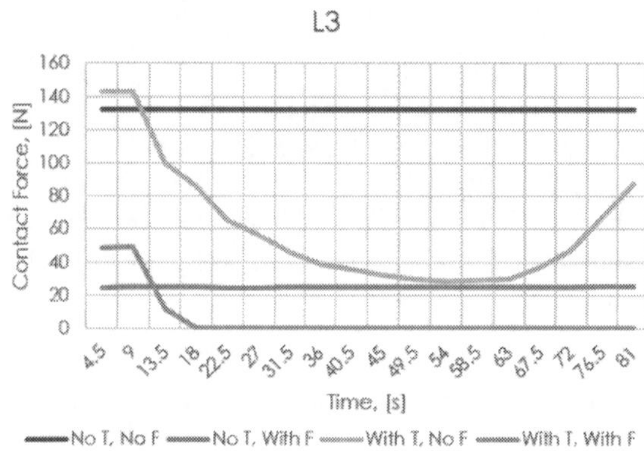

Figure 13. Contact forces for locator L3

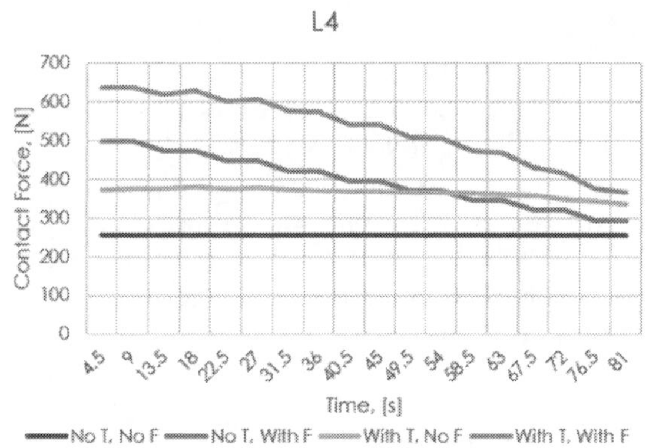

Figure 14. Contact forces for locator L4

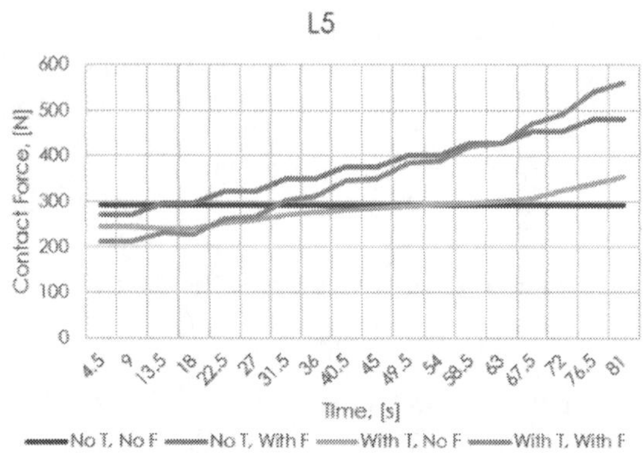

Figure 15. Contact forces for locator L5

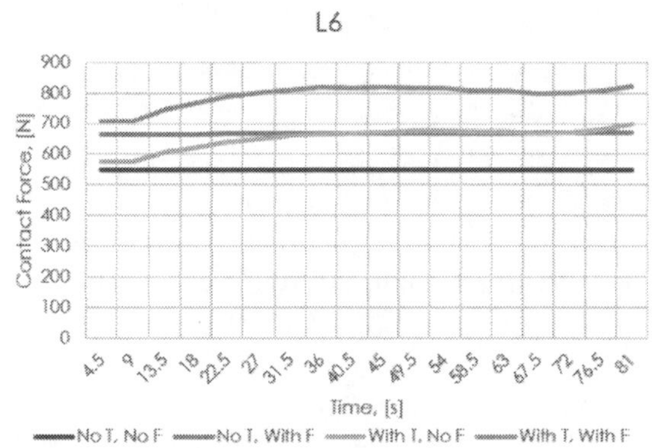

Figure 16. Contact forces for locator L6

Figure 17 shows the variation of the maximum deformation of the evaluated edge for the four loading cases, and Figure 18 for case I and II. It is important to take into account the maximum deformation of the evaluated edge because the edge is the reference element for measuring for the position dimension of the canal.

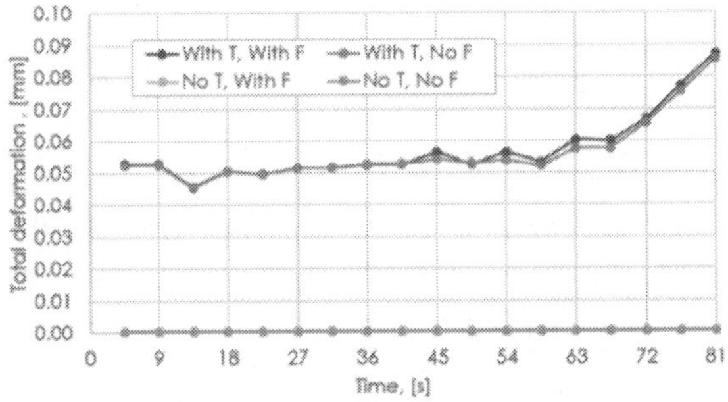

Figure 17. The maximum total deformation of the evaluated edge for Case I, II, III and IV

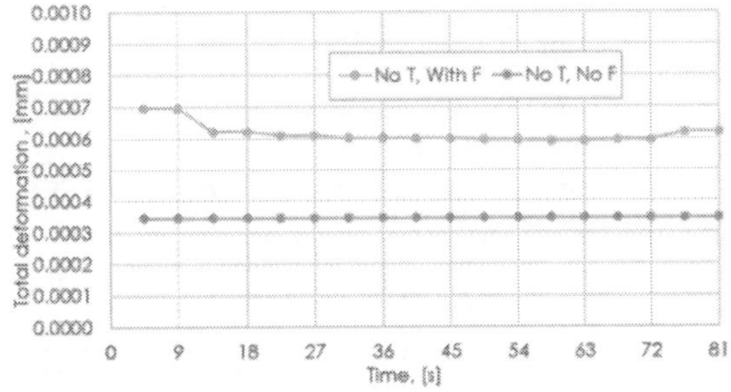

Figure 18. The maximum total deformation of the evaluated edge for Case I and II

4. Conclusions

From the graphical representations from above, it can certainly be inferred that the heat generated during cutting influences the size of the contact forces between the locators of the fixture and the workpiece subjected to machining.

This influence can be manifested in two ways:

- To determine an increase in the contact forces, leading to the destruction of the surfaces of the workpiece by the occurrence of permanent plastic deformations, especially in the case of locators with point and linear contact. This can be seen in Figures 11, 14, 16, when the contact forces between locators L1, L4 and L6 and the workpiece have higher values by approximately 25-30% than those obtained without taking into account the influence of the cutting temperature.
- To determine a decrease of the contact forces, resulting in the slipping of the workpiece on the locators or even in the loss of contact of the workpiece with the locators, which would lead to a compromise of the orientation scheme, with negative consequences on the machining precision. This situation can be seen in Figure 13, when the contact force corresponding to the locator L3 is almost equal to zero.

Clearly, both situations must be avoided.

It can be seen from Figures 17 and 18 that the total maximum deformation increases in the situation when the influence of the cutting temperature is taken into consideration, compared to the situations when it is ignored. If the deformation exceeds the size of the tolerance field of the quota, this could lead to the rejection of the workpiece.

In conclusion, following the simulation it was found that the heat released during the cutting process influences both the size of the contact forces between the locators of the fixture and the workpiece subjected to machining, as well as the machining precision. This is especially important when machining materials with a high conductivity coefficient (aluminium alloys, for example), with cutting regimes which can cause high cutting temperatures.

References

[1] Ivanov V, Mital D, Karpus V, Dehtiarov I, Zajac J, Pavlenko I and Hatala M 2017 Numerical simulation of the system “fixture–workpiece” for lever machining, *Int J Adv Manuf Technol* **91**(1) 79-90

[2] Ardelean M, Ardelean E, Heput T and Socalici A 2009 *Simulation results regarding to*

cinematic parameters of mechanical systems, Proceedings of the 13th WSEAS International Conference on Systems: Recent Advances in Systems: Mathematics and Computers in Science and Engineering, pp 254-259

[3] Miklos I Z, Alic C I and Miklos C C 2011 Experimental tests on indexing device with spherical elements (balls), *Annals of Faculty Engineering Hunedoara – International Journal of Engineering* **9**(4) 167-170

[4] Tirian G O 2011 Fuzzy system for control of a CNC cutting machine, *Annals of DAAAM & Proceedings of the 22th International Symposium* **22**(1) 601-602

[5] Cioata V G and Kiss I 2009, The machining error due to contact deformation of workpiece-fixture system, *Acta Tech Bull Eng* **2**(1) 33-36

[6] Cioată V G, Kiss I, Alexa V and Rațiu S A 2017 The optimization of the position and the magnitude of the clamping forces in machining fixtures, *IOP Conf. Ser.: Mater. Sci. Eng.* **200** 012015

[7] Miklos I Z, Miklos C C and Alic C I 2013 Temperature and relative speed influence on friction force from kinematic joints, *Annals of Faculty Engineering Hunedoara – International Journal of Engineering* **11**(3) 223-226

[8] Ardelean M, Ardelean E, Popa E, Josan A and Socalici A 2015 Research regarding to behavior on advanced plastic from rolling mills equipment, *IOP Conf. Ser.: Mater. Sci. Eng.* **106** 012035

[9] Wu B, Cui D, He X, Zhang D and Tang K 2016 Cutting tool temperature prediction method using analytical model for end milling, *Chinese Journal of Aeronautics* **29**(6) 1788-1794

[10] Pittalà G M and Monno M 2011 A new approach to the prediction of temperature of the workpiece of face milling operations of Ti-6Al-4V, *Applied Thermal Engineering* **31**(2-3) 173-180

[11] Rai J K and Xirouchakis P 2009 FEM-based prediction of workpiece transient temperature distribution and deformations during milling, *Int J Adv Manuf Technol* **42** (5-6) 429-449

[12] Stețiu G, Lăzărescu I D, Oprean C and Stețiu M 1994 *The theory and practice of cutting tools, vol. I (in romanian)*, Ed. Universității Sibiu, Romania

[13] Jonhson K L 1987 *Contact mechanics*, Cambridge University Press, Cambridge, USA

[14] Sanchez H T, Estrems M and Faura F 2006 Fixturing analysis methods for calculating the contact load distribution and the valid clamping regions in machining processes, *Int J Adv Manuf Technol* **29**(5) 426–435

IOP Conf. Series: Materials Science and Engineering **294** (2017) 012073 doi:10.1088/1757-899X/294/1/012073

Adaptive neuro fuzzy system for modelling and prediction of distance pantograph catenary in railway transportation

M Panoiu, C Panoiu and I L Lihaciu

Politehnica University of Timisoara, Electrical Engineering Department, Vasile Parvan str., no. 1-2, 300223 Timisoara, Romania

E-mail: manuela.panoiu@fih.upt.ro

Abstract. This research presents an adaptive neuro-fuzzy system which is used in the prediction of the distance between the pantograph and contact line of the electrical locomotives used in railway transportation. In railway transportation any incident that occurs in the electrical system can have major negative effects: traffic interrupts, equipment destroying. Therefore, a prediction as good as possible of such situations is very useful. In the paper was analyzing the possibility of modeling and prediction the variation of the distance between the pantograph and the contact line using intelligent techniques

1. Introduction

The problem of modeling is an important step in any activity that requires the identification of an engineering solution to a real problem. Solving a new problem involves: identifying and developing a model; using techniques to find an optimal solution based on the developed model (this can be done by simulation); using the optimal solution to the needs of the real problem

Modeling and simulation are the use of statistical or dynamic (timely) models to obtain data and decision making information. Modeling and simulation help reduce costs and increase product and system quality. The modeling and simulation technique is cheaper and more secure than real-world experiments in many practical situations. It is also sometimes closer to reality than traditional experiments and allowing the configuration of environmental parameters to the values encountered in the application domain. Simulation is (often) much faster than a real process, allowing for an efficient analysis of different situations (the most unfavorable case.) Once is identified and developed a model, this can be used many times.

The present paper uses modeling and simulation techniques based on fuzzy logic and neural calculus having the prediction of a time series, namely the variation in time of the catenary pantograph distance to electric locomotives. For the validation of the identified model, results of measurements of electrical quantities performed on a laboratory model were used. The catenary pantograph system has a dynamic oscillatory behavior and can be disturbed by certain turbulences during operation. It has been observed that at speeds of over 180 km / h the contact between the pantograph and the catenary can be interrupted and electric arcs can occur. There is no adequate mathematical model describing this behavior. Figure 1 shows the pantograph and figure 2 shows the aliasing of this movement.

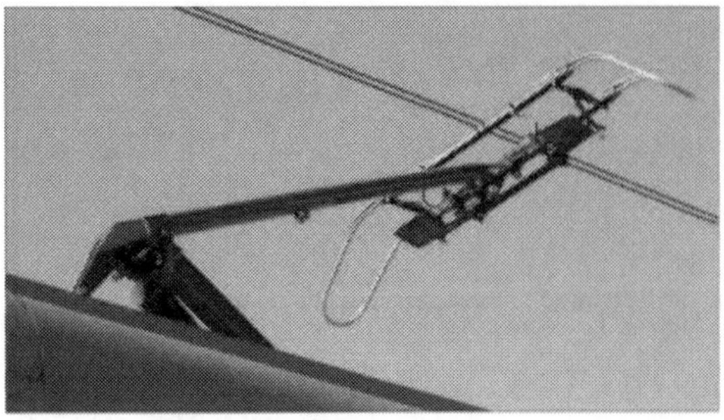

Figure 1. The pantograph

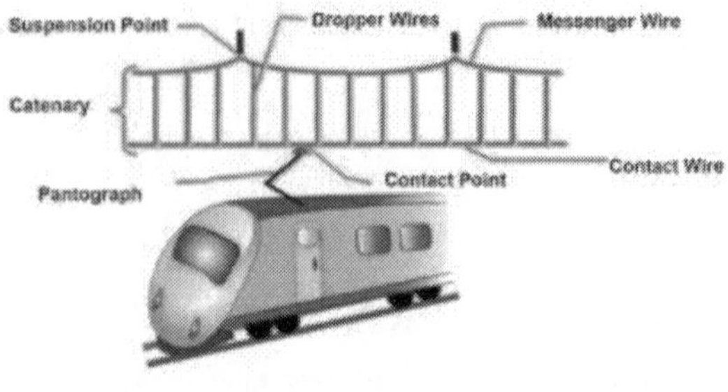

Figure 2. The aliasing of the pantograph catenary movement

The researches were made on a laboratory model for the pantograph realized at the scale 1:2 and controlled by a pneumatic system. The pantograph-contact wire assembly describes a trajectory in horizontal plan (zigzag) and another trajectory in vertical plan [1-3]. Measurements of the distance were made using a NI device by using optical tracking the pantograph motion and for acquiring the trajectory coordinates for statistics. Figure 3 shows the laboratory model of the pantograph used for tests.

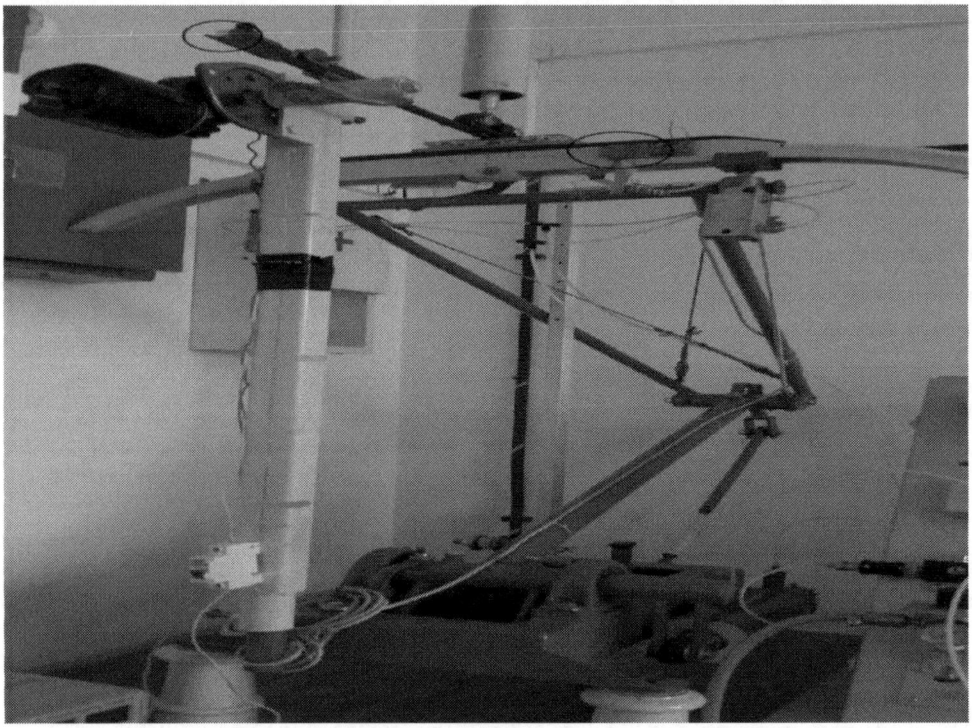

Figure 3. The laboratory model of the pantograph made at 1:2 scale

Folowing the measurements presented in [1-3] a variation of the pantograph – catenary distance in time was obtained ans figure 4 shows this variation.

2. Adaptive Neuro-Fuzzy Inference Systems

ANFIS hybrid neuro-fuzzy systems are equivalent adaptive neural networks from the functional point of view with Tagaki Sugeno fuzzy systems [4]. Compared to the classical fuzzy systems, ANFIS neuro-fuzzy systems have the capability to adapt during a learning process. In this way, by using an optimization method, the fuzzy membership functions can be adapted [5]. In order to explain the structure of a hybrid neuro-fuzzy system of ANFIS type it is considered a fuzzy system of Tagaki and Sugeno type of first order, which has two input values, *x* and *y*, and an output value, *f*. Fuzzy system rules base is considered made of two rules [5]:

- Rule 1: if x is A_1 and y is B_1 then $f_1 = p_1 \cdot x + q_1 \cdot y + r_1$
- Rule 2: if x is A_2 and y is B_2 then $f_2 = p_2 \cdot x + q_2 \cdot y + r_2$

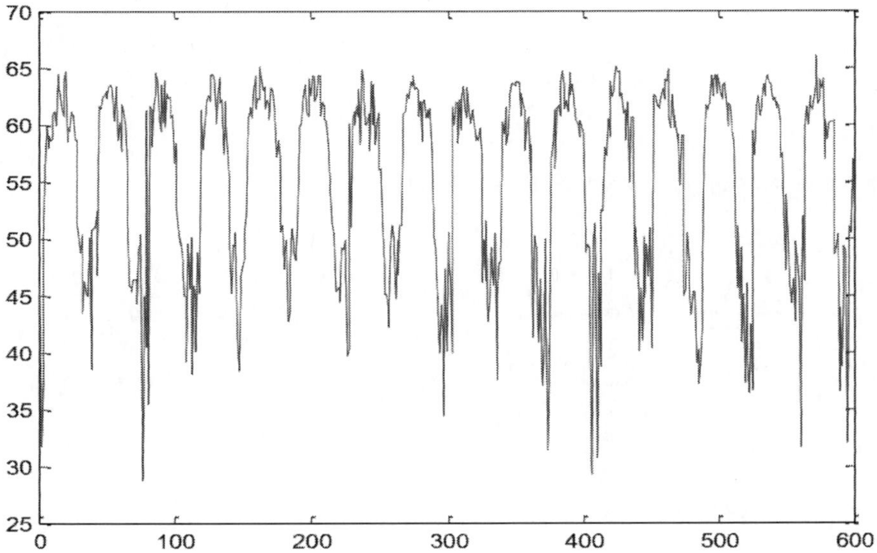

Figure 4. The variation in time of the pantograph – catenary distance

In Figure 5a is intuitively shown the way in which is made the reasoning in a fuzzy system of Tagaki and Sugeno type of first order, its output being computed as follow:

$$f = \frac{w_1 \cdot f_1 + w_2 \cdot f_2}{w_1 + w_2} = \overline{w_1} \cdot f_1 + \overline{w_2} \cdot f_2 \tag{1}$$

Figure 4b illustrates the structure of a considered neuro-fuzzy adaptive system. Equivalent ANFIS structure of the fussy system of first order is conditioned by 5 layers [4-8].

- Layer 1 has the output:

$$O_{1,i} = \mu_{A_i}(x), i = 1,2, \quad or \quad O_{1,i} = \mu_{B_{i-2}}(y), i = 3,4 \tag{2}$$

Parameters of this layer are called premise parameters.
If the Bell activation function is used, given by three parameters a_i, b_i, c_i then:

$$\mu_{A_i}(x) = \frac{1}{1 + \left[\left(\frac{x - c_i}{a_i}\right)^2\right]^{b_i}}, \tag{3}$$

or

$$\mu_{A_i}(x) = e^{-\left[\left(\frac{x - c_i}{a_i}\right)^2\right]^{b_i}} \tag{4}$$

where, $\{a_i, b_i, c_i\}$ is the parameter set of the membership functions referred to as premise parameters.

- Layer 2 has the output:

$$O_{2,k} = w_k = \mu_{A_i}(x) \times \mu_{B_j}(y), i = 1,2, j = 1,2, k = 1, \cdots 4 \tag{5}$$

Each node represents the activation value for a rule.

- Layer 3 has fixed nodes with the outputs:

$$O_{3,i} = \overline{w_i} = \frac{w_i}{w_1 + w_2}, i = 1,2 \tag{7}$$

- Layer 4 has adaptive nodes with the activation function

$$O_{4,i} = \overline{w_i} \cdot f_i = w_i(f_1 = p_1 \cdot x + q_1 \cdot y + r_1) \tag{8}$$

- Layer 5 is made of a fix node that computes the ANFIS overall output

$$f = O_{5i} = \sum_{i=1}^{2} \overline{w_i} \cdot f_i = \frac{\sum_{i=1}^{2} w_i \cdot f_i}{\sum_{i=1}^{2} w_i} \tag{9}$$

Learning within an ANFIS system is hybrid [3], [4].

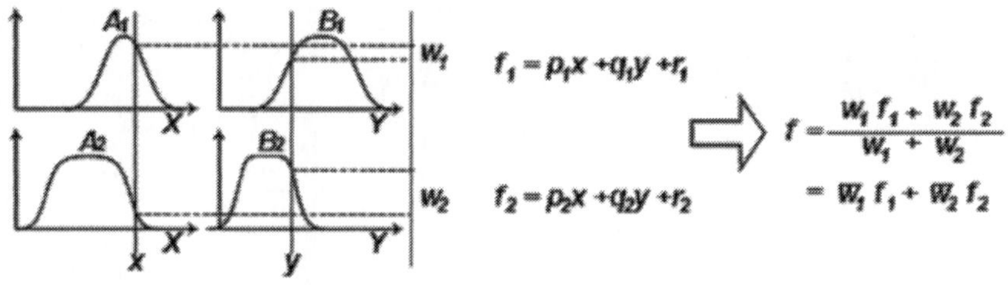

(a)

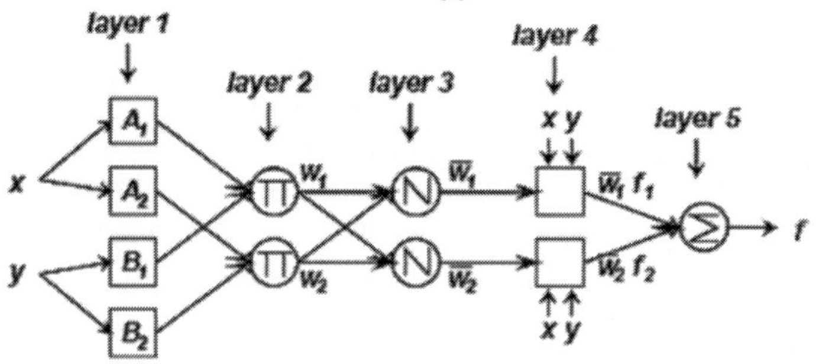

(b)

Figure 5. A two input first order Sugeno fuzzy model with two rules; (b) Equivalent ANFIS structure [3-7]

3. The ANFIS prediction of the pantograph catenary distance

In order to make a time series prediction, series values will be used up to the time T to predict the value at a time t + P [4-8]. The standard method for this type of prediction is to create a mapping from D samples, time-sampled, x (t - (D - 1)), ..., x (t - 1), x (t)) To predict a future value x (t + P).

For example, it can be choose D = 4 and Δ = P = 6. For each t, the input data for the anfis training will be represented by a vector with 4 dimensions:

$$W(t) = [x(t-18) \quad x(t-12) \quad x(t-6) \quad x(t)] \tag{10}$$

The Output data for training corresponds to the prediction of the trajectory:

$$S(t) = x(t+6) \tag{11}$$

For each t within the selected range, the input / output training data will be represented by a structure whose first component is the four-dimensional input, and the second component is the output s [4].

The structure of ANFIS system and the datasets used to train the system play a vital role in evaluating the performance of the system.

In order to generate the FIS (Fuzzy Inference System), three methods were used: grid partition, subtractive clustering, and FCM.

This paper is based on the design of Sugeno type ANFIS designed with all three methods and the usage of different datasets to train the system using MATLAB.

For training the ANFIS the measured data where used. The data were measured in the laboratory prototype [1].

3.1. ANFIS based on grid partition

If it is desired a prediction for the moment t= 4, samples d=3, sample time s=4:

trn= [x(t-8) x(t-4) x(t)]

and corresponds to the prediction for the output value y=x(t+4) .

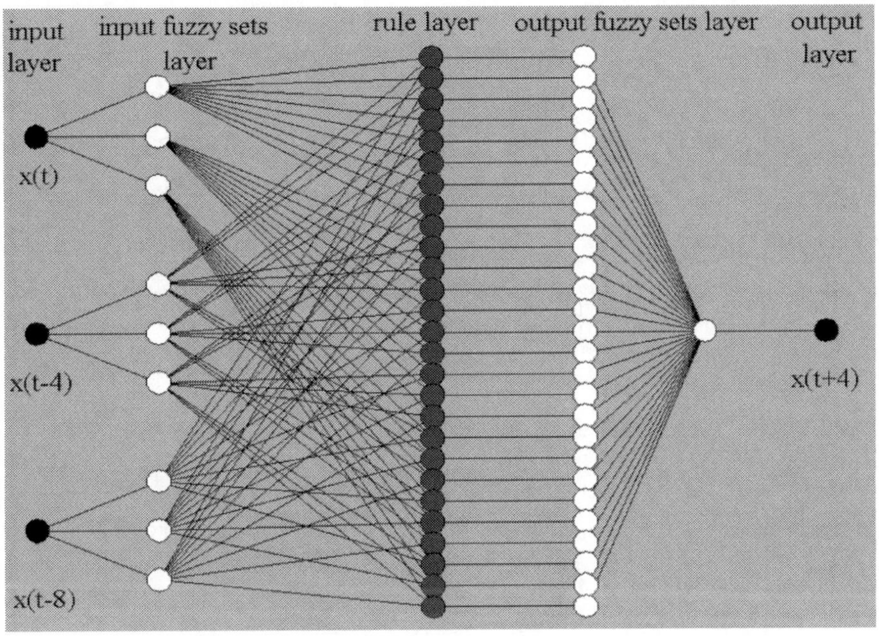

Figure 5. The ANFIS structure based on grid partition

3.1.1. ANFIS based on grid partition. Results. The grid partition method, is the default method used in Matlab for FIS (Fuzzy Inference System) generation. The method generates a Sugeno-type FIS structure used as initial conditions (initialization of the membership function parameters) [14].

The results obtained after the ANFIS is trained can be seen in the figure 6. Thus, in figure 6 a can be seen the output of the ANFIS when the input data are the training data. As mentioned from the 600 samples, 450 were used to train ANFIS and the other 150 samples were used for testing. Thus, Figure 6b shows the output of the ANFIS for the test data and Figure 6c shows the system output for the entire dataset, that is, the 600 samples. Figure 6d shows the regression coefficients of the ANFIS: for the trainnig dataset, for the testing dataset and also for the entire dataset.

3.2. ANFIS with subtractive clustering

3.2.1. Principles of Subtractive Clustering For Rules Selection. In a fuzzy inference system, the number of rules from a fuzzy inference system is decided by an expert familiar with the system to be modeled [9], [11]. The number of membership functions for the input variables is chosen empirically [10], [11].

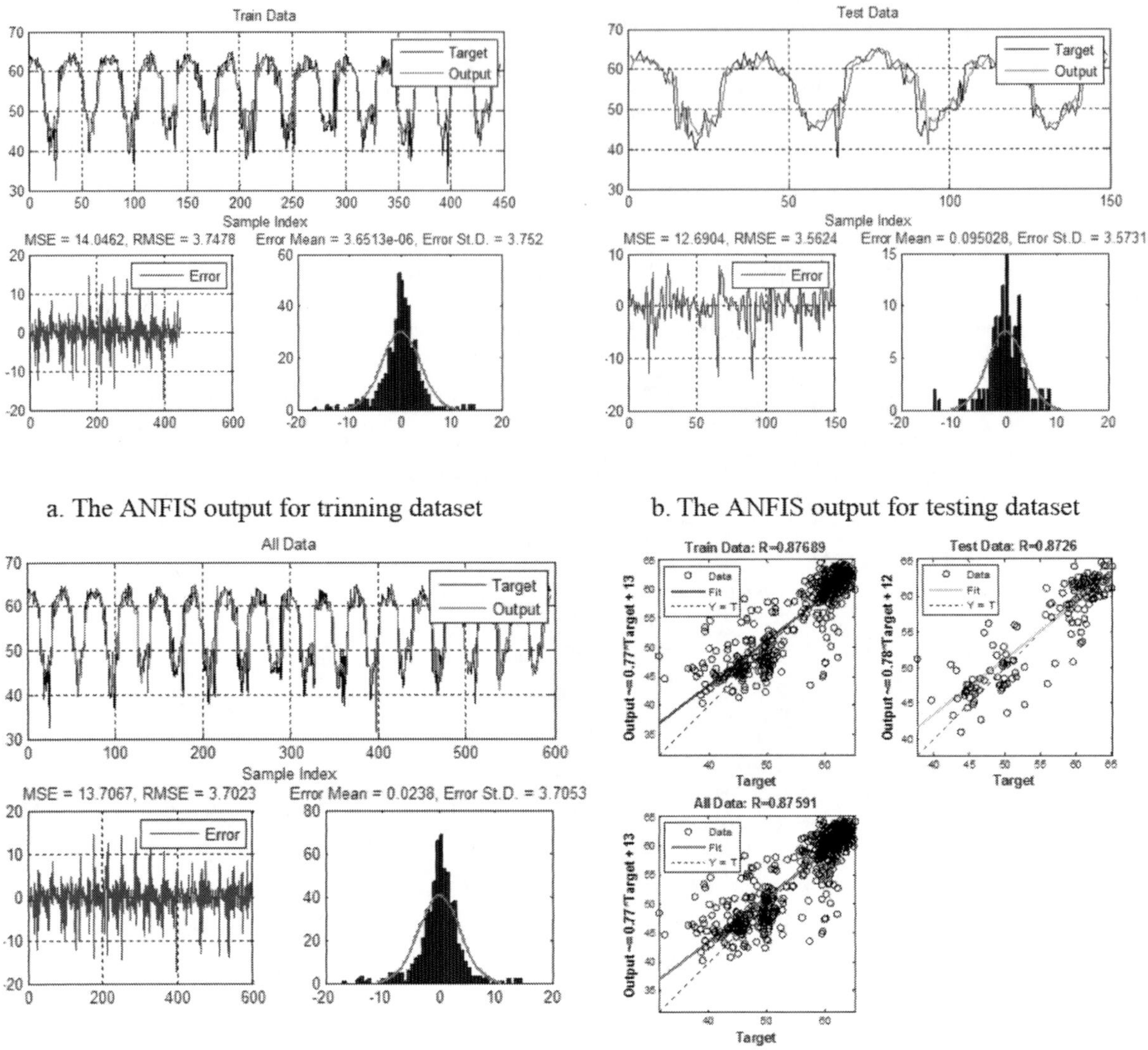

a. The ANFIS output for trinning dataset

b. The ANFIS output for testing dataset

c. The ANFIS output for all data

d. The regression coefficients

Figure 6. The results of ANFIS modelling and prediction with grid partition method

It is very difficult to establish the rules manually in order to obtain the precision with the minimized number of Membership Functions(MF), when the number of rules are larger than 3 [11]. The subtractive clustering algorithm is an approach to the synthesis of ANFIS, which estimates the cluster number and the cluster location automatically. In subtractive clustering algorithm, each sample point is seen as a potential cluster center [11].

3.2.2. ANFIS with subtractive clustering. Results. The results obtained after the ANFIS is trained can be seen in the figure 7. In figure 7a can be seen the output of the ANFIS for the training data. As in the above section, from the 600 samples, 450 were used to train ANFIS and the other 150 samples were used for testing. Figure 7b shows the output of the ANFIS for the test data and Figure 7c shows the system output for the entire dataset, the 600 samples. Figure 7d shows the regression coefficients of the ANFIS: for the trainnig dataset, for the testing dataset and also for the entire dataset.

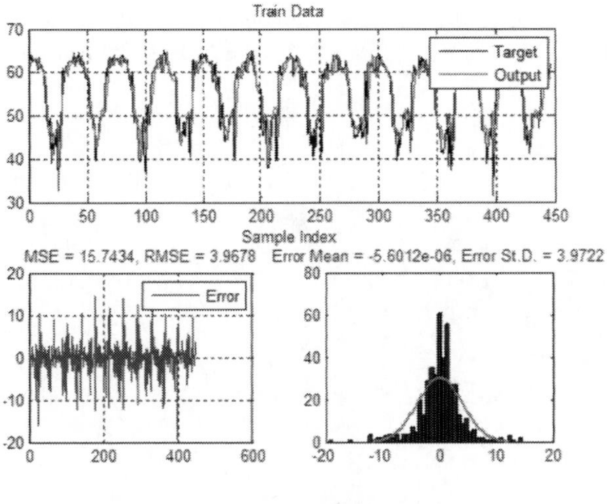

a. The ANFIS output for trinning dataset

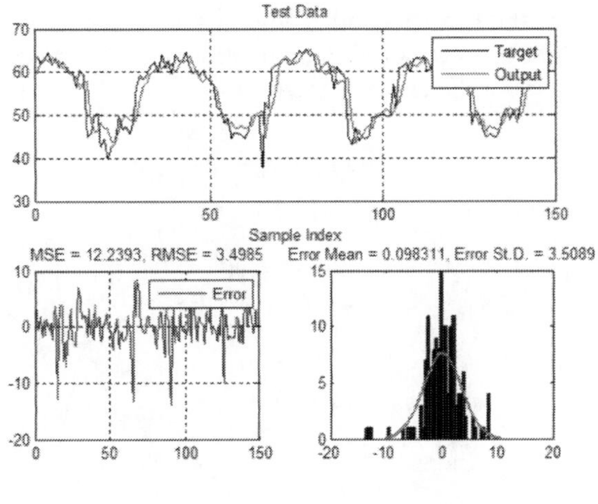

b. The ANFIS output for testing dataset

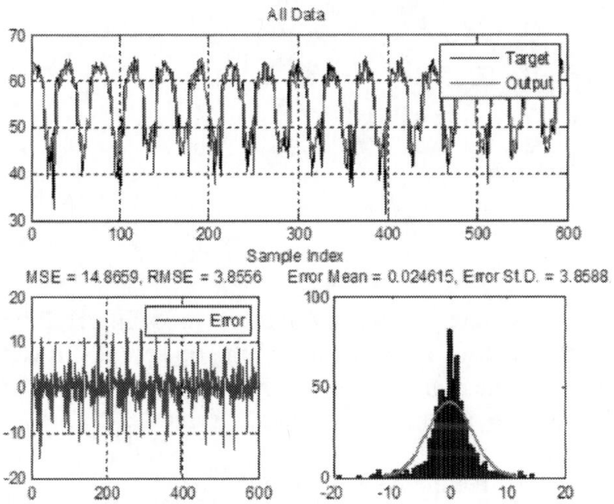

c. The ANFIS output for all data

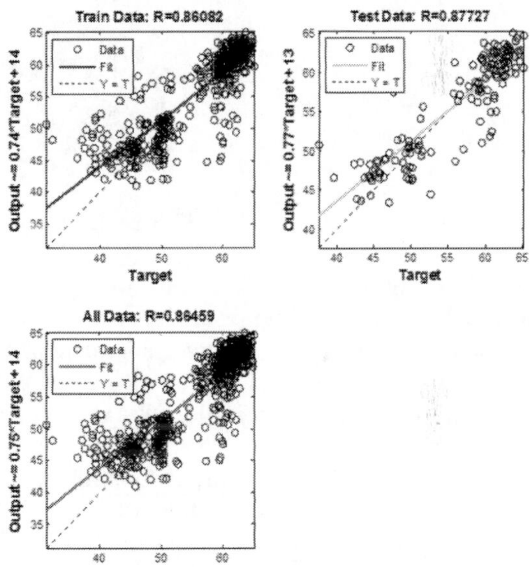

d. The regression coefficients

Figure 7. The results of ANFIS modelling and with subtractive clustering method

3.3. ANFIS with Fuzzy c–means clustering method

The FCM is a data clustering algorithm [12] in which each data point belongs to a cluster to a degree specified by a membership grade. FCM partitions a collection of n vector X_i , i= 1,2,...n , into C fuzzy groups, and finds a cluster center in each group such that a cost function of dissimilarity measure is minimized [13] as can be seen in Figure 8.

3.3.1. ANFIS with Fuzzy c–means clustering. Results. The results obtained after the ANFIS is trained can be seen in the figure 9. As in the other two methods of generating FIS, Figure 9a shows the output of the ANFIS for the training data. The 600 Data Samples were split in the same way like in the previous two methods, 450 for training and e 150 for testing. Figure 9b shows the output of the ANFIS for the test data and Figure 9c shows the system output for the entire dataset, the 600 samples. Figure 9d shows the regression coefficients of the ANFIS: for the training dataset, for the testing dataset and also for the entire dataset.

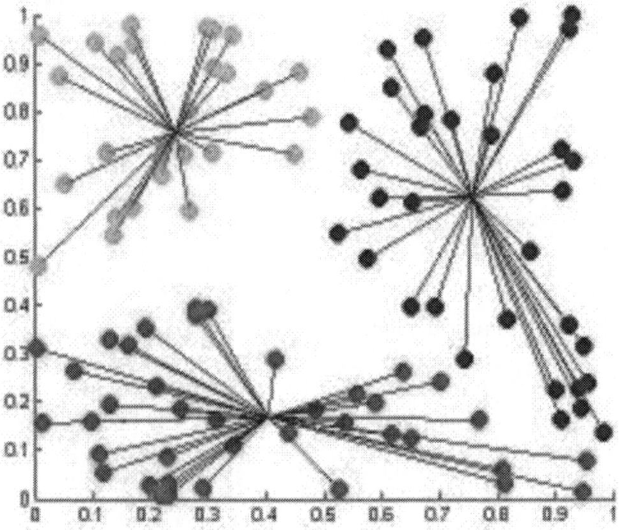

Figure 8. Assigning data points to clusters

4. Conclusions

From the analysis of the results obtained from the simulations we can find that the modeling and the prediction of the distance between the catenary and the pantograph was made with good results. It can be observed from Table 1 the Error Mean, the RMSE and the regression coefficients.

Analyzing the data in Table 1, it can be seen that all three methods offer good results.

Table 1. The results of simulation using the three methods of ANFIS design

ANFIS design method	Data set	RMSE	Error mean	Regression coefficients
Grid partition (3 Fuzzy sets)	Training set	3.7478	3.6513×10^{-6}	0.87689
	Testing set	3.5624	0.095028	0.8726
	All data	3.7023	0.0238	0.87591
Subtractive clustering	Training set	3.9678	-5.6×10^{-6}	0.86082
	Testing set	3.4985	0.098311	0.87727
	All data	3.8556	0.024615	0.86459
Fuzzy C-means (4 clusters)	Training set	3.4911	8.03×10^{-5}	0.89415
	Testing set	3.5501	0.30474	0.87716
	All data	3.506	0.076374	0.88972

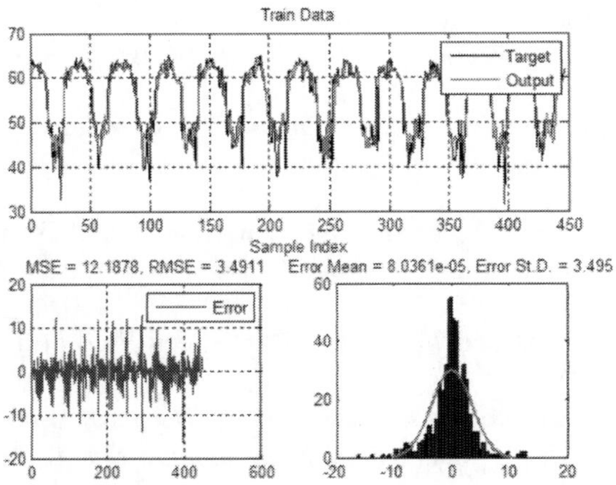

a. The ANFIS output for trinning dataset

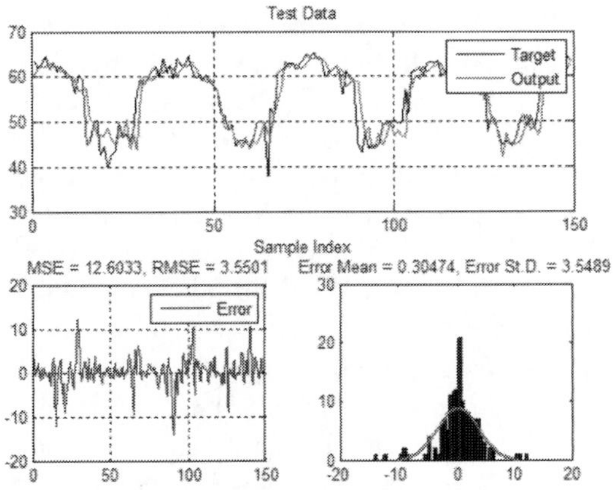

b. The ANFIS output for testing dataset

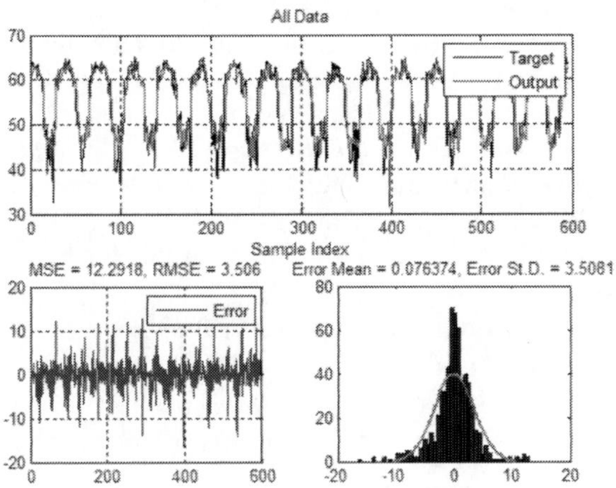

c. The ANFIS output for all data

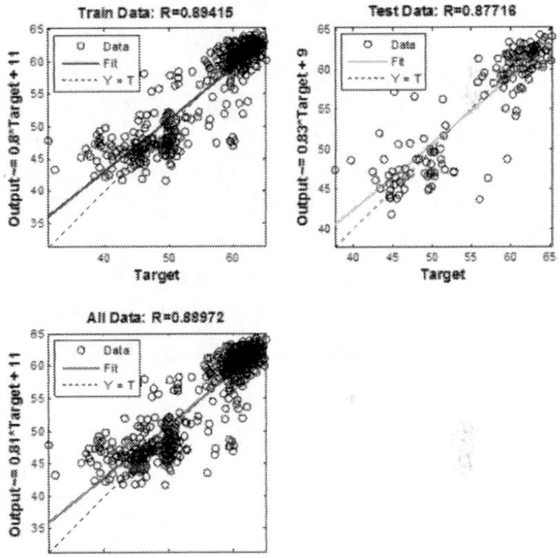

d. The regression coefficients

Figure 9. The results of ANFIS modelling and prediction with fuzzy C-means clustering method

Acknowledgement

This work was supported by a grant of the Romanian National Authority for Scientific Research and Innovation, CNCS/CCCDI – UEFISCDI, project number PN-III-P2-2.1-BG-2016-0473, within PNCDI III

References

[1] Rusu-Anghel S and Ene A 2017 Influence of typical faults over the dynamic behavior of pantograph-catenary contact force in electric rail transport, *IOP Conf. Ser.: Mater. Sci. Eng.* **200** 012056

[2] Rusu-Anghel S, Miklos C and Lihaciu IC 2014 Adaptive Control of Pantograph-Catenary Interaction Force, *WSEAS Transactions on Systems* **13** 708-719

[3] Rusu-Anghel S, Miklos C and Topor M 2012 *Chaos theory based control of contact force in electric railway transportation system*, 1st International Conference on Environment and

Electrical Engineering (EEEIC), 18-25 May, Venice, Italy

[4] Jang J S R 1993 Adaptive – Network – Based Fuzzy Inference System, *IEEE Transactions on Systems, Man and Cybernetics* **23**(3) 665-685

[5] Panoiu M, Ghiormez L and Panoiu C 2014 *Adaptive Neuro-Fuzzy System for Current prediction in Electric Arc Furnaces*, 6th International Workshop Soft Computing Applications SOFA 2014, Timisoara, Romania, July 24-26, pp 423-437

[6] Pérez E C, Algredo-Badillo I and Rodríguez V H G 2012 Performance Analysis of ANFIS in short term Wind Speed Prediction, *International Journal of Computer Science Issues* **9**(5) 94-102

[7] Valčić M, Antonić R and Tomas V 2011 ANFIS Based Model for Ship Speed Prediction *Brodogradnja* **62**(4) 373-382

[8] Soto J, Melin P and Castillo O 2013 *A new Approach for Time Series Prediction using Ensembles of ANFIS Models with Interval Type-2 and Type-1 Fuzzy Integrators*, IEEE Conference on ComputationalIntelligence for Financial Engineering & Economics, Singapore, April 16-19, pp 68-73

[9] Lim E A and Jayakumar Y 2008 A study of neuro-fuzzy system in approximation-based problems, *Matematika* **24**(2) 113-130

[10] Mandal S N, Choudhury and Bhadra Chaudhuri S R 2012 In Search of Suitable Fuzzy Membership Function in Prediction of Time Series Data, *International Journal of Computer Science Issues* **9**(3) 293-302

[11] Hiremath S M, Patra S C and Mishra A K 2012 *ANFIS with Subtractive Clustering-Based Extended Data Rate Prediction for Cognitive Radio*, 5th International Conference on Computers and Devices for Communication, Rourkela, Odisha, India

[12] Basir O, Zhu H and Karray F 2003 Fuzzy Based Image Segmentation, Fuzzy Filters for Image Processing, *Studies in Fuzziness and Soft Computing* **122**

[13] Moertini V S 2002 introduction to five data clustering algorithms, *Integral* **7**(2) 87-96

[14] ***https://www.mathworks.com/help/fuzzy/genfis1.html

IOP Conf. Series: Materials Science and Engineering **294** (2017) 012074 doi:10.1088/1757-899X/294/1/012074

Study of a phase-to-ground fault on a 400 kV overhead transmission line

A Iagăr, G N Popa and C M Diniş

Politehnica University of Timisoara, Department of Electrical Engineering and Industrial Informatics, 5 Revolution Street, Hunedoara, 331128, Romania

E-mail: angela.iagar@fih.upt.ro

Abstract. Power utilities need to supply their consumers at high power quality level. Because the faults that occur on High-Voltage and Extra-High-Voltage transmission lines can cause serious damages in underlying transmission and distribution systems, it is important to examine each fault in detail. In this work we studied a phase-to-ground fault (on phase 1) of 400 kV overhead transmission line Mintia-Arad. Indactic® 650 fault analyzing system was used to record the history of the fault. Signals (analog and digital) recorded by Indactic® 650 were visualized and analyzed by Focus program. Summary of fault report allowed evaluation of behavior of control and protection equipment and determination of cause and location of the fault.

1. Introduction

At present, power quality is an area of growing interest in electric utilities, due to continuous increase of complexity and interconnectivity degree between power systems, as a result of electricity markets globalization, electricity being traded at national or regional level. All of these aspects require more and more complex operating conditions, related to system control and mutual support [1].

Power quality refers to following electromagnetic phenomena that characterize the voltage and current: transients, voltage dips, voltage swells, interruptions, voltage fluctuations, imbalance, waveform distortion and power frequency variations. These events differ in duration, harmonic spectrum and waveform magnitude, the most severe being voltage interruptions and dips, because industrial processes are very sensitive to these type of disturbances [2], [3].

Protection relaying is very closely related to power quality. In the presence of harmonic distortion and power frequency variations, protective relays can trip incorrectly. On the other hand, protective relaying fault clearing result in voltage sags, that affect power quality [4-8].

The faults that occur on High-Voltage (HV) and Extra-High-Voltage (EHV) transmission lines can cause serious problems in underlying transmission and distribution systems. This imposes severe conditions for protection and control systems, which must instant isolate the faults to maintain the system stability [1].

On HV and EHV transmission lines the majority of faults are typically single-phase-to-ground and transient (e.g., flashover across an insulator), being cleared using automatic reclosing. In this case, the line is disconnected in order to allow the arc to extinguish, and after the dead time, necessary to sufficiently deionize the arc path, the line can be reclosed to restore normal service [9].

Auto-reclosing can be used, also, to cleared the semi-permanent faults (e.g., a small tree branch falling on the line), but with a delayed action. If the fault persists after the first trip and closure, is used double or triple-shot reclosing before the line will be put out of service [9]. Practice shows that the vast majority (almost 80%) of the faults are cleared after the first trip [1], [9]. 10% of the faults are cleared in the second reclosure, which is made after a time delay, 3% of the faults are cleared in the third reclosure and about 7% are permanent faults [9-11]. In the case of permanent fault, the reclosing relay is lockout, damaged part of the line being energized only after the fault has been removed.

2. Protections of 400 kV overhead transmission lines

In the case of 400 kV overhead transmission lines (OTL), basic protection against single-phase and poly-phase faults is provided at each terminal with two independent distance protections, which are of different construction types, completed with one communication channel to the opposite terminal, on separate paths [12], [13].

Each distance protection will send (simultaneously with the trigger signal for its own breaker), a signal through the communication channel, that will be received by the corresponding distance protection to perform one of the following functions: extending of zone 1 or accelerating of zone 2; fast triggering, with local control of the start and direction; in both cases, triggering will be selective on phases [13].

In the zone 1 extension method, zone 1 reach is normally set to cover 120% of line [14]. When fault occurs anywhere in the protected line, breakers at both terminals are tripped simultaneously without time delay and reclosed. Extending of zone 1 is done by means of the auto-reclosing relay under the following conditions:

- only for a certain type of short-circuit (typically only for single-phase short-circuits);
- only before or after the auto-reclose cycle (typically only before the auto-reclose cycle);
- coordinated with the auto-reclosing on the opposite terminal, for the fast clearing of short-circuits over the entire length of the protected line.

A contact of auto-reclosing relay is used to reset the zone 1 to 80% of line. If the fault is permanent, the breakers will be tripped according on the respective zone timers.

The major protections of analyzed 400 kV OTL are: line distance protection terminal REL 521 (group 1 of protections), distance protection relay LZ96a (group 2 of protections), auto-reclosing relay REXA 101 (group 2 of protections), distributed busbar protection REB 500 (group 1 of protections) and synchro-check relay SPAU 140 C.

2.1. Line distance protection terminal REL 521

REL 521 terminal has the following features [15]:

- fast and reliable fault detection, by simultaneous measurement of phase-to-phase and phase-to-earth loop impedances; numerical measuring elements are individual for each type of fault and each distance zones;
- minimum operating time is 13 ms;
- excellent possibilities for extensive combination of different options, based on multiprocessor design;
- numerical filtering and measuring technique ensures a correct performance during CT saturation and CVT transients;
- extensive self-supervision with fault diagnostics;
- detailed reports for the last ten disturbances (with up to 150 events for each disturbance).

Among the most important functions of REL 521 could be mentioned: distance protection (ZM); phase selection logic (PHS); power swing detection (PSD); power swing logic (PSL); pole slip protection (PSP); scheme communication logic for distance protection functions (ZCOM); current reversal and WEI logic for distance protection (ZCAL); automatic switch onto fault logic (SOTF); local acceleration logic (ZCLC); teleprotection; instantaneous overcurrent protection (IOC); time delayed overcurrent protection (TOC); thermal overload protection (THOL); stub protection (STUB); breaker failure protection (BFP); time delayed undervoltage (TUV) and time delayed overvoltge protection (TOV); disturbance report (DRP); disturbance recorder (DR); event recorder; fault locator

(FLOC); trip value recorder (TVR); autorecloser 1- and/or 3-phase, single or double circuit breakers [15].

REL521 terminal operates both for phase-to-phase and phase-to-earth fault, by distance protection function. It is possible single, two or three-phase tripping, through the separate phase selection function.

The distance protection zones can operate, independently of each other, in directional mode (forward - zones 1, 2, 4, or reverse - zone 3) or non-directional mode (zone 5). Distance protection zones 1, 2 and 3 can issue phase selective start and trip signals. Zone 5 can be used as a check zone, or as a time delayed zone, time delay being no more than 100 ms [15].

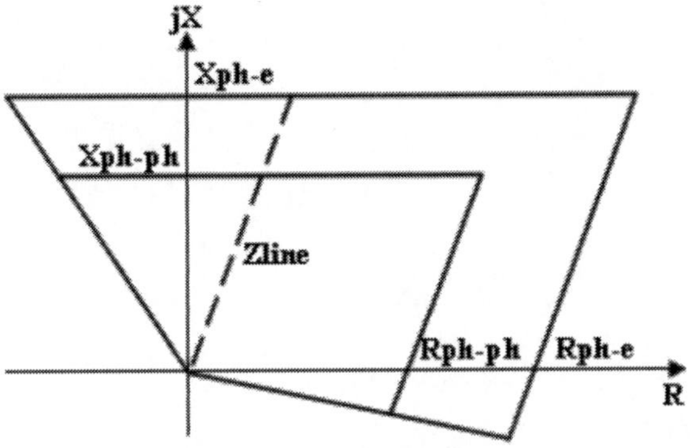

Figure 1. Operating characteristic for one distance protection zone (forward direction) [15]

Figure 1 presents the operating characteristic for one distance protection zone (in forward direction), where: Xph-e represents the reactive reach in the case of phase-to-ground (earth) faults; Xph-ph represents the reactive reach in the case of phase-to-phase faults; Rph-e represents the resistive reach in the case of phase-to-ground (earth) faults; Rph-ph represents the resistive reach in the case of phase-to-phase faults; Zline represents the line impedance [15].

Independent reactive reach setting for phase-to-phase and for phase-to-ground measurement secures high selectivity in networks. Simplified setting parameters reduce the complexity of necessary setting procedures.

2.2. Distance relay LZ96a

The distance relay type LZ96a is designed for the high-speed, selective protection of HV and EHV networks with effectively earthed neutral. LZ96a can be used to protect either overhead lines or cables, being ideal in networks with high short circuit power and critical stability conditions. It can detect all kinds of short circuit and offers important advantages: flexible and modular design, very short tripping time (20 ms) and very high accuracy [16].

The distance protection LZ96 uses the phase comparison principle for the measurement of the fault impedance. LZ96 has two measurement units with MHO characteristic. The first unit has three measurement zones. The first zone of this unit operates in permissive underreach trip transfer scheme and the second and third zones are for backup. The second unit has a measurement zone in permissive overreach trip transfer scheme that is used for the weak infeed scheme too [17].

LZ96a relay has two starting units: a main starting unit that operates simultaneously with the measurement units, and an auxiliary starting unit which operates with the second measurement unit as a component of the weak infeed scheme [16], [17].

The measurement principle also consider the system source impedance Zs, in order to detect ground faults of considerable impedance, where the characteristic inclines 14.4° approximately and expands towards the resistive axis [17].

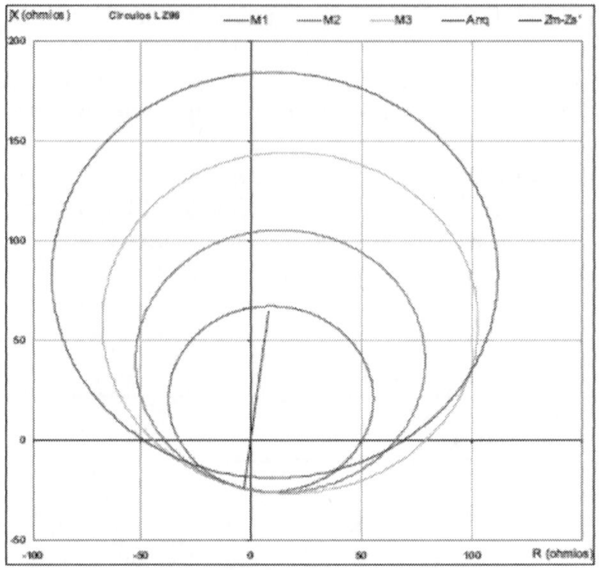

Figure 2. Operating characteristic for LZ96a [16]

2.3. Auto-reclosing relay REXA 101

REXA 101 auto-reclosing relay is used to restore the service of transmission lines when a circuit breaker is tripped single- or three-phase, due to a line fault [18].

The features of REXA 101 relay are: automatic high speed or delayed breaker reclosing; allows sequential reclosing of breakers in multibreaker arrangements; breaker position memory; the auto-reclosing function is performed by use of a pre-programmed single-chip microcomputer; allows extended shot 1 dead time when line protection communication link is out of service; prepared for coordination with synchronism and dead line check relay; reclaim time adjustable 10-180 s.

This relay allows setting three different time delays for dead time of the first single-phase reclosing, the first three-phase reclosing and the second three-phase reclosing [18]. Dead time settings are: single-phase, shot 1: 0.4-1.9 s (steps of 0.05 s); three-phase, shot 1: 0.25-60 s (steps of 0.05 to 1 s); three-phase, shot 2: 2-240 s (steps of 1 s) [18].

2.4. Distributed busbar protection REB 500

Distributed busbar protection REB 500 detects all bus and feeder faults for all type of system earthing. It operates selectively for all faults in the protected zone [19].

Among the main features of REB 500 could be mentioned: low-impedance busbar protection; stub and T-zone protection; high stability through fault, even in the case the current transformers are saturated; short tripping times; fully numerical signal processing; complete self-supervision; binary logic and timer in the bay unit; event recording; disturbance recording for power system currents [19].

Two independent measurement algorithms (stabilized differential current and directional current comparison) ensure a high functional reliability of REB 500, even in the case the busbar replica is not valid [19].

2.5. Bay control terminal REC 580

Bay control terminal REC 580 is used for the automation of complex HV substations [20]. REC 580 measures the instantaneous currents (I) and voltages (U) and computes the RMS values of U and I, active, reactive and apparent power (single and three-phase), energy values, power factor and frequency [20].

The most important functions of REC 580 are: control of circuit-breakers, isolators, earthing switches and fast earthing switches; switching between busbars; supervision of switchgear operating times; interlocking between bays; single and/or three-phase multi-shot auto-reclosing; synchrocheck; voltage check [20]. Other features of REC 580 consist of: event recorder (with resolution of 1 ms); oscillation suppression; circuit-breaker operations counter; self-monitoring [20].

2.6. Synchro-check relay SPAU 140 C

Synchro-check relay SPAU 140 C is an integrated microprocessor-based voltage measuring relay designed to be used for checking the conditions for circuit breaker closing [21].

SPAU 140 C incorporates two identical stages (which operate as independent units), with two functions: synchro-check function and voltage-check function.

The synchronism is checked before the protections send the closing signal to the circuit breaker. Synchro-check function finds the instant when the voltages on both sides of the circuit breaker are in synchronism, i.e. the voltages have the same frequency, are in phase and with similar magnitude. When the frequency, phase angle and voltage conditions are fulfilled, the duration of the synchronizing conditions is checked so as to ensure that they will still be met when the contacts of the circuit breaker close. The delay between the moment when the closing signal is issued and the circuit breaker is closed is determined on the basis of the frequency and phase difference measured, being about 50-250 ms [21].

The voltage-check function checks the energizing direction, if one or both sides of the circuit breaker to be closed are dead [21].

2.7. Indactic® 650 fault analyzing system

The basic functions of a station (or substation) automation system are monitoring, control, recording and protection.

Before the occurrence of a fault in the protected area of analyzed OTL, the fault recorder device, Indactic® 650, capture all the information (both analogical and digital: voltages, currents, signaling, alarms, etc.) related to the fault and permits its reconstruction and analysis. Indactic® 650 has a compact structure, with integrated current and voltage transformers. It provides impedance-based fault locating with a precision of under ±2% in relation to the line length and has a complete independence from other equipment, such as protections. Moreover, Indactic® 650 allows the investigation of combined events [22].

Focus for Windows program allows the visualization and analysis of the signals (analog and digital) recorded by Indactic® 650. The phasor diagrams of voltages and currents, and their harmonic analysis can be achieved using Focus program, also [23].

Further is analyzed a phase-to-ground fault, on phase 1, of 400 kV OTL Mintia-Arad, in Mintia station.

3. Summary of the phase-to-ground fault report

3.1. Values of pre-fault signals

Time period allocated to record the pre-fault signals was 99.13 ms. In normal operation of analyzed OTL (Figure 3, t=-57 ms), the RMS values of phase voltages and phase currents are close.

The phase angle deviation between the voltages is approximately 120°; also, the phase angle deviation between the currents is approximately 120°.

It is noted a symmetric loading on the three phases of the analyzed overhead transmission line. The homopolar voltage and the homopolar current have low RMS values: U0=2.415 kV, I0=13.69 A.

At t=-16 ms (Figure 4), phase angle deviation of the currents are very different from the normal operation (the phase angle deviation between IL2 and IL1 is only 20°).

It is noticed the decrease of the voltage and a very high increase of the current, on phase 1 of analzyed OTL. The current on phase 3 decreases compared to normal operation. Homopolar voltage and homopolar current have a huge increase compared to normal operation (U0=329.5 kV, I0=3.88 kA). All this indicates the incipient stage of a phase-to-ground fault on phase 1.

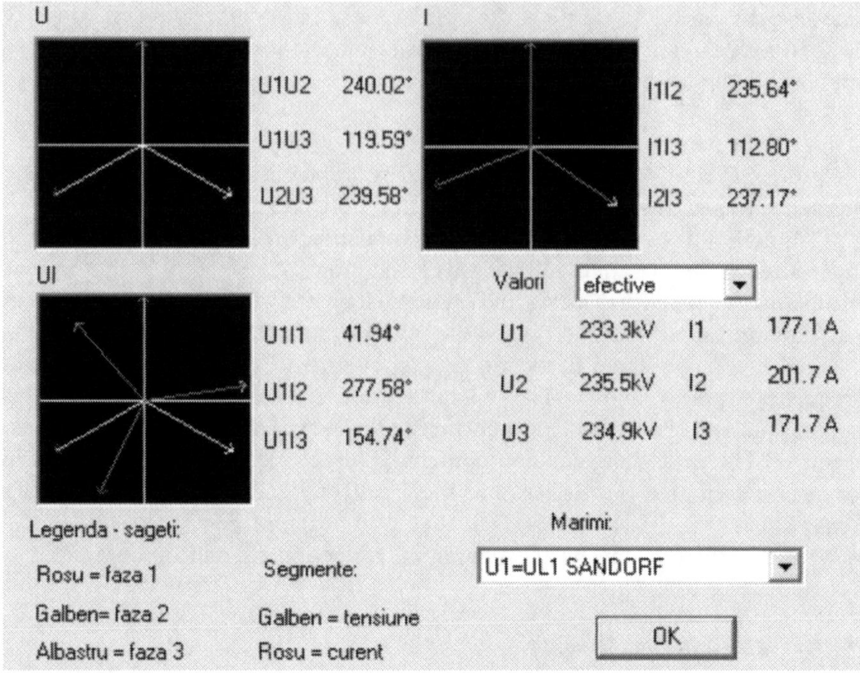

Figure 3. Marker 1: t=-57 ms (pre-fault). Phasor diagrams of voltages and currents

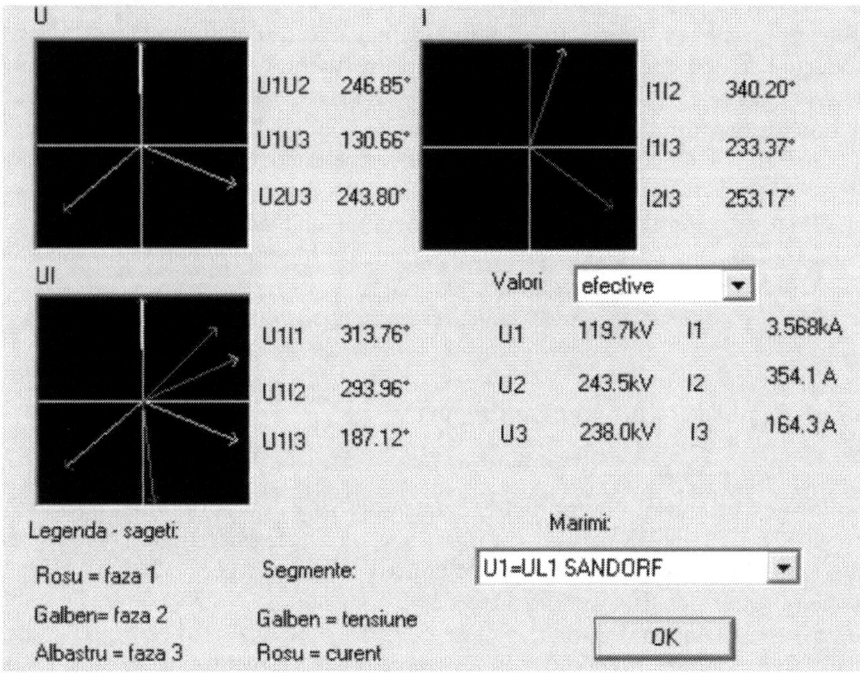

Figure 4. Marker 2: t=-16 ms (pre-fault). Phasor diagrams of voltages and currents

3.2. Values of fault signals

In fault operation (Figure 5, t=4 ms), one can notice a significant increase of the homopolar quantities (U0=400.4 kV, I0=5.245 kA). Phase angle deviations of the currents are very different from the normal operation.

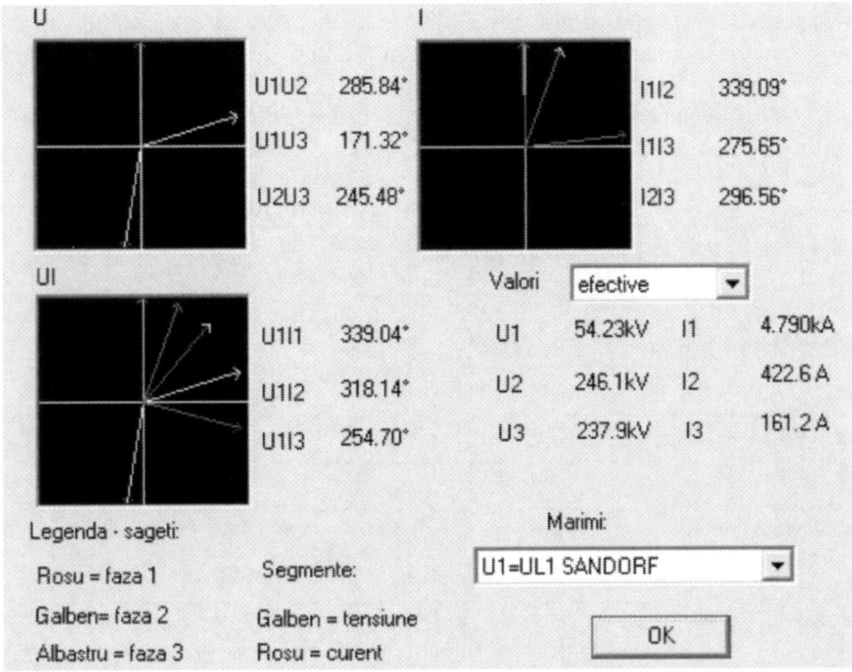

Figure 5. Marker 3: t=4 ms (fault). Phasor diagrams of voltages and currents

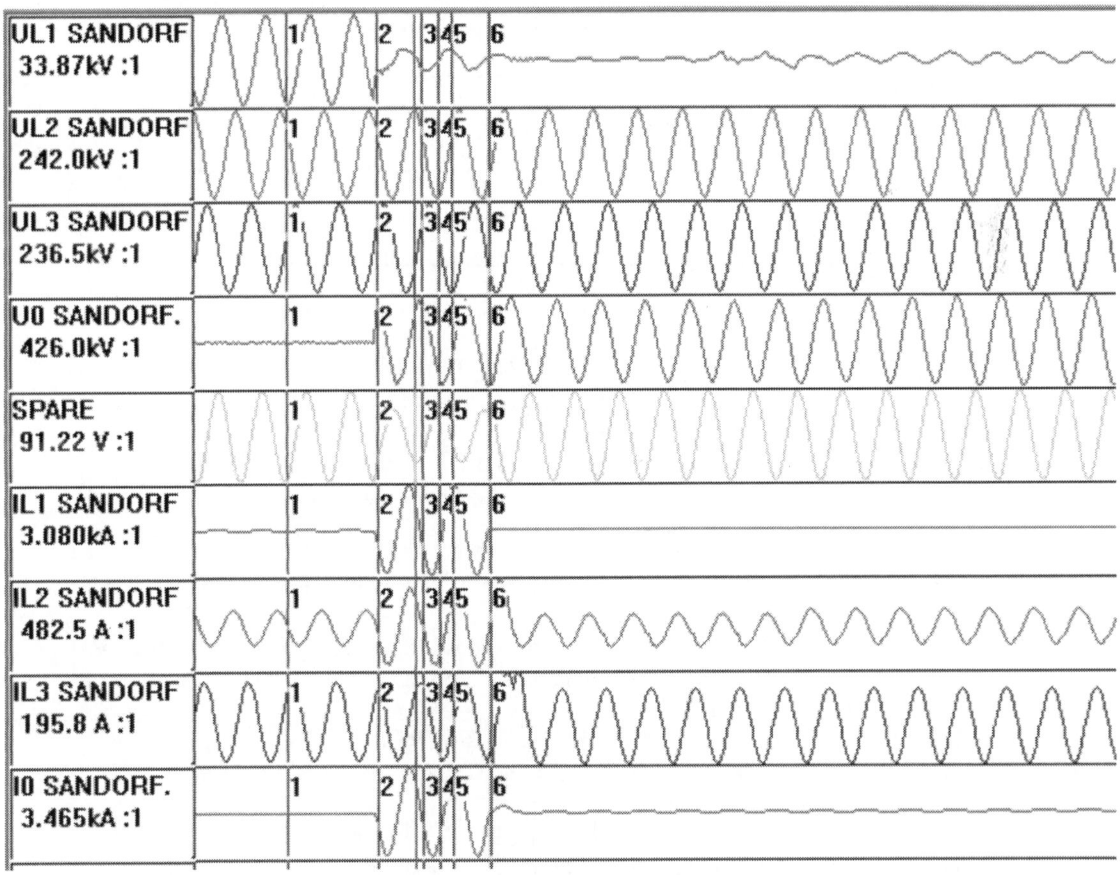

Figure 6. Analog signals recorded in the time period t=-99.24 ms...222 ms

At t=4 ms (Figure 5), voltage on phase 1 has a pronounced decrease and current has an even higher value than the previous moment. The impedance of OTL has decreased significantly from normal operation, which will cause the start of the distance protections.

Figure 6 and 7 show the analog and digital signals recorded in the time period t=-99.24 ms...222 ms (trigger at 16:20:07:769; number of samples: 1008; measuring time: 838 ms; marker 1: t=-57 ms, marker 2: t=-16 ms, marker 3: t=4 ms, marker 4: t=10 ms, marker 5: t=15 ms, marker 6: t=33 ms).

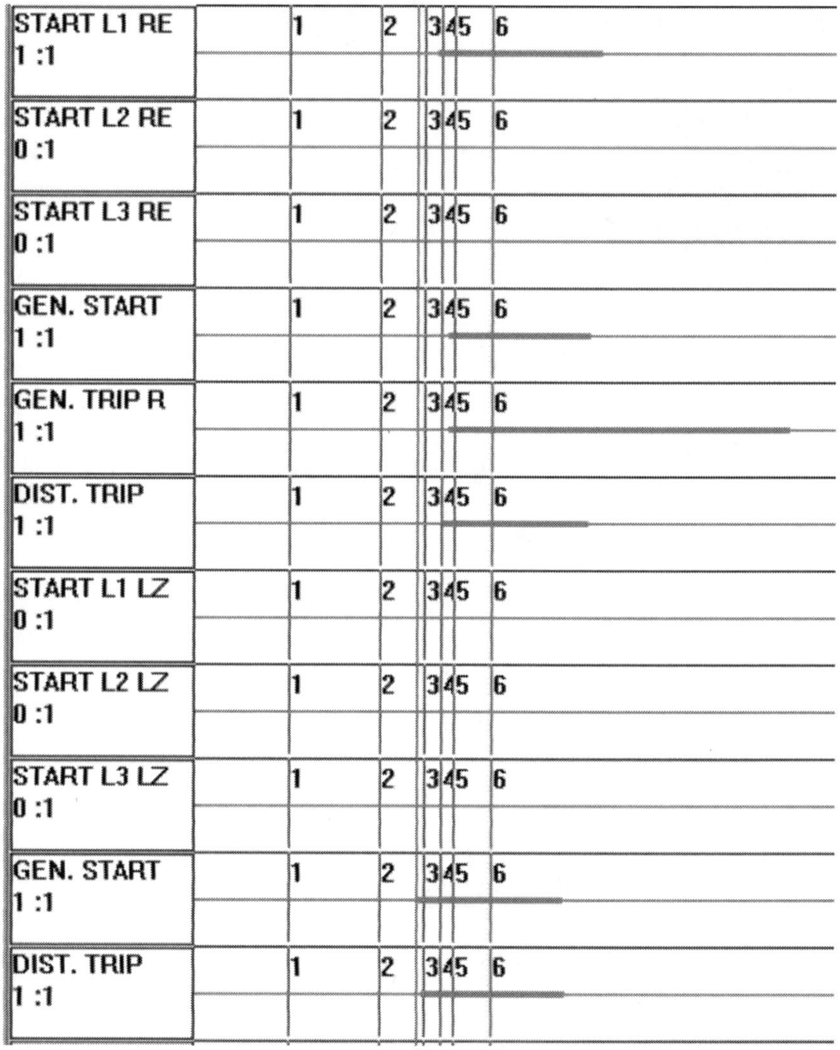

Figure 7. Digital signals recorded in the time period t=-99.24 ms…280 ms

At t=33 ms the faulted phase L1 was tripped. Figure 7 shows the start of distance protections:

- *START L1 REL* – start for group 1 of protections, through the line distance protection terminal REL 521, phase L1 (t=10 ms);
- *GEN. START REL* – general start of distance protection REL 521 (t=15 ms);
- *GEN. TRIP REL* – trigger pulse sent by distance protection REL 521 to the HV breaker of OTL, in Mintia station (t=15 ms);
- *DIST. TRIP REL* – trip of distance protection REL 521 (t=13 ms);
- *GEN. START LZ96a* – general start of digital relay LZ96a, group 2 of protections (t=2 ms);
- *DIST. TRIP LZ96a* – trip of digital relay LZ96a (t=4 ms).

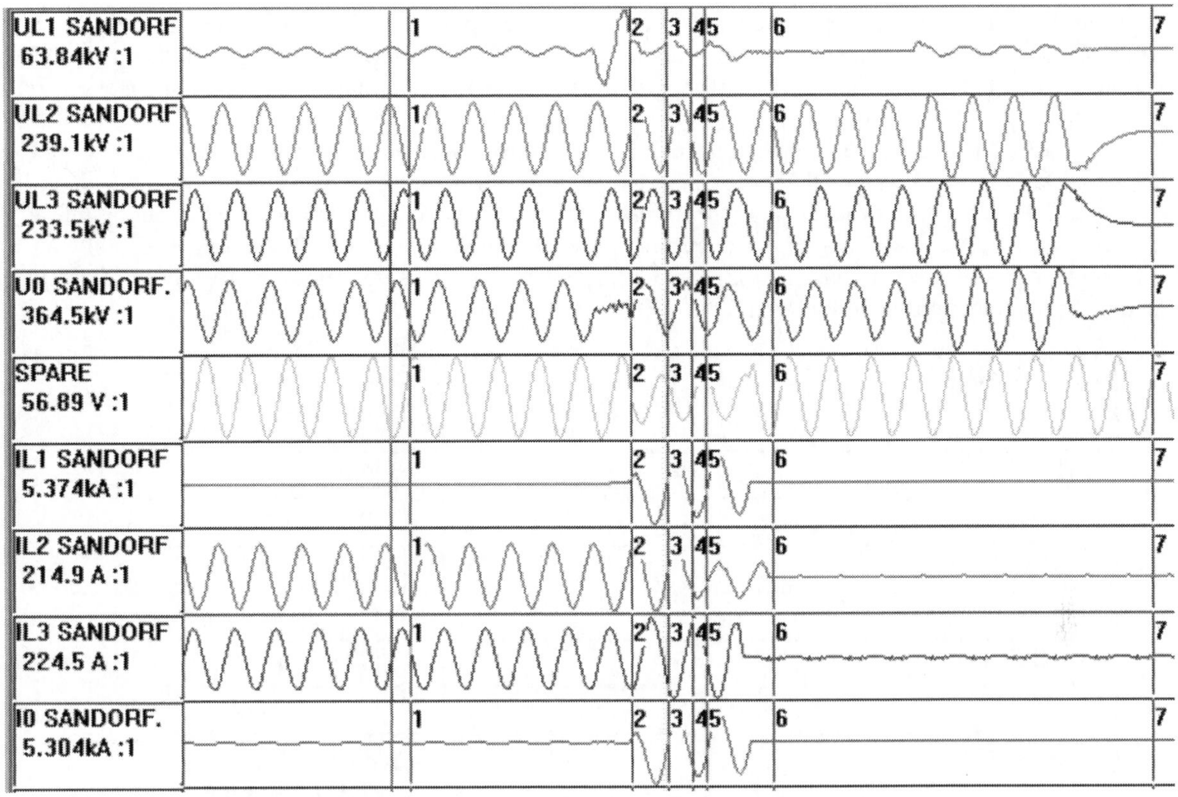

Figure 8. Analog signals recorded in the time period t=-99.29 ms...709 ms

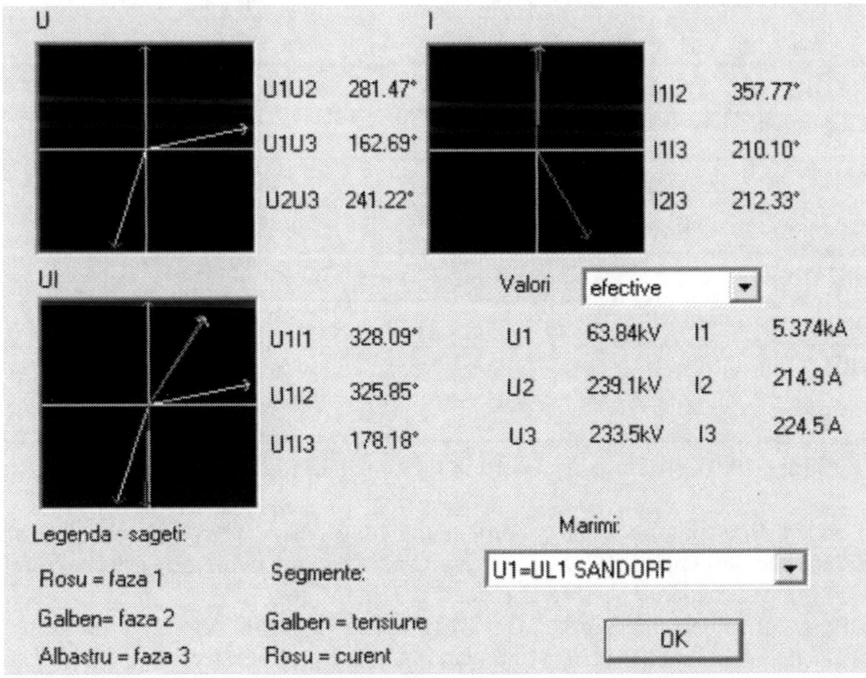

Figure 9. Marker 3: t=133 ms (fault, after the reclosing of breakers). Phasor diagrams of voltages and currents

Figure 8 shows the analog signals recorded in the time period t=-99.29 ms...709 ms (trigger at 16:20:07:769; number of samples: 1008; measuring time: 838 ms; marker 1: t=9 ms, marker 2: t=115 ms, marker 3: t=133 ms, marker 4: t=145 ms, marker 5: t=150 ms, marker 6: t=183 ms, marker 7: t=368 ms).

At t=115 ms phase L1 was connected (reclosed). Because the fault persisted, at t=173 ms phase L1 tripped. Markers 6 and 7 show the three-phase tripping of OTL.

After the reclosing of breakers (Figure 9), one can notice a very high value of the current on phase L1 (IL1=5.374 kA). Also, homopolar quantities have very high values (U0=364.5 kV, I0=5.304 kA). The voltage is still low on phase L1 (UL1=63.84 kV). The fault was not removed. Because the defect persists, the protections will start again.

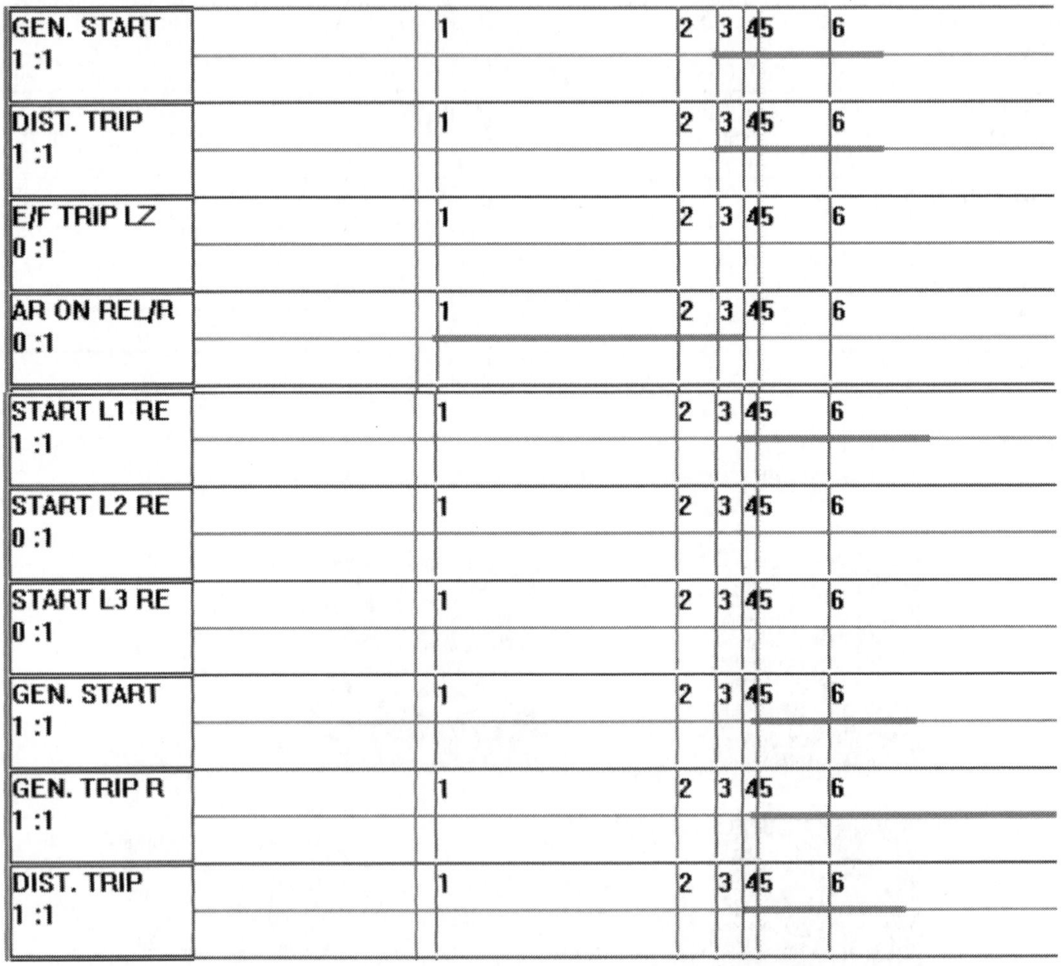

Figure 10. Digital signals record in the time period t=-99.29 ms...709 ms

Figure 10 shows the digital signals record in the time period t=-99.29 ms...709 ms (trigger at 16:20:08:769; number of samples: 972; measuring time: 808 ms; marker 1: t=9 ms, marker 2: t=115 ms, marker 3: t=133 ms, marker 4: t=145 ms, marker 5: t=150 ms).

When dead time (1 s) has elapsed, AR ON REL/REXA activates automatic reclosing, by sending reclosing signals to OTL breakers (in Mintia and Arad, by teleprotection). Auto-reclosing relay REXA requires a reclosing initiating signal from the line protection (REL 521). The breakers are reclosed when the reclose dead time delay has elapsed. Because the defect persists, the protections will start again, and OTL will three-phase tripped (markers 6 and 7).

IOP Conf. Series: Materials Science and Engineering **294** (2017) 012074 doi:10.1088/1757-899X/294/1/012074

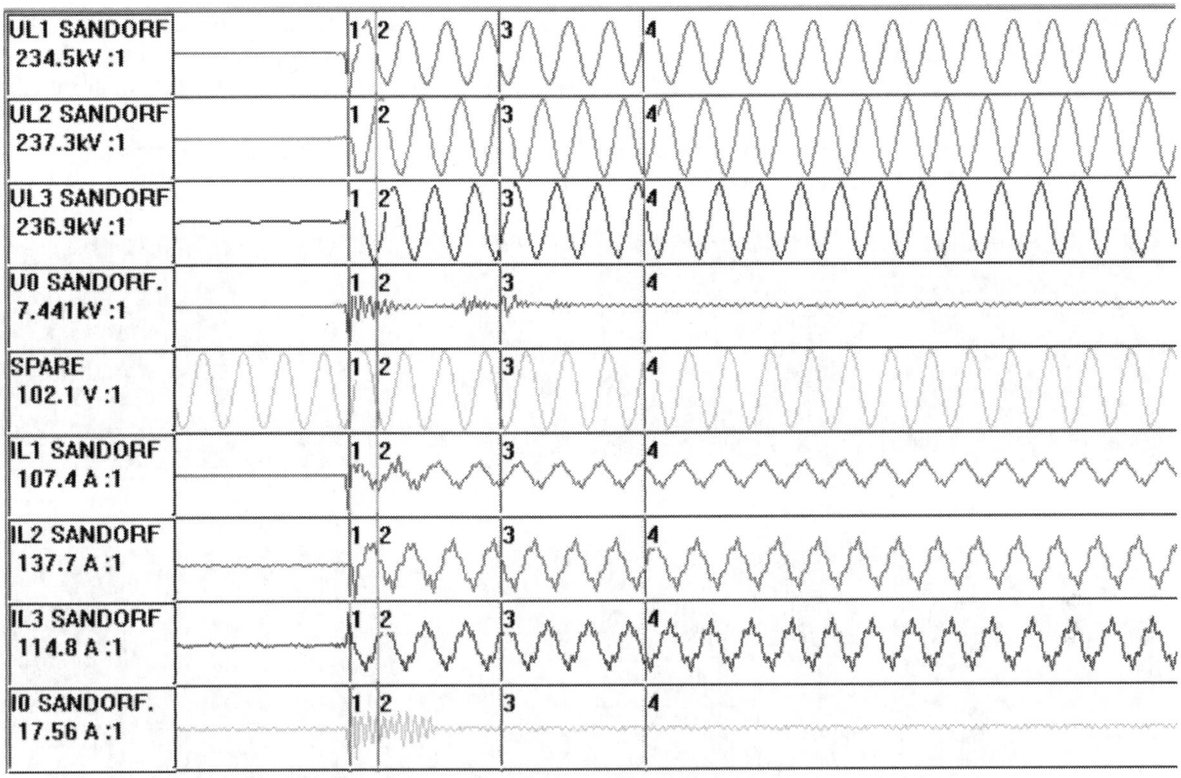

Figure 11. Analog signals recorded in the time period t=-98.29 ms...334 ms (marker 1: t=-14 ms, marker 2: t=0 ms, marker 3: t=61 ms, marker 4: t=131 ms)

3.3. Values of post-fault signals

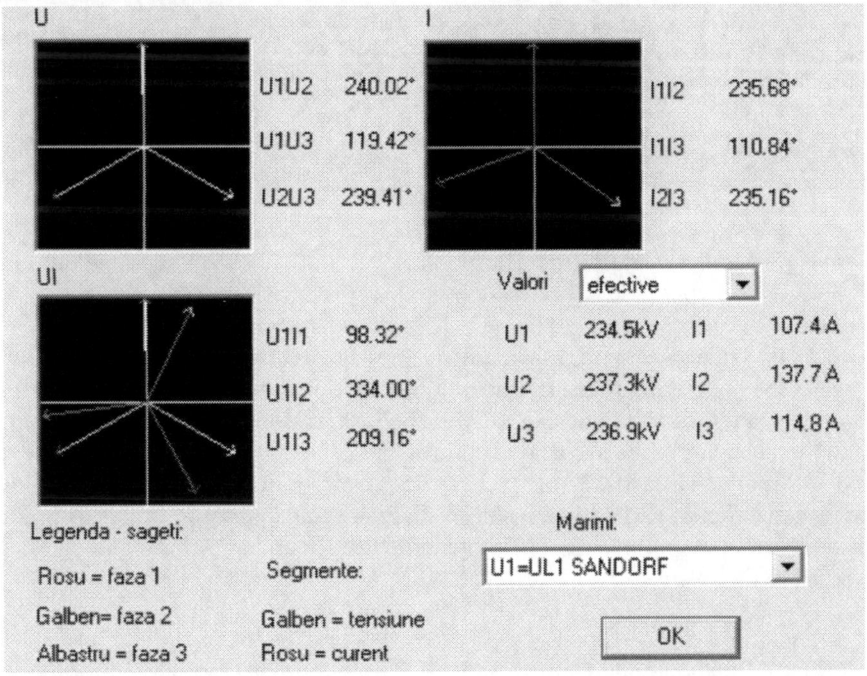

Figure 12. Marker 4: t=131 ms. Phasor diagrams of voltages and currents

After the fault has been removed (Figure 12), it is noted a symmetric loading on the three phases of the analyzed overhead transmission line. The homopolar voltage and the homopolar current have low RMS values: U0=10.77 kV , I0=17.56 A. At t = 131 (Figure 11, marker 4) analyzed OTL is in normal operation.

Focus program allows the harmonic analysis of the voltages and currents up to 10-th order. Figure 13 (marker 4: t=131 ms) shows the harmonic spectra of phase voltages and currents.

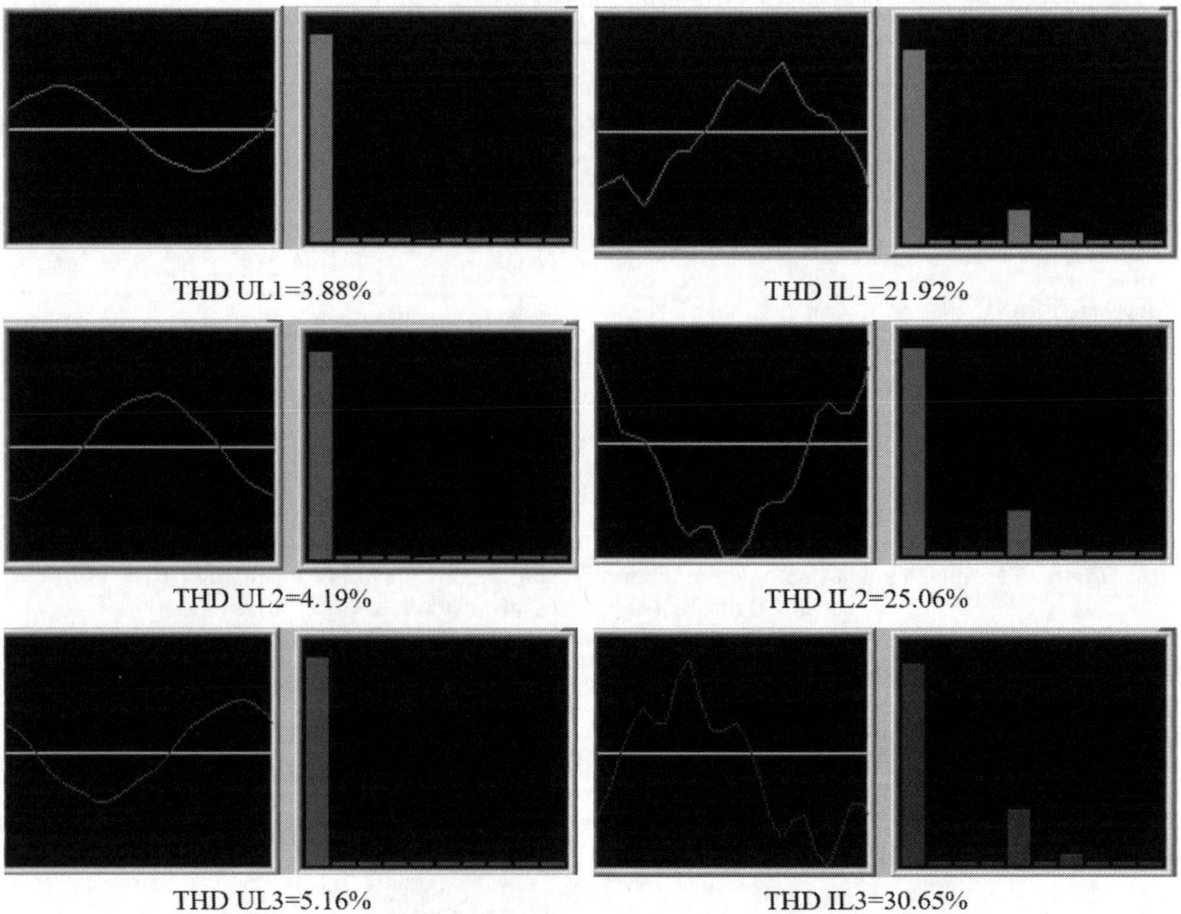

Figure 13. Marker 4: t=131 ms. Harmonic spectra of phase voltages and currents

4. Conclusions

Fast clearing of EHV faults is essential to maintain power quality in an area, because events on the EHV system affect the underlying transmission and distribution systems.

The cause and location of the fault can quickly be determined by analyzing the summary of the fault report, and the behavior of the associated control and protection equipment can be evaluated.

Are noticed especially the facilities offered by Focus program in analyzing of analog and digital quantities; the visualization of RMS values, phasor diagrams and harmonic analysis are in real time.

Fault analysis allows the improvement of protective designs and switching strategies, so that future faults will be prevented.

References

[1] Machowski J, Bialek J and Bumby J 2012 *Power System Dynamics: Stability and Control*, Second Edition, John Wiley & Sons Ltd, West Sussex, UK

[2] Baggini A 2008 *Handbook of Power Quality*, John Wiley & Sons Ltd, West Sussex, UK
[3] Dugan R C, McGranaghan M F, Santoso S and Beaty H W 2003 *Electrical Power Systems Quality*, Second Edition, McGraw-Hill, Columbus, OH USA
[4] Zamora I, Mazon A J, Valverde V, San Martin J I, Buigues G and Dyaeko A 2005 *Influence of power quality on the performance of digital protection relays*, IEEE Russia Power Tech, St. Petersburg, Russia, Jun 27-30, pp 1-7
[5] Sharif S, N A Latha and S Dawood 2011 Power Quality Disturbances and Protective Relays on Transmission Lines, *International Journal of Engineering Trends and Technology* **2**(2) 55-60
[6] Lee T M and Chan T W 1997 *An investigation on the effects of harmonic contents and phase shift on the performance of RCCB and over-current relays*, Proceedings of the International Power Engineering Conference (IPEC' 97), May 22 - 24, pp 775-779
[7] Stringer N T 1998 The effect of DC offset on current-operated relays, *IEEE Transactions on Industry Applications* **34**(1) 30 – 34
[8] Hart D G, Peterson W, Uy D, Schneider J, Novosel D and Wright R 2000 Tapping protective relays for power quality information, *IEEE Computer Applications in Power* **13**(1) 45 - 49
[9] ****Automatic Reclosing – Transmission Line Applications and considerations* (www.basler.com/downloads/autorecl.pdf)
[10] ***IEEE Power System Relaying Committee Working Group 1992 Single phase tripping and auto reclosing of transmission lines-IEEE Committee Report, *IEEE Transactions on Power Delivery* **7**(1) 182 -192
[11] Gamit V, Karode V, Mistry K, Parmar P and Chaudhari A 2015 Fault analysis on three phase system by auto reclosing mechanism, *International Journal of Research in Engineering and Technology* **4**(5) 292-298
[12] Ziegler G 2011 *Numerical Distance Protection: Principles and Applications*, 4th Edition, Publicis Publishing, Erlangen, Germany
[13] Schweitzer E O and Roberts J 2010 Distance Relay Element Design, *SEL Journal of Reliable Power* **1**(1) 4-14
[14] Kobet G et al. 2010 *Justifying pilot protection on transmission lines,* 63rd Annual Conference for Protective Relay Engineers, College Station, TX, March 29-April 1, pp 1 - 31
[15] ***Line distance protection terminal REL 521, ABB Technical brochure, 2003
[16] Bermúdez O E 2006 *Computational Tool for Advanced Analysis of Fault Recordings*, IEEE PES Transmission and Distribution Conference and Exposition Latin America, Caracas, Venezuela, August 15-18, pp 1-6
[17] Pineda H and Winkeljohann R 2006 *Computational Tool for Advanced Analysis of Fault Recordings*, IEEE PES Transmission and Distribution Conference and Exposition Latin America, Caracas, Venezuela, August 15-18, pp 1-5
[18] ***Auto-reclosing relay REXA 101, ABB Technical brochure, 1999
[19] ***Distributed busbar protection REB 500, ABB Technical brochure, 1999
[20] ***Bay control terminal REC 580, ABB Technical brochure, 1999
[21] ***Synchro-check relay SPAU 140 C, ABB User´s manual and Technical description, 2002
[22] ***Indactic® 650 fault analyzing system, ABB Power Automation, 1999
[23] ***Focus for Windows v2.0, TELECOMM, 2000

IOP Conf. Series: Materials Science and Engineering **294** (2017) 012075 doi:10.1088/1757-899X/294/1/012075

Study regarding the influence of environmental temperature and irradiation conditions on the performance of a photovoltaic solar module

C Hațiegan[1], N Gillich[1], C Popescu[2], E Răduca[1], L Cîndea[3], M F Predus[4] and I M Terfăloagă[1,5]

[1]Eftimie Murgu University of Resita, Department of Electric Engineering and Informatics, Traian Vuia str., no. 1-4, 320085 Resita, Romania
[2]"Constantin Brâncuşi" University of Târgu Jiu, Department of Electric Engineering, Eroilor str., no. 30, 210135 Targu Jiu, Romania
[3]Eftimie Murgu University of Resita, Department of Mechanical and Materials Engineering, Traian Vuia str., no.1-4, 320085 Resita, Romania
[4]"Politehnica" University of Bucuresti, Splaiul Independentei, nr.313, 060042, Bucuresti, Romania
[5]West University of Timisoara, Department of Computer Science, Vasile Parvan str., no. 4, 300223 Timisoara, Romania

E-mail: c.hatiegan@uem.ro

Abstract. In this paper the proposed study is verify to fit with the observed experimental data by choosing a correct mathematical model. The performance of the solar photovoltaic module provides a general view of the climate variables impacts and helps to find the efficiency of this module knowing the climatic parameters of a particular geographic area.The study was made when the climate is variable which is useful in developing an efficiency relationship of the photovoltaic solar module with major climatic parameters such as temperature, wind speed, humidity, dust, etc., so that the determined equations are in good correlation with measured data.The effectuate analysis shows that both variables: environmental temperature and irradiation can be used to optimize the efficiency of a photovoltaic solar module for different applications.

1. Introduction

Due to pollution and the realising of the limited fossil fuel reserves, regenerable resources for producing energy have risen constantly and have proven an adequate solution for humanity [1], [2].

Also, the technological dependency of the industrialised world on fossil fuel and the way in which these fuels have constantly degraded the terran environment in quite alarming.

Solar panels have become one of the most promising ways to meet the electrification demands of many isolated consumers around the world [3-5].

Power generation systems using photovoltaic systems located on the building surface can be combined with other functions of buildings that require electricity consumption [6-8].

Solar radiation (flux) is obviously a determining factor when it comes to studying the natural potential of solar energy as a source of renewable energy [9-10].

The efficiency of the solar photovoltaic module obtained during the day has different values. This fluctuation results from various factors that affect the performance of the photovoltaic solar module. These factors can be: latitude, altitude, temperature, humidity, wind speed, cloud cover, dust, impurities, etc., [11-15].

In this paper the average values of the efficiency, the surrounding environment temperature and the irradiation were determined experimentally. Also, the evolution of the average values of the efficiency, surrounding environment temperature and irradiation determined experimentally according to the months of the year 2014, as well as the evolution of the average values of the efficiency and of the irradiation in relation to the average values of the surrounding environment temperatures were presented graphically.

2. The system for monitoring the efficiency, irradiation and environmental temperature

The module performance provides an overview of the impacts of climate variables and helps to learn the efficiency of modules while knowing the climatic parameters of an area [16].

Solar radiation (flux) is obviously a determining factor when it comes to studying the natural potential of solar energy as a source of renewable energy. For tropical regions, on average, the potential for solar radiation is about 16.4 ± 1.2 MJ/ m2 per day [16], [17]. Solar flux is described by visible infrared radiation and near solar emissions, whereas the different spectra are described by their wavelength, which varies in a wide range.

Tracking of the equipment is done with a local monitoring system, but also at a distance, via the Internet. The Sunny WebBox (Figure 1) remote monitoring system acquires real-time data from the following components of the photovoltaic system [18]: from irradiation sensors, ambient temperature, photovoltaic panel temperature and wind speed; from the network inverter; from the battery inverter.

As the central communication interface, the Sunny WebBox connects the photovoltaic plant and its operator. The Sunny WebBox collects and documents all data of the connected devices, allowing continuous photovoltaic monitoring. Sunny WebBox provides the operator with all data recorded through an Internet connection.

The Flashview presentation software and the Sunny Portal site can be used to edit data or graphically display the stored data.

This means that operators can see returns both on the local network and via the Internet at any time using Flashview and Sunny Portal. The Sunny WebBox is also a powerful tool for operators when configuring cells or performing remote diagnostics via the computer. It allows continuous monitoring of photovoltaic installations and helps optimize photovoltaic cell yields [19].

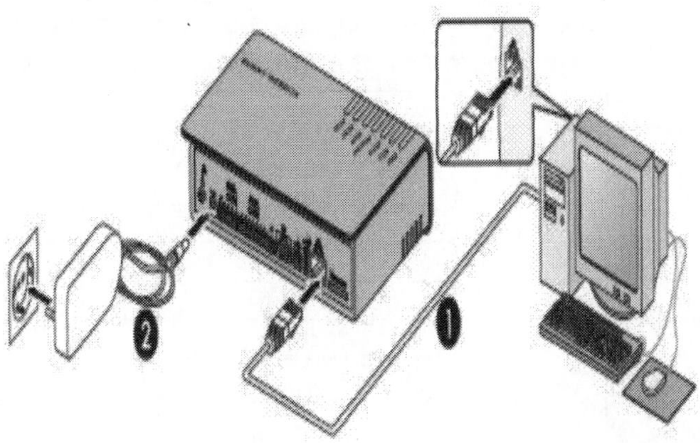

Figure 1. The e Sunny WebBox connecting way to a computer

Sunny SensorBox records the relevant environmental data to monitor the performance of photovoltaic cell systems.

For this purpose, the Sunny SensorBox has an integrated irradiation sensor as well as an external temperature sensor. The operator can also optionally be connected to an ambient temperature sensor and a wind sensor at the Sunny SensorBox.

3. Measurement results

The measurements were performed during the interval January 2014 - December 2014, in Resita, Romania, at various times of the day with clear or partly cloudy skies, averaging the values of efficiency, surrounding environment temperature and irradiation.

Table 1. Measured values in the period January- December 2014

Month	Efficiency average value	Surrounding environment temperature average value	Irradiation
January	8.53	3	535
February	10.79	5	801
March	9.12	10	881
April	10.36	15	892
May	10.00	20	978
June	10.12	38	928
July	10.16	39.4	910
August	10.66	33	954
September	10.83	22.3	978
October	9.36	14	697
November	8.83	9.6	600
December	7.90	20	473

The data presented in Table 1, refers to days with clear, cloudy or partly cloudy sky, averaging the values of efficiency, surrounding environment temperature and irradiation.

Figure 2 shows graphically the evolution of the average values of experimentally determined efficiency's according to the months of 2014.

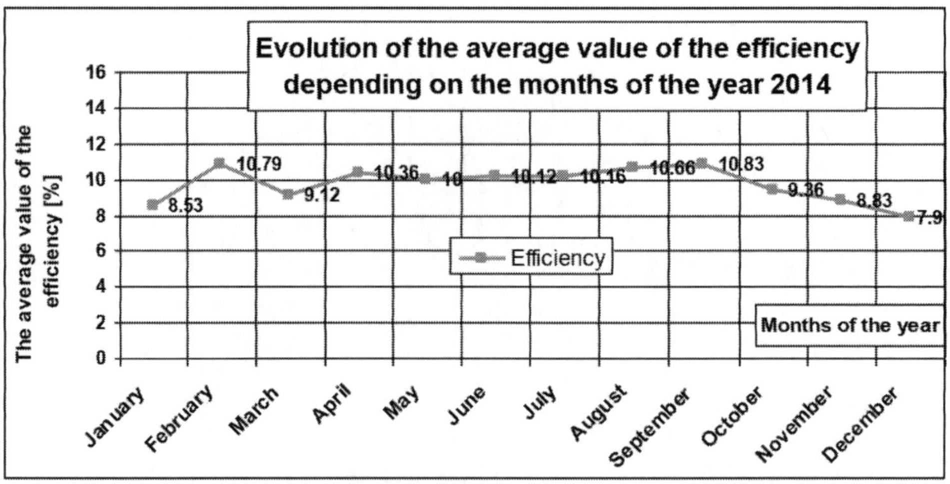

Figure 2. Evolution of the efficiency average value according the months of the year 2014

Figure 2 shows that the maximum value was determined in September and the minimum value in December.

Figure 3 shows graphically the evolution of the experimentally determined average temperature values in relation to the months of 2014.

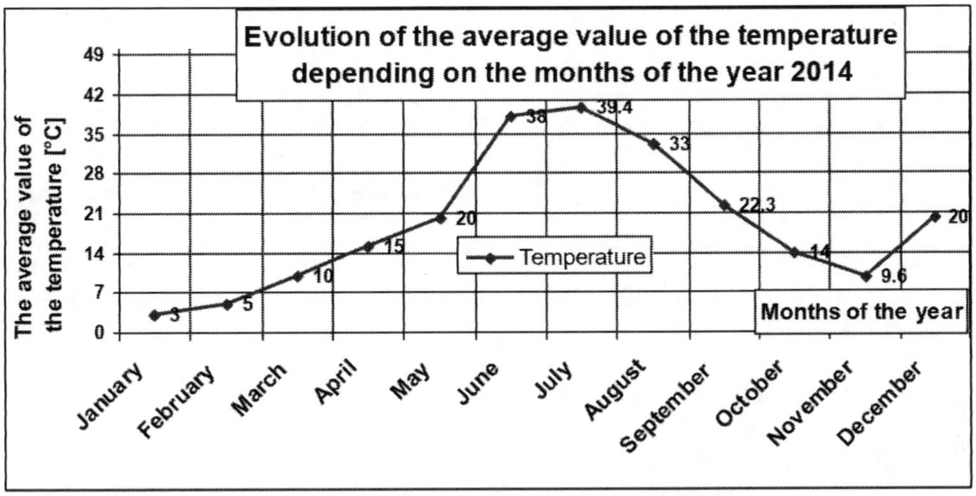

Figure 3. Evolution of the temperature average value according the months of the year 2014

The Figure shows that the maximum temperature was determined in July, and the minimum temperature in January. Figure 4 presents graphically the evolution of the average irradiation values determined experimentally according to the months of 2014.

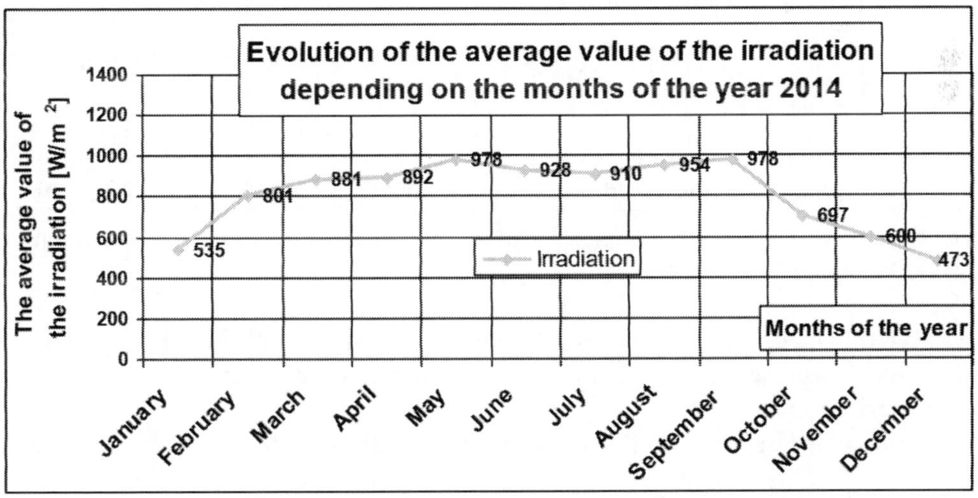

Figure 4. Evolution of the irradiation average value according the months of the year 2014

The Figure shows that the maximum irradiation value was determined in May and September, and the minimum irradiation value in December.

Figure 5 presents graphically the evolution of the average efficiency values according to the average temperature values determined in the months of 2014.

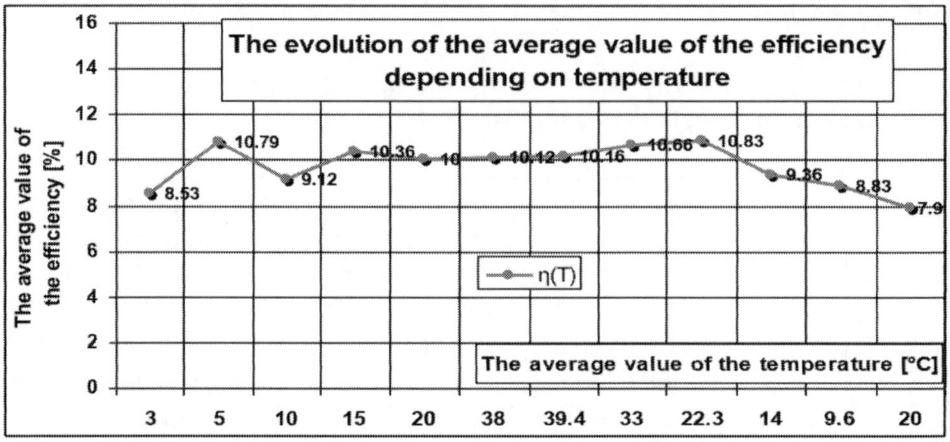

Figure 5. Evolution of the average efficiency value according to the temperature during the period January- December 2014

The Figure shows that the maximum yield of n = 10.83 was determined at 22.3 ° C and the minimum efficiency of n = 7.9 at 20 ° C.

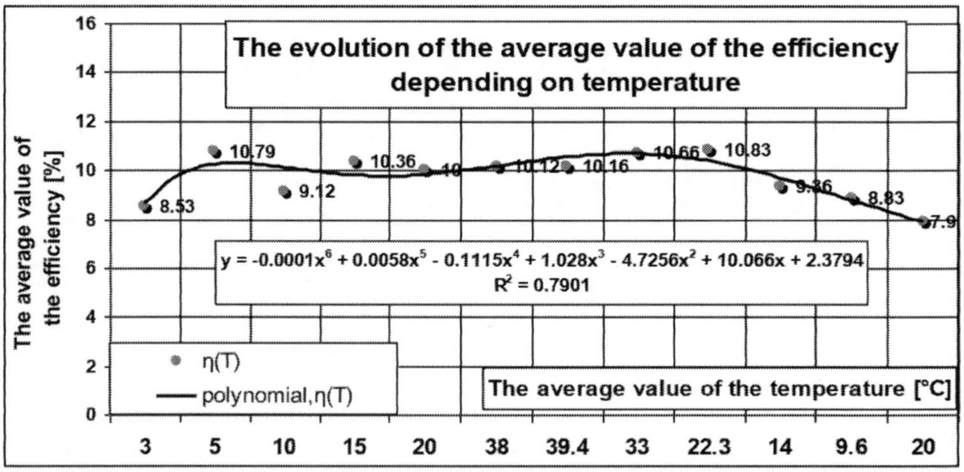

Figure 6. Evolution of the efficiency average value according to the temperature (January- December 2014)

The curve obtained in Figure 5 was interpolated in polynomial form in Figure 6 to obtain the mathematical representation of the average efficiency value in relation to the average temperature during January 2014 - December 2014.

The polynomial function of the efficiency according to temperature is given by the relation (1):

$$\eta(T) = -0.0001T^8 + 0.0058T^6 - 0.1115T^4 + 1.028T^3 - 4.7256T^2 + 10.066T + 2.3794,$$

with $R^2 = 0.7901$

Figure 7 shows the evolution of average irradiation values according to the average temperature values determined in the months of 2014.

IOP Conf. Series: Materials Science and Engineering **294** (2017) 012075 doi:10.1088/1757-899X/294/1/012075

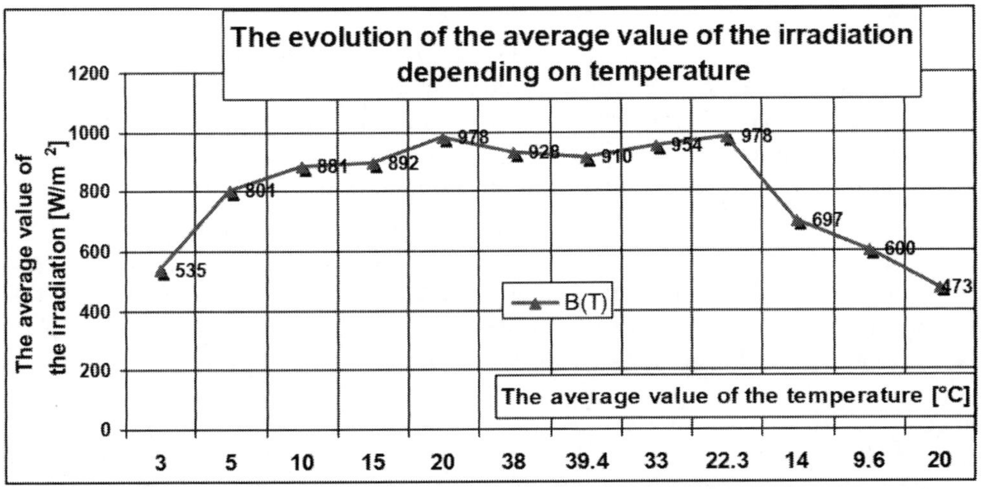

Figure 7. Evolution of the irradiation average value according to the temperature in the period January- December 2014

We can see from the Figure that the maximum irradiation B = 978 [W/m^2] was determined at temperatures of 20 °C and 22.3 °C and the minimum irradiation B = 473 [W/ m^2] at 20 °C.

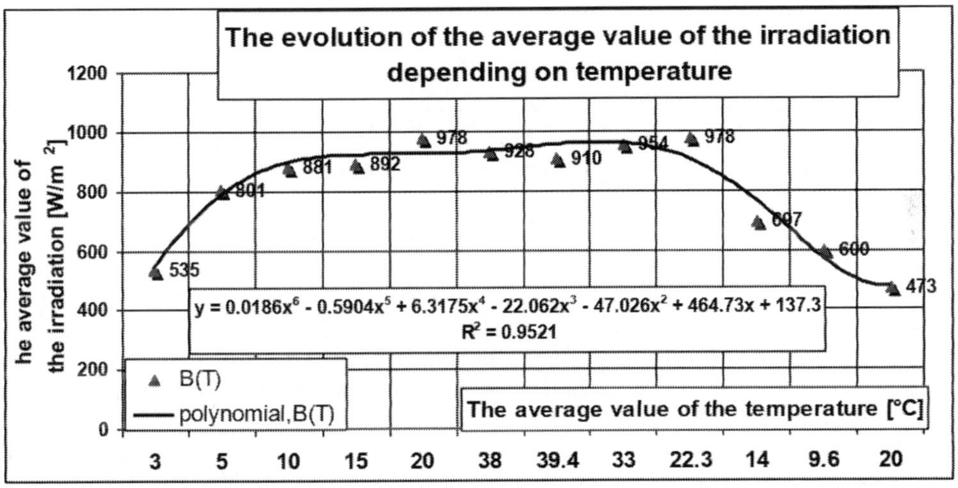

Figure 8. Evolution of the irradiation average value according to the temperature (January- December 2014)

The curve obtained in Figure 7 was interpolated in polynomial form in Figure 8 to obtain the mathematical representation of the average irradiation value in relation to the average temperature during January 2014 - December 2014.The polynomial irradiation function according to the temperature is given by the relation (2):

$$B(T) = 0.0186T^8 - 0.5904T^6 + 6.3175T^4 - 22.062T^3 - 47.026T^2 + 464.73T + 137.3,$$

with $R^2 = 0.9521$

4. Conclusions

The study was carried out when the climate is variable, fact which is useful in developing an efficiency relationship of the photovoltaic solar module with major climatic parameters such as temperature, wind speed, humidity; dust, etc., so that the determined equations are in good correlation with measured data.

The polynomial relations obtained through interpolation are approximated relations, but can be useful in order to determine the average efficiency value, respectively of the irradiation according to the average value of the ambient temperature.

The results of this analysis indicate that the measured parameters such as: the average values of the irradiation and ambient temperature can be used to optimize the efficiency of the photovoltaic module according to the geographic area.

References

[1] Chioncel C P, Tirian G O, Gillich N, Hatiegan C and Spunei E 2016 Overview of the wind energy market and renewable energy policy in Romania, *IOP Conf. Ser.: Mater. Sci. Eng.* **163** 012009

[2] Tokar A, Negoiţescu A, Hamat C and Roşu Ş 2016 The chemical and ecological state evaluation of a storage lake, *Revista de Chimie* **67**(9) 1860-1863

[3] Dubey S, Sarvaiya J N and Seshadri B 2013 Temperature Dependent Photovoltaic (PV) Efficiency and Its Effect on PV Production in the World A Review, *Energy Procedia* **33** 311-321

[4] Garcia A M C and Balenzategui J L 2004 Estimation ofphotovoltaic module yearly temperature and performance based on Nominal Operation Cell Temperature calculations, *Renewable Energy* **29** 1997-2010

[5] Negoiţescu A and Tokar A 2016 Solar Energy as an Alternative to Energy Saving and Pollutant Emissions Reduction, *Analele Universităţii "Eftimie Murgu" Reşiţa* **XXIII**(1) 195-202

[6] Bhattacharya T, Chakraborty A K and Pal K 2014 Effects of Ambient Temperature and Wind Speed on Performance of Monocrystalline Solar Photovoltaic Module in Tripura, *Journal of Solar Energy* **2014** 1-5

[7] Tokar A, Retezan A and Negoiţescu A 2017 Estimative analysis of investment payback for industrial spaces heated with radiant tubes, *Revista Română de Inginerie Civilă* **8**(1) 12-18

[8] Negoiţescu A, Tokar A and Hamat C 2016 The influence of the air thermal and rainfall regimes on storage lakes water turbidity, *Materiale Plastice* **53**(3) 542-545

[9] Luminosu I, Fara L, Pop N, Costache M and Fara S 2012 Exergy Analysis of the Air Solar Collector Based on Experimental Data, *Environmental Engineering and Management Journal* **11**(8) 1367-1374

[10] Sudhakar K and Srivastava T 2014 Energy and exergy analysis of 36 W solar photovoltaic module, *International Journal of Ambient Energy* **35**(1) 51-57

[11] Yao W, Li Z, Wang Y, Jiang F and Hu L 2014 Evaluation of global solar radiation models for Shanghai, *Energy Conversion and Management* **84** 597-612

[12] Enrique J M, Andujar J M and Bohorquez M A 2010 A reliable, fast and low cost maximum power point tracker for photovoltaic applications, *Solar Energy* **84**(1) 79-89

[13] Chioncel C P, Kohake D, Augustinov L, Chioncel P and Tiran G O 2010 Yield factors of a photovoltaic plant, *Acta Tech Corvin Bull Eng* **III**(2) 63-66

[14] Araneo R, Grasselli U and Celozzi S 2014 Assessment of a practical model to estimate the cell temperature of a photovoltaic module, *International Journal of Energy and Environmental Engineering* **5**(72) 1-16

[15] Popescu C, Haţiegan C, Racoceanu C and Bejinariu A C 2017 Aspects on the influence of starting the own services consumers of an energy group with unit of 330 mw on the power supply, *International Conference Knowledge-Based Organization* **23**(3) 84-89

[16] Chioncel C P, Chioncel P, Gillich N and Tirian G O 2009 Performance ratio of a photovoltaic plant, *Bulletin of Engineering, University Politehnica Timisoara* **II**(2) 55-58

[17] Shuklaa K N, Rangnekarb S and Sudhakarb S 2016 Mathematical modelling of solar radiation incident on tilted surface for photovoltaic application at Bhopal, *International Journal of Ambient Energy* **37**(6) 579-588

[18] Chioncel C P, Gillich N, Chioncel P and Gillich G R 2007 *CMS solutions in monitoring and real-time data transfer of photovoltaic plants,* XVth International Symposium on Electrical Apparatus and Technologies, SIELA 2007, Plovdiv, Bulgaria, May 31-June 1, Pp 14-17

[19] Ayompe L, Duffy A, McCormack S and Conlon M 2011 Measured Performance of a 1.72 kW Rooftop Grid Connected Photovoltaic System in Ireland, *Energy Conversion and Management* **52**(2) 816-825

IOP Conf. Series: Materials Science and Engineering **294** (2017) 012076 doi:10.1088/1757-899X/294/1/012076

Experimental measurements concerning the stator insulation partial discharge level of a hydro-generator of high power

C Haţiegan[1], E Răduca[1], C Popescu[2], C O Hamat[3], A Băra[3], D Anghel[4] and D A Pîrşan[3]

[1]Eftimie Murgu University of Resita, Department of Electric Engineering and Informatics, Traian Vuia str., no. 1-4, 320085 Resita, Romania
[2]"Constantin Brâncuşi" University of Târgu Jiu, Department of Electric Engineering, Eroilor str., no. 30, 210135 Targu Jiu, Romania
[3]Eftimie Murgu University of Resita, Department of Mechanical and Materials Engineering, Traian Vuia str., no.1-4, 320085 Resita, Romania
[4]"Politehnica" University of Bucuresti, Splaiul Independentei, nr.313, 060042, Bucuresti, Romania

E-mail: c.hatiegan@uem.ro

Abstract. The paper presents a determining and evaluating method from an experimental point of view of the partial discharges level from the insulation of the stator of a synchronous hydro- generator of high power that appear at different temperatures. The temperature is measured directly on the coil with placed thermo- resistant elements between the superior and inferior bars. Also, the level of partial discharges with the coil age in a cold state, at the temperature level of the surrounding environment from within the stator, but also in a warm state at different temperatures, is measured. Through this method there is created the possibility of highlighting some eventual degradations of the coil insulation of a hydro- generator.

1. Introduction

Partial downloads are generally consequently a local concentration of electrical insulation stress or on the surface of the insulation [1-3]. Generally such discharges occur in the form of pulse currents in the dielectric and outer circuit thereof with a duration much less than 1μs. Yet can occur and continuous forms downloads, such as the so-called low-intensity discharge ("pulse-less") in gaseous dielectrics [4-6].

Generally it can be considered that partial discharges in insulation systems are " symptoms" phenomena which is produced in that insulation and in most cases can be associated with insulation defects due to either the manufacturing technology, be of solicitation complex (mechanical, electrical, thermal, chemical etc.) on the material in which the insulation is achieved [7-18]. Though pierce from vacuole does not immediately lead to shorting dielectric, the phenomenon of partial discharge walls space respectively, finally leading to deterioration of the insulation [19-21].

Each discharge dissipates a small amount of energy, which will be located inside the cavity present in insulation.This energy spreads in all directions in the form of heat [22-25].

Apparent task analysis is important Is important to determine the corrective measures that need to be taken. The normal level of partial discharge is characterized by values up to 100 pC, it has a low impact on insulation but affects its lifetime equipment. The critical level appears after the threshold 1000 pC, at this stage the insulation is in an intense process of degradation. Values higher, 1000 pC affects considerably whether the action of the phenomenon takes a long time leaving carbonization traces in the insulation an effective method of assessment of the real state of the winding insulation stator is to monitor the level of partial discharge, highlighting any in homogeneities in the mass of the insulation, leading to serious damage of the insulation by shutting down the generator [8], [26-28].

In this paper the level of partial discharge for different voltages was determined and verified whether there was a change in temperature during the experimental determinations that could influence the parameters DP.

The temperature is measured directly on the thermal resistance coil placed between the top and the lower bar. It also measures the level of partial discharge the coil in the cold state with the ambient temperature inside the stator and at different temperatures in the hot state. The average values for the different levels of partial discharge voltages were represented graphically according to different temperatures and depending on the voltage at the generator terminals.

2. Partial Discharge Measurement System

The system used to measure the partial discharges shown in the Figure 1 is built of: coupling device, transmission and measurement devices.

Coupling device it is an integral part of the measuring system and a test circuit, its components being specially designed to achieve optimal sensitivity with a specific test circuit.

The coupling device (Figure 1b) is generally an active or passive network with four terminals (quadripole) and converts the input current into output voltage signals [8], [10].

These signals are transmitted to the measuring device DP from Figure 1a, through a transmission system. The frequency response of the coupling device, defined by the ratio of the output voltage to the input current It is selected so as to prevent the test voltage frequency and its harmonics to reach the measuring device.

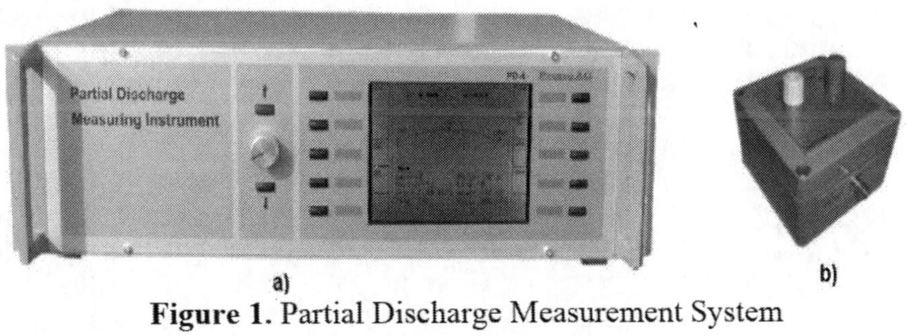

Figure 1. Partial Discharge Measurement System

Measuring device DP with coupling device, constitutes a measurement system DP broadband characterized by a transfer impedance $z(f)$ having frequencies lower limit values f_1 and an upper limit f_2 and adequate mitigation until f_1 and over f_2.

The response of this device to a non-oscillating current pulse due to a download part is generally a strongly damped oscillation. Both apparent charge and polarity Impulse current of DP can be determined from this response, and the pulse resolution time is very small. The wiring diagram of the plant for partial discharge measurement is shown in Figure 2, [8], [10].

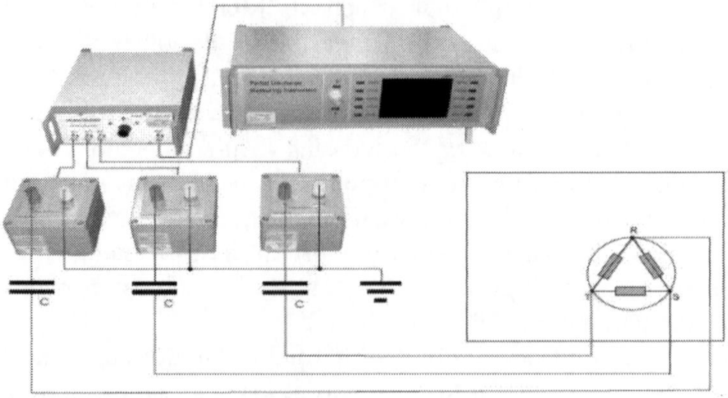

Figure 2. Installation wiring diagram of measuring partial discharges

3. Measurement results

Attempts have been made with the generator in idle functioning. Determining the level of partial discharges has been done at different values of the generator terminals: U_{line}=4kV, U_{line} =8,1 kV, U_{line} =12,3 kV, U_{line} =14 kV, U_{line} =15 kV, U_{line} =16kV.

For each voltage level partial discharges were measured on each phase, considering a number of 30 entries. From the average values measured, average values were determined for each phase. The results of the level of partial discharges for different voltages are given in the following Tables 1÷6:

Table 1. Results of the level of partial discharges for U_{line}=4 kV

No.	Phase A		Phase B		Phase C		No.	Phase A		Phase B		Phase C	
	U_A [kV]	PD_A [pC]	U_B [kV]	PD_B [pC]	U_B [kV]	PD_B [pC]		U_A [kV]	PD_A [pC]	U_B [kV]	PD_B [pC]	U_B [kV]	PD_B [pC]
1	2,10	398,4	2,20	2,20	2,20	281,3	16	2,10	398,4	2,10	2,10	2,10	318,8
2	2,10	407,8	2,10	2,10	2,10	337,5	17	2,10	365,6	2,10	2,10	2,10	300
3	2,10	407,8	2,10	2,10	2,10	328,1	18	2,10	403,1	2,10	2,10	2,10	393,8
4	2,10	398,4	2,20	2,20	2,20	309,4	19	2,10	393,8	2,10	2,10	2,10	323,4
5	2,10	407,8	2,10	2,10	2,10	356,3	20	2,10	407,8	2,10	2,10	2,10	318,8
6	2,10	281,3	2,10	2,10	2,10	332,8	21	2,10	398,4	2,10	2,10	2,10	332,8
7	2,10	323,4	2,20	2,20	2,20	365,6	22	2,10	403,1	2,10	2,10	2,10	346,9
8	2,10	398,4	2,20	2,20	2,20	337,5	23	2,10	300	2,10	2,10	2,10	337,5
9	2,10	403,1	2,10	2,10	2,10	337,5	24	2,10	407,8	2,10	2,10	2,10	332,8
10	2,10	407,8	2,10	2,10	2,10	360,9	25	2,10	351,6	2,10	2,10	2,10	426,6
11	2,10	351,6	2,10	2,10	2,10	304,7	26	2,10	398,4	2,10	2,10	2,10	346,9
12	2,10	412,5	2,10	2,10	2,10	328,1	27	2,10	393,8	2,10	2,10	2,10	332,8
13	2,10	370,3	2,10	2,10	2,10	332,8	28	2,10	398,4	2,10	2,10	2,10	304,7
14	2,10	290,6	2,10	2,10	2,10	281,3	29	2,10	398,4	2,10	2,10	2,10	276,6
15	2,10	435,9	2,10	2,10	2,10	328,1	30	2,10	356,3	2,10	2,10	2,10	351,6

Under the conditions of a nominal voltage de 4 kV to the generator terminals, we observe a very low level of partial discharges de 276,6 pC on the phase C, reaching to a value de 426,6 pC on the phase C.

Table 2. Results of the level of partial discharges for U_{line}=8,1kV

No.	Phase A		Phase B		Phase C		No.	Phase A		Phase B		Phase C	
	U_A [kV]	PD_A [pC]	U_B [kV]	PD_B [pC]	U_B [kV]	PD_B [pC]		U_A [kV]	PD_A [pC]	U_B [kV]	PD_B [pC]	U_B [kV]	PD_B [pC]
1	4,20	792,2	4,20	543,8	4,3	581,3	16	4,20	806,3	4,20	576,6	4,3	628,1
2	4,20	735,9	4,20	637,5	4,3	656,3	17	4,20	801,6	4,20	534,4	4,3	651,6
3	4,20	801,6	4,20	660,9	4,3	651,6	18	4,20	796,9	4,20	529,7	4,3	543,8
4	4,20	801,6	4,20	581,3	4,3	651,6	19	4,20	501,6	4,20	637,5	4,3	637,5
5	4,20	768,8	4,20	562,5	4,3	581,3	20	4,20	642,2	4,20	515,6	4,3	632,8
6	4,20	609,4	4,20	553,1	4,3	548,4	21	4,20	796,9	4,20	600	4,3	660,9
7	4,20	806,3	4,20	543,8	4,3	623,4	22	4,20	3796,9	4,20	628,1	4,3	646,9
8	4,20	764,1	4,20	604,7	4,3	642,2	23	4,20	890,6	4,20	534,4	4,3	726,6
9	4,20	796,9	4,20	656,3	4,3	590,6	24	4,20	750	4,20	604,7	4,3	656,3
10	4,20	806,3	4,20	632,8	4,3	590,6	25	4,20	792,2	4,20	642,2	4,3	656,3
11	4,20	600	4,20	562,5	4,3	632,8	26	4,20	801,6	4,20	534,4	4,3	567,2
12	4,20	801,6	4,20	557,8	4,3	651,6	27	4,20	478,1	4,20	618,8	4,3	651,6
13	4,20	782,8	4,20	590,6	4,3	614,1	28	4,20	464,1	4,20	618,8	4,3	628,1
14	4,20	787,5	4,20	632,8	4,3	632,8	29	4,20	787,5	4,20	534,4	4,3	581,3
15	4,20	670,3	4,20	600	4,3	656,3	30	4,20	806,3	4,20	628,1	4,3	637,5

The experimental results presented in Table 3 indicate a certain degree of imbalance of the partial phase discharge level. In conditions of a rated voltage of 8,1 kV the generator terminals, observe a low level of partial discharges de 464,1 pC on phase A, reaching to a value de 890,6 pC on phase A, the exception is for the case 22 where task apparent it is of 3796,9 pC on phase A.

Table 3. Results of the level of partial discharges for U_{line}=12,3kV

No.	Phase A		Phase B		Phase C		No.	Phase A		Phase B		Phase C	
	U_A [kV]	PD_A [pC]	U_B [kV]	PD_B [pC]	U_B [kV]	PD_B [pC]		U_A [kV]	PD_A [pC]	U_B [kV]	PD_B [pC]	U_B [kV]	PD_B [pC]
1	6,20	1171,9	6,20	946,9	6,30	843,8	16	6,20	1120,3	6,20	707,8	6,30	843,8
2	6,20	1125	6,20	768,8	6,30	890,6	17	6,20	731,3	6,20	900	6,30	937,5
3	6,20	1171,9	6,20	773,4	6,30	890,6	18	6,20	1017,2	6,20	1082,8	6,30	937,5
4	6,20	1110,9	6,20	778,1	6,30	843,8	19	6,20	792,2	6,20	735,9	6,30	1171,9
5	6,20	1134,4	6,20	895,3	6,30	937,5	20	6,20	1143,8	6,20	796,9	6,30	890,6
6	6,20	1021,9	6,20	773,4	6,30	750	21	6,20	745,3	6,20	754,7	6,30	843,8
7	6,20	1157,8	6,20	768,8	6,30	843,8	22	6,20	923,4	6,20	895,3	6,30	937,5
8	6,20	1195,3	6,20	942,2	6,30	890,6	23	6,20	1120,3	6,20	759,4	6,30	937,5
9	6,20	890,6	6,20	792,2	6,30	937,5	24	6,20	1007,8	6,20	890,6	6,30	843,8
10	6,20	703,1	6,20	735,9	6,30	937,5	25	6,20	1134,4	6,20	750	6,30	890,6
11	6,20	1167,2	6,20	759,4	6,30	937,5	26	6,20	956,3	6,20	890,6	6,30	890,6
12	6,20	1031,3	6,20	895,3	6,30	843,8	27	6,20	1129,7	6,20	656,3	6,30	890,6
13	6,20	885,9	6,20	895,3	6,30	890,6	28	6,20	1054,7	6,20	750	6,30	937,5
14	6,20	895,3	6,20	946,9	6,30	843,8	29	6,20	815,6	6,20	900	6,30	890,6
15	6,20	1101,6	6,20	778,1	6,30	937,5	30	6,20	1134,4	6,20	885,9	6,30	843,8

The experimental results presented in Table 3 indicate a certain degree of imbalance in the level of partial discharge phases. In conditions a rated voltage of 12,3 kV at the generator terminals, we notice a low level of partial discharge 703,1 pC on phase A, reaching value of 1171,9 pC on phase A respectively C.

Table 4. Results of the level of partial discharges for U_{line}=14kV

No.	Phase A		Phase B		Phase C		No.	Phase A		Phase B		Phase C	
	U_A [kV]	PD_A [pC]	U_B [kV]	PD_B [pC]	U_B [kV]	PD_B [pC]		U_A [kV]	PD_A [pC]	U_B [kV]	PD_B [pC]	U_B [kV]	PD_B [pC]
1	7,10	1359,4	7,20	1218,8	7,20	3093,8	16	7,10	1359,4	7,20	1312,5	7,20	1031,3
2	7,10	1265,6	7,20	1031,3	7,20	2812,5	17	7,10	843,8	7,20	1031,3	7,20	2953,1
3	7,10	1218,8	7,20	1031,3	7,20	1453,1	18	7,10	1453,1	7,20	1406,3	7,20	1875
4	7,10	1593,8	7,20	1031,3	7,20	1078,1	19	7,10	1031,3	7,10	1031,3	7,20	2343,8
5	7,10	1781,3	7,10	843,8	7,20	3046,9	20	7,20	7218,8	7,20	984,4	7,20	1406,3
6	7,10	1312,5	7,20	1125	7,20	2296,9	21	7,10	1312,5	7,20	1031,3	7,20	11953,1
7	7,10	1031,3	7,20	796,9	7,20	4921,9	22	7,10	1406,3	7,20	1031,3	7,20	2250
8	7,10	1312,5	7,20	1031,3	7,20	2109,4	23	7,10	1078,1	7,20	890,6	7,20	1687,5
9	7,10	1312,5	7,20	1031,3	7,20	1406,3	24	7,10	11953,1	7,10	937,5	7,20	2296,9
10	7,10	1265,6	7,20	1031,3	7,20	2437,5	25	7,10	1078,1	7,10	1031,3	7,20	3609,4
11	7,10	1453,1	7,20	843,8	7,20	1546,9	26	7,10	1312,5	7,10	1031,3	7,20	3000
12	7,10	1453,1	7,20	890,6	7,20	1687,5	27	7,10	843,8	7,10	890,6	7,20	3000
13	7,10	1312,5	7,20	890,6	7,20	2343,8	28	7,10	1078,1	7,10	1312,5	7,20	1640,6
14	7,10	1312,5	7,20	1078,1	7,20	2671,9	29	7,10	1312,5	7,20	937,5	7,20	1500
15	7,20	1640,6	7,20	937,5	7,20	1312,5	30	7,10	1031,3	7,10	1031,3	7,20	1828,1

The experimental results presented in Table 4 indicate a certain degree of imbalance in the level of partial discharge phases. In conditions a rated voltage of 14 kV at the generator terminals, we notice a low level of partial discharge 843,8 pC on phase A and B, reaching value of 11953,1 pC on phase C.

Table 5. Results of the level of partial discharges for U_{line}=15kV

No.	Phase A		Phase B		Phase C		No.	Phase A		Phase B		Phase C	
	U_A [kV]	PD_A [pC]	U_B [kV]	PD_B [pC]	U_B [kV]	PD_B [pC]		U_A [kV]	PD_A [pC]	U_B [kV]	PD_B [pC]	U_B [kV]	PD_B [pC]
1	7,60	5296,9	7,60	4125	7,70	11953,1	16	7,60	3375	7,60	3140,6	7,70	2625
2	7,60	11953,1	7,60	3843,8	7,70	6796,9	17	7,60	5484,4	7,60	3703,1	7,70	2484,4
3	7,60	4781,3	7,60	2203,1	7,70	2812,5	18	7,60	3421,9	7,60	3046,9	7,70	1359,4
4	7,60	6703,1	7,60	1687,5	7,70	2625	19	7,60	2859,4	7,60	3796,9	7,70	2250
5	7,60	5062,5	7,60	2531,3	7,70	1546,9	20	7,60	3656,3	7,60	1781,3	7,70	2156,3
6	7,60	5812,5	7,60	4453,1	7,70	2296,9	21	7,60	4031,3	7,60	11953,1	7,70	2812,5
7	7,60	4875	7,60	2296,9	7,70	3468,8	22	7,60	8062,5	7,60	3421,9	7,70	3375
8	7,60	11953,1	7,60	3656,3	7,70	1312,5	23	7,60	11953,1	7,60	1734,4	7,70	1453,1
9	7,60	3375	7,60	2859,4	7,70	3562,5	24	7,60	2953,1	7,60	11953,1	7,70	2250
10	7,60	3281,3	7,60	3140,6	7,70	2062,5	25	7,60	11953,1	7,60	11953,1	7,70	1640,6
11	7,60	6000	7,60	1312,5	7,70	3984,4	26	7,60	5109,4	7,60	11953,1	7,70	1265,6
12	7,60	9562,5	7,60	3187,5	7,70	2671,9	27	7,60	2437,5	7,60	2812,5	7,70	2250
13	7,60	5718,8	7,60	11953,1	7,70	1781,3	28	7,60	3562,5	7,60	5765,6	7,70	2578,1
14	7,60	4359,4	7,60	1640,6	7,70	1406,3	29	7,60	2625	7,60	2062,5	7,70	11953,1
15	7,60	11953,1	7,60	3140,6	7,70	11953,1	30	7,60	4828,1	7,60	4453,1	7,70	8203,1

The experimental results presented in Table 5 indicate a certain degree of imbalance in the level of partial discharge phases. In conditions a rated voltage of 15 kV at the generator terminals, we notice a low level of partial discharge 1265,6 pC on phase C, reaching value of 11953,1 pC on all three phases.

Table 6. Results of the level of partial discharges for U_{line}=16kV

No.	Phase A		Phase B		Phase C		No.	Phase A		Phase B		Phase C	
	U_A [kV]	PD_A [pC]	U_B [kV]	PD_B [pC]	U_B [kV]	PD_B [pC]		U_A [kV]	PD_A [pC]	U_B [kV]	PD_B [pC]	U_B [kV]	PD_B [pC]
1	8,00	11953,1	8,00	3703,1	8,10	2156,3	16	8,00	3468,8	8,00	3515,6	8,10	4218,8
2	8,00	3984,4	8,00	2578,1	8,10	3843,8	17	8,00	2578,1	8,00	2109,4	8,10	1968,8
3	8,00	11953,1	8,00	11953,1	8,10	4546,9	18	8,00	3468,8	8,00	2343,8	8,10	3375
4	8,00	1453,1	8,00	11953,1	8,10	5296,9	19	8,00	1593,8	8,00	2953,1	8,10	5250
5	8,00	11953,1	8,00	3468,8	8,10	2859,4	20	8,00	11953,1	8,00	3375	8,10	4359,4
6	8,00	8156,3	8,00	8250	8,10	4359,4	21	8,00	2578,1	8,00	5625	8,10	5484,4
7	8,00	11953,1	8,00	3421,9	8,10	2484,4	22	8,00	11953,1	8,00	4921,9	8,10	11953,1
8	8,00	11953,1	8,00	2578,1	8,10	2296,9	23	8,00	11953,1	8,00	3750	8,10	11953,1
9	8,00	6093,8	8,00	3046,9	8,10	2437,5	24	8,00	11953,1	8,00	2250	8,10	7875
10	8,00	11953,1	8,00	3890,6	8,10	4734,4	25	8,00	1828,1	8,00	3562,5	8,10	11953,1
11	8,00	11953,1	8,00	2718,8	8,10	3750	26	8,00	1453,1	8,00	1734,4	8,10	11953,1
12	8,00	11953,1	8,00	2203,1	8,10	4406,3	27	8,00	11953,1	8,00	1921,9	8,10	10781,3
13	8,00	11109,4	8,00	3750	8,10	2625	28	8,00	11953,1	8,00	4078,1	8,10	11953,1
14	8,00	3234,4	8,00	3187,5	8,10	2203,1	29	8,00	11953,1	8,00	3140,6	8,10	4921,9
15	8,00	2906,3	8,00	2671,9	8,10	2671,9	30	8,00	7593,8	8,00	3656,3	8,10	4453,1

The experimental results presented in Table 6 indicate a certain degree of imbalance in the level of partial discharge phases. In conditions a rated voltage of 16 kV at the generator terminals, we notice a very high level of partial discharge 1593,8 pC on phase C, reaching value of 11953,1 pC on phases A and C.

The measurement results of the level of partial discharges for the average values of voltage are given in the Table 7 and in Figure 3, where is presented the average values of partial discharges levels on the three phases for: U_{line} = 4 kV; U_{line} = 8,1 kV; U_{line} = 12,3 kV; U_{line} = 14 kV; U_{line} = 15 kV; U_{line} = 16 kV.

Table. 7 Average values for the partial discharges

Average values for U_{line} [kV	Phase A		Phase B		Phase C	
	U_A [kV]	DP_A [pC]	U_B [kV]	DP_B [pC]	U_C [kV]	DP_C [pC]
4	2,10	382,33	2,10	324,99	2,11	332,20
8.1	4,20	841,27	4,20	588,60	4,30	627,05
12.3	6,20	1019,69	6,20	826,87	6,30	896,88
14	7,11	1831,26	7,17	1023,46	7,20	2553,14
15	7,60	5900,01	7,60	4518,75	7,70	3629,69
16	8,00	8026,56	8,00	3943,75	8,10	5437,51

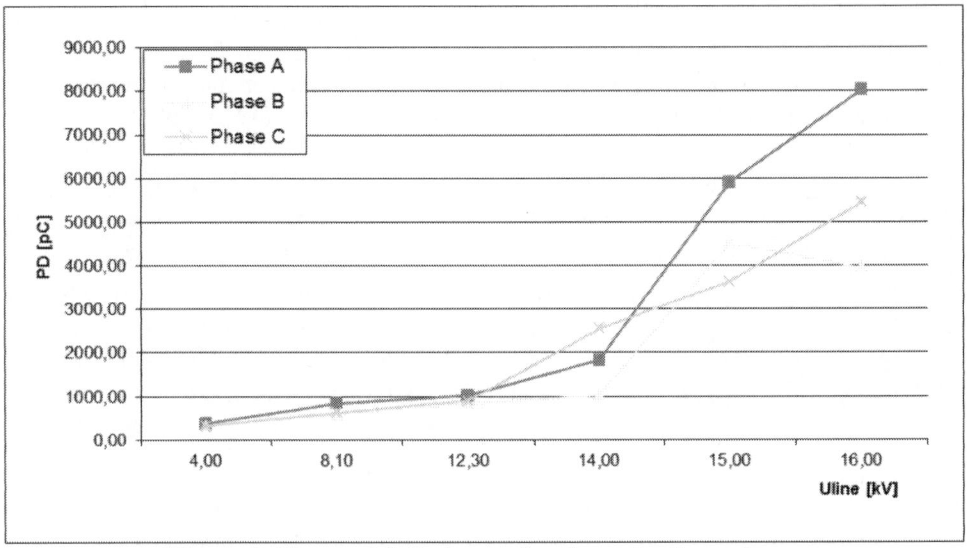

Figure 3. The partial discharges levels on the three phases

In Figure 3 we notice am increase in the level of partial discharge, according to the applied tension at the generator terminals, that is a sudden increase for the tension value at the generator terminals starting at U_{line}=14kV, especially on phase A and a slow increase of the tentsion at the generators terminals between U_{line}=4kV and U_{line}=14kV.

The results of the measurements presented in Figure 4 highlights degradations of the generators tested coiling, especially on phase A, in case in which the level of partial discharges surpasses a lot the recommended value of 1000 pC. This fact is due to the high degree of condense on the cooling conducts passing above the rotor.

For each different values of the line tension, there have been measured on the 3 phases of the hydro- generator: the partial discharges and the temperatures (upper radial bearing temperature, axial bearing temperature, inferior radial bearing temperature, turbine bearing temperature and ambient temperature), their values being inscribed in Table 8.

Table. 8 Average values of partial discharges and temperatures at different values of the line tensions

Uline [kV]	DPA [pC]	DPB [pC]	DPC [pC]	L.R.S. [^oC]	L.A.X. [^oC]	AIR [^oC]	L.R.I. [^oC]	L.T. [^oC]
4	382,33	324,99	332,20	51.87	63.66	27.8	49.8	45
8.1	841,27	588,60	627,05	52	63.8	27.5	50.2	46
12.3	1019,69	826,87	896,88	52	64	27.3	50.2	48.5
14	1831,26	1023,46	2553,14	52.25	64	27.7	50.4	50
15	5900,01	4518,75	3629,69	52	64	28.3	50.4	51
16	8026,56	3943,75	5437,51	52.37	64	28.7	50.4	51.5

For the measure of the coiling temperature, is placed between the bars of some of the coils thermoresistances, placed in the most requested thermal points.

For the measure of the bearing temperature, the sensors are placed in some of the segments, as well as in the bearing bath, thermoresistances. The transducer is part of plant monitoring of the boiler and is introduced into the oil bearing bath. In Figure 4, the average value of the partial discharge levels is represented graphically on the three phases according to the radial upper bearing temperature.

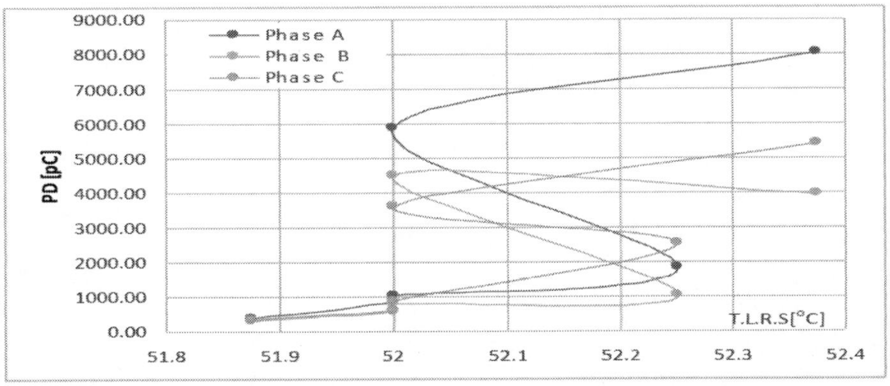

Figure 4. Radial upper bearing temperature

In Figure 5, the average value of the partial discharge levels is represented graphically on the three phases according to the axial bearing temperature.

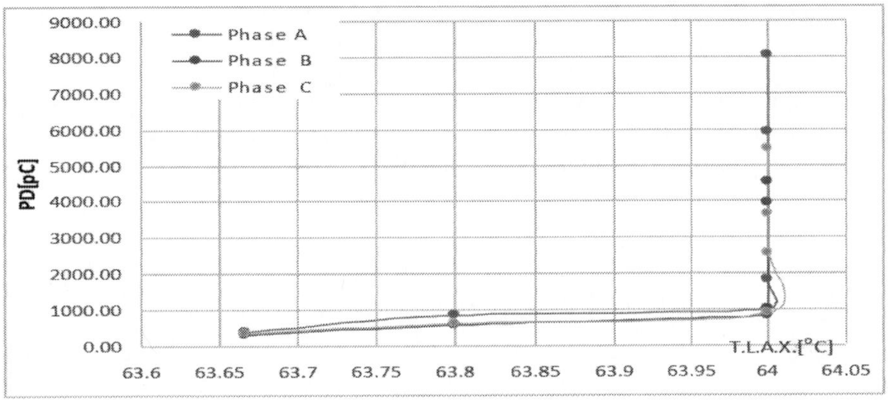

Figure 5. Axial bearing temperature

In Figure 6, the average value of the partial discharge levels is represented graphically on the three phases according to the inferior radial bearing temperature.

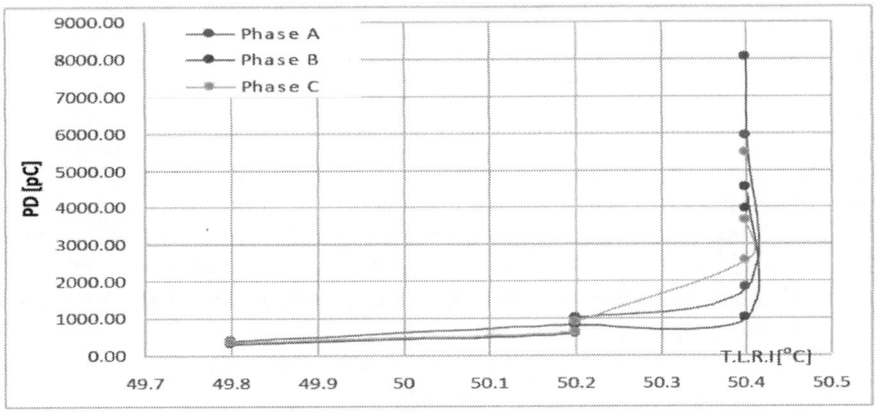

Figure 6. Inferior radial bearing temperature

In Figure 7, the average value of the partial discharge levels is represented graphically on the three phases according to the turbine bearing temperature.

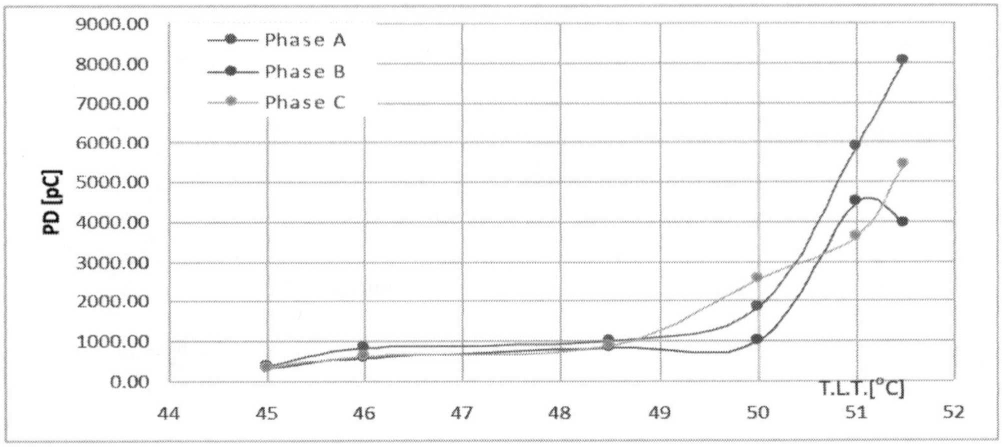

Figure 7. Turbine bearing temperature

To measure the ambient temperature, the sensor was mounted on a stem support at a height of 1.5 m in the generator fosse. In Figure 8, the average value of the partial discharge levels is represented graphically on the three phases according to the ambient temperature.

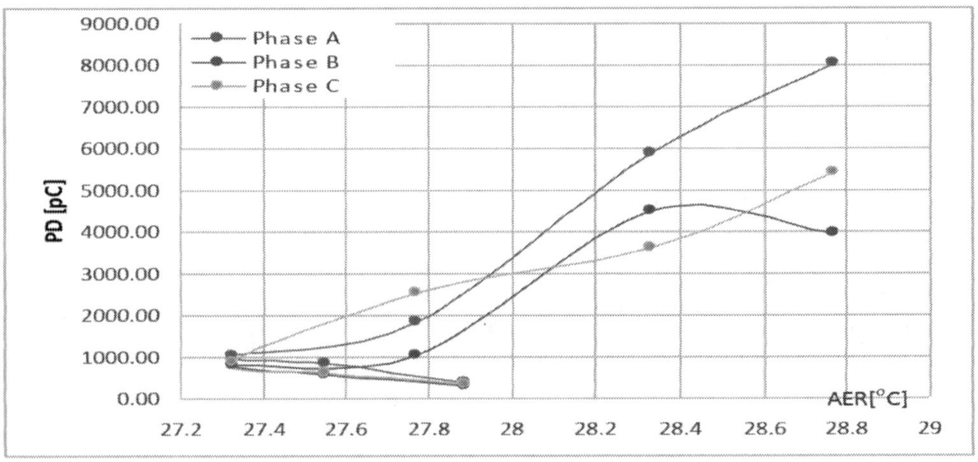

Figure 8. Ambient temperature

From Figure 8, it is observed that the average value of the partial discharge levels on the three phases increases approximately linearly, and the temperature variation is small.

From Figures 4,5,6 and 7, it is observed that the average value of the partial discharge levels on the three phases increases a lot, although the temperature variation on upper radial bearing, axial bearing, inferior radial bearing respectively turbine bearing is very low. This is because the voltage applied to the generator terminals is greater than 14 kV.

4. Conclusions

The Figure shows that for small values of line tension (up to 12,3kV), the partial discharges increase linearly with the temperature up to 1000 pC.

The Figure 3 shows that for high line tension (over 14 kV), the partial discharges increase sharply reaching a average value of 8026,56 pC. Note that although partial discharges have a sudden increase, the temperatures vary little.

One has observed, by measurement, that if the partial discharges increase linearly (for higher values of 1000 pC), depending on the temperature for different voltages, then the hydro- engine can be used safely.

It is noticed that the temperature in the cavity does not change, but there is a temperature gradient that increases with the partial discharge level, which means that the microcavity becomes a source of heat. Measurement of partial discharges found application as a test design, quality assurance test and condition evaluation test for high-voltage equipment.

References

[1] Paoletti G and Golubev A 1999 *Partial Discharge Theory and Applications to Electrical Systems*, Pulp and Paper, Industry Technical Conference Record of 1999 Annual, Seattle, WA, USA, pp 124-138

[2] Kannan M and Sreejaya P 2013 Partial discharge detection in solid dielectrics, *International Journal of Scientific & Engineering Research* **4**(8) 1-6

[3] Oliveira S C and Fontana E 2009 Optical Detection of Partial Discharges on Insulator Strings of High-Voltage Transmission Lines, *IEEE Transactions on Instrumentation and Measurement* **58**(7) 2328–2334

[4] Stone G C, Warren V, Sedding H G and Mcdermid W 2002 *Advances In Interpreting Partial Di From Scharge Test Results Motor And Generator Stator Windings*, Iris Power Engineering, 2-1 Westside Drive, Toronto, ON, Canada

[5] Li R, Pan L, Yan C, Li H and Hu B 2013 Condition Evaluation of Large Generator Stator Insulation Based on Partial Discharge Measurement, *Advances in Mechanical Engineering*

[6] Popescu C and Hatiegan C 2015 Aspects regarding the Calculation of the Dielectric Loss Angle Tangent between the Windings of a Rated 40 MVA Transformer, *Analele Universității "Eftimie Murgu"* **22**(2) 298-305

[7] Popescu C and Popescu L G 2014 *Aspects concerning the influence of the temperature on the ohmic resistance of the winding transformers with rated power 400 MVA*, Conference Proceedings SGEM 2014, Albena, Bulgaria, pp 19-24

[8] Hațiegan C, Padureanu I, Jurcu M R, Biriescu M, Răduca M and Dilertea F 2017 The evaluation of the insulation performances of the stator coil for the high power vertical synchronous hydro-generators by monitoring the level of partial discharges, *Electr Eng* **99**(3) 1013-1020

[9] Tokar A, Foris D, Negoițescu A and Foris T 2016 *The vulnerability of insolated communities and buildings regarding of the utilitties access*, 4rd International Multidisciplinary Scientific Conference on Social Sciences and Arts SGEM 2017, Vol. 2, pp 301-308

[10] Hațiegan C, Hălălae I, Popescu C, Gillich N, (Barboni) Hațiegan L, Răduca E and (Filip) Nedeloni L 2016 Evaluation insulation of the stator coil a hydro-generators through monitoring the level of partial discharges, *Annals of „Constantin Brâncuși" University of Târgu Jiu* **3** 57-62

[11] Hațiegan C, Chioncel C P, Răduca E, Popescu C, Pădureanu I, Jurcu M , Bordeașu D, Trocaru S, Dilertea F, Bădescu O, Terfăloagă I M, Băra A and (Barboni) Hațiegan L 2016 Determining the operating performance through electrical measurements of a hydro generator, *IOP Conf. Ser.: Mater. Sci. Eng.* **163** 012031

[12] Tokar A, Negoițescu A, Hamat C and Roșu Ș 2016 The chemical and ecological state evaluation of a storage lake, *Revista de Chimie* **67**(9) 1860-1863

[13] Pădureanu I, Jurcu M R, Augustinov L, Hațiegan C, Răduca E and Pădeanu L 2015 Optimisation of the Start-up and Operation Regimes of Cooling Water Pumps of a High-Power Hydro Generator, *Analele Universității "Eftimie Murgu", Fascicula de Inginerie* **XXII**(1) 345-358

[14] Tokar A, Negoiţescu A, Adam M and Tokar D 2015 Research on Mechanical Strength of Technological Fluid Storage Tank Made of Polyester Resin Reinforced with Fiberglass, *Revista de Materiale Plastice* **51**(3) 432-434

[15] Pădureanu I, Jurcu M, Augustinov L, Haţiegan C and Răduca E 2015 Implementation of an Automatic System for the Monitoring of Start-up and Operating Regimes of the Cooling Water Installations of a Hydro Generator, *Analele Universităţii "Eftimie Murgu", Fascicula de Inginerie* **XXII**(1) 359-368

[16] Jurcu M R, Padureanu I, Padeanu L, Augustinov L and Hatiegan C 2015 Tests Regarding the Transitory Regimes of Putting off Load of the Hydroagregate, *Analele Universităţii "Eftimie Murgu", Fascicula de Inginerie* **XXII**(2) 185-195

[17] Novăcescu F, Haţiegan C, Răduca M, Răduca E, Pop N and Nedeloni M 2012 A New Method for Testing the No-Load Work of an Electric One-Phase Transformer Using The Graphical Programming, *Scientific Bulletin of "Politehnica" University of Timişoara* **57**(71)2 54-67

[18] Haţiegan C, Răduca M, Frunzăverde D, Răduca E, Pop N and Gillich G R 2013 The modeling and simulation of the thermal analysis on the hydrogenerator stator winding insulation, *J. Therm Anal Calorim* **113**(3) 1217-1221

[19] Răduca M, Haţiegan C, Pop N, Răduca E and Gillich G R 2014 Finite element analysis of heat transfer in transformers from high voltage stations, *J Therm Anal Calorim* **118**(2) 1355-1360

[20] Popescu C, Popescu L and Popa R 2014 Aspects concerning the plastics used in the insulation of the conductors and the electrical cables, *Revista de materiale plastice* **51**(3) 309-312

[21] Tokar A, Retezan A and Negoiţescu A 2017 Estimative analysis of investment payback for industrial spaces heated with radiant tubes, *Revista Română de Inginerie Civilă* **8**(1) 12-18

[22] Negoiţescu A, Tokar A and Hamat C 2016 The influence of the air thermal and rainfall regimes on storage lakes water turbidity, *Materiale Plastice* **53**(3) 542-545

[23] Haţiegan C, Molnar M, Trocaru S, Pădureanu I, Jurcu M R and Ilie F 2016 *Modeling and Simulation of Thermal Analysis of a Teflon Coated Plate, International Conference Knowledge-Based Organization* **22** 639-643

[24] Răduca M, Răduca E, Haţiegan C and Ungureanu D 2011 Fuzzy controller for adjustment of liquid level in the tank, *Annals of the University of Craiova, Mathematics and Computer Science Series* **38**(4) 33-43

[25] Răduca M, Haţiegan C, Budai A M, Răduca E and Pop N 2012 *Optimization conditions of stator winding insulation of hydrogenerator*, National Symposium of Theoretical Electrical Engineering SNET'12, 14 December, Bucuresti, **3**(1) 316-321

[26] Lee S B, Younsi K and Kliman G B 2005 An online technique for monitoring the insulation condition of AC machine stator windings, *IEEE Transactions on Energy Conversion* **20**(4) 737-745

[27] Danikas M G and Prionistis F K 2004 Detection and recording of partial discharge below the so-called inception voltage, *Facta Univeritatis, ser. Elec. Energ.* **17** 99-110

[28] Popescu C, Haţiegan C, Racoceanu C and Bejinariu A C 2017 Aspects on the influence of starting the own services consumers of an energy group with unit of 330 mw on the power supply, *International Conference Knowledge-Based Organization* **23**(3) 84-89

Statistical research using the multiple regression analysis in areas of the cast hipereutectoid steel rolls manufacturing

I Kiss[1], V Alexa[1], S Serban[1], M Rackov[2] and M Čavić[2]

[1]University Politehnica Timisoara, Faculty of Engineering Hunedoara, Department of Engineering and Management, Hunedoara, Romania
[2]University of Novi Sad, Faculty of Technical Sciences, Novi Sad, Serbia

E–mail: imre.kiss@fih.upt.ro

Abstract. The cast hipereutectoid steel (usually named Adamite) is a roll manufacturing` destined material, having mechanical, chemical properties and Carbon [C] content of which stands between steelandiron, along–withitsalloyelements such as Nickel [Ni], Chrome [Cr], Molybdenum [Mo] and/or other alloy elements. Adamite Rolls are basically alloy steel rolls (a kind of high carbon steel) having hardness ranging from 40 to 55 degrees Shore C, with Carbon [C] percentage ranging from 1.35% until to 2% (usually between 1.2~2.3%), the extra Carbon [C] and the special alloying element giving an extra wear resistance and strength. First of all the Adamite roll`s prominent feature is the small variation in hardness of the working surface, and has a good abrasion resistance and bite performance. This paper reviews key aspects of roll material properties and presents an analysis of the influences of chemical composition upon the mechanical properties (hardness) of the cast hipereutectoid steel rolls (Adamite). Using the multiple regression analysis (the double and triple regression equations), some mathematical correlations between the cast hipereutectoid steel rolls` chemical composition and the obtained hardness are presented. In this work several results and evidence obtained by actual experiments are presented. Thus, several variation boundaries for the chemical composition of cast hipereutectoid steel rolls, in view the obtaining the proper values of the hardness, are revealed. For the multiple regression equations, correlation coefficients and graphical representations the software Matlab was used.

1. Introduction

The hot rolling is a highly complex engineering process, with interactions of the multiple process variables, which extremely dependents by the operating conditions. Within the framework of the hot rolling operations predominate the wear, it is necessary that rolls to have good hardness [1–3]. All of the rolls must have high exploiting qualities, which are determined by hardness, wear resistance and stability at high temperatures [1–3]. When the roll hardness is higher, the wear resistance is greater [1–3]. Since the properties of any cast alloy parts are determined by microstructure which are formed during solidification in the casting mould and under the influence of cooling rate, the criterion that determines the basic physic–mechanical properties of the rolls is the structure. Therefore, the composition of the alloy is an important parameter [1–3]. In case of Adamite hyper–eutectoid steel rolls the hardness depends on the degree of alloying elements and the special high temperature heat treatment process (annealing) followed by tempering cycle which determine the microstructure [4].

The intensification of rolling processes influences, directly, the durability of rolls, they being the most important components in the metal rolling [1–3]. The efficiency of rolling primarily depends by the quality of rolls, which durability in exploitation is determined by the features of material they are manufactured, the features of material that is laminated, and, not least, the exploitation conditions. Decreasing durability in exploitation of the rolls, which work in variable requests conditions due to the deformation process (requests that repeat cyclically, to certain intervals), represent one of the most actual problems met in the rolling sectors [1–3]. Therefore, the current work wants to answer to more of the problems linked by this aspect, having in focus the Adamite hyper–eutectoid steel rolls, creating a connection bridge between the casting manufacturing factors and the rolling exploitation conditions.

Figure 1. The cast hipereutectoid steel roll (usually named Adamite)

Prediction of the properties of the resulting cast alloy is a prerequisite for the management of metallurgy, especially in materials destined to the cast rolls (irons or steels). Required metallurgical quality of a particular type of alloys (irons and steels) and its properties are determined by chemical composition and a proper melting processing. Control of microstructure and high exploiting properties of cast alloy destined to rolling rolls is in perfect control of the melting process and the metallurgical melt processing [3], [5–12].

In the manufacturing rolls for hot rolling the roll makers are engaged in the production of various rolls including cast iron rolls and steel rolls (including Adamite steel). The best way for roll makers to achieve better rolls is to ensure that better materials and improved manufacturing processes are used and that roll users take account of rolling conditions and improved rolling processes. The problem of roll selection is complicated by the fact that mill conditions vary widely [3], [5–12]. In order to secure superior quality and durability characteristics, the main problem was and is the optimal chemical composition assurance. Therefore the roll grade is very important on roll properties. However, rolling conditions are important factors [1–3], [5–12].

2. Technological area of research: Adamite hyper–eutectoid steel rolls

Cast steel base rolls (or semi–steel base rolls) are classified into Adamite rolls and graphitic steel rolls. Cast steel rolls are available in wide of chemical compositions varying from hypo–eutectoid to hyper–eutectoid steels. Steel base rolls are heat treated in various ways and have no hardness drop [4].

The Adamite steel rolls typically have great strength, good thermal properties and high wear resistance. Unlike other metal rolls, these have uniform hardness from depth to the surface which makes them very special. Hardness of the Adamite steel rolls is maximum amongst the steel based rolls. Adamite steel rolls are stronger than cast iron rolls and harder than other steel rolls. The biggest advantage of these rolls is almost no hardness gradient in the working layer. These rolls show a low hardness drop from surface layer to the cross sectional centres and possess a high friction coefficient good for biting. Therefore, the Adamite hyper–eutectoid steel rolls are acknowledged for superior finish, durable life, wear resistance, strength and toughness [4], [13–15].

Adamite steel rolls are widely used for hot rolling applications in the different type of pre–roughing and roughing stands and also different positions in the section mills as well. Adamite hyper–eutectoid steel rolls are suitable for rolling medium and heavy sections where resistance is very important along with considerable amount of strength and toughness. Good biting is usually required for rolls for roughing stands, for which anti–heat cracking properties and high toughness are also required. These rolls are used in roughing mills and front stands of the hot strip mills. Also, they are used in roughing stands, intermediate stands and pre–finishing stands for bar mills [4], [13–15].

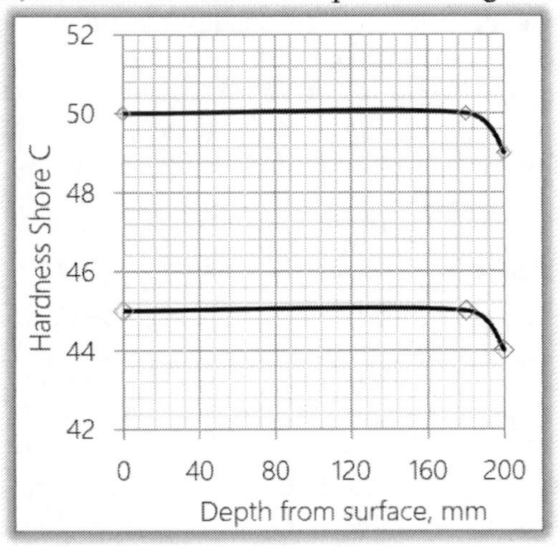

Figure 2. Hardness gradient in the working layer of Adamite hyper–eutectoid steel rolls (Hardness: 45–50 Shore / 308–344 Brinell)

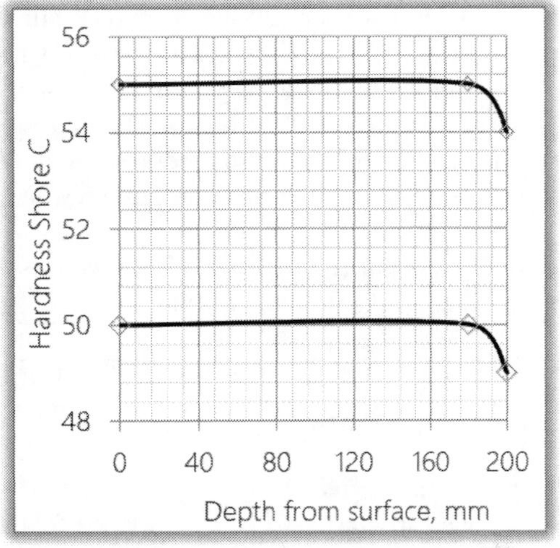

Figure 3. Hardness gradient in the working layer of Adamite hyper–eutectoid steel rolls (Hardness: 50–55 Shore / 345–381 Brinell)

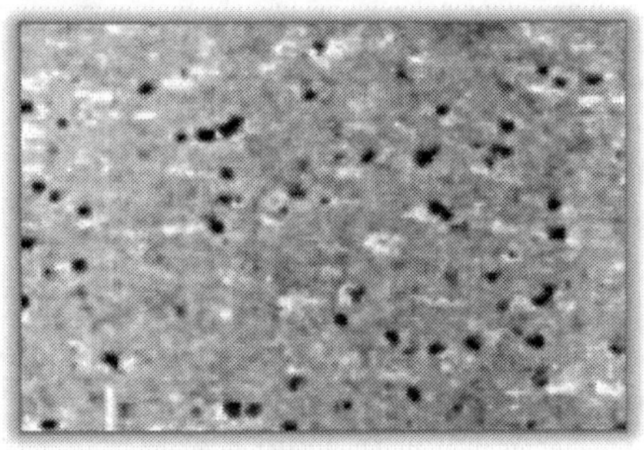

Figure 4. Microstructure of Adamite hyper–eutectoid steel rolls (Hardness: 45–50 Shore / 308–344 Brinell)

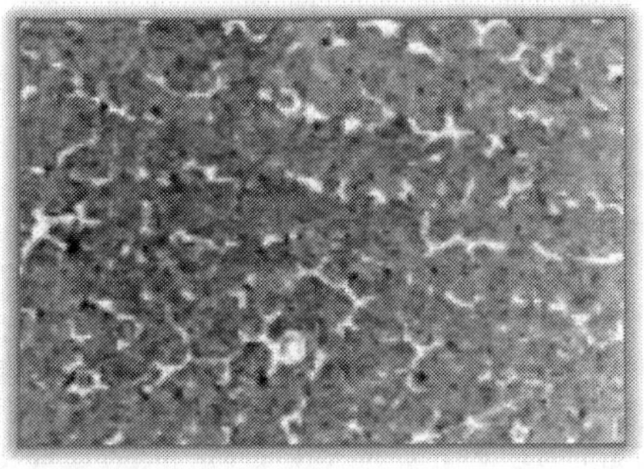

Figure 5. Microstructure of Adamite hyper–eutectoid steel rolls (Hardness: 50–55 Shore / 345–381 Brinell)

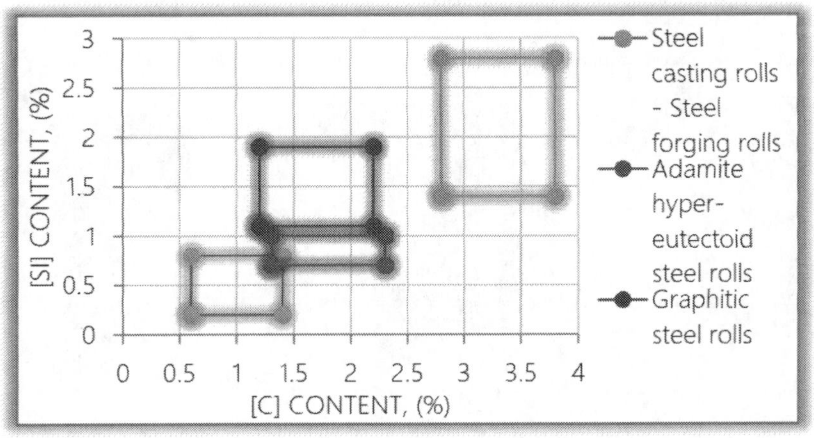

Figure 6. The technological domain of the Adamite hyper–eutectoid steel rolls

Adamite Rolls are special hyper–eutectoid steel rolls alloyed with Nickel, Chromium and Molybdenum. In fact, the Adamite is a material, mechanical properties and Carbon content of which stands between steel and iron [4], [13–15]. The extra Carbon – between 1.0 and 2.2% – and special alloys give it extra wear resistance and strength. Adamite steel rolls are mainly composed of fine pearlitic matrix with some cementite in the structure. The structure consists of well broken up and dispersed carbides in a pearlitic matrix complementing the hardness and wear resistance imparted by highly alloyed matrix [4], [13–15]. Depending on the primary requirements there are varieties of grades with a series of carbon content and alloying elements used considering the rolling mill operational conditions such as wear, strength, thermal cracks etc.

Table 1. Typical compositions of Adamite hyper–eutectoid steel rolls

Adamite hyper–eutectoid steel rolls	Class 1	Class 2	Class 3
Composition (%)			
Carbon [C]	1.40–1.60	1.50–1.80	1.60–2.00
Silicon [Si]	0.40–0.70	0.30–0.70	0.30–0.70
Manganese [Mn]	0.50–0.80	0.50–1.00	0.50–1.00
Nickel [Ni]	0.40–1.00	0.70–1.10	max 1.50
Chromium [Cr]	0.80–1.20	1.00–1.40	1.10–1.60
Molybdenum [Mo]	0.20–0.40	0.20–0.40	0.25–0.50
Mechanical Properties			
Hardness (Shore)	40–45	45–50	50–55
Hardness (Brinell)	270–308	309–344	345–381

Alloying combinations of different element like Nickel, Chromium and Molybdenum are normally used to achieve the desired physical and mechanical properties.In view of the demanding end use requirements these rolls cannot be used in an as cast condition. When a coarse grain pearlitic structure with cementite and ferrite in a continuous network is normally obtained. The rolls are subject to specially designed heat treatment cycles to arrive at the optimum microstructure configuration with fine pearlite and uniformly distributed suitable for the final application [4], [13–15].

Adamite hyper–eutectoid steel rolls are statically cast and undergo a special high temperature heat treatment process (annealing) followed by tempering cycle to give a microstructure consisting of fine pearlitic matrix with spherodised/broken carbide [4], [13–15]. This multi–stage heat treatment process followed by tempering phases are very important operations in order to develop the combination of hardness, toughness and wear resistance required for heavy duty wear applications [4], [13–15]. Owing to its alloy elements such as Nickel, Chromium and Molybdenum, there are some amounts of carbides in the matrix and because of certain special heat treatment technology, high wearability, good toughness and excellent resistance to fire crack is obtained [4], [13–15].

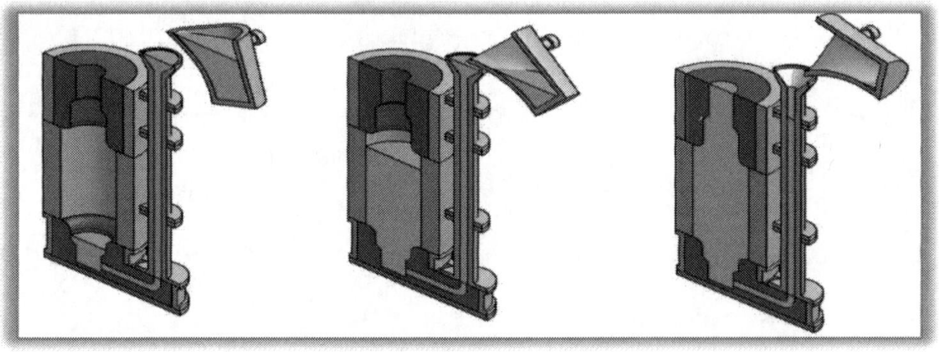

Figure 7. The static casting (bottom pour casting) of Adamite hyper–eutectoid steel rolls

The static casting is widely used to produce rolls for roughing mills, section mill, as well as universal mills [3]. The bottom pour casting is applied in the analysed Adamite hyper–eutectoid steel rolls production. By the control of the cooling speed techniquesby combined casting forms (sand for the necks and chill for the body) and the future heat treatment process, the static roll can get the performance requested on entirecast steel roll. The moulding practices, chill properties and casting specifications of the hyper–eutectoid steelare carefully controlled to achieve proper solidification rates across the full range of produced rolls. Exceptional care is needed during the casting process [3].

3. Methodology: the regression analysis

Modelling of cast rolls technologies by multivariate regression analysis are used to predict the rolls performance and to optimize the chemical composition components during the manufacturing process (i.e. melting and casting processes) and also to provide insights into control strategies that may improve the rolls exploitation properties [3], [5–12].

Many studies have been reported on mathematical modelling of cast rolls. Most of the studies reviewed above worked on modelling of cast iron rolls and their technological components which assure the proper hardness by the chemical composition variation, but not reviewed the cast steel–base rolls [3], [5–12]. We developed a general steady state mathematical model for cast iron rolls that can be applied to the hyper–eutectoid steel rolls too. The governing equations coupled with pertinent boundary conditions are solved by the correlation diagrams generation, in the Matlab environment. In all the above studies, the combined effect of basic chemical composition in addition to the proper alloying elements has been considered separate or simultaneously, in case of the cast iron rolls.

Here we have made an attempt to study the combined effects on the hardness, in case of the Adamite hyper–eutectoid steel rolls. We implemented into the research several lots of industrial data analyses, by the way of the statistical experiment, in two separate series. This analysis gathers within the framework of research, the cast Adamite rolls and aims to characteristics of the material (i.e. hyper–eutectoid steel). The analyses concern the influence of main basic and alloying elements in the chemical composition of the Adamite hyper–eutectoid steel, as well as the conditions under which the hardness is affected by the cumulative influence of the chemical components.

As results of the applied mathematical modelling, we described a number of multi–component regression equations, determined to the 3rd and 4th dimensions spaces. Also, we generated several regression surfaces and matrices of the level curves, which can be represented and interpreted by the roll makers and may be considered as correlation charts between the analyzed variables [3], [5–12]. For the multiple regression equations, correlation coefficients and graphical representations the software Matlab was used [3], [5–12].

4. Multicomponent relations determination

The variables and the hardness variation limits are presented in Table 2. The average values for the variables and the hardness, necessary for the calculation of the proper form of modelling are presented in Table 3.

Table 2. The variation domains of variables (basic and alloying elements)

Carbon, [C]		Silicon, [Si]		Manganese, [Ni]		Nickel, [Ni]		Chromium, [Cr]		Molybdenum, [Mo]		Hardness, [HB]	
min	max	min	Max	min	max	min	max	min	max	min	max	min	max
1.52	2.02	0.54	0.72	0.60	0.92	1.17	1.66	1.04	1.37	0.24	0.45	311	412

Table 3. The average values of variables

Carbon, [C]	Silicon, [Si]	Manganese, [Ni]	Nickel, [Ni]	Chromium, [Cr]	Molybdenum, [Mo]	Hardness, [HB]
Med	med	med	med	Med	med	med
1.73	0.62	0.76	1.31	1.22	0.32	356

The optimal form of modeling in the case of basic elements ([C], [Si], [Mn]) effect is given by the equation (1), where the correlation coefficient is $R^2 = 0.9322$. The modeling in the case of alloying elements ([Ni], [Cr], [Mo]) effect is given by the equation (2), where the correlation coefficient is $R^2 = 0.9330$. The polynomial coefficients have the values presented in Table 4.

$$[HB] = a_1[C]^2 + a_2[Si]^2 + a_3[Mn]^2 + a_4[C][Si] + a_5[Si][Mn] + a_6[Mn][C] + a_7[C] + a_8[Si] + a_9[Mn] + a_{10} \quad (1)$$

$$[HB] = a_1[Ni]^2 + a_2[Cr]^2 + a_3[Mo]^2 + a_4[Ni][Cr] + a_5[Ni][Mo] + a_6[Cr][Mo] + a_7[Ni] + a_8[Cr] + a_9[Mo] + a_{10} \quad (2)$$

Table 4. The polynomial coefficients of equations (1)–(2)

The polynomial coefficients	HB=f([C], [Si], [Mn])	HB=f([Ni],[Cr],[Mo])
the second–degree terms coefficients	$a_1 = 277.3852$	$a_1 = 69.9558$
	$a_2 = 1001.5596$	$a_2 = 133.2416$
	$a_3 = 318.0048$	$a_3 = -436.2808$
the product terms coefficients	$a_4 = -827.4239$	$a_4 = 433.7566$
	$a_5 = -324.6609$	$a_5 = 155.5801$
	$a_6 = -146.7932$	$a_6 = 923.9556$
the first–degree terms coefficients	$a_7 = -69.9044$	$a_7 = -669.3177$
	$a_8 = 300.0813$	$a_8 = -1331.4557$
	$a_9 = 285.3223$	$a_9 = -976.6554$
the constant term	$a_{10} = 56.4979$	$a_{10} = 1780.7491$

Because the two surfaces described by the equations (1)–(2) cannot be represented graphically, the independent variables were successively replaced with their average values ($[C]_{med}$, $[Si]_{med}$, $[Mn]_{med}$, $[Ni]_{med}$, $[Cr]_{med}$ and $[Mo]_{med}$). Therefore, in the case of basic elements correlation, the equations (3)–(5) were obtained. The correlation coefficient are R^2 at $[C]_{med} = 0.8195$, R^2 at $[Si]_{med} = 0.9122$ and R^2 at $[Mn]_{med} = 0.8940$. The polynomial coefficients of equations (3)–(5) in Table 5 are presented.

$$[HB] \text{ at } [C]_{med} = b_1[Si]^2 + b_2[Mn]^2 + b_3[Si][Mn] + b_4[Si] + b_5[Mn] + b_6 \quad (3)$$

$$[HB] \text{ at } [Si]_{med} = c_1[C]^2 + c_2[Mn]^2 + c_3[C][Mn] + c_4[C] + c_5[Mn] + c_6 \quad (4)$$

$$[HB] \text{ at } [Mn]_{med} = d_1[C]^2 + d_2[Si]^2 + d_3[C][Si] + d_4[C] + d_5[Si] + d_6 \quad (5)$$

Table 5. The polynomial coefficients of equations (3)–(5)

The polynomial coefficients	HB=f([C], [Si], [Mn])		
the second–degree terms coefficients	$b_1 = 1001.5595$	$c_1 = 277.3851$	$d_1 = 277.3851$
	$b_2 = 318.0048$	$c_2 = 318.0048$	$d_2 = 1001.5595$
the product terms coefficients	$b_3 = -146.7932$	$c_3 = -324.6609$	$d_3 = -827.4239$
the first–degree terms coefficients	$b_4 = -1135.3614$	$c_4 = -583.4589$	$d_4 = -316.8632$
	$b_5 = -277.9103$	$c_5 = 194.2126$	$d_5 = 188.4206$
the constant term	$b_6 = 770.0567$	$c_6 = 628.5762$	$d_6 = 457.5352$

Similarly, in the case of alloying elements correlation, the equations (6)–(8) were obtained. The correlation coefficient are R^2 at $[Ni]_{med} = 0.8358$, R^2 at $[Cr]_{med} = 0.8048$ and R^2 at $[Mo]_{med} = 0.9160$. The polynomial coefficients of equations (6)–(8) in Table 6 are presented.

$$[HB] \text{ at } [Ni]_{med} = b_1[Cr]^2 + b_2[Mo]^2 + b_3[Cr][Mo] + b_4[Cr] + b_5[Mo] + b_6 \quad (6)$$

$$[HB] \text{ at } [Cr]_{med} = c_1[Ni]^2 + c_2[Mo]^2 + c_3[Ni][Mo] + c_4[Ni] + c_5[Mo] + c_6 \quad (7)$$

$$[HB] \text{ at } [Mo]_{med} = d_1[Ni]^2 + d_2[Cr]^2 + d_3[Ni][Cr] + d_4[Ni] + d_5[Cr] + d_6 \quad (8)$$

Table 6. The polynomial coefficients of equations (6)–(8)

The polynomial coefficients	HB=f([Ni],[Cr],[Mo])		
the second–degree terms coefficients	$b_1 = 133.2416$ $b_2 = -436.2808$	$c_1 = 69.9558$ $c_2 = -436.2808$	$d_1 = 69.9558$ $d_2 = 133.2416$
the product terms coefficients	$b_3 = 923.9556$	$c_3 = 155.5801$	$d_3 = 433.7566$
the first–degree terms coefficients	$b_4 = -765.4032$ $b_5 = -773.6233$	$c_4 = -137.0983$ $c_5 = 157.0382$	$d_4 = -618.6764$ $d_5 = -1030.7081$
the constant term	$b_6 = 1026.4262$	$c_6 = 347.6522$	$d_6 = 1416.6237$

5. Generating the correlation diagrams

The regression surfaces described by equations (3)–(5), respectively equations (6)–(8) can be represented graphically, resulting several correlation diagrams.

Therefore, the correlation of twos independent variables can be made, so that the dependent variables can be obtained in between the requested limits. The correlation diagrams are presented in the Figures 8–13.

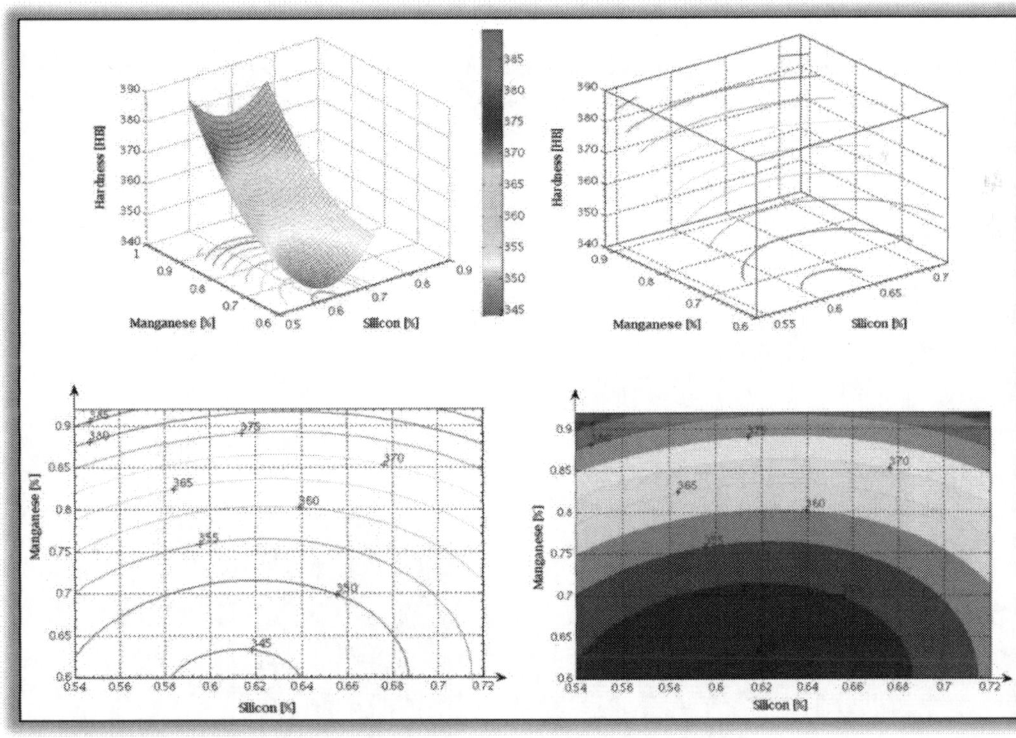

Figure 8. Correlation diagrams in case HB=f([C], [Si], [Mn]), when $[C]=[C]_{med}$

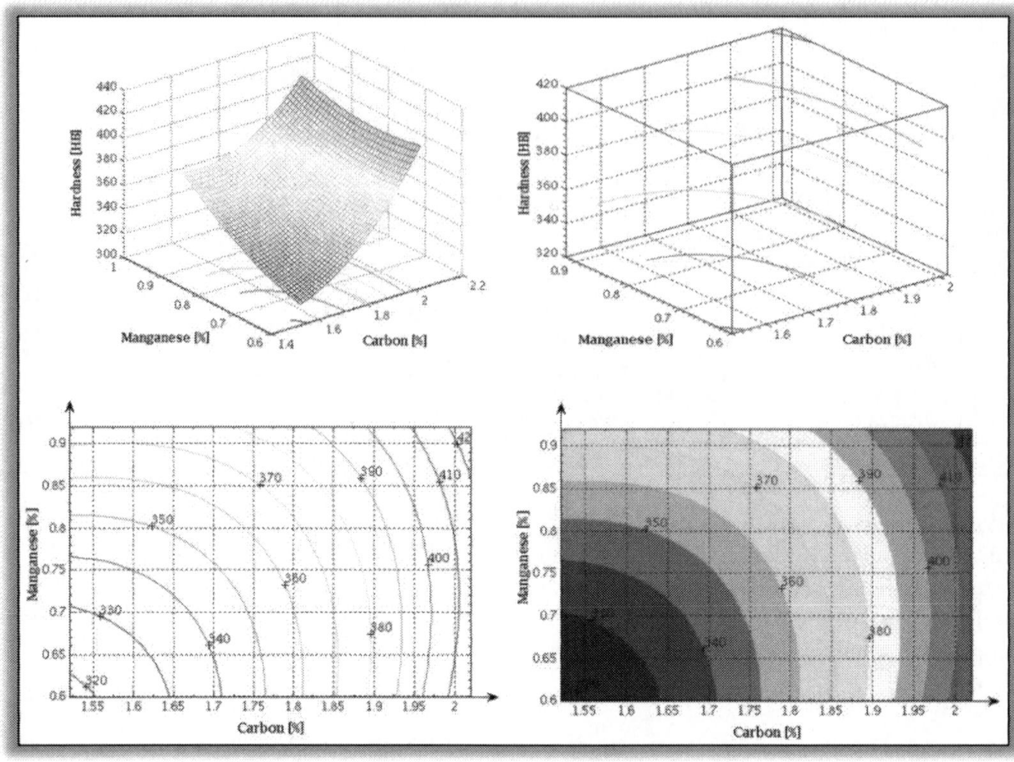

Figure 9. Correlation diagrams in case HB=f([C], [Si], [Mn]), when $[Si]=[Si]_{med}$

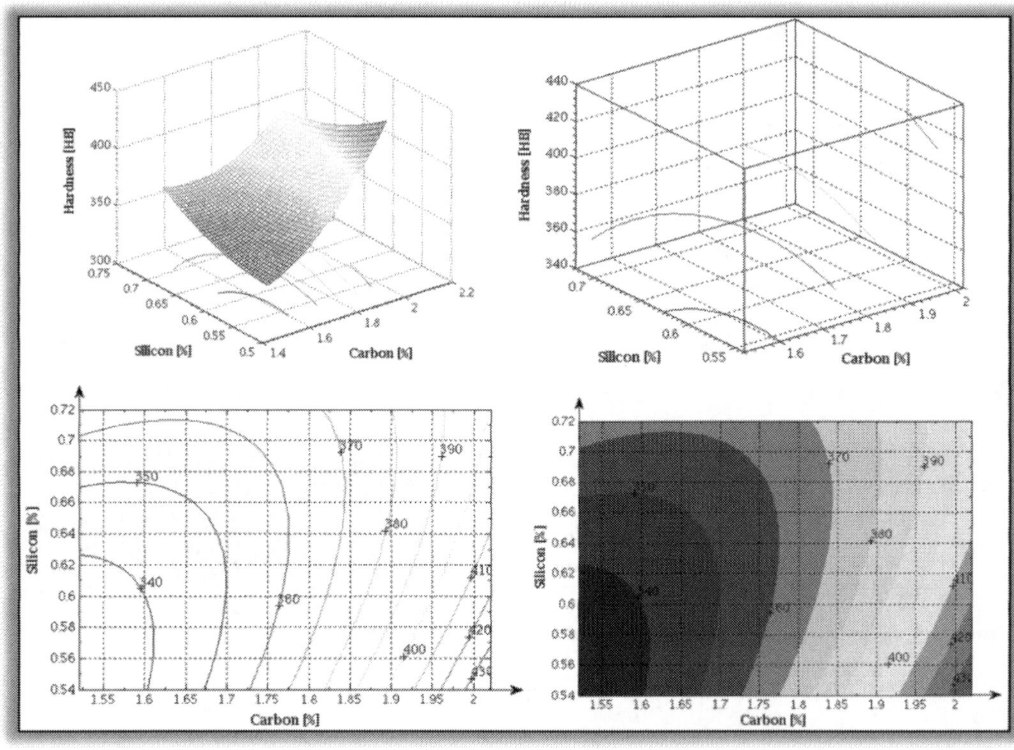

Figure 10. Correlation diagrams in case HB=f([C], [Si], [Mn]), when $[Mn]=[Mn]_{med}$

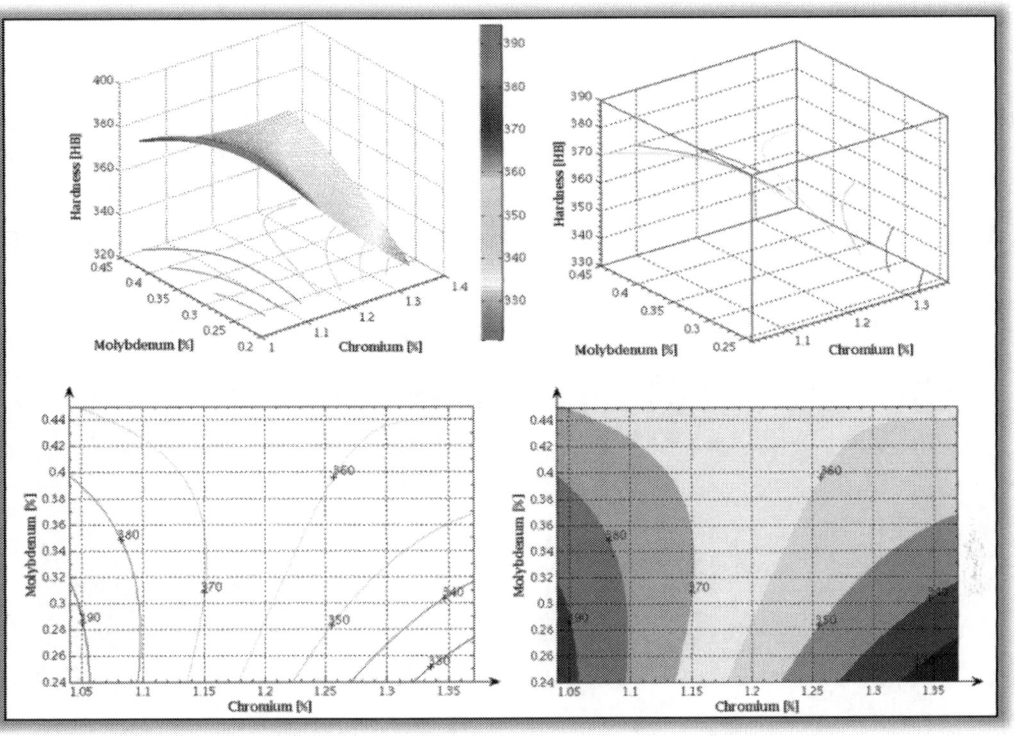

Figure 11. Correlation diagrams in case HB=f([Ni],[Cr],[Mo]), when $[Ni]=[Ni]_{med}$

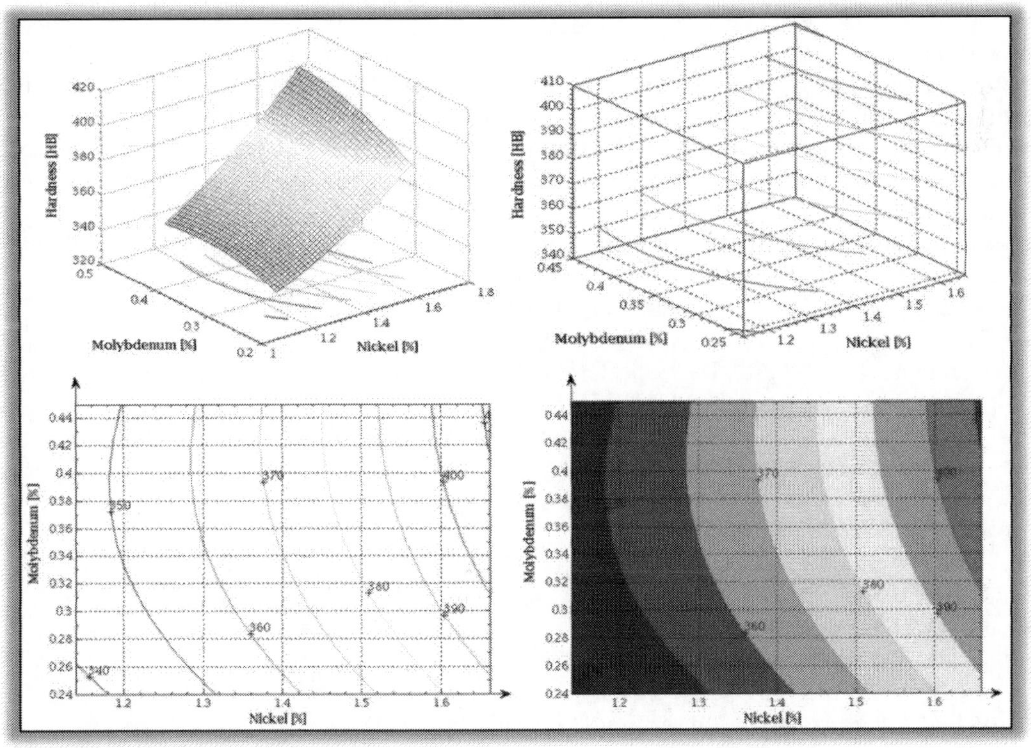

Figure 12. Correlation diagrams in case HB=f([Ni],[Cr],[Mo]), when $[Cr]=[Cr]_{med}$

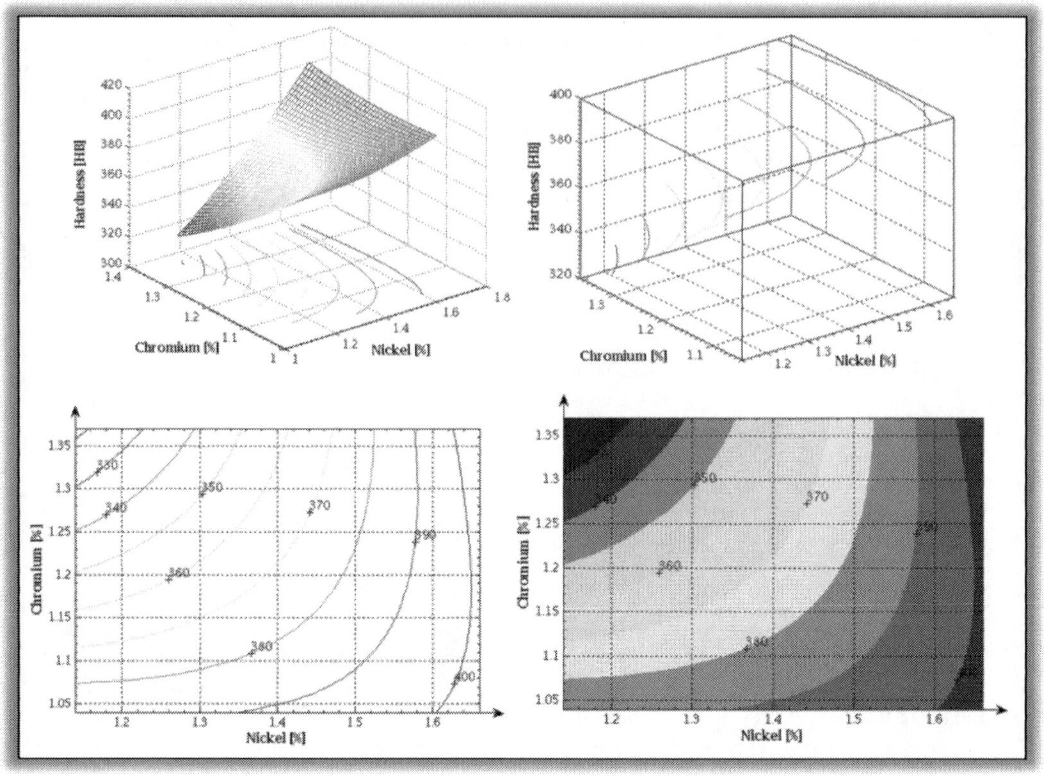

Figure 13. Correlation diagrams in case HB=f([Ni],[Cr],[Mo]), when $[Mo]=[Mo]_{med}$

6. Discussions

The optimal values of the chemical composition of hyper–eutectoid steel rolls are to be found on the correlation diagrams presented in Figures 8–10, respectively Figures 11–13. According to them, the optimal values in the concentration of each element can be noticed, values that can assure the adequate hardness of the rolls. Thus the optimal additions can be determined in these elements to assure the proper hardness`s.

If, according to a preferred material variant, the roll material has a composition:

» Carbon [C] = 1.5 to 2.0 %, preferably 1.7 to 2.0 %
» Silicon [Si] = 0.5 to 0.7 %, preferably 0.54 to 0.7 %
» Silicon [Si] = 0.7 to 2.0 %, preferably 1.0 to 2.0 %
» Chromium [Cr] = 0.6 to 1.5 %, preferably 0.7 to 1.4 %
» Nickel [Ni] = 0.8 to 1.6 %, preferably 1.0 to 2.0 %
» Molybdenum [Mo] = 0.2 to 0.6 %, preferably 0.25 to 0.4 %

with the remainder being iron and impurities.

According to the correlation diagrams, high quality assurance of the Adamite hyper–eutectoid steel rolls is achieved when the alloy contains:

» 1.7 to 1.8 % Carbon [C]
» 0.58 to 0.64 % Silicon [Si]
» 0.65 to 0.85 % Silicon [Si]
» 1.15 to 1.3 % Chromium [Cr]
» 1.3 to 1.5 % Nickel [Ni]
» 0.32 to 0.4 % Molybdenum [Mo]

with the remainder iron and impurities.

7. Conclusions

Hardness is strongly influenced by the roll surface conditions. And the variation of hardness in rolls is very important. For the roll makers the hardness assurance is the control of the roll manufacturing process. For the roll users, the hardness measurement is needed to avoid the roll failure due to surface fatigue. In fact, the hardness is the only property besides dimensions that can be tested by incoming inspections.

Careful investigations on the parameters given by roll users, combined with the vast experience of the roll makers in the roll materials area, often lead to improved solutions for finding a rolls which gives the best possible technically compromise. Besides the mechanical, physical and chemical properties of the roll materials the process technology and the parameters in the rolling mills are decisive for the roll`s performance.

Roll material is major factors affecting rolling performance and roll life. Therefore, a complex analyses of all the stages of manufacturing and the quality control process are needed. The performed research had in view to obtain correlations between the chemical composition and the hardness of cast steel–base rolls. Adamite hyper–eutectoid steel rolls are basically controlled under actual molding and casting conditions, therefore the desired hardness is obtained by the proper melting process, changing the chemical relationship between basic and alloying element, which is achieved by a close control of chemistry and process parameters. The proper exploitation properties requires extremely careful selection of melting metallic stock, closely controlled melting conditions (charging, temperatures regimes, time) and rigid control of chemical composition (alloying stages, corrections) to obtain the required hardness.

In the processing phase of the Adamite steel, the proper hardness is adjusted through the quality of the metallic charge and of the addition materials (especially the alloying elements addition, as well as through a proper leading of the melting and of the processing. The main chemical composition must be correlated with further addition of alloying elements, respecting the adequate proportions between nickel, chrome and molybdenum, besides an optimal ratio of carbon, silicon and manganese. Also, an optimal proportion between the silicon and the manganese contents is needed both from the basic metallic charges and from the ferro–alloy addition (FeSi, FeMn, SiMn).

References

[1] Schröder K H 2002 *Questions, answers, more questions, Twenty – five years of experience in discussing rolls and rolling technology*, ESW Handbook
[2] Schröder K H 1988 Rolling conditions in hot strip mills and their influence on the performance of work rolls, *Metallurgical Plant and Technology* **4** 44-56
[3] Kiss I 2008 *Rolling rolls – Approaches of quality in the multidisciplinary research*, Mirton Publishing House, Timisoara
[4] Ilca I, Kiss I, Alexa V and Ratiu S A 2016 Optimisation of the thermal treatment technologies for the cast hipereutectoid steel rolls, *Annals of F.E.H. – International Journal of Engineering*, **XIV**(3) 201-206
[5] Kiss I 2012 Investigations upon the indefinite rolls quality assurance in multiple regression analysis, *Revista de Metalurgia, CENIM*, **48**(2) 85-96
[6] Kiss I 2008 Research upon the quality assurance of the rolling–mill rolls and the variation boundaries of the chemical composition, *Revista de Metalurgia, CENIM*, **44**(4) 335-342
[7] Kiss I, Cioată V G and Alexa V 2010 Increasing the rolling–mill rolls quality in some multidisciplinary research, *Acta Technica Corviniensis – Bulletin of Engineering* **III**(2) 31-36
[8] Kiss I, Cioată V G, Alexa V and Ratiu S A 2011 Technological behaviour and interpretations in some multidisciplinary approaches, *Annals of F.E.H. – International Journal of Engineering* **IX**(4) 203-206
[9] Kiss I, Alexa V, Ratiu S A and Cioată V G 2014 The quality assurance in the rolling industry: Methods, approaches and tendencies, *Machine Design* **6**(4) 131-136

[10] Kiss I and Alexa V 2011 Statistical mathematic analysis and technological investigations in the area of the rolling mill rolls quality, *Machine Design* **3**(2) 79-84
[11] Kiss I and Alexa V 2012 The multivariate analysis in area of cast irons with nodular graphite, used in the rolling rolls manufacturing, *Machine Design* **4**(1) 43-48
[12] Kiss I, Alexa V and Cioată V G 2017 Statistical analysis using the multiple regression research in areas of the indefinite chilled cast–iron rolls manufacturing, *IOP Conf. Ser.: Mater. Sci. Eng.* **200** 012021
[13] Krawczyk J 2010 Influence of the microstructure and loads on tribological properties of cast steel, *Archives of Foundry Engineering* **10**(3) 39-44
[14] Hwang K C, Lee S and Lee H C 1998 Effects of alloying elements on microstructure and fracture properties of cast high speed steel rolls: Part I: Microstructural analysis, *Materials Science and Engineering: A* **254**(1–2) 282-295
[15] Noguchi H, Hiraoka H, Watanabe Y and Sayama Y 1988 Hardness and wear resistance of adamite for work rolls in hot rolling mill, *Transactions of the Iron and Steel Institute of Japan* **28**(6) 478-484

IOP Conf. Series: Materials Science and Engineering **294** (2017) 012078 doi:10.1088/1757-899X/294/1/012078

Statistical experiments using the multiple regression research for prediction of proper hardness in areas of phosphorus cast–iron brake shoes manufacturing

I Kiss[1], V G Cioată[1], S A Ratiu[1], M Rackov[2] and M Penčić[2]

[1]University Politehnica Timisoara, Faculty of Engineering Hunedoara, Department of Engineering and Management, Hunedoara, Romania
[2]University of Novi Sad, Faculty of Technical Sciences, Novi Sad, Serbia

E–mail: imre.kiss@fih.upt.ro

Abstract. Multivariate research is important in areas of cast–iron brake shoes manufacturing, because many variables interact with each other simultaneously. This article focuses on expressing the multiple linear regression model related to the hardness assurance by the chemical composition of the phosphorous cast irons destined to the brake shoes, having in view that the regression coefficients will illustrate the unrelated contributions of each independent variable towards predicting the dependent variable. In order to settle the multiple correlations between the hardness of the cast–iron brake shoes, and their chemical compositions several regression equations has been proposed. Is searched a mathematical solution which can determine the optimum chemical composition for the hardness desirable values. Starting from the above–mentioned affirmations two new statistical experiments are effectuated related to the values of Phosphorus [P], Manganese [Mn] and Silicon [Si]. Therefore, the regression equations, which describe the mathematical dependency between the above–mentioned elements and the hardness, are determined. As result, several correlation charts will be revealed.

1. Introductory notes

The best way for brake makers to achieve better shoes is to ensure that better materials and improved manufacturing processes are used and that shoes users take account of friction conditions and improved braking processes. The main problem in shoe's manufacturing processes is the optimal chemical composition assurance [1–5]. Therefore, the cast–iron brake shoes – with lamellar graphite and with a high content of phosphorus (0.8–1.1%) – need a special investigation [1–4], [6–8]. In order to establish the optimal condition for the cast–iron brake shoes we proposed a mathematical modelling study by using the statistical analysis and multiple regression equations [1–4]. Multivariate research is important in areas of cast–iron brake shoes manufacturing, because many variables interact with each other simultaneously [1–4], [8].

The properties of grey iron are primarily dependent on its composition, being influenced by the elements present in irons. The lower strength grades can be consistently produced by simply selecting the proper melting stock, but the higher strength grades require close control of their processing as well as their composition. Within a class of grey iron, hardness is a good indicator of its engineering properties and many of these properties are directly related to its hardness. The hardness is affected by the processing of the grey iron as well as the composition because these factors influence the microstructure [1–4].

The statistical analysis is useful for controlling the industrial production of braking shoes, due to the shoe's chemical composition is not specified by railway customers. Due to industrial data analysis, the improving of the braking shoes technological process of manufacturing, according to the specifications, is needed. Statistical evaluations are helpful for the brake shoe–makers in determining their actions to improve the shoe's quality [1–5]. Our studies on the multivariate analysis combine the statistical analysis techniques and the multivariate mapping for easier data interpretation.

Several correlations between the chemical components of the grey phosphorus cast–iron destined to the brake shoes – Carbon [C], Sulphur [S], Manganese [Mn], Silicon [Si] and Phosphorus [P] – and the obtained brake shoes hardness [HB] was presented in recent works related with [1–4]. Using the Matlab area, we proposed a multivariate statistical analysis related to:

- the general behavior of Carbon [C] content in relation with the rest of chemical elements, which have influence on the hardness of brake shoes [1], [3]. As result, correlation charts are determined (HB = f([C], [S]); HB = f([C], [Mn]), HB = f([C], [Si]) and HB = f([C], [P])).
- the combined behavior of all chemical elements of grey phosphorus cast iron on the hardness of brake shoes, in several bi–component correlations, with focus on the general behavior of Phosphorus [P] content in relation with the rest of chemical elements [1], [4]. As result, correlation charts are determined (HB = f([S], [P]); HB = f([Si], [P]), HB = f([C], [P]) and HB = f([Mn], [P]))
- the combined behavior of all chemical elements of grey phosphorus cast iron on the hardness of brake shoes, in several bi–component correlations, with focus on the general behavior of Manganese [Mn] content in relation with the rest of chemical elements [1], [4]. As result, correlation charts are determined (HB = f([Mn], [P]), HB = f([Mn], [C]), HB = f([Mn], [Si]) and HB = f([Mn], [S]))
- the relationship between the equivalent carbon content value (which is mathematically calculated on basis of the chemical composition) and the hardness of cast iron brake shoes [1], [4]. As result, correlation charts are determined for case HB = f([Cech])
- the cumulative effect of Carbon [C], Manganese [Mn] and Sulphur [S] upon the hardness of cast iron brake shoes [1], [2]. As result, correlation charts are determined for case HB = f([C], [S], [Mn])
- the cumulative effect of Carbon [C], Phosphorous [P] and Sulphur [S] upon the hardness of cast iron brake shoes [1], [2]. As result, correlation charts are determined for case HB = f([C], [S], [P])

The current performed research upon a number of 100 charges of phosphorous cast–iron brake shoes had in view to obtain multivariate three–dimensional correlations between the cast–iron brake shoes' hardness and its chemical composition, defined by several representative elements – Phosphorus [P], Manganese [Mn] and Silicon [Si] –, which determine the iron's microstructure. As result, several regression surfaces and correlation charts will be revealed and a comparative study will be made.

2. Methodology

The mathematical modeling establishes a methodology was described in several previous works [1–4]. In these studies we disposed industrial data from cast–iron brake shoes' manufacturing, therefore, the optimization model is based on these data [1–4]. The main goal of the used multiple regression analysis is to observe whether two or more variables co vary, and to quantify the strength of the relationship between these. The regression model expresses the relationship in the form of several regression equations, which predict the brake shoes behavior [1–4].

In the current statistical experiments we use the regression analyses for predict of proper hardness, having in view the interest related to:

- the cause–and–effect relationships, commonly used for investigating the relationship between two independent variables of the chemical composition and the hardness value. In this case, we try to determine the relationships – in form of several regression equations – between Manganese [Mn] and Silicon [Si], Manganese [Mn] and Phosphorus [P], respectively Silicon [Si] and Phosphorus [P], related to the hardness value, [HB]. The general technological scope is to identify if the variation of each element of the analyzed cast irons causes variation in hardness value of the final cast product (brake shoes) [1–4];

- the cumulative cause–and–effect relationships, commonly used for investigating the relationship between three independent variables of the chemical composition and the hardness value. In this case, we try to determine the relationship between three variables – Manganese [Mn], Silicon [Si] and Phosphorus [P], and the hardness value, [HB] [1–4];
- the associated variables (if exist), without necessarily inferring a cause–and–effect relationship, knowing that a strong association between two or more components results in an increase in the accuracy of the prediction of proper hardness [1–4].

Therefore, is searched a mathematical solution which can determine the proper chemical composition for phosphorous cast–iron brake shoes for the hardness desirable values [1–4]. Starting from the above–mentioned affirmations two new statistical experiments are effectuated related to the values of Phosphorus [P], Manganese [Mn] and Silicon [Si]. In fact, the obtained regression equations will describe the mathematical dependency between the above–mentioned elements and the hardness. Thus the optimal additions can be determined in these elements to assure the proper hardness [1–4].

3. Statistical experiments and results

3.1. The first statistical experiment

In the first statistical experiment, the cumulative effect on hardness of two elements – Manganese [Mn] and Silicon [Si] – is analyzed. The optimal form of modeling in the case of [HB] = f([Mn], [Si]) is given by the regression equation (1), where the correlation coefficient is $rf_{(1)} = 0.7372$.

$$[HB] = a_1[Mn]^2 + a_2[Si]^2 + a_3[Mn]\,[Si] + a_4[Mn] + a_5[Si] + a_6 \quad (1)$$

where the statistically determined values of the correlation coefficients are: $a_1 = 36.9005$; $a_2 = -189.9378$; $a_3 = -107.6425$; $a_4 = -70.7314$; $a_5 = 427.5707$; $a_6 = 158.5458$;

By the same methodology another two cumulative effects on hardness are analyzed, using the influence of Manganese [Mn] and Phosphorus [P], respectively the Silicon [Si] and Phosphorus [P] values. The regression equations are presented in (2)–(3), where the correlation coefficient are $rf_{(2)} = 0.8441$ and $rf_{(3)} = 0.8449$, representing the cause–and–effect relationships in the cases of [HB]= f([Mn], [P]) and [HB]= f([Si], [P]).

$$[HB] = a_1[Mn]^2 + a_2[P]^2 + a_3[Mn]\,[P] + a_4[Mn] + a_5[P] + a_6 \quad (2)$$

where the statistically determined values of the correlation coefficients are: $a_1 = -101.5218$; $a_2 = -57.3992$; $a_3 = -225.2918$; $a_4 = 313.5547$; $a_5 = 256.3256$; $a_6 = 14.3819$;

$$[HB] = a_1[Si]^2 + a_2[P]^2 + a_3[Si]\,[P] + a_4[Si] + a_5[P] + a_6 \quad (3)$$

where the statistically determined values of the correlation coefficients are: $a_1 = -67.2629$; $a_2 = 25.9138$; $a_3 = -70.6481$; $a_4 = 251.1173$; $a_5 = -29.5826$; $a_6 = 135.0779$;

3.2. The second statistical experiment

In the second statistical experiment, the cumulative effect of all three elements (Phosphorus [P], Manganese [Mn] and Silicon [Si]) upon the obtained hardness is analyzed. The optimal form of modeling in the case of [HB] = f([Mn], [Si], [P]) is given by the regression equation (4), where the correlation coefficient is $rf_{(4)} = 0.8024$.

$$HB = a_1[Si]^2 + a_2[Mn]^2 + a_3[P]^2 + a_4[Si][Mn] + a_5[Mn][P] + a_6[P][Si] + a_7[Si] + a_8[Mn] + a_9[P] + a_{10} \quad (4)$$

where the statistically determined values of the correlation coefficients are: $a_1 = 33.4696$; $a_2 = -125.8125$; $a_3 = -79.4532$; $a_4 = -72.4779$; $a_5 = 46.2496$; $a_6 = -280.5296$; $a_7 = -111.6077$; $a_8 = 521.8024$; $a_9 = 233.2149$; $a_{10} = 59.1273$

Because these areas, described by the equation (4), cannot be represented in the three–dimensional space, the independent variables ([Mn], [Si] and [P]) were replaced with their average values ($[Si]_{med}$, $[Mn]_{med}$ and $[P]_{med}$), successively. Therefore, in the case of correlation HB= f([Mn], [Si], [P]), the equations (5)–(7) were obtained, in which the correlation coefficient are $rf_{(5)} = 0.7356$, $rf_{(6)} = 0.8011$ and $rf_{(7)} = 0.7305$.

$$HB\,[Si]_{med} = b_1[Mn]^2 + b_2[P]^2 + b_3[Mn][P] + b_4[Mn] + b_5[P] + b_6 \qquad (5)$$

where the correlation coefficients have the following statistically determined values: $b_1 = -125.8125$; $b_2 = -79.4531$; $b_3 = -280.5296$; $b_4 = 391.1682$; $b_5 = 316.5751$; $b_6 = -33.3034$

$$HB\,[Mn]_{med} = c_1[Si]^2 + c_2[P]^2 + c_3[Si][P] + c_4[Si] + c_5[P] + c_6 \qquad (6)$$

where the correlation coefficients have the following statistically determined values: $c_1 = 33.4696$; $c_2 = -79.4531$; $c_3 = 46.2496$; $c_4 = -156.3266$; $c_5 = 60.1281$; $c_6 = 333.1839$

$$HB\,[P]_{med} = d_1[Si]^2 + d_2[Mn]^2 + d_3[Si][Mn] + d_4[Si] + d_5[Mn] + d_6 \qquad (7)$$

where the correlation coefficients have the following statistically determined values: $d_1 = 33.4696$; $d_2 = -125.8125$; $d_3 = -72.4779$; $d_4 = -74.0382$; $d_5 = 293.9225$; $d_6 = 196.1442$

4. Generation the variation domains

The equations (1)–(3) and (5)–(7), belonging to the three–dimensional space, are represented graphically, and several variation domains results, which can be interpreted as correlation charts [1–4]. In this way, the correlation of the values of the twos independent variables can be made, so that the dependent variables [HB] can be obtained between the requested limits [1–4].

The cumulative effect on hardness of Manganese [Mn] and Silicon [Si], corresponding to the first statistical experiment, is presented in Figure 1 and Figure 2. The cumulative effect on hardness of Manganese [Mn] and Phosphorus [P], respectively the cumulative effect on hardness of Silicon [Si] and Phosphorus [P] are presented in Figures 3–6, corresponding to the same statistical experiment.

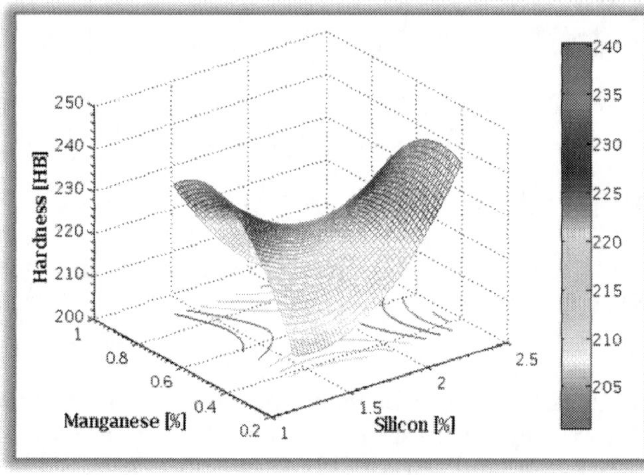

Figure 1. The regression surface of hardness in the [Mn]–[Si] correlation (according to the first statistical experiment)

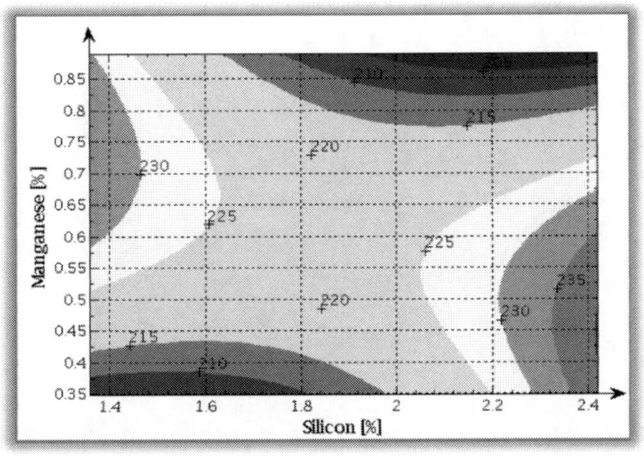

Figure 2. The hardness diagram in the [Mn]–[Si] correlation (according to the first statistical experiment)

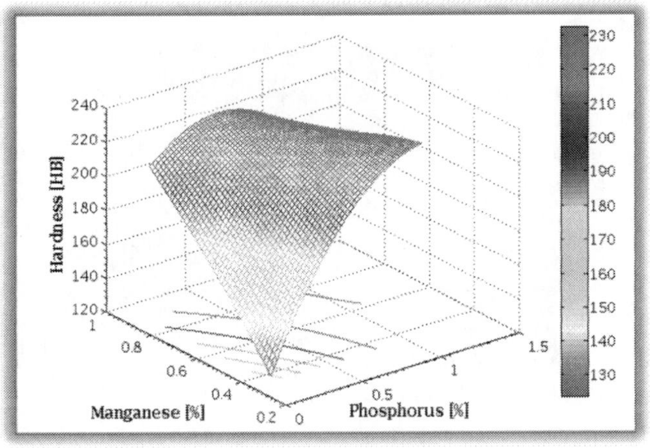

Figure 3. The regression surface of hardness in the [Mn]–[P] correlation (according to the first statistical experiment)

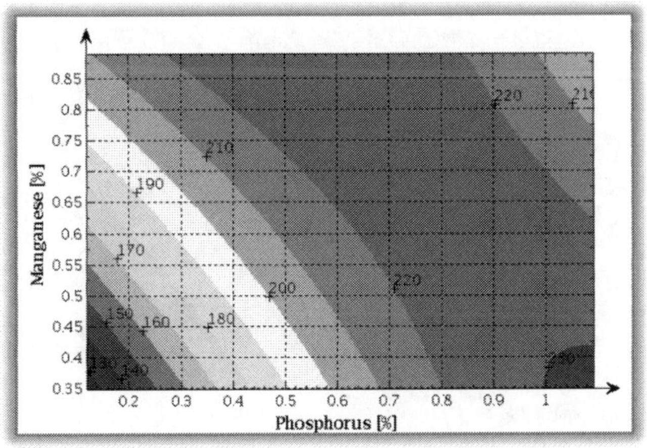

Figure 4. The hardness diagram in the [Mn]–[P] correlation (according to the first statistical experiment)

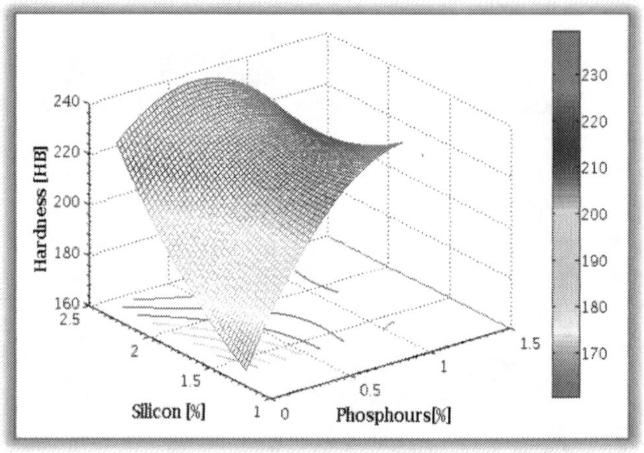

Figure 5. The regression surface of hardness in the [P]–[Si] correlation (according to the first statistical experiment)

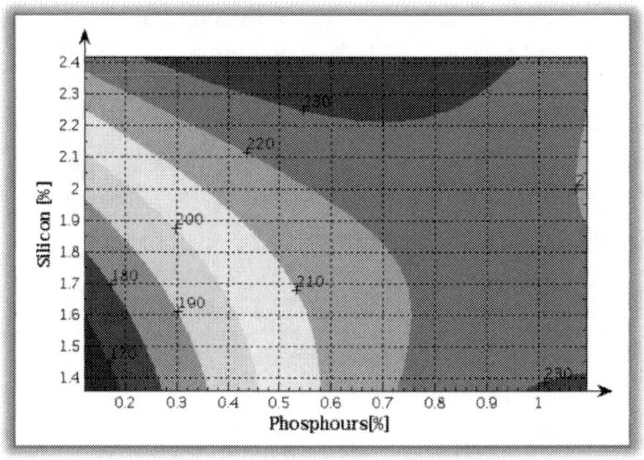

Figure 6. The hardness diagram in the [P]–[Si] correlation (according to the first statistical experiment)

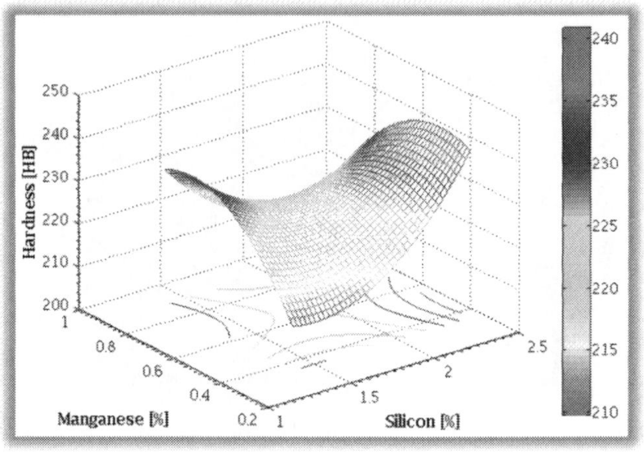

Figure 7. The regression surface of hardness in the [P]–[Si]–[Mn] correlation when [P] = $[P]_{med}$ (corresponding to the second statistical experiment)

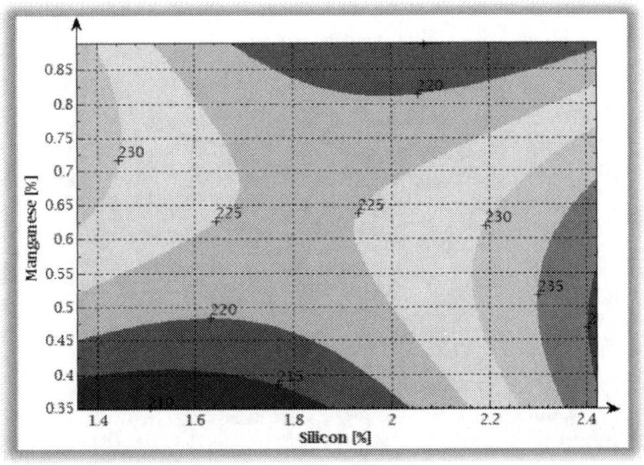

Figure 8. The hardness diagram in the [P]–[Si]–[Mn] correlation when [P] = $[P]_{med}$ (corresponding to the second statistical experiment)

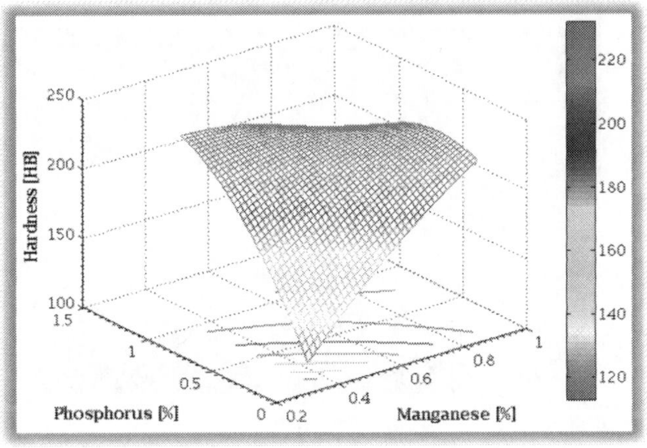

Figure 9. The regression surface of hardness in the [P]–[Si]–[Mn] correlation when [Si] = $[Si]_{med}$ (corresponding to the second statistical experiment)

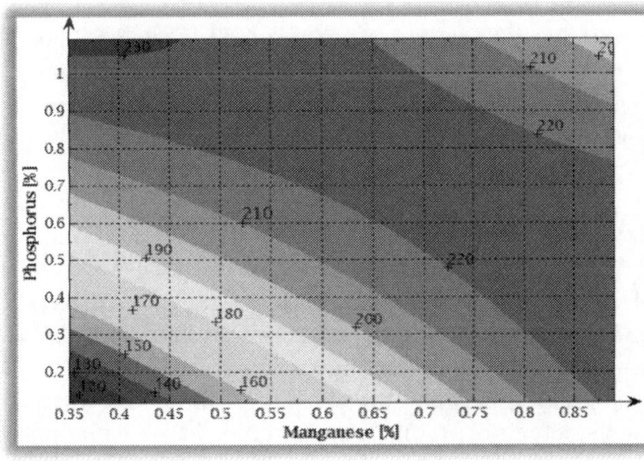

Figure 10. The hardness diagram in the [P]–[Si]–[Mn] correlation when [Si] = $[Si]_{med}$ (corresponding to the second statistical experiment)

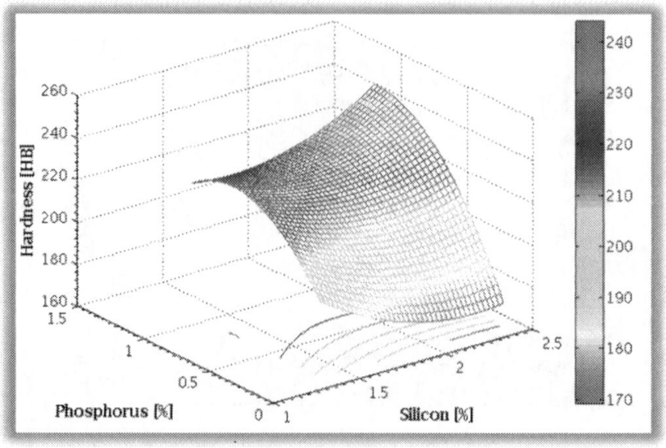

Figure 11. The regression surface of hardness in the [P]–[Si]–[Mn] correlation when [Mn] = $[Mn]_{med}$ (corresponding to the second statistical experiment)

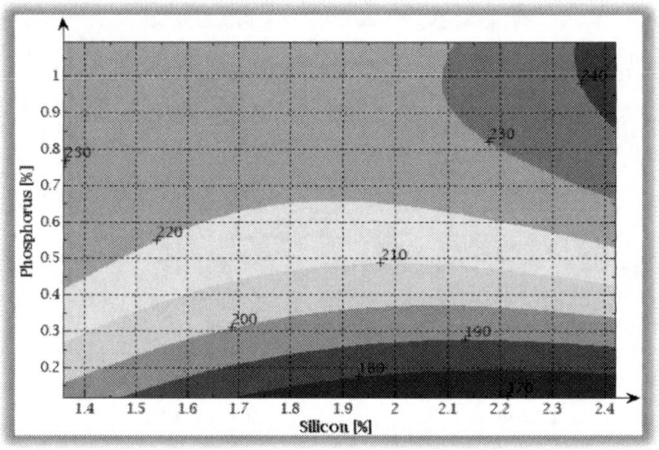

Figure 12. The hardness diagram in the [P]–[Si]–[Mn] correlation when [Mn] = $[Mn]_{med}$ (corresponding to the second statistical experiment)

The cumulative effects of Phosphorus [P], Manganese [Mn] and Silicon [Si] upon the obtained hardness, corresponding to the second statistical experiment, in the Figures 7–12 are presented. We make the mention that the analyzed statistical values were successively replaced with their average values ($[Si]_{med}$, $[Mn]_{med}$ and $[P]_{med}$), resulting new correlation diagrams.

5. Discussions

5.1. Discussion on the mathematical model

The mathematical model quantifies the influences of the independent variables ([Mn], [Si] and [P]) on the dependent variable ([HB]), and it was established by a statistical analysis. The real values of the process variables were chosen statistically, their limits of variation being given by the industrial data from the phosphorous cast–iron brake shoes manufacturing [1–4].

In order to settle the correlation between the [HB], and the above–mentioned parameters ([Mn], [Si] and [P]) the following model of polynomial equation type has been proposed:

$$Y = a_1[X_1]^2 + a_2[X_2]^2 + a_3[X_3]^2 + a_4[X_1][X_2] + a_5[X_1][X_3] + a_6[X_2][X_3] + a_7[X_1] + a_8[X_2] + a_9[X_3] + a_{10} \quad (9)$$

where $a_1, \ldots, a_{10}$ are the regression coefficients.

5.2. Discussion on the regression analysis

In order to estimate the independent variables overall influence on hardness [HB], the multiple correlation coefficients were calculated. The obtained values are: $rf_1 = 0.5773$ and $rf_2 = 0.6301$, respectively. Analyzing the values of regression coefficients one can conclude that the process performance it is noticeably influenced by all three parameters, in both statistical experiments. From the value of the determination coefficient resulted that the proposed mathematical model is adequate for a confidence interval of 57–63%. In other words, in the chosen intervals the independent variables influence the hardness in an extent of approximately 60%. The rest of approximately 40% can be attributed to the effect of other chemical elements or manufacturing factors ignored during these statistical experiments.

5.3. Discussion on the graphically addenda

For visualization of the regression surfaces and for the calculation correlation coefficients the software Matlab was used. The spatial representations of the response surfaces [HB] for constant values of variables reveal that all data are located inside the considered surface.

The three–dimensional regression surfaces are a graphical representation of the regression equations (1)–(3), according to the first statistical experiment, respectively equations (5)–(7), corresponding to the second statistical experiment. It is plotted to explain interaction of the variables and locate the optimal level of each variable for maximal response. Each regression surface represents the different combinations of two variables at one time while maintaining the other variable at the average value.

These regression surfaces obtained in the experiments (Figure 1, Figure 3 and Figure 5, respectively Figure 7, Figure 9 and Figure 11) and its respective correlation charts (Figure 2, Figure 4 and Figure 6, respectively Figure 8, Figure 10 and Figure 12) provide a visual interpretation of the interaction between two factors (in the bi–component analysis), and between three factors (when the third factor have an constant value). These regression surfaces facilitate the determination of proper experimental conditions. The point for which the response is optimized is the point called the stationary point. The stationary point may be a point of maximum response, minimum response or a saddle point (point of inflexion).

5.4. Comparative study

Comparative study can be made between:

- Figure 2 and Figure 8, which represents, in fact, the hardness diagram in the [Mn]–[Si] correlation.
- Figure 4 – Figure 10, which represents, in fact, the hardness diagram in the [Mn]–[P] correlation.
- Figure 6 – Figure 12, which represents, in fact, the hardness diagram in the [P]–[Si] correlation.

Table 1 present the comparative study's results.

Table 1. Comparative study (at [HB] = 220)

The [Mn]–[Si] correlation		The [Mn]–[P] correlation		The [P]–[Si] correlation	
The first statistical experiment	The second statistical experiment	The first statistical experiment	The second statistical experiment	The first statistical experiment	The second statistical experiment
[Si]		[P]		[P]	
1.62–2.04	1.68–1.96	0.7–0.9	0.5–0.85	0.75–1.6	0.65–1.1
[Mn]		[Mn]		[Si]	
0.5–0.7	0.48–0.8	0.5–0.8	0.72–0.8	1.7–2.2	1.4–2.1

Table 2. Technological domains according to the correlation diagrams (at [HB] = 220)

[Si]	[P]	[Mn]
1.7–1.96	0.75–0.85	0.5–0.7

Therefore, the comparative study is an optimisation process, in which we can determine the proper hardness [HB] that depends by the analysed variables – Phosphorus [P], Manganese [Mn] and Silicon [Si]. The technological domains for the analysed variables according to the correlation diagrams (at [HB] = 220), are presented in Table 2, as recommendation.

6. Conclusions

One of the main chapters of the statistics referring to the ability to predict. Although it is not find the perfect relations, by means of regression, can make statements of a variable, depending on the other values. The present researches are going to establish the influence of the Phosphorus [P], Manganese [Mn] and Silicon [Si] upon the hardness [HB] of the braking shoe's material (i.e. gray phosphorous cast iron with lamellar graphite).

Therefore, the realization of a proper chemical composition – Phosphorus [P], Manganese [Mn] and Silicon [Si], in this case – can constitute a technical efficient mode to assure the exploitation properties of the brake shoes. Having in view the previous works [1–4], presented as partial results in the complex statistical analyses related to the brake shoes manufacturing, we can conclude that the exploitation properties can be achieved by the manufacturing process, through a proper conduct of the iron's making processes.

References

[1] Kiss I, Cioată V G, Alexa V and Rațiu S A 2017 Multivariate research in areas of phosphorus cast iron brake shoes manufacturing using the statistical analysis and the multiple regression equations, *IOP Conf. Ser.: Mater. Sci. Eng.* **200** 012022

[2] Kiss I 2016 The chemical composition of phosphorous cast irons behavior in the manufacturing of brake shoes meant for the rolling stock, *Acta Technica Corviniensis – Bulletin of Engineering* **IX**(3) 77-84

[3] Kiss I, Cioata V, Alexa V and Ratiu S 2016 Investigations on the selection of friction materials destined to railway vehicles applications, *Annals of Faculty Engineering Hunedoara – International Journal of Engineering* **XIV**(4) 231-240

[4] Kiss I, Alexa V and Ratiu S 2016 Investigations on the phosphorous cast iron destined for brake shoes manufacturing, *Applied Engineering Letters* **1**(3) 61-66

[5] Pascu L 2015 *Research regarding increasing quality of brake shoes destined to railway vehicles applications*, University Politehnica Timisoara, Doctoral Thesis

[6] Pascu L, Socalici A, Popa E and Crisan E 2015 The influence of chemical composition upon the hardness of brake shoes meant for rolling stock, *AIP Conference Proceedings* **1648**(1), 12-18

[7] Pascu L, Stînea, E and Heputț T 2014 The influence of residual elements upon the hardness of the brake shoes made of sulfurous cast iron, *Annals of Faculty Engineering Hunedoara – International Journal of Engineering* **XII**(4) 157-160

[8] Socalici A, Pascu L, Ardelean E and Heput T 2016 The influence of Carbon, Manganese and Silicon content upon the hardness of phosphorus cast iron used in manufacturing rolling stock braking shoes, *Tehnički Vjesnik* **23**(6) 1643-1648

IOP Conf. Series: Materials Science and Engineering **294** (2017) 012079 doi:10.1088/1757-899X/294/1/012079

Social humanoid robot SARA: development of the wrist mechanism

M Penčić[1], M Rackov[1], M Čavić[1], I Kiss[2] and V G Cioată[2]

[1]University of Novi Sad, Faculty of Technical Sciences, Trg Dositeja Obradovića 6, 21000 Novi Sad, Serbia

[2]University Politehnica Timisoara, Faculty of Engineering Hunedoara, Revolutiei 5, 331128 Hunedoara, Romania

E-mail: mpencic@uns.ac.rs

Abstract. This paper presents the development of a wrist mechanism for humanoid robots. The research was conducted within the project which develops social humanoid robot Sara – a mobile anthropomorphic platform for researching the social behaviour of robots. There are two basic ways for the realization of humanoid wrist. The first one is based on biologically inspired structures that have variable stiffness, and the second one on low backlash mechanisms that have high stiffness. Our solution is low backlash differential mechanism that requires small actuators. Based on the kinematic-dynamic requirements, a dynamic model of the robot wrist is formed. A dynamic simulation for several hand positions was performed and the driving torques of the wrist mechanism were determined. The realized wrist has 2 DOFs and enables movements in the direction of flexion/extension 115°, ulnar/radial deviation ±45° and the combination of these two movements. It consists of a differential mechanism with three spur bevel gears, two of which are driving and identical, while the last one is the driven gear to which the robot hand is attached. Power transmission and motion from the actuator to the input links of the differential mechanism is realized with two parallel placed identical gear mechanisms. The wrist mechanism has high carrying capacity and reliability, high efficiency, a compact design and low backlash that provides high positioning accuracy and repeatability of movements, which is essential for motion control.

1. Introduction

In the near future, humanoid robots will be working in direct contact with humans, in an environment that is dynamic and unstructured, and will be performing numerous and complex tasks. One of expectations of modern society is the use of humanoids as an acceptable aid for people with special needs, regardless of whether they have a disability or have impaired mobility and motoric functions due to their old age. This kind of robots must be safe both for humans and for objects in their environment. Apart from this, humanoid joints – mechanisms of which are formed, must fulfil numerous requirements that exist during dynamic activity of the robot. Those are high carrying capacity and reliability, high efficiency, low noise, low vibration, low backlash which is essential for motion control, self-locking that provide the robot posture without actuator power supply, compact design etc. Therefore, special attention should be given to the design and realization of the wrist.

This paper presents the development of a wrist mechanism for humanoid robots. The research was conducted as part of the project which develops social humanoid robot Sara that should represent a mobile anthropomorphic platform for researching the social behavior of robots – Figure 1. The robot will be able to communicate verbally and nonverbally. For expressing facial expressions, biologically

IOP Conf. Series: Materials Science and Engineering **294** (2017) 012079 doi:10.1088/1757-899X/294/1/012079

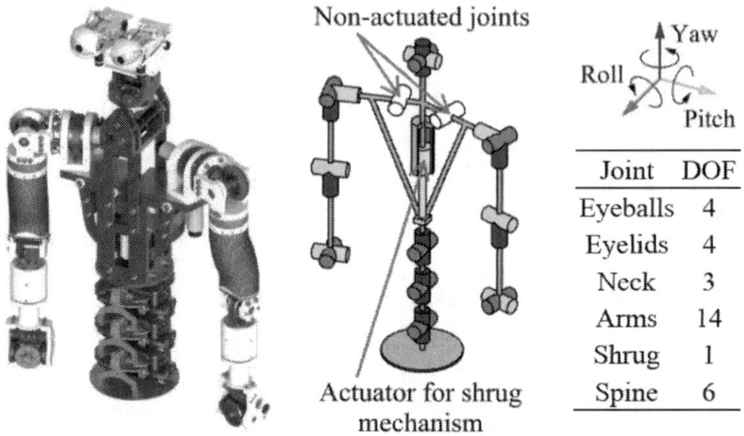

Joint	DOF
Eyeballs	4
Eyelids	4
Neck	3
Arms	14
Shrug	1
Spine	6

Figure 1. Social humanoid robot SARA: current prototype and its kinematic structure

inspired eyes are being developed with eyelids and eyelashes with total 8 DOFs [1], [2]. To extend the spectrum of nonverbal communication, the robot will be able to shrug if a question is confusing or it does not know what to answer [3-6]. In addition, Sara will have two anthropomorphic arms with 14 DOFs, a self-locking neck mechanism with 3 DOFs [7] and a self-locking multi-segment lumbar spine with 7 DOFs [8] to increase the mobility of the upper body without moving the lower body.

The paper is structured as follows: the first section shows the research motivation; the second section presents the analysis of existing solutions of the humanoid wrists; within the third section, a new wrist solution is proposed – dynamic analysis of the wrist was performed and proposed solution of wrist mechanism is presented in detail; the fourth section summarizes the paper contribution and outlooks future work.

1.1. Human wrist and its mobility

The wrist is a complex structure that connects the hand to the forearm. It allows the hand and fingers to function, it provides and controls the extension and contraction of the long flexors and extensors of fingers and it enables fine hand motoric. It consists of the radiocarpal joint that enables extension of the wrist, the mediocarpal joint for flexion movements of the wrist, the intercarpal joint consisting of a number of smaller joints and the radioulnar joint that enables forearm rotation. Large number of muscles and ligaments is involved in wrist movements. The wrist movements are extension – bending backward (Figure 2a), flexion – bending forward (Figure 2b), radial deviation – moving the palm to the left (Figure 2c) and ulnar deviation – moving the palm to the right (Figure 2d). The range of flexion movements is 80÷90°, while the range of extension movements is 70÷75°. Wrist movement in the direction of radial/ulnar deviation is approximately 60°, which is 15÷25° for movements of radial deviation, respectively 35÷45° for movements of ulnar deviation [9], [10].

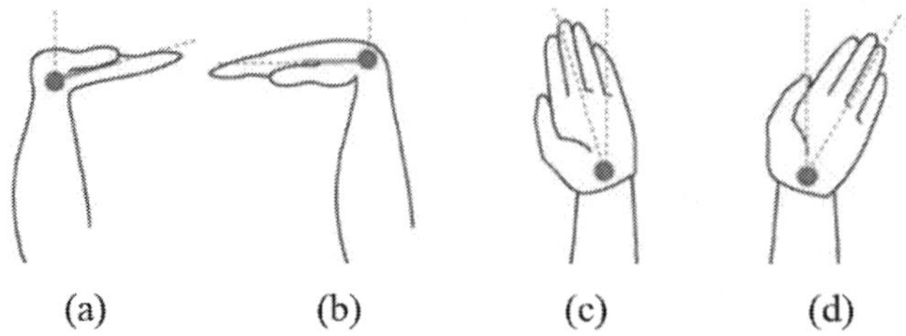

Figure 2. The movements of human wrist: (a) bending backward – extension, (b) bending forward – flexion, (c) palm motion to the left – radial deviation and (d) palm motion to the right – ulnar deviation

2. State of the art

There are two basic groups of robots capable to move the hand independently of the forearm by activating the wrist. The first group of robots has a wrist with 1, 2 or 3 by which each joint has 1 DOF – rigid structures, while the second group of robots has a wrist with 1 joint (condylar joint with 2 DOFs or ball-and-socket joint with 3 DOFs) – flexible structures.

Wrist with 1 DOF – movements about the yaw or pitch axis, are found in Affetto [11], Ever-1 [12], EveR-2 [13], CB^2 [14], ASIMO [15], HRP-4C [16] etc., wrist with 2 DOFs – movements about the yaw and pitch axes, are found in iCub [17], Robonaut 2 [18], Albert HUBO [19], AILA [20], TORO [21], KIBO [22], HUBO [23], WABIAN-2 [24], KOBIAN [25], ARMAR-III [26], SURALP [27] etc., while the wrist with 3 DOFs – movements about the yaw, roll and pitch axes, are found in James [28], Justin [29], Romeo [30], BERT2 [31], WE-4RII [32], HRP-4 [33] etc. The wrist mechanism of these robots usually consists of rigid and low backlash mechanisms that are interconnected – harmonic drive, cable-driven mechanism, spatial linkage mechanism, spindle drive, low backlash gears etc. In addition to having high carrying capacity and reliability, the advantage of these mechanisms is low backlash that provides high positioning accuracy that enables high accuracy and repeatability of movements, which is essential for motion control.

Biologically inspired wrist structures with 1 joint and 2-3 DOFs – movements about the yaw and pitch axes or yaw, roll and pitch axes, are found in Kenta [34], Kenji [35], Kotaro [36], Kojiro [37], Kenzoh [38], Kenshiro [39] and Kengoro [40]. The wrist of these robots consists of human-like bones and artificial muscles that are based on nonlinear spring units – NST, which has a nonlinear ratio between the tension and the spring constant of the tendon. Therefore wrist has variable stiffness which is possible to mechanically adjust. The advantage of NST is the simple realization and implementation by using only one spring and guided pulleys of tendon – wire.

3. Wrist mechanism

The primary requirement for the realization is adequate hand mobility about the roll and pitch axes. The wrist supposed to provide the position and orientation of the hand, because the hand must be in the appropriate position relative to the object with which intends to manipulate. Hand is position on the certain location by the arm, while the wrist provides the orientation. Therefore wrist must have more DOFs. Figure 3 shows a possible wrist design based on a differential mechanism. It consists of two driving gears G_1 and G_2, meshed with driven gear G_3 to which the hand of the robot is attached.

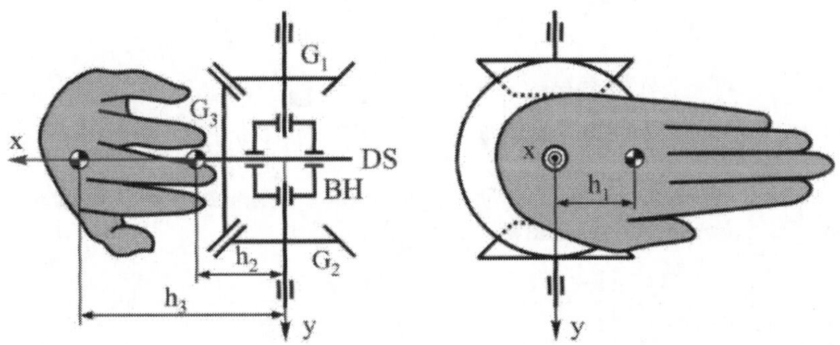

Figure 3. Kinematic scheme of the differential mechanism: G_1 and G_2 – driving bevel gears, G_3 – driven bevel gear to witch the robot hand is attached, DS – driven shaft and BH – bearing housing

3.1. Forces and torques

The hand of the robot is still in development wherefore is approximated with a 1 kg object whose centre of mass is located at distance h_1 relative to the wrist. Movements of the hand supposed to be as natural as possible and therefore is adopted that their duration is no longer than 1 s, which has certain dynamic effects as consequences. Based on the preliminary 3D model of the robot arm, a dynamic model of the wrist is formed and dynamic simulation is performed within which the driving torques of the wrist mechanism are determined. Three characteristic cases were examined, depending on the type of hand movements and the range of motion. The first one is in the direction of flexion/extension and the range of motion –90° to 15°, the total of 115° – Figure 3a, the second one is in the direction of ulnar/radial deviation and the range of motion –45° to 45°, the total of 90° – Figure 3b, and the third case includes simultaneous movements in the direction of flexion/extension –90° to 0° and ulnar/radial deviation –45° to 0° – Figure 3c. It should be noted that the forearm is immovable and that the hand is appropriately positioned.

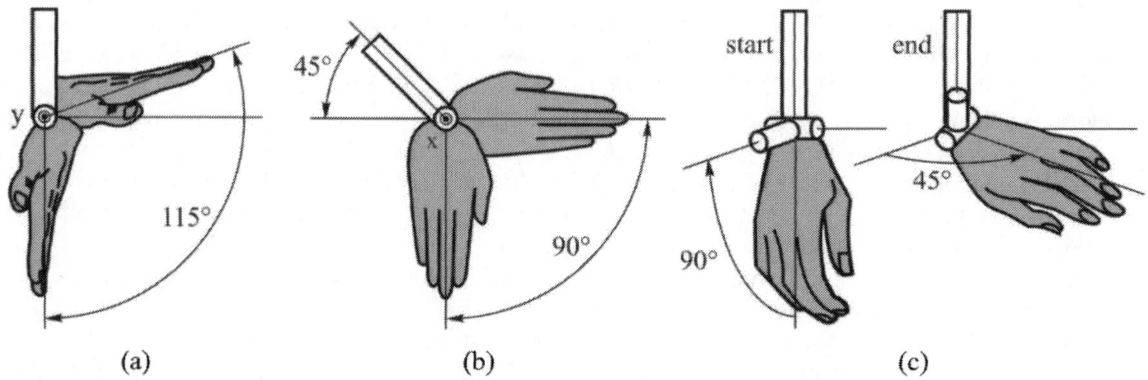

Figure 4. Dynamic simulation of the wrist mechanism: (a) flexion/extension of the hand from –90° to 15° – forearm is immovable and in vertical position, (b) ulnar/radial deviation of the hand from –45° to 45° – forearm is immovable and rotated by 45°, (c) simultaneous flexion/extension from –90° to 0° and ulnar/radial deviation from –45° do 0° – forearm is immovable and in vertical position

The adopted laws of motion of angle φ, angular velocity ω and angular acceleration ε are:

$$\varphi = H\left(\frac{t}{T}\right)^3\left[10-15\left(\frac{t}{T}\right)+6\left(\frac{t}{T}\right)^2\right] \tag{1}$$

$$\omega = \left(\frac{30H}{T}\right)\left(\frac{t}{T}\right)^2\left[1-2\left(\frac{t}{T}\right)+\left(\frac{t}{T}\right)^2\right] \tag{2}$$

$$\varepsilon = \left(\frac{60H}{T^2}\right)\left(\frac{t}{T}\right)\left[1-3\left(\frac{t}{T}\right)+2\left(\frac{t}{T}\right)^2\right] \tag{3}$$

where: T – duration of the movement and H – amplitude of the movement (maximum angle). The driving torques of the wrist about the x and y axes are defined as:

$$M_x = \left(m_H g\cos\varphi_y h_1 \sin\varphi_x\right) + \varepsilon_x\left(J_{Hx} + J_{Gx}\right) \tag{4}$$

$$M_y = \left(m_H g\sin\varphi_y h_1\cos\varphi_x + m_H g\cos\varphi_y h_2 + m_G g\cos\varphi_y h_3\right) + \varepsilon_y\left(J_{Hy} + J_{Gy}\right) \tag{5}$$

where: m_H – mass of the hand, g – gravitational acceleration, m_G – mass of shaft DS with gear G_3, J_{Hx}, J_{Hy} – moments of inertia of the hand about x and y axes, J_{Gx}, J_{Gy} – moments of inertia of the shaft DS with gear G_3 about x and y axes, φ_x, φ_y – rotation angles about x and y axes, $\varepsilon_x, \varepsilon_y$ – angular acceleration about x and y axes, h_1, h_2, h_3 – centres of mass coordinates of individual segments, seen Figure 3. Table 1 shows the values of the input parameters – segment masses, centres of mass and the appropriate moments of inertia.

Table 1. Input parameters

m_H [kg]	m_G [kg]	J_{Hx} 10^{-4}[kgm^2]	J_{Hy} 10^{-4}[kgm^2]	J_{Gx} 10^{-7}[kgm^2]	J_{Gy} 10^{-4}[kgm^2]	h_1 [m]	h_2 [m]	h_3 [m]
1	0.1	117	122	335	346	0.1	0.012	0.02

Given the Eqs. (1) – (5), dynamic simulation of the wrist mechanism has been performed and driving torques were determined – Figure 5. Based on the dynamic analysis, the maximum value of the driving torque for dimensioning of the wrist mechanism is 0.93 Nm.

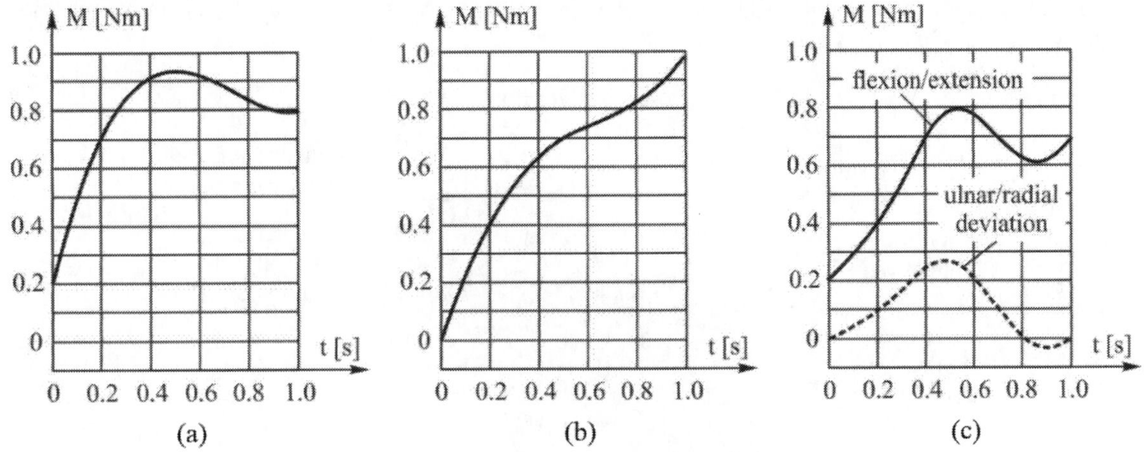

Figure 5. Time histories of the driving torque: (a) movement of flexion/extension (b) movement of ulnar/radial deviation and (c) simultaneous movement of flexion/extension and ulnar/radial deviation

3.2. Mechanical design

The realized wrist has 2 DOFs and enables hand movements in the direction of flexion/extension 115°, ulnar/radial deviation ±45° and the combination of these two movements – Figure 6. It consists of a differential mechanism with three spur bevel gears, two of which are driving and identical, while the last one is the driven gear. If driving gears have the same circumferential velocity and the same direction of rotation, hand moves about the pitch axis – flexion/extension movements, and for the opposite direction of rotation, hand moves about the roll axis – ulnar/radial deviation. If the circumferential velocities of driving gears are different, a combination of these two movements is obtained.

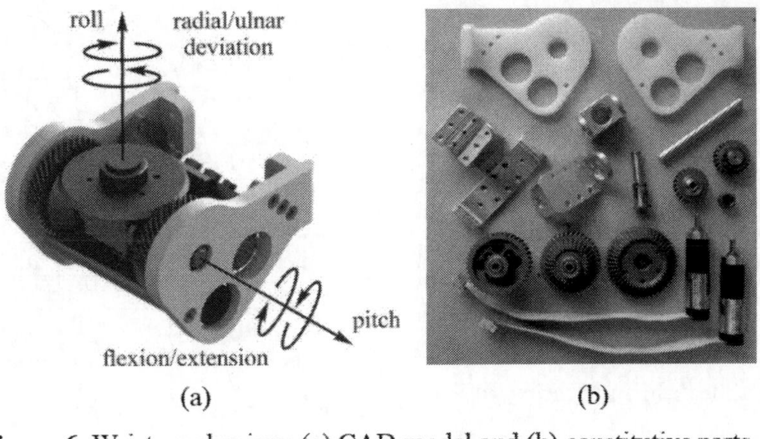

Figure 6. Wrist mechanism: (a) CAD model and (b) constitutive parts

Structure of the wrist mechanism is presented in Figure 7. It consists of three subassemblies, two of which are driving and identical, while the third one is driven to which robot hand is attached. Driving subassembly consists of an actuator and elements for power transmission and motion – driving shafts, helical gears, driving bevel gear, bearings etc. Actuator consists of a DC motor with graphite brushes, planetary gearhead with 3 stages and a magnetic encoder. Driving helical gear is directly attached to the gearhead shaft, while driving bevel gear is fixed to the driven helical gear that represent driving shaft. Driving subassembly consists of driven shaft, bearings, bearing housing and an output link of differential mechanism – driving bevel gear to which robot hand is attached. Low backlash in helical and bevel gear pairs is achieved by precise machining of gears, support structures and preloading – center distance is done with negative tolerance. Realized wrist mechanism is shown in Figure 8 and its configuration is given in Table 2. Hight of the wrist mechanism is 70 mm, its width is 80 mm, its length is 85 mm, while mass is 0.57 kg.

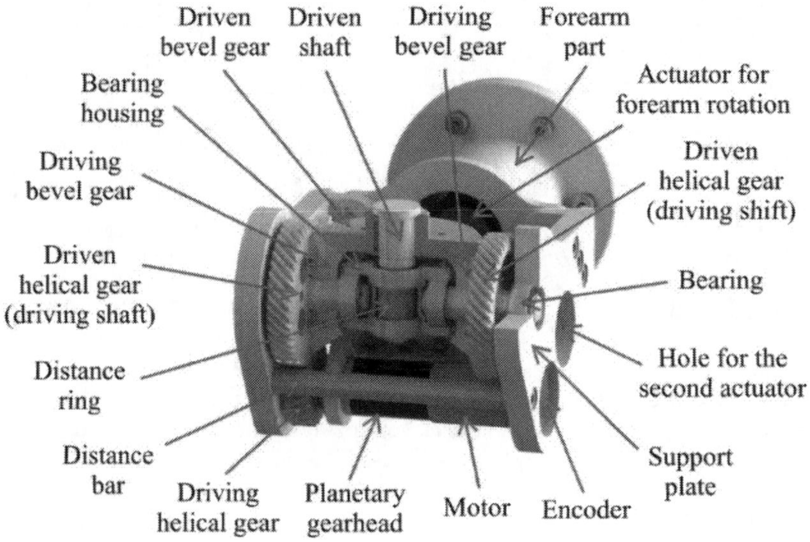

Figure 7. Wrist mechanism with details – cross section

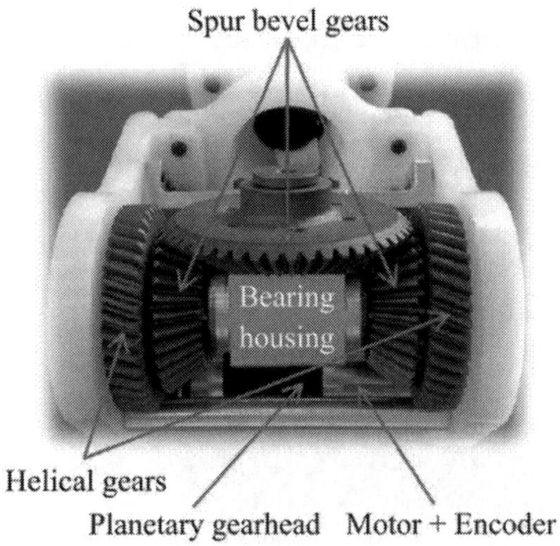

Figure 8. Wrist mechanism of the robot Sara

Table 2. Wrist mechanism configuration

Maxon motor RE-max 17 (2 pcs.)	Voltage [V]	24
	Power [W]	4
	Torque [mNm]	3.63
	Speed [rpm]	8520
Planetary gearhead GP 16-C (2 pcs.)	Reduction	157
	Torque [Nm]	0.4
	Efficiency [%]	73
Encoder MR-ML (2 pcs.)	Channels	3
	Counts per turn	512
Gearing with helical gears (2 pcs.)	Reduction	1.74
	Module [mm]	0.9
	Helix angle [°]	20
Gearing with bevel gears (2 driving & 1 driven)	Reduction	1.5
	Module [mm]	1
	Helix angle [°]	0

IOP Conf. Series: Materials Science and Engineering **294** (2017) 012079 doi:10.1088/1757-899X/294/1/012079

4. Conclusion

This paper presents the development of a wrist mechanism for humanoid robots. Considering that the hand is position on the certain location by the arm, while the wrist provides the orientation, basic requirement for the wrist realization is adequate mobility about pitch and roll axes. Having in mind the analysis of humanoid wrists, we proposed new wrist that require small actuators. Based on the kinematic-dynamic requirements, a dynamic model of the robot wrist is formed. A dynamic simulation for several hand positions was performed and the driving torques for the wrist mechanism were determined. The realized wrist has 2 DOFs and enables movements in the direction of flexion/ extension 115°, ulnar/radial deviation ±45° and the combination of these two movements. It consists of a differential mechanism with three spur bevel gears, two of which are driving and identical, while the third one is the driven gear to which the robot hand is attached. If driving gears have the same circumferential velocity and the same direction of rotation, then the movements about the pitch axis – flexion/extension of the hand are obtained, and for the opposite direction of rotation, movements about the roll axis – ulnar/radial deviation of the hand are obtained. If the circumferential velocities of driving gears are different, a combination of these two movements is obtained. Power transmission and motion from the actuators to the input links of the differential mechanism is realized with two identical helical gear pairs. Wrist mechanism has high carrying capacity and reliability, high efficiency, compact design and low backlash that provides high positioning accuracy that enables high accuracy and repeatability of movements, which is essential for motion control. Further work will be focused to the minimization of the overall dimensions as well as to the reduction of the backlash by using spiral bevel gears with small modules.

Acknowledgments
This work was funded by the Ministry of Education and Science of the Republic of Serbia under the contract III44008 and by the Provincial Secretariat for Science and Technological Development under the contract 114–451–2116/2011.

References

[1] Penčić M, Čavić M, Rackov M, Borovac B and Lu Z 2017 *Drive System of the Robot Eyeballs and Eyelids with 8 DOFs*, 12th IFToMM International Symposium on Science of Mechanisms and Machines SYROM 2017, Iasi, Romania, November 2-3, to be published

[2] Penčić M, Čavić M, Rackov M, Borovac B, Knežević I and Zlokolica M 2017 *Kinematic Analysis of the Robot Eyes Drive System with 7 DOFs*, 8th PSU-UNS International Conference on Engineering and Technology ICET 2017, Novi Sad, Serbia, June 8-10, pp PS-1.14-1–PS-1.14-5

[3] Penčić M, Čavić M, Savić S and Borovac B 2016 *Comparative Analysis of the Shrug Mechanisms for Humanoid Robots*, 3th International Conference on Electrical, Electronic and Computing Engineering (IcETRAN 2016), Zlatibor, Serbia, June 13-16, pp ROI1.3-1–ROI1.3-5

[4] Penčić M, Čavić M and Borovac B 2016, *Comparative Synthesis of the Shrug Mechanisms for Humanoid Robots*, 20th International Research/Expert Conference on Trends in the Development of Machinery and Associated Technology TMT 2016, Mediterranean Sea Cruising, September 24 - October 1, pp 249-252

[5] Penčić M, Čavić M and Borovac B 2017 Optimal Synthesis of the Worm-Lever Mechanism for Humanoid Robots Shrug, *Serbian Journal of Electrical Engineering* **14**(2) 245-256

[6] Penčić M, Čavić M, Rackov M, Borovac B and Lu Z 2017 *Kinematic-Dynamic Analysis of the Cam-Worm Mechanism for Humanoid Robots Shrug*, 12th IFToMM International Symposium on Science of Mechanisms and Machines SYROM 2017, Iasi, Romania, November 2-3, to be published

[7] Penčić M, Čavić M and Borovac B 2016 *Analysis of Mechanisms for Achieving Different Ways of Power Transmission and Motion of the Anthropomorphic Robots Upper Body*, 5th International Conference on Power Transmission BAPT 2017, Ohrid, Macedonia, October 5-8, pp 115-122

[8] Penčić M M, Borovac B A, Kovačević D I and Čavić M V 2016 Development of the Multi-

Segment Lumbar Spine for Humanoid Robots, *Thermal Science* **20**(Suppl. 2) S581–S590
[9] Nikolić D 2007 *Kinesiology*, College of Health Studies, Ćuprija, Serbia
[10] Opavsky P 1998 *Introduction to Sports Biomechanics*, Author's edition, Belgrade, Serbia
[11] Ishihara H and Asada M 2015 Design of 22-DOF Pneumatically Actuated Upper Body for Child android 'Affetto', *Advanced Robotics* **29**(18) 1151-1163
[12] Ahn H S, Lee D-W, Choi D, Lee D-Y, Lee H-G and Baeg M-H 2013 Development of an Incarnate Announcing Robot System Using Emotional Interaction with Humans, *International Journal of Humanoid Robotics* **10**(2) 1350017-1–1350017-24
[13] Ahn H S, Lee D-W, Choi D, Lee D Y, Hur M H, Lee H and Shon W H 2011 *Development of an Android for Singing with Facial Expression*, 37th Annual Conference of the IEEE Industrial Electronics Society IECON 2011, Melbourne, Australia, November 7-10, pp 104-109
[14] Minato T, Yoshikawa Y, Noda T, Ikemoto S, Ishiguro H and Asada H 2007 *CB^2: A Child Robot with Biomimetic Body for Cognitive Developmental Robotics*, 7th IEEE-RAS International Conference on Humanoid Robots HUMANOIDS 2007, Pittsburgh, PA, USA, November 29 - December 1, pp 557-562
[15] Sakagami Y, Watanabe R, Aoyama C, Matsunaga S, Higaki N and Fujimura K 2002 *The intelligent ASIMO: System Overview and Integration*, IEEE/RSJ International Conference on Intelligent Robots and Systems IROS 2002, Lausanne, Switzerland, September 30 - October 4, pp 2478-2483
[16] Kajita S, Kaneko K, Kaneiro F, Harada K, Morisawa M, Nakaoka S, Miura K, Fujiwara K, Neo E S, Hara I, Yokoi K and Hirukawa H 2011 *Cybernetic Human HRP-4C: A Humanoid Robot with Human-Like Proportions*, In book: Pradalier C, Siegwart R and Hirzinger G (Eds.) Robotics Research: 14th ISRR. STAR **70** 301–314, Springer, Heidelberg
[17] Parmiggiani A, Maggiali M, Natale L, Nori F, Schmitz A, Tsagarakis N, Santos-Victor J, Becchi F, Sandini G and Metta G 2012 The Design of the iCub Humanoid Robot, *International Journal of Humanoid Robotics* **9**(4) 1250027-1–1250027-24
[18] Diftler M A, Mehling J S, Abdallah M E, Radford N A, Bridgwater L B, Sanders A M, Askew R S, Linn D M, Yamokoski J D, Permenter F A, Hargrave B K and Platt R 2011 *Robonaut 2 – The First Humanoid Robot in Space*, IEEE International Conference on Robotics and Automation ICRA 2011, Shanghai, China, May 9-13, pp 2178-2183
[19] Park I-W, Kim J-Y, Cho B-K and Oh J-H 2008 Control Hardware Integration of a Biped Humanoid Robot with an Android Head, *Robotics and Autonomous Systems* **56**(1) 95-103
[20] Lemburg J, de Gea Fernández J, Eich M, Mronga D, Kampmann P, Vogt A, Aggarwal A, Shi Y and Kirchner F 2011 *AILA – Design of an Autonomous Mobile Dual-Arm Robot*, IEEE International Conference on Robotics and Automation ICRA 2011, Shanghai, China, May 9-13, pp 5147–5153
[21] Englsberger J, Werner A, Ott C, Henze B, Roa M A, Garofalo G, Burger R, Beyer A, Eiberger O, Schmid K and Albu-Schäffer A 2014 *Overview of the Torque-Controlled Humanoid Robot TORO*, 14th IEEE-RAS International Conference on Humanoid Robots HUMANOIDS 2014, Madrid, Spain, November 18-20, pp 916-923
[22] Lee S, Kim J-Y and Kim M 2013 Development and Walking Control of Emotional Humanoid Robot, KIBO, *International Journal of Humanoid Robotics* **10**(4) 1350024-1–1350024-35
[23] Park I-W, Kim J-Y, Lee J and Oh J-H 2007 Mechanical Design of the Humanoid Robot Platform, HUBO, *Advanced Robotics* **21**(11) 1305-1322
[24] Ogura Y, Aikawa H, Shimomura K, Kondo H, Morishima A, Lim H-O and Takanishi A 2006 *Development of a New Humanoid Robot WABIAN-2**, IEEE International Conference on Robotics and Automation ICRA 2006, Orlando, FL, USA, May 15-19, pp 76-81
[25] Endo N and Takanishi A 2011 Development of Whole-Body Emotional Expression Humanoid Robot for ADL-Assistive RT Services, *Journal of Robotics and Mechatronics* **23**(6) 969-977
[26] Asfour T, Regenstein K, Azad P, Schroder J, Bierbaum A, Vahrenkamp N and Dillmann R 2006 *ARMAR-III: An Integrated Humanoid Platform for Sensory-Motor Control*, 6th IEEE-

RAS International Conference on Humanoid Robots HUMANOIDS 2006, Genova, Italy, December 4-6, pp 169-175
[27] Erbatur K, Seven U, Taşkiran E, Koca Ö, Yilmaz M, Ünel M, Kiziltaş G, Sabanovic A and Onat A 2009 *SURALP: A New Full-Body Humanoid Robot Platform*, IEEE/RSJ International Conference on Intelligent Robots and Systems IROS 2009, St. Louis, MO, USA, October 11-15, pp 4949-4954
[28] Jamone L, Metta G, Nori F and Sandini G 2006 *James: A Humanoid Robot Acting over an Unstructured World*, 6th IEEE-RAS International Conference on Humanoid Robots, Genova HUMANOIDS 2006, Italy, December 4-6, pp 143-150
[29] Ott Ch, Eiberger O, Friedl W, Bauml B, Hillenbrand U, Borst Ch, Albu-Schaffer A, Brunner B, Hirschmuller H, Kielhofer S, Konietschke R, Suppa M, Wimbock T, Zacharias F and Hirzinger G 2006 *A Humanoid Two-Arm System for Dexterous Manipulation*, IEEE-RAS International Conference on Humanoid Robots HUMANOIDS 2006, Genova, Italy, December 4-6, pp 276-283
[30] Claudio G, Spindler F and Chaumette F 2016 *Vision-Based Manipulation with the Humanoid Robot Romeo*, IEEE-RAS International Conference on Humanoid Robots HUMANOIDS 2016, Cancun, Mexico, November 15-17, pp 289-293
[31] Lenz A, Skachek S, Hamann K, Steinwender J, Pipe A G and Melhuish C 2010 *The BERT2 Infrastructure: An Integrated System for the Study of Human-Robot Interaction*, IEEE-RAS International Conference on Humanoid Robots HUMANOIDS 2010, Nashville, TN, USA, December 6-8, pp 346-351
[32] Itoh K, Miwa H, Zecca M, Takanobu H, Roccella S, Carrozza M C, Dario P and Takanishi A 2006 *Mechanical Design of Emotion Expression Humanoid Robot WE-4RII*, In book: Zielinska T and Zielinski C (Eds.) ROMANSY 16: Robot Design, Dynamics and Control. CISM **487** 255-262, Springer, Vienna
[33] Kaneko K, Kanehiro F, Morisawa M, Akachi K, Miyamori G, Hayashi A and Kanehira N 2011 *Humanoid robot HRP-4 – Humanoid Robotics Platform with Lightweight and Slim Body*, IEEE/RSJ International Conference on Intelligent Robots and Systems IROS 2011, San Francisco, CA, USA, September 25-30, pp 4400-4407
[34] Inaba M, Mizuuchi I, Tajima R, Yoshikai T, Sato D, Nagashima K and Inoue H 2003 *Building Spined Muscle-Tendon Humanoid*, In book: Jarvis R A and Zelinsky A (Eds.) Robotics Research: 9th ISRR. STAR **6** 113-127, Springer, Heidelberg
[35] Mizuuchi I, Yoshikai T, Nakanishi Y and Inaba M 2005 *A Reinforceable-Muscle Flexible-Spine Humanoid "Kenji"*, IEEE/RSJ International Conference on Intelligent Robots and Systems IROS 2005, Edmonton, Canada, August 2-6, pp 4143-4148
[36] Mizuuchi I, Yoshikai T, Sodeyama Y, Nakanishi Y, Miyadera A, Yamamoto T, Niemela T, Hayashi M, Urata J, Namiki Y, Nishino T and Inaba M 2006 *Development of Musculoskeletal Humanoid Kotaro*, IEEE International Conference on Robotics and Automation ICRA 2006, Orlando, FL, USA, May 15-19, pp 82-87
[37] Mizuuchi I, Nakanishi Y, Sodeyama Y, Namiki Y, Nishino T, Muramatsu N, Urata J, Hongo K, Yoshikai T and Inaba M 2007 *An Advanced Musculoskeletal Humanoid Kojiro*, 7th IEEE-RAS International Conference on Humanoid Robots HUMANOIDS 2007, Pittsburgh, PA, USA, November 29 - December 1, pp 294-299
[38] Izawa T, Osada M, Ito N, Ohta S, Urata J and Inaba M 2011 *Development of Musculoskeletal Humanoid Kenzoh with Mechanical Compliance Changeable Tendons by Nonlinear Spring Unit*, IEEE International Conference on Robotics and Biomimetics ROBIO 2011, Phuket, Thailand, December 7-11, pp 2384-2389
[39] Asano Y, Kozuki T, Mizoguchi H, Motegi Y, Osada M, Shirai T, Urata J, Okada K and Inaba M 2012 *Design Concept of Detail Musculoskeletal Humanoid Kenshiro – Toward a Real Human Body Musculoskeletal Simulator*, 12th IEEE-RAS International Conference on Humanoid Robots HUMANOIDS 2012, Osaka, Japan, November 29 - December 1, pp 811-816
[40] Asano Y, Kozuki T, Ookubo S, Kawamura M, Nakashima S, Katayama T, Yanokura I, Hirose T,

Kawaharazuka K, Makino S, Kakiuchi Y, Okada K and Inaba M 2016 *Human Mimetic Musculoskeletal Humanoid Kengoro toward Real World Physically Interactive Actions*, 16th IEEE-RAS International Conference on Humanoid Robots HUMANOIDS 2016, Cancun, Mexico, November 15-17, pp 876-883

Instantaneous flywheel torque of IC engine grey-box identification

A Milašinović, D Knežević, Z Milovanović and J Škundrić

Faculty of Mechanical Engineering, University of Banja Luka, Stepe Stepanovića 71 78000 Banja Luka, Bosnia & Herzegovina

E-mail: aleksandar.milasinovic@unibl.rs

Abstract. In this paper a mathematical model developed for the identification of excitation torque acting on the IC engine flywheel is presented. The excitation torque gained through internal combustion of the fuel in the IC engine is transmitted from the flywheel to the transmission. The torque is not constant but variable and is a function of the crank angle. The verification of the mathematical model was done on a 4-cylinder 4-stroke diesel engine for which the in-cylinder pressure was measured in one cylinder and the instantaneous angular speed of the crankshaft at its free end. The research was conducted on a hydraulic engine brake. Inertial forces of all rotational parts, from flywheel to the turbine wheel of the engine brake, are acting on the flywheel due to the nonuniform motion of the flywheel. It is known from the theory of turbomachinery that the torque on the hydraulic brake is a quadratic function of angular speed. Due to that and the variable angular speed of the turbine wheel of the engine brake, the torque during one engine cycle is also variable. The motivation for this research was the idea (intention) to determine the instantaneous torque acting on the flywheel as a function of the crank angle with a mathematical model without any measuring and based on this to determine the quality of work of specific cylinders of the multi-cylinder engine. The crankshaft was considered elastic and also its torsional vibrations were taken into account.

1. Introduction

The internal combustion (IC) engine is still a dominant drive unit of automobiles. Due to economic and ecological aspects, the optimization and diagnostics of the quality of IC engine work is very important. There are two methods to determine the instantaneous torque acting on the flywheel of an IC engine: the direct method (by measurement) and the indirect method of parameter identification of a mathematical model. The direct method implies the installation of another complex device on the engine and its connection to the engine electronics. The indirect method of determination of the torque on the flywheel would be (imply) a software solution which would require a more powerful computer support.

The motion of the crankshaft is the result of excitation torques due to the combustion of fuel in the cylinder, inertial torques of the slider-crank mechanism, elastic deformations of the crankshaft and crankshaft motion resistance torque which is transmitted over the flywheel to the power consumers (engine brake). Under stationary engine operation all torques, acting on the flywheel, could be considered periodical and as such may be expressed in a Fourier series. The total motion of the crankshaft can be considered as a sum of the motions caused by individual harmonics.

It is known from theoretical dynamics that there are two approaches considering mechanical system motion: the first, where forces and torques acting on the system are known while the law of motion has to be determined and the second approach where the law of motion is known while the forces and torques causing this motion have to be determined. In this particular case the situation is more complex because the law of motion of the free end of the crankshaft is known (gained through measurement), but the law of motion of particular sections of the equivalent dynamical system of the crankshaft is not. The law of motion of the crankshaft depends both on excitation torques and dynamical characteristics of the crankshaft (reduced moment of inertia for the axis of rotation, torsional stiffness, friction, etc.). Accurate knowledge of these properties and a mathematical model which does not include assumptions that could lead to significant deviation between the actual and modeled system can lead to a good-quality simulation model of the dynamical behavior of the crankshaft. It is usually assumed that the torque on the brake is constant [1].

2. Crankshaft dynamics

Figure 1 shows an equivalent schematics of a dynamical model of elastic crankshaft. This approach of conversion of a real dynamical system into an equivalent system has proven to be a good solution in praxis [1-18].

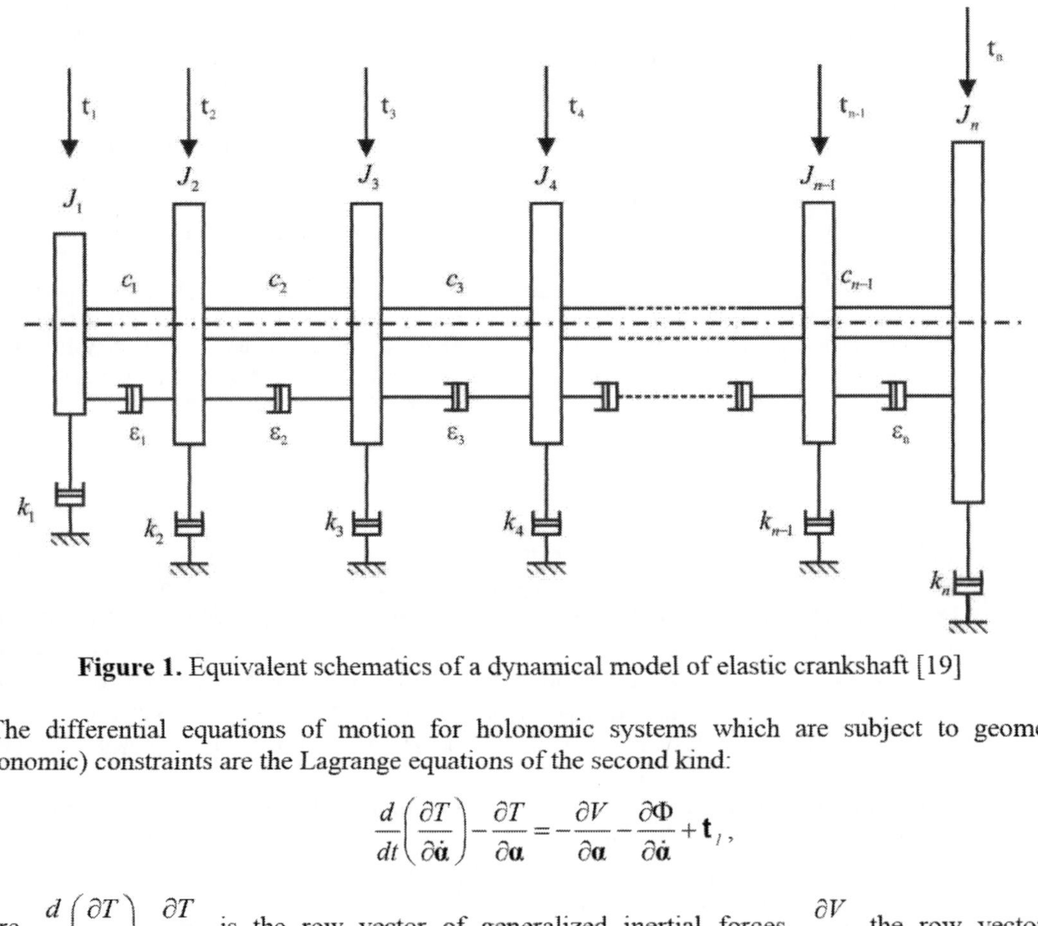

Figure 1. Equivalent schematics of a dynamical model of elastic crankshaft [19]

The differential equations of motion for holonomic systems which are subject to geometric (holonomic) constraints are the Lagrange equations of the second kind:

$$\frac{d}{dt}\left(\frac{\partial T}{\partial \dot{\boldsymbol{\alpha}}}\right)-\frac{\partial T}{\partial \boldsymbol{\alpha}}=-\frac{\partial V}{\partial \boldsymbol{\alpha}}-\frac{\partial \Phi}{\partial \dot{\boldsymbol{\alpha}}}+\mathbf{t}_l, \tag{1}$$

where $\frac{d}{dt}\left(\frac{\partial T}{\partial \dot{\boldsymbol{\alpha}}}\right)-\frac{\partial T}{\partial \boldsymbol{\alpha}}$ is the row vector of generalized inertial forces, $\frac{\partial V}{\partial \boldsymbol{\alpha}}$ the row vector of generalized conservative forces, $\frac{\partial \Phi}{\partial \dot{\boldsymbol{\alpha}}}$ the row vector of generalized dissipative forces and $\mathbf{t}_l$ the row vector of generalized external excitation forces. Here, external load refers to a vector comprising excitation forces on each lumped mass, regardless of externally applied force or forces of constraint. For example, if we observe the lumped mass model of the crankshaft, there is no external force acting on the first lumped mass - the flywheel, but there is a generalized force, which is the result of the constraint force acting between the flywheel and the brake. Frictional forces are not included in the vector of external excitation forces (they are included in the stiffness matrix through external damping) [20].

A crankshaft with n lumped masses, rotating at steady state operating conditions while the lumped masses are moving variably (nonuniformly) around the mean angular velocity, is observed. The angular position of each lumped mass is defined by a column vector of generalized coordinates $\boldsymbol{\alpha}=[\alpha_1,\alpha_2,...,\alpha_n]^T$.

$$T=\frac{\dot{\boldsymbol{\alpha}}^T\mathbf{J}(\boldsymbol{\alpha})\dot{\boldsymbol{\alpha}}}{2}, \tag{2}$$

where $\mathbf{J}(\boldsymbol{\alpha})$ is the moments of inertia matrix of the lumped mass model of the crankshaft.

The potential energy can be expressed as

$$V=\frac{\boldsymbol{\alpha}^T\mathbf{C}\boldsymbol{\alpha}}{2}, \tag{3}$$

where $\mathbf{C}$ is the stiffness matrix of the lumped mass model of the crankshaft.

If it is assumed that frictional resistance during the oscillation of the crankshaft is comprised of external and internal (resistance within the shaft), the dissipation function has the following form:

$$\Phi=\frac{\dot{\boldsymbol{\alpha}}^T\mathbf{K}\dot{\boldsymbol{\alpha}}}{2}, \tag{4}$$

where $\mathbf{K}$ is the damping matrix, both external and internal.

After substituting particular functional derivatives defined by (2)-(4) into the Lagrange equation (1), the equation of motion as a function of the crank angle is obtained:

$$\mathbf{J}(\boldsymbol{\alpha})\ddot{\boldsymbol{\alpha}}+\frac{1}{2}\frac{d\mathbf{J}(\boldsymbol{\alpha})}{d\boldsymbol{\alpha}}\dot{\boldsymbol{\alpha}}^2+\mathbf{K}\dot{\boldsymbol{\alpha}}+\mathbf{C}\boldsymbol{\alpha}=\mathbf{t}_l. \tag{5}$$

If the motion of the lumped mass model of the crankshaft is observed as being composed of the motion of all lumped masses with constant speed ω_0 together and the relative motion of every lumped mass in relation to the common crank angle of uniform motion $\omega_0 t$, and also by defining the row vector of angle of nonuniform motion $\boldsymbol{\varphi}$ of the lumped mass model of the crankshaft in relation to $\omega_0 t$, the following relation, regarding angular position of the lumped masses can be obtained:

$$\boldsymbol{\alpha}(t)=\boldsymbol{\omega}_0 t+\boldsymbol{\varphi}(t), \tag{6}$$

where $\boldsymbol{\omega}_0=[\omega_0,...,\omega_0]^T_{1\times n}$ is a column vector of mean angular speed, t is time (a scalar quantity) while $\boldsymbol{\varphi}(t)=[\varphi_1(t),...,\varphi_n(t)]^T_{1\times n}$ is a column vector of the angle of nonuniform motion of the lumped mass model of the crankshaft. Quantities which are functions of the vector of the angle of motion ($\mathbf{J}(\boldsymbol{\alpha})$, $\frac{d\mathbf{J}(\boldsymbol{\alpha})}{d\boldsymbol{\alpha}}$ and $\mathbf{t}_l(\boldsymbol{\alpha})$) may be approximated by a Taylor series [19].

After expanding into Taylor series and omitting $\dot{\boldsymbol{\varphi}}^2$, $\boldsymbol{\varphi}^2$, $\boldsymbol{\varphi}\dot{\boldsymbol{\varphi}}$ and $\boldsymbol{\varphi}\ddot{\boldsymbol{\varphi}}$ as lower-order terms, a system of differential equations describing the motion of an IC engine crankshaft in the form of a matrix is gained:

$$\mathbf{J}(\omega_0 t)\ddot{\varphi}+\mathbf{K}\dot{\varphi}(t)+\frac{d\mathbf{J}(\omega_0 t)}{d(\omega_0 t)}\omega_0\dot{\varphi}+\frac{1}{2}\frac{d^2\mathbf{J}(\omega_0 t)}{d(\omega_0 t)^2}\omega_0^2\varphi+\mathbf{C}\varphi=\mathrm{t}_l(\omega_0 t)-\frac{1}{2}\frac{d\mathbf{J}(\omega_0 t)}{d(\omega_0 t)}\omega_0^2-\mathbf{K}\omega_0, \quad (7)$$

After analyzing equation (7), it can be noticed that the excitation is composed of the excitation torque referring to external torque acting on all lumped masses $\mathrm{t}_l(\omega_0 t)$ and the torque due to the motion of the slider-crank mechanism $\frac{1}{2}\frac{d\mathbf{J}(\omega_0 t)}{d(\omega_0 t)}\omega_0^2$. The constant friction torque $\mathbf{K}\omega_0$ represents engine losses that occur as a result of absolute damping. The moment of inertia and its first and second derivative are periodic functions of the crank angle of uniform motion $\omega_0 t$. For the given system of differential equations, they are variable coefficients, i.e. the system is nonlinear. The term $\frac{1}{2}\frac{d^2\mathbf{J}(\omega_0 t)}{d(\omega_0 t)^2}\omega_0^2\varphi$ takes into consideration the excitation variation caused by the masses of the slider-crank mechanism, with the angle of nonuniform motion $\varphi(t)$ of the lumped mass model of the crankshaft. The equation (7) takes into consideration the variability of the moment of inertia as well as the variability of excitation caused by the motion of the slider-crank mechanism with the angle of nonuniform motion.

3. Parameter identification for the dynamical model of elastic crankshaft

The dynamic behavior of a multibody system can be described by a set of ordinary differential equations. A relative coordinate formulation is applied such that no algebraic equations due to kinematic joints arise [21]

$$\begin{aligned}&\mathbf{M}(\mathbf{q},\mathbf{b})\ddot{\mathbf{q}}=\mathbf{Q}^A(\mathbf{q},\dot{\mathbf{q}},\mathbf{b},t),\\&\mathbf{q}(t_0)=\mathbf{q}_0,\ \dot{\mathbf{q}}(t_0)=\dot{\mathbf{q}}_0.\end{aligned}, \quad (8)$$

In equation **Error! Reference source not found.** $\mathbf{M}$ is mass matrix, $\mathbf{Q}^A$ is generalized applied force, and $\mathbf{q}(t_0)$ and $\dot{\mathbf{q}}(t_0)$ are initial conditions. The column of generalized forces includes externally applied forces and torques.

The mathematical model of the dynamical system of equivalent crankshaft contains a large number of the unknown or approximately known parameters that significantly affect the dynamic motion of the crankshaft. Parameter identification of the unknown or approximately known parameters, from experimental tests of IC engine, is one of the key problems to achieve a good agreement of results obtained from the mathematical model and the experimental measurements. On the other hand, based on the identified parameters a better understanding can be obtained in those parts of the process where modeling is more or less formalized, and checks made assumptions done. The model parameters whose exact measurement in the working conditions of the engine impossible can be determined.

The goal of parameter identification is to determine the parameters of a fixed mathematical model so that the output from the model agrees with the results obtained by measuring. It should be noted that there is a clear distinction between the parameter identification (grey box model) and the system identification (black box model). Grey box model still has a number of unknown parameters which should be defined. The parameter identification calculates the mathematical model parameters that best fit a block of measured data. The least-squares parameter identification finds the parameter vector that minimizes the sum of squared prediction errors.

Equation **Error! Reference source not found.** represents differential equations of motion of the crankshaft. A common approach is to first determine all the parameters of the equivalent system (moments of inertia matrix $\mathbf{J}$, stiffness matrix $\mathbf{C}$, damping matrix $\mathbf{K}$, and torque vector t_l) and then solve differe7ntial equations – obtain vector φ. However, determining the parameters of the equivalent system is not simple. We can quite accurately determine some of the parameters such as moment of inertia (*Computer Aided Design*-CAD software), the gas-pressure torque (measuring the gas pressure in the cylinder) and the torque due to the motion of the slider-crank mechanism $\frac{1}{2}\frac{d\mathbf{J}(\omega_0 t)}{d(\omega_0 t)}\omega_0^2$. Determination of other parameters of the system such as the stiffness of particular parts of the crankshaft $\mathbf{C}$, damping matrix $\mathbf{K}$ and the reaction force acting between the flywheel and the brake. However, we can determine the boundaries within some of the unknown parameters range. In [15] a procedure for determining the stiffness crank of the crankshaft is given. It is important to determine as precisely as possible the border of the unknown parameter range. The mathematical model gives all the solutions. However, some of the solutions of the system are impossible from the standpoint of mechanics, e.g. negative stiffness crank of the crankshaft [15].

An integrated design approach for dynamic systems has to support all steps from problem formulation to problem solution by optimization, see Figure 1. Firstly, the technical system to be optimized has to be transformed to a mathematical model. Therefore, parameters of the mathematical model have to be classified either as design variables whose values can be chosen within given bounds or as system constants whose values remain fixed during optimization. In the presence of several criteria, multicriteria optimization strategies may be used to find optimal compromise solutions.

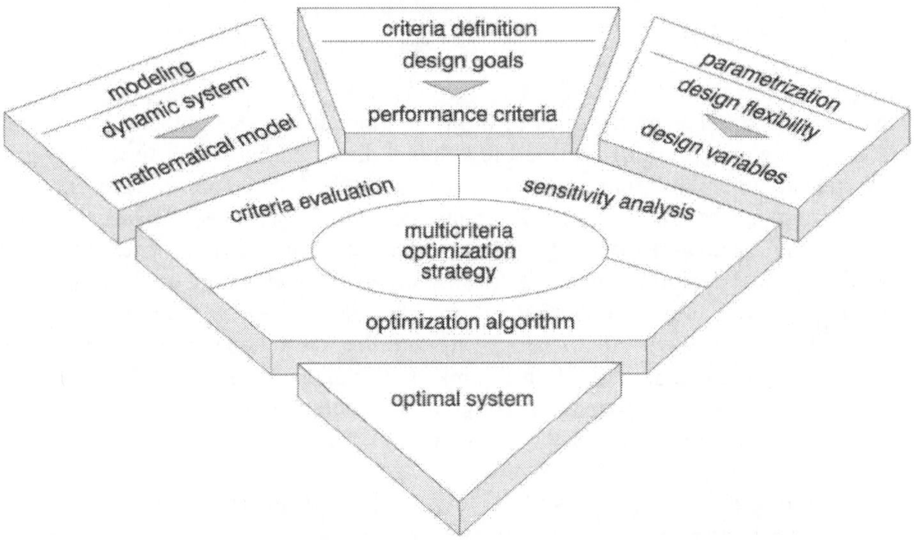

Figure 2. Components of an integrated design concept [22]

Typically in the parameter identification process [15], experimental data in the form $\{(t_i, y_i) | i = 1,...,n_m\}$ have been collected, and the design parameters $\mathbf{b} \in \mathbf{R}^p$ for the given model $\mathcal{M}(t,\mathbf{b})$ must be selected such that $y_m(t) \approx y_i$, where

$$y_m(t) = \mathcal{M}(t,\mathbf{b}), \tag{9}$$

Generally, there are far more data points than parameters $n_m > p$.

In equation **Error! Reference source not found.**, the mathematical model is an arbitrary nonlinear model. The system will include a vector of p unknown parameters $\mathbf{b}$. These parameters represent physical properties of the rigid body mechanism, such as masses, inertias, spring coefficients, damping coefficients. Note that any of the terms in equation **Error! Reference source not found.** can be a function of the unknown parameters $\mathbf{b}$.

The general approach to the solution of the parameter identification problem is to minimize a cost function of the model residuals, defined as follows [23-25]

$$r_i(\mathbf{b}) = y_i - y_m(t_i), \tag{10}$$

The cost function is given as the quadratic form

$$F(\mathbf{b}) = \frac{1}{2}\sum_{i=1}^{n_m} r_i^2(\mathbf{b}), \tag{11}$$

The estimate of the design parameters $\mathbf{b}$, based on n_m data points, is then defined as

$$\mathbf{b}^* = \min_{\mathbf{b}}\left[F(\mathbf{b})\right], \tag{12}$$

By assembling the model residuals into a vector, given as

$$\mathbf{r}(\mathbf{b}) = [r_1(\mathbf{b}),\ r_2(\mathbf{b}),\ldots.,r_{n_m}(\mathbf{b})]^T, \tag{13}$$

The design parameters must be chosen so that the residual vector is as small as possible, in some sense. The quadratic performance function $F(\mathbf{b})$ becomes

$$F(\mathbf{b}) = \frac{1}{2}\mathbf{r}^T(\mathbf{b})\mathbf{r}(\mathbf{b}), \tag{14}$$

4. Result analysis

In order to study the dynamic characteristics of the IC engine, an experimental setup was made whose scheme is shown in Figure 3. A turbocharged low-speed diesel engine with direct fuel injection was experimentally investigated. The main characteristics of the tested engine are presented in Table 1. The following parameters of interest for the engine operation have been measured: instantaneous angular velocity of the crankshaft free end, cylinder pressure in the cylinder next to the flywheel and the torque on the engine dynamometer. All these parameters were measured simultaneously. The measurements were carried out at different engine speeds and loads. The task of experimental measurements in the context of this paper is to get the data necessary to perform parameter identification of the dynamical model of the crankshaft. Since the parameter identification of the dynamical model is based on a comparison of actual and modelled angular speed of the free end of the crankshaft, the basic measurement item is the instantaneous angular speed of the pulley.

Table 1. Characteristics of the tested engine: TAM BF 4 L 515 C

Property	Symbol	Value	Unit
Piston head area	A_k	0.123	m^2
Crank radius	r	0.0725	m
Length of the connection rod	l	0.237	m
Firing order	---	1-3-4-2	
Reciprocating mass	m_{rec}	4.6	kg
Rotating mass	m_{rot}	2.6	kg
Cylinder capacity	V	7000	cm^3
Bore diameter	d	125	mm

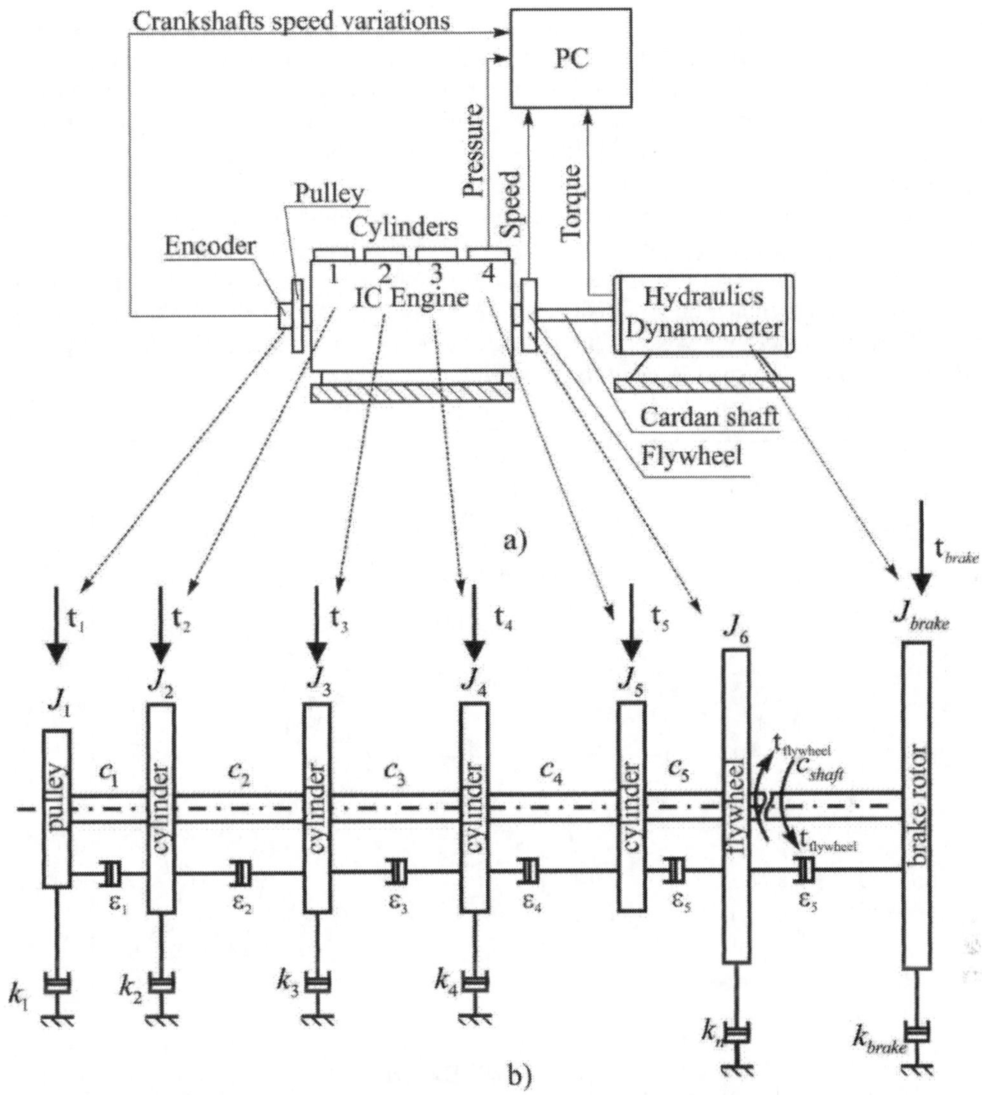

Figure 3. Experimental setup a) Schematic of measuring installations for measuring dynamic characteristics of the IC engine b) The lumped mass model

As function $\mathbf{J}(\boldsymbol{\alpha})$ and $\mathrm{t}_i(\boldsymbol{\alpha})$ that exist in the mathematical model **Error! Reference source not found.** are periodic -4π, we can decompose it on a Fourier basis. Harmonic analysis and synthesis is used to solve optimization problems. The first harmonic is a constant and it causes uniform movement of the crankshaft with a constant angular speed. For the test engine, the pressure is measured in the fourth cylinder. However, it was assumed that the curve of change of pressure as a function of the angle of the crankshaft in the other cylinders (the first, second and third) the same as the curve obtained by measuring the pressure in the fourth cylinder only is phase shifted. The aforementioned curve of change of pressure in the first, second and third cylinders is used as the initial iteration in the optimization process. When an engine is working properly, this assumption is possible. The torque due to motion of the crank-slider mechanism masses is the same for all cylinders, but is phase shifted as a function of the crankshaft angle position. Any possible error in such an assumption is negligible.

The moment of inertia is a function of both mass and position. Therefore, in the case of a crank-slider mechanism, where the geometry changes, the moment of inertia will also vary, hence the term varying inertia will be used for this. The moments of inertia a parts of crank-slider mechanism were determined using SolidWorks CAD (Computer Aided Design) package. Determination of the crankshaft torsional stiffness coefficient can be reduced to determining the torsional stiffness coefficient of simple cylindrical parts, and determining the torsional stiffness coefficient for a single crank of the crankshaft [15]. Torsional stiffness is determined by using CAE-Computer Aided Engineering software packages Catia and SolidWorks.

Figure 4 shows identified instantaneous torque that is a function of the crank angle. Figure 4 also shows identified torques of the cylinder in which the pressure is not measured (1-cyl., 2-cyl. and 3-cyl). Four peaks can be seen on the curve identified torque of the flywheel which are caused by the action of individual engine cylinders. Changes in torque at the flywheel during one cycle are significant. Moment of inertia brakes, angular acceleration of the parts of the brake and torque that is generated in the brake affect the torque at the flywheel.

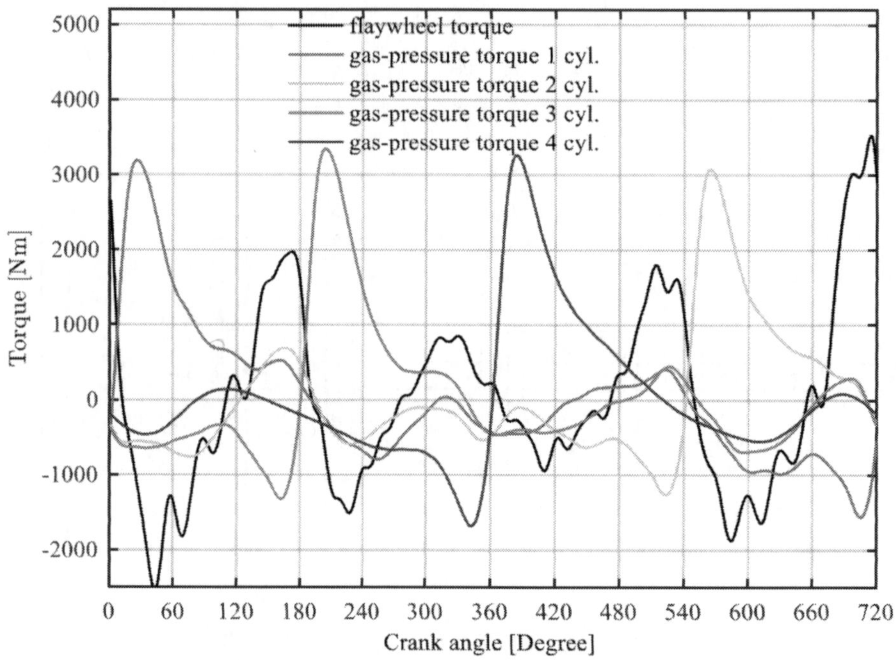

Figure 4. Identified torque of the flywheel and torques of the cylinder in which the pressure is not measured at 1000 rpm speed of rotation of the crankshaft.

Figure 5 shows the instantaneous calculated and measured angular velocity of the free end of the crankshaft at 1000 rpm speed of rotation of the crankshaft. It could be noticed that the simulated speed follows very closely the measured speed. Amplitude of the unevenness of rotation of the crankshaft is large, at 1000 rpm, and torsional oscillations are small.

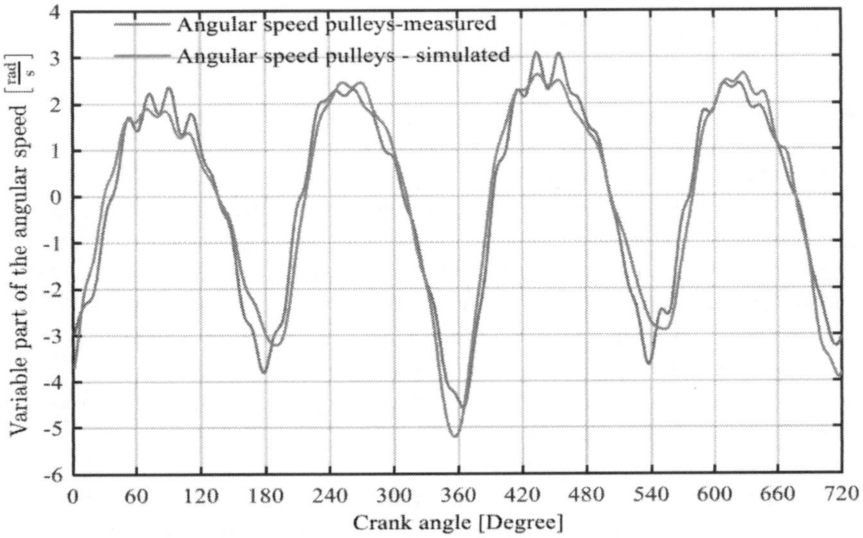

Figure 5. The instantaneous calculated and measured angular velocity of the free end of the crankshaft at 1000 rpm speed of rotation of the crankshaft

Parameter identification is done on the basis of harmonic analysis. Eight parameters are identified for a single harmonic: three amplitudes and three-phase pressure in the cylinders in which the pressure is not measured and amplitude and phase harmonica at the flywheel. "Trust-region methods" were used for the identification [26]. Changing the starting point of search for the optimal solution leads to a change in the number of iterations.

The given mathematical model of movement of the crankshaft, which is used to identify parameters, ends with flywheel. The impact of the brakes on the flywheel is replaced by reactive torque acting on the flywheel.

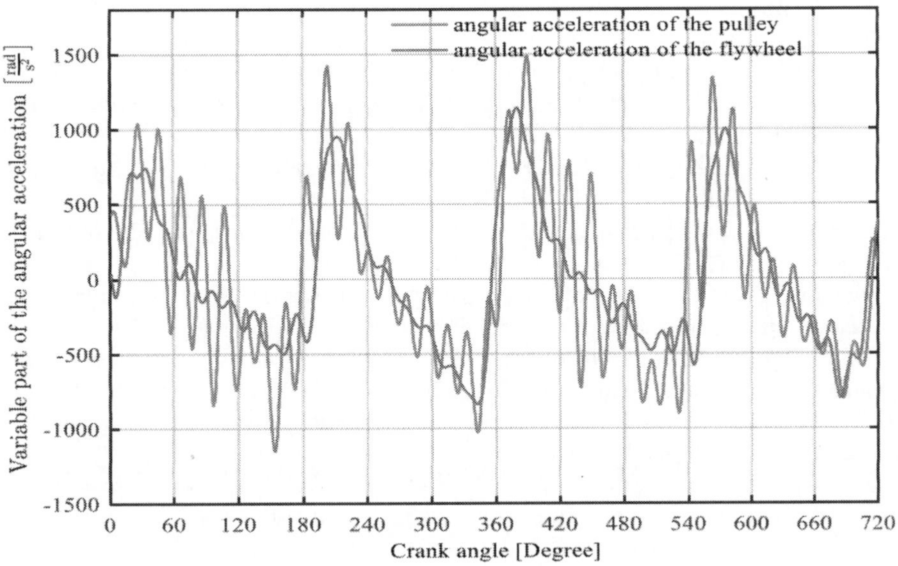

Figure 6. Angular acceleration of the pulley and flywheel in the function of the angle position of the crankshaft at 1000 rpm engine speed

Figure 6 shows the curve of change of angular acceleration of the flywheel and pulleys in the function of angle position of the crankshaft. Angular acceleration of the pulley has a dominant harmonics of higher orders. This phenomenon is caused by torsion vibration of the crankshaft.

5. Conclusions

Based on the established mathematical model and the analysis presented in this paper, the following conclusions can be made:

- The mathematical model must be as simple as possible and with minimum errors in determining the parameters of the model. Taking into account the variability moment of inertia of the system in the mathematical modelling of the dynamics of the crankshaft in the identification parameters of the dynamic system is not necessary. A greater mistake is made when estimating the friction or torsional stiffness in the system rather than neglecting the variability moment of inertia of the system.
- If some parameters of the mathematical model cannot be determined accurately then the segment can be determined in which the parameter ranges and the method of identifying parameters determine its real value.
- The instantaneous value of torque at the flywheel of IC engines is difficult to measure and such a sensor is not installed in the vehicle. By using the method of identifying parameters, it is possible to determine the value of the torque at the flywheel and this information can be used for diagnostic purposes.
- A common simplification that is created when the crankshaft is modelled as a rigid body leads to the inaccurate determination of excitation torque. The mathematical model of the elastic crankshaft allows the addition of detection of inadequate work of each cylinder and determines which cylinders are not working properly.

References

[1] Taraza D, Henein N A and Bryzik W 2001 The frequency analysis of the crankshaft's speed variation: a reliable tool for diesel engine diagnosis, *Journal of Engineering for Gas Turbines and Power* **123**(3) 428–431

[2] Kinkle U and Nielsen L 2005 *Automotive control system for engine, driveline and vehicle*, Springer-Verlag, Berlin

[3] Genta G 1993 *Vibrations of structures and machine*, Springer-Verlag, Berlin

[4] Hafner K E and Maass H 1985 *Torsionsschwingungen in der Verbrennungskraftmaschine*, Springer-Verlag, Berlin

[5] Taraza D 2001 *Quantifying relationships between the crankshaft's speed variation and the gas pressure torque*, SAE paper 2001-01-1007

[6] Taraza D 2002 *Accuracy Limits of IMEP Determination from Crankshaft Speed Measurements*, SAE Paper No. 2002-01-0331

[7] Taraza D 1993 *Estimation of the Mean Indicated Pressure from Measurements of the Crankshafts Angular Speed Variation*, SAE Paper No. 932413

[8] Taraza D 1993 *Possibilities to Reconstruct Indicator Diagrams by Analysis of the Angular Motion of the Crankshaft*, SAE Paper No. 932414

[9] Schagerberg S and McKelvey T 2003 *Instantaneous crankshaft torque measurements-modeling and validation,* SAE paper 2003-01-0713

[10] Taraza D, Henein N A and Bryzik W 1998 *Determination of the gas-pressure torque of a multicylinder engine from measurements of the crankshaft's speed variation*, SAE paper 980164

[11] Filipović I 2007 *IC engine dynamics and vibration*, Faculty of Mechanical Engineering, University of Sarajevo Sarajevo, Bosnia and Herzegovina

[12] Nestoridies E J 1958 BICERA Research Laboratory, *A handbook on torsional vibration*,

Cambridge University ACSS, Cambridge

[13] Bombek G, Milainović A, Filipović I and Hribernik A 2005 *Determination of torsional vibrations of a diesel engine crankshaft*, In Proceedings of the Conference on Innovative automotive technology (IAT'05), Bled, Slovenia, 21–22 April, pp 236-242 (University of Ljubljana, Ljubljana)

[14] Milašinović A, Filipović I and Hribernik A 2009 Contribution to the Definition of the Torsional Stiffness of the Crankshaft of a Diesel Engine Used in Heavy-Duty Vehicles, *Journal of Automobile Engineering* **223**(7) 921-930

[15] Rakopoulos C D, Giakoumis E G and Dimaratos A M 2007 *Evaluation of various dynamic issues during transient operation of turbocharged diesel engine with special reference to friction development*, SAE paper 2007-01-0136

[16] Brusa E, Delprete C and Genta G 1997 Torsional vibration of crankshafts: effects of non-constant moments of inertia, *J. Sound Vibr.* **205**(2) 135-150

[17] Chen S K and Chen S 1993 *Engine diagnostics by dynamic shaft measurement: a progress report*, SAE paper 932412

[18] Taraza D, Henein N A, Gade M J and Bryzik W 2005 *Cylinder Pressure Reconstruction From Crankshaft Speed Measurement in a Four-Stroke Single Cylinder Diesel Engine*, ASME 2005 Internal Combustion Engine Division Spring Technical Conference, Chicago, Illinois, USA, April 5–7, Paper No. ICES2005-1023, pp 387-395

[19] Milašinović A 2007 *Mathematical modeling and experimental investigation of nonlinear torsional vibrationof crankshaft of IC engine*, University of Banja Luka, Bosnia and Herzegovina, Doctoral Thesis

[20] Taraza D, Henein N and Bryzik W 2000 *Friction losses in multi-cylinder diesel engines*, SAE paper 2000-01-0921

[21] Etman L F P 1997 *Optimization of multibody systems using approximation concepts*, Eindhoven University of Technology, Doctoral Thesis

[22] Bestle D and Eberhard P 1994 *Integrated Modeling, Simulation, and Optimization of Multibody Systems*, in: Integrated Systems Engineering by Johannsen, G. (Ed.). Oxford: Pergamon, pp 35-40

[23] Serban R, Jeffrey S Freeman and Negrut D 1997 *Parameter Identification for Multibody Dynamic Systems*, Proc. of the 16th Biennial Conference on Mechanical Vibration and Noise, Sacramento

[24] Serban R 1998 *Dynamic and Sensitivity Analysis of Multibody Systems*, The University of Iowa, Doctoral Thesis

[25] Serban R and Freeman J S 2001 Identification and Identifiability of Unknown Parameters in Multibody Dynamic Systems, *Multibody System Dynamics Journal* **5**(4) 335-350

[26] Nocedal J and Wright S J 1999 *Numerical Optimization*, Springer

APPENDIX: SYMBOLS

T	- Kinetic energy of the lumped mass model of the crankshaft
V	- Potential energy of the lumped mass model of the crankshaft or cylinder capacity
Φ	- Dissipation function (Rayleigh's dissipation function)
$\boldsymbol{\alpha}$	- Column vector of generalized coordinates for each lumped mass
$\dot{\boldsymbol{\alpha}}$	- Column vector of generalized velocities of equivalent masses of crankshaft dynamic system
t	- Time
$\mathbf{t}_l$	- Row vector of excitation forces in generalized coordinates
$\mathbf{J}$	- Moments of inertia matrix of the lumped mass model of the crankshaft or Jacobian matrix
$\mathbf{C}$	- Stiffness matrix of the lumped mass model of the crankshaft
$\mathbf{K}$	- Damping matrix of the lumped mass model of the crankshaft

ω_0	- Mean angular speed of the crankshaft
$\boldsymbol{\varphi}$	- Column vector of the angle of nonuniform motion
$\mathbf{J}_{const}$	- Moments of inertia matrix of equivalent crankshaft system with constant terms not being functions of crankshaft position $\omega_0 t$
i	- Imaginary unit $\sqrt{-1}$
$\mathbf{i}$	- Unit column vector
$\mathbf{M}$	- Matrix of inertial coefficients of the system
$\mathbf{Q}^A$	- Column vector of generalized forces (all external forces, torques and Coriolis force)
$\mathbf{q}$	- Vector of generalized coordinates of global dynamic system
$\mathbf{b}$	- Vector of unknown parameters of global dynamic system
$\mathbf{d}$	- Step while searching for optimal solution
$\mathcal{M}(t,\mathbf{b})$	- Mathematical model of global dynamic system
p	- Number of unknown parameters of the system (parameters which have been identified)
y_m	- Exit from mathematical model (for instance, angular velocity of the pulley acquired by Calculation based on mathematical model $\mathcal{M}$)
y_i	- Results acquired by measuring (for instance, measured angular velocity of the pulley)
r_i	- Difference between measured and mathematically modelled value
$\mathbf{r}$	- Vector of differences between measured and mathematically modeled values
n_m	- Number of measured values
F	- Function of goal

Experimental model of a wind energy conversion system

C Vasar, C L Rat and O Prostean

Politehnica University of Timisoara, Department of Automation and Applied Informatics, Vasile Parvan str., no. 1-2, 300223 Timisoara, Romania

E-mail: cristian.vasar@upt.ro

Abstract. The renewable energy domain represents an important issue for the sustainable development of the mankind in the actual context of increasing demand for energy along with the increasing pollution that affect the environment. A significant quota of the clean energy is represented by the wind energy. As a consequence, the developing of wind energy conversion systems (WECS) in order to achieve high energetic performances (efficiency, stability, availability, competitive cost etc) represents a topic of permanent actuality. Testing and developing of an optimized control strategy for a WECS direct implemented on a real energetic site is quite difficult and not cost efficient. Thus a more convenient solution consists in a flexible laboratory setup which requires an experimental model of a WECS. Such approach would allow the simulation of various real conditions very similar with existing energetic sites. This paper presents a grid-connected wind turbine emulator. The wind turbine is implemented through a real-time Hardware-in-the-Loop (HIL) emulator, which will be analyzed extensively in the paper. The HIL system uses software implemented in the LabVIEW programming environment to control an ABB ACS800 electric drive. ACS800 has the task of driving an induction machine coupled to a permanent magnet synchronous generator. The power obtained from the synchronous generator is rectified, filtered and sent to the main grid through a controlled inverter. The control strategy is implemented on a NI CompactRIO (cRIO) platform.

1. Introduction

Due to the increase in energy demand worldwide and environmental concerns, the world is going increasingly green (renewable energies) in its energy consumption. Use of wind energy for electricity generation purposes is becoming an attractive energy source. But, in order for wind power to compete effectively with fossil fuels as an electrical power generator in the market, the production cost must be comparable to that of fossil fuels. The initial capital investment in wind power goes to the machine and the supporting infrastructure. Any factors that lead to a decrease in the cost of energy (turbine design, construction or operation) are vital in making wind power competitive as an alternative energy source.

Hardware-in-the-loop (HIL) systems which replicate the behavior of the wind turbine are the main alternatives for laboratory-scale feeders. These provide a more cost-effective alternative to field testing [1]. In a HIL structure only one part of the process is actually used, the other parts are simulated in a controller board [2]. The controller board contains the mathematical representation of the real plant and all its related dynamic systems, being able, therefore, to simulate the complexity of the plant. The impact of the "missing" turbine components into the testing system is dependent on the exactness of the mathematical model. These mathematical representations are referred to as the "plant simulation" [3]. These modeled sources can be explicitly controlled and are not the whims of the weather and also provide flexibility to change the characteristics of the sources as needed. By using such sources, the goal of this system is to use the given wind speed variations along with a software simulation of the turbine dynamics and related controllers to produce control signals that drive an appropriately sized motor and generator [4]. Wind, turbine and mechanical power train are emulated by a controlled induction motor to impose the dynamical behavior of the actual process on the generator shaft. Once connected to the feeder, the varying power produced by the generator will allow for a broader set of experiments dealing with the dynamic behavior of the wind turbine [4]. Such a methodology has been used in aeronautics, traction applications and electrical drives [2].

The advantages of this method are that it can provide a real world controller to be tested in many configurations without requiring the presence of the real plant.

This paper presents a Hardware-in-the-Loop (HIL) emulator for a wind turbine system, developed for the testing in a microgrid laboratory. The emulator includes: a LabVIEW real time model of the wind turbine running on a real time programmable automation controller, namely NI cRIO 9068, an induction machine (IM) drive with direct torque control (the wind turbine mechanical system equivalent), namely ACS 800, and coupled to an actual permanent magnet synchronous generator (PMSG) with the corresponding load [5].

2. Wind turbine

The wind turbine system is composed of a rotor that is connected to a permanent magnet synchronous generator [6]. These can be seen in Figure 1.

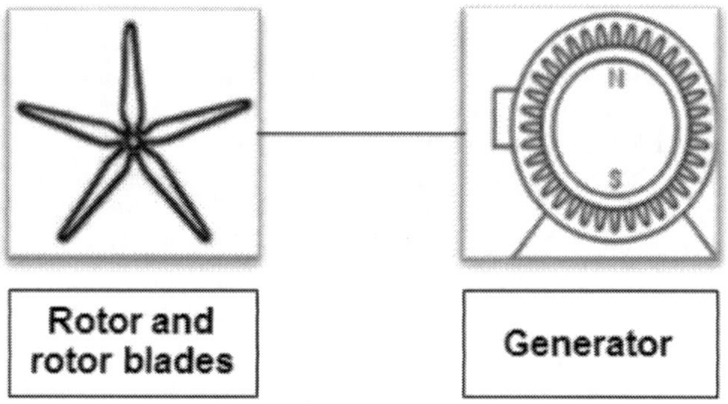

Figure 1. The wind turbine main components

The wind turbine output power varies with the wind speed:

$$P_{wt} = 0.5A\rho C_p(\lambda) * v^3 \quad (1)$$

where the speed ratio λ is defined as:

$$\lambda = \frac{\omega_T R}{v} \quad (2)$$

And the following denotations are considered:

- A : blade swept area [m^2];
- ρ : specific density of air [kg/m^3];
- v : wind speed [m/s];
- R : radius of the turbine blade [m];
- ω_r : rotating speed [rad/s];
- C_p : coefficient of power conversion.

IOP Conf. Series: Materials Science and Engineering **294** (2017) 012081 doi:10.1088/1757-899X/294/1/012081

The output power of a wind turbine is proportional with the cube of the wind speed. Larger rotors allow wind turbines to operate at higher wind speeds, increasing their output power. Because the area covered by the rotor is larger and so is the blades length, this leads to a large energy capture [7].

The mechanical part of the system is considered as a two-mass model, consisting of a large mass (corresponding to the wind turbine inertia J_{wt}) and a small mass (corresponding to the generator rotor inertia Jg). The dynamic equation is:

$$T_{wt} - T_g = J \cdot \frac{d\omega_r}{dt} \quad (3)$$

where:
T_{wt} : wind turbine torque [N•m];
T_g : generator torque [N•m];
J : inertia of the wind turbine system [kg•m^2]. This is composed of:

$$J = J_{wt} + J_g \quad (4)$$

where:
J_{wt}: inertia of the wind turbine [kg•m^2];
J_g: inertia of the PMSG [kg•m^2];

The corresponding power of the system can also be expressed as follows:

$$P_{wt} = T_{wt} * \omega_r \quad (5)$$

From (1) and (5), the wind turbine torque has the following equation:

$$T_{wt} = 0.5A\rho C_T(\lambda) * v^2 \quad (6)$$

where C_T is the wind turbine torque coefficient.

Expression (7) was obtained from the wind turbine test, where a, b, C_{T0} are the constants for the nominal tip-speed ratio λ_0.

$$C_T = C_{T0} + a\lambda - b\lambda^{2.5} \quad (7)$$

The parameters of the considered wind turbine are presented in Table1.

Table 1. The parameters of the considered wind turbine

Parameter	Value
Rated power	$P_n = 5.5\ [kW]$
Rated wind speed	$v_0 = 11\ [m/s]$
Maximum speed	$n = 126\ [rpm]$
Turbine inertia	$J_{wt} = 140\ [kg\ m^2]$

As it was mentioned, one key component of the wind energy conversion system is represented by the electric generator. Most used is the permanent magnet synchronous generator (PMSG) due to its certain advantages:

- The rotor has no winding, containing only permanent magnets, and thus has no electric loss or thermal dissipation, increasing the generator efficiency
- No need for electric circuit related to machine excitation system, reducing the control structure and increasing the overall reliability.

- Due to the absence of the mechanical components such as brushes and slip rings, the friction and wearing is reduced, with benefits on the maintenance and reliability, also it lead to a decreased physical size, and a low moment of inertia [8].

The main disadvantage is related to the cost and availability of the permanent magnet.

It can be mentioned also the fact that the control is quite simple because the permanent magnet assures a constant excitation, and the generator output voltage is variable and directly influenced by the rotor speed [9]. This is not a real limitation since proper developed control strategies can assure desired performances.

The parameters of the considered PMSG are presented in Table 2.

Table 2. The parameters of the considered PM generator turbine

Parameter	Value
Rated power	Sn = 5 [kVA]
Rated current	In = 12 [A]
Rated speed	nn = 120 [rpm]
Rated frequency	fn = 32 [Hz]
Number of stator slots	Nc =33
Pole pairs number	pp=16
PM flux	ΨPM = 1.32 [Wb]

The wind turbine characteristics, representing the captured power versus the rotating speed, at constant wind speeds, have the allures as depicted in Figure 2.

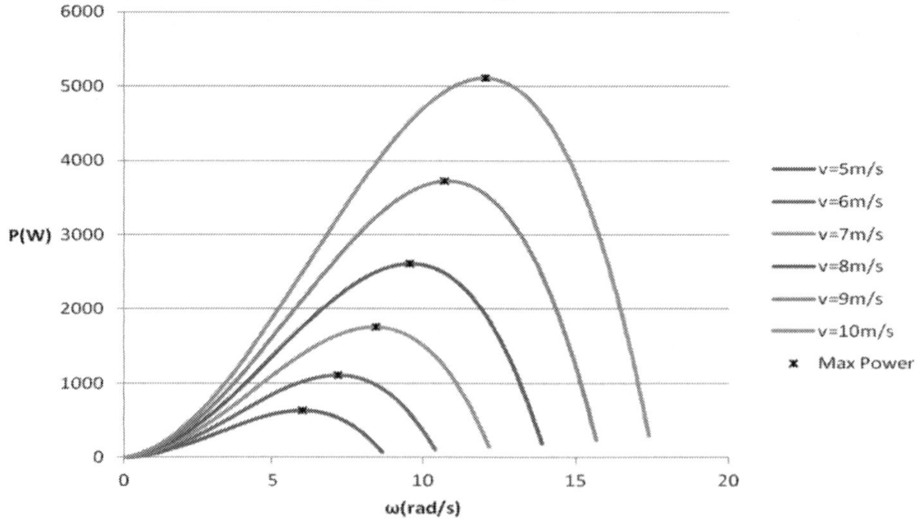

Figure 2. Wind turbine power characteristics

It can be noticed a maximum power point for every wind speed at an optimum rotor speed, highlighted in the picture with the symbol "*". Also the maximum torque point has its correspondent optimum value of the rotor speed as shown in Figure 3.

In order to provide the maximum power for the available wind condition, the wind turbine has to operate continuously in the maximum power point. This can be assured by using a maximum power point tracking algorithm (MPPT). As it can be noticed in Figure 2, for a constant wind speed, the provided power and angular speed increase until MPP is reached, and after that, the angular speed continues to increase while the provided power decreased based on the particular turbine specifications.

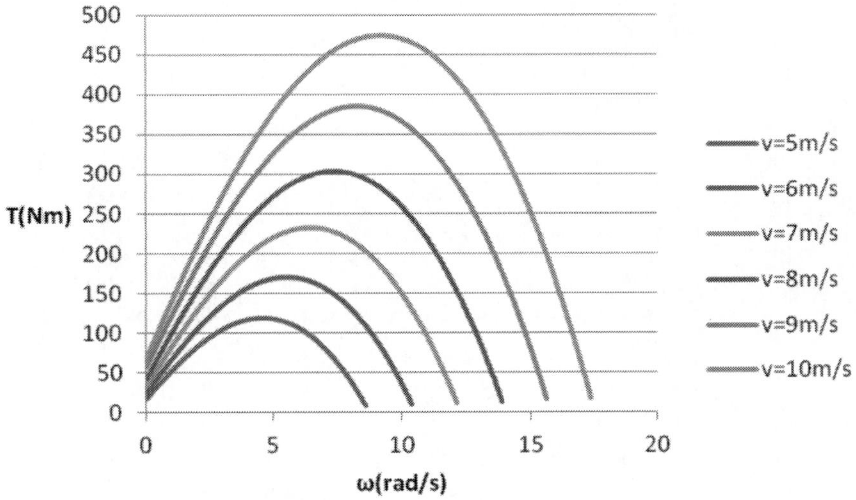

Figure 3. Wind turbine torque characteristics

3. Wind turbine emulator

The A wind turbine emulator consists essentially of two main components:

- **A software system** which implements the mathematical model of the wind turbine.
- **A hardware system** that provides the similar static and dynamic characteristics as the real studied system.

The software system includes a variable wind speed profile generator and a wind turbine aerodynamic model. The software system is deployed on the controller board, namely NI cRIO 9068. The wind turbine equations were coded in LabVIEW graphical development environment. The code is then deployed to the industrial NI CompactRIO hardware controller. The diagram of the program can be seen in Figure 4 and the front panel is depicted in Figure 5. The NI cRIO 9068 is represents a hardware/software platform for development of flexible applications of signal processing, data logging and industrial process control. The NI cRIO features 8 RIO (reconfigurable I/O) modules, however in this case, only 2 modules are used: one as an analogue input and another as an analogue output. CompactRIO also includes a dual core processor and an integrated FPGA, which provide the support for high-speed signal acquisition, assuring real time processing. The programming language is NI LabVIEW, but it can integrate also VHDL or C++ code by means of specialized wrappers. In [5] a similar system was programmed using a LabVIEW toolkit: Control Design and Simulation Module. This module permits the usage of a similar blockset to the model developed in Matlab Simulink without the need to implement a full scale solver for the model. The disadvantage of this, which the current paper wants to highlight, is that with the Control Design and Simulation Module the primary reason of using LabVIEW, namely its superior quality in acquiring, processing and displaying signals, is not taken full advantage of it. MATLAB and its LabVIEW equivalent, the Control Design and Simulation Module, are better suited for calculus and operations that require high computational effort, whereas LabVIEW's basic programming language G is more suited for real-time processing and control [10].

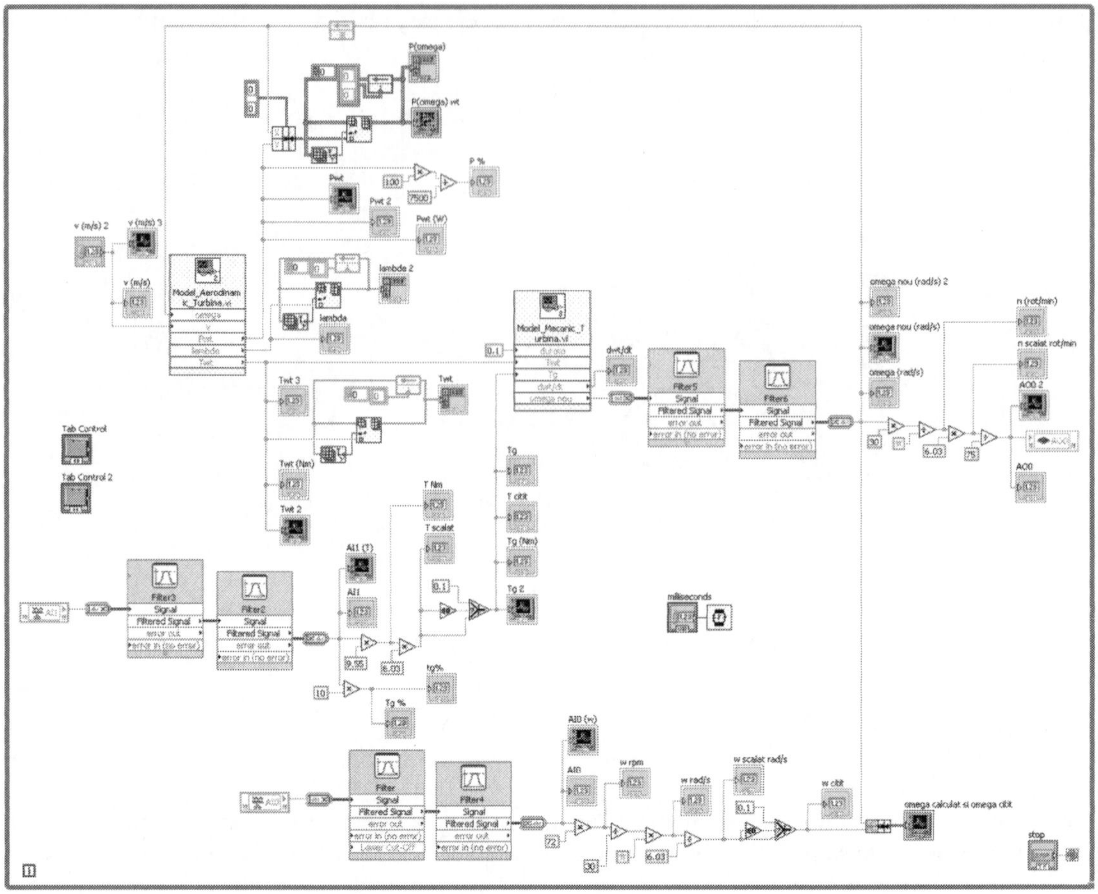

Figure 4. The LabVIEW software for the emulator - The block diagram of the .vi

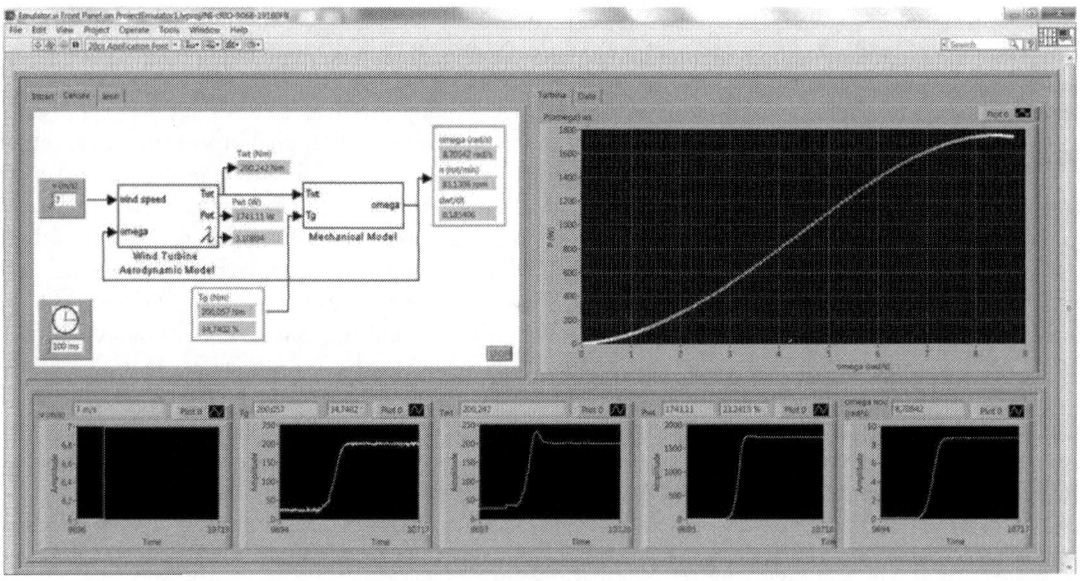

Figure 5. The LabVIEW software for the emulator - The front panel

The model diagram needs to be processed in a while loop. This loop actually calls the solver for the processing of the model and deploying the simulated system to run on the real time target [5].

The model is developed using 2 subroutines for calculating the parameters; one is based on the aerodynamic model of the turbine and another on the mechanical model. The program reads the torque estimated by ACS 800 and calculates the new rotating speed (ωr) based on the previous rotating speed and the torque calculated using the aerodynamic model of the turbine (6). The 1/6 gearbox is also accounted for in this step. Another method would be to read the rotating speed and calculate the torque, as can be seen in [6].

The mechanical component of the wind turbine is modeled as depicted in Figure 6.

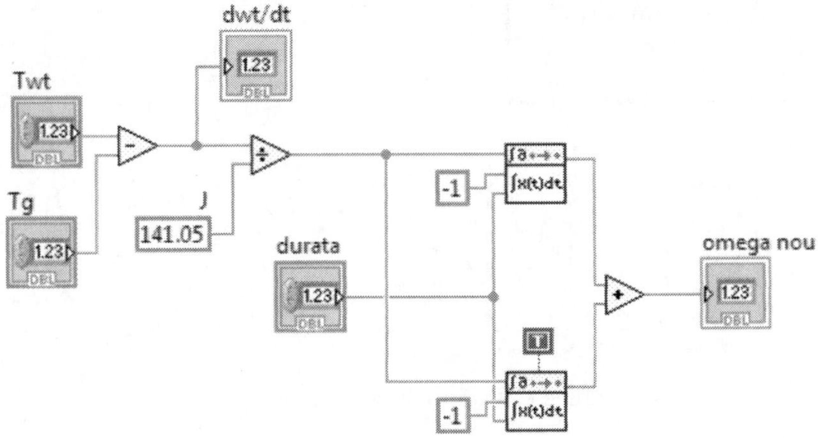

Figure 6. The mechanical model of the wind turbine

The considered aerodynamic model of the wind turbine is presented in Figure 7.

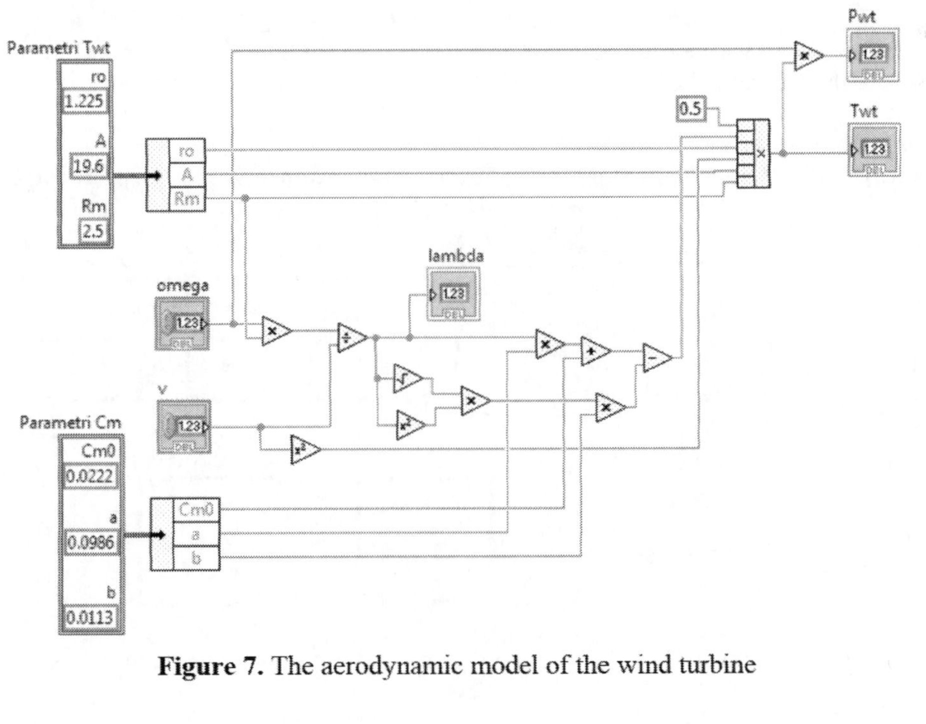

Figure 7. The aerodynamic model of the wind turbine

IOP Conf. Series: Materials Science and Engineering **294** (2017) 012081 doi:10.1088/1757-899X/294/1/012081

In order to operate correctly the loop need to be synchronized with a clock source. This way, the real time processing is enforced and the input and outputs of the model can be integrated with the other hardware models.

The hardware system contains an ABB ACS 800 drive and a 7.5 kW three phase squirrel cage induction motor (IM) with a gearbox (GB) (the wind turbine equivalent) and a permanent magnet synchronous generator (PMSG) [6]. The generator is the same one used in the real turbine system.

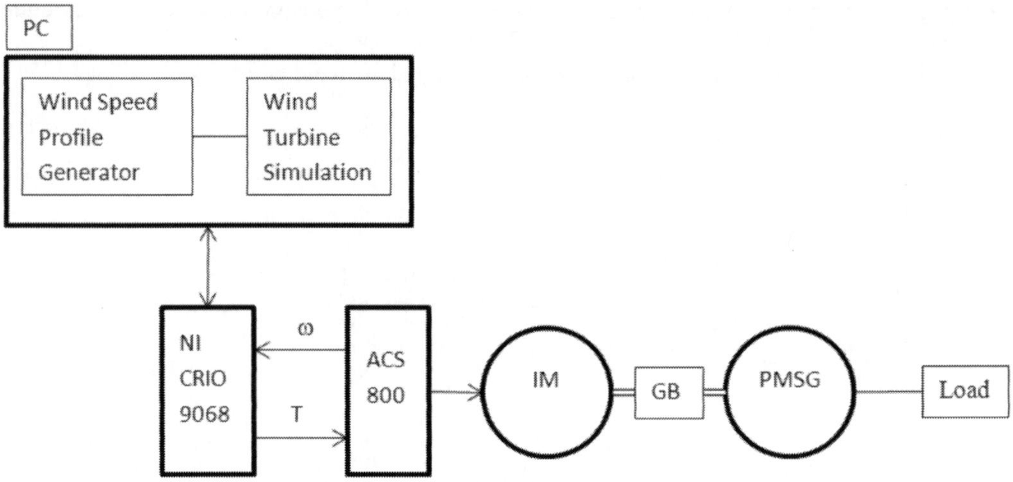

Figure 8. The emulator configuration

The emulator control was implemented in LabVIEW and it runs in real-time on the NI cRIO board, as shown in Figure 8. The software system sends reference signals to the physical system during the real time simulation. The control output signal (angular speed reference) is send to the voltage source inverter ABB ACS 800 [6]. The role of ACS 800 is to drive a 7.5 kW/720 rpm squirrel cage induction motor (IM). The IM receives the shaft speed as a control signal from the ACS 800. ACS 800, in turn, computes the estimated load torque obtained from the reference model of the induction machine and feeds it back to the controller board.

4. Experimental results

The structure of the experimental setup for wind energy conversion system is presented in Figure 9.

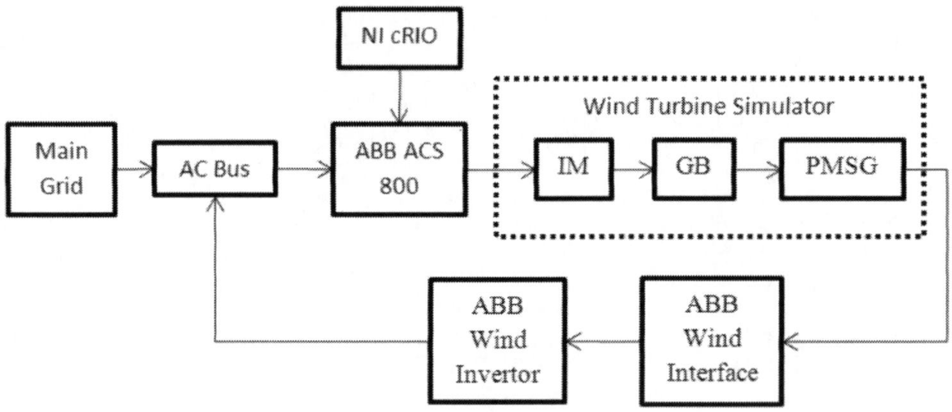

Figure 9. The complete structure of the experimental setup for a wind energy conversion system

In the presented experiments, the wind condition has a profile as depicted in Figure 10. It can be noticed that the wind speed had an impulse varying from 5 to 6 m/s. In Figure 11 is presented the angular speed.

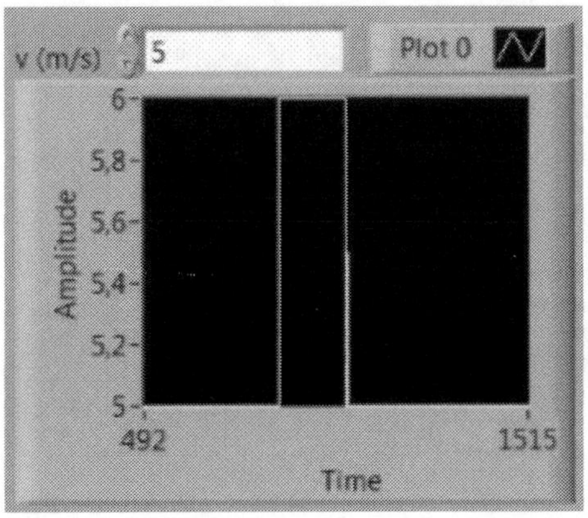

Figure 10. Wind speed

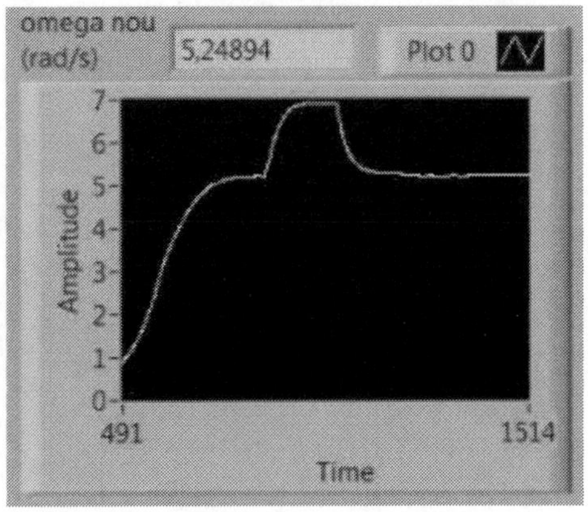

Figure 11. Angular speed

Measured torque variation is presented in Figure 12. As wind speed rise, the torque value increase, but due to the increasing load of the generator, as extracted power and angular speed increase, the torque decrease.

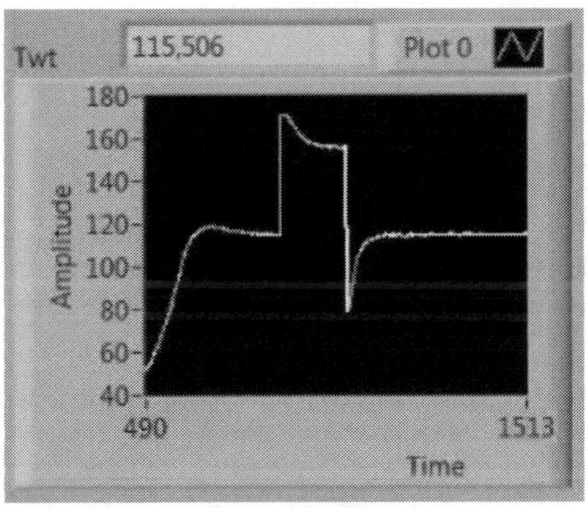

Figure 12. Torque variation

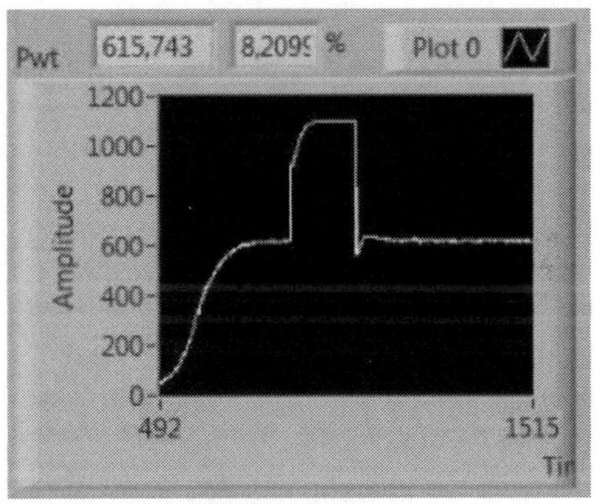

Figure 13. Power variation

The operating wind turbine power characteristic for the considered wind conditions is presented in Figure 13.

The wind turbine power characteristics achieved through simulations using the turbine model for various wind speed are presented in Figure 15.

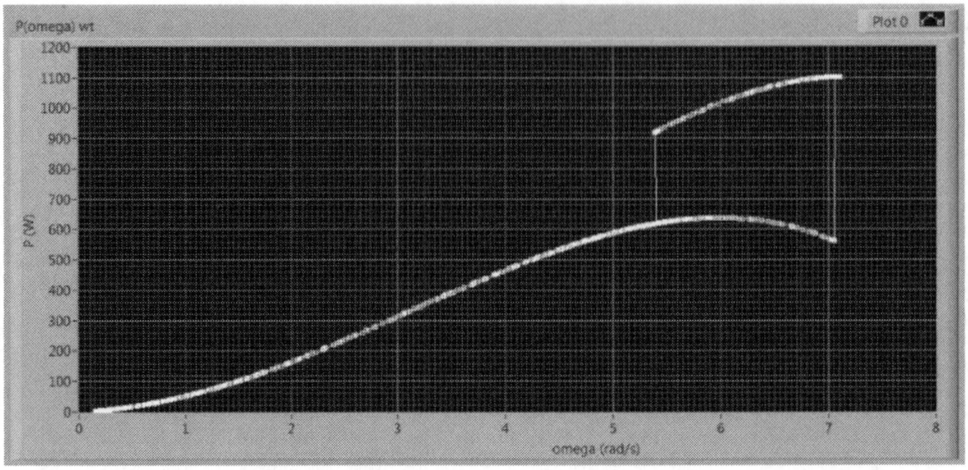

Figure 14. Response to an impulse variation of the wind speed from 5 to 6 m/s

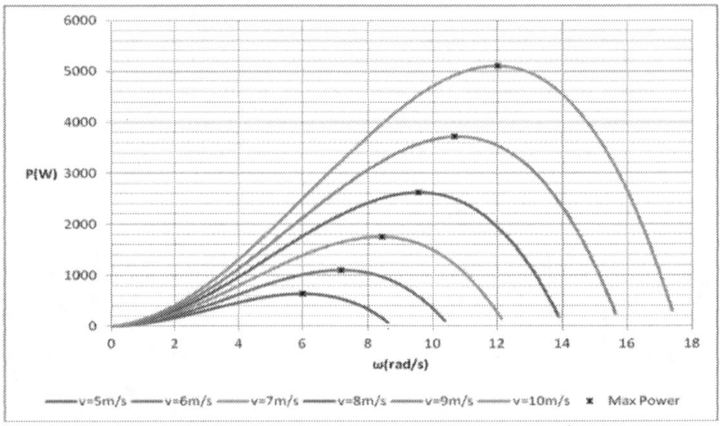

Figure 15. Wind turbine power characteristics P(ω), obtained by simulation

The wind turbine power characteristics obtained from emulator considering various wind speed are presented in Figure 16.

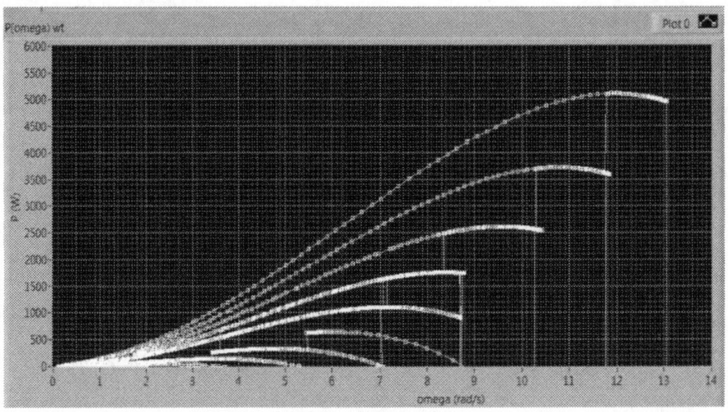

Figure 16. The wind turbine power characteristics obtained from emulator

The main parameters determined experimentally are presented in Table 3.

Table 3. Experimental determined parameters

Wind speed	5 m/s	6 m/s	7 m/s	8 m/s	9 m/s	10 m/s
Tg [Nm]	114.5	156	202	250	308	349
Twt [Nm]	115.7	157	202.6	252.3	309.1	350
P opt [W]	613.5	1101.5	1747.5	2575.6	3620	4752
ϖ opt [rad/s]	5.3	7	8.6	10.2	11.7	13.6
n opt [rpm]	50.6	67	82.4	97.5	112	130.2
Efficiency [%]	46.454	57.557	65.236	70.080	73.922	71.927

It can be noticed that the wind turbine efficiency varies with the wind speed and angular speed. There are significant power losses at low values of the angular speed mainly due to the mechanical mechanism the coupling the induction machine, gearbox and PMS generator.

5. Conclusions

The paper presents a powerful and flexible setup for wind energy laboratory, useful for study and testing real equipment with real control strategies, without the necessity of a real wind turbine.

HIL systems has certain advantages in comparison with software simulators, due to the fact that there are certain factors (as different losses) that cannot be very accurate included in software simulation, but can be achieved by the physical system of the emulator [4].

The main advantage is the system flexibility, the number of wind turbines types with or without gearboxes that can be implemented and tested using the emulator is virtually unlimited – the only limitation is the knowledge of a good mathematical model combined with the characteristics of the used asynchronous motor.

Turbines can be simulated using real-time sampled data obtained from existing turbines. Thus, there is no need to spend money for expensive wind turbines or aerodynamic tunnel tests.

A wide range of environmental conditions can be simulated: both regular wind conditions and special conditions (wind gust or extreme fluctuating wind), with any transient regime or successive steady-state regimes.

There is the possibility of testing different control loops, power converters, storage elements, electrical loads in real time.

All these facts can be used to optimize the turbine design and minimize the production costs, leading to decreased costs related to wind energy, and making it a viable energetic alternative [3].

The disadvantages are mainly related to the uncertainties regarding both the mathematical models and real parameters.

References

[1] Schkoda R, Fox C, Hadidi R, Gevorgian V, Wallen R and Lambert S 2016 *Hardware-in-the-Loop Testing of Utility-Scale Wind Turbine Generators* (No. NREL/TP-5000-64787), NREL- National Renewable Energy Laboratory, Golden, CO, United States

[2] Bouscayrol A, Guillaud X, Teodorescu R, Delarue P and Lhomme W 2006 *Hardware-in-the-loop simulation of different wind turbines using Energetic Macroscopic Representation*, IECON 2006-32nd Annual Conference on IEEE Industrial Electronics, Paris, France, November 6-10, pp 5338-5343

[3] Manyonge A W, Ochieng R M, Onyango F N and Shichikha J M 2012 Mathematical Modelling of Wind Turbine in a Wind Energy Conversion System: Power Coefficient Analysis, *Applied Mathematical Sciences* **6** 4527-4536

[4] Hardy T and Jewell W 2014 Hardware-in-the-loop wind turbine simulation platform for a laboratory feeder model, *IEEE Transactions on Sustainable Energy* **5**(3) 1003-1009

[5] Topor M, Muntean N and Sorandaru C 2014 Wind Turbine Emulator Development using NI cRIO 9068 and ABB ACS 800 Drive, *Journal of Electrical Engineering* 1-6

[6] Muntean N, Tutelea L, Petrila D and Pelan O 2011 *Hardware in the loop wind turbine emulator*, International Aegean Conference on Electrical Machines and Power Electronics and Electromotion, Joint Conference, Istanbul, Turkey, September 8-10, pp 53-58

[7] Petrila D 2013 *Energy Conversion And Storage Control For Small Wind Turbine Systems*, Timisoara, Romania, Doctoral Thesis

[8] Milivojevic N, Stamenkovic I, Schofield N and Emadi A 2008 *Electrical machines and power electronic drives for wind turbine applications,* Industrial Electronics, IECON 2008 34th Annual Conference of IEEE, Orlando, Florida, USA, November 10-13, pp 2326-2331

[9] Patrascu C, Muntean N and Hedes A 2016 *Microgrid Laboratory for Educational and Research Purposes*, IEEE International Energy Conference, Energycon, Leuven, Belgium, April 4-8

[10] Tasner T, Lovrec D, Tasner F and Edler J 2012 Comparison Of LabVIEW and MATLAB for scientific research, *Annals of the Faculty of Engineering Hunedoara* **10**(3) 389-394

Issues regarding the usage of MPPT techniques in micro grid systems

I Szeidert, I Filip, F Dragan and A Gal

Politehnica University of Timisoara, Department of Automation and Applied Informatics, 2 Vasile Parvan Bvd, Timisoara, 300223, Romania

E-mail: iosif.szeidert@aut.upt.ro

Abstract. The main objective of the control strategies applied at hybrid micro grid systems (wind/hydro/solar), that function based on maximum power point tracking (MPPT) techniques is to improve the conversion system's efficiency and to preserve the quality of the generated electrical energy (voltage and power factor). One of the main goals of maximum power point tracking strategy is to achieve the harvesting of the maximal possible energy within a certain time period. In order to implement the control strategies for micro grid, there are typically required specific transducers (sensor for wind speed, optical rotational transducers, etc.). In the technical literature, several variants of the MPPT techniques are presented and particularized at some applications (wind energy conversion systems, solar systems, hydro plants, micro grid hybrid systems). The maximum power point tracking implementations are mainly based on two-level architecture. The lower level controls the main variable and the superior level represents the MPPT control structure. The paper presents micro grid structures developed at Politehnica University Timisoara (PUT) within the frame of a research grant. The paper is focused on the application of MPPT strategies on hybrid micro grid systems. There are presented several structures and control strategies and are highlighted their advantages and disadvantages, together with practical implementation guidelines.

1. Introduction

The paper presents some issues regarding the usage of maximum power point tracking (MPPT) approaches within microgrid systems. The authors studied several types of microgrid structures. Some of them implemented on laboratory rigs at University Politehnica from Timisoara, other structures were only studied in dedicated modelling and simulation software such as Matlab-Simulink [1].

The usage of renewable energy resources has significantly increased in the past years at global level. It becomes obviously a domain of very high importance. The scientific community presents a great interest for the domain of renewable resources especially for wind and solar energy. The solar and wind energy have the advantage of being clean and free. However, one of the main drawback of the wind resources is that it has a sporadic character. Therefore, in a microgrid there are grouped several energy resources and supplementary there are included also energy storage systems (supercapacitors, batteries, other additional equipment) [2-5].

In the case of wind energy conversion system there is harvested the energy and converted into electrical energy. The output power of course depends on wind speed. The problem occurs due to the non-linear characteristic of the wind aggregate. There are developed several techniques that aim to extract the maximum amount of energy at any wind speed.

Similarly, the solar energy is a renewable energy resource which is converted into electrical energy by using photovoltaic (PV) cells. There are two main parameters - radiation and temperature – that disturb the output of the photovoltaic panels. In the case that irradiance increases then current increases and the variation of the voltage is smaller. In the case that temperature increases, open circuit voltage decreases, while if intensity of solar radiation increases, short circuit current increases. Therefore, the I-U and respectively the P-U characteristics change with temperature and irradiance variation, fact that leads to a shift of the maximum power point (MPP) on the characteristic. So, the most accurate MPPT techniques take into account these parameters [6-10].

Nowadays, in the technical literature there are used several MPPT technique types for solar and wind energy system. In the case of MPPT approaches for wind energy conversion systems there are widely spread three methods: perturb and observe (P&O); tip speed ratio (TSR) and respectively optimum-relation-based (ORB). In the case of MPPT strategies for solar energy conversion system there are mentioned: constant-voltage tracing; perturb and observe (P&O) and incremental conductance. Of course, this are only the most used techniques. In literature, there are many other related and variations of MPPT strategies presented and studied. Each of them presents advantages and disadvantages. The system engineers must perform studies on a certain microgrid implementation and choose the most suitable control strategy [11-13].

2. Issues on MPPT techniques applied at microgrid systems at PUT

The paper presents the proposed microgrid topology – at an experimental setup at Politehnica University Timisoara (PUT) laboratory [1].

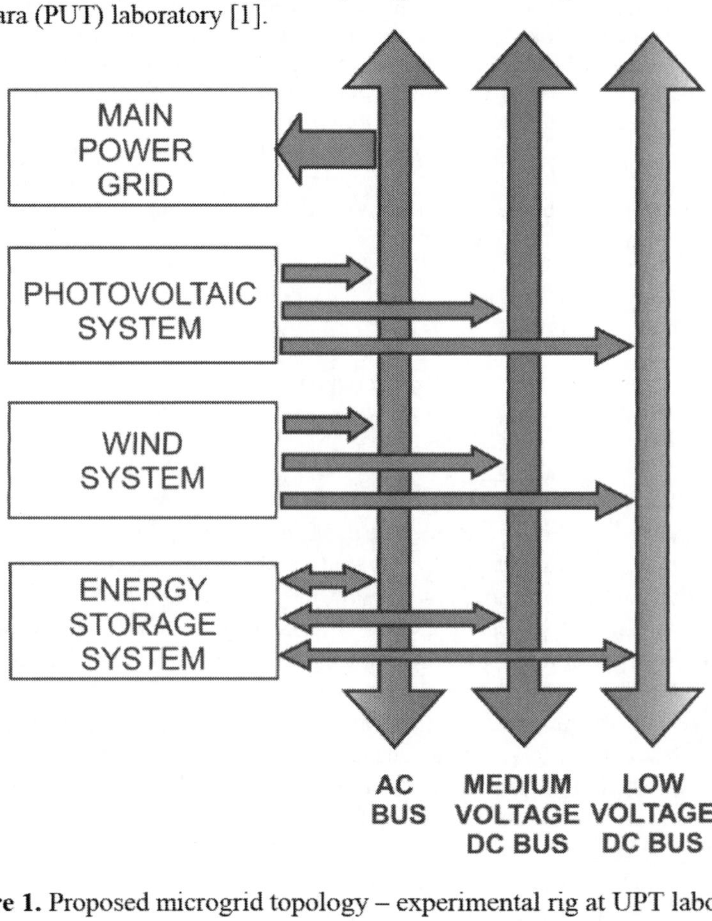

Figure 1. Proposed microgrid topology – experimental rig at UPT laboratory

In Figure 1, there is detailed the considered structure of the experimental microgrid. The main components are: photovoltaic subsystem; wind subsystem; energy storage (capacitor bank) and the interface with the power grid system. The system is based on three power buses: one for *ac* voltage and two for *dc* voltage (low and medium voltages). There were conducted several experiments and also different simulations of the microgrid components. In the case of wind energy subsystem, the experimental rig includes: asynchronous machine (squirrel cage), permanent magnet synchronous machine and a modified induction machine - doubly fed with α lagged windings [14].

In the case of the wind system there must be done some considerations regarding the mathematical modelling. The wind aggregate converts rotational mechanical energy into electrical energy. As already was stated the amount of electrical energy heavily depends on the availability of the wind. The variations of wind speed, lead to the increase or to the decrease of the electric energy production. Therefore, in the case of wind energy conversion systems the site selection is very important.

The mechanical power that is generated by the wind aggregate is given by relation (1):

$$P = 1/2 * C_p(\lambda, \beta) * \rho * A * V^3 \tag{1}$$

where: P – power; A – rotor swept area; ρ - air density; C_p -power coefficient; λ - tip speed ratio; β - blade pitch angle; V- wind speed.

Another issue is that the maximum value of C_p can be 0.48. There results that it cannot be converted all wind energy into electrical energy. In the technical literature is mentioned the Betz limit. There can be converted the wind energy only up to 48%.

For the wind energy conversion system, there must be known the specific wind turbine power characteristic (curve). The maximum power characteristics must be via off-line experiments in wind tunnels of via simulations on individual power turbines. Having this knowledge there can be customised the different MPPT approaches and strategies.

In the case of the solar energy conversion systems there are several mathematical models that model the photovoltaic cell. In the technical literature, the most used are: linear power model and the classical single diode model [15].

In this case, the MPPT is a technique that uses power inverters, solar battery chargers and related devices in order to obtain the maximum possible power from one or several photovoltaic cells/panels. The power output of a photovoltaic cell has a quite complex relationship that includes parameters such as: solar irradiation, temperature and total resistance. It is non-linear and can be studied based on the I-U (current-voltage) characteristic. The main goal of an MPPT system is to measure the output of the photovoltaic cells and to deliver the suitable resistance (load) in order to obtain the maximum possible power under any particular environmental conditions. MPPT equipment is usually embedded into an electric power converter that provides the required voltage or/and current conversion.

Figure 2 presents the conceptual diagram for generic MPPT methods. This structure is applicable both at solar and wind systems. There can be noticed that the MPPT approach is based on two hierarchic levels: superior and inferior. At the superior level, there are considered the human operator experience and some input from sensors and other measurement devices. The output at this level represents practically prescribed data for lower control system. The lower level (base level) practically implements the dedicated control system (there are PI controllers, fuzzy, etc). Its outputs represent actually the command variables (current, voltages, etc).

In Figure 3, is represented the MPPT strategy applied at a wind energy conversion system equipped with a modified induction machine – doubly fed induction generator. The generator has separated windings for excitation and load. The considered generator is presented in the research papers of the member of the research grants [14]. The structure has the wind turbine coupled via a gear box with the electrical machine. The output of the electrical machine is three-phased. The input in the MPPT controller is represented by the rotation speed, and the computed output is the prescribed optimal power (from the power characteristic – the maximum power point). The low-level controller acts on the electronic power converter achieving its required operating point.

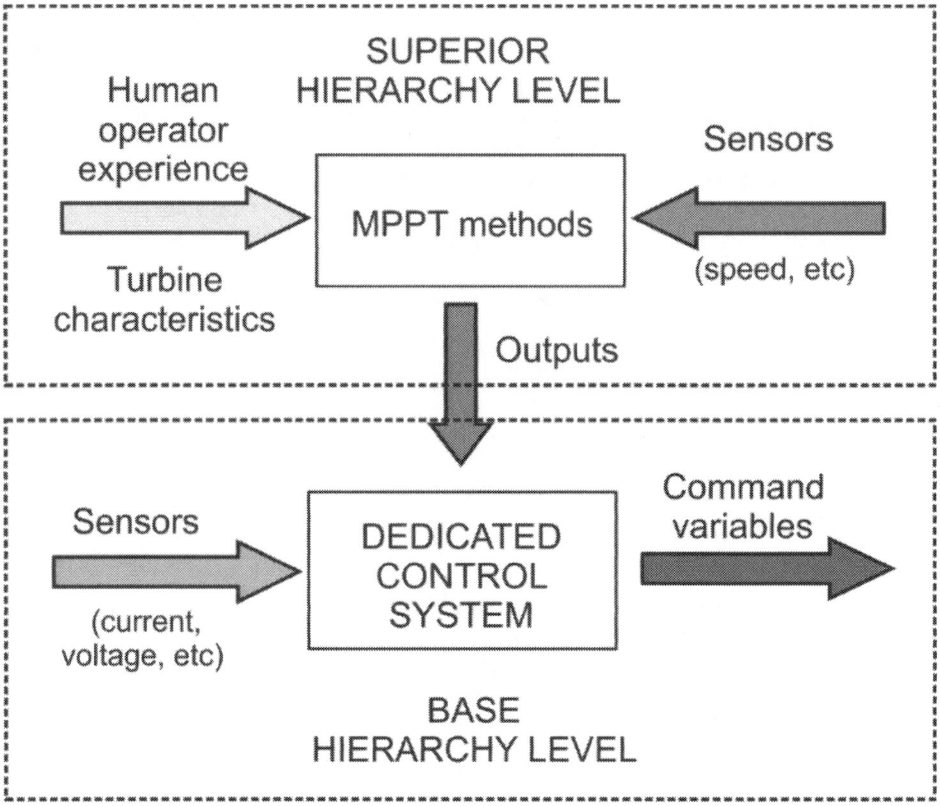

Figure 2. MPPT methods – conceptual diagram

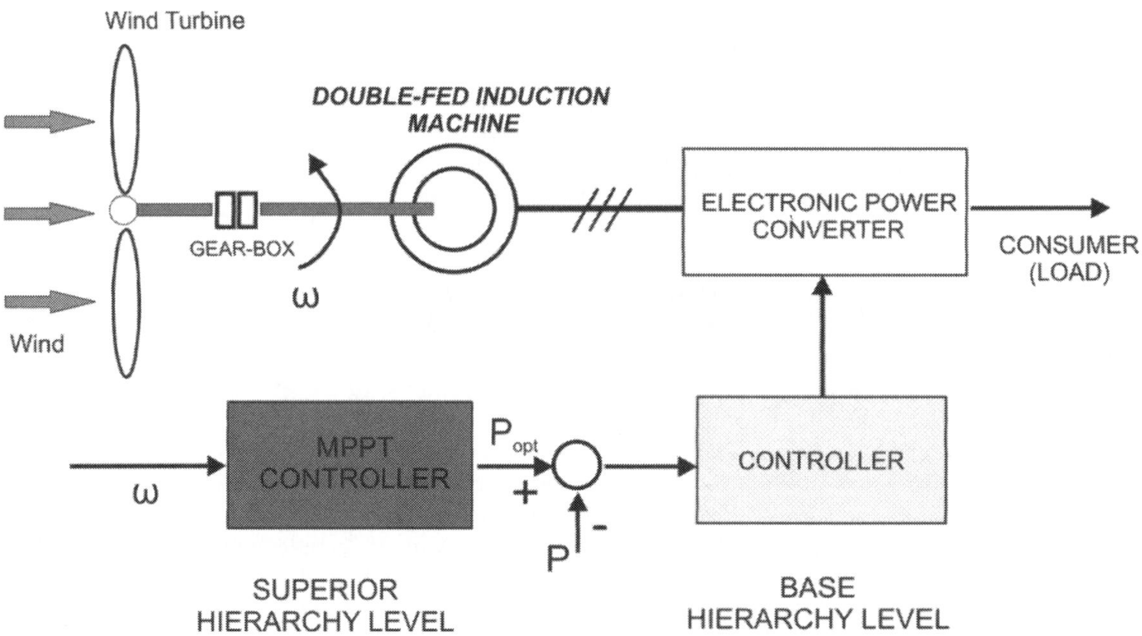

Figure 3. MPPT method applied at a wind energy conversion system

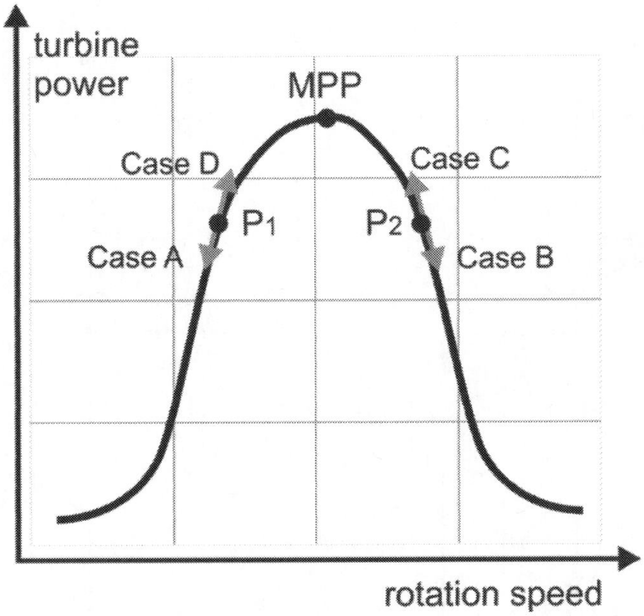

Figure 4. MPP – wind turbine characteristic

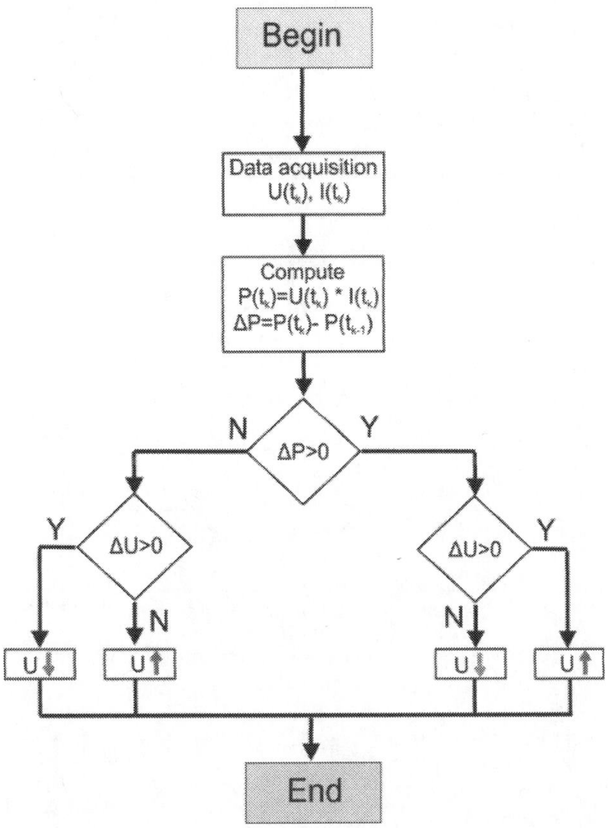

Figure 5. MPPT – Perturb and Observe algorithm diagram

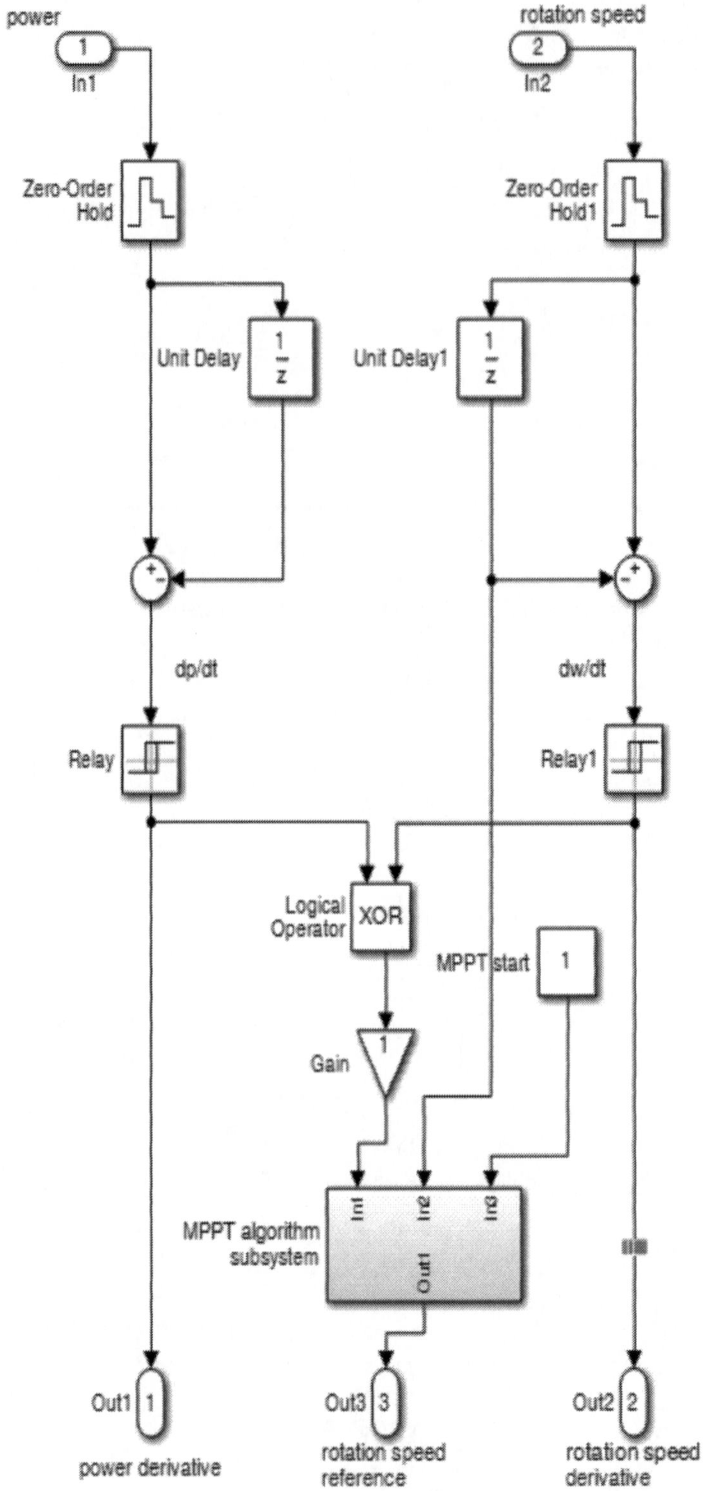

Figure 6. MPPT strategy – Simulink block diagram

In Figure 4, there is presented a typical wind turbine characteristic (turbine power vs rotation speed) with exemplification of the searching/tracking of the MPP (maximum power point). There are the four cases (A, B, C and D) regarding the two operating points below the MPP (P_1 and P_2). Practically, the operating points are shifted towards the MPP. The MPP tracking is further detailed in the Figure 5, with the notice that the figure 5 is elaborated for the case of a solar system. The same logic can be also applied in the case of wind energy conversion systems.

In Figure 5 there is presented the block diagram of the MPPT algorithm – Perturb and Observe in the case of a solar energy conversion system. In technical literature, there are presented different variants of this algorithm that consider the influence of irradiation and temperature. Maximum power point trackers use several types of logic to search for this point (the MPP) and consequently to allow the power converter to extract the maximum power available from a photovoltaic cell/array. The Perturb and Observe algorithm practically computes the power $P(t_k)$ by using the data acquired (voltage (Ut_k) and current (It_k)) and then it compares with the previous computed power $P(t_{k-1})$. The algorithm permanently perturbs (is continuously reiterated) the system; if the functioning point on the characteristic has positive variation and consequently the direction of the perturbation is reversed if it has negative variation. So, there is achieved in time the maximum power point on the characteristic of the energy conversion system. This approach is of the “hill climbing” type. Practically it depends on the characteristic curve of power versus voltage near the maximum power point.

The main issue of this approach is that there occur serious oscillations around the maximum power point. So, there must be considered a trade-off between a rapid tracking versus oscillations, and also there is the issue of user predefined constants. There are guidelines for choosing the time step and the method practically try to harvest a maximum possible amount of energy within a preset time period. As already stated, this algorithm can be used both in the case solar and wind energy conversion systems and is widely spread in microgrids implementations.

In figure 6, there is presented the MPPT strategy implementation – in dedicated simulation environment (Matlab-Simulink). The simulation studies performed by the authors used also this structure implemented in Simulink. There were considered different structures for the modular micro grid implementations [16-19].

3. Conclusions

In the technical literature, several variants of the MPPT techniques are presented and particularized at certain applications (wind energy conversion systems, solar systems, hydro plants, micro grid hybrid systems). The maximum power point tracking implementations are mainly based on two-level architecture. The first level controls the primary variables and a second one represents the MPPT control structure.

Paper presents micro grid structures developed at Politehnica University Timisoara. Also, the topic is focused on the application of MPPT strategies on hybrid micro grid systems. There are proposed structures and control strategies and are highlighted their advantages and disadvantages, together with practical implementation guidelines.

References

[1] Szeidert I, Filip I, Prostean O and Vasar C 2016 *Laboratory setup for microgrid study*, IEEE 20th Jubilee International Conference on Intelligent Engineering Systems (INES), Budapest, Hungary, June 30 - July 2

[2] Szeidert I, Prostean O, Filip I, Vasar C and Mihet-Popa L 2008 *Issues regarding the modeling and simulation of wind energy conversion system's components*, IEEE International Conference on Automation, Quality and Testing, Robotics, AQTR 2008, Cluj-Napoca, Romania, May 22-25

[3] Filip I, Szeidert I, Prostean G and Vasar C 2011 *Issues regarding mathematical modeling of wind aggregates*, IEEE 9th International Symposium on Applied Machine Intelligence and Informatics (SAMI), Smolenice, Slovakia, March 24

[4] Liu T, Zhu Y, Zhu H, Yu J, Yang Y and Ye F 2016 *Long-Term Renewable Energy Usage Maximization in a Microgrid*, 25th International Conference on Computer Communication and Networks (ICCCN), Waikoloa, USA, August 1-4

[5] Wang H 2016 *Microgrid generation planning considering renewable energy target*, IEEE International Conference on Power and Energy (PECon), Melaka, Malaysia, November 28-29

[6] Abdelsalam A K, Massoud A M, Ahmed S and Enjeti P N 2011 High-Performance Adaptive Perturb and Observe MPPT Technique for Photovoltaic-Based Microgrids, *IEEE Transactions on Power Electronics* **26**(4) 1010-1021

[7] Faranda R, Leva S and Maugeri V 2008 *MPPT techniques for PV Systems: Energetic and cost comparison*, IEEE Power and Energy Society General Meeting - Conversion and Delivery of Electrical Energy in the 21st Century, Pittsburgh, PA, USA, July 20-24

[8] Esram T and Chapman P L 2007 Comparison of photovoltaic array maximum power point tracking techniques, *IEEE Transactions on Energy Conversion* **22**(2) 439-449

[9] Alajmi B N and Alkandari F A 2016 Modified Perturbation and Observation Technique for Partially Shaded Photovoltaic Systems in Microgrids, *Journal of Clean Energy Technologies* **4**(1) 32-35

[10] Anurag A, Bal S, Sourav S and Nanda M 2016 A review of maximum power-point tracking techniques for photovoltaic systems, *International Journal of Sustainable Energy* **35**(5) 478-501

[11] Mali P S and Hans M R 2015 Wind and solar MPPT in hybrid microgrid, *International Journal of Innovations in Engineering Research and Technology (IJIERT)* **2**(6) 1-11

[12] Ishaque K and Salam Z 2013 A review of maximum power point tracking techniques of PV system for uniform insolation and partial shading condition, *Journal on Renewable and Sustainable Energy Reviews* **19** 475-488

[13] Mohamed A E and Zhengming Z 2013 MPPT techniques for photovoltaic applications, *Journal on Renewable and Sustainable Energy Reviews* **25** 793-813

[14] Budisan N, Prostean O, Boraci R, Szeidert I and Muller V 2010 *The dual induction generator for renewable energy conversion systems. Experimental results, problems and solutions*, International Joint Conference on Computational Cybernetics and Technical Informatics (ICCC-CONTI), Timisoara, Romania, June 21

[15] Koch-Ciobotaru C, Boraci R, Prostean O and Szeidert I 2013 *Voltage control in a microgrid using demand response*, IEEE 8th International Symposium on Applied Computational Intelligence and Informatics (SACI), Timisoara, Romania, May 23-25

[16] Saymote S and Atre Y R 2016 Simulation of Perturb &Observe (P & O) MPPT Technique for PV Micro - Grid System Using MATLAB/SIMULINKS *International Research Journal of Engineering and Technology (IRJET)* **03**(08) 1557-1560

[17] Shah K, Gaur V, Joshi S and Patel N 2015 Maximum Power Point Tracking Methods for Wind and Solar Conversion Systems for Standalone Generation PSIM based Perturb and Observe Method, *International Journal of Engineering Research and Development (IJERD)* 46-54

[18] Mihet-Popa L, Groza V, Prostean O and Szeidert I 2008 *Modeling and Design of a Grid Connection Control Mode for a Small Variable-Speed Wind Turbine System*, IEEE Instrumentation and Measurement Technology Conference Proceedings, IMTC 2008, Victoria, Canada, June 20

[19] Rob R and Rat C 2013 Mathematical model for studying the variation of the electrical parameters in functioning of nonlinear loads, *AIP Conference Proceedings* **1558** 1345-1348

The behaviour of a vehicle's suspension system on dynamic testing conditions

L Mihon, N Lontiş and S Deac

Politehnica University of Timisoara, Mechanical Machines, Equipment and Transportation Department, Mihai Viteazu blvd., no. 1, 300222 Timisoara, Romania

E-mail: liviu.mihon@upt.ro

Abstract. The paper presents a car suspension's behaviour on dynamic testing conditions through theoretical and mathematical simulation on specific model, on the single traction wheel, according to the real vehicle and by experiment on the test bench by reproducing the road's geometry and vehicle's speed and measuring the acceleration and damping response of the suspension system on that wheel. There are taking in consideration also the geometry and properties of the tyre-wheel model and physical wheel's properties. The results are important due to the suspension's model properties which allows to extend the theory and applications to the whole vehicle for improving the vehicle's dynamics.

1. Introduction

Road vehicle's movement on a dedicated surface, like concrete and asphalt, could be controlled by a driver in longitudinal and transvers directions as well as around the vertical axis (yaw rotation) within the limits imposed by the physical laws and the surfaces interaction through adherence. The transverse and yaw movement are very well connected each other.

In the vertical direction the movement of the vehicle follows the roadway, including all the bumps (ascending or descending) and declivities, without any action on the part of the driver. For comfort and a safety driving all these irregularities of the road should be minimized through the tires and suspension system of the vehicle.

2. Suspension System of the vehicle

All the external forces that act on a vehicle during his ride, with the exception of gravity and aerodynamic forces, are transmitted through the contact area between tire and road surface (tire patches). Thus, the suspension of a vehicle could be precept as a complex system which generate forces at the tire's contact patches and serve to transmit these to the vehicle: wheels and tires, brakes, suspension arms, hubs and nuts, steering system, springs and shock absorbers [1].

Due to his role of linking the vehicle and the road, the suspension system direct influence the handling dynamics and ride comfort and also affects space utilization, aerodynamics and costs. The suspension alone is not decisive in establishing handling dynamics and ride comfort of the vehicle, but the overall vehicle size, center of gravity height, weight distribution, wheelbase and truck and aerodynamics, at high speeds, influence also the vehicle's behavior.

Starting from this basic concept, where the vehicle's driving consists of moving a body within the atmosphere and the gravity field and in this process the forces (vertical, propulsion and braking, lateral

IOP Conf. Series: Materials Science and Engineering **294** (2017) 012083 doi:10.1088/1757-899X/294/1/012083

guiding) are transmitted to the road by tire contact patches, we will take in consideration all these forces for vehicle's dynamics analysis [2].

2.1. Longitudinal and vertical suspension forces

The major longitudinal forces transmitted from the road to the vehicle body are propulsion and braking forces. The primary task of propulsion and braking forces is to control longitudinal dynamics of the vehicles by acceleration and speed. Longitudinal acceleration also gives rise to pitching and vertical motions. Coupling depends on center of gravity position, wheelbase and suspension kinematics. Stabilizing factor are pitch and lift stiffness but also shock absorber damping. For a live axis the driveshaft torque reaction on the body must also be considered.

As a result of vehicle acceleration or braking [3] there are changing of the instantaneous normal forces at the front and rear axles as a dynamic weight transfer reaction. The elastokinematic toe changes due to the direct changes in tire tangential force and indirectly by the changes in wheel travel and the resulting change in side forces, generate side force components which affect vehicle handling. Toe changes on the other hand allow to improve the handling response of the vehicle's braking system as response to the load changes. Figure 1 [4] present the general problem of a vehicle under the loads and movements constraints and degrees of freedom (displacements, accelerations, rotations), with specific notations.

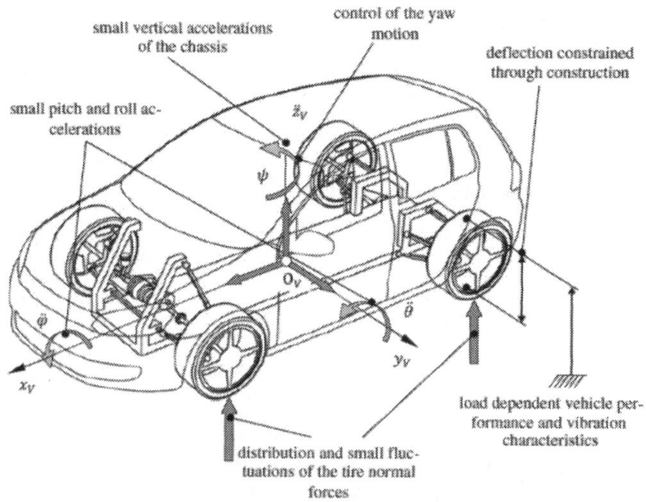

Figure 1. Vehicle model with applied forces

The simulation and analysis of vehicle models is influenced significantly by the type and the temporal behaviour of the applied forces. For a complete study one differentiates between different fundamental types of forces: (i) external and internal forces and torques, (ii) applied and reaction forces and torques and (iii) surface and volume forces. The first category (i) depends on the boundaries of the system in question. The outside forces and torques acting on said system are referred to as external forces. The forces acting within a system are thus referred to as the internal forces. Through a change in the system boundaries, the internal forces become external forces of the new system boundaries. The second category (ii) refers to the source of the forces. Applied forces can be described using the physical force laws with respect to position and velocity. The applied forces in vehicle systems are the forces due to gravity, the spring and damper forces, but also the gas forces in the cylinders of the combustion engine, the air forces and especially the tire forces. The reaction forces are the forces occurring inside the joints. If modeling the wheel and the road surface as rigid bodies, the normal forces from the road to the wheels are reaction forces. In the case of the modeling of the wheel with an elastic tire however, they are considered to be applied forces. The third category (iii) refers to the distribution of the forces. Surface forces are constrained on a surface, and the volume forces are distributed spatially. The surface forces in vehicles are the wind forces and the forces between the road surface and the thread of the tire. The volume forces in vehicles consist of all the forces that are due to gravity.

Another differentiation results from the temporal characteristic of the forces. One differentiates between forces with deterministic, and those with stochastic temporal characteristics. Deterministic forces are defined explicitly through their time functions. Stochastic forces do not possess a rule based temporal characteristic and need to be described using statistical methods.

For stability and driving comfort of a vehicle there is also important to reduce at maximum the tangential force differences between left and right sides, by appropriate choice of brake system and, for the driven axles, by use of a differential. In this situation the brake and slip control are similarly tuned. The yaw moment caused by unequal tangential forces may also be employed positively. Limited-slip and locking differentials generate a locking effect which tends to produce a correcting yaw moment, tending to bring the car back to a straight path. In order to prevent exaggerated understeer in acceleration, the differential locking effect is limited. In decelerating process, using engine braking, differential locking torque helps to compensate for oversteering moments caused by dynamic weight transfer. To improve the stability of the vehicle using this effect some vehicles use differentials with a higher limited slip factor while under engine braking than in acceleration. So, the vehicle stability control may be realized by means of differentials with controllable locking torque, which permit introduction of a correcting yaw moment.

Modern brake control systems use deliberate, controlled, asymmetric, driver-independent brake intervention to stabilize the vehicle at the handling limit.

Vertical forces, acting at the wheel road surface patch than transmitted and applied between suspension and body, consist mainly of spring and shock absorber (damping) forces. Their task is to support the vehicle's sprung mass on the wheels and impose limits on vehicle lift, roll and pitch motion relatively to the road surface. Through the extent limit permitted by spring travel, the body should be isolated against road surface irregularities and dynamic changes in wheel load minimized.

Tire normal forces, Figure 2, differ from the corresponding spring and shock absorber forces introduced to the body by the inertia of unsprung masses. The spring and shock absorber forces, Figure 3, through their vertical reaction forces from tire and tangential forces also affect the body, as do unsprung masses. These forces results from coupling of the tire contact patch longitudinal and transverse movement in response to vertical suspension motion (antidive, antisquat, sideforce reaction angle). These effects are used to reduce roll and pitch angles.

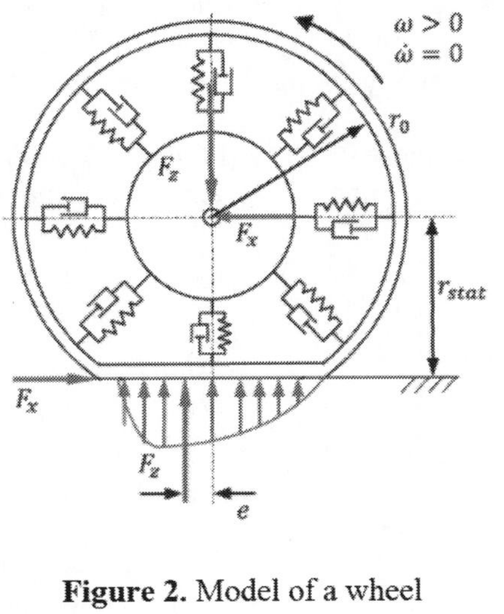

Figure 2. Model of a wheel

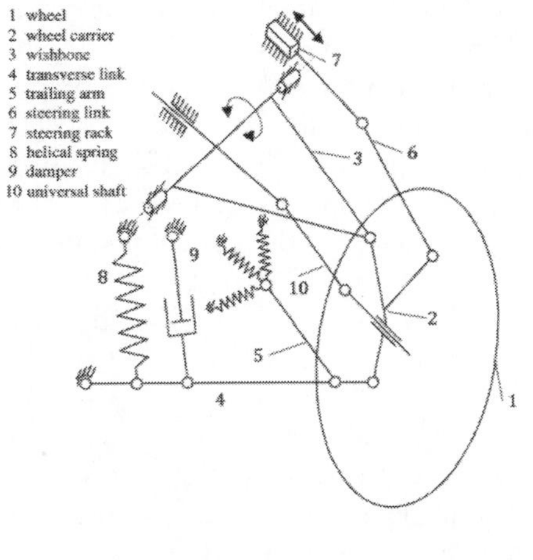

Figure 3. Scheme of an independent suspension

2.2. One Dimensional Quarter Vehicle Model

In order to be able to assess the possibilities and limits of mechanical systems, the use of theoretical mathematical techniques has proven to be of great value. By means of a suitable reduction of the model, the computing time and interpretability of the results can be optimized. Thus it was used a simple equivalent mass-spring system in order to theoretically analyze the vehicle vibrations in vertical direction. The influence of vehicle suspension and damping, wheel mass and tire elasticity on comfort and range of spring deflection can be determined simply by using quarter vehicle models.

The vehicle model hereby consists of a two-body-system, which is linked to a wheel model, Figure 4. This model reduction is acceptable for a passenger car because the coupling masses are usually very small compared to the car body mass. Because the influence on vibration's properties are reduced, for a correct analyze we neglect also the effects of pitch and roll oscillations in the evaluation of comfort.

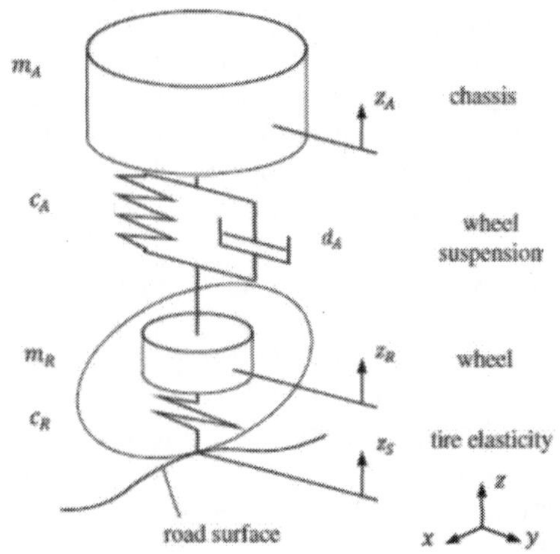

Figure 4. The quarter – vehicle model

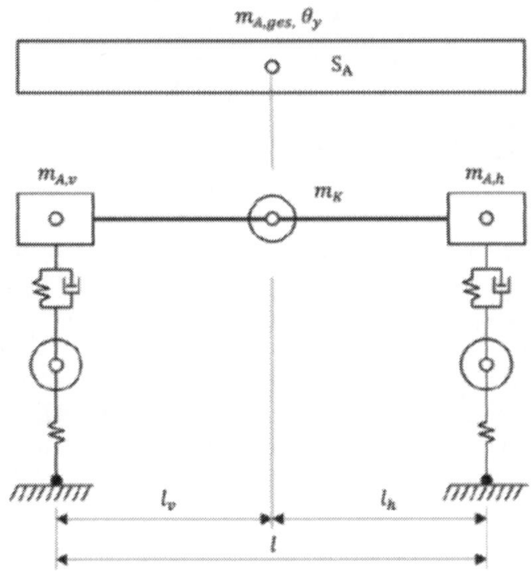

Figure 5. Derivation of the vehicle body mass

In Figure 4 [5] the notations are the following: x, y and z are the current position coordinate (and also Cartesian versors of axis), m_A is the body mass, expressed as ratio from the whole vehicle mass, m_R is the wheel assembly mass and respectively, c_A is the vehicle body spring stiffness, c_R is the wheel spring stiffness and d_A is the vehicle body suspension damper.

The contact to the road is established via the wheel (the patch with the road surface), which is modeled through the spring stiffness c_R. The equations of linear momentum for the system in z-direction, in accordance with Figure 4 notations are:

$$m_A\ddot{z}_A + d_A(\dot{z}_A - \dot{z}_R) + c_A(z_A - z_R) = 0 \quad (1)$$

$$m_R\ddot{z}_R - d_A(\dot{z}_A - \dot{z}_R) - c_A(z_A - z_R) + c_R z_R = c_R z_S \quad (2)$$

Equation 1 refer to the body (chassis) and equation 2 refer to the wheel.

In order to achieve better results from the quarter-vehicle model, the pro rata chassis mass can be determined by breaking down the half chassis mass onto three mass points $m_{A,v}$ (front axle), $m_{A,h}$ (rear axle) and m_K (coupling mass).

From the conditions for the preservation of the chassis mass $m_{A;ges}$, of the center of mass S_A and the mass moment of inertia $\theta_y = m_A i_y^2$, where i_y is the radius of gyration about y - axis

$$m_{A,y} + m_{A,h} + m_K = m_{A,ges} \tag{3}$$
$$m_{A,v}l_v - m_{A,h}l_h = 0 \tag{4}$$
$$m_{A,v}l_v^2 + m_{A,h}l_h^2 = m_A i_y^2 \tag{5}$$

one obtains three equivalent masses

$$m_{A,v} = m_{A,ges}\frac{i_y^2}{l_v l}, \; m_{A,h} = m_{A,ges}\frac{i_y^2}{l_h l}, \; m_K = m_{A,ges}\left(1 - \frac{i_y^2}{l_h l_v}\right) \tag{6}$$

For the wheel based, proportional body mass m_A of the quarter-vehicle model, m_A = ½ $m_{A;v}$ (front wheel) respectively m_A = ½ $m_{A;h}$ (rear wheel) is used.

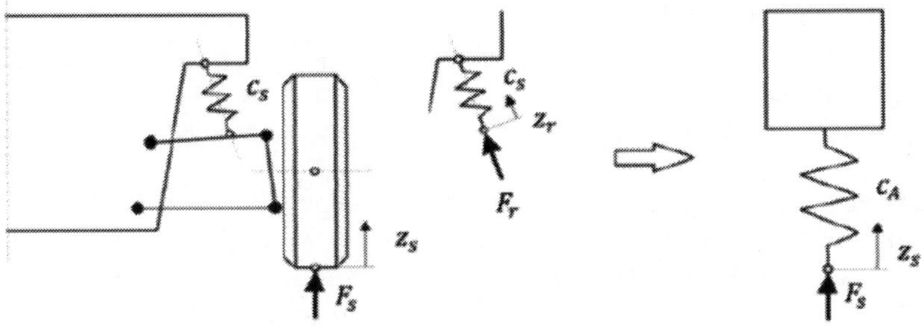

Figure 6. Kinematic transmission of the wheel suspension forces

Furthermore it is necessary to consider the kinematic transmission of the wheel suspension forces as shown in Figure 6. The kinematic transmission k results from the vertical deflection z_r of the wheel suspension as

$$k(z_s) = \frac{dz_r}{dz_s} = \frac{\dot{z}_r}{\dot{z}_s} \tag{7}$$

From the principle of virtual work, the relation between the force F_s at the tire-road contact point (in the contact patch) and the elastic force F_r can be derived as follows:

$$F_s\delta z_s = F_r\delta z_r = F_r\frac{dz_r}{dz_s}\delta z_s \tag{8}$$

then

$$F_s = \frac{dz_r}{dz_s}F_r = kF_r \tag{9}$$

Based on equations (7) to (9) one calculate the spring stiffness related to vertical displacement z_s of the tire contact point (in the contact patch) as follows:

$$c_s = \frac{dF_s}{dz_s} = \frac{d(kF_r)}{dz_s} = \frac{dk}{dz_s}F_r + k\frac{dF_r}{dz_s} = \frac{dk}{dz_s}F_r + k\frac{dF_r}{dz_r}\frac{dz_r}{dz_s} \tag{10}$$

Because $\frac{dF_r}{dz_r} = c_s$ and $\frac{dz_r}{dz_s} = k$ one obtain

$$c_A = \frac{dk}{dz_s}F_r + k^2 c_s \tag{11}$$

IOP Conf. Series: Materials Science and Engineering **294** (2017) 012083 doi:10.1088/1757-899X/294/1/012083

2.3. Tire model

The modeling of the tire forces requires special attention and care especially when along with the stationary behaviour the instationary behaviour has to be covered. There are three types of tire models such as mathematical models, physical models and a combination of the two. For the further processing one can either be accomplished with the help of an approximation through an algebraic function (Magic Formula Tire Model [6, 7]) or through an interpolation. This art of modeling is for the simulation of driving maneuvers mostly sufficient as the excitation frequencies in these cases are well below the Eigen frequencies of the belts. Problems can however occur, when a large number of influencing variables have to be considered, as in this case a considerable characteristic curve needs to be saved and evaluated. And it is practically impossible, to change individual parameters, without recreating the entire characteristic curves. That is why, for investigation and modeling of the comfort and vibration analysis are often use these models only to estimate the deformation of the tire.

From the point of view of a reasonable computation time, such models are still not used in the vehicle dynamics simulation in spite of the massive increase in computation power available today. In stationary models only the rim will be considered to have mass and inertia, and the belt will not be modeled as a separate body. These models are similar to the characteristic curve models mostly used in the simulation of stationary driving maneuvers. In order to extend the application area of these models, very often a first order time element (PT_1) is used to represent the delayed building up of the tangential forces.

However if excitations in the region of the belts eigenfrequency, between 30 Hz and 50 Hz, were to occur, such as through the use of elastic components in the wheel suspension or through the pulsation of the brake pressure during an ABS controlled braking procedure, and these have an effect on the characteristics of the vehicle behaviour, then the eigendynamics of the belt have to be considered as well. Thus the belt will be modeled as a rigid ring with mass. This will be applied at low speeds (below 10 m/s) highly instationary transport processes in the contact surface between the tire and the road (the contact patch), that require a special attention.

A kinematic model for the wheel-road contact [5] will respect the following: (i) the contact geometry tire/road are described through a representative mechanism and can be calculated using simple expressions, (ii) simple tire models, in which the longitudinal forces and the lateral forces can be calculated, without considering the eigendynamics of the belts, as a function of the longitudinal slip and the lateral slip, can access these variables, (iii) for more complex tire models, that need to represent the instationary processes, the patch can be discretized. The required kinematic values are available for these models also.

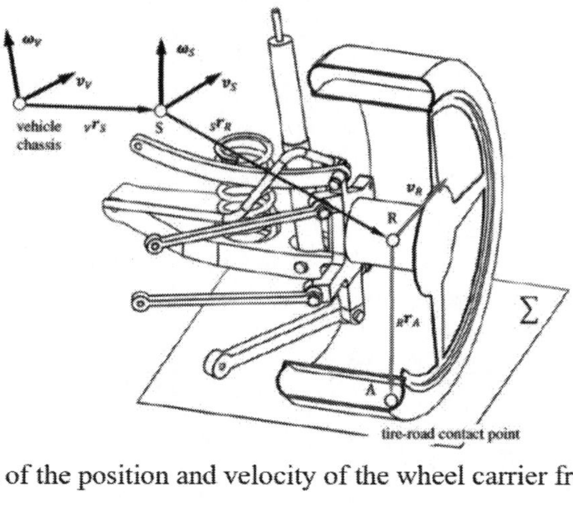

Figure 7. Calculation of the position and velocity of the wheel carrier from the vehicle chassis

It shall be assumed that the position and orientation of the wheel in space is known. For this one measures the position and velocity of the wheel from the vehicle chassis, see Figure 7. The position of the wheel carrier can be described through the wheel center point R and through the introduced wheel carrier fixed coordinate system according to Figure 3 and 4 notations. These values are known through the vehicle kinematics.

3. Modeling and Experimental results

Starting from the technical equipment that consist the core of the Vehicle Dynamics Laboratory of the Politehnica University Timisoara and also the MSC ADAMS package software for modelation and evaluation of the experiments involved in vehicle dynamics were obtained interesting satisfactory results of the proposed models.

Figure 8. MSC ADAMS Car model for the suspension analysis

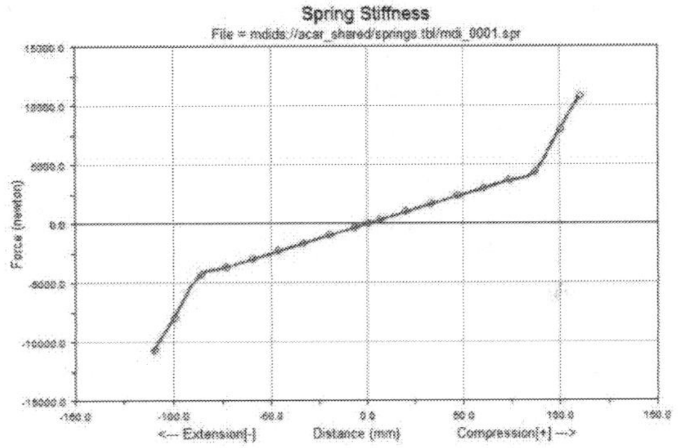

Figure 9. MSC ADAMS quarter-car model for the force loading through measurement

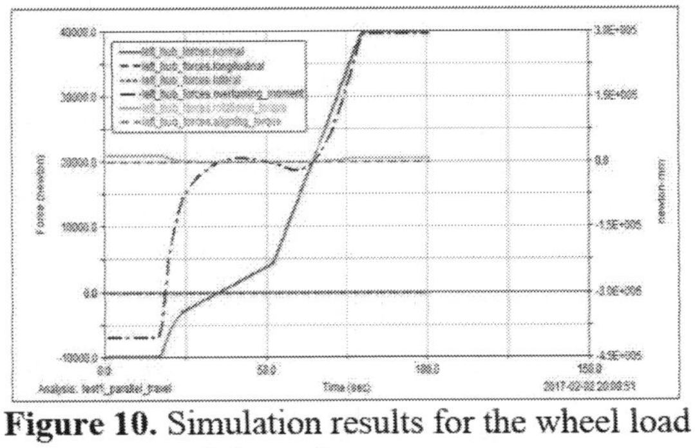

Figure 10. Simulation results for the wheel load and through measurement

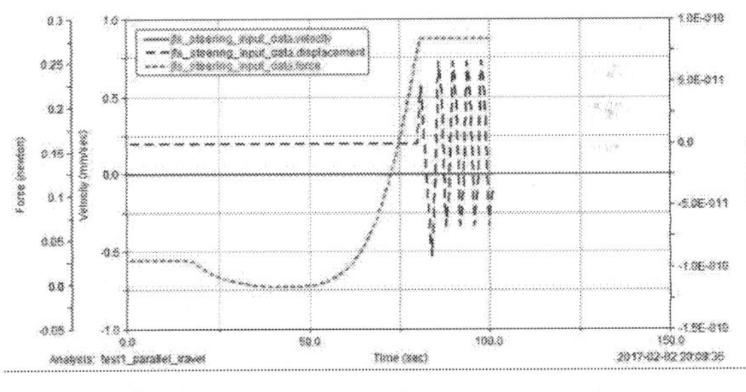

Figure 11. Simulation results for the suspension travel and through measurement

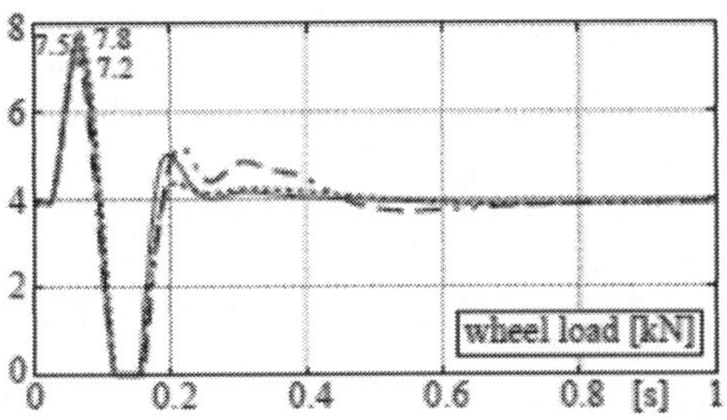

Figure 12. Compared simulation with experiment for the wheel load

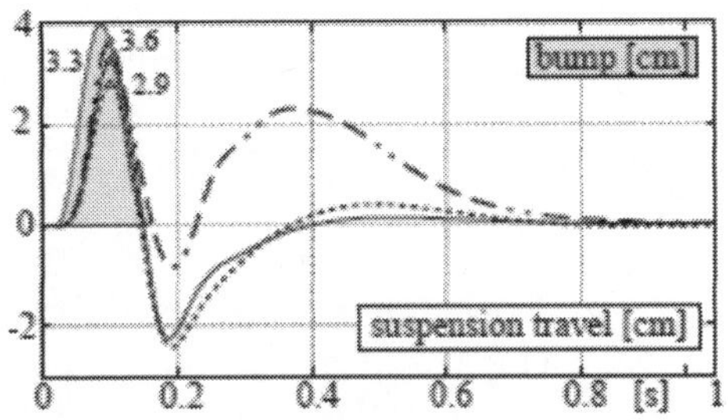

Figure 13. Compared simulation with experiment for the suspension travel

Thus, there were used the following equipment and software: manufactured suspension test bench, electric actuator for applied forces on wheel-spring-damper assembly, electronic equipment for electric actuator control, NI USB-6251 data acquisition and control device, LabView 2015 software [8], MSC Adams 2017 software [9], ASM R1K-L10 linear motion transducer, SACHS 1755017000 – 11 – 05 – 00, 19 – 04 – 00 and 19 – 09 – 01 acceleration transducers.

For the model and test bench was chosen a conventional MacPherson suspension assembly reproduced in Figure 8 after the modelation in MSC ADAMS Car Software. For the complete analysis in the model was included also the steering assembly that is missing on the suspension test bench.

For the quarter car model and through experiment (based on the manufactured test bench) there where used the following important data: sprung mass – 320 kg, spring diameter – 14 mm, spring coil diameter – 128 mm, shear modulus of steel – 85 kN/mm^2, upper arm length – 270 mm, lower arm length – 280 mm, tire 195/65 R15, wheel mass – 42 kg, wheel spring stiffness – 225000 N/m, damper value for the suspension test rig – 2750 Ns/m. Figure 9 expose the measured values for the loading force that was used in MSC ADAMS simulation.

Thus, from the experiment and simulation analysis we can compare and calibrate the values for the load forces that act on the tire – wheel – damper assembly of a quarter-car model. Based on the results and taking in consideration the equations and input data we will be able to obtain the comfort behaviour of the suspension assembly of the current classic MacPherson suspension system through variation of different values of tire pressure, tire model/manufacturer, tire damping, spring stiffness, absorber damper, etc.

Figures 10 and 11 present the simulation results made in MSC ADAMS starting from the measured elements (forces, displacements, accelerations) and time dependent working process, and in Figures 12 and 13 there are presented the compared values measured through data acquisition chain versus modeled values for the wheel load and suspension travel of the assembly on a bump-speed rolling sequence.

4. Conclusions

The studies developed for these models was first conceived and tested like a theoretical system and then simulated on specific environment. The results was compared with experimental data, obtained on a specialized test rig through data acquisition values and the results are very favourable, proving the good approximations and initial values conditions for the simulation and experiment. The models for the tires and suspension components could be modified due to the simulation and the time and costs of testing experiments could be preserved and the testing conditions could be varied on demands.

References

[1] Blundell M and Harty D 2004 *The Multibody System Approach to Vehicle Dynamics*, Elsevier Butterworth-Heinemann Publications
[2] Kiencke U and Nielsen L 2000 *Automotive Control Systems,* Springer
[3] Genta G 1997 *Motor Vehicle Dynamics. Modeling and Simulation*, World Scientific Publishing, Singapore
[4] Schramm D, Hiller M and Bardini R 2014 *Vehicle Dynamics. Modeling and Simulation*, Springer
[5] Rill G 1012 *Road Vehicle Dynamics. Fundamentals and Modeling*, CRC Press
[6] Pacejka H B 2006 *Tyre and vehicle dynamics. 2nd (ed.)*, Butterworth-Heinemann, Oxford
[7] Pacejka H B and Besselink I J M 1993 *Magic formula tyre model with transient properties*, Proceedings of the Berlin Tyre Colloquium, Vehicle System Dynamics Supplement 27, 145–155, Swets&Zeitlinger, Lisse, Berlin
[8] ***Labview/National Instruments – 2015 User's manual
[9] ***MSC ADAMS Car – 2017 User's manual

Determination of the unsulfonated color concentration from D&C Yellow No. 10 by the derivative spectrophotometry

A D Berdie and S Jitian

Politehnica University of Timisoara, Department of Electrical Engineering and Industrial Informatics, 5 Revolution Street, Hunedoara, 331128, Romania

E-mail: jitian_s@yahoo.com

Abstract. The method that we used is based on the measurement of the first derivative of the mixture of the two colorants at the wavelength for which one of them has the first derivative equal to zero.
The Code of Federal Regulations (21 CFR 74.1710) specifies for D&C Yellow No. 10 the maximum permitted levels of an unsulfonated subsidiary color and of diethyl ether-soluble matter other than that specified.
In the proposed method a color additive sample is dissolved in water and the unsulfonated subsidiary color are extracted from this solution with dichloromethane. The analysts in dichloromethane solution are determined by spectrophotometry.
The unsulfonated subsidiary colors determined are:
- D&C Yellow No. 11 [2-(2-Quinolinyl)-1H-indene-1,3(2H)-dione] (Y11), from which D&C Yellow No. 10 is manufactured by sulfonating and
- 1,5-Naphthyridinequinophthalone (1,5-NQ).

Another compound soluble in water and dichloromethane (which I called S) is present in dichloromethane solution after extraction together with the other two colors and can affect the correct determination of the concentrations. The dichloromethane-soluble matter other than specified is a mixture consisting mostly of chlorinated derivatives of the unsulfonated subsidiary color. Because the S color is present both in aqueous and in dichloromethane solutions, the spectra of calibration solutions should be corrected. The applied correction does not affect the determination of the unsulfonated subsidiary colors concentrations.
D&C Yellow No. 11 and 1,5-NQ are used as standard for unsulfonated subsidiary colors.

1. Introduction

D&C Yellow # 10 (2-(2-Quinolyl)-1,3-indandione disulfonic acid disodium salt) is a U.S. Food and Drug Administration (FDA)-approved colorant commonly used to color dose forms such as tablets and capsules. The FDA states that D&C Yellow No. 10 may be safely used as a colorant. D&C Yellow No. 10 is approved for use in drugs and cosmetics but is not approved as a colorant for food [1], [2]. The structure of D&C Yellow # 10 is presented in Figure 1.

D&C Yellow No. 10 is obtained by sulfonating of D&C Yellow No. 11. D&C Yellow No. 11 is manufactured by condensing quinaldine with phthalic anhydride to give the unsulfonated dye.

The color additive D&C Yellow No. 10 is a mixture of disulfonates (principally), monosulfonates and trisulfonates of 2-(2-quinolyl)indan-1,3-dione consisting principally of the sodium salts of: 2-(2,3-dihydro-1,3-dioxo-1H-indene-2-yl)-6-quinolinesulfonic acid and 2-(2,3-dihydro-1,3-dioxo-1H-indene-

IOP Conf. Series: Materials Science and Engineering **294** (2017) 012084 doi:10.1088/1757-899X/294/1/012084

2-yl)-8-quinolinesulfonic acid with lesser amounts of the disodium salts of the disulfonic acids of 2-(2-quinolinyl)-1H-indene-1,3(2H)-dione.

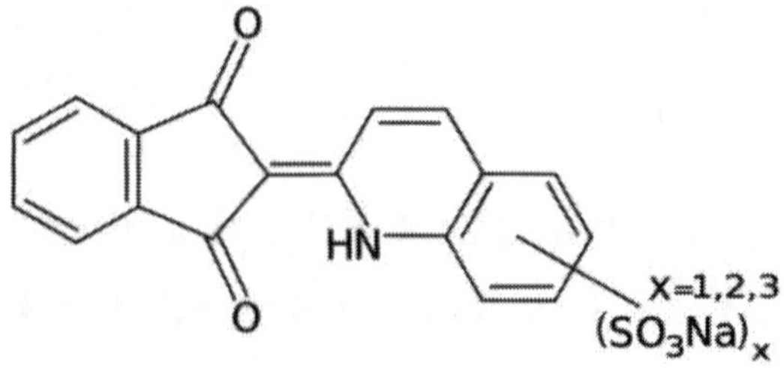

Figure 1. The structure of D&C Yellow # 10

In the manufacture of Y10 remain as impurities, especially Y11 unreacted and 1,5-Naphthyridinequinophthalone.

The Code of Federal Regulation (21 CFR 74.1710) specifies for D&C Yellow No. 10 minimum concentration 85% and the maximum permitted levels 4ppm for D&C Yellow No. 11. The maximum concentration of diethyl ether soluble matter other than that specified, must be 2 parts per million, using added 2-(2-quinolinyl)-1*H* -indene-1,3 (2*H*)-dione for calibration [3].

The unsulfonated subsidiary colors determined are:

- D&C Yellow No. 11 [2-(2-Quinolinyl)-1H-indene-1,3(2H)-dione] (Y11) [4], from which D&C Yellow No. 10 is manufactured by sulfonating and
- 1,5-Naphthyridinequinophthalone (1,5-NQ).

Another compound soluble in water and dichloromethane (called S) is present in dichloromethane solution after extraction together with the other two colors and can affect the correct determination of the concentrations. The four compounds present in D & C Yellow No.10 are shown in Figure 2.

Water soluble			
D&C Yellow No. 10	**S**	**1,5-NQ**	**D&C Yellow No. 11**
85% Min.	2ppm Max.		4ppm Max.
	Dichloromethane soluble		

Figure 2. The compounds water soluble and dichloromethane soluble

The dichloromethane-soluble matter other than specified is a mixture consisting mostly of chlorinated derivatives of the unsulfonated subsidiary color [5].

The D&C Yellow No. 11 is used as the standard for the unsulfonated subsidiary color an as a surrogate standard for the other ether-soluble matter.

The structure of D&C Yellow # 11 and 1,5-Naphthyridinequinophthalone are presented in Figure 3.

The UV-vis absorption spectra of the two sulfonated compounds that are present in the dichloromethane phase are shown in Figure 4. The absorption bands of the two compounds are very close, making it difficult to spectrophotometrically determine their concentrations.

For simultaneous determination of the concentrations from a mixture of two or more compounds, derivative spectrophotometry can be successfully used [6], [7]. Derivatization of the spectrum is especially useful when the absorption bands of the compounds in the mixture are very close and overlap. The overlapping signals may be thus separated and background may be eliminated by the presence of other compounds in the sample [8].

IOP Conf. Series: Materials Science and Engineering **294** (2017) 012084 doi:10.1088/1757-899X/294/1/012084

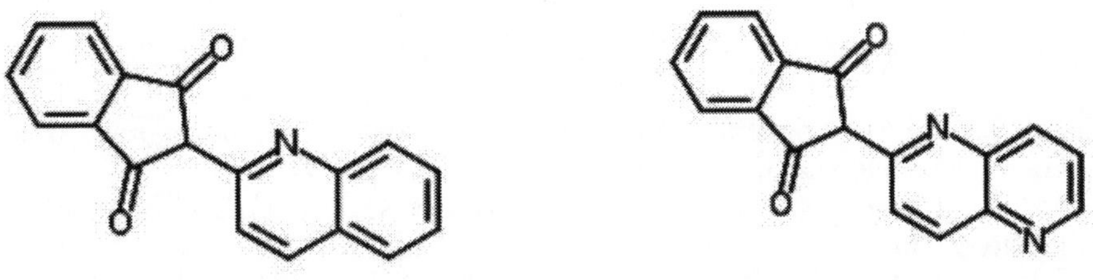

a) b)

Figure 3. The structures of: a) D&C Yellow No. 11 [2-(2-Quinolinyl)-1H-indene-1,3(2H)-dione] and b) 1,5-Naphthyridinequinophthalone

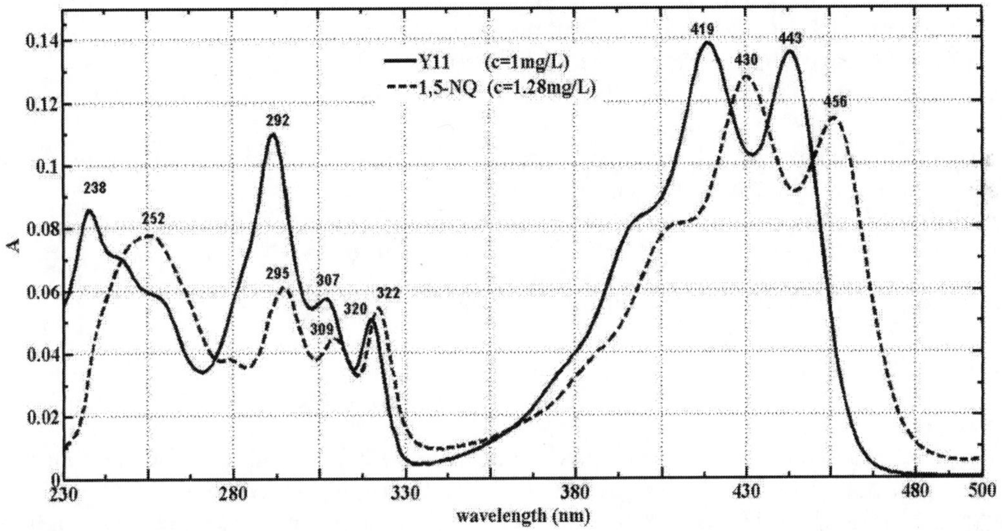

Figure 4. The UV-viz spectra of D&C Yellow No.11 and 1,5-Naphthyridinequinophthalone from the dicloromethane extract of D&C Yellow No. 10 batches

The derivative spectrophotometry is based on the measurement of the first derivative of the mixture of the two colorants at the wavelength for which one of them has the first derivative equal to zero.

For each colorant on a specific concentration domain, absorbance shows a linear dependence on the concentration of the form:

$$A=\varepsilon\cdot d\cdot c+n=m\cdot c+n \qquad (1)$$

where:

A = measured absorbance of the color solutions at the wavelength λ, corrected for the absorbance of the blank at the same wavelength;

ε = absorptivity value of the color at the wavelength λ;

d = cell path length (cm);

c = the concentration of color in solution (mg/L);

m = the slope of calibration line;

n = intercept of calibration line.

We denote with D^1 the first derivative of absorbance with respect to wavelength:

$$D^1 = \frac{dA}{d\lambda} = \frac{d\varepsilon}{d\lambda} \cdot c \cdot d + \frac{dn}{d\lambda} = m^1 \cdot c + n^1 \quad (2)$$

The first order derivative depends linearly on the concentration. The slope m^1 and the intercept n^1 are dependent on the wavelength.

For a mixture of two dyes with the concentrations c_1 and c_2 respectively, the absorbance of the mixture is additive:

$$A = A_1 + A_2 = \varepsilon_1 \cdot d \cdot c_1 + n_1 + \varepsilon_2 \cdot d \cdot c_2 + n_2 = m_1 \cdot c_1 + n_1 + m_2 \cdot c_2 + n_2 \quad (3)$$

where:

A = the measured absorbance of the mixture solutions at λ;
A_1 = the measured absorbance of the first color solution at λ;
A_2 = the measured absorbance of the second color solutions at λ;
m_1 = the slope of the calibration line for the first color;
m_2 = the slope of the calibration line for the second color;
n_1 = the intercept of the calibration line for the first color;
n_2 = the intercept of the calibration line for the second color.

It follows that in a solution containing a mixture of two dyes (Y11 and 1,5-NQ) the first derivative of the absorbance in relation to the wavelength, is additive.

$$D^1 = D^1_{Y11} + D^1_{NQ} = m^1_{Y11} \cdot c_{Y11} + n^1_{Y11} + m^1_{YNQ} \cdot c_{NQ} + n^1_{NQ} \quad (4)$$

The additivity is also valid for the higher order derivatives [9].

2. Experimental

2.1. Materials and instrumentation

All of the spectrophotometric measurements were made with a scanning UV-visible PerkinElmer Lambda25 spectrometer provided with quartz windows cells, 1 cm path length, 0.4 mL.

For all samples UV-VIZ spectra was recorded in 230÷500 nm, scan speed 120nm/min, data interval 1nm, split 1nm. Use the dichloromethane as blank.

The samples were weighed using microbalance, accurate to 0.01 mg and analytical balance, accurate to 0.1mg.

Micro-syringe 500 μL Gastight® and 50 μL Hamilton Co. Microliter® (Reno, Nevada, USA) was used to prepare the standard solutions and calibration solutions.

The solutions were prepared in silanized vials. The glassware surface deactivating reagent (Sylon CT, 5% dimethyldichlorosilane in toluene) was purchased from Supelco, (Bellefonte, PA, USA).

A hot plate magnetic stirrer with the speed controlled, IKA® Works Inc. (Wilmington, NC, USA) (model IKAMAG RT10 power) was used to stir the samples.

2.2. Reagents

- Water, HPLC grade, deionized with a Milli-Q water system from Millipore (Bedford, MA, USA).
- Dichloromethane, anhydrous (≥99.8%) obtained from Sigma-Aldrich (Milwaukee, WI, USA).
- Standards: D&C Yellow No.11 and 1,5-NQ, standards for the unsulfonated subsidiary color and for the other diethyl ether-soluble matter.

2.3. Glassware treatment

All the glassware used for analysis of Y11 and 1,5-NQ in the color mixture Y10 was silanized prior to use, as described previously [10]. The volumetric flasks and the vials used were filled with 5%

IOP Conf. Series: Materials Science and Engineering **294** (2017) 012084 doi:10.1088/1757-899X/294/1/012084

solution of dimethyldichlorosilane (DMDS) in toluene (Silon-CT). After 12 - 16 h they were rinsed once with toluene and once with methanol followed by oven drying at 150^{0}C for 1 h.

2.4. Preparation of solutions

2.4.1. The stock-assay standard solutions. The stock solutions of Y11 (80 mg/L) and 1,5-NQ (40 mg/L) were prepared by dissolving each of them in methanol chloride in 25 mL volumetric flask. The stock solutions are kept in the refrigerator. They are weighed each time and kept constant mass by adding CH_2Cl_2 (to compensate for the solvent evaporated in time).

2.4.2. The calibration solutions. Calibration is performed in the absence of D&C Yellow No. 10 The D&C Yellow No. 10 (Y10) which does not contain D&C Yellow No. 11 and 1,5-NQ are used for preparing of the calibration solutions.

Using an analytical balance, weigh 1 g (±0.001g) of D&C Yellow No. 10 (Y10) into six tarred 20 mL vials.

Add 10 mL water to the sample. Add a magnetic stirrer in vial and heat the mixture on a hot plate magnetic stirrer at 40^{0}C until the sample dissolves, then allow the solution to cool to room temperature.

Add with a micro-syringe 12.5 µL, 25 µL, 50 µL, 75 µL, 100 µL, 125 µL from stock-assay standard solution of Y11 (80 mg/L). Are obtained thus mixtures of Y10 and Y11 dyes in which **Y11** has the concentrations: **1ppm, 2ppm, 4ppm, 6ppm, 8ppm and 10ppm.**

Add with a micro-syringe 50 µL, 100 µL, 150 µL, 200 µL, 250 µL from stock-assay standard solution of 1,5-NQ (40 mg/L). Are obtained thus mixtures of Y10 and 1,5-NQ dyes in which **1,5-NQ** has the concentrations: **2ppm, 4ppm, 6ppm, 8ppm and 10ppm.**

Add 3mL dichloromethane in each vial, stopper and shake to mix, during 30 min at 1000 rpm on the hot plate magnetic stirrer, at room temperature, for extraction of unsulfonated subsidiary colors from aqueous solution.

Only the non-aqueous phase is transferred to other silanized glass vials with Pasteur pipette.

2.4.3. The reference solutions. Using an analytical balance, weigh 1 g (±0.001g) of D&C Yellow No. 10 (Y10) into six tarred 20 mL vials.

Add 10 mL water to the sample. Add a magnetic stirrer in vial and heat the mixture on a hot plate magnetic stirrer at 40^{0}C until the sample dissolves, then allow the solution to cool to room temperature.

Add 3mL dichloromethane in each vial, stopper and shake to mix, during 30 min at 1000 rpm on the hot plate magnetic stirrer, at room temperature, for extraction of unsulfonated subsidiary colors from aqueous solution.

Only the non-aqueous phase is transferred to other silanized glass vials with Pasteur pipette.

The reference solution is prepared from a color that does not contain the compounds: Y11 and 1,5-NQ. Unfortunately, the reference solution will also contain S-soluble compound in dichloromethane.

3. Result and discussion

3.1. The absorption spectra

The spectrum of the mixture of D&C Yellow No. 10 and S (water-soluble) is shown in Figure 5. It was measured using distilled water as a reference.

The spectrum of reference solution, corresponding to the color that is present in D&C Yellow No. 10 (soluble in both water and CH_2Cl_2), is diferent from that of the unsulfonated colors: Y11 and 1,5-NQ. Also it is different from that of D&C Yellow No. 10 due to the presence of S compound.

The spectrum of S compound (present in D&C Yellow No. 10) soluble both in water and dichloromethane is shown in Figure 6. Dichloromethane was used as the blank.

IOP Conf. Series: Materials Science and Engineering **294** (2017) 012084 doi:10.1088/1757-899X/294/1/012084

The spectrum of S compound is the average obtained from three spectra corresponding to three reference solutions.

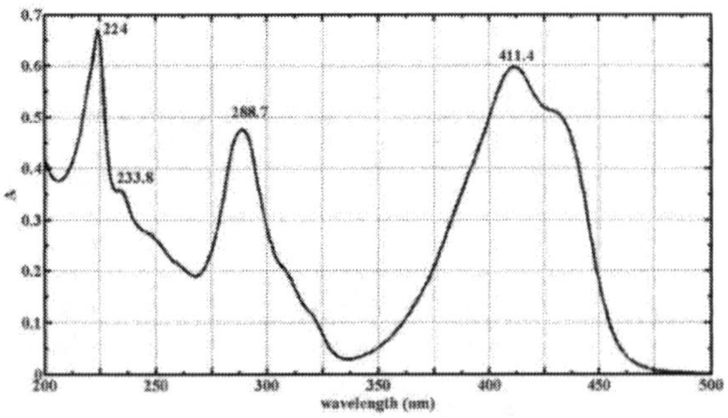

Figure 5. The spectrum of water soluble compound mixture from D&C Yellow No. 10

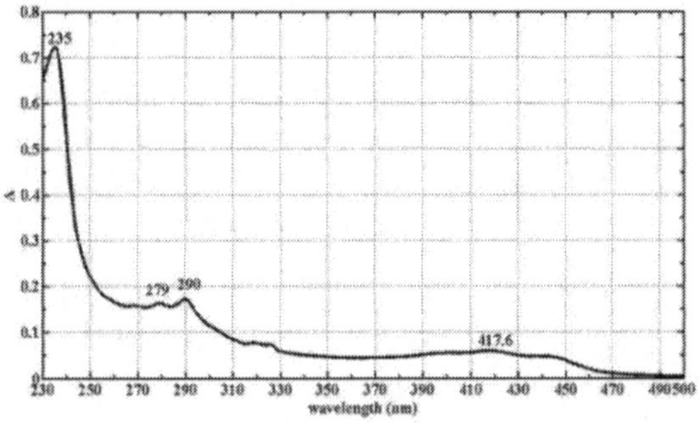

Figure 6. The spectrum of S compound, soluble both in water and dichloromethane

Because the S compound is present both in aqueous and in dichloromethane solutions, the spectra of calibration solutions should be corrected. The spectrum of the calibration solution containing Y11 color (c=4ppm) without correction and with the corresponding correction of S compound is presented in Figure 7.

We can use two methods to obtain the spectra of the corrected calibration solutions.

- Use the reference solution as blank. The spectra are obtained directly from calibration solutions.
- Use the dichloromethane as blank. The uncorrected spectra of calibration solutions are obtained. These are mathematically corrected by subtracting the spectrum of reference solution from spectrum of the calibration solutions.

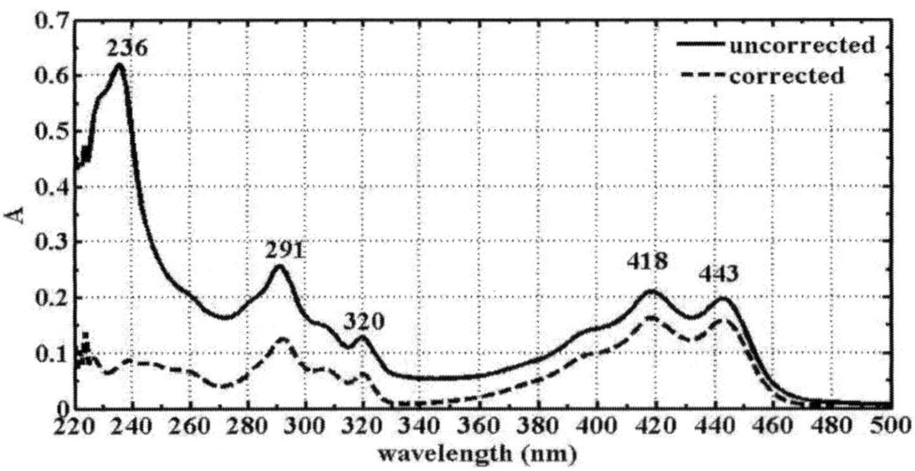

Figure 7. The spectra of the calibration solution containing D&C Yellow No.11 (c=4ppm)

The reference solution is prepared in both cases from a color that does not contain the compounds: Y11 and 1,5-NQ.

The spectra of the calibration solutions, obtained by the two methods are shown in Figure 8. The applied correction by both processes affect only the absorption bands positions below 295nm and does not affect the determination of the unsulfonated subsidiary colors concentrations.

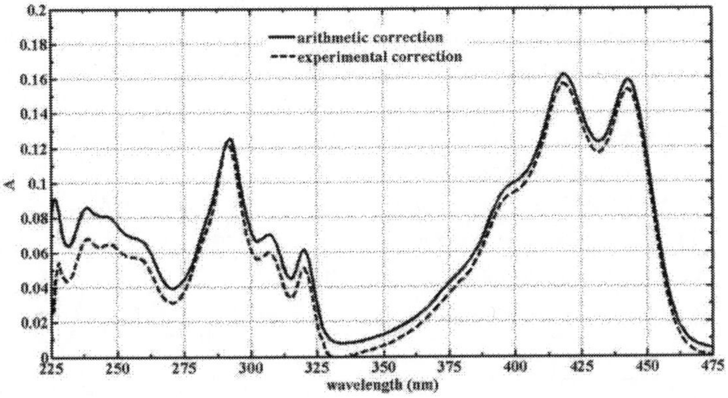

Figure 8. The spectra of the calibration solution containing D&C Yellow No. 11 (c=4ppm)

The corrected spectra of the calibration solutions are presented in Figures 9 and 10.

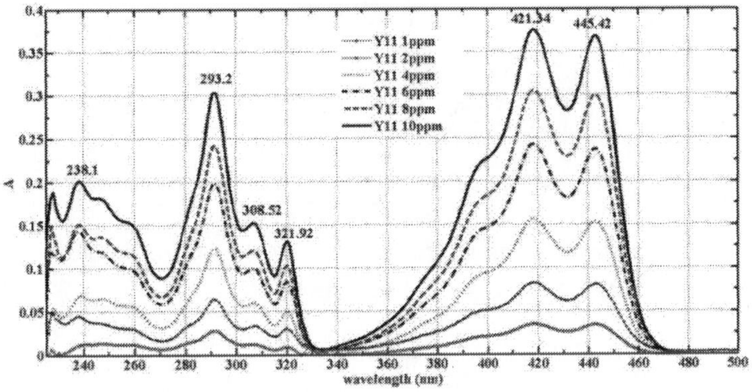

Figure 9. The corrected spectra of Y11 calbration solutions (from Y10 matrix)

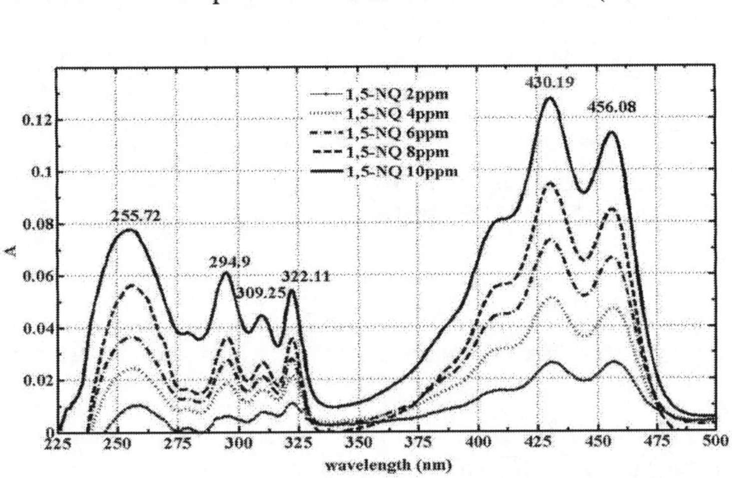

Figure 10. The corrected spectra of 1,5-NQ calbration solutions (from Y10 matrix)

3.2. The derivative spectra

Figure 11 shows the 1st order derivative spectra of the calibration solutions, obtained by the two methods, presented in Figure 7, where the concentration of the D & C Yellow No. 11 dye is 4ppm.

The differences corresponding to the two processes are of the order of 0.02 for Y11 and 0.0001 for 1,5-NQ.

The 1st order derivative spectra of D & C Yellow No. 11 and 1,5-Naphthyridinequinophthalone will be draw on the same graph, as in Figure 12.

The wavelengths at which the 1st order derivative of 1,5-NQ dye is zero and the 1st order derivative of D & C Yellow No. 11 dye is nonzero will be identify. The values λ = **294.75** nm and λ = **322.1** nm are found. The most appropriate value is λ = **294.75** nm.

For D&C Yellow No. 11 color, plot the calibration line $D^1_{Y11} = f(c)$ at this wavelength (Figure 13a).

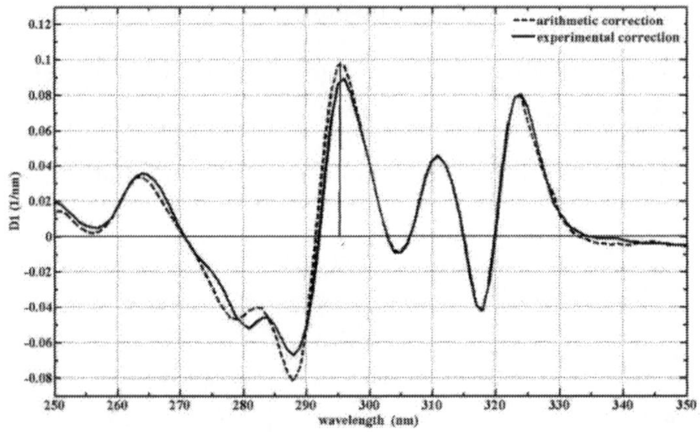

Figure 11. The first derivative spectra of the calibration solution containing D&C Yellow No. 11 (c=4ppm)

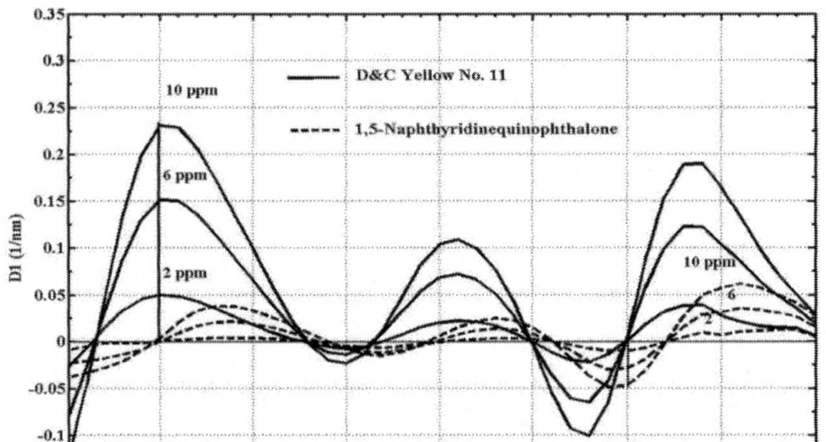

Figure 12. The first derivative spectra of calibration solutions for D&C Yellow No.11 and 1,5-Naphthyridinequinophthalone

The calibration line for unsulfonated subsidiary color from the calibration data using least-squares linear regression is expressed as:

$$y = mx + n \tag{5}$$

where:

- y = the first derivative of absorbance at maximum values (nm^{-1});
- x = the analyte concentration ($mg \cdot L^{-1}$);
- m = calculated slope of the calibration line($L \cdot mg^{-1} \cdot nm^{-1}$);
- n = calculated intercept of the calibration line (nm^{-1});
- r = correlation coefficient of data.

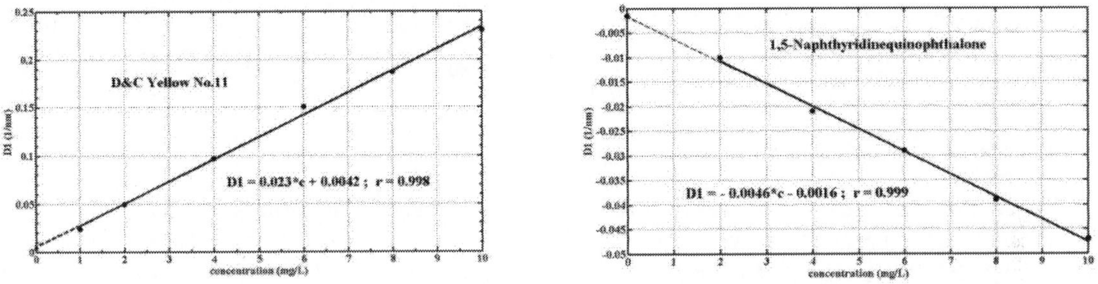

a) b)

Figure 13. The calibration line from first derivative spectra of unsulfonated subsidiary colors: a) D&C Yellow No.11; b) 1,5-Naphthyridinequinophthalone

The line parameters are calculated: the slope m^1_{Y11} and the intercept n^1_{Y11}. These values presented in Table 1 are used to determine the concentration of the D & C Yellow No. 11 dye in the analyzed samples.

Table 1. The parameters of the calibration line for D&C Yellow No.11.

m^1_{Y11} ($L \cdot mg^{-1} \cdot nm^{-1}$)	n^1_{Y11} (nm^{-1})	r_{Y11}
0.023	**0.0042**	**0.9984**

At the wavelength **λ = 294.75** nm:

$$D^1_{NQ} = 0 \text{ and } D^1 = D^1_{Y11} = m^1_{Y11} \cdot c_{Y11} + n^1_{Y11} \tag{6}$$

The D&C Yellow No.11 color concentration will be:

$$c_{Y11} = \frac{\left(D^1_{Y11} - n^1_{Y11}\right)}{m^1_{Y11}} \tag{7}$$

The wavelengths at which the 1st order derivative of D & C Yellow No. 11 dye is zero and the 1st order derivative of 1,5-NQ dye is nonzero will be identify. The value **λ = 319.95** nm are found.

For 1,5-NQ color, plot the calibration line $D^1_{NQ} = f(c)$ at this wavelength (Figure 13b).

The line parameters are calculated: the slope m^1_{NQ} and the intercept n^1_{NQ}. These values presented in Table 2 are used to determine the concentration of the 1,5-NQ dye in the analyzed samples.

Table 2. The parameters of the calibration line for 1,5-Naphthyridinequinophthalone.

m^1_{NQ} ($L \cdot mg^{-1} \cdot nm^{-1}$)	n^1_{NQ} (nm^{-1})	r_{NQ}
-0.0046	**-0.0016**	**-0.9986**

At the wavelength **λ = 319.95 nm**:

$$D^1_{Y11} = 0 \text{ and } D^1 = D^1_{NQ} = m^1_{YNQ} \cdot c_{NQ} + n^1_{NQ} \tag{8}$$

The 1,5-NQ color concentration will be:

$$c_{NQ} = \frac{\left(D^1_{NQ} - n^1_{NQ}\right)}{m^1_{NQ}} \tag{9}$$

4. Conclusions

The determination of the concentrations of the unsulphonated subsidiary colors in D & C Yellow No. 10 batches is very important because of the restrictions on their maximum admissible concentrations imposed by the Code of Federal Regulations. Small concentrations (of the order of ppm) of the compounds require the use of high precision and fineness methods, such as spectrophotometry.

The concentrations of compounds in a mixture can't be determined spectrophotometrically when their absorption bands are very close position.

Derivative spectrophotometry allows the determination of their concentrations using first order or higher order derivatives. For two-component mixtures, derivative spectrophotometry is based on the measurement of the first derivative of the mixture of two colorants at the wavelength for which one of them has the first derivative equal to zero.

The presence of the third soluble compound in both water and dichloromethane does not alter the determination of the concentrations of the unsulphonated compounds D&C Yellow No.11 and 1,5-Naphthyridinequinophthalone. This compound does not have absorption in the spectral region used to determine the concentrations of the two unsulphonated compounds.

If the concentrations of D & C Yellow No. 11 and 1,5-NQ from the D & C Yellow No. 10 analyzed certified batches do not fall within the required restrictions then those batches are rejected.

Aknowledgements

This project was supported in part by an appointment to the Research Participation Program at the Center for Food Safety and Applied Nutrition administered by the Oak Ridge Institute for Science and Education through an interagency agreement between the U.S. Department of Energy and the U.S. Food and Drug Administration.

References

[1] ***https://www.ecfr.gov/cgi-bin/text-idx?SID=3ee286332416f26a91d9e6d786a604ab&mc=true&tpl=/ecfrbrowse/Title21/21tab_02.tpl 2017 *Electronic Code of Federal Regulations (e-CFR). Title 21: Food and Drugs. Part 74. Listing of color additives subject to certification. Subpart B. Drugs*
[2] ***http://www.fda.gov/ForIndustry/ColorAdditives/RegulatoryProcessHistoricalPerspectives/default.htm 2011 *U.S. Food and Drug Administration. Color Additives: FDA's Regulatory Process and Historical Perspectives*
[3] ****Code of Federal Regulation* 2013 CFR 21, 74.1710 D&C Yellow No. 10 419
[4] ****Code of Federal Regulation* 2013 CFR 21, 74.1711 D&C Yellow No. 11 420
[5] Bell S 1983 *Internal Progress Report*, Office of Cosmetics and Colors, Food and Drug

Administration
[6] Sánchez R F and Bosch O C 2009 Recent development in derivative ultraviolet/visible absorption spectrophotometry: 2004–2008. A review, *Analytica Chimica Acta* **635** 22-44
[7] Kara D 2005 Spectrophotometric determination of Tartrazine, Riboflavine and Carmoisine in drinks by zero-order spectrophotometric method using determinant calculation and first derivative spectrophotometric method, *Asian Journal of Chemistry* **17**(2) 743-754
[8] Karpinska J, Mikołuc B and Piotrowska-Jastrzebska J 1998 Application of derivative spectrophotometry for determination of coenzyme Q_{10} in pharmaceuticals and plasma, *Journal of Pharmaceutical and Biomedical Analysis* **17** 1345-1350
[9] Karpinska J 2004 Derivative spectrophotometry-recent applications and directions of developments, *Talanta* **64** 801-822
[10] ***http://www.sigmaaldrich.com/Graphics/Supelco/objects/4800/4749.pdf Supelco, 5% DMDCS in Toluene Sylon-CT, Product Specification

IOP Conf. Series: Materials Science and Engineering **294** (2017) 012085 doi:10.1088/1757-899X/294/1/012085

Security of bottle to fill in a high pressure air

M Todic[1], T Latinovic[1], V Golubovic-Bugarski[1] and A Majstorovic[2]

[1]University of Banjaluka, Faculty of Mechanical Enginering, Stepe Stepanovica 71, 78000 Banjaluka, Bosnia and Herzegovina
[2]SC. Mech. Eng., Banja Luka, City Banja Luka, City Hall, Department for civil protection and professional territorial fire brigades

E-mail: mladen.todic@mf.unibl.org

Abstract: Charging the bottle of high pressure air isolation devices is performed by a high-pressure compressor. The charging time is in function of the compressor capacity and the intensity of the nominal pressure of the air in the bottle. However, in accident situations this time is long and therefore high-pressure accumulators are used where the filling time of the bottle of isolation apparatus has been drastically reduced. Due to the short filling time of the bottle through the air flow, there is a thermodynamic load of bottle material that could endanger the safety of users and other participants in the area. It is therefore *necessary to determine the critical parameters of the rapid charge and their intensity*.

1. Introduction

In fire-fighting interventions, firefighters use isolation apparatus for personal protection of the respiratory system. One of the essential modules of the insulating device is a tank (tank-bottle) so-called. medical air. When the reservoir is full, the pressure in it is 30 MPa, and during the use of the insulation device, this pressure decreases in the user's load function. Consumption of air ranges from 10 l / min to 100 l / min. in the function of airflow, ie. user load. Filling the bottle with medical air is most often done by high pressure compressors [1]. The charging time is in the function of the air flow (compressor capacity) and the nominal pressure, this charging in practice of a bottle of 6,8 l usually takes from 10 to 15 minutes. In accidents, time is a very important factor, because it is easier to overcome the fire when the zone where it came down is smaller, so the bottling time is very long. The user at the maximum load can intervene for 20 to 25 minutes and afterwards must withdraw from the accident zone in order to replace the bottle of air or its recharging with medical air. In order to overcome these problems i.e. shortening the charging time, charging is performed from the medical air accumulator (air banks). The battery usually has a volume of 100 dm3 to 300 dm3. The time to fill the bottle of insulation devices is about 1 minutes if it is charged from the medical air accumulator. In doing so, it is not necessary to dismount a bottle from the insulator carrier and remove the insulating device from the back of the user. So, the time to re-fire the fire is shortened. However, the question arises as to how quickly it is possible to quickly fill the bottle without the need for a new accidents during the existing one. Rapidly increasing the pressure in the bottle leads to a rapid change in the stress state in the bottle material. The voltage states are mechanical and thermal. Due to the flow of air through the valve of the bottle, there is the appearance of heat that causes temperature stresses in the bottle material. These temperature stresses are not evenly distributed, i.e. they appear in certain local places.

IOP Conf. Series: Materials Science and Engineering **294** (2017) 012085 doi:10.1088/1757-899X/294/1/012085

2. Filing time of the bottle

Bottles of insulating devices are made of steel, semi-composite and composite materials, Figures 1 and 2. The bottle material has no effect on the rate of its loading, but it has a schedule and intensity of stress conditions in the wall of the bottle. Also, the charging rate affects the arrangement and intensity of the strain-deformation state in the bottle material. Therefore, account must be taken, in particular for semi-composite and composite bottles, of possible damaged walls that can cause an accidental situation with severe consequences for the user and other participants in the bottling zone of the bottle with medical air. Due to the above mentioned research problems, they are focused on the rate of increase of pressure over time.

Figure 1. Bottles for compressed medical, steel material

Figure 2. Bottles for compressed medical, composite material

The amount of air in the bottle can be calculated over the equation

$$V_1 = V_o \frac{P_1}{p_o} \quad (1)$$

where is:

V_1- the quantity of air in the bottle, m^3
V_2 - bottle volume, m^3
P_1 - end pressure in the bottle, MPa
P_O- initial pressure in the bottle, MPa.

The mean value of the airflow volume when the bottle is filled, if the high pressure compressor charge is calculated according to the equation

$$Q = \frac{V_1}{t} \quad (2)$$

where is:

V_1- the quantity of air in the bottle, m3
Q - amount of flow in unit time, m3 / min
t - the time for which the pressure in the bottle increased to the desired (required, nominal) pressure, min.

2.1. Filling the bottle with a multistage compressor

The bottling time of the bottle in a standard way, i.e. compressors with high pressure, is in function of the volume of the bottle, the nominal pressure in the bottle and the capacity of the compressor. The nominal compressor flow in the experimental research is 250 lit./min. Table 1 shows the values of the pressure increase and the amount of air in the bottle in the function of time.

Table 1. Increase of pressure and quantity of air in the function of time-steel bottle

Time sec.	Pressure (MPa)	Quantity of air m3	Quantity of air l
0	0	0	0
60	0,2	0.012	120
120	3	0.18	180
180	7.7	0.462	462
240	11.2	0.672	672
300	15	0.9	900
360	18,6	1.116	1116
420	22.15	1.329	1329
480	25.9	1.554	1554
540	29.2	1.752	1752
600	32.7	1.962	1962

Table 2. Increase of pressure and amount of air in function of time-composite bottle

Time sec.	Pressure (MPa)	Quantity of air m3	Quantity of air l
0	0	0	0
60	0,2	0.010	100
120	4	0.272	272
180	7.2	0.4896	489
240	10.6	0.72	720
300	13.8	0.93	930
360	16.9	1.149	1140
420	19.9	1.353	1353
480	223.2	1.577	1577
540	26.2	1.781	1781
600	29.3	1.992	1992
660	32.3	2.196	2196

The real increase in air volume is about 0.210 m3 / minute, while the rated flow of the compressor is 0.25 m3 / min. The charging time of the 6 liter bottle is 10 minutes. The dependence of the increase in air pressure in the function of time is given in Figure 3.

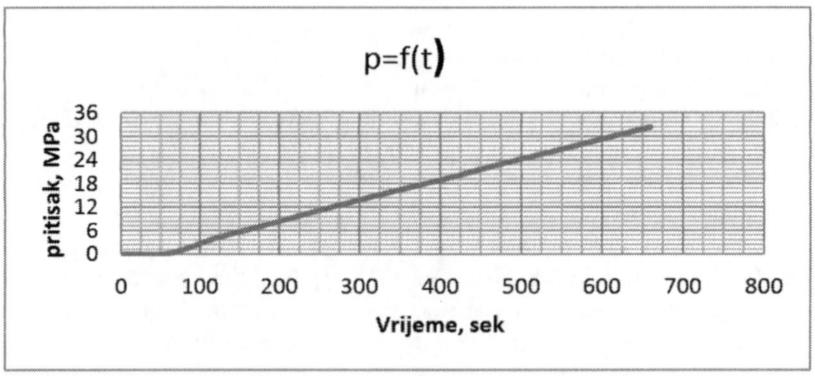

Figure 3. Pressure boost in function of time

When air flows due to air flow and friction, there is a rise in temperature in the segments of the bottle [1-4]. The recording was performed using a thermocouple, Figures 4, 5 and 6.

Figure 4. Temperature intensity at valve, comp. bottle

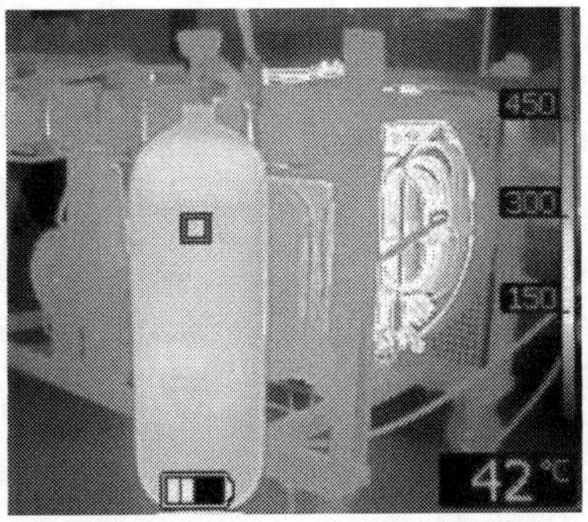

Figure 5. Temperature intensity in the upper half of the bottle, comp. bottle

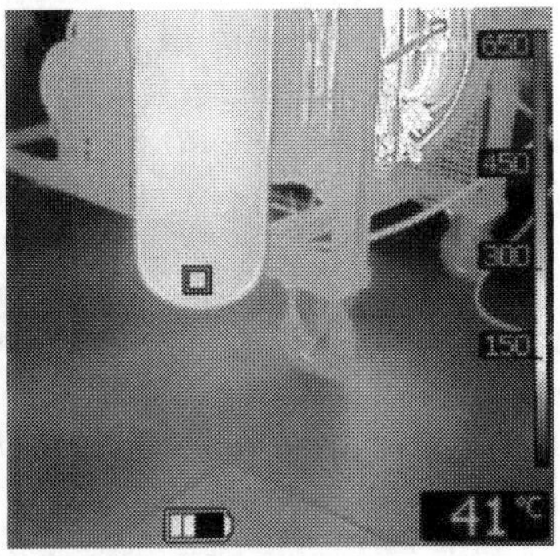

Figure 6. Temperature intensity at the bottom of the bottle, comp. bottle

There is no significant difference in the temperature of the temperature of the fillers from the steel or composite material at standard charge. These temperatures can not manifest significant temperature stresses in the walls of the bottle [3].

2.2. Filling the bottle from the medical air accumulator

Fast-loading the bottle of insulation devices is done from the battery (air bank). Banks can be stable and mobile. The bottling time at the nominal pressure (30 MPa) is up to one minute, according to the recommendations of the manufacturers of storage reservoirs for compressed air. This time can be regulated by the position of the flow valve, Figure 7. The capacity of the bank of air depends on the number and volume of individual bottles and the pressure of the air bank, Figure 8. [2].

Figure 7. Rapid charge control panel from the air bank

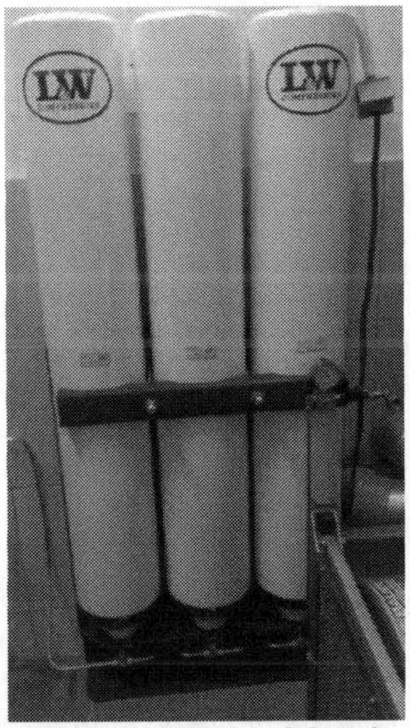

Figure 8. Bank of the air

The charging time may be shorter, as far as technical options are concerned (from 10 seconds to 60 seconds). However, there is a real danger of an accidental situation due to a dynamic (shock) change in the voltage state in the bottle material, where the value of the intensity of the voltage is far greater than the loading of the bottle with a slower increase in the pressure in the bottle. For this reason, research requires access to the identification of the intensity of the stress states in order to determine the parameters of the minimum time for fast bottling of the bottle with medical air that guarantees safety. The capacity of the air bank depends on the number and volume of individual bottles, and the pressure of the air bank. The dependence of the increase in pressure in fast charging in the function of time should be monitored with digital measuring equipment, as the bottling time of the bottle is quite short and there is a rapid increase in pressure. This rapid increase can cause an accident if the bottle is damaged during use, and the damaged has not been identified in time.

In addition, the air banks must be in conjunction with the compressor during their use, and the reason is that every bottle inflates the pressure in the air bank and the user bottle can not be filled to a nominal pressure of 30 MPa, Table 3.

Table 3. Drop of pressure and amount of air in the function of time-bank of air

Volume of the air bank l	Air pressure in the air bank (MPa)	Quantity of air m^3	Quantity of air l
150	30	4.5	45000
150	28.6	42.9	42900
150	27.2	4.08	40800
150	25.8	3.87	38700
150	24.4	3.66	36600
150	23	3.45	34500
150	21.6	3.24	32400
150	20.2	3.03	30300

The values in Table 3 are theoretical and, in real terms, a reduction of up to 10% can be counted as a result of losses resulting from manipulation during fast bottling. For this reason, it is necessary for the compressor to complement the air bank during interventions in order to be able to quickly flush the bottle at a nominal pressure.

3. Conclusion

When charging fast, it is necessary to take care of the correctness of the bottle, ie, that it was not damaged during interventions for the safety of its filling. In the event that the bottle is damaged, there may be severe consequences for the user. Fast charging should be done in one minute. Reduction of charging time leads to a reduction in safety. However, quick charging reduces the time up to 10 times in comparison to the classic bottling of the bottle. The disadvantage of the air bank is that it must be in conjunction with the compressor, if it is not continuously replenished, then it is impossible to fill more bottles at the nominal pressure as can be seen from the experimental testing of the pressure drop in the air bank in the case of quick bottling of several bottles.

References

[1] Švaic S 2006 *Equipment Under Preassure (monograph),* Faculty of Mechanical engineering and Shipbuilding, Zagreb, Croatia

[2] Majstorovic A and Todic M 2017 *The heat distribution at compression filling of bottles and expansion of medical air,* 13th International Conference of Accomplishments in Mechanical and Industrijal Engineering, University of Banja Luka, Faculty of Mechanical Engineering, Banja Luka, May 26-27, pp 863-868

[3] ***https://www.scubalife.hr/magazin/savjeti/ronilacke-boce/

[4] Majstorovic A 2015 *The Safety Using of Breathing Apparatus in Accident's Situations* Lambert Academic Publishing, Saarbrücken, Deutchland

Order management empowering entrepreneurial partnerships in the context of new technologies

M Tămăşilă, G Proştean and A Diaconescu

Politehnica University of Timisoara, Management Department, 14 Remus Street, Timisoara, Romania

E-mail: andra@publicist.com

Abstract. The expansiveness of latest generation technologies triggers manufacturers from different industry sectors more complex situations in order management with various loyal customers and occasional customers. More specifically, orders variations in logistics chain make it difficult to achieve entrepreneurial partnerships in the context of new technologies integrated into automotive and wind industry processes, which hinders getting major investments. Within this framework, the research team investigates the bottlenecks in the supply chain and indicates some rules and methods to solve the desynchronizations and fluctuations caused by the constraints of cutting-edge technologies. The paper aims to solve order management problems based on both an algorithm and an implementation in SAP. Also, in the paper, a conceptual model is created for the user whose basic task is the management of the entrepreneurial orders. Solutions identified based on the algorithm offers an order management plan by optimally adjusting inventories to deal with any kind of orders, thus achieving a profitable entrepreneurial approach between the two partners.

1. Introduction

Lately, a wide range of industries have focused their competitive strategies on mobilizing the skills and innovative capabilities that are found in the groups of customers and suppliers that make up their business supply chain. Considerable efforts have also been invested in quality management models, application of information technologies and procedural and organizational reengineering. Most of businesses have searched with confidence in supply chain management to provide new insights into new sources of competitive advantage [1]. A company that is in partnership with the supply chain can influence its performance in terms of the superiority of the new product through its entrepreneurial orientation. According to a complex study from the current literature, entrepreneurial thinking contributes positively to the superiority of the new product. Businesses can learn from their partners in the downstream supply chain, as they can directly increase performance and further strengthen the positive influence of entrepreneurial leadership on the superiority of the new product. It is worth noting that the merits of entrepreneurial guidance on the top-down of products diminish under three conditions: greater technological uncertainty, growing company age and reduced learning by supply chain partners [2]. The entrepreneurial orientation of a distributor positively mitigates the relationships between the co-operative factors and the acquisition of knowledge of a producer, implying that the strengthening of the entrepreneurial orientation of the distributor can improve the efficiency of the co-operation [3]. In academic research, within this complex field an extensive literature on partner selection criteria have been revised to show that adequate research from entrepreneurial firms is lacking. There are also various recommendations for business firms in choosing companies established as alliance partners [4]. It can be argued that firms with specialized resources, entrepreneurial orientation and local collaborations of partners gain competitive post-privatization advantage [5].

Taking into account all aspects of the extensive research study, the research team has sought to find feasible solutions for preventing variations in logistics chain orders so that it can achieve entrepreneurial partnerships.

The expansiveness of cutting-edge technologies of composite materials used in manufacturing parts with special technical characteristics both in the automotive and wind sectors has generated major changes in quality standards and parameters in delivering entrepreneurial contracts between main suppliers, intermediary suppliers and customers. Countering divergences between logistics chain links has influenced a new approach by which any chain link optimizes investment to ensure product quality and profit maximization, being challenged to make decisions by which the stocks held are properly dimensioned. The impulse offered by new technologies has enabled product variants to be diversified, offering consumers a much wider choice of products, also influencing unexpected transit of loyal customers [6]. This situation causes chain links to face increasingly sophisticated supply issues, such that making entrepreneurial partnerships creates an optimal flow of orders in the context of new technologies and offers new solutions for synchronizing variations between chain links. Cutting-edge technologies, current market requirements, and sudden changes in consumer preferences have prompted the development of ever-more complex collaborative relationships between logistics chain links for entrepreneurial order management [7].

Regardless of the challenges imposed by market changes, consumer buying behaviour and the emergence of state-of-the-art technologies, chain links in any logistics chain need to deliver a continuous flow of supply to manage a much larger order run. In this respect, the realization of the entrepreneurial partnerships in the logistic chain implies certain rules and methods, which control the variations imposed by challenges mentioned above.

2. Solving entrepreneurial orders issues

Speeding strategies of the new global economic environment have increasingly integrated more scientific concepts to solve the problems of managing entrepreneurial orders. Industrial organizations provide their competitive advantage through low costs and a high level of customer service. Basically, the necessary materials are planned to ensure the availability of the material, which is used to purchase or produce the necessary quantities in time, both for internal purposes and for sales and distribution. This process involves stocks monitoring and in particular, the creation of automatic procurement proposals for production. The realization of this process presupposes the support of the complete computerized link between production and distribution [8].

The study conceived in this paper shows how order management based on an algorithm using DBR (Drum-Buffer-Rope) philosophy and an implementation in the SAP system (Systems, Applications and Products for data processing) can manage production and processing logic to produce and assemble composite material parts needed both in the automotive and wind domain. More specifically, ERP systems are designed to connect integrally the departments of a business organization, namely to automate the flow of information from the purchasing, production, warehouse, delivery departments. ERP systems allow the procurement of raw materials to be planned in connection with production orders and deliveries available in warehouses, so that customer deliveries are made on a regular basis. Thus, the process of purchasing, storing and redistribution of raw materials or finished products is simplified. For example, in order to complete the manufacturing process, the SAP system, through Bill of Material (BOM), ensures that raw materials from the last level are purchased for the purpose of manufacturing the necessary subassemblies, then the other necessary procurement assures the final assembly of the finished product (Figure 1).

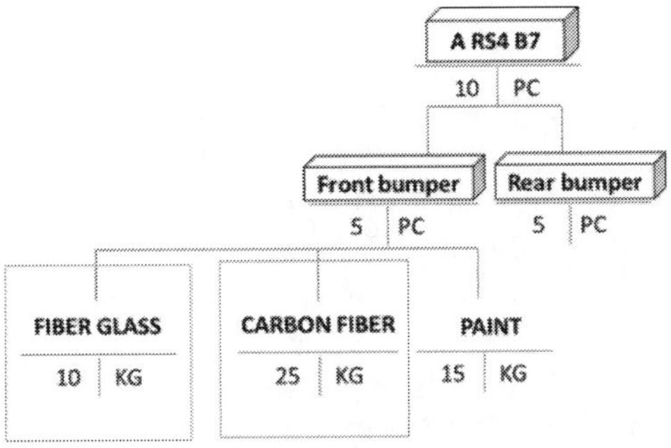

Figure 1. BOM for 10 pieces A RS4 B7

The acquisition process in SAP ERP is more efficient than traditional due to traceability and improved step automation. This facilitates purchase within the required plans, complete inventory management and reporting of all activities. In addition to facilitating access to vendor management tools, the system allows the selection and evaluation of suppliers and contracts negotiation [8].

3. Leveling entrepreneurial orders

DBR leveled entrepreneurial order along the logistics chain through safety stocks that are placed and dimensioned to protect the rhythm of each individual link in the supply chain. The purpose of using the DBR solution for dimensioning the intermediary stock is to protect the chain links against orders variations within the supply chain, provided the same rhythm is maintained at every link in the logistics chain. To illustrate this, the research team proposes a calculation model based on the DBR philosophy described by Goldratt (2003) and Cox et al (2003) through safety stock target (Tb) and safety stock status (Sb) [9-11].

The research team based on the study conducted between May and October 2016, data on the number of pieces produced by **Link X** for **Client A, Client B, Client C** and **Client D** were collected, these being current customers within a company that produces pieces from latest generation of composite materials (Figure 2).

For balancing the continuous flow of composite materials required for the production of the parts, for *Current Customers* and to ensure a logistic flow of composite materials in the production of the parts, for *Occasional Customers*, it is applied a sizing formalism by placing a safety stock at each supplier, respectively **Supplier Y, Supplier Z, Supplier W**.

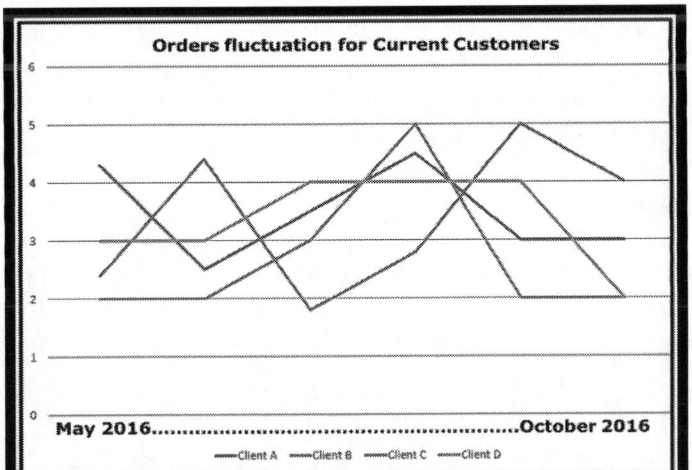

Figure 2. Orders fluctuation for *Current Customers*

For Current Customers placement of composite material orders is processed once a week. The replenishment of composite material stocks is performed at the same time from the three suppliers because the pieces manufactured by the **Chain Link X** requires combinations of composite material to obtain necesarry piece technical configuration (Table 1- Table 6). The tables presented in this paper have been adapted [12].

Table 1. Glass Fiber order from Supplier Y for *Current Customers*

Time phases for placing glass fiber order													
Min. 5 rolls	**Max. 20 rolls**												
Delivery time 1 week													
Week	**0**	**1**	**2**	**3**	**4**	**5**	**6**	**7**	**8**	**9**	**10**	**11**	**12**
Consumption	0	0	0	3	4	2	1	0	3	3	4	3	5
Stock held	10	10	10	7	3	1	20	20	17	14	10	17	12
Receive order	0	0	0	0	0	20	0	0	0	0	10	0	0
Placing new order	0	0	0	0	20	0	0	0	0	10	0	0	0

In week 3 for Client B, Customer C and Client D, 3 glass fiber rolls and 3 carbon fiber rolls were consumed to produce 10 pieces of A RS4 B7. Glass fiber requires a 100m^2 surface veil roll of 35g / m^2 and the carbon fiber weighs 200g / m^2, width 100cm / 100ml, thickness 0.32mm, and twill fabric 2/2 (Table 1, Table 2).

Table 2. Carbon Fiber order from Supplier Z for Current Customers

Time phases for placing carbon fiber order													
Min. 10 rolls	**Max. 35 rolls**												
Delivery time 1 week													
Week	**0**	**1**	**2**	**3**	**4**	**5**	**6**	**7**	**8**	**9**	**10**	**11**	**12**
Consumption	0	0	0	3	9	6	3	2	5	7	4	3	2
Stock held	15	15	15	12	3	17	14	12	7	25	21	18	16
Receive order	0	0	0	0	20	0	0	0	25	0	0	0	0
Placing new order	0	0	0	20	0	0	0	25	0	0	0	0	0

Table 3. Glass Fiber order from Supplier Z for Current Customers

Time phases for placing glass fiber order													
Min. 15 rolls	**Max. 25 rolls**												
Delivery time 1 week													
Week	**0**	**1**	**2**	**3**	**4**	**5**	**6**	**7**	**8**	**9**	**10**	**11**	**12**
Consumption	0	0	0	3	4	0	3	5	3	3	4	0	0
Stock held	10	10	10	7	3	3	0	15	12	9	5	15	15
Receive order	0	0	0	0	0	0	20	0	0	0	10	0	0
Placing new order	0	0	0	0	0	20	0	0	0	10	0	0	0

Table 4. Carbon Fiber order from Supplier W for Current Customers

Time phases for placing carbon fiber order													
Min. Max.													
10 rolls 25 rolls													
Delivery time 1 week													
Week	**0**	**1**	**2**	**3**	**4**	**5**	**6**	**7**	**8**	**9**	**10**	**11**	**12**
Consumption	0	3	4	3	5	6	3	9	2	3	2	2	4
Stock held	25	22	18	15	10	4	1	17	15	12	10	8	24
Receive order	0	0	0	0	0	0	25	0	0	0	0	20	0
Placing new order	0	0	0	0	0	25	0	0	0	0	20	0	0

Table 5. Glass Fiber order from Supplier W for Current Customers

Time phases for placing glass fiber order													
Min. Max.													
10 rolls 25 rolls													
Delivery time 1 week													
Week	**0**	**1**	**2**	**3**	**4**	**5**	**6**	**7**	**8**	**9**	**10**	**11**	**12**
Consumption	0	0	0	3	2	0	8	0	5	0	4	3	3
Stock held	18	18	18	15	13	13	5	25	20	20	16	13	25
Receive order	0	0	0	0	0	0	20	0	0	0	0	0	0
Placing new order	0	0	0	0	0	0	0	0	0	0	0	15	0

Table 6. Kevlar order from Supplier W for Current Customers

Time phases for placing kevlar order													
Min. Max.													
10 rolls 20 rolls													
Delivery time 1 week													
Week	**0**	**1**	**2**	**3**	**4**	**5**	**6**	**7**	**8**	**9**	**10**	**11**	**12**
Consumption	0	3	4	3	5	5	3	0	5	3	2	0	1
Stock held	20	17	13	10	5	0	12	12	6	3	11	11	10
Receive order	0	0	0	0	0	15	0	0	0	10	0	0	10
Placing new order	0	0	0	0	15	0	0	0	10	0	0	10	0

Chain Link X decide on the replenishment time for safety stock and takes into account following the three notifications:

-The **green signal** does not create another order.

- The **yellow signal** creates the order and sends it to the suppliers.
- At the **red signal** the safety stock has already been replenished.

The replenishment time for the safety stock is calculated through **Safety Stock Target (Tb)** and **Safety Stock Status (Sb)** [9], [10]. The algorithm for dimensioning the safety stock is based on the formula:

$$\boldsymbol{Tb = DLT \times \frac{MO}{U^{DLT}} \times SP} \tag{1}$$

Safety Stock Target Supplier Y (Tb_Y) for Glass fiber (G)

$$\mathrm{Tb}^{YG} = 1\,week \frac{25}{1\,week} \times 150\% = 37{,}5$$

Safety Stock Target Supplier Z (Tb_Z) for Carbon fiber (C) and Glass fiber (G).

$$Tb^{ZC} = 1\,week \frac{30}{1\,week} \times 150\% = 45$$

$$Tb^{ZG} = 1\,week \frac{20}{1\,week} \times 150\% = 30$$

Safety Stock Target Supplier W (Tb_W) for Carbon fiber (C), Glass fiber (G) and Kevlar (K).

$$Tb^{WC} = 1\,week \frac{15}{1\,week} \times 150\% = 22{,}5$$

$$Tb^{WG} = 1\,week \frac{30}{1\,week} \times 150\% = 45$$

$$Tb^{WK} = 1\,week \frac{25}{1\,week} \times 150\% = 37{,}5$$

Safety Stock Status (Sb) it is refers to the level of security stock at one time and is based on the formulas:

$$\boldsymbol{Percentage}\,\% \times \mathbf{Sb}^{yellow} = \frac{MO}{Tb} \times \mathbf{100} \tag{2}$$

$$\mathbf{MO}^{daily} = \frac{MO}{7} \tag{3}$$

$$\boldsymbol{Percentage}\,\% \times \mathbf{Sb}^{red} = \frac{\mathbf{MO}^{daily}}{Tb} \times \mathbf{100} \tag{4}$$

It is calculated

• the Stock Status for yellow color (Sbyellow) (formula 2),

• the Stock Status for **red** color (Sbred) (formula 4) which is calculated according to the average daily order according to formula 3.

Table 7. Sizing the safety stock for Supplier Y for Current Customers

Supplier Y	MO	DLT	SP	Tb	Sb_{yellow}	Consumption rhythm	Sb_{red}
Glass fiber	25	1 week	150%	37,5	**66,66%**	3,57/zi	**9,52%**

Table 8. Sizing the safety stock for Supplier Z for Current Customers

Supplier Z	MO	DLT	SP	Tb	Sb_{yellow}	Consumption rhythm	Sb_{red}
Carbon fiber	30	1 week	150%	45	**66,66%**	4,28/zi	**9,51%**
Glass fiber	20	1 week	150%	30	**66,66%**	2,85/zi	**9,5%**

Table 9. Sizing the safety stock for Supplier W for Current Customers

Supplier W	MO	DLT	SP	Tb	Sb_{yellow}	Consumption rhythm	Sb_{red}
Carbon fiber	15	1 week	150%	22,5	**66,66%**	2,14/zi	**9,51%**
Glass fiber	30	1 week	150%	45	**66,66%**	4,28/zi	**9,51%**
Kevlar	25	1 week	150%	37,5	**66,66%**	3,57/zi	**9,52%**

For *Occasional Customers*, placement of composite material orders is processed every 2 weeks. As with *Current Customers* orders, the replenishment of composite materials is made at the same time from the three suppliers because the parts manufactured by **Chain Link X** requires combinations of composite material to obtain the required technical configuration. Placement of orders every two weeks does not increase the investment and the materials are not kept in stock for too long. (Table 10 – Table 15)

Table 10. Glass Fiber order from Supplier Y for Occasional Customers

Time phases for placing glass fiber order													
Min. 5 rolls / Max. 20 rolls													
Delivery time 2 weeks													
Week	0	1	2	3	4	5	6	7	8	9	10	11	12
Consumption	0	0	0	3	4	2	1	0	3	3	4	3	5
Consumption for Occasional Customers								5					
Stock held	10	10	10	7	3	1	20	30	22	19	25	22	17
							30						
Receive order	0	0	0	0	0	20	0	0	0	0	10	0	0
						10							
Placing new order	0	0	0	0	20	0	0	0	0	10	0	0	0
Placing occasional order				10									

Table 11. Carbon Fiber order from Supplier Z for Occasional Customers

Time phases for placing carbon fiber order													
Min. 10 rolls / Max. 35 rolls													
Delivery time 2 weeks													
Week	0	1	2	3	4	5	6	7	8	9	10	11	12
Consumption	0	0	0	3	9	6	3	2	5	7	4	3	2
Consumption for Occasional Customers							3			5			
Stock held	15	15	15	12	10	24	18	16	11	24	25	22	20
Receive order	0	0	0	0	20	0	0	0	25	0	0	0	0
				7						5			
Placing new order	0	0	0	20	0	0	0	25	0	0	0	0	0
Placing occasional order		7						5					

Table 12. Glass Fiber order from Supplier Z for Occasional Customers

Time phases for placing glass fiber order													
Min. 15 rolls / Max. 25 rolls													
Delivery time 2 weeks													
Week	0	1	2	3	4	5	6	7	8	9	10	11	12
Consumption	0	0	0	3	4	0	3	5	3	3	4	0	0
Consumption for Occasional Customers						5				3			
Stock held	10	10	10	7	3	3	0	15	12	9	5	15	20
Receive order	0	0	0	0	0	0	20	0	0	0	10	0	0

						15						**5**	
Placing new order	0	0	0	0	0	20	0	0	0	10	0	0	0
Placing occasional order				**15**						**5**			

Table 13. Carbon Fiber order from Supplier W for Occasional Customers

Time phases for placing carbon fiber order													
Min. 10 rolls	**Max.** 25 rolls												
Delivery time 2 weeks													
Week	**0**	**1**	**2**	**3**	**4**	**5**	**6**	**7**	**8**	**9**	**10**	**11**	**12**
Consumption	0	3	4	3	5	6	3	9	2	3	2	2	4
Consumption for Occasional Customers											**7**	**15**	
Stock held	25	22	18	15	10	4	1	17	15	12	**23**	**6**	**22**
Receive order	0	0	0	0	0	0	25	0	0	0 **20**	0	20	0
Placing new order	0	0	0	0	0	25	0	0	0	0	20	0	0
Placing occasional order								**20**					

Table 14. Glass Fiber order from Supplier W for Occasional Customers

Time phases for placing glass fiber order													
Min. 10 rolls	**Max.** 25 rolls												
Delivery time 2 weeks													
Week	**0**	**1**	**2**	**3**	**4**	**5**	**6**	**7**	**8**	**9**	**10**	**11**	**12**
Consumption	0	0	0	3	2	0	8	0	5	0	4	3	3
Consumption for Occasional Customers										**12**			
Stock held	18	18	18	15	13	13	5	25	20	**20**	**16**	**13**	**25**
Receive order	0	0	0	0	0	0	20	0	0 **12**	0	0	0	0
Placing new order	0	0	0	0	0	0	0	0	0	0	0	15	0
Placing occasional order							**12**						

Table 15. Kevlar order from Supplier W for Occasional Customers

Time phases for placing kevlar order													
Min. 10 rolls	**Max.** 20 rolls												
Delivery time 2 weeks													
Week	**0**	**1**	**2**	**3**	**4**	**5**	**6**	**7**	**8**	**9**	**10**	**11**	**12**
Consumption	0	3	4	3	5	5	3	0	5	3	2	0	1
Consumption								**5**		**5**			

for Occasional Customers													
Stock held	20	17	13	10	5	0	12	**17**	**12**	**4**	**12**	**12**	**11**
Receive order	0	0	0	0	0	15	0 **10**	0	0	10	0	0	10
Placing new order	0	0	0	0	15	0	0	0	10	0	0	10	0
Placing occasional order					**10**								

Placement of composite material orders is processed every 2 weeks because this is the way material is optimally rolled out and Chain Link X has no excess inventory for a long time. The replenishment time for the safety stock for occasional order is calculated through **Safety Stock Target (Tb)** and **Safety Stock Status (Sb)** [9], [10]. For this case are used the formulas for replenishment time of safety stock also applied for Current Customers (Formulas 1, 2, 3, 4).

Table 16. Sizing the safety stock for Supplier Y for Occasional Customers

Supplier Y	**MO**	**DLT**	**SP**	**Tb**	**Sb_{yellow}**	**Consumption rhythm**	**Sb_{red}**
Glass fiber	25	2 weeks	150%	37,5	**66,66%**	1,78/zi	**14,26%**

Table 17. Sizing the safety stock for Supplier Z for Occasional Customers

Supplier Z	**MO**	**DLT**	**SP**	**Tb**	**Sb_{yellow}**	**Consumption rhythm**	**Sb_{red}**
Carbon fiber	30	2 weeks	150%	45	**66,66%**	2,14/zi	**14,26%**
Glass fiber	20	2 weeks	150%	30	**66,66%**	1,42/zi	**14,23%**

Table 18. Sizing the safety stock for Supplier W for Occasional Customers

Supplier W	**MO**	**DLT**	**SP**	**Tb**	**Sb_{yellow}**	**Consumption rhythm**	**Sb_{red}**
Carbon fiber	15	2 weeks	150%	22,5	**66,66%**	1,07/zi	**14,26%**
Glass fiber	30	2 weeks	150%	45	**66,66%**	2,14/zi	**14,26%**
Kevlar	25	2 weeks	150%	37,5	**66,66%**	1,78/zi	**14,26%**

For Supplier W dimensioning the safety stock for ***Kevlar,*** with a Tb of 25 orders, notify that yellow stock is 66.66%. The signal recorded for red stock at 14.26% requires emergency to refill the stock. Specifically, reaching 14.26% is the moment when the rope valve is pulled and acceleration plans for the material reception needs to be executed immediately. (Table 18) Orders for suppliers are sized with an average value, taking into account the average consumption and frequency set, for placing the order. In this way, through this control mechanism, the material is available at the right time for ***Occasional Customers***, while offering material availability for ***Current Customers***.

The dimensioning the safety stock minimizes "Customer Risk" by providing the right mix of composite material at the right time for both ***Current Customers*** and ***Occasional Customers***.

4. Conclusions and future work

The algorithm provides a synchronized process regarding the frequency of composite materials used, through which the intermediary supplier **Chain Link X** can offer at any moment the most appropriate technical combination. The applicability of the proposed solution offers a customized model of collaboration between the three selected suppliers and the intermediary supplier **Chain Link X** for ***Current Customers and Occasional Customers***.

IOP Conf. Series: Materials Science and Engineering **294** (2017) 012086 doi:10.1088/1757-899X/294/1/012086

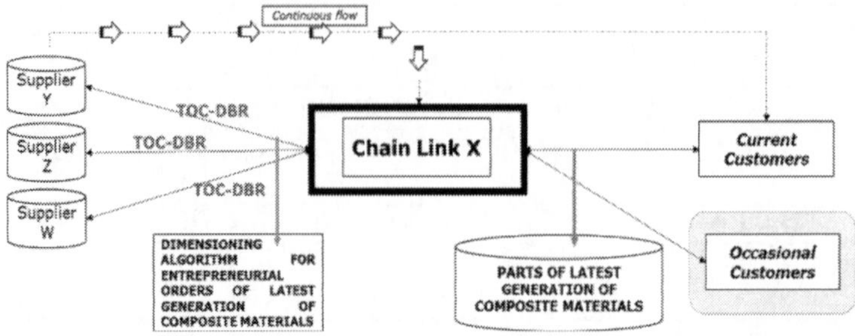

Figure 3. Dimensionig model for Chain Link X

The algorithm solved the dimensioning of certain safety stocks of the necessary materials for the production of the parts, ensuring a continuous flow of special raw materials for ***Current customers*** and ***Occasional Customers*** and contributed to the choice of the optimal combination of composite material in making a specific piece for automotive industry (Figure 3). The result of using the algorithm ensures material rolling, reduces storage costs and minimizes product risk.

References

[1] Ross D F 2016 *Introduction to e-Supply Chain Management: Engaging Technology to Build Market-Winning Business Partnerships*, CRC Press

[2] Bouncken R B, Plüschke B D and Pesch R N 2014 *Entrepreneurial Orientation in Supply Chain Partnerships - A Driving Force for Innovation*

[3] Yuan li, Yi Liu and Heng Liu 2011 Co-opetition, distributor's entrepreneurial orientation and manufacturer's knowledge acquisition, *Journal of Operations Management* **29**(1-2) 128-142

[4] Das T K and He I Y 2006 Entrepreneurial firms in search of established partners: review and recommendations, *International Journal of Entrepreneurial Behavior & Research* **12**(3) 114-143

[5] Doh J P 2000 Entrepreneurial Privatization Strategies: Order of Entry and Local Partner Collaboration as Sources of Competitive Advantage, *The Academy of Management Review* **25**(3) 551-571

[6] ***https://www.accenture.com/t20150523T042739__w__/hun/_acnmedia/Accenture/Conversion-Assets/DotCom/Documents/Global/PDF/Industries_2/Accenture-Next-Generation-Collaboration-New-Frontier-Shopper-Value-Industry-Growth.pdf

[7] ***https://www.accenture.com/t20160919T053639__w__/us-en/_acnmedia/PDF-21/Accenture-Digital-Collaboration-Retailers-Manufacturers-Accenture-Analytics.pdf

[8] Prostean G, Vasar C and Badea A 2016 *Logistics scenario for wind turbine assembly based on ERP*, *Advances in Intelligent Systems and Computing* **634** 194-200

[9] Cox J F, Blackstone J H and Schleier J G 2003 *Managing Operations: A Focus on Excellence*, Éditeur North River Press, Vol 2

[10] Goldratt E M 2003 *Production the TOC way* (revised edition), North River Press

[11] Goldratt E M and Goldratt A 2003 *TOC insights into operations*, Goldratt's Marketing Group

[12] Badea A 2017 *Considerations regarding supply chain collaboration. The case of renewable energy resources projects, in isolated regime*, Politehnica University Timişoara, Doctoral Thesis

Analysis of whole-body vibration on rheological models for tissues

A Neamţu, D Simoiu, E Nyaguly, I Crastiu and L Bereteu

Politehnica University Timisoara, Mechanical and Strength Materials Department, Mihai Viteazul str., no. 1, 300222 Timisoara, Romania

E-mail: liviu.bereteu@yahoo.com

Abstract. Whole body vibrations have become a very popular method in recent years, both in physical therapy and in sports. This popularity is due to the fact that, as a result of analyzing the groups of subjects, the effects of small amplitude vibration and low frequency vibration, it was found an increase in the force developed by the feet, a hardening of bone strength or an increase in bone density. In this paper we propose to give a possible explanation of the stress relieving in muscle and/or bone after whole body vibration treatment. To do this we consider some rheological models which after whole body vibrations and after the analysis of their response lead to various experiments.

1. Introduction

The results of the research on the influence of vibrations on the stress relaxation are controversial because there are no conclusive measurement methods whose effects can be demonstrated. Applying high frequency harmonic strain has been shown to stimulate bone resistance in different animals. However, the effects of whole body vibration on the human skeleton have rarely been studied. The effect of vibration on the whole body is difficult to quantify because the mechanical modelling of vibration behaviour of the whole body is not unitary but also offers many possibilities [1].

In another paper [2], it was hypothesized that through a non-contact vibration device, qualitative changes could be detected in a periodontal ligament and alveolar bone, identifying the mechanical parameters of the system: tooth, periodontal ligament and alveolar bone. In this case, the mechanical modeling of the analyzed system was the starting point of the study.

Whole-body vibration is a neuromuscular training method that recently has been developed and which increase strength compared with resistance training [3].

As a biological material, muscle tissue and bone tissue exhibit a complex microstructure and micro components, which result in complicated mechanical properties. Because of the viscoelastic nature of collagen fibers in the bone matrix, bone itself has remarkable viscoelasticity [4-6].

The vast majority of research has an experimental character and has not led to a clear view of the possibility of using vibrations to stress relaxation. The possibilities of reducing this stress are described in the paper [7] from the stress and deformation dependence, a characteristic of the material. Stress relaxation is based on the reinforcement of the material, meaning that by the influence of vibrations, the flow limit can be exceeded.

In this paper we propose a rheological model describing the elasto-plastic behavior with linear reinforcement. By generalization some elasto-plastic models of solid bodies will be analyzed, which show rigidity characteristics, following the effect of stress relief under the influence of vibrations.

2. Description of elasto-plastic behaviour of solid bodies

After reaching the flow limit, in many solid bodies there is a consolidation (stiffening) followed by increase the resistance. This observation is also valid for bone tissue. This means that the stress-strain diagram $\sigma-\varepsilon$ can be considered to consist of linear sections in both elastic behaviour and plastic behaviour. The slope of these segments is equal to the modulus of elasticity E for the first domain, equal to that of the plasticity E_p in the second domain (Figure 1). In the case of the monotone stress increase, for the elastic domain, this can be written analytically by the relation:

$$\sigma=\varepsilon E,\ \text{for}\ \sigma\leq\sigma_c\,. \tag{1}$$

In the equation 1, σ_c is the value of the stress to which the rheological model has a purely elastic behaviour. For the domain in which the rheological model has plastic behaviour it can be written:

$$\sigma=E_p\varepsilon+\frac{E-E_p}{E}\sigma_c,\ \text{for}\ \sigma>\sigma_c\,. \tag{2}$$

If we denote with σ_1 the maximum stress reached, then the return will be elastic and can be represented by the following relation:

$$\sigma=\sigma_1-E\left(\varepsilon_1-\varepsilon_2\right). \tag{3}$$

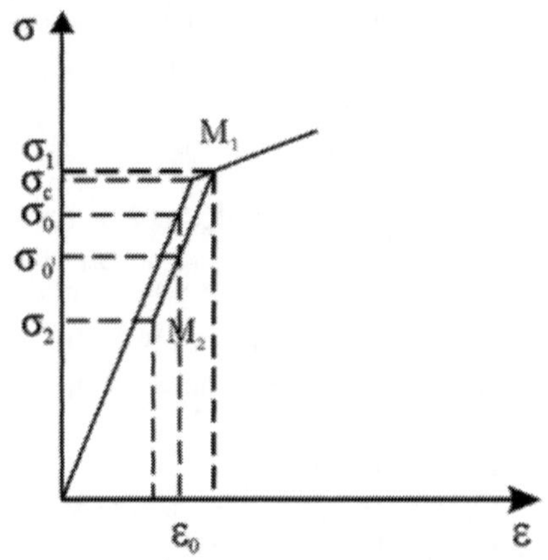

Figure 1. Elastic feature with stiffening

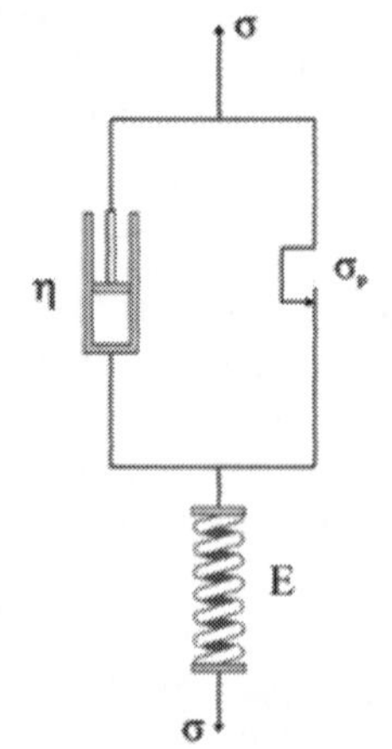

Figure 2. Elasto-plastic model with linear stiffening

The rheological model of elasto-plastic type with linear reinforcement is represented in Figure 2 and consists of two Hooke elements with E_1 and E_2 constants, as well as the Saint-Venant plastic element.

For the elastic domain in which $\sigma\leq\sigma_c$ the element E_2 does not work, the behavior of the pattern is purely elastic and corresponds to the E_1 element requirement. If $\sigma>\sigma_c$ then the element E_2 is also in operation being connected in series with E_1. Since the elastic modulus of the elasto-plastic solid model is E_1, and in the plastic domain is E_p, resulting by serial binding of the elements E_1 and E_2, (Figure 3), i.e.

$$\frac{1}{E_1}+\frac{1}{E_2}=\frac{1}{E_p}. \tag{4}$$

It can be concluded that the modulus of elasticity of the first element is:

$$E_1 = E, \tag{5}$$

and the second element is

$$\frac{1}{E_2}=\frac{1}{E_p}-\frac{1}{E}. \tag{6}$$

For verification, the constitutive equations can be written, having the following form:

$$\varepsilon=\frac{\sigma}{E_1}, \text{ for } \sigma \le \sigma_c, \tag{7}$$

$$\varepsilon=\frac{\sigma-\sigma_c}{E_2}+\frac{\sigma}{E_1}, \text{ for } \sigma > \sigma_c. \tag{8}$$

The conditions given by equations (5) and (6) must be justified in order to be in accordance with equations (1) and (2). Otherwise, meaning that the stress σ is decreasing, the tension of the Saint-Venant element would fall below the value of σ_c, which would lead to the deformation of the element E_2.

3. Elasto-plastic rheological models with viscous properties

The rheological model describing the elasto-plastic behaviour of viscous bodies is that of Goldsmith [8]. It is composed of two Hooke elements, a Newton element and a Saint-Venant element fused as in Figure 3. Until a stress is applied $\sigma \le \sigma_c$, the material behaves elastically, while at a higher stress it will behave as plastic.

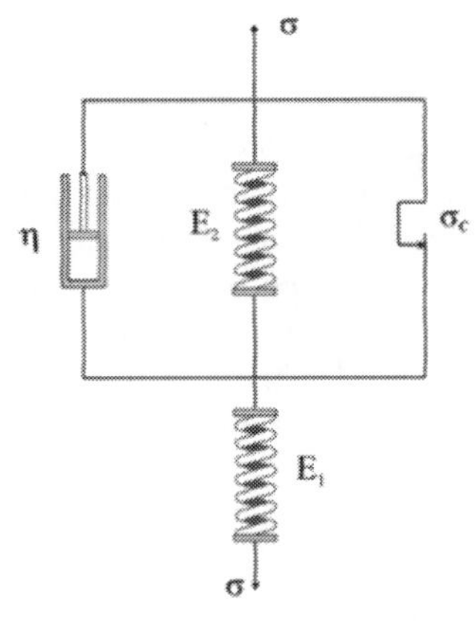

Figure 3. Goldsmith model of tissue

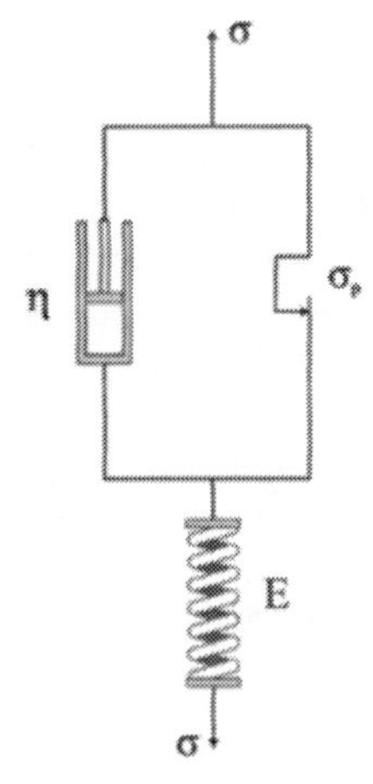

Figure 4. Bingham model of tissue

Another model describing a body with elasto-viscoplastic characteristics is that of Bingham as in Figure 4. To write the constitutive relationships we use a modified Bingham model, which takes into account the consolidation phenomenon.

In the domain of plastic deformations, the behaviour of the model can be written according to the relation:

$$\sigma_p = \sigma_c + E_p \varepsilon_p , \tag{9}$$

where σ_p is the value of the stress corresponding to the plastic deformation ε_p The constituent equations are:

$$\varepsilon = \frac{\sigma}{E}, \text{ for } \sigma \leq \sigma_c ; \tag{10}$$

$$\dot{\varepsilon} = \frac{\dot{\sigma}}{E} + \frac{\sigma - \sigma_c}{\eta}, \text{ for } \sigma > \sigma_c , \tag{11}$$

where η is the viscosity of damper element.
Plastic deformation can be obtained from total deformation if elastic deformation decreases

$$\varepsilon_p = \varepsilon - \frac{\sigma}{E}. \tag{12}$$

If this relationship is taken into account in the equation (10), it becomes the following:

$$\dot{\varepsilon} + \frac{Ep}{\eta} \cdot \varepsilon = \frac{\dot{\sigma}}{E} + \frac{1}{\eta}\left[\left(1 + \frac{Ep}{E}\right)\sigma - \sigma_c\right]. \tag{13}$$

4. Elasto-plastic rheological models with viscous properties

It is considered that the deformation following the application of the vibrations at a given point can be transcribed by the harmonic law:

$$\varepsilon = \varepsilon_0 + a \sin\omega.t , \tag{14}$$

where ε_0 represents the strain according to the initial state of the stress σ_0, and a is the small amplitude of the applied vibration and ω the circularly frequency ($\omega = 2\pi f$ is related to its frequency f). In the strain field of plastic deformation the rheological behaviour of the tissue can be described by equation (13) and the return will occur elastically to correspond to the experimental results. So the equation (13) and a harmonic variation of strain given by equation (14), can also be written as follows:

$$\dot{\sigma} + \frac{E + Ep}{\eta}\sigma = \frac{E.a}{\eta}\sqrt{E_p^{\,2} + \eta^2\omega^2}.\cos(\omega t - \varphi) + \frac{E_p\sigma_0 + E\sigma_c}{\eta}. \tag{15}$$

The general solution of this equation is:

$$\sigma(t) = e^{-\alpha t}\left\{C + \int_0^t \left[\sqrt{E_p^{\,2} + \eta^2\omega^2}.\cos(\omega t - \varphi) + \beta\right] e^{\alpha t} dt\right\}, \tag{16}$$

where the following notations were made

$$\alpha = \frac{E + Ep}{\eta}; \tag{17}$$

$$\beta = \frac{E_p.\sigma_0 + E.\sigma_c}{\eta};\tag{18}$$

$$tg\varphi = \frac{Ep}{\eta.\omega}.\tag{19}$$

Depending on these relationships the equation (15) may also be written as follows:

$$\sigma(t) = (C - \frac{\beta}{\alpha} - \frac{E}{\sqrt{Ep^2 + \eta^2.\omega^2}})e^{-\alpha t} + \frac{Ea\omega}{\eta(\omega^2 + \alpha^2)}\left[\sin(\omega t - \varphi) + \frac{\alpha}{\omega}\cos(\omega t - \varphi)\right] + \frac{\beta}{\alpha}.\tag{20}$$

The integration constant C is calculated for each cycle in the conditions of stress and strain continuity in the transition from the elastic to the plastic state. It is found that in the presence of vibrations the tension component decreases exponentially. It can be concluded that the model shown in Figure 4 shows a relaxation phenomenon.

After a large number of cycles, according to the relaxation time for this model, the maximum value of the tension will correspond to the maximum deformation value, as follows:

$$\sigma_{max} = \frac{E_p\sigma_0 + E.\sigma c}{E + Ep} + \frac{E.a\left[\eta^2\omega^2 + (E + E_p)E_p\right]}{\eta^2\omega^2 + (E_p + E)^2}.\tag{21}$$

On return, the process will be readily elastic along the straight line with the slope value equal to the elastic modulus E, therefore:

$$\frac{\sigma_{max} - \sigma_{min}}{2a} = E,\tag{22}$$

or

$$\sigma_{min} = \sigma_{max} - 2aE.\tag{23}$$

After stopping the vibration phenomenon, after several cycles, the voltage in the rheological model considered will stabilize in the average position of the extreme states of tension created by the subjection to harmonic vibrations, marked with $\sigma_0^{'}$

$$\sigma_0^{'} = \frac{\sigma_{max} + \sigma_{min}}{2},\tag{24}$$

or

$$\sigma_0^{'} = \frac{E_p\sigma_0 + E\sigma_c}{\eta} - \frac{E^2a(E_p + E)}{\eta^2\omega^2 + (E_p + E)^2}.\tag{25}$$

If it is assumed that the pretension of the viscoelatioplastic body tissue model is precisely at the limit of its elastic behaviour, meaning it corresponds to the limit of the plastic deformation input area $(\sigma_0 = \sigma_c)$ the end of the vibration process of tensing, the stress having the value given by the expression:

$$\sigma_0^{'} = \sigma_0 - \frac{E^2a(E_p + E)}{\eta^2\omega^2 + (E_p + E)^2} = \sigma_0 - \frac{E^2a(E_p + E)}{4\pi^2f^2\eta^2 + (E_p + E)^2} < \sigma_0.\tag{26}$$

The correct evaluation of the residual stress drop by applying to the studied model harmonic vibrations can be done only after identifying the parameters of the rheological model, a model that can be identified according to the procedure in the paper [9].

5. Whole body vibration system and method

Whole Body Vibration (WBV) is a technique used to increase neuromuscular performance, including muscle strength, power, movement velocity and flexibility, among various patient groups, including athletes [10-12]. The focus of other investigations was to determine the acute impact of a bout of whole-body vibration on athletic performance [13-15].

This technique has begun to be applied, with the long-term stay of cosmonauts in space, after which a decrease in muscle strength has been observed, but also a decrease in bone density [16].

The systems used for body vibration are based on the principles of acceleration training to stimulate the body's natural response to vibration. The WBV system creates vibrations of controlled frequencies and amplitudes, and in the case of law-shaped sinusoidal harmonic vibrations (14) they will produce amplitude of acceleration at the platform level given by the relationship:

$$A_{max} = \omega^2 a = 4\pi^2 f^2 a \approx 4 g f^2 a, \qquad (27)$$

where by A_{max} has been noted by the maximum acceleration, a is the amplitude of the vibration, f is the frequency, and g is the numerical value of the gravitational acceleration. The mechanical characteristics of the plate: the vibration amplitude and frequency can be changed, thus obtaining different levels of acceleration. Acceleration levels usually compare to the gravitational acceleration $g=9.81 m/s^2$. These accelerations will produce instability of the whole body, and while these vibrations transmit energy in waves, a variety of muscles contract and relax subconsciously throughout the body. This rapid, contraction and muscle relaxation cycle is what makes exercises with whole body vibration systems so effective.

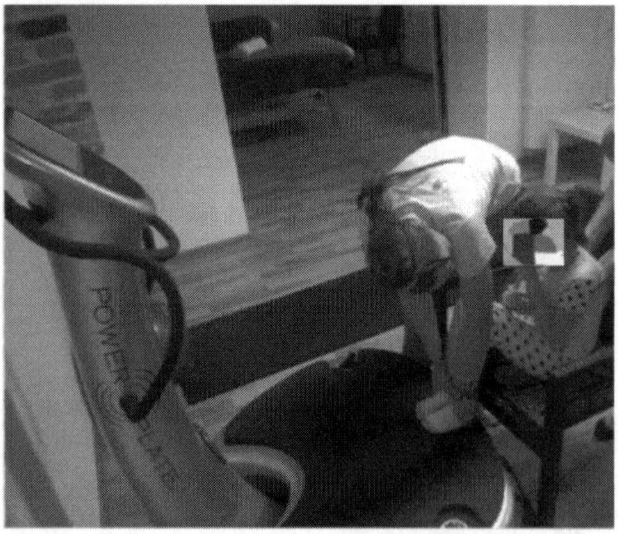

Figure 5. Exercise from the sitting position, keeping the heels on the platform

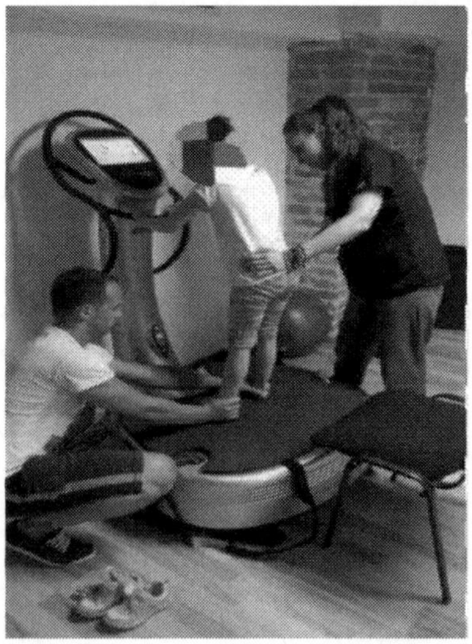

Figure 6. Exercise in orthostatic position and maintain balance through platform handles

A fraction of the kinetic energy introduced by the vibratory platform into the human body is found in other forms of body energy (kinetics or potential), and the other fraction will dissipate by heat. Dissipated energy is actually energy absorbed by the body as heat:

$$E_d = f\int_0^{\frac{2\pi}{\omega}} \sigma d\varepsilon = f\int_0^{\frac{2\pi}{\omega}} \sigma \frac{d\varepsilon}{dt} dt = f\int_0^{\frac{2\pi}{\omega}} \sigma \dot{\varepsilon} dt = \pi f E \varepsilon_1^2 \sin\varphi. \quad (28)$$

The value of the integral in equation (28) represents the energy dissipated over a single cycle, and the value of E_d expression represents the energy dissipated in one second.

From the analysis of the equation (27) it can be observed that the amplitudes of the acceleration at the level of the platform can be modified by the independent variables, the amplitude of the vibrations and the frequency. Whole body vibrations have been applied with amplitude between 2.5 mm and 5.5 mm, with a 0.5 mm step, and frequency between 5Hz and 30Hz with a 5Hz step. By modifying the two independent parameters, vibration amplitude and frequency could be worked in a range of accelerations ranging from 0.25g to 6.6g. Due to transmissibility, these accelerations are different at some body parts, and are harder to be highlighted by measurements. Given the objective of the investigations, it can be considered that the vibration level at the vibratory platform is the same as the vibration level at the toe.

A group of few children, insufficient to form a lot of statistical analysis, with problems walking on the toes of the legs was followed by the effect of physical therapy through light intensity exercises. For this purpose, the application of the vibration treatment was conducted using a Power Plate body vibration platform (Power Plate North America Inc., Northbrook, IL). For this group of children was conducted several types of exercises.

The first mode of vibration was from the seated position, maintaining the plantar vault on the surface of the vibrating platform for 30 seconds at a frequency of 5 Hz as can be seen in Figure 5. The exercise was repeated six times, with ten seconds pause between exercises, maintaining the amplitude of the platform vibration at 2.5 mm, but increasing the frequency with the 5 Hz step. In the following days, the exercise was repeated, maintaining the same amplitude, but increased by the 0.5 mm increments daily.

In the next exercise, the patient is in the seated position, flexes the leg and attempts to stretch the Achilles tendon so that the heel remains on the vibratory platform.

In another exercise, the patient is placed orthostatically on the vibratory platform and is maintained in this position with the support of the platform handles (Figure 6). Independent variables, exposure time, vibration amplitude and frequency remain the same as in the first exercise.

In other exercises the patient was in a sitting position to standing on the vibration platform, maintaining the calcaneus on the platform or in ventral cubitus and dorsal cubitus.

For all patients, there was an improvement in walking and other functional activities after only six weeks. In the case of a girl, who was at the beginning of her legs in a major adduction that made it impossible to walk, she found amazing progress, making her first steps without help.

6. Conclusions

Based on the analysis of the formula given by the equation (23), after the stopping of the vibration action, the rheological model remains tense at a lower stress value than the initial one. Therefore, the beneficial effect of vibrations on the whole body is not only a consequence of well-being or a psychological consequence. It should also be noted that the magnitude of the amplitude induces a decrease in the residual stress value while the vibration frequency must be small.

As can be seen from both the rheological models, based on the analytical calculations, and the results of the exercises on the vibratory platform, it is possible to obtain muscle stress relief and flexibility of the tendons.

Future research will focus on the statistical tracking of effects of independent variables, vibration amplitude and frequency, by walking analysis.

On the one hand, vibrations have been studied for a long time, for its dangerous effects on humans. On the other hand, recent works have suggested that mechanical stimulation of the low frequency and low amplitude of the human body is a safe and effective way to strengthen and relax the musculoskeletal structures.

References

[1] Verschueren S M P, Roelants M, Delecluse Ch, Swinnen S, Vanderschueren D and Boonen S 2004 Effect of 6-Month Whole Body Vibration Training on Hip Density, Muscle Strength, and Postural Control in Postmenopausal Women: A Randomized Controlled Pilot Study, *Journal of bone and mineral research* **19** 352-359

[2] Kobayashi C, Hayash M, Yamaoka M, Hashimoto K, Kato T, Komori N and Ogiso B 2012 Assessing qualitative changes in simulated periodontal ligament and alveolar bone using a non-contact electromagnetic vibration device, *Clin Oral Invest* **16** 1161-1169

[3] Delecluse C, Roelants M and Verschueren S 2003 Strength increase after whole-body vibration compared with resistance training, *Medicine & Science in Sports & Exercise* 1033-1041

[4] Bankoff A D P 2012 Biomechanical Characteristics of the Bone, *Human Musculoskeletal Biomechanics* 61-86

[5] Fan Z and Rho J Y 2003 Effects of viscoelasticity and time-dependent plasticity on nanoindentation measurements of human cortical bone, *J Biomed Mater Res* **67A** 208-214

[6] Wu Z, Ovaert T C and Niebur G L 2012 Viscoelastic Properties of Human Cortical Bone Tissue Depend on Gender and Elastic Modulus, *J Orthop Res* **30** 693-699

[7] Nicholson L, Maher C, Adams R and Phan-Thien N 2001 Stiffness properties of human lumbar spine: A lumped parameter model, *Clinical Biomechanics* **16** 285-291

[8] Sancaktar E 1987 Constitutive behaviour and testing of structural adhesives, *Appl. Mech. Rev.* **40** 15-27

[9] Perescu A, Suciu O, Neamtu-Popescu A, Nes C and Bereteu L 2015 Vascoelastic behavior simulation of cortical bone tissues using Burgers rheological model, *Applied Mechanics and Materials* **801** 267-272

[10] Cochrane DJ, Legg S J and Hooker M J 2004 The short-term effect of whole-body vibration training on vertical jump, sprint, and agility performance, *J. Strength Cond. Res.* **18** 828-832

[11] Issurin V B, Liebermann D G and Tenenbaum G 1994 Effect of vibratory stimulation training on maximal force and flexibility, *J. Sports Sci.* **12** 561-566

[12] Issurin V B and Tenenbaum G, 1999 Acute and residual effects of vibratory stimulation on explosive strength in elite and amateur athletes, *J. Sports Sci.* **17** 177-182

[13] Cormie P, Deane R S, Triplett N T and McBride J M 2006, Acute effect of whole-body vibration on muscle activity, strength and power, *J. Strength Cond. Res* **20**(2) 257-261

[14] McBride J M, Porcari J P and Scheunke M D 2004 Effect of vibration during fatiguing resistance exercise on subsequent muscle activity during maximal voluntary isometric contractions, *J. Strength Cond. Res.* **18** 777-781

[15] McBride J M, Triplett-McBride T, Davie D and Newton U 2002 The effect of heavy- vs. light-load jump squats on the development of strength, power, and speed, *J. Strength Cond. Res.* **18** 75-82

[16] Kaçoğlu C and Gürol B 2016 Effect of body position during whole body vibration on acute jumping performance, *Baltic Journal of Sport & Health Sciences* **3** 8-12

IOP Conf. Series: Materials Science and Engineering **294** (2017) 012088 doi:10.1088/1757-899X/294/1/012088

Determination of elastic mechanical characteristics of surface coatings from analysis of signals obtained by impulse excitation

E Nyaguly, I Craştiu, S Deac, C Gozman-Pop, G Drăgănescu and L Bereteu

Politehnica University Timisoara, Mechanical and Strength Materials Department, Mihai Viteazul str., no. 1, 300222 Timisoara, Romania

E-mail: liviu.bereteu@yahoo.com

Abstract. Most of the surface coatings are based on the synthetic polymers, which are substances composed from very large molecules that form tough, flexible, adhesive films when applied to surfaces. The other components of surface coverings materials are pigments that provide colour, opacity, gloss and other properties. Surface coatings are two-phase composite materials: constitute a polymer matrix on the one side, and on the other side of the pigments and additives dispersed in the matrix. Their role is not only aesthetically but also to ensure anticorrosive protection or even improve some mechanical properties of coated surfaces. In this paper it will follow, starting from the mechanical properties of the substrate, the metallic sheet in general, to determine the new properties of the assembly of substrate and the two coating layers, also the determination of mechanical properties of the layers. From the analysis of vibroacoustic signals obtained by the impulse excitation of the sample, one can determine the elasticity modulus. These results come to validate the results based on finite element analysis (FEA) of the same samples.

1. Introduction

Coatings provide a layer that changes the surface properties of piece to those of the material being applied. The piece becomes a composite material exhibiting properties generally not achievable by either material if used alone.

Coating metal surfaces is a challenge both for experienced paint professionals from various industries, and for those concerned with residential or commercial painting works. There are two types of metals: ferrous and non-ferrous. Ferrous metals will rust in contact with corrosive environments, and it is very important to cover the surfaces of these metals.

Knowing the metal nature of the coated surfaces is a vital issue because it influences the performance of the coatings. At the same time, in a paint-metal system, the primer layer plays an important role as a link between metal and paint and forms a suitable base for the next layer. Since cracking is the most frequent cause of failure in coating metal plates, it can be said that the different loads of the two components affect the coating. It is also appreciated that the impact of the rigidity of the adhesive film may be one of the vital reasons for preventing cracking. To prevent further corrosion of the metal surface, the rigidity of the coating should be equal to that of the metal used [1].

The coating of mechanical surfaces by metallization or painting is done for various purposes: from the aesthetic shape of the parts, through the given color, to avoiding their wear or to prevent their oxidation and corrosion. With the coating the elastic properties of the composite system change [2].

Several methods have been developed to determine the mechanical properties of materials, with a general classification in static methods and dynamic methods [3], [4]. Among the most widely used dynamic techniques are ultrasonic resonance spectroscopy [5-8] and resonant vibrational techniques [9-11]. Recent advances in identifying methods for the elastic characterization of materials and recent patents are reviewed in the papers [12-13].

In this paper the research on the determination of the mechanical characteristics of some materials by the impulse excitation technique is continued. The vibroacoustic signal resulting from the applied mechanical pulse was acquired by a condenser microphone or the movement of a sample point was recorded using a laser vibrometer [14-17].

The purpose of this paper is to determine the modulus of elasticity of some paints used for the protection of ordinary steel sheets used in the refrigeration industry. The elastic properties of the coating material can be determined by measuring the frequencies of the free vibrating bar excited by mechanical impulse, before and after coating.

2. Theoretical background

In the following, the samples are considered as homogeneous isotropic rectangular bar of constant section, consisting of a substrate and two coating layers, and no subject to external forces (Figure 1). To study the free vibration of the bar can consider different boundary conditions. The best known case in the literature is free–free conditions. It is assumed that the stiffness of the composite bar is equal to the sum of the stiffnesses of the component layers rigidities. Neglecting shear and rotary inertia effect, and assuming a perfect bonding at the interface the bending vibratory motion of the bar is governed by a differential equation with partial derivate Euler-Bernoulli [18]:

$$\left[(EI)_{1c}+(EI)_{s}+(EI)_{2c}\right]\frac{\partial^4 Y(x,t)}{\partial x^4}+\left[(\rho A)_{1c}+(\rho A)_{s}+(\rho A)_{2c}\right]\frac{\partial^2 Y(x,t)}{\partial t^2}=0 \quad (1)$$

where $Y(x, t)$ is the displacement of neutral axis of the beam at x distance to the left end, $(EI)_{1c}$ and $(EI)_{2c}$ are the stiffness's of the two coatings, $(\rho A)_{1c}$ and $(\rho A)_{2c}$ are the masses per unit of length of the two coating layers, $(EI)_s$ is the stiffness of the substrate beam and $(\rho A)_s$ is the mass per unit of length of substrate beam, E are the longitudinal elastic modulus, geometric moment of inertia of the cross section of the bar with respect to the neutral axis, the area of this section and material density while the subscripts s and c refer to the substrate and the coating, respectively.

This differential equation can be solved analytically and allows the determination of the shapes and the frequencies of the vibration modes. The resonance frequencies can be determined from the characteristic equation for free-free boundary conditions:

$$1-\mathrm{ch}X\cos X=0 \quad (2)$$

where

$$X=L\sqrt[4]{(2\pi f_{cb})^2\frac{(\rho A)_{1c}+(\rho A)_{s}+(\rho A)_{2c}}{(EI)_{1c}+(EI)_{s}+(EI)_{2c}}} \quad (3)$$

and f_{cb} is one of the bending resonance frequencies, L is the length of the composite bar.

The first two roots of the characteristic equation (2) are X_1=4.730, X_2=7.853, and for natural modes r>2, it is easy to show that:

$$X_r=\frac{(2r+1)\pi}{2} \quad (4)$$

It is obvious that for uncoated sample, i.e. the sample is only substrate, can be write a relationship similar to that given by the equation (3):

$$X = L\sqrt[4]{(2\pi f_s)^2 \frac{(\rho A)_s}{(EI)_s}} \tag{5}$$

where f_s is one of the bending resonance frequencies of the substrate beam.

From equations (3) and (5) it can obtain the relation between natural vibration frequencies f_{cb} of the composite beam, and natural vibration frequencies f_s of the substrate, assuming that the two coatings are of the same material and have the same thickness:

$$\frac{f_{cb}^2}{f_s^2} = \frac{2(EI)_c + (EI)_s}{(EI)_s} \frac{(\rho A)_s}{2(\rho A)_c + (\rho A)_s} \tag{6}$$

Figure 1 shows the design of the coated beam and how its dimensions are noted.

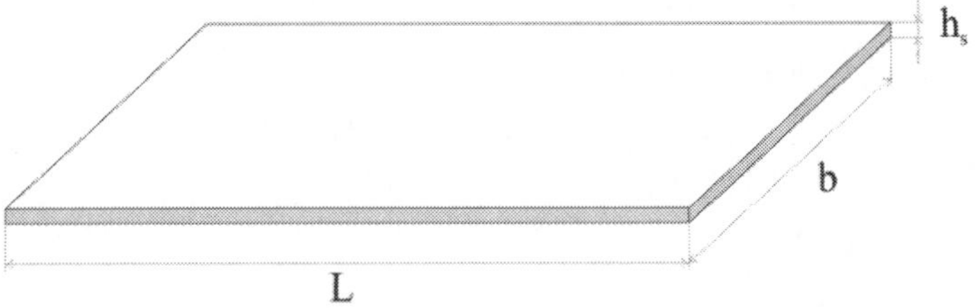

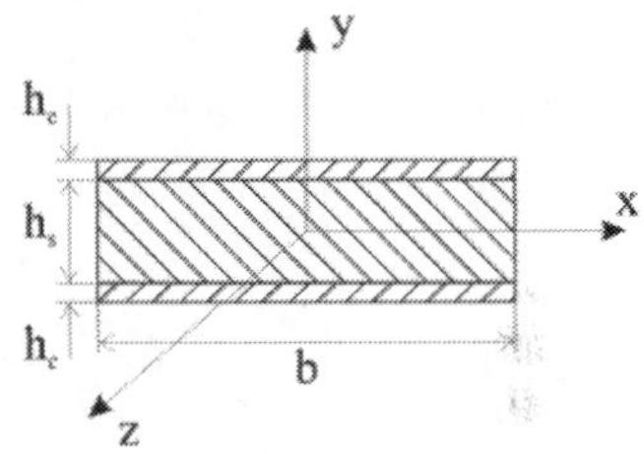

Figure 1. Coated beam and its noted dimensions

Figure 2. Beam cross section and its noted dimensions

For a composite beam with a rectangular cross section, Lopez has developed an analytical expression to determine the film Young's modulus [11]. In this model, it was considered a single coating layer and the simplification by which the neutral axis does not change after coating.

In the following, it will be considered to paint the bar with two symmetrical sides with respect to the substrate and with the same thickness. In this case, the neutral axis can be considered as not changing.

The geometrical moment of inertia of the substrate beam is

$$I_s = \int_{-h_s/2}^{h_s/2} by^2 dy = \frac{bh_s^3}{12} \tag{7}$$

and the moment of geometrical inertia of the two layers of coating, considering them of the same material and having the same thickness is

$$I_c = \int_{\frac{h_s}{2}}^{\frac{h_s}{2}+h_c} by^2 dy = \frac{(2h_c + h_s)^3}{24} - \frac{h_s^3}{24} = \frac{bh_s^2 h_c}{2} + bh_s h_c^2 + \frac{2bh_c^3}{3}. \tag{8}$$

The stiffness ratio of the bar after coating and the stiffness of the bar before to coating will depend on the thicknesses of the coating layer and the thickness of the substrate bar, as

$$\frac{2(EI)_c + (EI)_s}{(EI)_s} = \frac{E_c}{E_s}\left(\frac{2I_c}{I_s}\right) + 1 = \frac{E_c}{E_s}\left[\left(1 + 2\frac{h_c}{h_s}\right)^3 - 1\right] + 1 \tag{9}$$

The mass ratio per unit length of the coating layer and the substrate bar will depend on the ratio of the thicknesses of the coating layer and the thickness of the substrate bar and their densities

IOP Conf. Series: Materials Science and Engineering **294** (2017) 012088 doi:10.1088/1757-899X/294/1/012088

$$\frac{(\rho A)_s}{2(\rho A)_c + (\rho A)_s} = \frac{1}{1 + 2\dfrac{\rho_c}{\rho_s}\dfrac{h_c}{h_s}}. \tag{10}$$

By replacing equations (9) and (10) in equation (8), the ratio of the natural frequencies of the bar after coating and before coating is obtained. This ratio is a function of the thickness of the coating layer and the thickness of the bar.

$$\frac{f_{cb}^2}{f_s^2} = \frac{1 + \dfrac{E_c}{E_s}\left[6\left(\dfrac{h_c}{h_s}\right) + 12\left(\dfrac{h_c}{h_s}\right)^2 + 8\left(\dfrac{h_c}{h_s}\right)^3\right]}{1 + 2\left(\dfrac{\rho_c}{\rho_s}\right)\left(\dfrac{h_c}{h_s}\right)} \tag{11}$$

Then, the Young's modulus of the coating can be calculated as:

$$E_c = \frac{E_s}{6\left(\dfrac{h_c}{h_s}\right) + 12\left(\dfrac{h_c}{h_s}\right)^2 + 8\left(\dfrac{h_c}{h_s}\right)^3}\left[\left(\frac{f_{cb}}{f_s}\right)^2\left(1 + 2\left(\frac{\rho_c}{\rho_s}\right)\left(\frac{h_c}{h_s}\right)\right) - 1\right]. \tag{12}$$

Assuming that the thickness of the coating is much smaller than the thickness of the substrate, i.e. the ratio $h_c/h_s \ll 1$, then the square and the cube of this ratio can be neglected and the equation (12) takes the form:

$$E_c = \frac{E_s}{6\left(\dfrac{h_c}{h_s}\right)}\left[\left(\frac{f_{cb}}{f_s}\right)^2\left(1 + 2\left(\frac{\rho_c}{\rho_s}\right)\left(\frac{h_c}{h_s}\right)\right) - 1\right]. \tag{13}$$

This simplified model was given by Berry [19-20] in another way, but it can also be obtained from the Lopez model by developing Taylor series according to the ratio of the thicknesses of the coating and the substrate, and retaining only the terms of the first order.

Assuming that the thickness of the coating is much smaller than the thickness of the substrate, i.e. $h_c/h_s \ll 1$, and the density of the coating material is much lower than the substrate density then $\rho_c/\rho_s \ll 1$, then the equation (13) becomes

$$E_c = \frac{E_s}{6\left(\dfrac{h_c}{h_s}\right)}\left[\left(\frac{f_{cb}}{f_s}\right)^2 - 1\right]. \tag{14}$$

This model leads to a very quick and simple formula for determination of Young's modulus for the coating material into two very thin layers (on both sides) that have very small density compared with density of the substrate material. It can be used for thin, low density coatings (TLDC).

In the case of a free–free substrate beam with a rectangular cross-section, the Young's modulus can be determined by the following formula [21]:

$$E_s = k_1 f_s^2 \rho_s \frac{L^4}{h_s^2} T_f \tag{15}$$

where k_1=0.9465 and T_f is the geometrical correction factor introduced to take into account the shear and rotary inertia effects. For a ratio between the length of the bar and its thickness L/h_s>>20, this correction factor can be simplified into the following expression:

$$T_f = 1 + k_2\left(\frac{h_s}{L}\right)^2 \quad (16)$$

where k_2=6.585.

3. Numerical simulation

Figures 3 and 4 give the geometric model of the sample. It consists of three parts: the substrate bar and two layers of coating, symmetrical to the substrate and of the same thickness. The substrate is a rectangular cross section slender beam with a length L=250 mm, a width b=40 mm and a thickness h_s=2mm. It is made of steel S 235 JR, with material properties E_s=200GPa, ρ=7850 kg/m^3, μ=0.3.The painting was bonded to the substrate by electrostatic field having the possibility of thickness control. For the sample considered in the modal analysis, the thickness of the two layers assumed by equal thicknesses was h_c=0.0881mm after one deposition in the electrostatic field, respectively h_c=0.1882mm after two depositions.

For the free-free boundary conditions the modal analysis, were applied to beam from Figure 1, by ANSYS program. The 3D hexahedron element was used to mesh the structure. Mesh on the beam is generated automatically by ANSYS, while is used the spatial hexahedral elements as in Figure 3. The element is defined by 8 nodes while each node has three degrees of freedom. The solid hexahedral element has a quadratic shifting behavior and is suitable for modeling of the finite element regular mesh. The maximum size of the element is 4 mm. The mesh, in Figure 4, is created of 748 elements and of 5634 nodes.

To extract the resonance frequencies of the substrate beam and composite beam, the type of solver and the solution method in program ANSYS is selected automatically. For this modal analysis the direct solver including the block Lanczos method is used.

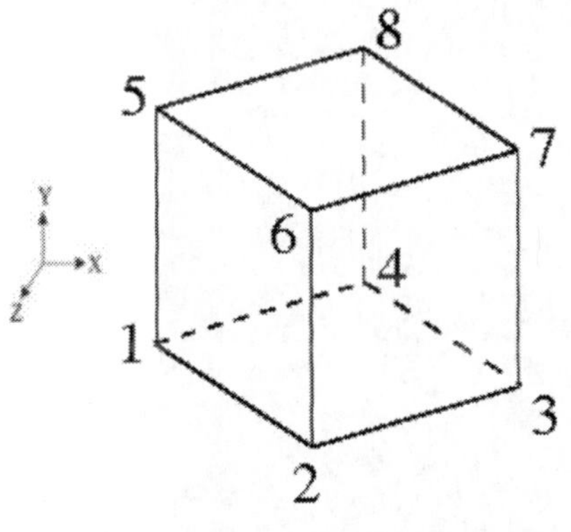

Figure 3. Spatial solid hexahedral element

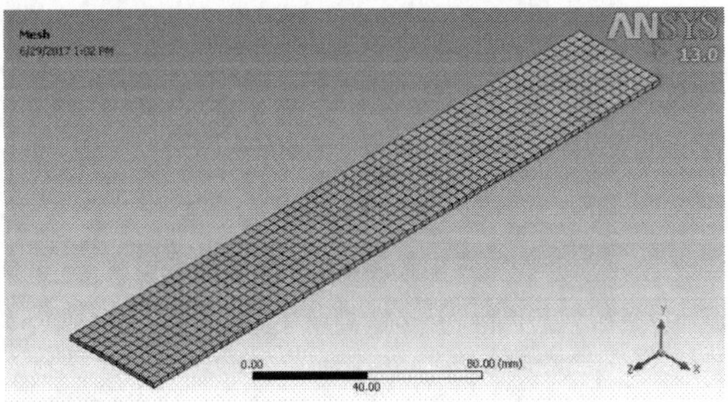

Figure 4. Mesh of finite elements on substrate beam

A total of twenty vibration modes have been extracted. The first six modes correspond to rigid body modes and have natural frequencies approximately equal to zero. The following two modes of elastic body correspond to bending modes in plane. The next one mode corresponds to the first torsion mode. In higher order modes, the first out of plane vibration mode appears.

The first mode of bending vibration (Figure 5) is useful for determining the modulus of elasticity of the substrate in equations (15), respectively for determining the modulus of elasticity of the covering material in the equations (12), (13) and (14).

The first mode of torsional vibration can be used to determine the shear modulus of elasticity.

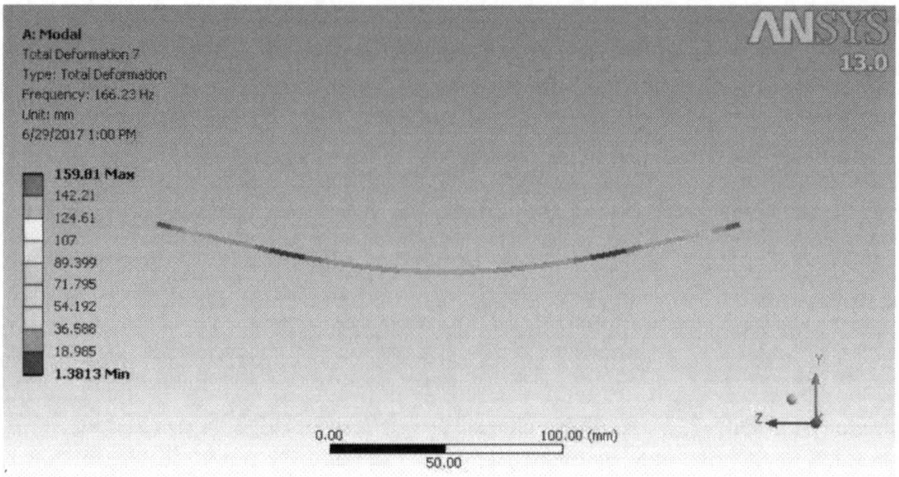

Figure 5. The shape of the first mode vibration beam with its vertical displacement

4. Experimental technique

To determine by measurements: the length L, width b and thickness h_s of the substrate beam, six measurements were made at six different points. To determine the thickness of coatings alleged equal, ten measurements were performed in six different points.

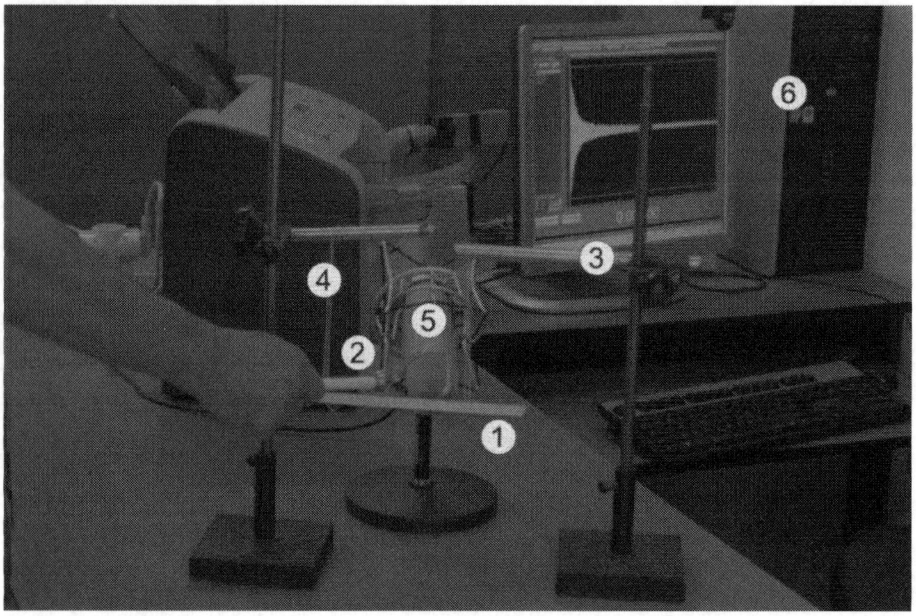

Figure 6. Experimental stand for vibroacoustic measurements

The substrate dimensions are its length L=250 mm, its width b=40 mm, and its thickness h_s=1.97mm mm. The coating has the same length and width as the substrate and a thickness of a single layer has h_c=0.0881mm after a single pass of the sample through the electrostatic field, respectively, h_c=0.1882mm, after two passes.

The purpose of the impulse excitation technique (IET) is to measure the resonance frequencies of a sample impacted by a small hammer impulse (Figure 6). A beam can be excited in different vibrations modes: longitudinal, flexural and torsional. Experimental stand used in the vibration lab of the department, is composed by: the sample **1**, which is the mechanical structure to be analyzed, impulsive mini hammer **2**, brackets to support the sample **3**, elastic threads for support of the beam in boundary conditions the free ends **4**, the acoustic sensor one condenser microphone **5**, and the computer with acquisition board **6**.

Before performing the test, the sample is placed on a support, elastic threads. The free limitation condition has been adopted for this procedure. The elastic threads are located at the nodal points of the sample, where the displacement is zero, i.e. for the first bending mode used in this study.

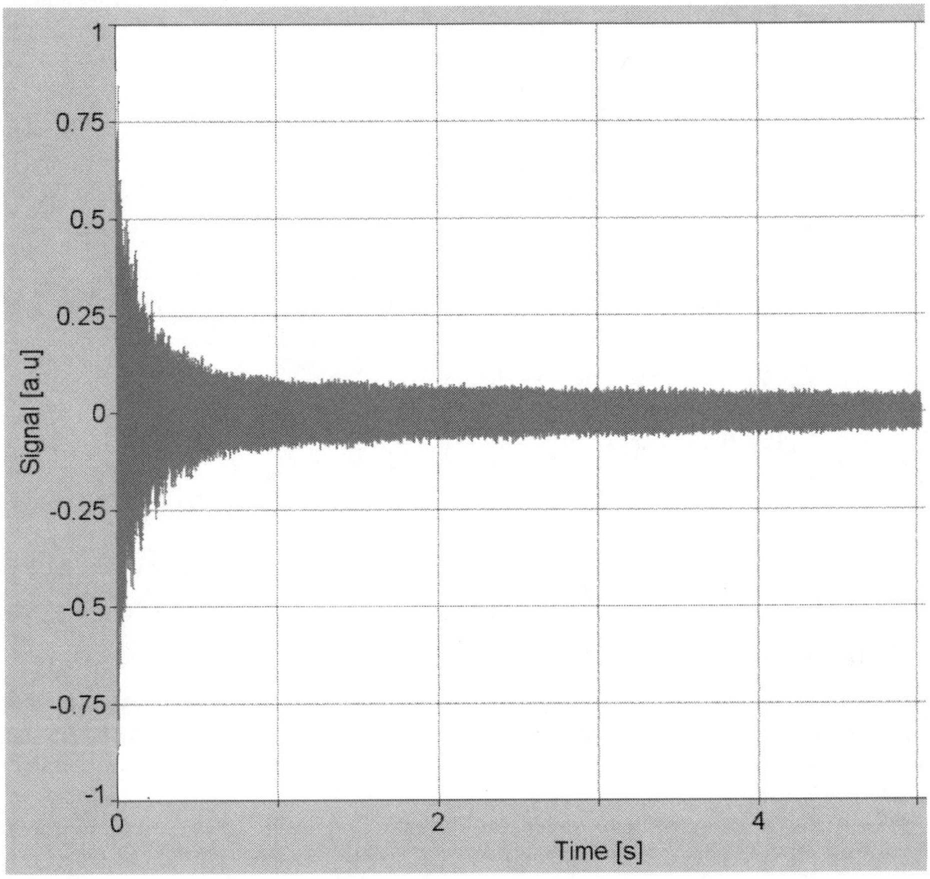

Figure 7. Recorded signal

These mechanical vibrations are detected by an acoustic microphone and transformed by it, which is actually a capacitive transducer, in an electrical signal that is represent as a time function (Figure 7). Fast Fourier Transform (FFT) is used to represent the signal as a frequency function and to extract the resonant frequencies of the sample.

For the purpose of this paper, it is sufficient to determine in the Fourier spectrum the fundamental bending frequencies, before coating the sample with paint and after covering the sample. For this reason the recorded signal is filtered, using a band pass filter with ±70 Hz from the central frequency, about which value is obtained information from the modal analysis. In the Fourier Spectrum represented in Figure 8, for the signal recorded from the sample not coated with paint, only the frequency corresponding to the first vibration mode was retained. This resonant frequency value, density or mass and sample sizes are used to determine the Young's modulus of elasticity of the uncovered sample.

To determine the elasticity modulus of the paint, which covers the substrate sample, the equations (12), (13) and (14) are used.

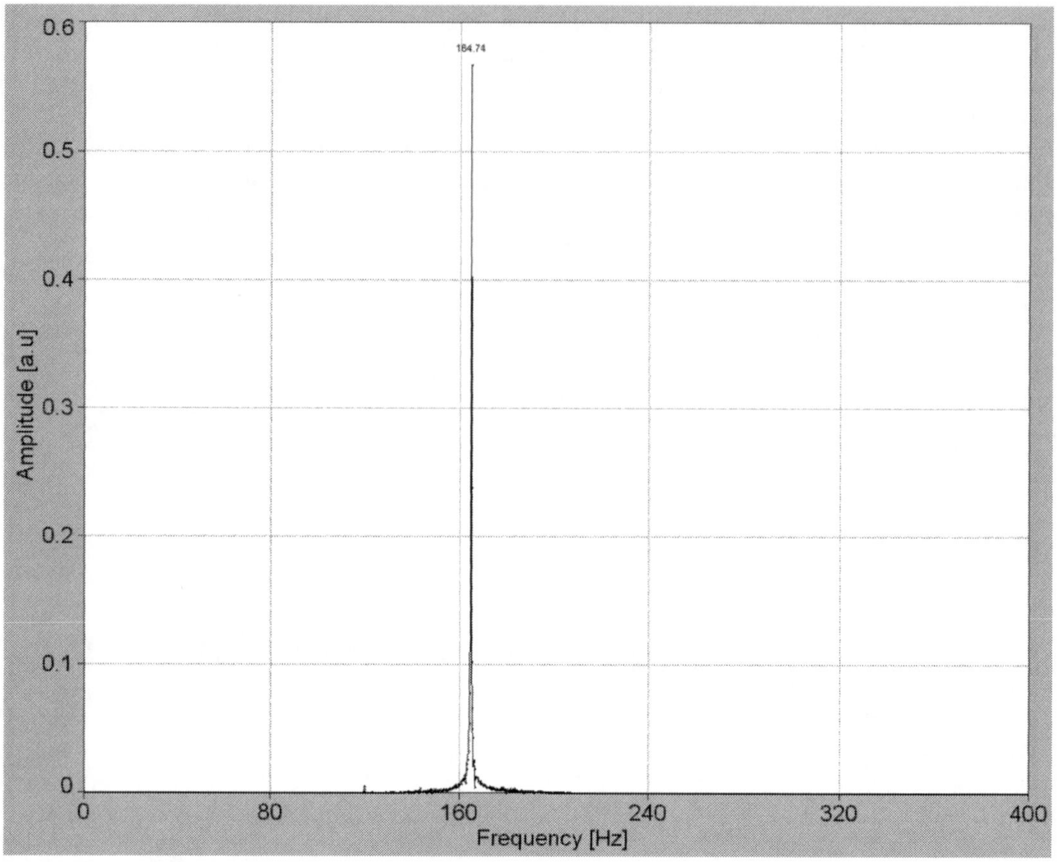

Figure 8. First resonant frequency from Fourier Spectrum for not coated beam

In these equations the frequency of the first bending mode of the composite beam must be known. From a recorded signal, similar to the signal in Figure 8, after filtering it and applying the Fast Fourier Transform, the Fourier Spectrum is obtained, containing only the first resonant frequency of the composite beam.

In the Figure 9 the Fourier Spectrum gives the frequency of the first bending mode for the painted beam bar through a single pass by the electrostatic field. For two passes, the thickness of the cover width increases and the frequency of the first mode increases to f = 166.83Hz.

The resonance frequency, density/ mass and sample sizes are used to calculate the elastic modulus of the paint material, and the results are shown in Table 2.

In a similar way, starting from the torsional vibration equation, relationships can be obtained for determining the shear modulus of the substrate and the paint layer, respectively. Between the three parameters that characterize the mechanical properties of a material: Young's module, shear modulus and Poisson's coefficient is a dependency relationship, and after experimental determination of the first two modules, Poisson's ratio can be determined. These issues are underway and are future research.

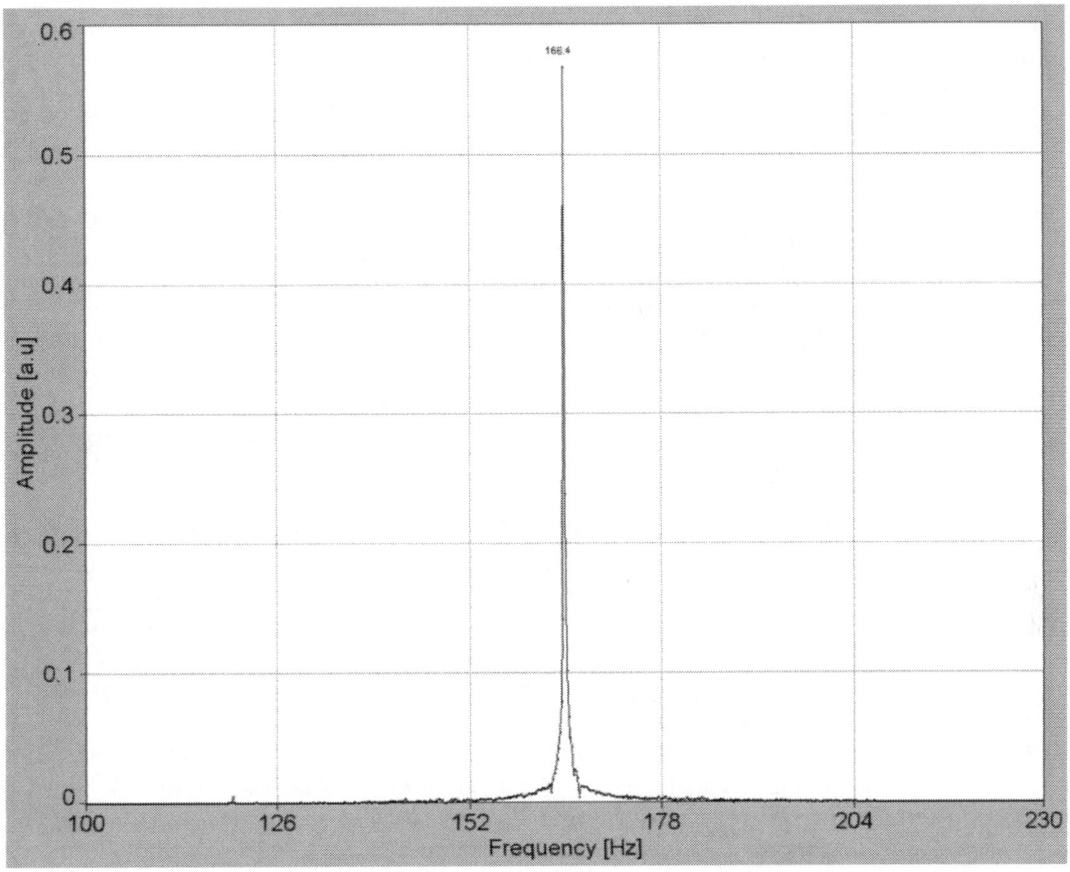

Figure 9. First resonant frequency from Fourier Spectrum for coated beam

5. Results and conclusions

To determine the elasticity modulus of the paint, the equations (12), (13) and (14) are used. The results obtained are shown in Table 1, in which experimental determinations were made after deposition of the paint through a single pass of the sample through the electrostatic field.

Table 1. Young's modulus of paint, after one pass of the sample through the electrostatic field

Frequency before coating (Hz)	Frequency after coating (Hz)	E_s (GPa) Young's modulus	E_c (GPa) TLDC method	E_c (GPa) Berry method	E_c (GPa) Lopez method
164.74	166.4	199.74	3.801	4.823	4.712

In Table 2 are presented the results of the elastic modulus of the paint determined experimentally from the sample covered by passing twice through the electrostatic field.

Table 1. Young's modulus of paint, after two passes of the sample through the electrostatic field

Frequency before coating (Hz)	Frequency after coating (Hz)	E_s (GPa) Young's modulus	E_c (GPa) TLDC method	E_c (GPa) Berry method	E_c (GPa) Lopez method
164.74	167.53	199.74	3.994	4.987	4.621

The values of the elastic modulus given in the paint catalogs are in the range of 4.2 GPa and 5.3 GPa. By comparing the results obtained experimentally with those given in the catalogs it is found that the method is valid and gives good results for determining the modulus of elasticity of coating materials.

References

[1] Lambourne R and Strivens T A 1999 *Paint and Surface Coatings. Theory and practice*, Second Edition, Woodhead Publishing Ltd, Cambridge, England
[2] Kubisztal M, Haneczok G, Chrobak A, Rams B, Miara D and Rasek J 2009, Elastic properties of epoxy system coatings on aluminum substrate, *Surface and Coatings Technology* **204** 120-124
[3] Bellan C and Dhers J 2004, Evaluation of Young modulus of CVD coatings by different techiches, *Thin Solid Films* **469-470** 214-220
[4] Radovic M, Lara-Curzio E and Riester L 2004 Comparison of different experimental techniques for determination of elastic properties of solids, *Mater. Sci. Eng.* **A 368** 56-70
[5] Thomasová M, Sedlák P, Seiner H, Janovská M, Kabla M, Shilo D and Landa M 2015 Young's moduli of sputter-deposited NiTi films determined by resonant ultrasound spectroscopy: Austenite, R-phase, and martensite, *Scripta Materialia* **101** 24-27
[6] Tan Y, Shyam A, Choi W B, Lara-Curzio E and Sampath S 2010 Anisotropic elastic properties of thermal spray coatings determined via resonant ultrasound spectroscopy, *Acta Materialia* **58** 5305-5315
[7] Sedmák P, Seiner H, Sedlák P, Landa R, Mušálek R and Matejícek J 2013 Application of resonant ultrasound spectroscopy to determine elastic constants of plasma-sprayed coatings with high internal friction, *Surface and Coatings Technology* **232**(1) 747-757
[8] Mazot P and Pautrot S 1998 Détermination du module d'young de dépôts par flexion dynamique: Application aux systèmes bicouche et tricouche, *Ann. Chim. Sci. Mat.* **23** 821-827
[9] Gadaud P and Pautrot S 2004 Application of the dynamical flexural resonance technique to industrial materials characterisation, *Mater. Sci. Eng.* **A 370** 422-426
[10] Gadaud P, Milhet X and Pautrot S 2009 Bulk and coated materials shear modulus determination by means of torsional resonant method, *Mater. Sci. Eng.* **A 521–522** 303-306
[11] López-Puerto A and Oliva A I 2014 A vibrational approach to determine the elastic modulus of individual thin films in multilayers, *Thin Solid Films* **565** 228-236
[12] Pagnotta L 2008 Recent progress in identification methods for the elastic characterization of materials, *International Journal of Mechanics* **4**(2) 129-140
[13] Alfano M and Pagnotta L 2009 Recent developments and selected patents on vibration based methods for elastic characterization of izotropic materials, *Recent Pat. on Mec. Eng.* **2** 130-143
[14] Suciu O, Bereteu L, Drăgănescu G and Ioanovici T 2013 Determination of the elastic modulus of hydroxiapatite doped with magnesium through nondestructive testing, *Advanced Materials Research* **814** 115-122
[15] Bereteu L, Vodă M, Pascu R and Nyaguly E 2015 Influence of some oxide coating lazers on the mechanical properties of steel plates, *Advanced Materials Research* **1111** 91-96
[16] Perescu A, Bereteu L, Simoiu D and Nyaguly E 2015 Nondestructive method for the determination of the elastic properties of welded aluminum plates, *Advanced Materials Research* **1111** 73-78
[17] Crastiu I, Bereteu L and Simoiu D 2015 Determination of the eccentricity of a rotor by impulse excitation technique of vibration, *Applied Mechanics and Materials* **801** 176-181
[18] Meirovitch L 2010 *Foundamentals of vibration*, Waveland Press, Inc.
[19] Berry B S and Pritchet W C 1975 Vibrating reed internal friction apparatus for films and foils, *IBM J Res Dev* **19** 334-343
[20] Berry B S,(Gupta D, and Ho P S Eds.) 1988 *Diffusion Phenomena in Thin Films and Microelectronic Materials*, Noyes, Park Ridge, New York, 73-141
[21] ***ASTM E 1876-15 2015 Standard test method for dynamic Young's modulus, shear modulus and Poisson's ratio by impulse excitation of vibration

A comparison between using distance sensors for measuring the pantograph vertically movement

R Rob, C Panoiu, S Rusu-Anghel and M Panoiu

Politehnica University of Timisoara, Electrical Engineering and Industrial Informatics Department, Revolutiei str., no. 5, 300223 Hunedoara, Romania

E-mail: raluca.rob@fih.upt.ro

Abstract. In railway transportation the most important problem to solve consists in assuring the safety traffic of people and freight. In this scope some of the geometrical parameters regarding the contact line must be measured. One of this parameter is the pantograph vertically movement, so it must use distance sensors. Present paper studies the performance of two kinds of distance sensors, an ultrasonic distance sensor and an infrared sensor. The performances are studied from the point of view of error distance measurement and the possibility of using a real time acquisition system. The researches were made on a laboratory model for the pantograph realized at the scale 1:2.

1. General description of the application

In order to measure the vertically movement of the laboratory pantograph, a hardware-software system is proposed. The system components are the followings and they are presented in Figure 1:
-Laptop equipped with LabVIEW Development System NI MyRIO
-Acquisition board NI MyRIO 1900, Xilinx processor, FPGA and Real time technology, 667MHz
-LabVIEW application for programming and monitoring
-Distance ultrasonic sensor Parallax Ping
-Distance IR sensor Pololu
-Temperature analogic sensor LM35

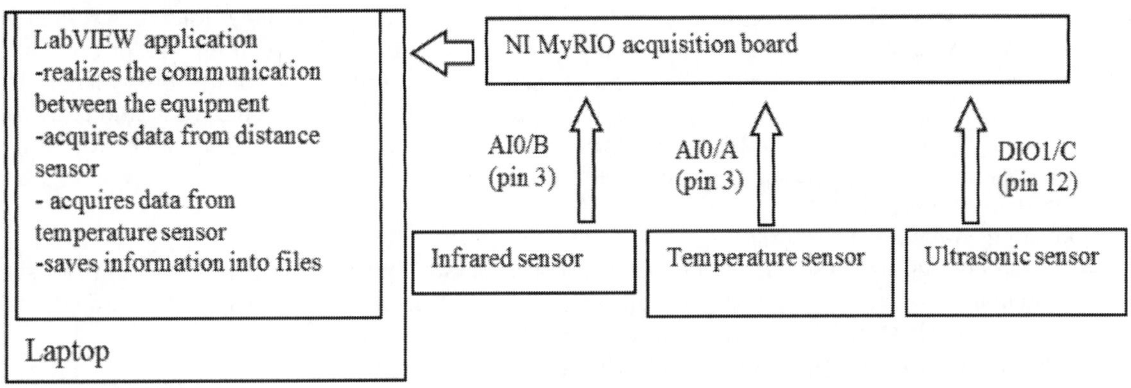

Figure 1. The block scheme of the application

IOP Conf. Series: Materials Science and Engineering **294** (2017) 012089 doi:10.1088/1757-899X/294/1/012089

2. Hardware application

2.1. NI MyRIO acquisition board

The myRIO Embedded Device features in/out pins on both sides of this device in MXP and MSP connectors. It includes analog inputs, analog outputs, digital I/O pins, LEDs, push button, a Xilinx FPGA, onboard accelerometer, and a dual- core processor. The user can program it with LabVIEW or C. With its onboard devices, this embedded device provides an affordable tool for students and educators.

2.2. Distance ultrasonic sensor Parallax Ping

The Parallax PING ultrasonic distance sensor provides accuracy; the range of distance is varying between 2 cm and 3 meters.

The PING sensor works by transmitting an ultrasonic burst and it provides an output pulse that is assigned to the time required for the burst echo to return to the sensor. By measuring the echo pulse width, the distance to target can easily be calculated.

Technical characteristics

-Range: 2 cm to 3 m (0.8 in to 3.3 yd)

-Burst indicator LED shows sensor activity

-Bidirectional TTL pulse interface on a single I/O pin can communicate with 5 V TTL or 3.3 V CMOS microcontrollers

-Input trigger: positive TTL pulse, 2 μs min, 5 μs typ.

-Echo pulse: positive TTL pulse, 115 μs minimum to 18.5 ms maximum.

-Supply voltage: +5 VDC

-Supply current: 30 mA typ; 35 mA max

-Communication: Positive TTL pulse. Package: 3-pin SIP, 0.1" spacing (ground, power, signal)

-Operating temperature: 0 – 70° C.

Communication Protocol

The PING sensor detects obstacles by emitting an ultrasonic burst and then the device waits to receive the echo. Controlled by a host microcontroller (trigger pulse), the sensor is emitting an ultrasonic burst with frequency of 40 kHz. The burst travels into the air, reaches an object and then it bounces back to the sensor. The PING sensor generates an output pulse to the microcontroller that will be finished when the echo is detected, hence the width of this pulse corresponds to the distance to the object. Figure 2 presents the time diagram of the sensor functioning.

2.3. Distance IR sensor Pololu

The presented distance sensors represent a popular choice for many projects which require accurate distance measurements. This type ofIR sensor is more economically than the sonar rangefinders, it provides much better performance than other IR alternatives. Interfacing to most microcontrollers is like in following: the single analogue output can be connected to an analogue to digital converter for taking distance measurements, or the output can be connected to comparator circuit for detection the threshold. The range for detection of this version is approximately 10 cm to 80 cm (4inch to 32inch). The GP2Y0A21 uses a 3-pin JST PH connector that works with our 3-pin JST PH cables for Sharp distance sensors. The time diagram is presented in the Figure 3.

Technical characteristics

-Operating voltage: 4.5 V to 5.5 V

-Average current consumption: 30 mA (note: present sensor draws current in wide, short bursts, and the manufacturer recommends connecting a 10 μF capacitor or larger across power and ground close to the sensor to stabilize the power supply line)

-Distance measuring range: 10 cm to 80 cm (4″ to 32″)

-Output voltage differential over distance range: 1.9 V (typical)

-Update period: 38 ± 10 ms

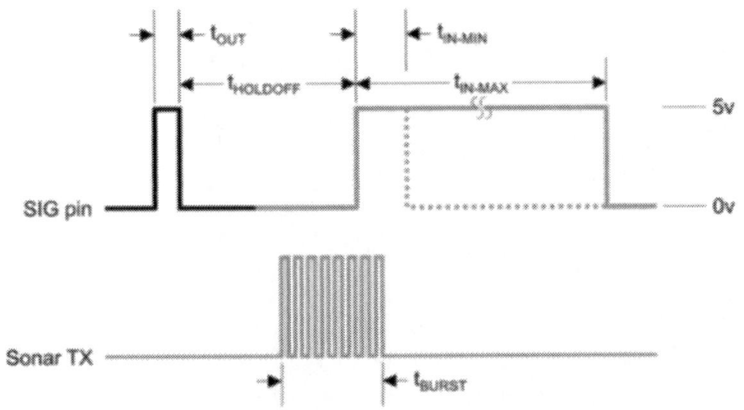

▬ (black)	Host Device	Input Trigger Pulse	t_{OUT}	2 µs (min), 5 µs typical
▬ (grey)	PING))) Sensor	Echo Holdoff	$t_{HOLDOFF}$	750 µs
		Burst Frequency	t_{BURST}	200 µs @ 40 kHz
		Echo Return Pulse Minimum	t_{IN-MIN}	115 µs
		Echo Return Pulse Maximum	t_{IN-MAX}	18.5 ms
		Delay before next measurement		200 µs

Figure 2. Parallax Ping ultrasonic sensor time diagram

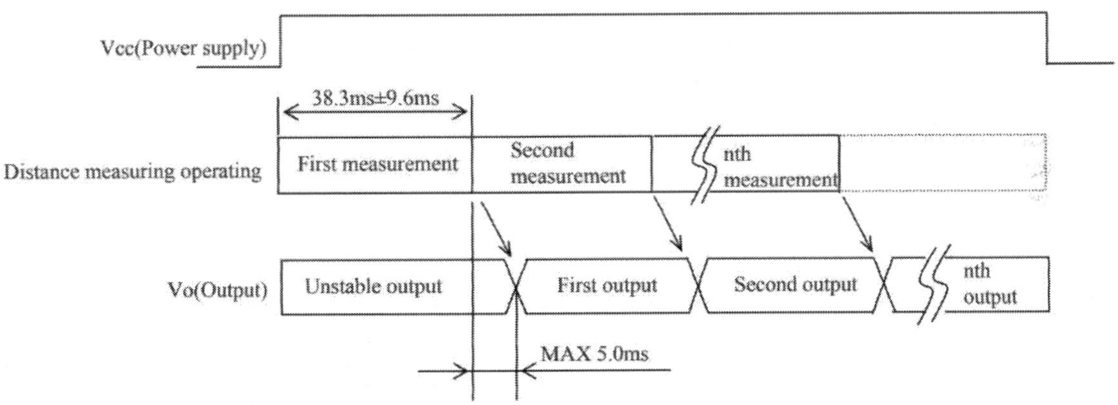

Figure 3. IR Pololu distance sensor time diagram

2.4. *Temperature sensor LM35*

The analogue temperature sensor is necessary due to the ultrasonic functioning.

The analogue sensor LM35 is rated to operate over a −55°C to 150°C temperature range. The LM35 sensor has the advantage over linear temperature sensors which are calibrated in Kelvin, so the user doesn't have to subtract a large constant voltage from the output to obtain convenient Centigrade scaling, so is presents a linear scale factor, + 10 mV/°C. The LM35 is connected at pin AI0 of B connector of NI MyRIO controller.

The relation (1) permits the computing of speed of ultrasonic burst in the air in function of temperature:

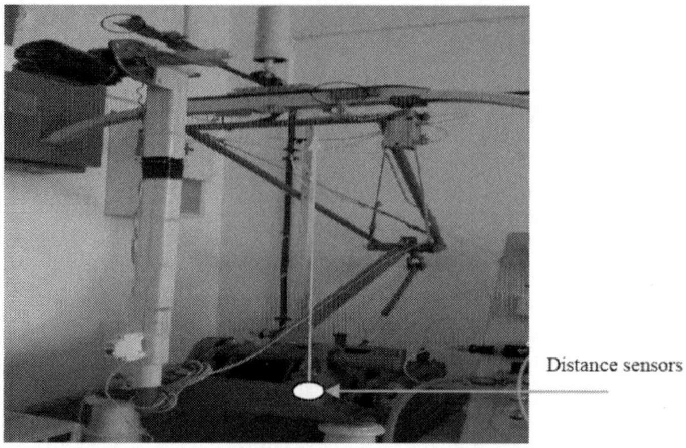

Figure 4. Figure with short caption (caption centred)

$$v_{air}[m/s] = 331.5 + 0.6 \cdot T[°C] \tag{1}$$

The distance measuring system is tested on a laboratory pantograph presented in Figure 4. The pantograph is made on 1/2 scales and it is commanded by a pneumatic system. By functioning of this mechanism, the pantograph-contact line assembly describes a trajectory in the horizontally plan (zigzag) and another trajectory in vertically plan. The distance sensor will acquire in real time the vertically distance [1], [2].

3. Software application for acquiring the ultrasonic sensor

LabVIEW MyRIO programming environment permits a reliable communication between hardware and software elements. The front panel of the application for measuring the distance using the ultrasonic sensor is presented in Figure 5. The real time communication ensures the acquiring of the temperature signal. A numerical filter is designed in order to cancel the acquisition noise. Therefore, the optimal value for the temperature window length is 100 samples. The application shows in real time the temperature value in Celsius, Fahrenheit and Kelvin degrees. The source code of the application is accomplished into a While structure for continuous acquisition. The block scheme of the project application is presented in Figure 6 and the block scheme of the Virtual instrument in Figure 7.

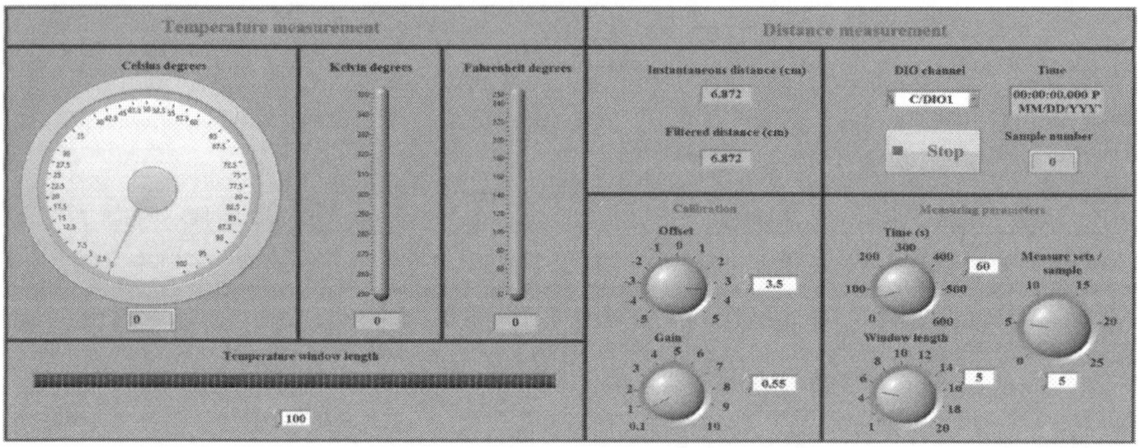

Figure 5. Front panel of the ultrasonic sensor application

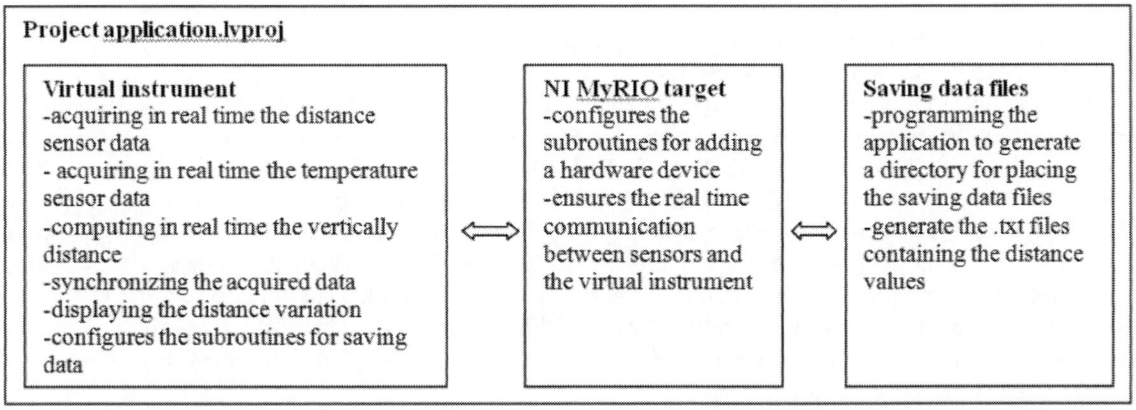

Figure 6. Block scheme of the project application

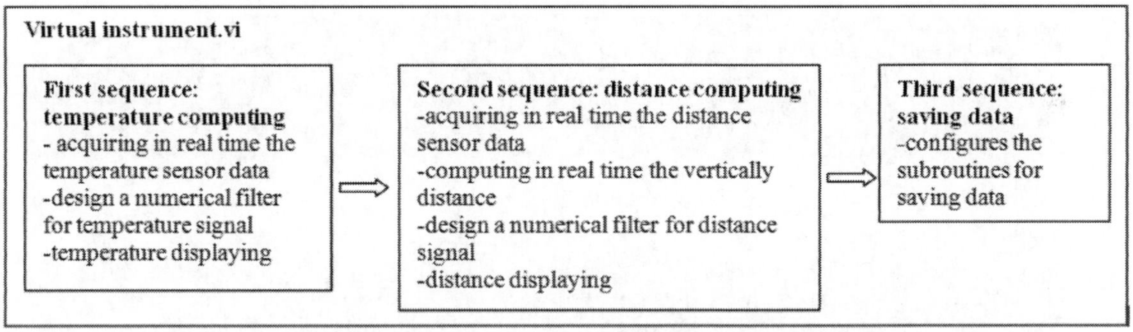

Figure 7. Block scheme of the Virtual instrument

Software application is realized as a LabVIEW project that is able to group and call together LabVIEW files or files not specific to LabVIEW, also it creates build specifications, and deploy or download files to real time targets. The project application calls in this case a virtual instrument which deploys acquiring data from distance and temperature sensors, NI MyRIO target which ensures the real time communication between sensors and the virtual instrument and an auto-populating folder which serves to data saving [3], [4].

The first sequence is realized for acquiring the temperature using the analogue input AI0 of NI My RIO.

The second sequence of the virtual instrument contains subroutines for acquiring data from distance sensor. The communication between the distance sensor and the MyRIO controller implies a single wire which is connected on DIO1 pin of the C connector. The I/O pin is configured to work properly using a structure with six cases presented in the followings:

Initialize – Initializes the session for acquiring the distance sensor. The DIO1 pin is getting ready to open the session using Smart Open.vi subroutine available in MyRIO library.

Trigger pulse – Generates the start pulse to the distance sensor. Using a sequence structure, the application waits 2μs, and then generates a 5μs pulse with Write.vi subroutine.

Rising edge – First, the application waits for the echo hold off for 750μs, then reads the information from the sensor

Falling edge – In function of the duration of the sensor pulse (115μs – 18.5ms), the application is ready to calculate the distance

Calculating the distance – The application uses the temperature information provided by the first sequence and the distance can be computed and displayed on the front panel.

Stop – MyRIO session must be closed.

The second structure uses the temperature acquired into the first sequence. The first sequence is placed into a While structure whose timing is set to 1000μs. In order to cancel the acquisitioning noise, a numerical filter is designed. The window length of this filter can be adjusted from the front panel, but it was observed the 100 samples ensure the maximum quality for the temperature signal. If the window length is increased, the measuring time is reduced, but the temperature measuring error increases due to the increased noise amplitude. The user is able to adjust the number of measuring set/sample. For example, the application can take 5 sets of measurement into 20ms and the average is considered to be the result into an interval of 100ms. The measurements are perturbed by the acquisitioning noise, so a smoothing filter is designed. This filter takes into consideration also the previous measuring results [5].

In Figures 8 and 9 the distance variation when the window length is l=10 samples and 5 samples is presented.

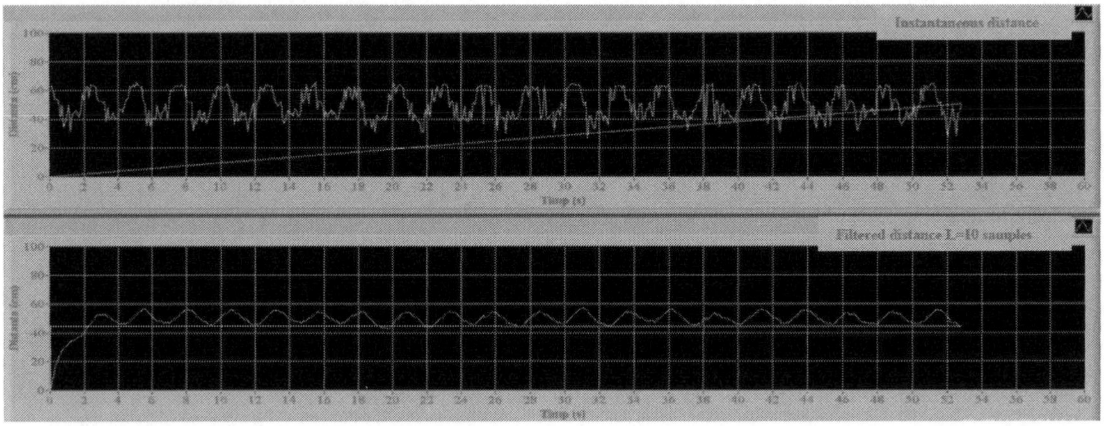

Figure 8. Variation of the unfiltered and filtered instantaneous distance with window length L = 10 samples using ultrasonic sensor

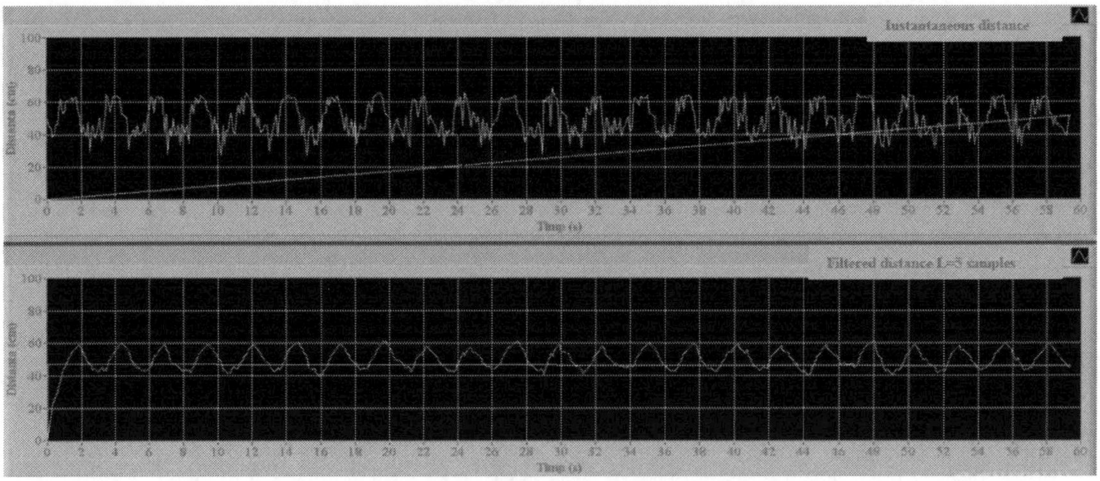

Figure 9. Variation of the unfiltered and filtered instantaneous distance with window length L = 5 samples using ultrasonic sensor

4. Software application for acquiring the IR sensor

Data used for generation the voltage-distance characteristic can be reached into a matrix which can generate the variation on the front panel of the IR sensor application (Figure 10). By choosing the factor and the interpolation type, the appropriated matrix can be generated. It realizes the interpolated characteristic. The matrix for interpolated data is presented in Figure 11.

This application is able to run in simulated and also in measuring mode. The functioning regime can be chosen from the front panel of the application. When the measuring mode is chosen, the application can acquire the signal from the IR sensor and to display the distance variation on the front panel. When the simulated mode is chosen, the application can display the variation by selecting the simulated voltage from the appropriated rotary button. In Figure 12 the variation of the distance in the measuring mode is displayed. The same numerical filter which has been presented in the previous application is used in order to cancel the acquisition noise.

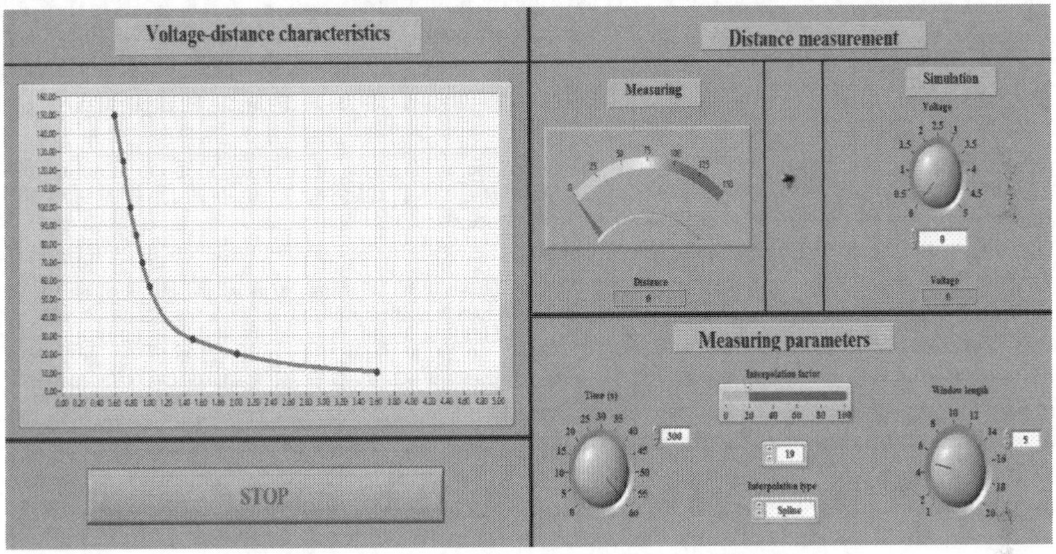

Figure 10. The front panel of the IR sensor application

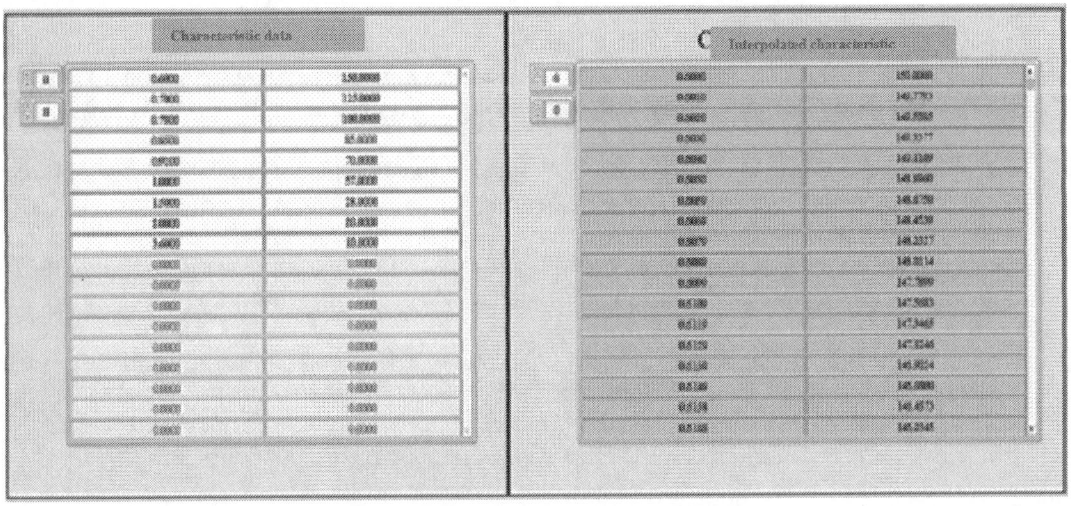

Figure 11. Matrix with interpolated data

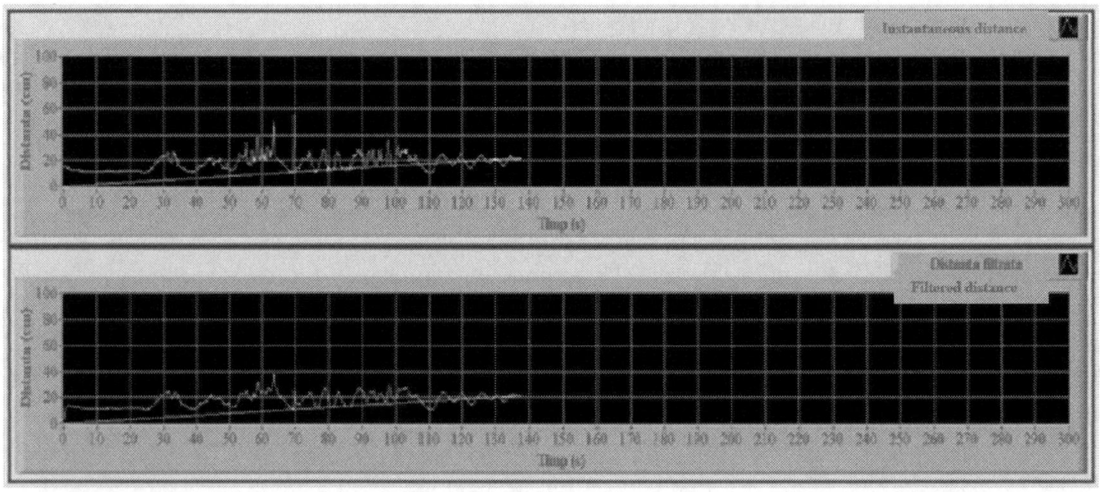

Figure 12. Variation of the unfiltered and filtered instantaneous distance using IR sensor

5. Conclusions

This paper describes two software applications which use two types of distance sensors. From the distance variations, the following conclusions can be taken:

The ultrasonic sensor presents a linear characteristic which depends on the ambient temperature. The update period is 20ms. The IR sensor has a nonlinear characteristic which is not dependent on the ambient temperature. It presents a better update period: 16.5 ± 4 ms.

The unfiltered variation of the distance in both cases presents acquisition noise. The filtered variations which uses the window length L = 5 samples presents a better accuracy, the lower limit 40cm and the higher limit 60cm are reached to each movement period. Increasing the window length, it can be observed that the error is also increasing due to the increased noise amplitude.

Acknowledgement. This work was supported by a grant of the Romanian National Authority for Scientific Research and Innovation, CNCS/CCCDI-UEFISCDI, project number 59BG/2016 within PNCDI III.

References

[1] Patankar A and Nikoubin T 2016 *Wearable System for Obstacle Detection and Human Assistance Using Ultrasonic Sensor Array*, Proceedings of the 7th International Conference on Computing Communication and Networking Technologies, Dallas, USA, July 2016, pp 60-63

[2] Migchelbrink M, White W N, Gorentz L, Wagner J and Blankenau B 2015 *Design, build, and test of an autonomous myRIO based Segbot*, American Control Conference (ACC), July 2015, pp 2783-2788

[3] Luo Y, Li C G, Zhang F and Wang K 2011 *The real-time monitor system based on LabVIEW*, International Conference on Computer Science and Network Technology, vol 2, December 2011, pp 848-851

[4] Popa G N, Diniş C M, lagăr A and Popa I 2017 An experimental comparative analyse between three low capacity PLCs, *IOP Conf. Ser.: Mater. Sci. Eng.* **163** 012034

[5] Iagar A, Popa G N and Dinis C M 2010 *Modern Equipment for Recording and Analyzing the Events in Electric Stations*, Latest Trends on Systems, January 2010, pp 222-227

Modeling and simulation of cars in frontal collision

S C Deac, A Perescu, D Simoiu, E Nyaguly, I Crâştiu and L Bereteu

Politehnica University Timisoara, Mechanical and Strength Materials Department, Mihai Viteazul str., no. 1, 300222 Timisoara, Romania

E-mail: liviu.bereteu@upt.ro

Abstract. Protection of cars, mainly drivers and passengers in a collision are very important issues worldwide. Statistics given by "World Health Organization" are alarming rate of increase in the number of road accidents, most claiming with serious injury, human and material loss. For these reasons has been a continuous development of protection systems, especially car causing three quarters of all accidents. Mathematical modeling and simulation of a car behavior during a frontal collision leads to new solutions in the development of protective systems. This paper presents several structural models of a vehicle during a frontal collision and its behavior is analyzed by numerical simulation using Simulink.

1. Introduction

Modeling the behavior of a car during an accident is one of the most important issues in strength design. Impact tests are complex and difficult to achieve in practical terms, requiring qualified personnel, sophisticated equipment, time and money. For these reasons was developed a mathematical model for the analysis of a collision, which can give by provided results an answer as exactly over behavior of the car, response that can replace even the experimental results. Over the years, were developed two main approaches on the behavior of the car during the collision: focus parameters method known as Lamped Mass-Spring (LMS) and finite element method (FEM). Analysis with finite element method (FEM) requires a complete and detailed description of the component geometry and material properties associated, in contrast, analysis of focus parameters method (LMS) is based on several discrete masses and springs of materials whose mechanical properties are determined by dynamic tests. The vehicles structure design process is based on determining deceleration-time history, called pulse accident, history provided by both methods.

First model, relatively simple but powerful to simulate the response of a car following a frontal impact was developed by Kamal [1] in 1970. This model, know as Lumped Mass-Spring (LMS) has become widely used by accident engineers due to its simplicity and relative accuracy. The vehicle is approximated one-dimensional and includes mass-spring system, a simplification which is quite acceptable for basic modeling of some characteristics in frontal impact collision. Because of the simplistic representation of an accident, the LMS model requires a user with knowledge, understanding and experience in dynamic testing of structural materials. Several versions of the LMS models were successfully used to simulate the impact of frontal, side and rear vehicle crashes [2-4].

The purpose of this paper is to present the dynamic behavior of a car in a collision, starting from simple models with concentrated parameters (LMS) with one, two and three degrees of freedom, by numerical simulation using Simulink.

2. Lumped mass-spring models

2.1. Mass-elastic element collision model

Since the car's mechanical design phase, the focus is on passenger safety.

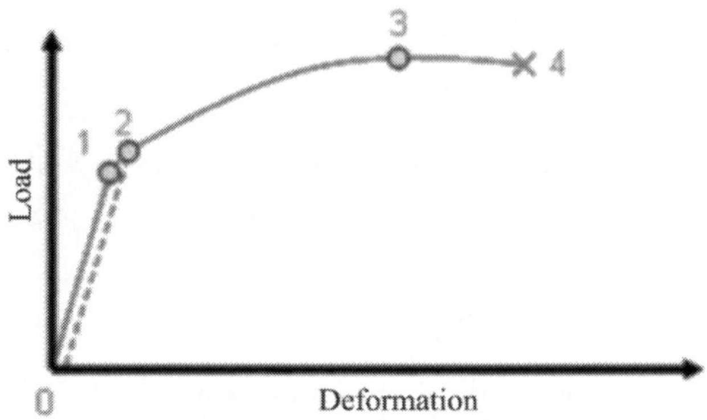

Figure 1. Load-deformation diagram

The car body is designed with special areas dedicated shock absorbing resulting from collisions. Deformation characteristics of the car structure by the collision, it can exemplify using load – deformation curve.

Depending on the speed and mechanical characteristics of vehicle and the object with which the collision occurs, the collision can be in character elastic, elasto-plastic or plastic.

Elastic collision (Figure 1, zone 0-1) usually occurs at very low speeds, the car has a low weight. In these cases, most of the time, activation of safety systems is not required.

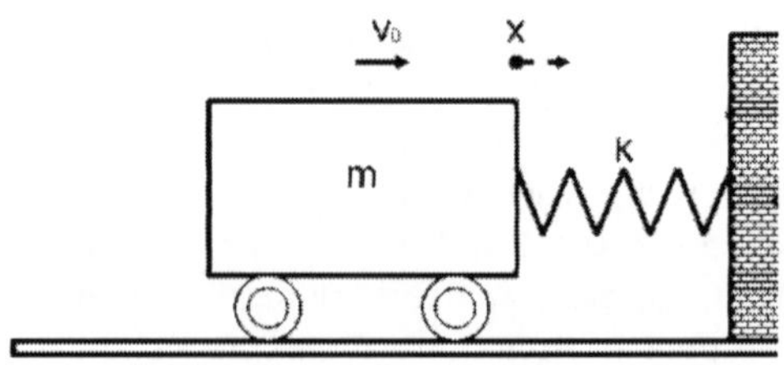

Figure 2. Mass-spring model

For the model in Figure 2 can write the differential equation valid for a short period of collision:

$$m\ddot{x} = -kx \quad (1)$$

2.2. Mass-elastic damping element collision model

Most vehicle accidents are elasto-plastic (Figure 1, zone 1-2), where, due to forces acting on passenger protection systems must be enabled.

Plastic collision happens at high speeds, and the vehicle having significant weight (Figure 1, zone 2-3-4). Often in this situation the ability to protect passengers the systems offered insufficient protection, it is usually about accidents with fatal results. Taking account of the elasto-plastic collision, we find a similar behavior with mass-spring type models or elastic element wider sense. Due to its shape and the material it is made, spring turns (after deformation) the mechanical work (W) in potential energy (E_p).

An important factor in the study of the collision, of analyzed mechanical systems is the behavior of the kinetic energy dissipation by friction, also called dissipative systems. Basic mechanical subassemblies can be equated with dissipative systems, and by adding these dissipative subsystems behaviors can describe dissipative behavior of the vehicle.

Dissipative systems connected to form a set of structural and functional elements, characterized by a number of parameters. Function of the dissipative system, like any dynamic system is to transform input into output sizes, useful technical (motion, force, energy). This transformation is achieved by dissipative system structure, i.e. the components which interact through processes of friction, which at the same time appearing not only friction loss, and tears.

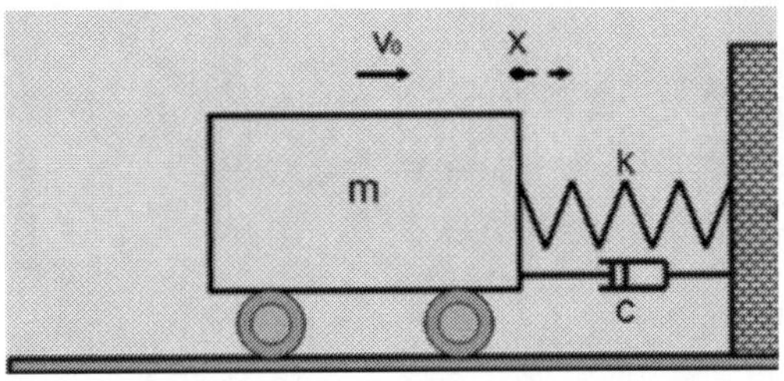

Figure 3. Mass-spring-damper model

For the model in Figure 3 can write the differential equation governing the motion for collision interval:

$$m\ddot{x} = -c\dot{x} - kx \tag{2}$$

3. Lumped mass-spring-damping models for collision of two cars

3.1. The behind crush simulation of two cars

In case of collision of two cars, there are two situations: the behind crash (Figure 4) and the front crash (Figure 5). To simulate dynamic behavior during collision using the SIMULINK module in MATLAB, the differential equations of motion will be determined. To determine these equations, Lagrange equations are used.

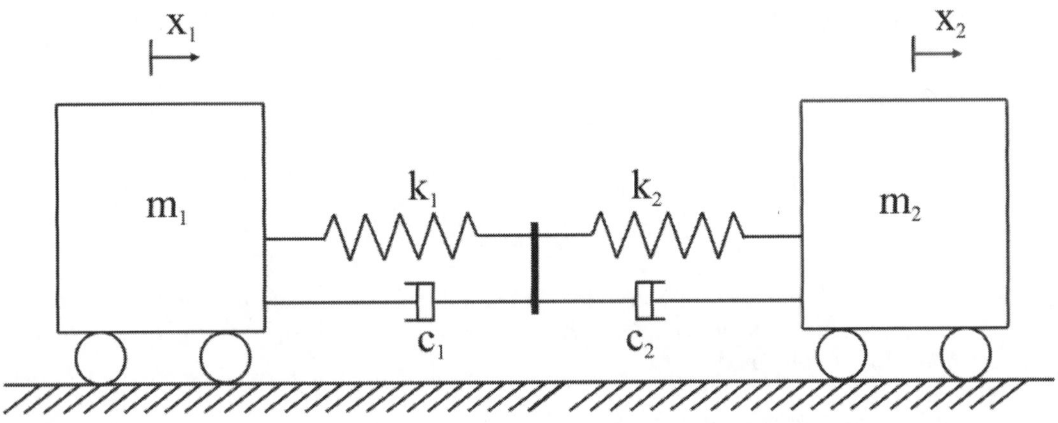

Figure 4. Behind collision of two cars

The kinetic energy of the system is:

$$E_c = \frac{m_1\dot{x}_1^2}{2} + \frac{m_2\dot{x}_2^2}{2} \tag{3}$$

where m_1 and m_2 are the masses of the two cars, $\dot{x}_1$ and $\dot{x}_2$ are velocities of cars, x_1 and x_2 are their elastic deformations. The two elastic elements with the constants k_1 and k_2 will be considered in series. Also in the series, the two elements of amortization, having the c_1 and c_2 damping coefficients, will be considered connected. By k_e, it is noted the equivalent elastic constant of the two elastic elements, and with c_e is the equivalent damping constant of the two damping elements. With these notations the expression of the potential energy, is determined with the expression:

$$E_p = \frac{k_e(x_1 - x_2)^2}{2}; k_e = \frac{k_1 k_2}{k_1 + k_2}; \tag{4}$$

and the expression of the energy of dissipation is:

$$E_d = \frac{c_e(\dot{x}_1 - \dot{x}_2)^2}{2}; c_e = \frac{c_1 c_2}{c_1 + c_2} \tag{5}$$

Lagrange's two equations have the following form:

$$\frac{d}{dt}\left(\frac{\partial E_c}{\partial \dot{x}_1}\right) - \frac{\partial E_c}{\partial x_1} = -\frac{\partial E_d}{\partial \dot{x}_1} - \frac{\partial E_p}{\partial x_1}; \tag{6}$$

$$\frac{d}{dt}\left(\frac{\partial E_c}{\partial \dot{x}_2}\right) - \frac{\partial E_c}{\partial x_2} = -\frac{\partial E_d}{\partial \dot{x}_2} - \frac{\partial E_p}{\partial x_2}. \tag{7}$$

By replacing Lagrange's equations with expressions of kinetic, deformation and dissipation energies, two differential equations of the second order that govern the movement of the two bodies for the time interval they are in contact are obtained.

$$m_1\ddot{x}_1 = -k_e(x_1 - x_2) - c_e(\dot{x}_1 - \dot{x}_2); \tag{8}$$

$$m_2\ddot{x}_2 = k_e(x_1 - x_2) + c_e(\dot{x}_1 - \dot{x}_2) \tag{9}$$

Given that simulation in Simulink aims to integrate the two differential equations, it explains from each equation the accelerations of the two bodies during the collision.

$$\ddot{x}_1 = \frac{-c_e(\dot{x}_1 - \dot{x}_2) - k_e(x_1 - x_2)}{m_1}; \tag{10}$$

$$\ddot{x}_2 = \frac{c_e(\dot{x}_1 - \dot{x}_2) + k_e(x_1 - x_2)}{m_2}. \tag{11}$$

These two equations can be entered into the Simulink block diagrams, with multiplication, integration and introduction of the initial conditions, based on which the deformation, velocity and acceleration diagrams are obtained.

3.2. The frontal crush simulation of two cars

For the second case considered, the two cars collide frontal in opposite directions, making the deformation of the elastic elements equal to the sum of the individual deformations of the two machines, and similarly the speed taken into account in the expression of dissipation energy is equal to the sum of the speeds of the two machines for the time interval of the collision.

With these observations, kinetic, deformation and dissipation energies will have the following expressions:

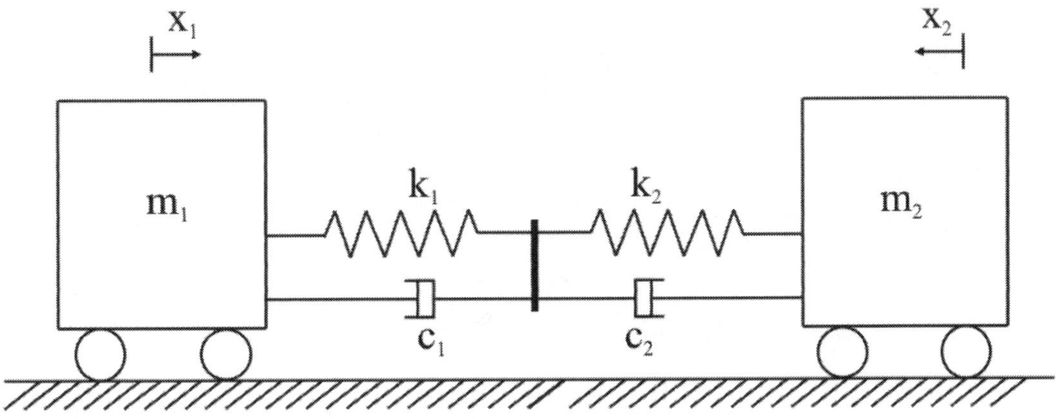

Figure 5. Frontal collision of two cars

$$E_c = \frac{m_1\dot{x}_1^2}{2} + \frac{m_2\dot{x}_2^2}{2} \tag{12}$$

$$E_p = \frac{k_e(x_1 + x_2)^2}{2} \tag{13}$$

$$E_d = \frac{c_e(\dot{x}_1 + \dot{x}_2)^2}{2} \tag{14}$$

Using the Lagrange equations that will have the same form as given in equations (6) and (7), are obtain the differential equations of the frontal collision of the two cars from opposite directions:

$$m_1\ddot{x}_1 = -c_e(\dot{x}_1 + \dot{x}_2) - k_e(x_1 + x_2) \tag{15}$$

$$m_2\ddot{x}_2 = -c_e(\dot{x}_1 + \dot{x}_2) - k_e(x_1 + x_2) \tag{16}$$

For the integration of the differential equations (15) and (16), the accelerations of the two cars must be explained during the collision, and based on them the model will be built in Simulink.

$$\ddot{x}_1 = \frac{-c_e(\dot{x}_1 + \dot{x}_2) - k_e(x_1 + x_2)}{m_1} \tag{17}$$

$$\ddot{x}_2 = \frac{-c_e(\dot{x}_1 + \dot{x}_2) - k_e(x_1 + x_2)}{m_2} \tag{18}$$

4. Simulations and discussions

4.1. Simulation car to barrier collision

For simulation was considerate weight of one medium class vehicle, m=900kg, damper coefficient, c=1,500Nm/s and stiffness constant, k=170,000N/m, v=25m/s.

These two models of collision were realized in numerical simulation software, Simulink. The scheme of these models is shown in Figure 6. The model was made following equation (1) and (2) of the system, for the first system, the damping ratio is zero, c=0.

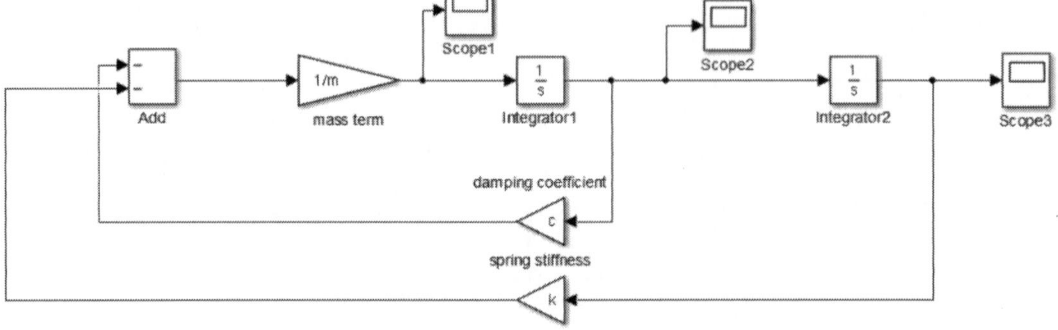

Figure 6. Mass-spring-damper model in Simulink

The simulation results can be seen in the diagrams of Figures 7a, 7b and 7c obtained on the scopes in the above scheme. In the figures are presented: deformation, velocity and acceleration during collision.

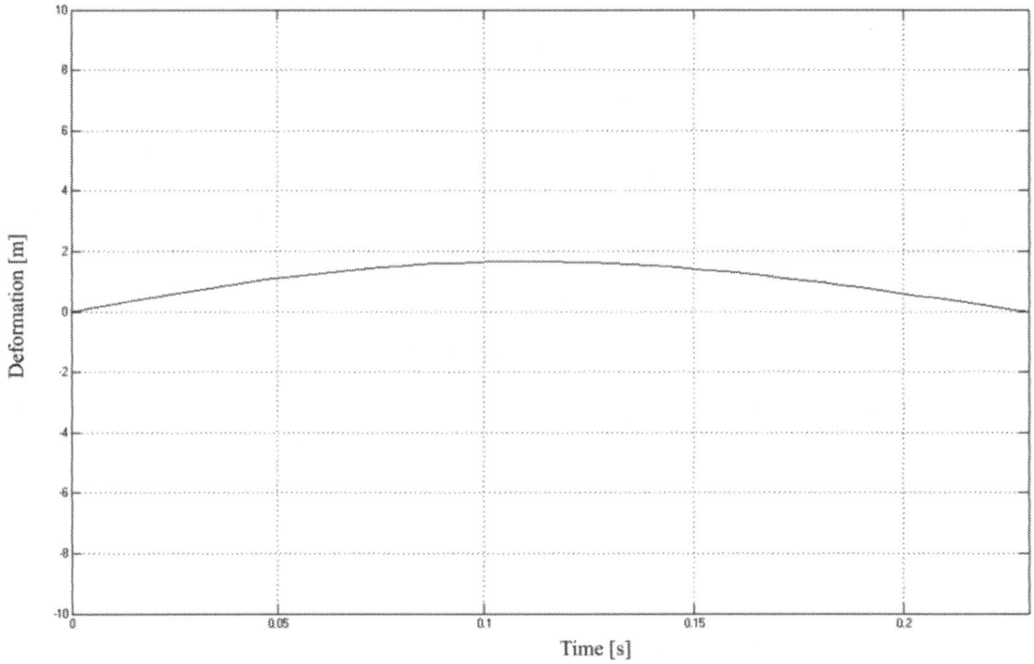

Figure 7a. Deformation of mass-spring-damper model

In the collision model of a vehicle with a barrier, in the absence of damping, the elastic collision ends after 0.16 s, and in the presence of the damping, with the values presented above, the collision ends after 0.232 s. When the velocity becomes zero, that is, after the time t = 0.116, the deformation phase is finished and follows the recovery phase, as can be seen in Figures 7a and 7b.

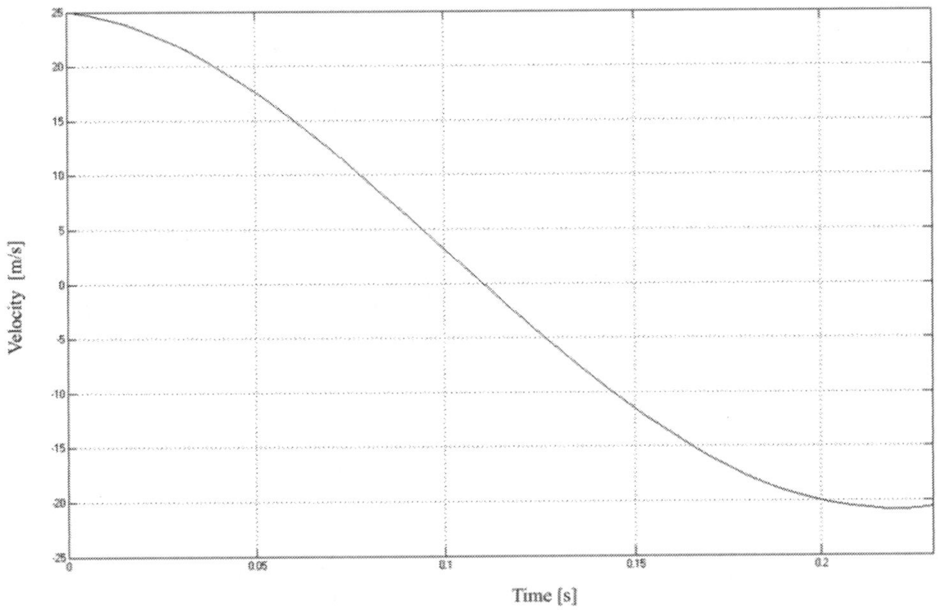

Figure 7b. Velocity of mass-spring-damper model

Therefore, the collision will have two phases. In the compression phase, the car deforms until the speed is canceled. In the second, deformation phase, a part of the deformation is recovered and held until the acceleration becomes zero. This happens after a time t = 0.232s.

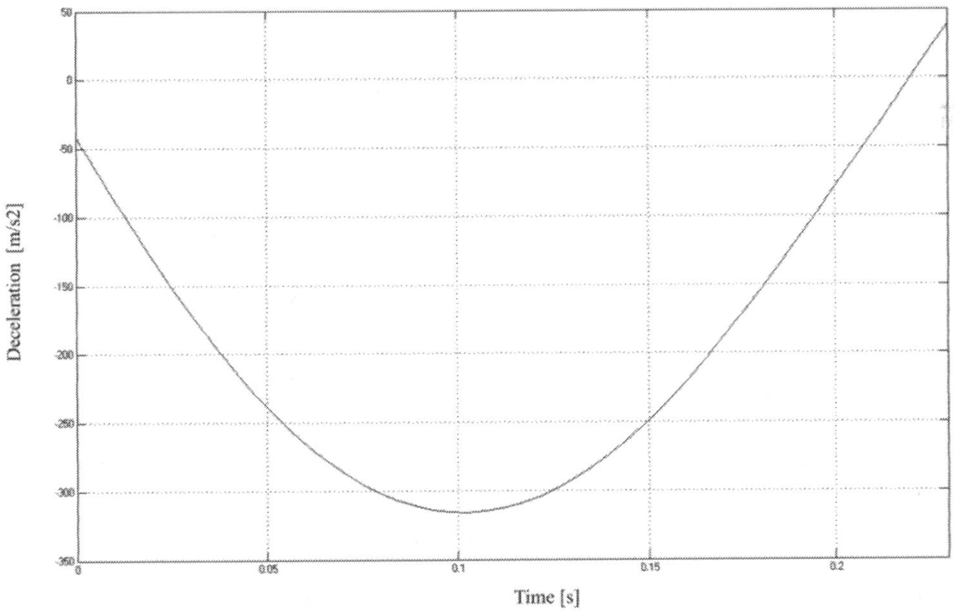

Figure 7c. Deceleration of mass-spring-damper model

It is found that in both cases there has been a sharp deceleration in 0.117 seconds for the first model and 0.1 seconds for second case. These decelerations are critical for vehicle passengers. Therefore, is the desire to increase the both of passive and active safety for new cars.

4.2. Simulation behind collision of two cars

Equations 10 and 11 are numerically resolved and plotted using the Simulink module in MATLAB, following the construction of the bloc diagram in Figure 8. In the simulation, the masses of two cars were considered m_1=1,200kg and m_2=1,400kg. For the two elastic elements, the following values for the elastic constants were considered as k_1=170,000N/m respectively for k_2=210,000N/m, and for the damping elements were considered the following values of the damping coefficients c_1=1,500Ns/m, respectively c_2=2,000Ns/m. The speeds of the two cars in the moment of collision were considered to be v_1=25 m/s, respectively v_2=20 m/s.

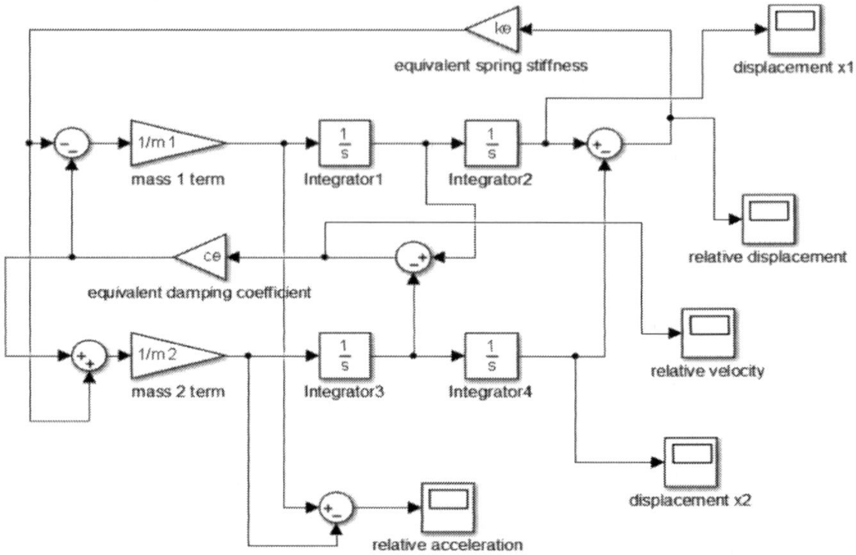

Figure 8. The behind collision model of two cars in Simulink

Diagrams in Figure 9 are the relative deformation, relative velocity and relative acceleration. For reasons of the order of magnitude, relative deformation and relative velocity are multiplied by 20 and 10. They are represented in overlapping graphs to observe the two phases of the collision.

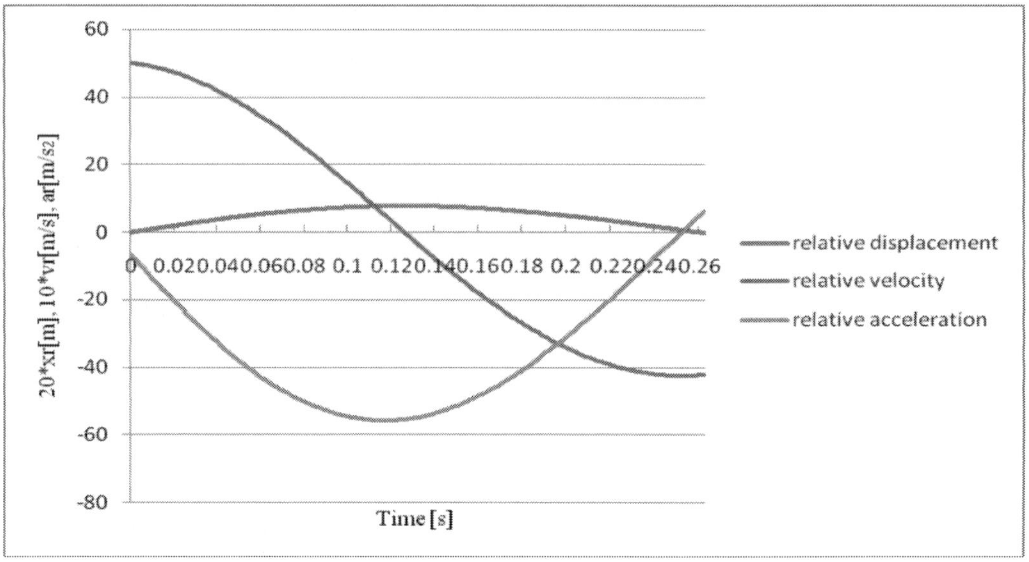

Figure 9. Relative displacement, velocity and acceleration diagrams, in behind collision of two cars

4.3. Simulation frontal collision of two cars

In case of front collision, the characteristic elements of the two cars are the same as in the previous case, only the speeds are v_1=25 m/s and v_2=10m/s. The frontal collision of two cars in Simulink is presented in Figure 10.

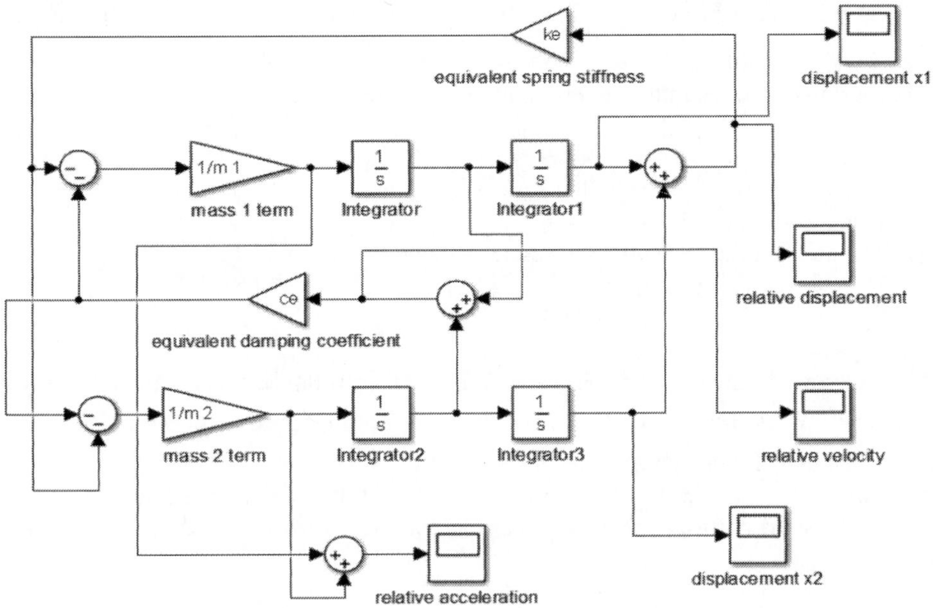

Figure 10. The frontal collision model of two cars in Simulink

The three diagrams representing relative displacement, relative speed and relative acceleration in the frontal collision of two cars, are represented in Figure 11 by overlapping to observe the two phases of the collision.

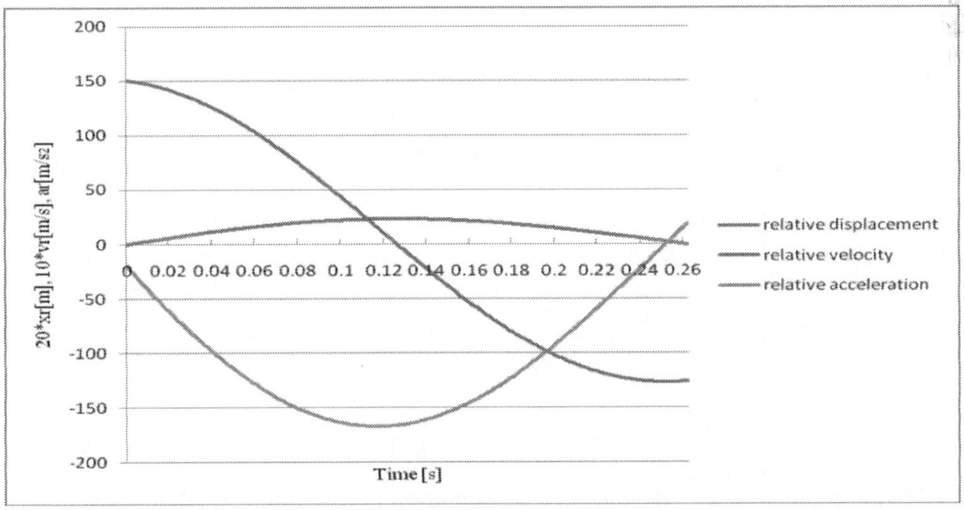

Figure 11. Relative displacement, velocity and acceleration diagrams, in frontal collision of two cars

4.4. Coefficient of restitution

Coefficient of restitution C_R is a parameter used to estimate the lost energy during a collision. It is a measure of the elasticity of collision. There are three definitions for this parameter [5].

IOP Conf. Series: Materials Science and Engineering **294** (2017) 012090 doi:10.1088/1757-899X/294/1/012090

Kinematic definition of the restitution coefficient, given by Newton:

$$C_R = \frac{P_c}{P_r} = \frac{v_2^{'} - v_1^{'}}{v_1 - v_2} \tag{19}$$

where P_c is compression impulse, P_r is restitution impulse, $v_2^{'}$ and $v_1^{'}$ are the velocities after impact of two cars, v_1 and v_2 are the velocities before the impact.

Kinetic definition of the restitution coefficient, given by Poisson:

$$C_R = \frac{P_c}{P_r} \tag{20}$$

where P_c and P_r are as defined above.

Energetic definition of the restitution coefficient, given by Stronge [6]:

$$C_R^2 = -\frac{W_r}{W_c} \tag{21}$$

where W_r is the sum of the work done by both normal impulsive forces during the restitution the collision, and W_c is the sum of the work done by both normal impulsive forces during the compression phase of the collision.

The three expressions are similar if the impact is collinear, the collision is centered, and the friction is neglected for the short collision time [7]. These hypotheses were assumed in the modeling made in this paper.

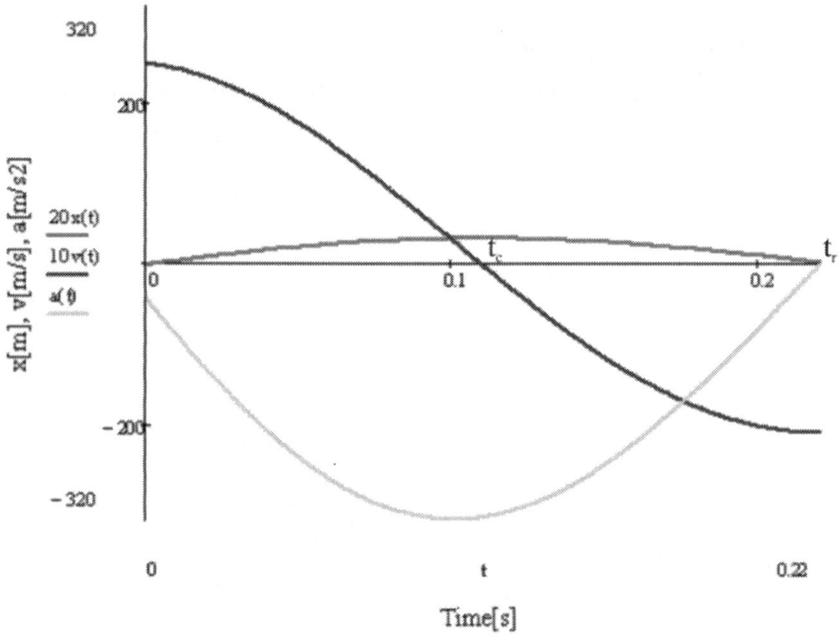

Figure 12 Deformation, velocity and acceleration of the mass, spring, damped model for v=25 m/s

Taking into account the order of magnitude of the three kinematic elements of motion: displacement, velocity and acceleration, graphically represented in the diagrams in Figure 12 and to observe the behavior of the car during collision with a barrier, the displacement was multiplied by 20 and the speed was multiplied by 10.

In order to discuss and conclude on the three simulations presented in Simulink, the coefficient of restitution was calculated in the case of a collision of a car by a barrier. It is considered the model of Figure 3 whose equation (2) was written in the form:

$$\ddot{x} + 2\sigma\dot{x} + \omega_n^2 x = 0 \tag{22}$$

where ω_n, p, σ, and ς are the circular frequency, the damped circular frequency, the damping factor and the damping ratio, having the following expressions:

$$\omega_n = \sqrt{\frac{k}{m}}; p = \omega_n\sqrt{1-\varsigma^2}; \sigma = \frac{c}{2m}; \varsigma = \frac{c}{2\sqrt{km}} \tag{23}$$

Considering the initial conditions given by the moment of the collision with the barrier, when the car has the speed v_1, the three kinematic elements during collision, deformation, velocity and acceleration have the following analytical expressions:

$$x(t) = \frac{v_1}{p}e^{-\sigma t}\sin pt \tag{24}$$

$$v(t) = v_1 e^{-\sigma t}(\cos pt - \frac{\sigma}{p}\sin pt) \tag{25}$$

$$a(t) = v_1 e^{-\sigma t}[-2\sigma\cos pt + (\frac{\sigma^2}{p} - p)\sin pt] \tag{26}$$

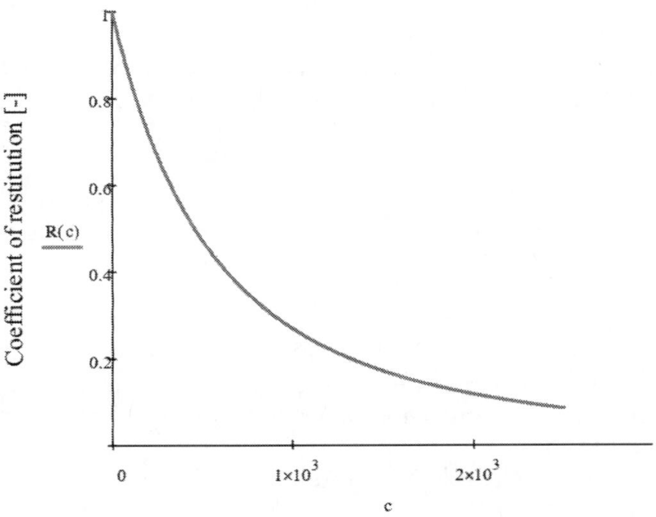

Figure 13. Coefficient of restitution versus damping coefficient

The compression phase lasts until the speed becomes zero this time is t_c, and the restitution phase lasts until the resulting force is zero, so the acceleration is canceled, this time being t_r (Figure 12). It can calculate the restitution coefficient in the collision of a car with a barrier, giving the expression:

$$C_R = \frac{v(t_r)}{v_1} = e^{-\frac{c}{\sqrt{4km-c^2}}\arccos\frac{c^2-2km}{2km}} \tag{27}$$

An analysis of this formula shows that the restitution coefficient depends on the three constituent parameters of the mechanical system mass m, stiffness k and damping c. By keeping constant the first two parameters m and k, and by modifying the damping coefficient c the dependence of the restitution coefficient is obtained as in Figure13.

It can be seen three important values for the coefficient of restitution:

1. For undamped system, c =0 and the coefficient of restitution is $C_R = 1$.The crash is perfectly elastic.
2. If the system has critical damping then, $c = 2\sqrt{km}$. After the elimination of the indetermination in the equation (27) the coefficient of restitution is: $C_R = e^{-2} = 0.13533$.
3. If the system is heavily over-damped then, $C_R = 0$. The collision is plastic.

5. Conclusions

To follow the collision of two cars in Simulink, the relative displacement, relative speed and relative acceleration are introduced. When the two velocities, before the collision, are in the same direction, the relative displacement is $x_r = x_1 - x_2$, and in the case of the front collision, the deformation is given by $x_r = x_1 - x_2$. Thus, practically the mechanical system can be considered with only one degree of freedom. In the diagrams given in Figures (9) and (11), it can be noticed that there is an identical behavior in the collision of two cars with the behavior of a collision of a car with a barrier. It should be noticed that here by t_c it can be understand the time after which the relative speed is zero, and the two cars arrive at an equal common velocity, followed by the restitution phase, until the moment t_r, when the relative acceleration is null. Similarly, it can be calculated the restitution coefficient that will have the same expression given by the relationship (27), where the mass m will be the equivalent mass m_e, the elastic constant k will be replaced by the equivalent elastic constant k_e and the damping coefficient c with the equivalent damping coefficient.

A structural analysis [8], that takes into account all the components of the front vehicle: torque box, front frame, drive line, sheet metal, wall radiator, fan and grille and transmission mount can bring a much better characterization of the behavior of a vehicle collision and sad simulation results close to real if the mechanical characteristics of these components are know.

These simulations help developed new technologies in vehicle crash safety, saving several resources. In simulation can bring a much better characterization of the behavior of a vehicle collision simulation results close to real if the mechanical characteristics of these components are know.

References

[1] Kamal M M 1970 *Analysis and Simulation of Vehicle to Barrier Impact*, SAE Paper No. 700414
[2] Magee C L 1988 *Design for Crash Energy Management- Present and Future developments*, The Seventh International Conference on Vehicle Structural Mechanics, Detroit, Michigan, USA, 2-4 April, 142-154
[3] Lim G G and Paluszny A 1988 *Side Impact Research*, XXII Fisita Congress- SAE Publication P-211, Paper No. 885055
[4] Green J E 1977 *Computer Simulation of Car-To-Car Collisions*, SAE Paper No. 770015
[5] Ahmad M, Ismail K A and Mati F 2016 Impact models and coefficient of restitution: A review, *ARPN Journal of Engineering and Applied Sciences* **11**(10) 6549-6555
[6] Stronge W J 2000 *Impact Mechanics*. Cambridge: Cambridge University Press
[7] Daily J, Strickland R and Daily J 2006, *Crush Analysis with Under-rides and the Coefficient of Restitution,* Institute of Police Technology and Management's 24 th Annual Special Problems in Traffic Crash Reconstruction
[8] Prasad P and Belwafa J E 2004 *Vehicle Crashworthiness and Occupant Protection*, American Iron and Steel Institute, Town Center, Southfield, Michigan

Determination of the wind power systems load to achieve operation in the maximum energy area

C P Chioncel[1], G O Tirian[2], E Spunei[1] and N Gillich[1]

[1]Eftimie Murgu University of Resita, Electrical Engineering and Informatics Department, Traian Vuia str., no. 3-4, 320085 Resita, Romania
[2]Politehnica University of Timisoara, Department of Electrical Engineering and Industrial Informatics, 5 Revolution Street, Hunedoara, 331128, Romania

E-mail: c.chioncel@uem.ro

Abstract. This paper analyses the operation of the wind turbine, WT, in the maximum power point, MPP, by linking the load of the Permanent Magnet Synchronous Generator, PMSG, with the wind speed value. The load control methods at wind power systems aiming an optimum performance in terms of energy are based on the fact that the energy captured by the wind turbine significantly depends on the mechanical angular speed of the wind turbine. The presented control method consists in determining the optimal mechanical angular speed, ω_{OPTIM}, using an auxiliary low power wind turbine, WT_{AUX}, operating without load, at maximum angular velocity, ω_{MAX}. The method relies on the fact that the ratio $\omega_{OPTIM}/\omega_{MAX}$ has a constant value for a given wind turbine and does not depend on the time variation of the wind speed values.

1. Introduction

In the current phase of harnessing wind energy, an optimum functioning is a fundamental requirement, considering the fact that the investments (in all stages: development, approval procedures, wind turbine acquisition, building & exploitation) are huge and optimum operating conditions under given wind characteristics, have to be ensured for maximizing the yield of wind power plants [1].

In the scientific literature the operation problem of the wind turbine (WT) in the maximum energy area is extensively treated, [2] but simplified, at constant wind speeds over time, using different mathematical models (MM - WT) given by the wind turbine producers and deduced under laboratory conditions, far different from those in operation. For that reason, the obtained electric energy has less value than the maximum possible provided in the Maximum Power Point (MPP) operation, obtainable at optimum angular mechanical velocity (ω_{OPTIM}). At time-varying wind speeds, the operation in MPP becomes a complex problem and not always timely solvable, due to large mechanical inertia and rapid changes in wind speed over time [3].

To reach the MPP area, currently the main used calculation procedure is the so-called small perturbation method; the obtained solutions introduce power swing in the generator and do not provide an operation in MPP. The main issue is to adapt and develop control algorithms and strategies that are based on the current wind speed value and the prescription of the optimum speed (RPM) [4].

The dependence of the WT power to mechanical angular velocity, ie the $P_{WT}(\omega)$ function, at a certain constant wind speed, has a maximum at ω_{OPTIM}. The optimum energy value is reached for ω_{OPTIM}, value at which through the wind turbine the captured power is a maxim, the MPP from the equivalent power characteristic. The important points from the power characteristic are the maximum power point (MPP) with ω_{OPTIM} and the zero power point with maximum angular velocity, ω_{MAX}. The correct determination of these points, in operating conditions, provides the operation in the MPP zone.

The maximum power point (MPP) coordinate, ω_{OPTIM} and P_{WT-MAX}, directly depend on the wind velocity variation in time. The correct determination of those depends on the correct wind speed information and the knowledge of their influence over the value of the MPP coordinates [5].

Therefore, achieving optimum performance for the wind energy system in terms of energy, two fundamental issues are distinguished: the determination of the MPP coordinates based on the estimated wind speed values.

In the present paper, the knowing of the wind speed time variation is done in two ways: through direct measurement or by measuring the mechanical velocity from the auxiliary wind turbine, WT_{AUX}, operating without a load. The determination of the MPP coordinates are based on the $\omega_{OPTIM}/\omega_{MAX}$ ratio has a constant value for a given wind turbine and does not depend on the wind speed variations.

2. Mathematical models

The simulations presented in the paper, are based on the classic mathematical models of the wind turbines and the Permanent Magnet Synchronous Generator [6], based on them, the optimum angular velocity, ω_{OPTIM}, is deduced.

2.1. Mathematic model of the wind turbine (MM-WT)

We envisage a classic model for the turbine model [7], which allows determination of the optimum angular velocity so that the captured energy is maximized.

The wind turbine power is computed based on the equation:

$$P_{WT} = \rho\pi R_b^2 C_p(\lambda) v^3 \tag{1}$$

where: ρ - air density, v – wind speed, Cp(λ) – power conversion coefficient, Rb - rotor blade radius [5].

The following relation determines Cp(λ), the power conversion coefficient:

$$C_p(\lambda) = c_1\left(\frac{c_2}{\Lambda} - c_3\right)e^{-\frac{c_4}{\lambda}} \tag{2}$$

where: $\lambda=\omega Rb/v$, $c_1 \div c_4$ – catalog constants and ω - angular mechanical velocity.

$$\frac{1}{\Lambda} = \frac{1}{\lambda} - 0.035 = \left(\frac{v}{1.5\omega} - 0.035\right) \tag{3}$$

Replacing equation (3) in (2), Cp(λ), the power conversion coefficient [5] results as:

$$C_p(\lambda) = c_1\left(c_2\left(\frac{v}{1.5\omega} - 0.035\right) - c_3\right)e^{-c_4\left(\frac{v}{1.5\omega} - 0.035\right)} \tag{4}$$

The wind turbine power is computed through (1) or

$$P_{WT} = k_1\left(k_2\left(\frac{v}{\omega} - 0.035R_p\right) - c_3\right)e^{-k_3\left(\frac{v}{\omega} - 0.035R_p\right)}v^3 \tag{5}$$

where: $k_1=1.225R_p^2 \cdot c_1$, $k_2=c_2/1.5$, $k_3=c_4/1.5$ [7].

The producers of wind turbines, WT, with Rp=1.5m, give the experimental power characteristics $P_{WT}(\omega,v)$ or torque $\tau_{WT}(\omega,v)$, called the experimental mechanical characteristics:

$$\tau_{WT}(\omega) = \frac{P_{WT}(\omega,v)}{\omega} \tag{6}$$

For the reference mechanical angular speed, ω_{ref}, we obtain, through derivation, the maximum of the P_{WT} function:

$$\frac{dP_{WT}(\omega,v)}{d\omega}=0 \qquad (7)$$

Solving the equation (7), we obtain

$$\omega_{OPTIM}=k_0\cdot v. \qquad (8)$$

The achieved result demonstrates the direct link between $\omega_{ref}=\omega_{OPTIM}$ and wind speed. Substituting the obtained results in the $P_{WT}(\omega,v)$ function, we obtain:

$$P_{WT}(\omega,v)=kp\cdot v^3. \qquad (9)$$

At an operation without load, correspond to the maximum angular mechanical velocity, obtained from

$$P_{WT}(\omega,V)=0, \qquad (10)$$

provides

$$\omega_{MAX}=k_{WL}\cdot v, \qquad (11)$$

where k_{WL} – represents the constant without load.

The ratio $\omega_{OPTIM}/\omega_{MAX}$ has the value

$$\omega_{OPTIM}/\omega_{MAX}=k_0\cdot v/\ k_{WL}\cdot v=k_0/\ k_{WL}, \qquad (12)$$

or

$$\omega_{OPTIM}/\omega_{MAX}=k_M, \qquad (13)$$

where k_M – is the constructive constant of the turbine.

The maximum power of the wind turbine, at a certain wind speed v, can be determined based on mechanical angular speed without load, ω_{MAX}, since replacing equation (11) in (9), results:

$$P_{WT\text{-}MAX}=k_{p\text{-}1}\cdot\omega^3{}_{MAX}, \qquad (14)$$

where $k_{P\text{-}1}$ – represent constructive turbine constant.

The value of the maximum mechanical angular speed, ω_{MAX}, is given from the operation, without load, of an auxiliary, low power, turbine, WT_{AUX}.

In cases where the wind speed varies significantly over time, the obtained result should be reconsidered and it imposes to define an equivalent wind speed [3].

2.2. A mathematic model of the permanent magnet synchronous generator (MM-PMSG)

To analyze at time-varying wind speeds the operation of the system (WT + PMSG), the orthogonal mathematical model for the permanent magnet synchronous generator is used, given through [8]:

$$\begin{cases} -U\sqrt{3}\sin\theta = R_1 I_d - \omega L_q I_q \\ U\sqrt{3}\sin\theta = R_1 I_q + \omega L_d I_d + \omega\Psi_{PM} \\ \tau_G = p_1\left(L_d - L_q\right)I_d I_q + I_q\Psi_{PM} \\ P_G = R\left(I_d^2 + I_q^2\right) \end{cases} \qquad (15)$$

where: I_d, I_q – stator currents, U – stator voltage, θ - load angle, τ_G – electromagnetic torque, R_1 – stator winding resistance, Ψ_{PM} – flux of the permanent magnet, L_d – own inductance of the stator winding on d axis, L_q – own inductance of the stator winding on q axis, p_1 – number of pole pairs. The two functions $P_G(R,\omega)$ the electric power debited by the generator and the electromagnetic torque, $\tau_G(R,\omega)$, depends on R – the load resistance and ω - mechanical angular velocity. At constant R, the torque [9] depends on linearly of ω, and the useful power depends on the quadratic ω.

From catalog data for a PN = 10 [kW] permanent magnet synchronous generators, following values for the parameters are obtained: R1= 1.6Ω, Ld = 0.07H, Lq = 0.08H, Ψ_{PM} = 1.3Wb, we obtained the torque, τ_G,

$$\tau_G=-845\,\omega(5R+8)\frac{4\omega^2+625R^2+2000R+1600}{\left(1250R^2+4000R+3200+7\omega^2\right)^2} \qquad (16)$$

and the generator power, PG,

$$P_G = 4225\,R\omega^2 \frac{4\omega^2 + 625\,R^2 + 2000\,R + 1600}{\left(1250\,R^2 + 4000\,R + 3200 + 7\omega^2\right)^2} \tag{17}$$

The transient analysis of the specific phenomena at variable wind speed is done by simulation, based on the movement equation:

$$J\frac{d\omega}{dt} = \tau_{WT} - \tau_{MPSG} \tag{18}$$

where: J – equivalent inertia moment, τ_{WT} – wind turbine torque and τ_{MPSG} – permanent magnet synchronous generator torque.

3. Control algorithm verification at different wind conditions

3.1. Analysis at constant wind speed

At a wind speed value of v=11[m/s], the power characteristics have the form as

$$P_{WT}(\omega,v)=1191.5\cdot(v/\omega-0.02)e^{-98.02(v/\omega)}\cdot v^3 \tag{19}$$

$P_{WT}(\omega,11)$, and by cancelling the derivative power,

$$\frac{dP_{WT}(\omega,v)}{d\omega} = 0 \tag{20}$$

we obtain ω_{OPTIM}=364.26[rad/s], the maximum power being $P_{WT}(\omega_{OPTIM},11)$=837.04[rad/s].

Considering that initially, the wind power system operates in point P_1, Figure1, at a mechanical angular speed of ω=469[rad/s], the power captured by the wind turbine results, from (19) as PWT=549.25[W].

Equaling the power from the wind turbine and the MPSG, PWT=PMPSG, at ω=469[rad/s], results

$$549 = 845\omega^2(5R+8)\frac{4\omega^2 + 625R^2 + 2000R + 1600}{\left(1250R^2 + 4000R + 3200 + 7\omega^2\right)^2} \tag{21}$$

and the resistance value from the point P_1 becomes R=673.64Ω.

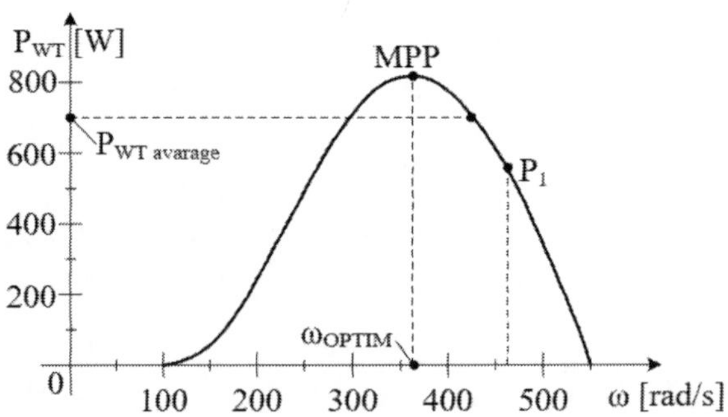

Figure 1. Power characteristic of the wind turbine, at v=11[m/s]

Remark 1. At a stable operation in point P_1, the power and equivalent load resistance of the PMSG is determined from measurements. To reach the maximum power point (MPP), in a imposed time interval, e.g $\Delta t = t_k - t_{k-1}$ = 2000[s], the PMSG has to take over, both, the captured wind energy by the wind turbine as well as the kinetic energy, $\Delta W_{KINETIC}$, of the rotating masses:

$$\Delta W_{KINETIC} = J\left(\frac{\omega_k^2 - \omega_{k-1}^2}{2}\right) = \left|45\frac{364.26^2 - 469^2}{2}\right| \tag{22}$$

where J – equivalent inertia moment, ω_{k-1} – mechanical angular speed at t_{k-1} and ω_k – mechanical angular speed at t_k.

The overtaken wind energy by the wind turbine in the Δt time interval is computed by integrating the power, under the form of:

$$W_{WT} = \int_{t_{k-1}}^{t_k} P_{WT} dt \tag{23}$$

and depends on the current angular mechanical velocity, ω, that changes in the time interval Δt in the field of $\omega_{OPTIM} \div \omega(P_1) = 364.26[rad/s] \div 469[rad/s]$.

For this reason, the power integration is not able to be computed and the captured wind energy is done approximating a linear power variation between the points P_1 and MPP, Figure 1.

Considering an average value for the wind turbine power,

$$P_{WT} = \frac{P_{WT-P1} + P_{WT-MPP}}{2} = \frac{594.25 + 837.04}{2} = 693.15W \tag{24}$$

we obtain the value of the captured wind energy through the wind turbine, in the given time interval Δt, as:

$$W_{WT} = \int_{t_{k-1}}^{t_k} P_{WT} dt = 693.15 \cdot 2000 = 1.386 \cdot 10^6 J \tag{25}$$

By summing the two values, we obtain the necessary energy, $W_{G\text{-necessary}}$, what the generator should produce in the time interval Δt:

$$W_{G\text{-necessary}} = W_{WT} + \Delta W_{KINETIC} = (1.3863 + 1.937) \cdot 10^6. \tag{26}$$

Reaching the optimum value of the mechanical angular speed is achieved in condition in that the power of the generator has the value $P_{G\text{-necessary}} = 3.35 \cdot 10^6/2000 = 1675[W]$, at an average value of the mechanical velocity of $\omega_{average} = (\omega_{OPTIM} + \omega_{P1})/2 = (364.26+469)/2 = 416.63[rad/s]$, resulting the value of the system load resistance, R, by replacing $\omega_{average}$ and PG-necessary in the PPMSG expression, as $R=168.64[\Omega]$.

Verification by solving the movement equation. At $R=168.64[\Omega]$ and $\omega(0)=469$, the movement equation becomes:

$$45 \frac{d\omega}{dt} \omega = 1191.5(11/\omega - 0.02) e^{-98.02(11/\omega)} 11^3 -$$

$$845\omega^2(5R+8) \frac{4\omega^2 + 625R^2 + 2000R + 1600}{(1250R^2 + 4000R + 3200 + 7\omega^2)^2}, \tag{27}$$

and the resulted value of the mechanical velocity at the end of the interval, becomes $\omega(2000)=372.95[rad/s]$, compared to the optimum, $\omega_{OPTIM}=364.26[rad/s]$, the recorded difference being worth 2.38%, therefore acceptable.

3.2. Analysis at variable wind speed

At a time variation of the wind speed [10] as in Figure 2 [5], the optimal values have been obtained, $P_{WT\text{-}MAX\text{-}1}=2690[W]$ at $\omega_{OPTIM}=579.7[rad/s]$.

It's considered that the wind power system operates initially in the point P_2, Figure 3, at the angular mechanical velocity of $\omega=669[rad/s]$. At this value of the angular mechanical velocity, the power of the wind turbine is given from the captured wind energy and has the value $P_{WT}=62849/35=1795.7[W]$.

Equaling the power from the MPSG and the wind turbine, PMPSG=P_{WT}, at $\omega=669[rad/s]$, results in the value of the load resistance in point P_2: $R=414.45[\Omega]$.

Remark 2. At a stable operation in P_2, the equivalent power and load resistance from the PMSG is determined from measurements. To reach the maximum power point (MPP), in an imposed time interval, e.g $\Delta t = t_k - t_{k-1} = 2000[s]$, the PMSG has to take over, both, the kinetic energy of the rotating masses (substituting the values in equation (22)), $\Delta W_{KINETIC}=2509\cdot10^6$[J] as well as the captured wind energy by the wind turbine.

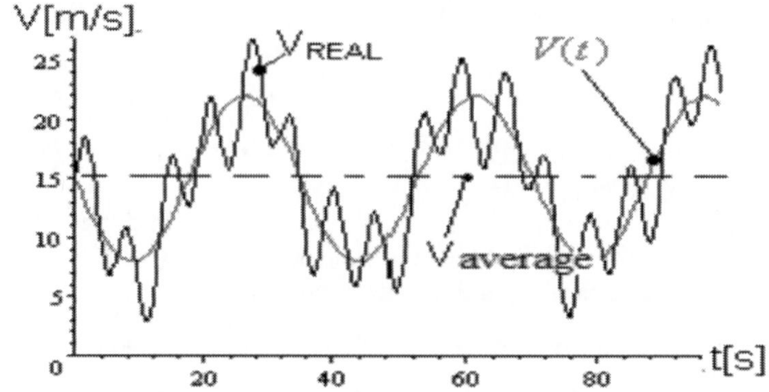

Figure 2. Time variation of the wind speed [5]

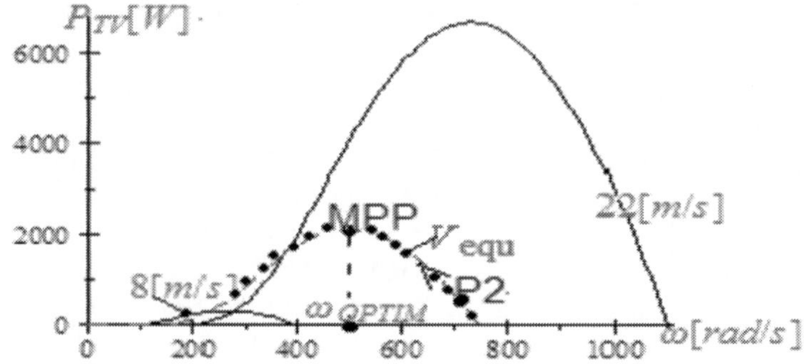

Figure 3. Power characteristics for wind speed [5]

The wind energy taken over by the wind turbine in this given time interval is computed based on the power integration and depends on the current mechanical angular speed, ω, that changes in Δt between $\omega_{OPTIM}\cdot\omega(P_2)$=579.7[rad/s]÷669[rad/s].

For this reason, the power integration is not possible to be computed, the captured wind energy is calculated, only approximate, in two ways:

- Considering a linear variation of ω between the points P_2 and MPP: $\omega=\omega(P2)-((\omega(P2)-\omega_{OPTIM})/2000)\cdot t=(669-0.04465\cdot t))$, we obtain $W_{WT}=3.9112\cdot10^6$J
- Considering a linear variation of the power between the points P_2 and MPP, Fig.3, and considering an average value for the power, $P_{WT\text{-}AVERAGE}=(P_{WT\text{-}P2}+P_{WT\text{-}MPP})/2=(1795.7+2060.9)/2=1928.3$[W].

Summing the two energy [11] values, $W_{WT}+\Delta W_{KINETIC}$, we get the necessary energy that the generator has to debit in the Δt time interval: $W_{G\text{-}necessary}=(3.356+2.509)\cdot10^6=6.36\cdot10^6$[J].

The optimum value of the mechanical angular speed is achieved when the generator power has the value $P_{G\text{-}necessary}=6.36\cdot10^6/2000=3182.8$[W], at an average of the angular mechanical velocity of $\omega_{average}=(\omega_{OPTIM}+\omega_{P2})/2=(579.7+669)/2=624.35$[rad/s]. The system load resistance, R, results through replacing $\omega_{average}$ and $P_{G\text{-}necessary}$ in the P_{PMSG} expression, as R=195.9[Ω].

Achieving the optimal mechanical angular velocity, based on the link between ω_{OPTIM} and ω_{MAX}, is done through measurements of the generators energy and mechanical velocity, in steps, in $\Delta t=t_k-t_{k-1}$ time intervals, based on the following control algorithm, synthetized in Figure 4 and reflected in Figure 5.

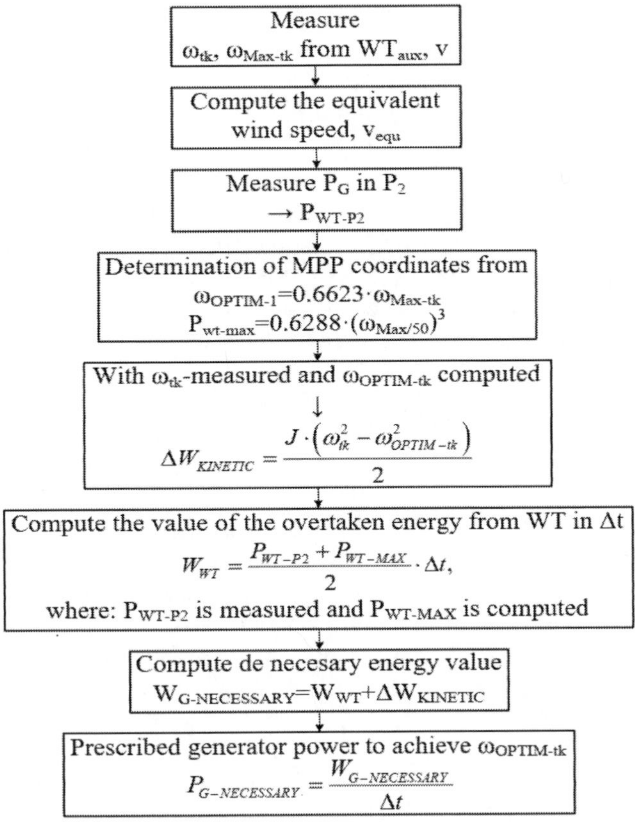

Figure 4. Proposed control algorithm

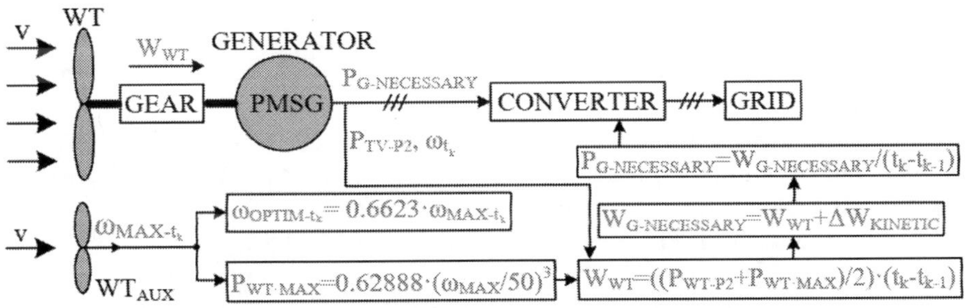

Figure 5. Optimal control of a wind system based on the link between ω_{OPTIM} and ω_{MAX}

4. Conclusions

Through measuring the mechanical angular speed ω_{MAX}, from the operation without the load of the auxiliary turbine, WT_{AUX}, we obtain the maximum power point coordinates: ω_{OPTIM}, using the value of the ratio $\omega_{OPTIM}/\omega_{MAX}=0.6623$ and the maximum power, $P_{WT\text{-}MAX}$, using the expression $P_{WT\text{-}MAX}=0.6288(\omega_{MAX}/50)^3$.

By analysing several cases we were able to determine the sizes that lead to an optimum performance in terms of energy and maintain the wind energy system in the MPP area, by measuring the angular mechanical velocity and the generator's energy. The correct determination of the maximum energy area is essential; the captured energy through the wind turbine has maximum values in the MPP area. The power of WT, for different values of the mechanical velocity, is estimated by computing the variations of the kinetic energies of the masses in the rotation and counting the debited electricity. By the presented simulation, we were able to identify the PMSG loads so the system reaches the optimum in terms of energy in the shortest time.

The proposed method is based on the determination of ω_{OPTIM} using an auxiliary operating WT, operating without load by the fact that ω_{OPTIM} is directly proportional with ω_{MAX}. This avoids the use of mathematical models in the control process, models that are valid only in certain circumstances. By measuring ω_{MAX} using an auxiliary WT, ω_{OPTIM} can be determined at any time, regardless the time variation of the wind speed. Calculating the variations of the kinetic energy and measuring the electricity debited by the generator, the value of the power is prescribed, a value which is achieved by the command of the power converter element, interposed between generator and network.

References

[1] Chioncel C P, Tirian G O, Gillich N, Hatiegan C and Spunei E 2017 Overview of the wind energy market and renewable energy policy in Romania, *IOP Conf. Ser.: Mater. Sci. Eng.* **163** 012009

[2] Koutroulis E and Kalaitzakis K 2006 Design of a Maximum Power Tracking System for Wind-Energy-Conversion Applications, *IEEE Trans. on Ind. Electr.* **52**(2) 486-494

[3] Balog F, Ciocarlie H, Babescu M and Erdodi G M 2016 *Equivalent speed and equivalent power of the wind systems that works at variable wind speed*, In: Balas V., Jain L., Kovačević B. (eds) Soft Computing Applications. Advances in Intelligent Systems and Computing, vol 357, Springer, Cham, pp 1325-1336

[4] Câmpeanu A, Enache M A, Augustinov L, Spunei E and Oprea G 2014 *Transitory Processes Determined by the Shunt-Reactors in the High Voltage Networks. Theoretical Synthesis and Simulation*, International Conference and Exposition on Electrical and Power Engineering, Iaşi, România, 16-18 October

[5] Chinchilla M, Arnaltes S and Burgos J C 2006 Control of permanent-magnet generators applied to variable-speed wind-energy systems connected to the grid, *IEEE Transaction on Energy Conversion* **21**(1) 130-133

[6] Chioncel C P, Bereteu L, Petrescu D I and Babescu M 2015 *Determination of Reference Mechanical Angular Speed for Wind Power Systems*, 6th International Workshop on Soft Computing Applications – SOFA, Vol. **357**, pp 1213-1222

[7] Chioncel C P, Spunei E and Babescu M 2016 *Limits of Mathematical Model used in Wind Turbine Descriptions*, 9th International Conference and Exposition on Electrical and Power Engineering (EPE), IEEE, Iasi, Romania, 20-22 October, pp 852-857

[8] Chioncel C P, Erdodi G M, Petrescu D I, Spunei E and Gillich N 2015 *Control of Wind Power Systems Imposing the Current in the Intermediate Circuit of the Converter at Variable Wind Speed*, 9th International Symposium on Advanced Topics in Electrical Engineering (ATEE), Bucharest, Romania, 7-9 May, pp 923-928

[9] Chioncel C P, Gillich N, Petrescu D I and Erdodi G M 2016 *Optimal Operation of an Wind Power System with Energy Storage in Electric Accumulators*, 9th International Conference and Exposition on Electrical and Power Engineering (EPE), Iasi, Romania, 20-22 October, pp 858-861

[10] Chioncel C P, Tirian G O, Gillich N and Raduca E 2016 Vector control structure of an asynchronous motor at maximum torque, *IOP Conf. Ser.: Mater. Sci. Eng.* **106** 012005

[11] Oguz Y, Güney I and Calik H 2013 Power Quality Control and Design of Power Converter for Variable-Speed Wind Energy Conversion System with Permanent-Magnet Synchronous Generator, *The Scientific World Journal* **2013** 783010

An evolutionary sensor approach for self-organizing production chains

M Mocan[1], E V Gillich[1], I C Mituletu[2] and Z I Korka[2]

[1]Politehnica University of Timisoara, Department of Management, Remus Str., no. 14, 300191 Timisoara, Romania
[2]"Eftimie Murgu" University of Resita, Department of Mechanics and Materials Science, P-ta Traian Vuia 1-4, Resita, 320085, Romania

E-mail: edwaldgillich@gmail.com

Abstract. Industry 4.0 is the actual great step in industrial progress. Convergence of industrial equipment with the power of advanced computing and analysis, low-cost sensing, and new connecting technologies are presumed to bring unexpected advancements in automation, flexibility, and efficiency. In this context, sensors ensure information regarding three essential areas: the number of processed elements, the quality of production and the condition of tools and equipment. To obtain this valuable information, the data resulted from a sensor has to be firstly processed and afterward used by the different stakeholders. If machines are linked together, this information can be employed to organize the production chain with few or without human intervention. We describe here the implementation of a sensor in a milling machine that is part of a simple production chain, capable of providing information regarding the number of manufactured pieces. It is used by the other machines in the production chain, in order to define the type and number of pieces to be manufactured by them and/or to set optimal parameters for their working regime. Secondly, the information achieved by monitoring the machine and manufactured piece dynamic behavior is used to evaluate the product quality. This information is used to warn about the need of maintenance, being transmitted to the specialized department. It is also transmitted to the central unit, in order to reorganize the production by involving other machines or by reconsidering the manufacturing regime of the existing machines. A special attention is drawn on analyzing and classifying the signals acquired via optical sensor from simulated processes.

1. Introduction

Nowadays, we face a trend in industrial development which is known as Industry 4.0. The idea is making equipment to self-organize and to collaborate with other machines and with people [1-5]. The actual state of implementing the principles in the industry of some developed countries in is found in [6]. According to these principles, the structure and organization for a factory is described in [7]. Here, the role of sensors and actuators collaborating via a central system and implemented algorithms is crucial and constitute a specific pillar of Industry 4.0. Thus, classical machines are equipped with sensors, which permit both machine state assessment and manufacturing process parameter extraction. In this way failure occurrence in machine parts and in manufactured pieces is observed. This makes classical machines to become intelligent and collaborative and reduces the need for major investments. Several attempts to integrate sensors in machines and to use the acquired data to gain complex information are described in [8-12], a significant amount of interesting results being presented.

In concordance to the actual legislation, machines should operate with low level of noise and vibration [13], [14]. Any deviation from the regular parameters indicates an anomalous behavior, indicating, beside machine or tool wear, a problem in the quality of machined pieces. The aim of manufacturing process monitoring is the observation of significant parameter changes. Usually, these parameters are acceleration, pressure or temperature, and are measured on the machine or on the manufactured piece. Manufacturing process monitoring consist of condition monitoring, used on machineries containing rotating or oscillating elements, and non-destructive testing techniques that is employed for stationary elements [15].

This paper aims presenting an example how optical sensors can be attached to a universal milling machine in order to supply data used to assess the manufactured pieces quality, to optimize the operation regime of machines linked in the production chain and to perform maintenance in due time.

2. The self-organizing concept for a simple production chain

The self-organizing process in an industrial structure is the process where the order arises from local interactions between elements of the disordered system. The process is spontaneous and need not to be controlled by any external entity, i.e. human intervention in the case analyzed in this paper. The self-organizing process is activated by random fluctuations and the resulting organization is completely decentralized, being distributed over all elements of the industrial structure. As such, the structure is able to self-repair substantial perturbation.

In this study, the perturbation consists of machine wear, tool wear or improper operational regime, simulated by different kind of actuation of the manufactured piece. The implemented optical sensor observes the perturbation and transmits the parameter changes to the central unit which takes action based on implemented algorithms.

2.1. The production chain

In this paper, by reason of simplicity, a production chain consisting of three machines is employed. The manufacturing process of a gas turbine palette, which has in the initial state the shape of a prismatic bar, is simulated. The operations are as following: (*i*) cutting to obtain the desired length, made on a cutting machine, (*ii*) milling the lateral surface in order to get the final shape, made on a universal milling machine, and (*iii*) welding the achieved elements on a rotor, made by a welding machine.

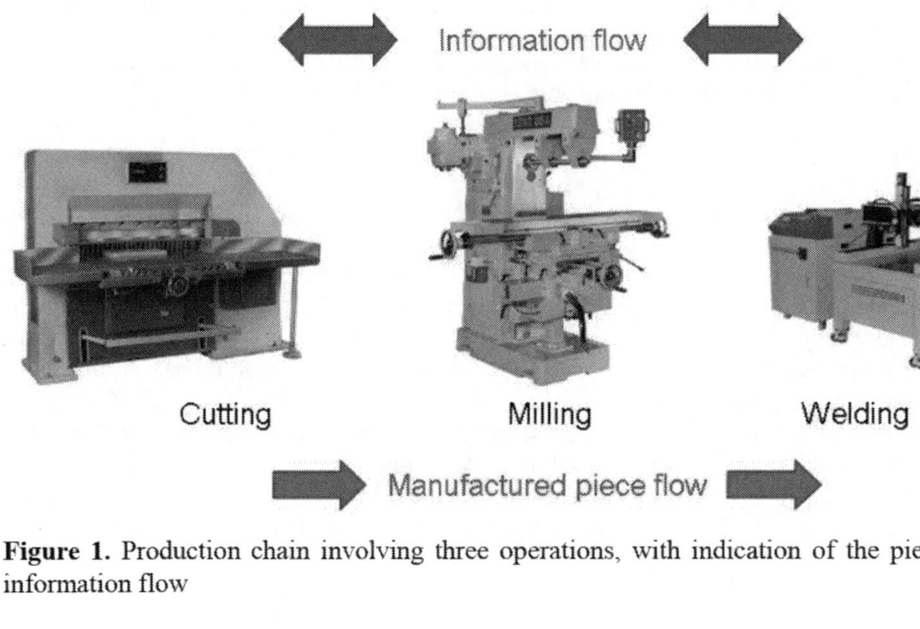

Figure 1. Production chain involving three operations, with indication of the piece respectively information flow

The chain is presented in Figure 1; one can observe in this figure the unidirectional flow for the manufactured piece, in the above-described succession of operation. On the other hand, the information flow follows two directions, to the cutting and welding machine which are placed before respectively after the milling machine in the production chain. Obviously, the information is transmitted to the staffs that are: the plant engineer, the accounting and sales representative, the quality assurance responsible and the maintenance team leader. The aim of transmitting data to staff members is for information purposes only, the human action being asked just for interventions that machines cannot handle. However, decision is taken by the central unit based on a priori developed algorithms.

3. Optical sensor implementation

The supervising process is exemplified on a universal milling machine which got attached an optical displacement sensor. This sensor acquires distances with high-precision, even for dynamic regimes. The responses to different operating scenarios are transmitted to the central unit, in this way a simple experimental stand being obtained. Since the aim of the paper is to demonstrate how sensors can be used to create self-organizing production chains, the study presented herein is limited to analyzing and classifying the signals transmitted by the sensor.

3.1. Sensor implementation

The optical sensor involved in the experiment is of type LG10A65PU; it can operate in analog and discrete mode as well. Figure 2 shows the experimental stand, where the machined piece is represented by a steel beam with dimensions 200x40x3 mm. The sensor produces a logic switching at the presence of an object, and analog voltage in relation with the distance till the pointed location signalized by red spot light is registered. Any distance variation determines a proportional voltage variation. The output voltage is analog–to-digital converted by the NI USB-6259 M series Multifunction DAQ (presented in Figure 3) and transmitted to the personal computer. Here, the signal is visualized and registered via the embedded Measurement & Automation application developed by National Instruments. Figure 3 also present the double power source, used to supply the optical with 0-3 A control and 0-30 V DC stabilized direct current voltage.

Figure 2. Experimental stand, consisting of a universal milling machine, machined piece and the optical sensor

Figure 3. Setup for the sensor calibration made before employing it in the operation monitoring process

The optical sensor (Laser Displacement Sensor) is a high-precision sensor that has a measurement range from 75 to 125 mm and the focal point at 180 mm. The analog resolution at 100 mm is as following: Fast: < 150 µm; Medium: < 50 µm; Slow: < 10 µm. It has one discrete PNP output (max. 100 mA) and output saturation DC voltage < 1.2 V at 10 mA respectively < 1.6 V at 100 mA.

Data acquisition module NI USB-6259 M series has 32 analog inputs (32 single ended or 16 differential), 4 analog outputs and 48 (32 clocked) digital I/O. The frequency range is 1.25 mega samples per second by 16 bits resolution, the data acquisition module being optimized for superior accuracy at fast sampling rates.

3.2. Sensor calibration

In order to acquire precise data, calibration is requested before the milling operation is monitored. As shown in Figure 3, the steel beam representing the machined piece is fixed on one end in a screw vice and has the other end free. The optical sensor is also fixed in a vise, the distance between sensor and piece being 100 mm. Both vices are rigid fixed on the table of the universal milling machine.

To check the precision of the optical sensor measurements, concerning the piece free end displacement, a standard dial gauge with 10 µm resolution is used. The dial gauge is placed on the beam free end, opposite to the Laser spot. The beam's free end displacement is accomplished by imposing, to the milling machine table, a displacement relative to the milling head. Calibration was made for the five displacement values indicated in Table 1, scaled to the relative voltage values.

Table 1. Example of calibration test values

Test No.	Displacement (mm)	Voltage (V DC)
1	0.1	0.02
2	0.2	0.04
3	0.3	0.06
4	0.5	0.1
5	0.85	0.17

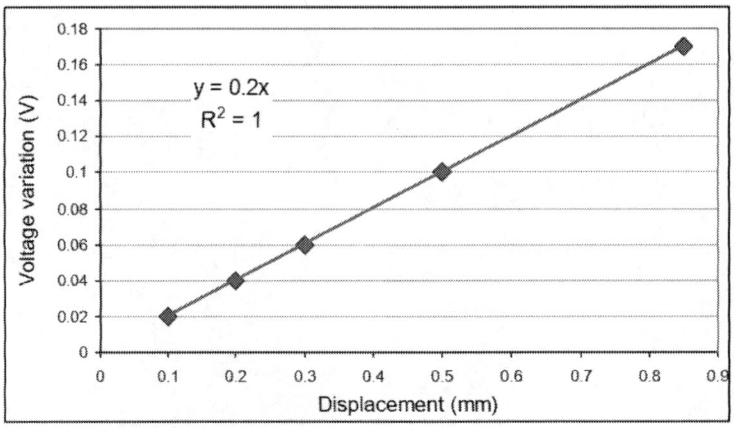

Figure 4. Displacement-voltage relationship

A sufficient DC voltage variation, i.e. 0.02 V, could be determined, even at a low displacement as 0.1 mm. The calibration results confirm the analog linearity indicated by the manufacturer (see Figure 4, namely +/0.2 mm from 75 to 125 mm; +/- 0.02 mm from 95 to 100 mm.

3.3. *Sensor output*

The Measurement & Automation permits a facile connection of the LG10A65PU sensor trough the NI USB-6259 Multifunction DAQ. The data acquired for a 0.85 mm displacement is shown in Figure 5. From this figure clearly results, on the horizontal axis, the time interval T for which the displacement of 0.85 mm is applied. The value can be also read from in the *HyperTrend Cursors* window, resulting T = 22.272 seconds.

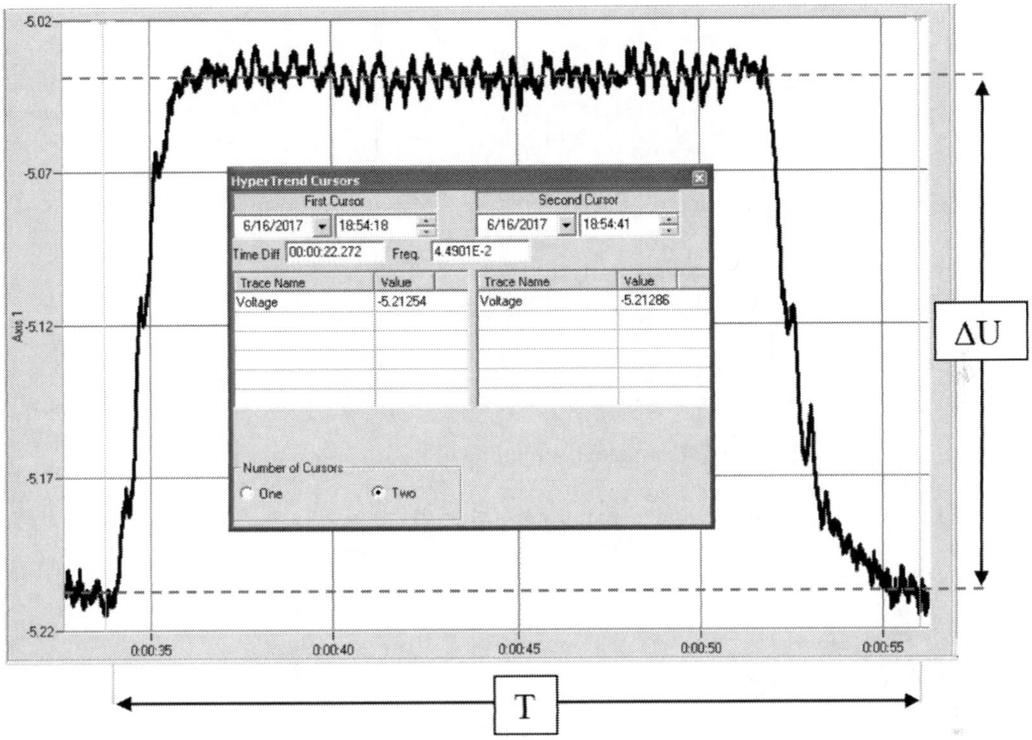

Figure 5. Data acquisition interface

Another parameter possible to be identified is the output voltage ΔU, proportional with the displacement of the free end due to the action of the milling head. This data can be also read from the *HyperTrend Cursors* window, if the horizontal cursors are adequately placed. The software can also compute the frequency that corresponds to the cycle framed between the vertical cursors, making possible an easy and fast quantification and adjustment of the operation cycles.

4. Results and discussions

Universal milling machines are not equipped, in the normal case, with any digital device for transmitting data to a central unit that can process it. By employing an optical sensor, the operation of the machine can be monitored and the gained information may be and used to plan production and maintenance. The decision can be taken by humans or by the central unit itself, in this latter case it being an algorithm-based management.

If long-term monitoring of the operation process is performed, the central unit recognizes the number of manufactured pieces. Figure 6 presents the acquired data in graphical form. Here, the optical sensor indicates 0 voltage output if the processed piece is not present in the vice. If the piece is loaded, the system indicates a voltage U proportional with the distance to the piece, that is U = -5.22 V in this example. Thus, the optical sensor's measurement output clearly indicates the process cycle time T used to manufacture one piece on the milling machine.

Another information resulted from Figure 6 is the time required for discharging and reloading the milling machine, i.e. the elapsed between two process cycles. If this time is in a predefined range, the process runs normally. Otherwise it indicates either a lack of raw material, or the need of maintenance. If this information is available in real time, a rapid reaction is possible to normalize the situation. In all cases, the stationary time over an analyzed time period can be found out.

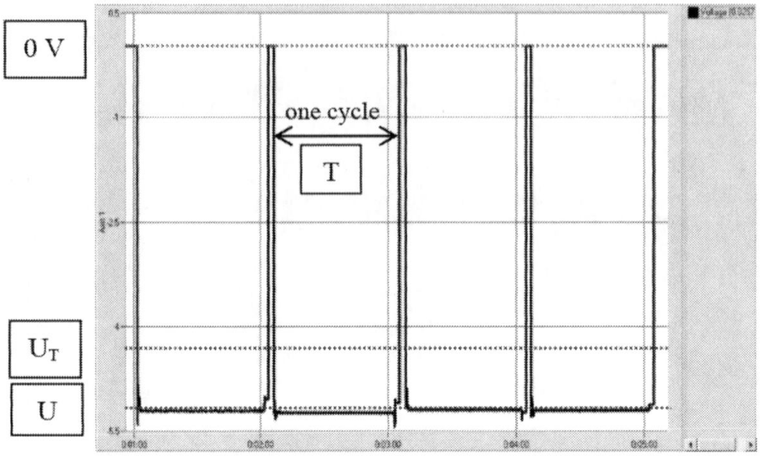

Figure 6. Long-term monitoring of the manufacturing process

From Figure 6 one can recognize that four pieces were processed, indicated by the four signal deviations from zero. The central unit automatically counts these pieces, by the algorithm:

if $U > U_T$
and $T_I < T < T_S$
than milled piece = milled piece + 1

The lower time limit T_I and upper time limit T_S are best visualized in Figure 7. This detailed view of a signal that comprises one process cycle shows that during the manufacturing process the signal amplitude varies in concordance with the loads applied to the bar and the resulted deformations. One can observe that three tasks are fulfilled during one process cycle. This variation of voltage is small, but the zoom on one task is better represented in Figure 8.

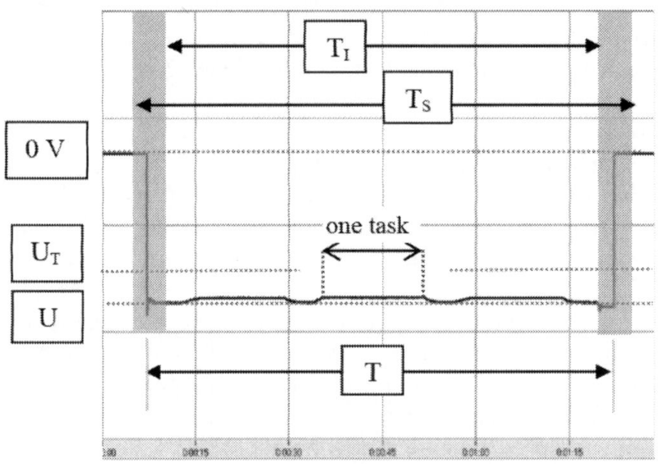

Figure 7. Zoom on one process cycle

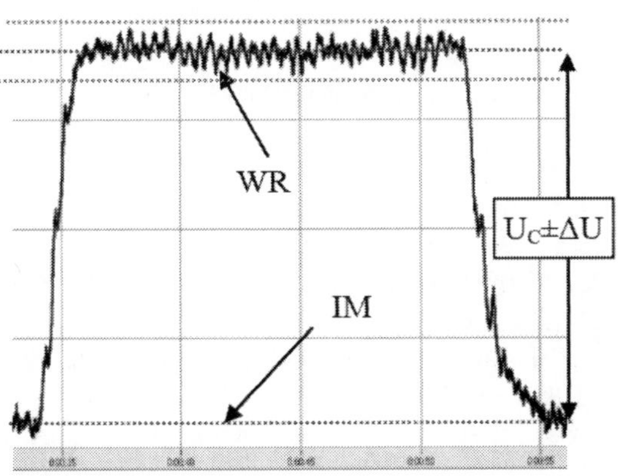

Figure 8. View on the signal portion corresponding to one task

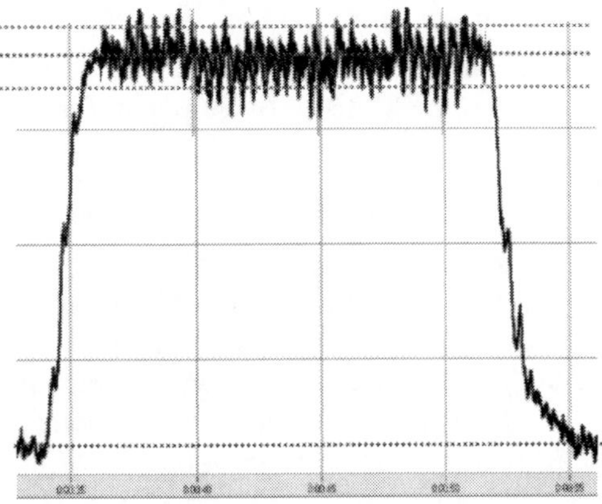

Figure 9. Signal portion representing a task with conditions indicating machine/tool wear

In Figure 8 the signal indicates the idle mode (IM) and the working regime (WR) of the milling machine. Obviously, during operation, due to the pressure exercised by the milling cutter, the bar is pushed closer to the sensor, this being indicated by a voltage decrease. One can observe the voltage alteration U_C due to the shift from the idle mode to the working regime.

The amplitudes $\pm\Delta U$ around the rest position WR, marked with red dotted lines in Figures 8 and 9, show if the machine respectively the tool is in good condition [16], [17]. Evidently, the higher amplitudes are associated with failure or machine/tool wear [18], [19].

if $|\pm\Delta U| < \Delta U$ threshold
than milled piece = milled piece + 1
else reject no. = reject no. +1
and !machine/tool alert!
and !supply raw material +1!

Thus, the information is transmitted to the cutting machine, in order to ensure the proper number of bars. Also, the quality assurance and maintenance departments are informed about the problem. If a longer term is requested to fix the problem, the information is transmitted to the cutting and welding machines in order to optimize the working regimes.

The above presented algorithm can be used, separately or simultaneously, for the idle mode as well. Numerous other scenarios can be imagined and algorithms implemented. The idea is to finally achieve a totally independent and self-organizing production chain.

5. Conclusions

This study shows how universal machines that are originally not foreseen with specialized monitoring systems can be equipped with sensors, in order to acquire and transmit digital data to a central unit. Algorithms implemented in the central unit convert this data in valuable information for the other machine composing the production chain and adjacent departments, e.g. maintenance and quality assurance. In this way, a self-organizing manufacturing process which quickly adjusts the malfunctions occurred due to perturbations is achieved, making the process as efficient as possible.

Simulations involving a simple production chain have shown that the developed concept, including a monitoring system and algorithms, provide adequate and trustful information, making it proper in industrial applications.

Acknowledgement
The work has been funded by the Sectoral Operational Programme Human Resources Development 2007-2013 of the Ministry of European Funds through the Agreement POSDRU/159/1.5/S/132395.

References

[1] Kagermann H, Wahlster W and Helbig J 2013 *Recommendations for implementing the strategic initiative Industrie 4.0,* Final report of the Industrie 4.0 Working Group
[2] Hermann M, Pentek T and Otto B 2015 *Design Principles for Industrie 4 Scenarios: A Literature Review*, Working paper 10, Universität Dortmund
[3] Bahrin M A K Othman M F, Nor Azli N H and Talib M F 2016 Industry 4.0: A Review on Industrial Automation and Robotic, *Journal Teknologi* **78**(6-13) 137-143
[4] Davies R 2015 *Industry 4.0 Digitalisation for productivity and growth*, European Parliamentary Research Service, PE 568.337
[5] Rüßmann M, Lorenz M, Gerbert P, Waldner M, Justus J, Engel P and Harnisch M 2015 *Industry 4.0: The Future of Productivity and Growth in Manufacturing Industries*, The Boston Consulting Group
[6] ****Global Industry 4.0 Survey: Industry 4.0: Building the digital enterprise* 2016 PwC
[7] Lucke D, Constantinescu C and Westkämper E 2008 *Smart Factory – A Step towards the Next Generation of Manufacturing*, in: Manufacturing Systems and Technologies for the New Frontier, Springer, London, pp 115-118
[8] Stavropoulos P, Chantzis D, Doukas C, Papacharalampopoulos A and Chryssolouris G 2013 Monitoring and control of manufacturing processes: A review, *Procedia CIRP* **8** 421-425
[9] Chung K T and Geddam A 2003 A multi-sensor approach to the monitoring of end milling operations, *Journal of Materials Processing Technology* **139** 15-20
[10] Zhao G 2011 Wireless Sensor Networks for Industrial Process Monitoring and Control: A Survey, *Network Protocols and Algorithms* **3**(1) 46-63
[11] Boud F and Gindy N Z 2003 Application of multi-sensor signals for monitoring tool/workpiece condition in broaching, *Journal of Materials Processing Technology* **139** 15-20
[12] Luo M, Liu L, and Luo H 2016 Real-Time Deflection Monitoring for Milling of a Thin-Walled Workpiece by Using PVDF Thin-Film Sensors with a Cantilevered Beam as a Case Study, *Sensors* **16** 1470
[13] Korka Z I, Gillich G R, Mituletu I C and Tufoi M 2015 Gearboxes Noise Reduction by Applying a Fluoropolymer Coating Procedure, *Environmental Engineering and Management Journal* **14**(6) 1433-1439
[14] Gillich G R, Gillich N, Chioncel C P and Cziple F 2008 Legal aspects concerning the evaluation of pollution effects due to vibrations in urban areas, *Journal of Environmental Protection and Ecology* **9**(2) 465-473
[15] Onchis D M and Gillich G R 2010 *Wavelet-type denoising for mechanical structures diagnosis*, Proceedings of the 3rd International Conference on Engineering Mechanics, Structures, Engineering Geology, Corfu Island, Greece, July 22-24, pp 200-203
[16] Gillich G R, Frunzaverde D, Gillich N and Amariei D 2010 The use of virtual instruments in engineering education, *Procedia-Social and Behavioral Sciences* **2**(2) 3806-3810
[17] Gillich G R, Mituletu I C, Negru I, Tufoi M, Iancu V and Muntean F 2015 A Method to Enhance Frequency Readability for Early Damage Detection, *Journal of Vibration Engineering & Technologies* **3**(5) 637-652
[18] Gillich G R, Praisach Z I, Abdel Wahab M, Gillich N, Mituletu I C and Nitescu C 2016 Free Vibration of a Perfectly Clamped-Free Beam with Stepwise Eccentric Distributed Masses, *Shock and Vibration* **2016** 2086274
[19] Nitescu C, Gillich G R, Abdel Wahab M, Manescu T and Korka Z I 2017 *Damage severity estimation from the global stiffness decrease,* 12th International Conference on Damage Assessment of Structures, Kitakyushu, Japan, 10-12 July, 012034

IOP Conf. Series: Materials Science and Engineering **294** (2017) 012093 doi:10.1088/1757-899X/294/1/012093

A few categories of electromagnetic field problems treated through Fuzzy Logic

M S Lolea and S Dzitac

University of Oradea, Faculty of Energy Engineering and Industrial Management, Department of Energy Engineering, Universităţii str., no. 1, 410087 Oradea, Romania

E-mail: mlolea@yahoo.com

Abstract. The paper deals with the problems of fuzzy logic applied in the field of electromagnetism. In the first part, there are presented some theoretical aspects regarding the characteristics and the application of the fuzzy logic in the general case. Are presented then, some categories of electromagnetic field problems treated by fuzzy logic. The accent is on the effects of exposure to the electromagnetic field on the human body. For this approach is dedicated a paragraph at the end of the paper. There is an application on how to treat by fuzzy logic the effects of electric field exposure. For this purpose, the fuzzy toolbox existing in the Matlab software and the results of some electric field strength measurements into a power substation are used. The results of the study and its conclusions are analyzed and exposed at the end of the paper.

1. Introduction

Practical research on the impact of electromagnetic field with technical systems is very numerous and has significant results. The study of the impact of the electromagnetic field on living organisms has led to more or less accurate results, and those related to human effects have generated great controversy.. This is due to the need for extensive time extension of research or the impossibility of direct exposure of humans to the electromagnetic field for scientific laboratory analysis due to legal and obviously ethical or spiritual constraints. In order to eliminate this shortcoming, it is necessary either to extrapolate animal results or to statistically - probabilistic analyzes or to create physic - mathematical models that will present reality as faithfully as possible.

In Romania, studies related to the influence of the electromagnetic field on living organisms are not numerous. The exact cause cannot be ascertained, but perhaps the subject has not raised much interest amongst the specialists. An almost generalized interest is manifested in other countries, such as in the northern European continent or in the USA.

There have been results that highlight the damaging effects of the electromagnetic field, linked of course, to certain factors or characteristic parameters of field quantities or generating sources.

Among the diseases caused by electromagnetic field exposure, we mention [1-4]: dizziness, loss of memory or lack of concentration, anxiety, sleep disturbance, infertility and impotence, decreased vitality, brain tumors or other forms of cancer etc. Many other studies in the light of the results obtained do not certify the harmful effects of the electromagnetic field. However, in our country, as in other countries, it was not possible to impose on the general population certainty about the harmful effects of exposure to electromagnetic fields.

2. Fuzzy logic. A short presentation

The fuzzy logic was enunciated in 1965 by L. Zadeh, putting its mathematical foundations. Modeling through fuzzy logic allows for the formalization of imprecisions due to the global knowledge of a very complex system and the expressions of its behavior by a word system. It also allows the standardization of a system description and the treatment of numerical data as if it were expressed in linguistic data [5-8].

The specificity of a fuzzy system is that it can simultaneously control numerical data and lexical knowledge. In a fuzzy system executes a non-linear transformation on the input data vector resulting in a scalar output. Fuzzy systems are considered a particular case of expert systems [6]. A fuzzy or vague crowd is a lot about which too many exact things are not known. There are several types of fuzzy systems according to the type of reasoning used [7], [9]: Mamdani, Larsen, Sugeno - Takagi, Tsukamoto etc. In fuzzy logic the domains correspond to the concepts of linguistic variables. The theory of fuzzy sets helps to transform qualitative human judgments into quantitative, discrete and deterministic numerical expressions.

The typical structure of fuzzy inference systems is a model that performs a certain correspondence: firm input value - input membership functions - inference rules - output characteristics - output membership functions - firm output value. The structure of a fuzzy system is shown in Figure 3, having the form in which it is found or described in other specialized papers [5], [7], [9], [10]. On the model, x_i are the input sizes and Y_i, the output ones.

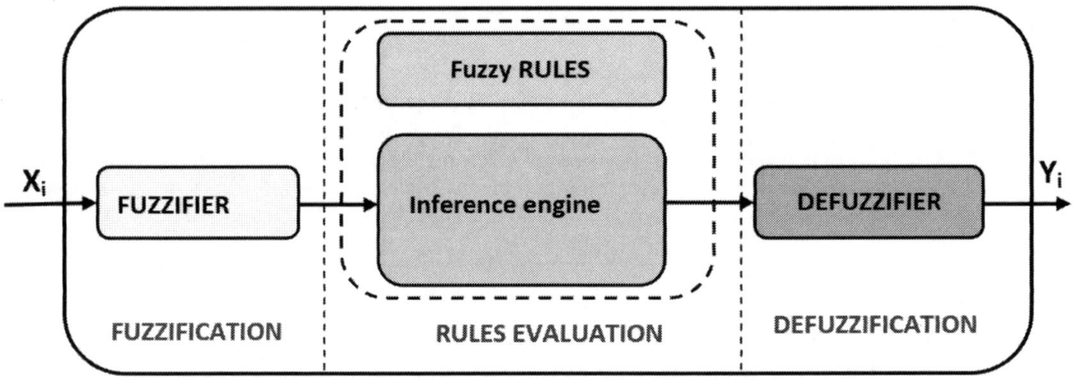

Figure 1. The structure of a fuzzy system and inference process [6]

Fuzzifier has the role of translating numerical expressions into fuzzy sets necessary for activating rules based on language variables and connectors.

The inference engine applies a transformation of the set of rules into fuzzy sets.

The defuzzifier executes the assignment of a real numerical value corresponding to a membership function(MF).

A fuzzy set is completely defined by the universe of values X and the function of belonging (MF) that describes the value / degree of belonging μ to the class of elements in the universe. Let X be the universe of the discourse, with elements notated x. A fuzzy set A of the discourse universe X is characterized by a membership function A (x) which associates each element x with a degree of belonging μA (x) to the set A [7-10]:

$$\mu_A(x): X \rightarrow [0,1] \tag{1}$$

To represent a fuzzy set, the function of membership must first be defined. In this case, a fuzzy set A, is completely defined by the set of doublets:

$$A = \{[x, \mu_A(x)] \mid x \in A\} \tag{2}$$

The membership function(MF) μ_A thus, assigns to each element x the degree of belonging to the fuzzy set A and the degree of belonging μ, expresses the extent to which one element belongs to a fuzzy set. There are several types of distinct membership functions: singleton, triangle, gamma, "S", "π", trapezoid, Gaussian etc. [5-7], [10].

Because it will be used in application from paragraph 5, here it's shown a MF by triangular type(trimf). This is defined by three parameters a, b, c, as follows:

$$\mu_A(x) = \text{triangle}(x;a,b,c,) = \begin{cases} 0, x \leq 0 \\ \frac{x-a}{b-a}, a \leq x \leq b \\ \frac{c-x}{c-b}, b \leq x \leq c \\ 0, c \leq x \end{cases} \quad (3)$$

Or, using the functions „mic" and „max"max, result:

$$\text{Triangle}(x;a,b,c,) = \max\left\{\min\left(\frac{x-a}{b-a}, 1, \frac{c-x}{c-b}\right), 0\right\}, \quad (4)$$

with x being the element of the fuzzy set A with the center of the membership function in point c.

The discourse (speech) universe is marked on the abscissa (axis X) of the orthogonal system, which is the range of all possible values applicable to a variable. On the order (Y-axis), the degree of belonging is represented. Like other types of functions on it, dilators and concentrators can be applied. Concentration of a fuzzy set results in a reduction of the membership by square-up of the membership function values. Fuzzy dilatation increases the degree of belonging of a value to the fuzzy set by applying the square root extractor. The phrases of the two operators are [10]:

a. Fuzzy concentrator: $$\mu_{CON(A)}(x) = (\mu_A(x))^2 \quad (5)$$

b. Fuzzy dilator : $$\mu_{DIL(A)}(x) = \sqrt{\mu_A(x)} \quad (6)$$

Figure 2 shows the graph of a normal or Gauss MF, and in Figure 3 a graph of a triangular MF.

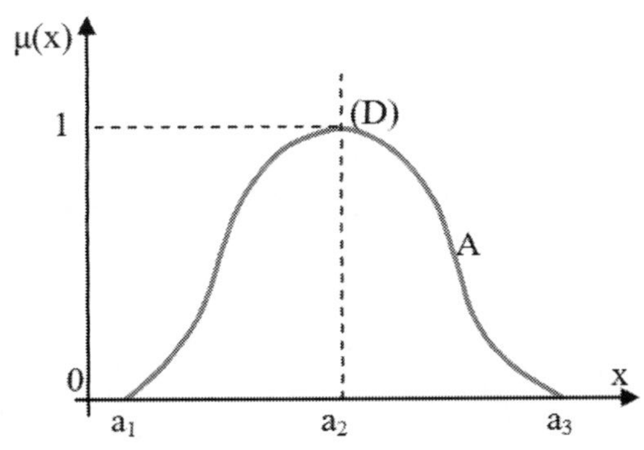

Figure 2. Gaussian MF, (D) – symetry axis

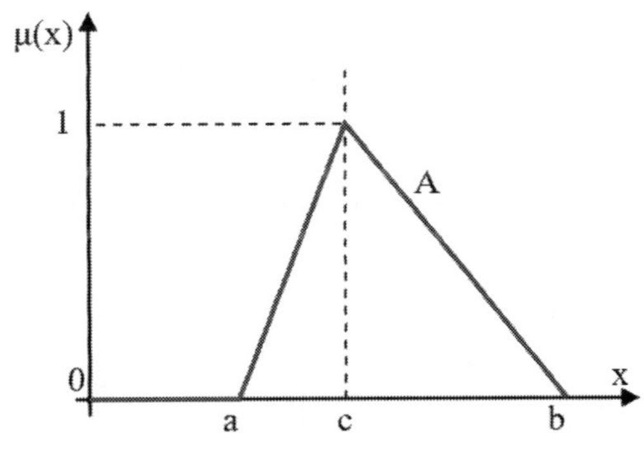

Figure 3. Triangular MF, (c) – set centre

In the paper application, an input variable in the fuzzy system is the age of man. Using this parameter will explain the difference between its treatment by fuzzy logic versus classical logic. In classical logic, the separation of levels is made after a fixed, deterministic value. Thus, a breakdown of age cycles and values in the years that define them are: childhood (0-14 years), adolescence (15-18 years), youth (18-35 years), maturity (36-60 years), old age > 60 years). Each of this range may have subdivisions with specific names [11].

In the graphs from Figure 4, the appreciation is random as intervals and was made by the authors. Age division has been conceived, however, in accordance with the active period of the man in which he is involved in the process of work in an organized form, in an economic institution.

In fuzzy modeling, the boundaries of each level will be merged by overlapping, because there is a different perception of people on them. For example, a 35-year-old person can be considered a young by a 50-year-old one, as well the same person, is considered old by a 19-year-old one. Age is an input variable in Matlab's fuzzy model that participates in assessing the effects of exposure, and the explanation for its perception is justified. Since every human being is unique in his own way, electromagnetic behavior can be considered different from person to person. This also suggests a different sensitivity to electromagnetic field exposure. So, the need to use fuzzy logic in the analytical and deliberative process is due to the assumption that the human perception is imprecise. For these reasons, the inference rules from the proposed application have also been formed.

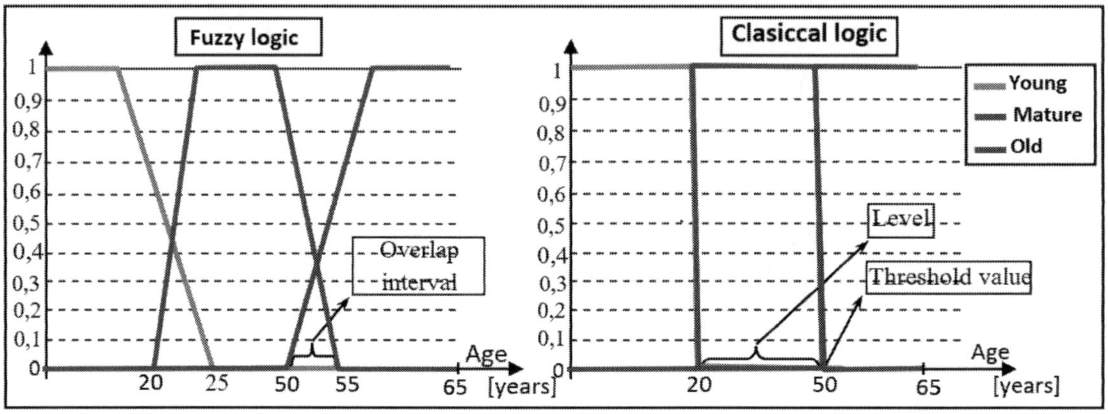

Figure 4. Active professional cycles of life in fuzzy logic and classical logic

3. Fuzzy Logic Toolbox from Matlab

For fuzzy modeling of applications in different fields, implicitly for electromagnetic field problems, it can use the dedicated toolbox, existing in the Matlab computer program.

This Toolbox has five components and functions as follows [5], [12], [13]:

a). The Editor of the fuzzy system, named „FIS Editor" (Fuzzy Inference System), allows the creation of inputs and outputs in the system;

b). The editor of the membership functions, .i.e. fuzzy sets over the universes of the speech inputs and output variables of the system with fuzzy logic, named „Membership Function Editor"(MFE);

c). The fuzzy rule editor, which allows the formulation of the set of rules of the fuzzy syste, named „Rule Editor";

d). Rule view window, where you can see the degree of activation of each rule and the result or effects of each rule, generating exit from the system for a certain pass-through value of the user-defined entry (Rule Viewer);

e). The Fuzzy Logic System Output Surface Viewing window, which graphically shows the dependence of the tangled output of the system with fuzzy logic on the fuzzy logic system inputs, determined by the fuzzy logic system operation; this dependence is generally known as the "system control surface," and the window appears under the name "Surface Viewer".

Toolbox components are dynamically coupled to each other, so a modification of the parameter on a work page automatically generates the modification of the other component components, provided they are connected to the set of rules. The graphical presentation of the working windows for each component of a fuzzy module is presented in point 5 of the paper, in which a case study will be presented.

4. Fuzzy approaches of some electromagnetic fields problems

It can be remembered a number of older and established applications of fuzzy logic in various fields of science: automatic control (temperature rules, subway speed control, autofocus of cameras), medicine (control of pacemakers), form recognition (fuzzy classification algorithms), Economy (fuzzy decision methods) or measurements (processing information provided by sensors).

Fuzzy logic is a deduction system that uses the concept of "degree of truth" or "calculus with words". In binary logic a sentence can be "true," with a value of 1 or a "false" value of 0. In fuzzy logic a sentence can be half true and / or half false, or 15% false and 85% true, etc. The human brain efficiently and continuously processes data with a wide uncertainty such as "probable," "almost impossible," "very difficult," "acceptable," situations that can be coupled by fuzzy logic modeling [9].

Due to the nonlinear and unpredictable character of electromagnetic field characteristics and the uncertainty of the effects produced by the impact with the technical and biological systems it is suitable to analyze these manifestations and phenomena by using fuzzy logic modeling. The main problem addressed in the paper is to determine the degree of health damage due to exposure of workers in the electromagnetic field.

For the correlation between the exposure parameters and the health damage level, different input sizes such as [14]: electromagnetic field frequency, electric field strength and magnetic induction, distance to source, exposure duration, age, gender, type of the source etc.

The application of fuzzy logic can be applied to other electromagnetic field problems, either solely or in combination with the technique of genetic algorithms, neural networks or binary logic. Examples of applications are [8], [15 -17]: estimation of the magnetic and magnetic field outside high and very high voltage overhead lines under normal and short-circuit conditions, electromagnetic influence produced by damaged power transmission lines on neighboring regions and induced tensions in pipelines, controlling the speed of an electric car by modifying electromagnetic parameters, the electromagnetic compatibility of an electrical device depending on the electrical and magnetic stresses occurring in the process, the identification of the short circuits in the electrical circuits of the industrial networks, the reactive power control and improving power factor etc.

For example, this latest approach is needed to establish the electromagnetic compatibility reduction of the apparatus since, after entering domestic or industrial use, the electromagnetic compatibility check is no longer carried out by users, although at least the degradation of the physico-chemical characteristics of the materials From the structure of the device seems obvious and does not survive over time.

5. Application

The application refers to fuzzy modeling of the effects of electric field exposure of personnel within an Power Substation(P.S.)

For the electric field strength, are used the values obtained by measurements at different distances from sources from the Remeţi Power Substation(PS), Bihor county, which serves a Hydro power Plant(HPP). The PS image is shown in Figure. 5. The measuring instrument used was SPECTRAN 5035, manufactured by Aaronia, Germany. Its image is shown in Figure 6. The age considered for the simulations is that of the electricians serving the HPP considered. Other input variables, set of rules, and degrees of severity are established based on norms recommendations or procedures used in other specialist papers [14].

Following the principle in paragraph 2, for the design of age classes specific to fuzzy logic, variation ranges for all input sizes are assigned in the Matlab Fuzzy Toolbox. Triangular memberships will be used.

Figure 5. Devices and Bus Bars on H.T. level from Power Substation Remeţi, Bihor

Figure 6. Field Detector SPECTRAN

The assignment of linguistic levels and variables is random and belongs to the authors. Similarly, the rules of inference are also formulated. Parameters required for simulation with the assignment of significance levels are presented in the Table 1.

Table 1. Simulation parameters

Input variables				Output variable
Electric field strenghtE[V/m]	**Ages [years]**	**Exposure time [min]**	**Distance from the source[cm]**	**Effect**
Low	young	short	small	Insignificant
medium	mature	medium	medium	Moderate
high	old	long	large	Significant

The Overlap Intervals(O.I.) of levels of the input parameters and the Allocated Intervals(A.I.) depending on the values of the delimitation thresholds(T.V.) chosen for the input variables, are shown in the following table:

Table 2. Assign intervals for input sizes

Field strenght[kV/m]			Ages[years]			Exposure time[min]			Distance[m]		
O.I.	**A.I.**	**T.V**	**O.I.**	**A.I.**	**T.V.**	**O.I.**	**A.I.**	**T.V**	**O.I.**	**A.I.**	**T.V**
3-5	0 - 5	5	20-25	0-25	0	40-60	0-60	0	1-2	0-2	0
8-10	5 - 10	10	50-55	20-55	20	240-300	60-300	60	3-5	2-5	2
10-12	10-15	15	50-55	55-65	50	300-360	300-480	480	5-7	5 - 10	5

To establish the ranges in Table 2, the reasoning in paragraph 2 of the paper was used. Allocation of age ranges took into account only the active age at work, 65 years being the retirement age, the period after this limit being out of context. For the determination of the electric field intervals, the delimitation thresholds are set according to the admissible value stipulated in the Romanian norms [18]: E_a = 10000 V / m. The effect or degree of harm refers to the negative influence on health. The inference rules are created with the operator "AND" and also with the conditional statements: "IF - THEN". All formed fuzzy rules are then inserted into the dedicated section from Toolbox. In order to solve the application, the options set in the fuzzy toolbox on analytics parameters are as follows: *document name*: "Exposure CELM", *type of fuzzy method*: mamdani; *number of input variables:* 4; *input variables type*: electric field strength, duration(exposute time), distance from the source and age; *inference rules number*: 20; *number of effect levels*: 3; *Types of affecting level:* insignificant, moderate, significant; *operator*: „AND"; *defuzzification*: maximum sampling method (Max); *fuzzy number allocated to MF*: triangular, by real variables. For input sizes, their values, variation ranges, implications in the set of rules, and the share of the final effects are assigned directly to the toolbox.

The application of the settings and the results obtained from the simulations are shown in Figures 7 ÷ 12. For the initial case of generating the cumulative effects, the operator's age is 30 years, the distance from the source is 200 cm, the field strength is 3000 V / m and exposure time is 120 minutes (Figure 10). To diversify the effects, the input parameters were changed as follows: age → 56 years, distance → 100 cm; duration → 400 min, field strength E → 7500 V/m(Figure 11). Other modifications are shown in Figure 12. Changes to mention parameters were made by moving the cursor marked with red line in the "Rule Viewer" window of the program, for the options chosen in each case.

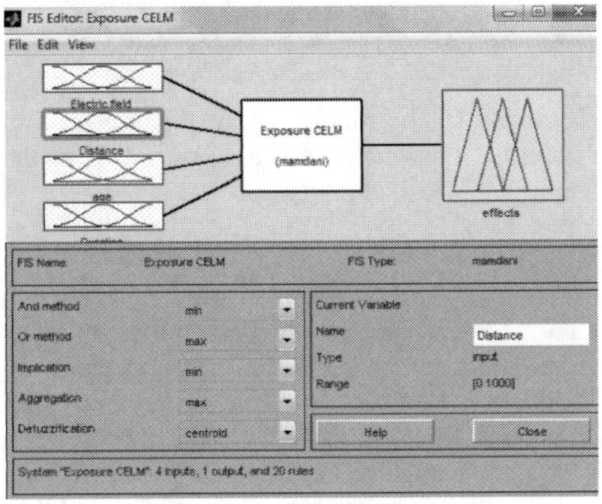

Figure 7. Inputs creation

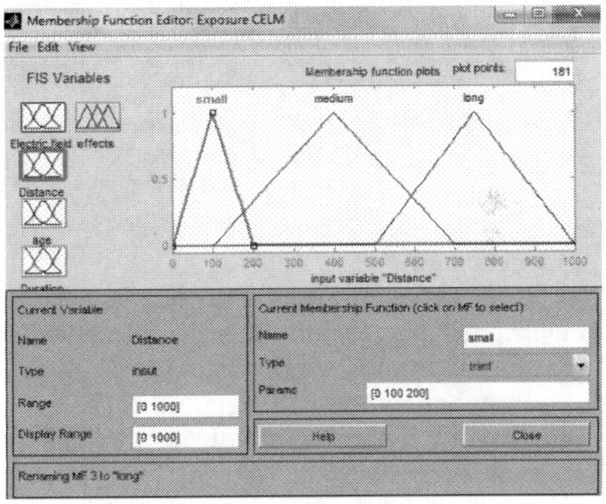

Figure 8. Assing input levels

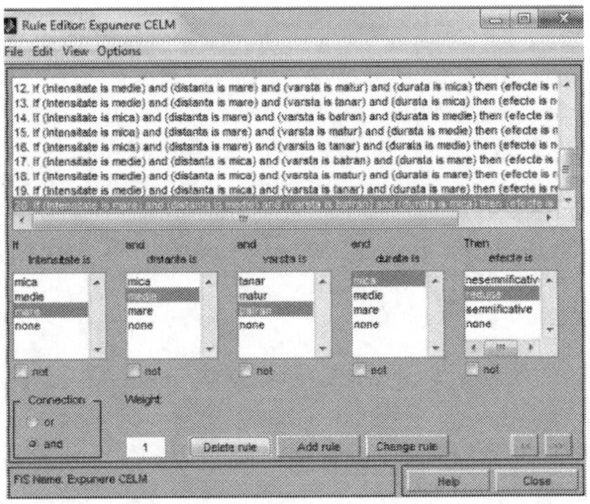

Figure 9. Rules set creating

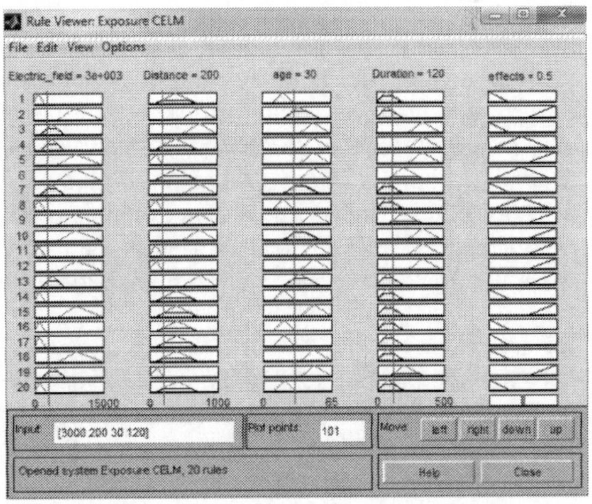

Figure 10. Generating cumulative effects

In Figure 11, all the values of the inputs originally set in Figure 10 have been modified. The change of input parameters, with the results of Figure 12, was performed by keeping the other complementary input parameters constant at the initial values from Figure 10.

IOP Conf. Series: Materials Science and Engineering **294** (2017) 012093 doi:10.1088/1757-899X/294/1/012093

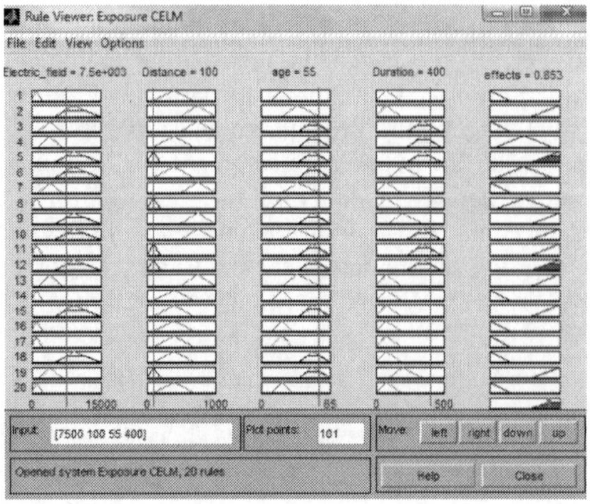

Figure 11. The modification of effects at changing of all input parameters

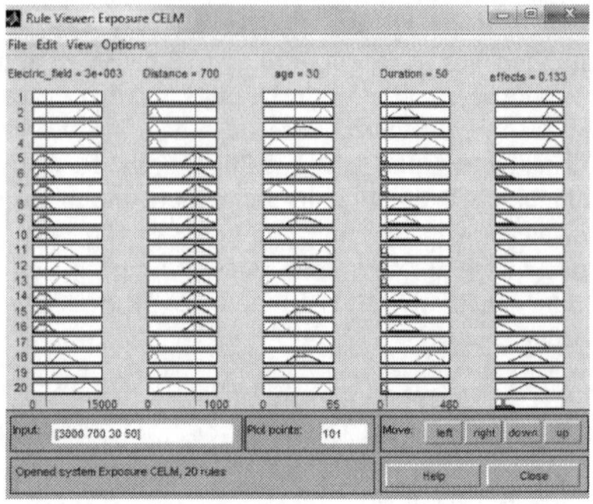

Figure 12. The modification of the effects at changing the length and distance values at the input

6. Conclusions

The implication of input parameters in generating effects can be highlighted in Matlab simulations provided they are comprised in the rules of inference. The biggest share has the "Insignificant" effects that resulted from 12 combinations of input sizes included in the rules of inference. "Significant" and "Moderate" effects hold the same share, each with four occurrences. Of course, the number of possible combinations of variables involved in writing inference rules was not exhausted, but the purpose of the simulation was reached. All three categories of effects were generated and the choice of different variants also took into account the results of other authors or recommendations from the dedicated literature. So, from the simulation results, the most favorable situations can be chosen to prevent the unwanted effects of exposure. This type of simulation is especially useful for workers from power grids.

According to different human perception of man, there are several ways of approach when it comes to defining thresholds. Fuzzy logic modeling uses this approach. The problem of exposure in the electric and magnetic field is to determine as accurately as possible the harmfulness thresholds, but also the magnitude of the harmful interval surrounding the source of electromagnetic pollution. An analysis technique based on the principles of fuzzy logic can lead to a better understanding of the effects of these fields. The precautionary principle must also take into account the future effects. An electromagnetic device sent to a person in the context in which there is the least chance of harming them should be canceled, even if the benefits of using it are obvious.

References

[1] Drăgulinescu A 2010 *Idolii fără fir. Telefonia mobilă și poluarea electromagnetică,* Editura Christiana, București, Romania
[2] Milham S 2012 *Dirty Electricity, Electrification and the Diseases of Civilizatio*n, Publisher: Universe Star, New York, U.S.A
[3] Morega M 1999 *Bioelectromagnetism*, Editura Matrix Rom , București, Romania
[4] Păunescu G 2010 *Câmpul electromagnetic. Studii asupra posibilelor efecte ale câmpului electromagnetic asupra sănătății*, www.infoscola.webgarden.ro
[5] Dzitac S 2009 *Fiabilitatea și disponibilitatea sistemelor de distribuție a energiei electrice. Modelare și simulare*, Editura Universității din Oradea

[6] Dosoftei C C 2009 *Utilizarea inteligenţei computaţionale în conducerea proceselor*, Universitatea Tehnică "Gheorghe Asachi"din Iaşi, Romania, Doctoral Thesis

[7] Preitl S and Precup R 1997 *Introducere în conducerea Fuzzy a proceselor*; Editura Tehnică; Bucureşti, Romania

[8] Mahmoud M 2013 *Aplicarea logicii fuzzy în reţele electrice de distribuţie, industrie şi securitatea muncii,* Universitatea „Transilvania" din Braşov, România, Doctoral Thesis

[9] Turcoane O 2012 *Logica fuzzy, fundament al deliberării*, din cercetări doctorale în domeniul: Democraţia digitală în societatea cunoaşterii, Bucureşti, Romania

[10] Văduva I and Albeanu G 2004 *Introducere în Modelarea Fuzzy*, Editura Universităţii din Bucureşti

[11] Jompan A 2012 De la medicina vârstelor la ciclul vieţii de familie, *Rev. Practica Medicală* **VII**(1)25 17-23

[12] Oltean G 20115 *Simularea sistemmelor cu logică fuzzy în Matlab. Prezentarea Toolbox-ului Fuzzy Logic*, Lucrări de laborator, Tehnici de inteligenţă computaţională în electronică, Universitatea Tehnică din Cluj Napoca, Departamentul de Bazele electrotehnicii, http://www.bel.utcluj.ro/rom/dce/goltean/tice/lab/2%20SimulareaSLF_Matlab.pdf

[13] Colhon M 2012 *Elemente de Logică Fuzzy*, disponibil on line la adresa URL următoare http://id.inf.ucv.ro/~ghindeanu/lab/sicc/carteb5.pdf, Craiova, Romania

[14] Boumaiza S, Bouharati S, Bouzidi A and Lalaoui M 2015 Predicting Health Effects of Electromagnetic Pollution in Algeria Using Fuzzy Logic, *International Journal of Public Health Research* **3**(6) 352-356

[15] Damousis I G, Satsios K J, Labridis D P and Dokopoulos P S 2002 Combined fuzzy logic and genetic algorithm techniques—application to an electromagnetic field problem, *Fuzzy Sets and Systems* **129** 371-386

[16] Lakshmi D and Sivakami P 2012 Computation of Magnetic Field Distribution by Using an Adaptive Neuro-Fuzzy Inference System, *International Journal of Engineering Innovation & Research* **1** 175-181

[17] Radulović J and Ranković V 2007 An ANFIS based approach to approximation of electromagnetic field around overhead power transmission lines, UDC 537.8:621.316.35(045)=111, *Facta Universitatis, Series: Mechanics, Automatic Control and Robotics* **6** 231-240

[18] ***H. G. Nr. 1136 din 30.08.2006 privind cerinţele minime de securitate şi sănătate referitoare la expunerea lucrătorilor la riscuri generate de câmpuri electromagnetice

IOP Conf. Series: Materials Science and Engineering **294** (2017) 012094 doi:10.1088/1757-899X/294/1/012094

Adjustment of gripping force by optical systems

C K Jalba and C Barz

Technical University Cluj Napoca, North University Center Baia Mare, Str. Doctor Victor Babeş 62A, Baia Mare 430083, Romania

E-mail: KlausJalba@t-online.de

Abstract. With increasing automation, robotics also requires ever more intelligent solutions in the handling of various tasks. In this context, many grippers must also be re-designed. For this, they must always be adapted for different requirements. The equipment of the gripper systems with sensors should help to make the gripping process more intelligent. In order to achieve such objectives, optical systems can also be used. This work analyzes how the gripping force can be adjusted by means of an optical recognition. The result of this work is the creation of a connection between optical recognition, tolerances, gripping force and real-time control. In this way, algorithms can be created, with the aid of which robot grippers as well as other gripping systems become more intelligent.

1. Introduction

In today's gripping technology, the grippers are increasingly specialized in specific gripping procedures.

In a gripping operation are involved various physical parameters, for example the gripping force, the weight of the gripped object, the friction between the gripper jaws and the gripped object, torques, etc. Regardless of the size of these forces there is a link between the distances between the gripper jaws and the object to be gripped.

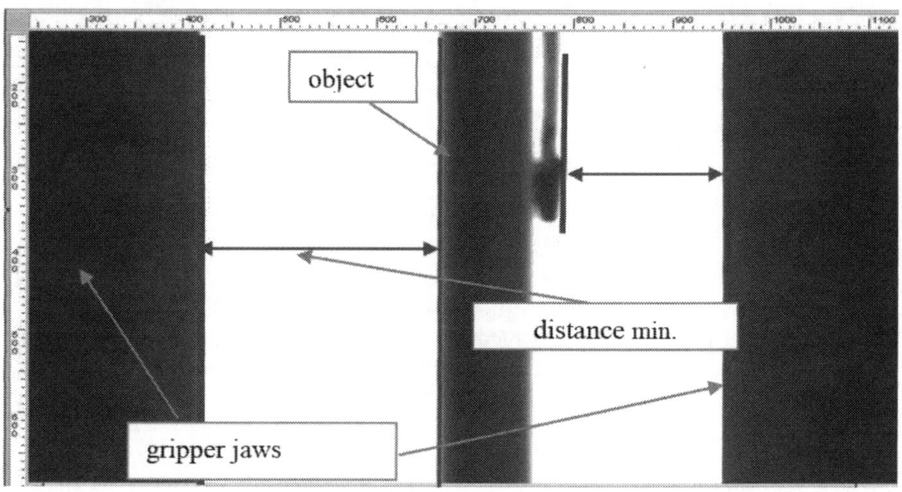

Figure 1. Magnification 5.5x

These distances tend to be zero, but they certainly do not reach this limit. For this reason, were made attempts to find out how these infinitesimal distances could be quantified. In the next two figures can be seen the same picture, by an enlargement of 550% (Figure 1) and 2500% (Figure 2).

In the first case, the contours from the object seems to be a clear well-defined line. But in the second picture it can be seen, that the contour is in fact a line with undefined accounts. The object contour becomes an optical field within certain boundaries. These boundaries can be viewed as a tolerance field.

A promising approach to measure this tolerance field was to make measurements in the tolerance system used in the mechanical engineering. In this system, the fits are divided into three categories: the clearance fit, the transition fit, and the press fit. If the tolerance is too great, the object falls between the fingers of the gripper, if it is too tight, then a softer object can be crushed.

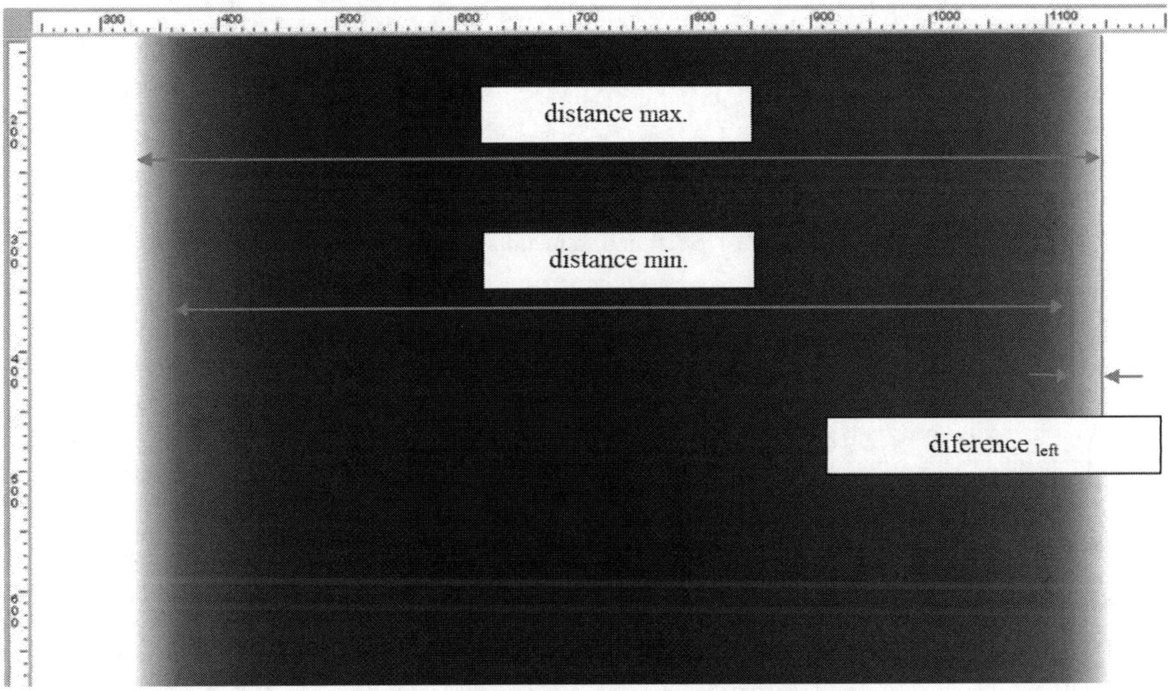

Figure 2. Magnification 20x

Only in the case of transitional adjustment does it seem that the objects are not falling through the jaws of the gripper, nor are they squeezed by excessive forces.

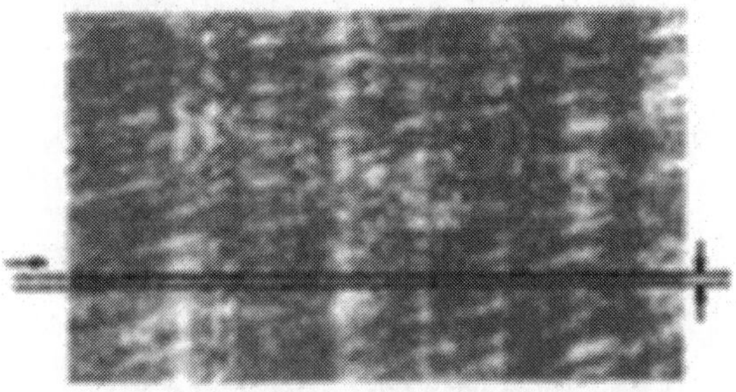

Figure 3. Metallic surface [1]

Because the tolerances of the transition fit are known from tables and they are located in the range of less than one millimeter, a measuring method must be found that can detect these tolerances. A method to find out this kind of tolerances is for example the measurement of material surfaces with roughness testers.

Figure 3 shows a metal surface machined by cutting. In the section, the surface of the metallic object has micron scattering (Figure 4) [1], [2].

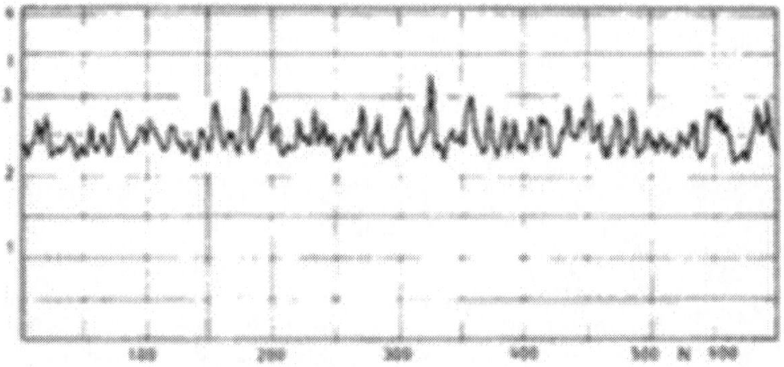

Figure 4. Section of a metallic surface [2]

Measurements with roughness testers are usually made with devices, which analyze surfaces with different mechanical sensors. The head with the touch sensor (Figure 6) slides on an air bed to avoid friction between the sensor and the object. Thus, the system becomes very sensitive and can measure by touching surfaces with accuracy of the order of the hundredth of mm (Figure 5). The whole assembly is positioned on a heavy polished granite plate. This plate gives a considerable inertia to the measurement system, being adapted to the dynamic forces of the translational movements [3].

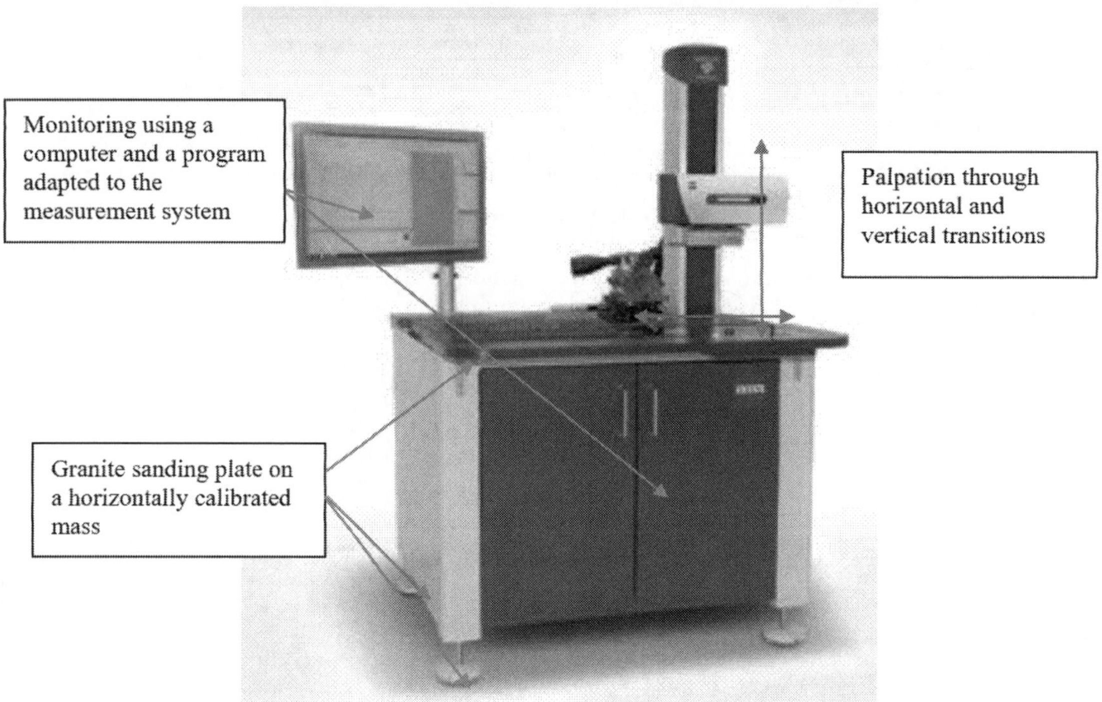

Figure 5. Roughness measurement with ZEISS SURFCOM [3]

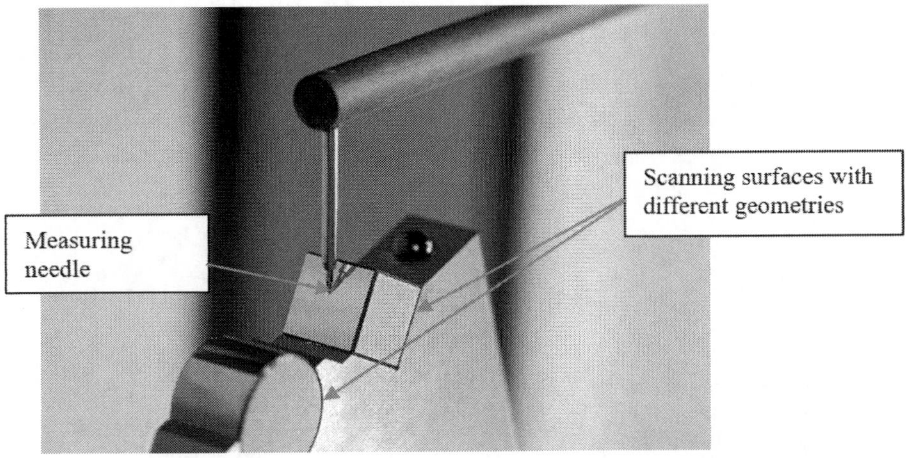

Figure 6. Section of a metallic surface [2]

For our measurements was used a ZEISS SURFCOM 1900 SD device (Figure 6) to pin out contours and asperities. The system can be equipped with tactile sensors as well as optical sensors. This system can reach a resolution of 0.016 mm with maximum sensitivity of 2 μm, 5 μm or 10 μm depending on the measuring head used (Figure 7).

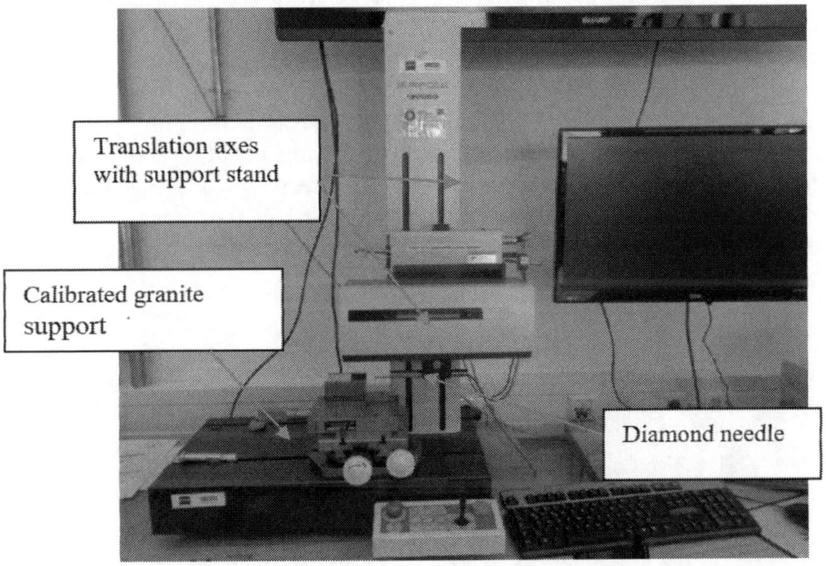

Figure 7. Workplace with ZEISS SURFCOM 1900 SD

With this system, measurements can be made for very fine rays, angles, phases or finishes, as measuring devices can be adapted to the requirements by changing the measuring heads. The possible track of measurement is 100 or 200 mm long.

2. Mathematical modeling

In the following example was measured the surface of a S235 steel shaft. In the graph (Figure 8) can be seen the roughness of the chipping process. The image was magnified vertically 2,000 times and horizontally 30.43 times.

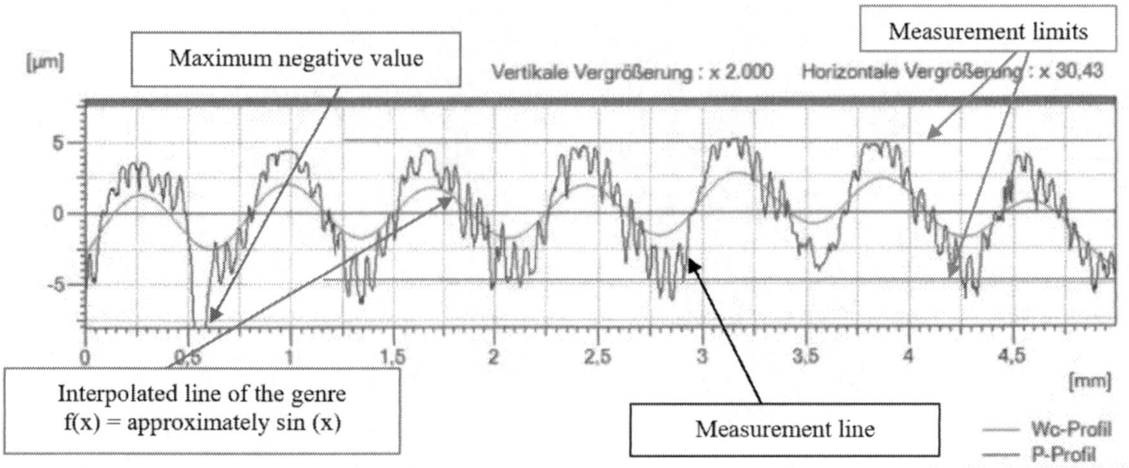

Figure 8. Roghness

According to this chart, the following values were obtained:

Ra = 1,7043 µm (average asperity)
Rz = 8,0779 µm (the average depth of asperities)
Rzmax = 12,1734 µm (maximum depth of asperity)
RSm = 275,0469 µm (average length of asperities)

The average of monitored asperities can then be considered as a function approximate to the formula f (x) = (sin x). This line does not show the exact asperities resulting from the cutting of the shaft, but in certain tolerances in the range of microns, can indicate the path traveled by the measuring head. In this way, extreme values can be eliminated and the trend of the function f(x) is visualized (Figure 9). For the above example, by mathematical modeling, the following function can be obtained for the entire interval: y = f (x)

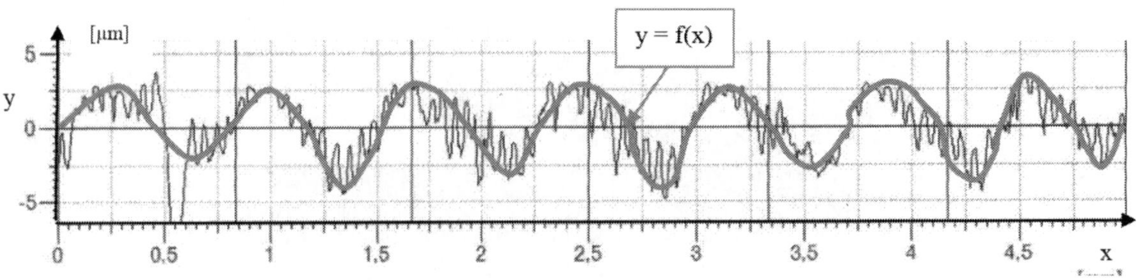

Figure 9. The trend of function f(x)

where

$$f(x) = f(x_1) + f(x_2) + f(x_3) + f(x_4) + f(x_5) + f(x_6) = \sum_{1}^{6} f(x) \tag{1}$$

and generally

$$f(x) = f(x_1) + ... + f(x_n) = \sum_{i=1}^{n} f(x) \tag{2}$$

By mathematical iteration, applying for example the Newton approximation, the function obtained is: f(x) = 3sin (7,5x)

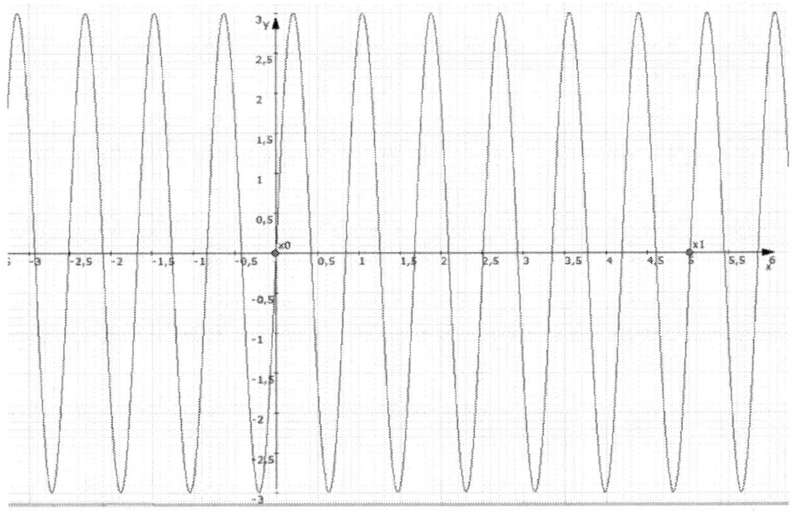

Figure 10. Approximate function

with the following values obtained by iteration (Table 1):

Table 1. Iteration resultes

Points	Bisection	Falsi Rule	Newton
P 1	(-1|-3,458589)	(-1|-3,458589)	(-1|-3,458589)
P 2	(2|1,060914)	(2,475096|0,725091)	(-0,163263|-,386907)
P 3	(0,5|-2,277188)	(1,872812|3,360299)	(0,237104|3,386907)
P 4	(1,25|-0,696104)	(0,457111|-1,292247)	(0,735508|-2,032976)
P 5	(1,625|-0,188456)	(1,326294|-2,480521)	(0,828179|0,327795)
P 6	(1,8125|3,442877)	(7,226111|-2,758128)	(0,815962|-0,000966)
P 7	(1,71875|2,167972)	(39,617143|-1,092724)	(0,815998|0)
P 8	(1,671875|1,057924)	(58,377015|-0,883749)	
P 9	(1,648438|0,441911)	(78,756683|-0,346201)	
P 10	(1,636719|0,127245)	(87,628283|2,267156)	
P 11	(1,630859|-0,030637)	(52,535186|2,371658)	
P 12	(1,633789|0,048317)	(30,757866|-3,282152)	
P 13	(1,632324|0,008841)	(621,530912|-3,183733)	
…...	……………………..		
P 23	(1,631997|0,000012)		
P 24	(1,631996|-0,000007)		
P 25	(1,631996|0,000002)		
P 26	(0,631996|-0,00002)		
P 27	(0,631996|0)		

If the function graph f (x) = 3sin (7.5x) is adjusted by scaling and expansion to the graph of the ZEISS SURFCOM 1900 SD (figure), then these two graphs become almost identical (Figure 11).

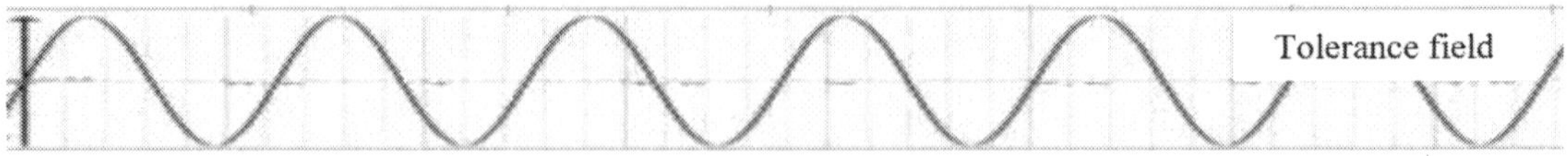

Figure 11. Adapted function graph

To compare mathematically the surface of the graph with surfaces obtained by other measuring methods, these surfaces are to be integrated according to the formula:

$$f(x = \int 3\sin(7{,}5x)dx = \int 3\sin(\frac{15}{2})xdx \rightarrow F(x) = -\frac{2\cos(\frac{15x}{2})}{5} + C \tag{3}$$

According to the ZEISS SURFCOM 1900 SD monitoring system presented in this paper, there were also used for optoelectronic measurements, [4-6] where the tolerance field had to be around 10 μm, an industrial processing system. The system has a camera, a lighting plate, an adjustable tripod and an image processing program (Figure 12).

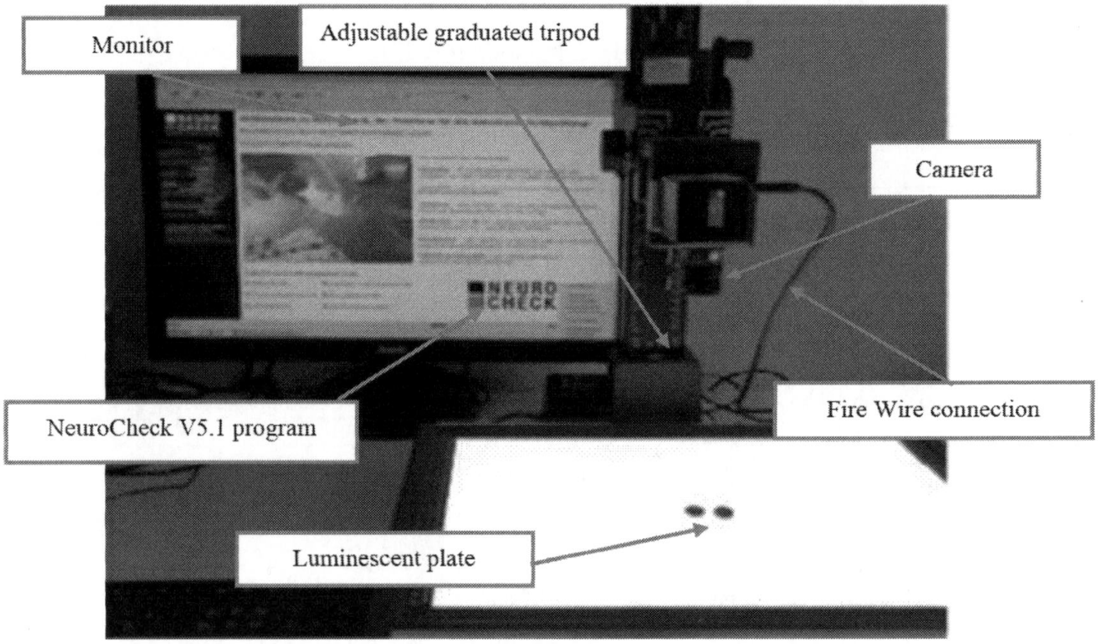

Figure 12. Work place for image recognition [6]

The data obtained from several series of simulations with objects of different shapes and sizes gave results similar to those obtained with the ZEISS measurement system. In the following figure can be seen the tolerances from the measurement of an object and in which tolerance field the measurements were framed (Figure 13).

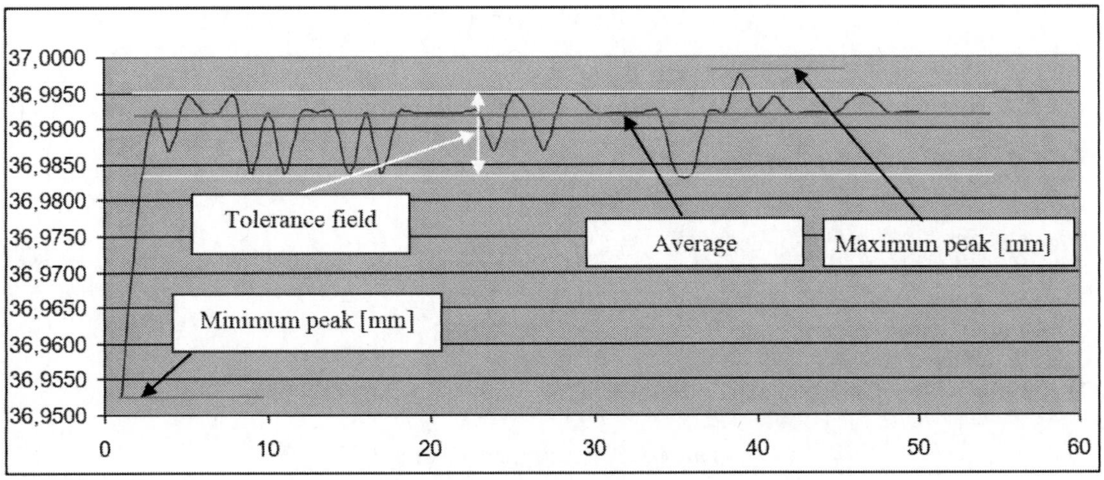

Figure 13. Measurements with image recognition

The measurements offered a series of representative results, from which the following information was extracted:

- The arithmetic means of the string according to the formula:

$$\bar{x} = \frac{1}{n}(x_1 + x_2 + x_3 + ... + x_n = \frac{1}{n}\sum_{i=1}^{n} x_i \tag{4}$$

where - $\bar{x}$ is the arithmetic mean,

x_i is an element of the string,

n is the number of the elements in the string,

with

$$\Delta x = \sigma_n \tag{5}$$

where: - Δx is the deviation for a single measurement,

- The variance according to the formula:

$$s^2 = \frac{1}{n-1}[(x_1 - \bar{x})^2 + (x_2 - \bar{x})^2 + ... + (x_n - \bar{x})^2] = \frac{1}{n-1}\sum_{i=1}^{n}(xi - \bar{x})^2 \tag{6}$$

- The standard deviation according to the formula:

$$\sigma_n = \sqrt{\frac{1}{n-1}\sum_{i=1}^{n}(x_i - \bar{x})} \tag{7}$$

where: - σ_n is the standard deviation,

$$\Delta\bar{x} = \frac{\sigma_n}{\sqrt{n}} \tag{8}$$

where: - $\Delta\bar{x}$ is the deviation from the arithmetic mean,

- The median, the value that divides the ordered statistical series into two equal subsets, the volumes being measured in the number of statistical units according to the formulas:

$$n = 2\,p + 1 \tag{9}$$

If the series has an odd number of values and

$$n = 2\,p \tag{10}$$

If the series has a just number of values,
- The variance coefficient v according to the formula:

$$v = \frac{s}{x} \cdot 100 \tag{11}$$

- Maximum deviations,
- Minimum deviations.

The results of this measurement can be seen in the following table (Table 2).

Table 2. Measuremets results

Temperature	23°C	4°C
Diameter max. [mm]	40.5064	40.4569
Diameter min. [mm]	40.4816	40.4242
Total	4049.1215	4044.3711
Math. average [mm]	40.491215	40.443711
Number of attempts *n*	100	100
Standard deviation	0.00901976	0.007228581
Median [mm]	40.4876	40.4433
Variance	8.1356E-05	5.22524E-05
Variance coefficient %	2.22758395	1.78731626

According to data obtained from the measurements we have chosen for (Figure 14), [7], [8]
- upper tolerance ES = 0.004 mm (for 0.0064 mm)
- lower tolerance EI = -0,016 mm (for 0,0284 mm).

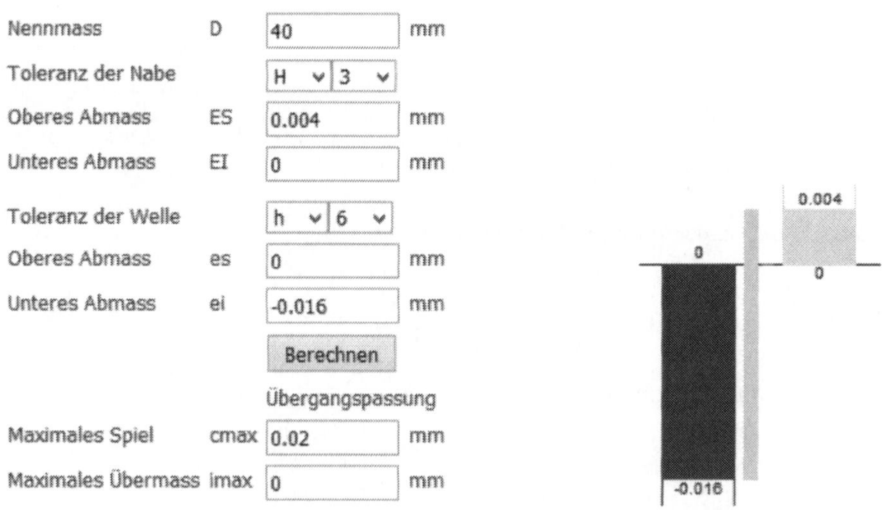

Figure 14. Calculation in Adjustment System [7]

These tolerances result in a combination of the H3 / h6 form, which means that the result of the measurements falls within the intermediate adjustment system (Figure 14).

3. Conclusions

As such, it has been shown that the resolution of the image recognition system is sufficiently precise for a smart intelligent gripping system to catch and pick up fragile objects without additional pressure sensors.

Numerical simulations helped determine contours resulting from surfaces. The results thus obtained can be processed in the algorithms for step-by-step motors programming for the exact operation of the gripper, respectively for the operation of the industrial robots used in the gripping process.

References

[1] Ahlers R J 1986 *Die Optische Rauheitsmessung in der Qualitätstechnik*, Springer-Verlag Berlin Heidelberg
[2] Burkhardt T, Feinäugle A, Fericean S and Forkl A 2004 *Lineare Weg- und Abstandssensoren: Berührungslose Messsysteme für den industriellen Einsatz*, Neuhausen
[3] ***2017 *ZEISS SURFCOM*, Carl Zeiss AG, internet site called 15.06.2017 https://www.zeiss.com/metrology/products/systems/form-and-surface/cnc-surface-measuring-instruments/surfcom-1500.html
[4] Demant C, Streicher-Abel B and Waszkewitz P 2011 *Industrielle Bildverarbeitung*, Third Edition, Springer-Verlag Berlin Heidelberg, New York
[5] Gonzalez R C and Woods R E 1992 *Digital Image Processing*, Addison-Wesley, Reading MA
[6] Jalba C K, Muminovic A, Epple S, Nasui V and Barz C 2016 Optical Gripper, *IOP Conf. Ser.: Mater. Sci. Eng.* **200** 012054
[7] ***2017 *MESYS AG*, Technoparkstrasse 1, 8005 Zürich, internet site called 20.03.2017 https://www.mesys.ch/?cat=5
[8] Schweizer A 2016 *Formelsammlung und Berechnungsprogramme für Anlagenbau*, Wasenöschstr. 11, DE 88048 Friedrichshafen

Innovation-Prototype. Making hydraulic and/or pneumatic plates using 3D printing technology

V Alexa[1], S A Raţiu[1], I Kiss[1], V G Cioată[1] and M Rackov[2]

[1]Politehnica University of Timisoara, Department of Engineering and Management, 5 Revolution Street, 331128, Hunedoara, Romania
[2]University of Novi Sad, Faculty of Technical Sciences, Trg Dositeja Obradovića 6, Novi Sad, Serbia

E-mail: vasile.alexa@fih.upt.ro

Abstract. Start 3D printing allows hydraulic and/or pneumatic plates to be obtained from a single technological process without requiring further post-production operations. 3D printing with innovative materials in a rich colour range has several advantages such as: time-saving, cost is the same for any type of plate and its reported just to volume, fast and impossible realization of hydraulic and/or pneumatic links compared to traditional and high accuracy technologies.

1. Innovation-prototyping. 3D printing

The 3D printing represents a technology not as new as it may be, but which has become more and more accessible in the last decade, stimulating the innovation and increasing the efficiency in many areas, including the hydraulic and pneumatic fields, due to the designing freedom, reduced time and low costs [1].

The 3D printing – regarded as a modern manufacturing process – has a number of advantages [2]:

- design optimization;
- multiple personalization possibilities;
- rendering to a high degree of complexity;
- decrease in production time;
- cost reduction;
- material saving;
- promoting the principle of sustainability.

The 3D printing is the ideal solution for the most demanding designers and engineers; therefore, through rapid prototyping, the organizations are helped to [3]:

- improve the communication modality in terms of developing new products;
- shorten the designing cycle;
- bring high-quality products to the market faster than the competition does;
- improve the precision of the patters;
- eliminate costly mistakes;
- bring innovation and high quality;
- optimize the collaboration between engineers, marketing and sales departments and management team, through elaborate presentations.

The application areas for 3D printing are many and various: mechanical and mechatronics engineering, engineering and management of technological systems, automation and computers, aerospace engineering, medical engineering, biotechnical systems engineering, machine building technology, electrical engineering, electronics and telecommunications, etc.

The benefits of 3D printing in engineering [4]:

- reduced costs of designing, building and marketing of a new product by verifying its shape and functionality at an early stage;
- highly efficient assessment of the ergonomics and aesthetics of the prototype;
- reducing and, often, eliminating the design changes of the manufacturing tools;
- improving the intercommunication among the members of the team, requiring a shorter time to launch a new product;
- reduced costs of machining tools by printing patterns or moulding components using 3D printed moulds.

The field of 3D printing is becoming more and more accessible, being a source of innovation similar to that of the Internet explosion.

2. Hydraulic/pneumatic blocks

The modular hydraulic/pneumatic blocks are standardized elements used for building hydraulic/pneumatic circuits with low consumption of ducts and connection elements. Constructively, they are parallelepiped-shaped.

The multiple positions in which they can be assembled have limited the number of modular board types. Typically, there is one type of board for each type of hydraulic/pneumatic device used in the control diagrams.

2.1. Role and constructive features

The hydraulic/pneumatic blocks include either embedded elements (i.e. cartridge elements) or applied elements (i.e. manifolds, valves), integrating all distribution and control elements of the hydraulic/pneumatic diagram.

The accelerated development of machine building requires the assembly of hydraulic/pneumatic equipment in a modular system.

The use of modular systems has some very important advantages:

- reduced size;
- low energy losses due to low consumption of ducts and connection elements;
- system flexibility, thanks to the possibility of modular device combinations;
- easy troubleshooting/maintenance.

The modular devices will be mounted on modular boards, enabling the building of the most diverse and complex hydraulic/pneumatic diagrams.

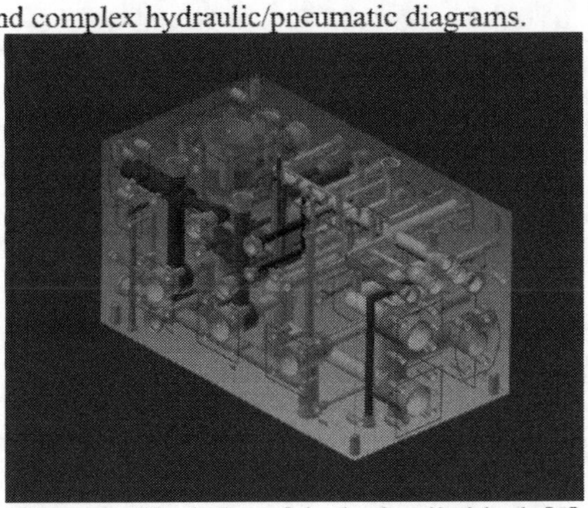

Figure 1. The holes of the hydraulic block [5]

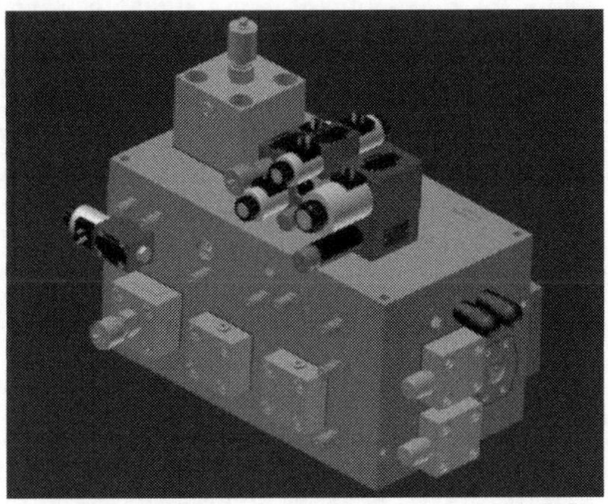

Figure 2. The hydraulic block assembly[5]

2.2. Hydraulic/pneumatic calculation of hydraulic/pneumatic blocks
Hydraulic/pneumatic connecting elements having the function of transporting the working agent require a hydraulic/pneumatic calculation to determine the minimum cross-sectional area of the holes through which the working fluid circulates [6].

2.2.1. Hydraulic/pneumatic calculation. It consists of setting the required nominal diameter (DN). For calculation, the relationship can be used:

$$d = 4{,}61\sqrt{\frac{Q}{v}} \quad (1)$$

in which:

Q - the flow of fluid flowing through the pipe, $\left[\frac{l}{min}\right]$;

v - the flow velocity of the fluid in the pipeline, $\left[\frac{m}{s}\right]$.

As it results from the relationship, the flow rate cannot be any size. The recommended values for these are:

$$v = 6...7\frac{m}{s},\ for\, l < 100 \cdot d$$
$$v = 3...4\frac{m}{s},\ for\, l > 100 \cdot d \quad (2)$$
$$v = 1{,}5...2{,}5\frac{m}{s},\ for\, the\, discharge\, pipe$$

The nominal pipe diameter, established with the previous relationship, is corrected to the normalized value, since the hydraulic apparatus is designed for DN nominal values with normalized values of 6; 8; 10; 13; 20 mm.

*2.2.2. Calculating the strength of the hydraulic block material.*Consider the hole in the block as a pressurized case subjected to a plane tension condition. Since the tension in the transverse direction is higher, the thickness of the wall can be expressed with the relation:

$$\sigma = \frac{p \cdot d}{2 \cdot \sigma_a} \quad (3)$$

σ_a - is the admissible mechanical strength of the material of the pipe material. The relationship (3) is valid for the calculation of thin-walled pipes ($16 \geq \delta d$).

For pipelines where $16 < \delta d$ behavior in terms of material resistance is described by relation (4):

$$\sigma_a = \frac{d^2 + 2\delta + 2\delta^2}{2(d - \delta)} \cdot p \quad (4)$$

resulting the thickness of the pipe wall:

$$\delta \geq \delta_{min} = \frac{d}{2}\left(\sqrt{\frac{\sigma_a + p}{\sigma_a - p}} - 1\right) \quad (5)$$

When designing, the pipe whose internal diameter corresponds to the required flow conditions is adopted, and then it is pre-dimensioned with one of the previous relations.

2.2.3. Linear load losses - hydraulic circuits. When flowing through real piping, two types of hydraulic losses occur:

- linear hydraulic losses h_{pd} (longitudinal or distributed), mathematically expressed by Darcy's formula:

$$h_{pd} = \lambda \frac{l}{d} \frac{v^2}{2g} \tag{6}$$

- local hydraulic losses h_{pl}, expressed by the formula of Weisbach:

$$h_{pl} = k \frac{v^2}{2g} \tag{7}$$

in which:

l - pipe length, $[m]$;

d - pipe diameter, $[m]$;

v - average fluid velocity per section, $\left[\frac{m}{s}\right]$;

g - gravitational acceleration, $\left[\frac{m}{s^2}\right]$;

λ - coefficient of linear hydraulic losses, $[-]$;

k - the local hydraulic losses coefficients corresponding to the different types of hydraulic resistances, $[-]$.

These coefficients k of local hydraulic loss are obtained from the experiments, and depending on the change of direction, the diameter of the pipes as well as the flow direction of the fluid may have the values given in figure 3.

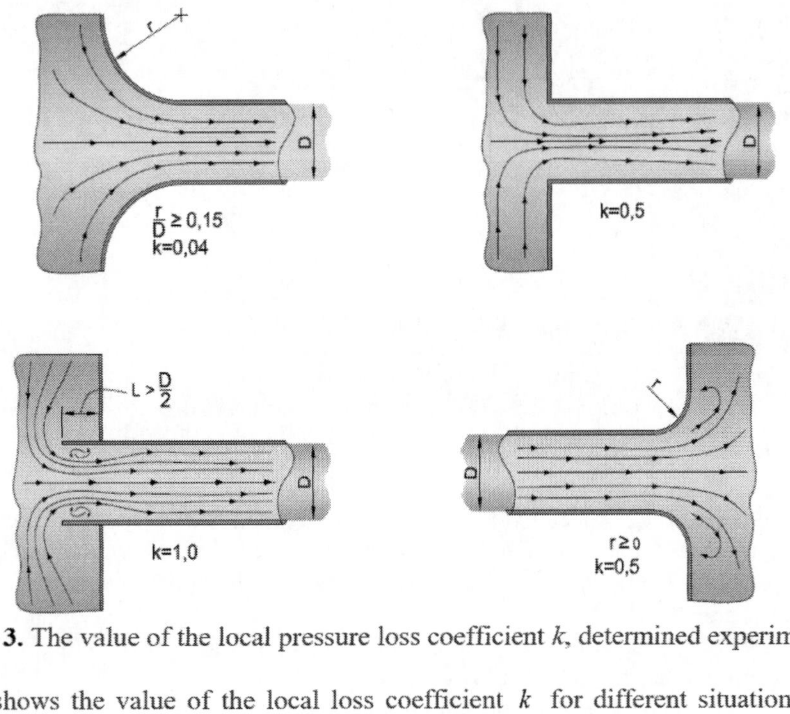

Figure 3. The value of the local pressure loss coefficient k, determined experimentally

The literature shows the value of the local loss coefficient k for different situations and elements found in practice.

Thus, Table 1 shows these values of the local hydraulic loss coefficient k :

Table 1. Values of the local hydraulic loss coefficient K - different types of elbows and joints

Types	Flanged	Threaded
Regular 45^0	0.2	0.4
Regular 90^0	0.3	1.5
Long radius 90^0	0.2	0.7
Return bend 180^0	0.2	1.5
Line flow	0.2	0.9
Branch flow	1	2
Union	0.08	0.09

3. Hydraulic block design. Modeling flow through holes

Using PLM software packages, users can simulate the full range of processes needed to develop a new product from the initial design phase to the design, analysis, manufacturing, and maintenance phases.

The CATIA V5 benefits from a sustained pace of development, and it is practically two to three months to come up with a new revision. At present, the CATIA V5 contains over 140 robust applications covering vast areas of assisted engineering [7]. Figure 4 shows the modelling of a hydraulic block with the specification tree, visualizing the holes [8].

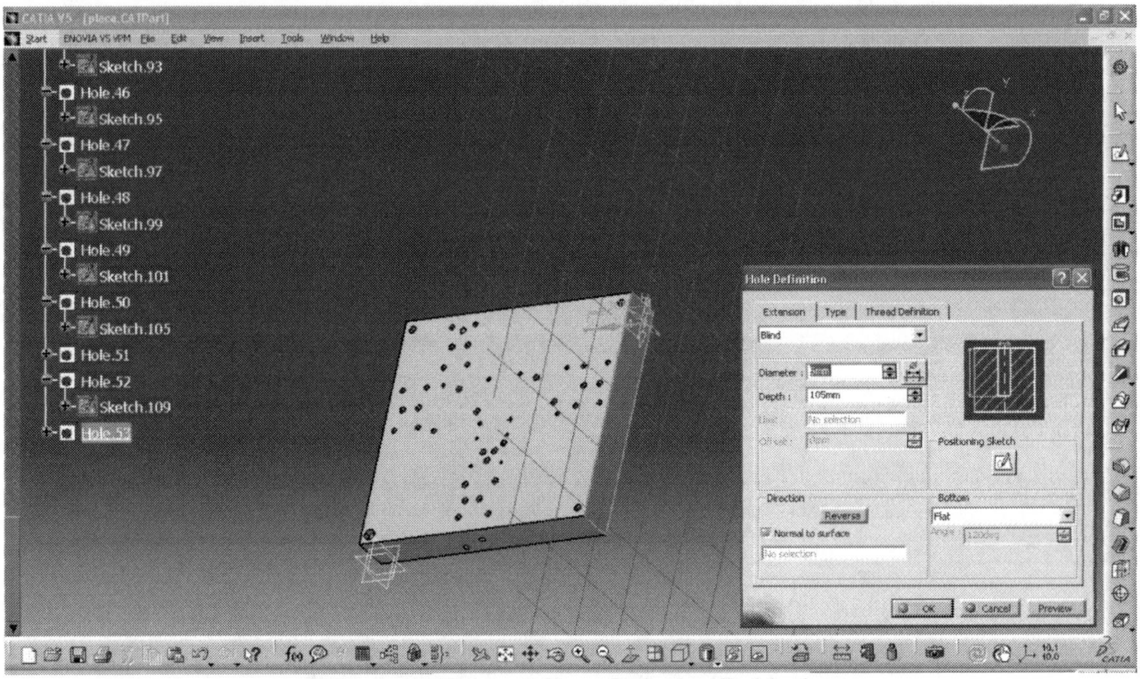

Figure 4. The modeling of a hydraulic block

Fluid mechanics equations, especially Newtonian fluid dynamics and Navier-Stokes non-stationary equations can be explained by specialized modeling and simulation software. Thus, a new branch emerged, namely CFD - Computational Fluid Dynamics, which complements the dynamics of the theoretical and experimental fluids.

This new branch of the fluid mechanics study, CFD, has a number of advantages compared to the dynamics of the experimental fluids, as follows:

- design and development time is substantially reduced;
- flow conditions that cannot be reproduced experimentally can be simulated;
- provide clear and detailed information (graphics and value);
- it is an economic method, with few material and energetic resources.

By modeling and simulation of CFD was followed by the suggestive presentation of the major energy differences of the flow of fluids through orifices processed by classical technology, respectively by the unconventional and innovative 3D printing technology.

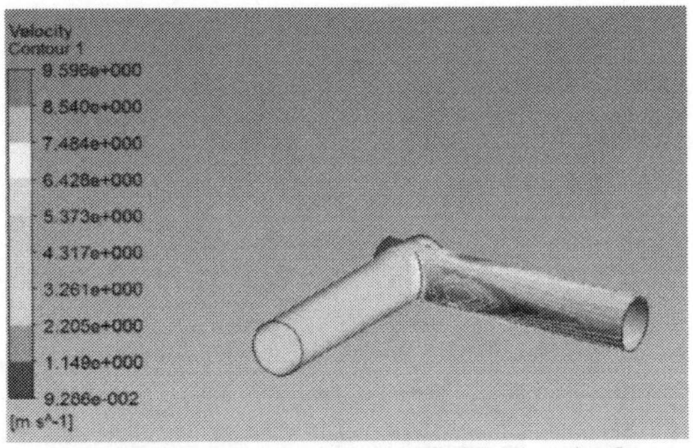

Figure 5. Fluid velocity field-classical mechanical processing

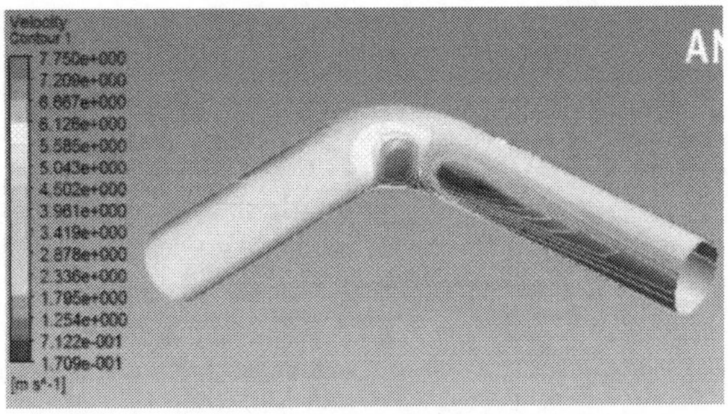

Figure 6. Fluid velocity field-3D printing processing

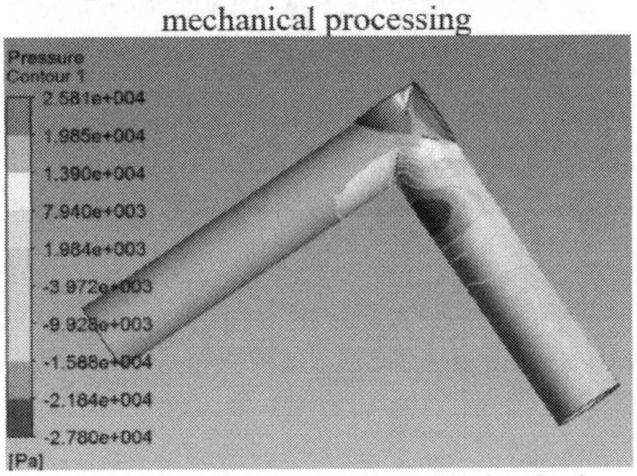

Figure 7. Fluid pressure field-classical mechanical processing

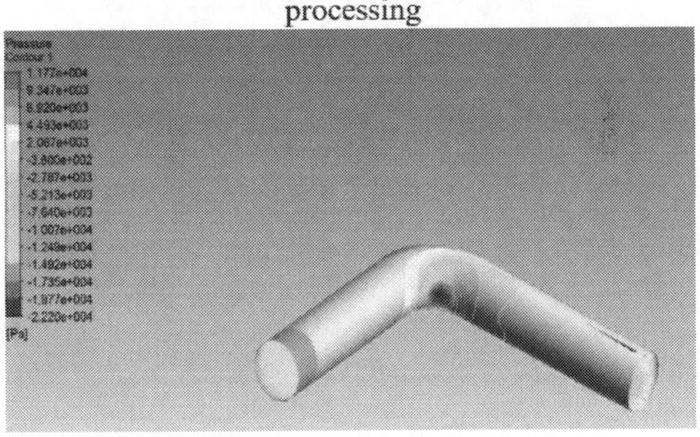

Figure 8. Fluid pressure field-3D printing processing

The differences between the two solutions are presented in Figures 5, 6, 7 and 8. The speeds have much higher values for the classic solution, Figure 5, which implicitly leads to a substantial increase of the local pressure losses, Figure 8. These major differences have made us physically achieve a hydraulic block with a 3D printer.

4. 3D printing of the designed hydraulic block

The 3D printer used uses the filament deposition technique. The printer model is Anet A6, bought as a kit and assembled. It is a printer working in Cartesian coordinates (X Y Z). The X and Y axes are driven by transmission belts, and the Z axis operates on the nut-screw principle.
Technical specification is presented in Table 2 [9].

Table 2. Technical specification

Specification	
model	A6
machine weight	7.6kg
gross weight	9.5kg
machine size	480x400x 400mm
packing size	450x 446x 215mm
LCD display	LCD12864
consumable material	ABS/PLA/HIPS/etc.
material diameter	1.75mm
framework	Acryllic

The extruder consists of a stepper motor, a filament pushing system, a filament heating and melting chamber, a nozzle, a thermocouple and an electrical resistance. The print pad is on the Y axis and has the option to be heated for better adhesion of more sensitive materials (ABS) [9].
Technical parameters are presented in Table 3.

Table 3. Technical parameters

Parameters	
print size	220x220x 250mm
printing speed	100 mm/s
nozzle diameter	0.4 mm
layer thickness	0.1-0.3 mm
XY position accuracy	0.012 mm
Z position accuracy	0.004 mm
extrusion head temperature	260°C
bed temperature	100°C

The 3D printer provides high-complexity pieces that cannot be achieved by known technology. Among the materials made available by producers, we chose the PLA (polylactic acid), which has the properties presented in Table 4.

PLA (polylactic acid) is a biopolymer, a biodegradable plastic. It is made of renewable raw materials, such as corn starch or sugar cane [10].

In addition to 3D printing, it is commonly used for packaging materials, plastic foils, plastic glasses and plastic water bottles.

Table 4. Polymer properties

Solid PLA	
print temperature	210-225°C
raft temperature	190°C
first layer temperature	200°C
print base material	MDF/Acrylic
raft required	Optional
filament diameter	3mm

With the elements given in the design phase, the actual building of the hydraulic block was completed. The hydraulic block was built in the Hydraulic and Pneumatic Hydraulic Driving Laboratory of the Faculty of Engineering in Hunedoara, Figure 9.

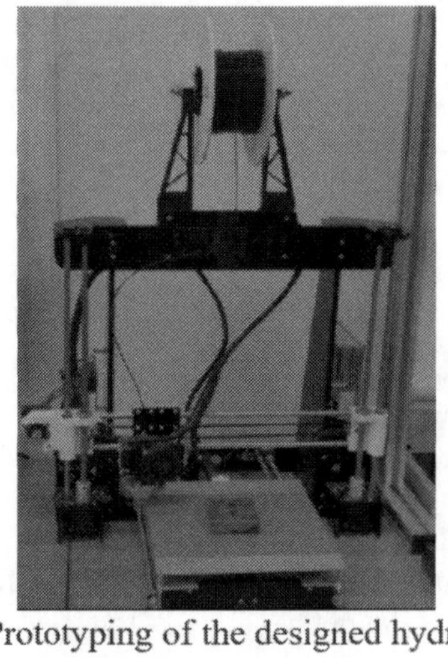

Figure 9. Prototyping of the designed hydraulic block

IOP Conf. Series: Materials Science and Engineering **294** (2017) 012095 doi:10.1088/1757-899X/294/1/012095

5. Conclusions

The hydraulic block obtained by this process combines the *technological* and *economic* advantages of the innovative design and 3D printing process.

From a *technical* point of view, the following have been highlighted:

- reduce the prototyping time of the product;
- eliminating mistakes and improving product quality;
- production of compact hydraulic blocks by eliminating the inconveniences generated by classical processing;

Economical, the advantages of 3D design and printing are given by:

- cost reduction;
- material saving;
- promoting the principle of sustainability.

The 3D printing represents a technology not as new as it may be, but which has become more and more accessible in the last decade, stimulating the innovation and increasing the efficiency in many areas, including the hydraulic and pneumatic fields, due to the designing freedom, reduced time and low costs.

References

[1] Bak D 2003 Rapid prototyping or rapid production? 3D printing processes move industry towards the latter, *Assembly Automation* **23**(4) 340-345
[2] ***http://www.printing3d.ro/avantaje/
[3] ***http://www.print3dbucuresti.ro/printare-3d/avantaje-imprimare-3d/
[4] ***http://www.zspotmedia.ro/blog/printare-3d/
[5] ***http://www.eurofluid.it/fr/conception
[6] Ţălu M D L 2014 *Calculul pierderilor de presiune distribuite în conducte hidraulice,* Editura Universitaria, Craiova
[7] Cioată V G 2009 *Proiectare asistată de calculator cu Catia V5*, Editura Mirton, Timişoara
[8] Alexa V, Sarosi J and Ratiu S 2014, *Processing simulation of hydraulic blocks NCMT,* Proceedings of 2014 International Conference of Hydraulics and Pneumatics – HERVEX, Călimanesti-Căciulata, Romania, November 5-7, pp 62-70
[9] ***https:// www.3dprint.wiki/reprap/anet/a6
[10] ***https://www.3dsystems.com/materials

Institute of Physics
Dirac House, Temple Back
Bristol BS1 6BE UK

ISSN: 1757-8981
ISBN 978-1-5108-5567-0

CURRAN ASSOCIATES INC.
proceedings.com